Jane's

MILITARY TRAINING AND SIMULATION SYSTEMS

Edited by Ian W Strachan

Seventh Edition

1994-95

ISBN 0 7106 1138 2
"Jane's" is a registered trade mark

In the USA and its dependencies
Jane's Information Group Inc, 1340 Braddock Place, Suite 300, Alexandria, VA 22314-1651, USA

I(T)P An International Thomson Publishing Company

British Library Cataloguing-in-Publication Data.
A catalogue record for this book is available from the British Library.

Printed and bound in Great Britain by Biddles Limited, Guildford and King's Lynn.

Contents

ADMINISTRATION

Publishing Director: Robert Hutchinson

Managing Editor: Charles LeMesurier

Yearbook Editorial Production Manager: Ruth Simmance

Production Editor: Diana Barrick

EDITORIAL OFFICES

Jane's Information Group Limited, Sentinel House,
163 Brighton Road, Coulsdon, Surrey CR5 2NH, United Kingdom

Tel: 0181 763 1030 International +44 181 763 1030
Telex: 916907 Janes G
Fax: 0181 763 1006 International +44 181 763 1006

SALES OFFICES

Send enquiries to International Sales Manager:
David Eaton-Jones (Trade, Middle East, Asia Pacific)
Fabiana Angelini (UK/MoD, Europe)
Jane's Information Group Limited, UK address as above

Send USA enquiries to:
Joe McHale, Senior Vice-President Product Sales,
Jane's Information Group Inc, 1340 Braddock Place, Suite 300,
Alexandria, VA 22314-1651

Tel: +1 703 683 3700
Telex: 6819193
Fax: +1 703 836 0029

ADVERTISEMENT SALES OFFICES

Advertisement Sales Manager: Sandie Palmer

Australia: Brendan Gullifer, Havre & Gullifer (PTY) Ltd, level 50, 101 Collins St., Melbourne 3000, Australia.

Tel: +61 3 6960288
Fax: +61 3 6966951

Benelux: Sandie Palmer, Jane's Information Group (see UK office)

Brazil: L Bilyk, Brazmedia International S/C Ltda, Alameda Gabriel Monterio da Silva, 366 CEP, 01442, São Paulo

Tel: +55 11 853 4133
Telex: 32836 BMED BR
Fax: +55 11 852 6485

France: Patrice Février, Jane's Information Group – France, BP 418, 35 avenue MacMahon, F-75824 Paris Cedex

Tel: +33 1 45 72 33 11
Fax: +33 1 45 72 17 95

Germany & Austria: Janet Scott, Jane's Information Group (see UK office)

Hong Kong: Jeremy Miller, Major Media Ltd, Room 1402, 14F Capitol Centre, 5-19 Jardine's Bazaar, Causeway Bay

Tel: +852 890 3110
Fax: +852 576 3397

Israel: Oreet Ben-Yaacov, Oreet International Media, 15 Kinneret Street, 51201 Bene-Berak

Tel: +972 3 570 6527
Fax: +972 3 570 6526

Italy and Switzerland: Ediconsult Internazionale Srl, Piazza Fontane Marose 3, I-16123 Genoa, Italy

Tel: +39 10 583684
Telex: 281197 EDINT I
Fax: +39 10 566578

Japan: Intermart/EAC Inc, 1-7 Akasaka 9-Chome, Minato-Ku, Tokyo 107, Tokyo.

Tel: +81 3 5474 7835
Fax: +81 3 5474 7837

Korea: Young Seoh Chinn, JES Media International, 6th Floor, Donghye Bldg, 47-16 Myungit Dong, Kangdong-Gu, Seoul 134-070

Tel: +82 2 481 3411
Fax: +82 2 481 3414

Scandinavia: Gillian Thompson, First Call International, 11 Chardmore Road, London N16 6JA, UK.

Tel: +44 181 806 2301/3538
Fax: + 44 181 806 8137

Singapore, Indonesia, Malaysia, Philippines, Taiwan and Thailand: Hoo Siew Sai, Major Media (Singapore) Pte Ltd, 6th Floor, 52 Chin Swee Road, Singapore 0316

Tel: +65 738 0122
Telex: RS 43370 AMPLS
Fax: +65 738 2108

Spain: Jesus Moran Iglesias, Varex SA, Modesto Lafuente 4, E-28010 Madrid

Tel: +34 1 448 7622
Fax: +34 1 446 0198

United States and Canada: Kimberly S Hanson, Director of Advertising Sales and Marketing, Jane's Information Group Inc, 1340 Braddock Place, Suite 300, Alexandria, VA 22314-1651

Tel: +1 703 683 3700
Telex: 6819193
Fax: +1 703 836 0029

USA: Jennifer Felix, Advertising Sales Representative (address as above)

United States and Canada: Maureen Nute, Advertising Production Manager
(see United States and Canada)

US West Coast: Ann Marie St. John-Brooks, Regional Advertising Sales Manager, Jane's Information Group, 1523 Rollins Road, Burlingame CA 94010

Tel: +1 415 259 9982
Fax: +1 415 259 9751

United Kingdom/Rest of World: Janet Scott, Jane's Information Group, Sentinel House, 163 Brighton Road, Coulsdon, Surrey CR5 2NH

Tel: 0181 763 1030 International +44 181 763 1030
Telex: 916907 Janes G
Fax: 0181 763 0643 International +44 181 763 0643

Administration: Karen Broomfield, Jane's Information Group (see UK office)

"Now we don't have to build all of our real-time 3D model data in-house."
REAL-TIME DATA
REAL FAST
Geoff Phillips,
Manager, Mission Simulations
Lockheed
Viewpoint has hundreds of pre-built Datasets for your real-time applications. And all are available with levels of detail and texture maps. We build custom Datasets too.
Call 1.800.DATASET (1.800.328.2738) for your free catalog.
Real-Time image capture features Viewpoint's SU-27B, AA-10 Missile and Refinery Datasets.
Viewpoint DataLabs
625 S. State Street • Orem, UT 84058
Ph 801.229.3000 • Fax 801.229.3300
VIEWPOINT
DATALABS

Classified list of advertisers

The companies listed, advertising in this publication, have informed us that they are involved in the fields of manufacture indicated below:

Aircraft and helicopter simulators and procedure trainers
CAE Electronics
ECC International
Reflectone UK

Aircrew trainers and simulators
ECC International

Air defence tactical training systems
CAE Electronics

Air defence training simulators
Comator Process

Air traffic control/radar/visual and fighter control training simulators
CelsiusTech Systems
Comator Process
RGB Spectrum

Air-to air combat simulators
ECC International

Anti-tank guided weapon trainers and simulators
CelsiusTech Systems

Anti-tank missile training systems
ECC International

Army tactical command and control trainers
CelsiusTech Systems
Comator Process

Artillery observation of fire trainer
Comator Process

Command information centre/ operations room simulators
CelsiusTech Systems
Comator Process
RGB Spectrum

Computer-based training systems
CelsiusTech Systems
Comator Process
ECC International
RGB Spectrum

Computer-generated imagery
CAE Electronics
CelsiusTech Systems
RGB Spectrum
Viewpoint

EW trainers and simulators
ECC International

Interactive video training systems
RGB Spectrum

Maintenance training systems
ECC International

Military training systems
CelsiusTech Systems
Comator Process
ECC International
FN Herstal
RGB Spectrum

Naval target systems
Comator Process

Part-task operator trainers
ECC International

Small arms marksmanship training systems
FN Herstal

Small arms range equipment and systems
FN Herstal

Submarine steering and depth control simulators
ECC International

Tank (and other AFV) crew tactical trainers
Comator Process
ECC International
FN Herstal

Tank (or direct fire) fire and fire-control trainers and simulators
Comator Process
ECC International

Weapon fire and effect simulators and systems
FN Herstal

Weapon operating trainers and simulators
ECC International
FN Herstal

Alphabetical list of advertisers

The Sogitec Super Puma Flight Simulator (see page 273)

Jane's

MILITARY TRAINING AND SIMULATION SYSTEMS

1994-95

Jane's Information Group Limited, Sentinel House, 163 Brighton Road, Coulsdon, Surrey CR5 2NH, UK
Jane's Information Group Inc, 1340 Braddock Place, Suite 300, Alexandria, VA 22314-1651, USA

Introduction

Note on the Jane's Group

Fred T Jane published his first edition of *Jane's All the World's Fighting Ships* (now *Jane's Fighting Ships*) in 1898, and the Jane's Group still publishes this title as a yearbook under the editorship of Captain Richard Sharpe RN. The next Jane's yearbook was *Jane's All the World's Airships, Aeroplanes and Dirigibles* (now *Jane's All the World's Aircraft*), first published in 1909. Other yearbooks and publications followed steadily over the years. In the 1980s it was decided that the yearbook *Jane's Weapons Systems* was becoming too big, and it was divided into several discrete sections such as Guided Weapons, Artillery, Armoured Fighting Vehicles and so forth. Currently the Jane's Information Group publishes the following yearbooks and binders as well as magazines such as *Jane's Defence Weekly, Jane's Intelligence Review, Jane's Airport Review, Janes Defence Systems Modernisation* and *International Defense Review*; they are listed in alphabetical order:

International ABC Aerospace Directory
International Defense Directory
Jane's AFV Retrofit Systems
Jane's Air Traffic Control
Jane's Air-Launched Weapons
Jane's Airports and Handling Agents
Jane's Airports, Equipment and Services
Jane's All the World's Aircraft
Jane's Ammunition Handbook
Jane's Armour and Artillery
Jane's Avionics
Jane's Battlefield Surveillance Systems
Jane's C⁴I Systems
Jane's Civil and Military Aircraft Upgrades
Jane's Containerisation Directory
Jane's Fighting Ships
Jane's High-Speed Marine Craft
Jane's Infantry Weapons
Jane's Land-Based Air Defence
Jane's Military Communications
Jane's Military Training and Simulation Systems
Jane's Military Vehicles and Logistics
Jane's Naval Weapon Systems
Jane's NBC Protection Equipment
Jane's Radar and EW Systems
Jane's Security and Co-In Equipment
Jane's Space Directory
Jane's Strategic Weapon Systems
Jane's Underwater Warfare Systems
Jane's Urban Transport Systems
Jane's World Airlines
Jane's World Railways

Your New Editor

Terry Gander decided to relinquish the editorship in order to concentrate on other editorial work within the Jane's group, and Ian Strachan was appointed in his place.

Ian was a Wing Commander in the Royal Air Force and retired in 1993. He has been involved in training since becoming a Qualified Flying Instructor (QFI) in 1965, achieving a Central Flying School A1 instructional category in 1969. He was also involved in the standardisation of other instructors. He then completed the ETPS course and served as a test pilot at Farnborough and Boscombe Down, commanding a test squadron during and after the Falklands (Malvinas) war when many extra trials had to be flown in order to support the conflict. He was awarded the Air Force Cross for test work on fast-jet night low flying using FLIR, LLTV and NVGs, and was the co-author of a definitive paper on this subject. He has flown some 8500 hours in over 100 types of aircraft, mainly fixed wing, but he is also a service-qualified helicopter pilot. Amongst other experience, he has over 900 air-to-air refuelling contacts in his flying logbook.

Between test pilot assignments, he was a lecturer in Guided Weapon technology at the RAF College of Air Warfare, and later became Senior GW Specialist at the Department of Air Warfare, RAF College, Cranwell. Some of his lecture subjects included military applications in the Infra-Red spectrum, and explosive and warhead technology. He was awarded the MBE for this work.

From 1985 he was in charge of the air systems training and simulation office in the Ministry of Defence in London. In this role he was Defence Staff sponsor for the Shorts Tucano training aircraft for the Royal Air Force, and for many flight simulators and training devices. Amongst other things he wrote the operational requirements for the RAF's Harrier, Tucano and Nimrod flight simulators and the Air Navigation Trainer, all now in service. The Harrier requirement was particularly demanding and involved not only a 6-DoF motion platform but a very wide field-of-view, solved by employing a dome combined with an eye-slaved projection system.

He has 'flown' over 50 flight simulators involving 35 types of aircraft from fighters to 747s and over 10 helicopter types. He is a member of the Royal Aeronautical Society Flight Simulation Group Committee and has been active in RAeS debates and symposia on simulation over many years. He has presented several papers to the RAeS, including one on Motion and Visual Cueing for Military Simulation and another on Military Helicopter Simulation. He admits to a particular interest in visual systems and motion platforms.

He continues to maintain a pilot's licence, instructs on gliders and motor gliders, and owns a part share of a Nimbus motor glider operating out of Lasham Gliding Centre some 70 km SW of London.

Foreword

The Training Scene

Reducing Defence Sector. The defence sector continues to reduce despite political uncertainties and many conflicts throughout the world. However, in many nations a plateau has been reached below which it is acknowledged even by the 'peace at all costs' lobby that it would be unwise to go. We can therefore see a glimmer of light at the end of the tunnel and some stabilisation of force structures and equipment programmes. The old adage 'if you want peace, prepare for war' still applies. I would adapt it to include the training factor and say: 'If you wish to avoid being threatened, maintain adequate and well-trained defence forces. Base long-term military planning on recent history, not altruism; human nature is flawed, particularly in stressful situations.' A shorter version might be: 'Read history, then learn from it.'
Capable Simulation Technology. Simulation technology is now very capable, and the training industry has nothing to hide in terms of costs, capabilities and timescales of delivery. In an uncertain military world we should have more synthetic training to back up the necessary real operation of the equipment in peacetime. This is particularly so with the availability of high-fidelity visual scenes to standard formats such as US DoD MIL STD 1821 (ex-Project 2851). This offers the capability for quick generation of a detailed real-world training scenario which can be practised by crews in a threat situation before they have to deploy to the area concerned.
Similarity between Military and Civil Simulation Equipment. One difficulty in producing a book with 'military' in its title is - which are military systems and which are civil? In the simulation industry there is often little difference in the technology employed. I have therefore erred on the side of including equipments and systems rather than excluding them. After all, the police (civil power) are involved in firearms training, the merchant marine in ship's bridge simulators, and the air forces of the world have transport aircraft, basic training aircraft and so forth which are also on the civil register. And the fire, police, and security services are now using sensors which in the past were exclusively military, such as intensifiers (NVGs and the like) and thermal imagers.
Lack of Investment in Modern High Quality Training Aids. One problem is that the Admirals, and Army and Air Force Generals and Air Marshals (and their financiers and political masters) still seem reluctant to make the investment in a modern and high quality range of training aids and simulators, with a few notable exceptions such as the US Army and the US Special Forces. A day at Fort Rucker is a stimulating experience, looking at the Apache CMS through to SIMNET. One can understand the policy of 'front line first', but the question is - what is the front line? A Full Mission Simulator (FMS) for front-line equipment, or, at the other end of the price scale, Computer-Based Training (CBT) or a Part-Task Trainer (PTT) for an activity such as war route familiarisation, are all capable of training for more combat scenarios than the combat vehicle itself and should be regarded as intimately related to front-line equipment. They are not, because 'you cannot fire them at the enemy'. They never figure in the Order of Battle. They should do, and there should be a category of 'aids to realistic combat training'. A potential enemy may well take them into account in his assessment of your capabilities. He should. A well-trained opposition is more of a threat than one, for instance, without modern training aids, which will be reduced to obtaining battle training 'on the job' with inevitable consequences in effectiveness and casualties.
Deterrence. The essence of armed forces is to act in the first place as a deterrent; conflict itself signifies a failure to deter and so a failure of the primary purpose of having armed forces in peacetime. The possession and proper use of good training aids, well advertised to the potential opposition, contributes significantly to the deterrent effect. And if, unhappily, deterrence fails and conflict becomes inevitable, those who have provided, funded and supported the training aids will know that they did their best to prepare their forces for conflict, maximise the chances of success, and minimise own casualties in both personnel and expensive military equipment.
Estimated Savings. Simulators cost much less to operate than the equipment itself except in the case of very simple equipment. They can run day and night, and can train for any scenario, any weather, any part of the world. A sensible split of training between simulation and the equipment itself will save money overall. This saving could be invested in other front-line projects in order to increase deterrence further. As an example of savings which have been calculated and declared, the US Army has estimated that its new range of aircraft simulators (its SFTS programme) will save $608.6 million per year or, with 131 cockpits simulated, $4.65 million per simulator cockpit. This is not because of parsimony in specifying the simulators since their SFTS programme includes state-of-the-art Combat Mission Simulators (CMS) utilising all the latest developments in motion, visual and computing systems.
Cheaper Training Aids than Full Mission Simulators (FMS). Much cheaper training aids are available than an FMS, some of which will achieve more than the equipment itself in peacetime. These include simple CBT involving PC-based systems for training such as familiarisation in possible combat terrain. The NATO DLMS or the US DoD MIL STD 1821 (ex-Project 2851) protocols for CGI, combined with satellite, mapping, and intelligence information, allow us to generate realistic scenes anywhere over the globe. In the days of the possible confrontation with the USSR and Warsaw Pact we used to call this 'war route training' but the concept of familiarisation with terrain, routes and targets over which you are not allowed to operate in peacetime or before the actual deployment, is always valid and should be recognised.
Regions of Conflict. Southern Iraq is relatively flat, former Yugoslavia is not. Many uninformed articles and letters have appeared in the press about the different responses by the UN and the Coalition Forces in Iraq and former Yugoslavia, but those experienced in military planning have only to look at the contours of the maps of the area of potential conflict to realise the fundamental difference in the military position. You cannot fight 'El Alamein' or 'Sinai' style Tank and Air Battles in terrain resembling the Alps, or even, crucially for the UN, provide a viable major response without very large and sophisticated forces equipped for fighting in rugged terrain. Politicians under attack seem not to have deployed this argument enough; have they really looked at the maps? Or are their briefers using maps without vertical contours? A good dose of real-world 3D CGI would sort this out for both politicians and the media! It is no coincidence that Dien Bien Phu and Que San were in mountainous terrain from which artillery could fire into the 'cauldron' below; shades of Sarejevo and elsewhere in the Balkans?
New Equipment - The Training Needs Analysis (TNA). With costly new front-line equipment, the concept of an early and thorough Training Needs Analysis (TNA) seems to be well accepted, and indeed imposed by financial planners before they will release the money for the overall equipment package. And guess what the TNA will inevitably recommend; a well-balanced suite of training aids and simulators ranging from CBT to, for complex military equipment, Full Mission Simulators; together with, of course, the optimum use of the military equipment itself in a training mode.
Financing a Good Standard of Simulator. Even the cynics on simulation agree that simulators are good for procedural training, and see the benefit of them as allowing sorties in the real equipment to concentrate on battle training rather than the routine. The argument should then be that savings on procedural training, which you no longer have to do in the real equipment, will easily finance a good suite of training aids within the overall budget for the equipment package. Once you have the money for good training aids, make sure that they are specified not only for the procedural work but also for battle scenarios not achievable with the real equipment in peacetime. These scenarios include: regular weapon firing; multiple engagements; force-on-force (blue on red) training with EW and attrition; tactics development for complex scenarios; and so forth.
Simulation of Equipment already in Service. Military equipment already in service often has to make do with inadequate training aids even if it is scheduled for front-line service for many years to come. I wonder how many systems have been retrofitted with Electro-Optical aids for night attack, and still have old simulators without visual systems?

Another factor is the refurbishment of old simulators in order to save small amounts of money, when the benefits of a new-build device using modern simulation technology are obvious. Some old simulators are just not worth saving, particularly those with old-fashioned motion systems or primitive hardware, computers, and general architecture. A breakthrough in motion platform technology happened in the early 1980s, when the second-generation 6-DoF synergistic platforms became available. The modern 'industry-standard' 6-DoF platform is a most impressive device. In my opinion it is equally suited to the simulation of agile fighter aircraft as well as transports and indeed AFVs, and is essential in the aircraft field if the customer requirement is fidelity of handling in addition to procedural training. Recent improvements in latencies (transport delay) of motion platforms can be seen in the table on motion platforms in the new section on Motion Cueing Devices.

Younger Generation. The saving grace for the future is that the younger generation will not accept shoddy training aids since not only have they come to expect good quality modern technology, but they were brought up on computer and arcade games and so have a different attitude from those of us who were not. Can the military training industry hang on until people with this outlook are in positions of power?

Industrial Scene. Looking at the industrial scene, Thomson Training and Simulation (TTS) is now the second force behind world leaders CAE and CAE-Link, having amalgamated the French Thomson Simulation interests with those of the ex Link-Miles, Rediffusion, and Burtek companies. It now has a formidable capability in the Sea, Land, and Air areas in both the military and civil fields. The big question is - can TTS maintain market share in the light of fierce competition, particularly from CAE who now 'go it alone' on visuals with its new MAXVUE™ system?

TTS has an initial difficulty. It has a 'surfeit of riches'; three good but different designs of 6-DoF motion platforms (Link-Miles, Rediffusion and Thomson), two CGI systems (VISA and IMAGE), two back-projected collimated display systems (WIDE and AWARDS), several different dome projection systems, and a plethora of different designs of collimated monitor displays, aside from various Instructor Station designs and different computing standards and simulator architectures. A good indication of TTS policy is its new SPACE™ CGI system, which takes the best from the previous VISA and IMAGE systems and should be formidable competition to market leaders such as Evans and Sutherland's ESIG series. No doubt we will see more standardisation of systems from TTS as time goes on, while the older systems will continue to be supported in order to keep customer goodwill. I understand that it has already decided to continue to offer both the flexible-mirror WIDE system and the glass-mirror AWARDS; this parallels other companies in this field which offer a choice of plastic or glass screens.

Many smaller companies have not survived, but those with both energy and good, cost-effective product lines are not only holding their own but going from strength to strength. One thinks of IVEX for CGI and SEOS for Display Systems.

Display Systems - Collimation and Vertical FoV. There is now a good choice of back-projected collimated wide-FoV display systems. First was Rediffusion's WIDE, then Link-Miles' AWARDS and McDonnell Douglas MULTIVIEW (now owned by FlightSafety International). Now we also have CAE MAXVUE and SEOS's PANORAMA. However, a serious deficiency for military application is not only the cost, but the limited vertical FoV. Good vertical FoV is needed for fighters and helicopters, and in many training exercises is more important than horizontal FoV (eg helicopter autorotations and engine-off landings). Personally I am not convinced that collimation is essential except where two pilots sit side by side and are widely separated; in an airliner cockpit, collimation provides the same perspective from laterally separated eye points. I have seen very effective visual scenes in military simulators which were based on direct projection in order to obtain vertical FoV. Under a training workload with rapidly moving scenes, I do not believe that the brain registers many alleged deficiencies in the scene as long as the cues are there for the exercise in hand. Research is needed to establish, for multi-pilot aircraft, what eye point separation can be sustained without adverse effects when using direct projection systems. This could also have profound effects on future civil regulatory rules.

Image Generators. IGs have steadily become more capable at less cost; just look at the polygon count per channel and the texture capability, even of low-cost systems. Evans and Sutherland and CompuScene (the latter now owned by Martin Marietta, soon to become Lockheed Martin) still dominate the high-capability market. But high- and even medium-cost systems had better look to their laurels with lower cost systems coming up from behind. However, customers should look at systems in the round. For example, in aircraft simulators, a data update rate of at least 25 Hz (greater for agile aircraft) is needed, irrespective of other features, if horizon-stepping during roll manoeuvres is to be avoided. And for all roles, a varied texture capability is often more fundamental than a high polygon count.

Virtual Reality (VR). Virtual Reality is a current buzzword. But we in the training and simulation world have used the concept of a realistic visual world and correct tactile cues for years; the difference is that our systems actually work as intended. Anyone who has seen a good quality simulator visual system must be awfully disappointed after donning a VR helmet. I wondered why users of VR do not suffer from disorientation or worse after experiencing such a wide artificial visual field without associated cues of motion; the times I have tried it lead me to believe that it is because the fidelity of the simulation is so poor that the brain is not convinced that it is real! However, as in all low-cost developments I look forward to an amalgamation of the best parts of existing systems; is it, for instance, possible to produce a display system of the power of CAE's FOHMD by combining a good VR helmet with an existing high quality CGI system? I suspect that you never get 'something for nothing' and the answer lies somewhere in the middle of the range of possibilities. But every sim lab should be researching using a FOHMD, VR, and dome projection, together with various levels of CGI, to quantify the trade-offs. That is if they can afford it all!

Finally, researchers and designers should bear in mind the inherent conservatism of operators and trainees, who may not want to don heavy and intrusive helmets but wish to use their normal equipment in a simulator. There is a significant difference between what you can get away with in research, and what is acceptable for training of front-line crews. An unhappy simulator trainee may soon become 'anti-simulator'; a happy trainee will become 'pro-simulator'.

Realism. What I sometimes call the 'ambience of realism' is often crucial in generating the right attitude in trainees, hence leading to good training effectiveness. Psychologists often say that realism is often overkill for the training required. What is often forgotten is the morale and motivation of the crew members under training; ask them and it quickly becomes apparent that there are intrinsic advantages in realism irrespective of the theoretical minimum standard to achieve the training. Experienced operators normally demand more realism than *ab initios,* because, being very familiar with the equipment and environment, they will not tolerate what they regard as a low quality of simulation. I am not arguing for 'realism at any cost'; the balance is always the cost-effectiveness of training. But additional realism almost always has benefits. Fortunately more 'realistic' systems are becoming cheaper. One only has to look at recent progress with textured CGI and in cheaper PC-based computing.

Editorial Aspects

New Data. I have tried to put in as much as seemed reasonable from the new data that was provided; if your company entry is not satisfactory please tell me straight away and send your update, in hard copy or preferably on PC-formatted 3.5″ disc in Word Perfect 5.1 or at least an ASCII-based system, together with good (preferably different) photos. The earlier I have the information the more time I have to format the entry for the 1995-96 edition.

Editing. Jane's publications are not simply for reprinting company publicity handouts or other company data; editorial judgement is applied in producing an even-handed approach to all contributions and concentrating on engineering and electronic characteristics rather than on opinion and the use of highly charged words. It is amazing how many companies seem to think that they produce the 'first', the 'best' or the 'most advanced' equipment in a given field. Can they all be right?

Abbreviations Section. The abbreviations section has been much expanded and a glossary of terms added. This section now contains brief explanations of terms such as APDS, BAI/CAS, CEP, CLOS guidance, HEAT, OCA/DCA and so forth. The section on Infra-Red now includes descriptions of the Near- and Far-IR, and terms such as intensifiers/NVGs, Thermal Imaging/FLIR, and Heat Seeking. Specialist terms used in the CGI field have also been included and comment is invited so that eventually an industry-standard consensus definition can be used wherever possible. Apparently straightforward terms such as 'resolution', 'field', 'frame', 'refresh rate' and 'update rate' can prove a minefield. The difference between Simulation, STimulation, and Emulation is also covered. Please look at the definitions provided and comment if you think they could be improved.

New Directory of Suppliers. At the end of the book before the indexes is a new section called 'Directory of Suppliers'. The use of word 'supplier' rather than 'manufacturer' is deliberate. This section includes not only manufacturers but also research organisations who are now looking for trade in the supply of services and expertise, indeed any organisation active in the field of training and simulation such as regulatory authorities.

New Diary of Events. A diary of future events related to training and simulation has been produced and will be continued in future editions if it is popular and receives support in terms of event organisers providing the data well in advance. It follows this section and shows the next two years, although I keep a database for as far forward as anyone wishes to give me provisional information. For instance I believe that I/ITSEC will be at Albuquerque in 1995 but return to Orlando for 1996 and maybe beyond.

***JAWA* Style.** I also do some editing for *Jane's All the World's Aircraft (JAWA)* and have followed its style by attempting to organise the text into headed paragraphs, each describing a discrete subject. For *JMTSS* these include System Components, Trainee's Station, Instructor/Operating Station (IOS), Visual System (Different paragraphs for Image Generation (IG) and Display System (DS)), Database(s), Computers, Motion Platform, and so forth, depending on the trainer or simulator being described. Also, the company data used in the Directory of Suppliers is modelled on the *JAWA* style for company information and includes a list of key personnel and a summary of company history and activities (where these details were provided by the company concerned).

Cost Data. I note that in *JMTSS* very little cost data is given whereas in *JAWA* almost every item has a non-contractual 'ball-park' cost. I have always been keen to know approximate 'orders of cost' of systems and will pursue this where possible, with the caveat that Jane's can only give an approximate scale and customers must negotiate real costs with suppliers. I believe that it is useful to know, for instance, that 6-DoF synergistic motion platforms listed in the new section on Motion Cueing cost between US$200 000 and US$800 000, whereas CGI systems refreshing at 25 Hz or above with texture capability vary from US$50 000 per channel to nearly US$1 000 000.

Standard Terms. Wherever possible I have used industry-standard terms such as PTT, CBT, CGI, FMS etc. I note some variance across the systems area; in aviation we only use the term Simulator for higher level devices with a full flight model and, in the case of civil devices, with a motion platform. Other devices are known as Trainers or Flight Training Devices (FTDs). There seems to be duplication in terms such as Computer Based Training (CBT) and Computer Aided Instruction (CAI); are these not the same? If so, CBT seems the most accepted term.

Exaggerated Terms. I note a trend by companies to call a simulator a 'Full Mission Simulator' (FMS) or a 'Combat Mission Simulator' (CMS) when, on analysis, it appears that its facilities do not merit this description. The generally accepted criterion for FMS/CMS (synonymous?) is that of being able to simulate all aspects of an operational mission with high fidelity, and on this criterion there are only a few real FMS/CMS about.

Next Year - the 1995-96 Edition

Change of Title. It has been decided to drop the word 'Military' from the title since many simulation systems are similar whether they are used in a civil or military role. Ship handling simulators, flight simulators, and truck driving simulators are examples. Also, simulation systems such as CGI, display, and motion, are common to civil, military, sea, land and air roles. The new title will be *Jane's Simulation and Training Systems (JSTS),* and will have new sections on simulators for trains, trams, vehicles generally, and other areas of the civil training scene. It is also intended to cover the more sophisticated Entertainment Simulators, and 'Virtual Reality' displays.

Explanations and Tables. I intend to include more explanations and tables or diagrams of subjects of interest such as regulatory criteria and terminology. The yearbook is designed not only for the training and simulation expert but also for those new to the field; do all readers understand terms such as collimated optics and the difference between designs employing the stimulation principle as opposed to emulation?

Wide Number of Training Aids. What is a training aid? The potential scope is very wide, from assessment of training with the real gun, AFV, ship or aircraft, up to completely synthetic devices such as full mission simulators. It has been said, 'All military activity is training except combat.' This illustrates the difficulty in compiling a training yearbook; the term simulation is more helpful. There has to be a pass-fail test for inclusion of a type of training equipment. I think one criterion for inclusion in *JSTS* should be that a significant amount of synthetic equipment (emulated or stimulated) is used in the particular training device described.

Possible Deletions. I doubt, for instance, whether equipment to calibrate or aim the firing of real rounds from guns should be included, and indeed have rejected some entries on this ground; they can appear in other Jane's Yearbooks such as *Jane's Infantry Weapons* and *Jane's Armour and Artillery.* I query whether any entry should be concerned with live, as opposed to synthetic, firing. Using lasers instead of rounds is appropriate in a book on simulation, but rounds as rounds is not. I wonder what customer reaction would be if we dropped the early sections on gun firing ranges? Training ammunition and pyrotechnics is another area which could well be transferred to the very popular *Jane's Ammunition Handbook.* Similarly with UAVs used as Air Targets which could go into the new binder *Jane's UAVs and Targets,* to be published for the first time next year.

Object of Equipment Yearbooks. The value of a yearbook such as *JMTSS* is that it can concentrate on a world overview of the more complex equipment in its field to an extent just not possible in articles or tables in the weekly or monthly press. It can also take an annual view instead of responding to short-term fluctuations in the perceived market or in technological trends. Inclusion of equipment which uses innovative or interesting design is particularly important (whether it seems to be selling or not), as is a comparison of the characteristics of many systems which is not often done in more frequently published periodicals.

Acknowledgements. My thanks to all those company contacts who have co-operated by sending useful material; hopefully you will recognise parts of your contribution in this edition; keep the information flowing. You may now see your name mentioned in the Directory of Suppliers. Thanks to Terry Gander for the first six years as editor and steering the yearbook on a steady course. Thanks also to Ted Hooton who has edited the Sea Systems section for many years, and this year did the first 85 or so pages of the Land Systems section. Diana Barrick has been the main contact point with the Jane's Yearbook Editorial team led by Ruth Simmance; Charles LeMesurier has been the Managing Editor for JMTSS and the contact with higher management and policy within Jane's. Thank you, Diana, Ruth and Charles for bearing with me in my first year in the hot seat. And to Fred T Jane for starting it all off with *Jane's All the World's Fighting Ships* in 1898!

Bentworth, Hampshire, UK   September 1994   Ian Strachan

Comments and Amendments

Any comments or proposed additions and alterations for next year would be greatly appreciated, please send as soon as convenient. Include any civil simulation material of which you are aware. We always like to change the photographs of existing equipment so please send some photos from different angles or of interesting technical detail. It would be helpful if proposed text for direct inclusion is provided in a form in which it can either be document scanned (the narrative in some brochures is not suitable due to heavy 'artistic' background) or sent on a PC-formatted (not Apple Mac) 3.5″ disc in Word Perfect 5.1 (preferable) or a well known system such as MS Word, Wordstar or AMI Pro; if the alphanumerics follow the ASCII or ANSI protocols then they are usable without retyping. Text will not necessarily be published verbatim, it will be edited to the *JSTS* house style.

The usual rules apply; all entries are published free of charge and without any obligation to your organisation, we in turn expect that the material that you provide is free of copyright and that there is no restriction on its use by Jane's. Publication is due in November and so material embargoed for release before that month may be sent for inclusion.

Please send to:

The Editor, *Jane's Simulation and Training Systems (JSTS),* Sentinel House, 163 Brighton Road, Coulsdon, Surrey CR5 2NH, United Kingdom.

Fax: +44 181 763 2582 (backed up by hard copy by post, please)

FUTURE WORLD EVENTS IN TRAINING AND SIMULATION

Note from Editor: As an experiment, this year we are publishing a table of future events in which there is an element of training and simulation. Events are not confined solely to those specialising in the training field but include general shows and exhibitions at which the training industry is likely to be represented. If the data is well received the intention is to continue in later editions. Would organisers be kind enough to send details, however provisional, of future events for as far ahead as possible. The organisers' own details will then appear in the Directory of Suppliers. I will keep records as far ahead as there is data, but publication in *JMTSS* will generally be confined to the two years following the edition in question. I am conscious that some military and naval conferences and exhibitions are not yet included, and that I have no information for some countries such as Russia; please send the information for next year's edition.

1994

Month	Dates	Title	Venue	Organiser (Details, Directory of Suppliers)	Remarks
Nov	16-17	Networking Simulators	RAeS London	RAeS Flight Simulation Group	Group AGM 17 Nov after conference
Nov/Dec	28-1	16th I/ITSEC Inter-Service/Industry Training Systems and Education Conference	Marriott Orlando World Centre, USA	I/ITSEC	

1995

Month	Dates	Title	Venue	Organiser (Details, Directory of Suppliers)	Remarks
Mar	19-23	IDEX '95	Abu Dhabi	IDEX, Abu Dhabi	Includes MILCON '95
Mar	21-26	Australian International Aerospace Expo	Melbourne	AirShows DownUnder	
Apr	25-27	ITEC/CATC. International Training Equipment Conference and Exhibition and Civil Aviation Training Conference	Netherlands Conference Centre, The Hague	ITEC, UK	
May	10-12	European Conference on Aviation Simulation (ECAS)	Salzburg, Austria	Salzburg Technology Centre	Emphasis is on Civil Aviation Training
May	17-18	Flight Simulation Technology - Capabilities and Benefits	RAeS London	RAeS Flight Simulation Group (FSG)	FSG 25th anniversary conference. Civil & Military papers
May	22-25	Flight Simulation - where are the challenges?	Braunschweig, Germany	AGARD	Military simulation within NATO
Jun	TBC	Flight Crew Training Conference	Paris	European Aviation Consultants (EAC)	During Paris Air Show
Jun	10-20	Paris Air Show	Paris	GIFAS	
Sep	26-28	ITEC-(Asia)/ACATC. International Training Equipment Conference - Asia, and Exhibition and Asian Civil Aviation Training Conference and Exhibition	Raffles City Convention Centre, Singapore	ITEC UK	
Oct/Nov	TBC	Flight Crew Training Conference	TBC, USA	European Aviation Consultants (EAC)	
Nov	TBC	Course, Introduction to Flight Simulation (1 week)	Cranfield University, UK	Cranfield University	Co-sponsored by RAeS
Nov	15-16	Engineering simulation devices used in training	RAeS London	RAeS Flight Simulation Group (FSG)	FSG AGM after conference
Nov	TBC	17th I/ITSEC Inter-Service/Industry Training Systems and Education Conference	Conference Centre, Albuquerque, New Mexico, USA	I/ITSEC	
Dec	5-10	Langkawi International Maritime & Aerospace (LIMA) Exhibition & Conference	Langkawi Island (N of Penang) Malaysia	Le Proton LIMA, Kuala Lumpur	Called the LIMA 95 Exhibition

1996

Month	Dates	Title	Venue	Organiser (Details, Directory of Suppliers)	Remarks
Jan	TBC	Aerotech 96	UK - NEC Birmingham	I Mech E and RAeS	
Apr	TBC	ITEC Europe. International Training Equipment Conference and Exhibition	TBC	ITEC, UK	
May	TBC	RAeS Spring Simulation Conference	RAeS London	RAeS Flight Simulation Group	
Jun	TBC	IMAGE VII Conference	Phoenix, USA	Image Society, Tempe, Arizona	
Sep	First Week	Farnborough Air Show	Farnborough, UK	SBAC	
Nov	TBC	RAeS Autumn Simulation Conference	RAeS London	RAeS Flight Simulation Group (FSG)	FSG AGM after conference
Nov/Dec	TBC	18th I/ITSEC Inter-Service/Industry Training Systems and Education Conference	Marriott Orlando World Centre, USA	I/ITSEC	

TBC = to be confirmed

GLOSSARY OF TERMS AND ABBREVIATIONS

This Glossary is particularly comprehensive because training and simulation systems embrace all areas of military endeavour on sea, land and in the air. In addition, the specialist knowledge of readers is unlikely to include all the technologies and disciplines mentioned in this book, and so more terms are included rather than less. The target audience is international and terms very familiar to some readers will not be to others; no apology is made for including them, with explanations where necessary. The editor invites corrections or useful additions for the next edition.

NUMERICAL

2IC Second in command

ALPHABETICAL

AAA Anti-Aircraft Artillery (ie Air Defence Guns)

AAC Army Air Corps. A term used to describe the air arm of the UK and other Armies. The UK Army Air Corps (AAC) was re-titled Army Aviation (AAv) in Sep 1994

AACS Air-to-Air Combat Simulator

AACT All Arms Classroom Trainer

AADTS Advanced Air Defence Training System

AAM Air-to-Air Missile

AAR Air-to-Air Refuelling

AAT All Arms Trainer

AATS Anti-Aircraft Training System

AAW Anti-Air Warfare

AAWS Advanced Anti-tank Weapon System

ABI Air Battle Instructor

ABM Anti-Ballistic Missile, part of Ballistic Missile Defence (BMD)

AC Alternating Current, or Air Combat

Acceleration-Onset Cueing This is a term for the motion cueing principle used by a simulator motion platform. The initial acceleration of the vehicle being simulated is replicated closely by the platform but, because of its limited throw, the continuing motion of the real vehicle cannot be continued in the simulator. Thus, after the initial acceleration, in the simulator the response is gradually decreased, eventually to zero (this is known as the washout phase). Then the motion platform is reset to the neutral position but at a rate below that of the sensory threshold of the simulator crew. Fortunately the body sensors (inner ear, muscle-and-joint feedback, etc) principally react to accelerations rather than steady-state motions, and have thresholds below which they do not react. Thus, the way the body signals motion to the brain coincides very well with acceleration-onset cueing in a simulator, and is the reason why well-designed modern low-latency motion platforms work so well for all aircraft from large transports to agile fighters.

ACEATM Aimed Controlled Effect Anti-Tank Mine

ACLOS Automatic Command to Line-of-Sight. Type of missile guidance, see under CLOS

ACM Air Combat Manoeuvre

ACMI Air Combat Manoeuvring Instrumentation. Normally used in referring to an instrumented Air Combat range.

ACMS Air Combat Manoeuvring Simulator

ACOVS Artillery Combat Observer Vehicle Simulator

ACS Air Combat Simulator

ACT Airborne Crew Trainer, or Air Combat Trainer/Training, or Aircrew Co-ordination Training

AD Air Defence, Assistant Director

ADF Automatic Direction Finding

ADI Australian Defence Industries

ADP Automatic Date Processing

ADLIPS Automatic Data LInk Plotting System

ADT³ Air Defence Tactical Training Theatre

ADV Air Defence Variant, eg of the Tornado aircraft; the Tornado ADV is the F3 mark

AEW Airborne Early Warning

AFCS Automatic Flight Control System

AFTA Advanced First Term Avionics

AFTS Airtech Firearms Training System, or Advanced Flight Training System

AFV Armoured Fighting Vehicle

AG AktienGesellschaft (German Company Designation)

AGE Advanced Graphics Engine, or Air/Ground Environment

AGES/AD Air-to-Ground Engagement/Air Defence

AI Artificial Intelligence, or Air Intercept (eg AI Radar)

AIFV Armoured Infantry Fighting Vehicle

AIO Action Information Organisation

AIM Air Intercept Missile. Used by the US DoD with a numerical designator, eg AIM-7 Sparrow MRAAM and AIM-9 Sidewinder SRAAM

AIT Air Intercept Trainer

AITAT Armoured Infantry Training and Advisory Team

ALERTS Airborne Electronic Warfare Receiver Training System

Aliasing, Anti-aliasing Aliasing is a term used in scanning technology to describe a number of undesirable effects produced by translating an original image into discrete pixels. These effects include jagged and/or crawling edges, gaps in thin polygons, and a tendency for small polygons to blink on and off. As an example of one of these effects, when a straight edge displayed in raster format is rotated to an angle diagonal to the raster lines, a jagged edge (sometimes known as 'jaggies') will be seen instead of a straight line; a technique known as 'anti-aliasing' is employed to soften the ragged edge effect and to preserve the presented image as a straight line. Anti-aliasing techniques embrace measures to reduce all aliasing effects and make calculations at sub-pixel level for colour, texture, shading, fading and translucency in order to smooth the edges concerned by averaging the characteristics of the pixels actually displayed in accordance with the sub-pixel calculations.

AMB Air-Mobile Brigade

AMCM Airborne Mine CounterMeasures

AMFCTS Artillery and Mortar Fire-Control Training Simulator

AMS Automatic Marking System

Ångstrom Name of Scandinavian scientist who gave his name to a unit of length. There are 10 000 Ångstroms in a Micron and 1 Million Microns in a Metre. Human visual acuity is from about 0.4 (violet) to 0.7 (red) Microns, or 4000 to 7000 Ångstroms. Today, the Ångstrom is less popular and the nanometre is often used instead, the visual region in this case being 400 to 700 nm.

ANT Air Navigation Trainer

Anti-Aliasing See Aliasing

Anti-G Suit/Trousers A special garment worn by crew of aircraft capable of pulling high G loadings in order to help the aircrew withstand G effects without losing consciousness ('blacking out'). Generally in the form of tightly-fitting trousers which fit either under or over (depending on the design) the flying suit. The trousers have inflatable bladders in them which, when pneumatically pressurised through a G-sensitive valve in the aircraft, press firmly on the abdomen and legs, thus restricting the draining of blood away from the brain during the period of high G. Anti-G suits/trousers are commonly (but inaccurately) known as 'G-Suits'. In addition, in some modern very high-G aircraft, the Anti-G suit effect is augmented by a small amount of pressure applied to the lungs (partial pressure breathing) which also enhances resistance to high G. The effects of Anti-G suits and partial pressure breathing are straightforward to replicate in a simulator, although the continuous G forces themselves can only be produced artificially in devices such as centrifuges.

AOI Area Of Interest. In simulator visual systems, the AOI means the centre of the area where the trainee is looking and so requiring high resolution. The AOI part of a display may be head-slaved or head- and eye-slaved, in order that expensive high-resolution imagery is not generated where the crew member is not looking.

AOV Artillery Observation Vehicle

AOSS Automatic Ordnance Scoring System

AP Armour-Piercing

APB Analysing Printer Base

APC Armoured Personnel Carrier

APDS Armour-Piercing Discarding Sabot. A type of anti-armour round in which the damage mechanism is the kinetic energy of a core of rotationally stabilised, sharply pointed mass of hard metal of high density, with the object of penetrating armour. Sometimes depleted uranium is used as the core since it is hard and dense and has no nuclear implications since it is principally the non-fissionable isotope U235. The softer outer shell (the 'sabot') is deliberately discarded in flight due to aerodynamic loads, leaving the metal core to impact with the target at high velocity.

APFSDS Armour-Piercing Fin-Stabilised Discarding Sabot. See also APDS above.

APS Adaptive Processing Sonar

APWT Annual Personal Weapon Test

AR&M Availability, Reliability and Maintainability

arc/arcmin/arc minute Implies angle measurement, in the case of minutes the letters 'arc' are sometimes added to clearly distinguish between minutes of time (1/60 hour) and minutes of angle (1/60 of a degree). Used in measurement of resolution of the human eye and of visual display systems eg 'arc-minutes per pixel' or arcminutes per Optical Line Pair. See under Resolution.

ARGS Advanced Resolution Gunnery Simulator

ARM Anti Radiation Missile. A guided weapon designed to home on and destroy a radiating transmitter, normally an enemy radar.

ARTBASS ARmy Training BAttle Simulation System

Arty Artillery

ASAT Avionics Situation Awareness Trainer

ASET Automatic Scoring Electronic Target

ASG Ausbildungsgerate Schuss/Getroffen

ASI AirSpeed Indicator

ASIC Application Specific Integrated Circuit

ASM Air-to-Surface Missile, Anti-Ship Missile

ASMI Airfield Surface Movement Indicator (for ATC Simulation)

AST Avionics Systems Trainer

ASTABS Automated Status Board, AutoSTABilisation

ASTT Action Speed Tactical Trainer

ASW, ASWT Anti-Submarine Warfare, Anti-Submarine Warfare Trainer

ATA Aerospace Technologies of Australia

ATACDIG Army TACtical Digital Image Generator. CGI system produced in the mid-1980s by CAE-Link, Binghamton, NY, USA. Fitted, amongst others, to the Apache Combat Mission Simulators. ATACDIG is superseded by more modern systems and is out of production

ATACO Air TActical Control Operator

ATC Air Traffic Control

ATCC Air Traffic Control Centre

ATCO Air Traffic Control Officer

ATCRU Air Traffic Control Radar Unit

ATE Automatic Test Equipment

ATEMS Advanced Threat EMitter Simulator

ATG Automatic Test Guide

ATGW Anti-Tank Guided Weapon. Now normally referred to as an ATM since current terminology uses the term 'Missile' for 'Guided Weapon', particularly in the USA

ATM Anti-Tank Missile

ATS Agile Target System, or Air Traffic System, or Aircrew Training System

ATU Aircraft Target Unit

AVDS Armoured Vehicle Driving Simulators

AWACS Airborne WArning and Control System, for instance the Boeing Sentry E3 aircraft, and equivalents

AWARDS Advanced Wide-Angle Reflective Display System, a proprietary collimated display system made by the ex-Link-Miles company in the UK, since taken over by the Thomson Training & Simulation (TTS) Division.

AWESS Automatic Weapons Effects Signature Simulator

BAI Battlefield Air Interdiction. Not Close Air Support (CAS), not long-range interdiction, but interdiction of supply routes and other forms of communications and support routes and assets which are immediately concerned with the local battlefield.

BATS Battle Area Tactical Simulation

BBGT Brigade and Battle Group Trainer

BBN Bolt, Baraneck and Newman. US Company which used to make the Delta Graphics product line of CGI systems used in the US Army SIMNET programme, amongst others. The BBN simulation interests were sold to Loral in 1993

BBS Brigade and Battalion Simulation

BCD Biological and Chemical Defence

BCS Battery Computer System

BCT Battlefield Command Trainer

BIRIS Battlefield Inoculation Remote Initiation System

BIT, Bit BIT = Built-In Test, Bit is an abbreviation of BInary digiT, ie one unit of magnetic memory used in computing, the small spot of material which is in one state or the other (ie in magnetic systems, either magnetised or not).

BITE Built-In Test Equipment

BMAC Boeing Military Aircraft Company (Huntsville, Alabama)

BMD Ballistic Missile Defence, such as by employing ABMs

BMTR Basic Marksmanship Training Range

Bn Billion, one thousand million. Ten to the power nine. eg $Bn

BPC Back-Projected, Collimated. A type of visual display system whereby the users in a simulator crew compartment (such as a flight deck) view an outside-world display by means of a large horizontal curved mirror mounted ahead of the crew compartment which reflects images of the outside world which are back-projected on a screen above (and invisible to) the crew. See the diagrams at the beginning of the section on Visual Display Systems. The mirror surface can be of glass, or flexible plastic (typically Mylar) with a reflective coating. The advantage of the system is that any occupant of the crew compartment in the simulator has a continuous outside-world view (limited only by the FoV of the system), collimated to an infinity or a distant focus in order to match the real world and to avoid anomalies with differing eye points within the crew compartment which would occur, for instance, with direct projection (ie non-collimated) display systems. BPC systems marketed include the original Rediffusion (now TTSL) WIDE series, TTSL (ex-Link-Miles) AWARDS, FSI MultiView, CAE (Montreal) MAXVUE, and SEOS PANORAMA.

BTT Basic Training Target

BUC Block Upgrade Configuration

Buffer RAM devices used for temporary storage of information such as before transfer to and from disk, to a printer, and so on. See Z Buffer.

BVR Beyond Visual Range. Applied to targets, particularly with respect to Air Defence Fighter Aircraft. In gunnery the equivalent term is Indirect Fire.

Byte A group of eight memory Bits (qv) which can be used to designate one character. This provides two to the power eight (256) combinations which can be used to specify letters, numbers, symbols and so on, in coding systems such as ASCII, ANSI, and so forth.

C2 Command and Control

C3 Command, Control & Communications

C3I Command, Control, Communications and Intelligence

C4 Command, Control, Communications & Computers

CAA Civil Aviation Authority. The name of the Civil Aviation Regulatory Authority in a number of countries including the UK

CAAC Civil Aviation Administration of China

CAAIS Computer-Aided Action Information System

CAF Canadian Armed Forces. See also CF and CFB

CAI Computer-Aided Instruction, similar to CBT

Calligraphic/Cursive Calligraphy and Cursive are words describing writing. The terms are used for a type of scan pattern, generally used on CRTs, where the electron beam traces an individually programmed shape rather than, for instance, a raster scan pattern which has a fixed form of horizontal lines. Calligraphic scan is often used in displays where bright features are required such as alphanumerics and symbols in HUDs for use in bright sunlight, or sharp light-points in simulations of runway night lighting. Calligraphy can obtain brighter results from the same phosphor when compared to raster scanning since the electron beam can be held on a point, or repeated several times over a symbol where high brightness is a requirement. In simulations, particularly of night scenes, display systems often interlace raster and calligraphic scans, the latter being used for the light-points in the scene and the former for the general background.

CAS Close Air Support, or Chief of the Air Staff

CAST Combined Arms Staff Trainer

CATIES Combined Arms Training Integrated Evaluation System

CATM Chemical Agent Training Mixture

CATS Combined Arms Training System/Simulator, or Comator Advanced Training System

CATTS Combined Arms Tactical Training Simulator

CBD Chemical and Biological Defence

CBI Computer-Based Instruction, same as CBT

CBT Computer-Based Trainer/Training, same as CBI

CC Close Combat

CCD Charge Coupled Device

CCR Computer-Controlled Range

CCT Combat Crew Trainer

CCV Computer (or Control) Configured Vehicle. See also FBW (Fly-By-Wire)

CEOT Communication Equipment Operator Trainer

CEP Circular Error Probable, normally to a 50% probability level unless stated otherwise (50% CEP), a radius within which 50% of rounds, shells, bombs or other weapons should fall if sufficient of them are fired or dropped in the defined conditions for which the CEP is quoted to be statistically significant. For CEP figures to be meaningful, the weapon impact pattern should be essentially circular. If not, other statistical terms should be used such as, for elliptical impact distributions, 50% RE (Range Error Probable) and 50% DE (Deflection Error Probable). The RE and DE figures essentially constitute the major axes of an ellipse in which 50% of impacts should occur if enough weapons are fired or dropped.

CEPT Cockpit Emergency Procedures Trainer, a Cockpit Procedures Trainer (CPT) configured for the practice of emergency and failure procedures including actions and switching.

CET Combat Engineering Tractor

CEV Combat Engineer Vehicle

CF Canadian Forces; CFB = CF Base (eg CFB Trenton).

CGCSS Commander and Gunner Crew Station Simulator

CGEN Character GENerator

CGI Computer-Generated Image/Imagery. Application of computer graphics techniques to generate images, generally for transmission of real-time imagery to a related but separate display system. In training applications the imagery is normally of real-world scenes which are sometimes generic but more often geospecific. The display system can vary in complexity from a single TV monitor to a large-FoV collimated display system or a projection dome. Large FoV display systems need several CGI channels each generating imagery of adjacent scenes from a common database.

CGS Crew Gunnery Simulator

CHAIR Control Handling Aid for Increased Range

Channel, Visual Channel When used with respect to CGI systems, implies one computing entity, often part of several, which processes image data for that part of the simulation presented visually through a display system. Such parts may be one of several elements of the outside visual scene, the radar, IR or NVG parts of the E-M Spectrum, or computer-generated instrument displays or controls (eg by touch-screen). When applied to the outside-world scene, the greater the number of IG channels, the larger the potential outside-world FoV provided by the display system. Typically three channels employed for an outside-world view and disposed horizontally to an even resolution could be displayed over a total horizontal FoV of between 120° to 150°, and five channels would give between 200° and 220° (depending on resolution). An IG channel may also be used for trainee station displays not concerned with the outside world such as HUD, HDD, EFIS and so forth.

CIC Combat Information Centre

CIC-PT Combat Information Centre Procedure Trainer

CIT California Institute of Technology

CIWS Close-In Weapon System

CLASS Closed-Loop Artillery Simulation System

CLOS Command to Line-of-Sight. Guidance principle used in guided missile systems where the missile follows the direct sight-line to the target without any predictions of target movement such as lead angles or the use of proportional navigation equations. The missile can be manually guided to the Line-of-Sight by direct operator commands (MCLOS) where the operator tracks both missile and target and guides one to the other. Guidance can be fully automatic (ACLOS) such as from radar returns which track both missile and target and generate guidance signals independent of the operator, or semi-automatically guided (SACLOS) where the operator tracks the target and signals are automatically generated to bring the missile on to the operator's tracking sight-line.

CMOS Complementary Metal Oxide Semi-conductor. Used, amongst other applications, in computer systems, particularly for ROMs, PROMs and EPROMs.

CMS Combat Mission Simulator. Synonymous with FMS (Full Mission Simulator). An example is the US Army Apache attack helicopter CMS at Fort Rucker and elsewhere.

COBRA Combat Outcome Based on Rules for Attrition (attrition calculation methodology).

CODLAG COmbined Diesel, eLectric And Gas

COFT Conduct Of Fire Trainer

COIN COunter-INsurgency

Collimation Derived from 'Co-linear' and implying parallel lines. In optics the 'parallel lines' constitute an infinity focus, and the term is used in astronomy to describe the focusing and adjustment of telescopes. In collimated visual display systems used in simulation, a focal distance of about 100 m is typical.

COMBAT-SIM COMputerised BATtle SIMulation

COTS Commercial Off-The-Shelf

COTT Control Of Transmission Test

CP Command Post

CPOP Command Post/Observation Post

cps/CPS characters per second, or Cardinal Points Specification

CPT Cockpit Procedures Trainer, also, for practice of emergency drills. See CEPT

CPU Central Processing Unit, eg of a computer system

CPX Control Post Exercise

CQB Close Quarter Battle

CQBR Close Quarter Battle Range

CQE Close Quarter Engagements

CREST Comprehensive Radar Effects Simulator Trainer

CRISP Computer Reconstructed Images from Scene Photographs

CRM Crew Resource Management (training subject used, amongst others, in multi-crew aircraft)

CRT Cathode-Ray Tube

CSG Contact Signature Generator

CSI Computer Synthesised Image/Imagery. A term describing a technique where a real-world background image is synthesised by scanning and then digitising photographic information, compared to conventional CGI where the image is created completely from computerised information. The CSI image is often overlaid with targets which can either be projected on the scene by laser or other target projectors, or targets created by normal CGI techniques can be integrated into the scene from the start.

CSRDF Crew Station Research and Development Facility
CST Crew Station Trainer
CSTS Combat Simulation Test System
CTRPS Cine Target Range Projection System
CTS Communications Training System
CTT Combat Tactics Trainer, or Command Tactical Trainer
Cursive Scan See Calligraphic
CW Chemical Warfare, Continuous Wave

DA Design Authority
DARKAS DARstellungsgerät KAnonenabSchuss
DARPA Defense Advanced Research Projects Agency (US DoD agency)
Database/ Management In CGI terms, a database is the geometrical and other information which the Image Generator will process to produce images in real time. At a minimum this will include an organised structure of polygons which are defined by the positions of their corners, and other data regarding individual polygons such as colour, texture, shading, priority, and so forth.
Database management is the process by which the real-time system in the Image Generator brings new portions of the database from storage into active use as the eye point moves around the gaming area.
DBGF DataBase Generation Facilities
DBWS DataBase Work Station
DC Direct Current
DCA Defensive Counter-Air, the potential for Air-to-Air combat with a view to defending friendly airspace, compared with OCA (Offensive Counter-Air) which is attacks on an enemy's air assets such as airfields and air defence radars
Defense Mapping Agency See under DMA
DE Deflection Error Probable. The deflection (lateral) element of a weapon impact distribution, normally to a 50% probability level unless stated otherwise. See RE and CEP
DEC Digital Equipment Corporation (Computer manufacturer)
DET Detonate
DEP Design Eye Point, of Visual Display systems
DF or D/F Direct Fire, Direction Finding
DFAD Digital Feature Analysis Data. Part of the DLMS system (qv)
DFWES Direct Fire Weapons Effects System
DIS Distributed Interactive Simulation
DLMS Digital Land Mass Simulation. The NATO system for digitising terrain (DTED) and feature (DFAD) information for mapping, targeting, simulation and other purposes. Although NATO-based, it is a series of digitising protocols applicable and used worldwide. So-called 'DLMS Level 1' data is about the equivalent content of a 1:250 000 topographical map and 'Level 2' to data contained on a 1:50 000 map. Superseded by the US DoD MIL STD 1821 (ex-Project 2851) protocols but DLMS formats are still in common use (qv under Project).
DMA Defense Mapping Agency. Normally refers to the US DoD organisation which deals, amongst other things, with DLMS and Project 2851 mapping data. See also DLMS and Project 2851 (MIL STD 1821).
DME Distance Measuring Equipment
DML Decision and Modelling Language
DMT Deep Mobile Target
DND Department of National Defence
DoD/DOD Department of Defense (Generally refers to the USA but some other countries such as Australia use the title also)
DOF/DoF Degrees of Freedom. There are six Degrees of Freedom or possible types of movements, of a freely suspended body. These are the three rotations Pitch, Roll and Yaw, and the three linear movements which are normally referred to in the training and simulation area as Surge (longitudinal movement), Sway (lateral movement) and Heave (vertical movement). The terminology is used for the movement of simulator motion platforms, for movements of targets, and so forth. See the table at the beginning of the section on Motion Platforms.
DORSI Dornier Radar Scoring Indication System
DRA Defence Research Agency. Specifically, the DRA UK (qv in Directory of Suppliers)
DRES Defense Research Establishment Suffield (Canada)
DRIU Damage Repair Instruction Unit
DRLMS Digital Radar Land Mass System. DLMS (qv) applied to radar prediction for navigation and targeting.
Drone An unarmed Unmanned Air Vehicle (UAV)
DS Display System
DTCS Drone Tracking and Control System
DTED Digital Terrain Elevation Data. The terrain part of the NATO DLMS system (qv)
Dusk/Night CGI System A CGI system not capable of producing full-colour day imagery. Limited colours may be able to be produced in the dusk and night scenes but image appearance will always be limited to not greater simulated brightness than dusk conditions. A dusk/night system employs light points and a palette of dark colours, whereas a day system has a palette of bright colours.
DVL Data/Voice Logger
DVVD Dual Vision Viewing Device

EC European Community
ECAN Etablissement des Constructions et Armes Navales
ECCM Electronic Counter-CounterMeasures, also referred to as Electronic Protection Measures (EPM)
ECM Electronic CounterMeasures
Edges When used in the context of CGI systems, an edge consists of the border where two sufaces or polygons intersect. The capability of a CGI system to generate models of objects is generally specified in terms of polygons, faces, or surfaces (interchangeable terms), but in some systems is specified in terms of the number of 'edges' which can be generated. Dividing the edge count by three gives the equivalent in terms of triangular polygons which are generally used for the purpose of polygon count in specifications.
EET Emergency Evacuation Trainer
EEWT Elbit Electronic Warfare Trainer
EFIS Electronic Flight Instrument System, colloquially called a 'glass cockpit'.
ELINT ELectronic INTelligence
EM, em, E-M ElectroMagnetic. The EM spectrum from short waves such as UV through the visual region, IR, and the radar and radio spectrum from Millimetric Wave, VHF, UHF, Short Wave to Long Wave. See also under UV and IR.
EMATT Expendable Mobile ASW Training Target
EMP ElectroMagnetic Pulse. Particularly associated with a nuclear explosion but can be caused by specially powerful RF emitters
Emissivity The ability of a surface to emit radition; generally used in the sense of thermal emissivity which is due to the temperature and other thermal characteristics of the surface and therefore pertains to the EM spectrum in the region of 10 microns (the far-IR). Observed thermal emissivity by appropriate IR sensors will vary with ambient temperature, the temperature and internal heat history of the surface (specific heat etc), sun angle, humidity, visibility, and so forth. In simulations all these effects should be modelled.
Emulation The simulation of equipment by entirely artificial means, by physical and software modelling. As opposed to simulation by STimulation of real equipment (such as radar, sonics, instruments etc) by artificially generated signals. See also Simulation and STimulation.
EO, E-O Electro-Optical. Electrical devices sensitive to the visual and IR spectrums such as TV cameras, contrast seekers such as EOGB homing heads, intensifiers and 'night sights', heat seekers and Thermal Imagers/FLIR. See also EOGB.
EOD Explosive Ordnance Disposal
EOGB Electro-Optical Guided (or Glide) Bomb. First used by US forces in Viet Nam, the homing head of an EOGB consists of a TV camera which produces a picture in the cockpit of the carriage aircraft while the bomb is on the aircraft weapon pylon. The operator places 'cross-hairs' on the target and the bomb is released when within ballistic range (in the 'basket'). The TV picture in the aircraft is then lost, but the guidance fins steer the weapon to the selected point which is 'locked on' within the TV homing head using a TV contrast tracking circuit which 'memorises' the part of the picture where the operator placed the cross-hairs. The homing head and guidance systems can be added at the front end of a conventional bomb, or be built in to a specific weapon such as Maverick in its E-O guidance variant. A complementary homing system to the LGB.
EOLs Engine-Off Landing(s). Generally used with respect to helicopter training. A potentially hazardous exercise involving the actual closure of the engine or engines before a steep autorotative descent to a forced landing. The term 'Autorotation' (as opposed to EOL) in helicopter training may be used for a less hazardous exercise where the collective lever is lowered fully, causing an autorotative state, but the engine or engines are not closed down and can be used to recover if the exercise goes wrong. The terms are relevant to flight simulation since these exercises can be safely simulated in a device with the appropriate cues such as a display system with a large vertical FoV
EPM Electronic Protection Measures, same as ECCM
EPROM Erasable Programmable Read Only Memory
ESM Electronic Support Measures. ie passive measures of signal analysis and surveillance, as opposed to ECM which can include active transmissions.
ESPRIT Eye-Slaved Projected Raster InseT. A proprietary very wide angle display system produced by CAE-Link of Binghamton, NY, USA
EST Electronic Support and Training
ETR Electric Target Range
EW Electronic Warfare, Early Warning
EWO Electronic Warfare Officer
EWTS Electronic Warfare Training System
Exit Pupil / Viewing Area or Volume In display systems which create an infinity or distant image focus, the exit pupil is the volume within which a viewer will see a substantially undistorted image. Sometimes also called the viewing area or viewing volume which is a more descriptive term.
Eye Point The point in space from which the Image Generator calculates images with correct perspective. Many CGI systems are capable of simultaneously making calculations for several eye points such as for different crew members and for sensor positions. For debrief purposes many eye points may be presented such as from enemy crew stations, threat sites, and the so-called 'God's eye-view' which is generally a plan view of the battlefield or engagement taken from height.

FAA Federal Aviation Administration. The civil aviation regulatory body of the USA
FAASV Field Artillery Ammunition Support Vehicle
Face, Faces When used in the context of specifications of CGI systems, face means the same as Polygon (qv) or Surface. Generally the term polygon is preferred because it is more descriptive and less ambiguous.
Fade Level of Detail See under LoD
FAST Fully Automatic Scoring Target
FATS FireArms Training System
FBW Fly-By-Wire; the words imply only an electrically signalled control system. However, it is generally used in the sense of computer-configured controls, where between the operator and the final control actuator or surface is interposed a computer system, frequently of full control authority, which modifies the operator's inputs in accordance with software programs carefully developed and validated in order to produce maximum operational effect without compromising safety. In aircraft, such systems are frequently quadruplexed (4 independent channels) in order to prevent disaster in the case of failure of one or two channels. Aircraft with FBW controls (also called CCVs or

Control-Configured Vehicles) may be deliberately designed to have low or even negative aerodynamic stability in some flight regimes, the rapid-reacting CCV controls compensating for the lack of natural stability. This also allows more manoeuvrability in combat aircraft since stability surfaces such as large horizontal and vertical stabilisers add weight and detract from manoeuvrability and performance.

Simulation of FBW Vehicles. Control responses for FBW/CCV platforms may be easier to simulate accurately since the control laws are man-made and precisely known, and do not involve the many variables due to changing height and speed which affect non-computerised control systems.

FCC Flight Control Computer

FCR Fire-Control Radar; the active radar in a radar-guided missile or AAA attack. Frequently closely tracks and then locks on to a target before firing, and frequently a different radar from the search device which will have different electronic parameters.

FCTS Flight Controller Training System

FD Full Development (Equipment Project Phase before Production)

FDDI Fibre Data Distribution Interface

FDS Functionally Distributed Simulation

FFS Full Flight Simulator. A flight simulator with all systems for training to the highest regulatory standard (FAA Level C/D, UK CAA Level 4). In the civil field this must include a good visual system with at least a 150° × 30° FoV, and a 6-DoF motion platform of high performance and low latency; civil devices without any motion platforms are not categorised as simulators but as Flight Training Devices (FTDs). See the introductory pages to the Aircraft Simulator section.

FIBUA Fighting In Built-Up Areas

Field / Field of View (FoV) / Field of Regard Field. When used in relation to display systems, a field is a single scan sequence across the whole screen area but in interlaced systems not to the full number of lines, for which a second field scan of the remaining lines is required.

Frames and Half-Rasters. In 2:1 interlaced systems, two fields (the odd and even rasters) are called a frame.

CGI Calculation Times and Data Updata Rates. Some CGI systems updata data for each field and are said to run at field rate; in this case the field time is also the CGI cycle time as well as the half-raster (odd or even) display time. Other CGI systems will output the same image for both odd and even raster fields while calculating the next image, and are said to run at frame rate. Some CGI systems draw the same image more than once while the updated data for the next image is being calculated. In terms of the performance of a CGI system, data Update Rate is the critical characteristic.

Refresh Rates. The term refresh rate can be applied to either the field or frame rates. In a 2:1 interlaced system the field refresh rate would be twice the frame refresh rate.

Field-of-View (FoV). The area of image produced by a display system, normally expressed as horizontal and vertical arcs with respect to the Design Eye Point (DEP) of the system. Where used in the sense of the visual field observed at any one time (ie with a fixed head and eye position) the more precise term Instantaneous FoV (IFoV) is used; for precise measurement it is assumed that the eye point is in the centre of the viewing volume (exit pupil) of the display system concerned.

Field-of-Regard (FoR). The total angular field of cover of a visual display system allowing for unlimited head and eye movement. In fixed-area display systems the FoR will be the same as FoV, but in head and head-and-eye slaved systems the FoR will be large and the FoV of the image presented at any one time will be much smaller. In slaved systems, the image FoV will follow the subject's head and/or eye movements; calculation and display of images in areas where the subject is not looking is therefore not required.

FII Frasca International Inc

FIST Facility for Infantry Situation Training

FL Flexible

Flicker In display systems, a distracting variation in brightness generally caused by the refresh rate of the display system being too slow for the desired brightness level. Night and dusk scenes can refresh at 30 Hz with little or no flicker whereas a bright day scene may require 50 or 60 Hz if flicker is to be avoided. Some people are more sensitive to this effect than others. Flicker is aggravated by the size of bright objects in the scene and also the size of the FoV; a system may be able to display a bright object if it is small enough (eg a light point), whereas a large bright object or the whole screen presented to that brightness, would be seen to flicker.

FLIR Forward Looking Infra-Red. Implies Thermal Imaging (TI) sensors operating at a wavelength of about 10 microns. See under IR.

FLOLS Fresnel Lens Optical Landing System

FLOPS, kFLOPS, MFLOPS, GFLOPS FLoating point OPerations per Second, k = thousand, M = Million, G = 1000 million.

FLS Force Level Simulation

FM Frequency Modulation, or Full Mission

FMS Full Mission Simulator, or Fire Marking System, or Foreign Military Support, or Flight Management System (The latter is an extension of Autopilot/Autothrottle system on aircraft with programming.)

FN Fabrique Nationale. Belgian armaments supplier

FNNH Fabrique Nationale Nouvelle Herstal. Belgian armaments supplier

FOC Full Operational Capability, see also IOC and ISD

FOD Foreign Object Damage (eg by objects ingested into engines)

FOGM Fibre Optic-Guided Missile

FOHMD Fibre Optic Helmet-Mounted Display. A proprietary very wide angle display system for simulation purposes, manufactured by CAE Electronics, Montreal, Canada

FOO Forward Observation Officers, or (financial) Forecast of Outturn

FoR Field-of-Regard. The total angular field of cover of a visual display system allowing for unlimited head and eye movement.

FoV Field-of-View. The area of image produced by a display system, normally expressed as horizontal and vertical arcs with respect to the Design Eye Point (DEP) of the system. Where used in the sense of the visual field observed at any one time (ie with a fixed head and eye position) the more precise term Instantaneous FoV (IFoV) is used.

Frame, Frame Rate See under Field.

FRCs Flight Reference Cards. Cards carrying lists of aircraft checks and actions. The cards are carried and used in flight, and in simulators for military aircraft, and are normally bound together so that they cannot be easily lost in the cockpit. In fighter aircraft the FRCs used to be designed to be carried in a flying suit pocket but in 'glass cockpit' aircraft the checks are often called up and read on one of the 'glass' screens. Cockpit Procedures Trainers (CPTs, CEPTs) are often used for basic training in FRC checks and drills

FS Feasibility Study

FSAT Full-Scale Aerial Target

FSD Full-Scale Development (same as FD)

FSI Flight Safety International

FSMTC Full-Size Moving Tank Target Carrier

FSU Former Soviet Union

FSUP Flight Simulator Update Program

FTA Frangible Training Ammunition

FTD Flight Training Device. A trainer not classed as a full simulator. See also under FSS, and the tables at the start of the Aircraft Simulators section

FWT FleetWork Trainer

FY Fiscal/Financial Year

***g*, G, G-Suit** *'g'* The small letter g is reserved for the relatively constant earth's gravitational acceleration of about 9.81 m/sec/sec or 32.2 ft/sec/sec.

'G' The large letter G is used to denote accelerations along the Normal or Z axis (the negative Zb axis) of a vehicle or platform, normally an aircraft or a motion platform. The term G is frequently used as an abbreviation for Gz, and units used are multiples of *g* (32.2 ft/sec/sec). The convention used in this book is that generally used in aviation where one G describes the situation at the earth's surface under conditions of no acceleration (other than gravity) along the Z axis, whereas, for instance, 2 G is the situation in an aircraft in a steady 60° banked turn and 4 G is generally necessary for a loop. Modern agile fighter aircraft can be flown in combat at G values up to 9, with the attendant risk of G-LOC (see below).

G-Suit. An abbreviation for the more correct term Anti-G Suit (qv).

GA Ground Attack, or General Aviation

Gaming Area The total geographical area over which a simulation is possible. This may be the same area as the database, but is often larger if database development has not yet extended to fill the total possible gaming area. Database density may vary over the gaming area, with particular areas for training being modelled to high resolution and other areas to low resolution or even simply relying on a generic database between training areas. An example is a commercial aircraft simulator where the gaming area will normally be specified to include the whole world but only key airports and their surrounding areas will be modelled in the database; this allows other airports to be added later as the operator's requirements change or as other customers use the simulator.

GAR Generic Airborne Radar

GB GigaByte, one thousand million bytes (1000 MB)

GCA Ground-Controlled Approach, eg of aircraft, a 'talk-down' by an Air Traffic Controller; see also PAR

GCI Ground-Controlled Intercept

giga 10 to the power 9, ie 1000 million. eg GHz (1 GHz = 1000 MHz), Gbyte

G-LOC G-induced Loss Of Consciousness. A highly dangerous situation in agile fighter aircraft where the pilot loses consciousness ('blacks out') as the blood drains away from the brain under high G loadings. Although the pilot will recover blood supply to the brain when G forces are relaxed, in the intervening interval (10-20 sec or so) he may have been shot down or flown into the ground. Recovery from G-LOC is often followed by a period of disorientation and unreality and this feature is also dangerous. Anti-G suits (commonly abbreviated to G-Suits, qv) can reduce, but not eliminate these effects.

GLONASS GLobal Orbital NAvigation Satellite System, the Russian GNSS system similar to the US GPS.

GmbH Gesellschaft mit beschränkter Haftung (German designation for a Limited Liability Company, the equivalent of 'Ltd' or 'Inc' after the company name).

GMOTS Gun Maintenance and Operational Training System

GMRT Ground-Mapping Radar Training

GNSS Global Navigation Satellite System (Generic term for all systems such as the Russian GLONASS and the US GPS)

GOLD General On-Line Diagnostic

GORU Gun Order Responder Unit

GP General Purpose

GPMG General Purpose Machine Gun

GPS Global Positioning System, the GNSS system controlled by the US DoD and Dept of Transportation but available for receivers worldwide

GPTB GunPell Target Board

GPTE General Purpose Test Equipment

GPWS Ground Proximity Warning System

GRDS Generic Radar Display System

GRP Glass-Reinforced Plastic (Fibreglass).

GTTC Ground Tactical Training Centre

GUI Graphical User Interface. Computer interface displays based on graphics as opposed to AlphaNumerics; Microsoft 'Windows'™ and the like.

GW Guided Weapon. Modern terminology now uses the term 'Missile' instead of 'Guided Weapon', particularly in the USA

GWST Gabriel Weapon System Trainer

HARM High-Speed Anti-Radiation Missile

HAST High Altitude Supersonic Target, or Harrier Avionics Systems Trainer

HAT Height Above Terrain. A capability in some CGI systems to calculate height of a test point (generally ownship position) above the polygon directly beneath it.

HCI Human/Computer Interface, ie MMI in the computer field

HCT Helicopter Control Trainer, or HOT Compact Turret

HDD Head-Down Display, normally refers to aircraft cockpit displays giving instrument or sensor information, which can only be viewed from sight-lines inside the cockpit and below the outside-window display of the outside world. The opposite is HUD, qv.

HDTM Heavy-Duty Tank Target Mechanism

HDTV High Definition TeleVision. New-generation commercial TV with about 1200 raster scan lines at between 50 and 60 Hz (about 1440 kPixels per screen refresh or up to 84 million processed pixels per second). Potentially of great benefit to simulation display systems which will be able to use the technology developed elsewhere at little extra cost except to ensure that feeder systems such as CGI have the capacity to produce data-rates to match the new display systems.

HE High Explosive

HEAT Hostile Expendable Aerial Target, or High Explosive Anti-Tank. The latter is a type of anti-armour round with a shaped HE charge, which, on detonation, uses the Monroe effect to focus the explosive force and to generate an intensely hot liquid metal slug which proceeds at high velocity and is able to penetrate armour.

HELIOS HELicopter Instrument and Operational procedures Simulator

HEP High Explosive Plastic

HESH High-Explosive Squash-Head. A type of anti-armour round which is designed to explode in contact with (the 'squash' in the title) armour such as the outer shell of an AFV. Transfer of energy across the sheet of armour then causes metal to detach on the inner side ('spalling') which is the damage mechanism inside the target.

HETA Harpoon Engagement Training Aid

HF High Frequency, or Human Factors

HFI, HFIP Human Factors Integration, Human Factors Integration Programme (as MANPRINT)

HHC Hand-Held Controller

HHCU Hand-Held Control Unit

HHLL Hand-Held Laser Locator

HIFPT Helicopter Instrument-Flying Procedures Trainer

HIP Howitzer Improvement Program

HITS High-definition Indoor Trainer for Small-arms

HITTS Harpoon Interactive Tactical Training System

HLL High-Level Language (Used in computer systems, eg Ada, FORTRAN, COBOL, C, and so forth.)

HMD Helmet-Mounted Display. The display signals are often taken to the helmet-mounted projector through a fibre optic bundle and displayed on a semi-transparent reflective surface in front of an operator's eye or eyes. Images can vary in complexity from simple aiming graticules (cross-hairs) to FLIR and other imagery. Used not only for simulation purposes but for the operational vehicle. An example of a specialised HMD for simulation only is CAE's FOHMD (qv). So-called 'Virtual Reality' displays are also included.

HMS Helmet-Mounted Sight

HOSAT Hands On Stick And Throttle, see HOTAS

Host, Host Computer The computer which drives the core simulator or simulation, and to which other systems (and, frequently, their own computers) are attached. A typical attached system would be a CGI system and associated displays. The interface with the Host is frequently a two-way Ethernet interface or similar.

HOT Haut subsonique Optiquement Téléguide (ATM system); also Harpoon Onboard Trainer.

HOTAS Hands On Throttle And Stick. Military aircraft concept where vital ancillary switching is immediately available to the pilot by being positioned on either the throttle or stick (control column) instead of elsewhere in the cockpit. Sometimes, in the USA, HOSAT (Hands On Stick and Throttle) is used

hp horse power

HTI Hughes Training Incorporated

HUD Head-Up Display. Normally refers to aircraft, where a collimated display is projected in front of a pilot above the cockpit instruments where the instrument display and real outside world can be observed simultaneously.

HUTTS Hayes Universal Tow Target System

Hz Hertz, the scientific unit for cycles per second

IAS Indicated Air Speed, the speed indicated on a pilot's Air Speed Indicator (ASI). The ASI is essentially a forward-facing pressure gauge, calibrated to the basic formula $0.5\ \rho\ V^2$ in units of speed and matching True Air Speed (TAS) only at Sea Level. At approximately 40 000 ft the air density ρ has dropped to about 1/4 that at sea level and so as root 1/4 is 1/2, the IAS will be half the TAS at this height. Since aerodynamic drag is approximately proportional to IAS, this explains why high-flying jet aircraft are able to cruise at high altitude both economically (relatively low IAS) and at high speed (high TAS).

iaw in accordance with

IC Integrated Circuits, in computers and related systems

ICV Infantry Combat Vehicle

IDF Israeli Defence Forces

IED Improvised Explosive Devices

IF/IFR Instrument Flying; Instrument Flying Rules. Flight without reference to external visual cues, sometimes colloquially known as 'blind flying' or 'cloud flying'. Body motion cues from the inner ear (vestibular apparatus) and muscle sensors become of greater importance than in visual flight. For aircraft, IFR are defined precisely in air law and regulatory rules. See also IMC and VMC.

IFCS Improved Fire-Control System

IFF Identification Friend or Foe. Electronic recognition system

IFoV Instantaneous Field-of-View. The visual field observed at any one time (ie with a fixed head and eye position). The human eye has an IFoV of an elliptical shape with major axes of about 160° horizontally and 140° vertically (±80° × ±70°), with high resolution acuity only in the centre field, resolution falling off rapidly away from the centre, the peripheral field being used to detect movement rather than the details of objects. Some visual display systems have more IFoV than this and some, less, with the appropriate consequences in terms of training and realism (depending on what aspect is being trained).

IFPT Instrument Flying Procedures Trainer

IFS Indirect Fire Simulator, or Instrument Flight Simulator, or Institute of Flight Safety

IFT Instrument Flying Trainer, Indirect Fire Trainer, or Instructor Flown Target

IFV Infantry Fighting Vehicle

IFWES Indirect Fire Weapon Effect Simulation

IG Image Generator/Generation. A system which generates visual images with a view to providing such data to a display system. Includes video disk systems based on photographs, model boards based on scale models and small TV cameras, and computer techniques such as CGI (qv). CGI is the most popular system.

IGE Instrumentation Graphics Environment

IKAT Interactive Keyboard And Terminal

ILS Integrated Logistic Support, Instrument Landing System (VHF-band aircraft landing aid)

IM Intra-Mural, or Insensitive Munition

Image Generator See under IG

IMC Instrument Meteorological Conditions (under which flight to IFR procedures is required). See also IFR, VFR and VMC.

IMTS Interactive Modular Training System

INS Inertial Navigation System. Navigational system using a set of (normally three) accurately calibrated mechanical or laser gyros

Interlaced Scan The characteristic of many display devices to draw half of the raster lines during one field scan, followed by the other half interleaved between the first set of lines on the following field scan. The interlaced fields are referred to as the odd and even fields. The human eye tends to merge these fields together into a single image. Until recently, most raster displays were of this type, including home TV sets. The trend of recent display systems is to move from interlaced scan to a complete screen of raster lines for each field scan.

IG aspects. Image Generators have to be capable of matching this scan pattern, but it should be borne in mind that such line-by-line interlacing needs two scans to produce the full screen update of computed information. In this case, the screen refresh rate will be twice the data update rate; typically a 50 or 60 Hz screen refresh rate and a 25 or 30 Hz data update rate. Also, raster/calligraphic systems nomally carry out a raster scan followed by a calligraphic phase.

I/O, I-O Input-Output system (generally referring to computers)

IOC Initial Operational Capability (date). A NATO term meaning the date at which an IOC is achieved or is forecast to be achieved by a new equipment being developed. See also ISD, which is the date of first service delivery rather than the development of a genuine operational capability which takes some time to work up. In the case of complex new equipment, a conversion unit may have to be set up and only after courses, personnel or units have passed through such a unit into the front line, and have worked up in their front-line role, will an IOC for the equipment genuinely exist and be formally declared to higher headquarters. Training aids and simulators should be in service (RFT, Ready for Training) when the first service unit is formed at or about ISD but some time before an IOC is declared.

IOS Instructor Operating/Operator Station. The control and instructor station of a trainer or simulator during a training exercise. Sometimes preset exercises are run from the IOS using a simulator operator rather than a qualified instructor.

IP Intellectual Property, or Initial Provisioning, or Initial Production

IPMS Integrated Platform Management System

IPR Intellectual Property Rights

IPTS International Pistol Target System

IR Infra-Red. That part of the electromagnetic spectrum with wavelengths between about 700 nm (the near IR) up to about 10,000 nm (10 microns or 10æ), the far IR. Heat-seeking missiles have sensors which generally operate in the band from 2 to 5 microns, depending on the active substance used in their homing head. Intensifiers are sensitive to reflections of ambient radiation from objects. This radiation occurs naturally together with light (from sun, moon, artificial lights) and is in the near IR between 700 nm and 1 micron. Thermal imagers (such as FLIR equipment and others) use the far IR at about 10 microns, and produce a picture of thermal emittance (temperature contrast) across a scene. Thermal imaging equipment is therefore not only passive, but is independent of reflected ambient light since it depends on 10 micron emissions from the sources themselves. Also, Thermal Imagers can be used to penetrate small amounts of smoke and cloud; this is due to the long wavelength used which has good transmittance characteristics in the presence of particles in the atmosphere smaller than the wavelength itself.

IRIS Integrated Raster Imaging System. A proprietary term for a range of graphics work stations by Silicon Graphics, Inc.

IRT Instrument Rating Test. A test conducted regularly (normally annually) on all commercial and military pilots, in order to ensure proficiency in Instrument Flying and IFR procedures.

IS Instructor Station. See also IOS.

ISD In-Service Date (for an equipment under development; see also IOC which is a later date), or Instructional Systems Development

ISO International Standards Organisation

IT Information Technology, or Intermediate Trainer

ITCS Integrated Target Control System

ITS Integrated Target System, or Interactive Training System

ITT Invitation to Tender (as RFQ)

IT^3 Interactive Tactical Team Trainer

IVD Interactive Video Disk

IVDU Intelligent Visual Display Unit

IVR In-Visual Range (opposite of BVR)

IW Individual Weapon
IWESS Infantry Weapons Effects Simulation System

JATO Jet-Assisted Take Off
JESS Joint Exercise Support System
JTIDS Joint Tactical Information Distribution System (US/NATO secure tactical information system, designed to aid control of the sea, land and air battle)

k kilo, one thousand. eg kb for kilobytes, $k for thousands of dollars, etc.
KD Known Distances
kN kiloNewtons. Unit of force, 1Kn = 224.8045lb force

LAH Light Attack Helicopter. If referring to the US Army programme, this has been retitled LH.
LAMPS Light Airborne Multi-Purpose System. Specifically, a USN Helicopter system
LAN Local Area Network. Of linked or networked computers
LANTIRN Low Altitude Navigation and Targeting Infra-Red system for Night (US DoD Programme and associated night fighting equipment)
LAR Light Artillery Rocket
Latency Dictionary definition: the time between stimulus and reaction. In simulation applications, the time delay, normally measured in milliseconds (1/1000 sec), between initial input and an output clearly discernible to the simulator trainee. Sometimes also called transport delay. Some authorities distinguish between latency and transport delay by using latency in the sense of the extra time delay of a system over and above the reaction time of the vehicle being simulated, but this requires a detailed knowledge of the vehicle dynamics and can be controversial. In this book the terms latency and transport delay are taken as identical, in accordance with the normal dictionary definition.

Importance of Motion and Visual Latencies. In simulators with both visual and motion systems, it is particularly important that the latency of the motion system is not greater than of the visual system , or symptoms of simulator sickness may result. This is because in the real world, motion cues are those of acceleration and are quickly transmitted to the brain, typically in less than 50 milliseconds; this is followed some milliseconds later by a perception of change in the visual scene. The visual scene change is essentially one of change of perspective and/or displacement of objects such as the horizon, which takes some time to build up to discernible amounts after the initial acceleration which caused the displacement. A simulator should therefore reflect the real-world situation by ensuring that the motion latency is equal to or less than that of the visual system and not the other way round.
LAV Light Armoured Vehicle
LCAC Landing Craft Air Cushion
LCC Life-Cycle Costs, the same as Through-Life Costs (TLC). The total costs of an equipment or system from inception through production, operational service to disposal
LCD Liquid Crystal Display
LDTM Light-Duty Tank Target Mechanism
LED Light Emitting Diode
LEOS Loral Electro-Optical Systems
Level of Detail See LoD
LFM Live Firing Monitor
LGB Laser Guided Bomb. First used by the US forces in Viet Nam, a laser seeker and guidance module is added to the nose of a conventional bomb, and aerodynamic guidance vanes are used to home the weapon to a laser spot on the ground which is placed and kept there by either a ground-based laser designator, or by an aircraft equipped with laser marking equipment.
LH Light Helicopter. When used specifically to refer to the US Army programme, refers to the intended new light attack helicopter.
Light Point, Light String When used with reference to CGI systems and their associated display systems, the computing and display of an essentially point source of light, such as a flare on the battlefield or an approach or runway light in aircraft simulation. The appearence of a light point will depend on its brightness and hue (colour) as modified by distance, atmospheric visibility, and directionality, rather than by perspective and size as with polygons. Light points will therefore be described in the database in terms of geographical position, fundamental brightness, colour and (where relevant) directionality of light. Calligraphic (qv) scanning is often used in display systems in order to produce intensely bright, sharp, light points, although display systems using a high number of raster lines often dispense with calligraphy and paint light points as a cluster of bright pixels as part of a normal raster scan.

A Light String is a data format for storing a large number of light points rather than defining each light point separately. Light Strings can be straight, curved, or produce a given density of lights randomly distributed over an area.
LHN Long-Haul Network (Long-range LAN)
LIG Laser Image Generation
List Priority CGI system feature; see under Priority
LKL Lochkegelleitwerk
LLTV Low-Light TeleVision. Similar to intensifiers, normally operates in the near-IR band using reflected energy between 0.7 and 1 micron (700-1000 nm)
LMOTS Launcher Maintenance and Operational Training System
LOCAT LOw Cost Artillery Trainer
LoD/LoD Management/ Fade LoD Level of Detail. When the term is used with respect to CGI systems, it implies scene management tools which ensure that from the visual database, which will normally be equally detailed over a large area, when a scene is running on-line in real time, a high LoD is included in the foreground, gracefully degrading to a low LoD in the distance. This prevents, amongst other things, so-called feature 'popping' where a cultural feature such a building or other object suddenly appears in the scene as if from nowhere as the eye point moves towards it. The mechanism is to increase the number of polygons and light points used for a feature image as the range of the feature from the eye point is decreased. As a feature is approached, more polygons will be called up to establish a more complex shape. In the case of buildings, windows, doors and finally surface detail such as bricks will be added as the eye point moves closer. Some CGI systems use their transparency feature to fade in new Levels of Detail and to fade old ones out; this is a management system known as 'Fade Level of Detail'.
LOFT Line-Orientated Flight Training
LoL Limit of Liability
LOMAH Location Of Miss And Hit
LOPT Lynx helicopter Observer Procedure Trainer
LOS/LoS Line-of-Sight
LP Local Purchase
LP/C Launch Pod/Container
LPO Local Purchase Order
LRATGW/LRATM Long-Range Anti-Tank Guided Weapon/Missile. Current terminology uses the term 'Missile' for 'Guided Weapon', particularly in the USA
LRF Laser RangeFinder
LRI, LRU Line-Replaceable Item/Unit
LSA Logistic Support Analysis
LSD Logistic Support Date
LSOT Landing Signal Officer Trainer
LST Laser-Spot Tracker
LSW Light Support Weapon
LTC Long-Term Costing
LTID Laser Target Interface Device
LTS Laser Training System

M Million. 10 to the power 6. eg Mb for MegaByte, $M for 1 000 000 dollars, etc
MAD Magnetic Anomaly Detector/Detection
MADS MPS Air Defence Simulator
MAGLAD Marksman And Gunnery LAser Device
MANPRINT MANpower and PeRsonnel INTegration; Human Factors Integration (HFI) within equipment and weapon systems
MARS Modular Adaptable Radar Simulator
MART Mean Active Repair Time
MAST Multi-Application Sonar Trainer
MASTER Modular Acoustic STimulator/EmulatoR
MATS Military Aircraft Target System, or Marconi Acoustic Training System
MAU Marine Amphibious Unit
MBT Main Battle Tank
MCAS Machinery Control And Surveillance, Marine Corps Air Station
M-CATT Multi-weapon Combined Arms Tactical Trainer
MCC Management, Command and Control, or Manual Control and Counter, or Mission Control Centre
MCFS MPS Coastal Fortress Simulator
MCLOS Manual Command to Line-of-Sight. Missile guidance principle, see under CLOS
MCM Mine CounterMeasures, Multi-Chip Module
MCMV Mine CounterMeasures Vessel
MCS Machinery Control System, Manager Control Station, Mission Control Station
MCT Maritime Crew Trainer. Used to denote the complicated tactical trainers used for the ASW crew of MPA, to distinguish the trainer from the pilot flight simulator which is different. In sophisticated systems the pilot flight simulator will be able to be coupled to the MCT in order to run integrated crew training.
MCTFIST Marine Corps Tank Full crew Interactive Simulator Trainer
MDI Miss Distance Indicator
MDTM Medium-Duty Tank Target Mechanism
MDTS McDonnell Douglas Training Systems, formerly the American Airlines Training Corporation
MET Marksmanship Expert Trainer
MGB Medium Girder Bridge, or Modular Glide Bomb
MiCoL Microprocessor-based Control Loading
MILAN Missile d'Infanterie Léger ANtichar, proprietary ATM system
MILES Multiple Integrated Laser Engagement System
MIL STD 1821 see Project 2851
MIP-map MIP comes from the Latin *Multum In Parvo,* (much in little (space)). The MIP-map technique used in CGI systems involves combining many features into one screen application. An example is combining multi-level texture descriptions in order to suppress flicker in distant views and to minimise unrealistic 'mosaic' effects in close-ups.
mips/MIPS Millions of Instructions Per Second. Data rate, such as in computer capability
MIS Management Information System. Often software-based
MISS Missile Intercept Scoring System
MITS Mobile Independent Target System
MKT Maschinen Kanonen Trainer (gunnery trainer)
MLI Mid-Life Improvement (same as MLU)
MLRS Multiple Launch Rocket System
MLS Microwave Landing System, for aircraft
MLU Mid-Life Update (same as MLI)
MMI Man/Machine Interface
MMSS MPS Multibeam Sonar Simulator
MMW MilliMetric Wave
MoD Ministry of Defence (term used in countries such as India, Israel and the UK. The US equivalent is DoD).
MOD/Mod Modification
MOD DIG MODular Digital Image Generator. CGI system produced by CAE-Link, Binghamton, NY, USA. Only a few units made, now out of production
Model/Modelling The term Model can be met in a number of contexts. Since a database creator is called a modeller, 'model' is sometimes used interchangeably with 'database'. It is also used to describe 3D cultural features

within a database (each of which requires detailed modelling). In specifications for CGI systems the system capacity in terms of the maximum number of 'moving models' will often be given; these could be aircraft, missiles, ships, or vehicles of all sorts on the ground.

MOST Mission-Orientated Simulator Training

MOU Memorandum of Understanding, normally between nations. A level of agreement below the formality of a treaty between governments

MOUT Mobile Operations in Urban Terrain

MPA Maritime Patrol Aircraft

MPFF MultiPurpose Fighter Facility

MPI Mean Point of Impact. The average point of impact, typically of a group of rounds or shells. See also CEP.

MRAAM Medium Range Air-to-Air Missile. For example the US AIM-9 Sparrow series, and the AMRAAM (Advanced MRAAM) programme

MRAD Multiple Range Alignment Device, or MilliRadian, 1/1000 of a radian. 1 radian = 360/2 =57.3°, so a Milliradian or Mil is 57.3/1000 or approximately 6/100 of a degree, or 3.44 seconds of arc.

MRS Mobile Reception System

MRT Mission Rehearsal Training

MS Mission Simulator

MSR Mobile Sea Range

MSS Mission Simulator System, or Mobile Subscriber equipment System

MST Microprocessor Simulation Technology

MT Missile Target, or Maintenance Trainer, or Motor Transport

MTBF Mean Time Between Failure/Fault

MTI Moving Target Indicator. A radar system facility to differentiate moving targets, normally through an electronic speed gate which suppresses targets or radar returns moving at less than the gated speed. This also reduces screen clutter.

MTM Moving Target Mechanism

MTM(WD) Moving Target Mechanism (Winch Driven)

MTOW Maximum Take Off Weight

MTS Moving Target Simulator, or Maintenance Training System

MTT Moving Turning Target

MTTR Mean Time To Repair

MUDAS Modular Universal Data Acquisition System

MUTES MUltiple Threat Emitter System

MVLT Morse/Voice Language Trainer

MWLD Man Worn Laser Detector

NALLADS Norwegian Army Low Level Air Defence System

NAMFI NATO Missile Firing Installation

NAS Naval Air Station

NASA National Aeronautics and Space Administration (USA)

NATO North Atlantic Treaty Organisation

NATS National Air Traffic System/Service

NAVTAG NAVal TActical Game

NAWST Night Attack Weapon Systems Trainer

NBC Nuclear, Biological and Chemical

NDB Non-Directional Beacon

NES Night Effects Simulator

NEWTS Naval Electronic Warfare Training System

Ni/Cd Nickel/Cadmium (battery, as opposed to lead/acid)

NLR Nationaal Lucht-en Ruimtevaartlaboratorium (Research centre, the Netherlands)

nm nanometre, sometime also nautical mile, although the latter should be abbreviated nml. There are 1000 nanometres in a micron and 1 million microns in a metre. The human visual range is from about 400 nm (violet) to 700 nm (red), over 700 nm being the IR part of the spectrum and less than 400, the UV.

Nautical Mile. A nautical mile is defined as one minute of latitude (ie 60 × 90 between the equator and pole), and is for civil measurement purposes also defined in both US and English law as 1852 metres exactly; this is the so-called 'International Nautical Mile'.

NoE/NOE Nap-Of-the-Earth. Ultra low flying, normally by helicopters in the tactical, Anti-Tank or Special Forces roles. Implies maximum use of intimate ground contours in order to avoid detection, operating at heights down to 1 metre below skids or gear, flight below the height of trees, and under high cables rather than over them. One of the ultimate challenges for flight simulation.

NPA Notification of Proposal to Amend, eg 'NPA procedure' in regulatory documents, giving forewarning of intended changes.

NTB National Test-Bed

NTC National Training Center

NTCT Naval Tactics and Command Trainer

NTDS Naval Tactical Distribution System

NVG/NVGs Night Vision Goggles. Use E-O Intensifiers working in the near IR

OA Operational Analysis

OAS Offensive Air Support

OAT Outside Air Temperature

OBEWS OnBoard Electronic Warfare Simulator

Object In the context of CGI systems, an object is a defined division of database elements, normally grouped together for the purposes of scene management. A specific number of polygons and/or light points or strings may be grouped in this way. A simple building may be included as a single object, whereas a complex model such as an aircraft might be modelled in the form of several objects such as one for each wing, one for the fuselage, and so forth.

Obturator A device for stopping a hole, particularly in gunnery where it may be used to stop the escape of gas until the propulsion phase is complete.

OBVACT OnBoard Visual Aimer Continuation Trainer

OCA Offensive Counter-Air, eg attack on the air assets of an opposition such as his airfields.

OCU Operational Conversion Unit. A training unit preparing crews for their front-line role before they join their operational front-line units.

Occultation/ Occulting Dictionary definition: to hide by interposing, to become temporarily invisible. In astronomy when the moon occults the sun, an eclipse occurs. On the Earth's surface, as you descend into a valley, the surrounding hills occult distant objects which were previously in view, and at a local level, 3D objects in the forground occult objects and terrain in the background. In CGI systems occultation has to be achieved through software and specific calculations are needed for correct lines-of-sight with respect to terrain and objects on the terrain. Incorrect occultation gives false visual effects such as revealing features on the reverse slope of ground which would not be seen in the real world. Some CGI systems are limited in the numbers of layers or levels of occulting, and occulting performance is related to the processing power and priority mechanisms within a CGI system. Also called hidden-surface removal, it can be achieved by various priority and buffering systems. See items under Priority and Z Buffer.

OFP Operational Flight Programme

OFT Operational Flight Trainer. Frequently intended for general flight training as opposed to specialist tactics and weapons training, for which see OFTT and WST. Often has a motion platform in order to enable realistic Instrument Flying (IF) and other procedural training.

OFTT Operational Flight and Tactics Trainer. A device more capable than an OFT but not as capable in weapons training as a WST or an MS/FMS.

OP Observation Post/Point/Position, eg for Artillery Observation by an AOV, manual observation, or an instrumentation/tracking

OR Operational Requirement by a customer, ie the military requirement, the basis for a Request for Proposals (RFP) or Quotations (RFQ); sometimes OR stands for Operational Research, but because of possible confusion with the term Operational Requirement, the term OA is generally used for this subject.

OTS Off-The-Shelf

OTT Operator Tactics Trainer, or Operations Team Trainer

Overload, Overload management A condition when a CGI system is asked to process more than its limits allow for. Overload can be caused by attempting to create too many polygons, light points, pixels, transparencies, occulting calculations, too many dynamic models, and so forth, or any combination of these. Overload management is the use of a number of methods to reduce the effect of overload on a user. Generally this will automatically reduce the number of polygons and/or lights in a given scene, although some systems respond by reducing the data update rate.

Ownship The vehicle being simulated, the principal eye point for which an image is created.

PAAT Passive Acoustics Analysis Trainer

PADT Point Air Defence Trainer

PAK Projector Alignment Kit

PAR Precision Approach Radar, for aircraft approaching to land in conditions of poor visibility or low cloud-base. A precision form of the so-called Ground-Controlled Approach (GCA) 'talk-down' system

PC Personal Computer. Normally implies the IBM-compatible type with processors such as the 286, 386, 486 or Pentium.

PCB Printed Circuit Board

PCU Portable Control Unit

PD Project Definition. A phase in an equipment project after a Feasibility Study and before Full Development and Production

PDD Player Detection Device

PDT Practice Delivery Torpedo

PERT Progress Evaluation and Review Technique. Often summarised in the form of a chart or charts showing critical and other paths of various project stages and activities

PGCT Precision Gunnery Crew Trainer

PGS Platoon Gunnery Simulator

PHIGS Programmer's Hierarchical Interactive Graphics System

Photographic Texture Production of Texture maps using photographic rather than mathematically generated techniques. Used to produce realistic visual effects such as areas of woodland, grassland, cornfields, generic urban backgrounds, and the like. In more elaborate applications photographs of more complex scenes such as the side of a typical apartment block can be digitised and called up in a scene in the form of a texture mat rather than by using many polygons to make the same visual effect.

Pixel /Pixel Fill Rate (PFR) The smallest PIcture ELement on a display screen or computed in a GCI system. Equivalent to EITHER: a single light point produced by screen phosphors when illuminated by an electron beam (CRT displays); OR a single liquid crystal (LCDs). Used to be called a 'picture-point' which is a good descriptive term. The smallest unit for which colour and intensity can be defined (CGI systems) or displayed (Display Systems).

Resolution and processing power of CGI systems and Display Systems is often quoted in their ability to process pixels, eg 625 × 625 (392 kPixels), 1024 × 1024 (1049 kPixels), 1280 × 1024 (1311 kPixels), or, with HDTV systems, up to 2000 × 2000 (4 MPixels). Resolution of raster display systems is often quoted in terms of minutes of arc subtended at the operator's eye per pixel displayed on the screen (arcmin/pixel); see under Resolution. Pixel Fill Rate (PFR). The rate at which a CGI system can recompute pixel data for transmission to a display system.

PLA People's Liberation Army, the Naval, Army and Air Forces of the People's Republic of China, capital Beijing/Peking

PLAT Pilot Landing Aid Television

PLRS Position Locating Reporting System

PLS Projectile Location System

PM Plastic Manoeuvre

PM TRADE Program Manager for Training Devices. US DoD Organisation with headquarters in Orlando, Florida.

Polygon A two-dimensional shape within a CGI system database which can be coloured, shaded and textured, and whose position is defined by the co-ordinates of its vertices (corners). The CGI system calls up the polygon structure of a scene and renders it in correct perspective ready for transmission of the processed data to a display system. Although the polygonal surface itself is inherently a two-dimensional structure, it may be placed in a visual scene in a three-dimensional orientation. Since a polygon can have many sides and so need many points to define it, for CGI specification purposes it is generally taken as a triangle and so is processed as three points in the x, y, and z axes, needing 9 geometrical descriptors plus coding for colour, brightness, shading, texture, NVG and IR characteristics etc. Because of the number of ways of specifying a polygon, when analysing the characteristics of an indidivual CGI system the exact definition of polygon count as it applies to that system, should be obtained.

Popping Popping or feature popping in CGI systems, see Level of Detail (LoD) management.

POTS Production Off-The-Shelf

PPI Plan Position Indicator. Normally applied to a radar presentation in mapping orientation (ie a vertical view of terrain).

PPS Pulses Per Second (in Pulsed Radars)

PRF/prf Pulse Repetition Frequency, of radars and other pulsed devices such as lasers used for rangefinding or coding

Priority When used in the context of CGI systems, the process which determines the relative priority of features in a scene. In the real world, occulting of one object by another and the visibility of one feature against another, happens naturally and automatically; whereas in a CGI system this process has to be handled mathematically by appropriate algorithms. One of two methods is generally used; a list priority method or a Z buffer (qv) or related approach such as a range (R) buffer.

List Priority. An approach to scene priority in which the first polygon or light string in the list has the highest priority in order to appear in the visual scene, followed by other features in the list in order. This method needs a mechanism to contantly adjust the priority list in accordance with movements of the ownship and principal eye point for which the scene is being created.

Project 2851 (MIL STD 1821) A US DoD project concerned with standard formatting of geographical data; a more sophisticated successor to the NATO DLMS system (qv under DLMS). Project 2851 protocols set the standard for formats for terrain and feature data for the future and are increasingly employed in CGI systems and database systems as a data format before detailed processing converts the data into the proprietary high-resolution, textured output which is then transmitted to the display system. Now designated MIL STD 1821.

PROM Programmable Read-Only Memory. A type of storage from which a computer can take data but which cannot be altered without reprogramming.

PSMC Pre-Selected Manual Control

PSSSU Power Supply and System Selector Unit

PST Propulsion Systems Trainer

PT Plastic Training

PTA Pilotless Target Aircraft

PTT Part-Task Trainer. A trainer designed for only one part of the overall training task, not as capable as a full simulator. For air equipments, see FTD

PVC Polyvinyl chloride. Plastic material.

QA Quality Assurance. Can be an organisation ' QA branch/section' or an activity 'QA testing'

qv *quod vide* (Latin), literally, 'which see', and means please see (look up) this quotation/abbreviation etc (the phrase, definition or abbreviation referred to)

R Buffer See under Z Buffer

RAC Royal Armoured Corps (UK Army)

R&D Research and Development

RADSIM RADar SIMulator

RAF Royal Air Force, the UK Air Force

RALACS Radar Altimeter Low Altitude Control System

R&M Reliability and Maintainability. See also AR&M

RAM Random Access Memory, used in a computer to store material while it is being worked on, but is wiped clean when the machine is switched off.

RAN Royal Australian Navy

Range Buffer See under Z Buffer

Raster A scanning pattern where a line is drawn, generally on a phosphor-coated screen by an electron gun, follwed by a rapid flick-back to draw the next line down. The speed at which the beam(s) can be modulated to change the colour and intensity, determines the number of pixels on each raster line. Many displays use an interlaced (qv) technique where only every other line is drawn during the first field scan, a second scan being needed to complete all of the information.

RATO Rocket-Assisted Take Off

RCC Rescue Control Centre (for SAR)

RCMAT Radio-Controlled Miniature Aerial Target

RCMDS Remote-Control Mine Disposal System

RCS Range Control Station, Radar Cross Section (Equivalent radar reflective area in the frequency band specified)

RCTC Rear Crew Trainer Cabin

RE Range Error Probable. The range element of a weapon impact distribution, normally to a 50% probability level unless stated otherwise. See DE and CEP

Real Time When used in the context of CGI, indicates that the system cycles at a rapid and stable rate. Due to inevitable latency (transport delay), no system can run in real time in the literal sense, and the term has come to imply a data update rate which eliminates image stepping and display flicker. For aircraft simulation this is typically 30 Hz for night and dusk scenes, and 50 Hz for day scenes.

REAS Realistic Electromagnetic Arena Simulator

Refresh Rate The rate at which an image is redrawn on a display device. In interlaced systems, the term can be applied to either the field (half-raster) or a full frame (complete raster). Therefore, in a 2:1 interlaced system the field refresh rate will be twice the frame refresh rate. See under Interlaced and Field. Data Update Rate. Generally the data update rate of a CGI system will be the same as the full frame refresh rate, however, some systems draw the same image more than once while the updated data for the next image is being calculated.

REILS Runway End Identification Lighting System

REME Royal Electrical and Mechanical Engineers

Resolution and Computed Pixels Dictionary definitions for resolution include the separation into components and the making visible of detail. In simulator Visual Systems it is the smallest level at which different visual features can be clearly differentiated. In Image Generation which outputs as raster signals to a display system, resolution is a function of the number of pixels processed, since, in principle, successive pixels can have different characteristics and so satisfy the general definition of resolution. Units: Scan patterns, eg 625 lines × 625, 1024 × 1024; or overall computed pixels eg, 500 kPixels, 1MPixels, etc. For example, computing for a display resolution of 512 × 512 per channel requires 262 000 computed pixels per scan. A typical high resolution system operating at a resolution of 1024 × 1024 per channel requires 1 049 000 computed pixels per scan.

Display System Resolution. Display system resolution is a measure of pixel size as displayed to the viewer. It is quoted in terms of the angle subtended at the viewer's eye in units of either minutes (of arc) per pixel, or per pair of scan lines (Optical Line Pairs or OLP) in raster-scanned systems. Since a pair of scan lines involves two sets of pixels, a resolution of, say, 3 arcmin/pixel will be the same as 6 arcmin/OLP. The pixel resolution figure is in more common use. However, pixel values for resolution are the 'best case'. In terms of setting up or calibrating a display system, a pattern of alternate black and white stripes is often used, the limiting resolution being reached when these are reduced progressivly in size until they just begin to merge into an even shade.

Resolution of the Human Eye. The best resolution of the human eye (6/6 vision or a Visual Acuity (VA) of 1.0) at its optical centre (the fovea) is about 1 arcmin/OLP or 0.5 arcmin/pixel, reducing rapidly away from the fovea to about 2 arcmin/pixel at 10° off the fovea, 5 arcmin/pixel at 25° and 10 at 45°.

Slaved Display Systems. This characteristic allows display systems which are slaved to head or head-and-eye movement to have a low resolution 'background' or peripheral vision channel, combined with a high resolution 'Area-of-Interest' channel for the foveal vision.

Database Resolution. This is a measure of how accurately geometric calculations are performed. Most systems use 32 bit processing and can provide a resolution down to 1/32 of a foot and up to 400 miles.

RESS Radar Environmental Simulator System

Reticle From the Latin meaning 'net', a network or grid of lines displayed in an optical instrument. Used for aiming and sighting. The minimum reticle consists of simple 'cross-hairs'.

RETS REmote Target System

RF Radio Frequency. The electromagnetic spectrum with wavelengths longer than the far Infra-Red. Generally taken to start at millimetric wavelengths.

RFCT Rapid Fire Crossing Target

RFI Request for Information

RFP Request for Proposals (to meet an Operational Requirement (OR)).

RFQ Request for Quotations (to meet an Operational Requirement (OR), a more definitive stage than RFP since firm cost quotations are required). Similar to ITT

RFS Return Fire Simulator

RFT Ready-For-Training (refers to training aids and simulators, normally associated with a date, eg 'RFT mid-1998'), or Rapid Fire Target

RGB or R,G,B Red, Green and Blue. The three primary colours which are used to make up a colour display. Implies full colour when applied to Image Generation or Display Systems.

RHUD Raster-HUD. RHUDS can display raster-format TV imagery, whereas HUDs which solely employ cursive/calligraphic scanning cannot display raster video information such as FLIR imagery.

RHWR Radar Homing and Warning Receiver, as RWR

RIO Radar Intercept Officer

RISC Reduced Instruction Set Computing

RM&T Reliability, Maintainability and Testability

RN Royal Navy, the Naval forces of the UK

ROM Read-Only Memory, a type of storage from which a computer can take data but which cannot be altered.

ROSIM Royal Ordnance SIMulation, UK company

ROST Radar Operator Training System

ROTA Royal Ordnance Training Ammunition

ROV Remotely Operating Vehicle

RPV Remotely Piloted Vehicle, a type of UAV

RRA Radar Reflective Area, as RCS

RRES Radar Return Environment Simulator

RSA Royal School of Artillery (Larkhill, UK)

RSDU Radar/Sonar Display Unit

R/T Radio Telephony, ie use of radios to transmit and receive

RTAD Remote Trigger Activator Device

RTS Robot Target System, or Release To Service (of equipment)

RUKL Reflectone UK Limited (ex-BAeSL company)

rv, RV Rendezvous, for instance between units at sea, on the battlefield, or in Air-to-Air Refuelling

RVR Runway Visual Range, for aircraft and ATC simulation. A particular way of measuring visibility at the runway landing threshold.

RWR Radar Warning Receiver

SA Semi-Active (homing principle), or South Africa

SAAF Small Arms Alignment Fixture, or South African Air Force

SABRES Small Arms Battlefield Realistic Engagement Simulator

SACLOS Semi-Automatic Command to Line-of-Sight. Missile guidance principle, see under CLOS

SAF Semi-Automated Forces

SAINT Sonar Analysis INitial Trainer

SAM Surface-to-Air Missile

SAR Search and Rescue, or Synthetic Aperture Radar

SART Self-Activating Reactive Target, or Semi-active Artificial Radar Target

SAT Synchronised Acoustic Transmitter, or Small Arms Transmitter

SAWE-RF Simulated Area Weapon Effects - Radio Frequency

SAWES Small Arms Weapons Effect Simulator

Scintillation In Visual Systems terminology, the same as Aliasing (qv)

SCS Submarine Control Simulator

SCTT Submarine Command Team Trainer

SDS Steering and Depth control Simulator

SEAD Suppression of Enemy Air Defence

SEFIS Simulated Electronic Flight Instruments System

SEMSA Simulateur d'Entrainement à la Maintenance Système d'Arme

Sensory Threshold When related to human perception, a threshold of sensory perception below which the brain is incapable of sensing the cue concerned. When related to motion in any of the possible 6 DoF, the fact that sensory thresholds exist is why it is essential that aircraft have blind-flying instruments; sustained instrument flight is not possible by 'seat-of-the-pants' cues alone since errors build up due to aircraft movements below the pilot's sensory threshold, ultimately leading to loss of control. In simulators, the motion sensory thresholds are utilised in the technique known as 'acceleration-onset cueing' whereby a motion platform is reset to approximately its neutral position after it has made an acceleration by being moved at a rate below the sensory threshold, and is then ready to respond to the next acceleration required.

SF Special Forces

SFTS Synthetic Flight Training System

Shading In the context of CGI systems, shading is the technique of varying the colour and intensity of a polygon in accordance with its angle to a defined light source which can frequently be a simulated sun. The orientation of the polygon to the light source is calculated and the less perpendicular the angle, the lower the intensity of polygon colour. At right angles to the light source, normally a defined minimum intensity is used rather than allowing the polygon to go completely black, which would look obtrusive and unrealistic except in night scenes. Several different types of shading may be met:
Fixed Shading. Where the intensity of a polygon does not vary with light source angle; shading may vary across the polygon.
Flat Shading. The entire polygon has the same intensity which is never varied.
Gouraud or Smooth Shading. A system where intensity can be varied across a polygon and adjacent polygons. It can be done in such a way that a flat polygon can give the visual illusion of being a curved surface, by using shading which mimics that which would appear on a genuinely curved surface.

SHINMACS SHipboard INtegrated MAchinery Control System

SHORGUT SHOrt Range GUn Training

SHS Ship Handling Simulator

SICOMORE SImulateur de Centre Operations MOdulaire et REconfigurable

SID Standard Instrument Departure (Aircraft Climb-out Procedure), see also STAR

SIDFAC Shore Integration and Demonstration FACility

Simulation To model, mimic or replicate artificially without using any or some of the equipment being simulated. There are two broad techniques of simulation, STimulation and Emulation (qv). STimulation uses artificial signals to STimulate real equipment, whereas Emulation uses completely artificial means such as software and hardware models, both in order to achieve the same result. With recent developments in computing, Emulation has become the predominant technique in the training area.

SINS Ships Inertial Navigation System

SIPS Software Intensive ProjectS. They may need special procedures, checks and balances, etc, due to the nature of the software concerned.

SIRAS Surveillance and IFF RAdar Simulation

SISMAS Sonar In-Situ Mode Assessment System

SKE Station Keeping Equipment (for formation flying of aircraft in IMC or at night, by using electronic cues for maintaining safe separation).

SLAMSHOT Shot Location And Monitoring - Shows Hit On Target

SLAT Supersonic Low Altitude Target

SLOTTS Surface Launch Organic Tactical Training System

SLS Single Lane System

SLUTT Ship-Launched Underwater Transponder Target

SMAS Ship Motion and Anemometer Simulator

SMS Space Mission Simulator

SOF/SOFATS Special Operations Forces/SOF Aircrew Training System

SOP Standard Operating Procedure(s)

SPAL Simulator Projectile Airburst Liquid chemical agent

SPAT Self-Propelled Acoustic Target

SRAAM Short-Range Air-to-Air Missile. For example the US AIM-9 Sidewinder series, and the ASRAAM (Advanced SRAAM) programme

SRBOC Super Rapid-Blooming Overboard Chaff

STA Surveillance and Target Acquisition

STANAG STANdardization AGreement. Specifically, NATO STANAG, normally given a unique number so that it can be quoted as a customer requirement in the same way as US MIL SPEC or MIL STAN

STAR STandard ARrival procedure, for aircraft; set three-dimensional descent pattern or patterns for approaching a busy airport in a standard and well-known way with the object of minimising congestion and collision risk. See also SID

STARS Surveillance Target Attack Radar System

STE Simulateur de Tir Embarqué. Gunnery trainer on board a ship

STEMS STandard EMitter Simulator

STimulation STimulation describes a type of simulation whereby artificially generated signals are fed to real equipment in order to STimulate it to produce the result required for training. The real equipment can be radar, sonics, instruments etc. In some cases the STimulation equipment can be carried in the real platform or carriage vehicle (ie the Ship, AFV or Aircraft), and be used for so-called 'embedded training' during its operation, by the generation of simulated scenarios which can be dealt with in a realistic manner by use of the normal controls and displays. In the overall definition of simulation, the alternative method is called Emulation, which is the simulation of equipment by entirely artificial means by physical and software modelling.

STORM Sector Tactical Operations Range Modules

STOL Short Take Off and Landing

STOVL Short Take Off and Vertical Landing

STPT Stinger Troop Proficiency Trainer

Sub-Pixel Processing involving elements which are smaller than a pixel. Used in anti-aliasing (qv) techniques where a number of sub-pixel elements are calculated in order to produce averaged data actually to be displayed by the appropriate pixel. This technique is effective in removing the harsh jagged edges otherwise produced when line features are displayed in a raster format at an angle to the raster lines except at 0° and 90°.

Surface In CGI technology, the same as Polygon (qv) and Face. Generally the term polygon is preferred because it is more descriptive and less ambiguous.

SWATT Simulator for Wire-guided Anti-tank Tactical Training

Synergistic The word Synergy is from the Greek *Syn,* together, and *Ergon,* work, and means combined or co-ordinated action. A 6-DoF synergistic motion platform has three mounting points on the floor under the simulator, from each of which two jacks displaced at any angle are connected to the simulator cockpit baseplate. By correct programming, movements in all six DoF can be obtained from such a platform design. Nomenclature for the six DoF are tabulated at the beginning of the Motion Platform section. It should be noted that, because of the disposition of the jacks, in a synergistic platform the amounts of movement possible will be greater if only one DoF is used at a time. However, frequently the movement of the vehicle being simulated requires, say, pitch, roll, heave and surge all to be applied together. Maximum displacement in a given DoF is reduced when, as is normal in real training operations, more than one DoF have to be used together.

TA Training Analysis, or Territorial Army

TACAN TACtical Air Navigation system. An aircraft navigational aid giving range (DME) and bearing from a beacon

TACDEW Tactical Advanced Combat Direction and Electronic Warfare trainer

TACPRO TACtical PROcedures performance analysis

TACSIM TACtical SIMulation

TACTS Tactical Aircrew Combat Training System

T&E Trial and Evaluation

TAIT Tornado Air Interception Trainer

TALAFIT TAnk Level Aiming and FIring Trainer

TALISSI TActical LIght Shot SImulator

TAS True Air Speed, or Test Attendant Station

TASC Touch-Activated Simulator Control

TASIG Target And Simulation Image Generator

TASWIT Tactical Anti-Submarine Warfare Interim Trainer

TCB Target Connection Box

TC&E Training Control and Evaluation

TCS Target Control Set

TCT Tank Crew Trainer

TDP Technology Demonstrator Programme

TDT Tank Driver Trainer

TEMPEST Transient ElectroMagnetic Pulse Emanation STandard. A standard of shielding of wires and equipment so that unwanted signals cannot be injected into the equipment, or sensitive signals from the equipment intercepted by outside agencies. An ECCM measure.

TES Tactical Environment Simulation, or Tactical Engagement Simulator/Simulation

Tessellate Dictionary definition: to mark like a mosaic. In CGI terminology, refers to texture maps, and also to some types of databases, and implies correct edge-matching. In the case of a mosaic, the squares or other shapes of which it is composed must properly match without visual anomalies; the same principle applies with tessellated boundaries of texture maps which, when the map is repeated in adjacent areas, must produce a plausible visual effect. A photo-texture map including objects such as trees in front of a wall, will tesselate in the horizontal but not in the vertical, whereas a texture map of brickwork will tessellate in all directions. Edge-matching is a more descriptive term.

Texel/Textel TEXture ELement. The smallest element or cell which makes up a texture pattern or texture map. The texture equivalent of Pixel. The general term in use is Texel (rather than Textel).

Texture A reticulated visual effect involving a contrast pattern within an otherwise evenly coloured or shaded polygon or surface. In CGI systems, texture is generated by mathematical or digitised photographic means and stored in the form of a library of texture maps. Texture effects, when displayed to an observer, trainee or operator of a CGI system, produce additional visual cues over and above the basic scene content using polygon modelling. This not only enhances realism but is particularly important in training tasks where the ownship vehicle or platform is moving with respect to the background scene being displayed. In these cases, the peripheral vision cue of visual vection or picture-flow is strong and gives important cues of speed, and, for aircraft, height and velocity vector independent of other cues involving conscious comparison with recognisable objects in the visual scene.
Dynamic Texture. Textures can be static or dynamic, the latter being produced by moving one or more texture-maps with respect to the surface being textured. This can be used to produce visual effects such as rotor-wash, sea states with wind, and corn or long grass being blown in a downwind direction. Richness of scene detail is a function of the combination of the number of polygons, textures, shading and special effects such as shadows.
Geospecific or Real-World Texture. Texture maps derived from real-world data such as from satellite and reconnaissance photography. Such texture maps do not tessellate at the edges but can be fitted together such that the joins are not visible (like joining paper maps).
Photo-Texture. Any texture map derived from a photographic source. Normally achieved by a scanning digitiser.

TF Terrain Following. Generally refers to an aircraft radar or a radar mode 'TF Radar', 'TF mode', or even, colloquially, 'TF-ing'.

TFD Terrain-Following Device

TFR Terrain-Following Radar

TFS Tactical Flight Simulator, or Trainer Fighter Simulator

TGMTS Tank Gunnery and Missile Target System

TI Thermal Imager, or Texas Instruments (US company). Thermal imagers scan a scene using sensor elements sensitive to the far Infra-Red at about 10 micron wavelength and produce a picture of natural heat contrast across the scene. Thermal imaging equipment is therefore not only passive, but is independent of reflected ambient light (unlike intensifiers) since it depends on 10 micron heat emissions from the sources themselves. Also, Thermal Imagers can be used to penetrate small amounts of smoke and cloud; this is due to the long wavelength used which has good transmittance characteristics in the presence of particles in the atmosphere smaller than the wavelength itself. See under IR.

TLC Through-Life Costs, the same as Life-Cycle Costs (LCC)

TLFS Tank Laser Firing Simulator

TMS Training Management System

TNA Training Needs Analysis

TOBA Terms Of Business Agreement

TOC Tactical Operations Center

TOL, ToL Take Off and Landing

TOM Training Outdoor Military

TORT Tactical Operational Readiness Trainer

TOTS Tower Operator Training System

TOW Tube-launched Optically tracked Wire-guided (proprietary ATM system), or Take Off Weight

TPGID Tank Precision Gunnery In-bore Device

TPT Troop Proficiency Trainer

Transport Delay See Latency. In this book, the two terms are treated as the same.

Transparency/ Translucency The capability of a CGI system to display all or part of a polygon as translucent (a completely transparent polygon would be of little value in the scene). Normally achieved through the medium of texture. Different levels of translucency will be defined in a CGI system architecture and in a visual database. The characteristics of the translucent polygon's pixels are mixed with those of the polygons behind it in the scene. The translucency facility is also used as a device in making Level of Detail (LoD, qv) changes by fading objects out as they become more distant and fading objects in as they get closer; a technique known as Fade Level of Detail.

TREE Transient Radiation Effect on Electronics

TRS Training Reception System

TRTG Tactical Radar Threat Generator

TSMT FES Trident Sonar Maintenance Trainer Front End Simulator

TSU Telescopic Sight Unit

TSV Through-Sight Video

TTC Tactical Training Centre

TTOMT Tank Turret Organisational Maintenance Trainer

TTT TeleType Trainer

TTTS Tanker Transport Training System

TUSSF Trident Unique Software Support Facility

TVT TeleVision Trainer

TWGSS Tank Weapon Gunnery Simulation System

TWT Travelling-Wave Tube (RF transmitter type)

UAH Up After Hit

UAV Unmanned Air Vehicle. The generic name for a range of air vehicles varying in complexity from small hand-launched air targets to large and sophisticated surveillance vehicles and armed Remotely Piloted Vehicles (RPVs) which can be ground-launched, ship-launched, or launched from aircraft. Unarmed UAVs are also sometimes called 'drones'. Also includes manned aircraft reconfigured for unmanned operation under remote-control, frequently obsolete aircraft now used as targets and also described as UnManned Aircraft (UMA).

UCQBR Urban Close Quarter Battle Range

UHF Ultra High Frequency, the RF frequency band above VHF at about 200-500 MHz

UIT Union Internationale de Tire (International Shooting Standard)

UMA UnManned Aircraft. Aircraft-type vehicle designed to be operated without an onboard pilot.

Update Rate The rate in Hz at which the complete scene data in a visual channel is replaced with newly computed information, eg: typically, for an agile fighter aircraft simulation, 60Hz. For less dynamic aircraft requirements, 25 Hz is generally considered the minimum since below this horizon-stepping can be noticed during manoeuvres, particularly when rolling. For non-aircraft applications, 15, 10 or even 5 Hz may be suitable, depending on the task. For off-line tasks such as data base modelling the update rate is much less critical than for real-time training tasks. The data update rate may not be the same as the display field or refresh/frame rates; see also field rate and interlaced scan.

USA United States of America, United States Army

USAFE United States Air Force Europe

USASTC US Army Signal Training Centre

USMC United States Marine Corps

UV Ultra-Violet. The part of the electromagnetic spectrum with wavelengths less than the violet end of the human visual capability. Wavelengths less than about 0.4 microns or 400 nm

VA Visual Acuity. A human VA of 1 is the same as 6/6 vision, and a VA of 0.5 is 6/12 vision. A VA of 1 is approximately equal to a resolving power of 0.5 arcmin/pixel or 1.0 arcmin per Optical Line Pair, and a VA of 0.5 to 1 arcmin/pixel or 2 arcmin/OLP.

VADS Vulcan Air Defense System

VAS Visual Augmentation System

Vestibular Apparatus/System The organ of the human Inner Ear sensitive to accelerations in all six-DoF. The organ consists of the semi-circular canals and otoliths. These are stimulated by accelerations, and the semi-circular canals behave as angle-rate transducers. There is, however, a sensory threshold below which the brain does not register the motion below a critical rate; this has important consequences in terms of motion simulation technology, in the blind-flying of aircraft, and in conditions leading to spatial disorientation. See also Acceleration-Onset Cueing and Sensory Threshold.

VDD Vehicle Detection Device

VDU Visual Display Unit

VEMS Versatile Exercise Mine System

Vertex/Vertice/ Vertices Dictionary definition: a meeting point of lines. In CGI terms, a point in a database having no volume but defined in space by co-ordinates with respect to the database datum. Normally Cartesian co-ordinates in the standard x, y and z axes are used. The smallest polygon is a triangle which is defined in a database by three vertices.

VFR Visual Flight Rules. For aircraft these are defined precisely in air law and regulatory rules. See also VMC, IFR and IMC

VHF Very High Frequency, in the RF spectrum the band between 10 and 200 MHz

VHIS Visual Hit Indicator System

Viewing Area or Volume In display systems which create an infinity or distant image focus, the viewing volume within which an observer will see a substantially undistorted image. Sometimes also called the exit pupil.

VIGS Video Interactive Gunnery Simulator

Visibility In meteorology, the distance at which an object can be clearly discerned; the international definition of fog is a visibility of less than 1 km; mist is a visibility of between 1 and 2 km and haze from 2 to 5 km. Fog and mist are generally assumed to be composed principally of water droplets, haze and smoke can be of smaller particle size; this has implications for sensors such as thermal imagers (TI/FLIR) operating in the far-IR at wavelengths of about 10 microns which are capable of penetrating haze and some smokes well because their particle size is smaller than the wavelength; the IR radiation is therefore not significantly deflected or absorbed by the particles.

CGI Aspects of Visibility. Visibility must be specifically calculated by CGI systems through appropriate algorithms, and then applied either by selection by an instructor or through a preset programme. Some aspects are:

Cloud. For simulation of ground vehicles and ships, a variety of cloud patterns can be created by texture maps. For aircraft simulation in which cloud has to be approached, flown through, and represented correctly both above and below it, much more elaborate processing is required. In this case cloud is treated as dense fog at specific height bands. The more capable aircraft-orientated CGI systems allow for patchy cloud in various amounts, and when the eye point is above cloud such systems show terrain above the cloud (such as high ground and mountains), and also show lower terrain through the gaps in the cloud. Multi-layers of cloud can also be represented. This requires substantial ingenuity and power in processing the scene and is not included in lower-cost CGI systems.

Ground Fog. What it says. The difficulty for aircraft is that since ground fog is shallow, typically being caused by night-cooling, it can frequently be clearly seen through from above, but not from shallow slant ranges. Since aircraft on instrument approaches typically use a 3° glidepath, on a CGI system it will have to be ensured that the ground can be seen from vertically above, but as the runway is approached and the aircraft descends into the shallow fog layer, the CGI visibility is reduced appropriately.

RVR. Runway Visual Range, expressed in metres. A measuring system is used at the side of a specific runway threshold when visibility is less than 1 km. The measuring device uses calibrated lights which are viewed by a light meter. RVR values for that runway are passed by Air Traffic to pilots about to land. In simulator terms, RVR implies modelling for low visibility but the visual system distance for revealing of runway and other lights should match the RVR set at the IOS. Normally runway lights will be seen at ranges somewhat in excess of RVR because of their high intensity, and sometimes high intensity flashing strobe lights are used just before the runway itself.

Scud. A variation of visibility when entering or leaving a cloud layer, intended to simulate an uneven bottom or top to the cloud. CGI systems may have different settings for the depth and intensity of the effect.

VISMOD VISual MODification, eg of old or cheaper combat vehicles or aircraft in order to mimic current vehicles or aircraft, for training and recognition purposes.

VIS-IDENT Visual Identification

VISTA Visual Imagery Simulation Training Aid

Visual System, Visuals A much abused term which can mean, depending on the context, any element of a visual system. Sometimes means the IG/CGI systems (with or without databases), sometimes the Display System, sometimes both systems, sometimes includes host integration, and a number of support issues such as facilities for database generation, record/replay, after-sortie debrief, and so forth. A term to be used with caution on its own.

VLAST Video-based Live Aircraft Stinger Trainer

VLSI Very Large Scale Integration

VM-AT Versatile Multi-Aimer Trainer

VMC Visual Meteorological Conditions. For aircraft these are defined precisely in air law and regulatory rules and flight to VFR procedures apply. See also IFR and IMC.

VMT Vehicle Manoeuvring Trainer

VOR VHF Omni Range, a type of radio beacon for aircraft transmitting in the VHF band

VR Visual Range, Virtual Reality. The latter is a term for realistic simulated visual scenes, frequently combined with a tactile environment by the use of an instrumented glove or gloves. Commercially available VR helmets are low-cost collimated displays of limited resolution and FoV which can be used with CGI databases.

VSS Video Signal Simulator

VSTT Variable Speed Training Target

VTOL Vertical Take Off and Landing

VTT Video Target Trainer

WAC Wide-Angle Collimated. Generally means collimated monitor display windows. Particularly: a term used by CAE-Link of Binghamton, USA, to denote its proprietary range of collimated monitor display systems. A 'WAC Window' is a single collimated monitor unit used for an out-of-window display in a simulator

WAN Wide Area Network (Long-range LAN, same as LHN)

WES Weapons Effect Simulation

WESS Weapons Effects Signature Simulator

WHDE Weapon Handling and Discharge Equipment

WIA Wounded In Action

WIDE™ Wide-angle Infinity Display Equipment (Proprietary name for collimated wide angle display invented by the UK Rediffusion company, now part of Thomson Training and Simulation (TTS).

WIFT Westland Indirect Fire Trainer

Window In Display System terminology, a single display unit such as a collimated monitor, normally fed by a matching channel of CGI and positioned in a simulator just outside the actual cabin window in the vehicle being simulated. Derived from 'outside-world window'. Hence, for instance, a 'three channel/three window' system consisting of three channels of CGI each driving a display window, or, for two side-by-side trainees, a 'three channel/four window' system where the centre two display windows are driven by the same IG channel and situated one in front of each trainee. See the diagrams at the beginning of the section on Visual Display Systems.

WMR War Maintenance Reserve

WSO Weapon System Officer

WST Weapon Systems Trainer. For aircraft, this type of trainer often has a complex terrain database, a wide-FoV high resolution visual system, and a high computer capacity, so that weapons work can be practised, recorded and debriefed. Because of the compromises necessary with a large FoV visual system, older WSTs tended not to have a motion platform, but with advances in visual display technology such as domes and partial domes compatible with motion platforms, and lightweight displays such as the CAE FOHMD, more WSTs are being fitted with motion platforms and so constitute Full Mission Simulators or Combat Mission Simulators (FMS/CMS)

WSTR Weapon System Training Rig

W/T Wireless Telegraphy, use of non-voice radio techniques such as the Morse code

WVR Within Visual Range

Z Buffer/Z Depth/ Depth Buffer/ Range (R) Buffer Refers to a priority (qv) technique in a CGI system whereby the Z distance (depth) of a polygon is compared at the pixel level to those features which have already been processed. Sometimes called a depth buffer. In hardware terms, the actual memory element where these values are stored.

Z Depth. The Z component of range from the computed eye point to a point in the scene (usually a pixel, sometimes a polygon).

Range (R) Buffer. A priority mechanism similar to a Z Buffer but using slant range instead of the Z component. The advantage of an R Buffer is for CGI systems which may have very large horizontal FoV, where a Z Buffer may run out of resolution.

ZFT Zero Flight Time. A regulatory standard of aircraft simulation allowing already experienced pilots to convert to an aircraft type of similar characteristics to that already operated, solely by the use of high-quality flight simulation. The pilot's first flights in the aircraft itself will be on revenue flights under the supervision of an already-qualified aircraft commander.

Seconds to decide,
and no second chance.

That's how
real combat training
should be.

Win or lose, TOM pushes soldiers to the limits. With advanced combat training for infantry, armour and heavy artillery that recreates the stress and shock of battle. The modular system imposes no fixed doctrine. You are free to set the scene with offensive or defensive mission profiles to suit your own chosen terrain, task force, and combat conditions. TOM's user friendly interface and interactive technology allow real time command and control. So you can set the pace. See the results. And really train for the future, today.

FN HERSTAL

GIAT *industries*

Full Force Partnership

LAND-BASED SYSTEMS

Small Arms Range Equipment
Full Calibre Gunnery Ranges
Small Arms Marksmanship Training Equipment
Sub-calibre Training Devices
Training Ammunition and Pyrotechnics
Tactical Engagement Simulation
Anti-tank Weapon Training Systems
Air Defence Training
Tank and Direct Fire Gun Trainers
AFV Crew and Tactical Trainers
Artillery and Mortar Fire-Control Trainers
Tank and other Driving Simulators
Maintenance Trainers
Command and Staff Trainers
Miscellaneous Land-based Trainers

SMALL ARMS RANGE EQUIPMENT

AUSTRALIA

ADI Ballistic Analysis System

The ADI Ballistic Analysis System uses patented projectile location technology to locate a projectile fired from large or small calibre weapons in the range 5.56 to 120 mm. The equipment comprises a main computer (IBM PC) and optional projectile detection units. The detection units are the open array, T-bar and 1 × 0.8 m box target. These devices can be used to measure a number of ballistic parameters, as listed below. A remote visual display unit can be fitted at the firing point, if required.

The system is controlled from the main computer. If the remote visual display unit is used, an optional limited control is available at the firing point. Shots will be displayed as they are fired over the system and are graphically imposed on the target image on the control computer and remote monitor. Once the shots have been fired, the control computer will calculate the smallest group possible and calculate a variety of parameters which describe this group. Data may be printed and/or stored for later statistical use.

Data collection includes the following: dispersion data; score values; time of flight; means and standard direction; group size; ammunition velocity; ammunition energy; extreme speed; rate of fire; and miss distance. General statistical information may be incorporated, that is weapon type, ammunition type, firer, and so on.

Status
Available.

Manufacturer
Australian Defence Industries Limited, Bondi Junction. (For details see Directory of Suppliers)

ADI Ballistic Analysis System

ADI Computer-Controlled Range (CCR)

The ADI Computer-Controlled Range (CCR) simplifies the programming and controlling of multi-lane, multi-bank ranges for all types of small arms. The target mechanisms involved can be all stationary or all moving, or in any combination.

The computer used to control the range functions uses high resolution graphics, normally in colour; all information used in programming the range is stored on a hard disk. A comprehensive on-line 'help' facility is integral to the system. If required the computer may be controlled remotely by a portable hand-held controller.

Simple and complex target exposure sequences can be programmed and stored on disk for use as often as necessary. The operator can revert to manual control at any time. Hard copies of scores are printed out in graphics or numerically, on request.

The CCR is fully self-testing. An automatic test is conducted at 'start up', and further tests may be undertaken as part of the comprehensive 'Maintenance Programme' which enables faults to be located quickly.

Status
In production.

Manufacturer
Australian Defence Industries Limited, Bondi Junction. (For details see Directory of Suppliers)

The controller's position on an ADI Computer-Controlled Range (CCR)

ADI Figure Target Lifting Mechanisms

ADI manufactures four types of stationary figure target lifting mechanisms, the Series 5, Series 6, Series 7 and Series 10. All are suitable for use with all types of small arms.

The Series 5 is both mains- and battery-powered and can be used on static and portable ranges. The Series 6 is mains-powered only and is used only on static ranges.

The Series 7 is battery-powered and has a hit counter, battery state and night effects driver functions built into the unit. The target raises and lowers over the unit permitting the unit to be installed in a shorter overall space.

The Series 10 can be battery- or mains-powered. The target mechanism is lightweight and has a sideways-acting target arm so when installed it requires less operating space.

The Multipurpose Portable Infantry Target System (MPITS) is a lightweight system. Weighing less than 15 kg it is one-man portable, battery-operated and requires minimal ballistic protection due to its low profile.

All mechanisms are able to present several types of targets singly, in pairs or as a triple presentation. The targets can be programmed to operate as follows: fall when hit; reappear after falling; appear on command; hold-up to accept a number of rounds per minute (rds/min) and count automatic fire (up to 1200 rpm with the inertia switch fitted or 2200 rds/min with PLS); and bob, that is, cycle up and down every 5 seconds until hit.

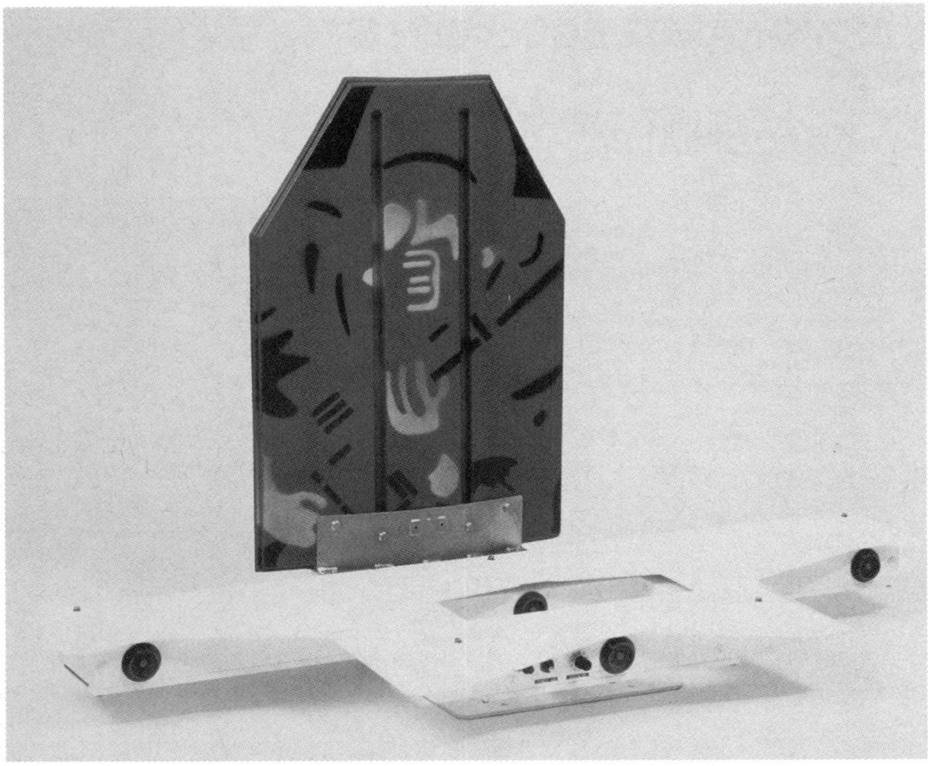

ADI Series 6 figure target lifting mechanism

There are four types of target arm which permit the target to be presented: towards the firer, sideways and from around corners; the fourth method exposes an edge-on target which remains upright. At night, target indication and target hit indication are provided by a night shoot attachment or a Night Effects Simulator (NES). The NES has an additional function of being able to silhouette the target with a blue haze effect. A retaliatory device uses detonators or squibs to simulate return fire from the target.

Control of the mechanisms may be by cable, radio, computer, automatic or remote. A manual control and counter uses cable to control the target functions, which are raise and lower, hold-up, detonate/illuminate, score, and reset scores.

Up After Hit (UAH) Generator

The ADI Up After Hit (UAH) Generator is a simple and inexpensive target reset control device fitted to the Series 5 target mechanism. The device is used instead of a receiver or transceiver to raise the target to the upright position automatically after it has been hit. A control switch on the UAH generator enables the period of time the target is to remain down to be preset and varied between 1 and 10 seconds approximately.

The target mechanism requires no modification as the UAH generator plugs directly into the receiver outlet on the mechanism or night light. The mechanism may be reconverted to radio control by unplugging the device and reconnecting the receiver or transceiver cable. The UAH generator can be used in conjunction with a manual hit counter.

ADI Proximity Detector

The ADI Proximity Detector comprises a detector assembly and reflector assembly and may be set up for use indoors and outdoors. The detector can be used with a single target or with multiple targets. When the firer breaks an infra-red beam between the detector and reflector, the target functions ie lifts up. The Proximity Detector is compatible with both the Series 5 and Two-way Turning Target mechanism (for details of the latter see following entry).

Status

In production. In widespread service.

Manufacturer

Australian Defence Industries Limited, Bondi Junction. (For details see Directory of Suppliers)

ADI Figure Targets

ADI manufactures a variety of figure targets suitable for all types of mechanical target systems. Plastic targets are available in various sizes: F and E type, 11 D, standing man; 13 D, weapon crew prone; and 15 D, prone man. For added realism a figure silhouette is painted on the target surface.

Depending on the shot concentration and calibre of the rounds used, ADI targets will accept between two and three thousand rounds of service ammunition before becoming unserviceable. All targets are ultraviolet protected and operate efficiently in all climates.

The targets are manufactured to fit three- and four-clamp target holders.

Status

In production.

Manufacturer

Australian Defence Industries Limited, Bondi Junction. (For details see Directory of Suppliers)

ADI International Pistol Target System (IPTS)

ADI produces a five-target international target pistol system which conforms to the UIT/ISU rules. ADI's system does not require a framework or bearing assemblies for the upper part of the target boards and therefore no metalwork is exposed to the field of fire. Consequently there is no need to provide protective armour for any upper bearings or target holders, the whole assembly is lighter, easier to assemble and requires virtually no maintenance, and by simply removing the target boards the firing range is unobscured for other types of target use. There is no need to move the IPTS frame away as its low profile is hidden behind a low protective wall. On some indoor ranges with suitable ceiling baffles the frame may be mounted on the ceiling.

The ADI International Pistol Target System (IPTS) uses mainly aluminium construction and is light in weight. It was designed to meet the standards required by the UIT/ISU regulations for the conduct of rapid fire, centre fire, standard and ladies' pistol matches. The system is also suitable for other disciplines such as service pistol matches and security and law enforcement officers' hand gun training.

The system comprises one or more target frames, an electronic range timing unit and an air supply.

Each target frame holds five targets which turn together. One pneumatic ram on the bottom of the frame turns the targets through 90° in less than 300 ms. The ram contains an adjustable cushion which dampens the shock at the end of each movement, thus preventing vibration and equipment damage. The target frames are manufactured from heavy gauge aluminium channel and require virtually no maintenance other than routine cleaning and target replacement.

The target boards are made of compressed chipboard or plywood and are from 6 to 16 mm thick (10 mm plywood is recommended). The centres of the target board can be cut out and removed to prolong the target board life. The target boards are capable of accepting a variety of paper target facings.

Up to three target frames may be controlled by one solid-state electronic range timing unit which requires a

ADI International Pistol Target System (IPTS) showing targets side on

ADI International Pistol Target System (IPTS)

240 V single phase supply or a 12 V DC supply from a battery or power pack. The timing unit enables an operator to select any one of 13 programmes, all of which meet UIT/ISU regulations. Alternatively the target exposures may be controlled manually or by remote-control, which can interrupt or reset an initiated programme.

For target ranges, or ranges which have a high throughput of firers, a range extender unit can be installed between the electronic range timing unit and target frames. This permits more than three target frames to be controlled simultaneously by one electronic range timing unit.

The air supply required must be from a clean and dry source at a pressure of 6 bar, and may be provided from an air bottle or a compressor. ADI can supply a suitable compressor appropriate to the number of frames required.

Status

In production.

Manufacturer

Australian Defence Industries Limited, Bondi Junction. (For details see Directory of Suppliers)

ADI Moving Figure Target Systems

Moving figure target systems produced by ADI include the Moving Target Mechanism Winch Driven (MTM (WD)) and the Moving Target Mechanism (MTM).

All systems can be fitted with the ADI Projectile Location System (PLS - see entry in this section) and can be used alone or in conjunction with the Computer-Controlled Range (CCR - see earlier entry). An alternative method of control is the ADI UHF Radio Hand-Held Controller (see entry in this section).

Moving Target Mechanism Winch Driven (MTM(WD))

The ADI MTM(WD) was designed to provide a cost-effective and versatile moving target system for small arms and some sub-calibre devices. With the system a figure target is moved in two directions on a straight run. Any three speeds between 1.3 and 13 km/h may be preset by the operator. The target will fall when hit and can be raised and lowered at any time during traverse. At night, target illumination is provided for target detection and hit indication. The MTM(WD) may be stopped and used as a stationary target.

The MTM(WD) is permanently installed and requires mains power. Crossing angles can be varied.

Moving Target Mechanism (MTM)

The Moving Target Mechanism (MTM) is a multipurpose target mechanism which accommodates figure targets and scaled vehicle targets. The mechanism may be mains-powered for use on permanent ranges or battery-powered for field firing and temporary ranges.

Any two speeds up to 13 km/h may be selected and preprogrammed. Up after hit, target hold and target score facilities are provided. The MTM may change speed and direction without stopping first, and auto-traverse indefinitely. Target presentation may be head-on or in profile. For night engagements the target is illuminated with a blue light and an orange light provides visual hit indication.

The MTM uses a monorail guidance system which requires little or no maintenance and is easily laid with minimum earth disturbance to follow ground contours and negotiate bends. The monorail may be of any length and several mechanisms can operate on the one rail as each is fitted with an anti-collision device. Alternatively, several monorail runs may be installed at one location. The choice of locating the target carrier on one of several runs prevents the firer from knowing in advance where a target will appear.

Status

In production and service.

Manufacturer

Australian Defence Industries Limited, Bondi Junction. (For details see Directory of Suppliers)

ADI Preselectable Manual Control System

The ADI Preselectable Manual Control System (PSMC) was produced to provide controlling and scoring for small arms practices, especially on 25 m ranges, as an alternative to the ADI Portable UHF Radio Control System (see later entry).

The PSMC comprises a console for 6 or 12 target mechanisms, a console cable, a six-way distribution box, and target mechanism cables. No additional power, other than that provided by the 12 V battery attached to the target mechanism, is required. A feature of the system is that no modifications are necessary to convert existing ADI Portable Radio Control systems as all cables plug into existing outlets on the target mechanisms.

The PSMC has three main functions: the target exposure sequence may be preselected and run automatically; manual override is possible at any time; and target hits will be scored. Three dials on the console enable preselection and setting of the time of exposure of the targets, the target down time between exposures, and the number of target exposures required. Targets are scored at 1 point/hit up to 1200 rds/min for all automatic and semi-automatic small arms. A 'hold' function enables automatic firing to be scored.

One person can operate the console and control the firers. The system can be operated after only a few minutes' instruction.

Status

In production.

Manufacturer

Australian Defence Industries Limited, Bondi Junction. (For details see Directory of Suppliers)

ADI Projectile Location System (PLS)

The ADI Projectile Location System (PLS) technology permits supersonic rounds to be marked automatically and scored whether or not the rounds hit or miss the target. The ADI PLS provides all the functions of a dedicated Location of Misses and Hits (LOMAH) system. PLS type ranges are normally purpose-built but standard classification and static electromechanical ranges can be converted and modified to accept a PLS.

There are two versions of the PLS; an open array of sensors and an enclosed array.

The open array uses a T-bar which is located below the target and protected by a mantlet. The T-bar may be positioned on a target lifting mechanism or mounted on a bracket with a permanently raised stationary target or aiming mark. The T-bar can also be mounted on a variety of moving target mechanisms. The T-bar contains an electronic package connected to four sensors which can locate supersonic rounds of all calibres passing over, or in the vicinity of, the sensors. The detection area is relative to the size of the round being used.

The enclosed array is used in a box target for precision or competition rifle shooting. The box target can be mounted on a moving target mechanism for sniper and counter-sniper training.

For each system the location of the round fired is displayed at the firing point on a Visual Display Unit (VDU), normally located next to the firer. Additional monitors may be available at a central control point.

The lanes of each firing point (except for some single lane systems) have a Lane Initiator or sensor positioned in front of the firer to identify the shots fired. Thus short rounds, ricochets and crossfires are identified but not allocated a score. When there is more than one firing point the VDUs and lane initiators can be moved between firing mounds as required.

PLS and VDU configurations can be varied to suit existing ranges or purpose built to suit training requirements and throughput. Configurations include single and multi-lane; single and multi-target banks; and single and multi-firing points or firing mounds.

Sequence of Events

The PLS is initiated by a shot fired at the firing point. Muzzle blast is detected by a lane initiator on the firing point (some systems do not require a lane initiator). The projectile passes over downrange detection sensors on a T-bar or box target. Signals from the sensors are processed and when requested are transmitted to the uprange display unit. Range communications may be via cable or radio. The downrange timing information is matched with information from the lane initiator. If a match is found, the sensor information is used to calculate the shot position, remaining velocity and time of flight for display and printout as required. The uprange visual display equipment allocates a status to score the shot, that is: Hit, Ricochet or Not In Target Area (NTA), and updates running totals. A graphical representation of the target and shots as they are fired is kept until the exercise is complete.

Programmable Functions

The ADI PLS system may be used to locate supersonic projectiles fired from small arms, including automatic weapons; AFV main armaments; anti-tank weapons; and air-to-ground guns and missiles. Some subsonic projectiles are also detected using the enclosed array.

Applications of PLS include grouping, zeroing, application of fire, precision and competition shooting, and ballistic analysis of ammunition and weapons. Software programs for PLS may be modified to suit particular applications and user requirements.

The grouping programme permits the firer or instructor to select the number of rounds to be used per group. At the completion of the group the extreme spread of the group is displayed. Shots may be displayed as they are fired or after the completion of the group. The size of acceptable groups may be specified and programmed for selection.

The zeroing programme calculates the mean point of impact and extreme spread of the group and displays the displacement of the mean point of impact from the point of aim. The ADI Zeroing Calculator can then be used to determine the sight adjustments to be applied to the weapon to move the point of impact to the correct zero position. Alternatively the sight adjustments to be applied to the weapon may be programmed to be displayed automatically at the completion of the group.

The application of fire programmes provides for a variety of scoring methods both on and off the target. Scores are totalled progressively and fire may be applied to both stationary and moving targets by day and by night.

The ballistic analysis programme provides a substantial amount of additional information on each round and group of rounds fired. Examples of this information could include horizontal and vertical spread, remaining velocity and energy, mean point of impact, measures of dispersion, cyclic rates of fire, and fastest and slowest rounds.

ADI T-bar

The ADI T-bar contains an electronic package connected to four sensors that can locate supersonic rounds of all calibres passing over or in the vicinity of it. The detection area is relative to the size of the round being used. For example, the detection area for a typical small arms round is a diameter of 5 m while the detection area for a typical large calibre AFV gun round is 20 m. As most targets are smaller than the detection area, hits and misses are distinguished and graded score values can be awarded to hits. Single rounds and automatic fire up to 6000 rds/min are accepted and all rounds are numbered in the order of arrival.

The T-bar needs to be aligned with the firing point

and provided with ballistic protection. It may be mounted on a target lifting mechanism that may be either stationary or moving. Alternatively, a permanently upright target may be used as an aiming mark. The T-bar does not need a target to function.

ADI Box Target System

The ADI Box Target is used when a high degree of accuracy is required for shot location. In competition shooting the size of the aiming mark can be quickly changed to match the type of shoot and range to target. Targets may be stationary for standard competitions and moving for specialist sniper training.

The Box Target consists of a frame with panels of self-sealing membranes secured on both faces, providing an enclosed volume of air. Within this compartment are situated three acoustic sensors. Supersonic projectiles entering the enclosed space produce a shock wave which is triangulated to give accurate positioning. At the base is an aluminium housing containing the sensor processing electronics. The housing is integral with the frame, giving air sealing and mounting rigidity. The target is insensitive to firing angles as the shock wave is contained within a thin air-slice. It is impervious to rain, snow, ice and wind effects.

The Box Target permits supersonic projectiles to be located with great accuracy and is therefore suitable for target rifle competition shooting. It can also be adapted to be used with certain calibres of subsonic ammunition. The system is compatible with ADI's Single Lane PLS Systems and Computer Controlled Ranges (CCR).

Box Targets can be mounted on existing manually operated target frames and recovered after use or they can be permanently mounted in containers which allow the target to be raised and lowered automatically as required. Several sizes are manufactured and the size of the aiming mark varied to suit particular applications. Replacement of the aiming mark is only necessary after several thousand rounds have hit the target.

ADI Single Lane System (SLS)

The ADI Single Lane System (SLS) is a low-cost, multipurpose target marking system. It is a variation of, and an extension to, the ADI projectile location technology.

The SLS may be permanently installed or portable. In the portable mode the system is easily transported to range areas and can be assembled and deployed ready to use within minutes. Rechargeable batteries or mains power may be used at the target and at the firing point. Communication between the target array and the firing point may be via land-line or UHF telemetry link.

Up to 10 T-bars, each located at a different distance, can operate to one Intelligent Visual Display Unit (IVDU). Each IVDU has an optional integral printer and is user-friendly. Several systems may be deployed to form a multi-lane, multi-bank facility which may be linked to a central computer.

The T-bar may be used with a target lifting mechanism of any type or used in the stand-alone mode with a target permanently on display. Up to 20 T-bars on target lifting mechanisms with all the uprange equipment and logistic support items may be deployed on a storage trailer to create a mobile and portable PLS for use with either small arms or AFV main and secondary armament.

The software used may be standard or designed to suit specific requirements.

An ADI PLS Single Lane System firing point

PLS Characteristics

Characteristics marked * apply only to T-bar or open arrays

Detection Area*: With the open array of sensors the detection area for small arms is a radius of 2.5 m from the target centre. The radial distance is increased to 20 m for AFV main armament weapons.

Accuracy: The mean point of impact of a group of shots located in the centre of the target is ±1 mm. Accuracy gradually diminishes towards the outer edge of the detection area and as the round velocity drops below 450 m/s.

Missed Rounds*: Rounds that miss the target but are within the detection area of the open array are located and displayed on the firing point VDU.

Wide Shots*: With the open array the position of shots that are within the detection area, but outside the VDU display area are indicated by an arrow pointing in the direction of the shot position.

Short Rounds: Shots which fail to arrive at the sensor array are identified when a lane initiator is used.

Ricochets*: Ricochets which pass through the detection area of the open array are identified as such but no position is displayed on the VDU.

Crossfire: Crossfire shots from other lanes are subjected to a high degree of discrimination and are indicated, rejected and not scored.

Minimum Remaining Velocity*: The open array will operate satisfactorily in conjunction with all conventional ammunition that is capable of producing a remaining velocity of at least 450 m/s at the target. Some subsonic projectiles are also detected using the enclosed array.

Firing Modes: All firing modes such as single shots, semi- and fully automatic fire may be used on this system providing that the ammunition complies with the previous paragraph.

Rounds numbered: Rounds are numbered in order of arrival alongside the shot position.

Last Shot: The last shot of a string 'flashes' on the VDU.

Rate of Fire: Up to a maximum of 2200 rds/min.

Weapon and Ammunition: The system can be programmed to operate with virtually all combinations of weapon and ammunition types, subject only to the limitations of remaining velocity and rate of fire.

Hit, Miss and Scoring*: The Hit/Miss decision is based on the comparison of the shot position against a mathematical hit registration device incorporated in the target mechanism. Therefore virtually any target material of any shape and size can be used with the open array.

Scoring Zones: Specified areas on and off the target may be identified for scoring purposes. These areas may be of any dimensional shape.

Target Mechanisms: The PLS may be used in conjunction with any type of target mechanism but special consideration must be made with regard to moving targets. A target lifting mechanism is not essential.

Zoom: A zoom capability permits a shot concentration on the target to be viewed in several different sizes on the VDU.

Status

In production and widespread service.

Manufacturer

Australian Defence Industries Limited, Bondi Junction.
(For details see Directory of Suppliers)

ADI Robot Target System (RTS)

The ADI Robot Target System (RTS) was developed as a training aid for counter-terrorist and special police forces. The RTS also has applications in military operations and urban training and in the close-quarter battle. In the urban environment the RTS has a role as an agile figure target and several RTS present very realistic targets on an ambush range.

The RTS comprises a low profile platform base on which stands a human form target. Rechargeable batteries, circuitry, an obstacle-avoidance system and drive mechanisms are contained within the base, which is protected by lightweight armour panels. The system uses a replaceable three-dimensional human form target which is realistic in appearance and can be dressed for all occasions. It has 'fatal' zones in the head and heart areas so a hit in either of these areas causes the platform to stop or the target to fall, depending on a switch setting. The machine can also be set to react to non-fatal hits with a stop or fall capability. Reactivation is achieved by radio control. If the target is set to fall it must be manually reinstated afterwards.

The target can accept all small arms calibres up to 0.45 in (11.43 mm); shotguns may be used but the target life is reduced. Once 'shot out' the target is quickly and easily replaced.

When the RTS is set for automatic movement it is capable of moving around a furnished room without colliding with the walls or furniture. It can turn away

ADI Robot Target System (RTS) in operation

from objects cleanly and with a minimum loss of speed. The platform turns in curves such that the target is always on the move and is therefore difficult to hit in a 'fatal' area. The RTS can also move at random. In this mode it behaves erratically whilst travelling between obstacles and is thus extremely difficult to 'kill' whilst moving at full speed. Sustained automatic fire can reduce this difficulty significantly but such a practice is considered unrealistic due to the potential effect of indiscriminate sustained automatic fire inflicting casualties to bystanders or hostages in real situations.

The RTS is radio-controlled via the UHF frequency range. A joystick governs the speed and direction of the platform while the obstacle-avoidance system may be left operating to avoid accidental collisions. When activated, the obstacle-avoidance system takes over from the operator, avoids the collision, and then returns control to the operator.

The RTS is supplied with replaceable, bullet-proof armour. The base is protected against low velocity rounds by the panels, but as the panels are easy to fit they can be changed according to the weapon calibre in use.

The RTS is designed to operate indoors and outdoors within certain surface and weather parameters. The onboard power supply is taken from 42 V rechargeable batteries which provide an operational life of between 60 to 100 operations at 30 s/operation. Maximum speed is 3 m/s.

Status

In production.

Manufacturer

Australian Defence Industries Limited, Bondi Junction.
(For details see Directory of Suppliers)

ADI Two-way Turning Target System

The ADI Two-way Turning Target System was developed to meet the requirements of training military and police personnel and security agents in the demanding skills of counter-terrorist and security operations. The system is designed for use on both indoor and outdoor ranges and may be permanently installed or used in a portable role.

The two-way turning target mechanisms are contained within their own carrying frame and can be fitted with all types of realistic full-size figure targets. From the edge-on position the mechanism can be turned through 90° to present either of two faces of the target. The turning time is 0.3 second.

The mechanisms offer a choice of hit functions such as waggle when hit, turn away, or turn away and return. Power is supplied by an internal 12 V battery with an in-built charging system.

Standard control consoles are available for systems of 6 or 12 mechanisms. The console unit contains a timer unit to allow exposure times to be varied and the number of exposures to be preset.

Friend and Foe Target System

The ADI Friend and Foe Target System is a price orientated system that can be used with all small arms. Each target mechanism is independently controlled and a system may comprise a single or any number of target mechanisms. An inexpensive three-core flexible cable of any length connected to a three-position toggle switch controls the target functions. From an edge-on start position the target will face either direction in less than one second. A variety of target types may be used. The basic system has no visual target hit indication but there is provision for a 'turn away when hit' function as an optional extra.

A target mechanism weighs 38 kg complete and is 370 mm high, 550 mm wide and 670 mm deep.

Status

In production.

Manufacturer

Australian Defence Industries Limited, Bondi Junction.
(For details see Directory of Suppliers)

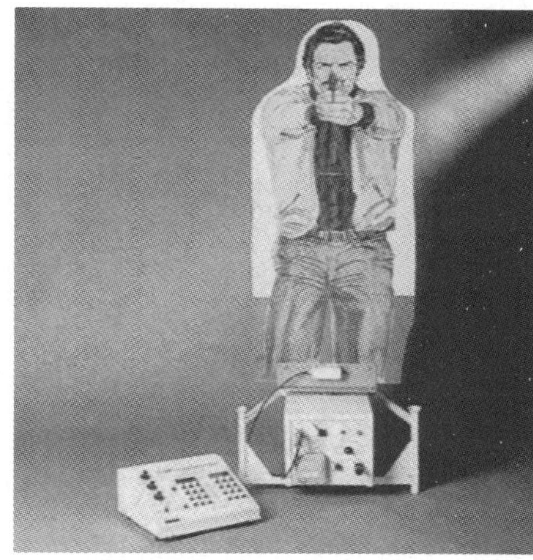

ADI Two-way Turning Target System

ADI UHF Radio Control System

The core of the ADI Radio Control System is a lightweight battery-powered Transceiver, or Hand-Held Controller (HHC) assembly which controls target mechanism assemblies. The target assemblies may be stationary or moving vehicle and figure targets and can be operated using single and multiple exposure combinations.

At the firing point the system comprises a programmable UHF HHC with a battery pack. At the target is a UHF transceiver to control the target assembly.

The HHC has a 32 push-button keyboard. Eight buttons are used to programme the target sequences and the remainder are used to operate and obtain scores from the targets. Control functions include: target up; target down; target up if not hit; target up after hit; target hold up; target detonate/illuminate; bank (of 10 targets); all targets; target score; clear score.

For moving targets the functions are: target move left; target move right; target auto-traverse; target stop.

The first three of the moving target functions are all at slow or fast speed.

Target commands and scores are displayed in a window above the keyboard. The display uses a 2 × 16-line liquid crystal display.

For programming the HHC uses a real-time sequencer which enables the operator to programme up to 500 events within 1 hour. One event equals one target command, for example target(s) up or target(s)

ADI UHF Radio Control System Hand-Held Controller (HHC) in use

down. Alternatively a computer may be used to programme, store and transfer target exposures sequences as required. All targets may be controlled individually, in groups of 10 targets (bank function), or all targets may be controlled simultaneously, for example all targets down.

The HHC can be used with all ADI target mechanisms. By using a headset with a microphone, two-way communications can be obtained between the deployed target assemblies and the HHC.

All range practices that require the target to 'fall when hit' can be conducted with the HHC. Using 30 figure target mechanisms deployed in three banks each of 10 targets, all small arms range practices normally conducted on known distance ranges by day or by night can be programmed and run repeatedly, and the scores obtained. Subsequently the targets can be rearranged to simulate a defended locality under assault or configured to enable firer moving and other assault practices to be conducted for individual and advanced collective training. Similarly tank main armament and anti-tank live firing practices may be conducted separately or in combination with all types of semi-automatic and automatic weapons.

The electronics of obsolescent in-service ADI equipment may be upgraded to ADI UHF standard simply and quickly as it requires only the transceiver, cable and antennas to be replaced and fitted to the target mechanisms.

The HHC weighs 2 kg with an antenna. The transceiver weighs 3.4 kg and the battery pack 3.6 kg.

Status

In production and widespread service.

Manufacturer

Australian Defence Industries Limited, Bondi Junction.
(For details see Directory of Suppliers)

BELGIUM

Browning Mobile Target Stands

There are three mobile target stands in the Browning range, the P1 and P2, and the B3. All three were produced primarily for use by law enforcement agencies but are suitable for other applications.

The P1 and P2 target stands are similar and differ only in the method of presenting the target. Both can present targets in a pop-up, swing-up or rotating mode, according to the positioning of the target stand unit. The P1 model allows a 90° target rotation clockwise or counter-clockwise according to the preselected position of an inverter unit inside the stand case. The P2 model can rotate through 90° either way without preset controls. Both target stands are supplied in the form of a stainless steel case. Both have dimensions of 200 × 280 × 300 mm and weigh 15 kg without their 12 V rechargeable battery.

The B3 target stand is larger and provides a combination of pop-up and friend or foe targets. For the latter, targets can be rotated through 180° when the target arm is in the down position. The B3 target stand is powered by compressed air provided from an outside source. The B3 weighs 25 kg and measures 660 × 500 × 700 mm.

Browning B3 mobile target stand

Browning P2 mobile target stand with target in rotating configuration; the P1 is essentially similar

All three target stands can be controlled by one of three methods: manual, timing or computer. In all three cases control signals may be sent to the target stands by wire, radio or via an electrical mains induction system. The P2 and B3 stands can be fitted with an impact detection system.

For manual control each target stand is controlled by one or more switches on a console located at a range control point. The same console can be used for the timing control mode when two timing systems are used. One timing system controls the exposure of a target after a timing sequence is initiated while the other determines the length of time a target is exposed.

Computer control is carried out using a standard personal computer with programmed sequences stored on a floppy disk. Once a programme has been stored the use of the system is simple and no prior knowledge of computer programming or operation is required. Up to 99 control programmes can be stored on one floppy disk.

Status

In production and widespread service.

Manufacturer

Browning SA, Herstal.
(For details see Directory of Suppliers)

Browning P2 mobile target stand with target in pop-up configuration

Control point for Browning mobile target stands showing manual or timing control console (left) with a radio control unit behind it, and a computer control on right

---

FN TOM Military Firing Ranges

Developed and produced by FN, the TOM (Training Outdoor Military) systems introduce new concepts in outdoor firing ranges, allowing valid and credible situations. TOM exists in two versions: TOM CC (Close Combat) and TOM FL (Flexible).

TOM CC is a fully computerised system based on an AT-compatible; while TOM FL is a field simulator, allowing tactical engagement.

TOM FL uses targetry to simulate the enemy threat and its evolution according to attrition or various user-programmed conditions until victory or defeat.

TOM FL can be either wire- or radio-controlled for use on permanent or unprepared ranges and relies on a novel power supply system eliminating the shortcomings of Ni/Cd batteries.

TOM FL is totally modular so can be readily matched to specific customer requirements while developments will, by design, be compatible automatically with the basic system. This concept enables the users to upgrade

TOM FL - fighting training system

A soldier engaging TOM FL targets under combat conditions

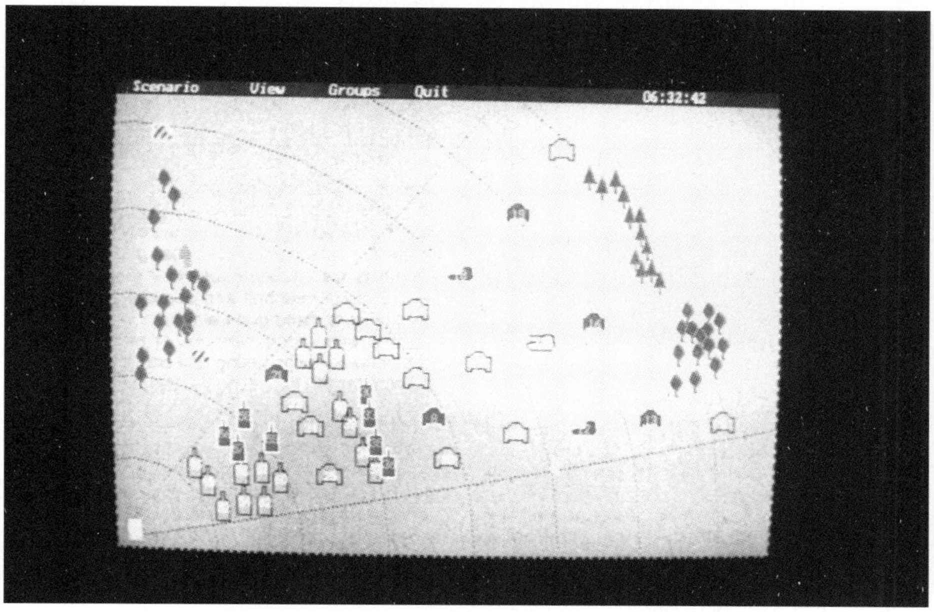

TOM FL computer monitor display showing range target locations

A TOM CC target unit showing a friend target

their training system according to the evolution of future threats and missions. Transport, installation and maintenance are simple and low cost. The TOM FL system can be used for all types of training, including defensive and battle firing, fire and movement, individual assault lanes, unit firing and instinctive firing. The useful range is up to 5000 m for calibres up to 120 mm while the target management capacity is up to 4000 targets on one unique frequency (radio version).

One of the main features of TOM FL is the system software which can be used by non-specialist personnel after a minimum of training. Although user-friendly the software is highly effective since it provides a simulated enemy with reactions dependent on the shooter's performance for the level of difficulty selected. The fictional enemy can win as well as lose, so the validity and credibility of the training are close to combat conditions, allowing the imposition of stress, incentives and penalties. Consequently the training value of TOM is extremely high. It has virtually unlimited scenario possibilities, plus the real-time command and control capability provided by the command unit with its fully topographic field representation using a colour monitor.

TOM software is further used to operate the system computer to check the correct functioning of all targets and also check the position of targets compared to a selected scenario. The computer monitor can be used to provide a complete view of a range and close-ups of specific areas on the range. The system computer and its associated hardware are fully militarised, with all major elements located in shock-proof metal racks. All the hardware can be transported in jeep-type vehicles without risk of damage.

The targets used with TOM FL are metallic and weather-proof and require little or no maintenance. Three types of pop-up infantry target are available; prone, kneeling and standing. Basic target units can be assembled into either a head-on or profile armoured vehicle target or a troop carrier target. All target actuating units are identical, using individual address plugs to guarantee electronic identification of each target. The address plugs allow the auto-checking of the exact position of targets at any time using the system computer monitor screen, without recourse to on-site inspection.

Among accessories available for use with TOM FL is a gas gun producing an explosion simulation, adjustable to simulate a small or medium calibre weapon report. It is available as a single gun or a pair of twin-gun units capable of simulating bursts. Effects produced are sound, shock wave and flame.

A special power system has been developed by FN to power TOM FL which uses constantly charged batteries. No daily battery maintenance is required, so operating and maintenance costs are low.

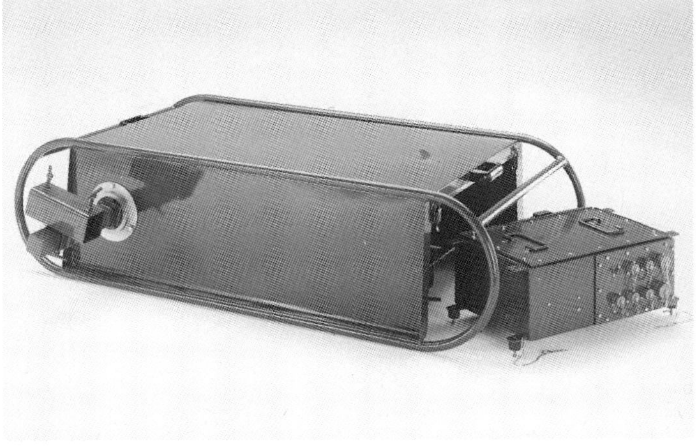

TOM FL tank target mechanism

The TOM FL system computer and monitor in use in a field environment

TOM CC is a close combat training system which features a wide variety of targets designed to provide close combat training situations and introduce discriminative firing in credible and valid scenarios. Available targets include pop-up, disappearing, running man and friend or foe. The system employs simple-to-use software and cables to provide close combat training, using weapons up to 7.62 mm in calibre and at ranges up to 500 m.

TOM CC is cabled. Its command unit manages up to four range racks redistributing messages to and from 100 targets each, thus allowing up to four different combat centres to be managed from one central unit. The real-time display of all sites appears on screen for full control. Operation is easy, in both full manual mode and sophisticated scenarios, which include random for rapid and simple programming of long duration shooting exercises.

TOM CC can incorporate synthetic sound simulation, using pyrotechnics or gas methods, for realistic stress generation.

Status

In production. In service with the Belgian Army.

Manufacturer

FN Herstal, Herstal.
(For details see Directory of Suppliers)

CHINA, PEOPLE'S REPUBLIC

BRC-1 Portable Programmable Radio Control System

The BRC-1 portable programmable radio control system can be used to control small arms range targets but has other applications in areas such as traffic control and so on.

The complete system comprises a command transmitter (plus one spare unit) and 36 receiver units. Every receiver can perform four functions. The transmitter has three preset frequencies with 12 receivers for each frequency. The effective control range is more than 3000 m in open terrain and not less than 1500 m in built-up areas.

The controller can use either random programming, manual or automatic preset control. Control commands can be written in EPROM for storage in the memory and use at any time.

Each receiver unit has four twin-pin output points with a capacity of 1 A at 27 V.

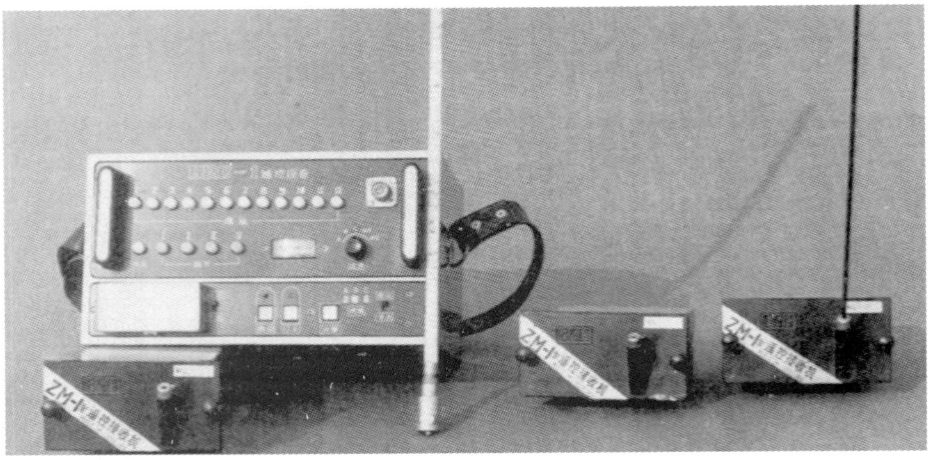

BRC-1 portable programmable radio control system

Transmission frequencies: 3 points around 154 MHz
Transmitter power: more than 2.5 W
RF modulation: FM
Weight:
(transmitter) 5.5 kg
(receiver) 0.85 kg
Dimensions:
(transmitter) 290 × 160 × 210 mm
(receiver) 160 × 110 × 70 mm

Status

In production. In service with the People's Liberation Army.

Developing Agency

Nanjing Research Institute on Simulation Technique.
(For details see Directory of Suppliers)

Target Systems for Infantry Training

This system includes a portable all-weather operation integrated system for infantry training consisting of pop-up targets, tank targets, sound units, smoke and pyrotechnic displays, a programmable radio control system, plus other range accessories for day and night training. It can be used for both tactical training at platoon or squad level, and firing practice with live ammunition for various light weapons.

The system involves a lightweight electric pop-up target mechanism using a plastic target board. The mechanism is driven by a 24 V permanent magneto motor and rechargeable batteries and can be used in wind speeds of up to 10 m/s.

Also included is a moving target system for light weapons consisting of a target, remote-control unit and rails. The rail length is 70 m and weight is 336 kg; the moving target weighs 26 kg together with the battery and receiver. Target speed is 2 to 3 m/s and it can climb an angle of more than 3°. Power is supplied by a 24 V 5 Ah Ni/Cd battery which can provide up to 1.5 hours of operation or 100 two-way movements.

Other similar target mechanisms powered by pneumatic systems have been developed as well as petrol engine-driven moving targets. These, plus the electrical

Electric pop-up infantry target with plastic target board

Moving target system for light weapons

systems, are in service with the People's Liberation Army.

Status
In production. In service with the People's Liberation Army.

Developing Agency
Nanjing Research Institute on Simulation Technique (NRIST).
(For details see Directory of Suppliers)

Electric pop-up infantry target unit with radio control unit (left) and portable control box (right)

FRANCE

Cegelec Soprano EM-10 Variable Speed Target System

This is a monorail system which carries one or more trolleys bearing falling targets. The rail can be of any length and the trolleys are moved by cable. The operation of the trolley is controlled by radio from a control point, which allows the instructor to raise and lower the target and move the trolleys backwards and forwards at varying speeds. The target falls automatically when hit and is re-erected by radio control. The system is adaptable to any service calibre weapon from 5.56 to 20 mm.

Trolley speed: (max) 5 m/s
Motor rating: 4 kW
Power supply: 220/280 V, 50 or 60 Hz (optional)
Target support power: 24 V DC battery
Battery life: 3000 operations
Radio transmitter: 150 MHz, 1 W, 9 V DC

Status
In production.

Manufacturer
Cegelec Soprano, Vaulx Milieu.
(For details see Directory of Suppliers)

Cegelec Soprano P-10/20/30 Multiple Position Target System

This system combines three electromechanical modules in the target holding mechanism which permit a variety of movements. The target can be made to rotate through ±90°, thus permitting the presentation of 'enemy' or 'friendly' aspects, as well as an intermediate 'edge-on' position; it can be commanded to rise and fall; the rising action can be combined with the rotating action; and the entire target unit can be revolved so as to add a third movement to the other two. Almost any conceivable target movement can thus be simulated, and the system provides a versatile training device for military or law enforcement personnel, as well as offering a wide selection of options for competition. It can be used with live or blank ammunition or with laser simulation systems. The target unit can be fitted with a hit recorder on request.

Power supply: 220 V 50 Hz (60 Hz optional)

Movement delay
Turn: 0.2 s
Rise: 0.6 s

Status
In production.

Manufacturer
Cegelec Soprano, Vaulx Milieu.
(For details see Directory of Suppliers)

Cegelec Soprano P101 Multi-function Remote-Controlled Target

This system was developed in collaboration with the French Army and Police as an improvement on the earlier P-10 system, and is intended for the training of infantry in small arms firing under realistic conditions. It can be deployed as a formal firing range or as an irregular tactical-type display, and can be used by day or night. It is fully waterproof.

The system consists of a remote-control unit with the possible use of any radio transmitter, and up to 100 target holders with receivers and batteries. The targets can be deployed at any distance from the firing point compatible with the type of radio in use.

On impact, the target automatically falls, thus indicating a hit; its return to the upright position can be either automatic or manual under the push-button control of the instructor. Lowering and raising the targets can be performed by remote-control, either individually or collectively up to the maximum number of targets deployed. In order to give a wide and varied range of exercises, and also to prevent the trainee from memorising sequences and timing, the target can hold four arms and execute five different movements.

The remote-control unit consists of an MSI fully integrated computer and radio command with the possible selection of any individual target, and the possible tape recording of various target movement scenarios (up to two hours). The system includes automatic feedback of the scoring of each individual target and display at the control unit. The MSI control box is equipped with user-friendly systems and a US-standard microchip terminal.

The P101A anti-personnel target is the only one able to execute five different movements: Eclipse to the rear; Pop-up, right and left; Rotation to the rear; Axial rotation (UIT); Rotation with friend or foe face display.

Liaison with the control unit can be carried out by radio or by wiring. The P101A target is designed for reflex and selective training. Rotation through 90° is possible in both directions in less than 0.2 second. This equipment can be used for UIT Olympic firing or for law enforcement training with the Friend or Foe silhouette.

Cyclic or unpredictable appearance of targets can be programmed, and the silhouette can be presented normally or obliquely. The results are fed back to the range officer who is informed of the scores of each trainee. Security and safety of personnel are thus improved, as there is no requirement to approach the targets at all.

The entire system is modular and can be configured as a permanent or temporary facility. The range of available equipment and accessories permits a wide variety of permutations.

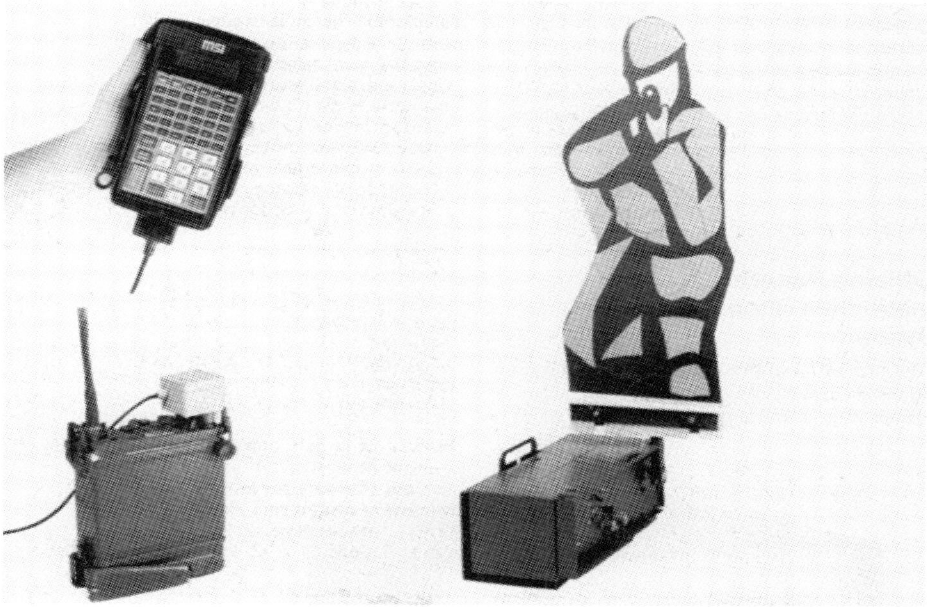

Elements of the Cegelec Soprano P101 target system

Target
Weight: 25 kg

Dimensions: 600 × 270 × 300 mm
Power supply: 24 V DC by incorporated batteries, or 120 V AC, 60 Hz or 50 Hz

Control unit
Weight: 0.5 kg
Dimensions: 90 × 170 × 30 mm
Power supply: 12 V DC battery
Temperature limits: −30 to +50°C

Status
In production. In service with the French Army and various police forces.

Manufacturer
Cegelec Soprano, Vaulx Milieu.
(For details see Directory of Suppliers)

Cegelec Soprano Programmable Control Desk

This is a P10 standard control desk with an interface for the connection of a low-cost magnetic tape recorder (300 Hz - 3 kHz bandpass) allowing the recording and/or execution of scenarios conceived by the user. No special programming knowledge is necessary. The instructor controls the targets in the usual way and can record the scenario in real time, of a duration limited only by the magnetic tape used. It is thus possible to automate training, to construct complex figures, and to repeat the same scenario as many times as desired. The programmable desk may be used with wire control for police targets, or with radio control for infantry targets.

Status
In production.

Manufacturer
Cegelec Soprano, Vaulx Milieu.
(For details see Directory of Suppliers)

GERMANY

Theissen Target Appliances

Theissen Fassaden- und Maschinenbau GmbH produces a wide range of target appliances for small arms and full calibre ranges. Small arms targets include infantry pop-up and turning target mechanisms with or without firing simulators that produce enemy rifle, automatic or anti-tank weapon fire.

Range installations may be radio- or cable-controlled. Radio-controlled ranges may be completely mobile and established as and when required using a radio control system using coded plugs for radio transmitters. The system allows the parallel operation of a large number of different radio transmitters of the same high frequency without mutual interference. Transmitter codes can be changed easily by changing the coded plug. The system operates in the 135 to 170 MHz band and by using digital and coded data transmission allows the control of up to 200 different receivers from one transmitter. Mobile installations are operated by Ni/Cd batteries for which Theissen can supply the charging and service units.

Cable-controlled installations involve a central control room which can contain computerised control units. On a typical Theissen cable-controlled range a protected data cable controls the targets. All commands are digital and coded and only one four-core data cable transmits commands to all units. On computerised ranges up to 40 moving targets and 500 other targets can be controlled simultaneously. The location of the targets and hit results are indicated on a monitor in the control room.

Theissen also produces a Precision Target System with which LOMAH (Location of Miss and Hit) sensors measure the exact position of a hit on a target. The system is computer-controlled and displays the hit location as a luminous dot on a monitor unit screen. Hard copy printouts of results can be produced.

Status
In production. In widespread service.

Manufacturer
Theissen Fassaden- und Maschinenbau GmbH, Düsseldorf.
(For details see Directory of Suppliers)

HUNGARY

IDEX Portable Programmable Target Controller

This portable programmable target controller can control up to 199 targets. An alternative frequency increases the total to 796. The controller can make targets appear or disappear as required and the controller can also initiate training pyrotechnics and smoke. The controller can be preprogrammed before an exercise but the programme can be interrupted and altered at any stage. It is stated that no special skills are required to operate the controller.

Number of targets controlled: 199 (up to 796 with frequency change-over)
Number of orders per target: up to 8
Number of programme steps: up to 999
Programme run time: up to 10 h
Power supply: 12 V DC, integral or external battery
Operating frequency band: 60 MHz
Range: approx 5000 m

Status
In production. In service with Hungarian armed forces.

Manufacturer
IDEX, Budapest.
(For details see Directory of Suppliers)

ITALY

Officine Panerai SAT 213 Instinctive Shooting Training System

The Officine Panerai SAT 213 instinctive shooting training system consists of a control unit and a number of dynamic targets. The targets may be of the turning or pop-up type and are intended to provide training in the instinctive and accurate live firing of small arms under day and night conditions.

The SAT 213 system enables instructors to present targets using a number of preprogrammed sequences. The programme can present friend or foe targets, vary the exposure time for each target from 0.1 to 10 seconds in 0.1 second intervals, vary the order of activation of the targets, and vary the number of target activations for each training sequence (up to 60).

The control unit consists of a personal computer with an 8086 processor, featuring 256 kbytes of random access memory and a floppy disk unit with 360 kbytes. The computer is housed in a special metal container

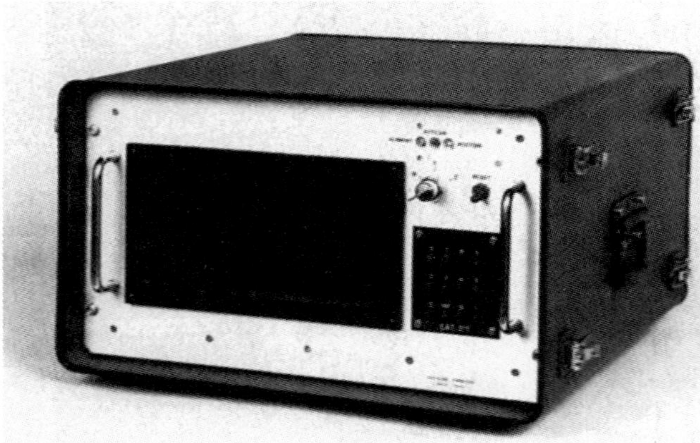

SAT 213 shooting system control unit

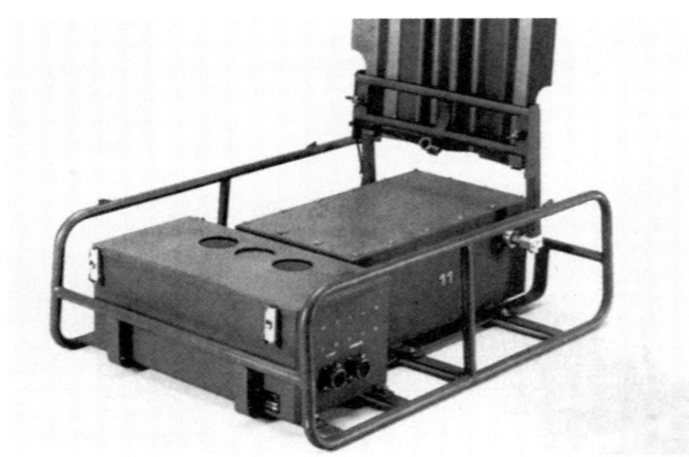

Detail of SAT 213 shooting system pop-up target

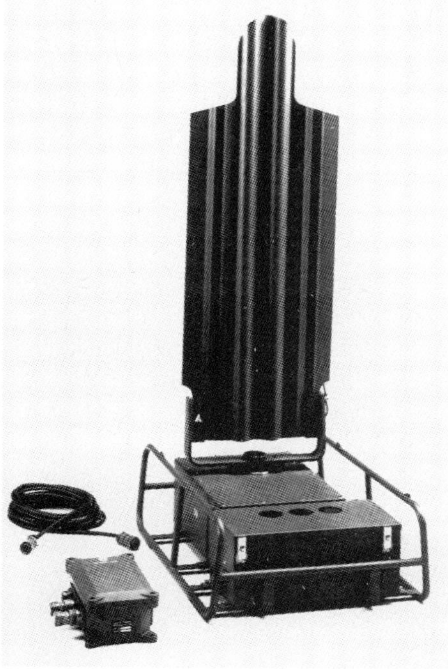

SAT 213 shooting system turning target

together with the interface electronics. There is provision for connecting a printer. An interface unit enclosed in a metal container can be supplied separately for connection to an IBM personal computer together with the user program on a floppy disk.

The control unit is used to programme, check and connect the operational parameters of a training sequence. The unit also contains in its memory the full sequence of 60 target activations, carries out its own self-test procedures, controls the programmed sequences, monitors the trainee's scores and subsequently evaluates the results.

Each target is controlled by its own microprocessor and is equipped with a three-colour lighting unit for use during night firing. Each target can be used either connected to the control unit or in a stand-alone mode using a random sequence of operations. Each hit on the target is detected by a hit sensor and recorded together with the length of time from the initiation of the target exposure. At the end of a sequence the control unit provides an indication of results by presenting a score of from 0 to 100, indicating target discrimination, accuracy, speed of reaction and economy of fire (number of rounds fired). The unit can also provide detailed results for each activation of each target involved. Two standard-sized targets are available. One is for a standing figure and measures 1270 × 466 mm overall. The other is for a kneeling figure and measures 980 × 466 mm overall. Each target can absorb several hundred hits before replacement is necessary.

The SAT 213 system can accommodate acoustic and infra-red sensors for target initiation in place of the control unit.

The SAT 213 system requires a 220 V AC supply (±10%), 50 or 60 Hz. All connectors are watertight.

Status
In production.

Manufacturer
Officine Panerai SpA, Florence.
(For details see Directory of Suppliers)

Scovati Italia Indoor Shooting Ranges

Scovati Italia has been producing indoor and outdoor range equipment on a 'turnkey' basis since 1980. Its indoor shooting ranges are constructed along traditional lines with an instructor booth located behind the firers and above the level of the gallery. The booth is sound-insulated and bullet-proofed and contains all the control and communications equipment required for the running of the range. Shooting stalls have side walls for ballistic protection of the students and there are safety bullet-retention panels in the firing area as well as protection panels along the length of the gallery. The end of the gallery has a bullet trap which may be of the anti-pollution 'Holdmatic' type produced by Scovati Italia.

Targets may be located in mobile target carriers and can be fixed/rotating, fixed/two-rotational, sliding/rotating, and sliding/two-rotational; friend or foe targets may be used.

General view of a typical Scovati Italia indoor range

If required, Scovati Italia can provide a variety of multi-stall combat ranges along which trainees can move towards the targets. Single stall combat ranges include a shoot-on-screen simulator with which firing causes the image projected onto the screen to stop and a blinking light denotes the position of the bullet strike; a cinerama screen installation is available, as is a library of training scenarios. Another type of installation involves pop-up targets; the targets are retained at floor level and pop up under electrical and compressed air control.

Scovati Italia can also provide all general range installation equipment such as lighting, cable and radio communications, environmental monitoring and conditioning systems.

Available indoor ranges include 100 m four-booth ranges prefabricated using modular elements and traditional shooting ranges up to 300 m long.

Status
Available.

Manufacturer
Scovati Italia, Turin.
(For details see Directory of Suppliers)

Scovati Italia Mobile Shooting Ranges

The basis of the Scovati Italia mobile shooting range is formed using two semi-trailers connected together to create a fully enclosed 10 or 15 m shooting gallery which is totally self-sufficient, providing its own power and carrying its own support services.

One end of the mobile shooting gallery contains a compartment located over the semi-trailer fifth wheel and housing an air-conditioning unit and a power generator. Located in an enclosed booth in front of the power compartment is the instructor's booth containing all the control and communication equipment necessary for the running of the range. The instructor's booth is located close behind the two-man firing position while at the other end of the range are the target holders and a venetian-blind bullet trap capable of stopping 9 mm Parabellum bullets. Targets may be of the conventional or projected type. The two semi-trailers are sealed together after their tail doors have been fully opened. The area behind the bullet trap can be used as a general support and service area with domestic accommodation and office facilities for range personnel.

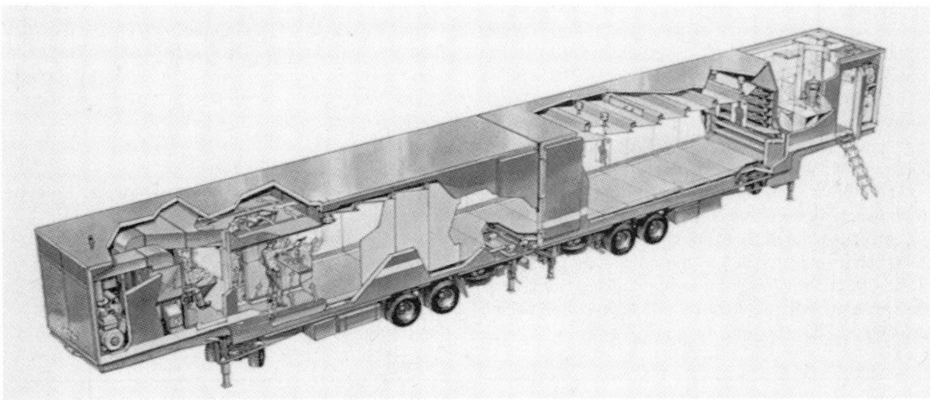

Artist's impression of a typical Scovati Italia mobile shooting range

The entire gallery is lined with baffles; steel and bullet-trap baffles in the firing area and steel baffles along the gallery sides and roof. Sound and environmental insulation is provided throughout and ventilation equipment along with fume monitoring and extraction systems are standard.

,Scovati Italia produces a generally similar range based on the use of one, two or three 44-foot/13.6 m ISO containers. This system is known as CONTAINE-RANGE and contains much the same range equipment as the mobile ranges but the internal service and support areas are not included, although an integral power-generating facility may be provided.

Status
Available. In service with Italian State Police Forces.

Manufacturer
Scovati Italia, Turin.
(For details see Directory of Suppliers)

NORWAY

HMV Rail System MAT-8 Moving Aircraft Target

The HMV Rail System MAT-8 Moving Aircraft Target was designed to provide a miniature air target for infantry training. The target is held above rails mounted 3 m above the ground and a capstan winch moves the target trolley in both directions using a stepless speed control.

The MAT-8 uses 0.25 m gauge vertically mounted rails which may be from 45 to 60 m long using 3 m sections. The rail supports are mounted on concrete and the target trolley can carry targets weighing up to 20 kg. The target is pulled by either a *Vossavinsjen El-Var 4F* 230 V AC capstan winch producing 4 hp or a *Vossavinsjen Ho-Var 11F* winch powered by an 11 hp petrol engine. In both cases a hydraulic drive system is involved and target speeds can vary from 5 to 30 km/h.

Status
In production. In service with the Norwegian Army.

Manufacturer
Hordaland Mek. Verkstad A/S, Voss.
(For details see Directory of Suppliers)

A typical HMV Rail System MAT-8 moving aircraft target system

SOUTH AFRICA

IST TARGET SYSTEMS

Developed to meet stringent requirements brought about by a South African need to produce rugged target systems capable of surviving arduous conditions while providing low unit costs, reliability and user-friendliness, the IST target system range comprises three main elements: a Portable Target System (PTS), a Mechanical Target System (MTS), and a Laterally Moving Target System (LMTS).

The PTS is an armour-protected, two-man portable target system combining a rotary (friend or foe) and pop-up action in one actuator. The battery-powered action can be remote-controlled by radio or infra-red sensor. Hits may be called up and displayed on the remote-control unit and added features such as flash-bang exist. Remote-control is via a Grinel DT309 Advanced Data Entry Terminal which can be interfaced with a client's HF/VHF/UHF radios. The Grinel militarised DP315 printer can provide immediate hard copy results on a score card. It is claimed that the PTS can achieve in excess of 4000 cycles before the battery needs to be recharged. Weight of a single PTS is 56 kg and dimensions are 440 × 440 × 380 mm. In the South African configuration, 12 PTS units are transported on a dedicated target trailer together with onboard battery charging facilities, spares, accessories and targets.

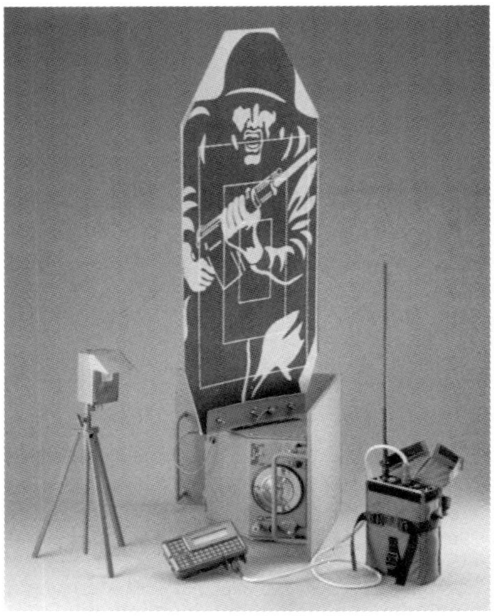

IST Portable Target System (PTS)

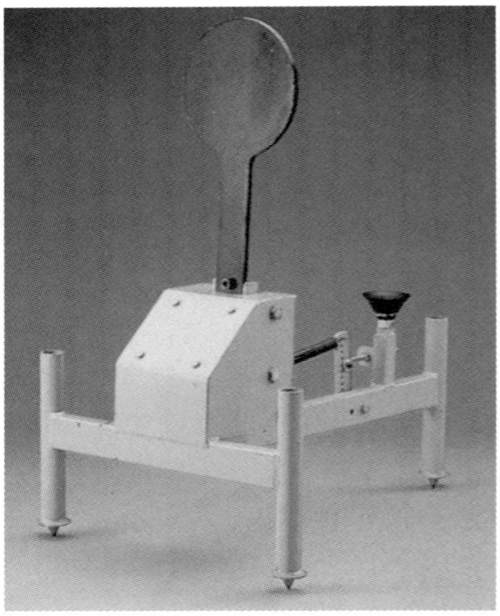

IST Mechanical Target System (MTS)

IST Laterally Moving Target System (LMTS)

The MTS is a purely mechanical pop-up target system which utilises bullet energy on the armoured expendable target plate to effect the pop-up action. The MTS can be used for all 5.56 mm small-arms firing from 15 to 200 m, and 7.62 mm weapons firing from 15 to 350 m. Weight of a single MTS is 15 kg and dimensions are 680 × 370 × 480 mm.

The LMTS presents a target moving at speed across the firer's immediate front for use either in the field or on urban warfare ranges. Power is provided by a petrol engine. Weight of a single PTS is 116 kg and dimensions are 1300 × 1050 × 770 mm.

Status
In production. In service with the South African Defence Forces.

Manufacturer
Integrators of System Technology (Pty) Limited (IST), Pretoria.
(For details see Directory of Suppliers)

SWEDEN

Saab Portable Target Systems

The Saab Portable Target System is used for field and combat training. The complete range includes infantry pop-up targets; tank pop-up targets; gun fire simulators; rifle fire simulators; combat noise simulator; and a radio control system.

The system is flexible and can be used in any combination for the most varied training situations.

BT 18AF Pop-Up Target

The BT 18AF pop-up target is used in fixed range installations. The BT 18AF is normally mounted on a solid vertical protective concrete or wooden wall. Power is derived from a single phase 110 to 240 V AC supply and the unit is controlled by a remote-control system via low voltage signal cables. There are two versions of the BT 18AF, one with and one without a lamp for hit scoring and target illumination. All the mechanisms are contained in a watertight box. The unit can be used for weapon calibres of up to 12.7 mm.

Dimensions: 320 × 320 × 190 mm
Rising and falling time: 1 s
Hit frequency: up to 1200 hits/min
Weight: 21 kg

BT 18BF Infantry Target

The BT 18BF pop-up infantry target is portable and used in flexible installations. A frame supports the mechanism box, battery and receiver box. The target is normally controlled by the BT 22FM radio system or the BT 24 range controller. The whole group is then controlled by one receiver.

There is a built-in lamp for scoring and target illumination for night shooting.

The box containing the electric motor and mechanism and the box containing the electronic circuit cards are both hermetically sealed.

The target disappears, reappears and is illuminated by radio control. The target is sensitive to hits in the raised position.

Dimensions: 830 × 620 × 240 mm
Power supply: Ni/Cd battery 24 V, 7 Ah
Capacity (number of cycles up and down): approx 4000/charge at approx 15 Nm load and 10°C
Weight: (excluding battery) approx 35 kg

BT 19F Sound Unit

The BT 19F sound unit is used to increase realism in field and combat training by adding sound effects. It can be used either for general combat noise or for machine gun sound. When used for general combat noise the sound is stored on a magnetic tape recorder. Sounds such as tank noise and helicopter landings can easily be reproduced. In machine gun simulation the BT 19F is connected as a slave to a pop-up target or controlled from a separate radio receiver. The machine gun sound is generated by high efficiency compression driver units controlled from an integrated electronic unit with sound characteristics stored in a non-erasable memory. The characteristics of the sound from machine guns have been analysed and broken down in a very special form that can be stored in the memory. This information is used to synthesise the machine gun sound without distorting the characteristics.

The simulator can also be used as a high power public address unit near or remote from the speaker. When used in this way it can be controlled from a cable connected microphone or from a walkie-talkie. The ability to use the simulator as a loudspeaker is extremely valuable when instructing soldiers in the field and in all other situations when large groups of people need to be addressed outdoors. The design principles guarantee excellent sound over very long distances.

The simulator unit is self-contained, stackable and easily portable by two men. It is powered from an electric rechargeable battery and is fully rainproof.

Dimensions: (horn down) 1070 × 580 × 400 mm
Weight: (excl battery) 47 kg
Battery: Ni/Cd 24 V
Capacity per charge: more than 15 000 shots

(Right) BT 18BF infantry remotely controlled target (foreground) BT 19F rifle fire simulator

BT 22FM radio control system

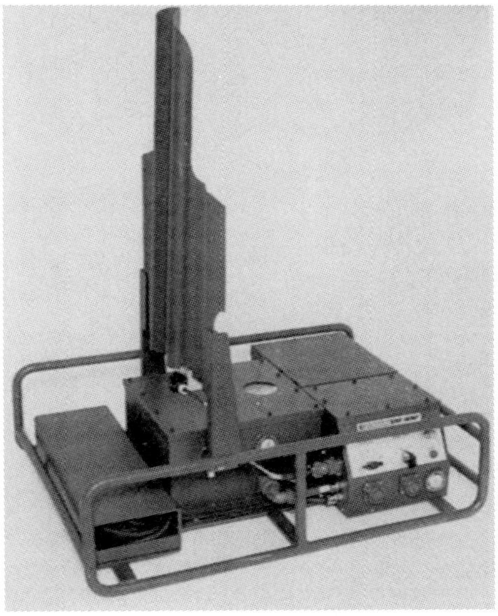

BT 18BF infantry target

BT 22FM Radio Control System

The radio control system BT 22FM consists of one transmitter and a number of receivers (from 1 to 40). Each transmitter offers the choice of any one of five different frequencies. On each frequency eight receivers with three different functions can be operated. The equipment is intended for the operation of stationary or mobile training equipment.

The transmitter is portable and is built into a rugged, rainproof housing. It is designed with CMOS circuits which require a minimum of stand-by power. The front panel has a keyboard with 27 push-buttons for operating the eight receivers, individually or as a group, a lamp for transmission and battery indication and a channel selector knob. This knob is removable and also functions as a key for the transmitter. It is supplied with an easily replaced Ni/Cd battery pack.

The receiver is housed in a diecast alloy metal cabinet which is both rugged and waterproof. It can be used together with various kinds of training equipment. It is designed for minimum current drain.

Frequency range: 143-165 MHz
Modulation: FM frequency shift 1.95-2.6 kHz
Number of functions: 3 - Fire, Up, Down
Temperature range: −40 to +70°C
Operating range: up to 5 km, open terrain

Transmitter
Output: 6 W
Supply: Ni/Cd battery 12 V, 1.8 Ah
Weight: 7.5 kg incl battery (0.95 kg)
Width: 320 mm
Height: 180 mm
Depth: 165 mm

Receiver
Width: 130 mm
Height: 140 mm
Depth: 160 mm
Weight: 1.1 kg

Saab Moving Infantry Target Trolley System BT 37C

The BT 37C target trolley system is a portable modular-built system with a track, trolley and driving unit for moving targets. It is intended for temporary, semi-permanent or permanent installations and requires a minimum of preparatory work. The modular design facilitates construction and minimises manpower

The Saab BT 37C moving infantry target trolley system

requirements for assembly. A 25 m long system can be assembled by two men in less than 1 hour.

The BT 37C system has three speeds: 4, 8 and 12 km/h. It is powered by a 115-230 V single phase supply or batteries. There are two types of supporting legs for the system. One type has fixed legs for permanent installations and the other has adjustable legs for use on uneven terrain.

The system can operate at speeds up to 20 km/h and speeds can be preprogrammed and controlled by the BT 24 or BT 22FM radio control systems.

Rail length: 3 m
Rail cross section: 40 × 40 × 2 mm
Total length: 200 m
Minimum track length: 14 m
Required protection height:
(maximum) 1 m
(minimum) 0.4 m

Status
In production. In service with several European armies.

Manufacturer
Saab Training Systems AB, Huskvarna.
(For details see Directory of Suppliers)

SWITZERLAND

DIXI AK-2 Target System

The DIXI AK-2 target outwardly resembles most similar devices, but there are some differences. It is designed to be fitted with its own armour, although this does not mean that the controls can be left above ground. The armouring is intended to prevent damage from ricochets and falling rubble. The other difference is that the target swings sideways instead of the more usual fore-and-aft movement.

For night shooting there is an electric flash and, instead of lighting the front surface of the target as do many others, the flash is reflected from steel mirrors at the side and directed forward by one which is just above ground level. This rather complex arrangement ensures that the source is well clear of any bullets and the only damage that can be done is to a small mirror. It also gives a clearer and brighter light, while the intensity can be controlled.

The swinging target can be of any size up to 1.5 m high and a fully charged battery will move this up and down for roughly 1000 cycles. Much attention has been paid to ease of maintenance and the manufacturer claims that work is restricted to cleaning off loose dirt and changing the battery.

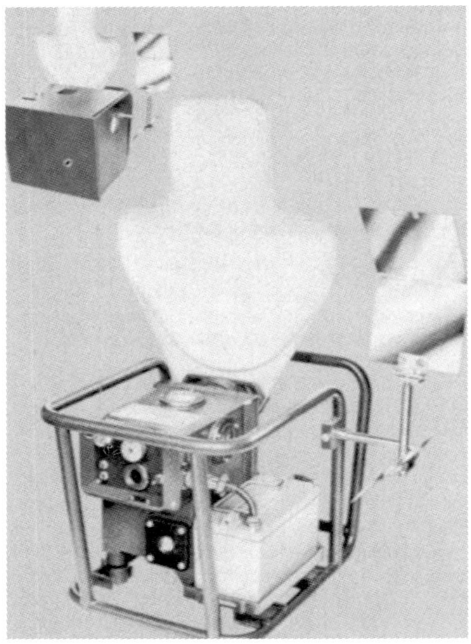

DIXI AK-2 side-swinging target

Target unit
Dimensions: 400 × 545 × 360 mm
Weights:
(unit alone) 20 kg
(battery) 7 kg
(reflector) 1 kg
(front plate) 14 kg
(full armour) 28 kg
Battery: fully sealed 12 V, 10 Ah
Temperature range: −25 to +55°C
Control unit
Dimensions: 280 × 174 × 166 mm
Weight: 6.5 kg
Number of channels: 8
Capacity: 12 targets/channel
Commands/target: (up, down) 2
Collective commands: (up, down) 2
Frequency ranges: 140, 160, 420 and 460 MHz band

Status
In production. In service with the Swiss and German armies and some Middle and Far East armies.

Manufacturer
DIXI SA, Le Locle.
(For details see Directory of Suppliers)

DIXI AK-30 Infantry Target System

The automatic AK-30 target can be radio-controlled or be connected to a range control station, can be operated at fixed ranges by programmable control units, or can be controlled with the DIXI portable computer remote-control unit. It is a motor-actuated pop-up type and can be fitted with single or multiple target screens. Compact, rugged and designed to meet the most severe environments, the AK-30 is reliable, cost-effective and almost free from maintenance costs. The self-contained mechanical elements have a minimum 15-year life expectancy, and a special feature is an electronic unit which prevents total discharge and possible reversal of polarity of the battery cells. There is also circuit-breaker protection against overload.

The shock detector system is sufficiently sensitive to register hits from small calibre ammunition, rubber bullets and plastic training ammunition as well as all types of service small arms ammunition. Optional equipment includes night illumination and flash hit indication, small arms return fire simulation, and armour plating to protect the operating unit from bullets up to 12.7 mm calibre.

Dimensions: 750 × 510 × 300 mm
Weight: 60 kg (48 kg without battery)
Target silhouette size: max 2 m^2
Power supply: 12 V DC lead acid or Ni/Cd battery
Battery capacity: 55 Ah
Frequency range: 68-87.5 MHz (i); 138-174 MHz (ii)
Number of channels: 8
Operating life: over 3500 cycles

Status
In production. In service with the British, German and Swiss armies and some Middle and Far East armies.

Manufacturer
DIXI SA, Le Locle.
(For details see Directory of Suppliers)

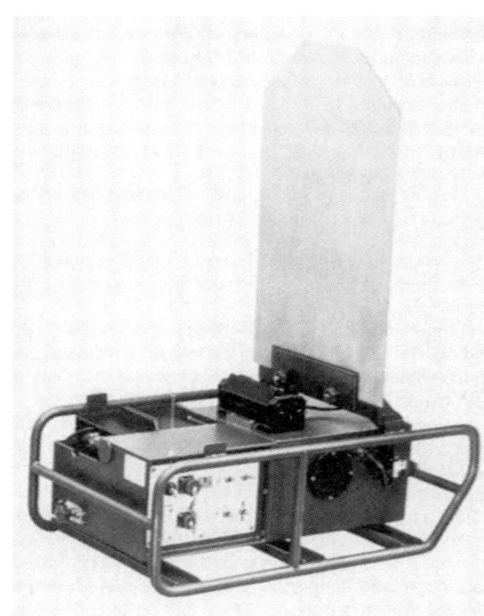

DIXI AK-30 target system

Polytronic Indoor Shooting Ranges

Polytronic International Limited has specialised for many years in shooting range equipment and range design and can supply and advise on a great variety of target installations and control systems and other aspects of indoor range design such as bullet traps, safety precautions, range illumination, sound insulation, ventilation, flooring, and so on.

Examples of Polytronic shooting and target installations are as follows:

TG 82. Pop-up targets (see separate entry).

TG 94/08. Monorail target system, available with intermediate distances.

TG 160. Stationary rotary target group.

TG 161. Mobile rotary target group, fitted to a ceiling.

TG 162. Single friend/foe rotary targets with 90° or 180° rotation (see separate entry).

TG 192. 10 m target transport installation for air weapon

TG 4000. Electronic hit scoring system (see separate entry).

Optical target installations can include slide, film or video combat equipment in combined or multi-lane installations, together with stationary or lifting target screens and light disturbance devices. Accessories include ear defenders, communications equipment, loudspeakers, weapon benches, barricades, and so on.

Depending on the size and purpose of the range the equipment can be controlled (individually and/or collectively) by the Computer RCS Range Control System (see following entry).

Bullet traps include Polytronic's steel lamellas trap. The rear wall of this trap has 6 mm thick steel plates while the lamellas, side and central supports are 10 mm thick. A splinter protection curtain is provided. Various types of rubber bullet traps are also provided according

Polytronic indoor range with SIANG sound insulation and SSB flooring

Polytronic steel lamellas bullet trap

Typical Polytronic indoor range

Polytronic video combat installation

to their suitability for the type of weapons and calibres to be used.

Range lighting may be diagonal to the shooting axis and can include spotlamps for target illumination. As an alternative, indirect illumination can be installed. In both cases the light intensity can be controlled.

Experience has demonstrated that SIANG is an efficient sound insulation system. With this, walls and ceilings are lined with sound-absorbing plates fitted onto a wooden grid with aluminium profiles. Compact foam mats can be glued onto concrete walls for extra sound insulation.

For flooring, SSB 29/30 rubber granulate safety flooring is laid in plates 30 mm thick.

Ventilation systems may be forced air or full air-conditioning.

Due to the complexity of indoor range requirements all Polytronic ranges are custom designed.

Status
All the above are in production and are in widespread use.

Manufacturer
Polytronic International Limited, Muri.
(For details see Directory of Suppliers)

Polytronic RCS Computer Control System for Target Installations and Indoor Ranges

The Polytronic RCS computer control system utilises a software package developed following years of shooting range and equipment experience. The system can control all types of target and target carrier.

Menus are presented on the monitor screen in the language of the user nation and are user-friendly. Due to the nature of the graphic and simple display of target installations an instructor can see the current stage of a

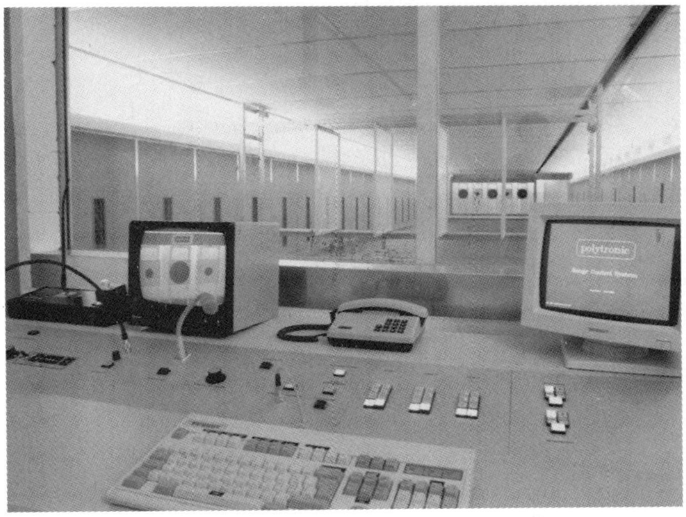

Typical Polytronic RCS computer control installation

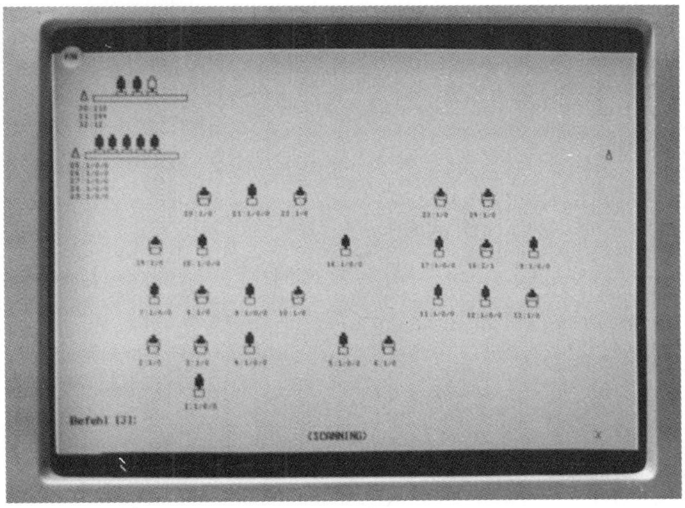

Typical display on a Polytronic RCS computer control monitor

shooting training programme at all times. The following information is shown on the monitor: target silhouette (pop-up, rotary, swing target or target mover); target number; target visible or invisible; target friend, foe or invisible; number of target rises or rotations; and number of hits on friend and foe. Further screen information can be added, such as the designation of shooter, range, line, and so on. Hits are stored and can be printed out in various forms at the end of a shooting programme.

An instructor can create, store and run a programme within a very short time. Range scenarios can be stored and recalled at any time. In addition to the automatic sequences, manual control of each individual unit can be carried out using the computer keyboard.

The RCS computer comprises an IBM or compatible computer, complete with a 20 Mb hard disk, a floppy disk drive, a waterproof keyboard, a colour card and a monitor. Also provided are a printer, an interface, a set of connection cables and the software package.

Special consideration was given to ease of service and maintenance at the design stage, with software maintenance available by telephone, either direct or via a modem.

Status
In production and in widespread use.

Manufacturer
Polytronic International Limited, Muri.
(For details see Directory of Suppliers)

Polytronic TG 82 AC Programmable Bi-directional Computer Radio Control Unit

The Polytronic TG 82 AC programmable bi-directional computer radio control unit can control from one to 99 target mechanisms, target trollies and accessories such as night light, laser sensors, muzzle flash or hostile fire simulators. The unit enables an instructor to select groupings, functions and timing of the various elements and programme the TG 82 AC accordingly. Prepared exercises can be called upon as required.

Target status and hit data are returned to the unit from the individual target device via a portable bi-directional radio link and scored on the LCD display; the display can be illuminated. The system has a range of 1000 m, or 4000 m with a booster option.

The transmitter is powered by a rechargeable 12 V battery but can be powered from a mains supply. Construction is modular to assist troubleshooting and maintenance. The unit housing is cast aluminium with all components made of a stainless or corrosion-resistant material.

The unit measures 230 × 280 × 90 mm and weighs 6 kg. Accessories include a battery charger, an antenna, a range booster and a leather case.

Status
In production.

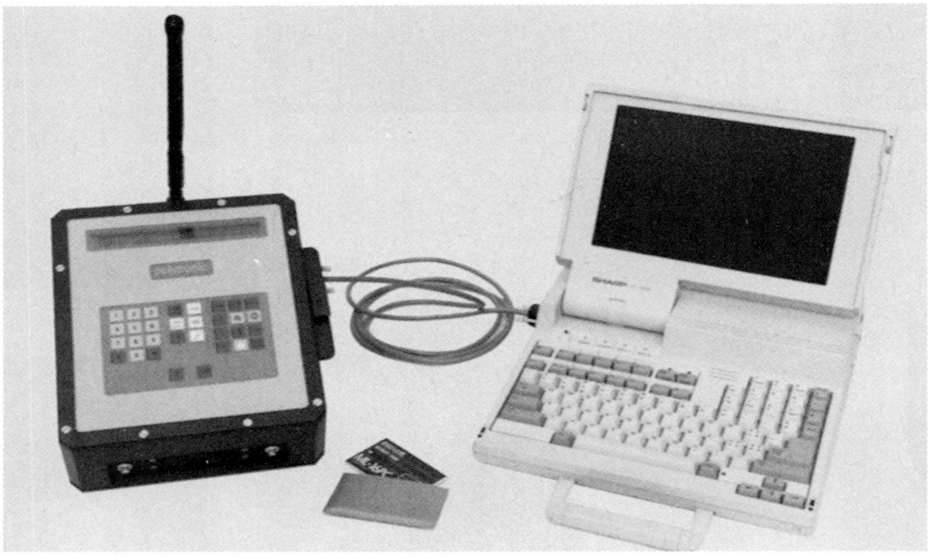

Polytronic TG 82 AC programmable bi-directional computer radio control unit

Manufacturer
Polytronic International Limited, Muri.
(For details see Directory of Suppliers)

Polytronic TG 82 Target System for Infantry

The Polytronic TG 82 pop-up target system for infantry is one component in the TG 82 target system, the other being a tank target system (see entry in *Full Calibre Gunnery Ranges* section for details).

The TG 82 system consists of single pop-up target mechanisms, tracks for moving targets, mono- and bi-directional wire or radio control, night lights and muzzle fire simulators, hostile fire simulators and hit indicators. The customer can select the components that meet his particular requirements and, as the system is modular, other components can be added to upgrade or alter an installation.

The radio-controlled TG 82-10 BES is portable and is powered by a 12 V battery, whereas the TG 82-10 NES is designed for fixed installations with mains power supplies; control can be via radio or via a data cable connected to a computer. The target is rotated by an electronically controlled 24 V motor. The system can be used with the LOMAH electronic hit scoring and evaluation system (see separate entry).

The system utilises electronics with an integrated processor, thereby allowing customer-specified software changes to the functions. The mechanism is fitted with ports for night light, laser sensor, muzzle flash or hostile fire simulators. The sensor electronics were developed so that reactions to near misses, muzzle bursts, wind, debris and splinters are filtered and suppressed.

Mechanism housings are made of cast aluminium with all components of stainless or corrosion-resistant materials.

Polytronic TG 82-10 BES remote-control pop-up infantry target mechanism

The TG 82-10 BES weighs 48 kg and measures 860 × 540 × 290 mm. The mains-powered TG 82-10 NES weighs 29 kg and measures 350 × 350 × 150 mm.

Status
In production and widespread service.

Manufacturer
Polytronic International Limited, Muri.
(For details see Directory of Suppliers)

Polytronic TG 162 Friend/Foe Rotary Target

The structure of the Polytronic TG 162 friend/foe rotary target is based on the main components of the TG 82 infantry target system (see previous entry). The target is available as a fixed hard-wired and mains-powered system or as a portable 24 V battery-powered target. The target is rotated by an electronically controlled 24 V motor.

The system utilises electronics with an integrated processor, thereby allowing customer-specified software changes to the functions. The mechanism is fitted with ports for night light, laser sensor, muzzle flash or hostile fire simulators. The sensor electronics were developed so that reactions to near misses, muzzle bursts, wind, debris and splinters are filtered and suppressed.

Mechanism housings are made of cast aluminium with all components of stainless or corrosion-resistant materials.

The TG 162 weighs 20 kg and measures 300 × 280 × 420 mm.

Status

In production and in widespread service.

Manufacturer

Polytronic International Limited, Muri.
(For details see Directory of Suppliers)

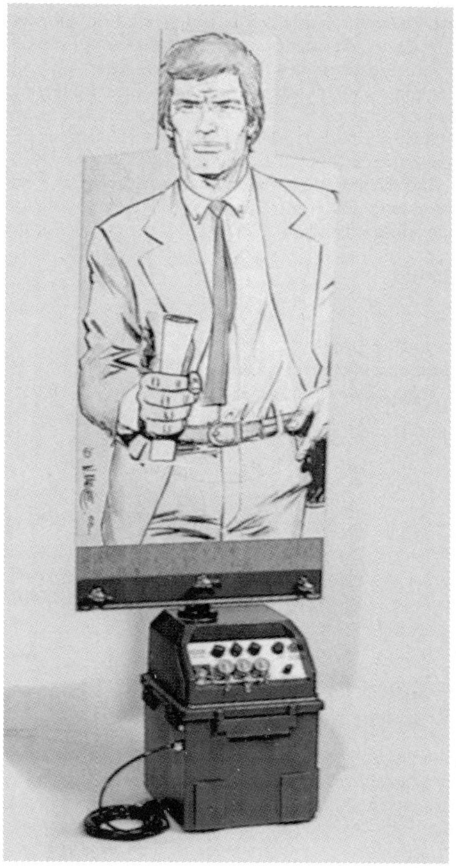

Polytronic TG 162 friend/foe rotary target

Polytronic TG 350 Infantry Target Mover System

The Polytronic TG 350 infantry target mover system is a multifunction system designed for sniper and machine gun training. The target trolley can carry up to three TG 82 or TG 162 infantry target mechanisms with weights of up to approximately 150 kg, and can travel at speeds of up to 15 km/h, adjustable in nine steps. The associated rail track length is selected by the user, with a maximum gradient of 5 per cent. Targets can appear and disappear at random.

The system is powered from a three-phase 380 V 50 Hz motor and controlled by either a TG 82 AC radio control unit or the RCS range control system.

Status

In production and in widespread service.

Manufacturer

Polytronic International Limited, Muri.
(For details see Directory of Suppliers)

Polytronic TG 350 infantry target mover system

Polytronic TG 3002 Electronic Hit Scoring System

The Polytronic TG 3002 electronic hit scoring system is used to provide small arms range users with details of their scores, target hits and shot-by-shot hit locations. It consists of three main components, a control unit, a target and a graphic monitor.

The target operates on the principle that a supersonic shock wave entering a sound chamber behind a targe' is picked up and measured by three microphones arranged at the base of the target. The time difference of the shock wave reaching the three microphones is used by the system computer to calculate the hit position of the round on the target. This information is transmitted to the graphic monitor where the hit location is shown on a 7-in (178 mm) cathode ray tube display. The monitor provides a display of shot location with the last fired shot flashing. Various zoom pictures can be selected and after the last shot in the programme a table is displayed of the total. The monitor features simple operation by membrane keyboard with user-programmable shot sequences while the system also has stray shot detectors. The system memory features 100 preprogrammed and 20 user-programmable shot sequences and the recording of the last 100 shots.

An optional extra is automatic data recording with the shooter's identification, shot sequence, weapon and location fed into the system by keyboard or barcode reader. Shot values, date, time, target number and fed-in data are transmitted via the network system to a personal computer for further processing. This computer can also be used as the central command unit for all TG 3002 systems installed in the range.

Status

The system is being marketed.

Manufacturer

Polytronic International Limited, Muri.
(For details see Directory of Suppliers)

Polytronic TG 4000 Electronic Hit Scoring System

The Polytronic TG 4000 electronic hit scoring system detects the position of a bullet as it strikes a target so that the position and score of the shot are shown on a monitor placed beside the shooter.

There are three different target detection systems available:

TG 4002 - for precision shooting using a closed target

TG 4010 - location of miss and hit (LOMAH) using an open target

TG 4020 - location of miss and hit (LOMAH) for angle shooting.

Acoustic detection outside the target allows the use of supersonic ammunition with no restrictions on calibre, from small arms to tank guns. A selection of up to 20 target pictures to suit different applications can be stored in the TG 4000 library; new targets can be created to suit customer requirements. The system can accommodate rates of fire of up to 60 rounds per second.

Up to 50 programmes, each with up to 20 sections,

can be compiled by the user and entered into the system. Each monitor can cover 15 targets and, for central control groups of TG 4000 units, can be linked to a host computer via a network. The internal memory retains set-up parameters and training programmes, even when the power is switched off or fails. If required, results can be printed out numerically or graphically.

The monitor is housed in a robust waterproof case measuring 350 × 322 × 430 mm; weight is 13.2 kg. A sun shield is fitted.

Status

In production and in widespread service.

Manufacturer

Polytronic International Limited, Muri.
(For details see Directory of Suppliers)

Monitor for the Polytronic TG 4000 electronic hit scoring system

Polytronic TG 4010 LOMAH Hit Scoring and Evaluation System

The Polytronic TG 4010 LOMAH (Location Of Miss And Hit) electronic hit scoring and evaluation system is described as an 'open' system which imposes no demands on the construction and design of the targets involved with the system. It can thus be used with any type of target including pop-up and moving targets.

The LOMAH system is based on an acoustic measuring principle in which a projectile travelling at supersonic speed produces a conical-shaped shock wave which can be detected by a piezoelectric crystal. The opening angle of the cone is determined by the ratio of projectile to sound velocity. The cone propagates perpendicularly to its surface at the speed of sound and its axis is identical to the trajectory of the projectile. The point of intersection of the cone axis with the target surface corresponds to the impact location of the projectile. The arrival and location of the supersonic cone is measured by three piezoelectric microphones and a further microphone located on a T-bar placed just in front of the target. The projectile velocity is measured by the central microphone on the T-bar and the front microphone. An additional device, consisting of a speaker, a resonance tube and a microphone, is used for measuring the momentary sound velocity. Using these measured values it is possible for the system computer to determine the exact position of the cone axis and consequently the projectile impact location. This information can be displayed on a monitor and recorded in the system computer.

Thus the central feature of the LOMAH system is the T-bar located in front of the target in use. The bar comprises a stable chassis made of aluminium sections on which four detector blocks and two spirit levels are installed. On the bottom, between the legs of the U-section, there is a metal housing containing the equipment for measuring the sound velocity and a socket for connecting the T-bar to the transmitter. The T-bar is mounted on a rack connecting it to the associated target unit by means of three rubber shock-absorbers. Also close to each target is a transmitter unit.

At each firing point are three components of the system: a sound detector, a monitor and an intercom unit. The sound detector is a special microphone on a support installed under the muzzle of the weapon involved. The sound detector records each shot fired from its firing point and should a shot from another firing point strike the wrong target, that shot information is recorded separately.

The monitor displays firing results in graphic and numeric form. Up to 60 shots in a series can be displayed graphically at the same time with the last shot fired denoted by a flashing point. The current shot number, the sector of impact, the corresponding score, the running total, as well as the projectile velocity, are all displayed graphically. At the end of a series the mean point of impact can be displayed. To restart a firing exercise a reset button is provided on the monitor to delete all stored data.

On a LOMAH-equipped firing range each lane has its own target system and monitor with the complete system controlled and monitored from a control desk located in a control tower or room close to the range. From this desk the target mechanisms are controlled and there is one control unit for each lane. Each control unit is responsible for data transmission from one lane and producing a video image for the monitor involved. The control unit can also operate a thermal printer to produce hard copies of firing results. The control desk has its own self-test programme and uses a simple menu-assisted programming system. The control desk also controls the intercom system and the instructor can select the monitor display for any particular lane.

A LOMAH-equipped firing range may be from 70 to 600 m long. Up to 20 shots/s can be registered over a rectangular area measuring 5 × 3 m around the geometric centre of each target. Each monitor can display up to 60 of the latest series of shots in each firing exercise.

Status

In production.

Manufacturer

Polytronic International Limited, Muri.
(For details see Directory of Suppliers)

SIUS-ASCOR SA-9001 Target System

The SA-9001 is an automatic scoring system. It provides the firer, trainer or scoring official with immediate shot-by-shot location and scores to an accuracy of 0.5 mm at the centre of the target.

The SA-9001 eliminates all requirements for Known Distances (KD) range equipment and accessories as currently known. Most importantly the SA-9001 eliminates all manpower, communication, paper targets, pasters-up, spotters and other associated requirements for operating KD target butts. Furthermore, the system eliminates the necessity for having score-keepers on the firing line, as a hard-copy printout for each shot fired, along with the total score, is provided.

The measuring instruments of this patented system detect the penetration of the tip of the projectile into the acoustic chamber. This chamber also contains a measuring system which records any variation of environmental parameters such as temperature and air humidity. These variations are then automatically

SIUS-ASCOR SA-9001 target system in use

compensated for each individual target; this unique feature maintains the high level of accuracy regardless of weather conditions. Rain, snow or hail do not affect the operation or accuracy of the system.

The SA-9001 target skin will accept between 20 000 and 50 000 shots before replacement is necessary.

A variation of the basic SA-9001 target system uses a T-bar mounting for microphones below the target. The point of strike of a projectile can be determined by the measurement of the time differences of the shock wave produced by the projectile on the microphones. SIUS manufactures the T-bars with lengths of from 1 to 6 m. Depending on the application, calibres from 5.56 to 205 mm can be located on targets from 2 up to 50 m². Three basic systems employing T-bars are produced. The SA-9001 TBS (the system mainly used for military purposes) has a T-bar length of 1 m while that for the SA-9001 TBM has a length of 1.3 m. These two systems can be used with weapons having calibres of from 5 to 30 mm. The SA-9001 TBB uses a T-bar 5.06 m long and is intended for use with weapons of from 12 to 200 mm. It can deal with tank-sized targets. All three systems can deal with rates of fire up to 30 shots/s. Results are displayed on the same unit as that used with the basic SA-9001 target system.

Data for basic SA-9001
Target frame
Dimensions: 1500 × 1600 × 40 mm
Weight: 30 kg

Target electronics
Dimensions: 350 × 150 × 70 mm
Weight: 6 kg

Monitor
Dimensions: 600 × 250 × 300 mm
Weight: 8 kg
Infra-red lane initiator weight: 2.5 kg
Target accuracy: ±1 mm
Bullet velocity: 250-1500 m/s
Temperature range: −20 to +50°C
Humidity: 95%
Rain and wind: no influence

Status
In production. In service with most European countries; over 7000 systems in use in Switzerland.

Manufacturer
SIUS AG, Effretikon.
(For details see Directory of Suppliers)

TIRAX Portable Target Systems

TIRAX refers to its target systems as combat simulators. They are target presentation device units that are handy and lightweight, can be used as rotary, turn or pop-up target mechanisms, are easy to install and operate, and require a minimum of maintenance.

The new version, available from October 1994, is TIRAX 2000, which is a modular, multipurpose, microprocessor-controlled combat simulator system for a variety of applications. It is based upon TI 92 and TR 101 units.

Each TIRAX 2000 unit consists of a small cylinder containing a power source, such as 1.5 V alkaline batteries or rechargeable batteries, together with a charge control and a target-turning drive mechanism. The basic version of T2000 is equipped with a programme for three different times for the target to return after a hit. An additional random programmed sequence can be set showing the target at certain times as friend or foe. The turning direction can be set according to the application. For turning targets the unit is placed on its base while for pop-up targets the unit is located on its side using a suitable bracket or other securing methods.

The basic version of the Type T2000 weighs 2.7 kg, it is 202 mm high and 150 mm in diameter. The properties of the basic T2000 can be expanded with interchangeable modules. These can be inserted into the body of the T2000 as a single upgrade or as a combination to meet the specific training requirements. Modules with the following functions are available:

Remote-control by radio to allow the synchronous use of unlimited targets over 12 channels in three different frequency groups. An input/output module makes it possible to use different release devices for the target or can send commands to other devices after a hit. A programming module is also available to set up a training sequence of eight different time cycles for showing or hiding the target, for friend/foe and for numbers of sequence repetitions.

The LAST module is for training without ammunition. The special target is hit by laser shots from devices used in commonly available tactical laser engagement systems (see entry in *Small Arms Marksmanship Training Equipment* section for details). The equipment containing the LAST module may also be used with live or training ammunition. An Urban Training module is available with T-Net for permanent installation in training villages or for temporary use for urban combat training. The different T2000 units are connected to a biaxial cable bus and the entire installation is programmable and controllable with the TIRAX software TRACOS. Other modules are being developed.

The TIRAX types TI 92 and TR 101 are partly compatible with the TIRAX 2000 system and are still being produced. Since they weigh only 2.7 kg or 3.3 kg, depending on the modules involved, they can be placed in any number of range configurations and are suitable for indoor and outdoor use under any weather conditions.

The Type TI 92 is controlled by a TC 4 control unit connected to the unit by 100 m of biaxial cable carried on a TCR reel. This combination allows up to 50 target units to be controlled but the TI 92 can also be controlled by radio using a TR 101 'master' unit as well as by the T2000 with the input/output module. An integrated sensor registers hits and causes the targets to turn. The battery life of both units permits up to 20 000 target turns.

The Type TR 101 weighs 3.3 kg, is 220 mm high and 145 mm in diameter. It has the same target functions as the TI 92, plus a friend/foe presentation mode, and is radio-controlled using a TT 100 control unit weighing 2.3 kg. A single TR 101 can act as the master unit for up to 50 TI 92 units (the similar TR 100 unit lacks this 'master' function). The battery life of the TR 101 permits up to 5000 target turns.

Status
In production. More than 12 000 units are in service with the armed forces and police units in more than 40 countries including Austria, Belgium, Canada, Chile, France, Germany, Greece, Hong Kong, Indonesia, Iraq, Italy, Japan, Kuwait, Malaysia, Morocco, Mozambique, Norway, Qatar, Singapore, Spain, Sri Lanka, Sweden,

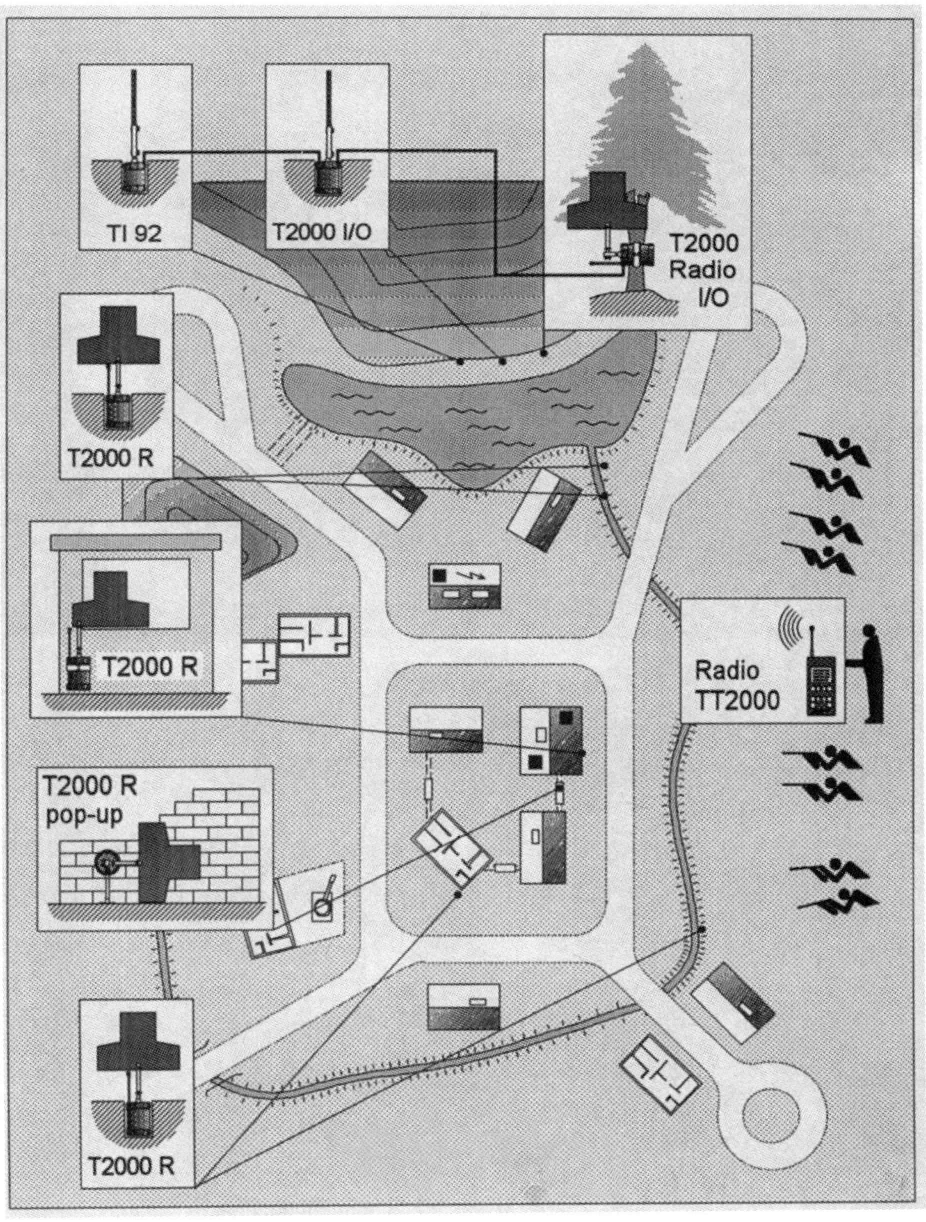

A diagram showing a typical application of the TIRAX 2000 system

Switzerland, Tunisia, United Kingdom and the United States (US Army and the US Marine Corps). The Swiss Army uses more than 5000 units for basic training.

Manufacturer
TIRAX AG, Dietikon.
(For details see Directory of Suppliers)

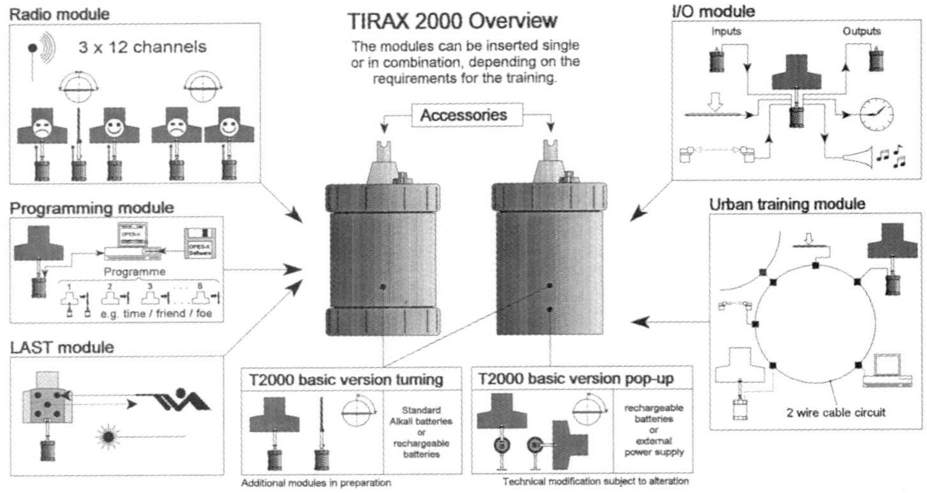

A drawing of the TIRAX 2000 system elements

UNITED KINGDOM

Gunpell Bullet Traps and Target Systems

Gunpell & Company Limited specialises in the manufacture and supply of a variety of shooting range equipment and materials. Gunpell portable electronic targets (MINIs) are in use with several armies and law enforcement groups worldwide. Gunpell Close Quarter Encounter (CQE) fully programmable, computer-controlled targets are in service with police and special forces. Gunpell also produces a cost-efficient bullet trap system for all types of shooting ranges.

Gunpell 'Soft-Wall' Bullet Trap

The 'Soft-Wall' bullet trap is for use in new and existing indoor shooting ranges for all handgun and rifle calibres. It will stop and encapsulate projectiles from 0.22 in rimfire through 9 mm to 5.56 mm and 7.62 mm calibres.

The trap is of simple construction and consists of a 'soft wall' of Gunpell-board fixed at a short distance from a range wall with the space between filled with Gunpell-granules. Only 30 cm depth of trap is required for hand-guns and 50 cm for rifles. Gunpell-board is a unique material which is completely self-healing and will allow hundreds of pistol and thousands of rifle projectiles to pass through it without producing any bullet holes. Gunpell-granules are specially compounded to be anti-pollutant and progressively to slow and stop all firearms ammunition.

In most conventional shooting ranges the targets are located at fixed positions so the majority of projectiles strike the bullet trap in small concentrated areas behind the targets. Under these severe conditions even Gunpell-board will eventually be damaged. For this reason a special feature of the Gunpell trap is the incorporation of High Impact Trap (HIT) Modules. These can be built into the trap in any positions behind the targets and consist of cylinders of Gunpell-board filled with Gunpell-granules. The HIT Modules are open in the front so that all bullets hitting the target near the aiming marks go directly into the granules where they are encapsulated without touching the Gunpell-board. The HIT Modules are also available as separate items for low-cost upgrading of existing ranges as well as for off-range safe discharging (clearing) of weapons.

Routine maintenance is easy and quick. The HIT Modules are simply withdrawn from the 'Soft Wall' and replaced with fresh ones. The spent bullets in the used HIT Modules can be separated at any convenient time from the surrounding granules and removed. The granules can then be refilled into a HIT Module ready for the next change. The materials form a bullet trap which is cost effective, safe against all ammunition natures and non-polluting.

Gunpell Outdoor Bullet Trap

This has all the features of the 'Soft-Wall' bullet trap and is covered against rain, snow and sand for use in all temperatures. The trap is quickly erected and installed

A selection of target pictures from the Gunpell 'Silhouette' target board range

Gunpell bullet trap installed in a wooden shed

Gunpell MINI target systems

in any outdoor location including flat and sloping ground, steep slopes and even vertical mountain faces. A special feature of the system is that it may be installed in existing ranges with very little ground preparation or disturbance to equipment. Because the trap will stop and encapsulate all firearms projectiles and prevent them entering sand butts or gravel beds there is no danger of bullet ricochet from buried bullets.

Gunpell 'Silhouette' Target Boards
These are designed for long lasting use with all rifle and pistol ammunition and will accept 10 000 hits from 0.5 in, 7.62 mm and 5.56 mm calibre rifles without producing bullet holes. They are designed for rough handling and are rigid enough to be used as full-size pop-up targets and to activate hit sensors. They may also be shot 'on edge' without damage to the target.

The boards may be designed for specific requirements in any shape and with any pattern or colour. They have permanent target patterns whose aiming marks will not get shot away and can be used with or without paper facings.

MINI Target Systems
The Gunpell MINI target systems are compact, rugged and portable target units that combine turning, pop-up and swing-out targets in one unit. They can be set and taken down within minutes and will continue to operate with 0.1 second precision for extended periods. They can be radio-controlled, line-controlled or self-operating.

Each target can be converted for use as a turning, pop-up or swing-out target by changing target arms; the targets turn through 90°. Each target mechanism weighs approximately 10 kg and comes fitted with a 12 V rechargeable battery (together with a battery charger), a hit sensor and turning and pop-up arms. The targets have integral circuitry for use as rapid fire targets. They are weather-resistant and can be used indoors or outdoors. If installed on a permanent range the units can operate off a mains supply.

A hand-held MINI Radio Controller can operate up to 15 MINI units, individually or in groups. The operating range is at least 1000 m.

MINI Rapid Fire Target
The MINI Rapid Fire Target (MINI-RFT) is a stand-alone target unit that will instantly turn or drop when hit and rebound for the next shot. It does not require any controller or cables. The time dropped or turned can be selected by a dial from between 1 and 30 seconds.

MINI Rapid Fire Crossing Target
The MINI Rapid Fire Crossing Target (MINI-RFCT) consists of a robust, flat-topped, self-driven wheeled trolley with a MINI Rapid Fire Target placed on top. Each time the target is hit it will swivel and return to face to indicate a hit. The trolley is self-propelled and operates off an integral rechargeable 12 V battery. It is guided by a wooden strip fixed to the range floor; speed is 1 or 2 m/s. The trolley is operated by a MINI target controller.

Close Quarter Encounter
Close Quarter Encounter (CQE) is an interactive three-dimensional, live fire, two-way turning target system that provides challenging scenarios for advanced police and special forces firearm training. It can also be used for basic and intermediate levels of shooting skills.

With CQE the targets are all double zone targets. There is a Centre of Mass portion and a Wound portion comprising the rest of the target body. A bullet strike in the Centre of Mass puts the target out of action. A strike in any other portion 'wounds' the target. When wounded the target turns to edge for 2 seconds and then refaces for the next shot. This is repeated three times. If the target is not hit within the facing time allocated by the selected programme, the target will Retaliatory Fire by electronic means. If the target is hit while in a Friend mode a derogatory buzz will sound. The total number of Hits, Wounds and Errors will be shown on the controller together with the shooting time to 0.1 second accuracy.

MINI Target System unit in pop-up mode

Status
All the above items are available and in worldwide service.

Manufacturer
Gunpell & Co Limited, Fleet.
(For details see Directory of Suppliers)

McQueen Target Equipment

The McQueen company has been manufacturing target equipment since 1890 and its extensive range covers conventional military paper targets in all the NATO standard patterns for use on miniature and full-sized ranges; butt equipment ranging from paper patches and paste to target frames; police and security force 'terrorist' and 'friendly' targets, complete with a wide range of accessory pictures which allow the target appearance to be changed frequently; pistol, rifle, and 'practical' targets for competition; landscape targets for classroom instruction; and a variety of training aids covering every aspect of small arms training.

The latest McQueen targets are special training targets printed in full colour and depicting various hostile or friendly figures to which a range of 'accessories' such as weapons and innocuous articles such as cameras can be added as required for use in decision making and reaction training. The older style targets were approximately ⅞th full size and fitted standard turning mechanisms. The new series targets are life size.

Examples from the range of McQueen targets

Status
In production. In service with military and police forces worldwide.

Manufacturer
McQueen Limited, Galashiels.
(For details see Directory of Suppliers)

MS Instruments Automatic Scoring Electronic Target (ASET)

ASET is a product for use during range marksmanship training. MS Instruments plc developed ASET from its proven Acoustic Target System, one of its extensive range of ballistic instrumentation products.

ASET comprises two elements which are the Firing Point Display Unit and the Target Sensor. The Firing Point Display Unit is a ruggedised, weather-proof display, with integral processor and controls, that allows the firer to examine and process the displayed shots at will. The Target Sensor accurately detects all supersonic projectiles passing through the detection area. ASET therefore provides accurate feedback to the firer on the firing point. Accuracy of the system is to within ±5 mm and it can accommodate projectiles with calibres between 5.56 and 120 mm. Target sizes may be 1 × 1 m, 2 × 2 m and 6 × 3 m.

The basic Acoustic Target system utilises two banks of pressure bars which provide X and Y co-ordinates for hit location assessment. For ballistic testing purposes the two bars may be set at right-angles to each other but for ASET the two bars are arranged in a 'V' configuration to enable them to remain below the line of fire.

Further development of the Acoustic Target and ASET system has enabled MS Instruments to adopt a new technique of pressure wave sensing by arranging the pressure wave sensors in two arrays, one each side of the target and below target level. Each individual

MS Instruments Automatic Scoring Electronic Target (ASET) units

array holds three pressure sensors in a triangular (delta) formation so that each individual sensor picks up the oncoming supersonic pressure wave at a different time from the others. The same pressure wave is also sensed by the sensor array on the other side of the target. By combining the data from the two arrays, and therefore all six sensors, the projectile co-ordinates can be assessed to a high degree of accuracy.

The results can be displayed on the Firing Point Display Unit, as with the basic ASET, or on a printer, with the data being provided in a form to meet customer requirements.

Status
In production and widespread service.

Manufacturer
MS Instruments plc, Bromley.
(For details see Directory of Suppliers)

Data processing unit used with ASET

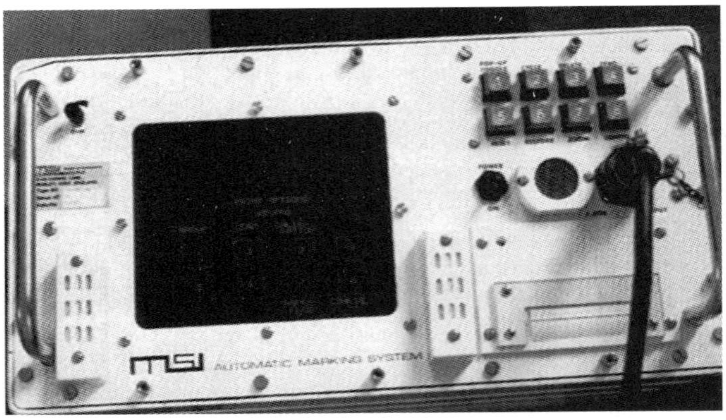

ASET firing point display unit

OTEC Facility for Infantry Situation Training (FIST)

One of the biggest challenges for armed forces is the growing demand to provide support for internal security operations. These require exacting and demanding skills, which evolve from thorough training covering the full spectrum of small arms infantry skills, culminating in live fire exercises in simulated urban environments.

The conventional close quarter battle urban environment is a permanent construction in which façades made of anti-ricochet concrete are erected in an architectural and building style and layout similar to the likely operational environment. It can be used for both foot and vehicle operations. All actions and commands of the trainees are closely monitored by the training officer through carefully positioned cameras and recorded for analysis and replay. The commander can develop the situation to exploit any weakness or errors in procedures through remotely controlled targets and battle effects, recording his comments upon tape.

This facility can be further enhanced by the addition of entry rooms and exit routes to enable hot pursuit exercises to be undertaken. Special moving targets can be added within the urban complex, or as a special range, for fire-control and fire discipline exercises for sniper training. A marksmanship-under-fire facility engages the trainees with live fire under strict control from a variety of weapons to exercise the trainee in identifying the source of aggressive fire.

The permanent close quarter battle urban facility and its associated enhancements provide comprehensive and thorough training but they do represent a significant investment which may be difficult to justify if the task and the obligations are unpredictable.

Building upon British armed forces' experience, OTEC can now offer the FIST (Facility for Infantry Situation Training) system. This consists of a number of prefabricated modular units which can be positioned quickly and easily by crane in any configuration or layout and 'dressed' to provide different architectural styles. This means the facility can be configured to match town, village or camp situations in different countries and allows troops to exercise patrolling techniques and incident reaction drills using live fire. Each FIST unit compares in size to a 6 m (20 ft) ISO container and can be transported and handled accordingly. The units can be stacked three high. Using FIST modules, a smaller initial investment is required to provide a facility for urban close quarter battle and this can be progressively expanded depending upon the requirement or funding.

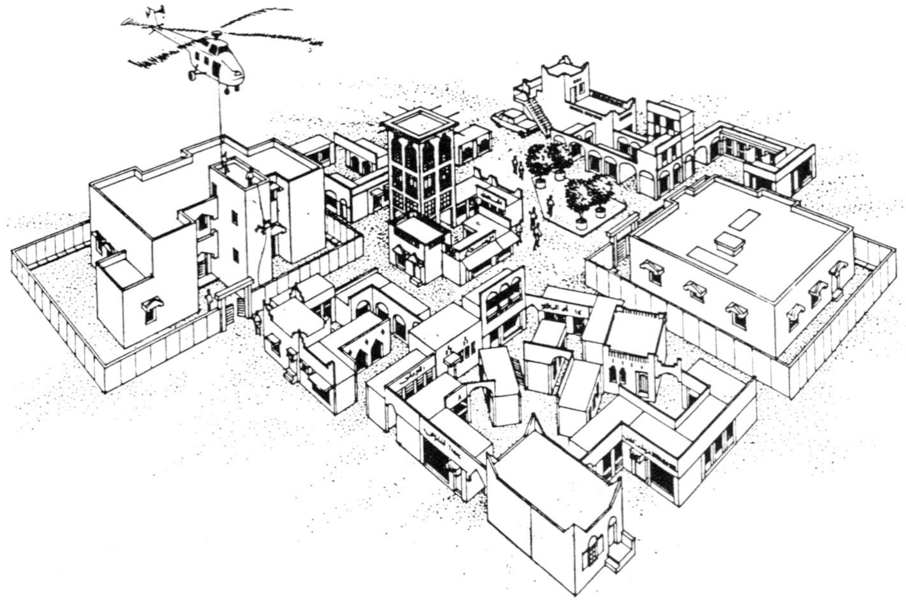

Artist's impression of FIST 2 and FIST 3 modules (with façades erected) in a mixed layout to form an urban range

There are two basic FIST units. The FIST 2 unit is of an 'H' construction with top and bottom floor and roof slabs. Both sides of the FIST 2 unit can form the sides of roads or alleyways and can be faced with low-cost local materials to create whatever building style is required. Windows and doors can be provided and may be opened mechanically to reveal targets.

The FIST 3 unit is a box section with end walls in place. The sides are open and units can be placed alongside each other to form rooms of different sizes. The exposed side walls are covered with a selection of panels including windows and doors. The resultant rooms can be used for hot pursuit, in-room and house-clearing exercises.

FIST 2 and FIST 3 units are constructed of anti-ricochet concrete and permit the retrofitting of monitoring and control systems when required. No specially prepared base is necessary for the units other than for it to be level and firm.

Status
Available and in service.

Manufacturer
OTEC Support Services Limited, Godalming.
(For details see Directory of Suppliers)

OTEC Range Systems

OTEC Support Services are specialists in fitting out ranges and have developed a number of systems which can be used in existing conventional indoor ranges to enable them to be converted for advanced training when required. These systems include the following:

Low Pollutant Bullet Traps
Bullet traps are one of the main causes of both acoustic and atmospheric pollution in ranges. The kinetic energy generated in stopping bullets with conventional metal traps causes some lead in the rounds to vaporise and to contaminate the atmosphere. The residue is difficult to collect and remove. The noise of the round hitting the metal adds considerably to the noise level within the range which is transferred to other parts of the building through the structure. OTEC has developed a series of soft bullet traps which progressively absorb the kinetic energy producing no noise or lead pollution. The rounds can be collected almost undamaged.

Acoustic and Anti-Ricochet Panels
Depending upon the materials used in the construction of the building and the calibre and volume of ammunition to be fired, OTEC can undertake to convert any range or building to permit firing through 360°, covering also floor and ceiling panels. Different grades of panels are fitted which provide acoustic and ballistic

protection meeting international health and safety regulations.

Night Light Simulation
A series of electroluminescent panels are supported on a frame suspended from the range ceiling. The panels are directed upwards and their glow is reflected downwards from the ceiling. By changing the intensity of the light of individual panels, all stages of night lighting conditions can be simulated. Additional sodium and mercury lights can be added as required to represent different street lighting conditions, as well as spot lights to represent car headlights and strobes for explosive effects.

The OTEC acoustic and anti-ricochet panels

Sector Partition System
A series of panels can be rapidly arranged to form diverse environments such as offices, rooms, shops and corridors for single and multiple room entry and for clearing drills for individual and team operations. Training for VIP protection duties and individual fire and movement can also be undertaken. The system includes doors and windows which can be moved rapidly into positions to change the configuration required.

CCTV Monitoring and Recording Systems
Video cameras are located overhead in key positions to provide full visual coverage of all trainees in the live fire zones. Special, non flare-out, solid-state, miniature cameras are used, with each camera position supported by a two-way intercom, emergency light and infra-red flood for night operations. All activities are monitored and recorded in a control room using either automatic tracking, with movement sensors, or quad splitters for continuous observation. In addition, for added safety control a matrix of the layout can be provided to indicate the position of each trainee using infra-red sensors.

Status
All these products and associated equipment are available and are in service.

Manufacturer
OTEC Support Services Limited, Godalming.
(For details see Directory of Suppliers)

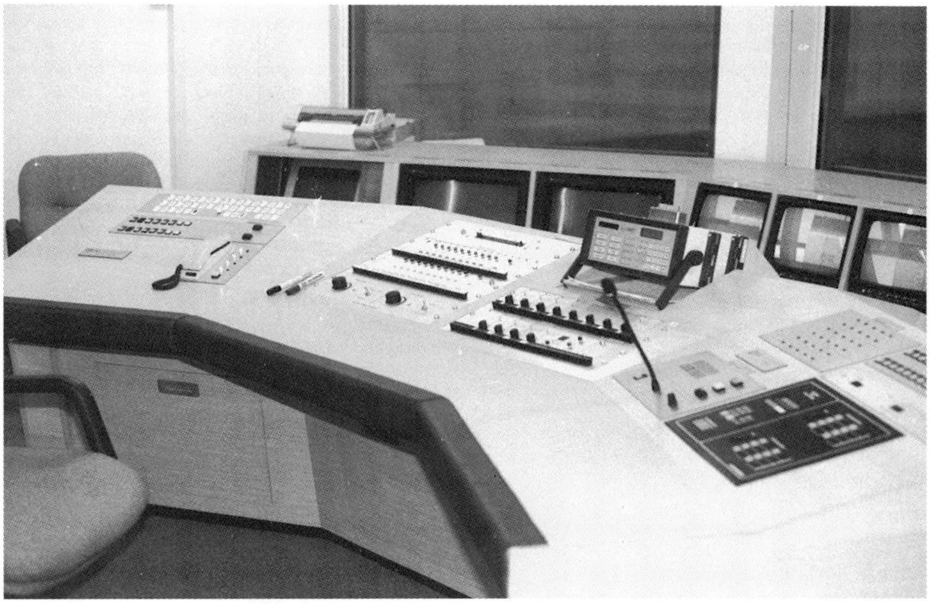

The OTEC CCTV monitoring and recording system

The CCTV miniature camera system

OTEC Universal Target Auto Scoring System (UTASS)

The Universal Target Auto Scoring System (UTASS) accurately locates projectiles of any calibre or velocity irrespective of the type of target used. It is contactless and can be used with ply or foam targets, pop-up, turning, slicing or vertical rising targets, UIT frames or any type of video, cine or slide projection screen.

The sensors can be located behind existing range baffles with simple fixings and the only maintenance required is occasional cleaning. A single set of UTASS sensors can locate all shots on video or slide projection screens up to 6 m wide. For sensing wider than five lanes or 6 m, additional sensors are required. Typically a single set of UTASS sensors can be used to locate all hits and misses for up to five lanes of a pistol range. It will simultaneously provide individual scoring for each lane and produce the results at range control or, alternatively, at each firing position. A VDU shows the precise location within ±5 mm of each shot within the detected area whether on or off the target. Each shot is numbered with the last one flashing. The mean point of impact and the score are shown and updated automatically with each shot. Group size can be shown as well as standard deviation if required. Hard copy printout with full firer's details, firing conditions and instructor comment can be stored for future reference.

Status
The UTASS is available and is in service.

Manufacturer
OTEC Support Services Limited, Godalming.
(For details see Directory of Suppliers)

RX Range and Target Systems

PX336 Target Mechanism
The RX PX336 is a portable dedicated pop-up target for infantry field use. The main features are: a smooth action; a built-in night light facility; an externally adjustable hit sensor; and 2000 operations from a 12 Ah battery pack. Control is by multiplexed radio or with stand-alone modules. A programmable radio system is available for the PX336 and up to 60 target units may be separately controlled; up to 10 sub-groups may be created. Hit responses can be selected for number of hits and mode of action.

PX236 Target Mechanism
The PX236 is a pop-up mechanism designed for permanent installations. It features buffered movement and a rapid action, with power taken from a mains supply. Control is dependent on operator choice.

PX95 Moving Target System
This is a track-based infantry training system which may be used in up to six lanes. It is intended for non-reactive targets but versions carrying pop-up and other reactive targets are produced. Four preset speeds up to 4 m/s are available. The PX95 is in service with the United Kingdom Ministry of Defence.

PX93 Radio-controlled Moving Target
This system is battery-controlled and moves along a quick-fit track carrying a PX180 or other target mechanism, and may be installed as a quick-installation tactical training system. A heavy-duty version produced for the United Kingdom Ministry of Defence can carry a pop-up target mechanism. Two preset speeds up to 3 m/s can be selected.

PX42 Milton Target Trolley System
The PX42 is a protected trolley driven by its own motor and powered by a self-contained rechargeable battery which also powers the target mechanism. The trolley

RX two-way turning and multipurpose target mechanism PX180/B

moves along the floor of a range guided by a wooden strip which also controls ground loop control wires. From the firing point the backward and forward movement of the trolley can be controlled to accurate distances together with the operation of the target mechanism. Control can be from a central console or by lane from each firing point.

PX43 Turning Target Mechanism

The PX43 is a single turning target rotating through 90° and is intended for use with vertical rotating targets weighing up to 10 kg. The mechanism operates using a 12 V battery and various hit functions are available. A modified heavy use version in a stackable frame is available.

PX180/B Two-Way Turning and Multipurpose Target Mechanisms

The PX180/B two-way turning and multipurpose target mechanisms use one mechanism and various brackets to rotate a target left or right from a central edge position, swing a target around cover (left or right), work as a pop-up target and have a slicing action (sideways pop-up). The PX180/R operates under line control.

The equipment is self-contained with an internal rechargeable battery giving at least one day's shooting. Weight is 17 kg, sufficient to provide stability yet light enough to be carried by one man. Dimensions are small enough to make the equipment easy to conceal in simulated tactical training situations.

Control is by one of three methods:

(a) line control mainly for indoor ranges offering a selection of controllers up to computerised operation

(b) local control for Close Quarter Battle (CQB) self-initiation, walk-through type ranges

(c) radio link for CQB and other outdoor applications.

The system offers features such as remote grouping and channel selection by radio; double-tap hit sensing options, accessory and trigger systems for operating the equipment, including infra-red.

PX200 Vertical Rising Target Mechanism

This target mechanism makes a target rise and fall in a vertical plane and lifts a 5 kg target 450 mm vertically in about 0.5 second. It can be controlled by all the usual control methods and is compatible with the PX180 system.

GX2003 Target

This target moves along the ground following a plastic rail which can be formed into curves as required. The guide rail is fixed in an area where firers do not normally have to move about. It has a simple start command and thereafter moves automatically to the other end of the track, waits and then returns. It can carry target mechanisms and there is a built-in delay after starting at each end to expose the target.

AJAX II General Purpose Turning Target Frame

This system is constructed from steel and weighs 65 kg. It has a turn time of less than 0.3 second and is battery-operated.

Self-Activating Reactive Target System (SARTS)

This is a range of equipment designed for use in a CQB or stress-type range, using the actions of a shooter or team to determine the actions of the equipment. It is a flexible system of 'triggers and responses' with the shooter or team activating the triggers and other devices producing the responses.

Examples of triggers are door switches, pressure mats, light or infra-red beams and movement detectors. Signals from the triggers are sent back to a PX205

PX180/B radio link control unit

master control box and directed to operate the required target or targets. Once a trigger has been operated it cannot be used again until reset. Eight target outputs are standard in the PX205 controller but some of the device outputs can operate certain targets if required. Up to 15 device outputs are standard. Devices include initiators for pyrotechnics such as stun grenades, smoke, machine gun fire simulators, dazzling or flashing lights and tape recorders or other sound sources.

SARTS uses the PX180/B and PX200 target mechanisms and it is also possible to use the PX180/R in the line control mode. Hits on targets can be used as triggers to operate other targets or device outputs.

SARTS can be presented as a package with enough parts to equip a CQB range and can include a training video covering installation, operation and maintenance.

Status

In production. RX equipments are in service in over 35 countries.

Manufacturer

RX Products Limited, Caerphilly.
(For details see Directory of Suppliers)

TTS Video Target Trainer (VTT)

The TTS (Trenton Technical Services) Video Target Trainer is a compact live fire training system complete with a movable Target Screen Unit. Moving targets or timed still targets from either supplied or user-generated video media, including computer-generated imagery (CGI), are projected onto the screen unit from a mounted projector (fixed installations) or a portable projector when a mobile system is required.

The VTT can operate in a manual or automatic mode with control exercised through a small hand-held control unit. When operating in the automatic mode a shot detected by the local or remote shot sensor puts the video projector into the pause mode to permit target hit assessment. The assessment time can be varied and at the end of the assessment period a clean target area is automatically restored on the screen. A monitor and cueing system enables firearms instructor lesson control and a voice-over sound system is standard.

The Trenton HAFT, a dry fire laser-based training device for teaching basic Aim, Hold and Fire techniques, is complementary to the VTT as is the Trenton MaSTER system. MaSTER is a series of add-on modules which enhances the VTT by the addition of interactive video, graphic target generation, and a scoring, record and analysis system. In addition to live fire with conventional ammunition, the VTT is in service with air pistols, paint ball and low velocity training ammunition.

Target screens can be selected from a variety of configurations to accommodate from one to four firers; a 180° panoramic target screen is in service. The screens utilise a specialised white/white opaque paper. Shot identification is either by continuous or flashing white light, with an infra-red option.

TTS Video Target Trainer (VTT)

Status

In production and service.

Manufacturer

Trenton Technical Services Limited, Southampton.
(For details see Directory of Suppliers)

Utilex Urban Close Quarter Battle Range

Utilex is a proprietary structural lightweight soft concrete which can be cast into large one-piece wall panels, and moulded, shaped and coloured to represent a wide range of building materials and styles. It also has the exceptional characteristic of being able to absorb small arms ammunition at a narrow angle without allowing the bullet to ricochet. This makes the material ideal for constructing military and anti-terrorist live ammunition urban warfare training ranges.

Ranges of this type can be built life-sized, to represent typical streets or buildings from any part of the world or in any desired style. The streets are then suitably decorated with cars, dummies, post-boxes, lampposts and telephone booths, the shops have window displays, and the whole is painted and signed to give the utmost degree of realism. Overseeing the whole area is a protected control tower from which it is possible to cause dummies to move, grenades to be thrown, weapons to be fired on fixed lines, and many other effects. The area is monitored by closed-circuit TV cameras.

Trainees enter the range and carry out various assigned duties. It should be emphasised that the buildings are not all mere façades, but some are fully built and furnished, so that full search procedures can be followed. Whilst the trainees move around the area, the controller may operate retaliatory fires, booby-traps, pop-up targets and other effects, to which the trainees can respond using live ammunition. At the end of the training period, the video record which has been made by the closed-circuit TV cameras can be replayed to the trainees and the instructor can conduct a critique of the operation, showing where and when any mistakes were made.

The principal advantage of this type of range is that, due to the controlled retaliatory fire and other hazards, the trainee is placed under stress throughout the exercise, a stress which leads to errors he is less likely to make on an unstressed firing range.

Utilex constructed an Urban Close Quarter Battle Range (UCQBR) in Hong Kong for use by the Hong Kong police as well as the British Army. The company has also built ranges to suit special applications, such as airport interiors for training in counter-terrorist operations.

Status

In service with the British Army and several police forces throughout the world.

Manufacturer

Utilex (UK) Limited, London.
(For details see Directory of Suppliers)

Street level view of the Utilex UCQBR in Hong Kong

Controller's view of the Hong Kong Utilex UCQBR

UNITED STATES OF AMERICA

Caswell Gamma Combat Video Range

The Caswell Gamma Combat Video Range is an optical live fire use-of-force scene simulator consisting of a video cassette player, a projector, an armoured screen assembly and a control unit. The system is portable and must be operated in a darkened, ballistically secure range.

The standard projected image is 1.2 m high and 1.8 m wide; larger screen sizes are available to special order. The screen is a multilayer heavy white paper. The trainee fires a weapon at the screen in response to the video scenario. At the sound of the shot the video image is frozen and automatic lighting shines through the bullet holes to identify the shot placement. Shots are erased by advancing the paper roll.

The user can make his own video scenarios and play them on the universal video machine. An optional camera/recorder is available to make scenarios and also to record the trainee on the range to provide a permanent record.

Status

In production. In service with US police, military and government agencies.

Manufacturer

Caswell International Corporation, Minneapolis.
(For details see Directory of Suppliers)

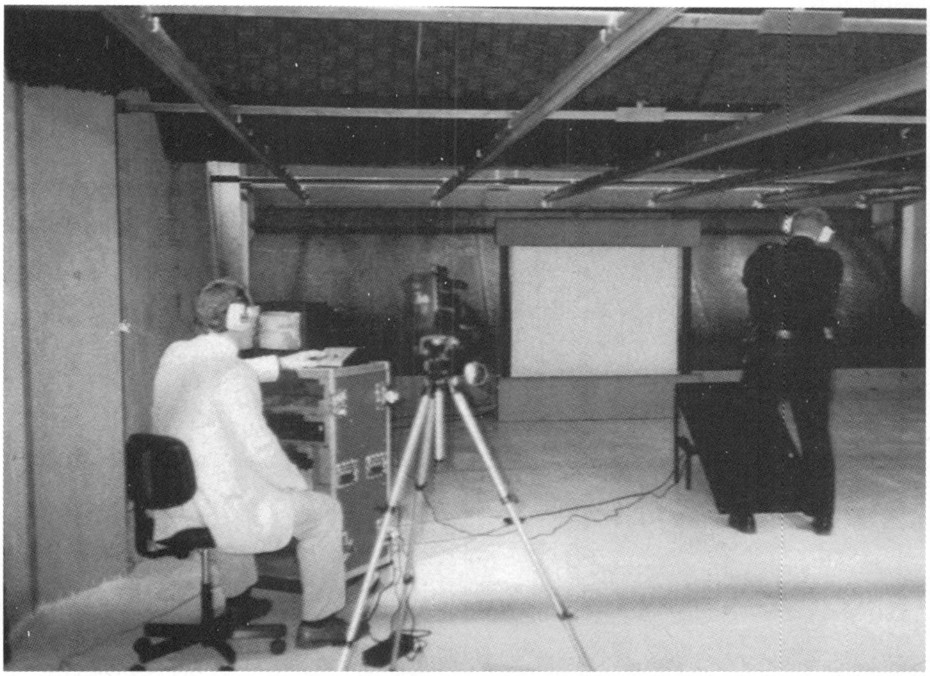

Caswell Gamma Combat Video Range

Caswell Mobile Shooting Range

The Caswell 'Road Range' is a fully functional training range contained in a trailer. It is intended for all military and police weapon training, except high velocity rifles, and contains three shooting points. It can be taken anywhere that training is required, and is invariably less expensive than transporting personnel from dispersed points to a central shooting range.

The mobile range conforms with US road traffic regulations and also complies with the environmental and health requirements that usually apply to shooting ranges. It has its own electric power generator and ventilation. The noise attenuation enables it to be used close to other activities without annoyance.

The standard mobile range contains three firing points. The bullets are arrested by a special bullet trap which allows use of all regular service hand-guns, shotguns and sub-machine guns. Automatic target carriers are used since these can withstand exceptional service and can be readily maintained wherever the trailer may be. The light dimming system enables combat situations to be simulated, and an audio system with speakers built into the ear protectors allows the range officer to communicate with the shooters.

Most standard police training courses can be conducted, even though the trailer is only 14.63 m long, since the targets are scaled to simulate shooting at all the usual distances.

Status

In production. In use by several US police and government agencies.

Manufacturer

Caswell International Corporation, Minneapolis.
(For details see Directory of Suppliers)

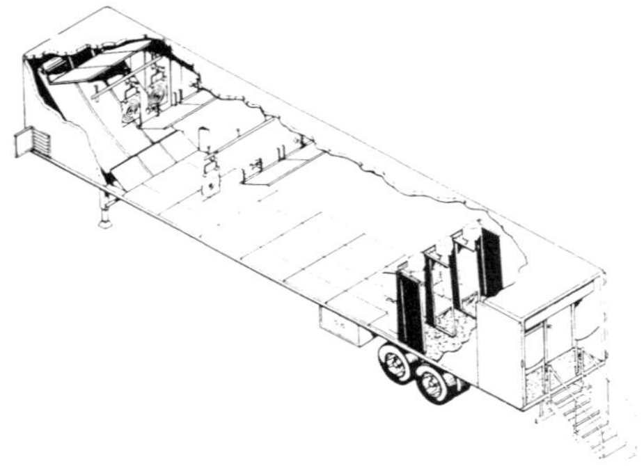

Cutaway view of Caswell Road Range

Firing points inside Caswell Road Range

Caswell Range Equipments and Systems

A typical Caswell outdoor shooting range

Since its foundation in 1926, Caswell International Corporation has provided shooting range equipment for firearms training. Included in the Caswell product range is all the necessary equipment for indoor and outdoor ranges including target retrieval systems, bullet traps for pistols and rifles, decision and stress target systems, remote radio or cable controls, ballistically secure shooting houses for assault and counter-terrorist training, and custom-engineered products to meet specialised mission requirements. The company also provides a consulting and design service as well as installation and training on a worldwide basis.

Target Retrieval Systems

Target retrieval systems allow shooters to remain on the firing line and have their targets transported to the desired distance. This method results in the greatest amount of safety and permits the largest number of shooters during a given time period. Caswell's systems range from manually or electrically powered simple cable and pulley arrangements to sophisticated monorail turning target carriers that allow friend-or-foe decision shooting. Microprocessor controls are available to preprogramme training courses, including expose/conceal times and distance selections.

Interior Fittings

Caswell can protect any structure from bullet damage with its range of baffles, guards and other shields. Special steel-lined modules are available to transform any interior space into a ballistically secure firing range. Also provided are lighting and ventilation systems.

Bullet Traps

Steel escalator bullet traps for hand-guns and rifles safely terminate shots and allow lead recovery and salvage. For close shooting and panoramic video projection, Caswell can provide a soft rubber bullet trap. Outdoor bullet traps are available to prevent soil contamination and possible pollution.

Full Calibre Gunnery Ranges and Systems

Caswell furnishes cable-controlled, fixed emplacement targetry as well as battery-powered, radio-controlled equipment for portable applications on all types of outdoor ranges.

The Caswell product line includes infantry running and pop-up targets, plus full-size side and front view tank and vehicle targets, either moving or pop-up. Audio, light and pyrotechnic hostile fire and battle effect simulators are available to enhance realism on the range. All mechanisms can be fitted with laser receptors.

Caswell manufactures its own microprocessor-based radio and fixed controllers and has developed software and interface packages for personal computers. All scenarios can be preprogrammed or run manually, and post-exercise report generation is available.

Infantry Target Mechanism

The Caswell Infantry Target Mechanism is an electro-mechanical actuator of the pop-up type for US Army E, F and similar personnel targets. The mechanism is remotely controlled and provides hit sensing and night

Caswell infantry target mechanism

A typical Caswell running man target system installation

A Caswell outdoor range bullet trap

illumination. It responds to the following commands: target up (expose); target down (conceal); target down when hit; target remains up when hit; target conceals and re-exposes when hit; repeat cycle; and target night illumination. The mechanism is manportable and suitable for permanent or temporary installation in a sheltered emplacement; an armoured version is available as an option. The mechanism is qualified for operation in severe climates and may be powered by rechargeable batteries or from a mains supply.

As an option, the mechanism can be equipped for return fire simulation. This includes a built-in strobe light that flashes on the target.

Running Man Target System

The Running Man Target System simulates a man moving rapidly across an open area. The system utilises a US Army E or F type silhouette target mounted on up to three pop-up mechanisms attached to the top of the target carrier. The target is carried laterally along a track, driven by an electrically powered drive unit located at one end of the track. The target system is installed behind a ballistically secure area with only the target exposed to the firer.

Status

In production. In service with military, security and police services worldwide.

Manufacturer

Caswell International Corporation, Minneapolis.
(For details see Directory of Suppliers)

Unisys Remote Target System (RETS)

The Unisys Remote Target System (RETS) is a system of modular target devices and communications equipment which can be used in a variety of live fire configurations for advanced infantry and armour marksmanship training. Both stationary and moving infantry and armour targets pop up on command from concealed positions to provide a realistic simulation of modern battlefield conditions, where precise and quick response is essential for successful engagement of hostile forces. In addition, the use of modular hardware allows individual training installations throughout the world to set up training exercises which meet their unique needs, from fixed lane infantry marksmanship training to a real-terrain battlefield through which troops sweep to engage up to 512 stationary and moving targets positioned throughout an area extending over several miles.

Realism is further enhanced by the use of three-dimensional infantry targets (moulded out of self healing plastic), the simulation of hostile infantry fire, and the use of smoke clouds to indicate a successful hit on an armour target. In addition, targets can be programmed to fall down after a specified number of hits (from one to fourteen), or after a specified exposure time if the target is not hit.

Two types of communication control are available, both of which use a microprocessor-based range controller. This system provides full duplex communication with all target emplacements by means of either buried cable or radio frequency links.

Unisys has introduced a new command and control system which replaces a system-unique computer with a standard personal computer and an upgraded software package. This Enhanced RETS (E-RETS) system allows the range operator to display and monitor downrange targetry graphically on the video display terminal. In addition, commands can be entered via the keyboard, a trackball or a unique voice recognition system. This allows the range operator to make real-time inputs to the training scenario in response to the ongoing training exercise events.

RETS has been type-classified by the US Army, having met all system specifications and passed all environmental, reliability and operational tests. Both the infantry and armour targets can be used in severe winter as well as desert environments. Prototype ranges have been operational at Fort Benning, Georgia, since 1978 and at Fort Riley, Kansas, since 1983. The system is in full-scale production with a large number of ranges operational.

The RETS ranges are supported by an integrated logistics support package, which includes operation and maintenance manuals, new equipment training, and complete spares support. Maintenance of the ranges is enhanced by a programme which emphasises the substitution of modular units in the field, with repairs performed at base workshop. This logistics philosophy optimises system availability so that malfunctioning hardware has no impact on training exercises.

Status

In production. In service with the US Army and Marine Corps (140 ordered) in USA, Germany, Hawaii and Korea, and the Egyptian Army (3).

Manufacturer

Unisys Corporation, Shipboard & Ground Systems Group, Huntsville.
(For details see Directory of Suppliers)

Unisys RETS armour moving target carrier

Unisys RETS infantry moving target carrier

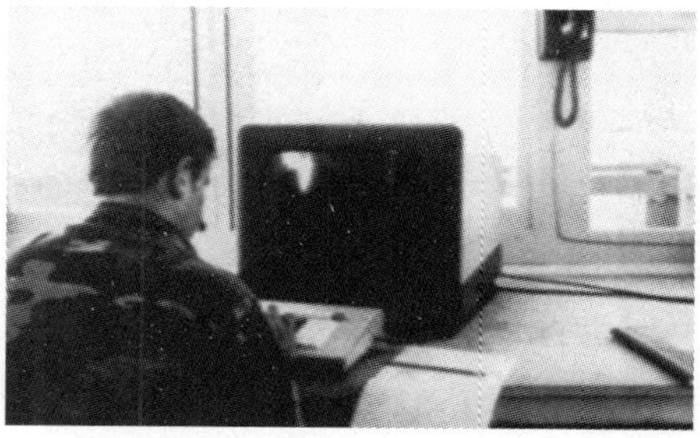

RETS Range Control Station with operator

YUGOSLAVIA (SERBIA AND MONTENEGRO)

Parachutist Target Launching Device M80

The Parachutist Target Launching Device M80, also known as the UIMP-80, is an unusual training aid intended to provide soldiers with targets simulating descending parachutists, both in groups and as individual targets. The targets can be used by small arms and light anti-aircraft weapons.

The M80 launching device resembles a large three-barrelled mortar carried on a trailer. Once at a launching site the trailer is stabilised using telescopic jacks and the three barrels are lowered to the ground from the rear of the trailer and onto a baseplate. The barrels are used to launch target carriers which resemble large mortar bombs. A propane/air mixture is ignited in a combustion chamber at the base of each barrel to act as the propellant. The muzzle-loaded target carriers are reusable.

The target carriers are fired using a hand-held control unit at 4- to 10-second intervals. At an altitude of over 350 m, controlled by a pneumatic regulator fitted to the fins of the target carrier, the carrier body opens and the airstream pulls out a small parachute and the target. Both remain attached to the carrier body and descend to the ground at a rate of from 4 to 6 m/s. The target simulates a full-size parachutist but the parachute and harness are on a reduced scale.

The M80 launching device is manned by a crew of two and can be set up in about 10 minutes. When travelling, the M80 carrying trailer can be towed at road speeds of up to 70 km/h.

Parachutist Target Launching Device M80 ready to fire

Status

In service with the armed forces.

Contractor

Holding Company YUGOIMPORT-SDPR Ltd, Belgrade.
(For details see Directory of Suppliers)

FULL CALIBRE GUNNERY RANGES

AUSTRALIA

ADI Vehicle Target System

The ADI Vehicle Target System comprises both stationary and moving target mechanisms which operate with a Combat Effects Simulator. The mechanisms are:

Stationary
- Light Duty Target Mechanism (LDTM)
- Medium Duty Target Mechanism (MDTM)
- Heavy Duty Target Mechanism (HDTM)

Moving
- Full-Size Moving Target Carrier (FSMTC)

All target mechanisms are controlled by the ADI UHF Hand-Held Controller and may be fitted with the ADI Projectile Location System (PLS - see entry under *Small Arms Range Equipment*). A range may comprise any combinations of stationary and moving, figure and vehicle target mechanisms.

Except for the LDTM, target boards are common and may be interchanged and rearranged to depict a variety of vehicle targets and crew-served weapons. The rail for the moving target mechanisms is permanently installed, but the stationary target mechanisms are transportable on a battery charger storage trailer. The FSMTC is also transported and stored on a trailer and can be moved easily between several monorail sites and its support area. All stationary target mechanisms may be staged (partially exposed) during raising and lowering.

Light Duty Target Mechanism (LDTM)
The Light Duty Target Mechanism is the smallest of the ADI range of vehicle target mechanisms. It lifts a target representative of a hull-down stationary AFV, and may be engaged by main and secondary armaments. The target board is of a suitable size to represent a half-scale vehicle target and may be fitted with a second target arm which presents up to five figure targets.

The LDTM is controlled by the ADI UHF Hand-Held Controller (HHC) and raises and lowers a target in 8 seconds. It is powered by a 12 V battery and 18 complete mechanisms can be transported and stored on a storage trailer which incorporates a 20-way battery charger. When the stabilising frame is folded for storage and transport the LDTM is compact and can be carried by two persons.

Medium Duty Target Mechanism (MDTM)
The Medium Duty Target Mechanism raises a full-size, head-on, tracks-up silhouette of an AFV. The robust construction of the electrohydraulic mechanism enables it to be used either permanently installed or, in the portable role, in all climates or terrain.

The MDTM is controlled by the ADI UHF HHC and requires 24 V to power the system. The batteries and transceiver are packaged in a carrying frame.

Heavy Duty Target Mechanism (HDTM)
The Heavy Duty Target Mechanism lifts a full-size silhouette of a MBT. Due to its size and weight it is most likely to be used as a permanent installation and used with anti-tank weapons. The HDTM can use two lifting shafts to expose up to seven target boards. Alternatively several groupings of target boards could be exposed to present multiple, head-on, tracks-up targets. As with the MDTM, the HDTM is controlled by the ADI UHF HHC and requires 24 V to power the system. The batteries and transceivers are packaged in a carrying frame.

Full-Size Moving Target Carrier (FSMTC)
The Full-Size Moving Target Carrier (FSMTC) is a self-propelled system which uses a monorail guidance system. It is capable of presenting a full-size vehicle target in profile, obliquely or head-on, at any two preselected speeds up to 26 km/h. When the target is hit the FSMTC will stop and discharge coloured smoke.

The FSMTC is hydraulically driven and is powered by an air-cooled petrol engine carrying sufficient fuel for 4 hours of continuous running. For situations requiring approaching and/or withdrawing targets, the FSMTC is fitted with a transverse target holder to represent an AFV, tracks-up. The normal target boards can be substituted for soft target boards when live rounds need to be fired (for example MILAN).

The monorail guidance system offers a number of advantages over conventional track. Monorail is more easily laid and is better able to follow ground contours and negotiate bends. The earth disturbance required to install monorail is small.

Combat Effects Simulator (CES)
The Combat Effects Simulator provides night illumination, pyrotechnic gunfire and target hit indication, and is used with the ADI Vehicle Target System.

In total 48 pyrotechnic capsules are mounted in a frame with a night light system. Half the capsules (gunfire) simulate main armament firing and the other half are (normally) orange smoke to simulate a target hit. The simulated return fire from the main armament is initiated from the firing point when the target is raised by using the ADI Hand-Held Controller. The coloured smoke is initiated when the target is hit.

Two lamps, one blue and one orange, comprise the night lights. When the target is raised, a soft blue light is projected onto the target to produce a realistic silhouette effect when viewed through gun sights at night. When the target is hit the orange lamp flashes.

The CES carrying frames are manportable, stackable and can be quickly connected to or disconnected from their parent target mechanism. When used with the FSMTC the CES is configured in a different form.

Status
In production. In widespread service.

Manufacturer
Australian Defence Industries Limited, Bondi Junction.
(For details see Directory of Suppliers)

Light duty tank target mechanism (LDTM)

Heavy duty tank target mechanism (HDTM) in permanent installation

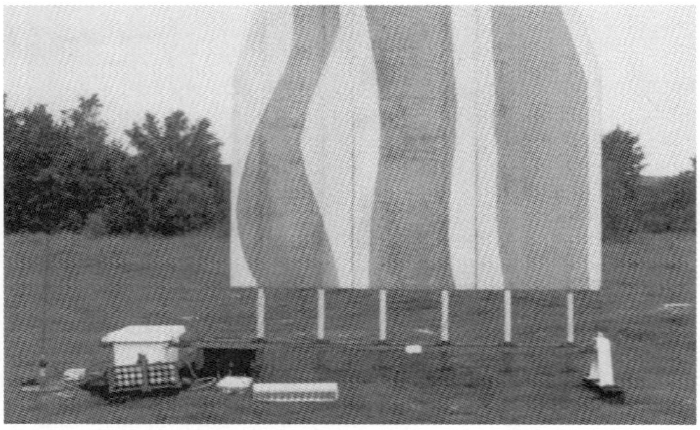

Medium duty tank target mechanism (MDTM) fitted with visual hit indicator system and gunfire simulator

Full-size moving target carrier (FSMTC)

CANADA

Boeing Canada Badger

The Badger tank target is based on the chassis of an Argo 6 × 6 all-terrain vehicle and represents a one-half or three-quarter scale mobile tank target. The Badger can reach speeds of up to 50 km/h over prepared terrain and is remotely operated at line-of-sight ranges of up to 6 km. The target can be equipped with miss-distance indicators, infra-red emitters and passive radar augmentation to simulate a full-size tank. The standard control station allows two targets to be operated at the same time with optional expansion for more. Multiple tank threats can be simulated by towing up to three tank targets behind the prime mover. The prime mover can be lightly armoured to protect it from sub-calibre weapon and small arms fire.

The Badger is 4.6 m long overall (3.45 m over the hull), 1.15 m high and 1.8 m wide.

Status

In production.

Manufacturer

Boeing Canada Technology Limited, Winnipeg.
(For details see Directory of Suppliers)

Badger half-scale tank target

FRANCE

Giat Self-Propelled Tank Target

This mobile tank target is propelled by an integral petrol engine coupled to an adjustable speed belt-driven transmission. The target can move along rails in either direction and the target can be raised and lowered as required.

The target moves along tracks 1.44 m wide arranged in a straight or loop configuration. The targets are made of plywood or wire mesh and the hydraulic target lifting mechanism is powered from a circuit on the carriage. The target may measure up to 4.6 × 2.3 m. A programmable external light is provided for night illumination.

The target may be controlled by radio from a distance of up to 4000 m using six frequencies in the 150 MHz band. The sensitivity of the hit sensors may be varied according to the type of ammunition in use.

The target carrier weighs 1200 kg and measures 4.93 × 1.608 × 1.29 m. Speeds are adjustable between 5 and 40 km/h and acceleration is from 0 to 40 km/h within 50 m. Payload of the carrier is 500 kg. One filling of the fuel tank will enable the carrier to run for up to 8 hours.

Status

In service throughout the world.

Manufacturers

Giat Industries, Tarbes.
Cegelec Soprano, Vaulx Milieu.
(For details see Directory of Suppliers)

Self-propelled tank target

GERMANY

Krupp Fast Moving Target System

The Fast Moving Target System is used for the training of crews of tanks and armoured vehicles. The development is based upon the NATO requirement for gunnery training systems which meet the high technical standards of modern armoured vehicles. It provides systematic and realistic training of tank crews in fields such as recognition, identification and hitting moving targets with live ammunition.

Computer-controlled target carriers, with automatic hit recognition, travel upon a monorail system which can be exactly adapted to the training requirements and skills of the troops through stored or programmable training programmes. The constant monitoring of results and the repeatability of certain training sections makes it possible to monitor training closely and to compare performances.

The target carrier weighs 800-900 kg and is 2.60 m long and 67 cm wide. It can travel at a maximum of 50 km/h (30 mph).

Status

In service.

Manufacturer

Krupp Fördertechnik GmbH, Duisburg.
(For details see Directory of Suppliers)

The Krupp Fördertechnik Fast Moving Target System

Theissen Tank Target Equipment

Theissen Fassaden- und Maschinenbau GmbH produces a tank target system which incorporates target presentation, hostile fire simulation and hit registration for moving and static target systems simulating day and night combat.

The target system can hold any tank target silhouette up to full size and raise and lower it on command from a controller even in strong winds. If the target system is installed on a moving target mechanism which changes the direction of travel, a target revolving device can be added. This device turns the target so that it faces the main direction of firing all the time. Thus the firer will constantly see the full target silhouette and there is no need to employ three-dimensional targets.

A flashlight is integrated into the target system to simulate the muzzle flash of an enemy tank gun and is activated by the controller. The flashlight is installed in such a way that its light will be reflected by the target silhouette and is strong enough to be observed under all weather conditions.

The system can provide a thermal target by using control electronics which control target heating and a low-cost heating foil on the target silhouette. The heating of the foil is controlled to ensure that the temperature difference between the target and the environment is constant at all times. The temperature difference can be selected either by the control electronics or the control unit.

The system uses an integrated LOMAH (Location of Miss and Hit) consisting of two sensor arrays installed in front of the target. Locations of hits or near misses are displayed on a monitor and a hard copy of results can be provided.

Status
In service.

Manufacturer
Theissen Fassaden- und Maschinenbau GmbH, Düsseldorf.
(For details see Directory of Suppliers)

Theissen tank target with target revolving device

HUNGARY

IDEX Moving Target Equipment

Industrialexport Hungary produces a range of moving target equipment for use on training ranges. These include infantry-type targets but also promoted is a range of moving target equipment suitable for use with tank targets. Equipment produced includes single- and double-rail systems and equipment not confined to travelling on a fixed track. Target carriers can be equipped with pyrotechnic devices producing white or red smoke. Also produced are moving target carrying systems running on overhead cable arrays.

Tracked carriers can run on tracks up to 300 m long and carriers can carry three or four targets with either single or double layers. The systems operate on three-phase 380 V, 50 Hz electrical supplies. Power demand is from 2 to 11 kW.

Included in the Industrialexport Hungary range is a pop-up helicopter target.

Status
In production. In service with the Hungarian armed forces.

Manufacturer
IDEX Budapest.
(For details see Directory of Suppliers)

Two-rail tank target carrier with pyrotechnic firing device

Pop-up helicopter target

Turning tank target carrier on two-track system

NORWAY

HMV Ground-Based Moving Target Systems

Hordaland Mek. Verkstad A/S (HMV) is a pioneer in the field of moving targetry based on winches utilising steel ropes or cables. On its ground-based winch-driven systems targets are mounted on a sledge-type target carrier which is pulled around a track by a system of pulleys and steel cables. These systems use a capstan winch with outputs of from 0.5 to 178 hp, although power from batteries, a mains supply or generators may be involved.

The winches and other equipment used with these systems can be operated over rough terrain conditions and are available to meet any target size and speed requirement. Track lengths of from 5 to 2500 m are available and target speeds may vary from 5 to 55 km/h.

On the sub-calibre HMV El-Dc 0,5 system the target is mounted on a small target sledge pulled by a capstan winch (*Vossavinsjen El-Dc 0,5*). This miniature system is portable and is intended for use with sub-calibre training weapons with the 24 V power supply for the winch being taken from the gunner's vehicle; the manually operated winch is normally located by the gunner's vehicle. The target runs in a straight line for distances of from 5 to 50 m with a speed of up to 10 km/h. Target weight may be up to 10 kg and the system can operate with two targets.

The 178 hp diesel-powered winch unit used with the HMV Di-Var 178 Agile Target System (ATS) with a target sledge in the background

Target sledge for an HMV El-Var 4 ground-based target system

On the medium ground-based systems, a variety of winches can be used. Target runs may vary from 5 to 150 m with speeds from 5 to 30 km/h. Targets may weigh up to 30 kg and it is possible to use this system with two targets. This system is normally used with sub-calibre weapons. The manually controlled, air-cooled, petrol-driven *Vossavinsjen Ho-Var 11M* and the *Vossavinsjen Ho-Var 11F* may be remotely controlled by cable or radio.

The HMV El-Var 4 follows the same general lines as the HMV Ho-Var 11 but uses a 230 V mains supply to power a 4 hp winch. In other respects the two systems are similar and use either manual or remotely controlled winches.

With the portable HMV Di-Var 70 system the targets are driven by a *Vossavinsjen Di-Var 70* diesel-engined winch producing 70 hp to power a hydraulic drive system. The target sledge uses ground-compensating skis and can carry full-size targets weighing up to 100 kg. Target runs may be from 100 to 500 m with target speeds of from 6 to 36 km/h. The trailer-mounted winch may be remotely controlled using various types of standard radio.

The largest system in the ground-based HMV range is the Di-Var 178 Agile Target System (ATS). This full-size target system is powered by a *Vossavinsjen Di-Var 178* diesel-powered capstan winch producing 178 hp to power a hydraulic drive. Up to two target sledges, each with a load of 200 kg, may be used with this system, with target runs of from 100 to 2500 m over rough terrain; target speeds are from 5 to 55 km/h. Target tracks can vary in shape from a straight line to almost a full circle. The system is remotely controlled by radio and it is possible to design complete programmes which can run automatically.

Status

In production. In service with the British, German and Norwegian armies.

Manufacturer

Hordaland Mek. Verkstad A/S, Voss.
(For details see Directory of Suppliers)

The sub-calibre HMV El-Dc 0,5 ground-based moving target system in use by a Norwegian Army M24 Chaffee tank

Target sledge for an HMV Ho-Var 11 ground-based target system

HMV Rail-Based Moving Target Systems

In addition to its ground-based moving target systems (see previous entry) Hordaland Mek. Verkstad A/S (HMV) also produces a number of rail-based moving target systems with the targets mounted on cable-drawn trolleys. These rail systems derive their tractive power from the same winches as those used for the ground-based systems.

Rail track gauges vary from 0.2 to 1.2 m with rail lengths of from 5 to 1500 m. Target speeds can vary from 5 to 55 km/h.

The HMV Rail System V-30 is a sub-calibre system providing a gunner with a 1:20 scale moving target. This system is powered by a *Vossavinsjen El-Fro 0,5* 220 V AC 0.5 hp capstan winch producing speeds of from 0.5 to 5 km/h. The target trolley runs on rails set either horizontally or vertically within protective grooves and uses a frequency control system that can operate several moving targets in conjunction with static pop-up targets.

The HMV Rail System S-100 uses horizontally mounted 1 m gauge rails along which a target trolley weighing up to 100 kg can move for distances of up to 300 m; straight or curved rails may be involved in sections 3 m long. The target is pulled by either a *Vossavinsjen El-Var 4F* 230 V AC capstan winch producing 4 hp or a *Vossavinsjen Ho-Var 11F* winch powered by an 11 hp petrol engine. In both cases a hydraulic drive system is involved and target speeds can vary from 5 to 30 km/h. Rail heights may vary from 0.4 to 0.75 m and curved sections have a radius of 22.9 m.

The HMV Rail System S-120 is intended for heavy gunnery training and uses rails mounted on standard railway sleepers. The rail gauge is 1.2 m and is used in lengths from 200 to 1000 m long. Each section is 6 m long and curved sections have a radius of 45.8 m; rail height is from 0.7 to 1.6 m. Power for the S-120 is provided by a *Vossavinsjen Di-Var 50* winch powered by a 50 hp diesel engine. Target speeds vary from 5 to 50 km/h and remote-control may be via cable or radio.

Status

In production. In service with the Norwegian Army.

Manufacturer

Hordaland Mek. Verkstad A/S, Voss.
(For details see Directory of Suppliers)

The HMV Rail System S-100

A typical HMV Rail System S-120 with Di-Var 50 winch

HMV Ropeway-Based Moving Target Systems

In addition to its ground- and rail-based moving target systems (see previous two entries) Hordaland Mek. Verkstad A/S (HMV) also produces moving target systems with which the target is pulled along steel rope tracks on an overhead trolley pulled by a capstan winch. These simple and relatively low-priced systems can be arranged to suit almost any terrain and use endless haul rope arrangements. Winches may be electrically or internal combustion powered. The track ropes may have diameters of from 5 to 15 mm and run lengths may be up to 1500 m.

An HMV Ropeway System TOW target

The HMV Ropeway System LPV uses an arrangement whereby the ropeway is supported by a number of adjustable pylons at a height of between 4 and 6 m above ground level; ground fixings for soil and rock can be supplied. The target size can vary according to customer requirements and can weigh up to 100 kg. The target is pulled by either a *Vossavinsjen El-Var 4F* 230 V AC capstan winch producing 4 hp or a *Vossavinsjen Ho-Var 11F* winch powered by an 11 hp petrol engine. In both cases a hydraulic drive system is involved.

The HMV Ropeway System TOW can be used with anti-tank guided missiles such as MILAN or TOW. With this system the ropeway is mounted between two adjustable pylons 6 to 12 m high and sufficient steel rope is supplied for a 1500 m run. The target size can vary according to customer requirements and can weigh up to 250 kg, with target speed from 5 to 40 km/h. Power to move the target is provided by a *Vossavinsjen Di-Var 17F* capstan winch powered by a 17 hp diesel engine. The winch can be remotely controlled by cable or radio.

Status

In production. In service with the Norwegian Army.

One of the pylons used with the HMV Ropeway System TOW

Manufacturer

Hordaland Mek. Verkstad A/S, Voss.
(For details see Directory of Suppliers)

SWEDEN

Saab BT 18SC Heavy Duty Tank Target

This is a pop-up target driven by an electric motor. It is powered by a Ni/Cd battery and controlled by the BT 22 FM radio system or the BT 24 range control system. The target consists of an electrohydraulic operating unit and one target holder, which is made in two parts from tubular steel. Each part of the holder has a clamp for securing the target holder and is connected to each side of the driving shaft. When transporting or storing the unit, each part with its clamp is easily fastened to each side of the target unit. The target is self-indicating and the signal voltage for hit indication is created by a sensor which is clamped to the lower edge of the target board. If the sensors or the cable to the sensors are damaged by a projectile, the target board disappears and the target cannot be operated by the radio, indicating that the hit sensing system is not intact. The number of hits and time to first hit can be transmitted to the BT 24 range controller where it may be displayed or printed out in hard copy form. Firing from the target can be simulated by connecting the unit with a BT 19 gunfire simulator, which can also be used as a hit indicator. The target board, which should be made of a rigid material such as hardboard, can be of silhouettes of a tank or a part of the tank.

Weight:
(torque mechanism less battery) 95 kg
(Ni/Cd battery): 21 kg
(lead battery): 14 kg
Turning angle: 85°
Capacity per charge: more than 200 cycles using target weighing 65 kg at 15 m/s wind speed
Rising and falling speed: 3-6 s/operation

Status

In production. In service with several armies.

Manufacturer

Saab Training Systems AB, Huskvarna.
(For details see Directory of Suppliers)

Saab BT 18SC heavy-duty tank target and BT 19G gunfire simulator

Saab BT 24 Range Control System

The Saab BT 24 range control system is a two-way radio control system. The communication system can link a maximum of 3375 stationary or moving targets on a single frequency with a radio transceiver located downrange. Each radio transceiver can control and also feed back hit data from 15 targets individually. The BT 24 programmable controller can direct targets with preprogrammed scenarios or with keyboard inputs from a range officer.

Before an exercise an operator can create programme scenarios which are then stored in a cassette. Targets can be programmed to fall after any number of hits; to remain exposed for a certain period of time; to ignite pyrotechnics at set times of target exposure or in response to hits; to link the actions of targets to firing results; and to vary speeds or appearances of moving targets at certain times or locations. The BT 24 programmable controller can direct preprogrammed tactical responses of target groups that vary according to the performance of the firing unit. The controller display permits an instructor to preview, select or reprogramme a scenario before an exercise commences.

Before commencing a scenario it is possible to test targets, batteries and other equipment and obtain a complete printout of the readiness status of all equipment. The programmed scenario can then be altered to operate around faulty equipment. Range equipment can also be re-tested following an exercise.

Saab BT 24 programmable range controller

During an exercise using the preprogrammed mode the BT 24 system automatically controls all targets on the range. This mode permits an instructor to observe a unit's response to a developing scenario rather than control targets. The instructor can interrupt a scenario at any point and manually operate targets. The scenario may be restarted at any programme line. An emergency key on the BT 24 programmable controller enables the instructor to lower all targets.

In the manual operation mode the instructor can control all targets in real time by using the keyboard on the controller. During an exercise the BT 24 system automatically collects hit data from all targets.

Hit data can be printed out for after-action review or can be stored on the removable cassette. The cassette provides data for an exercise replay or a permanent record for individuals or units. The BT 24 system provides information on the number of hits on each target, target exposure time before the first hit, the average exposure time of all targets, the percentage of hits, and other data to provide immediate feedback on performance.

Frequency range: 138 to 170 MHz
Frequency separation: 25 kHz
Range: 4000 m

Status

In production. In service with several armies including the British Army and the US Army.

Manufacturer

Saab Training Systems AB, Huskvarna.
(For details see Directory of Suppliers)

Saab BT 37 SD Moving Target System

The Saab BT 37 SD moving target system is a rail track operated and radio-controlled mobile target for tank and anti-tank gunnery training. The system consists of standardised rail-track sections, a self-propelled trolley with a target that can be raised and lowered, end sections with buffers, and a combined parking/charging station. Electric motors enable the target trolley to climb gradients of up to 8 per cent. Target hits are indicated by means of pyrotechnic cartridges or by lowering the target. The target may be controlled by the Saab BT 22 FM remote-control system or the Saab BT 24 range controller (see previous entry).

Power for the system is supplied by Ni/Cd batteries on the target trolley which provide a trolley range of 8 km, including 40 accelerations from standing to maximum speed. This capacity can be increased to about 100 km with an optional onboard generator.

The system can be equipped with accessories for simulation of fire from the target using pyrotechnics.

The rail track is built in modules with three main types of section: end section with buffers, straight section and curved sections with 5 or 15° change of direction. A track to suit any terrain can be made by combining the different sections. An embankment, 1.3 m high, is needed to protect the installation. Alternatively, the rail track can be laid in a trench of corresponding depth.

The target trolley has a maximum speed of 50 km/h, accelerating from standing to 30 km/h within 30 m. All functions and commands are controlled by a built-in microcomputer. The all-welded chassis contains the batteries, electronics and hydraulics. All four wheels have chain drive. There is also an anti-skid device which guarantees traction and optimum power output even under difficult conditions. Radio control of the trolley is achieved either with two BT 22 FM receivers for target control and trolley control, or with a BT 24 radio control system enabling data collection from the trolley.

The batteries are charged each time the trolley reaches the combined parking/charging station. For safety the battery temperature is monitored during the charging period. The charging power is switched off after a preset time (under the control of a timer) or when charging has been completed. The power source can be either 220 or 110 V mains or a portable engine-driven generator.

Safety features include automatic braking when the trolley approaches an end stop, automatic lowering of the target at the end stops, and shock-absorbers on the trolley and buffers at each end stop that permit impact at a speed of up to 35 km/h without damage.

Accessories include a pyrotechnic hostile fire simulator and a pyrotechnic hit indicator, both holding 15 cartridges. These can be carried on and used from the trolley.

Trolley
Weight:
(without batteries) 420 kg
(with batteries) 520 kg
Length: 2.2 m
(with target holder) 3.11 m
Height: 0.73 m
Width: 1.25 m
Speed: 0-50 km/h variable
Acceleration: 0-30 km/h within 30 m
Braking: 30-0 km/h within 20 m
Range: 8 km, including 40 accelerations from 0 to max speed; this can be increased to about 100 km by using an onboard generator
Target weight: up to 150 kg
Temperature limits: −25 to +40°C

Saab BT 37 SD moving target system

Track
Section length: 5.8 m
Track gauge: 1 m
Track height: 0.65 m
Max gradient: 8%
Recommended height of embankment: 1.3 m

Status

In production. In service with the Swedish and other armed forces, including the British Army.

Manufacturer

Saab Training Systems AB, Huskvarna.
(For details see Directory of Suppliers)

SWITZERLAND

DIXI CB-2 Tank Target System

This resembles the AK-2 target system (see entry under *Small Arms Range Equipment*) in general principles, but is larger and accommodates a tank target for gunnery training with tanks or anti-tank weapons. The mechanism is designed to be capable of raising and lowering a target 2 × 3 m in area in a wind of up to 60 km/h. It is normally used in a fixed position, but it can be put on a trolley so as to provide a moving target. It may be fitted with a gunfire simulator carrying 20 firing units, so that return fire may be simulated. Alternatively, smoke cartridges can be used to indicate hits or to simulate a tank smoke-screen.

Target Unit
Dimensions: 1800 × 650 × 450 mm
Weights:
(complete unit) 120 kg
(operating and electronic unit) 65 kg
Drive: electromechanical
Power source: 24 V DC Ni/Cd battery
Battery life: about 500 up or down movements in 24 h
Calibres accepted: 7.65-155 mm

Command Box
Dimensions: 280 × 174 × 166 mm
Weight: 6.5 kg
Power source: built-in battery or external
Capacity: 12 targets per HF channel

HF channels: 8
Frequency ranges: 140, 160, 420 or 460 MHz bands

Ni/Cd Battery Charger
Dimensions: 500 × 300 × 430 mm
Weight: 39 kg
Power supply: 110 or 220 V AC, 50/60 Hz
Target batteries: 6
Command box batteries: 2

Gunfire Simulator
Dimensions: 1020 × 350 × 300 mm
Weight: 59 kg
Number of rounds: 20
Sound pressure: 175 dB at 3 m distance

Status
In production. In service with the German Army.

Manufacturer
DIXI SA, Le Locle.
(For details see Directory of Suppliers)

DIXI CB/25 Target System

The DIXI CB/25 is an automatic tank target system. It will raise and lower full-scale plywood, plastic or glass fibre tank or armoured vehicle silhouette targets with areas of 3 × 2 m or more in winds of up to 100 km/h. The operation can be controlled from a central range console by land-line or radio, or by the DIXI portable computer remote-control unit, and the target unit also has its own controlled timer to raise and lower the target. There is also a moving target carrier system available onto which the CB/25 can be installed, thus giving all possible types of target presentation.

The CB/25 will withstand hits from TOW, MILAN, Dragon and similar missiles, gunfire from helicopters and all calibres up to and including 120 mm. A gunfire simulator can be connected to give realistic return fire.

Dimensions: 3030 × 505 × 730 mm
Weight: 220 kg (190 kg without battery)
Rise and fall time: 3.5-5.5 s
Power supply: 12 V or 24 V DC, 15 Ah or 55 Ah
Frequency range: 68-87.5 MHz (i); 138-174 MHz (ii)
Number of channels: 8
Operating life:
(24 V 15 Ah Ni/Cd battery) 250 cycles
(24 V 55 Ah torque starter type battery) 600 cycles
Delay timer action: 5-120 s

Status
In production. In service with the British, German, Swiss, Middle and Far Eastern armies.

Manufacturer
DIXI SA, Le Locle.
(For details see Directory of Suppliers)

DIXI CB/25 tank target system

Polytronic TG 82 Tank Target System

The Polytronic TG 82-30 tank target system is associated with the TG 82-10 infantry target system (refer to entry under *Small Arms Range Equipment* for details) and uses many components and accessories employed with that system, including the use of the TG 82 AC bi-directional radio control unit and the RCS computer control system (qv).

The TG 82-30 BES pop-up target is a radio-controlled unit for raising and lowering full-size tank targets and can be operated as a stationary target or (modified) on a moving platform. Depending on the model the mechanism is either mains (380 V AC, 50 Hz) or battery (12 V) powered, and may be controlled by radio, or a computer via cable. The electronics software can be varied to alter system functions. There are ports for night light, laser sensor, muzzle flash and hostile fire simulators.

While the TG 82-30 pop-up tank target system can accommodate the full range of Polytronic TG 82-associated accessories, items likely to be used include the HF-11 and HF-11C hostile fire cartridge devices (see entry under *Training ammunition and pyrotechnics*).

Status
In production and in widespread service.

Manufacturer
Polytronic International Limited, Muri.
(For details see Directory of Suppliers)

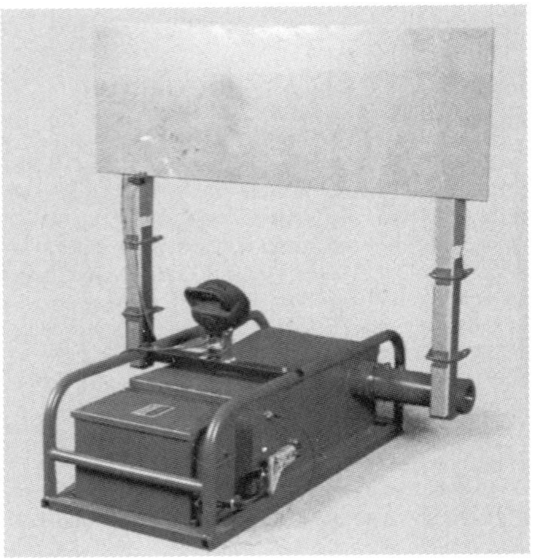

TG 82-30 tank target mechanism

Polytronic TG 300 Moving Target System

The Polytronic TG 300 moving tank target system permits training with tank and anti-tank weapons under realistic conditions.

A typical installation consists of a rail construction and a trolley carrying a combination of Polytronic mechanisms. The TG 300/1 is hauled by cable while the TG 300/2 is self-propelled. The system moves longitudinally, transversely or diagonally as required. The target trolley is either mains or battery powered and moves along two rails, with the rail length selected by the user. The speed may be selected in steps (maximum speed 50 km/h). To accelerate from 0 to 50 km/h a travel distance of 50 m is required. Gradients up to 8 per cent can be included.

The drive unit can be controlled by either the TG 82 AC bi-directional computerised radio control unit or the RCS range control system via cables. Targets may appear and disappear at random or according to a set programme.

The system construction can withstand rugged use under diverse environmental conditions.

Status
In production and in widespread service.

Manufacturer
Polytronic International Limited, Muri.
(For details see Directory of Suppliers)

Panzerzielfahrzeug 68 Target Tank

The Panzerzielfahrzeug 68, or PzZielfz 68, is a special target tank version of the Pz 68 main battle tank (for full details of the Pz 68 refer to *Jane's Armour and Artillery*). This vehicle was purpose-built for the target tank role and is used for training anti-tank weapon crews. It is fitted with extra armour for the hull, a heavily armoured turret and a dummy gun. The PzZielfz 68 weighs 38 000 kg and has a crew of two, the vehicle commander and a driver.

Status
In service with the Swiss Army only.

Manufacturer
Swiss Federal Armament Works, Thun.
(For details see Directory of Suppliers)

PzZielfz 68 target tank

UNITED KINGDOM

Chemring Modular Armoured vehicle Training Target System (MATTS)

The Modular Armoured vehicle Training Target System (MATTS) has been developed to meet the requirements for training troops to fire weapons with or without thermal imaging or infra-red (IR) sights. The use of IR sights on weapons launchers to guide missiles during night operations has highlighted the necessity for special IR training to aid vehicle recognition. Chemring has developed a training target system which meets UK MoD performance specifications for a Penetrable Anti-Armour Thermal Target System (PAATTS).

MATTS consists of a number of individual target panels of tubular construction which are 2.5 m high, 1.2 m wide and 25 cm deep. The panels are provided in a range of different density materials to suit environmental conditions and training scenarios and seven of them will make a main battle tank (which is 2.5 m high and 8.4 m wide) while five will make an armoured personnel carrier. A visual image is mounted to each target panel and if an IR signature is required it will be supplied as part of the image. The visual and IR signatures can be designed to meet specific customer requirements. Each panel is mounted into a panel holder secured to a target support beam which interfaces with conventional target moving systems.

Status

Available and in service.

Manufacturer

Chemring Limited, Portsmouth.
(For details see Directory of Suppliers)

Miltrain Tarquin Tank Targets

Miltrain Tarquin moving targets for tank and missile ranges are in service with the British Army and a number of other armies. Tarquin uses a radio-controlled, diesel-powered winch to tow up to three simple full-sized tank targets along the ground by means of a wire cable towed between pulleys. Target, cable and pulley layouts can be infinitely varied to suit the needs of those under training and systems can be transported and re-installed on a new site within a few hours, or returned to base to clear a range.

Targets used with Tarquin can be supplied by the manufacturer or can be locally produced. The Tarquin winch will tow a wide variety of targets so target size, shape, materials and construction can be varied to suit local conditions.

Winch operation is controlled from the firing point so that no range staff need to be located in bunkers down-range. There is a feedback on the radio control system to show the speed and location of targets on the range.

It is possible to combine the Tarquin system with other static or pop-up targets to add to range realism. In some countries Tarquin is used to provide targets for air-to-ground fire from helicopters and the target track is used to move the target between buildings, vehicles and other features to increase the training value.

Miltrain Tarquin tank target

The Tarquin system has been fully proven under active conditions at the RAC Gunnery Ranges at Castlemartin in South Wales.

Status

In service with the British Army and some other armies.

Manufacturer

Miltrain Limited, Godalming.
(For details see Directory of Suppliers)

UNITED STATES OF AMERICA

Caswell Full Calibre Gunnery Ranges and Systems

Moving Tank Target Mechanism

This is a rail track guided target for live firing training with tank or anti-tank weapons. The target-carrying vehicle is a self-propelled carriage moving on modular rail track sections at selectable speeds. The vehicle carries a target which represents a full-sized side view silhouette of a tank. The target can be raised and lowered and is equipped with hit sensors which can be commanded to lower the target or explode a pyrotechnic charge when hit. The vehicle is remotely controlled and responds to commands to move forward or backward, select speed, stop, raise or lower the target, retain the target down when hit, hold target when hit, night illumination, and pyrotechnic illumination. The mechanism is qualified for operation in severe environments. A protective emplacement is required to protect the installation.

Tank Target Mechanism

This is an electromechanical actuator of the pop-up type for frontal and side view silhouette targets. The mechanism is remotely controlled and provides hit sensing, night illumination, and an initiating interface for the optional pyrotechnic simulator. It is suitable for permanent or temporary installation in a sheltered emplacement and is qualified for operation in severe environments. The control mechanism responds to the usual range of commands and may be powered from rechargeable batteries or from a mains supply.

Battle Effects Simulator

Both hostile fire and battle effects are created with pyrotechnic charges fired by this simulator. The hostile fire charge simulates enemy return fire. The battle effects charge simulates a steel-on-steel hit constituting a kill. Standard 50 mm cartridges are used.

The simulator is a self-contained unit. It has a battery of 30 mortar tubes for the charges and a case containing the electronics to synchronise the firing. Charges can be fired either manually or by programme. Hostile fire simulation is triggered from the master control whereas battle effects simulation is usually triggered automatically by a hit on the target. The unit is qualified for operation in severe environments.

Status

In production. In service with the US and Middle East armed forces.

Manufacturer

Caswell International Corporation, Minneapolis.
(For details see Directory of Suppliers)

Caswell Moving Tank Target Mechanism

Caswell tank target mechanisms in fixed emplacements

CTA Integrated Thermal Target

CTA integrated thermal target

The CTA thermal target has its thermal image 'integrated' into the target surface. The thermal signature coatings are powered by a 12 V AC or DC transformer/rectifier system that converts 110/220 V AC, although they are also available for 110 or 220 V AC systems. Power is conducted through the target surface from wide busses that are situated well below the target area, thereby making it nearly impossible for hits on the target area to knock out the thermal signature until the target is destroyed. The thermal image is automatically activated to a temperature of at least 5°F above ambient in approximately 10 seconds after the target has been raised to the vertical position.

Hits on the CTA thermal target, from calibres of 5.56 mm up to 120 mm, typically hole 'cleanly' and do not disrupt the thermal signature except by the amount of impact surface removed by the hole left by the projectile, so even a large number of hits will not significantly degrade the thermal image. The targets are designed to fit most target-lifting mechanisms and will accommodate hit scoring sensors. The non-reflective surface of the targets allows them to be used with laser rangefinders.

CTA thermal targets may be two- or three-dimensional and in full, half, quarter or one-fifth scale. Targets produced to date are for the T-80 and T-72 MBTs, the ZSU-23-4 self-propelled air defence system, the BMP-2 IFV, the 122 mm 2S1 self-propelled howitzer, the BRDM-2 amphibious scout car, the BTR-70 APC, the 100 mm T-12 anti-tank gun, the Ural-375 truck, the BMD airborne combat vehicle and the Mil Mi-24 'Hind' helicopter.

Status
In production.

Manufacturer
Custom Training Aids (CTA) Inc, Swansea.
(For details see Directory of Suppliers)

CTA Laser Gunnery Target

CTA laser gunnery tank target

CTA has developed an inexpensive multi-spectral laser gunnery system utilising sets of lightweight, two-dimensional panels depicting full-scale visual and thermal images of NATO and former Warsaw Pact armoured vehicles for laser gunnery training. The panels may also be used for vehicle recognition training.

The system utilises a simplified version of CTA's decoy framing system with various high-fidelity visual images printed on durable fabric by the Milliken process. Integrated into the panels are thermal signature devices powered by 1.5 kW electrical generators providing 110 V AC.

The panels are designed to be placed in front of trainees manning thermal and optical sighting systems for the detection, identification, acquisition and engagement of threat vehicles in realistic settings. Targets may be placed around or along a trail for trainees to detect and 'engage' as they proceed around the course. As the panels are easily transported it is possible to vary training scenarios quickly. Target panels may be stabilised using inconspicuous guy ropes, and thermal images may be altered to depict vehicles in varying operational modes.

Status
In production.

Manufacturer
Custom Training Aids (CTA) Inc, Swansea.
(For details see Directory of Suppliers)

TVI High Resolution Multi-Spectral Targets

TVI high resolution multi-spectral targets are two-dimensional targets suitable for use in the field with laser-based gunnery simulators. Each target combines computer graphics visual technology with a thermal signature modelling capability. They can be used for stationary vehicle recognition and identification training and for employment during tactical exercises where the targets can be employed to constitute or augment an opposing force.

The targets accurately simulate the visual and thermal imager appearance of any vehicle likely to be encountered on a battlefield. The target vehicle can be produced in any configuration or camouflage paint scheme and can be presented from any aspect, including hull down, oblique, and side on. Whatever aspect is required, the corresponding thermal imager appearance will be accurately reproduced. The targets may be used as stand-alone fixed targets, on target lifting devices or mounted on light vehicles to produce movement. A Velcro strip on the target may be used to attach a MILES harness.

All the targets involved are capable of operating with only solar loading 'like the real tank' which presents a thermal signature of high fidelity. The simulants are four colours of a continuous tone that present a high-fidelity visual image of any threat or friendly vehicle. The result is that a two-dimensional target will appear to the human eye as three-dimensional. The illusion is more pronounced if an observer uses binoculars or other magnifying devices. The targets will continue to operate in adverse weather conditions such as rain, snow, ice and fog that are beyond the weather 'envelope' of AFV thermal sights.

Target is one-man portable and can be set up in minutes. No special tools, training or follow-on support is required. Each target is supplied fully assembled ready for immediate use and all targets are compatible with existing range equipments. They can be taken down and stored for reuse without difficulty. All systems fit into a soft container of duffel bag-type design fitted with a shoulder strap, zip fastener and snap hooks for securing to various vehicle mounting points. The maximum weight of a complete target is about 11.3 kg.

Target configurations produced include: T-72, T-62, T-54/55, BTR-70, BTR-60, SA-8 Gecko, ZSU-23-4, 122 mm 2S1, BMP, BRDM, Hind-D and Hind-E, M1 Abrams, M60A3, M2/M3 Bradley, M113, M901, M109, FISTV, HMMWV, Jeep, Chieftain, Centurion, Merkava, Leopard, AH-1S Cobra, and OH-58 Kiowa.

Status
In production. In service with the US Army and the Israeli armed forces.

Manufacturer
TVI Corporation, Beltsville.
(For details see Directory of Suppliers)

TVI Thermal Signature Targets

TVI thermal signature targets are used to provide training with armoured fighting vehicle thermal sights for main guns and secondary armament, and anti-tank guided missile systems. They can perform effectively under any weather conditions, and are durable, lightweight, easy to install and maintain.

The targets provide accurate thermal signature cues such as tank road wheels, tracks and engines, to the extent that some cues (tracks and engine exhausts for example) are 'hotter' than others. The signatures are provided by a TVI patented flexible heating element that uses a conductive coating rather than resistive wires. Since there are no wires involved the elements can be cut and shaped to simulate any vehicle 'cue', while hits on target will pass through the coated area without causing the target to cease operating. The targets are of modular design made up from several sections (known as target modules) with each module simulating a particular signature cue. Damage to one module will have no impact on the operation of the rest of the target. Certain critical parts of the overall target such as hull and turret sections have power leads at each end to form redundant power circuits, so that a hit on a power lead will not cause module failure.

In use the target modules are nailed or stapled to standard plywood or pressboard targets. There are no restrictions as to where nails or staples may be placed on the target. No special tools are required to set up, operate or maintain the targets. The targets use electrical power from batteries (via a power control interface), portable generators or mains supplies. Plug connectors joining the modules to a power supply can only be attached in one way and are of a simple push/pull type.

The targets do not emit any RF radiation and are adaptable to target raising devices, moving target ranges and computer-controlled ranges.

A typical TVI thermal target will absorb an average of 8-10 hits before repair or replacement is required. Repairing the targets is quick, simple and inexpensive. Familiarisation for range personnel with all aspects of setting up, operating and supporting the system can be accomplished in one to two days of on-site training.

A typical front or flank view thermal target weighs under 2.3 kg.

Status

In production. In service with the US Army.

Manufacturer

TVI Corporation, Beltsville.

(For details see Directory of Suppliers)

SMALL ARMS MARKSMANSHIP TRAINING EQUIPMENT

AUSTRALIA

ADI Marksman Trainer

The ADI Marksman Trainer was developed as a calibrated training aid to record, to recall and to display the pressures applied to a rifle by the firer at the moment of trigger operation. This allows faults in firing technique to be detected and corrected before they become habitual and permit higher standards of marksmanship. It consists of an instrumented service rifle which is connected by electrical cables to a display unit.

The rifle has strain gauges fitted to it to monitor the way the rifle is held when fired. Pressures at the firer-rifle interface points are monitored by the trainer to provide analysis on the consistency of weapon aiming, holding and firing techniques. A printout option is available for debriefing and analysis. The display memory can be used to recall up to 20 shots. The display unit can differentiate, record and display variations in the pressures applied.

Status
Available.

Manufacturer
Australian Defence Industries Limited, Bondi Junction.
(For details see Directory of Suppliers)

The ADI Marksman Trainer

BELGIUM

Browning Video Shooting System (VSS)

First produced during 1989 as a training aid for law enforcement agencies, the Browning Video Shooting System (VSS) has been further developed for military training applications. It is a video-based shooting and situation reaction training system which is both flexible and versatile and can be tailored to meet specific customer requirements.

The VSS can be set up in an indoor area with minimum dimensions of 10 × 5 m. Within that area up to four trainees can carry out laser or live shooting at targets projected by a video system onto a screen measuring 3 × 2 m. The target images may be projected from prerecorded video disk or tape and can provide targets at ranges of from 5 to 900 m. The trainee weapons involved may include hand-guns and rifles and can include larger infantry weapons. Targets can vary from the usual form of range targets to dynamic 'shoot/no-shoot' decision making situations. It is possible to produce interactive scenarios using video branching techniques.

As each laser or live round is fired the action on the screen is arrested and a computer-generated indicator denotes the shot impact on the target area; all shots fired are indicated. The system can take into account variables such as ammunition ballistics, wind speed and direction. The system can provide targets under virtually all lighting conditions. It is possible for the customer to produce their own video-based training scenarios.

Trainees can use the system by firing real ammunition, blanks, primed cartridge cases, compressed air or laser beam only. Hostile ripostes can be generated. For laser firing the system can produce computer-generated sound and recoil simulation can be provided using compressed air.

The VSS makes use of readily available equipment modules which may be combined to suit customer requirements. A basic system consists of two video U-MATIC players, a video projector, a personal computer with special software, a screen control monitor, a video image/text printer, a digital system for image processing, an interface unit for system control, a film camera, a firing detection system and a target screen. All equipment is stored off the floor in a rack and the screen may be folded away when not in use.

In use, the instructor is provided with simple menus to select a training sequence which is then projected onto the screen. Two video players permit the rapid linking of sequences to be projected. As the trainee fires, the system stops the picture which is digitised and projected onto the screen together with a point of impact position determined by the computer in the shape of a cross. Also projected onto the screen are the trainee's name and the reaction time. After each sequence the system automatically erases any live ammunition bullet impact from the screen by advancing the target screen paper. Digitised pictures can be stored in the system (up to 1000 frames can be stored) and results can be transferred to magnetic disks for later use.

It is possible to provide multichannel systems with up to three video projectors in use with a split screen layout, all under the control of a single control point.

Status
In production. In service with Belgian police and Gendarmerie units (15) and the Moroccan Ministry of Defence (1).

Manufacturer
Browning SA (Police Department), Herstal.
(For details see Directory of Suppliers)

Artist's impression of a Browning Video Shooting System (VSS) in use

CHINA, PEOPLE'S REPUBLIC

QSM-1 Basic Laser Aiming Simulator for Light Weapons

The QSM-1 laser aiming simulator for light weapons is intended to train new recruits in the basic steps of small arms marksmanship. The system involves a helium-neon laser mounted on a weapon and a fibre optics target.

The target is a reduced standard chest-type target scaled so that a range of 100 m can be simulated at 5 m. The system can indicate the scoring rings and sector for each shot with results displayed either shot-by-shot or in groups of 10 shots shown together.

Effective range: 5 m
Light power emitted: more than 0.3 mW
Sensitivity: less than 0.25 mW
Power consumption:
(transmitting) less than 50 VA
(stand-by) less than 2 VA
Continuous operation: 8 h
Operating temperature range: −10 to +40°C

Status

In production.

Development Agency

Nanjing Research Institute on Simulation Technique (NRIST), Nanjing.
(For details see Directory of Suppliers)

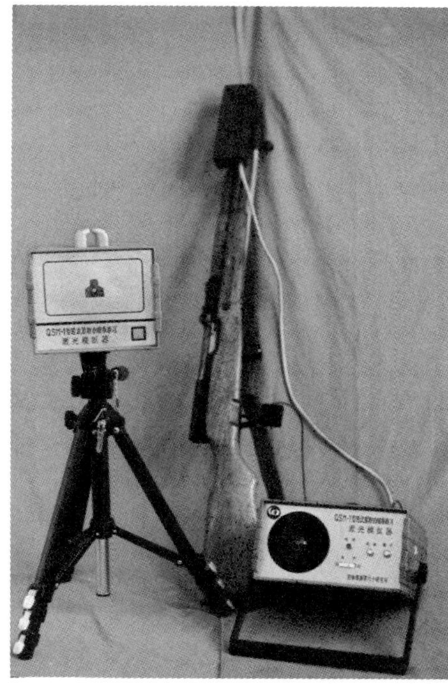

Elements of QSM-1 laser aiming simulator for light weapons

CROATIA

LAIR M83 Orthoscope Infantry Aiming Trainer

The M83 Orthoscope is a teaching aid designed to provide aiming and gunnery training exercises and their evaluation. Its operation is based on the characteristic of a semi-transparent mirror, attached to the frame of a weapon, to reflect 30 per cent and transmit 70 per cent of light. A trainee can practise aiming undisturbed while an instructor, at the same time, can inspect his aiming.

The device may be used indoors or outdoors and is very effective in the initial phase of training in marksmanship. By a simple adaptation the device may be attached to any type of weapon from rifles to heavy machine guns and certain types of artillery and mortar sight.

A set consists of seven holders and two frames with mirrors, stored in a carrying case. A complete set, in its case, weighs 5 kg.

Status

In production. In service with the armed forces.

Manufacturer

LAIR Croatia, Zagreb.
(For details see Directory of Suppliers)

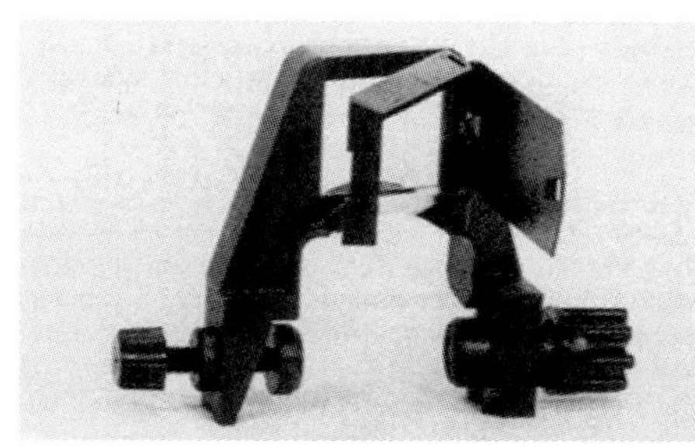

M83 Orthoscope infantry aiming trainer

FINLAND

Noptel ST-11 Shooting Trainer

The Noptel ST-11 is a laser device intended for personal marksmanship training and is attached to the barrel of a weapon by a universal adaptor; no modifications to the weapon itself are required. The ST-11 contains both the transmitter and receiver and no wires, electronic equipment or external power supply are required in the target area. The target, made of reflective material, can be of any size or shape and several firers can use the same target simultaneously.

The ST-11 can be operated either in the aiming or shooting mode. In the aiming mode the ST-11 emits laser pulses continuously and indicates successful aiming with a sound or light signal; this mode is used to practise aiming and holding. In the shooting mode the ST-11 emits a short laser pulse as the trigger is activated. If a hit is scored the pulse will be reflected back from the target and recorded by the receiver to provide a hit indication signal. Training can be carried out indoors or outdoors at ranges from 3 to 500 m.

The operating time with one 9 V rechargeable battery is 100 hours. Weight of the ST-11 is 380 g and the Class 1 device is shock- and water-resistant.

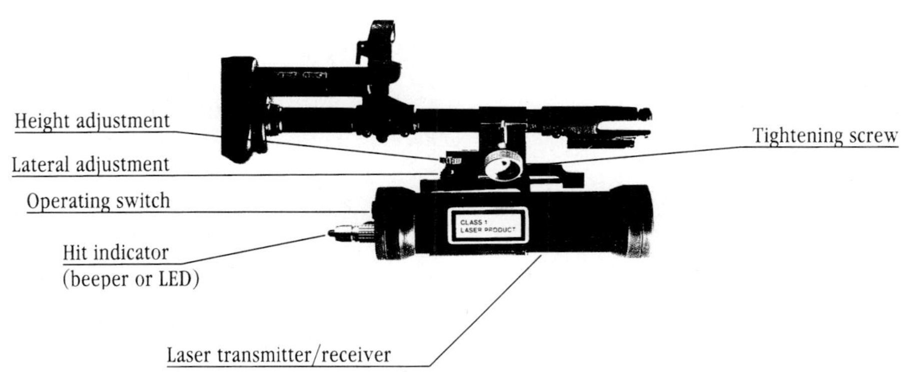

Main components of the Noptel ST-11 shooting trainer

Status

In production. Approved by several countries.

Manufacturer

Noptel Oy, Oulu.
(For details see Directory of Suppliers)

Noptel ST-15 Combat Trainer

The Noptel ST-15 Combat Trainer consists of a laser transmitter/receiver unit inserted into the barrel of a pistol plus an electronics unit mounted in the space normally occupied by the pistol magazine. No alterations to the pistol itself are required and the ST-15 can be replaced with the original parts for full operational use of the pistol. The ST-15 can be installed in most service pistols and was developed from the earlier ST-10 Shooting Trainer.

The target is covered with reflective material and can be of any size or shape, stationary or moving. As the pistol is fired the ST-15 unit transmits an invisible and eye-safe short laser pulse. If it hits a target the pulse is reflected back and the receiver indicates the hit by a sound signal. The operational range is from 3 to 50 m.

Weight of the ST-15 Class 1 laser device is 170 g and

it can be used for 30 hours of continuous operation with rechargeable 6 V batteries.

Status
In production. In service in 16 countries.

Manufacturer
Noptel Oy, Oulu.
(For details see Directory of Suppliers)

Components of the Noptel ST-15 combat trainer for pistols

Noptel ST-1000 PC Marksmanship Training and Analysis System

The ST-1000 PC marksmanship training and analysis system represents the second generation of laser training systems developed by Noptel, the first being the ST-1000 system. The ST-1000 PC provides computerised monitoring and analysis of the entire shooting process to enable an individual to improve and maintain marksmanship skills.

The ST-1000 PC can be used with rifles and pistols. The system scales a target automatically for different weapons, distance and projectile calibres. A trainee can utilise his own weapon without modification as the system utilises three components: a laser transmitter attached to the barrel of the weapon, an optical target and a computer connection.

As a trainee takes aim a narrow beam projected by the laser transmitter hits the optically sensitive target. The hit position is electronically measured every millisecond.

The ST-1000 PC operates with any IBM-compatible personal computer, offering a wide variety of freely selectable functions. The programme includes several display modes, data on shooting performance and printout facilities.

The Class 1 laser transmitter weighs 150 g and is shock- and water-resistant.

Status
In production. In service with 35 countries.

Manufacturer
Noptel Oy, Oulu.
(For details see Directory of Suppliers)

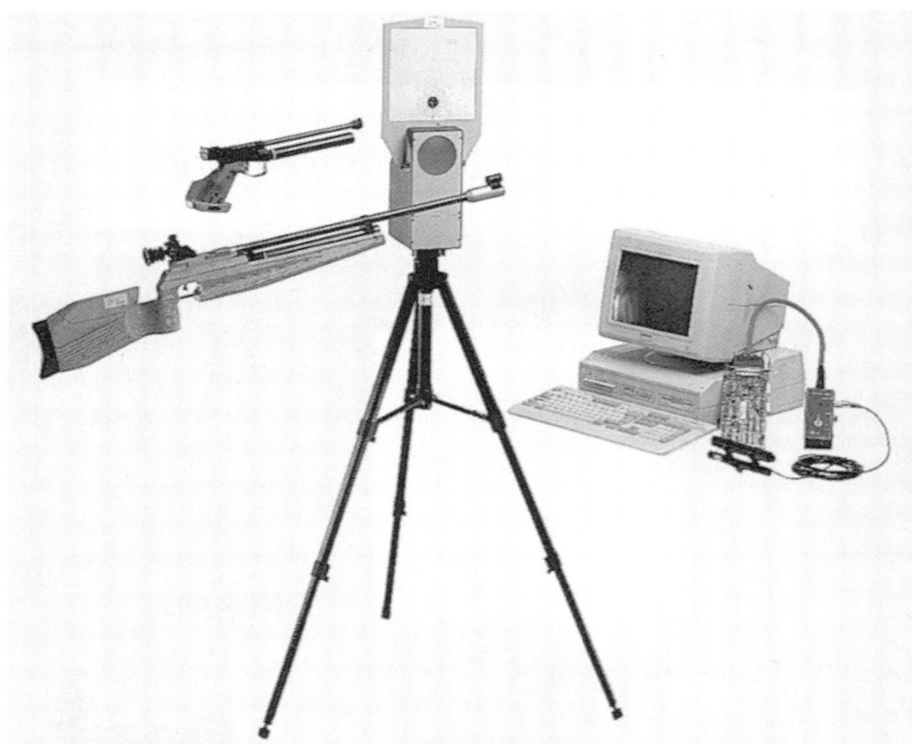

Components of the Noptel ST-1000 PC marksmanship training and analysis system

Noptel ST-2000 Marksmanship Training and Analysis System

The Noptel ST-2000 is a new generation of marksmanship training equipment which allows the marksmanship of one to eight or even more people to be monitored and analysed simultaneously. The system consists of interface electronics, cable and prism elements and a gun unit (transmitter/receiver/controller). Normal shooting targets are used with prism elements fixed to them.

The program is a graphic application and all functions can be controlled with a mouse which eases programming. The single-user program operates on notebook computers while multi-user programs use Intel 486-type personal computers. The software for the system, which permits training indoors or outdoors, includes advanced shot analysis and provides feedback on basic features.

The Noptel ST-2000 range in use

Status
In production during 1994.

Manufacturer
Noptel Oy, Oulu.
(For details see Directory of Suppliers)

FRANCE

Giravions Dorand DX 406 SITTAL Light Weapons Simulator

The Giravions Dorand DX 406 SITTAL light weapons simulator was developed to meet a potential French armed forces requirement. Designed for indoor use, the DX 406 provides training in the three traditional phases of firearms training: basic, technical and tactical. Besides providing individual and collective training, the DX 406 can also be used to provide training for squad leaders using realistic operational scenarios. The system can provide individual or group-based evaluation of results obtained during exercises.

Status
Prototypes.

Manufacturer
Giravions Dorand Industries, Suresnes.
(For details see Directory of Suppliers)

General view of a Giravions Dorand DX 406 SITTAL light weapons simulator

INTERNATIONAL

AirteQ Firearms Training System (AFTS)

The AirteQ Firearms Training Systems (AFTS) is a development of the FATS (Firearms Training System - originally intended for police training applications) and was specifically developed for military and police firearms training. The AFTS is a joint development by AirteQ Firearms Training Systems of the United Kingdom and Firearms Training Systems Inc of Norcross, Georgia, USA.

In late 1992 a derivative of the AFTS known as the Small Arms Trainer (SAT) was ordered by the British Army. The order was for 100 units to be delivered in four-lane (40) and 10-lane versions (60). A similar version was also ordered by the Dutch armed forces; this order was for 45 units.

The SAT provides British Army training establishments and line infantry units with a flexible and effective weapons trainer which can be operated in a classroom environment. The SAT provides a realistic simulation of SA80 Individual Weapon (IW) and Light Support Weapon (LSW) physical and operating characteristics and realistic representations of the ranges where live firing takes place. A converted Browning HP Pistol has been developed to simulate the in-service pistol. Both the four- and 10-lane SAT variants allow the instructor to control squad level firings or permit the individual firer to progress at his own pace through the standard army courses of live firing training. The SAT also permits the instructor to design unique range practices which replicate battle shooting situations based on a library of scenarios supplied with each SAT.

SAT is based on the AFTS trainer and both can be used at all skill levels from the recruit to the trained marksman. Training is software-driven and target imagery is provided from both software and laser disk. Images are projected onto large portable screens from floor-mounted projectors. Video scenarios may be used, giving the trainee the opportunity to practise marksmanship skills against life-size targets and tactical 'shoot/no-shoot' situations. Any number of weapons from pistols to manportable anti-armour weapons can be operated with the systems (the Dutch system can accommodate 11 types of weapon) but no live ammunition is fired. Weapons may be fired 'dry' or by using blank plastic cartridges containing a primer charge to provide a noise signature. The British Army reproduces battle and weapon noise effects through individual headsets and loudspeakers.

Weapons used with the simulator project a laser beam at the instant of firing. Both the AFTS and SAT have the capacity to provide trainees with detailed coaching feedback of each shot fired. The systems can also provide feedback on scores and reaction time, in addition to information on shot release. System computers allow data to be compiled on 3.5 in diskettes and trainees can be given hard copies of their results.

A variety of tactical and other scenarios may be used, from elementary range work to combat situations in rural or urban areas. The library of available situations is continually being enlarged and scenarios can be tailor-made to suit specific needs. Targets can be moving or static, single or multiple, and can be presented at ranges from 5 to 600 m. Trainees can be practised in zeroing their weapons or this function can be carried out using the system computer. Both SAT and AFTS can simulate weapon malfunctions to enable trainees to practise immediate action drills.

AFTS and SAT require a 240 V single-phase mains electrical supply. AFTS can be set up in a room measuring 3 × 6.1 m while the ten-lane SAT requires a room measuring 7 × 12.5 m. AFTS can be installed and removed in under 30 minutes. Both systems are highly portable.

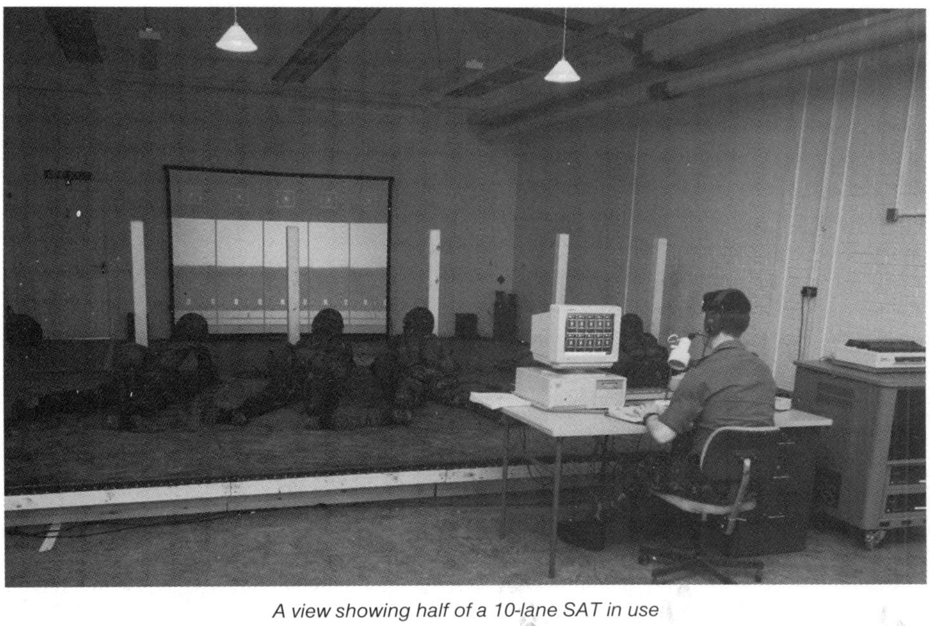

A view showing half of a 10-lane SAT in use

Status
In production. The original FATS are in service with over 600 customers in the United States (including police forces and the US Army Military Police). The AFTS has been ordered by the US National Guard, the US Marine Corps, the US Air Force Security Police, and the Royal Air Force Regiment. The SAT has been ordered by the British Army (100) and Dutch armed forces (45).

Manufacturers
AirteQ Training & Simulation Limited, Milton Keynes, UK
Firearms Training Systems Inc, Norcross, USA.
(For details see Directory of Suppliers)

ISRAEL

SIMTECH HOT HIT Marksman Trainer

The SIMTECH HOT HIT trainer is intended for use during the training of military and paramilitary marksmen. It involves the use of an optical sight fitted to a normal sniping rifle or a similar weapon but the sight is fitted with an image splitter. This device transmits the sight picture to a remote central control screen via a video cable (fibre optic cable or video data links are optional) where the sight picture is both displayed on a video monitor and recorded on videotape for subsequent analysis. The system can be used with most types of small arms and can be mounted on vehicles.

Status

Available.

Manufacturer

SIMTECH Advanced Training and Simulation Systems Limited, Holon.
(For details see Directory of Suppliers)

The SIMTECH HOT HIT marksman training system in use on a small arms range

SWITZERLAND

TIRA-LAST Laser Shooting System

TIRA-LAST consists of a low energy laser unit directly integrated into pistol, revolver or sub-machine barrels; no alteration to the weapon involved is required. The weapon is aimed and fired at a target in the normal way but the laser unit on the weapon emits a coded laser beam. The target area consists of a special synthetic and photosensitive material which converts the laser light into an electronic impulse. This pulse activates and controls the TIRAX TI 92 target turning unit (combat simulator unit - see entry in *Small Arms Range Equipment* section) on which the target is mounted. Thus, typically, a laser hit on the target will cause the target to turn side-on.

The TIRA-LAST unit mounted on the weapon is known as the LAST-Laser TLS. The LAST Sensitive Unit, the target, is the TLT and the Target Support is the TLE.

Status

Available.

Manufacturer

TIRAX AG, Dietikon.
(For details see Directory of Suppliers)

The TIRA-LAST laser shooting system showing main components

TIRAX LAST 2000 Laser Sensitive Targetry

The TIRAX LAST system consists of an electronic module which is inserted in the TIRAX 2000 target turning unit (see entry in *Small Arms Range Equipment* section) and a special laser-sensitive target or target cover. A weapon equipped with a laser unit used with a conventional tactical engagement system is aimed at the target in the normal way and the unit emits a coded laser beam. The LAST target contains special photosensitive material which converts the laser light into an electronic impulse. This pulse activates and controls the laser hit on the target which causes the target to turn side-on.

The TIRAX LAST provides optional reactive laser counter-fire. In cases where the equipment is integrated into tactical laser engagement training and the target is not hit, it will shoot a laser beam against the soldier. If he does not take cover in time he will be 'dead' by means of a corresponding signal of the tactical system.

The LAST system can be coded for most of the tactical laser engagement systems on the market. It can

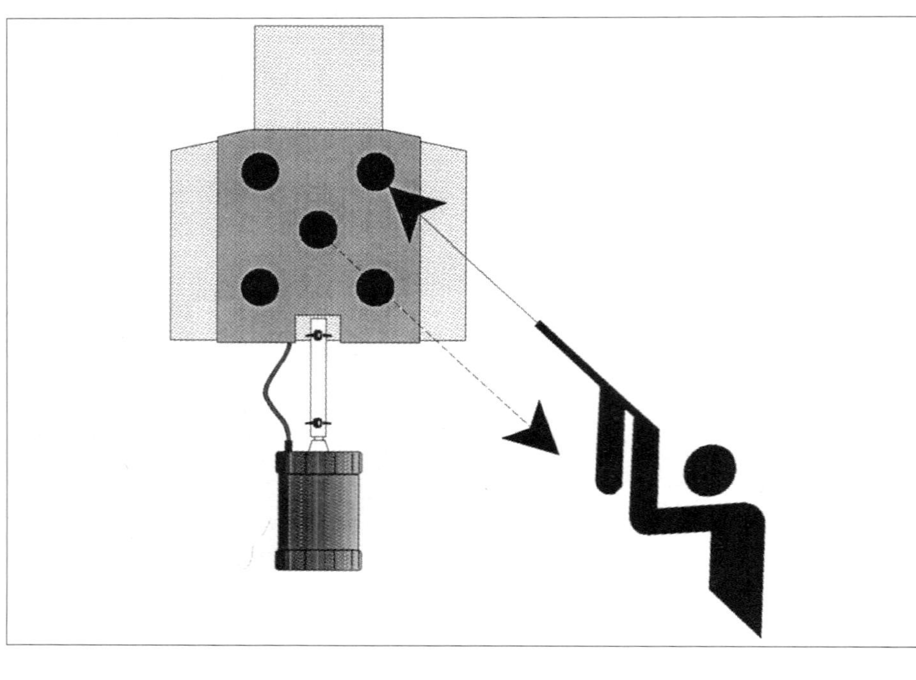

A diagram of the TIRAX LAST 2000 laser sensitive targetry

also be integrated into the new TIRAX battle training system, TRACOS. This special software-supported system makes possible combined battle training for infantry including all types of weapons such as small arms, anti-armour weapons and hand grenades equipped with laser units.

Status

Available from October 1994.

Manufacturer

TIRAX AG, Dietikon.
(For details see Directory of Suppliers)

UNITED KINGDOM

Marconi SMART Small Arms Trainer

The Marconi SMART Small Arms Trainer has been described as the most advanced rifle trainer in the world and consists of up to 10 trainee positions each comprising a modified in-service rifle, a high resolution monitor, ear defenders and a centralised instructor console.

The weapons used with SMART are adapted service rifles fitted with a light pen, handling sensors and a recoil simulator. Trainees may fire from the prone, kneeling or standing positions. The trainees' targetry is generated by computer graphics and displayed on a high resolution monitor approximately 700 mm from the muzzle of the rifle. The target types involved include point of aim, zeroing, static, snap and moving. Firing exercises encompass holding, aiming and firing, point of aim, grouping, zeroing, gallery range, electric target range, moving target range, close quarter combat range, application of fire, annual weapon test practice and competition shooting practice.

The point of aim target allows both the instructor and the student to track the latter's point of aim before, during and after firing and acts as an excellent diagnostic facility. All trainee actions and performances are monitored by the instructor from the central console from which he initiates and controls all range practices. The instructor has access to records of points of impact, mean point of impact, grouping circles, hit/miss analysis and scores for each trainee.

As each student engages a target the light pen detects the point of aim and the fall-of-shot is immediately displayed on the instructor's exercise monitor. This allows for detailed supervision of up to 10 trainees as each shot is fired. Recoil and weapon jump are simulated, and a realistic rifle report is reproduced through the ear defenders, which also provide a one-way audio link between instructor and student. The system can also simulate wind effects and allows for time of flight of a projectile. On completion of each practice the instructor can replay student performances and make group sizes and scores available to all trainees on their own monitors.

Other facilities within the system allow the instructor to monitor safety and ammunition state of the rifle, the student's hold of the weapon, trigger operation and the wind speed and direction at any time. The instructor can also check that correct immediate action drills are carried out. Weapon handling and safety drills can also be conducted.

SMART allows for training at basic, intermediate and advanced levels, providing realistic training for both recruits and trained soldiers. The system is simple to operate and no previous computer experience is needed. SMART provides detailed analysis of trainee performance, leading to effective fault diagnosis within a totally safe environment. SMART is designed for maximum participation which makes it efficient and cost-effective at all training levels.

SMART has been adapted to operate with a wide range of weapons including the SA80 weapon system, American 5.56 mm M16-series rifles, the Heckler & Koch G3 rifle and the French 5.56 mm FAMAS.

A SMART small arms trainer system in service at the Guards Depot in the UK

Status

In production. In service with the British Army and Royal Air Force Police, Denmark, the French Army, some Middle Eastern countries, and in South East Asia.

Manufacturer

Marconi Simulation and Training, Cwmbran.
(For details see Directory of Suppliers)

Hendry High definition Indoor Trainer for Small arms (HITS)

The Hendry High definition Indoor Trainer for Small arms (HITS) was developed for small arms training using the experience gained from the development of indoor training simulators such as the Hendry All Arms Classroom Trainer/Simulator (AACT). HITS was selected by the Royal Air Force for training recruits on the SA80 IW and LSW weapons and the general-purpose machine gun (GPMG), all with compressed air-driven recoil systems for both single shot and automatic fire.

HITS can depict targets to a range of 1000 m and shotfall accuracy is 10 mm at 100 m.

With HITS, each trainee is presented with a large screen and accurately scaled picture which includes the target. The screen is 1.2 m wide, 0.9 m high, and 3 m from the student. A realistic training weapon is used to fire an infra-red beam at the target and the system records the results, showing the point of impact, aim point trace, rounds fired, weapon drill errors and scores, on both an individual lane monitor and an instructor's monitor. The instructor has full access to control and to interrogate any lane. The system includes audio effects and lane control options, with results printout, disk storage of results and a summary printout.

Each firing lane has a projector station with two

The Hendry High definition Indoor Trainer for Small arms (HITS)

random access computer-controlled projectors containing up to 80 slides, a high resolution charge coupled device (CCD) camera, and a mirror-driven unit. The scene, including the target, is projected from one projector. The correct slide is automatically selected depending on the exercise involved with the picture presented to the student as an accurately scaled photograph of an actual firing range. If the exercise involves targets which fall when hit, the second projector is controlled to be ready with a slide depicting the same scene but with no target - if the target is hit the system fades rapidly from one projector to the other. When multiple targets are shown, the second projector's random access facility is used to select the scene with the appropriate target(s) missing. Moving targets are achieved by projecting the scene from one projector and the target from the other. A mirror is rotated by a stepper motor so that it is in the beam of the target projector. It is then driven at controlled speeds to deflect the projected beam, thus moving the target across the screen.

The high resolution CCD camera is mounted so that it covers the target area and detects on the screen the position of an infra-red light beam emitted from the weapon. Filters are fitted to the weapon, projectors and camera so that only the weapon's emissions are detected. Thus the output of the camera defines accurately the weapon's point of aim which is fed to a lane control subsystem for comparison with known target positions, range and dimensions.

The lane control subsystem for each firing lane contains electronic circuitry for slide projector control, processing of the CCD camera output, mirror motor drive and audio amplification. It also provides communication with the instructor's computer and provides the student with his own display and keypad for individual control of an exercise. Communication with the instructor's computer is performed via a bi-directional serial communications link. A student's display is provided on a 9-in/228 mm television monitor. The normal display mode shows an expanded view of the target with shot positions and mean point of impact marked. Each student wears a headphone/microphone set for communications from the instructor and for sound effects.

HITS can cater for any small arms weapon from pistols to light machine guns. Ballistic information is stored in the system computer at the instructor's station. Weapon cocking and recoil simulation are achieved pneumatically. The weapon incorporates an infra-red emitter assembly which is collimated with the sights.

The instructor's control station is where all exercises are controlled, ballistic calculations are performed, weapons are zeroed, effects such as crosswinds are introduced, student performance is monitored, and individual control is exercised. The control station is based on an IBM-compatible computer with hard disk and full colour graphics compatibility. Up to 10 lanes can be controlled from one station.

Status
In production. In service with the Royal Air Force.

Manufacturer
D Hendry Electronics Limited, Arundel.
(For details see Directory of Suppliers)

OTEC Ultima Small Arms Trainer

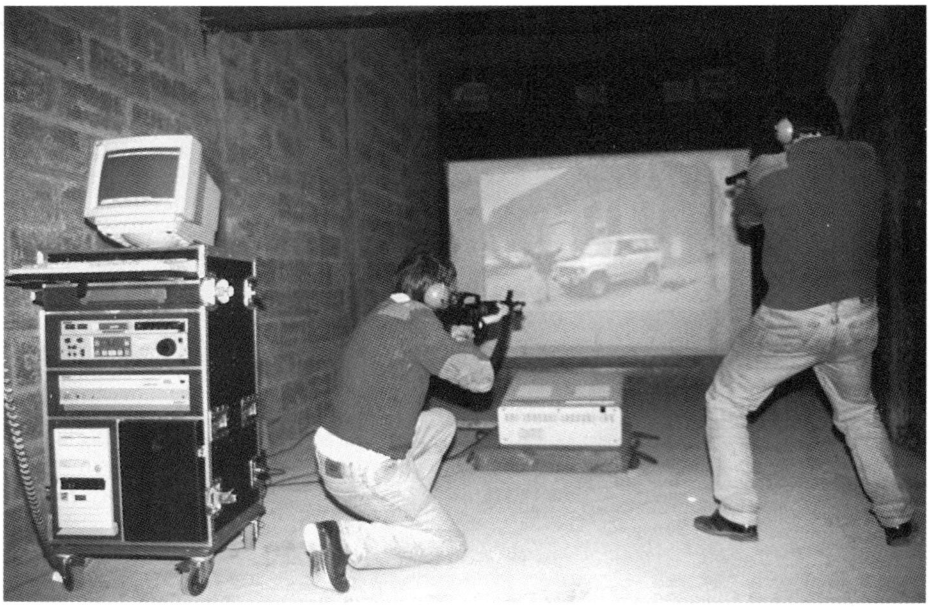

The Ultima Small Arms Trainer

The Ultima Small Arms Training System uses the latest multi-media technology to provide a highly flexible multi-task trainer for small arms. It is a modular system which can be expanded from a basic aim trainer to real-time duelling for force-on-force incident training. The system uses off-the-shelf hardware and six main levels of training can be provided in any combination:

Marksman Trainer
This uses laser-modified operational weapons and provides effective analysis of weapon and breathing control, aiming, firing and grouping and target engagement using any conventional target provided by the customer.

Interactive CGI System
This uses computer generation and digitised pictures taken from conventional videotape to provide an inexhaustible range of static and moving targets. It may be used for both conventional firearms training, simulating different ranges with pop-up and moving targets, or for more advanced operational practices involving fire discipline and target recognition. Material from hand-held video cameras may be used so that operational briefings and training exercises may be undertaken. Automatic scoring and marking of shots is provided, while the targets can be made to respond to the shooter's performance with reactions to the shot or time.

Slide Projection System
This is a computer-controlled slide projection system which provides high definition targets into which time and judgement elements can be introduced. The system can be provided with 140° or 180° screens and is ideally suited for long-distance, high definition targets. Customers can build up their own library of slides.

Interactive Videotape
Here sophisticated software provides a visual indication of the effect of target engagement using conventional videotape. The target will fall when hit and many different outcomes of each scenario can be presented. The customer can create and programme his own video films for the system to incorporate computer-generated images which will interact with shooting performance. The instructor can override the computer to simulate targets wearing body armour. The system can be used to train sniper teams in reporting and communication procedures and in sniper controlled volleys.

Interactive and Branching Video Disk
In this version, realistic video scenarios of tactical situations are stored on video disk for advanced situation training. Many different outcomes of each scenario can be produced, each depending upon the trainee's reaction.

Real-Time Duelling
Live video images are transposed from the 'incident' room into the 'attack' room and vice versa. Each shot is marked for replay and debriefing at the end of the exercise. Because of the 'real-time' nature of the system the scenarios acted out during a training session can relate to specific situations. The system can also be used as two individual shooting rooms for separate training.

Status
In service worldwide.

Manufacturer
OTEC Special Products Limited, Godalming.
(For details see Directory of Suppliers)

Reflectone Small Arms Trainer (SAT)

The Reflectone (formerly British Aerospace) Small Arms Trainer (SAT) is a computerised indoor trainer for small arms which simulates, under instructor control, live firing on outdoor ranges from basic firing lessons through to advanced marksmanship. In addition to firing ranges the SAT includes training for combat exercises with tactical and field situations, fighting in built-up areas, shoot/no-shoot guard duties and group training.

The SAT is a multi-lane system which is installed in training centres, barracks or gun clubs. There are different versions available, specifically tailored for military and police users. Typical configurations provide between four and ten firing posts with either a single large screen at which the trainees aim or with an individual screen per trainee. In both cases each trainee can operate either independently or together with others. The firers fire from the standing, kneeling, sitting or prone positions, at a distance of 4.5 m from the screen. They are able to practise all typical standard drills as if they were on an actual live range. The firing of single rounds and bursts is incorporated.

The weapons are either accurate replicas or demilitarised versions of real weapons modified for use with the SAT. They can include most types of personal weapon such as assault rifles or sniping rifles, pistols, machine guns, and anti-armour weapons. Ordinary iron or magnifying optical sights can be used. For the majority of rifle and pistol types the weapon can be offered with no cables or pipes connecting the weapon to the trainer so the firer is then free to hold the weapon and carry it around as they would with a live weapon.

Imagery is either video or digitised photographs of actual targets, ranges and training areas. During an exercise the relevant scenes and targets are processed in real time by the lane computers to present the scenes in accordance with the exercise parameters set by the instructor and the actions of the firer. The fall-of-shot and full scoring and debrief facilities can be called up

General view of a Reflectone Small Arms Trainer (SAT) individual screen version showing the instructor's station

General view of a Reflectone Small Arms Trainer (SAT) large screen version with four trainees

from the lane and the instructor station. Each SAT system has a facility allowing the instructor to create and to edit exercises and store them as a library ready for use. This also allows customers to prepare and to use their own videos and photographs.

Status
Entering production during 1994.

Manufacturer
Reflectone UK Limited, Bristol.
(For details see Directory of Suppliers)

Shorts S-1 Multi-Arms Trainer

The S-1 Multi-Arms Trainer is a high accuracy, video-based system for training military and law enforcement personnel in all aspects of the use of weapon systems ranging from small arms through to anti-armour weapons. The wide screen system offers both marksmanship (how to shoot) and scenario (when to shoot) training in a flexible, modular package which can be configured to meet a wide variety of training needs.

The system is available in four, five, eight and 10-lane configurations, all of which are controlled by a single instructor station. The instructor, as well as having the capability to run, pause and replay any practice, can also view the performance of any of the trainees on the instructor station monitor. Coaching can then be achieved by means of an intercom system allowing communication between the instructor and the trainees.

The system is noteworthy for the accuracy of the aim point detection system and the quality of the graphics depicting both background and target imagery. Other key features include the weapon simulation and the ease of use of the operator software. Weapon functions such as recoil, loading, unloading, reaction to stoppage and safety catch operations accurately simulate the live fire weapon. The instructor station interface is extremely user-friendly and allows operators to run the system with a minimum of training.

A wide variety of different environmental conditions can also be simulated allowing trainees to experience different lighting levels, wind and fog conditions. The comprehensive database system included in the S-1 means that performance data can be analysed under the varying conditions. Performance data can also be printed allowing hard-copy records to be retained by the students.

Manufacturers
Short Brothers plc (Defence Systems Division), Belfast.
(For details see Directory of Suppliers)

General view of a Shorts S-1 Multi-Arms Trainer showing instructor station

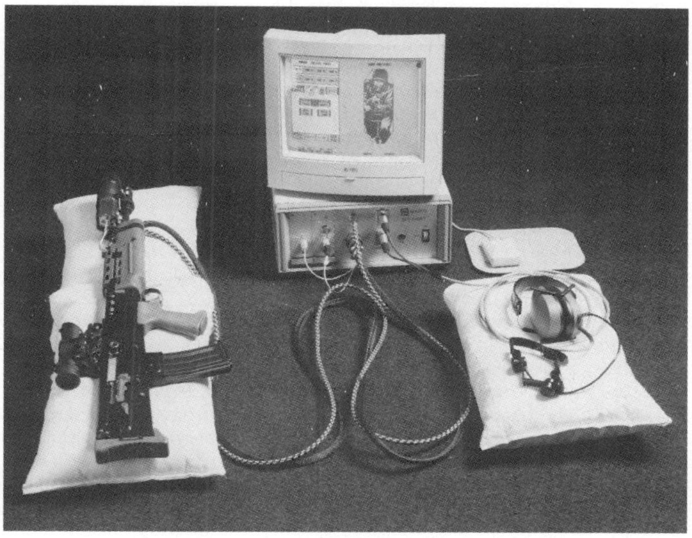

Components used at a student station on a Shorts S-1 Multi-Arms Trainer

TTSL Infantry Weapon Training Simulator (IWTS)

Development of the Thomson (formerly Link-Miles) Infantry Weapon Training Simulator (IWTS) began in early 1992. It provides a comprehensive training environment for all aspects of training in the use of small arms and marksmanship. The IWTS provides for platoon and individual training on ranges and in tactical scenarios; accurate simulation for both new recruits and experienced marksmen; realistic weapon handling and recoil; progressive training with result recording; and accurate real-time ballistic calculations (including wind effects). The system can accommodate handguns, rifles, machine guns and anti-armour weapons.

The IWTS features a collimated wide-angle visual scene display with the detail and perspective to provide comprehensive training in target acquisition, weapon engagement over a wide arc, fire-control procedures, and shoot/no-shoot decisions. The system can also be used for anti-armour weapons and for mortar fire control.

The wide-angle collimating mirror screen display is formed by moulding a thin plastic film coated in a reflective material over a suitably curved set of formers by a low-power vacuum pump. The resultant display ensures that all scenario and target images appear at the correct distance with the correct perspective for all trainees. The depth of field allows optical sights and image intensifiers to be used without modification.

Real infantry weapons, deactivated and instrumented, are used with the IWTS. Realistic recoil and automatic operation are provided using low pressure air. The balance and weight of the IWTS weapon are matched to those of the live weapon.

The main modification to the IWTS weapons is the installation of a small infra-red transmitter device either under or in the barrel of the weapon. The aim point is designated by the infra-red spot produced as the trigger is actuated; the spot is invisible to the eye and infra-red sights yet provides accurate aim point data through an array of sensors.

Computer-generated synthesised imagery is used with the IWTS to provide detailed targets superimposed on panoramic, photographic background scenes. Both range scenes and complex tactical scenarios can be displayed, either in lanes or as an overall wide-angle display. The system can also provide data on variable aspects of weapon control and marksmanship such as the aim point trace, butt and trigger pressure, weapon cant, and so on, and record them either for subsequent analysis at the instructor's station or for call-up display for each weapon station. Headphones and microphones at each firing station are used to provide battle, firing and target noise and to provide two-way communication between instructor and trainee.

The basic IWTS configuration is a five-lane system with a panoramic 72° horizontal field-of-view visual display. Two five-lane units can be combined, linked to one instructor station.

The IWTS is based on standard personal computers in a networked configuration (one for each firing position and one for the instructor). The system has facilities to enable instructors to prepare and present their own training exercises and scenarios.

Status

In production for Denmark and Germany.

Manufacturer

Thomson Training & Simulation Limited (TTSL) Crawley.
(For details see Directory of Suppliers)

Overview of an individual trainee using a Thomson Infantry Weapon Training Simulator (IWTS) and showing instructor station

Thomson Infantry Weapon Training Simulator (IWTS); the two-dimensional picture shown here cannot give a correct impression of the depth of the collimated display

Trenton FORCE ON FORCE Live Fire Training System

The Trenton FORCE ON FORCE training system was produced to satisfy the specific training needs of armed confrontation and hostage situations. It was designed and manufactured as a direct result of discussions with both military and police specialist firearms training instructors.

The system allows training in the imperative skills of armed confrontation and negotiations to be completed within a controlled training environment, fully monitored by training and instruction staff. The team allows specialist individual/team members to train together to build confidence in each other and eliminate possible tactical errors at the scene of an incident. With the use of modern projection and recording equipments, situation and judgmental training scenarios can be presented and completed for review and analysis. Training programmes are completed in real time.

The system allows for other visual media to be used on the target screen unit, thus extending the use of the equipments for other training requirements. The FORCE ON FORCE system is not associated with the concept of duelling houses or rooms. All firing is against one wall of each room only and the targets are displayed against a target screen unit.

The system was designed for the following training: training individuals/teams in negotiating skills under threat conditions; shoot/no-shoot situations; judgmental exercises; and operational approach procedures. The system is compatible with full bore or plastic ammunition or with sub-calibre devices, thus allowing for installation in existing live fire training complexes or uprated classrooms. Users are encouraged to prepare their own training programmes to suit local environments and no specialist programme facilities are required. Trainees can train with their own issued weapons.

The system has two distinct areas, the buildings and training rooms and the training equipments.

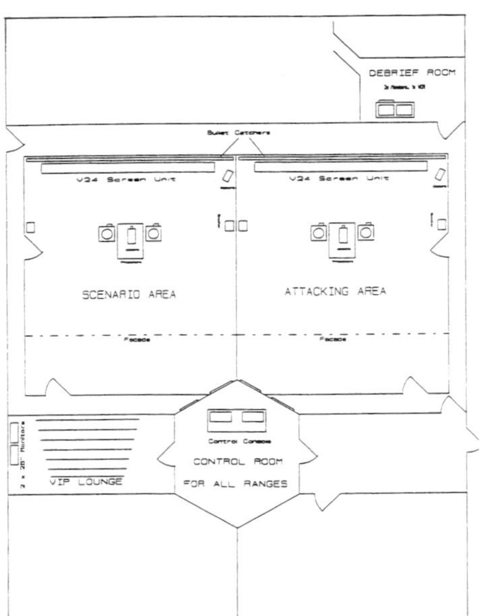

General layout of a typical Trenton FORCE ON FORCE live fire training system

There are two separate live fire training rooms, preferably adjacent but they can be as far as 25 m apart. A bullet trap is positioned at one end of the room with the other walls and the floor protected with anti-ricochet tiles and sound insulation. Because of the nature of the training the azimuth danger area is slightly increased due to the movement of personnel within the room. It is recommended that the bullet catcher specification is made to accept up to 0.44 Magnum ammunition.

The rear wall of each live fire room can represent a façade of whatever the customer requires, such as windows, doors and so on. Range room entry/exit doors are positioned other than on the front or back walls and have active lighting for emergency evacuation. It is recommended, as a guide, that the safety precautions adopted in the construction of 25 m indoor ranges be used as a basis for the construction parameters of the live firing training rooms.

The specialist training equipments are dispersed throughout the live firing training rooms and the debriefing rooms. Each live fire room is fitted with a ceiling-mounted combination video and slide projection system linked to video cameras in the opposite room for scenario display, recording and analysis. Thus each individual/team is able to see the other when in view and the instructors have complete visual control over the training.

A Type 24 target screen unit, designed and built to withstand live fire ammunition up to 0.44 Magnum, is positioned in front of the bullet catcher. The screen lighting system displays the projectile holes and a clear screen unit is restored either automatically or by the instructor at the correct time.

Each room is equipped with microphones and speakers situated at suitable positions to allow for audio interaction between the two range rooms.

The projectors, cameras and audio systems are all controlled from a central system console located in a control room. The control console records everything in real time and the recording and playback facility can be branched to other rooms for range control officers or VIPs in order to monitor progress of individuals or teams.

The modular construction of the FORCE ON FORCE system permits many variations to be supplied and installed.

Status

In production.

Manufacturer

Trenton Technical Services Limited, Southampton.
(For details see Directory of Suppliers)

UNITED STATES OF AMERICA

Firearms Training Systems (FATS) Marksmanship Trainers

Firearms Training Systems (FATS) is a world leader in the production of complete small arms marksmanship simulation systems. FATS simulators are not partial trainers. They use the latest advances in laser and interactive video technology to train military and law enforcement personnel in all aspects of marksmanship training from basic individual fundamentals to advanced collective combat skills. Up to 12 individual firers can train simultaneously using life-size projected video images on large screen displays.

FATS military systems can employ from one student up to squad-size units. Students can assume any position immediately, with no system reconfiguration, and simultaneously fire rifles, machine guns, grenade launchers and light anti-tank weapons in co-ordinated live action firing range and battlefield scenarios. The trainers offer numerous special features such as target box exercises and target trace analysis, immediate replay and feedback of both hits and misses, slow motion, stop action, instant search and threat identification, with printed results if required.

FATS law enforcement systems can be used to train students in target range and judgmental shoot/no shoot police skills in a variety of life-size scenarios ranging from ordinary patrol encounters to special operations and SWAT situations.

The FATS system is rugged, portable, user-friendly and does not require specially trained operators.

Variants of the basic FATS have been ordered by the British Army (100 units) and the Dutch armed forces (45 units). See under *International* for details of these variants.

Status

In production. In service with military and police forces throughout the United States (over 600 customers), the Americas, Europe, the Middle and Far East and the Pacific Rim. See also entry under *International*.

Manufacturer

Firearms Training Systems Inc, Norcross.
(For details see Directory of Suppliers)

FATS Light Anti-tank Weapon (LAW) training system in use

FATS Squad Engagement Training System in use

Simulaser MTS1000 Marksmanship Training System

The MTS1000 marksmanship training system is used with a laser-based weapon simulation system similar to MILES and is compatible with all fielded US Army MILES laser transmitters.

The usual configuration has a laser target and a computer/controller combined in a rugged transit case. The top half containing the target (which is similar to the US Army standard 25 m zeroing target) is separated and positioned 25 m from the firing line. The computer/control console is connected to a 110 V power supply and to a cathode-ray tube monitor.

In use, the student performs an alignment routine using the MTS1000's built-in BORESIGHT mode which provides a continuous display of the location of the laser beam on the monitor. After switching to the MARKSMAN mode the student fires shot groups of three rounds each at the target. The monitor displays the exact location of each 'bullet' and recommends a sight correction. The MTS1000 scores the student's performance and can provide a printout of the target on an optional dot matrix printer.

The MTS1000 can be configured for multiple lane operation. An additional feature is that in the BORESIGHT mode the MTS1000 can be used as an alignment system for any MILES transmitter.

Status

In production.

Manufacturer

Simulaser Corporation, California.
(For details see Directory of Suppliers)

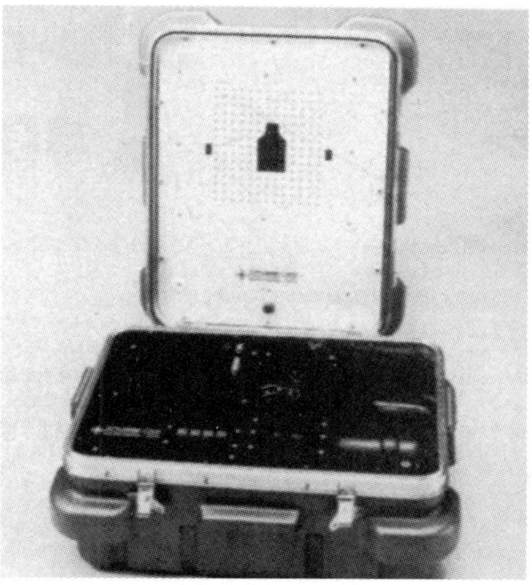

MTS1000 marksmanship training system

Spartanics Weaponeer M66 and M70 Marksmanship Training Systems

Weaponeer is an indoor simulator used for training small arms marksmanship. It comes complete with weapons, targets, firing positions and performance feedback in both video and printed media. To become fully operational the user need only provide electricity and a space of 6 × 2 m. No bullets or other expendable items are used in the simulation.

Features which distinguish Weaponeer are its accuracy, realistic weapon recoil and mechanical functions, the ability to operate in very bright or very limited lighting, unique performance feedback, training software and configuration.

Weaponeer M70 components include the aim sensing, recoil, sound, video, printer and data management systems. The infra-red sensing system is accurate to ±0.8 cm per 100 m. This is not a resolution figure but the actual accuracy taking into account all potential parallax, drift and alignment errors. This detection system samples data four or eight times faster than video rates so that trigger jerk and other critical firing errors will be detected. Adjustable net-force recoil is produced at strengths and cycle rates equal to live weapons. The recoil force is generated by a recoil rod attached to the front of the weapon. This counterbalanced system is below the firer's field-of-view so they neither feel nor see it. The recoil rod also controls the distance between the weapon and targets to prevent target scaling errors. All mechanical functions of the weapon are duplicated while realistic, high-volume, shot sound is generated in headphones. A video feedback monitor and printer are integral to each lane. All firing positions are built-in. The Weaponeer data management system permits an unlimited number of rosters to be created, edited, stored, transferred or printed.

The design of the M70 Weaponeer is modular and it readily adapts to specific customer requirements. The original M66 Weaponeer can be upgraded to M70 standards.

Status

In production and with 600 systems in service worldwide including North and Central America, Europe, the Middle East and Asia.

Manufacturer

Spartanics Limited, Rolling Meadows.
(For details see Directory of Suppliers)

Four M70 Weaponeer marksmanship training systems being used by US Army riflemen

YUGOSLAVIA (SERBIA AND MONTENEGRO)

Laser Infantry Weapons Simulator

This laser infantry weapons simulator consists of a gallium arsenide laser projector clamped to the muzzle of an infantry weapon for firing at a fixed target. The target has a laser detector so hits sensed by the target detector are relayed back to the firing point via a tripod-mounted laser-based connection unit to operate a scoring and control unit which logs results. The target may be placed at any range between 20 and 100 m from the firer.

Up to 10 firers may use one target at any one time. Hits are signalled by a flashing light (for about 2 seconds) and the control unit automatically provides a printout of results.

The laser projector mounted on the infantry weapon weighs 0.7 kg and emits a laser beam with a wavelength of 904 nm, a pulse power of 7 W and a beam divergence of 1.5 mrad. The control side of the system is powered by rechargeable Ni/Cd 12 V 56 Ah batteries.

The same basic laser projector can be used in conjunction with a helmet-mounted detector system, smoke cartridges and an umpire gun to convert the system into a tactical engagement simulation system. With this system the helmet carries four detectors to provide a full 360° coverage and smoke devices, also mounted on the helmet, provide an indication of incoming 'enemy fire'. Red smoke denotes a 'kill' and yellow smoke indicates 'wounded'. The weight of a helmet with the laser detectors is 1.75 kg. The tactical engagement version of the system can be used at ranges between 20 and 400 m.

Status

In service with the armed forces.

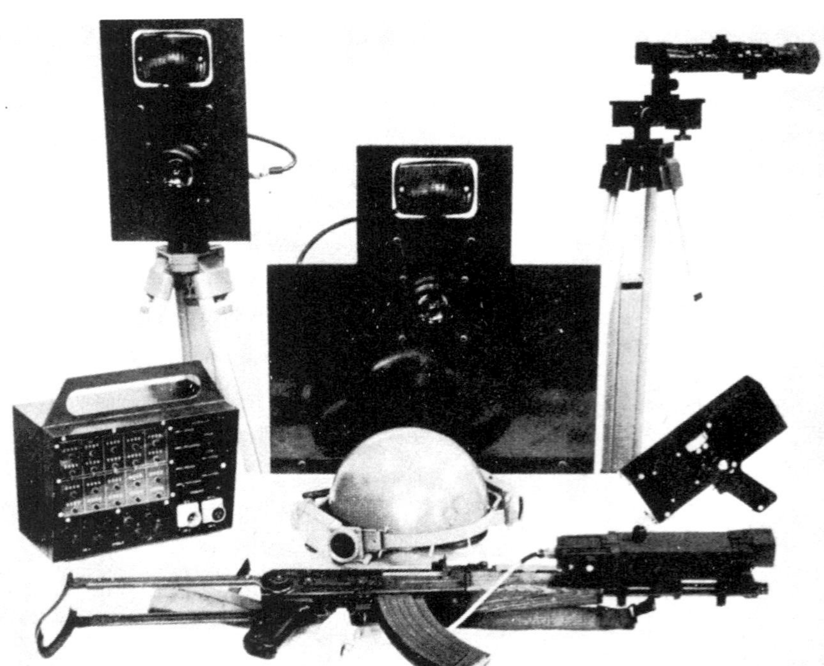

Components of laser infantry weapons simulator complete with helmet detector and umpire gun used with the tactical engagement simulation version

Contractor

Holding Company YUGOIMPORT-SDPR Ltd, Belgrade.
(For details see Directory of Suppliers)

SUB-CALIBRE TRAINING DEVICES

BELGIUM

MECAR TD-20-90 Sub-calibre Training Device

MECAR produces a sub-calibre training device for use with a range of 90 mm vehicle-mounted guns including the MECAR 90/28, the Cockerill Mark II and III, the ENGESA EC-90 and the DEFA F1. The device has the general appearance, dimensions and weight of an operational round of 90 mm ammunition. The device is fitted with a barrel 700 mm long, firing a 20 mm sub-calibre tracer round out to a range of 1000 m. Firing is carried out using the normal gun firing mechanism. Realistic training can thus be carried out with the system covering the complete range of gun command and control drills, loading, aiming and firing, all at a lower cost than with conventional ammunition.

The types of training devices and ammunition are as follows:

90/28 TD - M635
90/28 TP-T - M643
DEFA F1 TD - M648
DEFA F1 TP-T - M649
Cockerill Mk II-Mk III TD - M640
Cockerill Mk II-Mk III TP-T - M634.

Status
In production. In widespread service.

Manufacturer
MECAR SA.
(For details see Directory of Suppliers)

BRAZIL

Britanite Sub-calibre Ammunition System

The Britanite sub-calibre ammunition system is used for live firing training using mortars and anti-tank rocket launchers. The sub-calibre ammunition consists of a cartridge common to all weapon types which is loaded through the centre of an inert projectile. In the case of the mortar projectile the force of firing the cartridge is sufficient to drive the inert projectile from the mortar barrel and allow it to drop to the ground in front of the muzzle; the inert projectile can be reloaded with a fresh sub-calibre cartridge and reused. In the case of the rocket launcher the inert projectile remains inside the barrel. In both cases the sub-calibre cartridge fires a small projectile which lands within an area measuring 200 × 30 m, producing light, smoke and an explosion.

Inert projectiles and their common cartridge are available for 60 mm, 81 mm, 120 mm and 4.2 in (107 mm) mortars, and 3.5 in (88.9 mm) M20 (Bazooka) rocket launchers. A similar system is produced for 105 mm howitzers. In each case the inert projectile is handled, loaded and fired in exactly the same way as a real projectile.

Status
In production.

Manufacturer
Britanite SA Indústrias Quimicas, Paraná.
(For details see Directory of Suppliers)

CANADA

Diemaco 0.50 Calibre Blank Firing Adaptor

The Diemaco 0.50/12.7 mm Calibre Blank Firing Adaptor (BFA) is an easily fitted and removed device which mounts on the M2 HB machine gun. It was designed for and developed with the Canadian armed forces. The BFA body completely encloses the machine gun barrel in a rigid jacket. The muzzle piece of the BFA is adjustable to direct more or less power to the gun mechanism to correct for differing types of ammunition or other conditions which can affect reliable blank ammunition functioning. The muzzle piece also directs the exhaust gases forward and downward for safe operation. The system includes a live ammunition discriminator which fits directly into the feed tray of the weapon to prevent accidental feeding and firing of live ammunition, and a setting gauge to provide consistent initial positioning of the BFA muzzle piece. Full documentation is provided with the kit.

The BFA is fully compatible with all types of 0.50/12.7 mm blank ammunition including the US M1 and M1A2, the Canadian C48, and equivalents, including plastic types.

The BFA is 775 mm long and weighs 3.82 kg. The rate of fire is adjustable from 450 to 600 rds/min.

Status
In production.

The Diemaco 0.50/12.7 mm Blank Firing Adaptor (BFA)

Manufacturer
Diemaco, Kitchener.
(For details see Directory of Suppliers)

Training Rifle C10

The C10 Training Rifle is a 0.22 LR calibre weapon designed to meet a training requirement for the Canadian armed forces with whom it will be used as a cadet training rifle. Development of the C10 was funded under contract to the Canadian armed forces. It is an exact facsimile of the Canadian C7/C8 (American M16) combat rifle and has the same handling and shooting characteristics apart from firing on semi-automatic only. The C10 has approximately 80 per cent parts commonality with the C7 rifle and the 0.22 calibre barrel has

Training rifle C10

1:16 in rifling. The locking mechanism is of the inertial breech-block type firing from a closed breech. The breech mechanism can be removed as a unit from the weapon for separate storage if required. No special tools are required for cleaning or maintenance purposes.

The magazine capacity of the C10 is 10 rounds and the barrel is 518 mm long. Muzzle velocity is approximately 365 m/s. The weight of the C10 with magazine and sling is 3.44 kg.

Status
Pre-production.

Manufacturer
Diemaco, Kitchener.
(For details see Directory of Suppliers)

FRANCE

Giat 155 mm Artillery Training Aids

155 GCT Training Turret
The *Tourelle d'Instruction du 155 GCT* was produced to train 155 mm GCT self-propelled gun turret crews in gun drills and emergency procedures and also provide some maintenance and repair training for service personnel. The turret trainer is an open framework representation of the 155 GCT turret mounted on a static steel frame. The main approach is from the rear from an open platform to which access is provided by a metal stairway. Inside the frame are all the various items of equipment normally used in an operational 155 GCT turret, including the main breech, breech mechanism, ammunition stowage and handling system and the associated control equipment. Full and complete drills can be carried out by a turret crew and misfires and other emergency procedures can be introduced by an instructor manning a control panel on a movable simulator control system outside the turret. The training procedures can include the actual loading of training rounds. Once loaded, the breech closes as on the operational gun and a hydroelectrical conveyor not only simulates the gun recoil (a sound generator simulates the firing noise) but moves the projectile along the barrel until it falls out into a hopper under the trainer barrel assembly.

For maintenance training the simulator control panel can be used to introduce faults into parts of the turret systems. All the operating components used in the turret are standard production items. Platforms around the turret trainer can be used by the instructor for checking purposes or by students observing drills and procedures.

The 155 GCT training turret is 9 m long, 4.4 m wide and 3.66 m high. It weighs approximately 7500 kg.

TR Dummy Loader
This equipment was produced to enable gun crews of the 155 mm TR towed gun to carry out loading and rapid loading drills. Limited repair and maintenance procedure training can also be carried out. The equipment consists of a simulated TR breech and loading mechanism mounted on a carriage. The mechanisms are presented at the same height from the ground as with the real weapon and the angle of elevation can be varied. Dummy projectiles and charges loaded into the equipment are carried through the device and are ejected from the front. There is no barrel.

GCT Breech Demonstration Model
In order to provide a working model of the 155 GCT breech operation and mechanism, Giat produced a special demonstration mounting of a complete breech mounted on a tubular steel stand. Operation of the demonstration mounting *(Montage Dynamique de Fonctionnement de la Culasse)* is powered by an on-mount electrically driven hydraulic pump operating at a pressure of 1450 psi (100 bar). Using this system the operation of the breech can be slowed down from the normal operating speed to a rate slow enough for every operation to be demonstrated.

GCT Breech Cross-section Model
To demonstrate the inner workings and assemblies of the 155 GCT gun breech, Giat produces a static cross-section model of the breech mechanism mounted on a steel handling frame. The model includes the method of fitting the gun barrel to the breech. The model is known as a *canon coupe.*

Status
In production. In service with the French Army.

Manufacturer
Giat Industries, Versailles.
(For details see Directory of Suppliers)

Dummy loader for 155 mm towed gun

FAMAS à Plombs Training Rifle

The FAMAS à Plombs is a sub-calibre weapon intended for individual training of the soldier in elementary marksmanship and weapon handling. It resembles the standard 5.56 mm FAMAS F1 French Army service rifle but fires a 4.5 mm air rifle pellet by means of a 12 g CO_2 gas cartridge. It will fire shots in the semi-automatic mode, using ammunition fed from a vertical magazine. The shape, weight and balance are as close to the characteristics of the service F1 rifle as possible and it is issued complete with integral bipod and a sling.

This training weapon is produced at the Manufacture Nationale d'Armes de Saint-Étienne.

Calibre: 4.5 mm
Weight: (without magazine) 2.8 kg
Length: 765 mm
Length of barrel: 270 mm
Length of line-of-sight: 325 mm
Muzzle velocity: 130 m/s
Accuracy: 10 rds in 50 mm circle at 12 m
Range: gallery practice 25 m; moving targets, cine targets 15 m
Gas supply: sufficient for minimum of 70 rds at 20°C

FAMAS à Plombs training rifle

Status
In production.

Manufacturer
Giat Industries, Versailles.
(For details see Directory of Suppliers)

Losfeld Sub-calibre Mortar Training System

The Losfeld mortar training system has been in existence for over 25 years. It consists of a special sub-calibre barrel inserted into a standard mortar barrel and used to fire small, cheap and recoverable training rounds. The equipment fits all types of mortar and closely reproduces the actual firing of the full-size ammunition. There are different sub-calibre barrels, but all fire the same 30 mm ammunition and all rounds accept the same series of propellant charges.

The sub-calibre barrel is in two parts. For muzzle-loading mortars there is a fixed breech-block and barrel joined together to make one component. The barrel is 670 mm long, and the breech-block is screwed into it. This assembly is common to all sizes of mortar. For 60 mm mortars the breech rests directly on the bottom of the tube, while for other calibres it is held away from the bottom by an extension tube. For the 60 mm, 81 mm and 4.2 in mortars the firing pin is fixed; for the 120 mm calibre it is a trigger-operated mechanism.

The extension tube varies according to the calibre of the mortar and is so arranged that the end of the barrel is always level with the muzzle of the mortar. The sub-calibre projectile then has the same length of 30 mm barrel for all mortars, and the same length of shot travel. In the 120 mm mortar the extension tube is so long that it carries a third centering piece to ensure that it is always in the centre of the mortar tube.

The 30 mm sub-calibre projectiles are loaded in exactly the same way as with the full-size bombs and slide down the barrel to be fired using the firing pin. Thus the loading and firing drills are identical to those using real ammunition.

For breech-loading 60 mm mortars fitted to armoured vehicles there is a slightly different arrangement in which the 30 mm projectiles are loaded through the breech and fired using the 60 mm mechanism. A different sub-calibre barrel has to be used, but apart from this the general principle is the same and the loading and firing drills for the crew are unchanged.

The ammunition system consists of a 30 mm projectile with replaceable propellant cartridges, smoke generators and secondary charges.

Projectiles may be reusable or disposable. The reusable projectile consists of a machined-steel shell body onto which is screwed a truncated conical nose and a set of fins. The body contains the smoke generator and is pierced with four holes to release the smoke. The fin assembly accepts the primary cartridge in the central column and secondary charges are placed between the fins. There are five different increments, from 0 to 4. Charge 0 is the primary alone while Charge 4 is the primary and four secondaries. When the projectile strikes the ground at the target it ignites the smoke cartridge and releases a cloud of white smoke for several seconds. The colour of the smoke may be altered to suit the nature of the exercise if so desired.

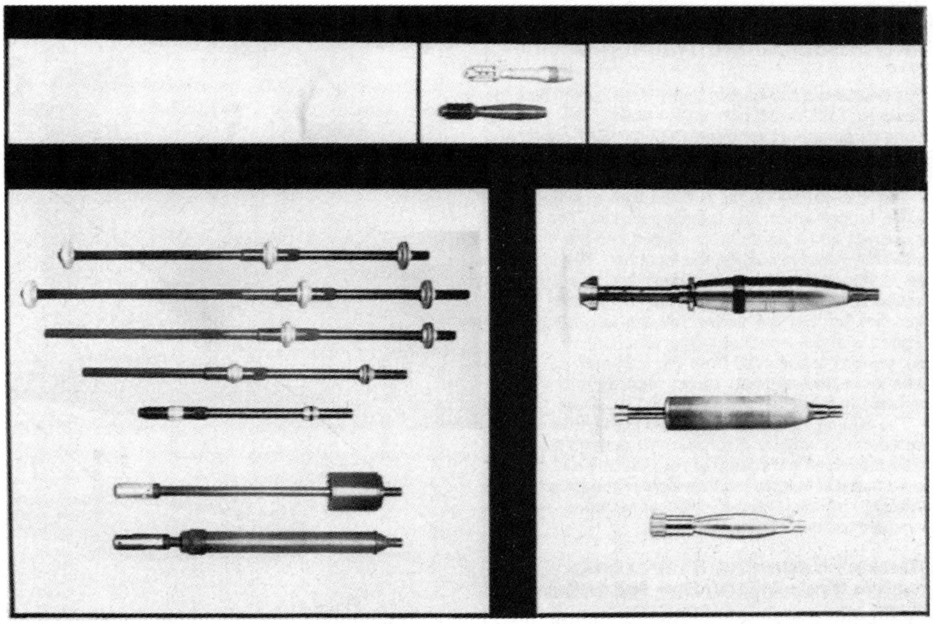

Main components of the Losfeld sub-calibre mortar training system. At the top are the 30 mm projectiles with the disposable model above; left are various sub-calibre barrels while on the right are three simulated sabot projectiles - from the top, 120 mm rifled PPR 120, 4.2 in PP 4.2 and 81 mm PP 81

Disposable projectiles are identical in operation to the reusable projectiles and carry the same payload but appear to be manufactured from a plastic material. They also have a different body shape.

When using the 30 mm projectiles the setting-up drills and all laying and firing drills are the same as with the full-size mortar (assuming that the sub-calibre barrel has been inserted beforehand). The mortar fire controller uses a special set of ballistic tables to calculate the fall-of-shot, and to make the corrections from the fire orders. Misfire drill is the same as for full-size ammunition and, though the rounds do not explode, the crew should carry out all normal safety drills at all times.

For recovering the reusable projectiles there is a special tool which is thrust into the ground to pull the projectiles free. The exact location of buried rounds is indicated by smoke stains on the surface.

Recharging the reusable projectiles is straightforward. The fin assembly is unscrewed using a special spanner and the empty case removed and replaced. The fin assembly is then screwed back on. The entire process takes only a minute or two.

A recent introduction is a range of what are described as simulated sabot projectiles that are sub-calibre barrels resembling a full-sized mortar bomb. These devices are loaded with a 30 mm sub-calibre projectile and are loaded into the mortar in the usual manner. The sub-calibre projectile is then fired and the simulated sabot projectile is recovered from the barrel using the usual recovery methods and drills. The device can then be reloaded with a 30 mm sub-calibre projectile and used again as often as required. This device is available for 81 mm (PP 81), 4.2 in (PP 4.2) and 120 mm calibres (for smooth-bored and rifled mortars - the rifled model is known as the PPR 120 and the smooth-bored the PPL 120).

Calibre:
(mortars) 60 mm, 81 mm, 120 mm, 4.2 in, and 60 mm breech-loading
(sub-projectile) 30 mm
30 mm projectile:
(length) 191 mm
(weight) 500 g
Primary cartridge:
(diameter) 12 mm
(length) 47.5 mm
Smoke generator:
(diameter) 18 mm
(length) 66.5 mm
Minimum range:
(Charge 0) 40 m
(Charge 4) 160 m
Maximum range:
(Charge 0) 140 m
(Charge 4) 300 m

Status
In production. In service with over 30 countries.

Manufacturer
Losfeld-Industries, Charenton.
(For details see Directory of Suppliers)

GERMANY

Dynamit Nobel 14.5 mm Tank Gunnery Training System

To allow full-size tanks to fire their guns on a small scale range, Dynamit Nobel produced its 14.5 mm tank gunnery training system (at one time known as the PS tank gunnery trainer). This system can be used on ranges of from 50 to 125 m; for safety reasons the range should be at least 600 m long. Firing at moving and stationary targets at a scale of 1:10 or 1:20, low-cost tank gunnery training can be carried out using a special 14.5 mm sub-calibre firing device located in the chamber of the tank main gun. A firing solenoid can be connected to the existing gun firing mechanism.

Sighting is carried out using the sight reticle grid applicable to HESH-T or HEP-T full calibre rounds with the range calculated at a scale of 1:20. Thus targets at 75 m on a scale of 1:20 are sighted using the 1500 m

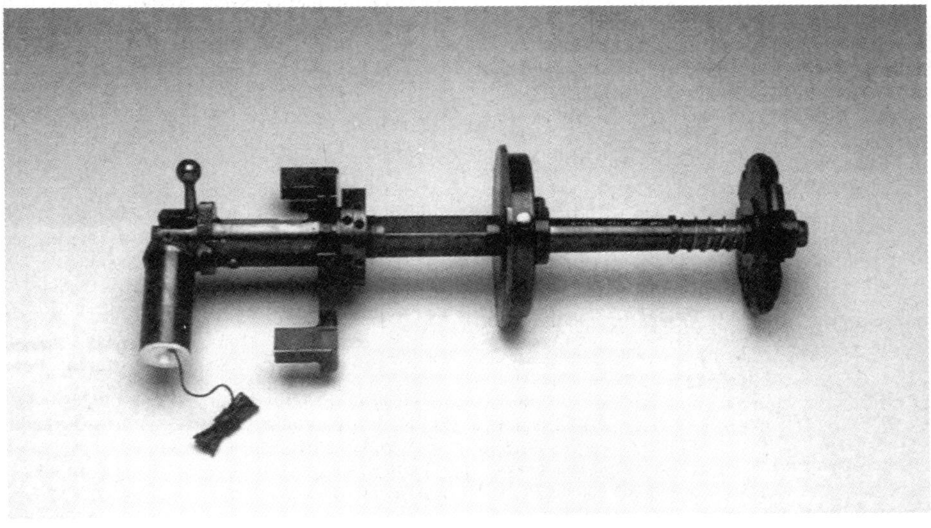

14.5 mm sub-calibre training device for Dynamit Nobel 14.5 mm tank gunnery training system

mark on the HESH-T/HEP-T grid. A prism inserted into the objective path of the sighting telescope compensates for parallax between the barrel and optical axes at the short ranges employed. As an alternative the normal sighting device can be adjusted to suit the system.

The training system fires the same 14.5 mm ammunition as the Dynamit Nobel 14.5 mm Artillery Training System (see entry in *Artillery and Mortar Fire-Control Trainers* section) but for tank gunnery training three types of ammunition are used. One has an impact fuze to simulate full calibre HE projectiles, one has an impact fuze and tracer to simulate HEAT-T, HESH-T and HEP-T, and one has a tracer only to simulate AP-T or APDS-T. The impact fuze version emits a smoke charge on impact and there is no other explosive or fragmentation effect.

Status

In production.

Manufacturer

Dynamit Nobel AG, Troisdorf.
(For details see Directory of Suppliers)

Dynamit Nobel 25 mm Mortar Training System

The Dynamit Nobel 25 mm mortar training system is a low-cost method of training mortar crews and fire controllers. The complete system consists of a training device, 25 mm calibre practice ammunition and range tables.

The training device consists of a training bomb with a 25 mm barrel running through its centreline. A 25 mm practice ammunition cartridge is loaded into the base of this barrel in the tail of the bomb. The bomb is loaded into the full-sized mortar where it drops to the bottom of the barrel and fires in the normal way. The propellant in the 25 mm cartridge fires the smaller projectile up the barrel and out of the muzzle. It also forces the bomb out of the mortar barrel, though at a low velocity so that it just clears the muzzle and falls to the ground a few metres in front of the mortar. The sub-projectile is shaped like a small mortar bomb and flies for a distance of 70 to 520 m depending on the charge used. It gives a true trajectory with good consistency and on landing produces a small cloud of marker smoke and a loud report. These can be used to correct the fire of the mortar and, due to the accuracy of the training system, the trajectory can be altered in the same way as for full-sized rounds, but using the training system range tables.

The training device can be quickly reloaded with another sub-calibre round and used continually since it does not suffer any damage. The sub-projectiles are harmless and do not produce fragments on impact.

The system is available for 60 mm, 81 mm, 120 mm and 4.2 in mortars. The minimum area required for the system measures 600 × 100 m with a maximum height of 300 m.

Status

In production. In service with some NATO armies (including Canada) and some other armies.

Manufacturer

Dynamit Nobel AG, Troisdorf.
(For details see Directory of Suppliers)

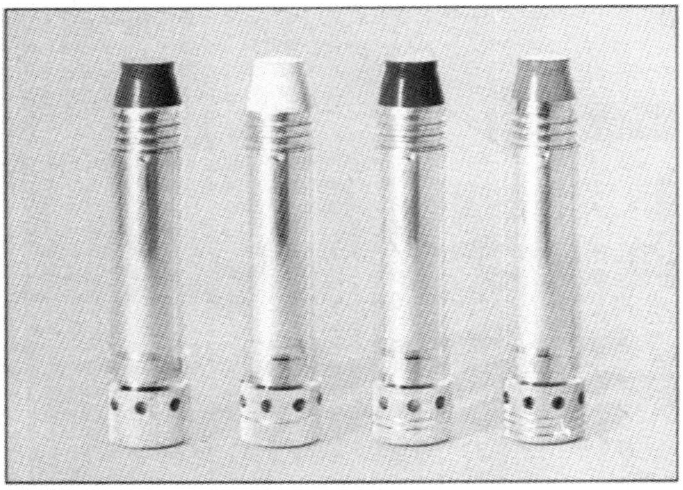

25 mm practice ammunition for the Dynamit Nobel mortar training system

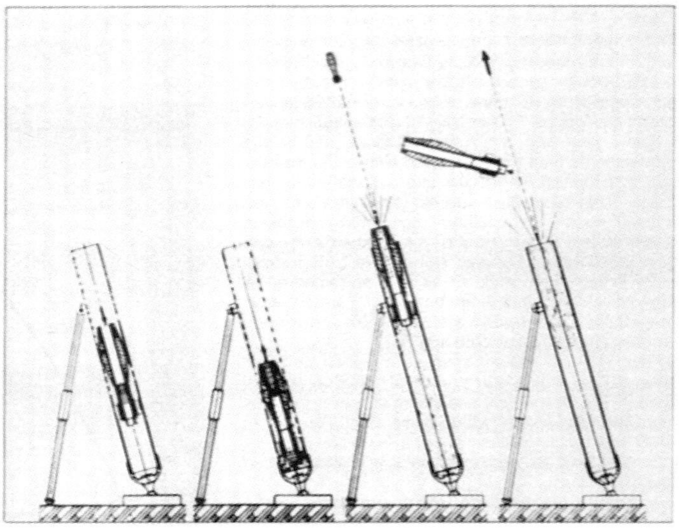

General operating principles of the Dynamit Nobel 25 mm mortar training system

Dynamit Nobel 110 mm Sub-calibre Training System for MLRS

The Dynamit Nobel 110 mm sub-calibre training system for the American Multiple Launch Rocket System (MLRS) was developed to allow realistic training in the full range of MLRS crew tasks, from loaded launch pod/containers (LP/C) handling to rocket launching, at a lower cost and with reduced range and safety requirements than those involved with using the full-size 227 mm MLRS rockets and associated components. The sub-calibre rockets employed with the training system are 110 mm Light Artillery Rockets (LAR) already in service with the German Army and also produced by Dynamit Nobel. The LAR rockets have a maximum range of 14 000 m compared to approximately 40 000 m for the MLRS. For full details of the LAR rocket system refer to *Jane's Armour and Artillery 1993-94* pages 605 and 606.

110 mm sub-calibre training system rocket being fired from MLRS launcher

The 110 mm system employs a launch pod/container (LP/C) dimensionally identical to that of the standard MLRS. The weight and centre of gravity are also identical to those of the MLRS component. The system LP/C has six LAR launch tubes (external diameter 126 mm) and the necessary cabling is incorporated into the LP/C.

In order to convert the MLRS programming and firing signals given by the fire-control system into LAR recognisable signals a supplementary electronic unit is attached to the rear part of the launcher and is connected to the fire-control system by cables. The unit accepts a power supply and fire-control signals and the output circuits transmit firing, fuze setting and ignition circuit testing signals to the two LP/Cs carried on the launcher. The electronic unit can be easily attached and removed. It has its own self-test capability and can be used as a test set for LP/S wiring. No other modifications to the MLRS launcher are required other than adapting the ballistic fire commands for MLRS to LAR ballistics using the unit's fire-control computer. An integration of LAR ballistic data software into the MLRS fire-control computer is possible.

The LAR rockets used with the training system are the DM 21 fragmentation rocket with a DM 54 proximity fuze, the DM 25/35 smoke rocket with DM 53/73 MTSQ fuze, and the DM 18/28/38/48 practice rocket. The LAR rockets have a minimum range of 6000 m and a maximum range of 14 000 m. Weight of a LAR rocket is 35 kg. Length is 2.263 m. The maximum velocity after two seconds of rocket burning time is approximately 640 m/s. There is a possibility of eliminating the supplementary electronic unit when using LAR practice rockets fitted with impact fuzes.

Status

In production for the German Army.

Manufacturer

Dynamit Nobel AG, Troisdorf.
(For details see Directory of Suppliers)

Heckler and Koch MP5PT Training Sub-machine Gun

The Heckler and Koch MP5PT is a specially designed training sub-machine gun which is based on the well-known Heckler and Koch MP5 and designed to fire PT plastic training ammunition only. The weapon allows single-shot fire, continuous fire or three-round bursts to be selected, and all handling and operating functions are identical to those of the service weapon. Two versions are manufactured; the MP5A4/PT has a fixed buttstock, while the MP5A5/PT has a retractable stock. To distinguish the PT versions from the service weapons they are identified by a blue cocking handle and the inscription 'Plastic Training Only' on each side of the receiver.

The bolt is identical to that of the service MP5 but does not have locking rollers; in order to provide the necessary impulse to the bolt to produce automatic operation, a floating chamber is fitted.

The maximum trajectory range of the 9 mm PT bullet is about 125 m, thus requiring a safety area of about 170 m. At 25 m range the projectile energy has dropped to below 10 joules. The sub-machine gun will group to a circle of 100 mm at a range of 8 m.

It is therefore possible to use this weapon in training on outdoor ranges with restricted safety areas or on indoor ranges, as well as in natural terrain for realistic tactical training. In an emergency the weapon can be used as a combat weapon for short-range engagements.

The MP5PT must never be used for firing ammunition other than the PT plastic ammunition. The use of any service ammunition is likely to damage the weapon, though there will be no danger to the firer.

Specifications

Cartridge: 9 mm PT plastic ammunition only
Operation: blowback, amplified by floating chamber
Type of fire: single-shot, automatic and 3-round bursts
Length:
(fixed butt) 680 mm
(extended butt) 660 mm
(retracted butt) 340 mm
(barrel) 225 mm
Weight:
(empty, fixed butt) 2.54 kg
(retractable butt) 2.88 kg
Feed: 15- or 30-round magazines
Rate of fire: 700 rds/min
Muzzle velocity: 400 m/s
Muzzle energy: 210 J

Status

In production. In service with German Special Forces.

Manufacturer

Heckler and Koch GmbH, Oberndorf-Neckar.
(For details see Directory of Suppliers)

Heckler and Koch MP5A4/PT training sub-machine gun

Heckler and Koch P7PT8 Training Pistol

The Heckler and Koch P7PT8 training pistol is a special semi-automatic pistol for firing 9 mm PT plastic training cartridges. All handling and service operations are identical to those of the P7M8 pistol.

The P7PT8 is similar to the standard P7M8 pistol. It also has a synthetic material insert at the top of the trigger guard for heat protection. In addition the rear sight is screwed into position and the grip is provided with a lanyard loop. In order to distinguish clearly between the training and service weapons, the P7PT8 is marked with a blue spot on each side of the slide and the blue inscription 'Plastic Training Only'. The slide is approximately 50 g lighter than that of the P7M8 and has no gas retarding piston. Moreover the P7PT8 has a different recoil spring and a longitudinally floating chamber to increase the recoil force on the slide.

When used with this pistol, PT training ammunition has a maximum range of about 125 m, which necessitates a safety area of about 170 m. At a distance of 25 m the bullet energy has dropped to less than 10 joules. Depending upon the ammunition lot, the pistol should group in a 100 mm circle at a range of 8 m.

In an emergency the P7PT8 may be used as a combat pistol for short-range firing. The special plastic projectile reduces the probability of injuring bystanders by ricochet, and the risk of damaging vital components in areas such as aircraft interiors is minimal.

The P7PT8 must never be used with any other ammunition than the PT plastic training cartridge. Firing service rounds may cause damage to the weapon, though the safety of the firer is assured.

Specifications

Cartridge: 9 mm PT plastic ammunition only
Operation: blowback, amplified by floating chamber
Feed: 8-round box magazine
Length: 170 mm
Barrel length: 105 mm
Weight empty: 790 g
Muzzle velocity: 400 m/s
Muzzle energy: 210 J

Status

In production. In service with German Special Forces.

Manufacturer

Heckler and Koch GmbH, Oberndorf-Neckar.
(For details see Directory of Suppliers)

Heckler and Koch P7PT8 training pistol

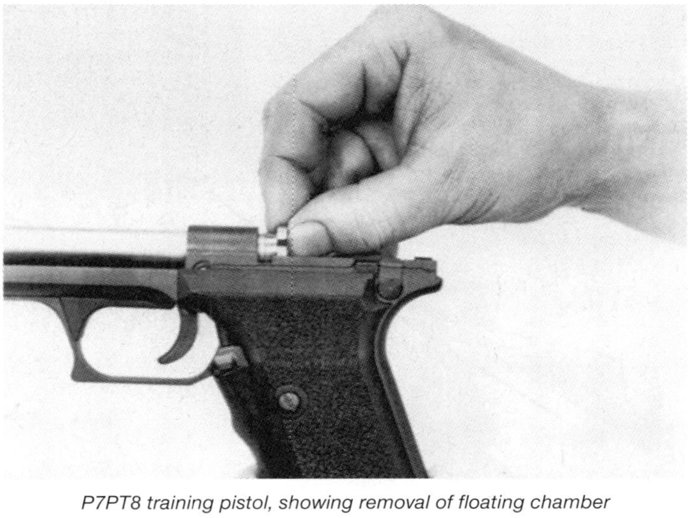

P7PT8 training pistol, showing removal of floating chamber

Mauser-Wegmann 35 mm Sub-calibre Training Device for 105 mm and 120 mm Tank Guns

Produced jointly by Mauser-Werke and Wegmann and Company, the 35 mm sub-calibre training device is intended for use with 105 or 120 mm tank guns. The device consists of a 35 mm barrel insert with clamping elements and an adaptor cartridge for 35 × 228 mm ammunition. In addition there is a breech opening device with a drive unit, control electronics and some accessories. The adaptor can be fitted to a tank gun by the tank crew in less than 1 hour. Once installed, the 35 mm ammunition is loaded into the gun in simulated full-bore cartridges. On firing, the 35 mm round produces about 30 mm of recoil which is used as a control pulse for the device's breech opening mechanism. The simulated cartridge, together with the spent 35 mm case, is then ejected for subsequent reloading (which takes approximately 2-3 minutes). Any 35 × 228 mm full calibre round can be used with the system although to improve performance two extremely accurate types of 35 mm ammunition have been developed to simulate HE and KE tank gun rounds. The tank's normal fire-control and other aiming devices can be used.

Specifications

Barrel insert
Calibre: 35 mm
Length: 2.95 m
Weight: approx 88 kg

Protective tube
Length:
(105 mm) 2.1 m
(120 mm) 1.961 m
Weight:
(105 mm) 7.6 kg
(120 mm) 6.5 kg

Breech opening device
Length: 500 mm
Diameter: 190 mm
Weight: approx 25 kg
Supply voltage: 21.5-30 V
Max power input: 190 W
Breech opening time: approx 0.3 s
Rate of fire: 5-8 rds/min

Adaptor cartridge with 105 mm TP-T
Length: approx 596 mm
Weight: approx 18 kg

Adaptor cartridge with 120 mm TP-T
Length: approx 596 mm
Weight: approx 19 kg

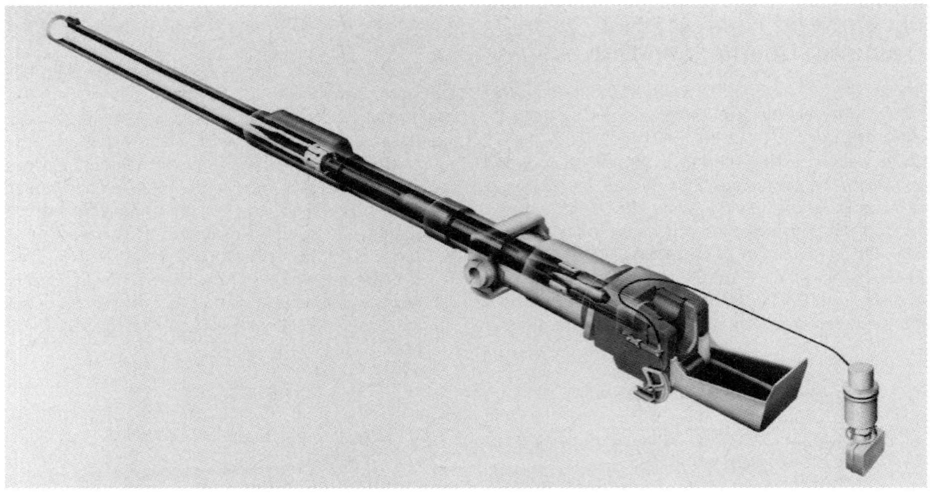

Outline illustration of the Mauser-Wegmann 35 mm sub-calibre training device for 105 mm and 120 mm tank guns seen here in the 120 mm gun configuration

Status

In production. In 1989 19 units were procured by US Army, Europe, as an off-the-shelf purchase for use in the 120 mm main gun of the M1A1 Abrams MBT.

Manufacturers

Mauser-Werke Oberndorf GmbH, Oberndorf-Neckar.
Wegmann and Company GmbH, Kassel.
(For details see Directory of Suppliers)

Nico-Pyrotechnik Anti-armour Sub-calibre Training Devices

In most of the Nico anti-armour sub-calibre training devices a standard sub-calibre barrel of 22 mm is used and in each case a small projectile is fired from the launcher and flies along the same trajectory as the full-sized round. Ignition can be electrical or using a percussion cap. The sub-calibre barrel is fitted into a standard launcher and the system can be used over and over again.

3.5 in Rocket Launcher

For this weapon the sub-calibre barrel is the same external shape as the live round and it is loaded in exactly the same way, thus giving the crew realistic handling practice. The sub-calibre round has a tracer element and can be followed visually throughout its flight.

Nico Light Anti-armour Weapon Sub-calibre Devices

Nico-Pyrotechnik produces sub-calibre devices for the Panzerfaust II (Lanze), Panzerfaust 3, Armbrust C, Carl Gustav, and M72 launchers. The Lanze sub-calibre round is matched to the real weapon in its flight ballistics and handling, this having been the prime requirement of the Bundeswehr. The devices are recoilless and have a target range matching that of the live rounds. After firing, the flight path of the 18 mm projectile is marked by a tracer. The sabot itself remains in the launcher. The main recoil counterbalance is obtained in the same manner as in the parent weapon, by dispersion of fine particles to the rear.

Specifications

Panzerfaust 3 sub-calibre DM 38
Dimensions: 18 × 96 mm
Weight of complete cartridge: 255 g
Weight of projectile: 120 g
Target range: max 400 m
Muzzle velocity: 250 m/s

Nico Training System for FW40

This is a sub-calibre training system for use with the FW40 anti-tank rocket launcher. It consists of a sabot which can be loaded into the launcher in the same way as the service munition. The sabot contains a sub-calibre smooth barrel which can be pre-loaded with a special 18 mm tracer round which is ballistically matched to the trajectory of the service rocket. It is fired by activation of the normal firing trigger system.

Specifications

Length of dummy: 932 mm
Weight, empty: 2.3 kg
Cartridge: 18 × 96 mm tracer
Range: 300 m

Status

All the above units are in production.

Manufacturer

Nico-Pyrotechnik, Trittau.
(For details see Directory of Suppliers)

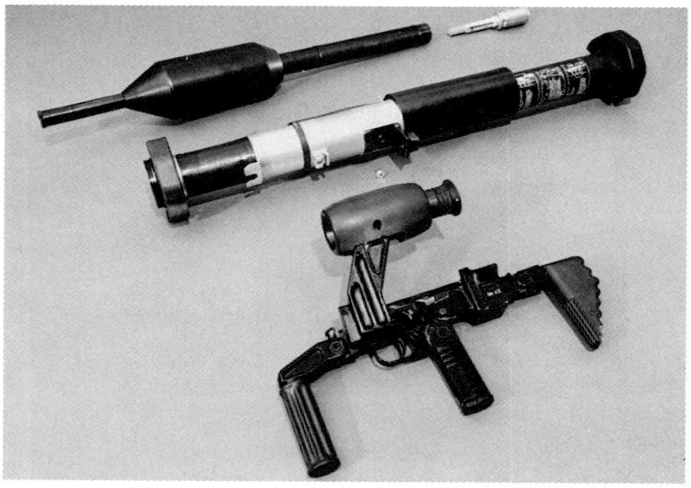

Nico sub-calibre training system for Panzerfaust 3

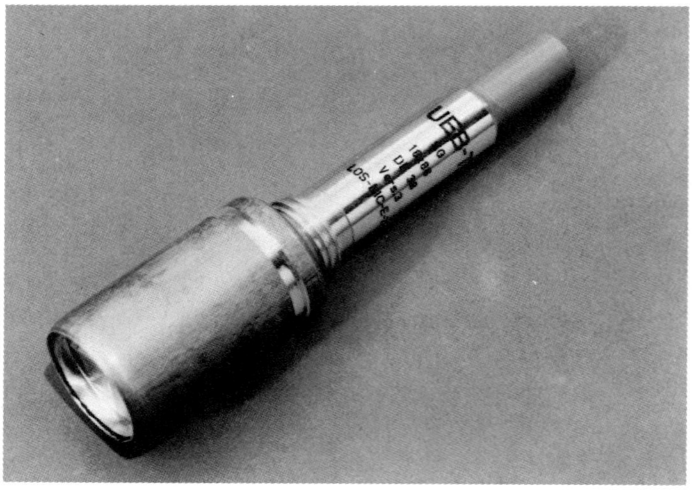

Nico sub-calibre cartridge for Panzerfaust 3

Nico-Pyrotechnik Mortar Training System

The Nico-Pyrotechnik Mortar Training System is produced for mortars with calibres of 60, 81/82, 107 (4.2 in), and 120 mm (rifled and smooth-bored). The system utilises a series of realistically shaped and weighted dummy mortar bombs (sabots) incorporating a 22 mm sub-calibre cartridge.

The 'loaded' sabots are handled, loaded and fired in the same way as live mortar rounds. The 22 mm projectile fired by the ejection charge in the cartridge is used to simulate the trajectory of the full calibre round on a scale selected by the user; the most usual scale requested is 1:10 with four charge variants. The projectile contains a charge to produce a target impact

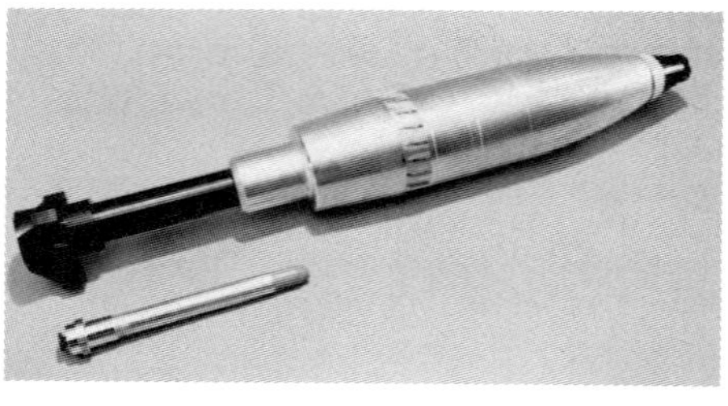

Nico sub-calibre training sabot for 120 mm rifled mortar

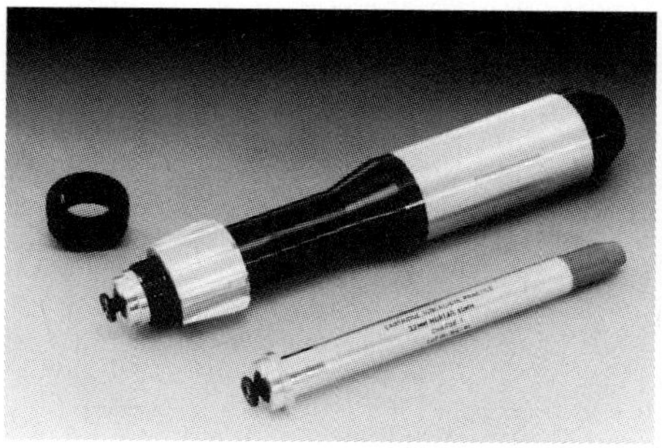

Nico sub-calibre training sabot for 60 mm mortar

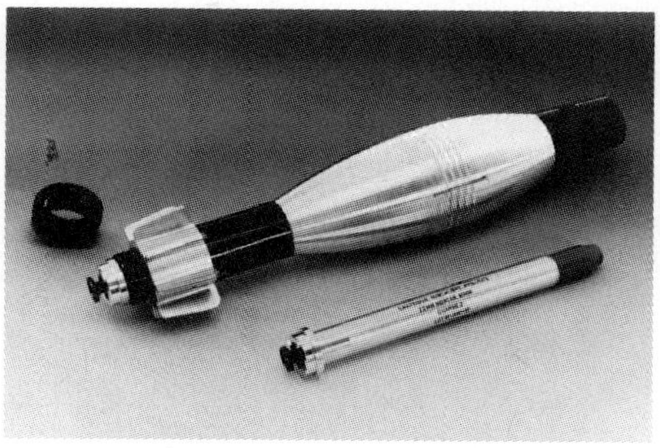

Nico sub-calibre training sabot for 81 mm mortar

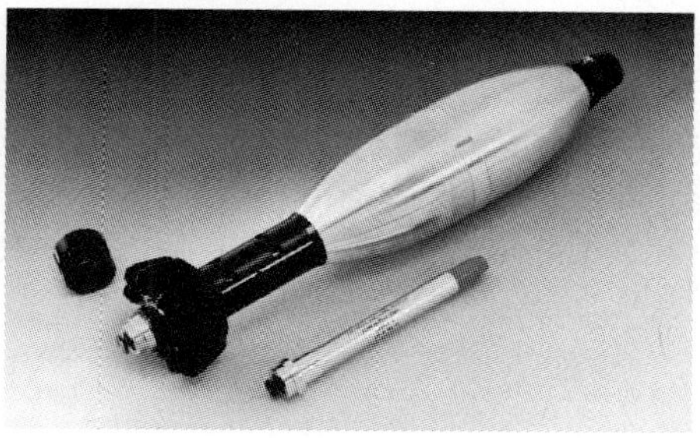

Nico sub-calibre training sabot for 120 mm mortar

signature large enough to provide visual and aural observation by day and night. The cartridge ejection charge is sufficient to propel the sabot from the mortar barrel, after which it falls to the ground after a few metres of travel. It can then be recovered and reloaded for further use.

The inherent accuracy of the fin-stabilised projectile permits realistic fire adjustment on miniature firing ranges and for training mortar fire controllers and forward observers.

No modifications are required to be made to the mortar involved with the system. If required, the sabot can be used to fire a blank (manoeuvre) cartridge which is sufficient to propel the sabot from the mortar barrel without the projectile being fired.

Status
In production. In service with Belgium, France, Germany, India, Indonesia, Peru, South Korea, Switzerland, the US Army and other armies.

Specifications

Calibre	**60 mm**	**81 mm**	**107 mm (4.2 in)**	**120 mm (smooth)**	**120 mm (rifled)**
Sabot					
(loaded)	3.4 kg	4.3 kg	10.2 kg	9.43 kg	15 kg
(unloaded)	3 kg	3.9 kg	9.7 kg	9 kg	14.3 kg
(length)	322 mm	397 mm	616 mm	595 mm	800 mm
Cartridge					
(calibre)	22 mm	22 mm	22 mm	22 mm	22 mm
(length)	244 mm	244 mm	253 mm	250 mm	253 mm
(charges)	4	4	9	4	4
(weight)	397 g	397 g	397 g	397 g	397 g
(max range)	550 m	550 m	530 m	540 m	540 m
(min range)	70 m	70 m	195 m	70 m	140 m

Manufacturer
Nico-Pyrotechnik, Trittau.
(For details see Directory of Suppliers)

HUNGARY

IDEX 14.5 mm Sub-calibre Device for 2A46 Tank Gun

The Hungarian defence industry has developed a 14.5 mm sub-calibre training device to be fitted into the 125 mm 2A46 tank gun used as the main armament of the Hungarian Army's T-72 tanks. The device is semi-automatic and is located inside the bore of the main gun, using a protective tube. It can be installed by the tank crew. Aiming and firing is carried out using the tank's usual instruments and controls, with the firing solenoid using the vehicle's 22-29 V DC power supply. The operational range of the device is between 1000 and 2000 m. Rate of fire is 6 rds/min.

The device has six main assemblies: the barrel; a cradle; the lock mechanism; a magazine; an electrical supply unit; and a control box. The complete weight is 119 kg. Maximum operational life of the device is stated to be at least 4500 firings.

The ammunition is the CIS 14.5 mm × 114 heavy machine gun round and the muzzle velocity quoted is 950 m/s.

A 12.7 mm version of this device has been proposed.

Status
In service with the Hungarian Army.

Agency
IDEX, Budapest.
(For details see Directory of Suppliers)

ITALY

Armi Adler 0.22 Rifles

The Armi Adler company manufactures a number of 0.22 calibre rifles for the commercial market which have obvious applications as training equipment since they faithfully duplicate the appearance and handling of well-known military rifles. They are not capable of automatic fire, but in other respects their operation is similar to the full-calibre rifles, with such controls as safety catches, cocking handles and magazine releases in the appropriate places and with sights which duplicate those of the parent weapons. The following designs are currently available:

AP 74 Reproduction of the American M16/AR-15; 515 mm barrel, weight 2.9 kg; also available in 7.65 mm ACP calibre.

AP 74/L Similar to AP 74 but with wooden furniture in place of plastic. Weight 3.5 kg.

AP 80 Reproduction of the AKM; 453 mm barrel, weight 2.8 kg. Folding stock version available. Also available chambered for 0.22 Magnum RF calibre.

AP 84 Reproduction of the Israeli Galil rifle; 453 mm barrel, weight 2.8 kg. Folding stock version available. Also available in 0.22 Magnum RF calibre.

AP 85 Reproduction of the French FAMAS rifle; 453 mm barrel, weight 2.8 kg.

All 0.22 RF models have 15- and 30-round magazines; those chambered for 0.22 Magnum RF have 10-round magazines. A telescope mount can be fitted to all models and slings and cleaning accessories are available. A nylon folding butt is available for the AP 80 and AP 84 rifles.

A bipod is available for rifles having barrels from 16 to 20 mm in diameter. It permits horizontal and vertical movement and allows weapons to be truly vertical when the ground is uneven. The legs, when folded, are 220 mm long and unfold to a maximum of 330 mm.

Status
In production.

Manufacturer
Armi Adler, Loano.
(For details see Directory of Suppliers)

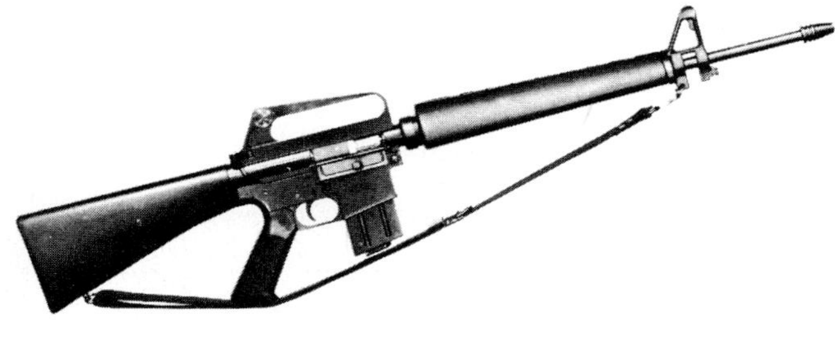

Armi Adler Mod. AP 74 rifle, based on the M16/AR-15

Armi Adler Mod. AP 80, based on the AKM

NETHERLANDS

Van Halteren M109 Full Turret Artillery Crew Trainer

The M109 Full Turret Artillery Crew Trainer was originally developed by Van Halteren Metaal bv and the Royal Netherlands Army in 1984. The trainers operate in exactly the same way as a real M109 turret and use a look-alike M109 turret with a simulated 155 mm howitzer breech mounted on a superstructure frame. All stages of ammunition preparation and loading plus aiming and firing can be simulated. The trainer interior is fully duplicated to the real vehicle dimensions. The trainer can use the original American design rammer, semi-automatic rammers or 'flick' rammers, all using the normal rammer velocities; maximum firing frequency is 3 rounds in 15 seconds. The turret can traverse over 1800 mils each side and the barrel elevates from −50 to +1280 mils. The full trainer can be canted over 100 mils over two axes.

After loading and aiming the howitzer a lanyard can be pulled so that the barrel produces a realistic recoil in length and velocity depending on the charge used. A firing shock is also generated.

The trainee crew is monitored on an instructor's station which shows sight bubble levels, installed azimuth and elevation angles, used ammunition and fuze settings. The instructor can remotely control the gun and introduce malfunctions. The trainers can be coupled easily to forward observer trainers and fire-control centres.

M109 Full Turret Artillery Crew Trainer

Status

The Netherlands Army has bought 15 and the Swiss Army has bought 18. One trailer-mounted system is available for demonstration.

Manufacturer

Van Halteren Metaal bv, Bunschoten.
(For details see Directory of Suppliers)

NORWAY

Raufoss M72-S 21 mm Sub-calibre Training Device

The M72 light anti-armour rocket launcher is a one-shot throw-away munition and the full-size round does not easily lend itself to being used for training. Raufoss, which has the contract to manufacture the M72 for NATO, developed a training sub-calibre rocket which closely simulates the firing of the full-sized round, yet which is both cheap and safe.

The sub-calibre training device uses the same launcher as the full-sized rocket with an additional insert clamped inside the tube. From this is fired a 21 mm rocket which gives the noise, flash and flight trajectory of the tactical round. The launcher can be a discarded item from a previous range practice with tactical ammunition and there is no need to procure special launchers.

There are two training systems, one for the M72A series and the other for the upgraded M72E series. The two systems are identical in function and differ only in dimensions and performance.

The insert consists of a smooth-bore barrel with a calibre of 21.45 mm. The rear end is fitted with a chamber which acts as a barrel support. At the front end is a muzzle sleeve and between the two is a supporting disc.

The 21 mm sub-calibre training rocket consists of a steel head and a steel rod connecting the head and the nozzle at the rear. The nozzle is formed as a tail ring which also helps to stabilise the rocket in flight, as well as holding the tracer element which burns up to 200 m. The igniter has the same primer block as that used for the full-size rocket and is connected in the same way. The propellant charge consists of six tubular sticks of the same material as used in the full-size rocket, with each stick attached to a plate at the head.

All launcher functions are the same as for the full-size round and the firer goes through the same sequence of operations to fire. The sub-calibre rocket is as accurate as the full-size round and can be used against armoured vehicles without harming them.

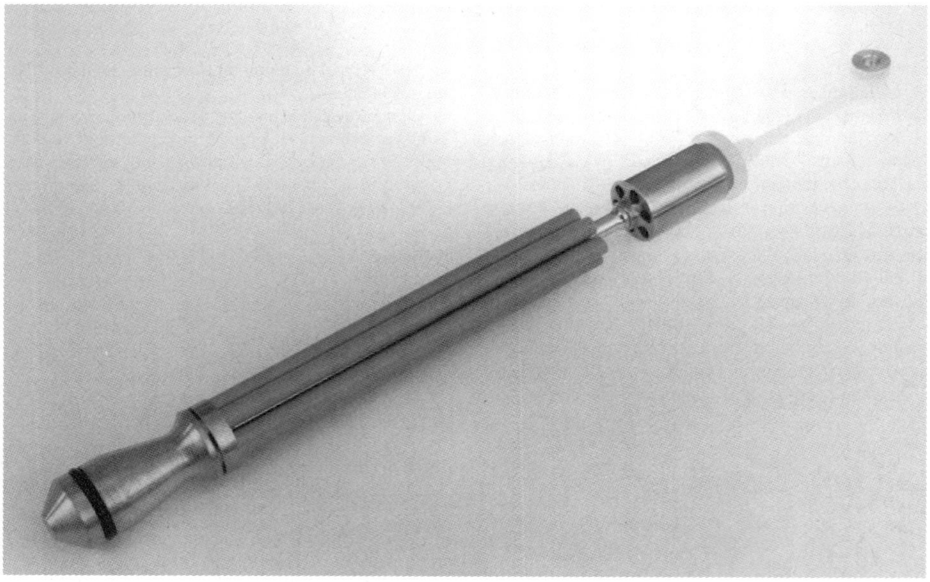

M72-S sub-calibre rocket

Specifications

	A-series	E-series
Sub-calibre barrel		
Calibre	21.45 mm	21.45 mm
Weight	1 kg	1.4 kg
Length	625 mm	729 mm
Complete assembly		
Weight	2.6 kg	3.5 kg
Length		
(closed)	664 mm	786 mm
(extended)	893 mm	982 mm
Sub-calibre rocket		
Calibre	21 mm	21 mm
Weight	195 g	170 g
Muzzle velocity	150 m/s	200 m/s
Maximum range	900 m	1200 m
Charge weight	16 g	18.5 g
Time of flight		
(to 100 m)	0.7 s	0.5 s
(to 200 m)	1.6 s	1.05 s
Tracer range	200 m	220 m

Status

In production. In service with the armies of Australia, Belgium, Canada, Denmark, Finland, Netherlands, New Zealand, Norway, Portugal, Turkey, and the United Kingdom.

Manufacturers

Raufoss A/S, Raufoss.
(For details see Directory of Suppliers)

POLAND

Ośrodek 23 ALW Sub-calibre Device for 2A46 Tank Gun

The 23 ALW-125 is a sub-calibre training device intended to be fitted into the breech and bore of the 125 mm 2A46 tank gun fitted to T-72 tanks to provide low-cost firing training. Using standard 23 × 152B mm ammunition, the device is issued as a kit which is assembled and inserted into the breech of the 125 mm gun once the breech-block has been removed. Once installed, the device utilises a 10-round magazine which feeds five rounds into a loading device under the control of the gun's automatic ammunition feed system; the system requires a 24 V electrical power supply for loading, elevation and traverse, as with the full calibre gun system.

The complete kit weighs 40 kg and is supplied complete with installation tools and spares in a canvas roll. The device has a cyclic rate of fire of 10 rds/min although the practical rate of fire is 5 rds/min. It can be used over a range of from 100 to 1100 m; maximum range is 2000 m. Muzzle velocity of the 23 mm ammunition is 970 to 1000 m/s.

Status

In production.

Manufacturer

Ośrodek Badawczo-Rozwojowy Sprzętu Mechanicznego, Tarnow.
(For details see Directory of Suppliers)

PORTUGAL

INDEP A2 Sub-calibre Mortar Trainer

The INDEP A2 sub-calibre mortar trainer was developed to be adapted for use with the 60, 81, 107 and 120 mm mortars used by the Portuguese armed forces. It provides mortar crew training and the carrying out of all fire direction control operations by forward observers and observation posts.

The trainer is a pneumatically operated attachment designed to adapt a mortar to fire a 25 mm training projectile. All ranges for each type of mortar can be simulated in a limited space, such as a sports field—the normal maximum simulated ranges are 50, 25 or 12.5 m. The 25 mm training projectile has a chamber in the nose to receive a 0.22 blank cartridge which detonates on impact to produce an impact signature.

The trainer comprises: a compressed air bottle containing enough air for approximately 1000 firings; two pressure gauges coupled by a regulator valve; a hose to connect the regulator to the barrel and ejection valve assembly; the barrel and ejector valve assembly composed of an air tube and a pressure chamber; three inner adaptors (for 81, 107 and 120 mm mortars respectively—for 60 mm mortars the outside diameter of the pressure chamber corresponds to the inside diameter of the barrel); four outer adaptors for the muzzles of 60, 81, 107 and 120 mm mortars respectively; twenty 25 mm training projectiles (which can be used repeatedly); a set of firing tables; a technical manual; and accessories and tools.

To prepare the trainer for use with a mortar the regulator valve is connected to the compressed air bottle valve and the hose is used to connect this valve to the barrel and ejection valve assembly. An inner adaptor is then installed, followed by an outer adaptor. The barrel and valve assembly is then inserted into the mortar barrel and secured at the top using thumbscrews.

For firing, trainee forward observers are provided with binoculars. The observation posts are provided with firing tables to convert real ranges into simulated ranges. After a fire-control plan has been calculated it is then only necessary to know the corrections given by the forward observers to estimate alterations using the firing tables. After knowing the range settings for the mortar, transmitted by the observation post, the gun crew makes the necessary alterations in azimuth and elevation. After this the commander can set the air pressure for the 'charge' as indicated by the observation post. The required pressure is fed into the trainer and retained inside a pressure chamber until the training projectile is loaded into the barrel. As the projectile strikes the bottom of the barrel a disc valve opens to propel the projectile.

Status

In production. In service with the Portuguese armed forces.

Manufacturer

INDEP - Indùstrias e Participações de Defesa SA.
(For details see Directory of Suppliers)

INDEP A2 sub-calibre mortar trainer fitted to 81 mm mortar

INDEP Sub-calibre Adaptor for 106 mm Recoilless Gun

This sub-calibre adaptor permits the training of 106 mm recoilless gun crews under simulated combat conditions by firing standard 7.62 mm NATO tracer rounds as opposed to full-scale 106 mm rounds.

The adaptor resembles the cartridge case of a 106 mm round and uses a centrally mounted 7.62 mm barrel provided with 40 × 4 mm holes drilled into it to reduce the muzzle velocity of the rifle projectile involved to 600 m/s. The projectile then has a 700 m range with a maximum deflection of about 0.5 m. The adaptor is aligned with the normal bore of the gun within the recoilless gun barrel and using elevating and traversing pins—a laser zeroing device is used for this operation. The device is retained within the barrel as the breech-block is opened and closed—the 7.62 mm cartridge is loaded and extracted manually. Both 7.62 mm NATO ball and tracer cartridges can be used.

The adaptor weighs 8.5 kg and is 665 mm long.

Status

In production. In service with the Portuguese armed forces.

Manufacturer

INDEP - Indùstrias e Participações de Defesa SA.
(For details see Directory of Suppliers)

INDEP sub-calibre adaptor for 106 mm recoilless gun

RUSSIA

2KH51 Turret Crew Trainer for 2C19 Howitzer

The 2KH51 Turret Crew Trainer for the 152 mm 2C19 self-propelled howitzer is intended for the crew training of all turret-located personnel. The full trainer consists of a replica 2C19 howitzer turret mounted on a fixed stand, an instructor's panel, a power supply complete with wiring harnesses, and inert projectiles and propellant cases.

The replica turret is formed over a light framework covered by a thick fabric. The turret interior is fully replicated, complete with all instruments and controls, and all functional components are either installed using original parts or close copies. Trainees can thus become fully familiarised with the layout of the 2C19 turret interior. The turret can be traversed through 360° and the dummy gun barrel and breech can be elevated over the full elevation range of from −3° to +68°.

To add further realism to the turret movement the full range of gun-laying drills and actions can be carried out while under instructor control. The gun-layer's sight views a drum placed over the replica turret, inside which a full photographic panorama can be observed. Standard fire-control orders can be transmitted to the trainee turret crew and results can be checked from the instructor panel. The same panel can be used to introduce equipment and control system malfunctions and for monitoring and recording trainee reactions and results.

Ammunition handling and loading can also be carried out using the 2KH51 trainer, including the actions of the two crew members who normally stand outside the turret to feed ammunition into the turret. Up to six or seven rounds can be 'fired' in a minute using the trainer's ammunition system. The inert projectiles and propellant cases have a service life of at least 10 000 loading cycles.

The complete 2KH51 trainer weighs 8890 kg, with the replica turret alone weighing 6934 kg. The trainer can be used continually for up to 8 hours at any one time, after which an interval of at least 1 hour is required for battery charging and other maintenance.

Status

In service with the Russian Army.

Manufacturer

State factories.

RPG-7 Sub-calibre Training Device

This device is a sub-calibre barrel insert which fires the standard 7.62 × 39 mm tracer bullet. The insert is made to represent the normal PG7-M HEAT rocket and is inserted into the launcher so as to resemble a loaded RPG-7 launcher in appearance, balance and weight.

The sub-calibre barrel is fitted with a simple bolt-action breech which has a folding handle. The breech is opened and a cartridge is loaded into the chamber when the insert is outside the launcher. The bolt handle is folded down into a recess and the complete loaded unit is then inserted into the muzzle of the launcher so that the striker release is automatically aligned with the launcher trigger mechanism. After firing, the unit is removed from the launcher, the empty case extracted, a fresh round loaded, and the unit replaced in the launcher.

The sub-calibre barrel is centred in the nose-cone of the device by adjusting screws; this permits the barrel to be moved as necessary to zero the strike of the bullet so that it simulates the trajectory of the rocket at the selected training range. The nose-cone can be removed to permit this adjustment.

Status

In service with the Russian and former Warsaw Pact armed forces.

RPG-7 launcher with PG7-M rocket and sub-calibre training device below

Nose-cone of the RPG-7 sub-calibre device removed, showing the method of adjusting barrel alignment

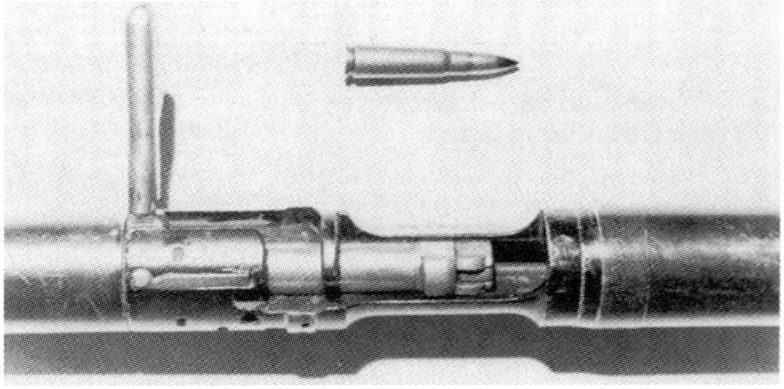

Breech of the RPG-7 sub-calibre device, with the bolt handle in operating position and breech open. The striker release is below the bolt handle

SPAIN

ECIA 25 mm Sub-calibre Mortar Training Round

In order to facilitate mortar training, and particularly to reduce the need for extensive ranges and danger areas, the firm of Esperanza y Cia (ECIA) produced the ECIA 25 mm sub-calibre training round. This round is fired from a full-sized mortar using a special bomb fitted with a sub-calibre barrel. The special bomb is loaded in the usual way and all the usual drills have to be carried out by the crew. When the bomb drops to the bottom of the barrel it fires the propelling charge in the tail of the sub-calibre round and is fired up the barrel in the carrier bomb. The sub-calibre round leaves the mortar on a trajectory identical to that followed by a full-size bomb fired at the same elevation. Due to the low power of the charge the sub-calibre round does not travel more than a few hundred metres—the maximum range is 820 m (minimum range 100 m).

The carrier bomb is also expelled by the same propelling charge, but it does not travel far from the muzzle and lands about 10 m in front of the mortar.

The sub-calibre round is a small fin-stabilised projectile fitted with a bright tracer so its ballistic path can be followed by eye throughout the trajectory. On striking the ground a small signalling charge detonates and a cloud of smoke is ejected, clearly marking the point of impact.

Range alterations are made by changes in elevation, and only one propelling charge is needed.

The manufacturer claims the following advantages over other similar systems: simplicity in range alteration due to the one charge; better visibility of the projectile trajectory due to the bright tracer; a clear indication of the point of impact due to the combined bang and smoke; and simplicity of operation using the projectile. There is no fuze since the terminal effects are initiated by the tracer burning through to the composition; safety is thus assured since there are no splinters.

The system is produced for 60, 81 and 120 mm mortars.

Status
In production.

Manufacturer
Esperanza y Cia SA (ECIA), Vizcaya.
(For details see Directory of Suppliers)

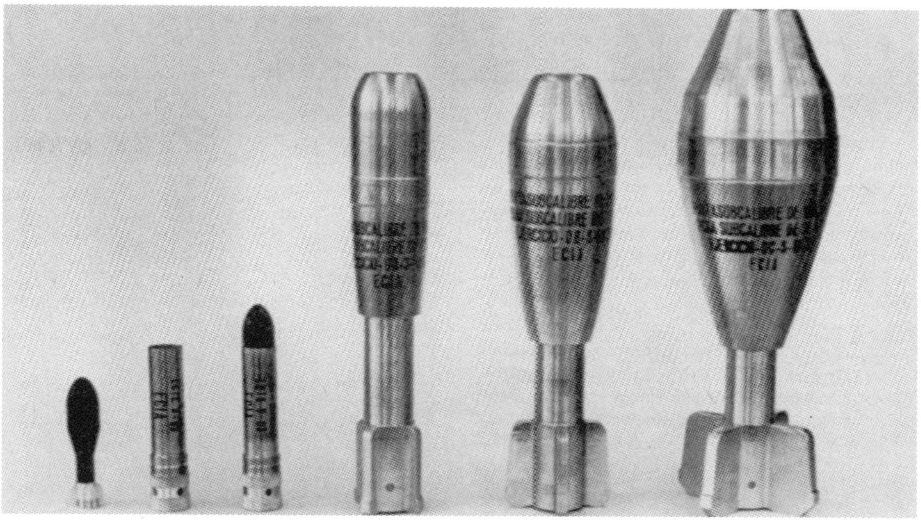

ECIA sub-calibre mortar training rounds with sub-projectile components on left

SWEDEN

Weibull Mortar Firing Simulator with Short-Range Ammunition

The Weibull mortar firing simulator with short-range ammunition is intended for the realistic training of mortar fire-control officers, fire-control sections, forward observers and mortar crews, all with a minimum of environmental disturbance.

In use the normal mortar barrel is replaced on the usual carriage by a simulated mortar barrel. Special mortar ammunition consisting of a dummy projectile and a short-range projectile is used. The short-range ammunition is loaded into the dummy projectile ready for loading into the simulated mortar barrel in the normal way. The short-range projectile is launched using the mortar's own ignition system acting on a primary charge on the short-range projectile so that the short-range projectile travels from the dummy projectile to a range of between 60 and 700 m, according to the elevation angle and loading. As the steel short-range projectile hits the ground it uses a nose fuze and pyrotechnic pellet to create smoke and flash for spotting purposes. Range measurements can be made using a laser rangefinder. After firing, the dummy projectile is removed from the simulated barrel via a hinged hatch and can then be reloaded with a new short-range projectile for further use.

If required, the dummy projectile and the short-range projectile can be fired from standard barrels but the dummy projectile will then have to be removed after firing using a special tool.

Status
In production.

Manufacturer
J L Weibull AB, Åkarp.
(For details see Directory of Suppliers)

Weibull Training Weapons

J L Weibull AB has produced a number of training weapons since 1953 when it produced 20 mm and 40 mm gun trainers, amongst others. Most training weapons produced concentrate on one aspect or another of weapon operation, like loading, and many bear only a passing visual resemblance to the real weapons involved. The training weapons outlined below are only a small selection from the range produced.

Loading Trainer for 105 mm Howitzer m/39 and m/40
Development of this trainer began in 1961 with the first deliveries being made to the Swedish Army the following year. Training ammunition with the same handling characteristics as live ammunition is used. When 'fired' the simulator recoils and the projectile drops out in front of the trainer. The crew then loads a new round, and the projectiles and cases are reassembled and used again. The main purpose of this simulator is for training in the fastest and safest way to fire and reload the howitzer, training in passing orders between the commander and the howitzer crew and the training of m/39 and m/40 crews to increase the firing rate of the howitzer. Optional equipment includes a sound simulator.

Loading Trainer for 105 mm Howitzer m/4140
This equipment follows the same general lines as the trainer for the 105 mm howitzer m/39 and m/40. The trainer for the 105 mm howitzer m/4140 was developed during 1962 and the first examples were issued for service in 1963. The practice rounds loaded into the breech are driven along the dummy barrel until they are ejected from the muzzle for reuse. An electric motor over the barrel drives the practice rounds through the trainer and also provides power for the breech mechanism.

120 mm Mortar Simulator
The 120 mm mortar simulator is a loading trainer intended to teach the drills for loading and serving the 120 mm mortar without firing the weapon. The difficulty in performing drills with muzzle-loading mortars is that bombs loaded by dropping down the tube need to be removed fairly frequently, thus interrupting drills with an unrealistic operation. In this simulator the dummy mortar consists of a barrel tube in three sections: the forward section which resembles the muzzle of a service mortar, the centre section which is hinged and can swivel away from alignment with the rest of the tube, and the bottom section which contains the firing mechanism and other mechanical devices for the operation of the trainer. The barrel unit rests on the service baseplate and is supported by the service bipod, so that the remainder of the mortar is of standard pattern and the correct drills can be carried out in laying and aiming.

When the simulator is in use, the soldier loads a dummy bomb into the muzzle. It drops, slowed by a damper which thus simulates the normal descent time, and comes to rest in the centre section of the tube. On operating the firing lever, the centre section is unlocked and, by means of a balance weight, the upper end swings forward and the rear end back. The rear end thus aligns with a bomb chute permanently attached to the lower section of the barrel; a trip is automatically released and the dummy bomb falls into the chute, from where it can be removed. The swinging section closes and the mortar is ready to be reloaded. It is possible for the instructor to lock the swinging centre section so that it does not function and thus simulates a misfire.

Status
All the above equipments are in service with the Swedish Army.

Manufacturer
J L Weibull AB, Åkarp.
(For details see Directory of Suppliers)

Loading trainer for 105 mm m/39 and m/40

Loading trainer for 105 mm m/4140

UNITED KINGDOM

Ensign Cadet Rifle

The Ensign Cadet rifle was designed as a training weapon for use by cadet and similar organisations and is calibered for conventional 5.56 mm ammunition, although a 0.22 adaptor kit is available. A low power 5.56 mm training round is also available.

The Ensign Cadet rifle has many components in common with the Enfield Weapon System Individual Weapon (IW), used by the British armed forces as the L85A1. However, the Ensign Cadet is a hand-operated single-shot weapon and uses a 10-round box magazine, although the 20- and 30-round magazines used with the IW are compatible. The Ensign Cadet uses only 'iron' sights but they are adjustable. Recoil energy is 4.5 J.

Muzzle velocity is 900 m/s and the weapon weighs 4.1 kg complete with sights and a loaded magazine. The barrel is 495 mm long and overall length is 755 mm.

Status
Available.

Manufacturer
Royal Ordnance Guns & Vehicles Division, Nottingham. (For details see Directory of Suppliers)

The 5.56 mm Ensign Cadet rifle

UNITED STATES OF AMERICA

Combat Engineer Vehicle Sub-caliber Training Device

In order to reduce Combat Engineer Vehicle gunner training costs the US Army Engineer School at Fort Belvoir, Virginia, supervised the development of the Combat Engineer Vehicle Sub-caliber Training Device. The reloadable device resembles a conventional round of ammunition for the M135 165 mm demolition gun fitted to the M728 Combat Engineer Vehicle (CEV). The device is handled, loaded and fired in the same manner as a conventional 165 mm round but contains a breech mechanism and barrel capable of firing 40 mm grenades, the same grenades fired from the MK-19 automatic grenade launcher. These 40 mm grenades have a maximum effective range of 1600 m but are more likely to be used at shorter ranges for the sub-calibre training role.

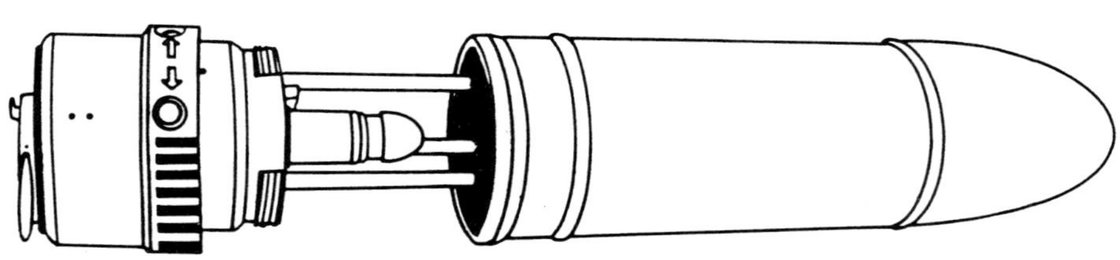

Combat Engineer Vehicle Sub-caliber Training Device

Status
In service.

M261 Conversion Kit

The 0.22 conversion unit that was to become the M261 was developed for the US Armament Corporation. Following standardisation, M261 production was carried out by the Saco Defense Systems Division of the Maremont Corporation of Saco, Maine. Over 63 000 M261 conversion units were produced under a contract worth $4.9 million.

The M261 conversion kit provides the M16, M16A1 and M16A2 5.56 mm rifles with the ability to fire 0.22 long-range ammunition. The rifle can be converted in less than one minute without the need for any tools. The outward appearance of the rifle and its handling and characteristics are unchanged, but by firing 0.22 LR

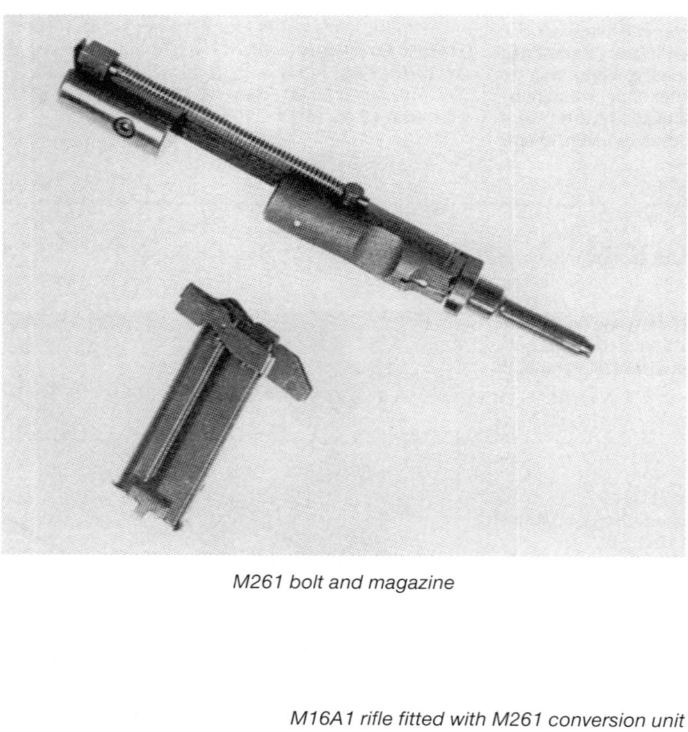

M261 bolt and magazine

M16A1 rifle fitted with M261 conversion unit

ammunition it can be used safely indoors and on ranges which are limited to low power rounds, while still providing realistic training.

Ammunition: 0.22 LR
Operation: blowback
Mode of fire: semi-automatic
Magazine: 10-round special box

Status
In service with the US armed forces and some other armies.

Red Gun Training Weapons

Armament Systems and Procedures produces a range of inert synthetic training guns. The weapons are dimensionally identical to actual firearms to allow a precise fit in holsters for weapon retention purposes. The entire surface of each weapon is coloured red to identify the units for training purposes. The overall surface texture, smooth edges and the light, non-metallic configuration prevent possible injuries during training drills.

Red Gun training weapons currently available include the following: Beretta 92; Glock 9 mm, 10 mm, 0.45; SIG P226; Smith & Wesson 9 mm, 0.40, 10 mm, 0.45, and K-frame revolver; Remington 870 shotgun; Heckler & Koch MP5; and the M16 rifle. Other weapons can be produced on request.

Status
In production.

Manufacturer
Armament Systems and Procedures Inc, Appleton.
(For details see Directory of Suppliers)

Examples of 'long' Red Gun training weapons: (top) Remington 870 shotgun, and (bottom) Heckler & Koch MP5 sub-machine gun

Part of the range of Red Gun training pistols

Tank Precision Gunnery Inbore Device (TPGID)

A development requirement has been issued for a sub-calibre device that can be fitted into the bore of the M256 120 mm main gun used on the M1A1 Abrams main battle tank. The device, known as the Tank Precision Gunnery Inbore Device (TPGID), will have to simulate the performance of service 120 mm projectiles. The requirement was issued following an increase in tank ammunition costs, limited range facilities, a general lack of associated resources and the need to reduce the effects of gun firing noise. The requirement has been requested by the US Army Armor School at Fort Knox, Kentucky.

In 1989 19 examples of the Mauser-Wegmann 35 mm sub-calibre training device (for details see entry in this section) were procured by US Army, Europe, as an off-the-shelf purchase for use in the 120 mm main gun of the M1A1 Abrams MBT.

Status
Development.

TRAINING AMMUNITION AND PYROTECHNICS

In this section the terms *training ammunition and pyrotechnics* will describe only those munitions produced specifically for the training role. It does not include ammunition types such as TP or TP-T that have the same range as their service equivalents but does include munitions that have no function other than employment for training purposes at reduced ranges.

AUSTRALIA

ADI Sound Effects Simulators

Combat Effects Simulator

The ADI Combat Effects Simulator provides night illumination, pyrotechnic gunfire and target hit indication when used with the ADI vehicle target system (qv).

A total of 48 pyrotechnic capsules is mounted in a frame with a night light system used to illuminate a target and produce evidence of a hit. Half the capsules (gunfire) simulate the firing of a tank's main armament. The other half are (normally) orange smoke to simulate a target hit. Simulated return fire from the main armament is initiated from the firing point when a target is raised; coloured smoke is initiated when the target is hit.

Battle Effects Simulator

This simulator uses an electronic sound synthesiser to simulate the noise signature of any weapon. The system is battery-powered and two-man portable. Within a carrying frame are incorporated the batteries, a strobe light to simulate muzzle flash and the electronic interface unit to a control unit. The simulator may be used as a stand-alone unit or as a peripheral to a target mechanism of any type.

Status

In production.

Manufacturer

Australian Defence Industries Limited, Bondi Junction.
For details see Directory of Suppliers)

AUSTRIA

Intertechnik SHORGUT Short-Range Disintegrating Training Projectile

The SHORGUT (Short-Range Gun Training) short-range disintegrating training projectiles are intended for use with guns and howitzers that have to fire in extremely limited safety areas. Applications could include live firing training from fortifications located in built-up areas, and functional or proof firing by maintenance personnel in limited areas.

The SHORGUT projectile consists of a shell of specially selected plastic which is filled with metal particles and closed at the top by a lid. When fired, the lid is detached from the shell and the particles are scattered about the nearby area. The plastic parts of the shell have a low cross-sectional density and high aerodynamic drag. After a short flight, which typically ranges from a maximum of 60 to 120 m, they fall to the ground. The metal parts have a maximum range of about 40 m. The large plastic pieces remaining can be gathered and recycled while the metal particles decompose into harmless metal oxides.

SHORGUT projectiles are in production for 105 mm L7 and M68 tank guns, 105 mm M2 and M101A1 light field howitzers, 106 mm M40 recoilless rifles, and 155 mm guns and howitzers. Other calibres, such as 90 mm, are available on request.

Status

In production—see text.

Manufacturer

Intertechnik, Linz.
(For details see Directory of Suppliers)

SHORGUT short-range disintegrating training projectiles

BELGIUM

MECAR Training Rifle Grenades

MECAR produces rifle grenades designed for use at all stages of training. The PRAC (HE) RFL-35 BTU M265 is the training grenade for the HE-RFL-35 BTU M262 service grenade, and the PRAC (ARP) RFL-40 BTU is the training grenade for the ARP RFL-40 BTU M260 service grenade. The training grenades are identical in weight, length and range to the service model but they contain no explosive components.

Status

In production.

Manufacturer

MECAR SA.
(For details see Directory of Suppliers)

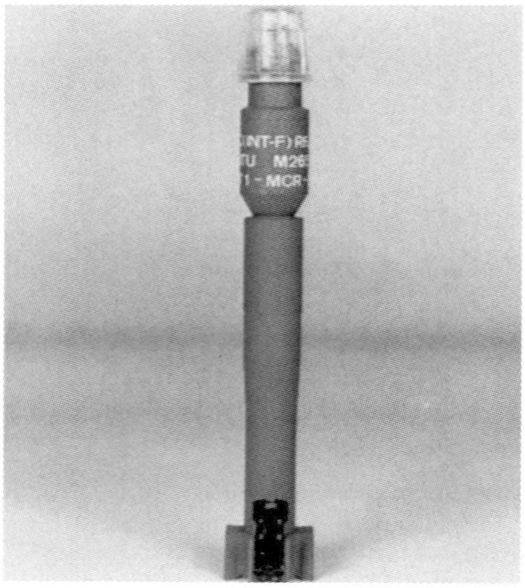

MECAR M265 training rifle grenade

CANADA

Simunition Cartridge 105 mm TP/FSDS-T C71

The C71 Target Practice round for Fin-Stabilised Discarding Sabot Tracer (TP/FSDS-T) is a tank target practice munition for use in 105 mm L7 and M68 tank guns. The shot consists of a sub-calibre tubular projectile with a discarding sabot pusher base, a driving band and obturator. After firing, the tubular projectile is a ballistic match for the APFSDS C76 projectile up to 3000 m but becomes aerodynamically unstable beyond that range and falls to the ground. The resultant safety template is 8 km along the line of fire and 2 km either side at a gun elevation of +10°.

Specifications

Cartridge length: 785 mm
Cartridge weight: 19 kg
Shot weight: 4 kg
Sub-projectile weight: 2.18 kg
Propellant type: M6 NH/P
Tracer: C3A3
Primer: C16
Cartridge case: C4
Muzzle velocity: 1590 m/s

Status

In production for the Canadian armed forces.

Manufacturer

Simunition, Québec.
(For details see Directory of Suppliers)

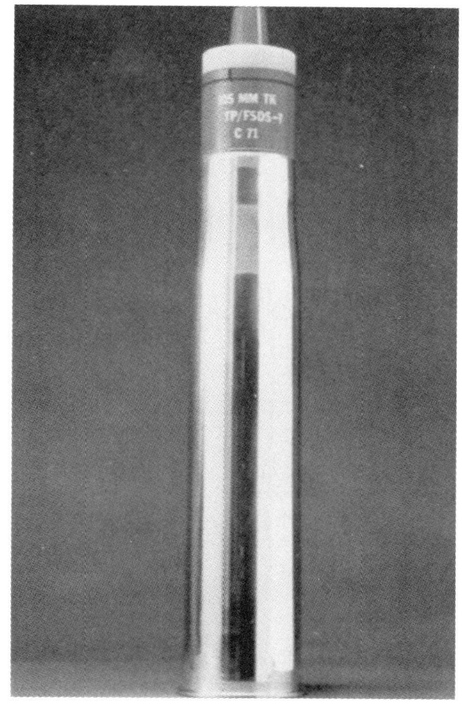

Cartridge 105 mm TP/FSDS-T C71 for L7 and M68 105 mm tank guns

Simunition Frangible Lead-free Training Ammunition

This lead-free training ammunition is available for 7.62 mm, 9 mm, 0.38 Special and 5.56 mm military weapons and was designed to match closely the ballistic and accuracy performances of conventional rounds at training ranges. They can be handled, loaded and fired in exactly the same manner as conventional rounds.

For the 7.62 mm Short Stop round the maximum effective range is 100 m, while the maximum range possible is 600 m as it utilises a non-contaminating, lead-free frangible projectile with finned nose. The 9 mm and 0.38 Special frangible rounds have a conventionally configured projectile with a maximum effective range of 250 m but maximum possible range of 1000 m. The figures for the 5.56 mm round are 250 m and 2000 m respectively. All 9 mm, 0.38 Special and 5.56 mm rounds use lead-free primers and are sold under Greenshield.

The projectiles fired by these cartridges will all break up on impact with steel plates as thin as 6.35 mm (mild steel for 0.38 Special, 9 mm and 7.62 mm, armoured steel for 5.56 mm), even at ranges as close as 10 m. All projectiles are made from a copper-polymer compound.

Status

In production.

Manufacturer

Simunition, Québec.
(For details see Directory of Suppliers)

Simunition frangible training rounds

Simunition FX Red Marking Cartridges

The FX Special Effects system was developed for close-quarter combat training in collaboration with police and special forces. The system utilises plastic projectiles that can be fired from normal service weapons, after modifications, and enables live firing to be carried out at close ranges in safety. Hits are denoted by a detergent-based inert fluorescent red marker which can be removed by washing with water. FX cartridges are available for 0.38/0.357 revolver and 9 mm pistols and sub-machine guns.

FX cartridges can be fired from the user's own weapon after the installation of a simple conversion kit that reduces the possibility of accidental discharge of live ammunition. The conversion kit involves a user-installed barrel for automatic pistols and a bolt for sub-machine guns and rifles. No modifications are required for revolvers.

FX rounds for 0.38/0.357 revolvers use an aluminium cartridge case and a full calibre projectile. The 9 mm cartridges use attenuated brass cases and hold their projectiles in a plastic sabot. Both cartridges have a muzzle velocity of 125 m/s and the maximum effective range for training is 15 m.

These cartridges have to be stored under cool and dry conditions and have a shelf life of one year.

FX is a registered trade mark.

Status

In production.

Manufacturer

Simunition, Québec.
(For details see Directory of Suppliers)

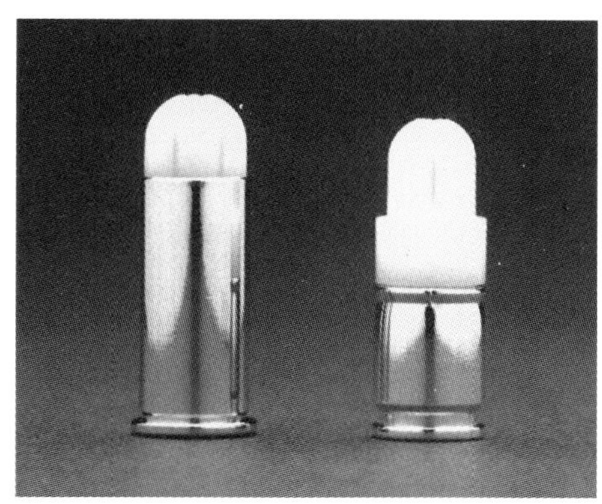

FX red marking cartridges; from left, 0.38/0.357 revolver and 9 mm

CHINA, PEOPLE'S REPUBLIC

Tactical Sound Effects Simulator

By means of discretion processing of a continuous signal, this simulator can create all manner of battlefield sounds which are first sampled, quantified and encoded in an EPROM memory. The sounds can be accessed and combined to produce a realistic training environment during exercises using either manual control, automatic random processing or computer program control.

Two types of sound unit are available, fixed or portable. High power hi-fi amplifiers can be used with this system and the power output can be altered to suit user requirements.

Status
In production. In service with the People's Liberation Army.

Developing Agency
Nanjing Research Institute on Simulation Technique (NRIST).
(For details see Directory of Suppliers)

Tactical sound effects simulator

FRANCE

Etienne Lacroix Plastic Practice Grenade Mle F1

The Mle F1 plastic practice grenade consists of three components: a blue plastic body filled with chalk dust or talc; a plastic sheath containing the explosive charge; and an igniter set. In operation the safety pin is extracted as the operating lever is held against the body of the grenade. The grenade is then thrown and after a 5 second delay the charge explodes with a loud detonation, and causes the grenade body to burst and scatter the contents. No fragments are produced.

This grenade can be used with any type of igniter, including the British L25.

The PC1 practice grenade produced by Royal Ordnance in the United Kingdom is very similar to the Mle F1 and was developed in association with Etienne Lacroix SA.

Status
In production. In service with the French Army.

Manufacturer
Etienne Lacroix SA, Muret.
(For details see Directory of Suppliers)

Plastic practice grenades Mle F1 fitted with various igniters

Ruggieri Pyrotechnic Devices

Pétard Mle F1

This device, full title *Pétard Simulateur de Feu d'Infanterie Modèle F1,* is used to simulate the firing or the projectile explosion of an anti-armour weapon, a mortar or any other similar piece of equipment. It consists of a cardboard cylinder with plastic end caps. The body contains a friction igniter, a 4-5 second fuze, a propellant charge and the actual explosive charge itself. In use the propellant blows the main charge out of the tube after a 4-5 second delay and it explodes in the open about 1.5 m away from the tube, accompanied by either a yellow or white cloud of smoke. The French Army uses the white smoke to show the firing of a weapon and the yellow to designate the arrival of a projectile.

Pétard Mle F1

Specifications
Length: 257 mm
Diameter: 35 mm
Weight: 110 g
Delay: 4-5 s

Salvo Simulator Mle F1

Salvo Simulator Mle F1

This simulator, made up from four charges in one pack, is fired electrically. Each charge is contained in a spherical container with its own fuze, igniter and wire. A set of four charges is strung up between supports about 2 m apart and connected to a central firing box. When fired, the four explode at short intervals to simulate the arrival of a salvo of artillery or mortar rounds. The delay between detonations can be 0, 2, 4 or 6 seconds, with the various delays denoted by the colour of the supports between the charges. A safety distance of 30 m has to be maintained when this simulator is detonated.

The full designation of this simulator is *Artifice de Simulation de Salve Percutante Modèle F1.*

Specifications
Diameter of one charge: 80 mm
Length of pack of four charges: 600 mm
Weight of four charges in pack: 1.5 kg

Atomic Explosion Simulator F1

This simulator is a large cylindrical container which is buried 0.5 m deep into the ground and initiated electrically using a supply of at least 60 V. When fired, it gives an intense blaze of light, a violent explosion and 60 seconds of smoke which rises to about 100 m above the ground, forming the familiar mushroom-shaped cloud. A smokescreen remains at the point of burst.

The full designation of this simulator is *Artifice Simulateur d'Explosion Atomique Modèle F1*.

Specifications
Height: 650 mm
Diameter: 426 mm
Weight: approx 53 kg
Length of ignition wire: 100 m

Status
All the above are in production. In service with the French Army.

Manufacturer
Ruggieri, Paris.
(For details see Directory of Suppliers)

Complete atomic explosion simulator F1

Ruggieri Tank Combat Simulator Mle F1 Type 1600/419

The tank combat simulator Mle F1 is a series of pyrotechnic devices intended to provide tank crews with some measure of tank gun firing experience relatively cheaply. The system has a pyrotechnic firing device, the Type 1600, and a further device fitted to the target, the Type 1601. Both fire smoke cartridges to simulate tank gun fire and the subsequent hit.

The Type 1600 is a 10-cartridge projector mounted above the firing tank gun barrel. It can be fitted to AMX-13 or AMX-30 tanks, and once fitted can be raised to 30° above the barrel elevation. A deflector over the gun barrel protects the barrel from the firing flash. When fired, each cartridge emits a report equivalent to that produced by firing a 105 mm gun and a cloud of brown smoke is emitted. The control box for the projector is carried inside the tank turret. The cartridges used are the *Cartouches Simulatrices de feu de Canon Mle F1*. They have a plastic body, light alloy base and are fired electrically.

The Type 1601 is fitted to the side of the target tank turret and is electrically controlled. The device can be set off by the tank crew or by an exercise umpire from outside the tank and, when fired, the Type 1601 fires a smoke pot that produces red or orange coloured smoke. The smoke pot involved is the *Cartouche Simulatrice de coup au But Type 138*. It has a light alloy body and is fired electrically.

Status
In production. In service with the French Army.

Manufacturer
Ruggieri, Paris.
(For details see Directory of Suppliers)

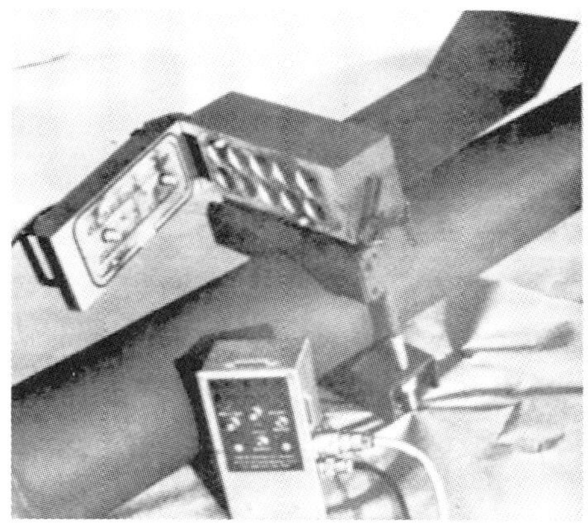

Tank combat simulator Mle F1 fitted above gun barrel with control box below

GERMANY

Comet Pyrotechnics

Comet Thunderflash I and III
These are the usual type of pyrotechnic grenade or explosion simulators for use in training. Manufactured from crush-resistant cardboard, there is no danger from flying fragments upon detonation. The two devices are similar in operation but the Thunderflash III is longer and provides a louder report.

Specifications
Data for Thunderflash I
Dimensions: 155 × 25 mm
Weight: 45 g
Ignition: friction/pull-wire igniter
Delay: 5 s minimum
Report: 165 dB at 3 m

Grenade Simulator
The Comet grenade simulator represents a standard hand grenade. It provides optical and acoustic effects during infantry training.

Specifications
Dimensions: 235 × 53 mm
Weight: 50 g
Ignition: friction igniter
Delay: 5 s
Report: 170 dB at 3 m

Impact Simulator
This pyrotechnic device, which resembles a stick grenade, is designed to give absolute safety to the user. Even if the simulator explodes while still being held in the hand of the operator, it will not cause injury. It is generally used during exercises to simulate grenades or other battlefield explosions.

Specifications
Dimensions: 165 × 40 mm (styropor head)
Weight: 45 g
Ignition: friction pull igniter
Delay: 4 s
Report: 131 dB at a distance of 3 m

Comet Hit Simulator
This is a pyrotechnic device which gives off a red flare and orange smoke visible up to 2000 m. It is for use in conjunction with various types of tank and gunfire simulators, such as those attached to targets on gunnery ranges or to provide simulation of hits by APDS and other anti-armour projectiles. It can also be manually fired on instructions from an umpire.

Comet has also developed Hit Simulators which indicate the specific signatures of APDS, TPDS and HEAT rounds. Using these simulators a tank kill could be indicated by a two minute cloud of black smoke.

A Cannon Fire Simulator, with flash, bang and smoke, audible and visible to 2000 m is also available.

Specifications
Dimensions: 150 × 50 mm
Weight: 150 g
Ignition: electric, 3 mW/s/Ohm
Colour of flare: red
Light intensity: 80 000 candela
Burning time: 1.5 s
Colour of smoke: orange
Smoke duration: 8 s

Comet Thunderflash I Nr.2209

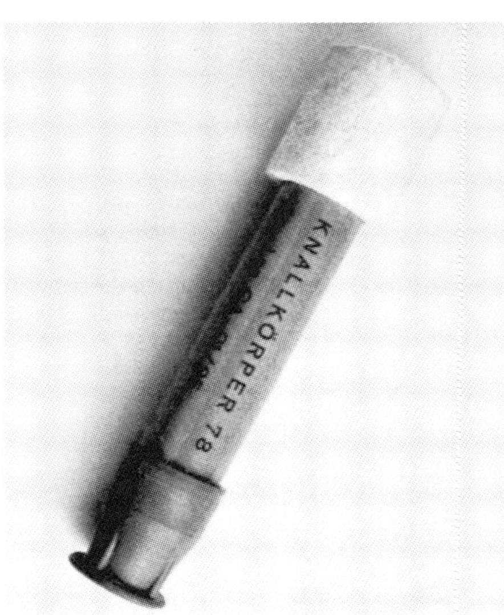

Comet Thunderflash III Nr.2215

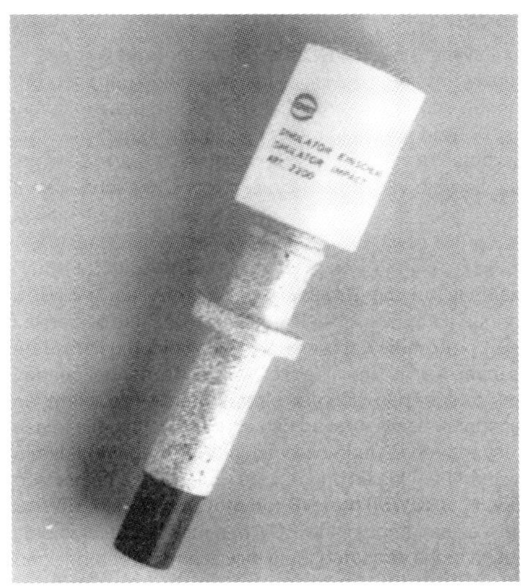

Comet impact simulator Nr.2200

Artillery fire Simulator

This provides audible and visual simulation of artillery fire with flash, smoke and report audible at 5000 m. It is also available as an artillery impact simulator.

Specifications

Dimensions: 280 × 76 mm
Weight: 750 g
Ignition: electric, 3 mW/s/Ohm
Report: 175 dB at 3 m

Status

All the above items are in production.

Manufacturer

Comet GmbH Pyrotechnik Apparatebau, Bremerhaven.
(For details see Directory of Suppliers)

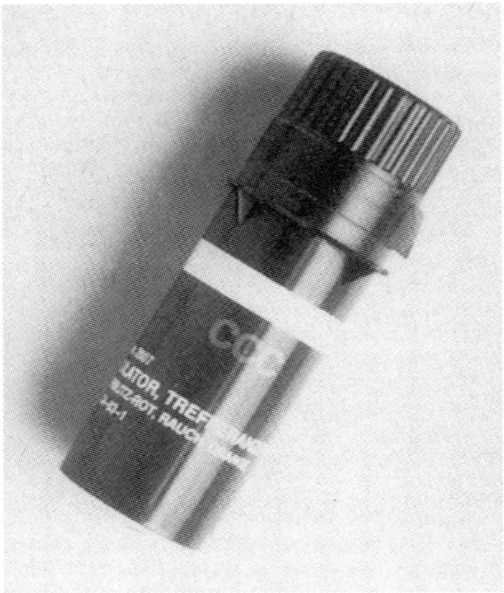

Comet hit simulator Nr.2007

Comet artillery fire simulator Nr.2040

DIEHL DARKAS 9/17 Tank Gunfire Simulator

The DIEHL DARKAS *(Darstellungsgerät Kanonenabschuss)* was developed by Hoffmann-Werke GmbH and may be used as a stand-alone tank gunfire simulator to reproduce the flash and report of a tank main gun being fired or a tank being hit by enemy weapons. It may also be used in conjunction with laser-based tactical engagement simulators.

The DARKAS 17 simulator holds 17 rounds of simulation ammunition (a 9-barrel arrangement, the DARKAS 9, is also available) in a twin horizontal layer magazine arrangement mounted directly on a tank turret or attached to a gun barrel with a calibre of 76 to 105 mm. The equipment can be fitted to any tank using special adaptors and safety clamps. Installation is simple, requiring no special tools, and can be carried out by the tank crew. Special attention has been given to safety when using the simulator to the extent that all electrical circuits, including the detonation circuit, are automatically broken if any faulty operating or handling procedures are carried out on critical components. Loading is only possible after a safety guard locked by two independent bolts and electrical circuits has been unfolded. Automatic self-test equipment is incorporated in the simulator control unit.

After loading, the safety catch can be released and the detonation cable plug on the ammunition rounds is inserted in the socket provided on the ammunition tray. Only when the safety guard has been locked can the simulator be fired. The cradle is connected by a cable to the control unit located inside the tank. As an additional indication that a hit has been scored (apart from the detonation of hit-simulating ammunition), a flashing lamp can also be connected.

DIEHL DARKAS 9 tank gunfire simulator fitted to Leopard 1 MBT

The control unit is fixed to its mounting inside the tank by a screw clamp. It is connected by a cable to the gun triggering circuit, the ammunition tray and the tank's main power supply. An interface is provided in the control unit for connecting laser-type combat simulators or sensors.

In use the simulator is fired using the normal gun fire-controls and reproduces the same flash and report as produced when firing live ammunition. The report and flash are visible over a combat range of up to 2000 m. The DARKAS 17 simulator capacity of 17 rounds is the same as that of the normal ready-use ammunition capacity of most tanks and the simulator cannot be fired again until a lapse of 6 seconds has occurred since the last firing, this interval being considered the minimum main gun reloading time.

The simulator can be used in two basic modes. The 'Without Hit' mode involves the loading of 17 simulation firing rounds (9 with DARKAS 9). The 'With Hit' mode involves the loading of 16 simulation firing rounds and one hit simulation round (8 and 1 for DARKAS 9). If the exercise umpire presses the control unit hit button or the laser beam of a combat simulator contacts the tank, the hit simulation round will detonate to produce a bright red flash, a report and a cloud of smoke.

The simulator ammunition is produced by Comet GmbH, a member of the DIEHL group.

Status

Several thousand systems are in service with the Belgian, Danish, German (1700), Italian, Norwegian and US armed forces.

Manufacturer

DIEHL GmbH & Co, Nürnberg 30.
(For details see Directory of Suppliers)

DIEHL DARKAS 60 Gunfire Simulator

The DIEHL DARKAS 60 *(Darstellungsgerät Kanonenabschuss)* was developed to simulate single or continuous firing of automatic weapons such as the MK-20 and MK-30. It may be used as a stand-alone tank gunfire simulator to reproduce the flash and report of a combat vehicle main gun being fired or a combat vehicle being hit by enemy weapons. It may be used in conjunction with laser-based tactical engagement simulators.

The DIEHL DARKAS 60 simulator holds 60 rounds of ammunition in a magazine arrangement mounted directly on a turret (another arrangement with less than 60 rounds is also possible). The equipment can be fitted to any vehicle using special adaptors. Installation is simple, requiring no special tools, and can be carried out by the tank crew. Special attention has been given to safety when using the simulator to the extent that all electrical circuits, including the detonation circuit, are automatically broken if any faulty operating or handling procedures are carried out on critical components.

When loading the magazine the ignition circuit is interrupted by a sensor as the magazine is opened. After loading and closing the magazine, the safety catch can be released and the detonation plug is inserted in a socket on the ammunition tray. Only when the ammunition arrangement has been completely locked can the simulator be fired. The cradle is connected by a cable to the control unit located inside the vehicle. As an additional indication that a hit has been scored on the carrier vehicle a flashing lamp can be included.

The control unit is fixed to a mounting inside the carrier vehicle by a screw clamp. It is connected by a cable to the gun triggering circuit, the ammunition tray and the vehicle's main power supply. An interface is provided in the control unit for connecting laser-type combat simulators or sensors.

In use the simulator is fired using the normal fire controls and produces the same report as when using live ammunition, producing little flash or smoke. For continuous firing the firing rate is limited to 5 rounds in 0.5 second; single rounds are also possible.

The simulator can be used in two modes. The 'Without Hit' mode involves the loading of 60 simulation firing rounds. The 'With Hit' mode involves the loading of 60 simulation firing rounds and one hit simulator round.

If an exercise umpire presses the control unit hit button or the laser beam of a combat simulator contacts the carrier vehicle, the hit simulator round will detonate to produce a bright red flash, a report and a cloud of smoke.

The simulator ammunition is produced by Comet GmbH, a member of the DIEHL group.

Status
Field testing with the German armed forces.

Manufacturer
DIEHL GmbH & Co, Nürnberg 30.
(For details see Directory of Suppliers)

DIEHL DARKAS 60 gunfire simulator fitted to Marder 1 A3 IFV

DIEHL Pop-up Target Gunfire Simulator (ASG)

The ASG *(Ausbildungsgerät Schuss/Getroffen)* was developed to fire pyrotechnics in connection with a pop-up tank target to demonstrate both the firing of a tank's main armament or a tank being hit. The ASG fires 7, 15, 20 or 30 seconds after raising the target or by remote-control from an instructor.

The ASG initiates a hit signal via the host target mechanism's hit sensor(s) or via a laser sensor on the target. Up to 10 ASGs can be connected in series. A safety guard on the unit cuts the detonation line when the unit is opened.

The ASG holds 10 rounds which may be 10 firing or 10 hit pyrotechnics or five of each. The rounds are inserted in 10 vertically arranged barrels with the angle of the barrels being 15° from the vertical.

For installation the connection cable to the target mechanism is installed. An ammunition select switch is positioned to either 10 firing, 10 hit or five of each and the delay time is adjusted via a separate switch. The safety guard is then closed, covering the detonator cable plug. As the guard is closed a 2-minute delay period is initiated to allow personnel to withdraw. During the last 30 seconds of the delay period a signal tone is emitted. The ASG is then ready to fire.

ASG pyrotechnics are produced by Comet GmbH, a member of the DIEHL group.

Status
Development complete. Accepted by the German Army.

Manufacturer
DIEHL GmbH & Co, Nürnberg.
(For details see Directory of Suppliers)

The DIEHL pop-up target gunfire simulator (ASG)

Dynamit Nobel KB Short-Range Ammunition

Dynamit Nobel KB short-range ammunition is a type of cartridge with a special bullet. Its weights and dimensions generally correspond to those of live ammunition and they can be handled in the same way. Without any changes to the weapon and without the use of any particular attachments, both single-shot and fully automatic fire can be carried out from the weapons concerned in the same way as live ammunition, irrespective of whether these weapons operate by gas pressure or the recoil principle.

The KB bullet consists of an axially drilled metal projectile body and a plastic nose cap. On firing, the plastic nose cap separates from the projectile body while still in the barrel. Immediately after leaving the barrel the plastic nose cap falls to the ground while the short-range bullet continues along its trajectory. Due to its flattened nose the metal body is strongly retarded in flight due to increased wind resistance, thus achieving a reduction in maximum range to about 50 per cent or less of the range of live ammunition. For example the normal range of a 7.62 mm bullet is reduced from 4000 m to around 1150 m. The normal range of a 25 × 137 mm projectile is reduced from 8000 m to approximately 3500 m.

KB projectiles can be fitted with tracer. They are available for 7.62 mm NATO, 12.7 × 99 mm, 20 × 139 mm and 25 × 137 mm.

Specifications

KB Cartridge	7.62 mm	12.7 mm	12.7 mm tracer	20 mm tracer	20 mm P-T	25 mm tracer
Bullet weight	5.55 g	38 g	40 g	120 g	122 g	180 g
Muzzle velocity	870 m/s	905 m/s	890 m/s	1040 m/s	1050 m/s	1100 m/s
Max training range	150 m	600 m	600 m	800 m	1200 m	1000 m
Max range	1150 m	3500 m	3500 m	3500 m	3700 m	3500 m

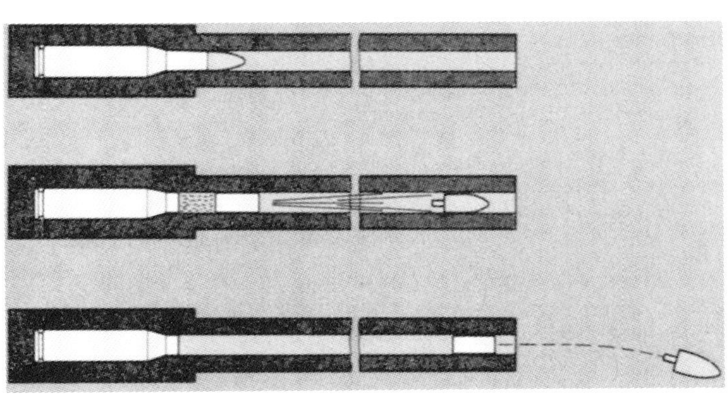

Principles of functioning of Dynamit Nobel KB short-range ammunition

Status
All the above are in production and are in service with the German armed forces and some NATO armed forces, including those of the USA.

Manufacturer
Dynamit Nobel AG, Troisdorf.
(For details see Directory of Suppliers)

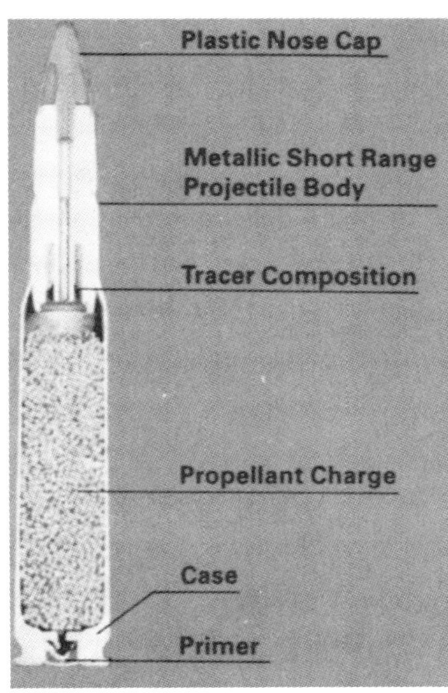

Main components of a Dynamit Nobel KB cartridge

Dynamit Nobel Training Ammunition

Dynamit Nobel PM Blank Ammunition

Dynamit Nobel manufactures a wide range of plastic manoeuvre (PM) ammunition for all calibres of small arms. The rounds are made in plastic and exactly simulate acoustically and functionally the equivalent live round. For automatic operation of the weapons a blank-firing muzzle attachment is necessary to reduce the gas flow out of the muzzle and build up sufficient internal pressure to drive the mechanism.

The PM cartridge has its plastic section made of injection-moulded polyethylene. This section includes the case and the bullet part—the 9 mm Parabellum cartridge has a metal case. The propellant charge is nickel/cadmium adjusted in quantity to suit the particular calibre in use and there is a metal base accommodating the propellant and primer cap. The primer cap uses SINOXID priming composition which is non-corrosive and non-erosive.

Specifications

Calibre	Cartridge weight	Muzzle safety distance
9 mm Parabellum	4.7 g	3 m
5.56 mm × 45	2.8 g	3 m
7.62 mm × 51 NATO	5.4 g	4 m
12.7 mm × 99	27.1 g	12 m

Dynamit Nobel PT Training Ammunition

Plastic training (PT) ammunition is intended for use at short range when there is no full-size range space available. It allows training in the use of the weapon and simulates exactly the functions of loading and firing and the sound and flash of using full-calibre rounds. However, the effective range is very short and the maximum range is little more than the length of a football pitch. The bullets are entirely plastic so that they lose velocity very quickly, will not ricochet and have an extremely short danger area. The ammunition can be used indoors and also for training in urban and street fighting exercises.

For automatic operation of these cartridges in the M16-series rifle and M2 HB 0.50/12.7 mm machine gun, special training breech bolts (M16) and recoil amplifying barrels (M2 HB) are available. The 5.56 mm and 0.50/12.7 mm cartridges have been type classified by the US Army as 5.56 mm Plastic Practice M862 and 0.50/12.7 mm Plastic Practice M858 Ball and M860 Tracer.

Heckler and Koch GmbH produces special versions of its P7M8 pistol and MP5 sub-machine gun to fire PT ammunition (refer to the entries in the *Sub-calibre Training Devices* section).

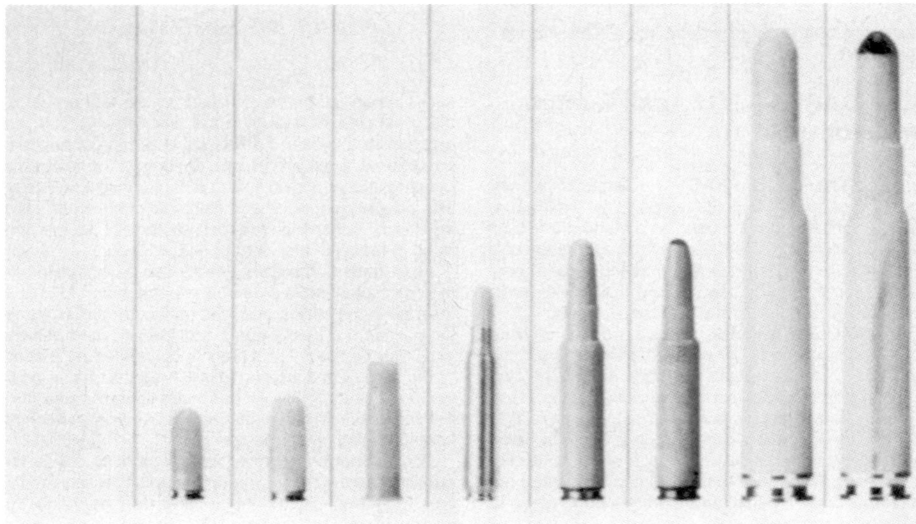

Varieties of Dynamit Nobel PT ammunition

Specifications

PT Cartridge	7.65 mm	9 mm Para	.38 Spec	5.56 mm	7.62 mm NATO PT	7.62 mm NATO PT-T	12.7 mm PT	12.7 mm PT-T
Bullet weight	0.22 g	0.42 g	0.49 g	0.2 g	0.7 g	1 g	3.2 g	3.2 g
Velocity	350 m/s	410 m/s	280 m/s	1000 m/s	1100 m/s	1100 m/s	1075 m/s	1045 m/s
Max training range	12 m	8-15 m	8-12 m	30 m	50 m	50 m	150 m	150 m
Max range	90 m	125 m	100 m	250 m	300 m	300 m	700 m	700 m

Status

All the above are in production and are in service with the German armed forces and some NATO armed forces.

Manufacturer

Dynamit Nobel AG, Troisdorf.
(For details see Directory of Suppliers)

Nico-Pyrotechnik Training Cartridges for Weapon Effect Signature Simulators

Nico-Pyrotechnik training cartridges for Weapon Effect Signature Simulators (WESS) have been tested under field conditions based on MIL-STD-810 D and feature a waterproof construction and primer initiation. On firing, the resultant flash and smoke are visible to the naked eye at 3000 m. For most applications the audio cue produced by the cartridge is limited to 140 dB at 20 m, although this can be modified for special requirements.

Status

In production.

Manufacturer

Nico-Pyrotechnik, Trittau.
(For details see Directory of Suppliers)

Nico-Pyrotechnik training cartridges for weapon effect signature simulators

Nico-Pyrotechnik Training Mine System

The Nico training mine system is a variable form of training device that can be adapted to suit virtually any type of operational land mine. In its simplest form it is a circular reusable mine body that can be fitted with an array of triggering devices such as pressure plates, tilt rods and trip-wire devices and when actuated produces a cloud of orange smoke. If required, the system can be incorporated into service mine bodies or the training mine body may be weighted to simulate a service mine. If required, a special body simulating an in-service mine can be supplied. Anti-tank and anti-personnel mines can be simulated.

Status
In production.

Manufacturer
Nico-Pyrotechnik, Trittau.
(For details see Directory of Suppliers)

Nico-Pyrotechnik training mine system in action

Piepenbrock Practice Demolition Charges

To provide training in firing demolition charges and correctly laying firing wires Piepenbrock has developed practice charges which are available in cubic and cylindrical form. Ignition is by all customary devices but without the blasting cap and the charges then produce a non-corrosive brown smoke which has a duration of some 3 seconds. The cubic charge, designated DM28 by the German Army and with the company Article Code 30001, is 40 × 50 × 70 mm and weighs about 53 g. The cylindrical charge, which has the company Article Code 30008, is 100 mm long, 30 mm in diameter and weighs 27 g.

Status
Both are in production.

Manufacturer
Piepenbrock Pyrotechnik GmbH, Göllheim.
(For details see Directory of Suppliers)

The Piepenbrock DM28 cubic practice demolition charge

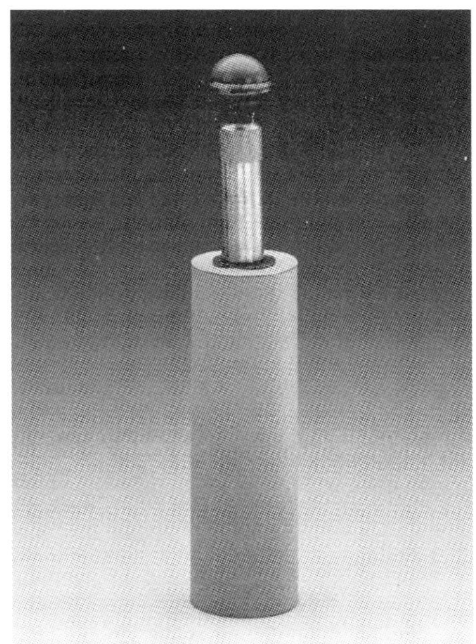

The Piepenbrock cylindrical practice demolition charge

Piepenbrock Simulator DM12

The Piepenbrock DM12 is a pyrotechnic simulator which represents a stick grenade and is operated in the same way, by a friction igniter fired by means of a pull-cord in the handle. It produces a loud report, a flash, and a grey smoke puff lasting approximately 3 seconds.

Specifications
Length: 240 mm
Diameter: 54 mm
Weight: 48 g
Delay time: 5 s nominal
Noise level: approx 175 dB

Status
In production. In service with the German Army.

Manufacturer
Piepenbrock Pyrotechnik GmbH, Göllheim.
(For details see Directory of Suppliers)

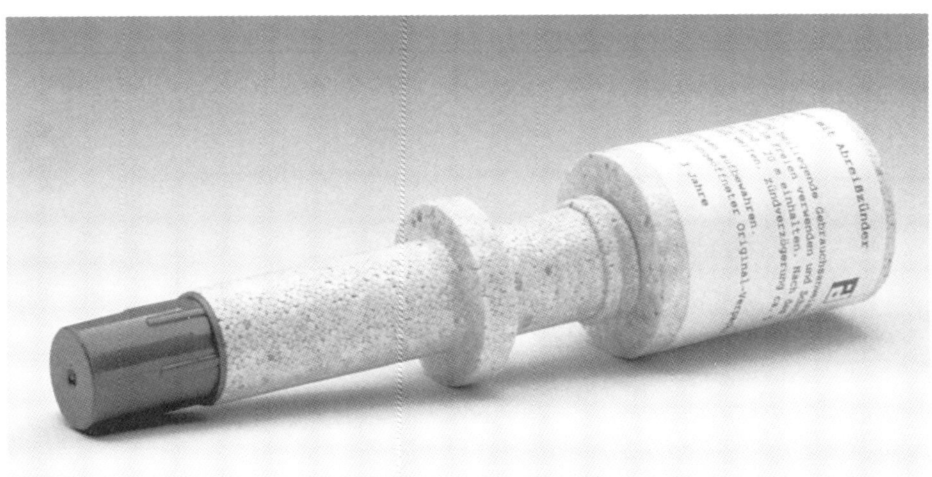

Piepenbrock simulator DM12

Piepenbrock Simulator DM48

The DM48 is a charge for practice hand grenades and may be used for training in close-quarter battle operations within an urban environment because after it explodes small splinters are created. It enables the practice hand grenade to be used in the same way as a live weapon but the ignition head is reusable. The sound pressure of the report is at least 156 dB in a 1 m high, 3 m radius area.

Specifications

Length: 96 mm
Delay time: 4.5 s
Noise level: approx 156 dB

Status

In service with the German Army.

Manufacturer

Piepenbrock Pyrotechnik GmbH, Göllheim.
(For details see Directory of Suppliers)

The elements of the Piepenbrock Simulator DM48

Rapp Rifle and Machine Gun Fire Simulator

The Rapp rifle and machine gun fire simulator is a modernised version of the German Second World War Zielfeuer Gerät 38. It is a recoil-operated weapon designed to fire blank cartridges only and is chambered for NATO standard 7.62 × 51 mm blank. The simulator resembles a sub-machine gun in some respects and is firmly anchored to the ground by ground spikes driven through three legs, though it can also be bolted onto target operating mechanism bases. Once emplaced, it is fired by means of a simple pull-line; this releases the bolt, which flies forward, loads and fires a cartridge. A pressure valve on the barrel restricts the flow of gas and thus ensures sufficient chamber pressure to operate the weapon in the normal blowback mode. If the pull-line has been released, then the bolt will be held back, ready for the next shot. If the pull-line is kept pulled, the weapon will continue to fire until either the line is released or the magazine is empty. The device can also be fired by a trip-wire.

The simulator can, according to the customer's requirements, be supplied with electrical remote-control, or mounted on a truck or armoured vehicle with electrical control.

Rapp rifle and machine gun fire simulator

Specifications

Calibre: 7.62 × 51 mm, blank only
Magazine capacity: 75 rounds
Rate of fire: (max) 500 rds/min

Status

In production. In service with the German Army.

Manufacturer

Theodor Rapp GmbH & Co KG, Haslach.
(For details see Directory of Suppliers)

Rheinmetall LKL Tank Practice Ammunition

Rheinmetall Wehrtechnik devised a novel approach to the design of kinetic energy training and practice ammunition by combining the aerodynamic device known as the Lochkegelleitwerk (LKL) with a kinetic energy penetrator rod. The LKL is a cone base attached to the projectile rod tail and this base is pierced with nine holes that run from front to rear. The hole size is accurately machined to allow a particular aerodynamic law to take effect. The holes in the cone will be aerodynamically closed when the projectile in flight slows down to a certain Mach number, which leads to an increase in air resistance caused by eddies formed behind the LKL cone base. The eddies effectively cause drag and reduce the velocity of the projectile rapidly, and hence the range.

In appearance and handling an LKL round resembles a conventional APFSDS round. As it is fired the sabot falls away but once in flight the LKL effect comes into play after about 2000 m. Up to that range the projectile behaves as a normal projectile. After about 2000 m the LKL effect reduces the projectile velocity to the extent that an APDS-simulated projectile has a range of about 7500 m instead of the usual 30 000 m plus. A lateral safety zone about 2100 m wide at 7500 m will be required for a tank gun elevated to over 10°.

Rheinmetall has applied the LKL principle to two types of tank gun ammunition, 105 mm and 120 mm. The DM128 105 mm is used in L7-series and M68 tank guns and the DM38 120 mm in the 120 mm Rheinmetall smooth-bore/M256 guns. Both calibres are identical in action and use but the LKL cone for the 105 mm round is solid (apart from the holes) while the 120 mm cone is hollow. Both have near-identical range performances.

In the United States, Alliant Techsystems developed the LKL concept for the 120 mm M865 training round used with the 120 mm M256 gun installed in the M1A1 Abrams main battle tank. This round is used as the training equivalent for the M827 and M829 APFSDS-T rounds.

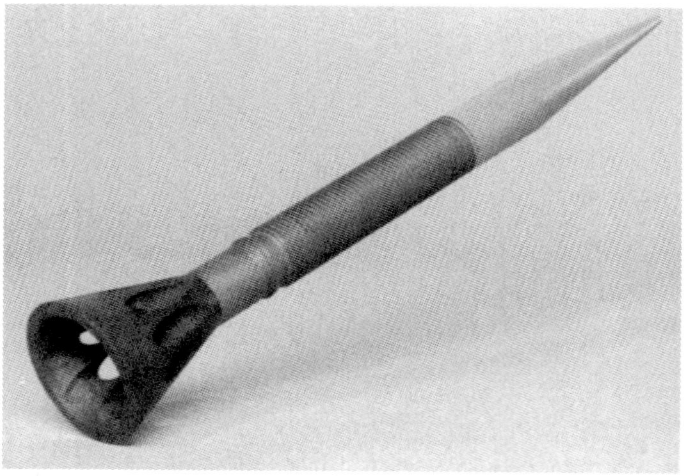

Rheinmetall 120 mm LKL projectile

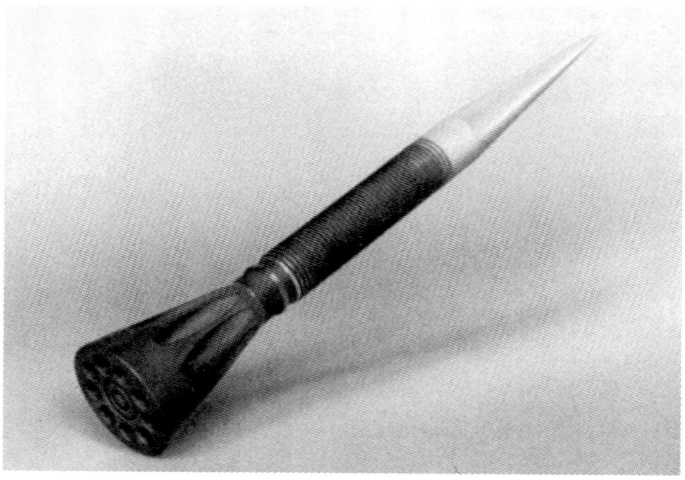

Rheinmetall 105 mm LKL projectile

Specifications

Calibre	**105 mm**	**120 mm**
Weight		
complete round	16.8 kg	18.4 kg
projectile/sabot	4.5 kg	5.9 kg
projectile	2.6 kg	3.2 kg
propellant	6±0.3 kg	8.2±0.3 kg
Diameter		
projectile	38 mm	38 mm
tail cone	80 mm	85 mm
Muzzle velocity	1640 m/s	1700 m/s
Max range		
(10° elevation)	7500 m	7500 m

Status

In production. In service with the German Army and the US Army and Marine Corp.

Manufacturer

Rheinmetall GmbH, Ratingen.

(For details see Directory of Suppliers)

NETHERLANDS

NWM De Kruithoorn Break-Up Training Ammunition

NWM De Kruithoorn is an independent subsidiary of Rheinmetall of Düsseldorf, Germany and manufactures break-up ammunition for a wide range of calibres including 40 mm (Bofors L/60 and L/70), 35 mm, 25 mm and 20 mm. The break-up projectiles disintegrate immediately after leaving the muzzle of the weapon and fall harmlessly to the ground within a very short distance. The size of safety area in front of the gun muzzle depends on the calibre but typical ranges are 20 mm 50 m, 40 mm 80 m, and 25 mm and 35 mm 100 m.

The handling, feeding and firing of break-up ammunition are the same as for conventional rounds, while the firing forces on the gun are the same as those produced when firing live rounds. The break-up rounds produce sound, flash and report identical to live ammunition and any aircraft can be used as a simulated target. Aircraft attack trajectories can be flown just as they would be in combat.

An area in front of the muzzle of 50 to 100 m long, depending on the calibre involved, and a maximum of 25 m wide is required as a safety area.

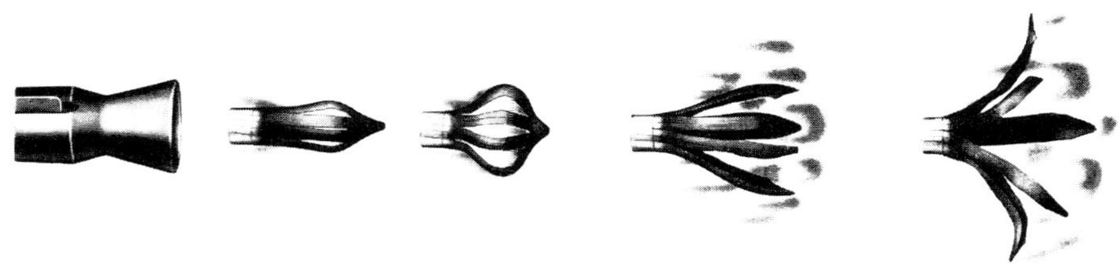

The break-up sequence of De Kruithoorn break-up ammunition; the sequence takes 0.00016 second

Types of break-up ammunition are as follows:

40 mm MN19 for Bofors L/70.
40 mm DM 18A1 for Bofors L/60.
35 mm MN16 for Oerlikon guns.
25 mm MN14 for Oerlikon guns.
20 mm DM78A2 for Hispano-Suiza HS820 and Rheinmetall MK20 Rh 202 guns.

Status

In production. In service worldwide.

Manufacturer

NWM De Kruithoorn BV, Hertogenbosch.

(For details see Directory of Suppliers)

NEW ZEALAND

Marine-Air Systems Battlefield Inoculation Remote Initiation System

The Battlefield Inoculation Remote Initiation System (BIRIS) was designed to assist the planning and conduct of realistic battlefield exercises and special operations. The equipment overcomes the limitations of cable-based and optical systems by using a secure UHF link.

The system comprises a transmitter and a number of receivers. The receiver can detonate up to 10 lines, each of which may be individually selected. Each line can fire up to six detonators depending on the length of cable connecting them. The transmitter contains a keyboard to permit selection of the receiver and line to be fired. A sequence of firing may also be preprogrammed through the keyboard. The transmitter maintains in memory a list of receiver/lines that have been fired. An illuminated display permits night operation.

This system is one of a family of remote initiation equipment available from Marine-Air Systems.

Status

In production.

Manufacturer

Marine-Air Systems Limited, Wellington.

(For details see Directory of Suppliers)

Battlefield Inoculation Remote Initiation System (BIRIS) transmitter

Battlefield Inoculation Remote Initiation System (BIRIS) receiver

POLAND

ISA Tank Gun Training Ammunition

ISA tank gun training ammunition centres around the use of plastic projectiles filled with water or an anti-freeze solution. As the rounds are fired the tank gun produces its characteristic flash and sound signature but the plastic body of the projectile is fractured by centrifugal forces as it travels down the barrel. As it leaves the muzzle the plastic body of the projectile fragments to release its liquid contents within an arc measuring 10°. The maximum fragment travel is 100 to 120 m.

The ISA training rounds are produced for two calibres: 125 and 100 mm. The 125 mm rounds are used with the D-81 tank guns fitted to Polish T-72 tanks while the 100 mm rounds are used with the D-10 guns fitted to T-54/55 tanks.

For the 125 mm ISA round the training projectile is 670 mm long and weighs 6 kg ±0.5 kg. It is fired using the 4Z40 propelling charge weighing approximately 5.85 kg.

Shelf life of the 125 mm projectile is 10 years, or 3 years when stored within a tank magazine. Shelf life of the 100 mm round is 5 years.

Status

In production. In service with the Polish armed forces.

Manufacturer

Zaklady Tworzyw Sztucznych 'PRONIT', Zakladowa 7.
(For details see Directory of Suppliers)

Polish tank gun training ammunition, with a complete 100 mm round (centre), 125 mm projectile (right) and 125 mm propelling charge (left)

RUSSIA

Russian Training Mines

The armed forces of the former Soviet Union use a wide range of training mines. Most are of the inert drill type and provide personnel with training in the handling and laying of minefields. Others are fitted with pressure or other fuzes which set off a smoke charge when activated.

One specialised training mine is the UITM-60 which, in appearance, resembles the TM-46 anti-tank mine. The UITM-60 can be laid by hand or mechanically. When a tank passes over a mine the lid is depressed until the bottom of the lid comes into contact with the top of the fuze and initiates the firing chain. After a delay of 3-4 seconds a signal cartridge is expelled, firing two green and two red flares 10 to 40 m into the air. This training mine is reusable and can also be fired electrically using a blasting cap.

Specifications

Weight: 8.7 kg
Diameter: 316 mm
Height: 125 mm
Operating force: 200-600 kg

Status

In service with the armed forces of the former Soviet Union.

Manufacturers

State factories.

SINGAPORE

Singapore Technologies Non-electric Thunderflash

This thunderflash is intended to simulate the sound of bombs, artillery projectiles, grenades and other explosive devices. In use a plastic end cap is removed, a striker surface on the cap is rubbed against the thunderflash fuze and the device is thrown. After a delay of 6.5 seconds the device detonates.

This waterproof thunderflash weighs 21 g, is 140 mm long and has a diameter of 24 mm.

Status

In production.

Manufacturer

Singapore Technologies, Chartered Industries of Singapore.
Marketed by: Unicorn International.
(For details see Directory of Suppliers)

Non-electric thunderflashes, showing packaging

Singapore Technologies Simulator Artillery TNT

This device is intended to simulate artillery fire for training or demonstration purposes and contains 370 g of high grade TNT flakes which can be detonated by detonating cord or an electrical or non-electrical detonator. Each simulator is encased in an anti-static plastic canister measuring 135 mm in height and 68 mm in diameter with a total weight of 438 g. When the simulator is detonated no lethal fragments are produced.

Status
In production.

Manufacturer
Singapore Technologies, Chartered Industries of Singapore.
Marketed by: Unicorn International.
(For details see Directory of Suppliers)

Simulator Artillery TNT

SOUTH AFRICA

Swartklip Training Rocket

Swartklip Products produces a range of hand-held distress rockets for maritime and other uses and has introduced a training rocket to the range. The training rocket is activated in the same manner as an operational distress rocket by removing a safety pin and pushing a lever against the side of the container. This releases a firing pin which ignites the rocket's solid fuel motor. The motor is contained in an aluminium housing which is ejected from the outer container and is propelled by the motor to a height of 300 m. At that height an inert payload is ejected, in place of the usual flare and parachute, without initiating a false distress signal.

The training rocket can be used for training in normal operating procedures, taking into account wind direction and velocity, and can also be used as a pyrotechnic to reproduce or simulate the effect of a missile being launched.

Status
In production.

Manufacturer
Enquiries to Armscor, Pretoria.
(For details see Directory of Suppliers)

SPAIN

PO Plastic Short-range Cartridges

PO (Plasticas Oramil) plastic cartridges are training rounds for use in areas where there is not sufficient space for a full-bore range and its corresponding safety area. The ammunition provides comparable accuracy to the full-size round at short range, but the velocity soon falls away and within 250 m the plastic bullet is virtually harmless. The muzzle velocity is very high, providing a flat trajectory for the first 50 m of flight, but the loss of kinetic energy is rapid. There is no danger from ricochets since the bullet crushes and disintegrates if it hits a solid object.

The ammunition is shaped to allow it to be loaded through a magazine in the normal way. The base of the cartridge is made of brass, but the walls of the case and the entire bullet are moulded plastic. The charge is gunpowder. The whole round, except for the brass base, is coloured blue to distinguish it from other types of round.

Types of PO plastic short-range cartridge produced include 7.62 mm NATO, 0.38 Special, 9 mm Parabellum, 9 mm Short, and 5.56 mm.

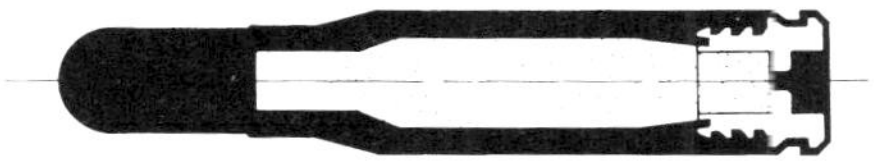

7.62 mm PO short-range cartridge

Specifications
Data for 7.62 mm short-range cartridge
Calibre: 7.62 × 51 mm
Weight:
(bullet) 0.7 g
(charge) 0.75 g
Max range: 300 m

Status
In production. In service with the Spanish armed forces.

Manufacturer
Plasticas Oramil (PO) SA, San Sebastian.
(For details see Directory of Suppliers)

SWEDEN

Bofors 105/50 TPDS-T Training Round

During 1982 Bofors introduced a new TPDS-T round for use with any 105 mm tank gun of the L7/M68 family. The round is known as the *Fläkten* (Fan) or the 105/50 TPDS-T. The projectile used with the round provides a flat trajectory out to a range of just over 2000 m, but after that range the ballistic performance is degraded by the use of fixed fins on the all-steel projectile body. The fins reduce the projectile maximum range to less than 7500 m.

The 105/50 projectile is held in a driving sabot which starts to disintegrate directly after leaving the gun muzzle. The prefragmented sabot shatters into small splinters within a space cone defined by a 20° included angle and a range of 150 m. The sabot base has a stable trajectory with a maximum flight distance of 1000 m. The barrel weight of the projectile is low so the round requires only a relatively small propellant load, resulting in low barrel wear.

Specifications
Calibre: 105 mm
Weight:
(complete round) 13.4 kg
(projectile/sabot assembly) 3.1 kg
(cartridge case) 5.6 kg
(propellant, nominal) 4.4 kg
(primer) 0.1 kg
(wear reducing liner) 0.2 kg
Muzzle velocity: (at +15°C) 1535 m/s
Tracer burning time: 2.5 s

Status
In production. In service with the Swedish Army.

Manufacturer
Bofors AB, Karlskoga.
(For details see Directory of Suppliers)

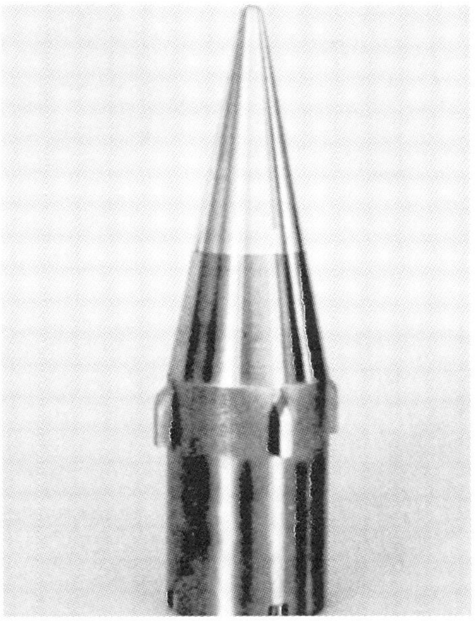

Bofors 105/50 Fläkten *TPDS-T projectile*

Bofors 105 mm APDS Training Round

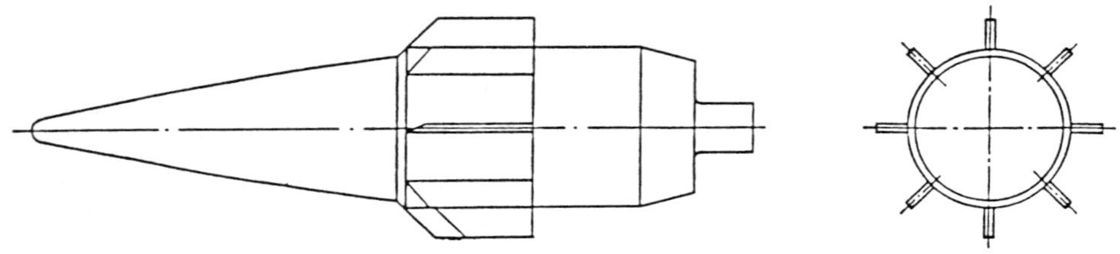

Drawings of converted Bofors APDS training sub-projectile

In view of the existing large stocks of Armour Piercing Discarding Sabot (APDS) tank gun rounds rendered obsolescent by the introduction of APFSDS projectiles, Lindesbergs Industri AB (LIAB - a Bofors subsidiary) proposed a modification programme that can convert APDS rounds into low-cost training rounds.

The programme involves the technology used with the Bofors 105/50 TPDS-T training round (see previous entry) in that a rotation decelerator on the sub-calibre projectile involved brings about gyroscopic instability that causes the projectile to tumble at around a predetermined range and so effect a powerful retardation in the flight trajectory. With the Bofors APDS training round the projectile behaves as normal over operational ranges of up to around 2500 m. After that range strakes on the projectile body gradually act to retard the projectile spin aerodynamically so that it becomes ballistically unstable and tumbles to the ground after travelling a distance less than a third of the maximum range of conventional APDS. Thus for 105 mm tank gun APDS training rounds a safety range can be 15 to 20 km instead of around the 80 km that a conventional APDS can achieve.

The conversion programme involves taking a conventional APDS round and removing the saboted tungsten penetrator. Mild steel vertical strakes (or fins) are added to the sub-calibre projectile which is then reinstalled into its sabot and replaced in the cartridge case. The training round is then handled in exactly the same way as a conventional APDS round and can be aimed and fired using existing gun sight APDS graticules.

Firing trials of converted 105 mm APDS rounds have taken place.

Status

Development.

Manufacturer

Bofors AB, Karlskoga.
(For details see Directory of Suppliers)

Norabel Pyrotechnic Simulators

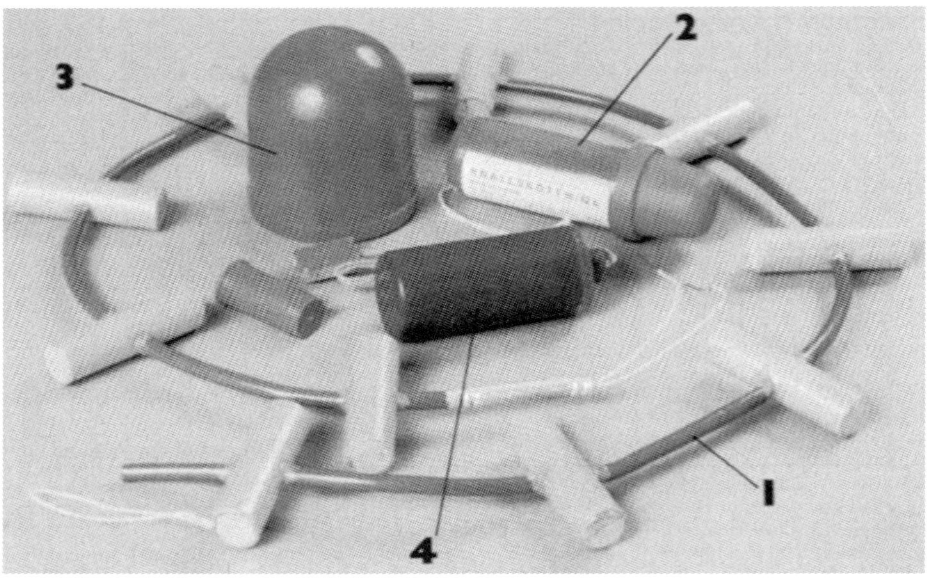

A selection of Norabel pyrotechnic simulators: (1) battlefield simulator; (2) Thunderflash; (3) air burst simulator; (4) gunfire simulator

Norabel, part of the Nobel Industries Group, provides a range of pyrotechnic simulators for all the usual battlefield effects. As well as standard models, special devices can be fabricated to customer requirements. The standard models are listed below.

Small Arms Fire Simulator

Also referred to as a Battlefield Simulator, this simulator produces a succession of sharp reports and flashes to simulate the firing of 5.56 or 7.62 mm rifles or automatic weapons. The reports are produced by firecrackers contained in cardboard cylinders connected by wires. The simulators may be for 5 or 10 shots and are ignited by a 0.7 A electrical current. A 10-shot simulator is 810 mm long while a 5-shot version is 720 mm long. A single-shot simulator is available.

Thunderflash

This simulates an explosion. The device uses a plastic casing with a waterproof friction igniter.

Bullet Hit Simulator

This device produces a crack similar to that produced by a bullet hitting the ground. It consists of an electric igniter filled with a pyrotechnic charge and in use is placed on the ground and covered lightly with dust or sand. For firing, the device is connected to a 0.7 A electrical supply. The simulator weighs 20 g and is 42.5 mm long and 7 mm in diameter.

Air Burst Simulator

These simulators produce a puff of smoke and burst of 10 firecrackers at an altitude of about 50 m. They are used with a special launcher and two versions are available, one for use at night, the other for daylight. Both are identical in dimensions (119 × 50 mm) but the night use version is heavier at 147 g (day version 126 g).

Ground Burst Simulators

These simulators produce a very loud report combined with a bright flash and are used to simulate explosions from weapons such as land mines, mortars and artillery fire. The simulators are supplied in polythene containers complete with electrical igniters and are produced in three sizes: light, medium and heavy. The light version weighs 32 g and contains 13 g of pyrotechnic composition. The medium version weighs 202 g and contains 150 g of pyrotechnics. The heavy version weighs 357 g and contains 300 g of pyrotechnics. A safety distance of 30 m has to be maintained for the heavy version.

Fireburst Simulators

These simulators produce a fierce ball of flame of short duration combined with a dense cloud of black smoke, and are used to simulate napalm or petrol bomb attacks. Two types are produced, medium and heavy, and both are supplied in polythene containers together with electrical igniters. The medium version weighs 303 g and contains 250 g of pyrotechnic material. The heavy version weighs 1.272 kg and contains 1.1 kg of pyrotechnics.

Gunfire Simulators

These devices simulate the firing of large calibre guns. They can be used with the Saab BT 19A gunfire simulator equipment.

Status

All the above items are in production and are in service with the Swedish armed forces.

Manufacturer

Norabel AB.
(For details see Directory of Suppliers)

SWITZERLAND

Polytronic HF-11 and HF-11C Hostile Fire Simulators

The Polytronic HF-11 and HF-11C hostile fire simulators can be used as smoke hit indicators, muzzle blast or enemy return fire simulators. Different types of cartridge may be used to create various effects such as coloured smokes, blast, flash/blast, flash/smoke, artillery explosion and illumination. The HF-11 has a capacity of 20 cartridges; the HF-11C has 40.

The units are portable and are easy to install and operate. During loading a security switch prevents accidental discharges.

These simulators are normally used with the Polytronic TG 82 target systems but may be used independently and can be controlled by radio via an additional 12 V battery receiver box. The control electronics are built into a waterproof aluminium housing. The main body of the unit is made of corrosion-resistant steel.

The HF-11 weighs 40 kg and the HF-11C 28 kg. Cartridges with diameters of up to 50 mm can be used.

Status
In production and widespread service.

Manufacturer
Polytronic International Limited, Muri.
(For details see Directory of Suppliers)

Polytronic HF-11 hostile fire simulator

UNITED KINGDOM

Aardvark Land Mine Simulators

The Aardvark mine simulators were developed in conjunction with the Aardvark Joint Services Flail Unit - for details of this equipment refer to *Jane's Military Vehicles and Logistics*. The simulators were originally conceived as flail training aids but have since evolved into a range of simulators.

The mine simulators consist of an electrical pressure switch which attaches to and initiates any of four types of simulator. The battery-operated electrical pressure switch is a circular moulded plastic case 150 mm in diameter and 43 mm deep. It is fitted with a safety key and has two external outlets and a battery compartment. The unit has 360° copper contacts and, for safety purposes, is remotely armed. The power source is fitted with a booster to ensure instantaneous initiation when contact is made.

A range of four simulators for use with the pressure switch is available: No 1 Classroom Demonstrator; No 2 Anti-personnel; No 3 Anti-vehicle; No 4 Anti-tank.

The No 1 Classroom Demonstrator is suitable for use within a confined space and produces a loud report, flash and a small amount of smoke; no explosive is used with this model. The No 2 Anti-personnel simulator produces the sound and flash effects of a small anti-personnel mine and was designed for surface use only. The effects are a loud report, flash and small amount of smoke. The No 3 Anti-vehicle simulator produces the effects of a large anti-vehicle mine. It can be surface laid or buried at a shallow depth. The effects are a large explosive noise with a vivid red flash and white smoke. The No 4 Anti-tank simulates the effects of a large anti-tank mine. It can be surface laid or buried at a shallow depth. The effects are a very loud explosion with a bright flash and a large white smoke cloud.

All simulators are supplied with a 5 m cable weighing 130 g and are initiated by an integral electric match head. On contact, initiation is instantaneous.

A safety test set and an arming kit are available.

Status
In production.

Manufacturer
Aardvark Clear Mine Limited.
(For details see Directory of Suppliers)

The full-range of Aardvark land mine simulators showing the pressure switch in the centre

Aardvark No 4 anti-tank mine simulator

Brocks Training Simulators

The scope of Brocks simulators is very wide, covering practically all battlefield effects. The following are some examples of the available range.

L20A1 Explosion Simulator
With a smoke and noise burst this device produces the realistic effect of a bomb explosion. Each simulator is 66 mm high and has a diameter of 57 mm.

Thunderflashes
Brocks produces a range of thunderflashes, including the standard Thunderflash CD in use throughout the world. This device uses friction ignition and has a delay of 5 to 7 seconds. It is 229 mm long, is 28 mm in diameter and weighs 45 g. A smaller version weighs 8.5 g, is 120 mm long and has a diameter of 13 mm. A small electric version ignited by a 4 to 12 V DC supply weighs 9.3 g and is 102 mm long. The No 6 Ground Flash is also electrically ignited and weighs 80 g. Length is 143 mm and diameter 30 mm.

EOD Simulators
These are designed to simulate the functioning of a terrorist bomb when personnel are being trained in Explosive Ordnance Disposal (EOD) and disarming Improvised Explosive Devices (IEDs).

Brocks L3A1 EOD simulator

The sound unit EOD electric L1A1 is a paper tube containing an electric fuze and pyrotechnic composition. It is designed primarily for indoor applications.

The sound unit EOD electric L2A1 is of similar construction but contains more composition and is designed for outdoor applications.

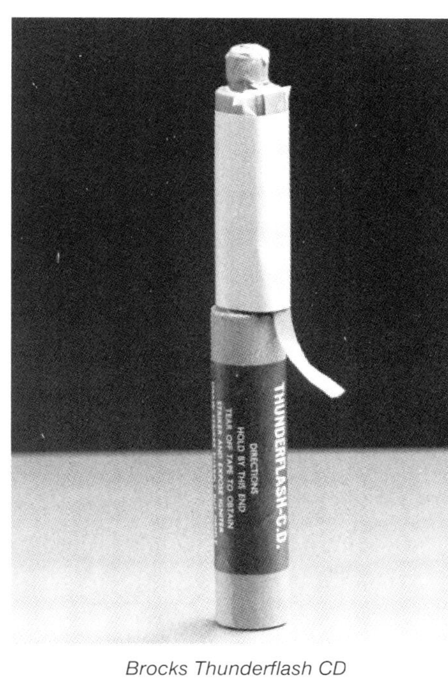

Brocks Thunderflash CD

The smoke unit EOD electric L3A1 is a chipboard tube containing an electric fuze and smoke composition which gives a visual smoke puff. It is designed for use alongside a simulated IED but may be built into it.

None of these units produces any particles which could injure personnel under instruction.

L21A1 Small Arms Strike Simulator

The L21A1 is a training device which is electrically initiated and, immediately after a sharp report, ejects an inert non-metallic projectile which creates a noise on striking a metal plate or other surface, similar to the impact of a bullet.

Machine Gun Simulator

The machine gun simulator consists of a length of very fast burning fuze with 10 equally spaced sound units which, after ignition, produce 10 loud reports in rapid succession, typifying the sound of a machine gun in operation. Two types of simulator are made. The electrical model is fired by means of a small electric current, while the friction model has a hand-operated friction igniter at the end of the fuze. The units are both about 1 m long.

Simfire System

Brocks manufactures pyrotechnic units for incorporation into the Simfire simulation system. The Simfire smoke generator is used to indicate that a tank has been hit by generating orange smoke for 8-12 seconds. The Simfire flash generator cartridge provides the visual and aural effects of a main tank gun in action.

Status

All the above items are in production and in service with the British Army.

Manufacturer

Brocks Explosives Limited.
(For details see Directory of Suppliers)

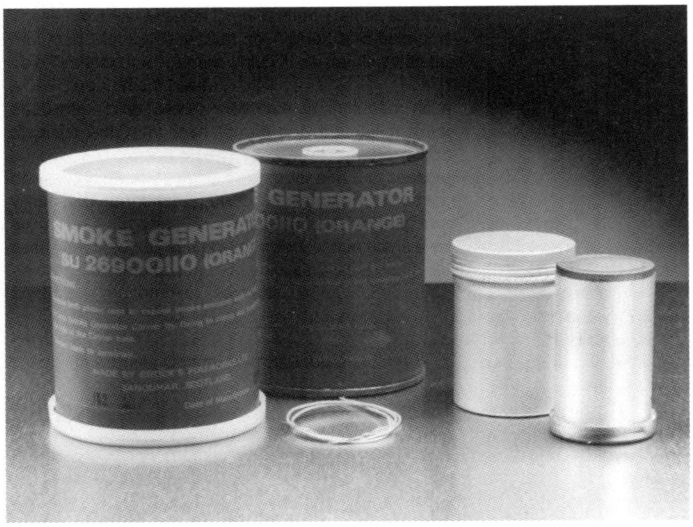

Pyrotechnic simulators produced by Brocks for the Simfire system

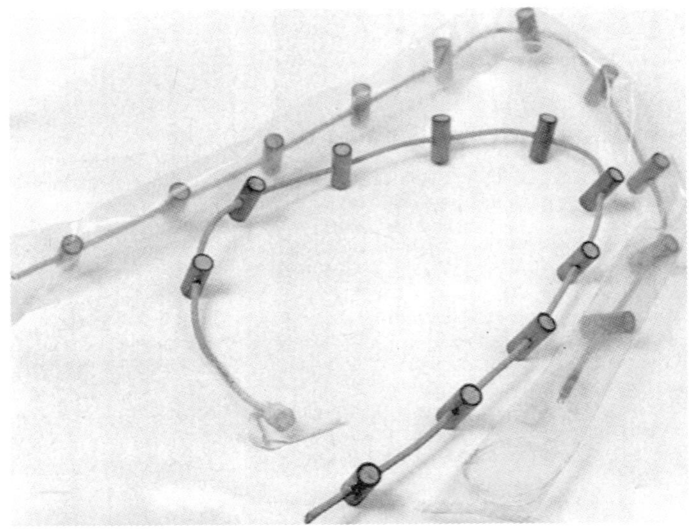

Brocks machine gun simulators

Haley & Weller Pyrotechnic Simulators

Thunderflash Mark 8 S150

The Thunderflash Mark 8 is for use as a training device to provide realistic simulation of battle noise such as a shell, bomb or other missile. The unit is initiated by a hand striker which, after a delay of 6 to 12 seconds, results in a bright flash and a loud bang. The Thunderflash Mark 8 is fully waterproof.

Specifications

Length: 216 mm
Diameter: 29 mm
Weight: 60 g
Delay time: 6-12 s (nominal)

Thunderflash N2 and N5

The S120 (N2) and S125 (N5) Thunderflash are used for battle noise simulation and are specifically designed for maritime use—the N5 may be used as a diver recall device. Both units produce a loud report but the N5 is smaller and so has a smaller report. There is minimum hazard from flying debris.

Specifications

Model	**S120**	**S125**
Length	151 mm	105 mm
Diameter	18 mm	13 mm
Weight	28 g	11 g

Sound Unit Training Electric (Indoor) S401
Sound Unit Training Electric (Outdoor) S402

The S401 and S402 electrically initiated sound units were designed for use in training personnel in the disarming and making safe of Improvised Explosive Devices (IEDs). When fitted to a training IED they indicate errors in drills by producing a loud report and bright flash without producing any hazardous fragments. The S402 is slightly longer (50 mm) than the S401 (40 mm) and both have a diameter of 8 mm. Weight is approximately 3 g.

BATTLEMASTER Weapon Effect Simulation

The BATTLEMASTER approach to weapon effect simulation is now standard equipment with armed forces worldwide and provides an easy-to-use system

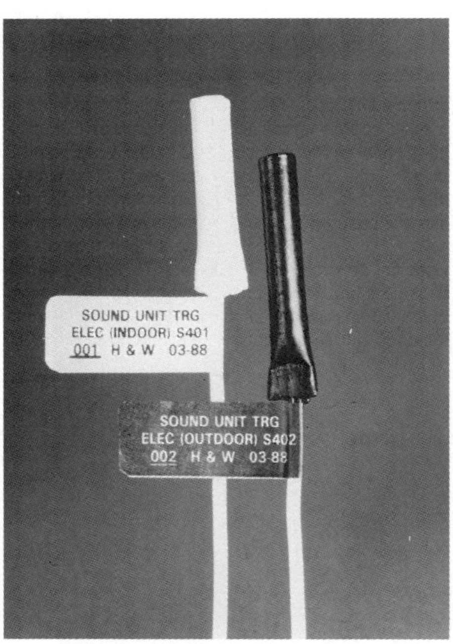

Sound Unit Training Electric (Indoor) S401 (left) and Sound Unit Training Electric (Outdoor) S402 (right)

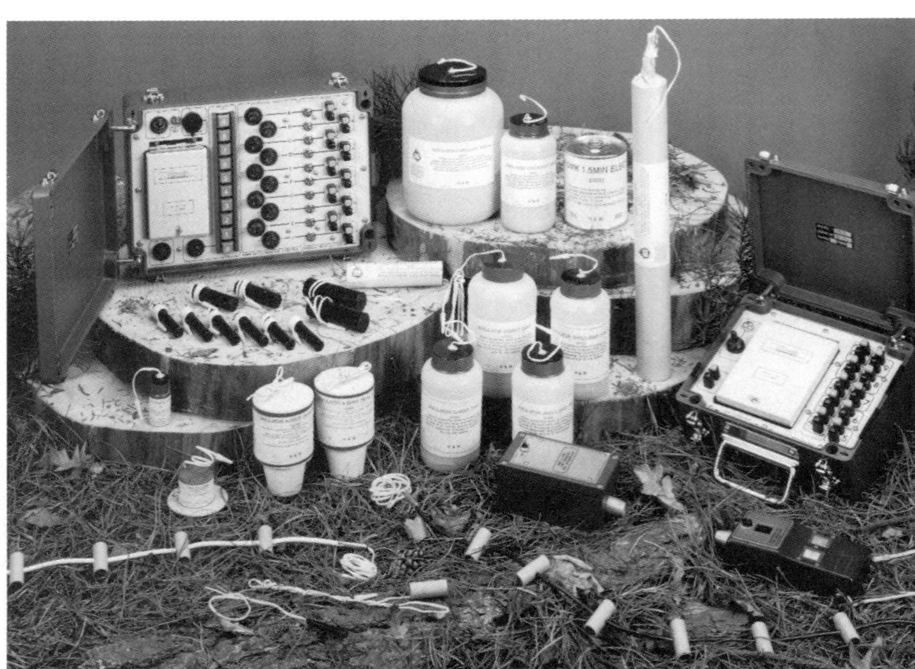

A selection of BATTLEMASTER pyrotechnics and control units

Thunderflash Mark 8 S150

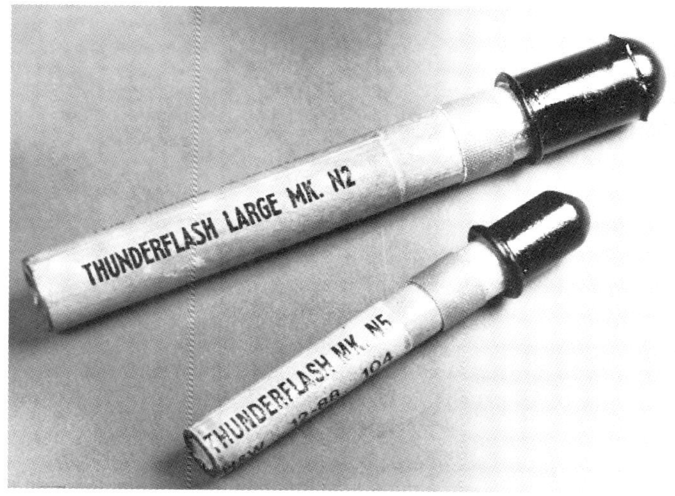

Thunderflash N2 and N5

requiring the minimum of manpower and training in setting up and use. It provides a combination of surprise, realism, safety and low cost (usually one per cent of live rounds).

The pyrotechnics involved with BATTLEMASTER are designed to provide realistic representation of a wide range of stores including small arms and machine gun fire, grenades, anti-personnel and anti-tank mines, bomb blasts, air bursts, mortar, light and heavy artillery shell bursts and a Thunderflash to simulate hand grenades and similar requirements. Also available are reduced effect pyrotechnics to simulate Close Quarter Battle (CQB) effects using a hand-held CQB controller.

Status

All the above are in production.

Manufacturer

Haley & Weller Limited.
(For details see Directory of Suppliers)

Unwin Pyrotechnic Battlefield Training Simulators

Unwin Pyrotechnics Limited specialises in the manufacture of a wide range of pyrotechnic weapon-effect simulators for battlefield training. The purpose of these simulators is to increase the realism and involvement of participants in field training and thereby enable instructors to bring their forces to a very much higher level of confidence and readiness under fire. The key factors of this training method are surprise, realism, safety (even when in close proximity to personnel, because of the absence of any dangerous fragmentation) and low cost (for example as little as 2 per cent of the cost of live rounds).

The pyrotechnics are designed to provide realistic representations of a wide range of stores including small arms and machine gun fire, grenade, mine, mortar, light and heavy artillery shellbursts, a powerful rocket for simulating guided missiles, and an effective and reliable Thunderflash. At the opposite end of the scale, reduced effect pyrotechnics are available to simulate booby traps, car bombs and similar explosions.

Used in conjunction with the Unwin BATTLEMASTER electronic firing system, up to 100 simulator effects can be fired as required over a range up to 3.5 km from the central control system, thereby giving flexibility over a large battlefield sector. For urban and other close quarter battle training situations, a hand-held unit is also available.

No specialised skills or maintenance are required in the use of the simulators or BATTLEMASTER.

Unwin shellburst simulator in use

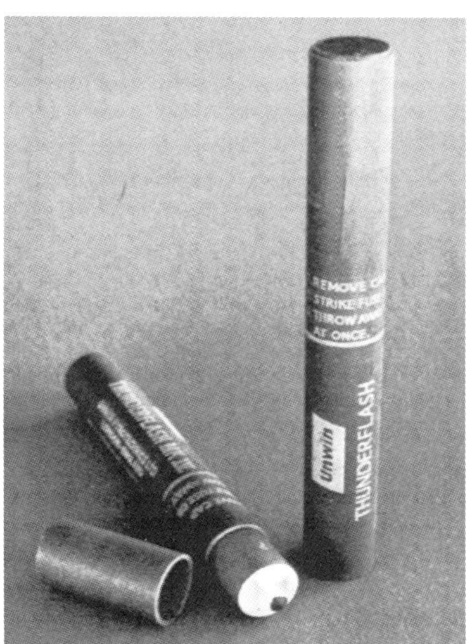

Thunderflash RU Mark IV

Airburst Simulators

These may be used for training aircrew in the effects of anti-aircraft gunfire or for acclimatising ground-based forces to airburst shelling. They reproduce accurately the burst of an anti-aircraft shell by giving a very loud report combined with dense smoke. The burst height is approximately 100 m.

Groundburst Simulators

These simulators are designed to simulate realistically land mine bursts, light and heavy mortar and artillery bursts and similar types of ordnance and explosive devices. They are suitable for use in battlefield exercises in conjunction with both infantry and vehicles.

Three sizes of groundburst simulator are available. They produce a very loud report combined with a bright flash and smoke cloud.

Effects of an Unwin groundburst simulator

Shellburst Simulators

These simulators may be used for infantry or paramilitary training under simulated battlefield conditions, for vehicle handling training under fire or for simulated land mine detonation. They reproduce the explosion effects of a wide range of shells from light mortar to artillery shell by producing a loud report combined with dense smoke and vivid flash.

Fireburst Simulator

This simulator may be used for the effective simulation of napalm attacks and shell or missile hits on buildings, vehicles and installations, as well as for infantry familiarisation with battlefield conditions. They produce a large and fierce ball of flame of short duration, combined with a dense cloud of black smoke. The burst effect size for the 110 mm type is approximately 1.5 m and for the 196 mm type, 4 m in diameter.

Thunderflash RU Mark IV

This is a hand-thrown training aid to simulate small explosive devices, including hand grenades. The device produces a brilliant white flash and a loud report. A special diver recall signal variant can also be supplied.

The charge is housed in a cardboard tube, part of which serves as a throwing handle. A unique shaped fuze head ensures particularly easy, reliable and positive ignition. The fuze is enclosed within a safety transit cap which incorporates the friction striker.

Small Arms Fire Simulators

These small arms fire simulators are used during the training of troops or anti-terrorist forces to accustom personnel to the conditions likely to be encountered during operational activities in battlefield and urban situations. These simulators are safe in use. No harmful fragmentation is produced. They may be placed in trees, bushes and old buildings to be fired remotely to simulate sniper or retaliatory fire.

Two types of simulator are available: a 5-shot rifle fire and a 10-shot machine gun. Both simulators produce a succession of sharp reports and bright flashes almost identical in duration to the shots produced by a 5.56 or 7.62 mm rifle or automatic weapon. The 5-shot rifle fire simulator incorporates a 3-5 seconds delay between effects and the 10-shot machine gun device simulates a rapid burst. Additional increments of 10 shots can be added to obtain longer bursts.

IED Simulator

Any military or paramilitary organisation must have the capability to make safe Improvised Explosive Devices (IEDs) such as those encountered during terrorist campaigns. For this purpose, members of Explosive Ordnance Disposal (EOD) teams have to be trained to recognise and disarm safely certain designs and variations of commonly used terrorist bombs.

Whilst undergoing training it is desirable that there is no risk to personnel, and yet a degree of realism is also required. Should the trainee make a mistake the IED device will go off, thus implying that an injury has occurred as a result of the trainee's last action. Alternatively, certain of these devices may be used in conjunction with remotely controlled EOD equipment so that familiarity with the robot may be gained.

Several types of these devices are available, all producing a sharp report, flash and a small amount of smoke. All these devices contain a small quantity of flash composition enclosed within a non-fragmenting case. The flash effect is initiated by an integral electric match head.

Petrol Bomb Simulator

This simulator was designed specifically for crowd control and internal security forces training. When thrown, the petrol bomb simulator closely resembles a real petrol bomb even to the point of a distinctly burning external fuze. After a short delay, the simulator bursts with a loud report and spreads a cloud of sand and sawdust over a radius of approximately 1.5 m to represent the area covered by burning petrol. There is no dangerous fragmentation.

Smoke Generators

Typical applications for these long duration smoke generators include fire-fighting training in vehicles, buildings or ships and low-cost screening smoke for infantry training in battlefield and urban situations. They generate clouds of dense, opaque white smoke of 1.5 or 3 minutes duration. A total area screening smoke of 4 minutes duration can also be supplied.

Small Arms Fire Simulators

Haley and Weller also produces a range of small arms fire simulators.

Status

All the above are in production. They have been sold to 19 nations.

Manufacturer

Haley and Weller Ltd.
(For details see Directory of Suppliers)

OTEC Frangible Training Ammunition (FTA)

OTEC Frangible Training Ammunition (FTA) was designed for use on indoor short-range combat and other training ranges where the use of conventional live ammunition could create safety problems, or where the normal range safety templates preclude the use of conventional full calibre ammunition.

The FTA projectile is manufactured using a plastic-based compound matrix into which metal powders and lubricants are embodied. The resultant projectile is combined with a conventional cartridge case and propellant load so that FTA rounds can be handled, loaded and fired in exactly the same way as ordinary ammunition. Recoil forces and firing signatures are the same as those produced by conventional ball ammunition. When the FTA projectile strikes a hard surface it disintegrates into powder that is completely free from lead particles; there is no splash-back and there are no ricochets. FTA projectiles cannot penetrate more than 2 mm of mild steel before disintegrating. Ballistic performances are virtually identical to the corresponding ball ammunition at training ranges but, compared to normal ammunition, range safety templates can be reduced by as much as 75 per cent.

OTEC FTA is available in 5.56 mm, 9 mm Parabellum, 0.38 Special and 0.357 Magnum calibres. Under development are 7.62 mm NATO and 0.50/12.7 mm Browning FTA rounds. The 5.56 mm FTA round can be fired from any type of 5.56 mm weapon, irrespective of rifling twist. The 9 mm FTA rounds can be fired from pistols and sub-machine guns, including full burst firing where appropriate. Any weapon can fire FTA without modification.

Status

In production. In service in the Americas, Europe and the Middle and Far East.

Manufacturer

OTEC Special Products Limited, Godalming.
(For details see Directory of Suppliers)

OTEC Frangible Training Ammunition (FTA) for 5.56 mm, 0.38 Special and 9 mm Parabellum

Pyramid Battlefield Simulation and Target Indication System

Pyramid is a radio-controlled initiation system which provides a flexible, versatile and economic means of firing pyrotechnic devices for field training. The system is simple to set up and dismantle and, when operated in accordance with the correct procedures, is safe to use on any training area.

Pyramid has four main purposes: to provide realistic battlefield simulation on field exercises; to mark targets on live firing ranges; to reduce live ammunition costs in 'adjustment of fire' training; and to add drama and provide special effects at demonstrations and displays.

The key features of the system are: radio control at ranges up to 5 km using exclusive digital messages; advanced safety interlocks; and modular design allowing control of up to 4800 pyrotechnic firings at 800 separate locations from one command station. Unauthorised use of the system is prevented by the provision of non-interchangeable keys and a unique access code which must be inserted before the encoder will operate.

Pyrotechnic simulators that can be used with Pyramid include shell, mortar and airbursts, machine gun fire, smoke, grenades and illuminants. NBC simulators can also be used.

Status

In production.

Manufacturer

Marketed by: Miltrain Limited.
(For details see Directory of Suppliers)

Pyramid in action

Richmond Training Bomblet

The Richmond Training Bomblet was designed to enable realistic, thorough and regular Explosive Ordnance Disposal (EOD) training to take place against bomblets which look and react exactly like the latest NATO and former Warsaw Pact versions.

The bomblet contains a programmable electronic control module (powered by one PP3 battery) featuring a time delay in 5-minute increments up to 64 hours, and an anti-disturbance initiator which can be set either at intermittent or continuous mode. A stainless steel pyrotechnic chamber contains a replaceable pyrotechnic effects module which, when initiated, produces a safe flash-bang-smoke effect. There is also a frangible nose/fuze module which, if removed dynamically with a de-arming tool, will neutralise the bomblet. The polythene nose-cone is a replaceable item and the body is aluminium.

The training bomblet weighs 3.2 kg without the pyrotechnics effect module. It is 330 mm long, 90 mm in diameter and is 127 mm wide across the fins.

A version has been developed which can be initiated remotely by a laser projector.

Status

In production.

Manufacturer

Richmond Electronics & Engineering International Limited.
(For details see Directory of Suppliers)

The laser-initiated Richmond Training Bomblet

Royal Ordnance Generator Smoke Training N5 Mark 1 and N6 Mark 1

The N5 and N6 smoke generators were designed to simulate adverse operational conditions and for training in fire-fighting operations. The generators were designed so that in a room of 43 m³ (N5) or 85 m³ (N6) with normal artificial lighting, objects are indistinguishable at 5 m distance.

The generator comprises pellets of a smoke-producing composition housed within an insulated aluminium container. Initiation of the generator is by a cocked striker system and a pyrotechnic delay. The unit is protected by a waterproof plastic cap which also serves as a means of removing the generator safety cap via an attached lanyard. On functioning of the striker mechanism, there is a delay of 2-4 seconds before the generation of smoke.

The smoke produced by the generators is of proven low toxicity but is a good obscurant and maintains realism, being a mild lachrymatory. These stores have an exceptionally low explosion and flame risk.

Specifications

Model	**N5**	**N6**
Weight	250 g	300 g
Height	105 mm	105 mm
Diameter	66 mm	66 mm
Contents (net)	44 g	88 g
Delay time	2-4 s	2-4 s
Time to full generation	10 s	30 s
Duration of smoke emission	30 s	30 s

Status

In production.

Manufacturer

Royal Ordnance plc, Chorley.
(For details see Directory of Suppliers)

Generator Smoke Training N5 (left) and N6 (right)

Royal Ordance Grenade, Hand, Smoke Screening, Training, L72A1

This training hand grenade is produced for use by forces undergoing training when individual soldiers are exposed to screening smoke and may inhale the smoke. The grenade comprises a pellet of smoke-producing composition housed in a tin-plate container. Initiation of the grenade is by a conventional fly-off lever and striker system. On functioning of the striker mechanism, ignition of the smoke pellet is immediate but there is a short delay (nominally 8 seconds) before production of effective smoke.

The smoke produced is of low toxicity and the safety of the grenade has been demonstrated during testing by the United Kingdom Ordnance Board and acceptance testing by the British Army.

Specifications

Weight: 450 g
Height: 120 mm
Diameter: 70 mm
Delay time to ignition: nil
Time to effective smoke production: 8 s
Duration of smoke production: 80 s

Status

In production.

Manufacturer

Royal Ordnance plc, Chorley.
(For details see Directory of Suppliers)

Grenade, Hand, Smoke Screening, Training, L72A1

Royal Ordnance Practice Flash Artillery Projectiles

These training projectiles were developed to provide cost-effective and completely realistic training for gun crew, command post and forward observers. The projectiles look and handle exactly as the high explosive shell they represent and are ballistically matched This enables standard charges and firing tables to be utilised.

An impact indicator in the base of the projectile produces a clear indication of the fall-of-shot out to a range of 4000 m and has been proven to function on all types of terrain, including meadows, sandy desert, peat bogs and deep snow. The indicator incorporates safety devices to ensure in-bore safety and no arming until the projectile is a safe distance from the gun.

Practice Flash Projectiles are more environmentally friendly than the equivalent high explosive shells they represent as they produce no fragments, craters or toxic fumes.

The cost of these projectiles is expected to be less than half that of the equivalent high explosive shell once they are in volume production. Versions in service include training projectiles for the 105 mm Light Gun and the 155 mm FH-70, M109 and AS90. Versions to match other 105 and 155 mm artillery systems are scheduled.

Status
In production.

Manufacturer
Royal Ordnance plc, Chorley.

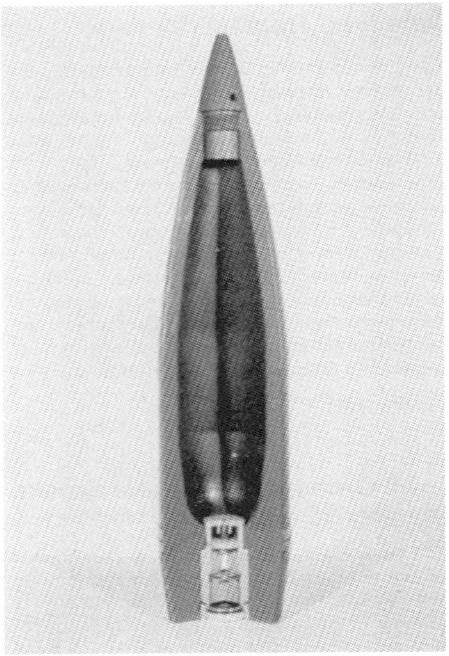

Cross-section of a Royal Ordnance 155 mm practice flash artillery projectile

Royal Ordnance 30 mm RARDEN Reduced Range Training Round

The Royal Ordnance RO 275 30 mm RARDEN Reduced Range Training Round (RRTR) was specially developed for use with the 30 mm RARDEN gun and utilises a tubular projectile. The trajectory is matched to the associated 30 mm APDS round to a range of 2000 m. Aerodynamic choking effects then limit the safety trace to a range of approximately 4500 m. The projectile incorporates a tracer which is effective to a range of 1500 m.

The technology involved in the RRTR is suitable for transfer to other gun-launched projectiles.

Specifications
Weight:
(round) 760 g
(projectile) 270 g
(propellant) 140 g
(cartridge case) 365 g
Muzzle velocity: 1300 m/s

Status
Development.

Manufacturer
Royal Ordnance plc, Chorley.
(For details see Directory of Suppliers)

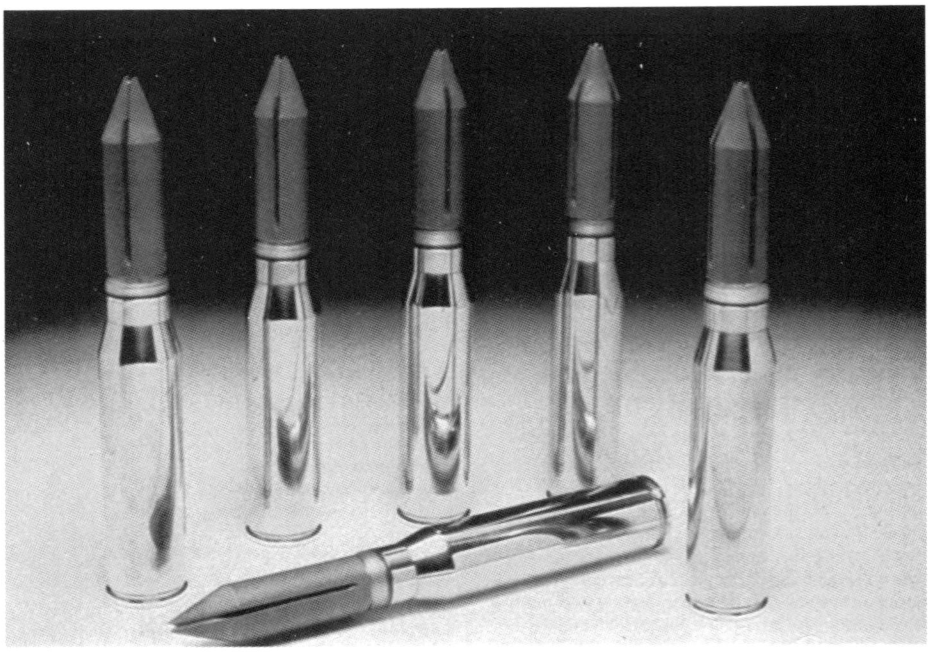

30 mm RARDEN Reduced Range Training Rounds (RRTR)

Royal Ordnance Training Ammunition (ROTA)

Royal Ordnance Training Ammunition (ROTA) was at one time during its development known as SPARTEN (Special Practice Ammunition for Realistic Training ENfield). It is produced in 5.56 mm calibre and looks and functions exactly like conventional live ammunition. The bullet is formed of a non-toxic metallic/plastic composition and the propellant load is less than that used for Royal Ordnance ball ammunition. The combination of these two factors means that ROTA can be handled and loaded in the same manner as fully operational ammunition and fired from unmodified weapons, yet the light projectile lacks kinetic energy and velocity rapidly falls away once realistic combat ranges have been reached. The bullets retain enough energy to activate automatic targets at ranges out to 150 m yet the overall safety template is only 850 m long, as opposed to the 4000 m long template for a conventional 5.56 mm ball round.

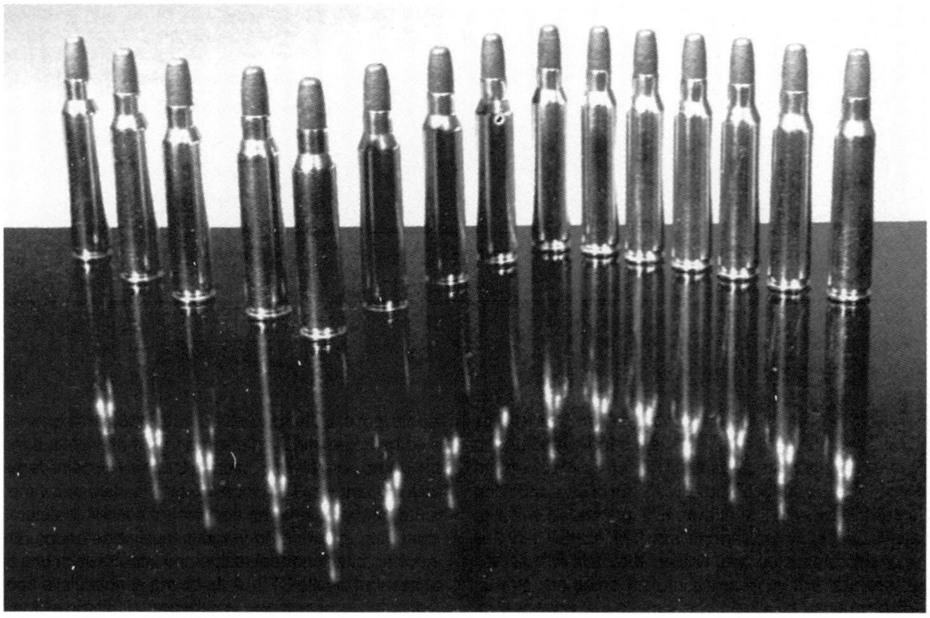

5.56 mm Royal Ordnance Training Ammunition (ROTA)

The bullet accuracy of ROTA matches that of a conventional NATO 5.56 mm ball round at 150 to 200 m. The bullet will not ricochet or splash back on impact with hard surfaces so ROTA is safe to use on indoor, Close Quarter Battle (CQB) and urban training ranges.

As an example of the kinetic energy loss of a 5.56 mm ROTA bullet, a bullet fired at a muzzle velocity of 1020 m/s has a muzzle energy of 0.744 kJ. At 150 m and 0.288 second later, the muzzle velocity will have dropped to about 325 m/s and the kinetic energy will have dropped to 0.075 kJ.

A ROTA 5.56 mm round weighs 10 g and the projectile 1.43 g.

Status
In production.

Manufacturer
Royal Ordnance plc, Chorley.
(For details see Directory of Suppliers)

Schermuly Simfire Pyrotechnics

Developed for use with the Loral Solartron Simfire system, these Schermuly pyrotechnics have been approved by the United Kingdom Ministry of Defence for use with Simfire systems and have undergone stringent testing. Two types of pyrotechnics are used: a flash cartridge and a smoke generator.

The flash cartridge (Schermuly reference Solartron SU 27580020; UK MoD L18A2; NSN 1370-99-966-6633; Item No 56000) simulates the flash and report of a tank's main armament. It consists of a weatherproof metal canister containing the smoke composition and is fitted with an electrical igniter. The cartridge is 70 mm long and has a diameter of 38 mm; weight is 80 g of which 12 g is explosive. The sound produced is audible at 2000 m.

The smoke generator (Schermuly reference Solartron SU 26900110; UK MoD L15A2; NSN 1370-99-966-6632; Item No 53885) is used to indicate a 'hit' by emitting orange-coloured smoke. It consists of a weatherproof metal canister containing the smoke composition and is fitted with an electrical igniter. The generator is 110 mm long and has a diameter of 89 mm; weight is 335 g, of which 170 g is the smoke composition. Burning time is from 15 to 18 seconds.

Status
In production and in widespread service.

Manufacturer
Schermuly (Pains-Wessex Limited), Salisbury.
(For details see Directory of Suppliers)

Schermuly Smoke Generators

Schermuly RPV Target Smoke Generator

The Schermuly RPV target smoke generator was developed to provide visual enhancement for Remotely Piloted Vehicles for simulation and training purposes.

The generator is ignited by the controller sending a firing signal from his RPV control box, after which a large volume of vivid orange smoke is emitted for 40 to 50 seconds. This enables the RPV to be located at ranges of from 6 to 10 km in fair weather conditions. An infra-red smoke generator is also produced.

The lightweight and simple design ensures that a maximum number of stores will fit easily into existing RPV pods. The design also ensures that Miss-Distance Indicators (MDI), if fitted, are free from interference. In addition, the environmental sealing remains intact during flight unless the generator is ignited, thus avoiding any possible ingress of moisture before firing and also allowing a return to storage if the generator is not used.

The Schermuly Item No for Orange Smoke is 53980.

Specifications
Length: 100 mm
Weight: 280 g
Diameter: 57 mm
Delay: instantaneous
Burn time:
(orange smoke) 40 to 50 s
(infra-red smoke) 20 s
Maximum surface temperature: +210°C
Operating voltage: 1.5 V minimum
Pulse length: 50 ms

Schermuly Generator Smoke Towed Target A25 (IR)

Schermuly Generator Smoke Towed Target A25 (IR)

Schermuly expanded their range of target smoke generators with their Generator Smoke Towed Target A25 (IR) for simulation/training purposes on Rusl-ton or similar targets. The A25's payload consists of Ir fra-Red (IR) smoke which emits radiation in the 8-14 and 3-5μ wavebands to enhance the target signature for missile guidance and homing. The generator is ignited electrically from the towing aircraft or remotely if used on RPVs. Radiation hazard protection is maximised by an electrical fuze via a radio frequency interference filter. The operation altitude and speed parameters are within those of the towed target envelope.

The Schermuly Item No for this item is A25 (IR) 54080. Other Generator Towed Target Enhancement Signals available in the same range are as follows:

A21 - red smoke - burn time 60 seconds - Item No 83000

A22 - visual flare - burn time 60-85 seconds - Item No 82020

A23 - infra-red flare - burn time 18-28 seconds - Item No 82050

A24 - red smoke - burn time 40 seconds - Item No 83040.

Specifications
Length: 508 mm
Diameter: 45 mm
Weight: 1.4 kg
Burn time: 20-25 s
Detection range: up to 15 km
Operational altitude and speed: as for towed target

Schermuly RPV Target Smoke Generator

Schermuly Low Toxicity Screening Smoke Grenade

Schermuly Low Toxicity Screening Smoke Grenade

The Schermuly Low Toxicity Screening Smoke Grenade was developed to meet the need for a training smoke which can be used without posing a health hazard to personnel or creating environmental problems. The smoke-producing composition is a widely accepted low toxicity and non-carcinogenic formulation which produces a greyish-white smoke to provide maximum obscuration. It is cool burning (having passed the United Kingdom Ministry of Defence dry straw test) and will not set fire to, or contaminate, ranges.

The operation of the grenade is similar to conventional hand grenades. It has a fly-off lever ignition system with specially formulated smoke pellets. Smoke emission is through apertures at the top of the grenade body to improve the rapid build-up of the screening effect.

The Schermuly Item No is 53740.

Specifications
Length: 138 mm
Diameter: 61 mm
Weight: 475 g
Burn time: 60 s

Status
In production. In service with the United Kingdom armed forces and with some Middle East countries.

Manufacturer
Schermuly (Pains-Wessex Limited), Salisbury.
(For details see Directory of Suppliers)

Schermuly Training Grenade

This training grenade is a small compact facsimile of a standard hand grenade, for use on both training exercises and operations. The grenade is constructed from moulded plastic and has a simple fly-off lever action. After the grenade is thrown the lever is released and a delay train of approximately 2 seconds duration is ignited. The contents are then ejected through the base plate and in less than 1 second explode with a very loud report and bright flash. As the contents have no case or covering to fragment there are no harmful particles and there is minimal smoke discharge.

The grenade can be used on training exercises to simulate air or ground burst explosions. After operation the outer case can be used for dry training.

The Schermuly Item No is 57940.

Specifications

Length: 110 mm
Diameter: 70 mm
Total weight: 233 g
Explosive weight: 33 g
Delay to ejection: 2 s minimum
Ejection to operation: 1 m
Sound level: up to 170 dB at 1 m

Status

In production. In service with some Middle East armed forces.

Manufacturer

Schermuly (Pains-Wessex Limited), Salisbury.
(For details see Directory of Suppliers)

Schermuly training grenade

ML Wallop NT Training Smoke

Although introduced primarily for training, this smoke has obvious operational uses. The smoke composition was developed by Nico-Pyrotechnik of Hamburg, and requires no casing, having sufficient mechanical strength to withstand handling.

The British Army purchased a quantity of the ML Wallop training smoke NT90 in place of the Smoke Generator No 8. The smoke block is 80 mm in diameter and 65 mm in height and weighs 450 g. It is ignited by windproof matches supplied with the blocks, or by ordinary matches, cigarette lighters or even by the application of a lighted cigarette.

The generator produces a large volume of white, low-toxicity screening smoke and burns for a nominal 90 seconds. The advantage of this store is that it has no casing, is weatherproof, and is completely consumed, leaving no litter or debris.

Using the same basic material, ML Wallop developed the training smoke grenade NT30 which is intended to replace the current Grenade No 80, at least in the training role where the white phosphorus content of the latter is hazardous.

The NT30 grenade body is 100 mm long and has a diameter of 53 mm. It is ignited by a conventional fly-off lever igniter set which fits into a plastic housing. The grenade burns for a nominal 30 seconds, discharging a large volume of smoke from the instant of ignition, 2 seconds after throwing. It is almost completely consumed, only the fly-off lever and striker housing remaining.

ML Wallop also produces a range of battlefield simulators from rifle fire through to heavy calibre shells in ground/airburst forms with day and night versions.

Specifications

NT30
Weight: 450 g
Diameter: 53 mm
Length overall: 130 mm
Length of body: 100 mm
Ignition delay: 2 s nominal
Burning time: 30 s nominal; 15 s also available

Status

In production. In service with the British Army.

Manufacturer

ML Wallop Defence Systems Limited.
(For details see Directory of Suppliers)

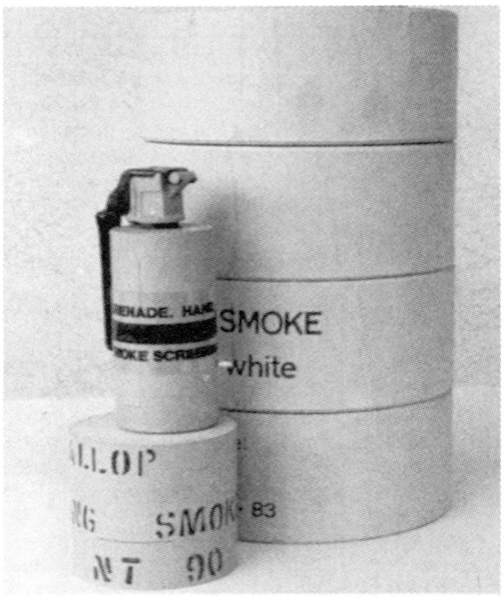

ML Wallop training smoke NT90 and grenade NT30

UNITED STATES OF AMERICA

CCI Blazer Lead-Free Training Ammunition

Blazer lead-free ammunition is intended for training use on indoor pistol ranges and employs a priming mixture that contains no lead or barium compounds. Non-reloadable alloy cartridge cases are used while the bullets are completely encased in pure copper. The rounds function in exactly the same way as conventional service ammunition and have identical accuracy and other performance, including full muzzle flash and recoil. The rounds can be used with automatic loading systems.

Blazer ammunition is available in a range of 28 loads, from 0.25 Auto up to 0.45 Colt, including the widely used 9 mm Parabellum and 0.38 Special.

Status

In production and in widespread service with military and law enforcement agencies worldwide.

Manufacturer

CCI, Omark Industries, Lewiston.
(For details see Directory of Suppliers)

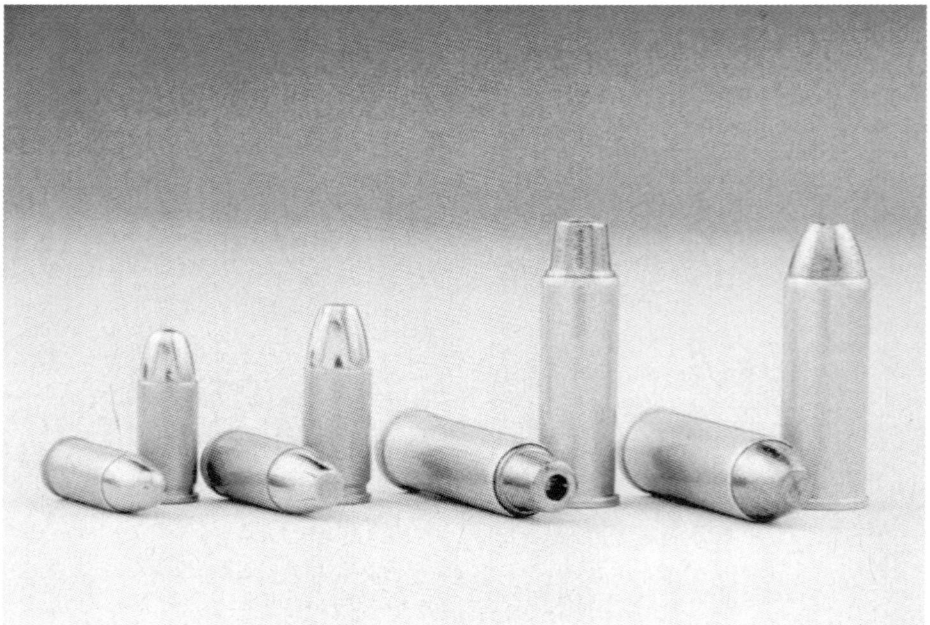

Examples of Blazer lead-free training ammunition

CTA Mine Training Kit

The CTA Mine Training Kit is a kit of completely inert land mine replicas and associated accessories contained in a robust carrying case. All the land mine types included in the kit are in service with the US Army.

The kit contains full-size replicas of M14, M16A2 and M18A1 anti-personnel mines and M15, M19 and M21 anti-tank mines. All the mines are exact in every visual and working detail apart from lacking any form of explosive filling. The training mines are supplied complete with replica fuzes, associated wiring and arming tools, and anti-handling devices where applicable. Each replica mine is moulded in a hard light blue plastic-based material and can be used for the complete simulation of testing, arming, deploying and disarming procedures and actions. Arming plugs may be set to armed and safe positions, while shipping plugs, anti-handling devices and fuze assemblies can be fitted and removed. In the case of the M18A1 (Claymore) the training mine is supplied complete with a test set, a blasting cap assembly and a carrying bandolier. Detonation of the M18A1 training mine is signalled by an electronic buzzer.

Status

In production. In service with the US Army.

Manufacturer

Custom Training Aids Inc (CTA), South Carolina.
(For details see Directory of Suppliers)

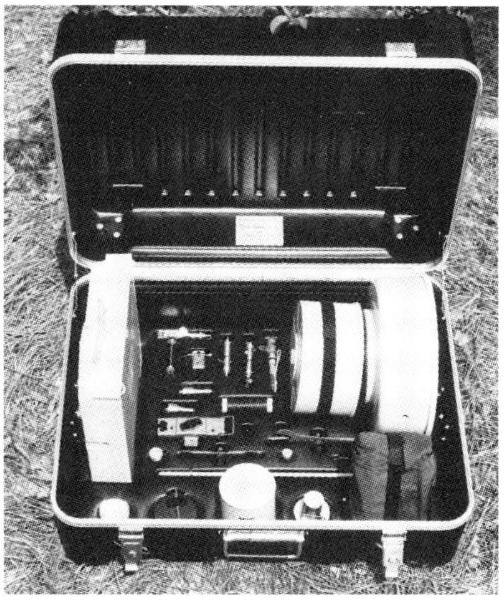

A complete CTA Mine Training Kit packed in its carrying case

CTA OXA Training Device

The CTA OXA Training Device is an expendable, movement-sensitive simulation of an unexploded land mine or submunition, designed for demolition by explosives during EOD training exercises. The OXA device's olive green body is constructed of non-toxic biodegradable materials which are nominally weather resistant.

The cylindrical CTA OXA training device contains connections for a 9 V battery (not included), a trembler switch easily adjustable for movement sensitivity, an electronic circuit board and components, and an on/off switch. When the 9 V battery is connected the device will detect slight movements and thereafter activate an integral buzzer audible at 50 m. A 40-second delay allows trembler switch movement to cease between device activation and circuit arming.

The device will remain watertight and functional for at least 24 hours in inclement weather conditions. The functioning temperature range is from −30 to +50°C. If not subjected to demolition the device is reusable. The instructor can adjust the level of movement sensitivity before initial use. There is sufficient metal within the device to activate a mine detector.

The OXA device weighs approximately 350 g, is 160 mm high and 100 mm in diameter.

Status

In production.

Manufacturer

Custom Training Aids Inc (CTA), South Carolina.
(For details see Directory of Suppliers)

CTA OXA training device

RamTech Long Shot Ballistic Impact Indicator

The Long Shot ballistic impact indicator is a long-range visual indicator which offers rapid confirmation of projectile impact on a target. The Long Shot senses the occurrence of projectile impact and rapidly signals back to the firer by means of a double flash from a high intensity strobe. The flash can be seen out to 1.6 km, even under bright light conditions.

Hit indication can be selected from two options: rapid (within 1.5 seconds of impact) or delayed (approximately 4-5 seconds from impact). An impact sensor can be attached to any location on the target perimeter via a 10 ft/3.05 m cable. It is a contact vibration sensor and is externally adjustable for sensitivity. The unit is self-monitoring and any functional damage of the sensor system is indicated by continual flashing of the unit for 8 seconds. The target sensor acts as a 'power up/power down' switch, so when an on-target impact is sensed the sensor's signal powers the strobe power supply, produces a discharge and then powers down the strobe supply to conserve battery power.

The unit is powered by a rechargeable 12 V battery. Dimensions are 109.5 × 178 × 203 mm.

Status

In production.

Manufacturer

RamTech Enterprises Inc, Texas.
(For details see Directory of Suppliers)

The Long Shot ballistic impact indicator

YUGOSLAVIA (SERBIA AND MONTENEGRO)

ITVM-M78 Underwater Burst Simulator

The ITVM-M78 underwater burst simulator is intended to simulate the subsurface detonation of artillery projectiles with calibres of 76 to 100 mm at sea or in lakes and rivers. In use the device is initiated with a friction fuze and dropped overboard or thrown using the fuze assembly as a throwing handle. The delay can be up to 5 minutes when the device detonates to produce flash and sound, a plume of spray and a cloud of smoke.

Specifications

Height:
(with fuze) 190 mm
(without fuze) 140 mm
Diameter: 96 mm
Weight: 0.8 kg
Weight of charge: 0.5 kg

Status

In service with the former Yugoslav armed forces.

Contractor

Holding Company YUGOIMPORT—SDPR Ltd, Belgrade.
(For details see Directory of Suppliers)

Effect of ITVM-M78 underwater burst simulator

Light Anti-aircraft Weapon Fire Simulator

This simulator is used during the training of anti-aircraft gun crews to simulate the sound of their guns firing during 'dry' firing exercises. The guns involved are usually the 20 mm M75-series anti-aircraft guns but the simulator could be used with other similar weapons.

The simulator can be mounted on the gun or placed nearby and creates the firing noise by mixing together propane or butane with air. The two gases are contained under pressure in cylinders and passed through pressure reduction valves before being mixed together and ignited in a compression chamber using a spark plug. The resultant sound is in the 110 to 130 dB range and the degree of sound is adjustable. The simulator will continue to simulate burst firing for as long as the weapon firing trigger is kept activated and the firing frequency is adjustable between 120 and 250 shots/min.

The simulator has provision for 0.6 kg of propane or butane and 4 kg of compressed air. The weight of the device packed in its case is 60 kg.

Status

In service with the former Yugoslav armed forces.

Contractor

Holding Company YUGOIMPORT–SDPR Ltd, Belgrade.
(For details see Directory of Suppliers)

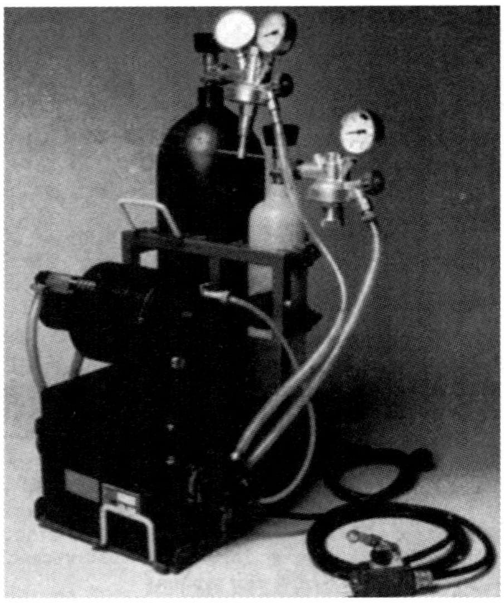

Light anti-aircraft weapon fire simulator

Nuclear Blast Simulator INE-M78

The nuclear blast simulator INE-M78 is normally placed in a prepared hole 350 mm in diameter and 600 mm deep. The device consists of a body formed from a watertight cardboard tube 327 mm in diameter and 1.2 m long. It contains a smoke charge assembly, a plastic vessel containing the smoke composition, and an explosive charge assembly, containing a plastic bottle filled with approximately 4 kg of amonal. An electric detonator with an electric lead 10 m long completes the simulator.

To prepare the INE-M78 for use after it has been placed in its prepared position, the electric detonator is removed from the tube and the smoke charge assembly is armed. Petrol and diesel fuel are poured around the tube to about two-thirds of the tube height; this is usually around 80 litres of fuel. The explosive charge is placed about 10 m from the device and the detonator is prepared and connected to the explosive charge, the smoke charge, and a blasting machine or exploder.

When detonated, the simulator produces a column of smoke 100 m high plus a fire ball 10 m in diameter, and lasting from 1 to 3 seconds. The sound effect is audible up to 3 km away.

When the INE-M78 is used the safety zone for men in the prone position or located in foxholes or trenches is 150 m. For men standing upright it is 250 m.

Status

In service with the former Yugoslav armed forces.

Contractor

Holding Company YUGOIMPORT–SDPR Ltd, Belgrade.
(For details see Directory of Suppliers)

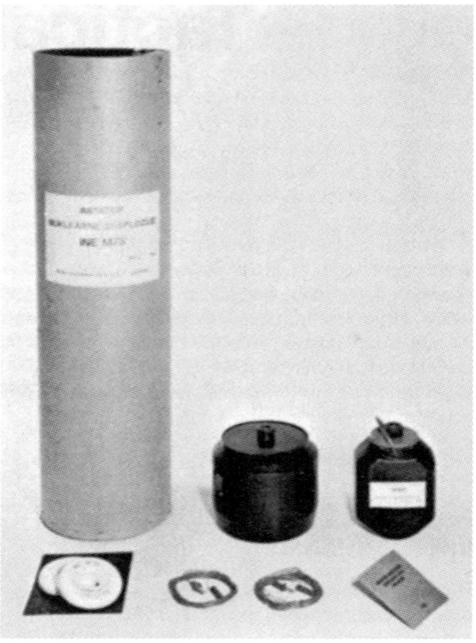

Nuclear blast simulator INE-M78

TACTICAL ENGAGEMENT SIMULATION

BELGIUM

Leentjens-Boës Tactical Engagement Simulators

Leentjens-Boës was formed in 1935 and in 1980 began the research and development for a series of laser-based tactical engagement simulators. The first of these was the Midra system for application with armoured fighting vehicles, anti-tank missiles and air-to-ground applications. This was followed by the Minidra system for hand-held weapons and sub-machine guns and, in 1988, by the Minidra S system for assault rifles and machine guns. All three systems are based on the use of laser transmitters and laser receiver/detector units.

Midra

The Midra simulation system is intended for use by armoured fighting vehicles and direct-fire anti-tank missile systems. It can also be used in an air-to-ground firing application by aircraft and helicopters. The basis of the system is a laser transmitter housed in a cylinder which can be fitted into a tank gun barrel or the launch tube of an anti-tank missile launcher. It can also be installed in aircraft or helicopter machine gun or rocket pods. Installing the laser transmitter involves a minimum of modification to the weapon system involved and the weapon can be aimed and fired using the normal controls. The laser transmitter contains its own power supply and emits a laser beam in a pulse-modulated form for identification and coding purposes.

CPU. Simulating the firing of guided missiles involves the use of a Central Processing Unit (CPU) to control the laser transmitter. The CPU is a self-contained, battery-powered waterproof unit on which an instructor selects the type of missile to be simulated and defines the firing range, as well as selecting the use of pyrotechnics or other firing effects involved.

Target Vehicles. Target vehicles carry a laser receiver known as a Mobile Reception System (MRS). When a target is hit by a laser beam simulating a projectile, a flashing lamp and an optional smoke discharger are initiated. The lamp will continue to flash until it is remotely reset by an instructor. The MRS has a 60° coverage in elevation and a 360° coverage in azimuth.

TRS System. A variation of the MRS system for short-range or simulated sub-calibre applications is known as the Training Reception System (TRS). This involves the use of a small target moving at variable speeds along a fixed rail. If the target is hit by the laser beam it comes to a stop and can then be restarted by a remote-control unit.

Air-to-Ground. For air-to-ground training applications the laser transmitter is fired against an Aircraft Target Unit (ATU). The ATU has a 60° coverage in elevation and azimuth.

Laser Transmitters. The Series 100 laser transmitter is for the simulation of short-range weapons with a maximum range of 500 m. The Series 500 transmitter is for longer range weapons with ranges of 500 to 3000 m. Three models are available in each series: the Model SM has a fixed laser source; the Model MP has an adjustable laser source; the Model MR also has an adjustable laser source operated by a remote-control unit.

Minidra

The Minidra simulation system is intended for use with hand-held weapons (such as pistols) and sub-machine guns. The basic components of the system are a laser transmitter and a detector unit worn as a body harness. Optional equipment includes an umpire's pistol and items such as battery chargers and transit cases.

The Minidra tactical laser simulation system in use

The Minidra S tactical laser simulation system in use

Laser Transmitter. The laser transmitter is housed in a small unit that is normally fitted under the butt of a pistol or secured under the barrel of a sub-machine gun by a clamp; adaptors can be supplied for each type of weapon. Power is provided by a 9 V Ni/Cd battery. The weapon involved is normally, but not necessarily, modified to operate with blank ammunition. Each time the trigger is pulled, the laser transmitter transmits an eye-safe coded laser beam along the line of fire.

The gallium arsenide laser involved has a wavelength of 904 nm and a pulse width of 200 ns. There are five laser beam diameters available with the diameter of the beam at 25 m being as follows: XS, 10 mm; S, 50 mm; M, 200 mm; L, 400 mm; XL, 800 mm. The XS and S versions are intended for precision shooting training while the M, L and XL are intended for tactical training. The transmitter measures 66 × 44 × 47 mm and weighs approximately 250 g. The laser has an life of about 15 000 'shots'.

Detector Unit. The body-worn laser detector unit carries eight laser detectors on a nylon textile supporting harness (four at the front, four at the rear) together with an electronic decoder unit. If a detector is hit by a laser beam a 'kill' signal is provided by an audible buzzer which can only be switched off by an umpire using an umpire pistol, a push-button on the decoder unit or a special key. A 'hit' light indicator mounted on the harness and powered by its own battery is available as an optional accessory. The detector unit is powered by a 9 V Ni/Cd battery which maintains the unit in operation for about 8 hours. When the battery is fitted the complete detector unit weighs 1.15 kg. Laser detectors may also be fitted to infantry range targets and target systems.

Minidra S

The Minidra S simulation system follows the same general lines as the Minidra system, is intended for use with assault rifles and machine guns and is configured accordingly.

Laser Transmitter. The Minidra S laser transmitter is contained in a unit measuring 135 × 45 × 65 mm clamped under the muzzle of a weapon and which emits a laser beam each time a blank cartridge is fired from the weapon; the system can be adapted to operate without firing blanks. It uses a gallium arsenide laser with a wavelength of 904 nm. At a range of 300 m the laser beam produces a 'kill' zone with a diameter of 50 mm and a 'near miss' zone with a diameter of 2 m. The transmitter is powered by a 9 V Ni/Cd battery which provides an operational life of about 10 000 'shots' for each charging. The laser transmitter weighs 600 g with the battery installed.

Detector Unit. The Minidra S laser receiver/detector unit carries 10 laser detectors on a nylon textile body harness: four detectors are on the back and four on the front with the other two worn one on each arm. A helmet harness carrying a further five detectors is available. The unit involves an electronic decoder unit which triggers a continuous alarm sound once one or more detectors have been hit by a laser beam; a 'near miss' produces a brief alarm sound. The 'kill' alarm can be silenced by the wearer lying on his back and the alarm can only be deactivated by an umpire using a special key or an umpire's pistol. A 'hit' light indicator powered by an extra battery is available.

The laser receiver/detector unit is powered by a 6 V accumulator providing power for about 24 hours. The unit weighs 1.25 kg.

Interoperability. The Minidra S simulation system is interoperable with other similar tactical engagement systems, including those in service with the US armed forces. It is in service with the Belgian Air Force.

Status

Available. The Minidra S system is in service with the Belgian Air Force, Army and the Belgian Ministry of the Interior. It is also in service with the French Army, NATO Headquarters Security Service and the EEC Ministers' Close Protection Service.

Manufacturer

Leentjens-Boës, Herstal.
(For details see Directory of Suppliers)

CHINA, PEOPLE'S REPUBLIC

QJM-2 and QJM-3 Laser Engagement Simulation Systems for Light Weapons

Laser-based weapon effect simulators were first developed in the People's Republic of China during the early 1980s and the first generation of laser engagement simulators for light weapons has been in service with the People's Liberation Army (PLA) for some years. In recent years a second generation of laser-based engagement simulation devices has been developed and is in production for use as an engagement simulator at infantry squad, platoon and company level.

QJM-2

The QJM-2 system for light weapons may be used with six types of weapons: rifles, sub-machine guns, light machine guns, machine guns, flame-throwers and pistols. A laser transmitter is located on or near the weapon muzzle, laser detectors on a body harness and helmet, and on a trigger mechanism unit. A buzzer set is located on the wearer's harness and a smoke emitting set on the helmet. There are five laser detectors on the helmet, three at the front and two at the back. Two laser detectors are located on the front of the harness and two at the back. The system is rain and humidity protected and is stated to have a low power consumption.

Laser Transmitter. When the weapon involved is fired, a gallium arsenide laser is actuated with a wavelength of 890 nm. The system employs coding and decoding of the laser pulses based on a single compact microprocessor board.

Hit and Miss Distance Assessment and Effects. Hit and near miss indications are calculated by the system. When a hit is indicated, a buzzer and smoke unit are both actuated. A near miss only produces a short duration sound signal from the buzzer. A hit will also cause the casualty's laser transmitter to be locked in an inoperable condition, thereby rendering the casualty tactically non-effective. A two-way induction system removes the need for cables between the weapon and body harness and weapon and helmet.

The cost of a QJM-2 set is 2500 Yuan (approximately US$290).

Specifications

Effective range:
(pistol, flame-thrower) 50 m
(rifle, sub-machine gun, light machine gun) 400 m
(machine gun) 800 m
Area of laser spot:
(pistol at 25 m) 400 × 400 mm
(flame-thrower at 25 m) 400 × 700 mm
(rifle and sub-machine gun at 250 m) 700 × 400 mm
(machine gun at 400 m) 1000 × 700 mm
Laser power emitted:
(pistol, flame-thrower) hit 0.15 W; near miss 0.3 W
(rifle, sub-machine gun, light machine gun) hit 0.3 W, near miss 0.4 W
Laser detector receiving angle: more than 100°
Casualty indication:
(hit) continuous buzz and smoke
(near miss) short buzz
Clock frequency of trigger pulse: 3 kHz
Power consumption:
(laser transmitter) less than 2 mA
(helmet receivers) less than 2 mA
(harness receivers) less than 3 mA
Effective audio control range: longer than 200 mm
Effective infra-red control range: longer than 1 m
Operational temperature range: −10 to +60°C

QJM-3 (Type III)

The QJM-3 (Type III) system is similar to the QJM-2 except that a target board has been added for marksmanship training and the output of the laser transmitter is increased, resulting in longer ranges.

The cost per set is 30 000 Yuan (approximately US$3450).

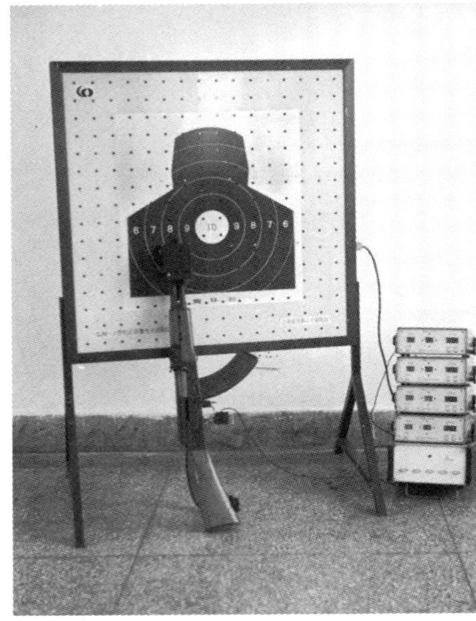

QJM-3 Type III laser engagement simulation system with its laser target board and recording equipment on the right

Specifications

Effective range:
(sub-machine gun, light machine gun) more than 400 m
(rifle) more than 1000 m
(pistol) more than 150 m
Laser power emitted:
(sub-machine gun, light machine gun) hit 0.2 W; near miss 0.3 W
(rifle) hit 0.25 W; near miss 0.3 W
(pistol) hit 0.15 W; near miss 0.25 W
Laser detector receiving angle: more than 90°
Effective audio control range: more than 400 mm
Effective infra-red control range: 1 to 2 m
Continuous operation: 10 h

Status

In production.

Development/Design Agency

Nanjing Research Institute on Simulation Technique (NRIST).
(For details see Directory of Suppliers)

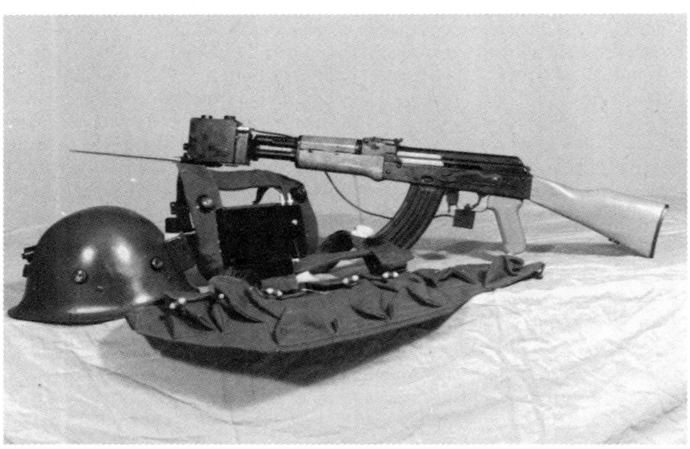

QJM-2 laser engagement simulation system for light weapons seen here in use with the 7.62 mm Type 56 assault rifle

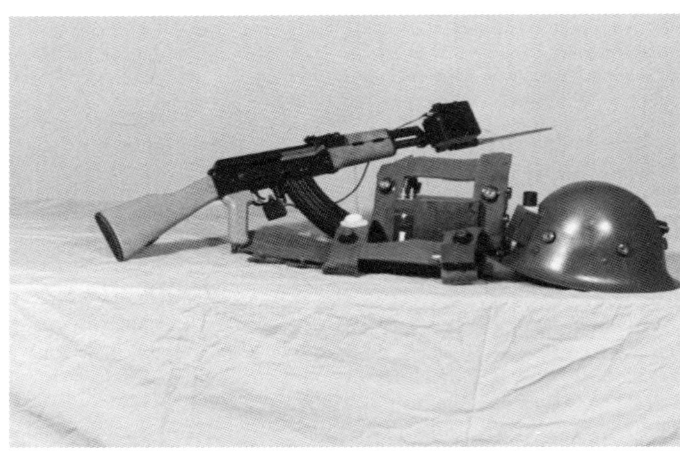

QJM-2 laser engagement simulation system showing the body harness between the helmet and rifle

CROATIA

LAIR Anti-tank Direct Fire Simulator

Developed to simulate direct fire from and at anti-tank guns, this system is produced in Croatia at the Laser and Atomic Research and Development Ruder Bošković Institute and is a laser system using projectors and receivers. The anti-tank gun component is part of a laser simulation system used by the Croatian Army and

M79PT anti-tank direct fire simulator fitted to 100 mm anti-tank gun T-12

is known as the M79PT; other components are the M76T used on AFVs and the KOGS-L used by infantry.
Components. The anti-tank gun fire simulator M79PT is issued as a kit and is used with 76 and 100 mm anti-tank guns. On both types of gun the laser transmitter is secured over the barrel just forward of the gun shield with a laser receiver unit mounted on top of the shield together with a flashing lamp and a smoke cartridge holder; similar units are mounted on the target AFV. If incoming laser signals are received by the laser receiver on the anti-tank gun, the lamp will start to flash and the smoke cartridge will emit coloured smoke.

Specifications. The M79PT is set to the combat ranges of the gun involved. For a 100 mm gun this is 1200 m and for a 76 mm gun 900 m. A 12 V power source is required. The laser employed has a power output of 7 W. Total power consumption of the system when switched on is 60 W and in standby 5 W. A hit counter is incorporated. The reception angle of the laser receiver is 90° in azimuth and 40° in elevation.

Status
In service with the Croatian Army.

Manufacturer
LAIR Croatia, Zagreb.
(For details see Directory of Suppliers)

FRANCE

Giravions Dorand DX 175 Tactical Engagement Simulators

The DX 175 tactical engagement simulators are a range of equipment designed to evaluate the effects of direct fire weapons during simulated combat. Each of the simulators comprises an optical unit equipped with a safe low power laser projector and a laser detector system. The laser transmits in place of firing a normal round of ammunition and the detectors mounted on a target sense and transmit the results for computed evaluation.

The DX 175 can be fitted on any tank or armoured vehicle, all light portable weapons, and helicopters or other vehicles equipped with missile weapon systems.

There are four types of DX 175.

DX 175/Tank
The DX 175/Tank combat firing simulator is used to evaluate and monitor the effects of direct shots fired from tanks or other armoured vehicles at real targets equipped with the appropriate detection systems.
Components. The main assemblies of the DX 175/Tank are an optical unit equipped with an eye-safe, low-powered laser projector, so-called 'beacon' units equipped with electro-optical detection and reflection systems, a microprocessor-based computer into which a specific data-carrying cassette for the carrier and its weapon are inserted, and a display and control unit.

Other components of the system include an umpire gun for the remote monitoring of the system operation and for umpire intervention, an operating unit for entering data into the carrier and weapon characteristics cassette, and pyrotechnics to simulate gunfire and direct hits.

On tanks equipped with automatic fire-control systems, an additional electronics unit enables the simulator rangefinder to be used in place of the operational rangefinder.

Special beacons for mounting on static targets or wheeled general-purpose vehicles are available.

The DX 175/Tank is in production for the French Army.

DX 175/Missile
The DX 175 Missile carries out the same training function as the DX 175/Tank and is completely compatible with the tank gun simulator. It can be configured to suit various short- or medium-range anti-tank missiles. Its MILAN configuration comprises an assembly housed in the missile tube which contains a laser projector similar to that used on the DX 175/Tank, along with computing and recording modules. The simulator is powered by batteries or the battery pack of the DX 143 MILAN weapon simulator. Other components (umpire gun, etc) are the same as for the infantry laser simulators. The system's target function can be performed by reflector beacons mounted on a target vehicle and/or sensor harnesses carried by the missile launcher crew.

DX 175/Helicopter
The DX 175/Helicopter follows the same general design as the DX 175/Tank system other than that the control unit located in the helicopter cockpit is of reduced size. Intended for the Gazelle armed with HOT or 20 mm cannon-armed, and the AATCP helicopter.

DX 175/STCAL
The DX 175/STCAL *(Simulateur de Tir de Combat pour les Armes Légères d'Infantrie)* is intended for use by infantry or other personnel using small arms. It uses a low-powered laser projector mounted on the weapon involved, the user wearing a sensor-equipped body harness which also carries the system computing

DX 175/Tank tactical engagement simulator fitted to a Panhard ERC 90 F4 Sagaie armoured car

DX 175/STCAL small arms tactical engagement simulator fitted to 5.56 mm FAMAS rifle

DX 175/Missile tactical engagement simulator configured as a MILAN anti-tank missile launcher

Gazelle helicopter equipped with DX 175/Helicopter tactical engagement simulator for HOT anti-tank missiles

equipment. It may be initiated by firing blanks. A boresighting kit is used to align the laser projector with the weapon's line-of-sight. Projectors are programmed at the beginning of an exercise by a downloading and readout unit which is also used to read results. An analysis unit uses data from the readout unit to summarise results achieved by all participants in an exercise.

The DX 175/STCAL is in production for the French Army. A total of 3500 systems has been ordered for delivery over a five-year period for initial in-service use during 1994.

Status
See text.

Manufacturer
Giravions Dorand Industries.
(For details see Directory of Suppliers)

GERMANY

Eichweber TALISSI Weapon Fire Simulator

The TALISSI *(TAktischer LIchtSchuss SImulator* - Tactical Light Shot Simulator) was designed by Kurt Eichweber in 1963 and demonstrated in 1965 with a ruby laser. Subsequent research and development, some on behalf of the Federal German Army, led to the present family of systems which employs gallium arsenide semiconductor lasers. After extensive trials 100 systems were delivered to the Federal German Army in 1977. This and subsequent deliveries total well over 1000 systems in service on Leopard 1, Leopard 2 and M48A2GA2 MBTs.

Components. The TALISSI comprises a laser transceiver which is normally fitted in the gun barrel, a central control unit located in the turret, optical receivers, a display unit, a monitoring unit for use by an umpire, and various cables and items of mounting hardware.

Sequence of Operation. The TALISSI 2 system sequence of operation on the Leopard 1 is as follows: the commander first observes the target and orders the gunner to engage it. The range to the target is measured (or estimated) and the gun aimed, and the range and type of ammunition being used are entered into the central control unit by the crew. When the gunner depresses the firing trigger, a laser pulse is transmitted and at the same time a bang and smoke are produced by the muzzle flash simulator. Using the range already entered and stored by the simulator, the gun is automatically aimed directly at the target when the ballistic compensation facility is provided. The divergence of the beam fired by the laser at the target corresponds to the weapon dispersion. Provided that the aim is correct, the laser pulse is detected by the optical receivers at the target. The pulse is reflected back to the transceiver by the reflectors mounted on the target and the elapsed time is used to calculate the target range. The result is accurate to within 5 m. The measured range is compared with that already in the computer memory. If the discrepancy lies within certain limits dependent on the type of round and the range, a second laser pulse is emitted which contains the additional 'hit' information and causes the target's TALISSI equipment to be switched off and a 'hit' to be displayed on the attacker's equipment. If the difference in range lies outside the limits, 'short' or 'over' is displayed on the attacker's equipment and 'under attack' is displayed in the target vehicle.

System Characteristics. Characteristics of TALISSI can be summarised as follows:

Its range of 3000 m corresponds to the tactical combat range

Laser beam divergence corresponds to the weapon's dispersion

Different properties of various types of ammunition are taken into account

Various types of ammunition can be simulated

Other functions of the weapon are not compromised

The laser is guaranteed for 100 000 shots

A hit is indicated at once and the result transmitted immediately to the umpire.

There are a number of TALISSI systems making up a family of interoperable simulators. These are as follows:

DWS-1. This is the simplest variation and is a one-way laser device. It is used as an umpire gun enabling an exercise umpire to interrogate a TALISSI-equipped vehicle to confirm that the system is operational; it can be used to 'kill' a vehicle if it enters a 'no go' area and in some cases is used to reactivate and replenish the 'ammunition' in an AFV.

The basic DWS-1 system was used to evaluate the concept of aircraft ground attack simulation. Two different systems were tested, one for small calibre helicopter-borne weapons, and the second for the 27 mm Mauser cannon of the Alpha-Jet.

DWS-2. This system, which is widely deployed by the German Army, incorporates a rangefinder which indicates the engagement result on the control unit as 'short', 'target' or 'over'. The system will score a 'hit' only if the gunner has laid the line-of-sight and determined the range accurately before firing. The actual target hit area is calculated from the fall angle ballistics and the range.

DWS-3. This simulator is similar in concept to DWS-2 but employs ballistic compensation according to different ammunition characteristics. With this system, the gunner is required to take into account the quadrant elevation as well as range and direction. The DWS-3 did not enter volume production.

DWS-4. This system is for Main Battle Tanks (MBTs) fitted with laser rangefinders and certain types of ballistic computer fire-control systems. Developed for the Leopard 2 fitted with the KAE (Hughes) fire-control, the DWS-4 may also be fitted to MBTs with fire-control systems that are similar to the Hughes model, such as M60A3 and Leopard 1/SABCA. As an option, data associated with the last 40 shots fired may be stored in the central processor for post-exercise analysis.

DWS-5. This was designed to operate with any MBT/ballistic fire-control system combination as no information needs to be received from or given to the ballistic computer. The overall system can be exchanged between vehicles with different fire-control systems and main armament calibres greater than 75 mm. Fall-of-shot data is stored in the central processor.

TALISSI for Dragon. This is a weapon effect simulator specifically for the Dragon anti-tank weapon system. The laser transceiver is built into a dummy tracker unit, and the central processor and launch/hit pyrotechnics are located in a dummy weapon tube. A monitor unit and a target control unit complete the overall system. The monitor unit is used for initial aiming training; this does not require a mobile target vehicle, and may be used with a single reflector located typically at 100 to 300 m from the training position. The target control unit is located in a mobile target vehicle. This unit displays and stores comprehensive data on the performance of the gunner, location of hit on vehicle, kill probability and aggressor identification. This data can be printed out later for analysis and assessment purposes.

TALISSI for MILAN. This was designed specifically to simulate the effects of the MILAN anti-tank weapon system. The laser transceiver is mounted on the localiser housing and may be used with the night sight fitted. The central processor and pyrotechnics are located in a dummy missile tube. All other characteristics, display unit, and so forth are as described for the Dragon simulator.

Loading a TALISSI 4 projector into the 105 mm gun barrel of a Leopard 1 MBT

TALISSI for TOW. This development for the German Army simulates tripod-mounted and vehicle-borne TOW anti-tank weapon systems. A version, first produced in 1983, has the laser transceiver located in the sight housing and the electronics and pyrotechnics in a dummy missile tube. This system, suitable for TOW, ITOW and TOW-2, offers the general characteristics as detailed for the Dragon simulator.

A variant for the TOW Armoured Launching Turret (ALT) was produced.

TALISSI EDDA. The development of TALISSI EDDA *(Echzeit Duell Daten Auswertegerät* - real-time battlefield event recorder) began in 1983. It is an additional facility for use with the TALISSI tactical and gunnery simulator system for MBTs, fire support systems, anti-tank weapons, machine guns and anti-tank helicopters. By using EDDA the events and results of a training exercise are registered in real time for subsequent assessment and analysis purposes.

When used with MBTs and field artillery, a TALISSI EDDA radio data transmitter is fitted to any weapon system equipped with a shot and kill effect simulator. The transmitter is usually fitted close to the hit and kill effect simulator. Data from the EDDA transmitter is sent to an

TOW infantry launcher fitted with TALISSI

Effect of a TALISSI 'hit' on a Leopard 2 MBT

EDDA radio receiver which acts as a recorder and control unit. In the receiver the EDDA digital codes are displayed for red and blue contestants, totalled and a data readout presents results for analysis. The receiver also acts as a counter for ammunition fired and for the number of hits on each combatant. The receiver may be set up in the field or carried on a vehicle.

Pyrotechnics

Associated with the various TALISSI systems are special pyrotechnics and firing effect equipment. The DG 60/3 flash/smoke unit mounted on the weapon system carries three smoke 'hit' pyrotechnics and three disposable blocks each containing 20 main armament muzzle flash and smoke devices. Electronic circuitry automatically sequences the firing of the 60 devices, thus enabling the reuse of partly exhausted blocks. The total weight of the flash/smoke unit is 4.2 kg and it is powered by the vehicle's 24 V supply or by the TALISSI system. A 5 m safety distance is recommended in the direction of firing. For the guided weapon simulators, the launch signature pyrotechnics are mounted in an annular assembly around the orange smoke 'hit' indicator pyrotechnic. This whole assembly, which is also disposable, is mounted on the rear of the dummy weapon tube.

Status

In service with the German Army, Jordan, the Netherlands and Switzerland.

Manufacturer

Kurt Eichweber Präzionsgerätewerk, Hamburg.
This company is understood to be no longer trading. Its last address is given in the Directory of Suppliers.

ITALY

Galileo SECAT Tactical Engagement Simulation System

SECAT is a laser-based system designed to simulate gunnery training procedures and armed vehicle engagements. The system is based on an eye-safe laser system and live ammunition is not fired.

The SECAT system has two main subassemblies, an attack subassembly and a target subassembly.

Attack Subassembly. The attack subassembly consists of a laser transceiver, a system computer and a control panel. It is mounted on an armoured vehicle, typically a tank, and both the commander and the gunner control the vehicle's fire-control system normally. Fire is simulated by means of a modulated laser beam aimed at the target subassembly which is installed on another armoured vehicle. Vehicles may be equipped with both attack and target subassemblies to provide both fire and target facilities for tactical engagement training.

On the attack subassembly the laser transceiver and system computer are interconnected. The laser uses a wavelength of 960 nm and both the transmitter and receiver are located on a tilting platform. The platform is controlled by a stepper motor to introduce a preprogrammed displacement of the laser beam, both in azimuth and elevation. On a typical installation the transceiver and its platform would protrude from the vehicle gun muzzle while the system computer would be installed inside the barrel.

The attack subassembly is completed by the system computer. It has two sets of controls, one for the instructor and the other for the trainee gunner/operator. The instructor can enter, under key control, parameters such as ammunition loading, a reset for shots on target, automatic or manual mode for the horizontal lead angle, an optional application for missile firing, the display or blanking of fire data (optional), and a selector for the automatic reset of the SECAT spinning alarm lamp (see below). Operator controls include self-test, ammunition selection, a 'fire' control to start the firing programme, and horizontal lead angle manual set and reset. Information displayed on the control unit includes ammunition selected, operating mode (self-test, ready to fire, shot fired), the actual horizontal lead angle, target under attack (left, right, up, down) hit on the target (up, down), score data, reaction time, and available ammunition.

Laser transceiver unit for SECAT tactical engagement system inserted in 105 mm tank gun muzzle

Target Subassembly. The target subassembly has four laser detectors, an optical-acoustic alarm and laser reflectors. The four detectors, two on the upper and two on the lower part of the target, are connected to the optical-acoustic alarm unit. This unit includes the relevant electronic circuitry, a spinning lamp and a buzzer. If the target subassembly is 'hit' by an attacking laser beam the spinning lamp comes on, thereby denoting the position and 'hit' state of the target, and the buzzer sounds inside the target vehicle. At the same time the target's attack subassembly is inhibited and cannot be operated until the locked control on the control panel has been operated.

The target subassembly is completed by a number of laser reflectors, usually four to a vehicle. Each consists of a spherical crystal with a field of operation over 180°; four reflectors provide a full 360° coverage for a vehicle.

Operation. When SECAT is initiated by depressing the firing button on the control panel the fire-control procedures are the same as those carried out during a normal engagement. A laser beam is transmitted and the tilting of the transceiver platform produces a path equivalent to the trajectory of live ammunition for the range concerned. If the beam hits the target, the signal is returned by the reflectors and is detected by the laser receiver in the attack subassembly. At the same time the optical-acoustic alarm unit on the target produces 'under attack' data. The time delay between two reflected signals is detected and processed by the laser transceiver circuits in order to provide range, elevation and target data with respect to the attacker. This data is compared with other fire control parameters, such as ammunition type selected, and calculations produce information such as hit on target (up or down) or target missed, (up, down, left, right). The transceiver continues to scan to allow the laser beam to hit the target's lower sensors to provide the target with the hit data. If the lower part of the target is shielded, the target data will be confined to 'under attack', but if the lower part of the target can receive hit data then a hit condition will be signalled.

Rounds Simulated. SECAT can be used to simulate the firing of HESH, HEAT and APDS rounds, and can also be used to simulate the firing of MILAN anti-tank missiles. For HESH the operational ranges are from 50 to 1500 m, for HEAT from 200 to 2200 m and for APDS from 300 to 2200 m. The MILAN simulated operational range is up to 2500 m. In all cases the maximum target speed is 15 m/s. The SECAT system has a storage limitation of 30 rounds for all three ammunition types.

The SECAT system computer can be converted to allow training with different ballistic or missile weapon systems. The equipment involved can be installed by first-line personnel.

Status

Available. In use in the Italian Army.

Manufacturer

Officine Galileo SpA, Firenze.
(For details see Directory of Suppliers)

NEW ZEALAND

Oscmar Infantry Weapons Effects Simulation Systems

The Oscmar Infantry Weapons Effects Simulation Systems (Oscmar-IWESS) are a range of tactical engagement simulators intended for use by infantry and which use eye-safe laser projectors to simulate fire of rounds. The basic system can be programmed in a number of ways, being compatible with existing systems such as MILES (Multiple Integrated Laser Engagement System). The system can also be identity coded and exercise computer evaluation is available.

Oscmar-IWESS is microprocessor driven and can be umpire programmed to match the characteristics required for an exercise. For example, the umpire can programme the standard eye-safe laser projector employed to match the range and ballistic characteristics of the weapon involved; an assault rifle could have a range of 400 m while a tripod-mounted machine gun could have a range of 1500 m. Equipment can also be umpire programmed to fire and receive the codes from other manufacturer's equipments, allowing international exercises to be conducted with existing equipment. Oscmar-IWESS uses 16-bit codes to allow for future code expansions.

Oscmar-IWESS is offered in a basic form but all systems are available with computer evaluation facilities. Using this facility, all projectors fire pulses coded uniquely for that projector, while all target systems have a time memory which records activity against a real-time clock. After an exercise, or periodically during an exercise as required, umpires can use the controller gun to interrogate target systems and transfer their recorded data to an Analysing Printer Base (APB). The APB sorts data by combatant and time and provides a hard copy printout to give combatants feedback on their exercise performance and to preserve the data for future analysis.

All data transfer from target systems through controller guns to the APB is by means of optical data linkages that do not require plugs or cables. To transfer data, the umpire selects the appropriate function on his controller gun, places the port against that on the equipment and pulls the trigger. Diodes then transfer data optically between equipments in a few milliseconds.

Controller/Umpire Gun. The controller gun allows an umpire to inflict hits, kills and near misses, and reset targets which have been killed, all from a distance of up to 800 m. The controller gun is also used to set target time clocks, programme infantry projectors for weapon type and range, enter ammunition stocks into direct fire weapon projectors, functionally test targets and projector operation and collect post-exercise data for transfer to the APB. The controller gun operates off a standard 9 V commercial PP3 battery and weighs 0.8 kg. Dimensions are 200 × 80 × 50 mm.

Laser Projector. The Infantry Individual Weapon Projector is used to fire the laser pulses at targets and is a standard unit that can be fitted to almost any infantry weapon using a two cam lever system. A variety of brackets is available for unusual weapon types. The laser-transmitted codes for kill and near miss are indicated by the firing of a blank cartridge; a sensitivity adjustment is provided to cover a variety and difference in quality of blank ammunition. A dry fire mode is provided but this is used only in conjunction with a target alignment kit.

The laser projector can be programmed by the controller gun to simulate the range and ballistic characteristics of the weapon to which it is fitted. Options available are assault rifle (400 m), light support weapon (M249 Minimi - 600 m), sniper rifle (800 m) and tripod-mounted sustained fire machine gun (1500 m). Ranges can be changed and additional weapons added to meet customer requirements. The laser beam divergence is set to 0.5 m for a kill and 1 m for a near miss. The projector is powered by a standard 9 V PP3 battery that provides approximately 15 000 simulated rounds of fire. Dimensions are 100 × 78 × 43 mm and the weight is 450 g (0.99 lb).

Soldier Harness. The Soldier Harness detects laser beams falling on or near the head and torso throughout a 360° azimuth. The harness has four sensors on the front of the body, four on the back, one on each upper arm facing to the side and four evenly spaced around the head. Each sensor has an angle of sensitivity of 120°. The harness indicates to the wearer that he has been engaged by emitting an audible alarm. A near miss will sound the alarm for one second while a kill pulse will activate the alarm until the wearer lies on his back. If he attempts to rise again the alarm will restart, until the harness is reset by an umpire using his controller gun. All harness and target memory electronics are contained in a chest-mounted box and the harness itself is fully adjustable to suit the different sizes of wearers and type of combat webbing worn. A self-test sequence is initiated when a battery is first inserted. Weight of the harness is 1.5 kg (3.3 lb).

Printer/Computer. The Analysing Printer Base (APB) is a portable data sorting computer which is ruggedised to meet military requirements in the field. A printer is included and there are cables to allow operations from an internal battery, a 24 V DC vehicle power source or a mains power supply. The function of the APB is to take and sort the data downloaded into it by an umpire using his controller gun. The APB correlates projector and target activity and presents a printout of each combatant's activity, showing data such as who he has killed, hit or near-missed and when, what type of weapon he was using, how many shots he fired, who killed, hit or near-missed him and when and what types of weapon they were using, how many times and when he was reset and by what controller gun, any controller gun kills, hits, and if he removed his battery to deactivate his harness. Printout presentation can be tailored to meet customer requirements. The ability to sort reports on a Red/Blue Force basis, and the ability to sort data in terms of time from H-hour is available. Analysis can then determine at what part of a plan a force sustained its casualties. The APB weighs 12.5 kg (27.6 lb).

A New Zealand Army exercise umpire using the Oscmar-IWESS controller gun to programme a squad's individual harness sets

Alignment. Oscmar-IWESS is completed by a Projector Alignment Kit (PAK) which is used to align the axis of the laser projector with the Line-of-Sight (LoS) of the weapon sights. The differing offset between the projector and the LoS for various weapons is compensated for by the use of a different aiming screen for each weapon type - a full set of screens is provided with each PAK. In use, a soldier connects a read-out unit to the PAK and withdraws until the 25 m cable between the two is fully laid out. The soldier is then the correct distance from the PAK and he can select the alignment mode on his projector, take aim at the screen cross-hair and dry fire. The read-out unit then tells him which way to adjust his projector by setting the adjusting knobs. The PAK can also double as an indoor marksmanship trainer.

Hand Grenade and Mine Simulation. Other Oscmar-IWESS equipment includes simulated hand grenades and Claymore mines.

Status

In production. In service with the New Zealand Army (640 sets). Ordered by the Australian Army (1200 sets), the Royal Thai Army (1400 sets), and Denmark.

Manufacturer

Oscmar Products Limited.

(For details see Directory of Suppliers)

PORTUGAL

INDEP SITPUL Tactical Engagement Simulator

The INDEP SITPUL Tactical Engagement Simulator (TES) contains codes for up to 8000 firers in tactical exercises, and can be integrated with other tactical simulation systems. The laser projector can be fitted to any type of direct fire small arms and is capable of 10 000 shots before a battery change is required.

The SITPUL TES uses laser detectors located around the helmet and on a body harness, all controlled by microprocessor. The sensors give a complete all-round coverage and can supply precise indications of a 'kill' or 'near miss' from an incoming laser beam. It can also be supplied with an event reader which allows exercise data to be readily available, and allows data storage for later analysis. The event reader can be used at any time in the field as well as at initialisation and exercise completion.

Characteristics. The SITPUL TES was designed to comply with the following requirements:

To display realistic indications of a small arms 'kill' or 'near miss' with an accuracy similar to the weapons involved

To ensure eye and other physical safety

To be able to read all events related to an individual user or weapon

To be compatible with other laser simulation systems by simple software exchange

To be capable of day and night operation

To be installed rapidly and easily on the weapons to be used

Not to interfere with the normal sight-line of the weapon involved

Not to affect the balance or the handling of the weapon involved

To allow fitting of the receiver sensor harness over any type of combat clothing, including NBC, desert and arctic warfare clothing

Not to permit the user to 'kill' himself (exercise suicide)

Not to be sensitive to the report of blank ammunition fired nearby.

SITPUL TES - Description of Operation The SITPUL TES has four main components: a laser transmitter, a detector, a controller gun and an event reader.

Before the Exercise Begins. In operation the laser projector is installed on a weapon and calibrated using a zeroing device and the detectors on potential targets in the simulated exercise. Before each exercise a controller checks the operation of all equipments.

Laser Projector. The laser projector is fitted to the user's normal weapon in such a manner that the normal sight-line is not obstructed. The laser is activated either by a weapon trigger press or the firing of blank ammunition. The laser beam carries coded pulses which identify the firer. The projector can simulate the weapon rate of fire and takes account of the amount of ammunition in the weapon magazine. It can also introduce a delay in firing which simulates the time required to change a magazine. The system can operate for a maximum of 120 shots for automatic rifles and 250 for light machine guns, before resetting is required. This is done by the insertion of a key into the laser transmitter by an exercise umpire or controller.

Detector Units. The detector units consist of a body harness and a helmet harness. The body harness is made of nylon fabric and carries 10 detectors. There is also a detector unit with a built-in microprocessor to register the identity of the laser beam firer and the result of the engagement, which could be a 'kill' or 'near miss'. The detector also includes a horn that can emit a continuous note for a simulated 'kill' or an intermittent sound consisting of three pulses a second for a 'near miss'. The helmet harness carries five detectors connected to the detector microprocessor and can be fitted to any kind of helmet. In the 'kill' condition the horn can be switched off only by the intervention of a controller using his umpire gun, or by the wearer or an umpire inserting a laser transmitter key inserted into a deactivating device on the detector harness. This action deactivates the laser transmitter and removes the user from any further part in the exercise. The detector programme allows the 'near miss' condition to be changed to a 'kill' when six successive 'near miss' shots have been indicated.

Controller Gun. The controller gun is a laser transmitter used to check system functioning or impose 'kill'

SITPUL tactical engagement simulator in use

SITPUL tactical engagement simulator showing head and body receiver harnesses, and laser projector mounted on a Heckler & Koch 7.62 mm G3 rifle

and 'near miss' conditions. It can also be used to reverse casualty conditions and allow 'killed' exercise participants to come back into action again.

Event Reader. The event reader allows data collected in detectors to be stored for further computer processing. Data can also be displayed on site using a liquid crystal display (LCD). Information such as the number and identification of detectors is available, together with data such as the number of hits received, precise 'kill' or 'near miss' indications, plus the identification of the weapon involved.

Status
In production.

Manufacturer
INDEP, Lisbon.
(For details see Directory of Suppliers)

ROMANIA

Laser Firing and Training Simulator for Tank Guns

This laser-based system is intended for the training of tank gun crews in basic shooting skills and can also be used for tactical engagement simulation (TES). It is based on a laser projector in a unit mounted on a tank gun barrel. A dummy round is loaded into the tank gun breech as a normal piece of ammunition and only then can the laser projector be used to transmit at targets which carry reflector arrays which then transmit back data on accuracy of fire.

Components. The complete system comprises a transmitter/receiver unit, a receiver unit, a control unit, a starting and simulating unit, a pyrotechnics unit, a results display unit, dummy rounds that simulate high explosive or armour-piercing rounds, retro-reflectors, and connecting cables.

System Characteristics. The system may be used between ranges of 600 to 1800 m at rates of fire of 6-7 rds/min. The system measures and displays the range to the target with an error of ±10 m and provides firing accuracy indications such as hit, over, under, left and right. The number of shots fired is also recorded. Loading simulated high explosive or armour-piercing dummy rounds alters the firing characteristics of the simulator. Power for the system is derived from the carrier tank's electrical system.

Status
In service with the Romanian Army.

Manufacturer
Romanian state factories.

Elements of the Romanian laser firing and training simulator for tank guns

---

RUSSIA

Electronintorg 9F824 Small Arms Tactical Simulator

The 9F824 small arms tactical simulator was first demonstrated in 1990 and is now in production.

Components. The simulator consists of a laser projector and rifle mount, control unit, body harness with six laser detectors, helmet band with four detectors, two batteries, and appropriate cables. The laser projector fits onto the rifle barrel to the left of the centre line of the weapon. The simulator is used on AK-74 assault rifles. The location allows the firer to use the normal weapon sights but may alter his grip on the weapon; the lateral balance of the weapon is also affected. A cable runs from the projector either to a battery pack on the

9F824 small arms tactical simulator in use showing the laser projector mounted on an AK-74 rifle and the control unit on the back of the firer (Henry Dodds)

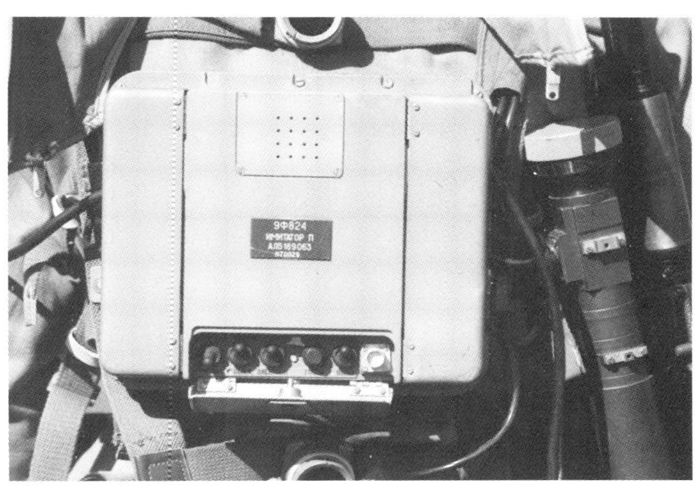

Close-up of 9F824 small arms tactical simulator control unit showing the lockable control panel flap in the open position (Henry Dodds)

firer's webbing harness or to the system control unit. A separate sight unit is used for system calibration.

Control Unit. The control unit is fitted to the body harness and is worn on the back. Four toggle switches, a push-button and a multi-pin connector are located under a lockable panel and are used to switch the system on and off and also into a training mode.

Target System. The target system consists of a total of 10 detectors mounted on the body and helmet harnesses, connected to the control unit by a cable.

The 9F824 can simulate firing over a range of 100 to 500 m and the system capacity before recharging is 128 electronic 'shots'. The battery delivers from 10.5 to 14.7 V. A complete equipment, packed, weighs 21 kg (46 lb).

Status

Offered for export.

Manufacturer

Electronintorg Limited, Moscow.
(For details see Directory of Suppliers).

Rear view of soldier wearing the 9F824 small arms tactical simulator showing detectors, the control unit and the laser projector mounted on the rifle; note the connecting cables (Henry Dodds)

Soldier wearing body and helmet harnesses for the 9F824 small arms tactical simulator (Henry Dodds)

Tank Gunnery Simulator

First seen in 1990, this tank gunnery simulator is mounted on main battle tanks or BMP IFVs and consists of a laser transmitter/receiver, a set of laser detectors and reflectors on a short mast with a flashing light on the top, a large control unit connected to a smoke grenade, three crew indicators (for the commander, driver and gunner), cables, a mounting kit, and green-tinted laser safety goggles; the provision of the latter may indicate that the laser and power used is not eye-safe. The system is mounted on a vehicle with the laser transceiver mounted in line with the main gun. The detector mast is mounted on the turret and the control unit can be on or inside the turret.

Operation. When the gun trigger is pressed, the transmitter unit sends a laser beam towards the target. If the target is equipped with deflectors/reflectors and the shot has been properly aimed, the laser beam is reflected back to the attacker and detected by a sensing system in the receiver module. This allows range and fall of shot relative to the detector/reflector mast to be calculated, thereby determining whether the shot fired was a hit or a miss. If a hit, then a smoke grenade is ignited on the outside of the target vehicle, a flashing red light is switched on and indicator units inside the target tank show the crew that their vehicle has been hit.

Dimensions. Approximate dimensions of the laser transceiver are 600 mm long, 300 mm wide and 250 mm high. The control unit is about 700 mm wide, 500 mm high and 300 mm deep.

Status

In service with Russian armed forces.

Manufacturer

State factories.

Components of the Russian tank gunnery simulator (Henry Dodds)

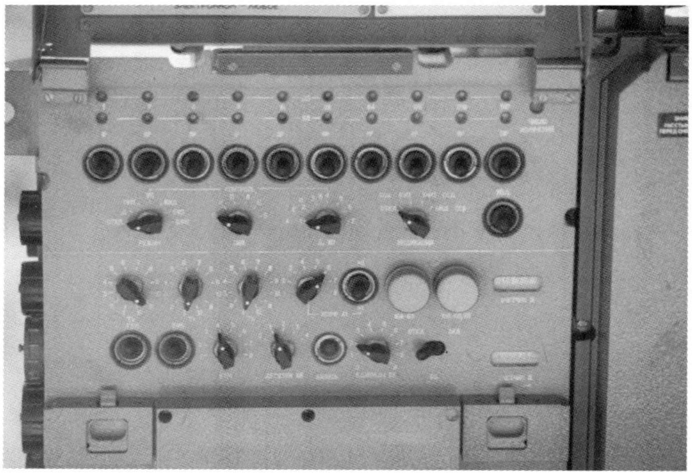

Close-up of the control unit panel used with the Russian tank gunnery simulator (Henry Dodds)

SINGAPORE

CEI Small Arms Battlefield Realistic Engagement Simulator (SABRES)

The Small Arms Battlefield Realistic Engagement Simulator (SABRES) was designed for use in operational tactical training to simulate small arms weapon fire and the effects of weapon fire on targets. SABRES also provides facilities for the control of tactical exercises by umpires from various suitably selected vantage points.

System Elements. The basic SABRES system consists of a laser transmitter, a head detector unit, a torso detector unit, an amplifier unit, a control unit and a quick-release clamp assembly. Optional equipment includes an umpire gun, an alignment fixture, a laser target locator and transit cases.

Transmitter. The laser transmitter is attached to a rifle, machine gun or other similar small arms weapon by means of a quick-release clamp assembly. Using the built-in zeroing mechanism and the optional alignment fixture, the laser transmitter can be adjusted in azimuth and elevation to align the laser beam to the sights of the weapon. The gallium arsenide laser transmitter comprises an optical laser tube, a battery compartment, a weapon key receptacle, a laser firing LED indicator and a 'dry firing' socket. The optional laser target locator allows the combatant a quick and effective method of zeroing the weapon.

Detector Units. The laser detector system comprises a head detector unit and a torso detector unit. These units are physically separated but are inductively coupled. A total of 13 laser detectors are used. The head detector unit has five silicon detectors, a sewn-in loop aerial and an amplifier unit or loop driver mounted on the head harness. The torso detector unit carries the remaining eight detectors, a buzzer, a key receptacle, a gravity switch, a loop aerial and a control unit. All these components are sewn onto the torso harness.

Amplifier Unit. The amplifier unit is powered by a commercially available 9 V battery, and drives the loop aerial on the helmet-worn head detector unit.

Control Unit. The control unit is driven by a 9 V battery. It decodes the data in the incoming laser pulses to indicate a 'kill' or 'near miss' condition by activating the buzzer. The system user can deactivate the buzzer by inserting a key taken from his own laser transmitter (thereby making his own weapon inoperable) or by lying down on his back.

Umpire Gun. The umpire gun, which is a customer option, tests the detector system and controls the tactical exercise.

Specifications

Weight:
(laser transmitter) 0.64 kg
(body harness) 1.08 kg
(head harness) 0.52 kg
(umpire gun) 0.555 kg

Dimensions:
(laser transmitter) 100 × 95 × 40 mm
(umpire gun) 110 × 95 × 40 mm (excluding pistol grip)

Range: typical weapon range simulated
(umpire gun) 400 m

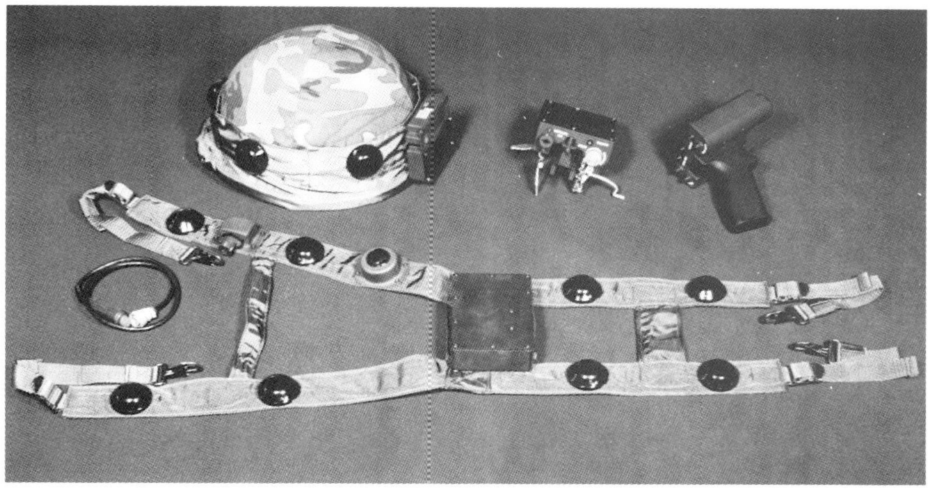

Main components of the SABRES system

Target size:
(near miss) 0.9 m diameter
(kill/hit) 0.45 m diameter

Operating temperature: 0 to +50°C

Storage temperature: −30 to +70°C

SABRES II

SABRES II was introduced in May 1993 and is generally similar to the earlier SABRES but has a smaller and lighter harness electronic control box, a built-in RS-232 interface and a modular design basis. The Motorola CATIES instrumented range system can be incorporated.

Detail design features of SABRES II include an improved anti-cheat system which indicates when a new battery is being loaded, event recording, deactivation of the audio cue which can come only when the user is lying down, and deactivation of the transmitter. A Player ID system is incorporated for after-action review and events received by the torso and head harness are automatically stored for downloading via the serial link. The electronics box on the torso harness has provision for an interface for communication with the CATIES instrumented range system to provide near real-time tracking of troop activities by a control station.

Status

In production. Offered for export.

Manufacturer

CEI Technologies Pte Limited, Republic of Singapore.
Marketed by Unicorn International Pte Limited.
(For details see Directory of Suppliers)

SWEDEN

Saab BT 41 Tank Combat Simulator

The BT 41 Tank Combat Simulator was developed by Saab Training Systems AB, (then Saab-Scania) between 1976 and 1978, to meet the requirements of the Swedish Army. It is a high precision weapon simulator which can be used for tactical and gunnery training at all levels, from basic to combat training. In addition, it can be used for the evolution of weapon systems and alternative tactics. It can be mounted on any type of tank and has been designed as an independent system which can be used either with or without the tank's own fire-control system.

System Components. The BT 41 system consists of a firing simulator, target simulator, display and interconnecting cables. Power is supplied from the tank's standard 24 V supply.

Firer Vehicle Units. The transmitter or firing simulator assembly consists of an electronic unit which can be placed anywhere on the vehicle, one or more tracer generators which are installed over the existing gunner's and/or commander's optics, and a laser transmitter unit which is placed in or on the gun barrel.

Laser Beam Characteristics. A low power laser beam is used to simulate the projectile. The trajectory is simulated in detail in real time with true speed, ballistic and gravity drop characteristics. Weapon movements after the moment of firing do not influence the simulation. The laser beam is also used to convey data between the firing unit and a target. Maximum range is about 3000 m, which can be exceeded with good visibility when laser transmittance is better. At a range of 1000 m, all misses within ±15 m in azimuth and ±3 m in elevation will be recorded. The accuracy in the calculation of the hit/miss location at this distance is better than ±0.2 m. Targets equipped with hit sensors and hit evaluation electronics can calculate damage caused by a hit. The target is divided into sub-areas of 0.25 × 0.25 m. The vulnerability in each sub-area for 24 aspect angles and up to eight projectile types is stored in a memory. A kill automatically stops the target tank. The standard version of the BT 41 is programmed for two trajectories but more can be added and both gun and guided missile trajectories can be simulated.

Components of the Saab BT 41 tank combat simulator arranged in front of a Leopard 2 MBT

Target Units. The target simulator assembly consists of a reference unit and an electronic unit. The reference unit contains corner reflectors and sensors, and is connected so that it can be mounted either with or separated from the electronic unit. It can also be divided into four quadrants to facilitate mounting on the vehicle.

Display Assembly. The display assembly consists of a panel unit and a printer unit. The panel has three sections, one showing the range of the target and point of impact, the second the effect of a hit, together with the recorded direction from which the shot came, and the third is used for system calibration and presetting, and includes a real-time clock and a test panel. The printer records all events on a paper, including each shot fired, each hit received or miss, and all settings and preliminary tests are recorded.

Training Modes. For basic training the target in its simplest form can be a remote-controlled tank silhouette target fitted with a reflector prism. A moving target can be provided by fitting a jeep with a number of reflectors to cover 360°. A more advanced target is provided by fitting the same jeep with the target simulation assembly of the BT 41, to give an evaluation of the effect of each hit. A variety of different hit indicators can be connected to the BT 41 target simulator, for example flash discharge lamps, pyrotechnic markers or smoke devices.

Miss Distance Calculations. The BT 41 provides three methods for determining the location of a hit or miss. Co-ordinates for the trajectory of the projectile relative to the target are shown on a display unit, recorded on a printed paper strip and also appear as a simulated trajectory in the gunner's and commander's optics. A brightly lit spot is shown in the gunner's normal sighting equipment. This follows the trajectory in real time and shows the projectile's track towards the target. The gunner should fire so that the light point falls on the target.

Target Vulnerability Settings. The BT 41 target simulator contains information on the vulnerability of the target vehicle. It describes the target from 12 different aspect angles, and models the sensitivity of the target to different types of ammunition. When the target is engaged, the co-ordinates of the firing unit are transmitted to the target, which then knows from which direction the shots are coming. The target simulator then evaluates whether the shot is a hit or a miss. If it is a hit the target simulator evaluates the effect at the point of impact. Every target is assigned a profile which is aspect angle-dependent in a realistic way. In other words, the effect of the hit is correctly related to the point of impact, distance, target aspect angle and the type of ammunition.

Status

In service with the Dutch, Finnish, German and Swedish armies.

Manufacturer

Saab Training Systems AB, Huskvarna.
(For details see Directory of Suppliers).

Saab BT 46 Anti-tank Weapon Simulator

In May 1990 it was announced that, following testing that lasted over a year, the Swedish Army had placed a substantial order with Saab Training Systems for a new anti-tank weapon simulator, the Saab BT 46. The BT 46 is a miniaturised, lightweight portable system based on the Saab BT 41 tank combat simulator (see previous entry).

The Saab BT 46 is interoperable with all other members of the Saab weapon effect simulator family and has the potential to simulate all known direct fire anti-tank weapons, including the main armament of battle tanks.

Other applications are under development.

Components. Component designations are BT 46G for the laser transmitter system, BT 46T target system, BT 46CP control panel and BT 46VI for the vehicle interface unit. Optional equipment includes BT 46HD hull detector unit and BT 46SD sight display unit. A control gun is integrated with a pair of binoculars, and a field computer and field printer complete the components. The computer is a peripheral unit which provides immediate feedback of the result of a shot in either alphanumeric or graphic form. In the graphic mode the field computer screen is capable of indicating the results of previous shots on a target profile. The computer can also be used to load weapons and target parameters. The field printer provides a hard copy of results.

Firing System. The BT 46G, BT 46CP and the BT 46VI (plus the optional BT 46SD) form the BT 46 firing system. The BT 46G consists of a compact cylindrically-shaped unit that performs the complete weapon effect simulation. The unit houses the eye-safe laser transceiver, the central microprocessor and a gyrosensor module. Target reflectors determine the position of shot strike with respect to the centre of mass with an accuracy equal or better than that of a live round. The scan pattern of the laser is large enough to simulate a hit at close range without having to aim directly at the target reflectors. The gyro module provides a stable reference to maintain the direction fire throughout the time of flight. During automatic fire the system performs simultaneous simulation of individual rounds. Application specific software is quickly and easily reloadable from the field computer. Information on the point of impact, firing system identity and ammunition type is transferred individually to every target within the sweeping area, which is particularly important when simulating high explosive rounds. The optional BT 46SD sight display unit incorporates a projectile tracer dot and impact burst into the gunner's sight to provide visual feedback of the fall-of-shot. The unit also acts as an interface to fire-control equipments to allow the simulation of laser rangefinding techniques. It is possible to connect pyrotechnic devices to the BT 46 firing system to provide a realistic firing signature.

Target Unit. The BT 46T is mounted on a potential target and consists of a microprocessor module and four reference units. The latter house prismatic reflectors, laser detectors and a high intensity strobe light, used to indicate a hit. The BT 46T may be installed as a cluster with all the elements mounted together as a single unit, or split with the reference units separately located. The target system simulates the effect of a hit on the target by comparing the relative lethality of the attacking ammunition with the vulnerability of the target at the point of strike. This determines whether the result is a kill, hit or an ineffective strike. The target is defined using grid net vulnerability models that cover the target from all required aspects.

Customers. The Swedish Army uses the BT 46 for a number of applications, including the RBS 56 BILL anti-tank missile, the 90 mm 1110 anti-tank gun and the one-shot AT-4 light anti-armour weapon.

The Finnish Department of Defence ordered the BT 46 in December 1990; the order also covered the modification and update of BT 41 simulators already in service. Deliveries commenced during the second half of 1991 and were completed during 1992.

In late 1992 the American Department of Defense (DoD) and the United Kingdom Ministry of Defence (MoD) both ordered large numbers, running into many thousands, of BT 46 systems for a number of applications. Deliveries will commence during 1994 and 1995. In the Spring of 1993 the Norwegian Army also ordered a number of BT 46 simulators.

Status

In production. In service with the Swedish and Finnish Armies and the US Department of Defense, the United Kingdom Ministry of Defence and the Norwegian Army.

Manufacturer

Saab Training Systems AB, Huskvarna.
(For details see Directory of Suppliers)

Elements of the Saab BT 46 anti-tank weapon simulator system

Loading a Saab BT 46G firing system into an 84 mm Carl Gustav anti-tank weapon

Saab METS II Area Weapons Effect Simulator

METS II is a system for the control and evaluation of combined arms team training involving the simulation of direct and indirect fire, minefields and NBC conditions. Its design involves a command and control system for the systems involved.

METS II can collect data from laser and radar simulators and GPS receivers and can simulate the effects of minefields and NBC warfare. Using a digitised map in a central control position, it is possible to designate minefields, areas exposed to indirect fire and areas affected by NBC conditions. The system can denote which areas are within the designated area and probable casualty rates are calculated using weapon effectiveness figures and a random number programme. METS II can also evaluate an exercise and record it for later evaluation.

Components. METS II consists of a central control position and a number of mobile units known as PRX-90s. The central control has two computers, two monitors, a four-colour printer and two PRX-90s operating on different frequencies. In the basic version, METS II can handle 256 mobile units but this number can be increased depending on the update rate which can be accepted.

Mobile Units - the PRX-90. The PRX-90 mobile unit has a built-in radio with a 5 W output, two processors and a real-time clock. It has a built-in 128 k memory which can be increased to 512 k. Each PRX-90 weighs 1.5 kg and measures 100 × 50 × 210 mm.

All PRX-90s are identical and are equipped with a number of programmes. The actual programme can be selected on command from the central control via the PRX-90's radio. The selection of programme depends on which function the PRX-90 has in a system. The basic functions are control position unit, player unit or relay unit. Each PRX-90 can operate in either a transparent or a compressed mode.

In the METS II system every PRX-90 has an identity (ID) consisting of an individual number and a group number. The PRX-90 has two bi-directional RS-232C ports. One bi-directional parallel port and two bi-directional RS-485 ports can be supplied as options. Each PRX-90 contains two standard programmes, one basic and one for radio control and the central mode. Up to a further 20 programmes can be stored.

A PRX-90 can be addressed in four different ways:

1. Challenge a single PRX-90 by individual and group number; for example, request for data or to give a command, or to give the result of damage due to indirect fire, minefields or exposure to NBC
2. Challenge all PRX-90s; for example, to set the time
3. Challenge the same ID number in each group; for example, unit leaders
4. Challenge all IDs in one group; for example, when a unit is in a minefield.

METS II can handle up to 4000 PRX-90s divided into 16 groups with 250 in each group. The update rate is 4 PRX-90s per second.

Central Computer. The central computer can be any equipped with an RS-232 interface. The central control PRX-90 connected to the central computer will automatically operate in the central mode (one of the PRX-90 standard programmes).

SOP. The Standard Operating Procedure (SOP) starts when the central control PRX-90 unit is provided with data on which of the PRX-90s are taking part in an exercise. This is done by the central computer which also supplies the functions of the PRX-90 units for the exercise, generally that of player or relay unit. After a start command from the central computer all PRX-90s will be called in sequence. If a PRX-90 does not respond the central unit will, after two recalls without response, ask one of the other PRX-90s if it has been able to read the non-responding unit. If it has, the central unit will request it to transmit this information. If no response is received no information will be lost as it is stored in the PRX-90 memory and transferred as soon as contact is established. During exercises in hilly terrain or over long distances it is usual to appoint one or more PRX-90s at strategic points to operate as relay units.

The entire call (polling) procedure is automatically controlled from the central unit. All collected data is stored in the central unit where evaluation and presentation in almost real time can be performed. Based on collected data or on command from the operator, data information can be transmitted from the central unit to one or more PRX-90s.

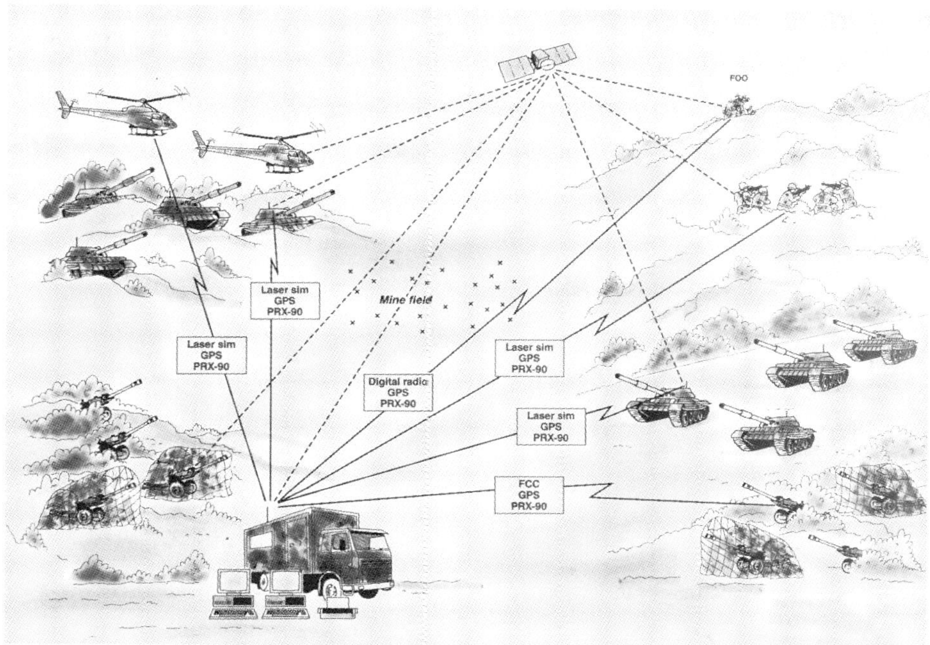

Overall operation of METS II area weapons effect simulator

Central Control Point. The central control point consists of two computers, one handling communications and processing of the firing data from laser simulators and positioning data from GPS receivers, and the other handling communications and the processing of information from targets and positioning data from GPS receivers. Information from the two computers is gathered and saved to provide the operator with data to evaluate an exercise.

Indoor Training. METS II can be used for indoor training. In this case a two-wire link is used to connect the PRX-90s in place of the radios. When a wire link is connected the radio is automatically disconnected

Artillery Applications. For artillery applications, guns, forward observers, troops and targets are equipped with PRX-90s. At the guns the PRX-90 is connected to a GPS receiver and a digital radio to provide firing data. The PRX-90 transmits this data to the METS II central control position where hit probability calculations are performed. Forward observers are equipped with a PRX-90 and a GPS receiver. Position and forward observer data are transmitted to the METS II central control where firing parameters are checked.

Direct Fire Units. Direct firing units using laser firing and target simulators are also equipped with a PRX-90 and a GPS receiver. Position, firing and hit data are transmitted to the METS II central control via the PRX-90.

Minefield and NBC Information. Information relating to minefields and NBC activities is registered on the METS II central control digitised maps and the central control PRX-90 transmits information to participating units if they are hit by indirect fire, enter a minefield or are subjected to NBC attack. The central control also transmits data regarding calculated indirect fire impact positions to forward observers.

Status

Available.

Manufacturer

Saab Training Systems AB, Solna.
(For details see Directory of Suppliers)

UNITED KINGDOM

Loral Solartron Simulators

Loral Solartron, formerly Weston Simfire, was involved in the development of laser-based Weapon Effect Simulation (WES) during the 1960s. The British Army adopted the SIMFIRE Mark II ER (Extended Range) for its MBTs and since the early 1970s 35 armed forces have purchased SIMFIRE simulators.

Loral Solartron is now owned by the Loral Corporation and, together with the Training and Simulation division of Loral Electro-Optical Systems (LEOS) in California, supply WES systems throughout the world. The direct fire weapons deployed by a modern battle group can be simulated to provide all or part of training needs.

Loral Solartron products cover basic to advanced system capability. Logistics support is provided. Current customers include the US Army and Marine Corps. (TOW and Dragon anti-tank weapon simulators), Canadian Forces (TOW under armour simulators), Austria, Denmark, Egypt and India (tank combat simulators), and Germany (infantry fire support simulators).

Levels of Training. The following levels can be covered: individual and crew training at platoon level and below; tactical and procedural training from platoon to battle group level and, when appropriate, to brigade level; command and staff training at battle group level and above.

M1 Abrams MBT fitted with SIMFIRE S Series precision gunnery trainer in firing mode

The SIMFIRE S Series Trajectory Display Unit provides an AFV gunner and commander with a visual image of a simulated projectile. This display indicates a hit

SIMFIRE exercises can be controlled by an umpire using the SIMFIRE controller's gun

The Loral Solartron range of laser-based WES systems are designed to meet the following training needs:

Tactical Training - includes weapon and troop deployment, weapon performance/capability, use of ground, fire and movement, all arms training, force-on-force, and command and control

Procedural Training - weapon handling, crew drills, aiming, tracking and firing techniques

Precision Gunnery - target acquisition, aiming, tracking, firing and missile guidance.

S-II for MBTs. The SIMFIRE S-II is adaptable for all direct fire anti-armour weapons and is interoperable with all known laser-based simulators. The addition of a GPS system allows it to integrate indirect fire and range instrumentation. It is modular and is software-programmable to enable it to meet diverse customer training and weapon requirements and to allow for future weapon and ammunition enhancements. The S-II system can interface with any computerised fire-control system or can operate independently. Trajectory display units display in the gunner's and commander's sights the simulated trajectory of a round. This enables precision gunnery skills to be practised. The system can accommodate 'slow' rounds, such as HESH or coaxial machine guns, and also rapid fire cannon.

S Series for IFSVs. The S Series simulates the firing cycle of a rapid-fire cannon and displays in the gunner's and commander's sights. It has a capability of over 1000 rounds. The system has been fitted to the Warrior, Bradley and Marder IFSVs.

S Series for Anti-armour Weapons. The SIMFIRE S Series can be configured to simulate a range of direct fire and wire-guided anti-armour weapons. The performance of the weapon simulated allows precision gunnery training as well as tactical force-on-force exercises.

PGTS. 1500 Precision Gunnery Training Systems (PGTS) for TOW and Dragon anti-tank weapon crews are being supplied to the US Army and Marine Corps. PGTS TOW units were shipped direct to the Persian Gulf for Operation Desert Storm. PGTS is configured for ITV, LAV-AT and ground-mounted TOW. Loral Solartron supplies Precision Gunnery systems for the TOW under armour vehicle for the Canadian Army.

SIMFIRE for Airborne Weapon Platforms. Simulation systems for armed anti-armour helicopters to simulate air-to-ground engagement scenarios are also available. Target systems can also be fitted to helicopters. Loral Solartron is a subcontractor for the US Army's Air to Ground Effects Simulation (AGES).

SIMFIRE Target Systems. Target systems for AFVs, B vehicles and ground positions are available. These can all be programmed with different vulnerability characteristics to allow for multiple targets.

Small Arms Weapon Effect Simulators. Small Arms Simulators are available which allow infantry to participate in tactical training exercises.

SIMFIRE Controller Gun. SIMFIRE simulators can be controlled by the controller gun which enables umpires to control tactical exercises. Umpires can interrogate, 'kill' and reactivate exercise participants.

Status

In production. Customers include the US, Canadian, German, Danish, Austrian and Indian armies. Potential future customers include several other European and Middle East armed forces.

Tank, Anti-Tank and Air Defence Trainers

Loral Solartron manufactures a range of laser and CGI-based weapon training systems for tank, anti-tank and air-defence instruction. This includes the PGTS/PGS laser-based trainer for TOW and Dragon wire-guided anti-tank missiles.

TOW Simulator. The anti-tank simulators for the TOW and Dragon guided missiles utilise a reconfigurable modular laser engagement system to replicate the engagement up to maximum range. The TOW simulator is in a replica Missile Guidance Set (MGS) and a modified TOW tube. The real tripod, trigger and traversing unit are utilised and simulated visual effects of firing, including obscuration, rocket flare and trace, are introduced into the operational day and thermal sighting systems. The audible firing signature is provided by initiation of an M80 blast cartridge. The system is configured for ground mount and vehicle launch platforms, including under armour.

Dragon Simulator. The Dragon simulator modules are in a complete missile replica. Power is supplied from a second missile tube, carried as a spare round. The real sights are used and the effects of firing and missile flight are simulated visually by injection of obscuration and flare into the sights. Audible effects are by initiation of an M80 blast cartridge, and physically, by an integral weight loss mechanism.

Status

In production. In service with the US Army and Marines and the Canadian armed forces.

Gunfire Simulator System

This simulator is used during the training of tank and anti-tank gun crews to simulate the sound, flash and smoke of their guns during dry firing gunnery or tactical training exercises. Variants are available replicating a number of weapon bores, and the degree of noise, flash and smoke are generated to match the weapon firing characteristics. The device is also suitable for

Loral Solartron Precision Gunnery System (PGS) for the Canadian TOW under armour vehicle

SIMFIRE S Series WES Precision Gunnery Training System (PGTS) for TOW for the US Army and USMC

deployment as a target hit indicator or fire marker.

Production of Effects. The effects of firing are generated by the safe, low pressure combustion of a propane/air mix. The combustion chamber is closed by a biodegradable diaphragm which is ruptured by the pressure of combustion, generating the noise of firing (130 to 160 dB) and releasing smoke material. Replacement of the diaphragm material is automatic, with a motorised feed system for multi-shot versions. The flash of combustion can be supplemented by electronic photo-flash if required. The simulator is self-contained, and has its own fuel supply and control electronics package. Refuelling provides 60 shots and is from a small disposable cartridge.

Loral Solartron Mobile CGI Gunnery Trainer

The Mobile CGI Trainer is a low cost trainer for weapon crews outside the constraints of a classroom or barracks setting. The system combines real video images with Computer Generated Imagery (CGI) of target and weapon effects. The combined display is then injected into the weapon sights of the operational weapon. High resolution colour displays give high realism of imagery, suitable for target recognition training at maximum engagement range. Gunners use their normal weapon controls to acquire and engage the artificial targets, which may be air, sea or land based. The computer imagery system generates simulated tracer and burst effects - no live rounds are fired. The size, power consumption and general portability of this system allows training to be conducted in front line situations without modification to the weapon system, without special provision of covered areas and with no need for off-weapon control consoles.

Artist's impression of AH-64A Apache helicopters using AGES in simulation exercises

Status

In development.

Manufacturer

Loral Solartron, Herts.
(For details see Directory of Suppliers)

Royal Ordnance Simulation Tactical Engagement Simulation

During 1989 Royal Ordnance Control Systems and Fuzes Division took over Centronic Limited Tactical Engagement Simulation (TES) and formed Royal Ordnance Simulation (ROSIM) which concentrates on the TES area.

ROSIM Laser Codes. MoD(UK) has a standard set of codes and the ROSIM Small Arms Weapon Effect Simulator (SAWES) (see later) functions on the appropriate MoD(UK) code. Battle Group exercises are carried out to a common code which attributes kill and miss distance data to attack systems, and allows for vulnerability of the target systems. The ROSIM Common Code is based on the TALISSI (TActical LIght Shot SImulator) PCM code format. ROSIM also offers equipment compatible with US Army MILES systems.

ROSIM SAWES in use with SA 80

ROSIM 100 Series Exercise Control Equipment

Controller kits. All the ROSIM controller kits include an Umpire Gun based on the SAWES ×4 optical projector. There are three types:

Type 101 - SAWES only (two modes)
Type 102 - ROSIM Common Code (six modes)
Type 103 - Common Code and RELACES (16 modes)

The umpire guns are used to check equipment and reset activated target systems. A battery test meter is fitted to the stock.

Indirect Fire Weapon Effect Simulation (IFWES). Pyrotechnics provide loud noise ('bangs') and are safe to personnel being trained but give the elements of surprise and disorientation that occur in real battle conditions. Kill conditions are signalled by laser codes that activate detectors on personnel or vehicles. Allowance is made for target signatures in the sensor spectrum of the attacking weapon (Radar/IR/Intensifiers/Visual).

Remote Laser Controlled Exploder System (RELACES). RELACES is deployed to create simulation of indirect fire at static targets such as dug-in positions or defined killing zones. The system comprises:

Type 103 - Controllers Gun
Type 131 - Receiver Control Unit (RCU)
Type 132 - Exploder Unit (EU)
Type 135 - IFWES Projector

Any mix of up to eight EUs and projectors are connected to one RCU which is activated by the Controller's Gun set as required at one of three modes: TEST, FIRE, and FIRE SALVO. Using the SALVO mode all RCU-connected units can be initiated simultaneously.

Mobile IFWES Kit. Where the use of RELACES is not appropriate, a vehicle-mounted combination of simultaneous bangs and kills may be used. The kit comprises the Type 104 Wide Angle Controller's Projector and the Type 137 Exploder Control Unit (ECU).

ROSIM Unexploded Ordnance Target System. The Type 120 unexploded ordnance target system has been supplied to RAF Regiment bomb disposal teams to assist in runway clearance training. The system uses existing SAWES subsystems configured as a bomb and compatible with the attack system. The Type 120 includes a flashing beacon and standard pyrotechnic for hit indication.

External Strike Internal Kill System (ESIKS). The Type 121 ESIKS simulates where shells or missiles attack a building or vehicle but, because a laser beam cannot penetrate brick or armour, the expected human casualties cannot be simulated in the same way as with man-mounted laser detector systems. ESIKS comprises an external detector belt which fits round the target structure, an external decoder box with battery supply, and an internal computer which calculates damage and kill probabilities and has a flashing beacon and pyrotechnic pot for external signalling of results.

The system detects the attacking round, activating the decoder which simultaneously initiates the visual indicators and a 360° 7 m 'kill' zone around the generator. Any personnel inside the building or vehicle wearing the SAWES man-detector system will be affected by the kill codes in accordance with casualty predictions set into the system.

ROSIM 200 SAWES

SAWES (Small Arms Weapon Effect Simulator) enables soldiers to aim and 'fire' their rifles or other guns during an exercise; junior commanders can witness the true consequences of their field decisions. The exercise controller or instructor's role has changed; due to the use of simulation, lessons can be largely self-taught as a result of individual simulated combat. There are various versions of SAWES equipment, as outlined below.

SAWES Man-Worn Detector System (MWDS). Each soldier is issued with a MWDS which comprises a ruggedised body harness with eight detectors and a headband supporting four detectors. The system is

The ROSIM SAWES Claymore mine

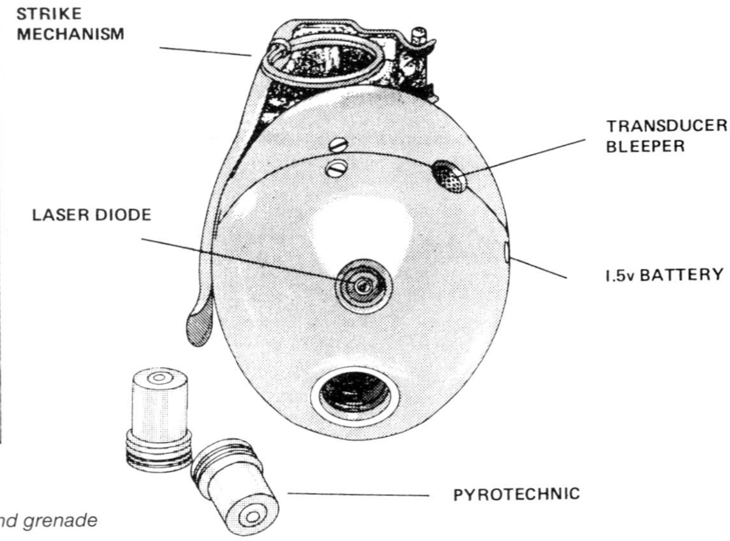

Component parts of the SAWES hand grenade

controlled by an electronics box positioned behind the wearer. The MWDS is activated by the insertion of a standard 9 V battery. The MWDS is designed so that an accurately aimed shot with a SAWES projector from any angle and from ranges between 5 and 800 m will guarantee a 'hit'. Longer ranges are possible with higher powered projectors such as those designed to simulate machine guns. The wearer recognises that he has been 'hit' by the initiation of a loud warbling alarm. This can only be stopped either by the wearer lying on his back or by intervention with the umpire gun. An optional facility is a 'near miss' indication which is given by a one second audio alarm.

SAWES ×4 Optical Laser Projectors. The Type 210 optical laser projector was developed for the British Army for laser simulation of fire and ease of use. The laser projector laser is integrated with an optical sight similar to the ×4 SUIT sight. The projector is fitted to the weapon without any form of alignment procedure and is ready for immediate use. The laser transmits on the firing of a blank cartridge and allows for target range. This avoids inaccuracies resulting from post-detonation gun jump and rewards a properly aimed shot. The projector can also be activated by a Trigger Switch Assembly (TSA) during dry firing. The equipment is powered by a 9 V battery.

SAWES Bore-aligned Laser Projectors. The Type 213/214 laser projectors can operate with infantry night vision sights. A SAWES is inherently accurate and does not require any form of field alignment during fitting. The bore-aligned projector uses miniaturised electronics to make the equipment compact, lightweight but still robust.

SAWES Claymore Mine. The SAWES Claymore mine is designed to represent the lethality zone of the M18-type anti-personnel mine. Initiation is by either the electrical or trip-wire method employed for the live mine. On actuation of the simulated mine, a pyrotechnic is initiated; simultaneously the SAWES kill code is transmitted in an approximate 60° fan up to a range of 50 m.

SAWES Hand Grenade Simulator. The SAWES hand grenade simulator is used in British Army FIBUA (Fighting In Built-Up Areas) exercises. The device incorporates a pyrotechnic which is initiated by a standard smoke grenade mechanism and with the normal ignition delay. Simultaneously the SAWES laser kill codes radiate up to a preset lethal range. Soldiers wearing the SAWES detector kit are 'killed' if they are within the lethal zone. An audio signal aids the location of the simulator which can then be reset for further use.

SAWES ×1 Optical Projector. The ×1 optical projector laser can be optimised for ranges up to 2000 m and is compatible with longer range weapons such as machine guns. The normal ammunition scaling for the particular weapon can be programmed so that a resupply period is introduced. The wide angle ×1 optical viewer ensures automatic zeroing and ease of use.

ROSIM 300 Series for Mechanised and Support Infantry

The 300 series is supplied with the appropriate Type 200 series SAWES equipment to permit battalion-level training with APCs and anti-tank support weapons. Type 100 series umpire guns are required, and RELACES can be supplied to simulate the mortars of the battalion.

Vehicle Target System. The Type 301 vehicle target system can be fitted to vehicles such as an M113 armoured personnel carrier or to a jeep-type vehicle. In each case the appropriate number of detector blocks is positioned round the vehicle and connected to a control unit. The exercise controller can set the correct target vulnerability (such as medium armour) and if a 'hit' is obtained from an attack system combined with the appropriate lethality calculations, an orange pyrotechnic and a flashing beacon are initiated.

AFV WES Cupola Kits. Kits have been supplied to simulate cupola machine guns of both 0.50/12.7 mm and 7.62 mm calibre. The ×1 laser projector is mounted on the weapon and programmed for the appropriate ammunition scaling. Instead of relying on blank rounds or pyrotechnics, an electronic flash/bang generator is supplied.

Anti-tank Weapons WES. ROSIM can offer simulators for the following anti-tank weapons: MILAN, LAW 80, TOW, 106 mm recoilless rifle, Dragon and 84 mm Carl Gustav. The systems are characterised by accurate representation of the weapon performance and require the same operator's skills as for the real weapon. When required, a result display panel is included for the benefit of the operator and the instructor. For MILAN, TOW and Dragon a vulnerability receiver unit can be included.

ROSIM LAW 80 simulator, part of the ROSIM 300 series

ROSIM 400 Series - Armour WES

The ROSIM 400 series includes infantry MICVs with main armament attack capability as well as a wide range of equipments for main battle tanks and cavalry units. The target system for these vehicles is based on the Type 301 system and can be programmed to represent heavy armour vulnerability in the DFWES coding system.

ROSIM 500 Series - Helicopter WES

A helicopter target system known as GATWES (Ground Attack Weapon Effect Simulator) has been supplied to the MoD(UK). The system permits land forces to simulate engaging attacking helicopters.

Status

The above equipments are in service with Canada (30 systems ordered in 1984), Jordan (922 systems ordered in 1984 followed by a repeat order for 500 in 1987), United Kingdom (1910 systems ordered in 1984 with more ordered in late 1992), and other armed forces, especially in the Middle East. ROSIM has produced a range of equipment for Malaysia, the US Army, UK Army, the Hong Kong Forces, and the US Marine Corps.

Manufacturer

Royal Ordnance Simulation (ROSIM), Blackburn.
(For details see Directory of Suppliers)

UNITED STATES OF AMERICA

ECC Close Combat Tactical Trainer (CCTT)

The CCTT is part of the Combined Arms Tactical Trainer (CATT) that trains crews up to battalion level in a realistic force-on-force, free play virtual battlefield environment.

Networked simulation modules can be programmed to replicate any armoured vehicle. Platoon, company or battalion exercises can be simulated. Exercises include those involving tactical, logistics, communications and support situations. Training for vehicle crews, commanders and support components is provided that would otherwise only be achieved by field exercises, and then with limitations imposed by the peacetime training environment.

Characteristics. Vehicle performance is simulated and includes available ammunition and results such as kill and miss distance probabilities. Crews are provided with the sights and sounds of armoured vehicle combat as well as the effects of combined arms against a semi-automated opposing force. Training performance is evaluated with an after-action review.

CCTT systems may be produced for any level of armour training; platoon, company or battalion. A wide variety of armoured vehicles can be simulated and networked.

Status
Available.

Manufacturer
ECC International Corp, Pennsylvania.
(Teamed with IBM and Evans and Sutherland, E&S supplying the Computer-Generated Imagery.)
(For details see Directory of Suppliers)

Example of CCTT Visual Display Showing Overhead Projectors and Five Visual Channels

Example of CCTT inside the AFV

Loral Multiple Integrated Laser Engagement System (MILES

MILES is a family of laser transmitters and laser detection devices that give tactical weapons training for field exercises involving large and diverse mobile forces. MILES uses low power gallium arsenide laser transmitters to fire simulated 'rounds' against silicon photodiode detectors. Weapon effects at firing are achieved by employing blank rounds. Simulated 'kills/hits/near misses' are denoted by audio and visual signals which can be observed at the targets. A 'kill/hit' results in the deactivation of the target weapon system. In the case of vehicles such as tanks, where a 'hit' may or may not cause a full kill, the logic in the detector system assesses the kill level based on the code message of the attacking weapon. Mobility and other levels of kill are registered. Unique codes are assigned to each type of weapon to give the capability of simulating a wide range of weapons.

Deliveries began in 1980 to the US Army. MILES has since been produced for use by numerous armies worldwide. Sales of MILES-based simulators total well over $750 million to date. The MILES family consists of infantry, armour and anti-tank simulator systems such as rifles, light and heavy machine guns, tank weapons (main gun and onboard machine guns), and anti-armour weapons such as TOW and Dragon.

MILES equipments also provide for air-to-ground engagement systems and air defence systems (AGES/AD I); the latter include Chaparral, Vulcan and Stinger. Other air-to-ground MILES systems include the OH-58A, UH-1H, and AH-1S helicopters. MILES equipment installed on AH-1S Cobra helicopters simulates the firing effects of TOW missiles, 20 mm cannon and 2.75-inch rockets. Upon receipt of a valid laser signal the Cobra's detection system will insert an audio tone in the intercom signal, thus activating both a light on the aircraft kill indicator in the cockpit and a flashing strobe light mounted on one skid. If the target is 'killed', a smoke signal collocated with the strobe light is activated.

Weapons are simulated by transmitters and the detection systems are configured for the various vehicles and helicopters. Lasers employed are eye-safe.

Support equipment includes boresight alignment aids, umpire control guns, a universal independent target system, detection for pop-up targets, audio/visual weapons effect simulation, and a full complement of maintenance test equipment.

US Army soldiers training with MILES equipment

MILES II and AGES II
This equipment provides full interoperability with basic MILES. Improvements include increased fidelity of weapons effects, event recording, provision to incorporate area weapons (such as artillery) and telemetry instrumentation. Weapon simulation for the latest aircraft weapon systems includes AH-64 Apache (Hellfire), OH-58D, CH-47 and UH-60.

The US Army employs MILES for home training as well as at the instrumented combat training centre systems at the National Training Center at Fort Irwin, California, the Joint Readiness Training Center at Fort Polk, Louisiana, and at the Combat Maneuver Training Centre at Hohenfels, Germany.

Current MILES Systems

Infantry Weapons/Systems. M16A1 and M16A2 rifles; Heckler & Koch G3 rifle; Galil rifle; BM59 rifle; AK-47 and AKM rifles; FN FAL and L1A1 rifles; M60 and M60E3 machine guns; M249 machine gun; M2 0.50/12.7 mm machine guns; M85 machine gun; MAG 58 machine gun; MG3 series machine guns; MG42/59 machine guns; M240 machine gun; M1919A4 machine gun; 12-gauge shotguns; 9 mm Uzi sub-machine gun; 7.62 mm Steyr rifle; 7.62 mm GPMG; ARMBRUST; 106 mm recoilless rifle; 84 mm Carl Gustav.

Missile and Rocket Systems. TOW I and II; LAW; Dragon; AT4; MILAN; Sagger; Stinger: Chaparral; SA-14; Shillelagh.

Tank and Vehicle Weapons Systems. M41; M48A1, M48A3 and M48A5; M60A1 and M60A3; M1 and M1A1 Abrams; M2 and M3 Bradley; M551 Sheridan; M551; Leopard 1A1 and 1A3; Leopard 2; Scorpion series; M113; AAV7A1; M901 ITV; M113 TOW; Chaparral; Vulcan. MILES systems are available for 20 mm, 25 mm, 73 mm, 90 mm, 105 mm, 120 mm and 152 mm guns and are also used on T-72, BRDM-2, BMP and M1974 VISMODs at the US Army's National Training Center at Fort Irwin, California.

Airborne Systems and Weapons. A-10; UH-1; UH-60 Black Hawk; AH-64 Apache; OH-58 Kiowa; Hughes 500MD; Hind-D (VISMOD); TOW and ITOW; 2.75-inch rocket; 57 mm rocket; 20 mm and 30 mm guns.

Compatible Systems. Small Arms Alignment Fixture (SAAF); Multiple Range Alignment Device (MRAD); Hand-Held Laser Locator (HHLL); Laser Target Interface Device Systems (LTIDS); Mobile Operations in Urban Terrain (MOUT).

Status
In production. In service with the US Army and Marine Corps. Over 250 000 vehicle systems and over 700 000 man systems supplied. In service with numerous other countries worldwide.

Manufacturer
Loral Electro-Optical Systems, Pasadena.
(For details see Directory of Suppliers)

Loral Precision Range Integrated Manoeuvre Exercise (PRIME)

PRIME is a MILES enhancement which provides gunnery range event-driven target control with simulated target shoot-back for MILES-based tactical/gunnery and force-on-force training for tanks, fighting vehicles and infantry units. The system is designed for platoon and company level exercises.

Characteristics. PRIME supports instrumented training with near real-time data collection and control of players and targets. The system is based on the US Army standard SAWE/MILES II player equipment. Post-simulation analysis is provided by combining time-tagged through-sight video with computer-generated reports and graphics.

PRIME allows tank and infantry to practise battle skills and measure proficiency. With a 250-infantry player and 150-vehicle player capability, PRIME supports MILES II-based force-on-force and computer-controlled target training.

Data Collected. PRIME data is collected via standard SAWE/MILES II player units and includes event time correlation; GPS-based position location; player identification; range of engagements; programmable player events including fire, weapon and ammunition type, hit zone and casualty assessment; and target status and shootback.

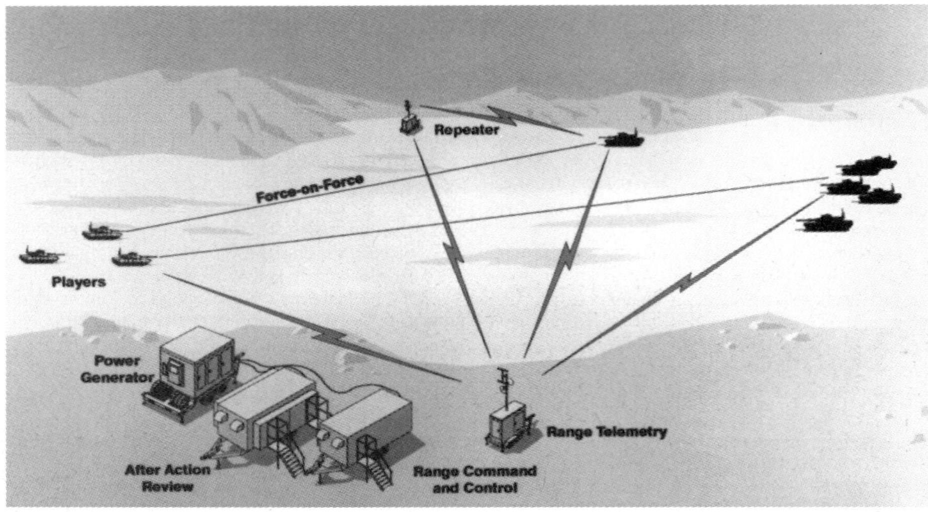

Schematic of a Precision Range Integrated Manoeuvre Exercise (PRIME)

Status

In production. First system due for operational readiness December 1994 at Fort Hood, Texas.

Manufacturer

Loral Electro-Optical Systems, Pasadena.
(For details see Directory of Suppliers)

Loral Simulated Area Weapon Effects (SAWE)

The system is for field manoeuvre force-on-force tactical training exercises and provides weapons effects for artillery, mortars, chemical weapons and mines.

SAWE utilises the Global Positioning System (GPS) to record accurate positions of units involved in the simulation. Each participating unit is equipped with a GPS receiver. A centralised mission planning/area weapons control computer provides the interface for artillery fire missions.

Operation. The computer controls a radio network which broadcasts artillery mission activity to the battlefield where an individual participant receives the artillery message. The radio message to each player provides point of impact, time of impact, and type of area weapon effect. The available position for each player is then compared with the artillery fire event to determine if the player is either in the lethal or near-miss area of the weapon. Weapon effect and appropriate casualty assessment decisions are based upon the nearness of the player to the weapon impact point. Audio visual effects can be included to provide cues on the battlefield.

Range of Simulations. SAWE simulates the effects of a broad range of area weapons from mortars to large calibre artillery and long-range missile delivery systems. High explosive conventional artillery, chemical, nuclear and minefields can be simulated. The system also enforces participants to take the proper countermeasures, such as the proper utilisation of respirators when under chemical attack.

Portability. Because of the use of GPS for position recording, the system requires minimal site preparation prior to installation and it can be moved to other training areas. Player equipment is rugged yet lightweight so demand on participants is minimised.

Compatibility. SAWE is compatible with MILES II and is an optional capability which can be added to MILES II infantry, vehicle and aircraft systems. The US Army has selected the SAWE/MILES II player engagement system as its standard for all its Combat Training Centers (CTC).

Status

In production. Deliveries in progress to the US Army Combat Training Centers.

Manufacturer

Loral Electro-Optical Systems, Pasadena.
(For details see Directory of Suppliers)

MILES II-equipped soldier fully equipped for a Simulated Area Weapon Effects (SAWE) exercise involving chemical weapons

Precision Range Integrated Manoeuvre Exercise (PRIME) Tank Player Units during Tactical Exercises

Motorola Combined Arms Training Integrated Evaluation System (CATIES)

CATIES (pronounced cat-eyes) is an area weapons effect tactical engagement simulation system designed to simulate indirect fire, NBC effects and landmine warfare effects. The system includes simulation of naval gunfire as well as other guns, rockets, missiles and mortars. CATIES is compatible with MILES and other similar systems.

Components. CATIES has three primary components: a master station, actuators and appliqués.

Master Station. The master station initiates and controls the system through the transmission of attacking weapons and timing data to selected actuators. The master station may be fixed or mobile and uses a PC with a 10 Mbyte hard disk, a colour display of topographical features, a Liquid Crystal Display (LCD) and an ink jet printer. The master station is powered by a 110 V AC generator.

The master station acts as a link with firing unit fire direction and fire support elements. Its processor captures and interprets messages and enters the data into the master station PC. Data recorded includes the target location, method of fire, the weapon/munition combination, time of flight and time of shot. The master station computes the data required to cause at least three coded, omnidirectional, RF energy pulses to intersect over the target location at the proper time. This intersection is achieved by transmitting the appropriate data to each of three or more actuators which retransmit the energy pulse. The maximum range of the master station from the actuators is 32 200 m, line-of-sight.

Relay/Actuators. The relay/actuators transmit directly, or as relays, weapon and timing data from the master station to appliqués in the target area. The system can handle up to 64 addressable actuators to cover large ranges and rugged terrain. The output power is 10 W and power is derived from an internal, solar-charged 12 V battery. The actuator is contained in a ruggedised carrying case measuring 228 × 305 × 457 mm. The weight of an actuator is 15 kg.

Appliqués. The appliqués may be placed on individual soldiers or pieces of equipment. They sense actuator transmissions, determine and indicate the effects on the targets, and propagate these lethal and suppressive effects through interfaces with the following:

MILES-type direct fire training equipment systems that provide limited audio indications of CATIES effects, appropriate disablement of weapons systems and vehicle mobility, and transfer of CATIES data to other instrumentation systems

Range instrumentation devices

Audio-visual devices that supply the flash/bang cues to soldiers affected by indirect artillery fire, NBC contamination or land mine warfare.

Appliqués are powered by an internal 6 to 9 V battery with an operational life of over 70 hours. Their weight is 450 g and each unit measures 76 × 32 × 114 mm. The appliqués can provide either 'hit' or 'near miss' information.

Diagram of CATIES operating system

Links. A link is provided to the command and control elements that provide the data necessary for CATIES to simulate the desired effects. These links must be digital to be compatible with fire direction systems such as TACFIRE and the Battery Computer System (BCS). Manual links are also provided to ensure redundancy as well as worldwide application.

Concept. The concept of CATIES was developed by LB & M Associates of Lawton, Oklahoma, and has been technically defined and supported by the Government Electronics Group of Motorola Inc. The first proof-of-concept contract was placed by the US Army Training and Doctrine Command in November 1986. A prototype system was first operated in April 1987 and was demonstrated at the US Army Field Artillery School in August 1987. A follow-on contract awarded in April 1988 covered two master stations, 10 actuators and 11 appliqués for evaluation at the National Training Center (NTC) at Fort Irwin, California. On 21 January 1989 a contract to provide production equipment for the NTC was awarded. This entailed four master stations, 32 actuators, 600 vehicle appliqué units and 600 audio-visual cues. The contract also covers the development and testing of 16 reduced-size 'player detection' devices (man-worn appliqués). Deliveries began in March 1990.

Specifications

CATIES Data

Weapon simulation: up to 16 000 different weapon-target combinations including field artillery (guns, howitzers, rockets and missiles), mortars, naval gunfire, NBC contamination and minefields

Targets: capable of 128 possible target types (both red and blue forces)

Effective radii of effects: selectable, eg 25, 50, 150 or 250 m

Simulations: up to 50 missions/s

Interfaces: MILES; range instrumentation systems; audio-visual cue devices; digital and voice command and control systems; unique audio-visual cueing for simulation of suppressive effects

Range:

(actuator to actuator) 20 miles/32.2 km line-of-sight

(actuator to target) 10 miles/16.1 km

(extended range through up to 3 relays) 60 miles/96.5 km

Resolution: 1 to 2 m

Accuracy for simulated impact areas: ±10 m using pulse compression techniques

Status

In production - see text.

Manufacturer

Motorola Inc, Arizona.

(For details see Directory of Suppliers)

ANTI-TANK WEAPON TRAINING SYSTEMS

CANADA

Istec Thru Sight Video Recorder for TOW Night Sights

The Thru Sight Video Recorder (TSV) is a system that allows the image from the AN/TAS-4A and AN/TAS-4C TOW night sight to be recorded on videotape and simultaneously be displayed on a remote video monitor. The normal operation of the sight can continue and the operator's image is unaffected. The system allows an instructor to assess the performance of individual TOW gunners, monitor missile performance, and use the videotape for record and debrief purposes.

The system consists of an optical splitter assembly that is mounted on the night sight and a remotely located video recorder and TV monitor screen. A video sync unit generates video timing and control signals to synchronise the video frame rate with the range of mechanical scan rates of the night sight. The optical splitter assembly contains a video camera and a prism and optics to send the appropriate images to the camera. The unit is mounted using the same threaded coupling that mounts the normal eye shield assembly. No modifications to the night sight are required. The system can operate from either 115 V AC 60 Hz or 24 V DC electrical supplies.

Status
In production.

Manufacturer
Istec Inc, Ontario.
(For details see Directory of Suppliers)

Thru Sight Video Recorder (TSV) fitted to TOW night sight

CHINA, PEOPLE'S REPUBLIC

CEIEC Anti-tank Simulator SHMX-11-11

The SHMX-11-11 is a three-piece portable unit designed to train anti-tank missile operators. It was developed initially for the Sagger MCLOS system but can be modified to other similar systems. The system comprises a computing unit, a CRT screen unit, and a joystick control unit, totalling 40.7 kg (89.7 lb). It can simulate various target characteristics and projectile trajectories in various conditions of operating and weather. It allows recruits to complete a programmed training course and can be deployed for field operation to attain a standard virtually guaranteeing a 90 per cent hit probability when the first live missiles are fired. An improved computer-controlled version is also in service.

Status
In service with the Chinese Army.

Agent
China National Electronics Import & Export Corporation (CEIEC), Beijing.
(For details see Directory of Suppliers).

CROATIA

LAIR Training Equipment for Anti Tank Guided Missile Systems (SIM-TREND)

The SIM-TREND system is for training missile gunners of ATM systems of first and second generation such as Sagger, Spigot or Konkurs. The training system can also be modified to other similar SACLOS guidance systems. SIM-TREND is based on a multiprocessor integrated with a low-cost PC-based platform. The simulations of anti-tank combat have a 6DOF missile flight model, and a 6DOF model of target motion with missile trajectories and target movement programmable by instructors. The models also include simulation of different weather and other environmental conditions. The trainer computer and interface units are housed in a ruggedised case. The procedure for installation and calibration is designed to be simple and user-friendly, and self-test software gives automatic detection of defects.

Scenarios and Cues Provided. The training scenarios are based on a visual environment which includes a 3-D background scene based on digitised photographs of the real world, and CGI 3-D models of various types of targets. Visual effects for missile-grounding and missile-target impact are also provided. Sound effects of rocket motor, impact on target or grounding, and battlefield sounds such as gunfire are also available.

Components. SIM-TREND uses real ATM launchers but with their sight units corrected for optical alignment in the simulator. It has interfaces for connecton to the training guidance equipment or to the real guidance

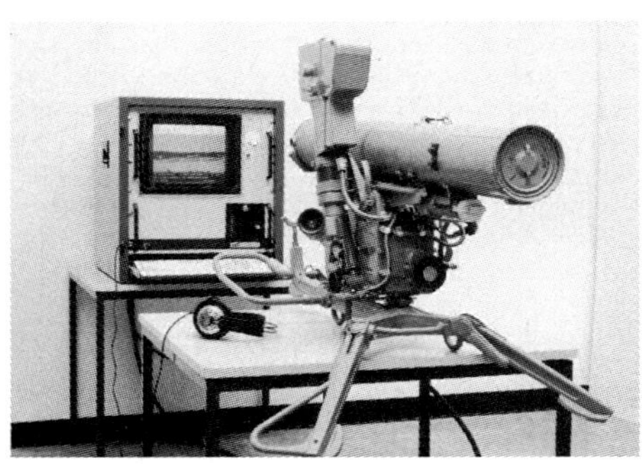

LAIR Croatia Sagger ATM Trainer

system. For Sagger it has the possibility of hardware-in-the-loop testing of the real guidance system.

Instructor Role. The instructor chooses scenarios and lesson plans from dialogue screens which are designed to be easy to understand. This equipment has a user-friendly software interface for control and mode selection based on MicroSoft Windows. On-line help is provided which shows the various mission options. Instructor selections are made through a keyboard or a mouse. During a training cycle, the instructor can change scenarios and other parameters such as types of targets and different landscapes. Replay of missile and target trajectories is provided for analysis and debrief. System software also provides automatic assessment of trainee actions and the results and assessments are saved automatically.

LAIR Croatia Spigot ATM Trainer

Status
In service with the Croatian Army (1994).

Manufacturer
LAIR Croatia, Zagreb.
(For details see Directory of Suppliers)

FRANCE

Giravions Dorand Weapon Simulators

Giravions Dorand Industries has produced a range of weapon simulators and training devices since first generation missile simulators were produced for the SS 11 wire-guided ATM. Most of the weapon-aiming simulators work by introducing simulated targets into the weapon sight or fire-control system of the weapon being trained. Search, track, fire and in-flight guidance to the simulated target can then be assessed by an instructor using a video monitoring unit. Some simpler trainers consist of a TV console and work station where the student is taught to use weapon controls or sights, compared to trainers which use a complete weapon installation.

DX 131. The DX 131 trains for HOT missiles fired from the UTM 800 turret. The equipment consists of an optical unit installed in front of the missile operator's sight, an electronics unit, a control unit and a video camera and monitor. This system can be used either against actual targets or against synthetic targets projected into the sight and viewed against the landscape. The video system can be used for instant playback. Test equipment is available for third line maintenance.

DX 132. As DX 131 but configured for HOT fired from the Lancelot turret.

DX 133. As DX 131 but configured for HOT fired from the VAB/Mephisto turret.

DX 136. Like the earlier DX 131, this is configured for HOT fired from the Lancelot turret but uses more recent technology. This simulator can also be used for training involving either synthetic or real targets. Components consist of an optical unit, electronics unit, control unit, and an instructor's monitor unit.

The optical unit generates synthetic targets and missile imagery which is superimposed on the real background seen by the trainee through the optical sight of the weapon.

The electronics unit generates missile trajectories and target characteristics and movement, both in the visual spectrum and also through a Castor thermal camera.

The control unit allows the instructor to modify and then store several target trajectories, initiate exercises and display results.

The instructor's monitor unit displays the image as seen by the trainee.

DX 137. The DX 137 is similar to the DX 136 but is configured for HOT missiles fired from the UTM 800.

DX 143. The DX 143 is used with the real MILAN launcher and consists of a dummy missile tube containing a computer unit and the 24 V power supply. An optical unit or video camera is mounted on top of the MILAN aiming unit and is used by the instructor to follow a student's performance. Training can be carried out against real or simulated targets which are projected into the aiming unit, and the full range of MILAN drills and actions can be practised.

The DX 143 is in production and use with several armies including the French Army.

Giravions Dorand DX 164 simulator for the TOW anti-tank missile

DX 147. The DX 147 is used for training operators of the HOT ATM in helicopters such as the Gazelle. As with the ground-based simulators described above, a simulated tank target is injected into the helicopter weapon sight. The DX 147 uses a simulated target at a range of 3500 m. The system can be used with the helicopter still on the ground if required. The system consists of an optical unit (BO-15) which is mounted on the helicopter sight in place of the IR missile tracker, a computer (CA-12), a control unit (BC-3) for use by the instructor, and a matching unit (MM-7) on the computer. The system weighs 21 to 55 kg (46.3 to 121.3 lb) plus the interconnecting cables.

The DX 147 is in production and service with the French armed forces.

DX 148. The DX 148 is configured for the Viviane sight fitted to the Gazelle helicopter armed with the HOT ATM. It is similar to the DX 136 and DX 137 and permits training using both the optical and thermal sighting systems.

DX 164 and DX 166. The DX 164 simulates TOW and the DX 166 simulates the ERYX short-range ATM. They can be used either in a classroom or outdoors, using real or dummy targets. The equipment consists of an optical unit, a missile canister containing the system electronics, and an instructor's control unit, which can be seen in the picture.

Status

In production and in service with over 50 countries.

Manufacturer

Giravions Dorand Industries, Suresnes.

(For details see Directory of Suppliers)

HUNGARY

IDEX IPR'90 Anti-tank Missile Firing Simulator

The IPR'90 anti-tank missile firing simulator is a development of the earlier IPR'88 version. It is based on an IBM PC with a 286 processor working at 12 MHz and 640 kbyte of RAM. The PC uses 3.5 inch disks on which missile firing and target data is stored. A colour monitor is used. Training is normally carried out indoors.

The basic IPR'90 is used to train the operators of the 9M14 and 9M14P1 'Maljutka' rockets used by the Hungarian armed forces (the Maljutka is a version of the Russian Sagger ATM). The IPR'90 consists of a mock-up of the missile operator's station used on the BRDM-2 amphibious scout car or the turret of a BMP IFV. The system can also be used (after some modifications) for training with the 9M111-2, 9M114 Fagot and Konkurs missiles. It can be expanded for training with the hand-controlled 9P133 Maljutka and for the Falanja rocket launched from helicopters, both of which use SACLOS guidance.

The instructor sits between two training stations, which may include full-size mock-ups of vehicle hulls and/or turrets or missile launchers. Landscapes are produced on the colour monitor and simple vehicle targets at various simulated ranges are then introduced. The trainee uses his launch unit sights and controls to acquire and track the target, and after simulated missile launch, the computer produces a representation of the missile exhaust and trajectory, and sounds of launch and other effects are produced. The operator uses his missile guidance controls in the normal way (MCLOS or SACLOS guidance depending on the missile being trained). The computer then assesses the accuracy of the shot and produces a hit/miss result.

A status panel in front of the instructor provides the instructor with the pre-launch and other states of the missile being simulated, and an intercom allows two-way conversations with the trainees. Firing data is stored on floppy disk. To switch from one missile type to another, the instructor changes the disk to one containing the new data and turns his monitor to face the other missile launch position.

Status

In service with the Hungarian armed forces.

Contractor

IDEX, Budapest.

(For details see Directory of Suppliers)

ISRAEL

Elbit SIMANTA Anti-Tank Missile Firing Trainer

SIMANTA is a computer-controlled simulator for training for ATMs such as TOW, MILAN or Swingfire. It consists of a number of cubicles for trainees, supervised by a single instructor. Each cubicle is about 8 x 3 m in area and accommodates a trainee missile gunner together with aiming and guidance equipment, a screen onto which a film strip is projected, a video display and recorder. Another trainee acts as a crew commander and as the station instructor (controlling the equipment within the cubicle).

Film is projected onto the screen. Targets may be either moving or stationary and more than one target may be projected. Under the command of the instructor/trainee commander (who may use binoculars if required) the trainee gunner selects a target, tracks, aims and fires the missile. A videotape recorder next to the launcher tube records and displays the results for subsequent analysis. As the missile is fired, the system simulates the firing flash, dust and smoke. The missile tail tracer is simulated in the normal way. Battlefield sounds such as gunfire and flashes may be introduced if required. The maximum screen time of any particular scenario is about 30 minutes. Target projection 'freezes' as the missile impacts.

The instructor has a central console which he can use to observe and record results on any particular cubicle, using selector switches to view individual sight pictures. He can also select the scenario film strip and the degree of battlefield effects to be used. Trainee gunners and commanders exchange roles when necessary. Trainees control and correct their actions using video playback, and if trainees are doing well, the role of the instructor reverts to that of system supervisor and troubleshooter.

Target range: up to 3000 m
Field-of-view or scene: 18°
Size of typical cubicle: 3 × 8 m
Distance of trainee from screen: 6 m approx
Display accuracy: 0.1 mrad
Display resolution: 0.1 mrad
Computed aiming accuracy: 0.1 mrad

Status
In service with the Israeli Army.

Manufacturer
Elbit Computers Limited, Haifa.
(For details see Directory of Suppliers)

SIMANTA training for the TOW ATM

SIMTECH Anti-tank Missile Gunnery Trainers

SIMTECH Anti-tank Missile Gunnery Trainers use real anti-tank guided missile launchers with the normal sight units replaced by a target injection sight unit. Target injection is based on images from video disk. Training is provided in pre-firing procedures, target search and track, missile firing and guidance, and post-fire procedures including actions following misfires and malfunctions. Subject ATMs include Dragon (see entry under USA in this section), TOW and the Israeli laser-guided MAPATS; several SIMTECH trainers for MAPATS have been sold in South America. SIMTECH can be operated both indoors and in the field.

Components. The system consists of a trainee station and an instructor station. The trainee station includes a real launcher and a synthetic sight unit containing a video monitor. Moving the launcher controls causes the image on the video monitor to move accordingly.

Instructor Station. A colour monitor displays the picture as seen by the trainee. Instructor control is through a touch-sensitive alphanumeric display and facilities are provided for training session management, planning, performance evaluation and scoring. Behind the front panel is a rack housing two monitors, the system computer, a video disk player and a printer. The computer contains the system application software, a sound and voice synthesizer, and a graphics module that combines real-world images from video disk with the synthetic images required for the situation such as target and missile effects.

Status
In production.

Manufacturer
SIMTECH, Holon.
(For details see Directory of Suppliers)

The SIMTECH ATM gunnery trainer configured for TOW, in service with the Israeli Defence Forces

SIMTECH ATM gunnery trainer configured for the MAPATS ATM

SWEDEN

Saab BT 61 Anti-tank Weapon Trainer

The Saab BT 61 anti-tank weapon trainer is for training gunners using direct fire weapons with optical sights. It is based on the use of computer-generated imagery and can be used on a wide range of weapon types. The weapons involved do not require modification other than the attachment of a weapon interface kit. It is also possible to use the equipment on vehicle-mounted weapon systems. Up to four trainees at one time can use the BT 61 as a troop firing trainer.

Components. A basic BT 61 training unit consists of a simulated sight with a small monochrome TV monitor which replaces the normal weapon sight, a weapon interface kit containing mechanical mountings, gyros and weapon sensors, and a computer which generates visual scenarios for trainees and also provides performance analysis and other functions. An evaluation unit is available as an option.

Operation. A trainee looking into the simulated weapon sight is presented with an image field-of-view approximately 45° wide and 20° high, depending on the magnification of the sight involved. The training scenario uses a three-dimensional database consisting of digitised photographs of real terrain. A training area up to 19 × 19 km can be used. Computer-generated targets can move freely in the gaming area and foreground objects are called up from a library. Up to four targets can be presented at one time, each being scaled and orientated according to range and position. The speed and direction of the targets is preprogrammed during the exercise definition stage and targets can execute evasive manoeuvres when fired upon. Weapon sight graticules are computer-generated.

Targets. Targets consist of three-dimensional models and have six Degrees-of-Freedom (DOF). At least five projectile models may be included in the programme. During the time of flight the projectiles may be visible, simulating tracers or missile flares. Muzzle flash and/or smoke obscuration are generated as the weapon involved is fired and results are represented by different types of burst. 'Killed' ground targets no longer move and air targets disappear. Also simulated are target sounds, firing effects, firing commands and so on, all presented to the trainee via a lightweight head set. Thermal image targets may also be presented.

Computer and Displays. The simulator computer unit is based on a 286 PC and is housed in a ruggedised case. The hardware is provided with diagnostic programmes for maintenance and repair down to board level. Two video monitors are connected to the computer with one monitor displaying the same image as the trainee's sight. The other monitor displays data and programme options for the instructor.

Weapon Interface Kit. The weapon interface kit is that part of the equipment attached to the weapon itself; the components involved are designed on a modular basis to permit transfer between weapon types. The main functions of the kit are to provide the visual and aural presentation of the training scenario to a trainee, to sense weapon movements and to sense weapon trigger and safety operations. A gyro unit is mounted on the barrel or launch tube to detect movements and gyro signals are processed by a small computer before they are sent to the main computer.

Status
Available.

Manufacturer
Saab Training Systems AB, Huskvarna.
(For details see Directory of Suppliers)

Saab BT 61 anti-tank weapon trainer components used with 84 mm Carl Gustav and TOW

Saab BT 61 anti-tank weapon trainer mounted on a TOW launcher

UNITED KINGDOM

Hendry All Arms Classroom Trainer/Simulator (AACT)

The Hendry All Arms Classroom Trainer/Simulator (AACT) is a portable trainer for most short-range direct fire anti-armour weapons, small arms, and ATMs. It is currently in service with the British Army for LAW 80 anti-armour weapon indoor training and has been adopted by other armed forces. Up to four firing lanes with static or pop-up targets can be provided with separate graphic and/or printed records of results. The system can be used for low light or night target image intensification with typical target and background signatures.

Weapon Fit. Weapons are fitted with an eye-safe 1 in/25.4 mm infra-red emitter tube, either inserted into or clipped onto the barrel. The insert is battery- or mains-powered with an acoustically-coupled trigger. Blanks or computer-generated sounds may be employed and recoil can be simulated on certain types of weapon.

Projected Scene. A 4 m wide background scene is projected onto a wall or screen. Ranges of 75 to 600 m can be simulated on the standard equipment and an extended range system extends the maximum range to 1200 m.

Targets. The Target Motion System comprises a 35 mm carousel target projector, an infra-red television camera and point-of-aim spot projector, all fitted on shock mountings within the transit case. A further background scene carousel projector is mounted on an upper platform of the case. The projectors are installed at least 6 m from the 4.25 × 2.1 m area of wall or screen involved. Targets may be moving or static with moving targets controlled by the instructor's joystick or preset to a programme. Target speeds may be up to 50 km/h. Up to eight target types are available with up to eight aspects available per target.

Control Unit. The Control Unit console has an instructor's joystick and other controls together with a 9 in/228 mm television monitor. The TV monitor provides the instructor and/or a class with a magnified view of a student's weapon sights and aim point, together with target data such as speed and range.

The Hendry All Arms Classroom Trainer/Simulator (AACT) in use with LAW 80

Performance Assessment and Scoring. An optional scoring facility is available which gives printouts of performance. Impact point and tracer are indicated by a screen light spot in the target vicinity or by a graphic display and readout on the monitor or printer. A recording/freeze facility shows the aim point prior to and on firing.

Transit Cases. The AACT may be packed into man-portable transit cases, the main packages being the Target Motion System and the Control Unit.

Status
In production. In service with the British Army and some other armed forces.

Manufacturer
D Hendry Electronics Limited, Arundel.
(For details see Directory of Suppliers)

Invertron Simulator for Wire-guided Anti-tank Tactical Training (SWATT)

SWATT is a computer-controlled system which simulates the characteristics of current ATM and is also designed to adapt to future missiles. The system can also be adapted for use by direct fire weapon systems such as tank guns. It is suitable for most anti-tank guided weapons such as HOT, TOW, Dragon, BILL and Hellfire.

General Characteristics. The trainee missile operator sees a projected view of terrain on which moving targets are either controlled by an instructor or by an automatic computer program. The trainee uses his own weapon sighting equipment (vehicle or ground based) and the simulator can be used for training not only in the guidance of the missile involved, but also for difficult conditions such as winds which are from the side or gusting. The simulator can also be used for training in target acquisition and identification. Two missile operators can use the system at one time and the system can be located in a classroom or in a hangar. In the latter case anti-tank guided missile launching vehicles complete with their fire-control systems can be used. If the student fails to 'gather' the missile once launched, it is destroyed, as in a real firing.

Sound Effects. SWATT incorporates an audio system that simulates sounds such as missile launch (including gyro run-up where applicable), in-flight noises from the missile, warhead detonation, moving target sounds and the effects of enemy gunfire being directed towards the firing position.

Terrain View and Targets. The terrain view can be produced using the customer's own slides and can represent an area over which troops might have to fight. Targets grow in size as they approach the firer, and vehicles will be screened as they pass behind obstacles in the line-of-sight. Target aspects are automatically changed, as when a side-on target turns towards the firing position. The targets automatically turn red and stop if a 'hit' is made and a 'target destroyed' message is flashed to the instructor on the computer's visual display unit. Battlefield visibility and lighting conditions, and up to eight independent and simultaneous moving targets can be projected onto the terrain view.

Invertron Simulator for Wire-guided Anti-tank Tactical Training (SWATT) configured for TOW

Record and Replay. A record is kept of the missile performance and a mission replay facility is provided. The mission replay can produce a 'freeze' of any part of the mission for debrief and analysis purposes.

Night Operations. Training for night targets uses infra-red projectors and special slides that can be projected in the dark. This is used in conjunction with thermal imaging sights.

Customers. SWATT was first delivered to the British Army in September 1982 and a further 17 systems were ordered, six for use in the United Kingdom and the remainder for BAOR. These are used by Swingfire Long-Range Anti-Tank Guided Weapon (LRATGW) units of the Royal Armoured Corps. The system was extended for use with MILAN, the 30 mm Rarden gun and the 76 mm gun fitted to Scorpion light reconnaissance vehicles.

The US Army purchased a version of SWATT referred to as M-CATT (Multi-weapon Combined Arms Tactical Trainer). It allows for the simultaneous training of two TOW or two Dragon anti-tank missile operators plus, in each case, a 40 mm MK 19 automatic grenade launcher. It also allows artillery observation of fire training. First deliveries were in May 1988.

Status

In service with the British Army (17 systems) and the US Army (one M-CATT).

Manufacturer

Invertron Simulated Systems Limited, West Sussex.
(For details see Directory of Suppliers)

UNITED STATES OF AMERICA

ECC Javelin Anti-Armour Weapons System Training Devices

Javelin training devices were developed by ECC International Corporation and are for use in the classroom and in the field. The training devices include the two major elements of the Javelin system, the Command Launch Unit (CLU) and the Launch Tube Assembly (LTA). The LTA is an expendable handling/launch tube that houses the missile, power pack attachment and CLU interface (see photo).

Javelin System - Background. The Javelin anti-tank weapons system itself was formerly known as the Advanced Anti-Tank Weapons System Medium (AAWS-M). It is a lightweight, one-man portable, 'fire and forget' anti-armour weapons system. Developed jointly by Texas Instruments and Martin Marietta, it is designed to overcome advanced armour designs and operate on the battlefield under a variety of visual conditions.

The Basic Skills Trainer (BST). The BST is designed for training in the classroom, aboard ship, or in garrison. Preprogrammed training scenarios are provided. Visual images are viewed through a colour LCD display in the simulated CLU. Other features are a lightweight target acquisition device that has an integrated day and night/thermal (IR) sight, launch controls and the gunner's eyepiece display. The BST provides initial training in field surveillance, target locating and acquisition, and fire control.

Javelin (ex AAWS-M) Basic Skills Trainer (BST) by EEC Corp

Field Tactical Trainer (FTT). The FTT is a MILES-compatible training device designed not only for training on weapon ranges but also for force-on-force tactical training. The FTT is used with the tactical version of the missile launch unit (CLU) and contains a MILES laser, AAWS-M Launch Effects Simulator (ALES); infra-red video is produced synthetically, simulating missile tracker video. The laser is capable up to a range of 3000 m. In addition, the laser is directed by the system tracker after lock-on. The ALES is a unit that provides simulation of both the flash and sound of a missile launch.

Missile Simulation Round (MSR). The MSR for the Javelin system is an inert round simulator for training in handling and transportaion.

Status

Available.

Manufacturer

ECC International Corporation, Pennsylvania.
(For details see Directory of Suppliers)

Perceptronics Dragon Precision Gunnery Training System

The Perceptronics Dragon Precision Gunnery Training System (PGTS) was developed in association with SIMTECH of Israel and is in service with the US Army and Marine Corps. The system consists of two stations, one for the student and the other for the instructor.

Student Station. Target scenarios are displayed with photographic realism in both day (colour) and night (monochromatic) sight channels. Scenarios are filmed in the field and stored on video disk. They feature actual combat vehicles engaged in realistic tactical manoeuvres and formations. The terrain and climatic conditions can be varied. Visual effects of missile firing, flight and hit/miss are generated by a computer and added to the scene as the student fires. Sound effects of firing are also created by the computer and presented to the student via earphones.

The student's station is a reconstruction of the Dragon launcher. The student uses either the day or night sight tracker for target identification and engagement. Each sight has focusing adjustment to avoid eye strain. On firing, the rubber-coated rear drop-off pod simulates the weight shift effect of a Dragon missile launch. The infra-red mirror tracking system provides feedback on aiming without any restriction on bipod movement. The tracking mirror is mounted in the student station carrying case for quick system set-up. The student station in its container weighs 60.3 kg.

Instructor's Station. The instructor's station contains the system computer as well as the floppy disk and video disk drives. A typical video disk contains 30 scenarios and two video disks are provided. Data display is at eye level, the drives are front loading and there is no keyboard; all input is via a 'mouse' controller. The instructor is free to observe student performance on a colour display and to provide immediate feedback either directly or through a headset. The instructor station in its container weighs 88.45 kg.

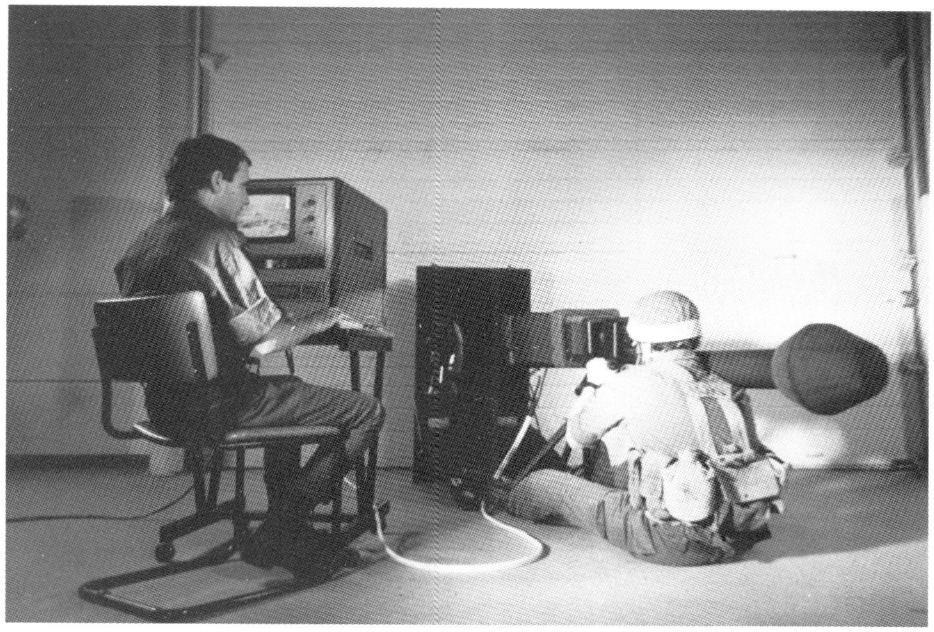

Perceptronics Dragon Precision Gunnery Training System

Analysis of Results. Evaluation of student performance is by automatic data recording and playback. While the student tracks and fires, the instructor views the target scene and the missile flight and hit/miss effects. After the mission ends, the student's performance can be assessed using playback at normal or slow speed. Scores, and computer-generated graphs of up-down and left-right aiming error are provided.

Training Sessions. Computer screens show the various mission options. These are selected by the instructor with the mouse. The planning process takes only minutes to complete.

Status

In production. In service with the US Army and Marine Corps.

Manufacturer

Perceptronics Inc, California.

(For details see Directory of Suppliers)

AIR DEFENCE TRAINING

CANADA

CAE ADATS Trainers

The Canadian armed forces operate the Oerlikon-Contraves Air Defense Anti-Tank Missile System (ADATS). Three trainers for this equipment are produced for the Canadian Forces by CAE Electronics of Montreal. These are a Full Mission Crew Station Trainer, a trailer-mounted Mobile Crew Station Trainer, and field-portable Tactical Unit Trainers which allow training with the ADATS equipment itself.

Features. The following features are included: visual displays with simulated terrain and cultural features such as roads and rivers; different atmospheric conditions, light level and weather; radar and E-O sensors; visual and sound cues; up to 30 stationary or moving air/ground targets; ECM; malfunctions; scoring and assessment; up to 100 prerecorded lesson plans available. Networking of separate units simulates battlefield conditions for complex exercises.

Status

In service with the Canadian armed forces.

Manufacturer

CAE Electronics Ltd, Québec.
(for details see Directory of Suppliers)

A CAE ADATS Crew Station Trainer at the Canadian Forces Base at Chatham, New Brunswick

CHINA, PEOPLE'S REPUBLIC

HN-5A Trainer

The HN-5A is the Chinese version of the Russian SA-7 'Grail' manportable SAM. A trainer has been produced to provide training on the handling, operating and firing sequences.

Components. Components of the trainer consist of a dummy missile and firing unit, a reusable battery, a monitoring and scoring unit operated by an instructor, and a moving target simulator which consists of a model aircraft moving along an overhead wire. The dummy missile and launcher are connected to the monitoring and scoring unit by a cable and all trainee operator actions are recorded for subsequent playback or hard-copy readout.

Status

In production. In service with the Chinese armed forces.

Agent

China Precision Machinery Import and Export Corporation, Beijing.
(For details see Directory of Suppliers)

CROATIA

LAIR Anti-aircraft Gun Simulator L70BF40

The L70BF40 simulator is for the training of operators of anti-aircraft guns equipped with the BOFI fire-control system. The simulator may be set for various levels of training and will operate indoors as well as in the open. Training exercises can be controlled either by an instructor or automatically using the system software. Performance evaluation results are produced and may be stored in files for each trainee.

Components. The L70BF40 components include a gunner control unit and control panel, and a personal computer (PC) with appropriate software and a colour TV monitor. The software supports all the operation modes of the BOFI fire-control system and training exercises can be reproduced for analysis. Other options available include: an additional monitor for an instructor; a motion platform with one or three DoF; laser spot-supported training with a special screen; and simulation and training in known terrain using images recorded on a magnetic disk or optical media.

Status

In service.

Manufacturer

LAIR Croatia, Zagreb.
(For details see Directory of Suppliers)

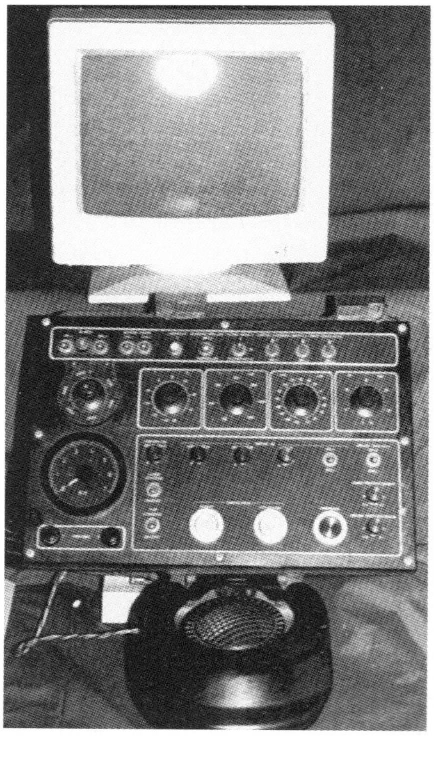

Anti-aircraft gun simulator L70BF40

LAIR Laser Quadrant Air/Ground Tracking and Aiming Simulator

This Air/Ground Tracking and Aiming Simulator provides training in weapon aiming and firing against aircraft targets. It can also be used for tracking ground targets. Evaluation of engagement results is through a panel display.

Fit to Weapons. The simulator can be fitted to all types of air defence guns. It can also be fitted to the larger direct fire weapons, including artillery, tank and anti-tank guns.

Components. Components include a laser transmitter/receiver unit, an instructor's or commander's control panel and unit, and a target reflector unit and various adaptors and cables.

Specifications

Weight: (without battery) approx 30 kg (66.1 lb)
Operational range: 20 to 400 m
Type of laser: GaAs laser diode; wavelength: 904 nm
Beam divergence: centre of beam 3 mrad; edge of beam about 12 mrad
Power supply: 12 or 24 V DC battery

Status

In service.

Manufacturer

LAIR Croatia, Zagreb.
(For details see Directory of Suppliers)

Air/Ground Tracking and Aiming Simulator fitted to 12.7 mm DShK-38/46 tank machine gun. The system control unit can be seen in front of the cupola at the end of the wire

FRANCE

Giravions Dorand DX 160 Anti-aircraft Gunnery Simulator

The Giravions Dorand DX 160 anti-aircraft gunnery simulator is intended for training gunners of 20 and 30 mm anti-aircraft towed or self-propelled guns, and operators of other short-range ground-to-air weapons. It can be configured to fit any particular weapon and no modification of the weapon is required.

Components. The DX 160 consists of an optronic unit which is placed on the sight bracket of the weapon involved. A control and processing unit operated by an instructor is connected by cable to the optronic unit. The control and processing unit consists of a control panel board, a video monitor displaying the same field-of-view observed by the trainee through his weapon sight, and a microprocessor-based computer.

Operation. The DX 160 uses the sight target injection system employing synthetic targets generated by the simulator. The targets are representational and may be in the form of generalised aircraft or helicopter shapes. The trainee views the targets through his normal weapon sight, and the target speed and course is controlled by the instructor. The target aspect and dimensions will change depending on their flight path, and targets may vanish behind simulated features such as trees or clouds. The sight display features weapon effects such as tracer trajectories and the results of hits and near misses.

Status

In production.

Giravions Dorand DX 160 simulator in use with a Cerbère 76T2 twin 20 mm automatic anti-aircraft gun

Manufacturer

Giravions Dorand Industries, Suresnes.
(For details see Directory of Suppliers)

Thomson-CSF New Generation Anti-aircraft Gunnery Simulators

Thomson-CSF, in co-operation with Standard Telephon und Radio AG, was chosen by the Swiss Ministry of Defence to supply 16 new generation anti-aircraft gunnery simulators. These simulators, delivered during 1992, provide both basic and advanced gunnery training on 20 and 35 mm anti-aircraft guns and manportable launchers. The simulators are designed for transportation and they can be used in field and base training environments.

System Description. The visual system produces computer-generated ground and air targets in colour in any direction. Eight targets are simultaneously available and may be aircraft, helicopters, armoured vehicles, troop carriers and trucks, as well as several types of thermal flare, shells and missiles. The system offers daylight, dusk and night visibility training conditions. A sound system simulates the combat environment and firing effects, as well as all messages given.

Facilities Available. Facilities available include ideal aiming, situation freeze, playback, automatic trainee evaluation and so on. Simulators can also be interconnected in a network.

Status

In service with Switzerland (16).

Manufacturer

Thomson Training and Simulation, Cergy-Pontoise.
(For details see Directory of Suppliers)

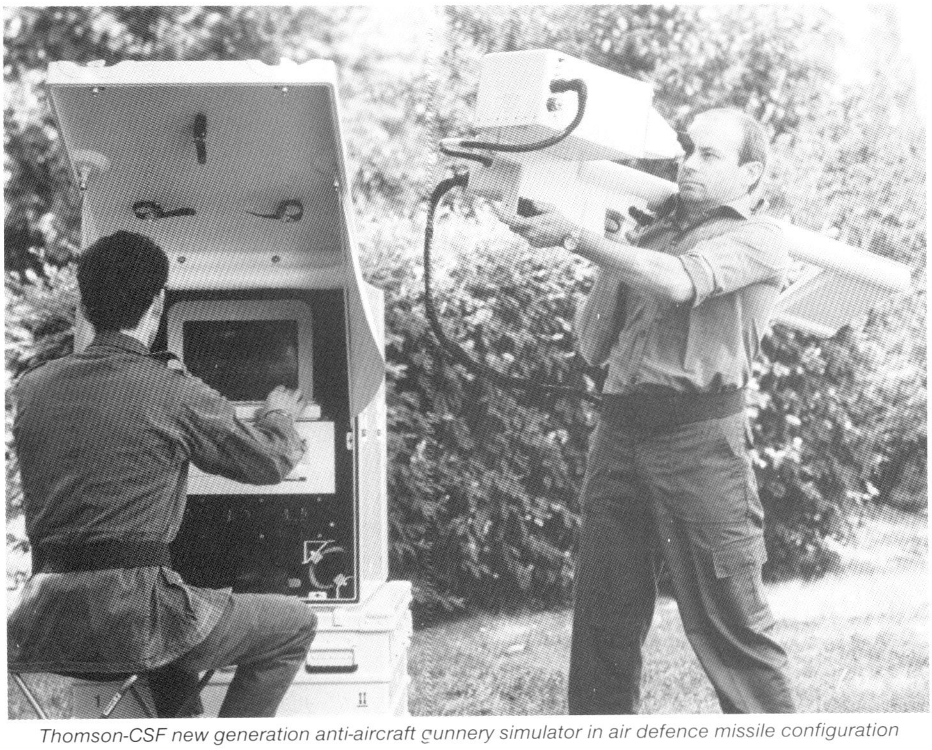

Thomson-CSF new generation anti-aircraft gunnery simulator in air defence missile configuration

Thomson-CSF Onboard Gunnery Simulator

Thomson-CSF has delivered 50 Onboard Gunnery Simulators to the French Army and exported a further 26. These simulators provide both basic and advanced training with the 20 mm guns on the AMX-30 MBT, AMX-10P MICV and AMX VCI ICV, and the 53T1 and 53T2 light anti-aircraft guns.

CGI System. A background scene with ground and air targets is produced by Thomson-CSF computer-generated imagery (CGI). The gunner sees tracer paths, impacts and the destruction of hit targets, and the CGI surfaces may be textured to enhance realism and improve judgement of distances.

Facilities Available. More than 200 target trajectories are available to an instructor and consist of images of aircraft, helicopters, armoured vehicles, personnel carriers and trucks. Instructor facilities include indication of the ideal aiming position, continuous display of aiming errors, and target speed and range. These can be displayed in the gunner's sight if required. In addition, situation freeze, playback and firing results are possible.

Design and Fit. The simulator is modular in design. It can be readily adapted to small and medium calibre guns on armoured vehicles, armoured personnel carriers, anti-aircraft guns and surface-to-air missiles; no modification of turrets is required.

Status

In service with the French Army (50 systems). They are in service with the anti-aircraft light artillery regiments and the AMX-30 and AMX-10P regiments. A total of 26 systems has been exported.

Thomson-CSF onboard gunnery simulator

Manufacturer

Thomson Training and Simulation, Cergy-Pontoise.
(For details see Directory of Suppliers)

GERMANY

Dornier Air Defence Simulator for Stinger

Dornier is the prime contractor for the licence production of the Stinger manportable air defence weapon in Europe. Dornier has developed the Fla-Simulator FIF2 which is designed for the training of Stinger teams which consist of commander and gunner. FIF2 is contained in a dome and utilises a computer-aided visual system which projects images on the inside of the dome surface. Maintenance is supported by BITE. The first example was delivered to the German armed forces in late 1991.

The Dome. The dome diameter is 20 m with a height at the centre of 6.5 m. Trainees, consisting of one or two Stinger teams, are located at the centre of the dome. The dome trainers were designed primarily for use with shoulder-launched Stingers but the design accommodates expansion to other systems such as helicopter-launched Stingers. The dome is therefore provided with large access doors measuring 4.5 × 4 m.

Targets. A target generator produces one or two targets using computer-generated imagery (CGI). Targets are projected onto the inner wall of the dome and separate optical projectors produce various background landscapes, clouds and sun. Images are produced over the complete 360° of azimuth and from 0 to +90° of elevation. Flight paths of fixed-wing aircraft and helicopters may be simulated, with target masking where appropriate. The Philips CGI target generator produces three-dimensional targets in shades of grey with representation of curved surfaces by shading. The target image update rate is 25 Hz and the image repetition rate 50 Hz. Resolution is 780 × 625 pixels and the targets are represented in their true size for ranges from 0.5 to 10 km. Speeds of up to Mach 1.2 may be simulated. A separate target generation unit (computer graphics station) allows the production of additional targets and flight profiles by the user.

The simulators for the German army have the capability for only one target at a time, but a second target generator may be added if required. The simulator for the Netherlands will produce two targets.

IR Simulation. Since Stinger is a passive homing missile using a heat-seeking sensor operating in the IR band at about 3-4 microns, as well as visual imagery, thermal conditions in this band are also simulated. Target IR signatures are calculated, and provision is made for artificial targets such as flares and natural false targets such as sun, reflections from clouds and background IR noise.

Gunner's Weapon and Engagement Procedures. The gunner's weapon is a Stinger aiming simulator without a seeker head and cooling gas unit. Seeker head behaviour and status is simulated throughout the engagement process. After firing, the missile flight path and indications of hit or miss are represented and displayed. Target noise, firing and warhead detonation sounds are simulated along with battlefield noise. Up to five engagement procedures with different target simulations and flight paths may be carried out within 10 minutes.

Instructor's Facilities. An instructor's console is in a separate room. The instructor starts an exercise and during the engagement process evaluation data is stored for later analysis, evaluation and production of statistics of team performance. Communication between the team commanders and instructor is via headsets.

Compatibility with other Systems. The system is compatible with the ROLAND classroom combat trainer, including the control panel, CGI assembly and MUDAS (Modular Universal Data Acquisition System). Because of the large doors to the dome, it is also possible to accommodate AAA weapons.

Status

In service with the German armed forces (5 units) and the Netherlands (1 unit).

Manufacturer

Dornier GmbH, Friedrichshafen.
(For details see Directory of Suppliers)

Dornier Anti-Aircraft Training System (AATS)

The Dornier Anti-Aircraft Training System (AATS) is an electronic training system for use by anti-aircraft batteries, enabling them to conduct dry firing exercises involving real targets. AATS is designed for military

Dornier Anti-Aircraft Training System (AATS) gun data terminal

transport and can be taken to wherever it might be required. The system has its own built-in test equipment. Up to 20 gun crews may be trained at one time although it is possible to involve more. The system can be used in a natural scenario using targets of opportunity as well as prearranged targets. AATS can be adapted for a number of light anti-aircraft weapons.

Components. The system has four main components: a central control station; one gun data terminal for each gun involved; an adaptor for each gun; and a target tracking system.

Target Tracking System. The target tracking system is used for following the flight paths of real aircraft or helicopters. Information on target position is then continuously transmitted by telemetry to the central control station.

Central Control Station. The central control station receives the target position data and transmits it to the guns. It also receives, stores and presents various other data and training results. It consists of monitoring devices with CRT displays, an electronic evaluation system, telemetry receivers and a floppy disk-based memory system.

Gun Data Terminals. A gun data terminal, contained in a water-resistant box, is connected to each gun and compares the actual target positions with the calculated trajectories of rounds. This is done by measuring the azimuth and elevation of each gun barrel and making appropriate trajectory calculations on trigger press. At each gun, hits, misses and the differences between the flight paths of target and rounds fired are displayed. Results of the simulated firings are presented on the screen of the terminal and are stored for up to ten training runs.

Status

In service with the German armed forces.

Manufacturer

Dornier GmbH, Friedrichshafen.
(For details see Directory of Suppliers)

Dornier ROLAND Classroom Combat Trainer

The classroom trainer developed for the ROLAND SAM system may be based in a vehicle-carried shelter or (more usually) a classroom. It employs no element of the actual ROLAND equipment. The classroom trainer consists of simulated commander's and gunner's compartments and an instructor's station. The classroom trainer is transportable. The trainer simulates all stages of an engagement cycle, including search, target tracking, identification, missile launch and interception.

Commander's Compartment. In the commander's compartment all weapon system functions are simulated, including visual and radar displays which are emulated electronically. A comms link to the gunner's compartment is also fitted for integrated training.

Gunner's Compartment. The gunner's compartment contains a simulated optical sight which gives 360° coverage. This can display a background of trees, vegetation and other terrain features, and up to three aircraft targets.

Instructor's Station. The instructor's station is a desk and console with an interactive colour display. The instructor can enter training sequences and simulated system malfunctions using a light pen. The system is capable of exercises from basic to advanced, with simulated hostile radar jamming if required. Various aircraft attack profiles and different aircraft types can be selected by the instructor. The trainer can also be used for basic maintenance training exercises. A monitor on the console can be used to follow exercises and display results. The system has its own built-in test equipment.

ROLAND classroom trainer showing instructor's station (left), commander's compartment (centre) and gunner's compartment (right)

Specifications. All subsystems are connected by a standardised serial bus data (MIL-STD-1553B). The central computer employs a 16 bit system with a memory expandable to 64 k words. The complete trainer weighs 2100 kg and requires a 220/380 V, 50/60 Hz, 4 kVA electrical supply. The environmental temperature range is from +10 to +40°C. Mean Time Between Failures (MTBF) for electronic units is stated to be up to 1000 hours and up to 300 hours for electromechanical components.

Status

In service with the French and German armies.

Manufacturer

Dornier GmbH, Friedrichshafen.
(For details see Directory of Suppliers)

DST AGG Training Simulator

Deutsche System-Technik GmbH (DST) has developed a gun turret trainer for the German-Dutch FlakPz 2 air defence vehicle. The system provides operational training of individual crews, tactical training programmes for the turret crews of up to seven networked vehicles taking part in battle exercises, and networking with other independent weapon systems via a computer highway.

Status

Development.

Manufacturer

Deutsche System-Technik GmbH (DST), Bremen.
(For details see Directory of Suppliers)

Honeywell Florett Anti-Aircraft Training System

Florett is a computer-controlled training system for anti-aircraft guns that was developed by the German Honeywell company to meet the requirements of the Swiss Army. Development was completed in 1975 and it is in service with the Swiss Army. The system permits simultaneous, individual training of light anti-aircraft gunners, including the co-ordination of a complete anti-aircraft unit.

Components. There can be up to 40 guns within 2000 m of the system computer centre. A terminal is assigned to each gun, and is connected to the computer centre by a simple two-wire pair. The terminal is equipped with a gun data transmission unit, a display unit, controls for identification and communication, and angle encoders for azimuth and elevation and firing switches. The display unit shows targets and indicates results and correction values.

Honeywell Florett anti-aircraft training system in use by Swiss Army gunners

A central terminal is used for supervising the entire exercise as well as monitoring the performance of the individual guns. The instructors can switch to any of the guns in the course of the exercise, or to issue orders to all the guns connected to it. Record/replay facilities are fitted for monitoring and debrief purposes.

The computer control centre consists of an air-conditioned module containing the main system computer, two magnetic tape units, an input and output typewriter and a printer. It can be removed from a truck if required by using its own winches.

The target-tracking radar is connected to the computer control centre and is used for monitoring the target situation. It can be either a Contraves Super-Fledermaus or Contraves Skyguard.

Operation. The central terminal controls the exercise and work at the guns. In the course of an exercise the process (main) computer system determines the relative position to the target for every actual or simulated burst of fire using data on target movement, ammunition ballistics and information on gun positions. From this data the target errors and correction data for individual gun sights are determined and stored. A maximum of one to two minutes elapse between firing and the display of the results, with simultaneous control of 40 guns. The processing capacity of the computer centre depends on the processing time and the number of guns handled; the processing time is reduced to about half to one second with only one gun. There are additional computer programs to cover recording and allocation of parameters for target practice, preparation of statistics and determination of the qualification ratings of the gunners.

Status

In service with the Swiss Army.

Manufacturer

Honeywell Geschäftsbereich Sondertechnik, Maintal.
(For details see Directory of Suppliers)

Honeywell Mobile Automatic Cannon Training System (MKT)

This system was produced for the German Army and is known as the Maschinen Kanonen Trainer (MKT). It is a computer-controlled system for training gun commanders and gunners of automatic cannon of all calibres. It is intended to be used in an air-conditioned shelter at unit level, but it is mobile. Up to 20 crews can be trained at once. Development of the MKT began in 1978 with the first training system delivered in July 1982.

Weapon Fit. Each weapon is equipped with an adaptor set which comprises a TV monitor, collimator and data acquisition system. There are also sensors for traverse, elevation, trigger movement and sight adjustment. A cable set completes the gun equipment and each gun commander has a monitor set.

Operation. The MKT feeds synthetic targets into the gun sight and the gunner can track the target and open fire; synthetic tracers then appear in the sight picture. Firing results can be displayed immediately or recorded in a computer for later printout.

Instructor's Station. A central instructor's station can oversee up to 20 equipments in use at any one time and an intercom is supplied to each gun station.

Status

In service with the German Army.

Manufacturer

Honeywell Geschäftsbereich Sondertechnik, Maintal.
(For details see Directory of Suppliers)

MKT instructor's station and console

Honeywell MKT on a 20 mm anti-aircraft gun

ISRAEL

SIMTECH Air Defence Gunnery Trainers

SIMTECH Advanced Training and Simulation Systems makes a series of air defence gunnery trainers for use with 20 mm, 23 mm and 20 mm Vulcan (Gatling) anti-aircraft guns. The trainers are in service with the Israeli Air Force and are known as the Advanced Air Defence Training System (AADTS).

General Characteristics. The trainers use real air defence guns with their normal sight units replaced by a target injection unit. Synthetic targets are produced from video disk, and real terrain backgrounds are also simulated. Missile trajectories are simulated, together with sound and visual imagery of the firing, guidance, hit and miss effects. Immediate feedback and performance evaluation is provided. AADTS allows trainees to practise pre-firing procedures, target search, detection, acquisition, tracking and firing, misfire and post-fire operational procedures.

Instructor Station. The instructor station has two positions, one to run the exercise and the other for post-exercise debriefing. Both are in a shelter carried on a suitable vehicle and are connected to the guns via cables. The instructor station uses colour monitors displaying the same picture as seen by the trainees. A touch-sensitive alphanumeric display is used for management, planning, performance evaluation and the trainee's score presentation. A computer contains a graphics module which combines the target video images with graphic effects, a sound and voice synthesizer and the system application software. The instructor station can control up to three guns at one time.

Status

In production. In service with the Israeli Air Force.

Manufacturer

SIMTECH Advanced Training and Simulation Systems Limited, Holon.
(For details see Directory of Suppliers)

A complete SIMTECH air defence gunnery training system in operation

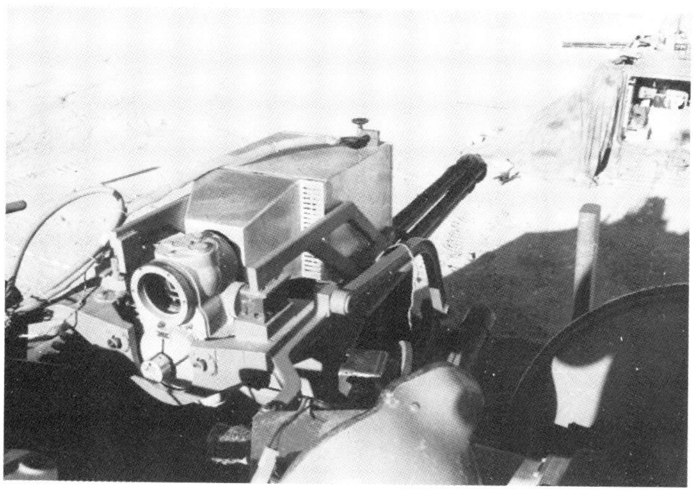

The trainee station of a SIMTECH air defence gunnery training system on a 20 mm self-propelled anti-aircraft gun

A SIMTECH air defence gunnery training system on a Vulcan 20 mm self-propelled anti-aircraft gun

SIMTECH Stinger Troop Proficiency Trainer (STPT)

The Stinger Troop Proficiency Trainer (STPT) was developed by SIMTECH as a subcontractor to the American Kollsman company and several hundred are in service with the US Army.

Launcher Unit. A simulated Stinger launcher is used with the normal sighting unit replaced by a simulated sight unit housing a small video monitor. STPT employs CGI targets superimposed on real landscapes, and sound cues are also provided. The system may be used indoors or outdoors.

Instructor. The instructor views all trainee actions on a colour monitor and can record all parts of any exercise for replay and debriefing purposes.

Status

In production. In service with the US Army.

Manufacturer

SIMTECH Advanced Training and Simulation Systems Limited, Holon.
(For details see Directory of Suppliers)

SIMTECH Stinger Troop Proficiency Trainer (STPT)

NETHERLANDS

NEDINSCO Video-based Live Aircraft Stinger Trainer (VLAST)

The Video-based Live Aircraft Stinger Trainer (VLAST) was developed by the Nationaal Lucht-en Ruimtevaartlaboratorium (NLR - National Air & Transport Laboratory) for the Royal Netherlands Air Force. It is manufactured by the NEDINSCO company. It takes video recordings of Stinger engagements using real aircraft in localities such as training ranges or where targets of opportunity are engaged.

General description. The VLAST system records video, audio and engagement data using a wide-angle video camera mounted on a Stinger Tracking Head Trainer. Missile seeker head position and weapon control positions are also recorded. Recordings are then used to debrief Stinger team commanders and gunners and may also be used to provide aircrew with visual evidence of their tactics and actions.

Control Unit and Power Supply. The system uses a control unit which may be up to 100 m from the missile launch site and camera. It operates from a vehicle battery but a 220 V AC mains supply can also be used.

Status

In service with the Royal Netherlands Air Force.

Manufacturer/Developer

Manufactured by Nederlandsche Instrumenten Compagnie 'NEDINSCO' BV, Venlo; developed by NLR (Nationaal Lucht-en Ruimtevaartlaboratorium), Amsterdam.
(For details see Directory of Suppliers)

Video-based Live Aircraft Stinger Trainer (VLAST)

NORWAY

MPS Air Defence Simulator (MADS)

The MPS Air Defence Simulator (MADS) provides training for land-based low level air defence systems. The system provides training for both individual operators and also the whole system team in overall co-operation and decision making. Training can include adjacent headquarters and Electronic Warfare (EW) sites. The simulator produces an automated evaluation of each student's performance.

Components. Each MADS has six instructor positions at three instructor consoles, six student positions at three radar operator consoles, 12 student positions for RBS 70 training, and 24 student positions for weapon terminal operators. Facilities also include a debrief console, a game preparation console and system control station, a laser printer, a plotter and four voice recorders. The console type is the MPS Multipurpose Tactical Console.

The MADS utilises real radar operator consoles, weapon terminals and missile equipment. Synthetic displays are based on digitised maps and high resolution digital terrain modules for the radar spectrum. Attacks are simulated by fixed-wing aircraft or helicopters, including use of weapons and EW equipment.

Computer. MADS uses distributed real-time computers with a local area network (using an Ethernet IEEE 803.2), full recording and playback/debrief facilities, and a voice communication system.

Service Operators. The Royal Norwegian Army has two MADS for operators of the Norwegian Army Low Level Air Defence System (NALLADS). One was delivered to the Royal Norwegian Army Training Establishment at Fredrickstad in October 1992 with the other delivered to Bardufoss in September 1993. The system provides operator training involving Ericsson Giraffe radars, weapons terminals and real RBS 70 missile equipment.

Status

Two systems in service with the Royal Norwegian Army.

Manufacturer

Micro Processor Systems A/S, Kongsberg.
(For details see Directory of Suppliers)

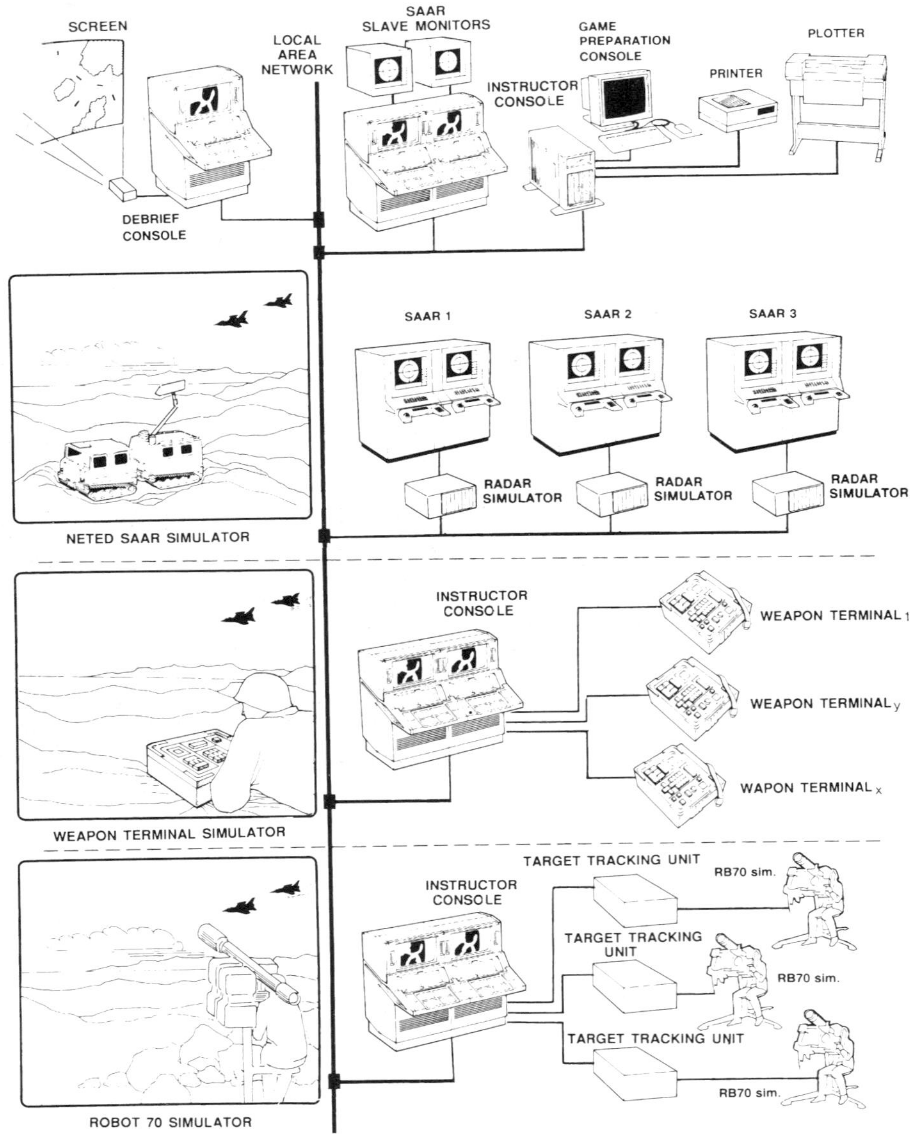

Schematic layout of an MPS Air Defence Simulator (MADS)

POLAND

Cenzin IGLICA Air Defence Missile Training System

The IGLICA Air Defence Missile Training System was developed to provide training for operators of short-range air defence missiles such as the Russian SA-7 'Grail' shoulder-launched SAM. Up to four missile operators can be trained at any one time and the system is portable and may be used in virtually any locality.

Components. The IGLICA consists of four replica SA-7 launchers complete with sighting systems and all controls. All four replica launchers are connected to a central control and recording unit via cables. Simple targets are injected into the sighting system but the IGLICA is intended for use as a sequence trainer rather than as an aiming trainer.

Operation. In operation the IGLICA launcher is held at the ready until a simple target outline is injected into the sight. The operator then tracks the target while carrying out weapon switching actions. These include switching on the launcher battery, moving the trigger lock, making the necessary elevation and lead angle corrections, and the actual launch. Sound cues such as gyro wind-up and launch are also produced. The IGLICA system monitors and records all operator actions and sequences and assesses launch accuracy over a series of up to 10 launches. Individual or group sequences can be recorded and printed out for recording and analysis.

Power Supply. Power is from either a 220 V AC mains supply or two 12 V batteries. The system can be utilised for up to 16 hours at a time provided a 1 hour break is taken after the first 8 hours.

IGLICA air defence missile training system for SA-7 Grail shoulder-launched SAM

Status

In service with the Polish armed forces.

Marketing Agency

Cenzin Company Limited, Warsaw.
(For details see Directory of Suppliers)

SOUTH AFRICA

Synertech ZUMSIM Air Defence Gun Training Simulator

During operations against various states bordering South Africa the South African Defence Forces acquired many twin 23 mm automatic anti-aircraft ZU-23 guns. The type was incorporated into the South African inventory in a self-propelled form known as the ZUMLAC. A training simulator for ZUMLAC has been developed and is known as ZUMSIM. Although ZUMSIM was developed for the 23 mm ZU-23 it can be adapted to suit virtually any air defence gun system between 20 and 35 mm calibre.

General. ZUMSIM produces a moving three-dimensional target, background images and flight path profiles which are injected into the weapon sights. No modification is required of the weapon involved other than connection to the simulator. Each simulator can provide images for up to three guns. The simulator can generate an unlimited number of target flight paths and scenario generation is stated to be simple and rapid.

Container. The ZUMSIM simulator is housed within a standard ISO shelter/container for easy transit and is connected to the guns involved via cables held on booms. The container has an integral auxiliary power unit for self-sufficiency. Installation of the simulator takes one hour using a step-by-step manual and does not require special tools.

Record/Replay. There is a playback facility for evaluation of tracking accuracy, a percentage scoring system for performance classification, a recording facility for debriefings and a printout of student results.

Status

In service with the South African Defence Forces.

Manufacturer

Synertech (Pty) Limited, Halfway House.
(For details see Directory of Suppliers)

SWEDEN

Saab BT 53 Rapid Fire Simulator

The Saab BT 53 rapid fire simulator is for field training. It can be used with any type of anti-aircraft gun, air defence missile or air defence armoured fighting vehicle with only minor changes to the system computer hardware. Round and guided missile trajectory and ballistic data is programmed into the system and it is easy to change data by reprogramming the BT 53 computer.

Components. The BT 53 simulator employs an eye-safe low power laser unit normally mounted on brackets over the barrel of the gun involved. Other components are a tracer unit, a computer unit, a display unit, an elevation sensor unit, a trigger device unit and a power supply unit. Reflectors for the laser are carried on target vehicles.

Operation. When a simulated round is fired a tracer is shown in the gunner's sight, simulating the first round of each burst (or the tail flare of a guided missile if that is being trained). The tracer is visible until the simulated round has hit or passed the target. It is then turned off and, for gun training, the next projectile tracer approaching the target is displayed, and so on. Thus the gunner can obtain an immediate feedback of his firing performance. He can judge the hit/miss distance relative to the target and can assess his gun laying procedures and accuracy. Target speed, range and direction are measured and recorded. The target hit area is defined in the simulator system, allowing simulation of direct impact or proximity-fused ammunition. No input to the simulator system is required other than a signal from the trigger device unit connected to the normal weapon trigger. The accuracy of the BT 53 simulator is stated to be the same as with live ammunition.

Recording of Results. The BT 53 provides a printout and displayed result of the time of opening fire, the range to the target at that instant, the speed of the target, the minimum perpendicular distance between the gun and the target flight path, the best round hit/miss position and the mean value of the burst. It also provides the number of fired rounds in the burst and the number of hits obtained in the target area.

Saab BT 53 on a Gepard twin 35 mm AAA system; note the laser unit fitted to the port-side barrel above the ladder

Air Targets. For RPV targets, miniature laser reflectors are used and special reflector pods are used on aircraft targets. The aircraft reflector pods are 2.46 m long and fit standard weapon pylons; weight is 44 kg. An alternative and relatively low-cost solution to providing fast-jet aircraft targets is the fitting of reflectors to civil light aircraft. These 'Standard' laser reflector pods are 370 mm long, weigh 1.2 kg and can be fitted to most light aircraft within minutes.

Status

Production complete. In service in Sweden, Germany and six other countries.

Manufacturer

Saab Training Systems AB, Huskvarna.
(For details see Directory of Suppliers).

Weibull Training Equipment

MANTEQ

The MANTEQ (MANportable Television EQuipment) is a lightweight television equipment weighing 22 kg (48.5 lb) and intended for outdoor use. The TV camera is mounted on the weapon involved so that the picture shown on the 127 mm screen is the same as that seen by the trainee aimer. Any aiming graticules involved are also recorded. Extra monitors may be connected.

The trainee may watch his results immediately after an exercise. By using a recorder known as the INFO unit, features such as date, time, mission number, target distance, firing, hits and miss-distance can be involved, and the instructor may add his own comments to the sound track. An optical laser rangefinder can be added to the system if required.

The MANTEQ is in service in Sweden configured for the RBS 17/Hellfire coastal defence missile system.

Pbv 302 Target Surveillance Unit

The Pbv 302 Target Surveillance Unit records the aiming of the turret gunner in the Pbv 302 APC. The equipment is mainly carried inside the Pbv 302 personnel compartment, and uses the vehicle's power supply. The main component is the monitor and recording unit, which consists of a television camera mounted over the vehicle gun outside the turret. A link with the gunner's sight is so arranged that the monitor can introduce reference and other information onto the picture. During live firings the tracers are also recorded. When not in use the monitor and record unit and camera are packed in a metal container.

The Pbv 302 is in service with the Swedish armed forces.

AMX TG 230 Anti-aircraft Trainer

This device is used for training the AMX gunner in target search, tracking, acquisition and firing. For indoor training the gunner fires against a target simulator and for outdoor training against real targets. A recording system is used for evaluating the results.

First deliveries were to France in 1978.

Surface/Air Target-track Trainer

This equipment simulates air and ground targets for anti-aircraft weapons or APCs, using small scale models of targets. Scales used are 1:200, 1:400, 1:600 and 1:1200. Targets can be moved with simulated speeds from 0 to 320 m/s. Depending on the space available, two or more weapons can engage the same target.

Components. The equipment consists of a dolly with a drive unit and a mast dolly. Targets have built-in photocells and move along a steel tape which is suspended between the two dollies. Light projectors for air and ground targets are accommodated in holders on the weapon. The light of each projector is modulated differently to ensure that the firing results from each weapon do not inadvertently mix. In the basic training mode, the gunner can see the light of the projector on the target. As training progresses, it is possible to work with other parts of the electromagnetic spectrum such as the near IR, using intensifiers.

Hit Indicator. The hit indicator registers hits or misses and has up to five channels, depending on the training system used. It allows the simulation of single shot and burst with various rates of fire. Further, a firing noise generator can be added to enhance the realism of the simulated combat situation.

Deliveries. First deliveries were in 1957. It is still in service with Denmark, France, Germany, Norway, Switzerland and Sweden.

TV Equipment for Training and Monitoring

The sight picture of a weapon is projected into the camera of this equipment via mirror. Playback is used for assessing results. The TV equipment is easy to transport.

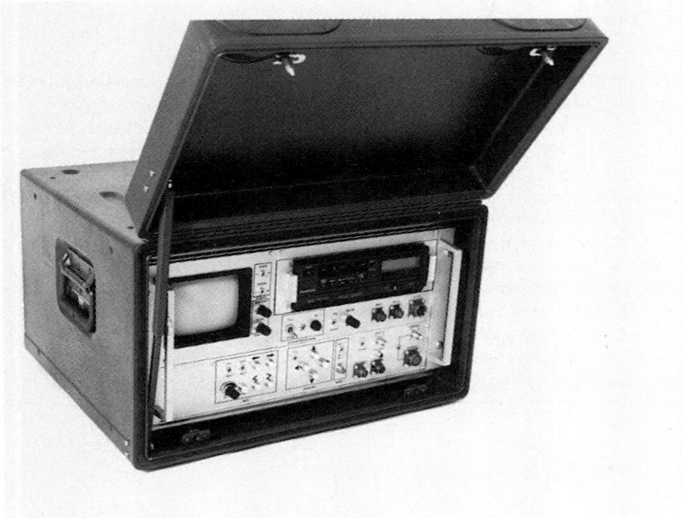

Monitor unit for the MANTEQ manportable television equipment

MANTEQ mounted on RBS 17/Hellfire coastal defence missile target designator

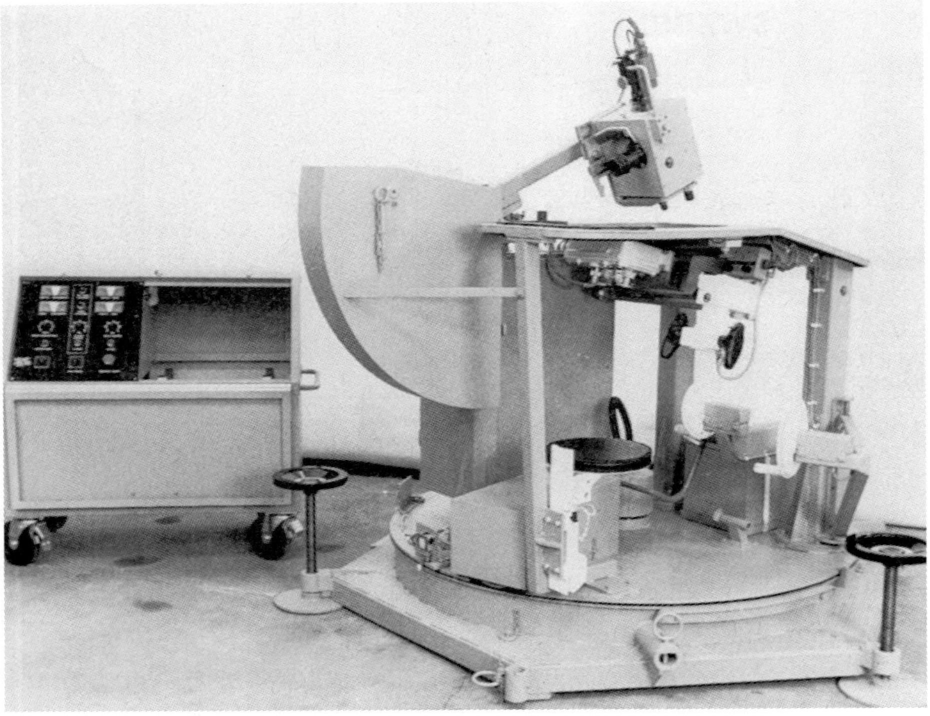

AMX TG 230 anti-aircraft trainer (right) with operator's console (left)

TV equipment for training and monitoring mounted on Pbv 302 APCs

It has been in service with France and Sweden since 1974.

Hit Indicator

A hit indicator indicates the number of rounds in the magazine and is connected to a light projector. The hit indicator permits a choice of number of rounds fired and the choice of different fire rates from single shots up to 2000 rds/min. The hit indicator has been designed for the training of one to five gunners at the same time. The latest model is intended for two weapons.

First deliveries were in 1965. It is still in service with Denmark, France, Germany, Norway, Sweden and Switzerland.

Sighting Trainer

This sighting trainer is used for basic training of AAA and other gunners. The unit gives the gunner an opportunity to assess various target situations and to aim accordingly. For supervision the instructor uses a light projector to monitor sight adjustments. Training is carried out indoors.

In production.

Status

See individual entries.

Manufacturer

J L Weibull AB, Möllegården.
(For details see Directory of Suppliers)

SWITZERLAND

Oerlikon-Contraves Air Defence Training and Simulation Systems

Oerlikon-Contraves has been producing air defence and associated training systems for many years. Several simulation systems concern air defence training requirements. The systems outlined below are currently available:

Firing Evaluation and Training System (FEATS)

The Firing Evaluation and Training System (FEATS) is a range system usually applied to 35 mm and other air defence guns. The system can provide continuous tracking of an air target and provide measurement of the trajectories or rounds for bursts lasting up to 60 rounds. The system can measure any type of projectile or guided missile and can cope with multiple bursts. Angles can be determined to an accuracy of less than 0.5 mrad. Results can be determined within seconds and the system can immediately inhibit the firing of any gun violating safety regulations. The system can also be used for 'dry' firing exercises involving guns not cleared for live firing.

Tactical Engagement Simulator and Trainer

The Tactical Engagement Simulator and Trainer (TEST) can simulate an air attack, and the air defences involved. It then calculates the hits scored by both the attackers and defenders. The system interacts with air defence command staffs and supports the staffs in real mission planning. Results of air defence engagements can be presented according to various scenarios and the results of various plans can be compared.

ETS Gunnery Training Simulators (GTS)

This gunnery training simulator family was designed for the Oerlikon-Contraves 35 mm air defence gun using the Gun King sight module. The system trains the gunner in the basic functions of the controls and displays at the gunner's position; provides all the operating sequences without the intervention of an instructor; and provides training in target acquisition, tracking and engagement of both air and ground targets. The system can provide the full range of gunner skills from basic to the networking of up to four trainee stations (the ETS/N). Computer-generated targets and weapon effects are employed and the simulated sight system can provide a full 360° coverage.

Skyguard Air Defence Training Simulator TS-2

The Air Defence Training Simulator TS-2 was designed for use with the Skyguard low level air defence radar. It provides training for personnel operating the Skyguard consoles and controls. It presents simulated aircraft targets on the commander's PPI display and on the target operator's TV monitor. ECM is also modelled. The TS-2 may be used in a field environment, and can be employed with air defence systems other than Skyguard.

The TS-2 has artificially generated background scene with the light intensity individually adjustable by computer control. Artificially generated targets are displayed on the TV monitor. The TS-2 is used in conjunction with an operational Skyguard system. The TS-2 operator's console corresponds approximately to that of a Skyguard and includes a TV screen, a PPI display and a computer keyboard. Training sequences are stored on tape cassette and include background data, targets and evaluation methodology for given exercises. The TV monitor can display up to four targets at a time.

Status

Available. In widespread service.

Manufacturer

Oerlikon-Contraves AG, Zürich.
(For details see Directory of Suppliers)

Four 35 mm Oerlikon air defence gun cabins connected to form an Engagement Training Simulator Network (ETS/N) together with the instructor station

UNITED KINGDOM

Marconi Air Defence Tactical Trainer (ADT2)

The Marconi Air Defence Tactical Trainer (ADT2) is a computer-based indoor training system derived from the latest version of the ADT3-C (see next entry). The system provides a portable indoor training system for manportable air defence systems such as Stinger, Javelin, RBS-70 and Mistral. The ADT2 provides training in all phases of an engagement.

Screen. The ADT2 employs a screen upon which both background and individual targets are projected. The curved screen provides for a minimum arc of 120° in azimuth and 45° in elevation.

CGI Target System. The computer-generated image system and projectors used by the ADT2 allow for the projection of up to four aircraft of various types, including helicopters, anywhere on the screen surface. Targets are generated in real time and can be provided by the end user or by Marconi. The range of target flight profiles is unlimited and is generated by the end user. The system allows for engagement ranges of less than 500 m out to ranges in excess of 14 000 m, in all visibility conditions ranging from full sunlight to night, and normal binoculars may be used to identify visual targets.

Instructor Station. The facilities provided to the instructor allow for the preprogramming of a number of differing exercises, the control and monitoring of an exercise including a readout of the gunner's view.

Inside a Marconi Air Defence Tactical Trainer (ADT2)

Analysis of an engagement is made with a hard-copy option for debrief purposes.

Status

Under development.

Manufacturer

Marconi Simulation and Training, Cwmbran.
(For details see Directory of Suppliers)

Marconi Air Defence Tactical Training Theatre (ADT3-C)

The Marconi Air Defence Tactical Training Theatre (ADT3-C) is a computer-based indoor training system for training air defence weapon operators and crews against high-speed low-flying aircraft. It provides training in all phases of engaging an air target - search, acquisition, identification, tracking and firing, with hit/miss indications and malfunctions.

Targets. Targets are produced by computer-generated imagery (CGI) and up to four target projectors are normally used; additional targets can be added. The system projects images of targets overflying or attacking from different directions. The aircraft are shown in colour, vary with aspect and their size and sound correlate with range.

Background Scenes. Scenes of background terrain are projected on the inside wall of the theatre and allow simulation of target obscuration.

Engagement Procedure. Operational or simulated weapons fitted with video-electronic adaptors provide the computer with the necessary weapon and target data. The sound of the weapon firing is simulated to provide realism and psychological training. For guided missile systems a target flare projector is used to simulate the missile from launch to impact. Simulation of electronics countermeasure flares is also available A successful engagement is indicated by the simulated destruction of the target.

Instructor Station. Exercises are controlled from the instructor's console by means of a visual display unit and keyboard. A television monitor displays the same sight picture as that seen through the trainee's weapon sight and also gives performance data on an engagement. An analysis of the engagement is stored for debrief purposes.

Naval Version. A naval version is available which includes a motion platform to simulate ship movement.

Customers. Several ADT3-C systems have been supplied to the Royal Air Force, the British Army, and the US Air Force. Further trainers have been sold to other

Rapier SAM crew training in a Marconi Air Defence Tactical Training Theatre

Royal Air Force Regiment 35 mm Oerlikon gun crew training in a Marconi Air Defence Tactical Training Theatre

overseas forces, including naval versions. Additional systems have been ordered by the British Army for its latest missiles systems, including the Starstreak high velocity missile.

Status

In production. In service with the British Army, Royal Air Force, US Air Force and others.

Manufacturer

Marconi Simulation and Training, Cwmbran.
(For details see Directory of Suppliers)

GEC-Marconi Anti-aircraft Gunner's Aids TA 500 and CT 550

TA 500

Designed primarily for use with the GEC-Marconi GSA series gun sight, the TA 500 is an add-on unit that provides a trainee gunner with an artificial target in the gun sight for tracking practice against a target travelling at operational speeds. It can also be used with a variety of other gun sights such as the ring and bead sights used on the 40 mm Bofors L/60, 20 mm Oerlikon guns and 20 mm Vulcan guns.

Components. The TA 500 is composed of two main units, the optical unit which is mounted on top of the GSA Mark 3 gun sight, and the input and control unit. The display unit weighs 7 kg and the input unit 8 kg.

The optical unit generates and projects a synthetic target image into the gunner's sight.

The input unit of the system is used to enter the target criteria into the system. It has controls for the initial range of the target (from 1000 to 3000 m in 500 m steps), pass distance azimuth and elevation (from +300 to −300 m in 100 m steps), target velocity (from 200 to 350 m/s in 50 m steps), and initial target bearing in both elevation and azimuth. An interconnecting harness is also supplied.

Operation. Rate gyros in the rate measurement unit measure the angular rates of the gun in both traverse and elevation. The output of these gyros is integrated to produce gun displacement angles which are then fed to the optical unit in order to stabilise the image of the synthetic target reticle as the gun moves. The gunner, looking through the optical unit, will see against the background the gun sight reticle (which moves with the gun) and the image of the synthetic target, which does not. A function generator causes the target image to move against the background and remaining independent of gun motion. The gunner's task is then to track the target image with the gun sight reticle. Closing of range is indicated to the gunner by a steady increase in the brilliance of the target image.

Assessment of Results. A video camera is built into the optical unit so that the instructor can assess the gunner's ability in tracking the synthetic target by observing his performance on the video monitor located beside the input unit. In addition, when the gunner presses his firing button, the aim error indicator (in traverse and elevation) indicates his tracking at the instant of firing. An additional unit, the Delta Analyzer, prints out the programme being operated, the miss-distance of a projectile and the range at the instant of firing.

Electrical Power. Power requirements for the TA 500 are one 24 V DC battery and the system may be operated from two 12 V car batteries.

Customers. A version of the TA 500, known as the TA 502, was sold to the Japanese Self-Defense Forces during 1987 for use with the 20 mm Vulcan Air Defence System. The TA 502 is also used on 20 mm guns operated by the Royal Brunei Armed Forces and by Chile (with 20 mm guns). A single system has been used by the Royal Auxiliary Air Force squadron operating 35 mm Oerlikon guns at RAF Waddington.

CT 550 Classroom Trainer

The CT 550 classroom trainer simulates a gun by mounting a TA 500 display unit, a video camera and a gun sight on an arm where the units can be counterbalanced as required. The arm is attached to a swivel, providing 340° movement in azimuth and 100° in elevation. An input unit and a video monitor are mounted on a table where they are used by the instructor. Four training programmes of graded difficulty can be used to assess the gunner's performance and provide comparison with other gunners.

Status

In production. In service with Brunei, Chile, Japan and the United Kingdom.

Manufacturer

GEC-Marconi Avionics Limited, Edinburgh.
(For details see Directory of Suppliers)

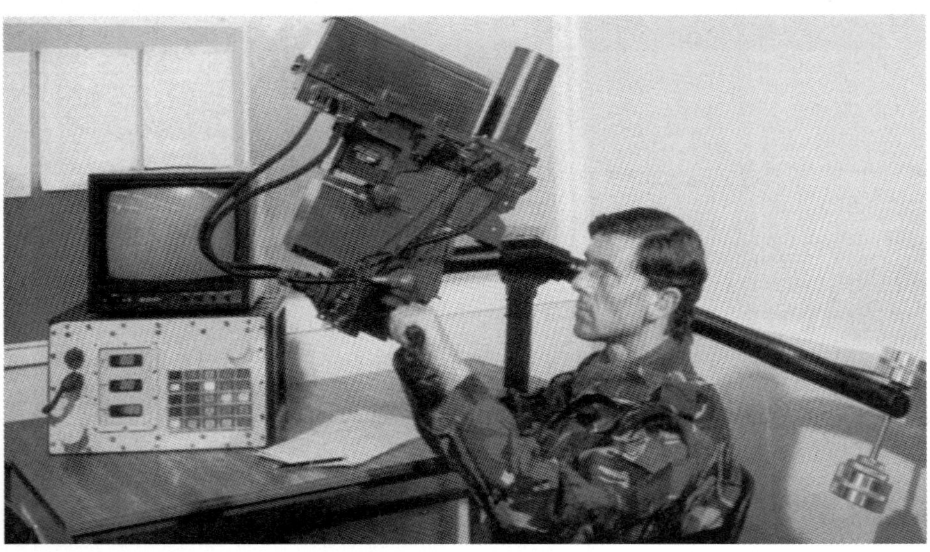

GEC-Marconi CT 550 classroom trainer

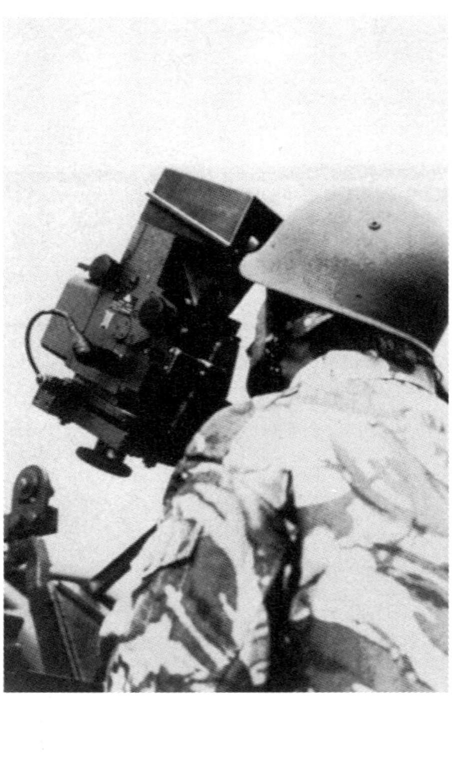

GEC-Marconi anti-aircraft gunner's aid TA 500

Invertron Skyfire Air Defence Training Systems

Skyfire System A

Skyfire System A is designed for training AAA and SAM operators in low level engagements. Simulation of the target aircraft, tracer trajectories and background terrain is achieved by injection of images into the weapon sight. The system can be fitted to a number of weapons for multiple operator training. It can also be used for optically sighted SAM.

With the weapons deployed in their firing positions, Skyfire System A can present either identical or differing target images to each weapon sight as required. The background can be either real or a digitised photograph of terrain, depending on the nature of the training.

Targets. Up to 16 target types can be stored in the system's image generation (IG) memory, and this capacity can be increased if required. The system produces up to five independent targets in each weapon sight. More targets can be made available by adding more IGs which are housed in the instructor's trailer. The target flight paths and formations can be controlled. Synthetic targets can be made to move at speeds from 0 to 600 m/s with a rate of change from 0 to 120°/s. The instructor can present targets through 360° of weapon traverse with elevation restricted by weapon limits only. Target destruction and tracer indication is simulated. Target images can be presented out to a range of 9000 m; visibility is variable from 0 to 9000 m. A cloud-base effect can be introduced (from 0 to 3000 m or clear sky) and glare from the sun can be injected into the weapon sight. The sun position and target attitudes realistically affect the shading of target surfaces.

Instructor's Station. The instructor's station is in an air-conditioned mobile caravan or semi-trailer containing the control station and operational equipment. It may be deployed either in the field or in a large building. From within the trailer the instructor can control all aspects of training, generate the required targets, and produce simulations of weapon and target sound effects. The instructor has immediate access to scoring and assessment facilities for individual trainees and groups of trainees. A mission replay facility and hard-copy student assessment printout are also available. Repeat monitors, showing the operator's sight view, are available at each weapon location for the use of the weapon commander.

Use of Real-World Aircraft and Background Scenes. An option is available for the engagement of real targets against a real terrain background using live fire-control data to determine tracking accuracy, hit probability, and so on. Another option is targets produced by computer-generated imagery (CGI) against a real background.

The terrain background is a digitised photograph providing target obscuration where appropriate. This task can be carried out by the user.

Naval Version. A shipborne version of Skyfire System A is also available.

Skyfire System A in use

Interior of a Skyfire System B air defence training system

Skyfire System B

Skyfire System B projects target and photographic terrain background on a large screen. The system was designed primarily for outdoor training with portable and shoulder-fired SAM such as Stinger, Javelin, RBS-70 and Mistral. It can also be used for the crew training of low level AAA guns and optically sighted SAM. Skyfire System B is suitable for employment in buildings or drill halls when used with shoulder-fired weapons. When used with larger weapon systems it is best deployed in larger buildings such as hangars in order to permit access by the weapon system's associated vehicles. Classrooms can also be used but need a ceiling height greater than 3.8 m. The system can simultaneously accommodate more than one weapon system and crew for training, provided there is sufficient space to locate the weapons. Power requirements are 220 to 240 V, 50 Hz, single phase.

Screen. The curved screen employed with Skyfire System B and provides a FoV of 120° in azimuth and 45° in elevation. When assembled the screen is approximately 10 m long and 4 m high. The screen can be built up in a modular fashion using large numbered blocks and, when complete, is free-standing.

Dome Version. The system may also be configured inside a dome, which is constructed of standard screen panels each covering a field-of-view of 20°. These are arranged in a circle, with extension panels around the top of the structure. The total usable screen area offered by the dome is 270° in azimuth and 115° in elevation (+90°, −15°). The azimuth restriction is due to the allowance made for an access doorway and the instructor's position.

The Skyfire System B dome was not designed to be weather-proof as it is intended to enclose the structure within a conventional building. This building can be supplied by the customer or by Invertron as part of a training package. The dimensions of the Invertron building housing the dome are 16 m long, 14 m wide and 6 m high (to the eaves). This building allows space for a small workshop or office area.

Projector. The projector is a monochrome TV type employing a lens and mirror system which gives correct target scaling and positioning on the screen. The complete system is controlled by a DEC PDP11-series computer.

Background Scene. The background scene is of photographic terrain which is selected by the instructor for the exercise. Terrain views are selected beforehand and photographed by the user using a conventional camera and wide-angle lens. The system can simulate range coverage from 300 to 9000 m with visibility conditions from sunlight to night conditions controlled by the instructor.

Targets. Target information is stored on videotape and there is no limit to the number of target types that can be stored. Attack profiles can be generated by entering new flight paths into the system computer by means of a graphic pad, or by manual flight path control by the instructor using a joystick control. Terrain background provides target obscuration behind ground, trees, buildings or other cover, where appropriate, and production of the terrain database can be carried out by the user. Weapon effects and target sounds are realistically simulated, including missile guidance flares. Simulation of intensifiers and IR thermal imagers can be provided where required. Target types and profiles are unlimited with up to nine simultaneous, independent targets displayed at any one time. Although each target is capable of 270° of movement in azimuth, the projector configuration and size of the weapon system involved may cause minor restrictions on the positioning of some images. However, with a four-target configuration it would be possible to position at least two targets at any given point on the screen and all four targets in the primary arc (120°).

Instructor's Station. The instructor's station includes a monitor showing the view through the weapon sight, a control visual display unit and keyboard, a target control box, a communications system, and a printer to provide a permanent record of trainee or crew performance. The instructor has the facility to preprogramme exercises of an unlimited duration (including typical 'battlefield days') and he can analyse and replay missions. The instructor has full monitoring facilities from target acquisition to termination of the engagement, including 'target hit/kill' indication.

Status

System A in service with the Austrian Army.

System B in service at Canadian Air Defence Artillery School, Chatham, New Brunswick.

Manufacturer

Invertron Simulated Systems Limited, West Sussex.

(For details see Directory of Suppliers)

Reflectone Tactical Air Defence Weapons Trainer

The Reflectone (ex-BAeSL) Air Defence Weapons Trainer provides realistic training for operators of low level air defence weapon systems. Originally developed for training Rapier SAM fire unit crews, the trainer can also be used for a wide range of SAMs such as Stinger, SA-7 'Grail', Mistral, Javelin and RBS-70, and also for AAA. The trainer can be configured to suit virtually any type of short-range air defence weapon system, and can be used for different types of missile and gun systems with minimum change-over times between training sessions.

Image Projection - Dome or Wide Screen. Image projection may either be on the inside of a number of domes, or on a wide-angle curved screen. A 12 m or 16 m diameter dome is used for ground-mounted weapon systems, and a 20 m dome for multiple ground and vehicle-mounted systems. The dome versions can be installed in a building or housed inside a weather-proof outer dome. A wide-screen version is also available which consists of curved dome segments, designed for installation in drill halls, hangars or in purpose-built buildings.

Naval Versions. Naval versions can include a motion platform to simulate ship movement.

Targets. Computer-generated targets are projected by a scanning laser onto the inside of the dome. Up to 10 independently-controlled targets can be projected simultaneously in the 20 m dome and four on the wide-screen version. The target images move in all aircraft axes and can also be assigned IR signatures. Targets are visible at simulated ranges from 200 to 14 000 m, and because of their high resolution can be viewed through magnification sights and hand-held binoculars. The brightness of the laser target images allows relatively high levels of background illumination to be used, representing normal daylight conditions. Target signatures can be adjusted to produce glint effects and simulate their appearance under poor visibility conditions. A separate red laser is used for simulating missile flare and tracer for guns. If the range of the target and missile/tracer coincide then a realistic impact effect occurs and the target aircraft is seen falling from the sky. The target library includes 50 different types of helicopters and fixed-wing aircraft, and additional types can be modelled on a work station and added to the library.

Background Scenes. The background panoramic scenes are produced from a bank of 35 mm slide projectors and are derived from photographs of the user's operational sites and/or training ranges.

Sound Effects. A multi-channel audio system relays realistic computer-controlled sound effects of the targets, weapon system operation and battlefield noises.

Simulated Radar Control of Firing Platforms. Elevation and azimuth signals can be produced to drive missile or gun platforms under simulated radar control and a trainee commander can be provided with a radar display and IFF information. This allows a commander to select targets for engagement either visually or from radar information.

Training Exercises. Training exercises are managed by an instructor seated at a console inside the dome. The system computer produces menu lists of prompts to guide the instructor through lesson preparation and exercise management. Targets are selected from the library and are assigned tracks with heights, waypoints and speeds. Use of correct performance data ensures that only realistic attack profiles are generated for a particular target type. A string of flight paths and events can be prepared with specific time intervals designed for different training tasks, ranging from basic operating procedures to complex tactical operations. Scenarios of up to 18 hours of combat operations can be prepared with realistic periods of inactivity followed by intense periods of battle. Pre-prepared exercises are stored in the computer system for recall when required.

During a training session, the instructor can introduce system faults and misfires, alter an IFF status, introduce radar jamming, cause targets to take evasive actions and dispense flares, and destroy a target to simulate a successful 'kill' by a neighbouring weapon. The instructor's displays enable him to monitor all aspects of the training, including the view through the weapon sights. Video, audio and student performance data from a training session can be recorded for later debriefing and analysis.

Reflectone Tactical Air Defence Weapons Trainer in a dome

Wide-screen version of the Reflectone Tactical Air Defence Weapons Trainer

Status

In production. Two trainers delivered to the US Air Force for use in Turkey. Two trainers are being supplied to a customer in the Far East, two more are being supplied to the French Navy and a further trainer is on order for the Royal Norwegian Navy (a joint programme with MPS of Norway.)

Manufacturer

Reflectone UK Limited, Bristol.
(For details see Directory of Suppliers)

Tector Visual Systems Defence Training Systems

The Tector Defence Training Systems were developed from the range of Tecstar computer-generated image targets (see entry under *CGI and Visual Display Systems* for details of Tecstar). The systems cover all aspects of air defence and anti-tank simulators and can be provided in individual or group configurations. The trainers vary from simple desktop part-task trainers for weapon aiming or similar functions to onboard trainers presenting simulated targets in missile and gun system sights. The trainers may be used in a field, classroom or dome training environment.

A Tector onboard trainer configured for the Rapier SAM system

Target Generation and Background Scenes. The Tecstar CGI system is used for generating target images, and a full-colour 360° photographic background scene is provided. When required, the background may be changed in about two to three minutes. The instructor may select targets or tracks and he has the capability to generate new tracks. A total of 20 dynamic CGI targets is provided from a library of 50 but the customer can purchase a target preparation package which will allow him to generate his own targets. All targets are three-dimensional and up to four targets can be presented on one channel. Target obscuration takes place when appropriate and targets can 'move' at speeds up to 2250 km/h. Up to 256 surfaces (polygons) are available for modelling and the image update rate is 50 Hz.

Status

In production and widespread service.

Manufacturer

Tector Visual Systems, Newark.
(For details see Directory of Suppliers)

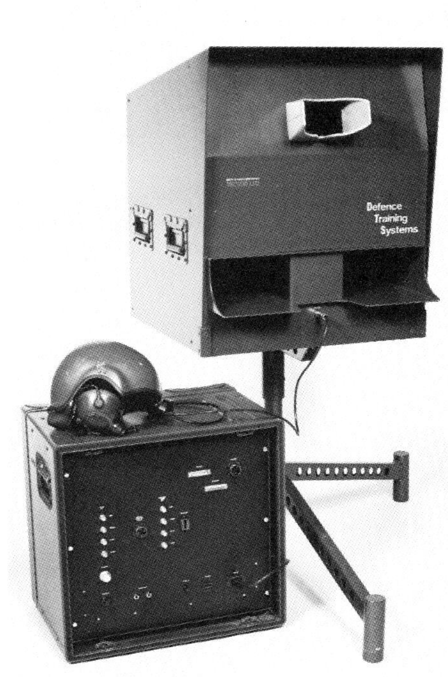

A Tector part-task trainer for classroom use

UNITED STATES OF AMERICA

AAI Corporation MTS-II Multiple Weapon Procedures and Tactics Trainer

The MTS-II (Moving Target Simulator II) is a second-generation version of AAI's M87 Redeye/Stinger simulator. It gives a wider selection of scenario options and training facilities. Gunners can be trained on a wide variety of weapon systems including Stinger, Vulcan, Javelin, Chaparral and so forth.

Housing. The MTS-II is housed in a 64 ft (19.5 m) or 20 ft (6.09 m) diameter dome. This serves as the 360° display screen for targets, background, countermeasures and infra-red images.

Scene and Target Generation. The MTS-II projects up to three independent aircraft images, each with its own infra-red signature and infra-red countermeasures. The target projector assembly is located on top of a gantry in the centre of the dome. Three video projectors are mounted on computer-driven gimbals, allowing independent target motion at realistic angular velocities. Located directly below the video projectors is the mount for the terrain projectors, which provide instructor-selectable background scenery. Sky colour projectors are located at the dome periphery, and three or more visual/infra-red cloud projectors are located as required. Facilities include different weather conditions, visual and infra-red images, directional audio effects, and strategically positioned infra-red emitters.

Assessment and Scoring. The trainee's performance is scored and evaluated by the MTS-II and data is provided for debriefing by the instructor.

Other Weapons Systems. Multiple weapon systems can be accommodated. Vehicles can be driven in through the 4.5 × 3.5 m door in the side of the dome. The 5 m gantry clearance allows vehicular weapon systems to be centred under the projectors while allowing full weapon manoeuvrability. For hand-held weapons systems, such as Stinger, the operators are allowed to move about on the floor.

Customers. Two 20 ft (6.09 m) models were built for the US Marine Corps and, under a separate contract, a larger 32 ft (9.75 m) drive-in model was supplied to the Turkish Land Forces. Two additional larger models are in service with Italy. In October 1990 the US Army Missile Command agreed to purchase five trainers for a total of approximately $10 million. The contract called for turnkey installations in Germany, and at four sites with n the USA. Work was completed during 1993.

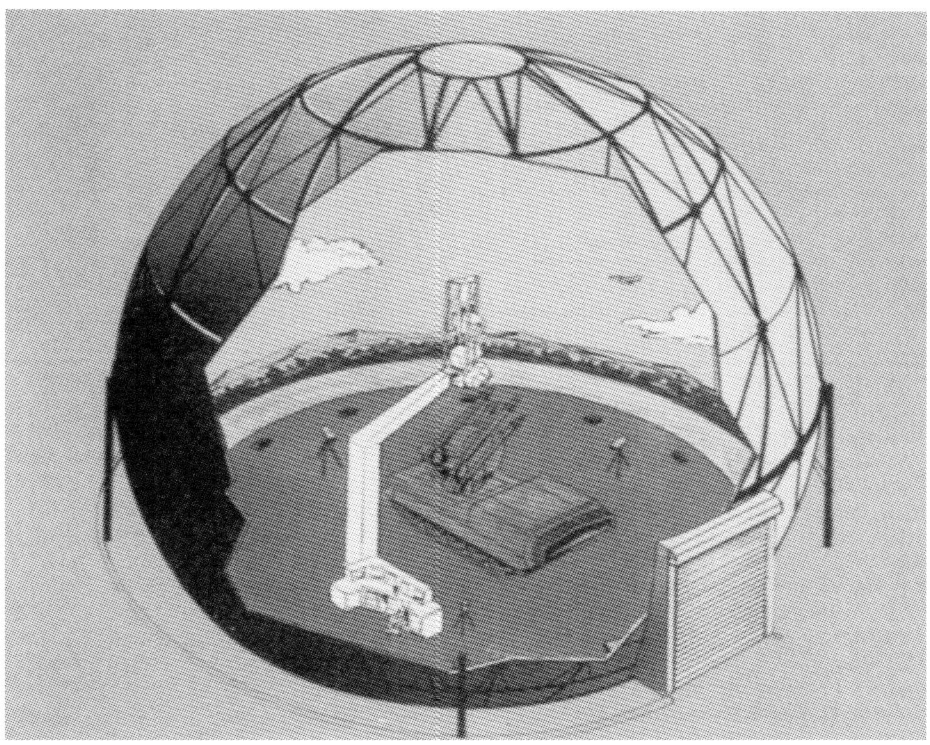

AAI Corporation MTS-II 32 ft (9.75 m) multiple weapon procedures and tactics trainer

Status

In service with the US Marine Corps (2), Italy (2), Japan (1), Turkey (1), and the US Army (5). In late 1993, AAI completed negotiations with the US DoD for four additional systems which includes three systems for the US Army and one for the USMC. Deliveries are scheduled to take place during 1995. Two additional systems have been purchased by the Japanese government. One system is currently operational at the Japanese Air Force Base in Hyakuri, Japan. Delivery for the second system is scheduled for 1995.

Manufacturer

AAI Corporation, Maryland.
(For details see Directory of Suppliers)

Patriot Missile System Operator Tactics Trainer

The Patriot Missile System Operator Tactics Trainer (OTT) enables one instructor to train eight students simultaneously in the operation of the Patriot SAM/ABM system. Students are seated at control consoles in a classroom environment and are presented with radar displays to which they must respond using the same controls and actions as they would in an operational Patriot system van.

Status

In service with the US Army (12), Germany (3), Netherlands (1), Saudi Arabia (1), and Japan (4).

Manufacturer

Lockheed Sanders, New Hampshire.
(For details see Directory of Suppliers)

A US Army Patriot SAM/ABM System Operator Tactics Trainer (OTT)

YUGOSLAVIA (SERBIA AND MONTENEGRO)

PAT Anti-aircraft Gun Trainer

The PAT anti-aircraft gun trainer is contained in a dome and is for training light anti-aircraft gun crews, mainly 20 mm calibre. Training provided is in the detection, tracking and engagement of moving aircraft targets. A light AAA gun or a gun simulator is placed inside the dome, and aircraft targets are simulated by film projected on the dome surface. A computer is used to control target speeds and attack profiles against a background scene which can also include a range of battlefield sounds and meteorological conditions.

Dome. The dome stands on a circular concrete foundation with a diameter of up to 14 m. The dome itself is 8 m in diameter. Erection time is 48 hours while 24 hours are needed for removal. Access doors are provided for weapons and other equipment. A three-phase 220 V 50 Hz electrical supply is required.

Components. The complete PAT trainer consists of the dome, a sighting system on a gun or a simulated gun, a film projector, an instructor's console, a sound effects generation system, a visual effects projector, various accessories and tools and manuals. The films used with the trainers are taken using aircraft or helicopter models. A microprocessor system controls the film projector servo-mechanisms to control target speeds and positions and also to synchronise the relevant sound effects. A video monitor for the instructor can be supplied.

Status

In service with the Serbian armed forces.

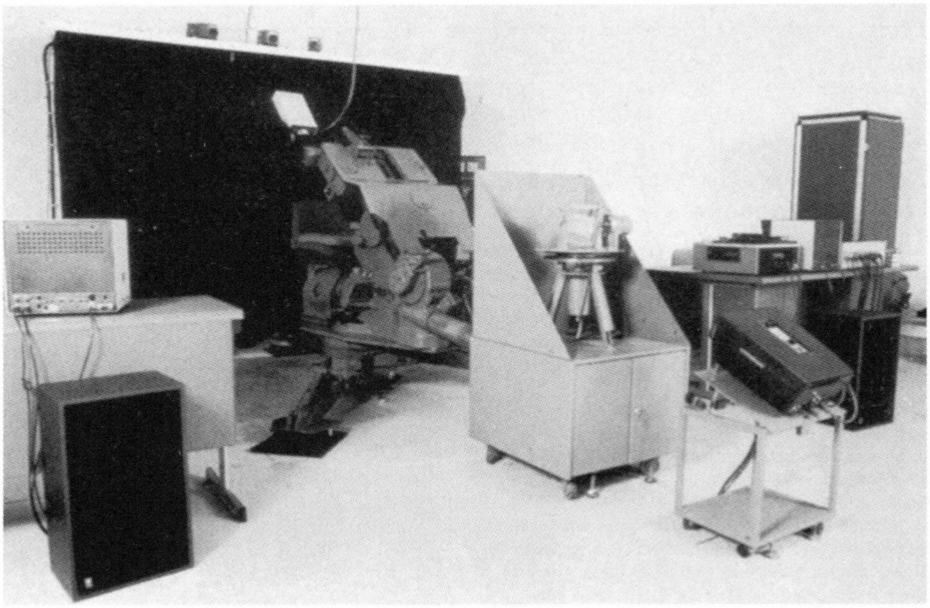

Main components of PAT anti-aircraft gun trainer, including a simulated 20/3 mm M55 anti-aircraft gun

Contractor

Holding Company YUGOIMPORT-SDPR Ltd, Belgrade.
(For details see Directory of Suppliers)

RO EI TV Launcher Effect Trainer 9M32M

The Launcher Effect Trainer 9M32M is a replica of an SA-7 'Grail' shoulder-launched SAM launcher. The former Yugoslav armed forces used the Russian designation 9M32M for this. As well as the SA-7 replica, the equipment consists of a monitoring and control system for an instructor. The trainer can be used in several ways from carrying out 'dry' aiming and launch sequences to the simulation of firing a missile using small rocket motors to eject a dummy rocket from the launcher.

General Description. The Launcher Effect Trainer can be used to carry out drills such as interrupting a target engagement sequence, dealing with unit failures and general safety procedures. Each Launcher Effect Trainer is issued with a case containing 15 ejection motors. The RT. S2M.110 is issued in a separate case, while the two monitoring units are packed in further cases along with the associated interconnecting cables. Other equipment issued with the trainer includes tools, accessories and documentation. The entire trainer weighs approximately 83 kg. Electrical power is from mains or batteries.

Launching Barrel Simulator. The launching barrel simulator exactly resembles an operational SA-7 and is used to fire the RT. S2M.110 training missile. The training missile can be used for a minimum of 15 times and is launched using small ejection motors known as the 15EM. A safety area measuring 10 × 80 m is required in order to fire the training missiles. The launching barrel simulator has an operational life of at least 10 000 firing sequences.

Electronic Unit. The Launcher Effect Trainer is controlled and monitored by an electronic unit known as the 9M58. This unit monitors a trainee's missile preparation, time reactions, correctness of operating sequence and aim. A printer in the unit can provide a hard copy of trainee actions and results. The system's thermal battery is simulated by the B17 unit which is used by the instructor to switch on and control a

Training Functional Set T. S2M.100 without training missile

training sequence. During training the instructor is located close enough to the launch point for verbal commands and instructions to be given to a trainee. Should the need arise, the instructor can immediately stop any training sequence by using a key on the battery simulator unit.

Training less Launch of Training Missile. A similar trainer that carries out all the functions of the Launcher Effect Trainer apart from the launch of the training missile, is known as the Training Functional Set 9M32M (T. S2M.100). This trainer uses the same components as the Launcher Effect Trainer apart from the Battery Simulator B17. A loudspeaker system is used to simulate the sounds of missile launch.

Status

In service with the Serbian armed forces.

Manufacturer

RO EI TV Elektronika, Vlahovića.
(For details see Directory of Suppliers)

Launcher Effect Trainer for SA-7 with training missile

TANK AND DIRECT FIRE GUN TRAINERS

CANADA

Atlantis Tank Gunnery Training Systems

Thru-Sight Video (TSV) System

The Atlantis Thru-Sight Video (TSV) system consists of a number of modular, compact, clip-on assemblies that permit real-time reviewing and recording of tank gunner performance during static and mobile firing. This allows instructors to observe precisely what a trainee gunner is viewing through his sights during firing and to assess the actual performance. The TSV system allows normal gunner and crew interaction with the system without interference with the equipment on which it is installed. No modification to the actual sight equipment involved is required.

Fit to AFVs and IFVs. The TSV is available off-the-shelf for the Leopard 1 and M1/M1A1 Abrams MBTs, the M2/M3 Bradley IFV, and the TOW ATM sight. It can be configured easily for virtually all sight systems, including the Leopard 2, M60, Challenger 1 and 2 MBTs. It can also be configured for rifle and similar weapon sights.

Television-based Tank Training System (T3S)

The Atlantis Television-based Tank Training System (T3S) consists of the Atlantis Thru-Sight Video (TSV - see above), a range observation video system and an instructor/debrief station. The system is designed to provide real-time recording and after-action review of student gunner performance, individually and by platoon, during static firing and battle runs. The system permits the automatic calculation of Mean Point of Impact (MPI) of round zeroing during static firing.

Status

The TVS and T3S are both in service with the Canadian armed forces.

Manufacturer

Atlantis Aerospace Corporation, Ontario.
(For details see Directory of Suppliers).

CHINA, PEOPLE'S REPUBLIC

TSM-2 Tank Gunnery Trainer

The TSM-2 Tank Gunnery Trainer, sometimes referred to as a Tank Firing Simulator, is used for gunnery training with the 100 mm gun mounted on the Type 59 (and T-54/T-54A) tank. With slight modification it can also be used for gunnery training with other tank and direct fire weapons. The simulator can be used to conduct low-cost and convenient indoor training which would otherwise have to be conducted using live ammunition on ranges.

Training Scenarios. The following basic training can be conducted: operating with the gun elevation and traversing mechanisms and with the gun electrical drive and stabilisation units; target search, acquisition, recognition, tracking, aiming, firing and fire correction against static and moving targets. The trainer can also be used for gunnery training under simulated day and night conditions, training in primary crew procedures and various combat firing drills.

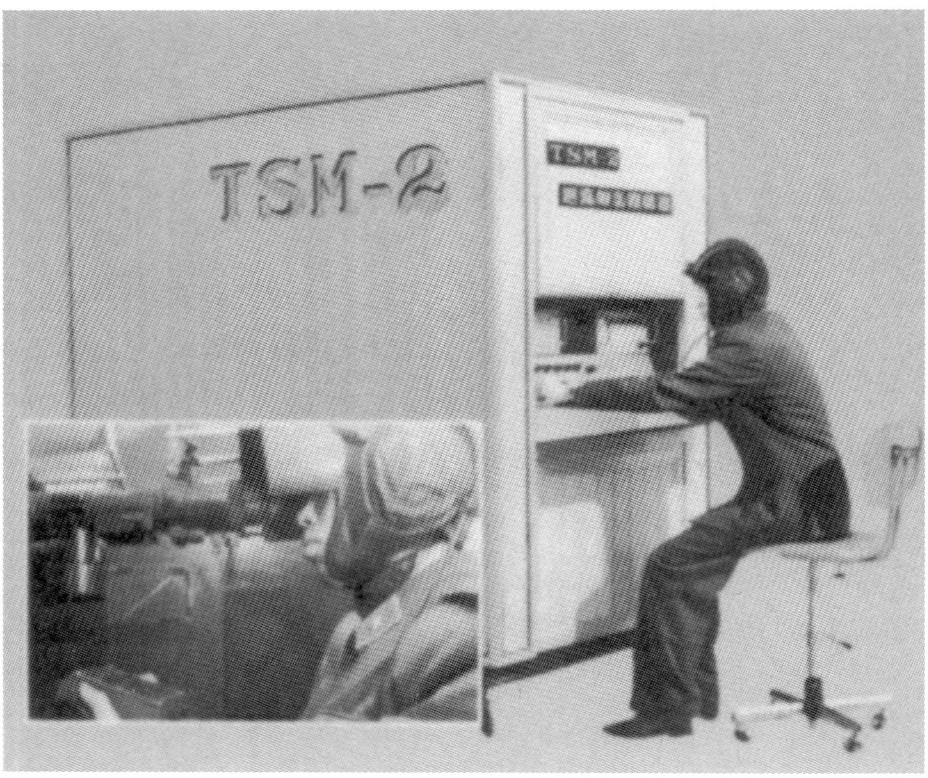

Components. The TSM-2 simulator is housed in a box-type structure to facilitate transport and installation in a classroom. The control position is located at the front with the trainee cabin occupying the interior. The control position is used to control the visual image systems of the simulator and includes helium and neon laser generators, an automatic slide projector, a target system, an optical system, the elevation and traverse drive system, right- and left-hand control units, the main control panel, power supply controls, and various DC supplies. The trainee cabin contains the simulated gunner's position of a Type 59 turret: a sighting telescope, elevation and traverse controls, a control box, a replica of the main gun, a vibration platform and mechanism, intercom devices, and various other components.

Trainee Cabin. The trainee cabin is configured for a Type 59 tank and the trainee is provided with real controls and a control box. The background image and target displays seen through the fire-control telescope. Displays are controlled by a microcomputer, which also controls the shot ballistics. Optical techniques such as fibre optics are used to produce images on the visual image system. The fall-of-shot is indicated by a red spot of neon laser light.

Specifications

Simulated firing range: 400 to 2000 m
Operational voltage: 220 V ±10 V 50 Hz
Power consumption: <3 kW
Weight: <900 kg (1984 lb)
Dimensions: 2.2 × 1.55 × 1.85 m

Status

In production. In service with the Chinese Army.

Agent for Manufacturer

China Xinxing Corporation, Beijing.
(For details see Directory of suppliers)

Photographs showing interior and exterior of TSM-2 tank gunnery trainer

CROATIA

LAIR Tank Gunnery Trainer for T-55 and T-72 (TGT1 T-55/T-72)

The TGT1 T-55/T-72 is a training aid designed to provide aiming and gunnery exercises for T-55 and T-72 MBT crews at short ranges. The system is suitable for initial training before using live ammunition and can be used for additional and advanced training.

General. The system provides simulated firing and can be used for both main and auxiliary weapons. Operating ranges correspond to the combat ranges of the weapon involved: 400 to 3000 m for a main gun and 400 to 800 m for machine guns. The trainer works best when used in hangars or demonstration rooms equipped with relief models. The trainer is directed towards the centre of a relief model at a maximum range of 10 m.

Computer Programmes. The system is computer-controlled and based upon three software support programmes—Automatic Operation, Control, and Rectification programmes.

Automatic Operation. The Automatic Operation programme is used for training with targets on the relief model. A command block moves the weapon sight over the model. When the student detects a target, simulated by a suitably scaled model, he checks necessary parameters for firing and then fires. As he fires, the system produces the appropriate sound. The computer then calculates the position of the fall-of-shot and the time the projectile took to get there, using sensor data. A marker indicates the position of the projectile fall so if this corresponds with the target a hit is produced. From his command position the instructor is able to alter a projectile trajectory due to changed conditions. A series of exercises can be observed through the sight eye-piece and also on a monitor which allows observers to take note of actions taken.

Rectification Programme. The Rectification programme is used during initial installation and whenever the Control programme does not function correctly. The programme aligns the optical head of the trainer with the optical sight involved, reads data for ballistic characteristic generation, and reads the range values of the front and rear of the relief model.

Control Programme. The Control programme checks the data readings during the Rectification mode so that the results can be used during automatic operations or during repetition of the Rectification programme.

Status

In service.

Manufacturer

LAIR Croatia, Zagreb.
(For details see Directory of Suppliers)

General view of the Tank Gunnery Trainer for T-55 and T-72 (TGT1 T-55/T-72)

LAIR Tank Gunnery Trainer TGT2

The TGT2 is a trainer designed to provide aiming and gunnery exercises for tank gunners engaging short range targets. The system is suitable for initial training before using live ammunition and can also be used for advanced and refresher training. The system is used at simulated ranges of between 400 and 3000 m, using a demonstration room equipped with landscape models. The room needs to be 10 × 5 m or larger, with the trainer directed towards the centre of the model at a distance of 4.7 m.

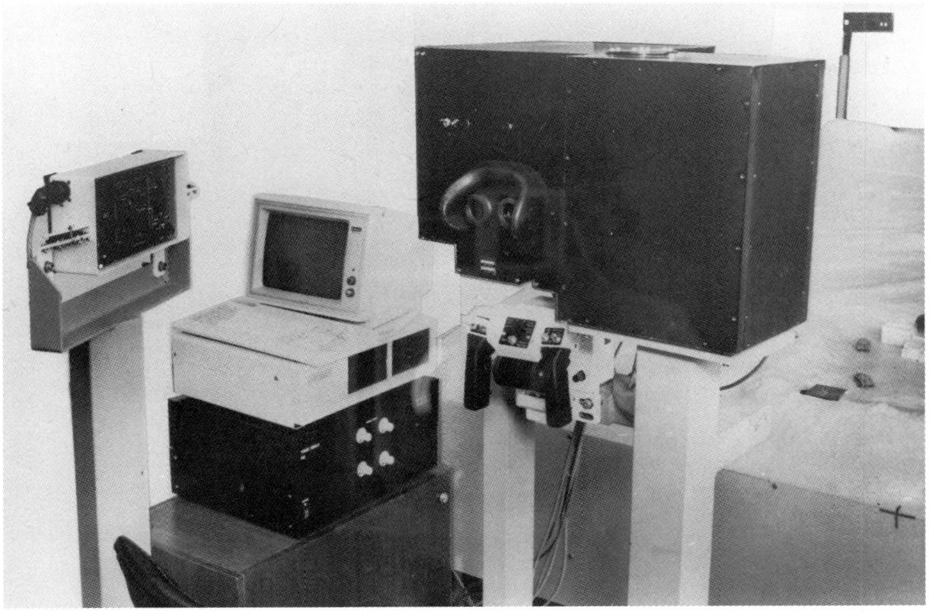

General view of the Tank Gunnery Trainer TGT2

System Operation. The trainee observes target models moving over the landscape model at varying ranges, using his gunnery controls to move the simulated sight. When a target is acquired, the normal controls are used to track the target, obtain the range, and fire. The system then calculates the fall-of-shot, which is then indicated in the sight field-of-view. A correct shot will cause a stationary target to fall or a moving target to stop.

Software. The software involved can provide: training in switching the fire-control system on and off; zeroing the fire-control system; data input for the control computer; aiming; tracking; range measurement; firing at stationary or moving targets; and more complex training.

Instructor's Role. The instructor selects the desired exercise on the main computer and observes the trainee's actions on a video monitor. The monitor displays the same scene as seen by the trainee through the sight. An additional monitor can be added for other observers.

Status

In service.

Manufacturer

LAIR Croatia, Zagreb.
(For details see Directory of Suppliers)

FRANCE

Giravions Dorand DX 200-Series Tank Gunnery Training Aids

The Giravions Dorand DX 200-series tank gunnery training aids are used for the basic training of tank gunners. They are classroom desktop equipments consisting of consoles with CRTs, and basic gunner's fire-control devices and controls. An instructor's control console and a computer are included. The CRTs present a simulated landscape and static or moving targets. Using a sighting circle the student can use the fire-control system controls to aim at the target and, as he does so, the computer assesses accuracy and aiming. As the trigger is pulled, the student is presented with results on the screen. The instructor can monitor actions and the computer can store results and scores.

Other Systems. The DX 200-series basic system can be adapted for other fire-control systems. A design for the APX 586 fire-control system exists but has not yet been ordered.

Status

The DX 201 (EVIC) is in production for the COTAC integrated fire-control system used in the AMX-30 B2 MBT and the AMX-10 RC reconnaissance vehicle. The DX 202 version is in production for the SOPTAC integrated fire-control system.

Manufacturer

Giravions Dorand Industries, Suresnes.
(For details see Directory of Suppliers)

Thomson-CSF Turret Gunnery Trainers

In November 1989 Thomson-CSF was selected by the French Ministry of Defence to design and develop a Turret Gunnery Trainer for the Leclerc MBT. The Turret Gunnery Trainer is being installed at regiment level to provide training for both the tank commander and the gunner and has been designed so that it does not require specialised personnel for its operation. The trainer is designed to be user-friendly; it does not require a special installation and may be set up in a mobile shelter.

Components. A turret crew station reproduces the gunner's and tank commander's stations, along with a processing and computing system. It can be used in two modes: a technical instruction mode (training in turret operation and firing) or a technical and tactical instruction mode. It is also possible to link up to six Turret Gunnery Trainers, all monitored by a single instructor, for training in platoon and squadron exercises.

Visual System. A colour visual display system is fitted which generates images in the sights at the commander's and gunner's stations. The visual system can generate up to five simultaneous targets of two different types as well as weapons and weather effects, alphanumeric insets, image intensifier and thermal imager presentations.

Gunner-only Training. The Turret Gunnery Trainer may be designed to train only the gunner. In this case the trainer features the gunner station along with the associated computing system and visual system.

Recent Customers. Eleven similar infantry combat simulators were delivered during 1993 to the Belgian Land Forces to train AIFV 25 mm gunners, either in separate or platoon mode.

Status

In production. In service with the French Army (for Leclerc MBT) and the Belgian Army (11 for AIFV).

Manufacturer

Thomson Training and Simulation, Cergy-Pontoise.
(For details see Directory of Suppliers)

Gunner's position on the Leclerc Turret Gunnery Trainer

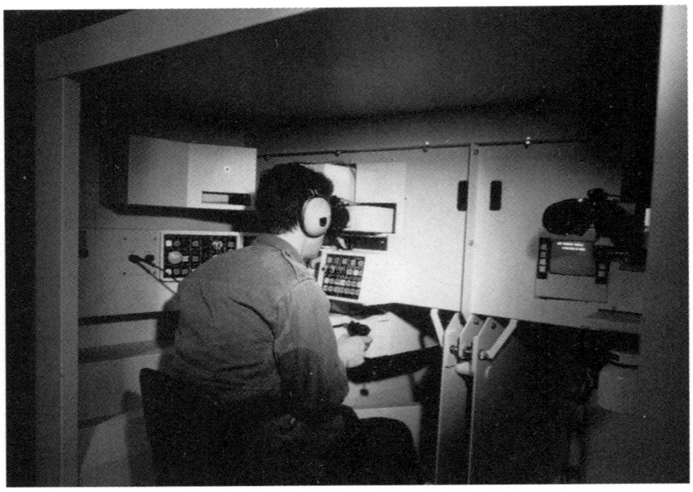

Leclerc MBT Turret Team Trainer

GERMANY

Atlas ASPT Gunnery Training Simulator

The ASPT Gunnery Training Simulator is a member of the TACOS family of gunnery and combat simulators for tanks. It utilises experience gained in the development and production of the ELSALEO and ASPA simulators (see following entry). It is used for gunnery training involving the commander and gunner of the Leopard 2 MBT.

Visual System. The ASPT uses a colour, textured Atlas DISI 4 CGI system observed through the commander's and gunner's sights and viewing devices.

Container. The ASPT is accommodated within one 2 × 2 × 4 m container which can be carried to the desired site by any suitable truck. The only requirement for start-up is connection to a suitable supply of electrical power. Part of the interior is taken up by the instructor's and crew stations while a further compartment, entered by a separate door, is used for maintenance access only.

Crew Stations. The controls at the commander's and gunner's stations use selective realism for some of the controls not involved in the gunnery procedures and these are in simplified form, replaced by pasted photographs or simply omitted. Real controls and display elements are used when required by the training objectives. These include the eye-piece assemblies and the gunner's and commander's control units.

Instructor Role. The instructor manages exercises using two monitors which show external views as seen by the commander and gunner. Lesson plans are controlled by menu and operation via a keyboard. The instructor selects exercises and starts them, communicating with the trainee crew via a voice link. The results of the exercises are automatically produced on a printer for debriefing. The computer automatically raises the level of exercise difficulty, depending on results. To prevent trainees from becoming accustomed to individual exercises the system is equipped with a randomiser. Data for central statistical evaluation is gathered automatically.

Autonomous Operation. The system can operate without an instructor. In this case the menu is injected into the commander's view monitor. The commander's station is equipped with a keyboard identical to that at the instructor's station.

Adaptation to other AFVs. The ASPT uses a modular construction which makes it easily adaptable to suit other types of armoured vehicle, such as the Leopard 1A5 MBT. The ASPT can also be adapted to meet other requirements, such as incorporating a three-channel visual system similar to that of the ASGM (the gunnery training simulator for the Marder MICV). If required, the ASPT can be equipped with additional functions such as playback or a vehicle status display, and can be extended to cover training in the handling of emergencies.

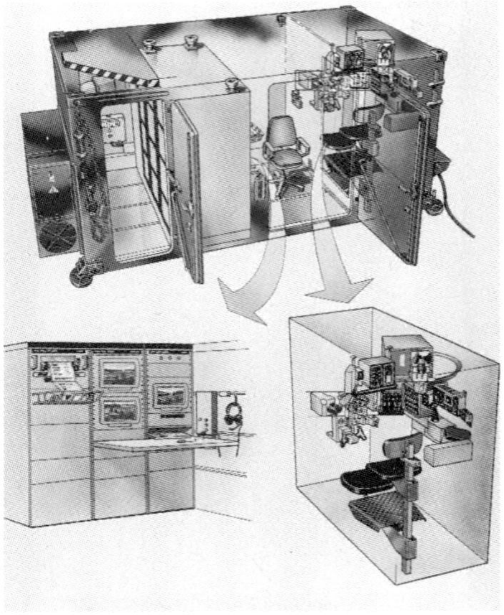

Artist's impression of the ASPT Gunnery Training Simulator

Status

In service with the German Army. First unit installed at the Combat Arms School 2, Munster, in January 1992, 35 units in service 1994.

Manufacturer

Atlas Elektronik GmbH, Bremen.
(For details see Directory of Suppliers)

Atlas ELSALEO Tank Gunnery Training System

The Atlas Elektronik ELSALEO (full designation, *ELektronische SchiessAusbildungsanlage für Panzer LEOpard)* is a tank gunnery training system and was developed for the Swiss Armaments Services Group. It is a member of the TACOS family of gunnery and combat simulators for tanks. The first example was delivered in 1988 for use on a Swiss Army recruit training course for crews training on the Panzer 87 Leopard (Leopard 2). ELSALEO also involves some participation from Swiss industry. Up to four crew compartment/turret simulators can be interconnected to permit subunit gunnery training up to platoon level.

Components. The system employs a single dummy turret, the interior of which is equipped with replicas of the commander's, gunner's and loader's positions. The simulator uses the DISI 3 CGI system developed by Atlas Elektronik. Using this system it is possible for the turret crew to identify and engage up to 20 moving targets at ranges of up to 2500 m.

ELSA Pz 68/88

Following the acceptance of ELSALEO by the Armaments Services Group it was decided that a similar system was to be procured as Phase 1 of the Electronic Gunnery Training System ELSA Pz 68/88. This system is based on the ELSALEO concept and provides training for Pz 68 MBT crews following its combat effectiveness enhancement programme. One ELSA Pz 68/88 system, consisting of four crew compartment/turret simulators, was procured with some production participation from Swiss industry.

ASPA
A further development of ELSALEO is known as ASPA. This is a tank platoon gunnery trainer which is accommodated in a training centre and consists of four tank crew compartment simulators each containing a realistic replica of a Leopard 2 MBT interior. The systems are controlled from an instructor's console. The system allows 'open hatch' observation by trainee tank commanders and involves a 180° visual display system. The first ASPA was installed at Combat Arms School 2 at Munster.

Status
In service with the German and Swiss armies.

Manufacturer
Atlas Elektronik GmbH, Bremen.
(For details see Directory of Suppliers)

Overall view of the ELSALEO tank gunnery training system

Wegmann TV-Monitoring and Gunnery Training System for Armoured Vehicles (FESAP)

The FESAP system is used as a gunnery training and monitoring system for armoured vehicles and other application which have optically sighted weapon systems. No modifications of the weapon system are required and adaptors allow quick and easy mounting and removal without impairment of the operator and system performances. The system can be employed at platoon level, in motion and over large distances. It is versatile and can be used for training and practising sighting and rangefinding as well as monitoring and recording combat firings. The robust construction of the system makes it suitable for operation in armoured vehicles under all operating conditions. The modular concept means it can be adapted quickly and easily to special requirements and to other systems.

Components. The FESAP equipment consists of: the transmitter unit made up of a camera with mechanical and optical adaptors; an electronics box with voltage supply; an electronic data preparation and transmitter; a cable harness and antenna; the receiver unit with electronic data preparation, receivers, monitors, a video recorder, a communication unit and antenna; the target observation unit with a control unit with direct and position memory control, monitors, communication data, a target observation camera with azimuth and elevation control, weather protection; and the cable set.

Camera System. The camera is mounted to the weapon sight and records the image as it is seen by the gunner without impairing his field-of-view. The camera has a low light level capability, automatic adaptation to the light level so there are no burnt out or low contrast images, a small volume, low weight, and a high shock, vibration and temperature resistance. It is suitable for connection with the existing vehicle night sights and thermal imagers. In an electronic unit, information relating to the fire-control data and operation mode is added to the image taken by the camera. The data package is then transmitted by a multichannel transmitter and a specially designed antenna. The image information is transferred to the multichannel receiver, and is provided with a fade-in time and an identification number. The resulting image can be seen directly on one of the two monitors and recorded by the integrated video recorder. A camera remote-control is also possible, as is control of the camera by a target position memory which has up to 15 positions. This means that the push of a button can align the camera to its programmed position. The camera control unit also has two monitors which show the images of the target observation camera and the optical sight of the weapon system to be fired. The control unit can be connected to the communication system. Playback is by 19 mm video tape recording with faded-in time and system data.

Target Images. The target image is controllable by a second monitor on the receiver unit. During playback after the firing, a clear target image will be produced which is not impaired by secondary effects of the firing.

Secondary Effects. Secondary effects during the firing such as blast and flash can interfere with the observation (monitoring) of the hit impact and the effect on the target. These secondary effects can be avoided by means of the target observation camera with its large focal length and precision azimuth and elevation controls, set up remotely from the vehicle. A sound sensing unit allows precise recording of the hit impact and hit effect.

Receiver Unit. The receiver unit allows radio communication with the operators of the weapon systems, which can be tape recorded. Additional language information can be recorded immediately or subsequently on a second sound track of the videotape. The receiver unit is suitable for semi-mobile or stationary operation on vehicles, in tents, and so on.

Status
In service with the Royal Netherlands Army and the German and Swiss armies. A TSV (Through Sight Video) version is part of a training system for NATO forces in Europe.

Manufacturer
Wegmann & Co GmbH, Kassel.
(For details see Directory of Suppliers)

Schematic of the major components of the Wegmann FESAP

HUNGARY

MARTA and MÁTRA Weapon Sight Training Equipments

The MARTA weapon sight training equipment can be used with virtually any direct fire weapon, towed or self-propelled, but is of particular value to self-propelled or towed artillery weapon crews where the gun barrel elevating and traversing operations might be carried out by two men. It can be used either in barracks or on a weapon range.

Operation. MARTA uses the normal direct fire sighting telescope which is aimed at a target board about 20 m away. Traced on the board are target tracks which the gunner(s) follows using the direct fire sight and sight controls. As the gun barrel moves, an electrical position sensor mounted on the gun muzzle follows a track on a much smaller indicator board. On the indicator board is the same target track as that on the large target board and any deviation of the sight from the required target track will result in an indication since the muzzle sensor will diverge from the track on the indicator board. Deviations and errors, plus their timing and duration, are recorded on an electronic monitor unit which has counters and an indication of the target track involved during the exercise. A light on the monitor target track shows what track position trainees have reached during an exercise. The time taken to complete an exercise is also recorded. A 220 V AC or 24 V DC electrical supply is required and the electrical consumption is 25 W.

Guns Used. The MARTA is used for 122 mm 2S1 self-propelled howitzers and can be used with other similar weapons.

The MÁTRA system is intended for use with tanks and uses the same method of operation. With MÁTRA a single recorder unit can record the movements of up to three tank gun barrels.

Status

In production. In service with the Hungarian Army.

Manufacturer

Struktura Szerezesi Vallalat.

Enquiries to IDEX, Budapest.

(For details see Directory of Suppliers)

The recorder unit for the MÁTRA tank weapon sight training equipment

The gun muzzle position sensor and indicator board used with the MÁTRA tank weapon sight training equipment

ISRAEL

Elbit GTS-2000 Tank Gunnery Training System

The Elbit GTS-2000 tank gunnery training system is available in two versions. The GTS-2000 is ruggedised and fits the M32 eye-piece for on-tank training; with modification it can be used with other eye-pieces. The GTS-2000C is used for indoor training.

Operation. In the GTS-2000 a scene is viewed by the trainee gunner through his normal tank gun sight and the scene is also displayed on a television monitor. An optical splitter mounted on the sight eye-piece enables a picture to be registered on a small solid-state television camera, thereby displaying the gunner's performance on a monitor screen. The instructor/tank commander can then follow the gunner's actions in real time. Factors such as target designation and aiming procedures may be followed and exercises are recorded on a video cassette recorder. The training system is integrated with the tank internal communication system.

Other Modes. The GTS-2000 can act as a tank commander's sight or can be adapted to the gunner's night-sight. The system can also be used as an image-capturing device for a video tracker, for motion detection, or can be incorporated into automatic target acquisition and cueing units. It can also serve as a battlefield briefing camera during field operations.

Status

In production.

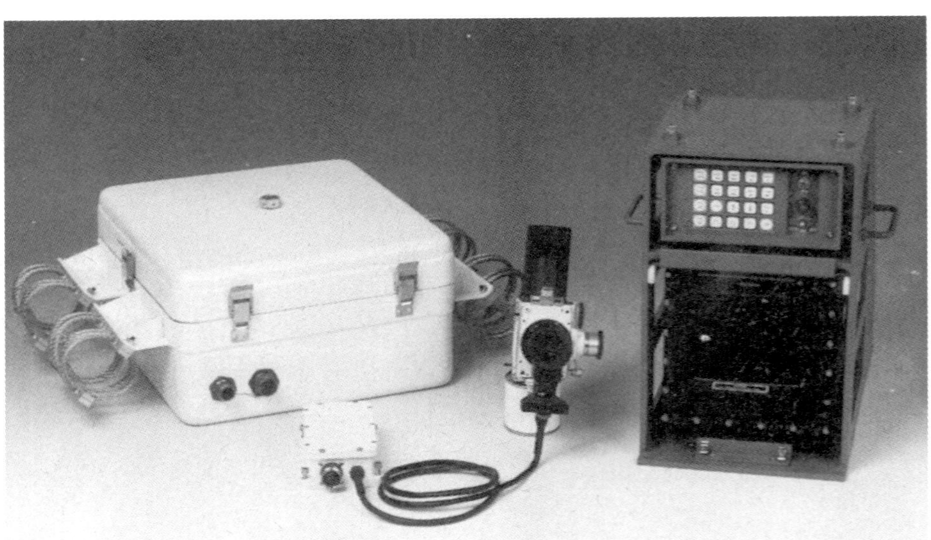

Main components of the Elbit GTS-2000 tank gunnery training system

Manufacturer

Elbit Limited, Haifa.

(For details see Directory of Suppliers)

Rafael SIMUGUN Tank Gunnery Trainer

SIMUGUN is a modular, computerised, tank-mounted, precision gunnery training system for training tank gunners and commanders in the basic procedures of tank gunnery, from single tanks up to platoon level. It was developed by Rafael in co-ordination with the Israeli Defence Forces (IDF). The number of simulated rounds that can be 'fired' in one hour of SIMUGUN training is between 150 and 200. The system can be adapted to all MBTs and integrated with modern fire-control systems. Using a SIMUGUN platoon gunnery trainer, up to three tank crews can train simultaneously. It can also be used to train forward artillery observers, anti-aircraft gun crews and firers of ATMs which use LoS guidance.

Components. The SIMUGUN system fits in a single vehicle and includes colour and monochrome TV displays, computer, printer, video disk player and a control unit. The training vehicle is placed in front of the trainee's tanks, and connecting cables and other fits are made in a few minutes. A colour display of the synthetic battlefield is set up in front of the gunner's sight.

Operation. The gunner uses his normal sight and locates a synthetic target on the colour TV display, positions the cross hairs, feeds in the firing data and 'fires' using the tank's normal gun controls. A firing flash and sound is then generated by the simulator together with gun jump and dust effects. The simulated round in flight then appears in the gunner's display and the impact is observed.

Results. The system is computer-controlled and provides feedback of gun position, digital output and the visual sight reticle display. Results are provided in real time and Record/Replay facilities are included. Printouts of data provide hard copy of results.

Visual Scenes. The SIMUGUN system makes use of a library of video disks showing targets and real landscapes. SIMUGUN can also be connected to a SIMUVISION unit, a real-time or off-line technology that creates a realistic scenario and firing effects. Custom-made scenarios may be generated and stored for future use.

Indoor version. A desktop version of SIMUGUN provides the same training options as the vehicle-based version but in an indoor environment.

Training Effectiveness. According to Rafael, tests have demonstrated that after 50 to 75 minutes of first-time training with SIMUGUN, both the tank commander and gunner can achieve a 60 to 100 per cent overall improvement in gunnery performance. Improvement levels for hit/miss identification show a 200 to 400 per cent improvement, laying accuracy is 25 to 50 per cent improved, and rapidity of response is from 50 to 150 per cent improved.

SIMUGUN tank gunnery trainers in use with IDF M60A1 MBTs

Status

In production. In service with the Israeli Defence Forces and some other defence forces, including some in South America.

Manufacturer

Rafael, Haifa.

(For details see Directory of Suppliers)

SIMTECH TP-series Table-top Tank Gunnery Trainers

The TP-series of table-top tank gunnery trainers has been developed by SIMTECH in association with the American Perceptronics company. These trainers can be used for basic training and continuation training and to provide training in target recognition, acquisition and engagement, gunnery skills, laser ranging and thermal sighting. The following models are made:

TP-2 - for M48/M60 tanks with advanced fire-control system

TP-2A - for fire-control systems involving laser and thermal (TI) rangefinders

TP-3 - for Centurion tanks

TP-3A - for fire-control systems involving laser and thermal (TI) rangefinders

TP-4 - for Merkava tanks

TP-5 - for M48/M60 and Centurion tanks with advanced fire-control systems

TP-6 - for M60 and Centurion tanks with highly advanced fire-control systems.

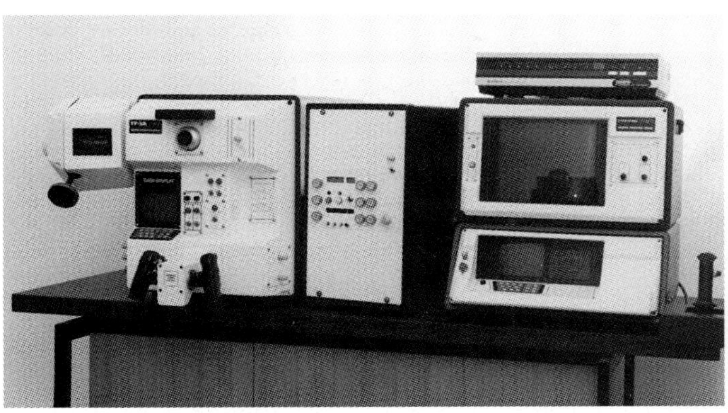

SIMTECH TP-3A table-top tank gunnery trainer for Centurion MBT with laser rangefinder

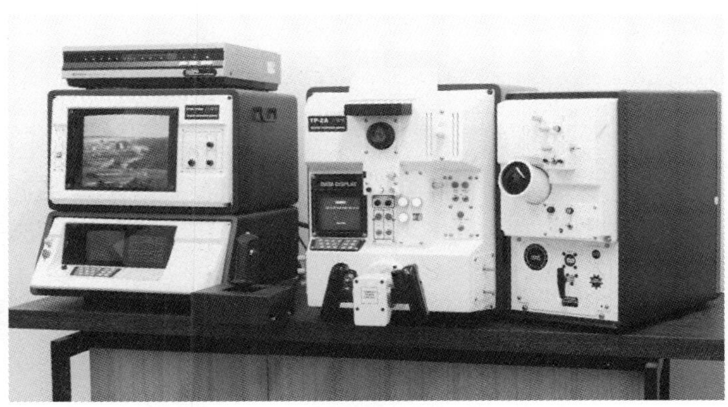

SIMTECH TP-2A table-top tank gunnery trainer for M60 MBT with advanced fire-control system plus laser and thermal rangefinder

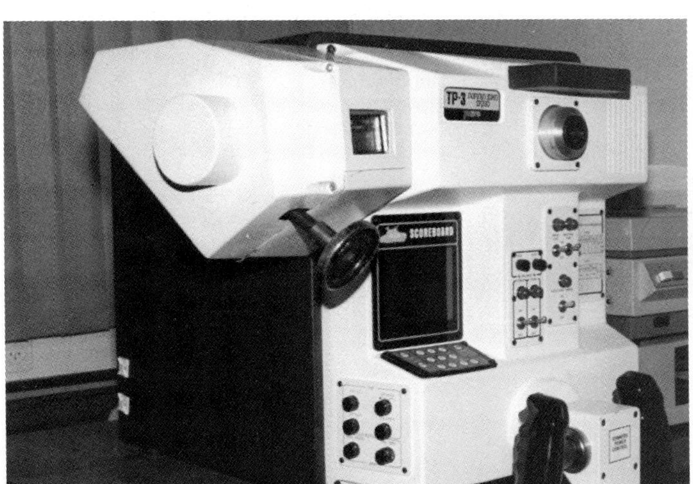

SIMTECH TP-3 table-top tank gunnery trainer for Centurion MBT

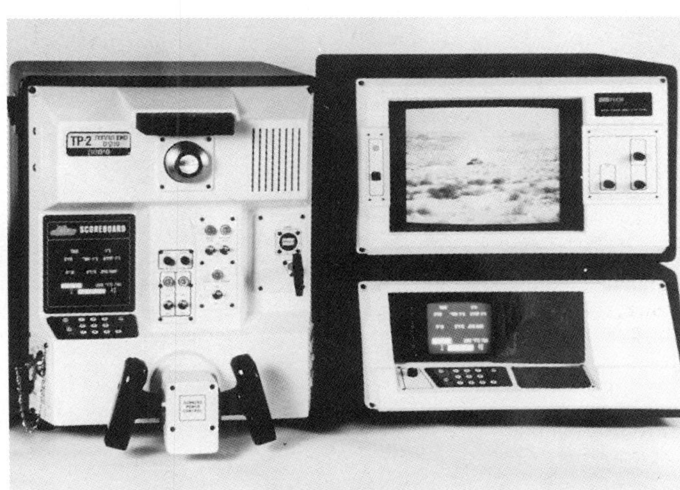

SIMTECH TP-2 table-top tank gunnery trainer for Centurion MBT

One further trainer incorporates a stabilised line-of-sight feature. Other types of TP-series trainer can be produced to meet virtually any requirement.

Components. Each trainer consists of a gunner's station console and a video disk player. Each has all the various gun controls and gun control panels associated with a particular fighting vehicle. There is also an eye-piece with a brow pad, and mounted on the trainer's front panel is a cathode ray tube together with its associated input keypad. A master key controls the on/off switch.

Operation. The type of exercise can be selected using the CRT keypad and the trainee looks through the eye-piece upon which is imposed a suitable sight graticule. He will see a training scenario derived from a video disk. Using the gun controls provided, the trainee then carries out the exercise selected. Targets may be single or multiple, short- or long-range, moving or stationary, and concealed or in the open. The lighting conditions may be day or night so the targets can also be thermal or optical. Training exercise factors can include variables such as mission type and sequencing, target type, ammunition type, ammunition allowance, loading intervals and induced errors.

Firing and Results. After any simulated firing, computer-generated tracer paths and hit or miss explosions will be simulated. The system computer will then calculate and record results such as actual and possible scores, time elapsed since exercise start, rounds used and rounds remaining, and elevation and deflection errors in mils.

Status
Available.

Manufacturer
SIMTECH, Holon.
(For details see Directory of Suppliers)

SIMTECH Onboard Tank Gunnery Trainer

This training system may be seen as an extension of the technology involved in the SIMTECH TP-series of table-top tank gunnery trainers (see previous entry) except that a real tank is used. A display monitor is installed opposite the periscope used by the tank fire-control system. Up to four tanks can be linked together for platoon training.

Gunner's Station. The trainee gunner sits inside the tank and operates the fire-control system in the usual manner. The picture he sees in his sight is the picture displayed on the monitor integrated with the rangefinder and other fire-control features. As well as depicting targets on a real video-recorded landscape the system integrates tracer trajectories, explosions and so on.

Instructor's Station. The instructor's station may be installed up to several metres away from the tank and includes a monitor and all the functions necessary for managing a training session.

Status
Available.

Manufacturer
SIMTECH, Holon.
(For details see Directory of Suppliers)

Montage of SIMTECH onboard tank gunnery trainer

POLAND

WCBKT BESKID-2 for T-72 Tank Gunner

The BESKID-2 simulator is designed for basic training of the T-72 tank gunner. It enables training in operating the sight (rangefinder TPD-K1), measuring target distance between ranges of 500-5000 m, aiming and firing at stationary and mobile targets within ranges of 500-4000 m (for the main gun) and 200-2000 m (for the machine gun), fixing advance (WYP) and range (CEL) corrections according to given target parameters and atmospheric conditions, and detection and recognition of targets (23 profiles).

Calculations are made according to the type of round, barrel wear, atmospheric conditions, and target conditions such as range and movement.

Training Methods. Three training methods are included. In the basic mode, the simulator indicates a succession of actions required for accomplishment of task. In the self-training mode, the simulator does not show actions required, but it verifies correctness of actions and measures time taken, indicating any errors. In the checking mode, the simulator checks for correctness of actions and measures time taken. It stores a record of errors as well as the advance (WYP) and range (CEL) corrections used.

Abbreviated Technical Data:

Electrical supply: mains at 220V/50Hz, max 600VA
Accuracy of target calculation: velocity ±10%; distance ±10m
Accuracy of projectile flight calculation: ±0.15 sec
Accuracy of fire range calculation: ±0.5mil
External dimensions: length: 2.0m; width: 0.76m; height: 1.8m
Weight: 250 kg (551 lb)

Status
In service with the Polish Army. Available.

Manufacturer
WCBKT, Bemowo; the marketing organisation is the Cenzin Company, Warsaw.
(For details, see Directory of Suppliers)

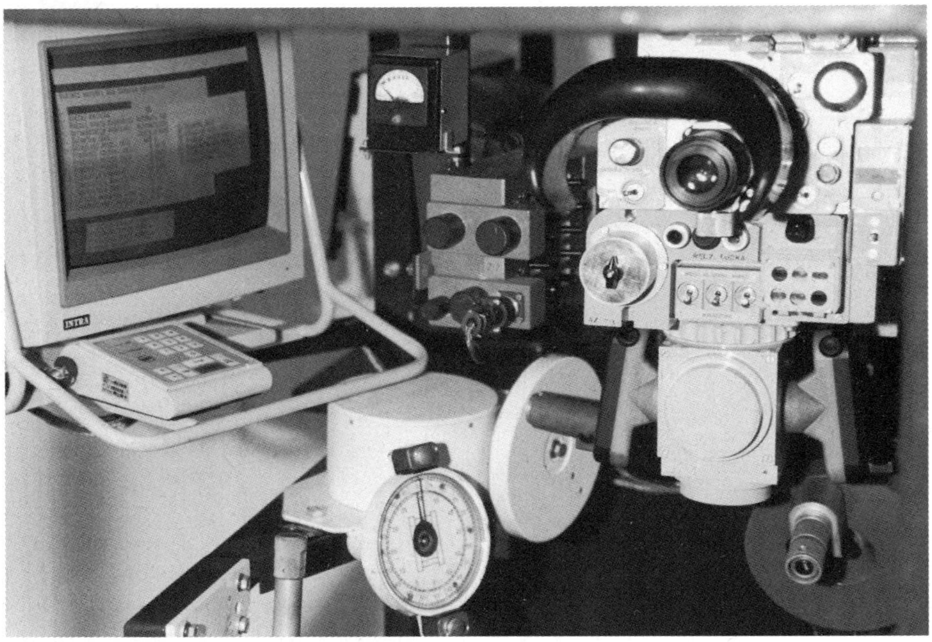

BESKID-2 T-72 Gunner Trainer

WCBKT ORTLES-3 for Initial Training of APC Gunner

The ORTLES-3 trainer is designed for initial training of APC BMP gunners. It provides basic training in operating the PN22M2 gun sight, recognition of targets (23 profiles), estimation of target distance and velocity, making corrections according to given firing conditions, tracking targets and firing the main gun (HEAT and HE projectiles simulated) and the machine gun (heavy projectiles), and judging the effect of firing and recognising methods of fire corrections.

Different target characteristics and round ballistics may be selected for different training exercises. Results of firing are presented for assessment on a chart during and after firing.

Abbreviated Technical Data:

Electrical supply: mains at 220V/50Hz, max 300VA
Accuracy of target calculation: velocity ±15%; distance ±7%
Accuracy of projectile flight calculation: ±0.1 sec
Accuracy of fire range calculation: ±25m
External dimensions: length: 1.8m; width: 0.65m; height: 1.65m
Weight: 200 kg (441 lb)

Status
In service with the Polish Army. Available.

Manufacturer
WCBKT, Bemowo; the marketing organisation is the Cenzin Company, Warsaw .
(For details see Directory of Suppliers)

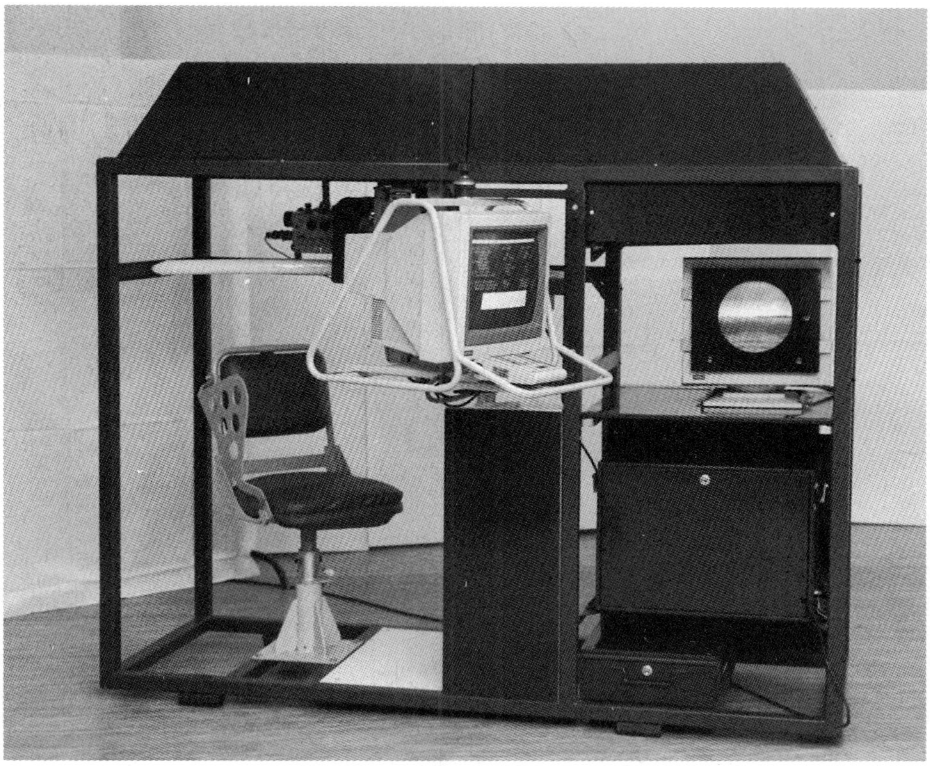

ORTLES-3 APC Gunner Basic Trainer

OBRUM SJ 08 Fire Training Stand

The SJ 08 Fire Training Stand is a live fire training device configured to provide gunnery training for Polish T-72M1 MBT gunners and commanders. It consists of a T-72M1 turret replica, complete with a full-scale mock-up of the 125 mm main gun, and a mock-up of the tank hull. It is mounted on T-72M1 road wheels for mobility.

Gunner's Station. The trainee gunner uses his normal gun controls and a modified TPD-K1 fire-control system to aim the main gun at scaled targets up to 100 m away which can represent stationary or moving targets at simulated ranges from 500 to 2500 m. The fire-control system is geared to convert these short ranges to the 'true' range as far as the fire-control system is concerned and the necessary adjustments are automatically made to the armament. Fire is provided by a coaxial 7.62 mm PKT machine gun. The device is provided with an intercom system for the turret crew.

Status
Available.

Manufacturer
OBRUM, Gliwice.
(For details see Directory of Suppliers)

SJ 08 Fire Training Stand

OBRUM SJ 02 Tank Gun Loading Automatic Dummy Training Stand

In order to provide training in the procedures required by the autoloader used with the 2A46 125 mm main guns fitted to T-72M1 tanks used by the Polish Army, a loading trainer has been produced by OBRUM. Automatic, semi-automatic and manual loading procedures may be carried out. The stand doubles as a training device to impart the required maintenance and trouble-shooting procedures. The stand contains a complete autoloader and gun mounted within a steel tube cage to allow view and supervision of trainee activities.

Status
In service with the Polish armed forces.

Manufacturer
OBRUM, Gliwice.
(For details see Directory of Suppliers)

SJ 02 Tank Gun Loading Stand. Note the rotary ammunition storage just above the baseplate. The dummy barrel is pointing away from the direction of the photo

RUSSIA

Tank Gunnery Training Equipment

The Russian and Soviet armed forces have been using gunnery simulators since before the Second World War. Tracking and firing on the move are simulated with the aid of rocking platforms for the gunner and his sight and for the vehicle. A typical individual rocking platform consists of a metal frame onto which an individual crew seat or AFV can be placed and rocked to simulate cross-country movement. To conserve ammunition, 23 mm sub-calibre ammunition is used rather than full-calibre rounds. It has been reported that Russian tank crews are limited to between 20 and 25 rounds of full-calibre ammunition a year.

Turret Trainers. Tank crew members are trained in a turret trainer which oscillates both horizontally and vertically (2-DoF, Pitch and Roll). This is used to teach conduct of fire against fixed and moving targets during daylight and darkness. In addition, it also facilitates the training of reconnaissance, radio communications, range determination, rules of fire and the loading and firing of both the main armament and the coaxial machine gun.

Status

Most of the illustrated equipments are believed to be still in service.

Manufacturer

State Armament Factories, details not known but information welcome.

KOP-RM optical control instrument for training tank commanders

UB-765 training turret for BMP-1 MICV, used for gunnery training with a machine gun for range training in place of the main 73 mm armament

TOPT-2 trainer used for classroom gunnery training

3T260 tank gun-layer trainer for classroom or field training which simulates targets in the layer's sight

SOUTH AFRICA

UEC Gunnery Training Simulators

The UEC Gunnery Training Simulator (GTS) is for training armoured vehicle gunners and commanders in gunnery. The GTS can be configured to simulate almost any armoured vehicle turret interior, and GTS has been produced for both the 20 mm and 90 mm versions of the Ratel IFV (GTS:Ratel) and for the Rooikat reconnaissance vehicle (GTS:76). Each GTS can be used as an individual unit or in a network of up to 20 units controlled from a central console. The GTS:Ratel weighs approximately 500 kg (1102 lb) and measures 1.26 × 1.67 × 1.7 m. Each GTS unit is mounted on swivel wheels for easy transport.

Components. The GTS consists of a lightweight metal cubicle which duplicates the internal dimensions, layout and controls of the AFV turret concerned. Provision is made for the commander and gunner. Gunnery visual effects are created by electronically manipulating digitised scenario images which move according to the movement of the hand controls. Fall-of-shot, tracer lines, dust and smoke are created using CGI. Sound effects are also simulated. Any type of target can be presented in a variety of background scenes. Up to 20 targets can be presented in an exercise scenario and targets can be active or passive.

Lesson Plan. The GTS can run a computer-based teaching module presented to the trainee on a video monitor located close to the trainee station. The module presents visual displays and sound effects; a voice-over can supplement the visuals.

Instructor Control. During an exercise the instructor can stand next to the GTS and control the exercise using a keypad and monitor display mounted on the GTS unit. The instructor also has his own menu-driven diagnostics and calibration routine. Exercises can be recorded for analysis and/or debriefing and there is a hard-copy printout facility.

Status

In production. In service with the South African Defence Forces.

Manufacturer

UEC Projects (Pty) Limited, Natal.

(For details see Directory of Suppliers)

Trainee gunner's position in a UEC Gunnery Training Simulator (GTS)

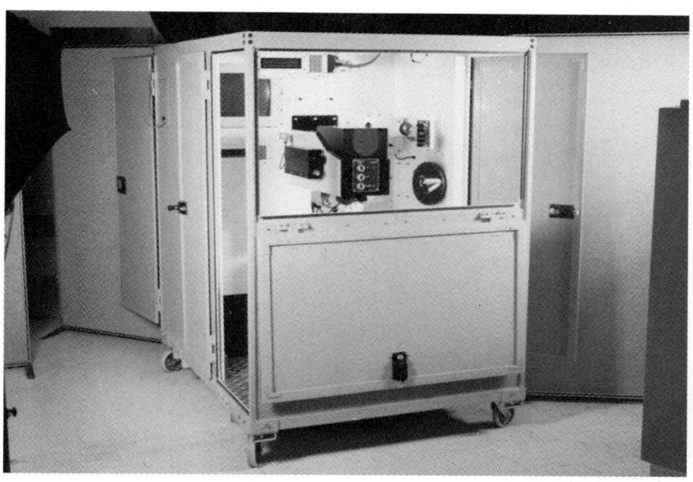

UEC Gunnery Training Simulator (GTS) for the Ratel IFV (GTS:Ratel)

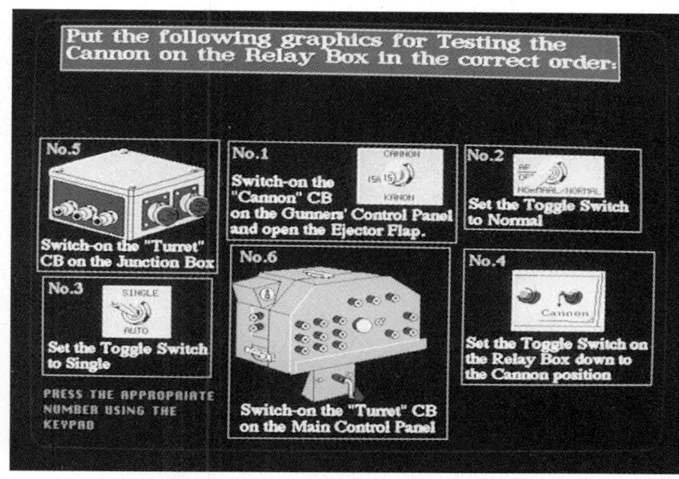

Instructional display in a UEC Gunnery Training Simulator (GTS)

SWEDEN

Weibull Tank Gunnery Training Equipment

S Tank Gunnery Trainer

This tank gunnery trainer entered service with the Swedish Army in 1970 and is used for training the gunner/driver or commander of the Bofors S tank. For indoor training the gunner fires against a target simulator. For outdoor training against real targets, the trainer is used with the Weibull optical surveillance system.

Ikv-91 Gunnery Simulator

This simulator entered service with the Swedish Army in 1975. It is used for training the Ikv-91 tank commander and gunner in target acquisition, aiming and firing. For indoor training the gunner fires against a target simulator. For outdoor training against real targets, the simulator is mounted on a 4 × 4 cross-country vehicle. The Weibull optical surveillance system is used for evaluating the results.

Pbv 302 Gunnery Simulator

This gunnery simulator entered service with the Swedish Army in 1966. It was also adopted by the Swiss Army. It is used to train gunners of the 20 mm gun turret

Weibull S tank gunnery trainer

Weibull Ikv-91 gunnery simulator from front

fitted to the Pbv 302 APCs of the Swedish Army. The Swiss Army use it for training gunners of the M113s fitted with the Swedish 20 mm gun turret called the Schützenpanzer 63/73. The simulator is used for indoor training against a target simulator, with hits being registered on the Weibull hit indicator.

Status
In service with the Swedish and Swiss armies.

Manufacturer
J L Weibull AB, Möllegården.
(For details see Directory of Suppliers)

Pbv 302 gunnery trainer

SWITZERLAND

SINTRO Gunnery Training Simulators

Thc SINTRO Gunnery Training Simulators have been designed for Individual Training in target acquisition, tracking and firing procedures. Modular design enables them to be configured to any gunnery system, whether for artillery, tanks, fighting vehicles, air-defence or for any other requirements. They are available in three versions: Desktop Version, Classroom Version and Shelter (Cabin) Version.

Visual Systems Available The SINTRO 2-D low-cost visual system is based on digitised backgrounds and superimposed targets and effects. It gives a balance between realism and cost. Software tools can be utilised to create individual exercises. The SINTRO 3-D visual system is also available for more advanced training scenarios involving multiple forces, and provides a CGI landcape, targets and special effects, together with the usual software tools.

User-Friendly Design. All of SINTRO's Gunnery Training Simulators are designed to be user-friendly and easy to operate without specific knowledge or training, and can be set up within minutes.

Status
Available.

Manufacturer
SINTRO Electronics AG, Interlaken.
(For details see Directory of Suppliers)

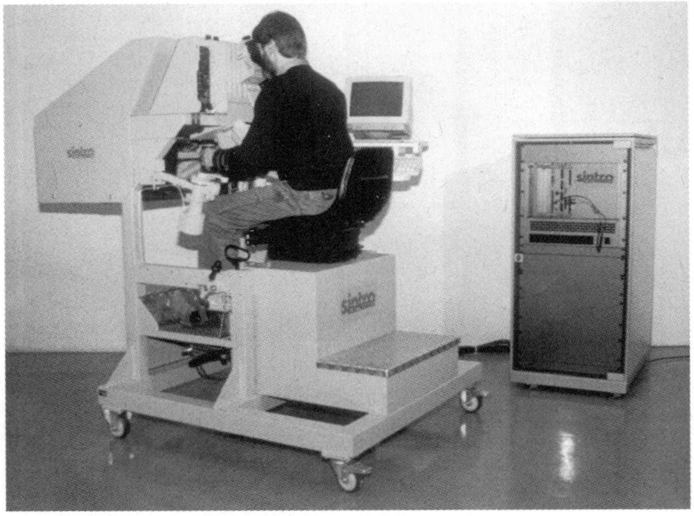

SINTRO Classroom Gunnery Trainer

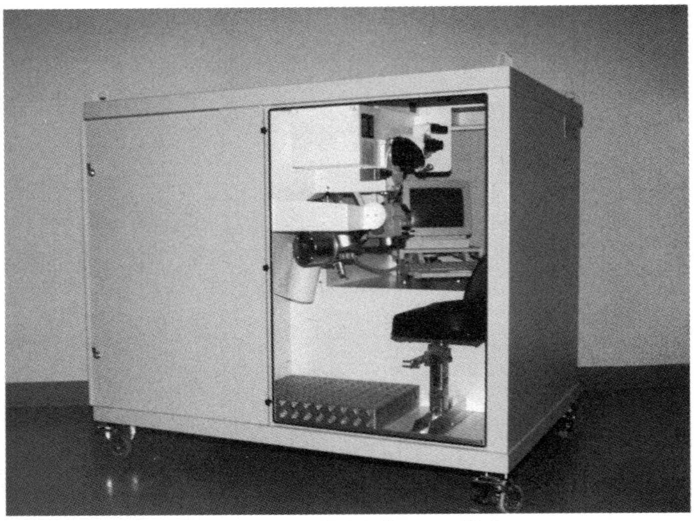

SINTRO Shelter/Cabin Gunnery Trainer

UNITED KINGDOM

Invertron Advanced Resolution Gunnery Simulator (ARGS)

The Invertron Advanced Resolution Gunnery Simulator (ARGS) is compatible with all tank fire-control systems, both visual and thermal imaging. Although initially configured as a tank gunnery simulator, the ARGS technology is equally applicable to other types of simulator such as guided missile and recognition simulators.

Visual System. The high resolution visual system employs a combination of digitised photographs, video disk and CGI technology and is PC-based. It has an angular resolution of 0.15 mrad. It has a 360° capability and allows for the simultaneous display of up to 32 targets. Of these 16 may be moving targets, including ground vehicles, or air targets such as helicopters. Target tracks are programmable by the instructor. Visual effects include tracer, fall-of-shot effects and overshoot or undershoot simulation. The resolution is such that the ARGS can simulate the passage of a projectile between a tank's gun and hull or turret without scoring a hit. The designated kill zone is precise, and may be varied depending on the type of target, its aspect and the type of ammunition involved.

View through Invertron ARGS simulated sight demonstrating a miss on a tank target

View through Invertron ARGS simulated sight demonstrating a hit on a truck target

Forward Observer Trainer. The visual system used with ARGS is also employed with the Invertron Deployable Forward Observer (DFO) Trainer. For details refer to the entry under *Artillery and Mortar Fire-Control Trainers and Simulators*.

Status
Available.

Manufacturer
Invertron Ltd, West Sussex.
(For details see Directory of Suppliers)

Invertron Desk Top Gunnery Trainer

The Invertron Desk Top Gunnery Trainer was originally developed as part of the Warrior Training Package but can be configured for any AFV. It is a low-cost system designed to produce basic gunnery training for trainee MBT and other AFV gunners. It is contained in two travelling box units, one of which becomes part of the trainer when in use.

General. The Trainer is based on a standard desktop PC, and uses simple CGI and graphics-based imagery. The student is provided with basic gun controls, elevation and traverse controls and a firing switch. With the Warrior training Package the trainer uses the sight graticule and controls for the 30 mm RARDEN gun. The trainee gunner views a simplified scene representative of day or night, including the weapon sight graticule. Terrain obscuration is incorporated for moving targets where applicable.

Student Exercises. The student's performance can be assessed on a duplicate monitor positioned by the instructor, and a Record/Replay facility is available. There is also a 'self teach' facility using preprogrammed exercises with the vehicle commander's voice produced for the trainee gunner by voice synthesis. At the end of each exercise a student is given a score based on many factors including the range to the target, time taken to engage and the number of hits.

Status

In service with the British Army's Armoured Infantry Battalions.

Alvis Limited, Aerospace Division, has adapted the Desk Top Trainer to support the Scorpion family of light vehicles.

Examples of the Scorpion Desk Top Trainer are in service with the Venezuelan Army and Nigeria.

Manufacturer

Invertron Ltd, West Sussex.
(For details see Directory of Suppliers)

An Invertron Desk Top Gunnery Trainer configured for the Warrior IFV with the 30 mm RARDEN gun

Invertron Intermediate Direct Fire Gunnery Trainer

The Invertron Intermediate Trainer (IT) was produced using the experience gained with the Direct Fire Trainers developed for the Warrior Training Package. The Intermediate Trainer uses a modular construction and a variety of gunnery controls can be installed. The Intermediate Trainer is a low-cost training aid for MBT gunners, enabling them to practise the use of their gunnery controls, plus target acquisition, tracking and engagement, using a variety of ammunition. The trainer consists of an instructor station and a modular student station unit into which any MBT gunnery controls can be installed.

Student Station. The student station is a metal box in which the trainee gunner is seated and is fitted with replica sights and all the usual manual and power controls in their correct respective positions for the crew position being simulated. Where the vehicle is equipped with stabilisation controls, a laser rangefinder, or a gunnery computer, these are also simulated.

Visual System. The trainee gunner views a simplified day or night terrain scene on which computer-generated targets manoeuvre. As targets pass behind terrain features such as buildings they are hidden or partially hidden as appropriate. Also produced are ammunition characteristics such as tracer and fall-of-shot representation for various types of projectile. The system enables trainee gunners to practise firing at targets whilst both the target and their own vehicle are moving.

Instructor Station. The instructor's control station consists of a desk into which is built the system computer, plus associated electronics and power supplies. The instructor has a VDU and keyboard plus a TV monitor which shows him the same view as the trainee sees through his sight. The instructor can either give the appropriate commander's orders to the trainee himself or these can be provided by the computer from synthesised recordings. From his station the instructor can control targets manually or replay complex target exercises, depending on the level required by individual students. The system displays an assessment of each firing and there is a Record/Replay facility.

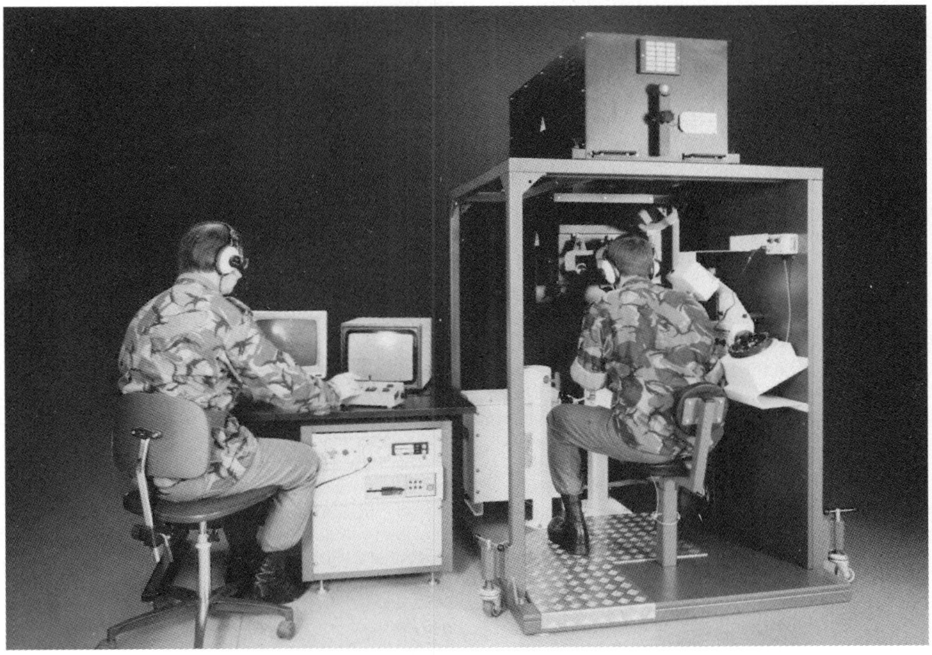

Invertron Intermediate Trainer for M60-series tank

Status

In service with the Austrian Army.

Manufacturer

Invertron Ltd, West Sussex.
(For details see Directory of Suppliers)

Marconi Tank Gunnery Trainer (TGT)

The Marconi Tank Gunnery Trainer (TGT) was designed to allow tank gunners and commanders to practise equipment and gunnery skills. The TGT provides a simulation of episcopes, sights and thermal imagers to permit detection, recognition and identification at realistic ranges in all weather conditions.

Databases and Modelling. The simulation computer models the dynamics of the vehicle concerned and its weapon systems, together with their interaction with a range of terrain conditions. Terrain databases encompass a variety of selected areas.

Instructor Station. The instructor's work station is designed to reduce workload by automation. Courseware is supplied by an automated student records system which captures results from individual students. Results control a trainee's progress through a series of exercises and a replay facility permits later analysis.

Status

Available for all armoured fighting vehicles.

Manufacturer

Marconi Simulation and Training, Dunfermline.
(For details see Directory of Suppliers)

Imagery from the Marconi Tank Gunnery Trainer (TGT)

Trenton Tank Gunnery and Missile Target System

The Trenton Tank Gunnery and Missile Target System (TGMTS) is designed for tank and missile gunnery training. It can be used in any building capable of providing access for the AFV concerned, but it needs less area when used for training with smaller systems such as mobile shoulder-launched SAM.

General Principle - Screen and Projector. Inside a suitable building a back-projection screen is installed 5.6 m from a 16 mm film or video projector. The tank is positioned on the opposite side of the screen with the main armament pivot point 5.6 m from the screen. When the tank commander and gunner are inside the turret and using their optical sights, this positioning in relation to the screen ensures a wide field-of-view. It also represents a normal scene in clear focus through the sights. Due to the back-projection, no obscuration of the scene can take place from tank parts such as the gun barrel. Normal rates of turret movement and normal angles of armament elevation and depression are used when tracking and engaging simulated targets.

The screen is not electronically connected to the remainder of the equipment but is aligned accurately with it. It is provided with the necessary height adjustment so that the screen centre can be aligned with the optical sights of different types of AFV.

Targets and Effects. A modified film target projector is positioned above the TGMTS console or alternatively the targets can be projected from videotape images. Films of moving targets are projected on the screen. Different scenes are available to show first simple target movement, progressing to more complex target movements, and finally multiple target scenes with smoke.

A projected target is engaged by following normal tank gunnery doctrine (precision or battlesight) as applied to laser or stereoscopic rangefinders, optical sights and ballistic/lead angle computers, or optical sights only. The TGMTS will permit training in Burst-on-Target procedures.

Laser rangefinders can be used normally with the TGMTS by means of an additional device known as the Laser Range Simulator.

Firing. On firing, a simulated gun flash is seen in the Primary and Secondary sights, a simulated weapon report is heard within the turret and simulated smoke briefly obscures the field-of-view. The projectile trace is observed and the target continues to move during the time of flight. At impact, the film image is stopped and the impact point projector is made to show the result of the fired round.

At the instant the target range equals the projectile range a brilliant point of eye-safe laser light is projected onto the screen from the impact point projector located inside the control console. This indicates the 95 per cent hit probability zone which may be adjusted to suit the user's weapon system.

Hit Probability Zones. The size of the 95 per cent hit probability zone may be varied from a diameter of 3 mm (0.5 mils at a distance of 5.6 m) to a diameter of 64 mm (11.5 mils at 5.6 m). If the zone intersects the target completely it is designated a kill. If half of the zone intersects part of the target, it is considered damaged and further shots will be required to produce a kill.

Results. The circle of laser light on the target will permit the tank commander, gunner and instructor to assess immediately the result for a preset number of seconds before the film and target resume motion. If an obvious miss occurs the instructor can extend the assessment time delay for as long as required to provide instruction. With skilled tank crews the assessment time can be set at zero. The impact point is briefly flashed onto the screen and a target scenario can be continued without interruption.

Instructor Control. The TGMTS is controlled by the instructor, who uses a single remote-control unit fitted with a wanderlead. As an alternative, the instructor can take the part of the tank commander and control the training from within the tank.

Projection unit used with Trenton tank gunnery and missile target system

Status

In service with M60A1/A3 and Leopard 1 MBTs.

Manufacturer

Trenton Technical Services Limited, Southampton.
(For details see Directory of Suppliers)

Vickers Tank Gunnery Part Task Trainer

The Vickers Tank Gunnery Part Task Trainer (VTGPTT) was developed in response to requirement for Challenger MBT training aids for the UK MoD. The contract was awarded to Vickers in May 1993 and the system will support Challenger 1 and Challenger 2 MBTs which are also produced by Vickers. The trainer allows gunnery students to undertake initial training and assessment under realistic conditions.

Components. The trainer consists of a single Instructor Control Station (ICS) connected to up to six student stations, each representing a tank gunner's position.

Student Stations. Each student station uses authentic hand controls coupled with high quality graphics. Each station is self-contained and can be moved by one person. Stations can be used in either a stand-alone mode or linked to the ICS, and two stations can be linked together to allow tank gunners and commanders to be trained together. Each station has its own visual system containing all databases, targets and exercises, and up to 16 targets are visible at any one time. Engagements to full battle distances can be undertaken in simulated day, night and limited visibility conditions. If required by the customer, visual performance may be enhanced.

Use for Other AFVs. The gunnery trainer can be configured for any AFV of any origin, and can be supplied with a full training package.

Status

On order for the United Kingdom and Oman. First deliveries to the British Army in 1994, followed by further deliveries to Oman.

Manufacturer

Vickers Defence Systems, Newcastle upon Tyne.
(For details see Directory of Suppliers)

The Vickers Tank Gunnery Part Task Trainer

UNITED STATES OF AMERICA

ECC Videodisc Gunnery Simulator (VIGS)

The Videodisc Gunnery Simulator (VIGS) is produced by the ECC International Corp (ECC). VIGS is used to train gunners for the M60A1, M60A3 and M1/M1A1 Abrams MBTs and for the M2/M3 Bradley IFV. Prototypes have also been developed for the Challenger MBT and the M48CH.

The VIGS is a portable, low-cost, desktop simulator replicating the gunner's station on an armoured fighting vehicle. It is a part-task trainer (PTT) used to enable student gunners to develop target acquisition and graticule aiming skills, and to maintain proficiency levels for experienced crewmen. VIGS exercises may include the identification of friendly and enemy targets. An alternative VIGS training mode is range determination practice, where a trainee estimates the distance to a target and is then graded on his ranging ability.

Components. Each desktop VIGS has a replica of a gunner's control panel appropriate for the vehicle and gun involved. On the front face of the VIGS console is the gunner's eye-piece, a brow pad, the gunner's control handles, a gunner's switch box panel and a stabilisation control selector panel.

There is also a 127 mm (5-in) Cathode Ray Tube (CRT) which displays a training exercise menu and then records and displays the gunner's performance for each exercise.

A videodisc player beside each VIGS supplies the exercise scenarios seen through the gunner's eye-piece. Each VIGS is supplied in a portable, protected carrying case.

Exercise Menu. Once the VIGS is turned on, a menu of training exercises is listed on the CRT. The student or instructor selects an exercise via a keypad under the CRT. After selection, the audio command - 'Gunner' - starts the trainer. An elapsed time meter is activated to evaluate gunner performance. The timer stops when a target is destroyed, when allotted ammunition is exhausted, or when allotted exercise time has run out.

Exercise Sequence. After an exercise is selected, an image will appear in the gunner's field-of-view. Tank commander fire commands will be heard. The gunner must then acquire, identify and make the necessary adjustments in aiming prior to firing. After the gun has been fired, a computer-generated tracer allows the gunner to see the projectile path. On impact, a computer-generated explosion occurs showing smoke and fire. The explosion and audio cues for a miss are different from those for a hit. When the target is hit, the exercise mission ends and the video image is frozen. If there is more than one target involved, the exercise will continue until all targets are hit or the end of the video scene is reached. The gunner can practise his proficiency with both the main gun and the coaxial machine gun. A kill zone from 5 to 300 per cent of the target size is selectable. Training exercises include single and multiple target missions in both short- and long-range scenarios. Stationary, moving, concealed and troop formation targets are also included.

When an exercise mission ends, the video image is frozen. All rounds fired are superimposed on the target by a dot in their hit or miss position relative to the target in the gunner's field-of-view. Simultaneously, the computer evaluates the firing performance and displays the detailed analysis on the CRT on the front of the console.

Two ECC Videodisc Gunnery Simulators in use

Status

In production. In service with the US Army, Reserves and National Guard (over 1200 units).

Manufacturer

ECC International Corporation, Pennsylvania.
(For details see Directory of Suppliers)

Perceptronics Gunnery Trainers

The Training and Simulation Systems Division of Perceptronics Inc developed, in association with SIMTECH of Israel, a series of gunnery trainers for use by the gunners of various armoured vehicles. The trainers involved are all part-task trainers which are portable and intended for use on desk tops. These trainers can be used for basic and continuation training. They can also be used to provide training in target recognition, acquisition and engagement, gunnery skills, laser ranging and thermal sighting.

Components. Each trainer consists of a gunner's station console and a video disk player. Each has all the various gun controls and gun control panels associated with a particular fighting vehicle. There is also an eye-piece with a brow pad, and mounted on the trainer's front panel is a cathode ray tube together with its associated input keypad. A master key controls the on/off switch.

Exercise Selection. The type of exercise can be selected using a CRT keypad and the trainee looks through the eye-piece, upon which is imposed a suitable sight graticule. He will see a training scenario derived from a video disk and then carries out the exercise selected. Training exercise factors can include mission type and sequencing, target type, ammunition type, ammunition allowance, loading intervals and induced errors.

Targets. Targets may be single or multiple, short- or long-range, moving or stationary, concealed or in the open. The lighting conditions may be day or night and the targets can also be optical or thermal (TI imagery). Firings will result in computer-generated tracer paths and hit or miss explosions.

Scoring. The system computer will calculate and record results such as actual and possible scores, time elapsed since exercise start, rounds used and rounds remaining, and elevation and deflection errors in mils.

Visual and Sound Effects. Visual and sound effects provided in most of these trainers include: an introductory target identification sequence; fire commands; firing flash; tracer graphics for the coaxial machine gun and main gun; coaxial machine gun strike graphics; 'hit but not kill', 'miss' and 'kill' explosion graphics; and an 'up' announcement indicating round reload. The MK-728 has an introductory animated tutorial on target engagement procedures.

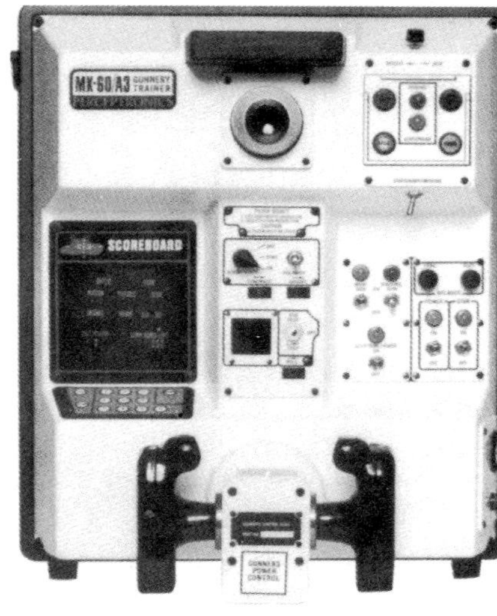

Perceptronics MK-60/A3 tank gunnery trainer

Trainers produced include the following:

MK-60/A3 and MK-60/A1 Tank Gunnery Trainers
Intended for the M60A3 and M60A1 main battle tanks respectively, these trainers are configured for 105 mm main tank guns and their 7.62 mm coaxial machine guns. Both models provide a gunner's field-of-view greater than 41° and there is a ×8 magnification of the target scene. Both trainers weigh 56.2 kg and measure 559 × 457 × 700 mm.

MK-48/A5 Tank Gunnery Trainer
This trainer is essentially similar to the MK-60/A1 trainer but is intended for use by M48A5 main battle tank gunners.

MK-728 CEV Gunnery Trainer
This trainer is intended for training M728 Combat Engineer Vehicle (CEV) gunners. The trainer is configured for the CEV's 165 mm demolition gun and its coaxial machine gun, so the simulated ballistics are limited to the HEP projectile for the main gun. Weights and dimensions are as for the MK-60/A3.

MK-2/3 BFV Gunnery Trainer
This trainer is intended for training M2/M3 Bradley Infantry Fighting Vehicle gunners, so is configured for the TOW anti-tank missile, the M252 25 mm Chain Gun and a 7.62 mm coaxial machine gun. The gunner's field-of-view is 54°. Weight is 77.3 kg and dimensions are 635 × 533 × 787 mm.

MK-13 Tank Gunnery Trainer
This trainer is configured for a 75 mm main gun and a coaxial 7.62 mm machine gun. Weight and dimensions are the same as for the MK-2/3.

All these trainers may be supplied with an optional instructor's console, a sight repeater module and a printer.

Status
Available.

Manufacturer
Perceptronics Inc, California.
(For details see Directory of Suppliers)

YUGOSLAVIA (SERBIA AND MONTENEGRO)

Control Training Equipment for M-84 Tank Gunnery Training

The CTE M-84 is a training aid intended to monitor tank gunnery training exercises involving the M-84 MBT, a derivative of the Russian T-72. The system may be used on indoor or miniature training ranges and in field conditions and is capable of both dry and live firing exercises.

Commander's TV Monitor. A video camera provides an image which is identical to the one seen by the trainee gunner through the sight eye-piece. The M7 monitor enables the tank commander to observe the trainee's actions and a video recorder can be used to play back results for later analysis. An external M44 monitor can be used by other observers.

System Operation. The system measures the time taken for the gunner to use the fire-control system; controls the training such as range measuring and the setting up of lead angles; uses the video recorder to record results; evaluates scoring; and aligns the optical sight reticule with an electronically generated reticule. Optical adaptors are fitted when training on indoor or miniature training ranges.

Status
In service.

Contractor
Holding Company YUGOIMPORT-SDPR Ltd, Belgrade.
(For details see Directory of Suppliers)

AFV CREW AND TACTICAL TRAINERS

CHINA, PEOPLE'S REPUBLIC

TCM-1 Tank Crew Training Simulator

The TCM-1 tank crew training simulator was designed and developed by a research institute for the Armour Headquarters of the PLA. Development began in September 1987 and the system entered service in May 1989. It is used for indoor gunnery training for Type 79 MBT turret crews consisting of commander, gunner and loader. With some modification it could also be used for other types of tank. Training carried out using the TCM-1 is claimed to be 1/400th of the cost of using a real vehicle.

Training Station. The Training Station is located inside a production Type 79 tank turret mounting a real 105 mm main gun. Included are the main gun elevation and traverse mechanisms, a control box, a coaxial machine gun, electronic equipment, a recoil movement hydraulic subsystem, a sound synthesis subsystem, what is described as a head-shell separating subsystem, and simulated ammunition. The simulated ammunition is full size and has the same weight and appearance as real ammunition. Simulated recoil movements and sound effects are stated to be realistic. Also included are three television monitors, one for each crew member.

Instructor Station. The instructor station is located next to the gunner's position. It consists of a computer and keyboard, a printer, a monitor, a video disk player, a sound synthesis subsystem, an electrical control subsystem, and power supplies. The monitor displays the same view through the gun sights as that seen by the trainee gunner. If necessary the instructor can act as the tank driver and can provide target and firing commands.

Visual System and Operating Sequence. The TCM-1 through-the-sight visual display is a recorded landscape presented from the video disk player. Computer-generated mobile targets are inserted in the display, together with an aiming point and a sight reticule. The trainee crew then loads the main gun and carries out complete target acquisition and firing sequences. The system computer can provide a complete printout of each crew member's performance.

Status

In service with the People's Liberation Army.

Manufacturer

State armament factories, for the Armour Headquarters of the PLA

FRANCE

Giravions Dorand DX 211 Turret Trainer

The Giravions Dorand DX 211 Turret Trainer has been produced to simulate LAV 25 commander and gunner stations.

Training Station. The training station comprises an 'open' turret model which reproduces the crew observation and control facilities. The system uses a video disk to provide landscape images seen through the crew observation and sighting devices. Targets and firing effects are overlaid on the landscape image.

Instructor's Station. The instructor's station has devices needed to select the preprogrammed scenarios, monitor trainee performance via video recording, and the reading and general recording of results.

Status

In production for the US Marine Corps.

Manufacturer

Giravions Dorand Industries, Suresnes.
(For details see Directory of Suppliers)

Giravions Dorand DX 211 Turret Trainer configured for a LAV 25

Thomson Leclerc Tank Crew Training Simulator

The Leclerc Crew Training Simulator was developed by Thomson Training and Simulation in co-operation with Giat Industries, and allows training of the entire crew. Previous individual training will have taken place for the driver on the Driver Trainer (described later under *Tank and Other Driving Simulators*) and on the Turret Team Trainer for the commander and gunner (described earlier under *Tank and Direct Fire Gun Trainers*).

Components. The Leclerc Crew Training Simulator incorporates two compartments mounted on correlated motion systems, one for the driver and one for the turret crew.

Visual System. The visual system uses the Thomson SPACE CGI system (see details later in the CGI section) which uses photographic textures to represent details of the terrain between 0.5 and 2.5 m in height. Thirty targets can be presented simultaneously with animation and gunnery effects. The terrain database covers 400 square kilometres in a format that is compatible with the CGI terrain and other protocols of the US DoD Project 2851.

Thomson SPACE CGI system - Illustration of battlefield scene on Leclerc Tank Crew Simulator

Instructor Station. The instructor monitors the trainees' work at a supervision and animation station which has been designed to be user-friendly. Intelligent software animates the movements of friendly and enemy forces based on commands entered by the instructor when preparing the exercise.
Networking. Four Crew Training Simulators can be networked at the same site to provide 'troop mode' operation. Networking between remote sites is also possible.

Status
In service with the French Army at the Training Centre at Carpiagne.

Manufacturer
Thomson Training and Simulation, Cergy-Pontoise.
(For details see Directory of Suppliers)

Thomson Leclerc Tank Crew Simulator

Thomson Training Systems Crew and Platoon Gunnery Simulator

The Thomson Crew and Platoon Gunnery Simulator provides both basic and advanced training for gunners, gunner/tank commander crews, and complete platoons. Twelve Crew Gunnery Simulators have been in service since 1985 at the French Army Advanced Gunnery Training Centre at Canjuers and the Armoured Officers and NCO School at Saumur. During 1988 Thomson upgraded these simulators so that four (instead of the former three) turret crews form a platoon. The equipment is modular in design, and can be adapted to all types of tank currently in service.

Visual Image Generation. Image generation is by the Thomson SPACE CGI system. This provides a large gaming area with realistic landscape features, targets and weapons effects. Up to 30 simultaneous moving targets may be shown. Target images shown include slow appearance; firing or observation under cover; movement at different speeds, paths and aspects; and masking effects produced by terrain, vegetation or other cover. For training from fixed firing positions, the simulator can also be fitted with a PC visual system which gives enhanced realism.
Training Exercises. The platoon commander and crews can be trained in all the problems involved in gunnery and fire control, such as: terrain observation; target detection, identification, selection and distribution; aiming, rangefinding and control.

Status
A total of 12 Crew Gunnery Simulators (equivalent to two Platoon Gunnery Simulators) in service with the French Army.

Manufacturer
Thomson Training and Simulation, Cergy-Pontoise.
(For details see Directory of Suppliers)

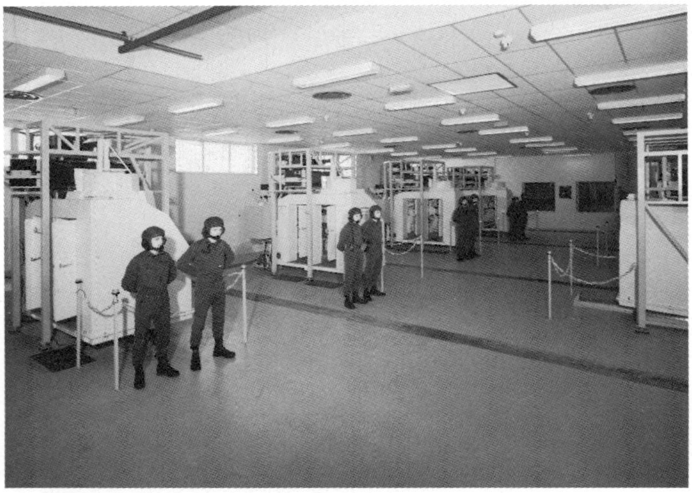

Student stations of the Platoon Gunnery Simulator at Canjuers

Instructor stations of the Platoon Gunnery Simulator at Canjuers

GERMANY

Atlas AGPG Mechanised Infantry Platoon Combat Simulator for the Marder ICV

The Mechanised Infantry Platoon Combat Simulator (AGPG) is for training Marder crews in combined-arms combat. It consists of three identical crew simulators housed in individual containers. A fourth container is for the instructor and control facilities, and also houses a debriefing theatre.

Crew Compartment. This contains workstations for the commander, gunner, driver and group leader which are realistic replications of those in the Marder 1A3 ICV. The optical system uses an ATLAS DISI 4 GCI system, which includes texture generation and photo-texture and a 20 x 30 km database. Dismounted combat training is also possible.

Instructor Station. The control and evaluation facility (instructor station) is normally operated by a principal instructor and three assistant instructors. Each has a monitor displaying the GCI view as seen by the trainee crew, and a multifunction monitor for exercise control. Record/replay facilities are also provided, and the AGPG can be networked with other simulators.

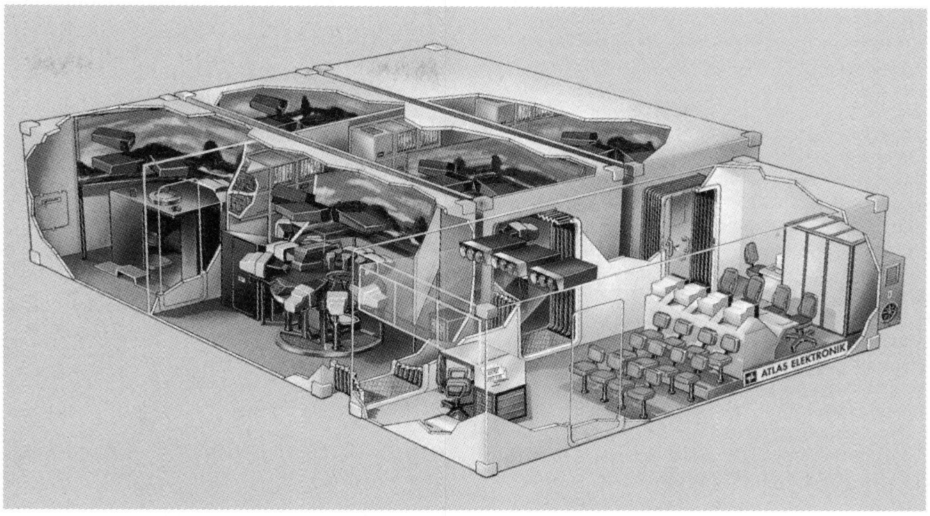

Atlas AGPG Platoon Combat Simulator

Status
Available. There are 22 units in service with the German Bundeswehr.

Manufacturer
Atlas Elektronik, Bremen.
(For details see Directory of Suppliers)

Atlas AGSM Gunnery Trainer for Marder

The Atlas AGSM is for gunnery training of the commander and gunner of the Marder AIFC 1A3. It is housed in a portable air-conditioned container.

Crew Compartment. The interior is a model turret with controls provided either as real units or working replicas. The optical system uses an ATLAS DISI 4 GCI system, which includes texture generation and photo-texture. Day, night and reduced visibility images are provided. A sound system produces realistic battlefield noises.

Instructor Station. The instructor has a monitor displaying the GCI view as seen by the trainee crew, and a multifunction monitor for exercise control. Record/replay facilities are also provided

Status
Available. In service with the German Bundeswehr.

Manufacturer
Atlas Elektronik, Bremen.
(For details see Directory of Suppliers)

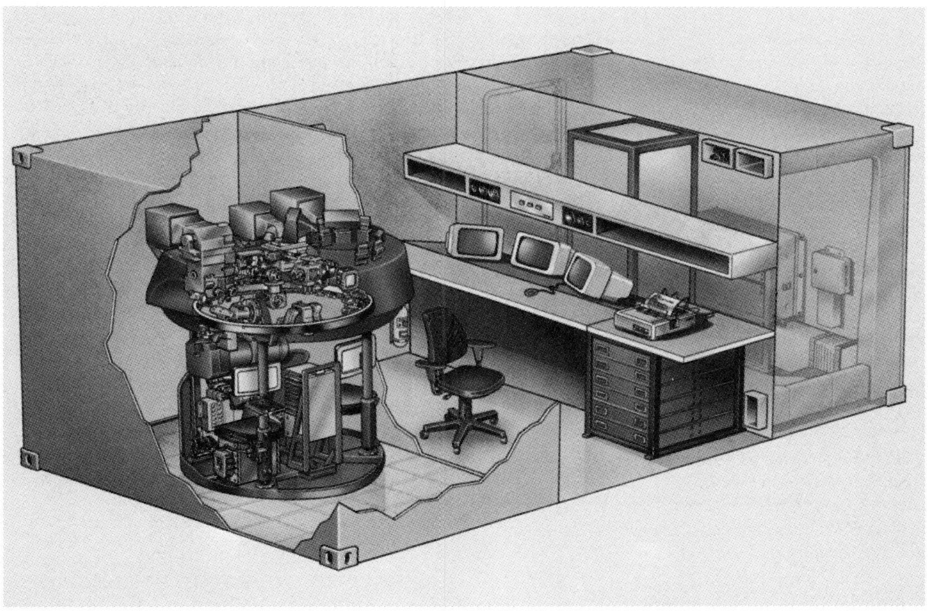

Atlas AGSM Gunnery Trainer for Marder

Honeywell Tank Crew Trainer (TCT)

The Honeywell Tank Crew Trainer (TCT) is a computer-controlled training system for full mission combat training of up to four tank commanders and gunners of MBTs. It is contained in an air-conditioned building measuring 38 × 21 × 7 m. The commander and gunner team can be trained in battlefield observation under a wide variation in lighting and visibility conditions, target detection, recognition and identification, target designation by the tank commander, ammunition type selection, rangefinding, coarse and fine gun-laying, firing from a stationary position or on the move, and shot sensing and correction.

Terrain Model Board. The centre of the trainer is a large terrain model board built to a scale of 1:160 which provides a simulated area of 2.5 × 2 km. The terrain model is provided with 56 moving and stationary targets. The trainer is equipped with a computer-controlled illumination system which has 35 floodlights, each with variable incident angles, height and brilliance.

Training Turrets. Around the terrain model are four training turrets. Each has a complete optical system with a ranging area of 90° in azimuth and 30° in elevation, and a possible trunnion tilt angle of ±15°. Each turret is complete with all its normal equipment, including a dummy gun. Each has its own instructor station equipped with video monitors and recorders, input data equipment and data output devices. Audio and tank noise simulation is available for each turret crew which can be supplied with hit or ricochet simulation in its sights, and the targets can be moved either manually or by preset sequences. As each turret is complete in all details it can also be used for procedural training, and troubleshooting can be practised.

Lesson Plans and Record/Replay. The central computer can accept either manually entered lesson plans for each exercise or a series of stored plans. Results are stored for later printout or longer-term storage.

Deployment. The first TCT training centre was ordered by the German Army; a Leopard 1A4 complex was specified. Operational evaluation of the prototype unit finished in March 1980. Delivery of three additional turrets for the Leopard 1A3 and 1A4 was carried out in the years 1983 to 1984. Three additional centres, each with four turrets and located in different parts of Germany, were delivered between 1986 and 1990.

General view of terrain model for a TCT

Status
In service with the German Army.

Manufacturer
Honeywell GmbH, Maintal.
(For details see Directory of Suppliers)

Wegmann AGPT Tactical Team Training and Gunnery Simulator

Drawing of a Leopard MBT tactical team training and gunnery simulator

The Tactical Team Training and Gunnery Simulator was developed to be a reasonably priced simulation system for tactical and gunnery training up to company level. A platoon configuration of this system for the Leopard 2 MBT was selected by the German Bundeswehr to support tactical training with embedded gunnery training at platoon level; this trainer is known as AGPT. Wegmann GmbH is the prime contractor responsible for overall system integration, BBN Systems and Technologies Corporation of Cambridge, Massachusetts, adapted features of the SIMNET programme, produced software and provided the CGI, and Perceptronics of Woodland Hills, California, designed the AGPT Leopard 2 crew cabins. The AGPT system has been optimised to meet the requirements of the German armoured units. Due to its modular design adaptations for other customer requirements are possible.

Components. The AGPT consists of three weapon system simulators for the Leopard 2 MBT and an Instructor Station known as the Training Control and Evaluation station (TC & E). In the weapon system simulators three full tank crews practise combat tasks in a 'closed hatch' environment. The system is in air-conditioned containers.

Design Principles. The following design principles apply:

Distributed simulation. Each simulator and Instructor Station has its own computer and power for simulation and image generation. There is no central simulation computer in the system.

Modular design allows use for other weapon system simulators without development risks, as well as easy software upgrades.

Selective fidelity. To achieve low hardware costs for the crew stations only those details which are essential for the training purposes are physically and functionally simulated.

Instructor Station. In the TC & E the exercise director and his assistant establish the initial tactical situation, and start and control the exercise by means of assignments to the exercising platoon and/or by controlling the enemy by means of semi-automated forces (SAF), and own flanking forces. They subsequently conduct the after-action review in the presence of the tank crews. Work stations and monitors include one plan-view display including a terrain map and the tactical situation for the exercise director and another for his assistant; two monitors representing the director's out-of-the-window view provided by an Invisible Director's Vehicle (ULF); a status display showing operating data for the simulators as well as information from BITE. A communication system provides a simulation of battlefield radio communications to the trainee crew members.

ULF is a two-channel computer image generator providing an out-of-the-window view into the battlefield or providing displays of the main optics of the simulators in order to monitor target acquisition and gun laying operations. ULF may be used during the run time of an exercise as well as during the after-action review.

Record/Replay. For the after-action review, all exercise activities, including radio traffic, is recorded by a Data/Voice Logger (DVL). The DVL may be used with the ULF for a full-scale after-action review.

Visual System. Each simulator has two BBN GT 120 CGI systems which display the 20 × 30 km terrain database as well as target models and special effects. The terrain area can be extended. Sixty simultaneous targets can be displayed. Images of the synthetic battlefield are displayed in the gunner's primary sight, the thermal image unit, the commander's independent viewing devices, the crew's vision blocks, and the driver's image intensifier. The present terrain database depicts an area near Wurzburg.

Conversion to Other Platforms. Image generation systems, the sound systems and combat station simulation are connected by interface electronics which, as they use modular software, can be easily converted for use in other weapon system simulators.

Containers. The TC & E and the weapon system simulators are enclosed in portable and climatically protected containers which can be moved when necessary. The containers can be handled by standard military material handling equipments and are carried on standard 5-tonne trucks. On arrival all that is required is connection to a suitable electrical power supply (380 V, 50 Hz) and the connection by cable of the four containers. Each container measures 5 × 2.2 × 2.45 m and weighs approximately 3600 kg when equipped.

Status

In production for the German Bundeswehr. Deployment completed by January 1994 with 15 platoon sets at separate locations throughout Germany. Adaptation for the Combat Improved Leopard 2 MBT for the German Bundeswehr and Swedish Army is taking place for deliveries 1994-96.

Manufacturer

Wegmann & Co GmbH, Kassel.

(For details see Directory of Suppliers)

Wegmann Turret Trainer for Leopard 2

Wegmann turret trainer for Leopard 2 MBT

The Wegmann Turret Trainer was designed to train Leopard 2 MBT turret crews in basic training in the turret, loader training for the main armament, and refresher drill training for turret crews using the armament and fire-control system. Several crews can be trained at one time.

Components. The turret trainer uses a cutaway configuration and includes all the components necessary to operate the turret system. Most components used are the same as those used in the production turret. To accommodate additional trainees, a platform with two tiers of seats for six persons is located on the commander/gunner's side; a seat for an instructor is located next to the electronics compartment. A permanently mounted ring-shaped step runs around the remaining part of the trainer.

Gun Trainer. The gun mount trainer allows training for the loader. As on the real equipment, the gun mount trainer is supported by two trunnion bearings and can be moved by means of a production line elevation mechanism. The use of semi-fixed inert practice ammunition in conjunction with the recoil mechanism and simulated case ejection ensures a sequence of operations identical to the real tank from the point of removing rounds from their ammunition racks to the loading operation and firing. To simulate barrel recoil movements, hydraulic cylinders are integrated in the recoil buffer housings. An unfired practice cartridge can be removed as a complete round. The entire operating sequence is controlled electro-hydraulically using control electronics located in a unit on the turret roof next to the instructor's seat.

Specifications

Weight: approx 9000 kg (19,842 lb)
Height: 2.24 m
Platform diameter: 3.5 m
Step diameter: 4.5 m
Elevation/depression: +20°/−9°
Traverse: n × 360°
Max number of trainees: 10 (inc instructor)
Mains supply: 3 × 380 V AC

Status

Available. In service with the German Army.

Manufacturer

Wegmann & Co GmbH, Kassel.
(For details see Directory of Suppliers)

ISRAEL

SIMTECH Tank Crew Trainers

These trainers involve mock-ups of tank turrets and/or interiors, usually involving the commander and gunner positions. The interiors include all the usual fire-control and other equipment found in a real tank. The mock-ups may be either free-standing or located in mobile shelters. Up to three or four mock-ups can be combined to create platoon trainers. SIMTECH is involved in two platoon trainer projects, one for Belgium and the other for Finland. Both involve units for two platoons with an option for a further two platoons. The system for Finland also has a 'gunner only' option based on video disk technology.

Training Station. Trainees have access to all their normal vision devices and can view moving scenes produced by digital graphic imagery or computer graphic imagery recorded on a video disk. The scenes depict the position and firing of nearby tanks, explosions, tracer trajectories and real world effects such as day/night, according to customer requirements.

Other Training Uses. The technology used can be applied to other weapon systems such as SAM launchers.

Status

Available.

Manufacturer

SIMTECH Limited, Holon.
(For details see Directory of Suppliers)

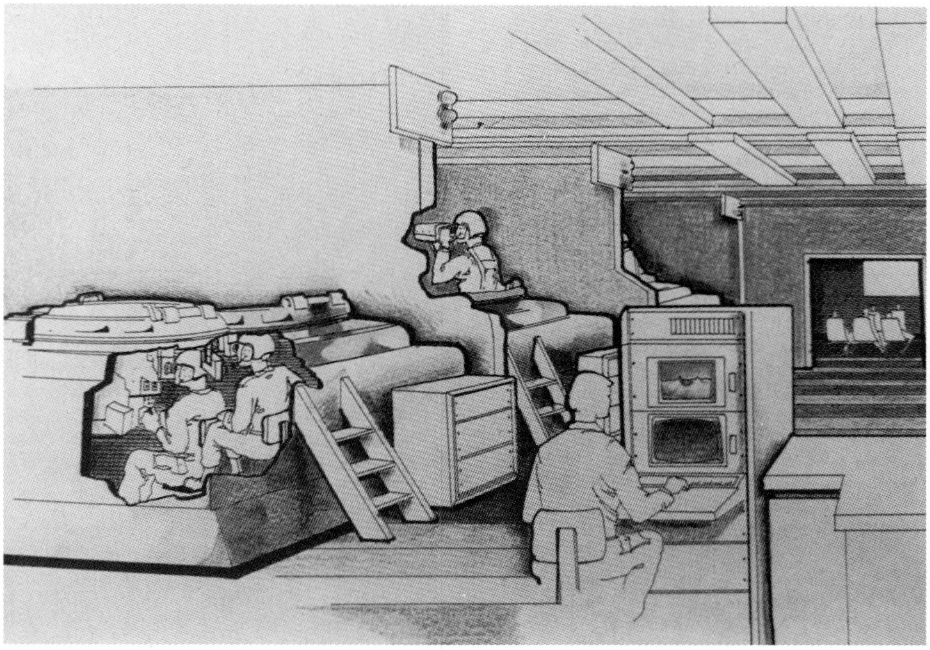

Drawing of SIMTECH Tank Crew Trainers arranged as a platoon trainer

SOUTH AFRICA

UEC Battle Training Simulators (BTS)

The UEC range of Battle Training Simulators (BTS) was designed to train AFV commanders in combat drills and tactics. BTS systems are available in a rugged console form or as a desktop classroom system. Multiple BTS systems may be configured in various linked modes, training up to 18 vehicle commanders under one instructor, thereby enabling training at both troop and squadron levels, including a force-on-force mode. Each BTS system caters for up to four students simultaneously under the guidance of a single instructor.

Training Station. Each student has a digitised map containing contour and terrain information. Numerous simulated actions are available. A dual-frequency radio net is simulated to provide inter-vehicle and vehicle-HQ communications.

Instructor Station. Features include the definition, activation, monitoring, evaluation and recording of a range of battle exercises against various enemy scenarios. BTS software provides all the functions necessary to define a simulated battle environment, and caters for accurate mobility, visibility and engagement modelling.

Vehicle Configuration. The BTS has been configured for Rooikat reconnaissance vehicle commanders and in this form is known as the BTS:76. This version contains up to three preprogrammed scenarios suitable for squadron training, and up to 50 preset exercises.

Status

In production. In service with the South African Defence Forces.

Manufacturer

UEC Projects (Pty) Limited, Natal.
(For details see Directory of Suppliers)

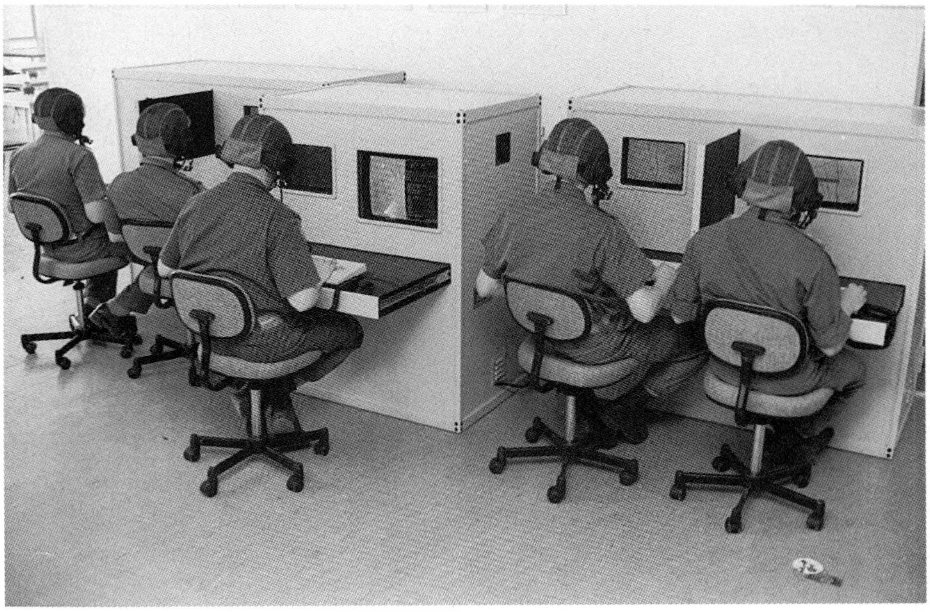

UEC Battle Training Simulator in BTS:76 form for Rooikat reconnaissance vehicle commanders

SWITZERLAND

SINTRO Tank Crew Trainer

The SINTRO Tank Crew Trainer is for training a full tank turret crew: commander, gunner and loader. Original turrets are utilised and modified by SINTRO; it is also possible to use full-scale replicas instead of original turrets. If required, openings are cut through the external walls of the turret to allow direct supervision by an instructor. Complete tank platoon trainers are also produced for training both crews and the platoon commander.

Training Station. Target areas and targets are presented to the tank commander and gunner by a high resolution, real-time, three-dimensional sighting system. A communication system allows crew-to-crew and instructor-to-crew communication. Training rounds are made from hard-wearing material and are identical to actual rounds in size, weight, centre of gravity, handling and appearance. A barrel recoil simulator replicates the recoil produced by a live round. The crew also hear realistic engine, hydraulic and firing noises in their headphones.

Instructor Station. The instructor controls the activities of the turret crew from a control console outside the turret. The instructor is able to monitor individual actions, check gun laying and monitor the target selected by the gunner. Weather conditions can be varied, including searchlight dazzle, fog and mist. Recording equipment, including a video recording system, enables trainee performance to be assessed. A plotter is provided for detailed analysis of results.

Training Exercises. The Tank Crew Trainer can be used for a wide range of training exercises including: basic target acquisition, identification, assignment and attack under varying external conditions; tank commander's briefings and crew teamwork; fire observation procedures including tracer, projectile impact and flash; and firing at stationary and moving targets at varying ranges.

SINTRO Tank Crew Trainer for Centurion MBT showing dummy rounds, main gun on left

Status
Available.

Manufacturer
SINTRO Electronics AG, Interlaken.
(For details see Directory of Suppliers)

UNITED KINGDOM

Invertron Challenger 2 Commander and Gunner Crew Station Simulator (CGCSS)

In early 1989, Invertron Simulated Systems Limited received a contract to provide a Commander and Gunner Crew Station Simulator (CGCSS) for the Challenger 2 MBT for the British Army.

Training Station. The CGCSS uses CGI for realistic terrain scenes, targets and firing effects. These are injected into the commander's and gunner's sights. A thermal image display is also produced in the Thermal Imager (TI) sight. The system software simulates the onboard fire-control computer, gun and sight control equipment, ballistic offsets and projectile trajectory. The trainee station provides a realistic environment for the Challenger 2 commander and gunner to perform target acquisition and engagement procedures, both when stationary and on the move.

Container. The CGCSS is housed in a purpose-built portable air-conditioned container.

Record/Replay. A detailed analysis is provided of each engagement.

Present Use. The system is used by Vickers Defence Systems to demonstrate use of the sights and fire-control equipment. The simulator also provides a tool for analysing the ergonomics of the turret layout.

Status
In service with Vickers Defence Systems.

Manufacturer
Invertron Simulated Systems Limited, West Sussex.
(For details see Directory of Suppliers)

Invertron Turret and Troop Trainers

The Invertron Turret and Troop Trainers are both part of the Warrior Training Package but can be configured to simulate any type of AFV.

Turret Trainer

The Turret Trainer comprises a dummy turret housing a commander and gunner. The crew is provided with both high and low magnification simulated sights capable of day and night operation. For the Warrior Training Package, Alvis supply the dummy turret and Invertron the training suite.

Training Station. Using the Turret Trainer, full target acquisition and gunnery skills can be practised against realistic computer-generated tactical targets. The graphic displays are similar to those used with the Invertron Desk Top Trainer (see under *Tank and Direct Fire Gun Trainers*) although more advanced computer-generated, video disk or computer-synthesised imagery systems are available. There is a choice of ammunition and weapon type, and target tracking using the sight graticule. Main and secondary armament facilities are provided and display the full ammunition characteristics of, for example, tracer and firing detonation. Warrior turret trainers are provided with a dummy 30 mm RARDEN gun and a real 7.62 mm Chain Gun using dummy ammunition. Weapon sound effects are transmitted to the crew during target engagement and simulated radio nets can be included in a tactical training scenario. The turret weapons can also be used for training in ammunition loading and stoppage clearing procedures.

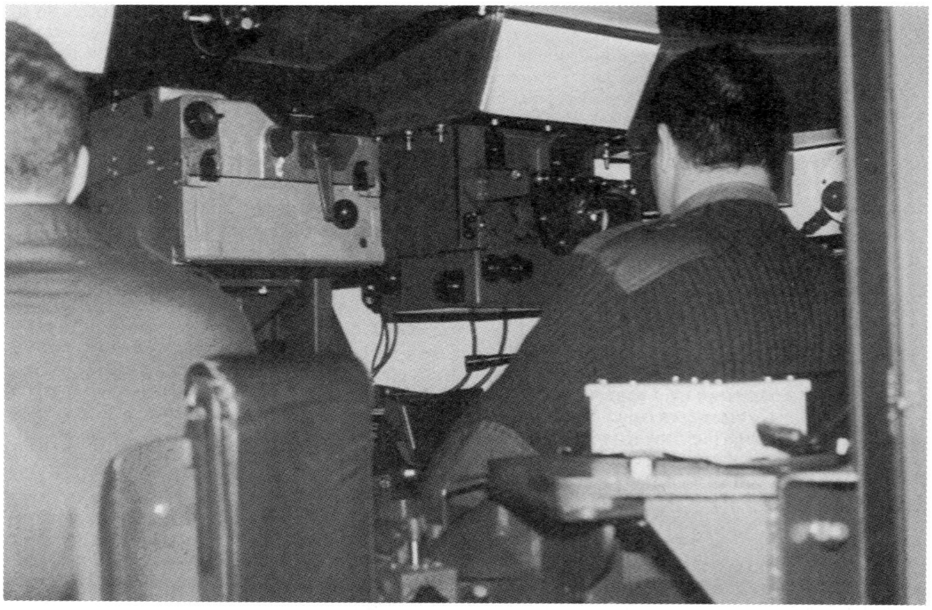

Interior of Invertron Turret Trainer configured for the Warrior ICV

Instructor Station. External to the trainer is the instructor's station where a full log of the trainee crew's performance is maintained by the computer. The instructor can replay preprogrammed target exercises or control individual targets manually. A hard-copy facility is included, with each crew receiving their own performance assessment from the computer printer. Two video monitors are mounted over the trainer to allow waiting or observing trainees to watch the actions and progress of the crew using the trainer.

Installation and Transportability. The dummy turret and instructor's station can be palletised and are transportable by truck, enabling a number of units to share the same training facility. The trainer can also be installed permanently in a building. The British Army scale of issue to Warrior units is one per armoured infantry company.
Troop Trainers. Troop Trainers are also referred to as Platoon Trainers and comprise three or four Turret Trainers each housed in a separate air-conditioned cabin and linked via computer data cables to a central instructor station.
Warrior Package - Trainee's Facilities. The Warrior Training Package consists of four Turret Trainers. All four turret crews view the same general scenario including common targets. However, each crew may be looking at a different part of the scene, depending on the direction the turret is facing and the elevation of the gun. This provides tactical training. The terrain views are produced by computer-synthesised imagery derived from pictures of real-world terrain. Targets are produced by imagery derived from scale models and are automatically occulted when passing behind terrain features or obstacles in the gunner's line-of-sight. Targets include a mix of friendly and enemy vehicles, both ground and air. In a typical troop trainer, 20 different target types are offered. Up to six static and six moving targets can be displayed on the terrain simultaneously. The fall-of-shot of adjacent vehicles is visible to each crew, correct for the previously determined vehicle locations at the start of the exercise. A jockeying facility is also included. Targets have the capability of returning fire once engaged by own troops.
Instructor Station. The instructor can record up to 120 target tracks on the computer disk over a total recording time of 120 minutes.
Networking. Several Troop Trainers can form a network. Each vehicle would see the same targetry and the fire effects of other vehicles. With additional viewing points, the networked vehicles could be separated into friendly and enemy forces, thus permitting force-on-force training.
Scorpion Version. Alvis Limited, Aerospace Division, has adapted the Turret Trainer to support the Scorpion family of light vehicles. Examples of the Scorpion Turret Trainer are in service with the Venezuelan Army and Nigeria.

Status
In service with the British Army's Armoured Infantry Battalions.

Manufacturer
Invertron Simulated Systems Limited, West Sussex.
(For details see Directory of Suppliers)

UNITED STATES OF AMERICA

CSC Tank Full-crew Interactive Simulator Trainer (T-FIST)

Computer Sciences Corporation's Tank Full-crew Interactive Simulator Trainer (T-FIST) was developed as a low-cost, interactive video disk-based training system for US Marine Corps M60A1 tank crews. It has been in use with US Marine Corps Reserve units since 1989 and was deployed on Operation Desert Shield/Desert Storm when it was used on board amphibious assault ships. The system is used to train an entire tank crew of commander, loader, gunner and driver in the actual vehicle they will use in combat. The system ensures that all crew members use and operate their own controls but the tank does not move, the main engine is not activated and no rounds are fired. T-FIST is available for M60A1, M60A3 and M1A1 Abrams tanks but can be modified for any armoured fighting vehicle. The system takes between 45 and 60 minutes to install and requires no special tools or computer skills.
Components. T-FIST has three basic components: an instructor/operator station for simulator control and the collection/display of crew performance data; optical assemblies to direct the video imagery into the tank gun sights and vision blocks; and a sensor subsystem that allows the crew to control the tank during simulation.
Training Stations. The commander can operate the rangefinder and other controls to acquire targets, determine range and fire the main gun and coaxial machine gun. The gunner can employ the primary sight, telescopic sight and gunner controls, select ammunition type and fire the main gun and coaxial machine gun. The driver can use his normal driver controls, tachometer, and speedometer to control the tank's simulated movement. The loader can select ammunition and simulate loading the main gun and set the safe/fire switches for the main gun and coaxial machine gun.
Visual Systems. The system is available with visual databases which use either CGI or high resolution video disk. The video disk-based system uses filmed terrain retrieved from a video disk. The CGI system employs phototexture to produce high quality exercise scenes. Both systems provide a wide range of realistic ground and air targets and weapons effects both in the visual and IR spectrum.

Tank Full-crew Interactive Simulator Trainer (T-FIST) installed on a US Marine Corps M1A1 Abrams tank

Training Exercises. T-FIST training exercises include a wide variety of gunnery tasks from appropriate tank combat tables, such as: engaging several types of stationary or moving targets; engaging single or multiple targets; employing the stabilised or unstabilised gun control modes; firing on the move or halted; using precision, battlesight or degraded gunnery techniques; moving across country and into and out of turret-down and hull-down positions; and training to meet performance standards under specified combat conditions.

Status
In service with the US Marine Corps reserves.

Manufacturer
Computer Sciences Corporation, Newport News.
(For details see Directory of Suppliers)

Martin Marietta Conduct of Fire Trainer (COFT)

The US DoD Project Manager for Training Devices (PM TRADE) in Orlando, Florida, asked for tenders for the development of a Conduct of Fire Trainer (COFT) on behalf of the US Army Material Development and Readiness Command. In September 1982 General Electric (GE) Simulation and Control Systems Division (Daytona Beach, Florida) was awarded the initial increment of a $610 million contract for COFT. The initial increment was for the production and start-up costs for four M1 Abrams COFTs and was worth $40.6 million. In 1993, GE sold their Simulation and Control Systems Division at Daytona Beach to Martin Marietta and the organisation now operates under the title of Martin Marietta Simulation and Automated Systems Division; for details see the Directory of Suppliers.
Deliveries from Daytona Facility. The first GE COFT was delivered in 1984. Since then more than 250 GE COFTs have been delivered to the US Army for battalion training. The COFTs are configured for the M1A1 and M60A3 MBTs and for M2/M3 Bradley Fighting Vehicles. GE COFTs have also been delivered to Egypt and Saudi Arabia.
Production in Israel. During 1987, as the result of competitive tendering bids, production of follow-on COFTs switched to Elbit Computers Limited of Israel. They produced M-COFTs (see next page) for the M60A3 MBT, with production for the US Army running until January 1991.
Components. The COFT is contained in a mobile shelter with training stations for the vehicle commander and the gunner, an instructor's/operator's station and console, a central computer which generates full-colour action scenes, a briefing/debriefing area, and some maintenance facilities.
Training Station. The training station is a replica of the vehicle interior involved and contains all the same equipment. Scenes are generated from a digital database covering a 7000 × 10 000 m range or area of operations. Within this area the training vehicle may move with perspective maintained all the time. Degraded visibility can be introduced along with special effects such as those involving thermal imaging sights. These images are produced on video screens and updated at 30 Hz. Sounds are provided such as track clatter, drive train whine, gun firing and breech-block noises and the clatter of spent cartridges. Tracers are introduced into the field-of-view along with gun flash.
Instructor Station. A series of over 200 preprogrammed exercises can be fed into each COFT, each with as many as four different versions including terrain and varying degrees of difficulty. Exercises are measured in three areas: target acquisition; systems management (selection of ammunition, proper weapon and so on); and accuracy of aim. An intercom allows the

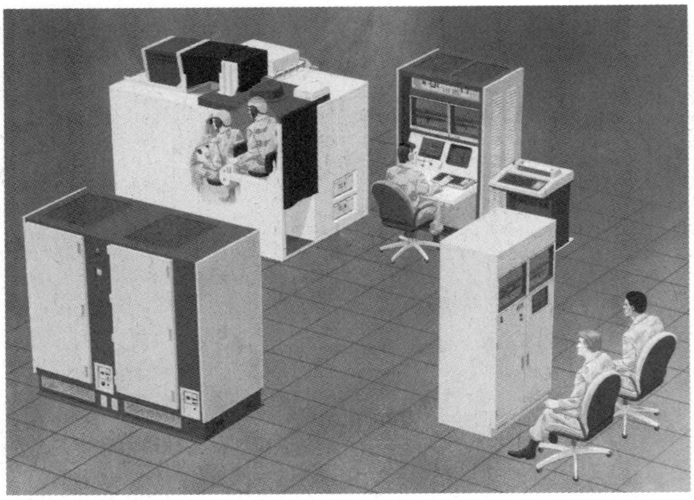

Drawing of a C-COFT2000 Crew Conduct of Fire Trainer in use

The instructor station for the Platoon Gunnery Trainer for the AMX-13SM1 in service in Singapore

instructor to communicate with the trainee crew, and accuracy and response times can be displayed and recorded.

Weapons Simulated. Weapons simulated include the 120 mm gun for the M1A1, the 105 mm main gun for the M1 and M60, 25 mm Chain Gun for the M2/M3, 7.62 mm machine gun, 0.50/12.7 mm machine gun, TOW and smoke grenades.

Basic Configurations - U-COFT and I-COFT. The basic COFT is the Unit Conduct of Fire Trainer (U-COFT). Variants of U-COFT include the Institutional Conduct of Fire Trainer (I-COFT), which is composed of four U-COFT modules, a master control function and an associated training package. The I-COFT is used for group instruction of entry level trainees in basic gunnery skills in a training school or centre environment.

M-COFT. The Mobile Conduct of Fire Trainer (M-COFT), normally consists of two shelters, each containing one U-COFT module, carried on a flatbed trailer. These trainers are used for training National Guard and Reserve units.

Compu-Scene PT2000 CGI System. In early 1991 a new family of armour training systems based on the COFT were announced, featuring the Compu–Scene PT2000 CGI system. The new family includes a Crew Conduct of Fire Trainer (C-COFT2000), a Platoon Conduct of Fire Trainer (P-COFT2000) and the TDT2000 Tank Driver Trainer (see entry in *Tank and other Driving Simulators* section).

C-COFT2000. The C-COFT2000 embodies all the functions of the original COFT plus the extra realism provided by the Compu-Scene PT2000 visual system. The C-COFT2000 can be configured as a fixed base shelter-contained Unit Trainer for battalion level training, as a Mobile Unit Trainer for Reserve units, or as an Institutional Trainer for classroom applications.

Version for Abrams MBT. A version of the P-COFT2000 for the M1A1 Abrams MBT was designed, built and certified ready for training by the US Army in less than 12 months from the time of contract award. This trainer is in service and was used for troops deployed for 'Desert Storm'. The P-COFT2000 applies lessons learned during the networking of standard US Army COFTs to create interactive Platoon Gunnery Trainers for the M1A1 tank and M2 Bradley. In addition to its role as an interactive platoon trainer, the P-COFT2000 can operate as an individual crew trainer.

Version for the AMX-13SM1 MBT. GE worked closely with the Singapore Armour Simulation Centre to develop a Platoon Gunnery Trainer for the AMX-13SM1 tank. This trainer was delivered to Singapore in July 1989.

Status

In production. In service with the US Army, Egypt and Saudi Arabia. A Platoon Gunnery Trainer for the AMX-13SM1 tank is in service in Singapore.

Manufacturers

Martin Marietta, Simulation and Automated Systems Division, Daytona Beach.

(For details see Directory of Suppliers)

ARTILLERY AND MORTAR FIRE-CONTROL TRAINERS

FRANCE

Thomson Training Fire Team Support Vehicle Simulator

Thomson Training produces a Fire Team Support Vehicle Simulator for the training of crews of Artillery Observation Vehicles (AOV). The simulator trains in target acquisition and firing, typically at enemy tanks, troop carriers or self-propelled guns.

AOV Crew Composition. The AOV crew of five consists of the driver together with an observation/liaison team made up of an observer, laser rangefinder operator, a ground navigator and radio operator.

AOV Mission. The AOV stays close to its front line units but at the same time keeping out of direct sight of the enemy. Before each move to the next Observation Point (OP), the crew has to select which is the best position for an OP and get there without being seen. Once at the selected observation point the AOV crew has to identify the enemy and his position and collect the necessary range data, including the accurate position of the AOV and the enemy, enemy movements and the type of fire required. This data is then relayed back to the fire direction centre, computations are conducted and firing initiated, followed by any necessary adjustment of fire by the AOV crew.

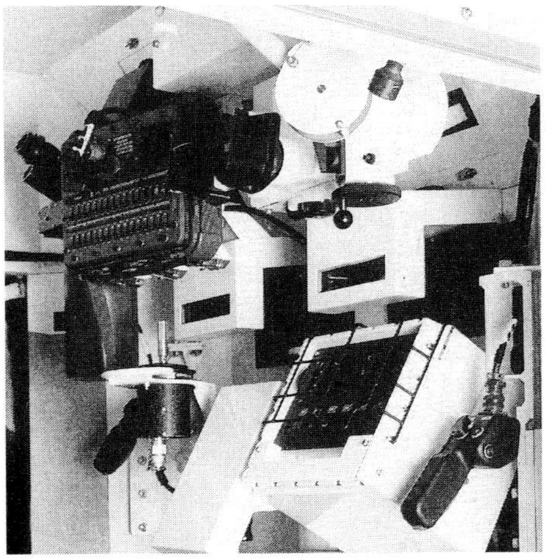

Interior of Thomson Fire Team Support Vehicle Simulator

External view of Thomson Fire Team Support Vehicle Simulator and the instructor station

Simulator Facilities. The Fire Team Support Vehicle Simulator provides a realistic mechanical, tactile, visual and aural simulated environment. The simulator allows the AOV team to move about in a gaming area of 75 km^2, with 70 prerecorded OPs. From each OP the crew has to deal with a new tactical situation where it has to define its new position and then that of the enemy forces which can be observed through the periscopes. It then identifies the enemy troops and vehicles, calls for fire support, initiates it and adjusts the firing.

Status

In service: 3 units delivered in 1991 to the French Army at Droguignan.

Manufacturer

Thomson Training and Simulation, Cergy-Pontoise.
(For details see Directory of Suppliers)

GERMANY

AS Artillery Trainer

The Dynamit Nobel AS Artillery Trainer consists of a small gun firing 14.5 mm rounds which can be used in a scaled-down environment to model artillery fire. To use the system fully a range 1350 m long and 800 to 1000 m wide is required, and the maximum range of the gun is 1150 m. Model scenarios including buildings and villages on a scale of 1:10 are recommended.

Tripod Mounting. The AS trainer is mounted on a tripod together with the dial sight system of the gun involved. The head of the tripod can be levelled and the trainer barrel then simulates the laying of the real gun.

Mounting in the Artillery Piece. If the full artillery piece and crew are to be involved, a special sub-calibre version of the trainer device can be inserted in the barrel or breech. Adaptors are available to suit virtually any type of known gun or howitzer.

Training Ammunition. Both types of trainer fire special 14.5 mm ammunition fitted with either impact or delay fuzes - the delay may be 3 or 6 seconds. The projectiles are lead-jacketed and contain a smoke charge which is ejected at the point of impact or after the required delay. The delay option can be used to simulate high bursting points or proximity-fused ammunition.

US Army Nomenclature. In US Army service this device is known as the 14.5 mm Field Artillery M-31 Indirect Fire Trainer.

Status

In production. In service with more than 30 countries including France, Germany, the United Kingdom and the USA.

Manufacturer

Dynamit Nobel AG, Troisdorf.
(For details see Directory of Suppliers)

AS 14.5 mm artillery trainer

ISRAEL

Rafael SIMUGUN-AR Artillery Fire-Control Simulators

The Rafael SIMUGUN-AR artillery fire-control simulators were developed to train artillery officers at all levels, from the technical level for the fire-control observer to the tactical level for support officers and battery commanders. The simulators make use of realistic images either displayed on monitors or projected on a screen placed in front of students. The images make use of actual films taken during field activities and are combined with realistic sound effects. Digital imaging techniques are used to create dynamic scenarios with multiple targets on different and wide landscapes.

There are four models in the SIMUGUN-AR series of simulators:

AR1 - classroom simulator for technical training
AR2 - individual simulator for technical training
AR3 - classroom simulator for group training
AR4 - tactical and technical simulator.

The field-of-view for the AR1 model is 68°; for the other three models it is 23°. The AR1 model has a display resolution of 3840 × 1024 pixels; the other models have a resolution of 1280 × 1082 pixels. The display can show an unlimited number of static targets and up to six moving targets, with a range of target selections. Up to 12 explosive effects (hit or miss) can be displayed simultaneously. The explosive effects and the sounds involved are real and recorded from field situations. Various atmospheric conditions can be reproduced. There is a rerun and playback facility.

The simulators can be used with realistic mock-ups of sighting devices and radio communications equipment. Options include thermal imaging simulation and laser range measurement.

Status

Available.

Manufacturer

Rafael Electronic Systems Division, Haifa.
(For details see Directory of Suppliers)

Rafael SIMGUN-AR Artillery Trainer

SWEDEN

Comator Advanced Training System Fire-Control Simulator

The Comator Advanced Training System (CATS) Fire-Control Simulator is also known by the Swedish acronym ESIM. It is intended for the training of forward observers for artillery and mortar batteries, including airborne observers, in the observation and adjustment of fire-control techniques and in fire planning procedures. The system employs a large projected terrain picture, measuring 6 × 1.5 m, on which target and other symbols are superimposed out to a simulated range of 2500 or 5000 m—longer ranges up to 10 000 m are optional. In addition to its primary function, the trainer can also be used for basic and advanced map-reading exercises, position sighting and terrain recognition, and exercise pre-planning. Direct firing simulation using rifles or anti-tank weapons is an option. The Swedish Army has indicated an interest in connecting several simulators together on a network and future trainers will have this facility. The system can also be upgraded to become a tactical trainer.

Components. The system comprises a computer, an operator station with two keyboards and two VDUs, a ceiling- or floor-mounted projection system with two projectors, and an audio system for the simulation of burst and target sounds. The equipment is normally installed in a classroom measuring about 7 × 10 m. Installation time is about 6 hours and a 110 V 35 A or 230 V 16 A 50 Hz single phase supply is required. Up to 30 trainees can be positioned to view the projected terrain scene. Other optional equipment includes specially calibrated binoculars.

Operation. When the simulator is used for basic training the trainee forward observer gives oral fire-control commands to the system operator, who then controls the production of further bursts and target symbols. For more advanced training a simulated radio network may be used and different types of laser rangefinders can be simulated. Trainees can report through a computer terminal and all commands can be stored for evaluation and feedback. A printer is provided to produce hard copies of results.

Terrain Scenes. Five terrain scenes are provided with the trainer, along with a computer terrain database. Each database contains terrain elevation data and visibility data as seen from an actual observation point. Further terrain scenes can be produced using a standard 35 mm camera with a wide-angle lens, and the system can store up to 100 pictures—more can be stored if required.

Targets. Moving and static targets are generated at the correct scale for their range. Air, impact and mixed artillery bursts can be simulated accurately to scale with respect to the observer burst range. The trainer also has the capability to simulate night illumination flare and smoke missions, with wind drift initiated by the instructor/operator. Target symbols are semi-three-dimensional and are shown at three different angles. A typical target is 10 pixels wide at an observation distance of 1.5 km. The update rate is 10 Hz when 10 targets are displayed and the targets move without sudden jumps or jerks. Up to 1000 target symbols may be stored. Burst symbols include covering burst (completely or partially visible), air bursts from time or proximity fuzes (completely or partially visible, with or without fragment fall), smoke bursts and flare bursts. Targets that are hit will stop and/or start burning for a preset time before being removed from the scene.

Sound Effects. Sound effects include projectile flight and impact, single shot and automatic weapon firing, tracked and wheeled vehicles, and helicopters and fixed-wing aircraft. All sounds are delayed and attenuated automatically according to range. Other sounds can be added to the system as an option.

Instructor Control. The time taken to start up the system is about 5 minutes. System operation is carried out using menus and menu trees. One of the instructor's displays shows a map displaying the current state of events in the projected terrain area and the instructor can select and enlarge any area for modification or examination. A mouse is used to 'draw' unit, target or other routes. Exercises are directed using the map only with all exercise definitions and movements shown on the map. Three kinds of symbol are used in the system: target, burst and map symbols. Only the first two are used with the terrain pictures as the map symbols are used by the instructor. The system can be put into a stand-by condition from which restarting takes about 10 seconds.

Status

In production. In service with the Swedish Army (8 systems), the Royal Norwegian Coast Artillery (2 systems) and a Far Eastern country (2 systems).

Manufacturer

Comator Process AB, Helsingborg.
(For details see Directory of Suppliers)
Marketed outside Sweden by CelsiusTech Systems AB, Järfälla.
(For details see Directory of Suppliers)

Overall view of a Comator Advanced Training System Fire-Control Simulator classroom showing the operator/instructor station

Saab BT 33 and BT 33ALM Artillery Fire-Control Simulators

BT 33

The BT 33 Artillery Fire-Control Simulator is an indoor simulator for training artillery and mortar fire controllers and consists of terrain projectors, moving target projectors, target projector, burst projectors and the main control unit.

Visual System. Terrain projectors are standard 70 × 70 mm colour slide units. One shows a view of a terrain area as seen by a fire controller from an observation post and the other simulates the effect of flare shell over different parts of the same terrain at night. Moving targets including tanks, landing craft and helicopters, can be projected onto the terrain view.

Burst projectors are used to represent ground burst, air burst, shrapnel impact, rising burst cloud, and concealed burst.

Sound Effects. The sound of the projectile during its trajectory is heard for a few seconds. The burst explosion is then heard, delayed in accordance with the observation range. Both the time and the intensity of the explosion can be varied.

Instructor Control. An instructor operates the main control unit which includes an analogue computer for calculating simulated projectile trajectories, a memory for the target terrain, and control units for the burst projectors and the sound system.

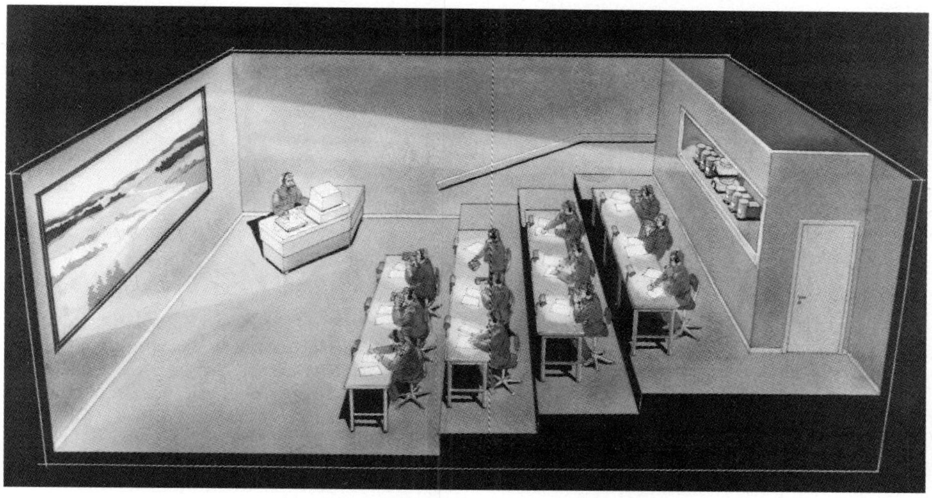

Artist's impression of Saab BT 33ALM artillery fire-control simulator in use

Summary of BT 33 Specifications

Target area
Screen: 6 × 1.5 m at 17 m between projector and screen
Battlefield scene: sector with 72° top rake angle
Target area: minimum 200 m, max 2500 m, topographic variations of ±100 m

Central unit and burst projectors
Firing units: 1 battalion of 3 batteries of 4 guns each
Rate of fire: max 12 shots/12 s

Central control panels (for each of the batteries A, B and C)
Distance: distance from OP graduated from 200 to 2500 m
Width: angle from azimuth line of target area to target, graduated from −700 to +700 mils
Gun selection: firing from 1 gun, 1 section, wing guns or all battery

Control panel (common to all batteries)
Angle of impact: 10, 20, 30 and 40°
Target angle: −800, −400, −200, −100, and 0 mils, +100, +200, +400 and +800 mils

BT 33ALM

The Saab BT 33ALM artillery fire-control simulator, also referred to as an artillery procedure trainer, was developed from the basic BT 33 simulators. The BT 33ALM is a modular system designed for use either as a stand-alone simulator or as an upgrading kit for an existing BT 33 installation. The BT 33ALM can be installed in existing BT 33 simulator classrooms or in ordinary rooms having mains electrical power and dimensions of at least 12 × 7 × 3.5 m, accommodating up to 25 students.

BT 33A Main Module. The BT 33A is the main module in the new system and is used to simulate indirect artillery fire missions.

BT 33L Laser Rangefinder Simulation. The BT 33L is a Laser RangeFinder (LRF) simulator designed to simulate a hand-held LRF that can be moved around to any part of a classroom. Up to five LRFs can be connected to the BT 33L simultaneously. The BT 33L is compatible with BT 33A modules and with upgraded BT 33 installations.

BT 33M Scenario Presentation System. The BT 33M is a presentation system for target scenarios that can be connected either to an existing BT 33 installation or to a new BT 33A module.

Instructor Aspects. Preset exercise plans are used which take the form of descriptive scenarios. They are produced by an instructor or by a centralised training headquarters. Communications during an exercise use both real hardware and digital message devices. The BT 33ALM can store and subsequently present the results of an exercise. Software is written in the Ada language.

Summary of Tactical Data
Firing units: 6 × 10 guns
Ammunition: optional; max 10 different types of ballistics
Range measurement: 5 hand-held or tripod-mounted LRF simulators
Field of observation: 70° angle up to 2500 m; expandable to 5000 m

Status

BT 33 in service with over 10 countries including Canada, Denmark, Ireland and Sweden.

BT 33ALM available.

Manufacturer

Saab Training Systems AB, Huskvarna.
(For details see Directory of Suppliers)

SWITZERLAND

SINTRO Artillery Simulators

SINTRO Electronics AG manufactures artillery simulators which are designed for stand-alone crew training, for training in subsystems (loader training, gunner aiming training, command training) and for training with a fire direction centre and with a forward observer. Systems for Field Howitzers and Mortars are included.

Training Station. Original turrets and weapon systems may be utilised and modified by SINTRO. Alternatively, it is possible to use replica equipment. If required, the turret can be delivered with cut-out openings to allow the instructor to directly supervise the crew's activities. Training is provided for gunners, assistant gunners and crew commanders in gun placing, laying, boresighting, pre-firing checks, loading and firing in both direct and indirect fire modes. Realistic noise simulation in the turret is provided, and a communication system allows crew-to-crew and instructor-to-crew communication.

Visual System. A 3-D visual system is fitted, which presents the turret commander and gunners with a realistic high resolution image of the target area and of targets themselves.

Drill Ammunition. Drill ammunition consists of primers, fuzes, rounds and charge, and models real rounds in size, weight, centre of gravity, handling and

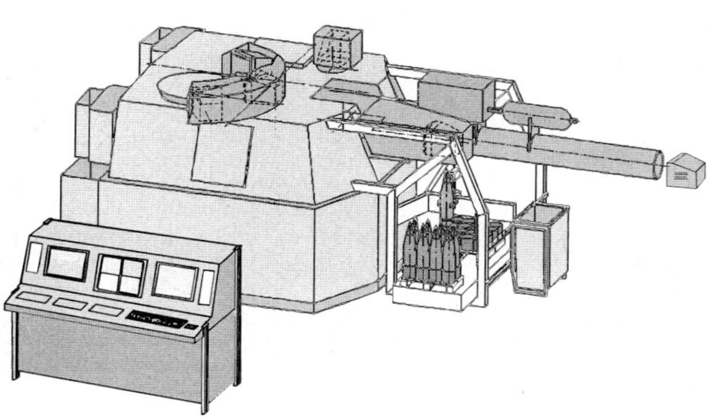

Diagram of SINTRO artillery simulator. Note palletised drill rounds, and visual system on top of turret

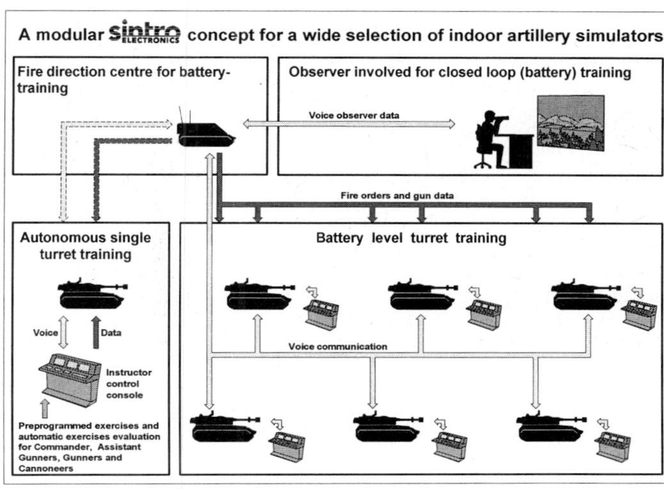

SINTRO modular artillery training concept

appearance. An ammunition identification system is used together with an automatic loading system (if applicable) and a barrel recoil system with forward ejection of the drill ammunition.

Instructor Aspects. The instructor controls the activities of the turret crew from the control console, located outside the turret. The instructor is able to monitor the individual actions of each of the turret crew, to check the laying of the main armament, and to view the target selected by the gunner on a monitor. It is also possible to install video cameras into the turret to record crew activity.

Status

Available.

Manufacturer

SINTRO Electronics AG, Interlaken.
(For details see Directory of Suppliers)

UNITED KINGDOM

Invertron Artillery and Mortar Fire-Control Training Simulator (AMFCTS)

The Artillery and Mortar Fire-Control Training Simulator (AMFCTS) provides artillery, infantry and army aviation personnel with initial, continuation and refresher training in the observation and adjustment of fire, and training for fire planning by Forward Observation Officers (FOO) and Battery Commanders (BC). The simulator may also be used for basic and advanced map-reading exercises, position siting and basic terrain recognition, and exercise pre-planning. The first unit for the UK MoD was installed at the Royal School of Artillery (RSA) in July 1979, other units are listed later under 'Status'.

Visual System. The equipment provides students with a large projected scene of a target area. Shell bursts, smoke-screens, enemy vehicles, and enemy fire effects are shown on the screen. The effect of wind on smoke development and flare illumination is represented, and the equipment can handle single and multi-flare night illumination patterns. Target area scenes including image-intensifier, radar and thermal-imager simulations can also be projected. The system can be fitted into existing classrooms. A screen 6 m long by 1.5 m high is adequate in a 12 m long room containing 30 students, and the same training can be achieved in a room 3 to 4 m shorter if necessary; the trainer has been set up in an area only 4 m^2, with provision for six students.

Communications. Communication among students, instructors and the operator is by three simulated tactical radio networks and an intercom system.

Optional Extras. Optional extras include: laser ranging and target markings; image-intensifier; radar and thermal-imager scenes; wind effect on illuminating shells; and the ability to reduce minimum range to 40 m.

Instructor Aspects. The instructor can control up to 80 different targets, including single-shot small arms fire, machine gunfire, anti-tank gunfire, troop concentrations, armoured vehicles, trucks and assault craft. The selection of a target automatically presents it on the screen at its preselected starting point and with the appropriate sound effects at the correct level for that range. The instructor can 'drive' moving targets such as tanks or landing craft by means of a joystick, and he can occult the target where it would be obscured by buildings or trees. The system can also simulate the development of a smoke-screen, which automatically expands slowly and drifts across the target area in accordance with the simulated wind speed and direction, according to instructor settings. All types of shell can be simulated, including ground bursts, smoke plumes, air bursts and air bursts with splinters. The size of the burst varies with range, and the distance between air bursts and splinters hitting the ground is a function of the fuzing set on the keyboard by the instructor following the input of commands from the observation point. Night scenes of the target area are included and the trainer will automatically handle single or multi-flare patterns, the illumination remaining available to students for a realistic period. It also drifts in accordance with whatever wind speed and direction the instructor has chosen.

The instructor/operator can set up to four batteries, each with a maximum of eight guns, in any desired battery layout configuration. Once set up, a particular exercise can be stored on a disk and replayed later. The four batteries can be fired simultaneously with differing rates of fire, which enables realistic fire planning exercises to be carried out.

Three tactical communications networks and a private telephone line are provided between the instructor and each of the student observation post positions. Static interference and fading can be provided as an option to increase realism.

The equipment is supplied with eight target areas representing observation posts of the customer's choice. Other targets may be added by the user. A programme is provided to prepare the database for any desired target area. The system can handle up to eight target areas if necessary, each with a range beyond 5000 m.

Sound Effects. The sound simulation includes the aerodynamic noise of shells which pass over the observation post, together with shell bursts and target sounds.

CPOP - Command Post Trainer. An optional extra to the basic trainer which converts it into a Command Post/Observation Post (CPOP) trainer. Orders from the student observer, instead of being sent direct to the computer operator, are sent to a Command Post which can be in another room or even in a vehicle parked nearby. The Command Post staff will have an input of meteorological data and the ballistic data for each gun in a battery and, using this and the orders from the observer, produce gun data using their normal method (computer, graphs and firing tables). This gun data, instead of being sent to the guns, is sent to the trainer's computer operator. He will have entered the same meteorological and ballistic data as the Command Post and can therefore enter their gun data to produce the appropriate burst symbols. If the Command Post staff has made an error, the burst symbols will appear in the wrong location, as in real life. Using the CPOP trainer, therefore, not only are the observers of a battery trained, but also the Command Post staff and the signallers. To convert the basic trainer into a CPOP trainer requires no additional hardware but a considerable amount of extra software. A number of users purchased the option, including the US Army.

CLASS - Artillery Simulation System. A further extension to the CPOP trainer has been introduced by the development of the Closed-Loop Artillery Simulation System (CLASS). Using this system, FOOs, battery command post crews and gun crews can all be trained at the same time. All the normal artillery procedures are used by the FOO to transmit target data to the Command Post where gun data is calculated. Data passed to the guns and then 'fired' by those guns, is transmitted to the computer controlling the AMFCTS. This trainer displays the effects of the rounds 'as fired' by each of the guns in the battery. Should any gun fire on incorrect data, the error will be seen.

CLASS involves a Gun Unit which is a fully ruggedised weather-proof module containing a two-axis gyro system and an inclinometer. The unit is mounted on top of a gun barrel using a mounting adaptor and quick release toggle clamps. Installation and removal do not require any modification to the gun. The purpose of the two-axis gyro is to supply the bearing and elevation of the gun barrel, while the inclinometer provides a reference to compensate for the effects of trunnion tilt. A 'FIRE' signal is provided by a small switch unit mounted on the gun. All the sensors are connected to the gun data recorder, a unit containing a microprocessor-based printed circuit board which converts signals sent from the gun unit into bearing and elevation angles. Each gun data recorder and gyro system is driven from a central interface control unit sited beside a Command Post. This unit transmits CLASS data over a VHF link to the AMFCTS which can be up to 8 km away. The data is then error-checked, processed and displayed on the instructor's monitor. The data is used to calculate the fall-of-shot which is then displayed on the projection screen at the correct time, according to the projectile time of flight.

Use with Laser Designator Simulators. The AMFCTS can be used in conjunction with the Invertron Ground/Vehicular Laser Locator Designator (G/VLLD) simulator. This can be used to provide operator designation and tracking training against stationary and moving targets for laser-guided weapons such as Copperhead. The G/VLLD can also be used for

The Invertron AMFCTS installed at the US Army Field Artillery School at Fort Sill, Oklahoma

Simulation of a six-gun 155 mm battery on an AMFCTS screen

rangefinding and position finding. In July 1987 the US Army ordered 80 G/VLLDs.

Status

In service with 25 countries including the United Kingdom (21 systems), Canada (8 systems), Egypt (5 systems), France (20 systems with 11 delivered in 1983 and the remaining 9 in 1984), Greece (2), Jordan (2), Singapore (2), South Korea (2), Portugal (2), Oman (2), Thailand (2), USA (87 systems) and other countries in Europe and the Middle and Far East. A total of 170 trainers have been delivered.

Manufacturer

Invertron Simulated Systems Limited, West Sussex.
(For details see Directory of Suppliers)

Invertron Deployable Forward Observer Trainer

The Invertron Deployable Forward Observer Trainer (DFO) incorporates the Invertron Advanced Resolution Gunnery Simulator (ARGS) visual system also being produced for a tank gunnery training system (see entry under *Tank and Direct Fire Gun Trainers).*

Visual System. Terrain views are derived from photographs taken of the customer's training area. They are then digitised and projected on a screen by high resolution television projectors. The screen area may be from 1.5 × 1.5 m to 6 × 1.5 m, depending on the number of projectors and the number of students to be trained. Up to 32 targets may be displayed at any one time, of which 16 may be moving and 16 static. Target images are presented in some detail. Fall-of-shot effects include high explosive, proximity and mechanical time fuse, improved conventional munitions, smoke and illuminating.

Fire Unit Training. Four batteries, each containing up to eight guns, can be deployed. Each battery can be split into two sections, giving a total of eight fire units. Fire units can comprise different weapon types, including 105 and 155 mm howitzers and 120 mm mortars.

Sound Effects. A sound system provides effects of vehicles, fall-of-shot and general battlefield background noises.

Record/Replay. Recording and subsequent replay of entire missions provides evaluations of student performance.

Laser Designators. Laser rangefinders and laser locators may be incorporated in the training.

Status

In production. Ordered by the Austrian Army (4) and the Royal Netherlands Army (2).

Manufacturer

Invertron Simulated Systems Limited, West Sussex.
(For details see Directory of Suppliers)

Marconi Simulation M-ARTY Mortar and Artillery Observer Trainer

The Marconi M-ARTY is a low-cost personal computer-based classroom trainer for the training of soldiers in directing mortar or artillery fire. All aspects of training can be covered and the system can also be used for training in tactics, the production of fire plans, for tank recognition and for the practice of laser rangefinding. It is designed to deal with up to four batteries of eight guns or nine mortars, each of which can be controlled independently, thus permitting four simultaneous engagements.

Visual System. A screen is used to display high quality colour pictures of target terrains to a class of 12 to 16 students.

Instructor Aspects. The instructor or simulator operator has a VDU and keyboard. He has control over all the parameters affected by the initial orders and corrections that an artillery or mortar forward observer would have to make. He also controls the display of moving targets, enemy fire and battlefield noises. The system can be controlled by an operator with a limited knowledge of gunnery, leaving a role-specialised instructor free to concentrate on training rather than running the simulator.

M-ARTY mortar and artillery observer trainer display

Status

Available. In January 1992 it was announced that the British Army had ordered 83 M-ARTY trainers.

Manufacturer

Marconi Simulation, Dunfermline.
(For details see Directory of Suppliers)

Phoenix Simulation Artilleryman Trainer

The Phoenix Artilleryman Trainer meets the training requirements of an entire artillery regiment. It also provides simultaneous training for forward observers, fire direction teams, battery commanders and gun crews. It is designed as a modular mobile system that can be installed and operated from a standard army vehicle. The modular design allows users to reconfigure the system to suit differing training requirements as they arise, including part-task training when required. Miscellaneous equipment for the system includes modified binoculars and map boards. Options include a high resolution colour system, a simulated laser rangefinder, customer terrain views, documentation in various languages as required, and additional training.

Visual System. The visual system is based on the use of high brightness video projection using video disks, coupled with CGI. Gun and sight deflection data use laser techniques and produce an accuracy of better than 1 mrad. Gun elevation determination is carried out using an inclinometer capable of measuring angles to an accuracy of better than 0.5 mrad. The laser module used in the sight unit is a low power helium/neon device and is eye-safe.

Montage of Phoenix Artilleryman Trainer

Instructor Facilities. All the facilities of the Artilleryman Trainer are controlled by a single instructor/operator using a keyboard and visual display unit. The system operates on a series of dedicated 'pages' and preprogrammed function keys, so previous experience of computers or computer-based equipment is not necessary. Functions available through the keyboard include the following: selectable exercise parameters such as the number of batteries (from one to four), the number of guns in each battery (from one to eight), the battery layout, wind and temperature conditions, dispersion factors, offsets and weapon types (guns or mortars—each of the four batteries may have a different calibre if required), ammunition types, up to 10 simultaneous moving targets, 20 individual missions stored on disk, time sequenced and with an on-line modification facility, laser rangefinder/designator options, miscellaneous facilities such as mission record/replay, storage of reference points, terrain database, airborne observer procedures and command post training.

Battery Training. There is also a battery training system for up to eight guns with direct measurement of actual gun deflection, actual gun quadrant elevation, sight direction and cross-level settings, plus direct sensing of the firing mechanisms. The battery involved can be deployed in its normal terrain positions with a possible live firing facility. Gun equipment is ruggedised and weather-proofed. The system ensures the correct display of fall-of-shot on the projected terrain, and any incorrect sight settings or firing data calculations will result in a corresponding incorrect fall-of-shot to be observed on the projection screen.

Computing. The computing power is distributed to the various points on the system where each of the individual tasks is being carried out. Each individual processor has the ability not only to carry out its own task, but also to communicate via a networking system with all the other processors, effectively producing a single large processor with capacity for expansion. The computing system provides an individual processor for each of the separate areas concerned, which are weapons, control and visuals. All computers are IBM-based.

Power Requirements. Power for the system can be a 24 V DC source or a 110/240 V AC supply.

Status

Available. In service with the Royal School of Artillery at Larkhill, Wiltshire.

Manufacturer

Phoenix Simulation, West Sussex.
(For details see Directory of Suppliers)

Westland Indirect Fire Trainer (WIFT)

The Westland Indirect Fire Trainer (WIFT) training system is for low-cost indirect fire training at unit level. It is designed for use by infantry, artillery and armoured units as well as airborne units responsible for directing field artillery.

Components. The system comprises a slide and video projector, a microcomputer, special binoculars, a laser rangefinder, and a screen. Full operator documentation, target areas and maps are supplied with each system.

Visual System. The ceiling-mounted video projector is used to superimpose computer-generated imagery for targets and shell bursts on slide-projected target areas. The screen is a standard video projection screen which may be folded away into its ceiling or wall housing. Binoculars used are specially calibrated 6 × 30 field binoculars and are employed for adjustment of simulated fall-of-shot.

Training Exercises. Engagements are modelled for high explosive, illuminating, proximity and delay fuze fire missions. The student can adjust fire along his observer target line or any direction he chooses. Battery commander fire planning can be exercised, including the deployment of smoke missions. Variable distribution, rates of fire and rounds fired are all simulated in accordance with the observer's orders. Training varies from basic map reading and adjustment procedures to multiple target missions using grid or polar data.

A British Army Westland Indirect Fire Trainer (WIFT) system

Student Self-Assessment. Although WIFT was designed primarily for supervised continuation training, the system can be operated by an unsupervised student to learn and practise individual skills and procedures without specialist support.

Instructor Aspects. WIFT is simple to set up, operates in semi-light conditions and does not require the use of a specialised or dedicated building. Following a simple procedure to calibrate the grid projection of the terrain area, the WIFT instructor/operator initialises the system for battery/mortar line locations, enemy target positions, and any ballistic errors due to meteorological conditions or incorrect settings. Operators require no previous knowledge of computers, but a short course is provided during system installation. The instructor/operator can vary training during the session through the selected use and rate of enemy targets being made active. He can also introduce random errors in the fall-of-shot positions. Dead ground conditions are simulated. At any time the session may be 'frozen' to allow instructional points to be made. A 'check fire' facility is also available. The system is time based so that preplanned fire missions may be exercised using recorded targets, of which up to 100 can be stored in the system. Direct fire training may also be exercised with the Westland Direct Fire Trainer software package, enabling rural and urban environments to be simulated.

Sound Effects. Sound and visual effects are modelled for different ammunition types, and the sounds of other gunfire, enemy fire and shells passing overhead are also simulated.

Installation and AR&M. WIFT is supplied as an installed turnkey system based upon commercially available hardware. Good AR&M is claimed for in-service systems.

Status

In service with the United Kingdom (7 systems), Canada (5 systems), Ireland (1 system) and New Zealand (5 systems).

Manufacturer

Westland System Assessment Limited, Yeovil.
(for details see Directory of Suppliers)

UNITED STATES OF AMERICA

Daedalean GUARDFIST II

The Daedalean GUARDFIST II (Guard Unit ARmory Device Full crew Interactive Simulation Trainer) is a transportable trainer for field artillery forward observers. The trainer superimposes computer-generated targets and munitions effects on real world background scene photographs stored on video disk. Any real TACFIRE-compatible digital device may also be used with the trainer. Artillery fire options include ICM, HE/Quick, HE/Time, HC Smoke and illumination rounds.

Components. The trainer consists of two stations: an Instructor/Operator Station (IOS) and a Training Station for a Forward Observer. The IOS consists of a personal computer with text monitor and keyboard, a video disk player, an audio mixer and amplifier, a stereo headset, and a full-colour scenario display, all housed in a computer console desk.

Visual System. Twenty colour background scenes are provided, representing the variety of terrain a forward observer may encounter around the world, including desert, farmland, rolling hills and mountains. A 1:50 000 military map is provided for each scene. All scenes may be displayed in day, night and illuminated conditions. Target and munition images are occulted, partially or completely, by terrain features. Moving targets appear and disappear behind trees and buildings; rounds will be 'lost' behind hills. The background scene library is stored on removable optical media to permit a virtually unlimited number of scenes to be added to the trainer.

Training Station. The forward observer station consists of a colour scenario display and stereo headset, a digital message device simulator that replicates the AN/PSG-2, and a trackball to simulate the use of the M19 binoculars and M2 compass. The forward observer may call for fire either verbally using the audio headset microphone or digitally using the AN/PSG-2

A GUARDFIST II trainer in use

A typical GUARDFIST II forward observer display showing an artillery projectile impact

digital message device simulator. Time of flight, flash-to-bang time delay and probability of kill are all accurately replicated.

Instructor Station. The system supports field artillery tasks such as area fire, co-ordinated illumination, and impact and time registration. The trainer has a library of 18 ready-to-run exercises. The instructor may create and save new exercises in the library by selecting a background scene, task, target, target location and target activity. Target options include tanks, personnel carriers, trucks, helicopters, bunkers, machine gun positions and fuel dumps. The system supports both manual and automated field artillery units.

Scoring. All exercises are scored both automatically and manually. When the forward observer completes an exercise, the trainer provides the instructor with a series of reports defining exactly where the targets were located, the actual spotting of each round fired, and the correction the forward observer gave to adjust the round. The instructor has the option of printing any of the reports while performing after-action reviews and debriefing following a session. The trainer also saves the reports in a non-line database so the instructor can quickly review a soldier's performance history and plan further training.

Status
In production for the US Army reserves and National Guard (over 400).

Manufacturers
Daedalean Inc, Maryland.
(For details see Directory of Suppliers)

In April 1993, Fidelity Technologies Corporation of Reading, Pennsylvania, was awarded a contract to produce an additional 55 GUARDFIST II simulators, with final deliveries by October 1994.

Institutional Fire Control System Trainer (IFCST) for M109A6 Paladin

The ECC Institutional Fire Control System Trainer (IFCST) for the M109A6 Paladin self-propelled howitzer is a desktop trainer.

Student's Presentation. The student has a full simulation of the Fire Control System Display Control Unit from inside the cab of a 155 mm M109A6 Paladin. A second element is an interface unit which is used for preprogrammed instructional information, battery control commands and fire direction centre data information.

Training Exercises. Operators may be trained in the tactical and procedural use of the section commander's display and control unit. It provides the artillery commander with the means to practise the engagement of multiple targets and mass fires when a fire mission demands. It also allows him to practise ammunition inventory control and land navigation exercises.

Status
In service with the US Army.

Manufacturer
ECC International Corporation, Pennsylvania.
(For details see Directory of Suppliers)

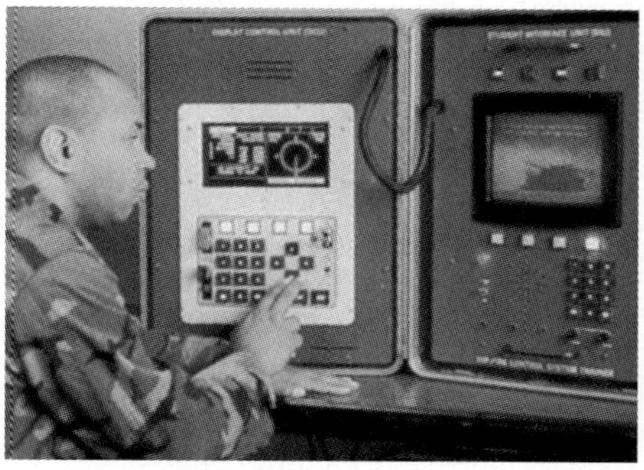

Paladin Howitzer Fire Control System Trainer

FATS Indirect Fire Training Simulator

The FATS Indirect Fire Training Simulator is a derivative of the FATS Marksmanship Trainer (see entry under *Small Arms Marksmanship Training Equipment and Systems* for details) and uses the same image projection system and computer hardware. Its software is reconfigured to produce landscapes and artillery-related images to enable forward observers to note the fall-of-shot from artillery and mortar units and make the necessary fire corrections. The system can be expanded to include a fire direction centre and a gun section with all three elements performing their normal combat functions.

Training Undertaken. The forward observer can identify targets by grid co-ordinates, polar plot or shift from a known point and send a call for fire to the fire direction centre. The fire direction centre can then interpret the call for fire and send the necessary data for the guns to set. Whatever data the guns set will then be displayed as actual shots for the forward observer to interpret.

System Equipment. The system makes use of special binoculars, maps and simulated communications equipment.

Status
Available.

Manufacturer
Firearms Training Systems Inc, Georgia.
(For details see Directory of Suppliers)

FATS Indirect Fire Training Simulator in use

YUGOSLAVIA (SERBIA AND MONTENEGRO)

Artillery Fire Simulator SAV-78

The Artillery Fire Simulator SAV-78 is a classroom simulator intended for training artillery observers and fire-control centre personnel. It is computer-controlled and enables trainees to train in target location and tracking, the issuing of fire commands, fire adjustment and tactical judgements.

Components. The SAV-78 comprises a computer unit, an instructor's console, a large screen display, a video terminal, an automatic slide projector, two slide projectors for projecting the terrain panorama, a projection screen, two 45 W loudspeaker units, up to 40 pairs of binoculars, and a set of interconnecting cables.

Classroom Environment. Up to 40 observers plus eight fire direction centre personnel can use the SAV-78 at one time. The classroom containing the equipment should be at least 16 × 8 × 3 m and equipped with desks and seating for the trainees. Some degree of blackout is required. A panoramic view of a battlefield is projected in colour onto a 6 × 1.5 m screen.

Method of Operation. Firing data produced by trainees is fed into the computer by an operator using a keyboard. The instructor checks the fire-control data given previously by the computer and makes any corrections. The simulator provides a choice of weapon and ammunition types and the presentation of different types of target on the screen. Terrain and visibility conditions are automatically taken into account by the system computer during fire and target simulation. The system computer can be used to alter exercise variables such as meteorological and ballistic data. All such exercise data is entered via the system keyboard.

Targets. Targets are generated by the system computer, using symbols representing enemy activity. Targets may be moving or fixed and graphic hit symbols produced by the computer can vary according to the type of ammunition used and other factors. Synthetic noise generation produces the sounds of shell bursts and flight, small arms and automatic fire, and vehicle noise.

Status
In service with the armed forces.

General view of an Artillery Fire Simulator SAV-78

Contractor
Holding Company YUGOIMPORT-SDPR Ltd, Belgrade.
(For details see Directory of Suppliers)

UVAG-M69 Artillery Trainer

The UVAG-M69 artillery trainer consists of a small gun which can be used in a scaled-down environment to model artillery fire, a similar principle to the Dynamit Nobel German AS artillery trainer (see entry under Germany in this section). The UVAG-M69 gun used is the M48 7.92 mm rifle firing practice rifle grenades. The trainer consists of a gun carriage mounted on a tripod and has provision for mounting panoramic sight carriers. Elevation controls are carried out using a DB-1 range quadrant as used on the 76 mm M48(B1) mountain gun.

Operational use. In use a gun crew can carry out 'dry' fire-control and laying procedures with all the actual fire-control settings being made using the sights on the UVAG-M69 trainer. The trainer fires the 40 mm M68 practice grenade to a maximum range of approximately 570 m so the trainer can be used on comparatively small ranges and with a safety area very much smaller than that for a full-size artillery weapon. As the grenade hits the target area it produces a small puff of smoke and a flash for observation and fire correction purposes.

Mortar Crew Training. The UVAG-M69 can be used for training with mortar crews. In this mode the trainer is fitted with an NSB-1 sight carrier.

Specifications

Weight in firing position: 80 kg
Barrel calibre: 7.92 mm
Range of practice grenade: approx 570 m
Muzzle velocity:
(practice grenade) approx 90 m/s
(7.92 mm round) approx 720 m/s
Max height of line of fire: approx 1.5 m
Elevation/depression: +75°/−10°
Traverse: 360°

Status

In service with the armed forces.

UVAG-M69 artillery trainer

Contractor

Holding Company YUGOIMPORT-SDPR Ltd, Belgrade.
(For details, see Directory of Suppliers)

TANK AND OTHER DRIVING SIMULATORS

FRANCE

Thomson-CSF Driving Simulators

Thomson-CSF has delivered over 160 driving simulators of various sorts, and French, German, Swiss, UK and USA AFVs have been simulated.

Germany - AFV Simulators. In association with Krauss-Maffei, the German Army was supplied with 24 driving simulators equipped with the Visa 4 computer-generated imagery visual system. Leopard 2, M109, Jaguar, and Marder vehicles were simulated. This followed an earlier contract in association with Krauss-Maffei, for 16 Leopard 1 simulators.

USA - M60 Simulators. In 1987 the American branch of Thomson Training, Burtek Inc, was awarded a $9.5 million contract by the US Army to supply five M60A3 tank driving simulators. These simulators are installed at the US Army Armor School at Fort Knox, Kentucky.

France - National Armoured Training Centre (CIABC) At Carpiagne, France formed a centre for training the drivers of all types of AFV. This is called the National Armoured Training Centre, CIABC being the initials in French. The centre has 22 simulators installed.

Leclerc Driving Simulator. The Leclerc driving simulator was delivered in 1993 as part of the Leclerc complete crew trainer (qv in Tank Crew Trainer section). It includes the SPACE CGI system (details in the CGI section), a three-periscope plus rear-view mirror vision system, a detailed exercise terrain, and a compact electric (no hydraulics) motion system developed by Thomson-CSF. A photo is included.

Truck Driving Simulators. Thomson Training has a truck driving simulator in-house demonstrator (see photo), and a contract was awarded in August 1993 by the French Army to develop and supply a driving simulator for a super-heavy truck to the Engineer Corps. The truck concerned is amphibious and enables MBTs to cross rivers; it is known as the Engin de Franchissement de l'Avant, or EFA, not to be confused with the aircraft which is now known as Eurofighter 2000.

The demonstrator has a high-resolution CGI system, images from which are displayed in real time by projection on a curved screen with a horizontal FoV (FoV(H)) of 140°. Simulation of all weather conditions by day and night is provided, as are different vehicle weights and centres of gravity including work with trailers. The design allows an instructor to manage up to eight driving simulators simultaneously, and customer options include an enhanced FoV(H) to 180° and the Thomson electric motion system as fitted to the Leclerc driving simulator (see photo).

Example of Engin de Franchissement de l'Avant (EFA) vehicle, showing its large size which leads to difficulties in real-road training and the need for a driving simulator

The EFA Truck Driving Simulator features a 3-DoF electric motion platform and a visual system based on CGI. The CGI images are displayed on a screen with 130° FoV(H) and also on two synthetic rear view mirrors. A digital sound system and user-friendly instructor station are included.

Thomson Training and Simulation - AFV driving simulator for the German Army

Status

AFV simulators - available and in service. Users include Belgium (Leopard 1, Scorpion, M113); France (including AMX-30, AMX-30B2, AMX-10P, AMX-10RC, Panhard AML, Leclerc); Germany (16 Leopard 1, 8 Leopard 2, 4 M109, 3 Jaguar, 9 Marder); Greece (AMX-30); Italy (Leopard 1); Norway (Leopard 1); Saudi Arabia (AMX-30, AMX-10P); Spain (AMX-30); Switzerland (Centurion, Pz 55, Pz 61 and Pz 68, M109, 4 Leopard 2 (Pz 87)); United Arab Emirates (AMX-30, AMX-10P); Venezuela (AMX-30); USA (5 M60A3).

Truck Driving Simulators - demonstrator available in-house. Super-heavy Truck (EFA) Driving Simulator in development for the French Army for delivery during 1995.

Manufacturer

Thomson Training and Simulation, Cergy-Pontoise.
(For details see Directory of Suppliers)

Thomson Training and Simulation - In-house demonstrator of truck driving simulator

Leclerc MBT driving simulator by Thomson Training and Simulation

GERMANY

Atlas LVTS Truck Driving Simulator

The Atlas truck driving simulator consists of a simulated cab and a projected four-channel visual system giving a 180 degree FoV(H). The cab can be changed for that of another vehicle if required. A training room of about 9 x 9 m is required to accommodate the simulator.

Driver's Station. The driver has functionally correct controls and the simulation allows for steering, gearbox, braking and torque effects. A sound system produces noise effects, and a motion system provides movement cues.

Status

Available.

Manufacturer

Atlas Elektronik, Bremen.
(For details see Directory of Suppliers)

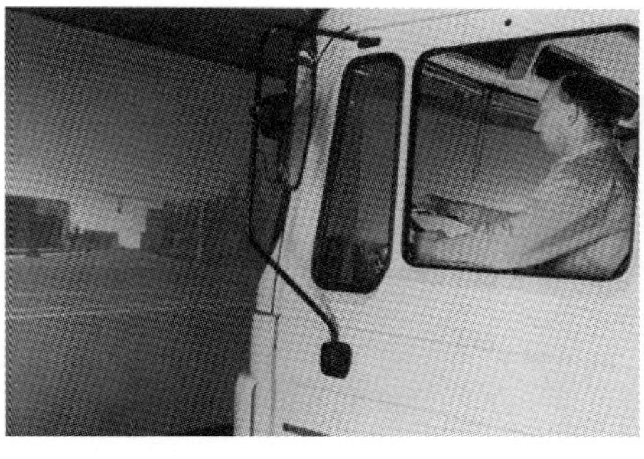

Atlas LVTS Truck Driving Simulator

INDIA

Bharat Simu-Drive B-Vehicle Driving Simulator

The Bharat Electronics Simu-Drive B-vehicle driving simulator is for training vehicle drivers with no previous driving experience and can also be used to improve experienced driver techniques.

Components. The simulator consists of a mock-up of a driver's cab with dashboard instruments, a steering wheel and column, a gearbox and foot controls. This is located in front of a curved wide-angle screen on which are projected images of driving situations. The simulator also includes an instructor's station and an engine sound system.

Visual System. The visual Display System uses projectors mounted over the trainee position. Image generation may be from CGI, video disk or a simpler shadow-based system.

Status

In production.

Manufacturer

Bharat Electronics Limited, Bangalore.
(For details see Directory of Suppliers)

Bharat Electronics driving simulator

POLAND

OBRUM Armoured Vehicle Driving Simulators

Jaguar 3 for the T-72

The Jaguar 3 driving simulator is configured for the T-72 MBT and is intended for training in basic driving skills before practising on a real vehicle.

Components. It consists of a mock-up of the driver's position. The mock-up is contained within a light metal housing on wheels to allow the simulator to be moved to wherever it is to be used. Inside the mock-up compartment all the controls are in their correct locations and closely resemble the real controls; in some instances the components are those used in the real vehicle.

Trainee's Video Monitor. In front of the driver's position is a video monitor secured to a boom extension. Lessons are displayed on it for the trainee to follow. The trainee may view the monitor either with the driver's hatch open or through the vehicle vision devices. The trainee follows the instructions provided at his own pace. If mistakes are made, the computer program will display a message explaining the error and repeat the exercise until the correct standard has been reached.

Instructor's Role. A monitor panel located externally at the rear of the simulator allows an instructor to monitor the exact position and state of all relevant controls at

Inside the Jaguar 3 driving simulator for the T-72 MBT

Jaguar 3 driving simulator for T-72 MBT - trainee viewing screen directly

Pancernik driving simulator with trainee viewing screen through optics

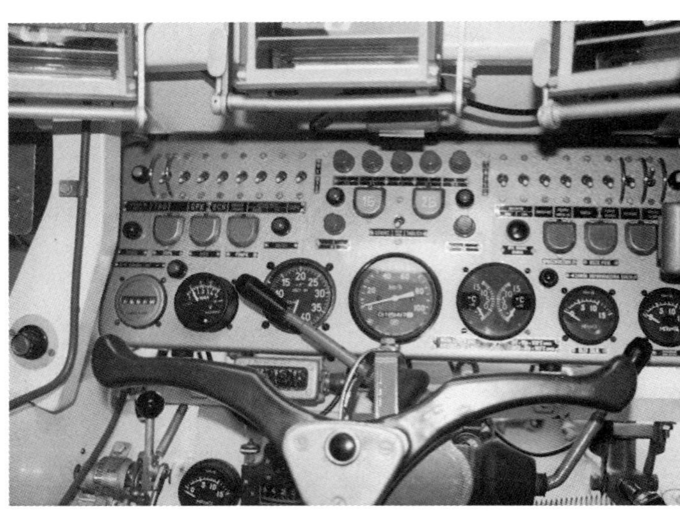

Pancernik driving simulator inside view

OBRUM Driver Training Stand for T-72 MBT

OBRUM SJ 09 Training Vehicle for T-72 MBT

any time. Typical training exercises include vehicle start-up, moving off, handling the vehicle on slopes, emergency braking, and so on.

Size, Weight, and Power Requirements. The Jaguar 3 weighs about 800 kg and is 3.145 m long, 1.5 m wide and 1.47 m high. It requires an electrical power supply of 110 to 220 V AC.

Pancernik-1 driving simulator for the BMP IFV

The OBRUM Pancernik-1 driving simulator is intended for the training of BMP IFV drivers. It is similar in design and specifications to the Jaguar 3 but is slightly lighter and smaller. It weighs about 700 kg, and is 2.98 m long, 1.43 m wide and 1.53 m high.

OBRUM SJ 03 Driver Training Stand for T-72 MBT

This static trainer is designed to familiarise drivers with the driver compartment of the T-72 MBT before progressing to other trainers or the T72 itself. It allows training in engaging and changing gears, turning, braking, fuel delivery system (control of injection pump), operation of louvres and operation of driver's and emergency hatches.

OBRUM SJ 09 Training Vehicle for T-72 MBT

This tracked vehicle trains drivers of the T-72 MBT in a cheaper environment than that of the T-72 itself. It may be obtained in full T-72 kit, or as a bare hull for the customer to fit out with T72 or other controls. It is also possible to disassemble a real T-72 turret and fit it to the SJ 09 chassis if required for training.

Status

In production. In service with the Polish armed forces.

Manufacturer

OBRUM, Gliwice.
(For details see Directory of Suppliers)

Marketing Agency

Cenzin Company Limited, Warsaw.
(For details see Directory of Suppliers)

RUSSIA

Basic Driver Trainer

This trainer has a generic driver station combined with a road pattern projected onto a screen from a disk located over the top of the driver's position. It is similar to the Serbian TOVT 55P tank driver trainer (qv).

Status

In service with the Russian armed forces.

Manufacturer

State factories.

Basic driver trainer in operation

BMP Driver Trainer

This trainer consists of a mock-up of a BMP driving compartment fitted with dials and controls taken from a real vehicle. The driver uses his periscope to view a scene projected by a cine-projector located behind the trainer and, by using the accelerator pedal, the trainee can control the rate at which the film is shown and hence the apparent speed. Instructions to turn left or right are given via headphones by an instructor sitting beside the projector. Realism is added by simulating engine noise and by moving the compartment up and down as the vehicle travels over rough terrain. The harder the accelerator is pressed the more violent the vertical movement becomes.

Status

In service with the Russian armed forces.

Manufacturer

State factories. Details requested.

BMP driver trainer in use (Henry Dodds)

Malysh T-72

The Malysh T-72 (Mite T-72) is a driver trainer for the T-72 MBT adapted from existing training components combined into a programmed sequence simulator for the basic training of tank drivers. The trainer is designed for use mainly in school environments but is portable and may be taken to field environments where some form of shelter is available.

Components. The equipment consists of an instructor's console, a T-72 driving simulator, a control panel, a small portable 'Malysh T-72' training unit, a programmer unit and the interconnecting cables and junction boxes.

Training Station. Each driver station is a standard T-72 driver simulator with the addition of starting controls and equipment and a trainee control panel powered from a 24 V DC source. This panel can be used to request further instruction and to initiate or terminate training programmes. The driver receives a task from a programmer unit, overseen by the instructor at his console. If the task is completed correctly a 'Completed Correctly' indication is given to the driver and instructor. In the early stages of training the correct sequences are indicated to the trainee by indicator lights. In the more advanced stages, any incorrect sequence will not complete the programme and the trainee can request the programmer unit to rerun the sequence to indicate the error point. The instructor may also advise the student verbally.

Instructor Station. The instructor's console has illuminated panels which are used to transmit data and instructions to the trainee driver, and indicators which show the position and state of each driver's controls. Each instructor's console can control up to three driver stations. A programmer unit has three sub-units: the programmer unit itself; an error-detection unit; and an indicator panel with control switches and lamps. The programmer unit has two basic programmes. The first is preparing the tank for start-up, starting up and switching off the engine. The other programme is for moving off, gear changing and braking. The unit provides 11 test sequences and contains 65 tasks.

Driver station of Malysh T-72 driver trainer with driver's control panel on left

Status
In service with the Russian armed forces.

Manufacturer
State factories.

T-54/T-55 Tank Driving Simulator

Russian tank driving schools make extensive use of various forms of simulators for training tank drivers, ranging from simple static stands to simulators employing film projection, engine noise, travelling vibrations, and pitch and roll simulations. Basic driving on roads and cross-country is carried out using light tracked and other vehicles.

Status
In service with the Russian and CIS armed forces.

Manufacturer
State factories.

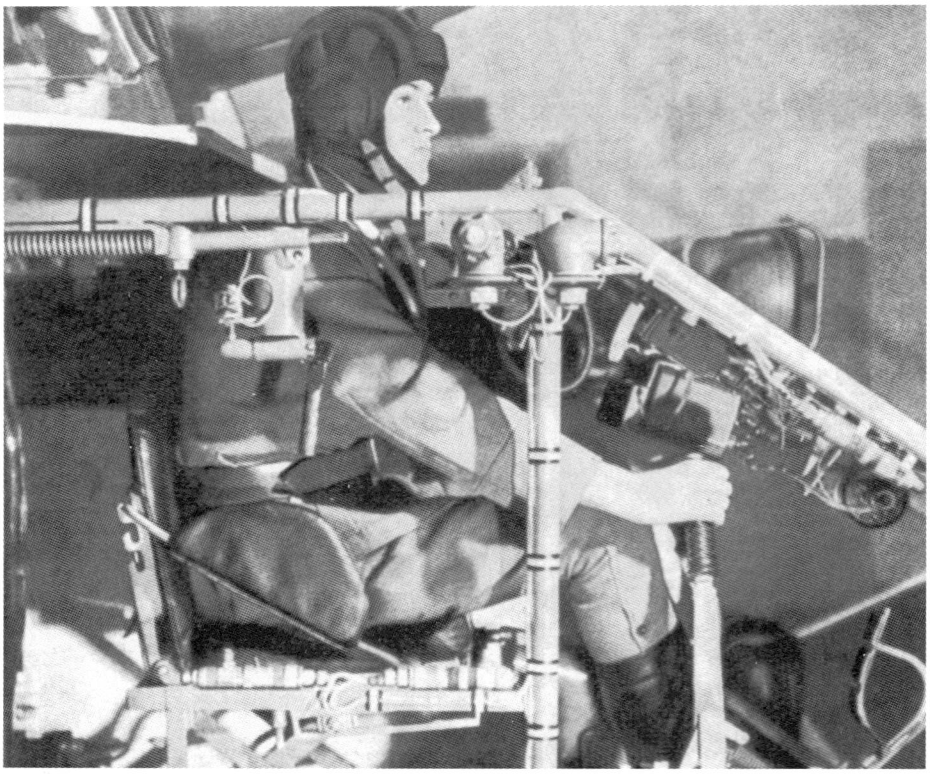

Trainee station on a T-54/T-55 tank driving simulator

SOUTH AFRICA

UEC Driver Training Simulators

UEC Driver Training Simulators (DTS) are for training AFV drivers. The DTS system can be used to present computer-controlled tutorials to a trainee driver and to test trainees in driving skills such as warning sign recognition.

Training Station. A simulated driving cab gives good functional fidelity, and moving visual scenarios are provided to the trainee driver on a video monitor. Authentic sound effects provide audio cues and enhance realism.

Instructor Control. The instructor controls exercises via a simple keypad and can have access to a number of preprogrammed exercises. He is presented with the same visual scene as the trainee on a separate monitor screen. He communicates with the trainee through a hatch in the side of the simulator or via a driver-commander intercom. A number of DTS units can be connected to a central instructor console. Assessment and recording facilities are included.

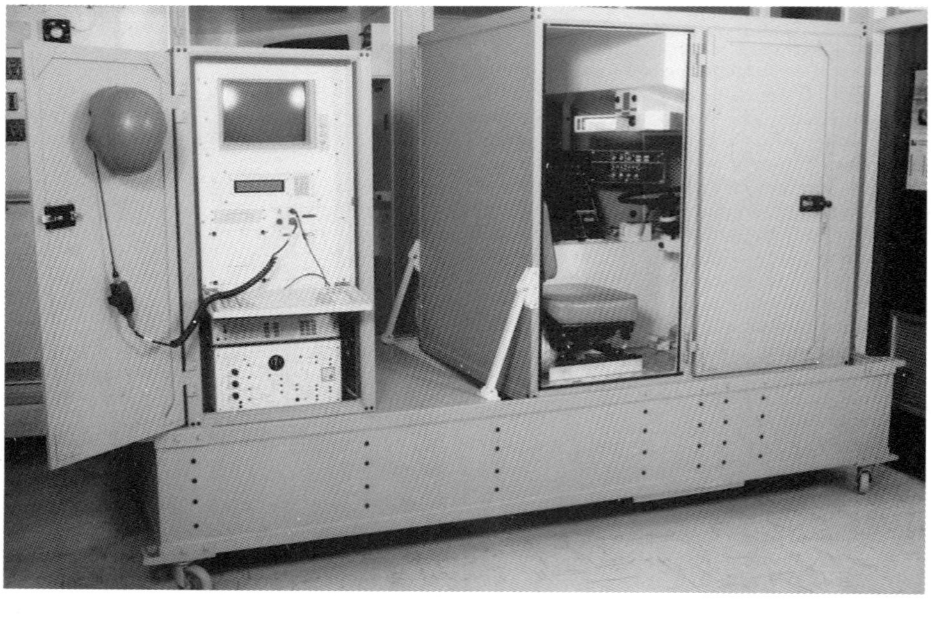

UEC Driver Training Simulator for Rooikat recce vehicle, instructor station left, trainee cubicle right

Truck Driver Simulator. UEC also produces a Truck Driver Training Simulator (TDS) for all aspects of truck driver training. The TDS can feature a single-axis motion platform.

Status
In production. In service with the South African Defence Forces.

Manufacturer
UEC Projects (Pty) Limited, Natal.
(For details see Directory of Suppliers)

SWITZERLAND

SINTRO Driver Trainers

SINTRO Electronics AG constructs driving simulators for a number of on-road and off-road vehicles.
Training Station. The trainee is presented with visual, motion and sound cues. Visual systems use monitors, screens, or, for AFVs with hatches down, images in the periscope. Where vehicles have rear-view mirrors, the appropriate images are simulated. Motion is simulated either through a special trainee seat or by a simple motion platform.
Instructor Aspects. Automatic progress control is provided together with user-friendly computer support. Monitoring and recording facilities are provided for exercise evaluation.
Adaptation to Other Vehicles. The design is modular and can be adapted to simulate the driving position of any vehicle, self-propelled howitzer or MBT.

Status
Available.

Manufacturer
SINTRO AG, Interlaken.
(For details see Directory of Suppliers)

SINTRO truck driver trainer. Note CGI visual system and rear-view mirror imagery on left side of cab

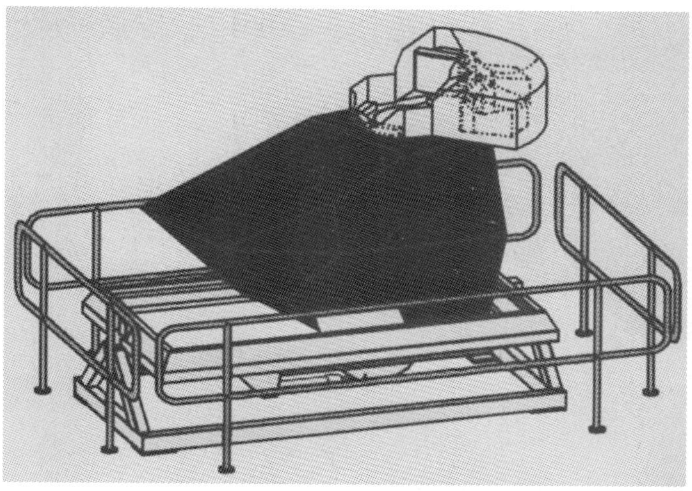

Example of SINTRO driver trainer for APCs, note motion platform for pitch and roll, and visual system on top of cab

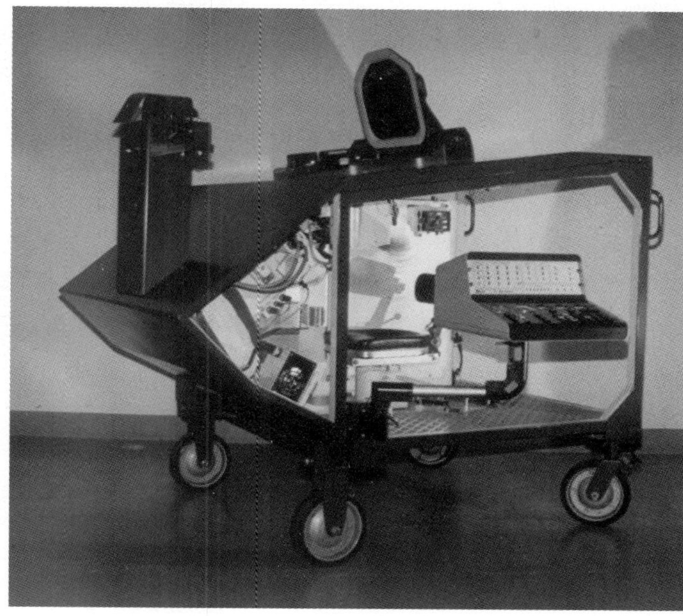

SINTRO driver trainer for the Pz 68/88 MBT

UNITED KINGDOM

Marconi Simulation Tank Driver Trainer

The Marconi Simulation Challenger 2 MBT Unit Driver Trainer (UDT) is designed as a complete driver training system which allows the practice of all equipment and operational procedures and driving skills a tank driver will require. In January 1992 it was announced that Marconi Simulation had been chosen by Vickers Defence Systems to supply five UDTs, plus four part-task trainers and a courseware package based on Marconi Simulation's 'Mandarin for Windows' computer-based training product. The complete package will be used to train British Army Challenger 2 MBT drivers.
Training Station. At the trainee's position, high-fidelity visual simulation is provided. Day and night periscopes view video monitor display units, and head-up operations are possible through the driver's hatch. Modelling includes the tank engine, transmission, steering, braking, suspension and motion dynamics and their interaction via the driver with a range of terrain conditions. The visual terrain database includes a variety of local areas including roads in built-up areas, hill slopes, soil conditions of mud, snow and ice, obstacles, and special items such as garages and low loaders. The driver's station may be mounted on a motion platform if required.
Instructor Station. The instructor's console facilities are designed to reduce the workload by automation. This gives the instructor the flexibility of interaction with the exercise but frees him to coach students unhampered by the equipment. The facilities also enable contractor-supplied courseware to be altered either by the instructor or the contract maintainer. Courseware is supplemented by an automated student record system which captures results from individual students, controls their progression through exercises, prints out results and allows for later analysis.

Status
Five being supplied for British Army's Challenger 2 MBT training programme.

Manufacturer
Marconi Simulation, Dunfermline.
(For details see Directory of Suppliers)

Thomson Training and Simulation (UK) AFV Driving Simulators

Thomson Training and Simulation took over Link-Miles in 1990. In 1964, the then Singer Link-Miles built an AFV simulator, for the British Army Chieftain Mark 3 MBT. This was followed by simulators for the FV432 APC. By 1990 Link-Miles had supplied 36 AFV simulators for 18 different vehicles to eight armed forces.

Motion Platforms. On the first AFV simulators the replicated driver's cab was mounted on a hydraulically operated 3-DoF motion platform. Later models were mounted on a 6-DoF aircraft-type motion platform, and this is available today.

Visual Image Generation - CGI systems. Since 1981 the IMAGE series of CGI systems has been available for AFV as well as aircraft simulation, and since Thomson Training took over Rediffusion in 1993, the new Thomson SPACE CGI system is available, combining the best features of IMAGE and Thomson's earlier VISA CGI system.

Visual Display System. Normally, three display monitors would be fitted, giving a 127° FoV(H) and a refresh rate of 50 Hz. Other configurations would be at customer request.

IOS. The Instructor Operating Station (IOS) includes student performance evaluation and such features as a touch-sensitive screen for exercise setting up and control. The instructor is able to monitor the student's progress, inject malfunctions and also drive the simulator to demonstrate particular techniques to students. An additional student console may also be incorporated to allow the rest of a class to follow a training exercise.

Status

In service with Denmark (Leopard 1 and Centurion), Egypt (T-54 and T-62), Italy (Leopard 1 (5) and M60A1 (2)), Netherlands (Leopard 1, Leopard 2, YPR 765, M113), India (Vijayanta), Iran (Chieftain Mark 3 and Mark 5) and United Kingdom.

Manufacturer

Thomson Training and Simulation (UK), Crawley.
(For details see Directory of Suppliers)

Link-Miles (now Thomson Training) AFV driving simulator for Leopard 2

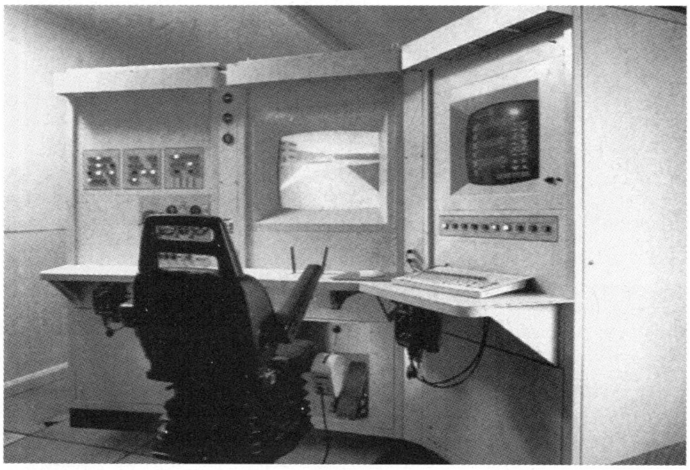

IOS for a Link-Miles (now Thomson Training) AFV driving simulator with monitor repeating the CGI visual scene

Example of CGI database for AFV driving simulator

UNITED STATES OF AMERICA

Doron Driver Training System

Doron Precision Systems Inc produces the L-300 driver trainer systems for enhancing the driving skills of new and experienced drivers. Doron can produce a complete range of driver training systems from standard ground vehicle simulation systems and portable desktop training systems to customised driver training systems.

Vehicle Manoeuvring Trainer (VMT). The VMT is a full training system with interactive capabilities. It allows trainees to practise the manoeuvring of articulated, oversized or tracked vehicles. The combination of VMT and L-300 simulation-aided training systems provides a comprehensive driver training and testing system.

3-DoF Motion Platform. Doron offers a hydraulic 3-DoF motion platform to enable other manufacturers who do not make their own motion platforms to add motion cues.

Single Doron driving simulator

A group of 12 Doron driving simulators

Status

Available, in production and in widespread service.

Manufacturer

Doron Precision Systems Inc, New York State.
(For details see Directory of Suppliers)

Martin Marietta TDT2000 Tank Driver Trainer

The US Army awarded a contract in 1989 to General Electric (GE) for the delivery of up to 20 TDT2000 tank driver trainer systems for the M1 Abrams. Subsequently, the GE facility at Daytona Beach was taken over by Martin Marietta and is now designated its Simulation and Automated Systems division.

Cabs and Motion Platforms. Each TDT2000 tank driver trainer has two independent trainee driver cabs, each mounted on a 6-DoF hydraulic motion platform. A system known as 'terrain feedback' produces motion and vibration effects of the surface texture and hardness of the terrain being traversed.

Visual System. A single-channel CompuScene PT2000 visual system is used to provide the visual display for two independent driver cabs using databases simulating actual driving environments. The visual system has the capability to simulate battlefield conditions such as smoke, fog, haze, incoming and friendly fire, plus weather conditions such as snow and rain.

Training Exercises. Trainee drivers are able to drive in convoy scenarios or alone in day, night, dawn or dusk conditions. The TDT2000 training area incorporates urban regions, plains, rolling hills, mountainous terrain and shoreline. Within this topography is an obstacle course, marked and unmarked minefields, water and vertical obstacles, rail and aircraft loading areas, a motor pool, and a basic driving course. The trainee can have either an open or closed hatch view of the surrounding terrain, with a 132° FoV(H). Malfunctions and emergencies can be simulated, including engine fires, loss of power, steering, brakes, tracks, engine oil or transmission fluid and engine overspeed.

IOS. Instruction features include computer-generated performance analysis, scene playback, selectable malfunctions and either preprogrammed or free run scenarios.

Networking. The trainer makes use of modular technology and has the capability to network with other training systems.

Status

In production for the US Army.

Manufacturer

Martin Marietta Corporation, Simulation and Automated Systems, Daytona Beach.

(For details see Directory of Suppliers)

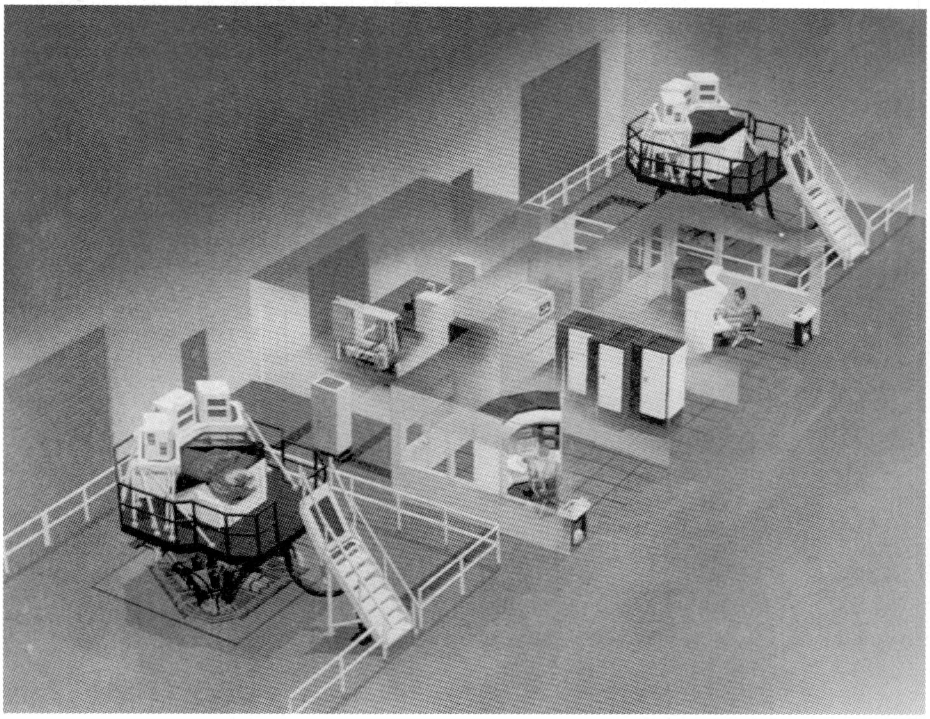

Artist's impression of TDT2000 tank driver trainer for M1 Abrams MBT

Instructor's station for TDT2000 tank driver trainer for M1 Abrams MBT

NHTSA Advanced Driving Simulator

The US National Highway Traffic Safety Administration (NHTSA) has announced contract awards of $1.25 million each to TRW Inc of San Diego, California, and to Contraves Inc of Tampa, Florida, for a design competition for an Advanced Driving Simulator. Although this is not a military simulator, basic driving and safety training is as relevant to military operations as to civil, and so this item is included for information. It is hoped to report on the technology employed (Motion, CGI, Display and Computer Systems) in later editions of *Jane's Military Training Simulation Systems* as the project develops.

Project Timescale. At the end of a 13-month design competition NHTSA will select a single contractor to build and test the simulator. It will be installed at the University of Iowa Research Park in Iowa City.

Intended Capabilities. The simulator will allow studies of drivers in realistic crash situations. Examples include obstruction, animals or people on the road ahead, skids on oil or glaze ice, or situations with out-of-control other vehicles. When the simulator becomes operational, NHTSA has said that it may be the most sophisticated driver simulator in use

Status

In design.

Sponsor

US Department Of Transportation, National Highway Traffic Safety Administration (NHTSA).

(For details see Directory of Suppliers)

YUGOSLAVIA (SERBIA AND MONTENEGRO)

TOVT 55P Tank Driver Trainer

The TOVT 55P is an indoor tank driver trainer used in a darkened room to allow stylised images of roads or tracks to be projected on a screen mounted in front of the trainer. A transparent circular disk carrying the outline of a road or path network is located over the top of the trainer. A projector displays the outline of the road or track on the screen and any movement of the controls by the trainee driver will result in a corresponding alteration of the projected image. The TOVT 55P measures 1.74 × 0.96 × 1.6 m and weighs approximately 200 kg.

Training Station. The trainee's station is built within a steel frame and is configured for the T-55 MBT. It contains representations of all the tank driving controls and a panel for controlling the projector and other sectors of the trainer. The frame is built onto a rocking platform to simulate travelling movement.

Training Exercises. The trainer can be used for preliminary and main driving training and also for driver proficiency testing. Training includes starting and stopping the engine, moving off and braking, driving over various types of ground with the associated gear changing, and general driving procedures.

Instructor Station. The TOVT 55P is normally located in its own compartment with up to eight trainers under the control of a single instructor station.

Car Driver Trainer Variant. A variant of this trainer intended for training car drivers and following the same general lines is known as the AS ELS82.

Status

In service with the armed forces.

Contractor

Holding Company YUGOIMPORT SDPR Ltd, Belgrade.

(For details see Directory of Suppliers)

TOVT 55P tank driver trainer

T-55 Tank Basic Driving Trainer

The T-55 basic driving trainer is a simplified version of the TOVT 55P and is used to provide basic training in driving seat adjustment, driving control location and feel, and instrument layout familiarisation. An instructor's control panel is provided to initiate and control various procedural exercises during which the instruments produce the appropriate readings and responses to trainee actions.

A 12 V DC supply is required; weight is approximately 200 kg.

Status

In service with the armed forces.

Contractor

Holding Company YUGOIMPORT SDPR Ltd, Belgrade.
(For details see Directory of Suppliers)

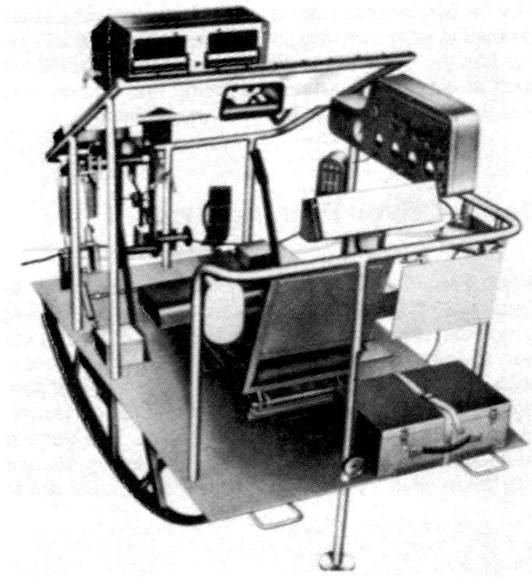

Trainer for basic training of T-55 tank drivers

MAINTENANCE TRAINERS

FRANCE

Giat Maintenance Training Aids

Although Giat is primarily a manufacturer of military equipment rather than training aids, it produces cutaway or other models and full-size replicas of its products. An example is its trainer for the AMX-10P infantry combat vehicle.

AMX-10H Trainer. The AMX-10H consists mainly of a wooden full-scale mock-up of the hull of the AMX-10P ICV mounted on a frame. The frame can be tilted 20° to either side to assist viewing and can be mounted on small wheels to allow the trainer to be moved around a classroom area. The right-hand side of the hull includes elements from the suspension while the hull interior contains a wooden mock-up of the main power unit fitted with some real components, such as the drain plugs, filling systems and dipsticks. The upper part of the hull is manufactured from a transparent material that permits observation of the interior including the driver's compartment, the water propulsion components, and the fuel and cooling circuits. The rear ramp on the hull can be opened or closed as required. An instructor can use a control console to illuminate parts of the interior or switch on lights to indicate the location of various key components. A footbridge over the trainer can be provided if required.

Other Models. Giat also produces many other similar mock-ups and models, both cutaway or complete, of complete vehicles, subassemblies or components. They also produce 'breadboard' models for electrical or hydraulic circuits, and operating switchboards which can be used with or without training mock-ups to assist maintenance and fault-finding instruction.

AMX-10H full-scale trainer for the AMX-10P infantry combat vehicle

Status
Trainers produced to customer requirements as required.

Manufacturer
Giat Industries, Versailles.
(For details see Directory of Suppliers)

GERMANY

Krauss-Maffei Gepard Maintenance Training Equipment

The Krauss-Maffei Gepard twin 35 mm self-propelled anti-aircraft gun system maintenance training equipment is produced as three classroom rigs: an electronic training rig; a weapon training rig; and a power supply training rig. All use original Gepard components and the test equipment used by the trainees is the same as used in the mobile workshops where they will work. Instructors can use the rigs to introduce faults into the various items of Gepard equipment and trainees can then be taught to diagnose, isolate and correct the faults.

Electronic Training Rig. The electronic training rig includes an operational Gepard fire-control unit and terrestrial navigation unit. A digital processor controls the various safety circuits involved with the simulation of turret and weapon controls, muzzle velocities, vehicle speeds and movements. Using the same processor the instructor can increase progressively the complexity of the faults introduced.

Weapon Training Rig. The weapon training rig comprises the gun itself and its associated hydraulic equipment, muzzle velocity measuring system and the gun electrics.

Power Supply Training Rig. The power supply training rig duplicates the electrical assemblies and circuitry involved on the Gepard. The circuits include those for the turret and weapon aiming drives, and the load, power supply and operation simulators.

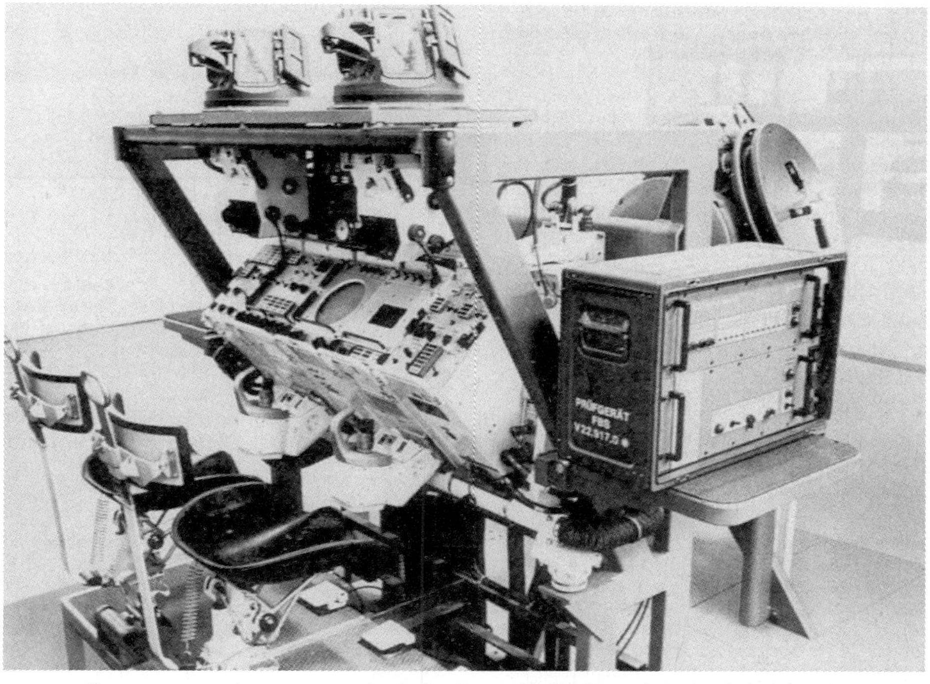

Commander and gunner consoles in the Krauss-Maffei Gepard electronic training rig

Status
In service with the German Army.

Manufacturer
Krauss-Maffei AG, Munich.
(For details see Directory of Suppliers)

Wegmann Trainer for Leopard 1 MBT

This Wegmann indoor trainer consists mainly of a modified Leopard 1 turret equipped with turret hydraulics, turret electrical system and optical components. The sidewalls and roof of the turret have been replaced by a tubular structure to make the interior visible from outside. Appropriate weights maintain the centre of gravity relative to the traverse axis. A removable platform for the instructor and 11 students gives a view of the turret interior and is mounted around the sides of the turret. Turret and platform traverse together during gun/turret control exercise. The seat pans are fitted with safety belts. The gun itself cannot be fired, but has the same weight, and functional and inertial data as the operational 105 mm gun.

Failure Input Unit (FIU). The failure input unit (FIU)

allows the introduction of preset failures from a control console. The FIU is integrated with the movement simulator control panel, making up the control console which is mounted on the rear of the training turret. A total of 255 malfunctions can be selected and introduced, of which up to eight can be simultaneous. The failures are grouped in three categories: non-hazardous, critical and hazardous.

Motion System. The motion system consists of a platform on which the turret is mounted, hydraulic power supply and a control box. It can be controlled either manually or automatically. In the automatic mode a sensor unit introduces and follows a terrain configuration. Although designed primarily for the Leopard 1 turret, the motion system can be used with other types of tank turret.

Specifications

Weight:
(without turret) 8000 kg
(with Leopard 1 turret) 19 000 kg
Traverse speed: 30°/s
Elevation speed: 15°/s
Elevation/depression: +17°/−17°

Status

Production as required. In service with the German Army.

Manufacturer

Wegmann & Co GmbH, Kassel.
(For details see Directory of Suppliers)

Wegmann Trainer for Leopard 2 MBT

This trainer for the Leopard 2 MBT is a gun and electronics maintenance training system. It permits real-time failure simulation in electrical and hydraulic turret components and, in particular, the fire-control systems and gun sights.

Components. The trainer consists of a two-axis motion system, a turret trainer, a failure input system and a television system.

Motion System. The motion system consists of a motion platform, a hydraulic system, and an electrical power supply and electronic control system complete with a control panel.

Turret Trainer. The turret trainer is a special welded turret fitted with real and modified components from the Leopard 2 MBT. The side walls and turret roof have been replaced by a tubular construction permitting observation of the turret interior from the outside. It is fitted with an operable main gun which is identical to the Leopard 2 main armament. The turret trainer mounts two television monitors which can be viewed by all trainees. Together with the television camera, video recorder and central electronic unit, they form the trainer's television system which is designed to monitor and record the gunner's laying operations.

Instructor Station and Platform for Trainees. The instructor station is located in the area normally occupied by the ammunition racks, along with the failure simulator and control panel for the motion platform. A seating platform for up to 10 trainees is located around the turret trainer, and each trainee is provided with a seat belt. A standing platform is located at the rear. The turret roof mounts the training display panel which is connected in parallel with the standard Tank Test System RPP 1-8 display panel, and enables all trainees to follow the readouts on the RPP 1-8. Also mounted on the roof are the trainer television monitors.

Failure Input System (FIS). The failure input system (FIS) is designed to introduce malfunctions into the electronic and electrical components of the hydraulic gun, and turret control system and functional turret components. A total of 255 different malfunctions can be entered, including up to eight malfunctions simultaneously.

Wegmann trainer for Leopard 2 MBT

Specifications

Weight:
(motion simulator without turret) 9500 kg
(motion simulator with turret) 18 500 kg
Traverse range: n × 360°
Traverse angle: max ±25°
Elevation/depression: +11°/−11°
Traverse rate: 7 rpm
Operating pressure: 200 bar
Mains supply voltage: 3 × 380 V AC

Status

In production. In service with the German and Swiss armies.

Manufacturer

Wegmann & Co GmbH, Kassel.
(For details see Directory of Suppliers)

POLAND

OBRUM T-72M1 Maintenance Trainers

OBRUM makes a wide range of maintenance trainers for the T-72M1 MBT used by the Polish armed forces. The range varies from a full-scale interactive mock-up of the main systems employed on the T-72M1 down to individual training devices devoted to secondary systems.

SJ 01 Training Stand. The largest trainer is the SJ 01 training stand which is a skeleton mock-up of a complete T-72M1 for indoor use. It incorporates all armament systems, electrics and communication systems, driver controls, and other systems. The stand involves numerous hatches cut into the main hull and turret for access and training purposes.

SJ 03S Stand. The SJ 03S stand is a cage-like mock-up of the T-72M1 hull which exposes all drive-train components. It is for the training of mechanics and the vehicle driver who is expected to carry out limited running maintenance and repairs.

SJ 01 training stand for the T-72M1 MBT

SJ 03S drive-train stand for the T-72M1 MBT

Other Trainers and Rigs. These smaller training stands and rigs for the T-72M1 are usually laid out on dedicated benches or as free-standing rigs. The parts are in a similar arrangement to their real position in the T-72M1 vehicle, and the trainers are designed to be installed and operated in a classroom. They include:

SJ 05 Engine Lubrication System Training Stand.

SJ 06 Fuel System Training Stand, for the left front fuel tank.

SJ 07 Engine Fuel Supply System Training Stand, for the engine fuel supply system and the thermic smoke generation system.

A fire extinguisher training rig.

Status
In service with the Polish armed forces.

Manufacturer
OBRUM, Gliwice.
(For details see Directory of Suppliers)

UNITED KINGDOM

ECC Challenger 2 Simulated Maintenance Trainers

ECC Simulation Limited, the European division of ECC International, is currently producing a family of maintenance trainers for the Challenger 2 MBT for Vickers Defence Systems. The package contains: a turret maintenance trainer (armament and instruments); a turret maintenance trainer (weapons); hull frame electrical layouts; turret frame electrical layouts; firing circuit panel electrical layout; and loader drills trainers.

ECC Simulation Limited has also produced a maintenance trainer for the Challenger armoured repair and recovery vehicle, plus gunnery and maintenance simulators for the Swiss Pz 68 MBT.

Status
In production for the British Army.

Manufacturer
ECC Simulation Limited, West Sussex.
(For details see Directory of Suppliers)

ECC simulated maintenance trainer for Challenger 2 MBT

Pennant Combat Engineer Tractor Electrical System Trainer

The Combat Engineer Tractor (CET) Electrical System Trainer was designed by Pennant International to provide training in fault identification and diagnosis, and the repair and maintenance of the electrical system.

Components. The trainer consists of a full-scale replica of the CET constructed in the form of an open framework into which the actual electrical system is incorporated, including driver and commander instrument panels, generator, starter, steering unit, lighting, fuel, lubrication, hydraulics, bilge pump, wiring and distribution panels, NBC and alarm systems. To assist in locating components and tracing cables, clear acrylic materials are used for the manufacture of bodywork and bulkhead panels.

Instructor Station. Control of the trainer facilities is provided from a separate console, from which the instructor can introduce a range of faults.

Status
In service with the Indian Army.

Manufacturer
Pennant International Limited, Cheltenham.
(For details see Directory of Suppliers)

Pennant Combat Engineer Tractor Electrical System Trainer

UNITED STATES OF AMERICA

ECC M1 Abrams Tank Turret Organisational Maintenance Trainer (TTOMT)

ECC M1 Abrams Tank Turret Organisational Maintenance Trainer (TTOMT)

ECC International Corporation make the M1 Abrams Tank Turret Organisational Maintenance Trainer (TTOMT). These are installed at the US Army Armor School at Fort Knox, Kentucky, and the USAEUR Armor School at Vilseck, Germany. They are used for fault analysis, utilising the STE/M1 test set, and for teaching the removal and replacement of electrical components, hydraulic accumulators and of the optical system.

Components. The trainer is a full-size mock-up of the M1 turret which rotates in a surrounding stand. The roof and sides are made from wire mesh allowing up to 15 trainees to view the operations inside the turret. The stand itself is a welded steel frame which also contains the turret rotary drive power. The interior contains M1 components, locations and attachments, and the 105 mm gun assembly can move to allow maintenance procedures to be carried out.

Instructor Station. An instructor station at the turret rear top is equipped with procedural check facilities, emergency shut-off and other control functions. The system's troubleshooting simulation model test board can be located close to the trainer turret.

Transportability. The trainer turret has oversized locking wheels to allow it to be moved into its operating position, where it is lowered onto pedestals.

Status
In service with the US Army.

Manufacturer
ECC International Corporation, Pennsylvania.
(For details see Directory of Suppliers)

ECC EC3 Maintenance Trainers

ECC International Corporation makes a series of EC3 maintenance trainers for second-level maintenance. Equipments included are the M1 Abrams and M60A3 MBT systems, and TOW/Cobra. All the EC3 trainers are basically similar, but the presentation panel for the M1 is slightly larger than the others.

Presentation to Trainees. The EC3 trainers present a diagrammatic display of a particular system. Each of the test sets used for the system concerned is simulated on the display panel, allowing the trainee to perform the same tests as he would on the actual equipment.

Instructor Aspects. The simulator programme shows both the normal operating condition and a variety of malfunctions which are entered from a keyboard. Different learning conditions can be selected from a training list on a TV monitor which also presents student performance. Demonstration, test, practice, and individual training modes may be selected. A keyboard is used to enter student information and a line printer to record performance.

M1 Trainer. The EC3 trainer for the M1 provides programmes for turret hydraulic and electrical troubleshooting, the ballistics computer, hull electrics, engine and transmission troubleshooting. An additional programme provides for troubleshooting on the laser gun sight.

M109 Trainer. The M109 HIP (Howitzer Improvement Program) fire-control and maintenance trainers use the actual weapon system and use the Ada programming language. The automatic fire-control trainer trains operators in the use of the Section Chief's Display and Control Unit (SCDCU). It allows engagement of multiple targets simultaneously, and mass firing when the mission demands. This allows crew chiefs to use the howitzer as a training aid for initial and refresher training in the gun park as well as during field exercises.

Other Trainers. Other similar maintenance trainers are produced for the M1A1 Abrams, the M109A6 Paladin, and the M2 Bradley (upgraded to the M2A1 configuration - see following entry).

Status
In production. In service with the US Army.

Manufacturer
ECC International Corporation, Pennsylvania.
(For details see Directory of Suppliers)

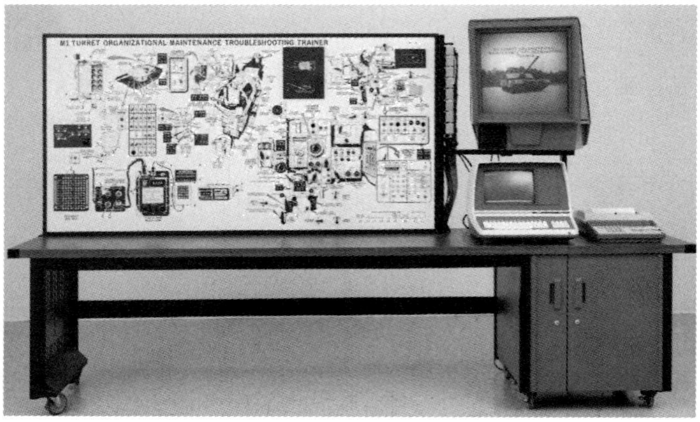

EC3 maintenance trainer for M1 Abrams MBT

EC3 maintenance trainer for M60A3 MBT

ECC M2 Bradley Turret Maintenance Simulator

The ECC Bradley M2 Fighting Vehicle turret maintenance simulator is a full-scale simulation of the vehicle's turret, and has been upgraded to the M2A1 configuration.

Components. The simulator consists of a turret and gun assembly. The turret rotates, the gun elevates and all replaceable units are the same weight and have the same centre of gravity as real parts.

Trainee Aspects. The trainer provides the trainee turret mechanic with component troubleshooting practice using the STE/M2 test set. This trainer also includes the TOW missile system as installed on the M2 Bradley. First and second level maintenance training are included. The student turret mechanic learns the use of simulated test sets, realistic component interactions, and to practise troubleshooting procedures.

Status

In service with the US Army.

Manufacturer

ECC International Corporation, Pennsylvania.
(For details see Directory of Suppliers)

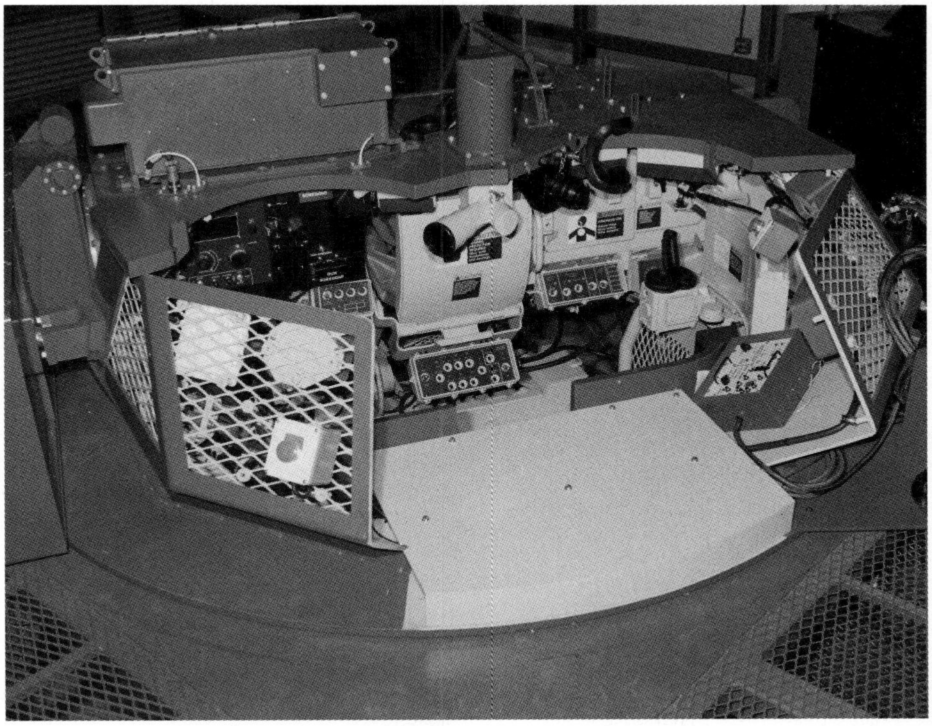

The M2 Bradley fighting vehicle turret maintenance trainer

COMMAND AND STAFF TRAINERS

CANADA

Crisis Simulations Auriga Low Intensity Conflict Simulator

Auriga is a command and control simulation designed to train all five functional staff branches of a divisional level headquarters in the management and co-ordination of counter-terrorism and civil disorder operations in a domestic emergency environment.

The system is similar to Polaris (qv) in that it is based on three stand-alone PCs. Depending upon the nature of the particular scenario being used, five different scales of simulation display may be employed.

Status

Operational in Canadian Army.

Manufacturer

Crisis Simulations Limited, Nova Scotia.
(For details see Directory of Suppliers)

Crisis Simulations Centaurus Urban Disaster Simulator

Centaurus is a computer-assisted command and staff trainer designed primarily to train municipal emergency management groups in the handling of natural or technological disasters. The system can be used to train military units who have a disaster relief role.

The Centaurus system was developed by adapting Polaris technology. The physical structure and major component elements are similar to those described for Polaris (qv).

Status

In use in several Canadian municipalities and academic institutions.

Manufacturer

Crisis Simulations Limited, Nova Scotia.
(For details see Directory of Suppliers)

Crisis Simulations Polaris Battalion Battle Simulator

The Polaris Battalion Battle Simulator is a computer-assisted command and control trainer. Its prime function is to train battalion-level operations and administrative staffs in the conduct of mid- to high-intensity operations. The system is designed to be user-operated and does not require full-time staff.

Components. The system includes training modules for system management, system operation and computer operation. Three IBM PS2 computers are used. Software reflects the equipment of the user and the enemy characteristics, and Turbo Pascal is used.

Portability. The control centre can be set up in any large room with dimensions of at least 20 × 15 m. Command groups practising on the system are in their regular command posts and link to the control centre through normal tactical radio and telephone nets. For transport the system is packed into five rigid cases, each measuring 1.22 × 0.914 × 0.914 m and weighing approximately 30 kg.

Terrain and Environment. Users specify a piece of ground 6 × 24 km, and this is modelled on a scale of 1:3345. As well as simulating day and night conditions, software is provided to portray winter, chemical and electronic warfare environments.

Training Capabilities. The system includes air reconnaissance, air strikes, indirect fire, direct fire and close combat. Fuel and ammunition consumption as well as vehicle and human battle casualties are also calculated. Supplementary manual routines portray vehicle breakdowns, disease and non-battle injury, reinforcement and other administrative functions.

A Polaris control centre in operation during an exercise

Status

Deployed to all regular force brigades and reserve force areas of the Canadian Army. English and French versions available.

Manufacturer

Crisis Simulations Limited, Nova Scotia.
(For details see Directory of Suppliers)

Crisis Simulations Sirius Brigade Battle Simulator

Sirius is the second in a series of related but independent command and control simulators developed for the Canadian Army. Sirius was designed to exercise brigade level operations and administrative staffs. The system retains the portable, user-operated, computer-assisted design features of the Polaris system (see previous entry), but is considerably more complex.

Components. Sirius includes both low and high resolution simulation display areas. The computer suite consists of six IBM PS2-based workstations linked to a 386 architecture file server by a Novell local area network. As with Polaris, Sirius is programmed in Turbo Pascal.

Portability. The system is packed in a total of eight transport cases similar to those used with the Polaris system.

Status

Operational prototype held by the Canadian Army.

Manufacturer

Crisis Simulations Limited, Nova Scotia.
(For details see Directory of Suppliers)

GERMANY

Interatom GESI Battle Simulation System

The initials GESI are derived from GEfechtsSImulation (battle simulation). GESI is a computer-based simulation system based on mathematical models. These describe and display numerous weapon systems and units in connection with graphical situation displays, based on military symbols. GESI simulates map exercises either one-sided or two-sided (force-on-force). The exercise setting is available in the form of a battle scenario and does not require additional data generation. The system can be configured to accommodate special command and control tasks to be performed or accomplished by individual staff officers. GESI makes use of a modular system architecture and is based on the use of a UNIX workstation.

Battlefield and Weapon Models. The battlefield is shown in the form of digitised terrain and represents the information content of normal military maps. Terrain-based fortifications and obstacles can be inserted by the user either before or during an exercise. Visual contact with the enemy is displayed by a graphic message, giving the operator the perceived location, direction of movement and type of weapon system of the enemy. Target allocation and engagement are conducted by weapon effectiveness and vulnerability models which determine the results of direct and indirect fire. Performance data is included on combat and combat support troops, airborne units and logistic support systems. This data can be changed by the user and so modifications, enhancements and future weapons can be inserted.

Display to Units. The display of subordinate units takes into account built-up areas and vegetation, weather conditions, time of day and other factors such as smoke.

Status
Available.

Manufacturer
Interatom GmbH, Bergisch Gladbach.
(For details see Directory of Suppliers)

Wegmann Tactical Team Gunnery Trainer

The Tactical Team Trainer consists of five interconnected containers, four of which contain tank crews under training and one is the Instructor Station. Each system has two Loral (ex BBN) GT 120 CGI systems, similar to those used in the US SIMNET equipment (qv). The monitor screens on which the CGI images are presented give realistic outside information to the gunner and commander. A sound system simulates battlefield noise and a Semi-Automated Forces (SAFOR) system represents own and enemy forces in the simulation.

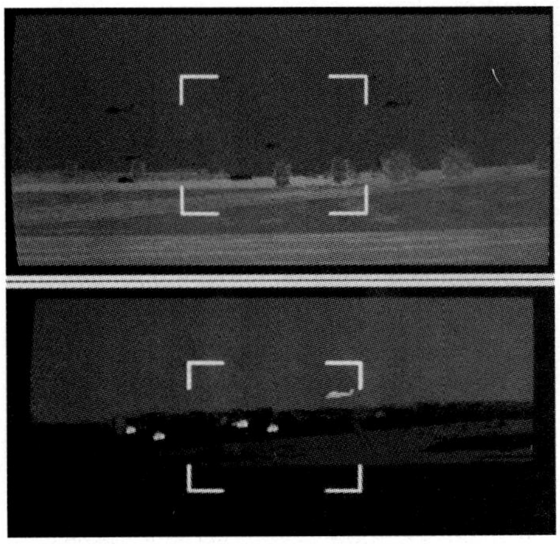

Wegmann Tactical Team Gunnery Trainer - Illustration of Thermal Imagery, black hot (top) and white hot (bottom)

Wegmann Tactical Team Gunnery Trainer

Status
Available.

Manufacturer
Wegmann & Co GmbH, Kassel.
(For details see Directory of Suppliers)

ISRAEL

Elbit Ground Tactical Training Centre (GTTC)

Elbit Computers Limited make the Ground Tactical Training Centre, or GTTC.

The GTTC is housed in a permanent one-storey building. It is programmed to run military exercises of all kinds for the commanders and command staffs of units up to brigade and corps level. The building houses a control centre, computer, a theatre for briefing and debriefing, and various command centre cubicles which may be fitted in the same manner as field command posts. If required, extra real field command post and control vehicles may be located outside the GTTC and connected to the computer and control system, or may even be used in place of the GTTC cubicles.

Exercise Control. A Chief Exercise Controller determines the form of the exercise, its duration and type, and the time in which the exercise is to take place. The basic exercise area is a simulated stretch of terrain measuring 120 × 120 km. An area of this terrain up to 60 × 60 km may be used as a particular 'battle run'. For detail the computer can break down the terrain into 500 × 500 m or 100 × 100 m blocks. Three forms of exercise can be carried out: basic, standard or extended. The time scale of an exercise may also vary as it may be in real time, double time or speeded up by a factor of three. Normally the lowest level involved will be the battalion. The computer can also be programmed to present results in changing climatic conditions and degrees of visibility. The central computer can simulate up to eight different types of weapon or vehicle, and special equipments can be programmed to suit any exercise. The exercise can be 'frozen' at any point and full replays can be made for analysis and correction at any time.

Ground Tactical Training Centre (GTTC) instructor console

Friendly and Enemy Forces. The Exercise Controllers may play the part of the 'enemy'. They also control a computer-generated display on which selected areas display geographical and tactical symbols generated by the computer. This display is on a master console which also displays totals of units, unit strengths, ammunition stocks, forces readiness, types of weapon involved, transport states, detection sensors, communication and electronic warfare states, and other such indications. The central display may also be projected onto a large screen in the GTTC theatre for debriefing or for command trainees.

Trainees. The trainee command staffs are normally located in their command post cubicles or vehicles and communicate via VHF and UHF networks. Trainees can practise the four basic skills of operations, intelligence, logistics and support, either separately or combined, and referees can intervene and oversee at any time, as in a conventional exercise. All traffic is recorded so that an order by a trainee will result in the appropriate movement on the central display.

Specifications
Total exercise area: 120 × 120 km
Exercise run area: 60 × 60 km
Element area partition: 500 × 500 m or 100 × 100 m
Exercise time rate: real-time, ×1, ×2, ×3
Command levels: battalion through to corps
Vehicle types simulated: 8
Weapon types simulated: 8
Special equipment simulated: definable for each system
Tactical communications: VHF and UHF separately controlled several networks in each band
Debriefing: communication-grid traffic two separate situation displays
Force types: up to 10

Status
In service with the Israeli Defence Forces.

Manufacturer
Elbit Computers Limited, Haifa.
(For details see Directory of Suppliers)

Rafael Army Unit and Battle Group Simulation (ABS) 2000

The Rafael ABS 2000 system is for training commanders and staff from battalion to corps level in combined arms command in realistic battle conditions. It is UNIX based and uses C++ with client-server technology. The diagrams show a battle group exercise but higher level exercises are possible, substituting, for instance, brigade commanders for company commanders in the simulation control centre.

Simulation Modules. The basic simulation modules cover terrain, order of battle, environment, movement, detection and identification, suppression and recovery, and direct fire.

Factors Included. Factors and forces included in the simulation are armoured, mechanised and infantry forces, fire support, air support, C^{3}I, engineering, air defence, logistic support (ordnance, supply, inventory and medical), chemical warfare, displacement and transport, civil defence, EW, personnel management, warfare in built-up areas, and finally evaluation and after-action review.

Status

In service with the Israeli Defence Forces.

Manufacturer

Rafael, Haifa.
(For details see Directory of Suppliers)

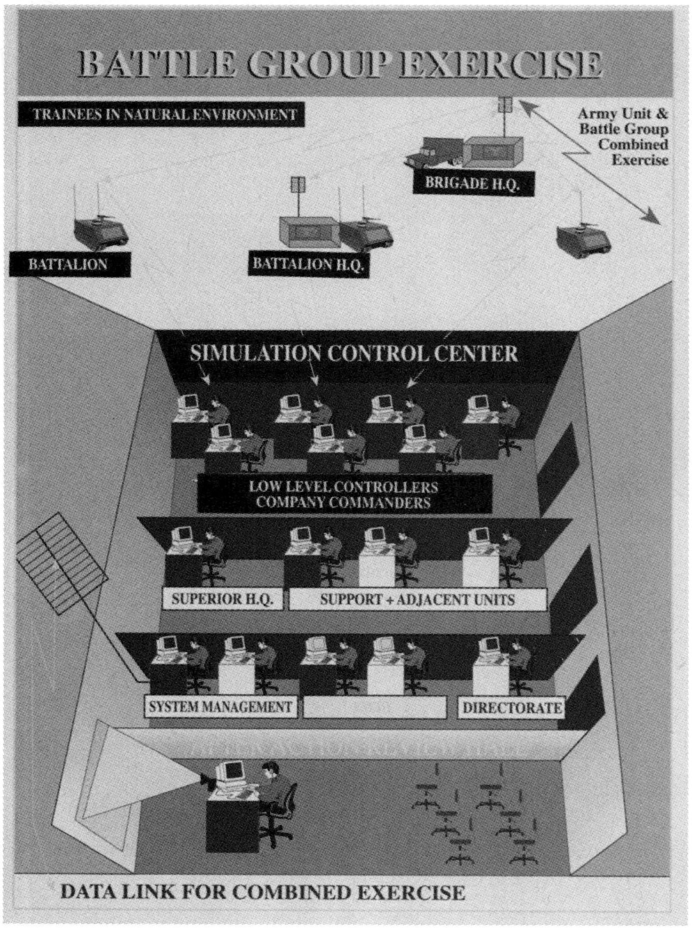

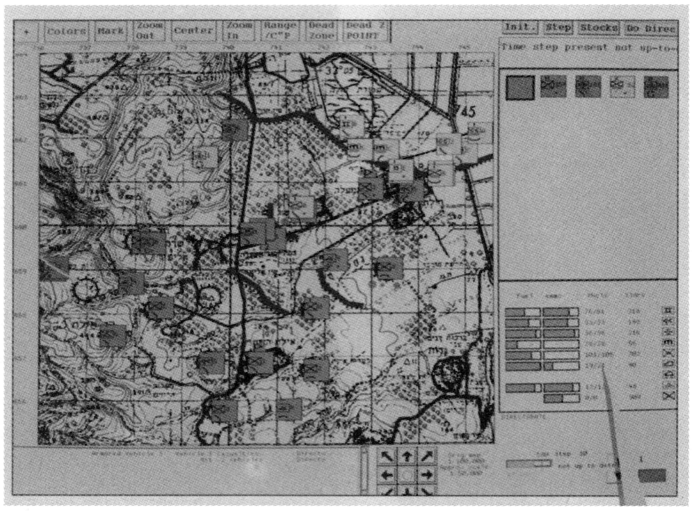

Typical display of Rafael ABS 2000 system

Rafael ABS 2000 Training System for Commanders and Staff

NETHERLANDS

KIBOWI

KIBOWI is a wargame system for staff officers at battalion and brigade level and is used at the Army's Infantry School and Staff College up to division level. It is also used for research purposes dealing with doctrine improvement and weapon procurement planning. The basis for the simulation is the representation of the combat environment including terrain, roads, bridges, water obstacles, slopes, and artificial obstacles such as minefields. Terrain features and elevation determine the firing zones and line-of-sight conditions. Both combat and supporting units are modelled together with their command structure. Computing includes detection, firing, attrition, fuel and ammunition consumption, logistics, and engineer activities.

Gaming Controls. 'Enemy Control' commands the exercise enemy units. Lower Controls command the

A typical KIBOWI work station

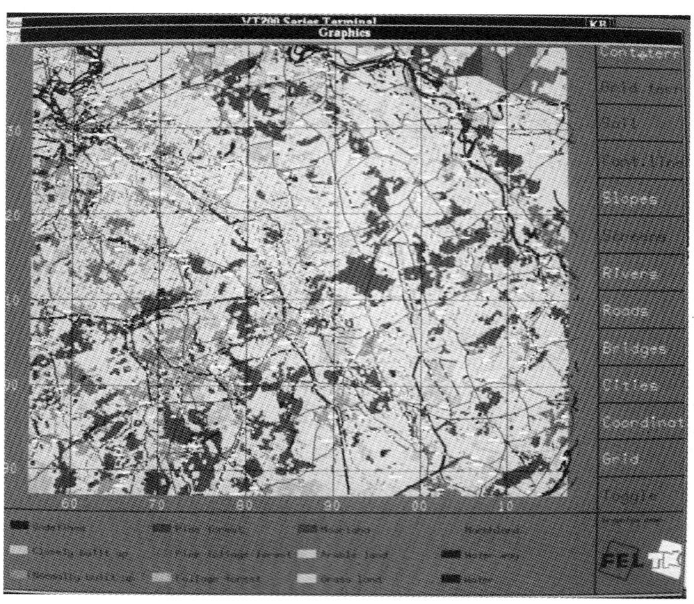

A typical KIBOWI work station screen display

units of the battalion, brigade or division to be trained and are the contacts between the player staff(s) and the KIBOWI system. Higher Control acts as the senior commander to the player staff(s). The player staff(s) to be trained carry out their staff duties as they would in reality and do not have direct interaction with the KIBOWI system. Monitor Control follows the battle to check whether the exercise serves its training purpose and can, if necessary, directly control elements and processes in the simulation.

Computer Aspects. KIBOWI software is written in Ada and runs on a DEC/VAX local area network or cluster. The number of work stations, printers and terminals depends on the size of the exercise supported. In December 1990 the Netherlands Army installed a DEC/VAX configuration consisting of a VAX 6410, over 30 colour work stations and over 60 numeric terminals and printers in a semi-permanent location.

Status
In service with the Royal Netherlands Army.

Development Agency
TNO Physics and Electronics Laboratory, The Hague.
(For details see Directory of Suppliers)

SWEDEN

CATS STRISIM Tactical Command Simulator

The Comator Advanced Training System (CATS) STRISIM (acronym in Swedish) is a tactical command simulator designed to train a tank battalion commander and his staff in communication, control and command (C^3) in a simulated battle.

General. The battalion commander is located in a mock-up of an armoured command vehicle together with his artillery commander. Images of the surrounding terrain are displayed in front of them on three screens. Friendly and enemy activities are shown as symbols on the projected images. Other personnel are seated at a total of 13 workstations. On each screen a map or picture of the corresponding terrain is shown. Symbols representing friendly or enemy activities can be shown on both types of display. A position is provided for the battalion second-in-command in a separate room, and in another room for the battalion quartermaster and a communications officer. Company commanders are seated in a gaming room. They control the battle within their companies, acting on orders from the commander. Smaller units within each company are simulated by the company commanders. Communication between players is via a simulated radio net.

Training Controller. A controller controls training in conjunction with the battalion commander and arbitrates the outcome of activities. An operator is located in the armoured command vehicle mock-up and guides the battalion commander during a game. He also acts as the 'driver' of the command vehicle. An artillery controller is located in the gaming room and monitors artillery activities. He also takes the role of forward observer to the company commanders and the battalion artillery commander's superior officer who is also responsible for the artillery units in the game.

Enemy Forces. A controller (enemy) controls the enemy units based on the type of training selected. In addition, he plays the roles of enemy intelligence units.

Workstations. Interface to the game operators is based on DEC Windows. A keyboard and mouse are used. Screens can show a map, a terrain picture or perform text input. By having several windows open at the same time the operator is able to change quickly between functions. The dialogue can be in any language. The system enables a workstation to change function, for instance the game controller can move to a workstation normally used by a company commander.

Visual Scenes. Five general scenes are provided with the system and maps and pictures supplied for each. At the start of a game a map covering 30 × 30 km is displayed automatically. The map shows the terrain with the units depicted as symbols. Each player's map shows only those units he would be able to observe in the real terrain except, that is, for the game controller, artillery controller, instructor and game controller (enemy), who can observe all the symbols on the map.

Three sizes of map are used in the system: 4 × 4 km, 10 × 10 km and 30 × 30 km. Three types of symbol are superimposed by the system: target and burst, used with the terrain pictures only, and map symbols. The size and shape of artillery bursts will be affected by the type of ammunition used, the terrain and the distance. Seven types of shell bursts can be produced: illuminating, smoke, visible air or ground bursts, hidden bursts, shrapnel effects following an airburst, and enemy muzzle flash. The system can reproduce eight types of weather condition, eight types of terrain and eight variations of ground cover and building density.

Targets and Symbols. Target symbol sizes vary with the viewing distance. They can move independently in the picture and can be occulted. About 50 target symbols are stored in the system and more are being added. They can be either silent or create a sound which will vary with range and intervening terrain or objects. Map symbols are squares and triangles of various colours. About 30 map systems are contained in the system and more are being added.

Mines. Both mine-laying and clearing can be accommodated within the system although a unit can lay mines only if it is 'equipped' to do so.

Capacity of System. Up to 10 artillery battalions, each with five batteries, can be defined within the system.

Noise Simulation. Sounds are also produced and vary according to point of origin and distance and are produced using four loudspeakers.

Computing. The system is based on an Ethernet network into which all workstations and other peripherals are connected; the computer and peripherals are manufactured by DEC. Data is presented on a standard monitor or on three 3 × 1.5 m screens. All monitors in the system use graphics with a maximum resolution of 1024 × 864 pixels. Software is written in the C language and runs under the UNIX operating system.

Record/Replay. All activities happening on the game controller's screen can be video recorded for playback and evaluation. Four channels from the radio net are also recorded on the tape.

CATS - All Arms Trainer

The CATS - All Arms Trainer (CATS-AAT) is based on the CATS and can be used as a training system for most small arms. CATS-AAT has most of the CATS function but has a 6 m wide video display on which the trainee views and hears targets and the effects after firing. CATS-AAT has been delivered to the Royal Norwegian Coast Artillery where it is used for troop training up to 84 mm Carl Gustav level.

Status
In production. In service with the Swedish Army.

Manufacturer
Comator Process AB, Helsingborg.
(For details see Directory of Suppliers)
Marketed outside Sweden by: CelsiusTech Systems AB, Järfälla.
(For details see Directory of Suppliers)

STRISIM in action showing the game controller on the left and the battalion commander on the right observing a terrain scene

STRISIM with the game controller's work station on the left, and the commander's vehicle mockup and visual screen on the right

CATS Search and Rescue C^3I Trainer

This Comator simulator is used during the training of Search And Rescue (SAR) control officers in conducting operations both on the scene and from Rescue Control Centres (RCC). SAR operations covered when using the simulator include: situation analysis; strategic and tactical decision making; formulating and issuing orders; communications; plotting; information; planning; and co-operation during joint actions.

Trainee Stations. Trainees work in an environment which resembles their normal RCC or vehicle as closely as possible. Vehicles can be parked close to the simulator screen and connected to the communication system. Trainees and operators in the game room have access to a continuous visual presentation of the incident site or area of operation. Results of actions ordered by trainees can be seen on the screens.

Exercise Control. Radio and telephone messages are used from 'exercise control' which has a number of operators situated in an adjoining room. Exercise control also represents external parties. The exercise controller can manipulate the visual presentation or give instructions to the operators to respond in a certain way to requests and orders given by the trainees. The game control is manned by a limited number of operators and an exercise director. Operators represent field units and personnel not being exercised. The operators work in positions equipped with computer terminals. On the display they have a personal plot of units under their 'command'. The plot shows in real time the movements of units with a realistic speed dependent on the choice of route, terrain and weather conditions. They also monitor fuel states and other consumables, likewise the work time of dispatched personnel. A log of the resources available for the operation is also shown.

Accommodation. The system can be built as a mobile unit in a container. This contains the computer, the game control with two operator and one exercise director positions, and a visual display. The container can be taken to wherever it is required and command vehicles can be connected into the system.

Rescue Control Centre (RCC) Training. When used for RCC training, the simulator contains five rooms furnished and equipped as different types of control centre resembling temporary centres established close to an incident. Exercises can then be conducted with the simultaneous participation of fire services, government SAR services, and other organisations such as police and medical services. Each control room can be equipped with a personal computer to be used by training units that normally have access to data support. The computers can be connected to a public telecommunications net. One room is used as an incident site terrain room, to which trainees have access to obtain visual impressions of the incident site. The scene is displayed on three large video screens and sound effects can be added. If trainees are operating from a command vehicle the vehicle is parked adjacent the terrain room and the scene is viewed by opening sliding doors.

Visual System. Visual presentations consist of two main groups, environments or backgrounds and incidents. The backgrounds are: pictures of locations such as city or urban areas with high buildings; suburban areas with small houses; industrial areas; motorways, roads and railways; oil and chemical storage areas; harbours and airports; rural areas; forests and wilderness areas; coastal areas and archipelagos and open sea. In these environments rescue units can be shown as symbols and moved about in a realistic manner. Different light, visibility and weather conditions can be applied. Environments may also be photographed interiors of areas such as apartments, factories, public places, and transport systems such as aircraft, ships, buses or trains. Fire, smoke, injured people and damage can be added to these backgrounds by using computer graphics. Incidents and incident sites are a combination of photographs and computer graphics. They may include fires, transport incidents, landslides and floods. The incident site displays can be altered gradually dependent on actions taken by the trainees.

Communications. For communications the control rooms are equipped with three to five contact points for standardised communication units. The exercise director position, the operator positions and the parking area for command vehicles have one contact point each. Two additional contact points are available on the outside. The communication units are identical and enable communication on 10 different frequencies. Availability, readability and range can be altered by the exercise director. There is also an internal telephone net connected to the public net. Messages can be sent to all rooms on a separate one-way communication net with receivers in all rooms. The transmitter is located in the game control.

SAR simulator terrain and environment display with SAR vehicle and personnel

Status

Delivered to the National Rescue Services in Sweden.

Manufacturer

Comator Process AB, Helsingborg.
(For details see Directory of Suppliers)

SAR simulator control room; note same visual scene as adjoining picture

UNITED KINGDOM

BAeSEMA Brigade and Battle Group Trainer (BBGT)

The British Army's Brigade and Battle Group Trainer (BBGT) was developed from earlier Battle Group Trainers, one of which remains in use in the United Kingdom at the Armoured Corps depot at Bovington, Dorset, UK. The later BBGTs are in service with the British Army at Sennelager (Germany) and Catterick (UK). BAeSEMA continues to support the existing systems and to enhance them as required.

Training Provided. The BBGT provides training for brigade and battle group commanders and their staffs. It enables the command, staff support and service functions within a headquarters to be exercised realistically at a static location.

Components. The BBGT consists of three basic areas: an Operations Headquarters; an Exercise Control area; and a System Control room. The system provides a representation of the standard communications systems deployed in the field, centrally controlled by a computer which also stores a digital terrain database. The main systems represented include combat net radio and Ptarmigan trunk communications. The system was designed to allow visiting units to operate their own in-service command information systems such as Wavell.

The staff located in the Operations Headquarters are housed in either replica command vehicles or in a reproduction of a headquarters building. As this environment is realistic it allows the exercising staff to derive the maximum benefit from their limited training staff.

The Brigade and Battle Group Trainer (BBGT)

Exercise Control. The Exercise Control area contains a map table on which the force dispositions are depicted. Around the side of the table, cells contain staff officers representing higher and lower headquarters as well as an overall Exercise Controller. A separate System Control room holds the equipment needed to sustain the communications networks. The Exercise Controller has the ability to inject EW effects on net communications. The communication links can also be monitored and recording made for later debriefing purposes.

Enemy Forces. The 'Orange' force (opposing the 'Blue' force) is represented by permanent BBGT-based staff.

Results. Results are processed using a combat evaluation software package which runs in near-real time and ensures that the flow of the simulated battle is not interrupted. Evaluation includes direct and indirect fire, and also the effects of close air support and minefield breaching.

Status

In service with the British Army.

Manufacturer

BAeSEMA Limited, Surrey.
(For details see Directory of Suppliers)

Defence Research Agency Wargaming and Tactical Trainers

The UK Defence Research Agency (DRA) have developed a number of land systems combat models and trainers as part of their research activity. Some of these are available for hire at normal commercial rates, subject to the DRA's commitments to other customers including their primary customer, the UK MoD.

UK DRA Close Action ENvironment (CAEN)

CAEN is a two-sided close combat model which has been developed by the DRA Operational Studies (Land) Department. The model represents close combat up to company strength. It has been used as an OA tool, and can also be used either as an interactive wargame or as a tactical training simulation.

Terrain. Digital terrain is provided from the DTED data of the NATO DLMS system. A simulated battle area 5 km by 6 km is provided at a resolution of 10 m. Seven types of rural culture and seven types of urban buildings are provided, together with roads, rives, field defences and obstacles.

Graphics. Graphics are provided which give variable magnification from x1 to x128, cross-section, and sensor view.

Forces Modelled. Forces up to company strength on both sides are modelled. Representation of individual men, crew-served weapons and AFVs are included, as are indirect fire support from artillery and mortars.

Models. Autonomous models allow units to move and interact. Models include: surveillance, suppression of units and suppressive fire, detection, acquisition and engagement, minefields, indirect fire, obscuration, weapon sharing, and fighting in buildings.

Player Facilities. Players input units and ammunition scales and select routes, arcs and activities. Preset activities by individuals and groups can be configured, as can recce of the battlefield using sensor view, square profile and cross-section views.

UK DRA JANUS Battle Group Wargame

The JANUS Battle Group Wargame is an interactive, two-sided, computer wargame of ground combat employing stochastic (statistic-based) computations. Play is closed, in that information can only be gained about an enemy force which can be gathered by friendly units.

Terrain Model. The gaming area is a digital terrain map of 600 by 600 cells which allows an area of 60 km square at 100 m resolution. Seven types of vegetation and seven types of urban terrain are provided each with its own properties and symbol. Rivers of specified widths and roads can also be represented.

Database. Menu-driven inputs using toolbars drive a structured database, split into system data for weapon capabilities and force data for unit and task force organisation. Assets such as aircraft, artillery and engineer equipment can also be included.

LoS Algorithm. A LoS algorithm determines whether LoS exists between elements on the terrain. A player is therefore able to explore the LoS opportunities at a given location.

Other Facilities. Other facilities include surveillance and target acquisition (STA), direct fire, personnel units (mounting, dismounting, going prone under fire, etc), indirect fire, obscurants, engineering systems, aircraft, fibre-optic guided missiles (FOGMs), and post-game analysis.

Operational Use. The JANUS system has already been used as an OA tool to examine weapon performance, tactics and doctrine at brigade and battle group level, to provide scenarios to drive simulations, and was used to give support for the Gulf War. Recent studies using JANUS include: the future UK recce vehicle; MLI for the Warrior ICV; new rounds for the RARDEN cannon; requirements for the Air Mobile Brigade (AMB), and the optimum mix of anti-armour weapons.

Status

Available at commercial rates subject to other bookings.

Supplier

DRA (UK), Battle Group Wargaming & Simulation, Fort Halstead, Kent.
(For details see Directory of Suppliers)

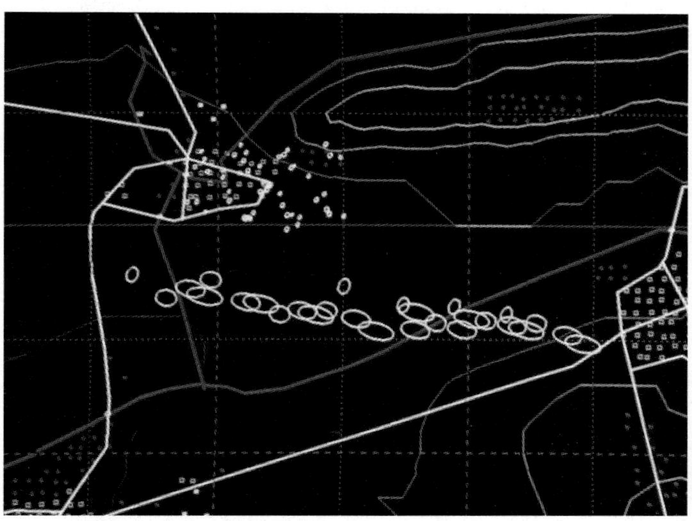

JANUS - plan view of obscurants landing on the battle area

CAEN CGI sensor view

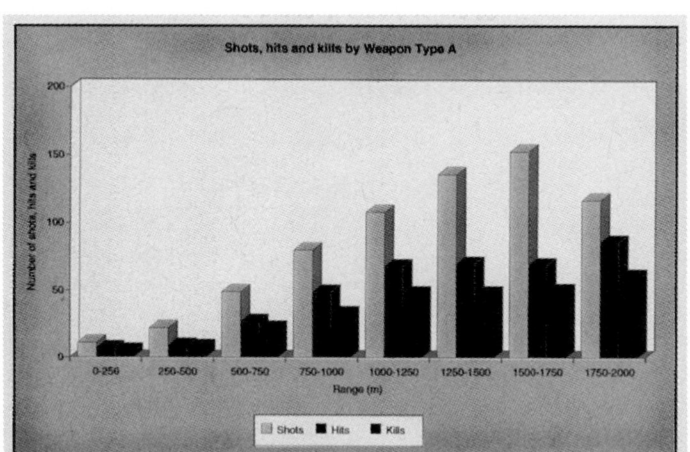

JANUS - example of post-action analysis

CAEN CGI showing a general area view

UNITED STATES OF AMERICA

CACI TAC Thunder

The CACI TAC Thunder package was developed from the earlier TAC Warrior programme. TAC Thunder simulates theatre-level air and ground warfare together with logistics to provide a tactical air campaign model. Its purpose is to assist the analyst studying the impact of policy decisions such as force location, alternative equipment choices and theatre-level air tactics.

Air Model. The model simulates the air war to the level of flight groups on missions while individual air bases servicing aircraft are also modelled. Air war planning is accomplished by means of generating mission orders from intelligence collected in the model and from air resources collected both dynamically and from input. When aircraft are available missions are flown.

Ground Model. The ground war is modelled with divisional units which engage each other causing the Forward Line Of Troops (FLOT) to move. The C^3I net and logistics are modelled to assess unit effectiveness and mobility features are also included. The purpose of the air operations package is to assess the effectiveness of operations upon the ground war and the targets which may be attacked include units, choke points, C^3I and the logistics network. The effect of air defence resources is also taken into account. Geographic databases can be prepared in three to six months, force structure databases in one or two weeks, while scenarios can be produced within a few hours.

Status

Available. In service with the US Air Force.

Manufacturer

CACI International Inc, Virginia.
(For details see Directory of Suppliers)

CAE-Link Army Training Battle Simulation System (ARTBASS)

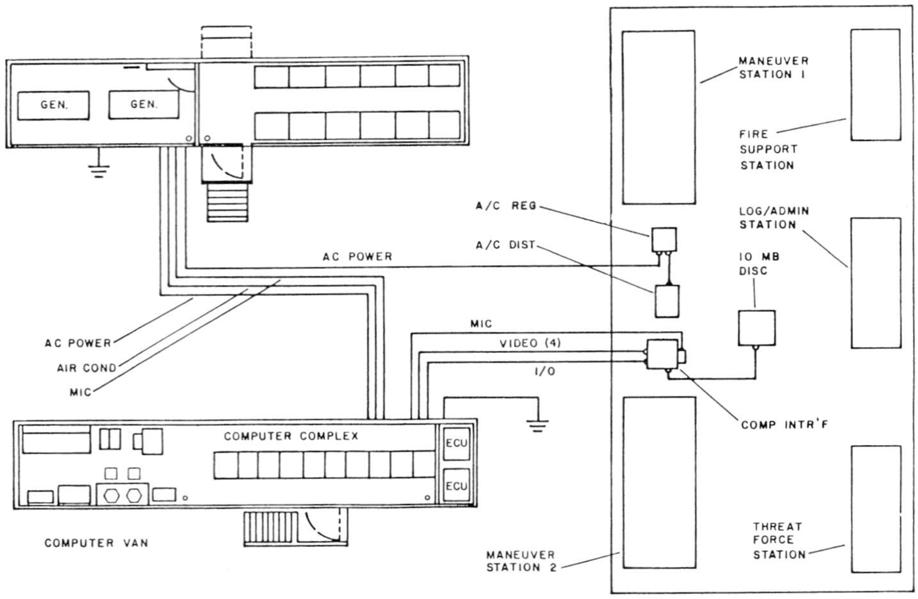

Schematic layout of an operating ARTBASS

The CAE-Link ARTBASS is designed to train battalion commanders and their staffs in the conduct of battlefield operations. It was developed from an earlier simulation system known as the Combined Arms Tactical Training Simulator (CATTS). It provides real-time, two-sided, interactive simulation of tactical operations. The system can be used for exercising new tactics and doctrines and to research the effectiveness of new command and control procedures. Other applications include New Organisation Team Training (NOTT), command group training and evaluations, and preparation prior to attending exercises at the National Training Center.

Components. ARTBASS is mobile and can be used at existing bases, utilising the user organisation's communications and command post equipment. The major subsystems are: a model of a ground combat environment; a control facility; a trailer-mounted central computer system; and a trailer-mounted power system. Two commercial-model semi-trailer vans are used, one for the computer system, while the second houses the power generation system and the control facility equipment. On arrival at a training site the semi-trailers are parked close to a suitable building and the control equipment is unloaded and installed in the building.

Terrain Model. The terrain model, based on digital DMA data, covers 5000 km² and airspace is included to an altitude of 10 km. Water features, weather, vegetation and forests, ground surface and soil types, wind and visibility are also included. Map displays are in 64 colours. Six scales are available from 48 × 36 km to 4 × 3 km. Weapon graphics can include tanks, anti-tank rockets and missiles, field artillery, mortars, air defence, air missions, impacting fire and preplanned targets. The graphic displays are updated once every minute for ground units, four times every minute for air units and upon repositioning of the map.

Units Simulated. The system can simulate a total of 200 units simultaneously during an exercise. These can include combat, combat support and combat service support units. Both air and ground units are included in the model.

Control Facility. The Control Facility equipment group is divided into six stations: Manoeuvre Stations 1 and 2; a Threat Force Station; a Fire Support Station; the Logistics/Administrative (Log/Admin) Station; and a Higher Command Station. The Manoeuvre Stations are operated by the simulator staff, with the company commanders (role players) collocated to fight the battle. The Log/Admin and Fire Support Stations are arranged to be manned by the staff and role players while the Threat Station is manned by staff only. Each station has Multi-Function KeyBoards (MFKBs), the cursor control bit pads, controller Input/Output (I/O) terminals, Terrain Map (TM) and Terrain Appreciation (TA) displays, and line printers.

Types of Training. Types of tactical operation that can be carried out include: movement to contact; attack; reconnaissance; exploit and pursuit; ambush; hasty and deliberate defence; delay and withdrawal. Battalion elements can include infantry, airborne infantry, air assault, mechanised, tank and cavalry. Targets are identified as static, moving, ground or air. Command and control inputs possible include: unit initial locations and missions; changed operational state; move units; control firing; establish control measures; significant event tagging; Army aviation; Air Force assets; and air defence. Special inputs include: manoeuvre; fire support; logistics; environment and weather; obstacles and mines. The following are modelled:

- Equipment types per unit - 30
- Ammunition types per unit - 30
- Equipment types overall - 200
- Ammunition types overall - 150
- Concurrent smoke missions - 150.

Air Aspects. Simulated aviation assets employed in support of the Blue battalion operations include utility, observation and cargo helicopters, helicopter gunships, close air support aircraft and reconnaissance and surveillance aircraft. The controllers are provided with the capability to assign air missions interactively and they can specify mission types, flight routes, number of aircraft, type of aircraft, types of equipment, targets, weapon or sensor load. Any of the air units specified can be designated as Red force units. Air defence parameters include maximum detection range, maximum and minimum horizontal engagement range and altitude, acquisition probability, rate of fire, probability of kill per round, and EW factors.

A typical ARTBASS manoeuvre station

A typical Battlefield Command Trainer control room

Damage Calculations. The effects of weapons on personnel and equipment are calculated. Results are provided in terms of personnel casualties and equipment damage. Factors considered are weapon and target characteristics, weapon factors, firing unit characteristics, minefield casualties, personnel casualties, equipment damages, environmental damage, and the suppression of unit strength. The value of suppression is used in the calculation of rate of movement, vulnerability, detectability, rate of fire, and weapon effectiveness.
Logistics. Each unit has a starting quantity of equipment and supplies allocated and, as the unit consumes these, or loses them in combat, that unit's allocation is reduced. Any resupply causes the unit's supply to be increased. Resupply will occur as a result of a command to transfer resources between units. A record is maintained continuously of the personnel, equipment and supply status. A minimum threshold can be specified by the user which, when reached, will produce an alert message at the responsible control station. The logistic elements must manage to supply volumes where they are needed when they are needed. The system also models maintenance activities to allow for the recovery and repair of partially damaged equipment, and for combat engineers to build and remove obstacles and minefields. Up to 20 critical repair equipment types can be designated for repair.
Record/Replay. Record, replay, freeze and restart facilities are included. At the end of an exercise the system generates user selectable hard-copy post-game summary reports on the computer line printer.

Status
In service with the US Army and National Guard.

Manufacturer
CAE-Link Corporation, New York State.
For UK version: (design of the controller station user interfaces and installation, maintenance and logistics support), Invertron Ltd, West Sussex, UK.
(For details see Directory of Suppliers)

The interior of a typical ARTBASS tactical operations centre

CAE-Link Battlefield Command Trainer (BCT)

Building on previous experience with the ARTBASS system (see previous entry), CAE-Link developed the Battlefield Command Trainer (BCT). It utilises improvements in computer techniques to enable a high level division/corps trainer to simultaneously have weapon-on-weapon combat resolution. It also incorporates detailed support models, and uses the NATO standard 25 metre data for terrain and feature information over a large gaming area.
Higher Level Staff. The HQ Staff to be trained is located in its tactical command post facilities in a field location and never actually sees the trainer. The HQ Staff operates as it would in combat, maintaining contact with subordinate, adjacent and superior units by normal combat radio nets, land-lines, or other appropriate tactical communication methods. All staff planning and control functions are carried out as in combat.
Role Players. The role players (called Unit Controllers) represent subordinate, adjacent and higher echelon units in the field. Subordinate unit commanders and higher headquarter staff members typically perform the Unit Controller (role player) function, providing valuable training to them as well as the HQ Staff being trained. Unit Controllers interface directly with the HQ Staff via normal tactical communications channels and with the computer simulation via the Unit Controller workstations.
Aggregation of Formations. Aggregate Control allows individual companies to be aggregated into battalions and individual battalions to be aggregated into brigades. Unit Controllers can choose to issue commands to and receive information from either the aggregated unit (the brigade) or any of the lower echelon units (battalions or companies).
Factors Included. The following factors are included in a similar way to ARTBASS but to a more sophisticated level: Manoeuvre; Intelligence/Electronic Warfare; Ground Combat (utilising stochastic models to resolve combat at the individual weapon system level); Artillery/Air Defence Artillery Naval Gunfire; Attack Helicopters; NBC; General Aviation; Combat Engineer; Maintenance; Supply; Personnel; Medical/Casualty; Amphibious Operations.
Terrain Data Base. Exercise areas may be up to 200 x 400 km to a 25 metre resolution. NATO Digital Terrain Elevation Data (DTED) and Digital Feature Analysis Data (DFAD) is utilized to generate Line-Of-Sight (LoS) and Cross-Country Movement (CCM) files. All LoS and CCM files are registered with display maps which are currently to scales of 1:50 000, 1:250 000, and 1:500 000. Scale maps are displayed using NATO standard Arc Standard Raster Product (ASRP) image data. Display map features include 4 X Zoom and Scroll at any map scale.
Map Generation. Maps may be generated for new exercise areas from standard ASRP, DTED and DFAD data (see above).
Computing. The configuration consists of a server computer and multiple Personal Computer (PC) based Unit Controller workstations, communicating over a modular, 'hub and spoke' type Ethernet local area network (LAN). Some 200 000 lines of code are devoted to modelling battlefield interactions. A Silicon Graphics (SGI) Challenge L is the main server. An Optical Data Systems (ODS) concentrator is used for the LAN and a Ciszo 4000 router for a WAN. The Unit Controller stations are configured with a DEC 486 PC, 17 or 19 inch colour monitor, ASCII terminal, dot matrix printer and a mouse. Standard systems are used such as Windows, Empress DataBase Management System (DBMS), MS-DOS, C, C + +, FORTRAN, and PC/TCP-IP Network software.
Developments. Work has begun to integrate BCT with Interactive Tactical Environment Management System (ITEMS), an entity level simulation. This will provide a virtual/constructive simulation interface, allowing full-range tactical training from the individual soldier level through to the Corps Staff level. Work has also begun to update CAE-Link's patented terrain data generation capability. This will provide BCT with the ability to supply detailed terrain data to customers with limited or no access to 25 metre terrain data. The following other features are also being explored: Automated and semi-automated threat forces; enhanced Air Force and Naval modelling; Special Operations Forces capabilities; multi-force conflicts (red vs blue vs green); civilian emergency management (disaster) training.

Status
Available for sale. A brigade level BCT is in service with the New Jersey National Guard (known as 'Enhanced ARTBASS'), and a division level BCT was sold to the Singapore MoD in 1989 (known as FULCOM). A corps level version will be delivered to the UK MoD in 1994 under their Higher Formation Trainer (HFT) programme, for use with the multinational ACE Rapid Reaction Corps (ARRC).

Manufacturer
CAE-Link Corporation, New York State.
(For details see Directory of Suppliers)

A CAE-Link Battlefield Command Trainer (BCT) manoeuvre station

CAE-Link Junior Leader Tactical Training Programme

A training-orientated Interactive Video Disk (IVD) programme has been produced for the US Army National Guard under the direction of the US Army Research Institute field unit at Boise, Idaho. It is known as the Junior Leader Tactical Training Programme.

Prototype Programme. The prototype programme consists of approximately 30 hours of courseware designed to present several sequential and progressive training modules for armour platoon leaders.

Trainee Cues. Simulation graphics, video motion, video stills, and digitised audio for voice and weapons effects, are used to create the exercises to train the platoon leader to manoeuvre the platoon, know the capabilities of enemy anti-tank weapon systems, to recognise enemy defensive tactics, and plan a movement-to-contact operation.

Record/Replay. The feedback facilities have graphic overlays and multiple video images used with digitised voice to analyse performance and provide teaching points. Hard-copy results are also provided.

Status

Initial fielding Spring 1991. Further courseware completed Spring 1992. Two modules (22 hours) address the execution of tactical operations in the offence; two modules (22 hours) train skills involved in the planning and execution of a hasty defence mission.

Manufacturer

CAE-Link Corporation, New York State.
(For details see Directory of Suppliers)

Combined Arms Staff Trainer

The Combined Arms Staff Trainer (CAST), or Device 16N8, is a US Marine Corps combat simulation device designed to train Marine Amphibious Unit (MAU), Marine Amphibious Brigade (MAB) and Marine Amphibious Force (MAF) staffs in supporting arms co-ordination and application. CAST will enable the US Marines to practise the control and co-ordination of artillery, air, naval gunfire and mortar fire support.

Terrain Model. The terrain subsystem uses both two- and three-dimensional terrain. Either type of surface may be used to depict the locations of both friendly and enemy forces and to display the impact of fire support missions on these units. The terrain surface can be as large as 12 ft × 32 ft (3.657 × 9.75 m). The three-dimensional terrain boards are designed in 4 ft × 8 ft (1.22 × 2.438 m) sections and depict a variety of geographical locations. The scale of the terrain varies from one exercise to another when two-dimensional surfaces are used. The three-dimensional surface is constructed at a scale of 1 in to 125 m.

Computer System. The computer subsystem is a series of networked PCs that provide for the entry of fire mission data. The same method is used for Forward Observer/Forward Air Controller (FO/FAC) requested missions. The computer applies predetermined delay factors and then executes the mission by illuminating the terrain surface with lasers at the appropriate time and for the appropriate duration. Fire mission requests, fire plans and notes in an exercise time-sequenced list are stored for post-exercise analysis and critiques.

Fire Marking System. The application of fire support from artillery, aircraft, naval gunfire and mortar assets is presented on the terrain subsystem through the use of a computer-controlled Fire Marking System (FMS). The FMS consists of eye-safe lasers that illuminate the terrain surface, displaying the impact points of fire support missions. The lasers display red and green spots for enemy and friendly fire as well as providing steady or pulsing illumination of high explosive shells and pulsing illumination for white phosphorous ordnance.

Communication System. Each player position can communicate with other network stations. The CAST also provides the controllers with the capability of monitoring, recording and the application of jamming techniques over any frequency. The controllers can communicate with each other during an exercise.

Housing. The CAST is housed in a permanent building with two floors. The centre of the building is occupied by a well in which the terrain boards are located. Various cells and rooms are located around the well, which extends to the roof, and there is space for exercise spectators as well as control staff and players. The fire marking subsystem occupies the ceiling area. Each cell or room is occupied by a particular staff function and space is provided for offices, a lecture/debriefing room and storage space.

Deployment. The CAST is operational with the US Marine Corps Air Ground Combat Center (MCAGCC) at Twentynine Palms, California, to support brigade and unit level Combined Arms Exercises (CAXs). CASTs are being installed at: Camp Lejeune, North Carolina; Camp Pendleton, California; Camp Butler, Japan; and MCAS Kaneohe Bay, Hawaii. The MCAS Kaneohe Bay CAST will support MAB and MAU level exercises while the Camp Lejeune, Camp Pendleton and Camp Butler CASTs will support up to MAF level exercises.

Manufacturer

CAE-Link Corporation, New York State.
(For details see Directory of Suppliers)

JPL Corps Battle Simulation (CBS)

Corps Battle Simulation (CBS) is a computerised battle simulation system designed to drive Joint Readiness exercises, Army corps and division command post exercises, and Battle Command Training Program War Fighter exercises. The system supports Army staffs down to brigade and battalion levels and Air Force staffs.

Capabilities. CBS simulates all major aspects of the air/land battle, including combat, combat support, and combat service support. CBS supports intelligence play via an interface with Tactical Simulation (TACSIM) and allows other external interfaces. Additional software has been developed to support scenario preparation, game initialisation and after-action review. CBS is operational at the following sites: I Corps, Fort Lewis; III Corps, Fort Hood; XVIII Corps Headquarters, Fort Bragg; Combined Arms Center, Fort Leavenworth; Joint Warfare Center, Hurlbery Field; 21st Theatre Area Command, Kaiserslautern; 32nd AADCOM, Darmstadt; 8th Army, Korea; 3rd Infantry Division, Schweinfurt; 8th Infantry Division, Krevtznach; 75th Maneuver Area Command, Houston. The current version is CBS 1.3 which is the fourth fielded version of the system. Current CBS development is by the Jet Propulsion Laboratory of CIT, sponsored by PM-TRADE.

Computers. A CBS suite consists of a central computer, normally a DEC VAX 6420 containing the battle simulation software. The 6480 is interconnected with MicroVAX computers which support up to 256 controller workstations.

Networking. Ethernet-based Local Area Networks (LAN) connect the CBS site workstations using the DECnet communication protocol. Wide Area Networks (WAN) permit worldwide exercises by linking CBS sites.

Displays and Terrain Models. Map images and graphic military symbols are displayed on colour monitors. A video disk player displays map images which may contain up to 54 000 discrete map frames. Existing terrain databases cover the Korean peninsula, Southwest Asia, Central America and Central Europe. Unit databases cover most US forces and many allied and OPFOR orders of battle.

Database Development. The Scenario Preparation Program (SPP) assembles simulation databases describing all model parameters and units. The Report and Scenario Verification Program (RSVP) then evaluates the database's internal consistency, thereby eliminating run time problems caused by data errors. The Terrain Database Preparation System (TDPS) develops a new terrain description. The US Army Combined Arms Center - Training (CAC-T) at Fort Leavenworth, Kansas, currently prepares CBS databases.

Events Modelled. The following military actions are modelled:

Ground movement - The impacts of congestion, march interval, road and terrain conditions, obstacles, military police and unit capability are considered.

Ground combat - The Combat Outcome Based on Rules for Attrition (COBRA) is an enhancement to Lancaster methodology for computing ground combat attrition. COBRA uses a rule-based expert system to evaluate combat situations and determine appropriate multipliers for simulation attrition calculations. COBRA describes the rules in each combat assessment in English text which is archived and can be reviewed. The simulation may be run with or without COBRA. A post-processor provides a capability to review COBRA rules. Individual weapon ranges, day/night, weather, unit posture, terrain, fortifications, NBC conditions, suppressions due to incoming artillery fire and direct fire, system crew capabilities, combat fatigue, infantry/armour mix, and engagement geometry are all considered. The COBRA system analyses each situation and determines the appropriate multipliers for the Lancaster equations.

Artillery - Ammunition types are HE, improved conventional munitions, scatterable mines, Copperhead, chemical and nuclear. Copperhead projectiles require a forward observer to designate a target.

AAA - This can utilise radar-guided missiles, radar-controlled guns, infra-red missiles, and optically sighted guns. Modes of operation are automatic or manual pairing, and self-defence. The units weapon status of Free, Tight or Hold will affect automatic engagements.

Engineers - Engineers can repair or destroy bridges, create or breach point obstacles and anti-tank ditches, install or breach minefields, fortify battle positions, and cut or repair runways.

NBC - Nuclear and chemical projectiles will cause casualties while equipment will have to undergo decontamination. Units can wear protective equipment but will suffer from degraded performance and heat casualties.

Supply - Supply consumption for each supply category can be daily, per person per day, or as used. All supply classes except II and VI are monitored. Convoys pick up supplies from supply centres and deliver to requesting units.

Medical Support - Wounded personnel are sent to medical units. Losses due to disease are included. Severe casualties are evacuated out of theatre.

Maintenance - Equipment items under repair are monitored. Workloads can be shifted between maintenance units. Losses due to non-battle damage are included along with air maintenance.

Radar and EW - Counter-battery radars are included and can be detected by electronic intelligence. AWACS acquisition radars on fixed-wing aircraft and ground-based radars can detect both enemy and friendly aircraft. Degradation of radars by airborne jammers is modelled.

Fixed-Wing Air Operations - Mission types represented are: defensive counter-air, close air support, battlefield air interdiction, air interdiction, offensive counter-air, suppression of enemy air defence, reconnaissance, airlift/airdrop, and air refuel. Air mission packages can be used to combine a variety of missions into a co-ordinated strike force. Ingress and egress routes can be specified.

Rotary-Wing Air Operations - Mission types represented are: attack, blocking, support, screening, and airlift/airdrop. Operations may be conducted from any ground combat unit, from a forward area rearming and refuelling point, or from any location within the gaming area. Ingress and egress routes and associated speed, altitude and mode of flight can be specified.

Airlift/airdrop - Both fixed- and rotary-wing aircraft can airlift/airdrop units and supplies. An automated load planner is available to assist the controller.

Enhancements. Scheduled product improvements include the following:

Special Operations Forces - These will be represented in the simulation along with missions specific to Special Forces and Ranger units. In addition, the ability to infiltrate these forces and other dismounted and infantry forces through enemy-occupied areas to perform reconnaissance missions and to engage in high resolution combat will be added.

COntroller ASsisT (COAST) for Infiltrate - Infiltrating the platoon elements of battalion-sized light infantry units is a labour-intensive task for a CBS controller. COAST for Infiltrate will automatically manage many aspects of the infiltration task for the controller.

Reconnaissance - The controller will be able to use the CBS graphics as a situation map by specifying perceived information to be displayed (for example minefields, enemy troops) in addition to the automatically displayed information.

Fire Support Modifications - Artillery and mortar units will be able to deliver fire as explicit indirect fire missions while engaged in combat. Artillery will be able to place smoke screens at specified co-ordinates.

Battlefield Smoke - Smoke and its effects on combat, observation, movement, and target designations will be modelled. Battlefield smoke will be deployed by artillery, smoke generators, or smoke pots.

Laser Designation - Ground units will be able to designate targets for RW and FW aircraft, and for artillery. Helicopters will be able to designate targets for themselves, artillery fire missions, FW aircraft, and other helicopters. Targets will include supply depots, bridges, runways, aircraft shelter clusters, missile launchers, convoys and units.

Supply Collections - Supplies for issue associated with a supply centre will now be considered a Supply Center Collection (SCC). Ground units and convoys will be able to create Ground Supply Collections (GSC). Convoys will be able to pick up from and deliver to SCCs and GSCs. Supply collections created by airlift are modelled and are called Airlift Cache Collections. Supplies within a supply collection or unit will be vulnerable to attrition, and supply collections will be subject to demolition missions. Specific supply types can be reduced by point impact weapons, bombs and artillery.

Air Defence Artillery Modifications - Enhanced aircraft recognition will be employed. Weapon deployment will be measured from the deploy radius of the unit. There will be an envelope defining the allowable delivery parameters of weapons from aircraft.

Army Aviation Modifications - It will be possible to divert returning missions to a new attack point, and to modify the egress route for a mission.

Task Organisation - OPFOR units will be able to split, merge, and cross-attach in a manner similar to BLUFOR units. The Senior Controller will be able to create OPFOR templates dynamically.

Workstation Display - Convoys, point obstacles and smoke screens will be displayed. The workstation will be able to display additional information on the graphics screen as specified by the controller.

The number of military order-of-battle units which can be represented in the simulation will be increased to 15 000. The possible number of unique workstations will be increased to over 300.

Status

In use by the US Army.

Development Agency

Jet Propulsion Laboratory, California.
(For details see Directory of Suppliers)

Perceptronics COMBAT-SIM

COMBAT-SIM is the acronym for COMputerised BATtle SIMulation and is a low-cost trainer for the command and control of military combat operations. It was developed in 1985 by Perceptronics Inc for the Australian Army. In 1986 it was selected by the US Army as the basis for their Brigade and Battalion Simulation (BBS) and 47 sets were ordered for fielding during 1988.

System Components. The Australian Army system consists of 10 IBM PC in a local area network. The US Army version runs on a 10-station DEC MicroVAX 11 suite. New modelling features may be incorporated, and a data editing capability has been added which allows the user to read the information presented in common language using a spreadsheet. For the Australian Army system this uses DBase 111 or Lotus 123, whereas in the US Army system a DEC on-screen editor is used.

Characteristics Modelled. Weapon and vehicle characteristics include range, actual firing results against various types of targets, detection, identification, ability to acquire a target and expected results given a hit. Units are modelled as a collection of personnel, equipment, weapons and supplies located within a geographic area. Models are included of fire support, close air support (CAS), helicopters and fixed-wing aircraft, engineers, air defence, and personnel and logistics. Red and Blue forces are also modelled. Three modes of play are used:

(1) Single echelon battalion staff exercises.

(2) Multi-echelon exercise for both brigade and battalion staffs.

(3) Single echelon brigade staff exercise.

Terrain Model. For terrain and environment simulation a video disk map display allows each workstation to pan and zoom independently through six levels of view ranging from a display of 3 × 4 km to 90 × 120 km. The model determines automatically when line-of-sight exists between units, incorporating the effect of foliage and terrain. The model also controls the unit speed over the various types of terrain. Provision is made for cross-sectional views, and a perspective or three-dimensional view from a specified observation point to a specified target location.

Weather and Visibility. Cloud cover, precipitation, visibility and light conditions are modelled.

Minefields and Obstacles. Minefields and other battlefield obstacles can be emplaced or cleared at any time during an exercise. Combat engineering assets are controlled and manoeuvred by the senior control station and can be field exercised in the Control Post Exercise (CPX) if desired.

Conduct of Exercises. Exercises may take place in a classroom. Another mode of operation is to conduct command post exercises with the commanders and their staffs in the field or in vehicles located in an appropriate outdoor area. Command post exercises can also be carried out with the commanders and their staffs in mock-ups of their command posts in a simulation facility in a suitable location such as a garrison area. For a brigade- or battalion-level exercise the control room needs to be large enough to accommodate 10 workstations, planning areas, a briefing and visitor area, administrative space for a permanent staff and the usual domestic facilities. A minimum of three permanent staff is required to conduct any COMBAT-SIM exercise: a manoeuvre controller, a threat controller and a systems manager. Other directing and controlling personnel can be drawn from existing unit strengths and assigned as required.

Modelling. The following are modelled:

Ammunition types - 200
Concurrent smoke missions - 120
Number of units - 2150
Minefields/obstacles - 100
Control measures - 100
Interactive workstations/consoles - 10
Data recording capability - 24 h
Digitised terrain area - 10 000 km^2
Equipment types per unit - 10
Weapon types - 300
Operational states - 16
Weather types - 8
Terrain types - 16
Terrain data resolution - 100 m
Preparation time for new game - 4-8 h
Restart/instant replay archive interval - 15 mins
Game cycle (time between updates) - 15 s

Status

Available. In service with the Australian Army and the US Army (47 sets). Under consideration by several other armies.

Manufacturer (under licence from the Australian Army)

Perceptronics Inc, California.
(For details see Directory of Suppliers)

Perceptronics SIMNET

The development of SIMNET began in 1982 as a research project sponsored by the Defense Advanced Research Projects Agency (DARPA) in partnership with the US Army. The objective of the initial programme was to develop the technology for networking large numbers of interactive combat vehicle simulators and their supporting elements. Fully manned platoon-, company- and battalion-level units can conduct force-on-force simulated engagements against opposing forces of similar composition. From 1995 the US Army will begin to equip each of its armoured divisions with SIMNET simulators.

Current Deployment. 252 individual SIMNET simulators have been installed at: Grafenwöhr, Fulda, Schweinfurt and Friedberg in Germany; Fort Knox, Kentucky; Fort Benning, Georgia; Fort Stewart, Georgia; and Camp McCain, Mississippi. Combat vehicles simulated to date include the M1 Abrams MBT and the M2 Bradley IFV; other combat vehicles are planned for inclusion, including the AH-64A Apache helicopter. An AIRNET complex for helicopters has been established at Fort Rucker, Alabama, and studies are under way to produce a JETNET to include US Air Force A-10 and F-15 aircraft simulators. A SIMNET installation at Fort Knox, Kentucky is being used to test new simulation

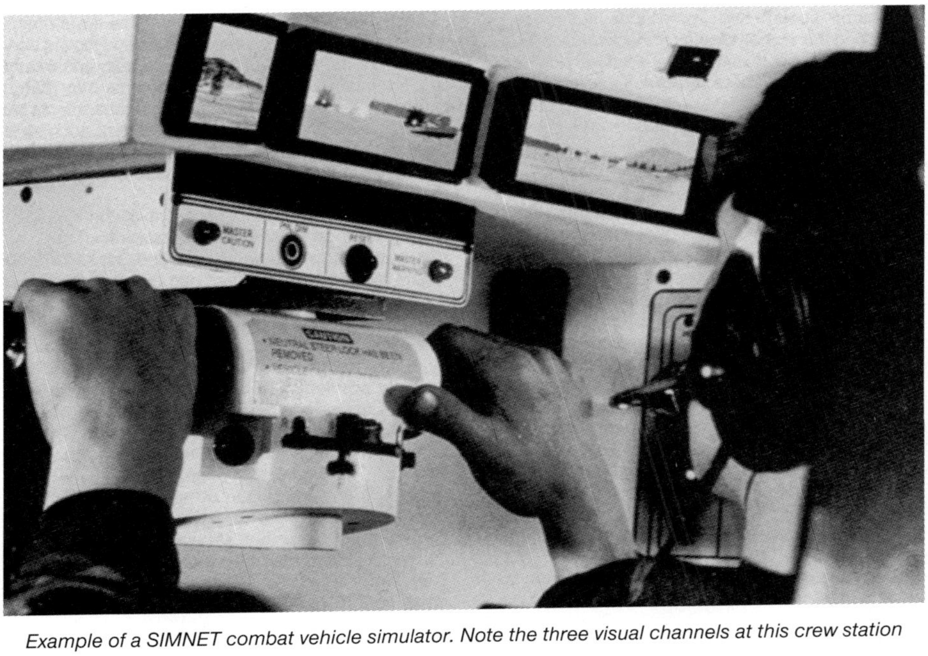

Example of a SIMNET combat vehicle simulator. Note the three visual channels at this crew station

concepts and system operators.

Simulator Components and Characteristics. Each simulator has a host microprocessor, a visual and sound system, controls and instruments, and a terrain database. It is fitted with facsimile controls for the gunner, commander and driver. Realistic delays are included for reloading the main gun and transferring ammunition inside the turret. Five different types of terrain are provided, and any attempt to cross an unfordable water obstacle will result in the vehicle bogging down. For fuel or ammunition resupply the simulated vehicle has to be driven to within a set distance (usually 20 m) from a resupply vehicle and an inactive appropriate resupply time is imposed. Maintenance parameters are also built into the system.

Visual System. Each simulation element has a low-cost multi window CGI system (based on the BBN Delta Graphics CGI system) that provides a view of the battlefield to each crew member. Through these screens the crews can see terrain constructed using US Defense Mapping Agency (DMA) data. The terrain representation is realistic enough for crews to navigate through it, recognising roads, rivers, hills, tree lines and other distinctive terrain features. In addition, they also see combat vehicles, combat support vehicles and other vehicles such as fuel trucks, helicopters and self-propelled howitzers. The actions of these vehicles reflect the control actions of other vehicle crews in other simulators elsewhere on the SIMNET network.

Support Aspects. The support vehicles are not fully crewed. They are controlled from consoles connected to the SIMNET Management, Command and Control (MCC) system. A Logistics Console is used to direct fuel and ammunition vehicles to rendezvous points where combat vehicles are resupplied. The Fire-Control Console is used by the battalion Fire Support Officer to direct artillery and mortar fire. A Close Air Support Console is used by the Air Liaison Officer to direct air strikes. The latter two consoles are collocated in the battalion Tactical Operations Center (TOC).

Networking. The simulators within each installation are modular and are able to function independently, but when forming a Local Area Network (LAN) they are connected by Ethernet. Each LAN requires one extra microprocessor and from two to six additional microprocessors for user interface. LANs can be connected together into what is described as a Long-Haul Network, or LHN, also described at times as a Wide Area Network, or WAN. Using an LHN allows units at remote sites to train together. Thus, units involved in the LAN at Fort Knox can 'engage' units involved in the LAN at Grafenwöhr in Germany. Use of the LHN will also enable any unit to engage the helicopters and attack aircraft based at the Fort Rucker AIRNET facility. Three types of LHN system are under evaluation: dedicated digital land-lines, wideband satellite, and microwave.

Work Share. Perceptronics Inc was responsible for the hardware design and development and the overall establishment of the SIMNET system. Others involved included Bolt Beranek and Newman Inc (BBN) and their subsidiary BBN Delta Graphics, who were responsible for CGI, data communication and distributed software. BBN has subsequently sold its simulation business to Loral (See under Loral Simulation in the Directory of Suppliers).

Status

In service with the US Army.

Manufacturers

Perceptronics Inc, California.
(For details, see Directory of Suppliers)
Loral Simulation Inc, Massachusetts.
(For details see Directory of Suppliers)

MISCELLANEOUS LAND-BASED TRAINERS

GERMANY

Dornier Radio Teletype (RTTY) Trainer

The RTTY trainer is a computer-based system which allows the simulation of various types of teletype transmissions between stations. A number of trainee stations are connected to an instructor station. This allows for sending messages from station to station or the broadcasting of messages to groups of stations.

Operation. Callsigns and routing indicators are allocated and message handling functions are programmed to enable training in the complete range of RTTY procedures. Messages are writtten on keyboards and are printed out as hard copy. The keyboards are modified in order to provide teletype 'machine functions'. Other teletype equipment may be provided such as tape punchers and tape readers.

Status

Available

Manufacturer

Dornier Luftfahrt, Bremen.

(For details see Directory of Suppliers)

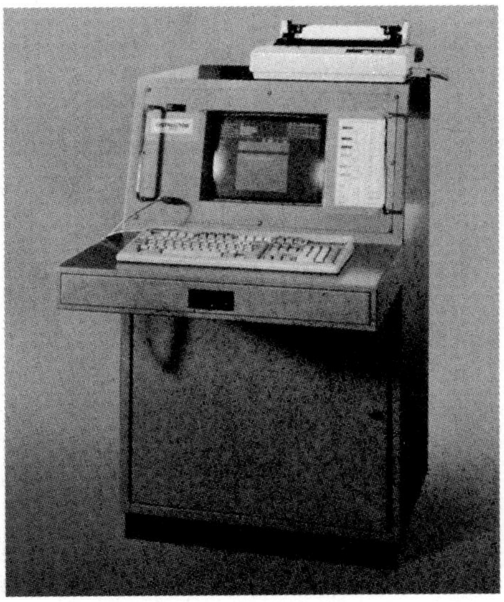

Dornier RTTY Teletype Trainer

SOUTH AFRICA

UEC Aerial Target Instructional System (ATIS)

The UEC Aerial Target Instructional System (ATIS) is a computer-based training aid which, although intended for the training of air defence gunners, can be adapted to impart training in many other different skills. ATIS is built as a free-standing and self-contained instructional system intended to impart the skills of aircraft and insignia recognition plus estimation of aircraft target approach speed, range, angle of approach and required aim-off. Other skills could be taught by the same system.

Operation. ATIS may operate as a single stand-alone unit or a number of ATIS stations may be connected to run as a networked system; the unit is programmable to either mode. Menus are selected by the student by operating a trackerball. The courseware was designed along the lines of an arcade game and includes gaming elements to stimulate the learning process.

Instructor Aspects. Functions available to the instructor include student registration, course assignment, messages, and generating student or lesson reports, among others. Student responses are constantly logged and recorded for later assessment and analysis.

Status

Available.

Manufacturer

UEC Projects (Pty) Limited, Natal.

(For details see Directory of Suppliers)

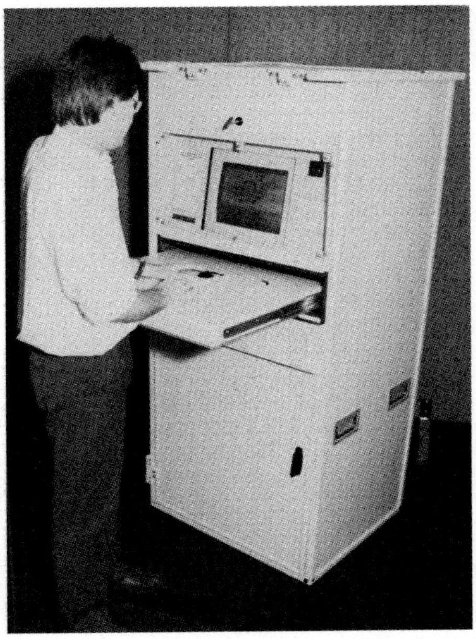

UEC Aerial Target Instructional System (ATIS) in use

SWEDEN

TELEANALYS Communications Training System (CTS)

CTS is an integrated communications training system comprising an instructor's console with a switchboard, and trainee tables with connection units. Up to 36 students can be instructed in cable telegraphy, radio communications, morse, light morse and message sequence training.

Telegraphy Training. For telegraphy training the equipment can be supplemented with a Character GENerator (CGEN), a PC-based device for the generation of telegraphy messages operating through Microsoft Windows. All data is stored on hard disk and can be accessed for lesson preparation. The same messages can be used in several lessons with different parameters. It is also possible to change the character relationship in a transmitted message which can then be used for training in the reception of unrhythmic transmissions.

Control of Transmission Test (COTT). The CGEN can be supplemented with equipment to monitor transmission tests by use of the Control Of Transmission Test, (COTT). This is a single card system which is connected to the CGEN and the trainee's telegraphy key. COTT was designed as a complement to CGEN and permits control of eight students simultaneously. When monitoring transmission tests a tolerance level of 0 to 50 per cent can be selected for each of the eight trainees.

Status

CTS, with CGEN and COTT, is in service with Swedish armed forces schools (Army, Navy and Air Force). FMV ordered 60 systems for delivery in 1995.

Manufacturer

AB TELEANALYS, Solna.

(For details see Directory of Suppliers)

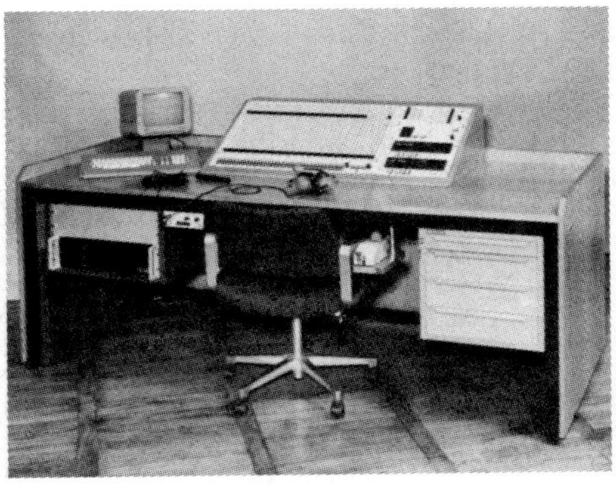

Instructor's console for the TELEANALYS Communications Training System (CTS)

TELEANALYS Simulated Communication Systems

AB TELEANALYS has been supplying communication and training systems to the Swedish armed forces since 1947. Two examples are provided below:

RF-based Simulated Radio System
This is a signal communication system for use by army battle training centres for battalion commanders and their staffs. The basic concept of the system permits a battalion commander and his staff to be trained in a realistic battle radio environment with access only to their normal field communications equipment.
Operation. RF signals are attenuated directly at the radio station output and fed into the system through double-shielded (Tempest standard) coaxial cables. All communications can be logged for later evaluation and analysis. The software-controlled simulation system provides realistic battlefield communication conditions by introducing range limits, interference and background noise.
Developments. It is intended to replace real radios with facsimiles and simulate radio communications by sound sources, complete with interference and background noise. This is intended to meet the communication requirements for future artillery and armoured forces.
Deployment. The first system installed at the Swedish Infantry Combat School is integrated with the NobelTech Electronics Army Battle Training Centre where the system can remotely control all parameters of the communication system to monitor the actual situation in a 'game'.

Audio Frequency Simulated Radio System
This is a radio simulator for the Swedish Air Force central ground training schools. The simulator is required to model all radio communications including air and ground radios. This includes base-to-aircraft, base-to-base, and base-to-ground. It uses audio frequencies and provides a realistic radio environment and system functions for the trainee operator.

Status
In service with the Swedish armed forces.

Manufacturer
AB TELEANALYS, Solna.
(For details see Directory of Supplies)

UNITED KINGDOM

AirteQ Target Identification and Recognition Trainer (TIRT)

The Target Identification and Recognition Trainer (TIRT) provides basic and advanced training in many aspects of target identification and recognition. A prototype TIRT was produced during 1989 and was used to support United Kingdom operations in the 1990 Gulf campaign. Hardware includes a PC and can include a time meter.
Operation. TIRT is basically a software program designed to produce target images on a personal computer monitor screen. The images may be in the visual or IR spectrum. The software can be configured to meet other recognition training needs.
Instructor Aspects. A TIRT system may be networked with a central Master Station for the instructor. Lessons can be self-paced and include time exposure presentations as well as exposures made according to varying levels of difficulty. Comparison lessons may be produced and the system records all student results for later analysis.

Status
In service with the United Kingdom MoD and the Canadian DoD.

Manufactur
AirteQ Limited, Milton Keynes.
(For details see Directory of Suppliers)
Marketing for TIRT: Invertron Ltd, West Sussex

IPL CREST Radio Simulator and Trainer

Innovative Projects Limited (IPL) make the CREST low-cost, multipurpose, combat net radio simulator and trainer. It can be used for basic voice and morse training and also advanced exercises, both in the classroom and in the field.
Components. It is a modular system using plug-in units which, in its basic form, can cater for a net of six students under the control of one instructor. By adding further modules the system's capability can be increased to 30 students with up to five instructors.
Operation. It is modelled on the Clansman system but can be styled to mimic any of a wide variety of radio systems currently in service. The system imitates HF and VHF transmissions, whilst variable signal strength (from good to not receiving), fading and jamming effects can all be determined by the instructor. Instructors also can communicate directly with individuals or groups without disrupting the others.

Status
In service with the British Army.

Manufacturer
Innovative Projects Limited, Chichester.
(For details see Directory of Suppliers)

The CREST radio simulator and trainer

Marconi Simulation Training Simulator for Gun Sound Ranging Equipment

This simulator is for use with Gun Sound Ranging Mark II equipments. It was originally developed by Marconi Instruments for the British Army.
Operation. The simulator is connected to sound ranging equipment fitted with Pen Recorder and the Power Supply and System Selector Unit (PS & SSU). When operated by an instructor, the simulator produces pen recorder signals which model live firing in a range of battlefield scenarios. Students can be trained either in a classroom or under field conditions. For the latter, the simulator may be quickly connected to a deployed sound ranging unit.
Components. The simulator is contained in a protective casing and is a portable, free-standing unit. The cable connecting the simulator to the PS & SSU is 15 m long, enabling it to be located in a nearby vehicle if necessary. The front panel of the simulator carries a keypad, an alphanumeric display unit, power and pen recorder connectors, status indicators and switches. Power supplies are taken from a 24 V DC 75 Ah vehicle battery while cables and other accessories are carried in a separate case.
Instruction Aspects. An instructor can prepare training or operational scenarios in advance. These can be retained in the simulator library and recalled later.

Specification
Some specification points for acoustic effects are as follows:

A 200 km^2 area of scenario
Gun signatures from typical weapon types including heavy, medium and light guns and mortars
Zero, one or two echoes from gun waves
Shell supersonic shock waves
Shell bursts
Shell splinter signatures
Microphone wind noise
Faulty microphones
Moving wheeled and tracked traffic
Aircraft and helicopters moving across the scenario.

All of the above effects may be placed at any point on the scenario using national or UTM grid co-ordinates. The simulated microphone locations and advance posts are similarly inserted and also W/V and air temperature. BITE is fitted.
Weights: simulator 37 kg; case 18 kg; accessories 7.3 kg; total 62.3 kg.

Marconi Simulation Training Simulator for Gun Sound Ranging Equipment in use (by the left operator in the picture)

Status

Available. In service with the British Army and some other armed forces.

Manufacturer

Marconi Simulation, Dunfermline.
(For details see Directory of Suppliers)

Miltra Military Training Aids

Miltra produces a wide variety of training equipment ranging from 1:100 scale models of vehicles and troops, through to full-scale training mock-ups of items of military equipment. The company can produce models and mock-ups to suit almost any requirement. Some examples follow:

Training Mine Replicas. A range of full-size land mine replicas in metal, plastics and wood are produced. Fuzes and arming mechanisms are reproduced to permit disarming drills to be carried out. Each can be produced either as an inert replica, a mine fitted for a smoke charge, or in cross-section form. A corresponding range of dummy fuzes is also produced. The range includes virtually all in-service land mines, both anti-tank and anti-personnel. They are in service in the armed forces of Canada, Germany, Norway and the USA.

EOD Training Devices. Miltra produces a range of replica bombs and bomblets for training Explosive Ordnance Disposal (EOD) operatives. These devices are detailed and accurate and can be used not only for identification purposes but also for practising disarming techniques. A corresponding range of bomb and bomblet fuzes in armed and unarmed states is also available.

Large Scale Vehicle Models. Detailed models of vehicles in scales from 1:35 to 1:5 are manufactured. Uses range from sand table exercises and logistic loading to image generation (CGI applications). Large scale models are also in use with radio control for laser gunnery trainers.

Modular Terrain Models. Modular terrain models are used to train forces in tactics, deployment, recognition, field exercises, the defence of airfields and other specific objectives. Scales are normally 1:10 000, 1:1000, 1:500 or 1:100, but for specific applications any scale can be modelled. Terrain is modelled from photographs of actual locations.

Terrain Dioramas and Systems Models. 1:100 scale terrain layouts are available, and can be used for purposes varying from the layout of communications and supply systems to the depiction of battles and historical events.

Field Recognition Set Mark III. Using 1:100 model vehicles and resin model buildings the Miltra Field Recognition Set Mark III is issued in standardised field containers to permit all aspects of armoured and other fighting vehicle recognition to be taught either in the classroom or in the open. The British Army obtained sets for use by the Territorial Army and sets have been adopted by the US Army in Europe.

Full-Size Mock-ups. Full-size mock-ups and training aids for a variety of purposes are largely purpose-built to meet customer requirements. Full-size mock-ups of combat vehicles have been produced.

MGB Training System. Miltra produces 1:12 scale models of the Williams Fairey Medium Girder Bridge (MGB) which reproduces the characteristics of the full-size bridge. The model can be used for all aspects of MGB training and training drills, and includes items such as the MGB Link Reinforcement Set, the Span Junction and the Pier Set. These sets are in use by the British Army, the US Army and Marine Corps, the Canadian armed forces, plus the armed forces of Australia, Italy and Iraq and some other nations.

Status

In production to suit customer requirements. In service with the US and UK armed forces and many other nations.

Manufacturer

Miltra Engineering Limited, Watford.
(For details see Directory of Suppliers)

Miltra 1:12 scale model of Williams Fairey Medium Girder Bridge (MGB) used with the Miltra MGB training system

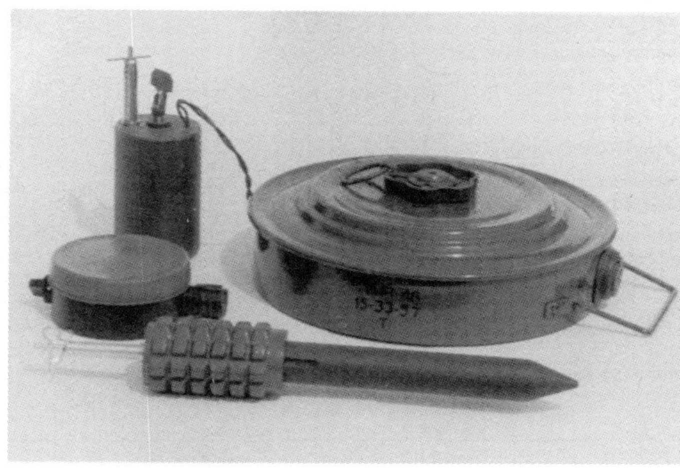

Miltra full-size land mine replicas

Racal Communications and Electronic Warfare Training Simulator COMSIM - SS2932

The SS2932 COMSIM provides a realistic training environment for personnel in the land battle concerned with radio communications. Personnel to be trained vary from junior radio operators to communication and Electronic Warfare (EW) managers and general staff.

Components. COMSIM comprises a Central Operations Control Board with an interactive map board, an Engineering/ECM Console (in which the system computer is located), two equipment racks and radio interfaces to which user radios are connected. Static and transportable installations are available.

Operation. Trainees use their current in-service radios which are connected to the simulator by coaxial cable, so avoiding stray RF radiation. Exercises are carried out based on any mapped exercise area; each participant has a copy of the map in use. A computer-controlled network of attenuators, known as a propagation matrix, represents the RF transmission paths between notional locations in the exercise area. Propagation losses dependent upon the positions and the intervening terrain are calculated by the main system computer which applies the appropriate settings to the attenuators. Each user experiences the effect of terrain-dependent communications.

Instructor Aspects. The exercise controller, who may be an instructor or a signals officer under training, first establishes the notional positions of each user with reference to the area map which is placed on the interactive map board. Each mapped location is entered by stylus contact with the control board and map. Similarly the exercise controller will determine and enter the positions of hostile ECM or ESM stations. Using this simple 'pointing' mechanism, other facilities such as antenna heights, path profiles and manual

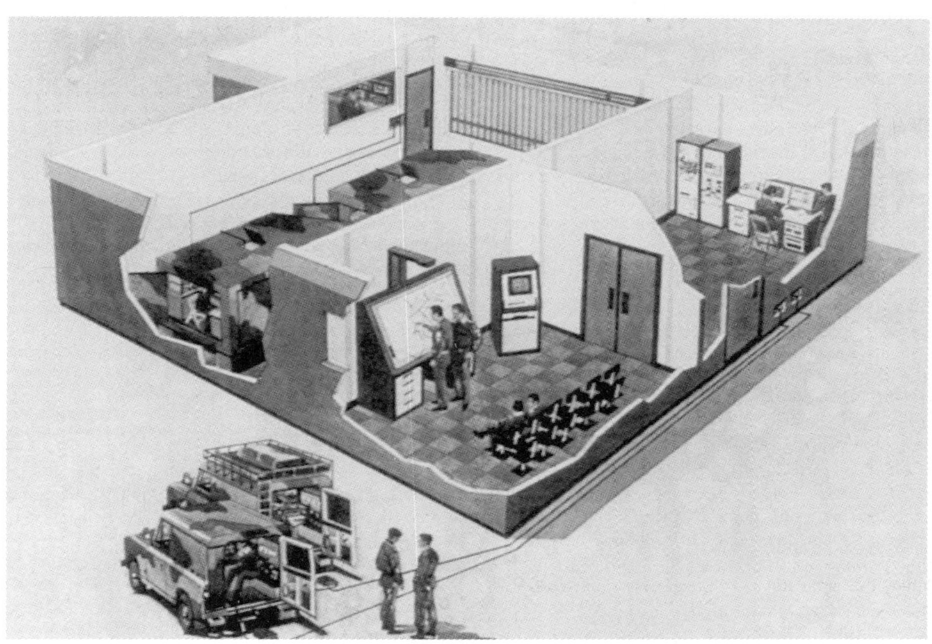

Layout of a permanent SS2932 COMSIM communications simulator installation

override can also be called up. The simulator then provides the interconnection between radios with the signal paths and various forms of interference or jamming that may be expected. The exercise controller has a direct private intercom with each separate radio station so that requests for changes of position can be passed between the trainee and the exercise controller.

Record/Debrief. Facilities for data logging and recording ensure that the current level of information is available for post-exercise debriefing and game analysis.

Specifications

Some specifications include:

Terrain exercise area: 40 × 40 km 1:50 000 maps

Number of independently positioned radio stations: (including simulated interference and jamming sources) 10 max

Max number of radios: 4 per radio interface, 40 max

Frequency range of radio equipment to be connected to simulator: 1.5-88 MHz

Max power output of radio equipment to be connected to simulator in a radio interface unit: 100 W any one radio; max of 250 W into one radio interface

Number of simulated interference/jamming sources: up to 8

Exercise control and display facilities: colour graphics display unit; visual display unit and control keyboard; audio monitoring unit; interactive map board with pointer system

Monitoring: switch-selectable split earphone operation enables listening to individual radio traffic

Intercom: 10-way intercom enabling all exercise participants and the exercise controller to converse without using the exercise radio links

Status

In production. In service with NATO armed forces, including Italy, and others worldwide.

Manufacturer

Racal Communications Limited, Reading.

(For details see Directory of Suppliers)

Exercise control board and VDU of a SS2932 COMSIM communications simulator

Reflectone EW RF Training Simulator

Reflectone UK Limited has been awarded a contract by the UK MoD for the supply of an Electronic Warfare (EW) Radio Frequency (RF) Training Simulator for use by the British Army. The equipment is an RF Talisman F version of the Talisman II Series of EW Training simulators produced by Reflectone UK Limited for training operators of Electronic Support Measures (ESM) and Electronic Intelligence equipment. The simulator will provide multiple radar emitters within a real-time environment which includes terrain screening and other features. It will be housed within a screened and air-conditioned transportable container. The simulator will drive two vehicle-mounted mobile ESM systems each receiving identical RF scenarios but with different simulated locations.

Status

In production for the British Army.

Manufacturer

Reflectone UK Limited, Bristol.

(For details see Directory of Suppliers)

Reflectone Nightowl Night Vision Goggle and Thermal Imaging Trainer

The Nightowl is a self-contained, low-cost trainer in the use of a variety of night vision devices including Night Vision Goggles (NVGs) and Thermal Imagers (TI). Both ground-based and airborne applications are included. The equipment can also be used for the set-up and testing of NVGs. The design is modular, enabling the trainer to be configured for other customer requirements.

Operation. A projector is used to display imagery derived from either a terrain model or from real world sensors. Images are displayed with contrast and other characteristics appropriate to the part of the spectrum being used (TI devices, thermal signatures of objects at about 10 microns; NVGs, reflected radiation in the near IR between 0.7 and 1 micron). Training can be by individuals or small classes. A series of programmable LEDs allows the training and testing of efficient scan techniques.

Instructor Aspects. A computerised instructor station is used for set up and for running exercises, and a debrief facility provides automatic student assessment.

Status

In service with the United Kingdom Ministry of Defence.

Manufacturer

Reflectone UK Limited, Bristol.

(For details see Directory of Suppliers).

Previously British Aerospace Simulation Limited (BAeSL).

UNITED STATES OF AMERICA

CTA VISMODs

Custom Training Aids (CTA) VISMODs are produced in the form of kits that can convert suitable vehicles into a replica of an Eastern Bloc AFV. The VISMOD (Visual Modification) kits contain all the necessary components to change the external appearance of the host vehicle, such as a jeep or CUCV, to match that required.

AFVs Modelled. VISMOD kits are available for the T-80 and T-72 MBTs, the ZSU-23-4 23 mm self-propelled anti-aircraft gun system, the BMP-2 IFV, the BTR-70 APC, the BRDM-2 MICV and the 2S1 122 mm self-propelled howitzer. Other kits can be produced on request.

Construction. The component parts are made of GRP and are bolted to the host structure with no modification necessary for the host vehicle. VISMOD components can absorb strikes from trees without suffering structural failure as the components utilise a special filler. Operational turrets, radar reflectivity and thermal image signature options are available.

Small Arms Replicas. CTA also produces solid plastic (urethane) full-scale replicas of small arms such as the M16A1 and AK-74 rifles. These replicas are the same weight as the real weapons and have the same point of balance. The barrels of the training rifles are metal.

Status

Available.

Manufacturer

Custom Training Aids (CTA) Inc, South Carolina.

(For details see Directory of Suppliers)

CTA M16A1 training rifle, a solid plastic replica

CTA VISMOD of CIS T-72 MBT

ECC Communications Network Simulator (CNS)

ECC International Corporation has developed a Communications Network Simulator (CNS) for training network managers and operators.

Equipment Simulated. The Mobile Subscriber Equipment (MSE) system simulated by the CNS is a digital tactical communications system for echelons below corps level. MSE links nearly every element of modern digital and cellular communications, including satellite communications, the Combat Radio Net, SINCGARS, NATO systems, and commercial telephones. The CNS will provide training on the Node Center Switching Shelter and the Node Center Operations Shelter. Multiple student stations and instructor stations interact with up to 42 simulated nodes and provide training across a wide range of realistic tactical scenarios. Each simulated shelter incorporates models of actual components and some actual equipment.

Status

In service with the US Army Signal Center at Fort Gordon, Georgia.

Manufacturer

ECC International Corporation, Pennsylvania.
(For details see Directory of Suppliers)

ESC Tactical Radar Threat Generator (TRTG)

The Electronic and Space Corporation's (ESC) Tactical Radar Threat Generator (TRTG) is a land-based, mobile, modular radar emitter based on the AN/APQ-159 airborne fire-control radar. It is used to train aircrew in a realistic electronic environment by stimulating the aircraft RWR. The TRTG emits radar signals that simulate signals emitted by AAA gun laying radars and SAM tracking radars. It is modular in construction and can be installed in various vehicles and in semi-permanent installations.

Type I Configuration. The basic TRTG configuration is the Type I, consisting of a radar, an indicator, a motorised turret/radome, and the controls needed to operate the system. The Type I provides a radar/beacon operation.

Type II Configuration. The Type II TRTG adds a television subsystem, consisting of a television camera, monitor screen and videotape recorder. This adds an electro-optic mode, a radar-assisted electro-optic mode, and the capability to record the video for the later evaluation of aircraft responses to RWR indications.

Type III Configuration. The Type III TRTG system adds digital scan conversion to the basic TRTG system. The Type III system displays both radar and video on the television monitor screen. It can also show mission data such as flight number and date and time, on the monitor screen.

Type IV Configuration. The Type IV TRTG system adds a score/instrumentation subsystem to the Type I system, and also records the aircraft responses.

Additions. The following capabilities can be added to the TRTG system: command guidance; millimetre wave, continuous wave; laser threats; additional missile threats; and helicopter air-to-air threat simulations.

Carriage Vehicles. The TRTG can be carried on the M880 (4 × 4) 1-ton truck or the M548 tracked cargo carrier (both housing the equipment in an S-250 shelter/container) and the M114 tracked vehicle (used by the US Air Force), although many other vehicles could be used.

Status

In service with the US Army and US Air Force.

Tactical Radar Threat Generator (TRTG) mounted on an M880 (4 × 4) 1-ton truck

Manufacturer

Electronics and Space Corporation, St Louis.
(For details see Directory of Suppliers)

Hughes Firefinder Training Devices for Weapon Locating Radars

There are three Firefinder Training Devices used to provide training for the operators and maintenance staff of AN/TPQ-36 and AN/TPQ-37 weapon locating radars. They are used at the US Army Field Artillery School at Fort Sill, Oklahoma.

A17E11 Operator Trainer

The A17E11 presents a trainee operator with a variety of simulated training scenarios. These scenarios require the trainees to format data to perform system start procedures, to locate hostile weapons such as artillery and mortars, and to initiate fire registration assistance for friendly batteries. The A17E11 consists of five units consisting of 15 individual pieces. The main units are a computer system, an instructor station, a main power distribution system, an operator console and a shelter power assembly. Total weight is 3470 kg.

A17E12 and A17E14 Maintenance Trainers

The A17E12 and A17E14 simulate various maintenance aspects of the AN/TPQ-36 and AN/TPQ-37 radar sets respectively They can provide malfunction indications through the simulation of diagnostic test results, or can be used for manual troubleshooting procedures for the Firefinder radar system. Removal and replacement procedures are simulated by the use of alphanumeric codes that identify the various maintainer-replaceable units. When the trainees have located the faulty unit they then enter the corresponding alphanumeric code via the simulator keyboard.

The A17E12 is basically an A17E11 operator trainer plus a signal processor, an AN/TPQ-36 trailer and a shelter power supply, and consists of 33 pieces weighing 6192 kg overall.

The A17E14 consists of an A17E11 plus a signal processor, an AN/TPQ-37 trailer and a shelter power supply, and consists of 33 pieces weighing 5458 kg overall.

Instructor Aspects. With all three trainers the instructor has the capability of selecting a set of initial radar conditions to be used in an exercise at the time of assigning the exercise to a student. A zeroing mode allows the instructor to return a simulation problem to the preset initial condition. The instructor also has the option of freezing, aborting or assigning problems at any time. The instructional workload can be shared between two instructor positions, each with the capability to control the training scenario at any or all of the student stations. On-line test diagnostics are provided to verify the operational status of the trainer. An off-line diagnostic test is provided to allow the isolation of any trainer malfunctions. The trainers perform student grading while student actions are being summarised on the instructor displays. At the end of each exercise the instructor can obtain a printed copy of student performance. Each trainer can support up to 12 students.

Status

In service with the US Army.

Manufacturer

Hughes Aircraft Company, Fullerton, California.
(For details see Directory of Suppliers)

Hughes Joint Surveillance Target Attack Radar System Trainer

The Joint Surveillance Target Attack Radar System (Joint STARS) provides commanders with a 'window on the battlefield', permitting US Army and Air Force decision makers to detect, track and engage enemy forces. The Hughes Joint STARS Ground Station Trainer is used to train soldiers and airmen in interpreting data and providing near real-time combat information.

Joint STARS Testbed. A Joint STARS Ground Station Testbed was developed to examine the impact of alternative technology, mission and man/machine interfaces. The testbed is used to simulate changes in enemy doctrine, enhanced operator aids, and integration of information from multiple sensors to predict and quantify improvements in system performance. The Joint STARS Project Office uses this data as a foundation for system improvement specifications for the operational equipment.

Joint STARS ground station trainer

Joint STARS Operator Trainers. The Joint STARS operator trainers are at the US Army Intelligence Center and School at Fort Huachuca, Arizona. Classroom lessons and trainer exercises are used to teach students how to use the Joint STARS Ground Station Modules. The trainer permits students to develop and refine skills that support the Joint STARS surveillance, target acquisition and target attack missions.

Scenarios. Scenario-based imagery depicting hypothetical battlefields is used. This capability is used in conjunction with worldwide field data collection to produce detailed scenarios depicting the activities of tens of thousands of individual vehicles, emitters and aircraft across the area of battlefield that Joint STARS will see. These scenarios, updated as perceptions of enemy activity evolve, provide imagery depicting how these battlefields would appear to a variety of sensors. This imagery is used in the trainer to permit simulation exercises and evaluations, potentially over several days.

Status

In service with the US Army and Air Force.

Manufacturer

Hughes Training Inc, Minneapolis.
(For details see Directory of Suppliers)

Hughes Position Location Reporting System (PLRS) Master Station Trainer

The Master Station Trainer has been developed to support the training of operators and maintainers of the PLRS system. PLRS is a tactical data system which provides ground, air and amphibious force commanders with automatic, near real-time positioning information for forces on the battlefield, regardless of terrain, weather, visibility or geographical location.

Components. The trainer has two main elements, the student stations and the instructor station. The five student stations provide either operator or maintenance training. Each station provides a simulation of the PLRS Master Station, an accurate emulation of the real-time tactical PLRS display, and on-line performance monitoring with audio and visual feedback. Wherever possible, commercial off-the-shelf equipment is used. Software provides accurate real-time responses to student actions. An interactive graphics system is used to emulate both the PLRS display console and an oscilloscope display. The trainer includes self-test system diagnostics which can be run on any student station without interfering with the operation of any of the other four student stations. The student stations are stand-alone devices so training can be conducted even during a failure of the instructor station.

Instructor Aspects. The instructor station for the trainer provides exercise monitoring, control, and computer-managed instructional features. A student progress summary, continually presented at the instructor terminal, permits the instructor to monitor multiple student stations. He may request a list of the student's previous switch actions and/or monitor in real time either of the student station displays. The instructor may also 'freeze' an exercise while he verbally tutors the student using intercom equipment, the exercise is then resumed at the point frozen or at other selected points in the exercise.

Status

In service with the US Army Signal Training Center (USASTC) at Fort Gordon, Georgia.

Manufacturer

Hughes Aircraft Company, Fullerton, California.
(For details see Directory of Suppliers)

National Training Center AFV VISMODs

Following their withdrawal from service with the US Army during 1978-79, a total of 330 M551 Sheridan light tanks/reconnaissance vehicles was assigned to the National Training Center (NTC) at Fort Irwin, California. There they were modified visually and in other ways to enable them to take part as 'aggressor' vehicles in the large scale tactical exercises conducted at the NTC. Most vehicles have had their original armament removed and their appearances have been altered by the addition of suitable superstructures and dummy armament to produce VISual MODifications (VISMODs) that resemble Eastern Bloc combat vehicles. In October 1986, 24 of these NTC vehicles were permanently transferred to Fort Knox, Kentucky.

AFVs Modelled. VISMODs in use at the NTC include the ZSU-23-4 23 mm self-propelled anti-aircraft gun system, the T-72 MBT, the BMP IFV, the BMD airborne combat vehicle, the ASU-85 airborne self-propelled anti-tank vehicle, and the 122 mm M-1974 (2S1) self-propelled howitzer.

Replacement. Most of these vehicles will require replacement in the near future. FMC Corporation has proposed its M113 OPFOR vehicle consisting of a lightweight M113 chassis and drive train with a turret assembly similar in outline to that of a BMP-2. Facsimile hulls for the BMP-2, ZSU-23, 2S1 and T-80 armoured vehicles could be built on to this chassis. CTA has proposed a VISMOD (see earlier entry) and General Motors of Canada has proposed a series of VISMODs based on the hull of its wheeled LAV.

Status

In service with the US Army at the National Training Center, Fort Irwin, California, and Fort Knox, Kentucky.

BMD airborne combat vehicle VISMOD based on the M551 Sheridan

T-72 MBT VISMOD based on the M551 Sheridan

TAS Radar Video Simulation (RVS)

Developed by the Technology Applications and Service Company (TAS), Radar Video Simulation (RVS) was designed to be used in training or testing applications in which a combat system requires radar video inputs. RVS provides the target, clutter, jamming and IFF/SIF video outputs with the associated trigger and timing signals. The main modes of all known two-dimensional radars can be generated while the majority of major US Navy radars are instantly selectable.

Three-dimensional simulation of the AN/SPS-48(C) and the AN/SPS-52(B) are also available. A mixed mode is also available in association with a live radar.

Status

In production.

Manufacturer

Technical Applications and Service Company (TAS), Maryland.
(For details see Directory of Suppliers)

NAVAL SYSTEMS

Ship Handling Simulators
Weapon Operating Trainers and Simulators
Operations Centre Simulators
Submarine Command Team Trainers
Tactical Training Simulators
Propulsion/Electrical Power Generation and Distribution System Simulators
Submarine Steering and Depth Control Simulators
Sensor and Electronic Warfare Simulators and Trainers
Naval Maintenance Trainers
Underwater Practice Weapons and Targets
Miscellaneous Naval Simulators and Trainers

SHIP HANDLING SIMULATORS

GERMANY

Atlas Elektronik Ship Handling Simulator (SHS)

A major manufacturer of naval electronics, Atlas Elektronik GmbH also produces the Ship Handling Simulator (SHS). This system includes a replica ship's bridge, an instructor's console, model data input stations, briefing/debriefing rooms, a computer room and (optionally) student cubicles each equipped with facsimiles of the necessary sensors and inputs. The SHS is designed to train watch officers as well as providing refresher courses and continuation training for holders of maritime certificates and ships' masters. The two naval systems are used to train junior officers in manoeuvring and navigation. The system can also be used for research into bridge layout and design, development of new ports and channels and for fleet training purposes with up to seven 'ships' (that is, students in cubicles) sailing independently.

Each system has a life-size replica of a ship's bridge with a full range of navigation aids and accessories mounted on an electrohydraulically controlled motion base which provides pitch and roll movements of between 5 and 12°. Computer-generated images are projected in colour onto a 250° panoramic screen. All forms of environmental and visibility conditions can be reproduced.

The simulator can be programmed to produce the performance and characteristics of various specific ships with the German Navy, one having the characteristics of the 'Bremen' class (Type 122) frigate, the Type 143 A Fast Attack Craft, the 'Lindau' class (Type 331) mine-hunter, a supply ship and merchant ships. Traffic situations involving up to 20 instructor-controlled ships can be conducted, with the exercises being recorded for playback.

Status

Five systems have been sold since 1982: two to Germany (one for naval use), two to Australia (one for naval use) and one to Taiwan. The Royal Australian Navy system, which was delivered to HMAS *Watson* in Sydney in March 1985, was the first purpose-built naval ship handling simulator. A similar system was supplied to the German Navy's Marine Training School at Mürwik, Flensburg, in October 1986. Both systems have 250° viewing arcs but the Australian system has four cubicles while the German system has six. In September 1993 the Royal Thai Navy awarded the company a contract for a system with a 240° viewing arc which was scheduled to be installed in the Royal Thai Naval Academy, Bangkok, in August 1994.

Manufacturer

Atlas Elektronik GmbH, Bremen.
(For details see Directory of Suppliers)

The instructor's station in the German Navy's ship handling simulator at Mürwik

The bridge section of the German Navy ship handling simulator

INDIA

Bharat Electronics Mariner

Mariner was designed to give an extremely realistic environment for ship handling training in normal and hazardous conditions, including under way replenishment, navigating through restricted channels and navigating in shallow waters.

The system consists of a computer networked to a graphics processor, a radar processor, a feedback processor and a DA processor. The graphics processor provides nine-channel realistic visual simulation, while the radar processor is used to simulate the navigation radar, and the DA processor is used to interface with the 'pilot house'.

Status

In service with the Indian Navy.

Manufacturer

Bharat Electronics Limited, Bangalore.
(For details see Directory of Suppliers)

NETHERLANDS

Maritime Simulation Centre Mermaid 500

The Mermaid 500, produced by the Maritime Simulation Centre the Netherlands (MSCN), is designed to meet the requirements of ships' officers, pilots and students who need to learn, or to augment their knowledge of ship behaviour, the effects of dynamic forces, working with tugs, controlling computerised systems and actions in emergencies. It is a modular system which allows for hardware or software upgrading and the software is in the Ada programming language. The system can simulate up to 10 fully controllable ships, and a motion system with two or three degrees-of-freedom may be added.

The system has eight subsystems: instructor, ship, bridge, outside view, radar, special effects, preparation and debriefing, which are all linked by a local area network. The instructor subsystem consists of a run selector, simulation manager, an area manager which monitors the exercise and a tug server which controls 'tugs' in the exercise.

The ship subsystem consists of own ships, targets (other traffic) and tugs. Own ship behaviour is based upon mathematical models of genuine ships and consists of three sub-components: autocon (which models autopilot functions); engine room (which supplies control settings); and matmod (a mathematical model of ship locations, angles and speeds). The targets sail along fixed routes and at various speeds, all of which the instructor can change. Tugs can be controlled by the instructor or they can be given own ship capabilities.

The bridge subsystem is designed to represent the bridge of a real ship and consists of consoles with instruments which are controlled by a front-end computer. These may be reconfigured to represent other ships. The outside view subsystem features up to 10 sectors covering a sector of 36°, each with a graphical computer and a projector which generates a visual image. One or two radar simulators of the radar subsystem are also in the bridge subsystem.

The special effects subsystem provides audio effects and special dynamic effects as well as controlling any motion system. The preparation subsystem features a central database, a preparation programme and a run time data structure generator. The debriefing system has a dedicated computer which is connected, not to the real-time local area network, but to a large screen projection system.

Status

In production and in service with the Royal Netherlands Navy.

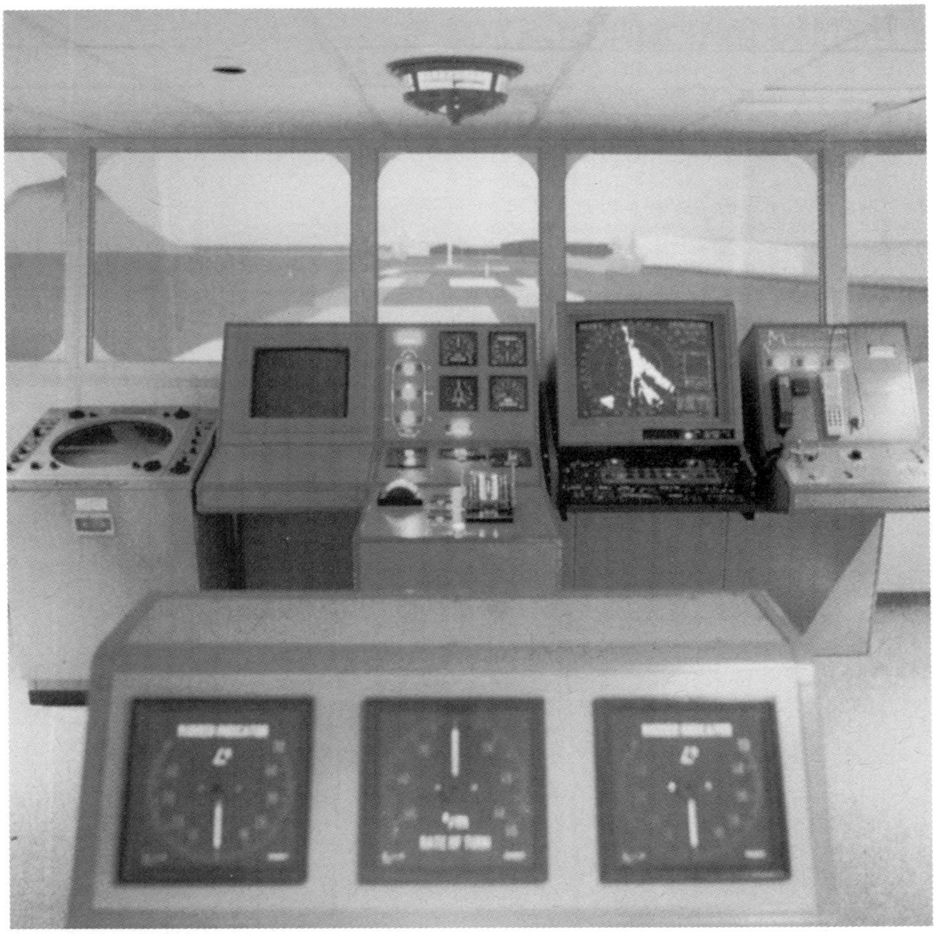

A typical MSCN Mermaid 500 bridge subsystem

Manufacturer
Maritime Simulation Centre the Netherlands, Wageningen.
(For details see Directory of Suppliers)

NORWAY

Norcontrol Ship Handling Simulators

Norcontrol is the world leader in maritime simulation with more than 300 simulators sold to more than 40 countries. Products range from part-task trainers to full mission ship handling simulators and propulsion plant trainers. The company is wholly owned by Norsk Forsvarsteknologi A/S (NFT), Norway's largest manufacturer of defence, aerospace and maritime products, which itself is government-owned through the Ministry of Industry. The Norcontrol Ship handling Simulators (NSS) series is a family of trainers configurable to meet any need and budget.

In its basic configuration the NSS consists of a mock-up bridge equipped with ship-like instrumentation, real radar displays, a visual generation system covering a field-of-view from 120 to 360° and an instructor station. The extremely flexible concept allows for almost any combination of bridge equipment to be installed initially or at a later stage.

Norcontrol has a large selection of panels/instruments including both simulated and real instrumentation. The library consists of more than 50 different panels/instruments and is constantly increased due to different customers' requirements. Typical instrumentation consists of a helm position, engine throttles, thruster control, anchor control, navigation receivers, logs and communications equipment.

There are four types of visual systems from Norcontrol for the ship handling simulators: NorView-1000, NorView-2000, NorView-3000 and NorView-4000. For details see the *Computer-generated Image and Visual Display Systems* section.

The Norcontrol ship handling simulator

Since the accuracy of the manoeuvring characteristics is of vital importance in ship handling simulators, Norcontrol has established a close co-operation with the Danish Maritime Institute. The institute has a 25-year record of specialising in ship manoeuvring and operates its own ship tank which helps it to provide Norcontrol with the most complex hydrodynamic models which cover three to six degrees-of-freedom.

The high-fidelity hydrodynamic models cover the complete range of speeds and manoeuvres which a ship experiences when subject to its own control forces and/or to external control forces from tugs, anchors and fenders. This allows training in all aspects of ship handling including berthing, shallow water manoeuvres, replenishment at sea, towing and fleet work.

The bridge may be mounted on a motion platform with one to six degrees-of-freedom. The motions of the platform are controlled from the hydrodynamic model and interfaced with the visual system.

Great care is taken with the design of the instructor station such that no specialised or programming knowledge is required by the operator. The instructor controls the scenario and environment, including traffic, all of which may be preprogrammed. Debrief facilities such as playback, X/Y plotters and printers are controlled from the instructor station or from a separate debriefing station.

The instructor stations may control more than one bridge. Several other fully equipped bridges and/or cubicles may be connected to the simulator. Cubicles may be linked to the bridge and thus play the role of an operations room when training ship teams.

In November 1992 the United Kingdom Ministry of Defence awarded a major contract to a consortium consisting of Norcontrol and Marconi Simulation. This is for a Bridge Simulator and Blind Pilotage Trainer (BS & BPT) for the Royal Navy. It will provide instruction in navigation and ship handling for a wide range of surface vessels as well as for 'Vanguard' class nuclear submarines.

The BS & BPT will consist of a representative 'Duke' class frigate bridge and three pilotage cubicles all controlled from a single instructor station. The bridge will be on a motion platform in a large projection theatre, creating a highly realistic visual scene with 270° field-of-view. The visual database will include major ports and more than 100 different models of ships, submarines, ferries, tenders, tugs and aircraft. The visual system, based on Marconi's real-time software, has already been supplied in periscope simulators delivered to, or ordered by, the Royal Navy, the Royal Netherlands Navy and the Canadian Navy.

Training on a wide range of vessels is achieved by reconfiguring the high-fidelity hydrodynamic model from the Danish Maritime Institute to simulate specific vessel types. The simulator will be installed in a purpose-built building at HMS *Dryad*, the Royal Navy's School of Maritime Operations at Fareham, Hampshire.

Status

The company has supplied ship handling simulators to more than 40 customers in Australia, Bulgaria, Canada, China, the Commonwealth of Independent States (CIS), Finland, France, Iran, South Korea, Netherlands Antilles, Norway, Poland, Singapore, Sweden, the USA, the United Kingdom and Venezuela.

Manufacturer

Norcontrol Systems AS, Horten.
(For details see Directory of Suppliers)

UNITED KINGDOM

Marconi Simulation Ship's Bridge Simulator

The Marconi Ship's Bridge Simulator consists of a facsimile of a specific ship's bridge or a generic reproduction. It features realistic operation of all required bridge equipment (including real radar displays), accurate and realistic presentation of the out-of-the-window visual scene, faithful representation of the behaviour of different classes of vessels and full instructor facilities. The creation of exercises and exercise areas as well as databases may be performed easily by the customer with minimum training.

The latest technology is used to display a panoramic view of the outside environment including three-dimensional sea waves in different sea states, wind and tide conditions, full time-of-day variations together with weather effects such as fog, clouds and rain. There are also realistic models of the ships, harbour approaches, coastal features, buoys and navigational aids.

The simulator includes comprehensive, easy-to-use facilities for the instructor. These allow him to concentrate upon student instruction rather than operating the simulator.

Status

A system has been ordered by the Royal Navy.

Manufacturer

Marconi Simulation, Dunfermline.
(For details see Directory of Suppliers)

The bridge subsystem of the Full Mission Ship handling Simulator (FMSS) training system

Reflectone Full Mission Ship handling Simulator (FMSS)

Reflectone UK Limited (ex-BAeSL) has developed a Full Mission Ship handling Simulator (FMSS) for both training and research purposes and also offers cost-effective ship handling simulators based upon powerful work stations as well as ship's bridge trainers.

An FMSS supplied to Singapore in association with Ship Analytics, features a visual generation system representing selected coastal areas and harbours in various visual conditions. It also has authentic controls and instrumentation as well as realistic ship handling simulation.

There is a separate tug/own ship station simulator which may be used either as a stand-alone system or for interactive training. Tug control facilities are provided by the instructor station and the system will include a classroom with remote monitoring, debriefing and feedback capabilities.

Status

In service with the Port of Singapore Authority.

Manufacturer

Reflectone UK Ltd (ex-BAeSL) - Sea Systems, Farnborough.
(For details see Directory of Suppliers)

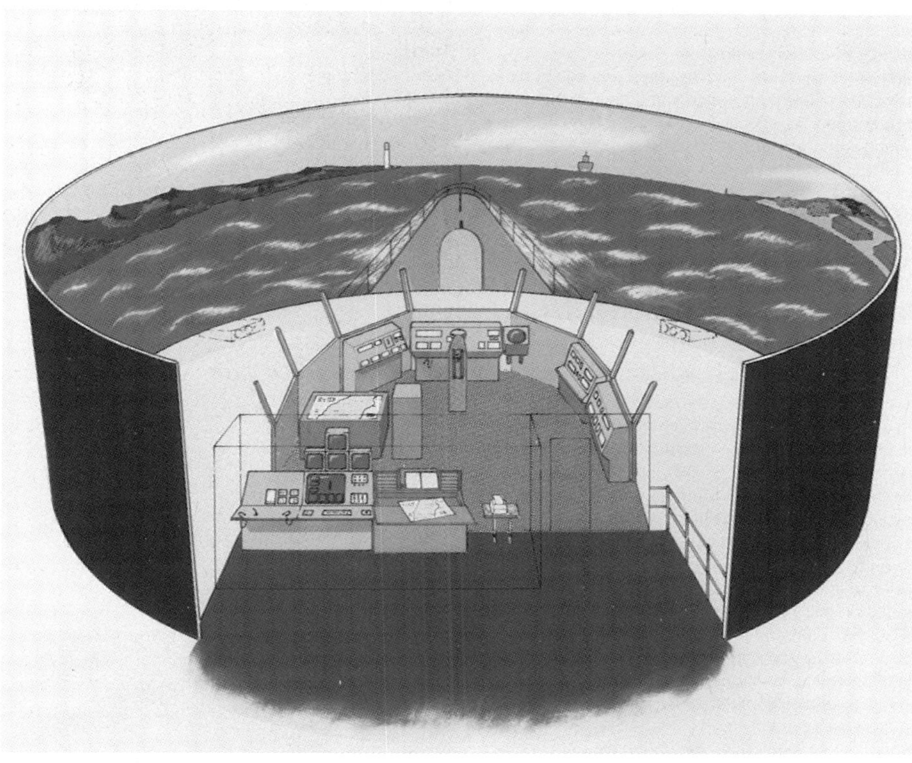

An artist's impression of the Marconi Simulation Bridge Simulator which has been ordered by the Royal Navy

Vickers Ship Motion Simulator

Vickers Shipbuilding and Engineering Limited (VSEL) designs and produces a number of simulators for shore-based Royal Navy integration facilities and trainers. These include ship motion simulators for submarines as well as ship motion and anemometer simulators for surface ships.

The Ship Motion Simulator is used to simulate submarine motion while the Ship Motion and Anemometer Simulator (SMAS) simulates the motion and wind sensors for a surface vessel. The company also manufactures a decoy simulator which simulates launcher and round deployment together with the characteristics and motion of chaff clouds.

Status

Five systems are in service supporting simulators for 'Upholder' class submarines and 'Duke' (Type 23) class frigate simulators. Another eight are being built.

Manufacturer

Vickers Shipbuilding and Engineering Limited, Barrow-in-Furness.

(For details see Directory of Suppliers)

UNITED STATES OF AMERICA

ECO CAPTAINS™ Ship Manoeuvring Training System

CAPTAINS™ has been developed to fulfil a training need for a cost effective, real-time ship simulator. The system is capable of replicating the handling and manoeuvring characteristics of a wide range of naval and merchant ships. The system also allows the simulation exercise to be performed 'within' any port and under any combination of environmental conditions during the 'day' or 'night'.

The CAPTAINS™ system is available in four configurations. The basic system functions in all four versions which are performed through the use of a ship console integrated with a minimum of one standard 486 microprocessor, together with an EGA graphics controller and a ship controller for one particular ship type and port layout. Also included is the software which provides ship selection type and port region: control of magnitude and direction for wind and current effects; the ability for autopilot control; the use of a variable range marker and an electronic bearing line; the selection of ship's head-up or North-up orientation for the radar display; and selection of radar range.

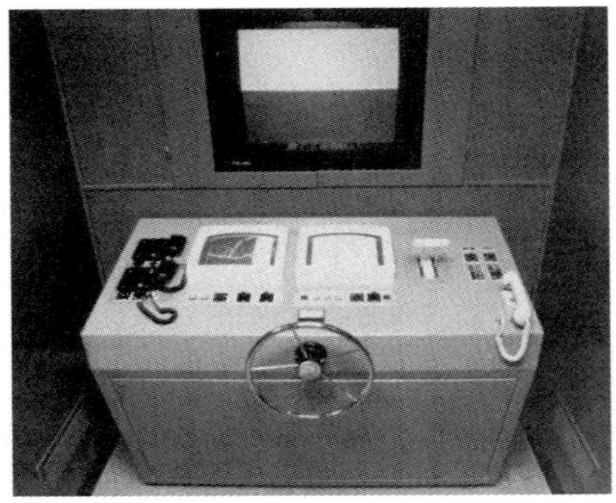

A CAPTAINS™ III unit installed at the US Naval Academy, Annapolis

The CAPTAINS™ system has three main displays. One is a situation display giving a plan view presentation of a port area and the controlled ship; the scaling and position of the display may be adjusted. The second is a radar display with the ship at the centre of the screen with the options of North-up or ship's head-up; range and bearing markers are available. The third is an instrument display. This shows the current settings of the rudder and engines with the course, rate of turn, speed and depth. Relative wind speed and direction and autopilot setting are also given.

The system makes use of unique manoeuvring equations developed over several years. Mathematical models incorporate the following hydrodynamic effects: acceleration and deceleration; propeller side-force effects; loss of speed in turn; single/twin-screw turning effect; wind effects; and current effects.

Status

The system has been selected by the US Navy for inclusion in the curriculum for seamanship training at the US Naval Academy in Annapolis, Maryland.

Manufacturer

Engineering Computer Optecnomics Inc (ECO), Maryland.

(For details see Directory of Suppliers)

Hughes Full Mission Trainer for the Landing Craft Air Cushion (LCAC) (Device 20G6)

Hughes Training's Landing Craft Air Cushion (LCAC) Full Mission Trainer is under development to act as the key element in an extensive training programme for the LCAC Craftmaster, Engineer, Navigator and Group Commander crew positions. The trainer will accurately simulate the performance, manoeuvring capabilities, craft systems, electronic equipment, control reactions and responses of an actual LCAC, and will simulate the operational environment, that is, over water, over land, and in LCAC support ship deck wells.

The craft dynamics module computes solutions to six-degrees-of-freedom equations. It also includes lift, thrust and inertial effects. Handling qualities and performance characteristics peculiar to the LCAC are modelled. Systems models provide complete simulation of the LCAC's communications and navigation suite (including radar), plus the electrical, fuel, hydraulic, craft control and auxiliary power systems. The functional simulation covers all appropriate operating modes and conditions.

Six software-controlled hydraulic actuators are configured to provide six degrees-of-freedom. This simulation of actual craft motion is co-ordinated with craft controls, instrumentation and visual imagery.

The instructor station is located on the motion system, adjacent to the simulated control cabin. This arrangement allows the instructor complete control of the training scenario, over-the-shoulder observation and a direct view of the visual scene.

Status

Two trainers have been supplied to the Naval Amphibious Schools at Coronado and Little Creek.

Manufacturer

Hughes Training Inc, Herndon, Virginia.

(For details see Directory of Suppliers)

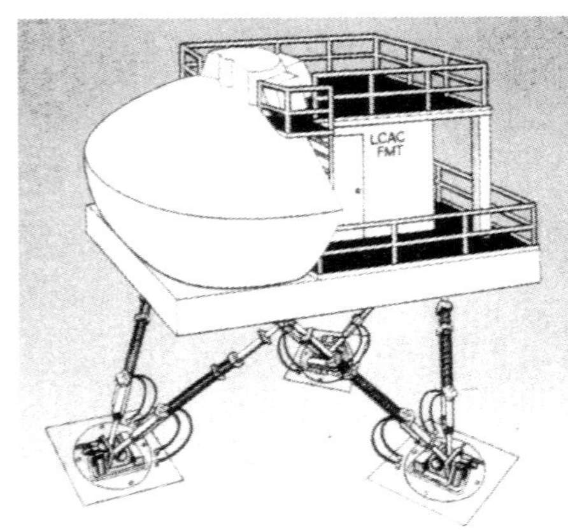

Artist's impression of the Hughes LCAC Full Mission Trainer

MarineSafety International (MSI) Ship handling Trainer

A subsidiary of FlightSafety International, MarineSafety International, has developed commercial and naval ship handling trainers which emulate the characteristics of most ship types in various hydrodynamic and weather conditions.

For naval officers the company has a facility, MSI Newport, which provides ship handling skill and navigation team training for prospective commanding officers and executive officers, for department heads and division officers and navigation teams. Courses are provided for aviation and submarine as well as surface warfare officers. Intermediate and advanced courses include docking and undocking, harbour transits, working in channels, anchoring, formation manoeuvring, under way replenishment, emergency ship handling and case studies.

Four interactive visual ship handling trainers, all featuring full-colour computer graphics, are used to enhance the refinement of decision making expertise. The facility has also conducted specialised training for

Observing a 'task force' from a visual ship handling trainer

the Military Sealift Command, the US Naval Reserve and foreign navies.

For naval training, the facility can reproduce the response characteristics of 18 US Navy vessels. These range in size from aircraft carriers such as the 'John F Kennedy' (CV 67) class to the 'Avenger' (MCM 1) class mine-countermeasures vessel. They include submarines, cruisers, destroyers, frigates, amphibious warfare and combat logistic vessels.

Geographic areas created include Boston, Norfolk, New London, San Francisco-Oakland, Los Angeles, San Diego, Charleston, the Panama Canal and Pearl Harbor. A wide range of forces including multiple currents, winds, passing ship interaction, mooring lines, dredging anchors and tugs is created, while there is also the full range of visual effects including haze, fog, sea swells, wakes and bow waves.

Status

The MarineSafety Newport Simulator Complex has four interactive ship handling simulators for which the company has contracts from the US Navy and Canadian Navy to provide tuition. The US Coast Guard uses the facility to qualify radar observers and also for bridge team management courses.

Manufacturer

MarineSafety International, Rhode Island.
(For details see Directory of Suppliers)

A navigation team from an 'Ohio' class submarine conducting team training on the full mission bridge

Ship Analytics Full Mission Ship handling Simulator (FMSS)

Ship Analytics Full Mission Ship handling Simulator (FMSS) systems are designed to provide training of naval personnel in basic and complex ship handling techniques such as: ship manoeuvring and tactics, collision avoidance, docking/undocking, transiting shallow waters, ship handling in various sea and weather conditions, navigation in confined/restricted waterways, entering/leaving harbours and ports, blind pilotage (radar only), use of navigation aids, and replenishment underway.

The FMSS is designed and built to replicate actual own ship bridge and equipment. This provides the most realistic environment in which training can be conducted. Naval vessels modelled include aircraft carriers, destroyers, frigates, tenders, and various auxiliary ship types. The simulation software allows for complex or simple training exercises/scenarios that cover the major ship handling techniques. Advanced hydrodynamics and state-of-the-art visual imagery enhances training effectiveness. Standard or custom bridge configurations are available, depending upon the user's specific training or research requirements.

Status

Over 12 FMSS systems have been supplied to US Navy, Coast Guard, and Maritime Schools, as well as to foreign commercial and military customers.

Manufacturer

Ship Analytics, Connecticut.
(For details see Directory of Suppliers)

A typical Ship Analytics FMSS bridge

WEAPON OPERATING TRAINERS AND SIMULATORS

FRANCE

DCN SIMPAP

The DCN has one establishment producing weapon operating simulators: DCN, F-16600 Ruelle, which produces simulators for the PAP 104 (Poisson Auto Propulsé) remotely operated vehicle and the F17 Mod 2 torpedo.

SIMPAP is a piloting simulator to train operators of the PAP 104 remotely operated mine disposal system. The student learns to approach objects using the sonar, to identify them with the television and then to neutralise them by means of an explosive charge or by cutting their cable.

The simulator consists of a PAP control panel, a sonar display, a television monitor served by a Sogitec GI500 computer-generated image system and an instructor's console with plotting table. Multiple parameters may be inserted including various sea bed conditions, currents, mines, visibility, other objects and even the mine explosion.

Status
A SIMPAP simulator was delivered to the French Navy in April 1987. A new version of SIMPAP, with the characteristics of the PAP Mark V, is under development.

Manufacturer
DCN, Paris.
(For details see Directory of Suppliers)

The DCN SIMPAP simulator

DCN F17 torpedo simulator

A simulator for the wire-guided F17 torpedo has been developed by DCN. It is designed as a complete, independent tactical system simulating the system used in the 'Rubis' class fleet nuclear submarines with all of the torpedo firing control functions. The system generates and maintains target elements, calculates firing parameters and controls the 'torpedo' to simulated impact. A precise mathematical model is used to reproduce the effect of the acoustic propagation conditions upon the performance characteristics of the torpedo homing head.

Status
One simulator was delivered to the French Navy in December 1987 and another in November 1988. They are at Lorient and Toulon.

Manufacturer
DCN, Paris.
(For details see Directory of Suppliers)

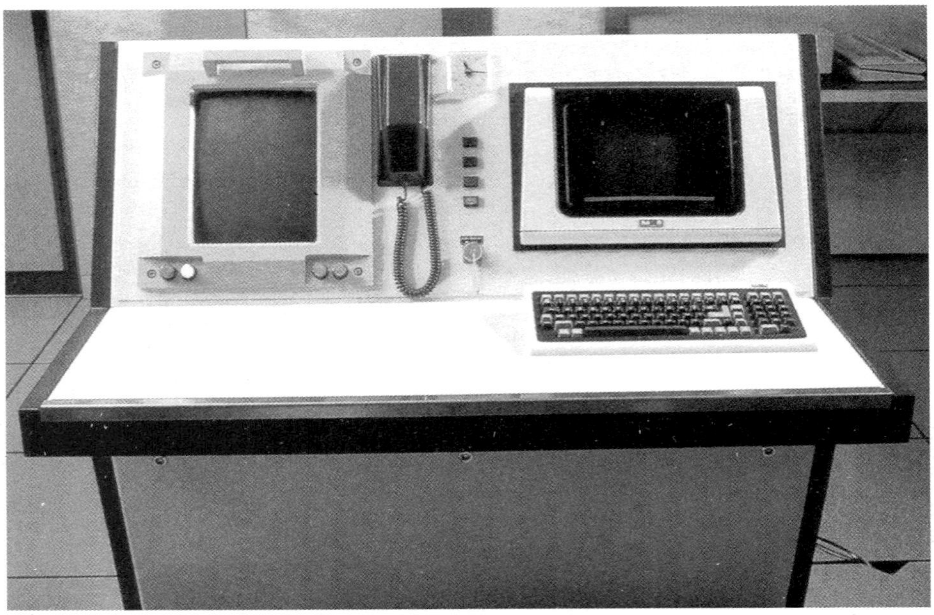

The instructor's station for the F17 trainer

ISRAEL

IAI Gabriel Weapon System Trainer (GWST)

Israel Aircraft Industries (IAI), the aircraft and missile manufacturer, designed and developed the Gabriel anti-ship missile and, to support it, produced a weapon operator trainer.

The Gabriel Weapon System Trainer (GWST) features an instructor's console and a number of simulated operator consoles and provides the student with familiarisation with the operations room target designator, the weapon selection and control units together with the missile selection unit. Programming and mid-exercise instructor intervention are conducted through a standard teleprinter. Operators are also given familiarisation courses with associated naval gunnery, missile launch and optical director units. A digital computer provides control and scenario simulation of target data, sea state and electronic warfare environment.

Status
In service with the Israeli Navy.

Manufacturer
Israel Aircraft Industries Limited, Lod.
(For details see Directory of Suppliers)

The IAI Gabriel Weapon System Trainer

NETHERLANDS

Hollandse Signaalapparaten Goalkeeper Operator Training System (OTS)

Goalkeeper is a Close-In Weapon System (CIWS) and the Operator Training System (OTS) is designed to train operators to use the manual override and monitoring facilities in what is otherwise an autonomous and automatic system. It is also used for training basic operating procedures within environmental and jamming conditions in realistic situations. The system consists of a real weapon control console with video simulators and dedicated processors for the simulation function.

Status

In production with one system known to have been purchased by the Royal Navy. This system is installed at the School of Maritime Operations (HMS *Dryad*).

Manufacturer

Hollandse Signaalapparaten BV, Hengelo.
(For details see Directory of Suppliers)

SWEDEN

Weibull Naval Air Defence Simulator

This simulator is designed to provide operators with an environment for realistic aiming exercises and simulated firing of a wide range of naval air defence weapons including guns and missiles.

It consists of an electrohydraulically operated platform capable of pitching 0 to ±15° or rolling 0 to ±25° for periods of 5 to 30 seconds. The weapon simulator is attached to the platform.

Status

In production and in service with the German, Portuguese, and Swedish navies.

Manufacturer

J L Weibull, Åkarp.
(For details see Directory of Suppliers)

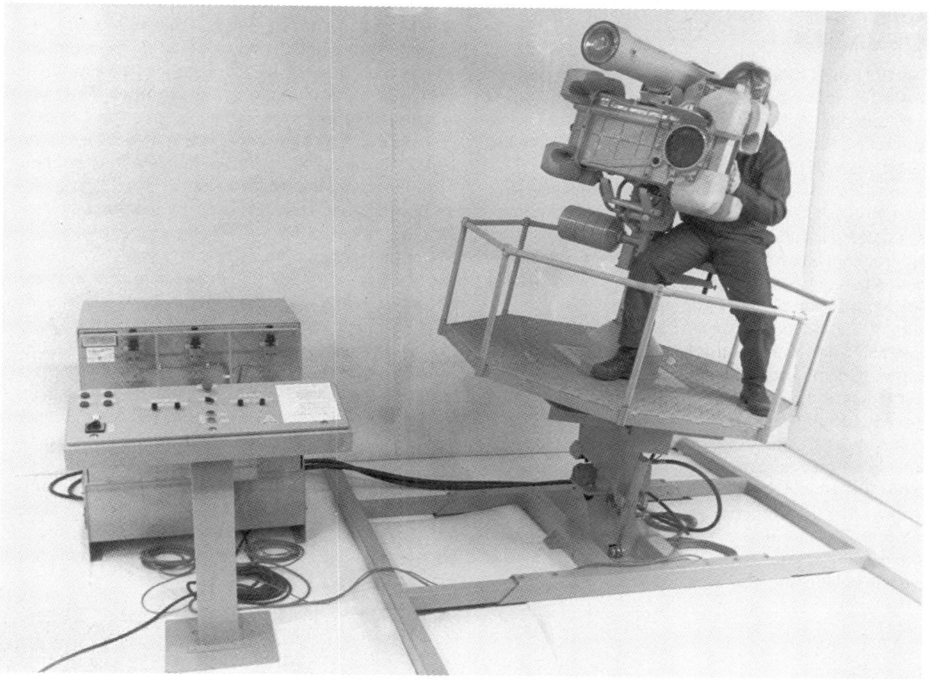

The Weibull naval air defence simulator pictured with an RBS 70 air defence missile simulator

UNITED KINGDOM

Data Sciences Harpoon Onboard Trainer (HOT)

Data Sciences (formerly Software Sciences) has developed the HOT (Harpoon Onboard Trainer) for use with the McDonnell Douglas RGM-84A Harpoon anti-ship missile. The system interfaces with the Harpoon missile system over an RS-232 link via the Harpoon Interface Translation Unit which is also supplied by Data Sciences.

The HOT system is based upon an IBM-compatible Compaq Portable II personal computer which includes a Motorola 80286 16-bit microprocessor and a Motorola 80287 floating point co-processor. In addition, there is a synchro unit which receives simulated ship motion data from the computer and then interfaces with the Harpoon weapon system. The HOT exchanges data through the weapon interface with the weapons control system and simulates the combat command system.

It has three modes of operation: set-up, play and analysis. In the first (set-up) a library of target platforms and scenarios for engagements is created. These may be carried out in real time or faster. In the second (play) the selected scenario is run and the HOT automatically responds to background and sector track requests as well as approving or rejecting plans. Part of the screen shows a map which displays either the true or the perceived state of the scenario while the remainder holds engagement data. In the last mode (analysis) the exercise is examined in 'snapshot', tabular or rerun modes.

Status

The Ministry of Defence has purchased 14 systems for use on board Royal Navy ships and in training establishments.

Manufacturer

Data Sciences Limited, Farnborough.
(For details see Directory of Suppliers)

The Data Sciences' HOT system

DSD Remotely Operating Vehicle Emulator

Digital Systems and Design Limited (DSD) is a United Kingdom-based defence electronics engineering company which provides a comprehensive range of compatible modular training facilities. The company's approach to the requirement for Simulation, Training, Education and Procedural Systems (STEP) is to provide maximum on-line training by means of generic training in the form of a number of Intelligent Procedure Trainers (IPT).

The Remotely Operating Vehicle (ROV) Emulator provides operators with a real-time, three-dimensional visual emulation of complex underwater scenarios. This allows them to be trained in different types of naval mine and the wide variety of environmental conditions in which the ROV may have to operate. The trainer may be either shore- or ship-based.

Status

Development complete.

Manufacturer

DSD Digital Systems and Design Limited, Eastleigh.
(For details see Directory of Suppliers)

EASAMS Shore Integration and Demonstration Facility (SIDFAC)

EASAMS has been awarded a contract by the United Kingdom Ministry of Defence to provide and to operate a Shore Integration and Demonstration FACility (SIDFAC) for the 'Swiftsure' and 'Trafalgar' class nuclear submarine tactical weapons system update programme.

The purpose of SIDFAC is to bring together the component parts of the tactical weapons system to allow it to be exercised in an arduous but realistic manner before it is refitted into submarines. This enables the refit to be completed earlier than might otherwise have been the case. SIDFAC is housed in EASAMS' purpose-built test block and the Ministry of Defence is providing the individual items of equipment which make up the tactical weapons system.

A further contract has been received for the supply of two system test facilities to enable the SIDFAC test procedures to be applied to refitting submarines. The system test facilities consist of proprietary and dedicated hardware and software which are embarked for alongside testing.

As part of EASAMS' weapon system integration authority contract for Stages 3 and 4 of the 'Swiftsure' and 'Trafalgar' class updates, EASAMS will extend the SIDFAC to provide a shore integration facility.

Status

In service.

Manufacturer

EASAMS Limited, Camberley.
(For details see Directory of Suppliers)

Loral Solartron On Board Visual Aimer Continuation Trainer (OBVACT)

The On Board Visual Aimer Continuation Trainer, or OBVACT, was developed and manufactured by Solartron Defence Systems, now Loral Solartron, in response to a Royal Navy requirement for a naval gunnery trainer. OBVACT is a system adaptable to a wide range of small calibre guns and uses Computer-Generated Imagery (CGI) to provide realistic weapon aimer training on board a ship without using live ammunition or target drones. The system also provides training in target recognition, improves trainee familiarisation with a weapon type and provides a comprehensive printout of student performance. OBVACT is used by the Royal Navy to provide training on all its surface target and anti-aircraft guns ranging from 20 to 40 mm in calibre. The system is used on all types of ship and at land-based gunnery training centres.

The high resolution CGI graphics used in OBVACT provide realistic seascape backgrounds and accurate three-dimensional scaled targets. The stabilised picture ensures that the synthetic horizon is correctly aligned and that the relative orientation of the targets is maintained. Air or surface targets can be selected and up to two targets at a time can be utilised.

OBVACT 'on gun' modules include a weapon interface kit, a target image unit and interface optics, with the latter two components mounted on the sight unit. The weapon interface unit contains gyros to provide the computer unit with angular rotation rates. The target image unit receives video signals from the computer unit and converts them into visual images on a cathode ray tube. The interface optics couple the cathode ray tube image and the weapon sight.

'On deck' modules are used by the instructor and include the computer unit which provides overall control of the system. The computer unit also processes the stabilisation of the synthetic picture and generates all CGI video signals including the scene background, targets, tracer paths and weapon and target noise. Other instructor units include a hand-held unit with a keypad and a liquid crystal graphics display. This unit interfaces with the computer unit and has a built-in test facility. The hand-held unit is used to initiate an exercise, has an exercise run facility, and is also used for target programming and replay control. There is also a printer and power supply 'on deck' unit which can be used to provide a hard copy of a trainee's performance and progress results. The display facility on the instructor's hand-held unit also provides an indication of trainee results.

OBVACT is manportable and can be fitted onto a gun within 15 minutes. No electrical interface with a gun platform or gyros is required. Once installed the system is self-contained. An instructor can select exercise general visibility conditions, the seascape and the sun position. The system can be programmed to suit different ballistic characteristics. A number of preset exercises can be selected for unsupervised training and progressively more difficult targets can be introduced as skill levels rise. For all exercises the trainee is provided with a visual simulation of target hits.

OBVACT requires a 115 V AC, 300 W supply with short circuit and over-voltage protection. Starting current is less than 8 A and maximum steady state current is less than 5 A. The gyroscope full-scale deflection rate for both slew/pitch and roll is 160°/s.

Status

In production. In service with the Royal Navy, the US Coast Guard and the Singapore Navy.

Manufacturer

Loral Solartron, Herts.
(For details see Directory of Suppliers)

Loral Solartron OBVACT installed on a 30 mm gun

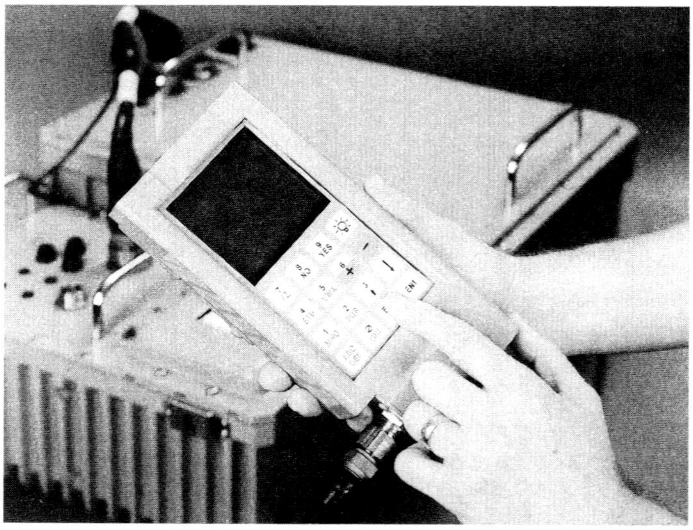

The instructor's hand-held unit used with the Loral Solartron OBVACT

Marconi Point Air Defence Trainers (PADT)

Marconi (formerly Ferranti International Simulation and Training) has produced a number of weapon trainers, the first being a Sea Dart Operator Trainer which used an FM1600B computer and was delivered to the Royal Navy School of Maritime Operations (HMS *Dryad*) at Fareham, Hampshire.

A Point Air Defence Trainer (PADT), using FM1600E and FM1600B to provide complete training for controllers of the Seawolf surface-to-air missile (GWS 25 Mod 0) and part-task training for missile directors, electronic warfare directors and air warfare officers was delivered to HMS *Dryad* in 1985 and became operational in the summer of 1986. This system features an instructor's annex with two consoles and a radar display with simultaneous monitoring of up to eight communications channels, a computer room and a student display room. The latter consists of actual Seawolf, Exocet, and electronic warfare operating equipment and a simulated CAAIS (Computer-Aided Action Information System). The company also supplied four containerised WSA 421 weapon control system trainers to HMS *Cambridge* in 1985.

Status

In service with the Royal Navy.

Manufacturer

Marconi International Simulation and Training, Cheadle Heath.
(For details see Directory of Suppliers)

Marconi Versatile Multi-Aimer Trainer (V-MAT)

As a major supplier of naval radar systems, Marconi developed the V-MAT (Versatile Multi-Aimer Trainer) missile operator trainer. This system is used to train up to eight gun or missile aimers simultaneously using simulated infra-red or low-light television sights.

The system is based upon the Tepigen (Television Picture Generator) to produce computer-generated images which create realistic 'targets'. The objects, either ships or aircraft, can move freely within the 'play area' under the instructor's control while scenery or light condition details can be changed easily.

Based upon the Marconi Locus 16 processor, one of which is used with the Tepigen and the other as the control computer, V-MAT can be used to simulate various defensive and offensive roles using missiles or guns simply by changing the software. Student progress is monitored from the Exercise Control Console which can be used to insert a wide variety of problems into the student's console. A printout of the student's progress can be made immediately from the Exercise Control Console which also incorporates record and replay facilities.

The Malaysian system features a rolling platform to simulate a ship's deck, two visual displays and a Bofors 40 mm L/70 gun. The system also provides computer-generated weapon-mounting sounds.

Status

One system for training missile operators was bought for Royal Navy use during the 1970s. Another system for gunnery training was purchased for the Royal Malaysian Navy.

Manufacturer

Marconi Radar Systems Limited, Chelmsford.
(For details see Directory of Suppliers)

Strachan & Henshaw Weapon Handling and Discharge Equipment (WHDE) Training Equipment

Strachan & Henshaw provides all the Royal Navy's submarine Weapon Handling and Discharge Equipment (WHDE) having been prime contractors to the Ministry of Defence for such equipment since 1972. Given the difficulties of moving a 1.5-ton torpedo safely, the company has also provided WHDE training equipment.

The 'Upholder' class weapons handling and discharge trainer is the latest such system supplied to the Royal Navy for installation in HMS *Dolphin* in Portsmouth. It was formally opened in June 1990.

This system consists of a facsimile weapon stowage compartment into which dummy weapons are embarked via the 'torpedo hatch' having been lowered onto the 'casing'. Within the 'compartment' they are lowered onto the centreline positions and may then be traversed and stowed in all the upper and lower weapon loads on the starboard side. A weapon may be loaded into No 3 tube, the system can be prepared for launch and a launch may then be simulated. A fault simulation system is available for the instructors to insert faults during the firing sequence.

Status

In addition to installing the 'Upholder' class WHD trainer in 1990, the company installed a similar system in HMS *Dolphin* for the 'Swiftsure'/'Trafalgar' class in 1981 and for the 'Vanguard' class in 1992. The latter has similar capabilities to those of the Upholder trainer and is controlled via a desktop personal computer. The company has also provided a similar facility at HMAS *Stirling* to train Royal Australian Navy personnel in the use of the WHDE of the 'Collins' class.

The latest Strachan & Henshaw WHDE training system

Manufacturer

Strachan & Henshaw, Bristol.
(For details see Directory of Suppliers)

Vosper Thornycroft Missile Compensation Trainer

In 1967 Vosper supplied a missile compensation trainer to simulate operations from 'Resolution' class ballistic missile submarines. This system is designed to provide handling simulation for submarine personnel and provides an indication of the change in trim and other handling characteristics once a Polaris missile has been fired. The system is installed at the Royal Navy Polaris School at the Faslane Naval Base (HMS *Neptune*).

Status

In service with the Royal Navy.

Manufacturer

Vosper Thornycroft (UK) Limited, Portsmouth.
(For details see Directory of Suppliers)

UNITED STATES OF AMERICA

Delex Systems Harpoon Engagement Training Aid (T0403)

The Harpoon Engagement Training Aid (HETA), which has the US Navy training device number T0403, is a single station tactical trainer for Harpoon anti-ship missiles whether they are air-, ship- or submarine-launched. It can be used to support either anti-surface warfare tactical training or Harpoon team training.

HETA can simulate the launch of an unlimited number of missiles in realistic tactical situations. The system uses commercial desktop computers with Delex-designed software. It provides scenario generation, target designation training, fire-control solution generation, simulated missile firing and an immediate result.

Status

Delex has developed 19 different programs for the US Navy as well as for navies and air forces of 18 foreign countries.

Manufacturer

Delex Systems Inc, Virginia.
(For details see Directory of Suppliers)

The Delex HETA uses desktop computers and Delex software

Indian Head Division, Naval Surface Warfare Center, Missile/Weapon Simulators

The Indian Head Division of the US Naval Surface Warfare Center (IHD/NSWC) serves as the US Naval Sea Systems Command centre of excellence for missile weapon simulators and training shapes. It provides total quality management of development, acquisition, manufacturing and life-cycle support in weapons simulation, emulation, training shapes and systems test equipment not only for the Department of Defense but also for allied nations. The devices simulate the physical and/or functional characteristics of missile weapons for use in weapon system installation and checking, operational readiness testing, field maintenance and the training of operational and maintenance personnel.

A wide range of pre- and post-launch missile as well as fire-control simulation systems is produced. They provide low-cost alternatives to firing for system and user training, for system and launcher diagnostics and for built-in simulator diagnostics. Missile simulators include the Mark 78 for Harpoon with Mark 124 (ASROC cable) and Mark 125 (canister cable) derivatives as well as the Mark 98 and Mark 105 simulators for submarine-launched Harpoon. There is the Mark 81 sea-launched Tomahawk digital weapon simulator which may be supplemented by a Link 11 interface parallel to serial data converter. A variety of simulators is produced for the Standard Missile (SM) in both Medium Range (MR) and Extended Range (ER) versions. Three missile simulators include the Mark 63 (SM-1MR/Harpoon), the Mark 89 (SM-1ER/SM-2ER) and the Mark 92 (SM-1MR/SM-2MR/Aegis) as well as two interface test sets. Mark 96 (Aegis Vertical Launch System/Tomahawk/Harpoon/ASROC/Seasparrow) and Mark 682 (Vertical Launch System/SM-2/ASROC) are also produced.

Dummy missiles, training rounds as well as subsystem training systems are provided for Harpoon, Tomahawk and Standard Missile. The Guided Missile Training Round (GMTR), which simulates US Navy surface-to-air missiles, is used aboard cruisers and destroyers for crew training and for daily fire-control system testing. The GMTR consists of two major components: the Guided Missile Dummy Round (GMDR) and the Guided Missile Simulator (GMS). Four GMTRs are stored in each ship's magazines (two forward and two aft) and are hoisted to the deck by the same mechanical system as the real missiles. GMTRs functionally check the operation of the launching system interfaces with the weapon systems and train personnel in launcher operation and missile firing. Other examples of weapon simulation include the Mark 48 Torpedo Launchable Vehicle, which is used as a final test gauge for torpedo tubes which are installed in submarines using the Mark 48 torpedo. To support the Mark 41 vertical launch system IHD/NSWC produces Missile Echo Units (MEUs) and Vertical launch system Operational Test Sets (VOTS). The MEU simulates the electrical/electronic characteristics of missiles and enables weapon control systems and the vertical launch control system to conduct missile inventories and to execute tactical launches. The VOTS supports troubleshooting of launcher problems when normal maintenance procedures have failed to isolate the fault. It is used to support land-based/shipyard testing of the vertical launch system during development, integration and installation.

Status

The missile simulators and other training systems are produced for the US Navy.

Manufacturer

Indian Head Division, Naval Surface Warfare Center, Maryland.

(For details see Directory of Suppliers)

Litton AN/SSQ-91 Combat Simulation Test System (CSTS)

Litton Data Systems' AN/SSQ-91 Combat Simulation Test System (CSTS) is in service with 'Wasp' (LHD 1) class amphibious assault ships as well as New Threat Upgrade (NTU) cruisers and destroyers. The system uses 62 Motorola 68040 microprocessors on VME cards with as many as eight in any one node and it can be linked via emerging distributed simulation technology to provide tactical training in port or at sea using the ship's equipment. It provides a gateway to NAVNET and the Tactical Combat Training System (TCTS) as well as to other defence training networks. It conforms to the interoperability standards of the Distributed Interactive Simulation (DIS) protocols which will allow it to communicate with other participants in Navy battle force tactical training and joint operational exercises.

Using a SAFENET local area network the AN/SSQ-91 simulates a complex, multi-threat tactical environment involving up to 4096 simultaneous tracks (including 12 active and 16 passive ECM events per sensor), with 1 second track update rate, to exercise the ship's weapons and sensors by providing entirely realistic, maximum stress combat scenarios. The combat system teams train in the major warfare areas using their own consoles to learn tracking, detect-to-engage track processing, surface/amphibious operations, Link 11, air traffic control, Link 4A/air intercept control, electronic warfare engagements and weapons firing. The CSTS monitors both operator performance and equipment.

The AN/SSQ-91 uses the ship's equipment and operating procedures and is flexible enough for easy configuration to specific ship combat systems. It simulates the ship's radars using RF injected at the wave guide to produce video/tracks at the operators' consoles. For electronic support measures equipment, emitter parametric data and bearing information is sent as digital stimuli to the computer.

CSTS features include a 7000 km^2 x 3000 m area, eight-day scenarios, meteorological and other environmental features, a 5000 vector land mass, littoral contour mapping and a real-world "game board".

Status

The system is operational in 'Wasp' (LHD 1) class amphibious assault ships and two NTU vessels, while 20 more ship systems have been ordered. Elements are also used for data link testing in the 'Ticonderoga' (CG 47) class cruisers.

Manufacturer

Litton Data Systems, Agoura Hills, California.

(For details see Directory of Suppliers)

The Litton AN/SSQ-91 CSTS tactical control console

Loral Librascope Submarine Combat Control System Mark I Training System

The Loral Librascope Corporation produces submarine fire-control systems and has developed a combat control system simulator to train operators. This is a land-based system designed to train tactical information and weapon control personnel in the use of the Submarine Combat Control System Mark I. It consists of a simulated weapons control console with appropriate communications and a computer to run the system and generate a variety of appropriate data which can be injected into the system.

Status

The SCCS Mark I has been purchased by Australia, Canada and India. In May 1987 Librascope (now Loral Librascope) announced the delivery of a Mod 151 training system to the Indian Navy for installation at a dedicated facility in Bombay. The two year programme cost $6.14 million.

Manufacturer

Loral Librascope, Glendale, California.

(For details see Directory of Suppliers)

The Librascope SCCS Mark I training system which has been delivered to the Indian Navy

United Defense Automated Training Aid (ATA)

The Automated Training Aid (ATA) uses an Intel 486 work station or Macintosh to provide computer-based training for shipboard operators of the Guided Missile Launching Systems Mark 13 and Mark 26. ATA enhances classroom instruction on the functional operation of these launching systems.

ATA aids the shipboard operator in understanding the Guided Missile Launching System functional operations of train, elevation, load, unload, system activation and missile selection/direction. It uses computer animation to describe the operation of the subassemblies and the student can use a mouse to observe the effects of hydraulic fluid flow and pressure, valve operation, piston motion and electrical power distribution.

Status

Under development with deployment to the Great Lakes Naval Training Center expected by the end of 1994.

Manufacturer

United Defense LP, Minneapolis.
(For details see Directory of Suppliers)

ATA lets the operator see how the launcher subsystem will respond during each cycle of operation

United Defense Launcher Maintenance and Operational Training System (LMOTS)

The Launcher Maintenance and Operational Training System (LMOTS) provides simulation training for shipboard operators of the US Navy's Guided Missile Launching System Mark 13 Mods 4, 5, 6 and 7 or the Guided Missile Launching System Mark 26. LMOTS training includes normal operation, casualty identification and fault isolation procedures.

An LMOTS system consists of the primary tactical guided missile launching system control system elements and a missile simulator, together with a Launcher Order Responder Unit (LORU). The control system elements include the power distribution panel, an operator's control console, a missile power supply, a launcher relay control box and a missile simulator. The use of the standard control system elements is a key feature of the LMOTS concept. They are identical to those with which the guided missile launcher system operator will be working aboard ship. When training, the operator receives the same visual indications and stimuli, and takes the same actions as when operating the shipboard guided missile launching system and overcoming problems which may be encountered.

The heart of the LMOTS is the LORU. The LORU consists of an instructor's console which duplicates the electrical, timing and response function of the launcher, the magazine and the jettison hydraulic mechanical assemblies. It also duplicates weapons control sufficiently to generate the proper guided missile launching system and missile orders in the time intervals encountered with the real launching system, as well as the electrical interface characteristics of missiles sufficiently to complete the necessary simulated firing responses. It inserts simulated malfunctions of the launching electrical system and operational conditions.

LMOTS systems are installed in shore-based training facilities which include visual and other instructional aids. They are used in the initial training courses for shipboard operators with 50 per cent of the time spent in classrooms, 30 per cent in operational training with LMOTS, and 20 per cent in hands-on training with the actual system.

Status

LMOTS Mark 13 and Mark 26 have been produced for the US Navy while LMOTS Mark 13 has also been produced for Taiwan. Other countries which have expressed interest in the system include Australia, France, Germany, Italy, Japan, the Netherlands and Spain.

Manufacturer

United Defense LP, Minneapolis.
(For details see Directory of Suppliers)

The power distribution, operator's control and display panels in the LMOTS

United Defense Gun Maintenance and Operational Training System (GMOTS)

The Gun Maintenance and Operational Training System (GMOTS) is similar in concept to the LMOTS (see previous entry). It consists of a power distribution panel, an operator's control panel, a display panel, an electronic fuze setter unit, an electronic fuze simulator and a Gun Order Responder Unit (GORU).

The GORU duplicates the electrical, timing and response functions of the Mark 45 Mod 1 or Mod 2 gun mount, its ready service loader, lower hoist and fuze setter. It simulates the ammunition selection, loading and unloading as well as strikedown operations, and duplicates weapon control orders and interfaces. Simulated malfunctions can be inserted in the 'gun mount' electrical circuits and operational conditions. It can also insert mechanical timing errors.

The GORU is also used during gun mount overhauls to test the Mark 45 control panels. A special test program is loaded into the GORU in order to accomplish this additional test capability.

Status

In production and in service with the United States and Turkey.

Manufacturer

United Defense LP, Minneapolis.
(For details see Directory of Suppliers)

OPERATIONS CENTRE SIMULATORS

AUSTRALIA

British Aerospace Australia Integrated Operations Team Training Facility (IOTTF)

British Aerospace Australia Limited (BAeA) won an Australian $9.5 million contract to supply a facility to train the operations team of the Royal Australian Navy's 'Perth' class missile destroyers. In the past, the Royal Australian Navy has relied upon training its personnel within ships but, with the modernisation of the destroyers, it has been decided that a simulator would provide the most cost-effective means of training personnel and the system will also provide training for crews of the 'Adelaide' class frigates.

The IOTTF consists of operations rooms, a monitor room and a debriefing auditorium. It features a distributed VME architecture using Motorola 68030 processors and Ada language software which includes more than 280 000 lines of code. The simulator has been designed by Australian Defence Industries Pty Ltd (ADI) and the Ada language software will be provided by SD-Scicon UK Limited. The operations rooms are reproductions of those in the destroyer and the frigate with facsimile consoles whose performance is emulated by the extensive use of commercial off-the-shelf equipment.

Testing of facsimile consoles for the destroyer operations room before they are handed over to the Royal Australian Navy

The monitor room is used by the instructors and uses work stations to create and to run the training scenarios. The instructors attack or support the 'ship' by moving 'platforms' in a computer-generated environment then monitor the performance of individual students. After the exercise debriefing can take place in the auditorium with the aid of a large screen projection system.

Status

Delivery of the first, destroyer, phase of the system to HMAS *Watson* in Sydney took place early in 1994. The second, frigate, phase will be delivered in mid-1995.

Manufacturer

British Aerospace Australia Limited, Canberra.
(For details see Directory of Suppliers)

CANADA

Litton Systems Canada Automatic Data Link Plotting System (ADLIPS)/CCS-280 Simulators

ADLIPS (Automatic Data Link Plotting System) is the control and tactical data communication system used by the Canadian Navy's frigates, while the CCS-280 (Command and Control System for the DDH-280 or 'Tribal' class) is a similar system for Canada's destroyers. Each replaces electromechanical plotting tables with associated radar displays with a horizontal display which processes data from shipboard sensors, as well as those of friendly ships and aircraft received by data link. The systems process the data and present it as a composite tactical display.

The simulators use the original horizontal display systems but they are operated by specialised computer systems controlled from the instructor station. The computer produces realistic responses of both 'own ship' sensors and weapon systems, as well as those of hostile ones, with the instructor monitoring the students' response and inserting complicatory factors.

Status

Three ADLIPS display systems were provided in the early 1980s for the Fleet Training School at Halifax, Nova Scotia, while a CCS-280 system was installed at the Canadian armed forces' Combined Support Division (CSD) at Dartmouth, Nova Scotia.

Manufacturer

Litton Systems Canada Limited, Ontario.
(For details see Directory of Suppliers)

Students operating a horizontal console of a CCS-280 system at the Combined Support Division facility at Dartmouth, Nova Scotia

GERMANY

Atlas Elektronik Corvette Combat Training Simulator

Atlas Elektronik has supplied the Republic of Singapore Navy with a CIC simulator to train command and specialist personnel in 'Victory' class corvettes. The system has been designed in collaboration with Singapore Electronic & Engineering Limited (SEEL).

The simulator uses both simulation and so-called 'stimulation' techniques to provide an authentic, integrated environment for individual and team training in all possible weather and warfare conditions. It includes facsimiles of the corvette's command and weapon control system, variable depth sonar, torpedo fire-control, Harpoon anti-ship missile, electronic warfare and communications facilities. The host and simulator computers are based upon Atlas Elektronik MPR processors.

Status

In service with the Republic of Singapore Navy from January 1993.

Manufacturer

Atlas Elektronik GmbH, Bremen.
(For details see Directory of Suppliers)

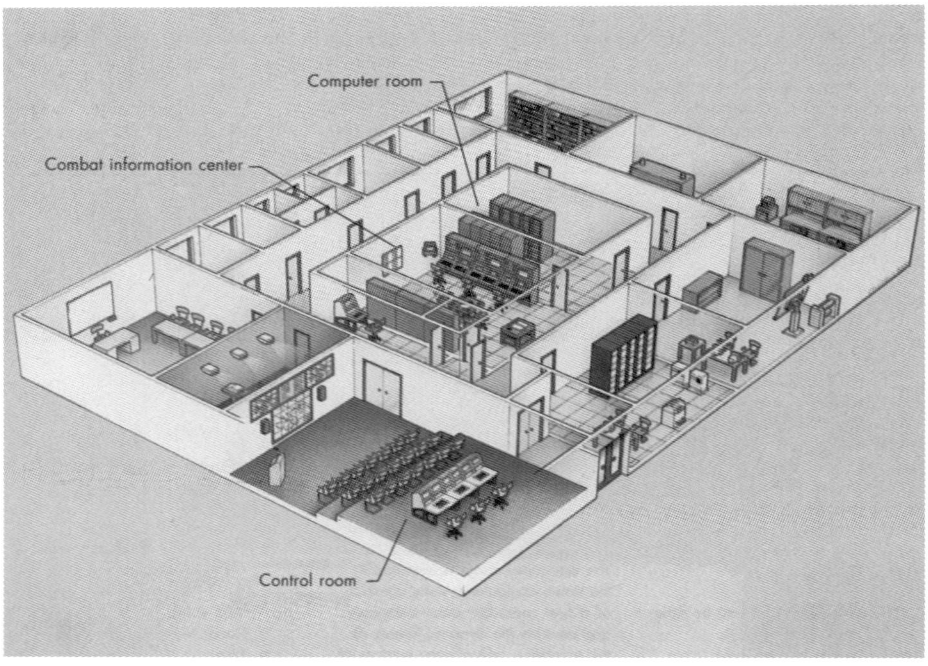

An artist's impression of the Singapore corvette training simulator

Atlas Elektronik Minehunting Command and Operator Trainer (MCOT)

The Minehunting Command and Operator Trainer (MCOT) is designed to assist operator and command team training in minehunters. It provides procedural and tactical training for minehunting as well as maintainer training and includes ship handling, detection and classification of underwater objects, the identification, neutralisation and documentation of mines as well as the control of the Remotely-Operated Vehicle (ROV).

MCOT consists of a CIC room, a maintenance room, a computer room and an auditorium for briefing and debriefing. The CIC room consists of the commander's station, tactical display and command equipment, a console for a minehunting sonar, a console for 'controlling' an ROV, a ship control console, a plotter and a printer as well as instructor's console. The maintenance room consists of a real ROV with associated electronics cabinets. The computer room features a host computer for central management of 'targets' and the 'exercise area', a digital sonar simulator, a sonar image generator (for the active identification sonar), a TV simulator for the ROV and the database generation station.

Status

An MCOT based upon the 'Frankenthal' (Type 332) minehunter was delivered to the German Navy in 1993. This features simulated DSQS-11M sonar and the Pinguin B3 ROV.

The instructor's station in the Atlas Elektronik Minehunting Command and Operator Trainer

Manufacturer

Atlas Elektronik GmbH, Bremen.
(For details see Directory of Suppliers)

DST Verfahrenstrainer-Marine Ortungsschule (VT-MOS) and Taktik und Verfahrenstrainer Marine (TVTM)

To meet a German Navy requirement for a tactics and procedural training system for personnel in Combat Information Centres (CIC), DST has supplied the VT-MOS and TVTM simulator systems. They simulate all existing and planned command and control systems and their operation takes into account environmental limitations which have an effect upon tactical scenarios. VT-MOS is designed to simulate procedural tasks within the CIC including weapon control, tactical navigation and communications. TVTM is a similar system which emphasises tactical procedures.

Each system features several simulated CICs, a control room and a 'bridge' with each CIC driven by a Concurrent Computer Corporation computer which has three 32-bit Motorola 68040 processors using software based upon a real-time Unix system which is coded in Ada. Each CIC is also supported by a DST GC-10 graphics system which can display up to 800 target tracks.

The CICs are linked by a DECNET local area network and feature a number of work stations or Multi-Function Command Consoles (MFCC) with VT-MOS having a total of 42 MFCC while TVTM has 57. Each MFCC features a 19-inch (483 mm) raster scan display, an alphanumeric display, a touch-screen, function key unit and qwerty keyboard for alphanumeric input. Collectively they simulate different ships' command systems and man/machine interfaces with between one and six representing one simulated unit such as a helicopter, a ship or a submarine. Within the simulator CIC they may be supplemented by peripherals such as printers, steering units and network terminals.

TVTM has a special feature, the tactical analysis system, which analyses recorded data from real situations acquired in exercises or by intelligence. This is inserted to provide a realistic background and assists analysis during debriefing. Each system has a debriefing display auditorium with a large screen display.

Status

The systems were ordered in 1989 and 1990 and are now in service at the German Navy's Operations School. They replace three operations centre simulators.

Manufacturer

DST Deutsche System-Technik GmbH, Kiel.
(For details see Directory of Suppliers)

MSG Mine Countermeasures Vessel Simulator

MSG Marine- und Sondertechnik GmbH (formerly the Marine and Special Products Division of MBB) developed and produced a mine countermeasures vessel simulator. This system uses a mine-location sonar into which are injected simulated scenarios. There are some 15 student cubicles with their own 'sonar system' and search and classification screens upon which computer-generated images are projected. In addition to realistic sonar performance, the simulator also displays the seabed and objects which may be encountered there. A synthetic image of a remotely operated vehicle is also displayed for mine classification.

Status

A system to train crews for the German Navy's 'Göttingen' class mine countermeasures vessels has been installed at the Naval Weapons College at Eckernförde.

Manufacturer

MSG Marine- und Sondertechnik GmbH, Bremen.
(For details see Directory of Suppliers)

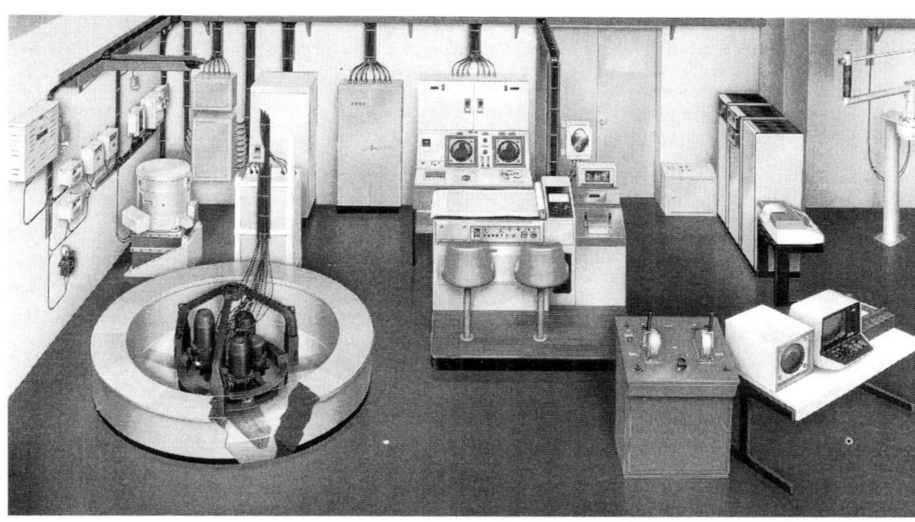

A model of the MCMV operations room produced by MSG and installed at the German Naval Weapons College

ISRAEL

Elbit Combat Information Centre Procedural Trainer (CIC-PT)

The Elbit Combat Information Centre Procedural Trainer (CIC-PT) was designed to train members of the operations centre. It features facsimile radar and command/control consoles, plotting tables, internal and 'external' communications.

Status

Believed to be in production. Reportedly a 'Wielingen' frigate CIC version is in service with the Belgian Navy.

Manufacturer

Elbit Computers Limited, Haifa.
(For details see Directory of Suppliers)

SOUTH AFRICA

Teklogic Genius Surface Ship/ Submarine Combat Team Trainers

Teklogic is offering two combat team trainers with realistic operational scenarios based upon its Genius (Generic Integration and Utility Simulator) range. These include a target controller capability and pre-programmable targets which behave autonomously.

A typical system for surface warships would consist of a simulated CIC with facsimile consoles, an instructor's station, a helmsman's station and a target controller station. The system would use computer-generated visual imagery.

For submarine command teams there is the similar Cabano system. This would include a periscope simulator using computer-generated imagery and a sonar simulator.

The Genius range includes emulators for sonars and underwater weapons, radars, electronic warfare systems, electro-optical systems, periscopes, weapons, communications and helmsmen. There are simulators for radar systems, radars and sonars.

Status
Available.

Manufacturer
Teklogic (Pty) Limited, Halfway House.
(For details see Directory of Suppliers)

SWEDEN

CelsiusTech 9LV200 Fire-Control Simulator

CelsiusTech has supplied many navies with the 9LV200 command and weapon control system for fast attack craft and larger warships. The company has recently developed, in association with Micro Processor Systems A/S (MPS) of Norway, a fire-control simulator for ships with the 9LV200/300 Mark 2 and Mark 2.5.

The system uses facsimiles of the 9LV200 consoles which are linked to an instructor's station based upon a personal computer. This generates realistic simulations from all the electronic and electro-optical sensors in a realistic combat training scenario. The system may be used also to superimpose simulated events upon real radar images.

Status
Available.

Manufacturer
CelsiusTech Systems AB, Järfälla.
(For details see Directory of Suppliers)

UNITED KINGDOM

Marconi Operations Room Simulators

Marconi (ex-Ferranti International Simulation and Training) supplies simulators to provide comprehensive training for operations room personnel at skill levels from track supervisor to principal warfare officer. The system allows for the training of individual operators, pre-commissioning training at team level, continuation training of an established team and, if required, multiple ship training. The operations room trainer consists of a central simulation suite, exercise control room and one or more operations rooms for a specific class of warship featuring a mix of simulated actual ship equipment, facsimiles and emulation to provide a realistic environment.

The trainer assists all aspects of operations room activities including tactical picture compilation, sensor control and supervision, weapons systems response and target tracking control, electronic warfare and communications. User-friendly, menu-prompted procedures and access to extensive data libraries from the instructor's workstation allow for the efficient generation of exercise scenarios. The instructor is able to control the tempo and complexity of the exercise to suit the ability of the students and is able to highlight points of specific interest either as the lesson progresses or as part of the debrief.

The Type 22 frigate operations room simulator in HMS Dryad

Seven operations room simulators have been supplied to the Combined Tactical Trainer at the Royal Navy's School of Maritime Operations (HMS *Dryad*) at Fareham, Hampshire. These systems represent 'County' class destroyers (1970), Ikara-equipped 'Leander' class frigates, CAAIS-equipped 'Leander' class frigates (1976), Type 42 destroyers (1976), Type 21 frigates (1979), and Type 22 frigates (1985). The 'County' class system was removed in 1984 while the Type 42 system was extended in 1979.

A two-operations room system for a general-purpose ship was delivered to an unidentified navy in 1977. In 1985 a frigate operations room for a 'Niteroi' class frigate was delivered to Brazil. In November 1987 the Canadian armed forces ordered a simulator of the Command and Control System (CCS) as part of the TRibal Update and Modernization Program (TRUMP). This system, which has five instructor consoles, is based upon the VME 68020 processor. A similar system has been ordered by Hollandse Signaalapparaten BV as the basis for the Portuguese Navy's 'Vasco da Gama' class (MEKO 200) frigate combat information centre simulator. It also features the Marconi CREST radar simulator and was delivered in 1990.

Status
In service (see text). In May 1991 a £17 million contract was awarded to update the Type 22 and Type 42 simulators to reflect upgrades to the ship weapon and sensor systems.

Manufacturer
Marconi International Simulation and Training, Cheadle Heath.
(For details see Directory of Suppliers)

Marconi Mine Countermeasures Vessel (MCMV) Operations Room Simulators

'Hunt' class

The Mine CounterMeasures Vessel (MCMV) operations room simulator for the 'Hunt' class is a modular system designed to provide training both for individual operators and command teams and features a Computer Aided Action Information System (CAAIS) computer, with two Argus 700 processors. Data for up to four hours of exercise time can be stored permitting a reconstruction of the exercise for debriefing purposes with facilities including freeze and fast run.

The system is housed in two semi-mobile trailers, one of which houses the computer and simulation hardware suite with two generators, while the second houses the CAAIS computer, instructor's and students' areas. The instructor's area consists of a two-seat station with displays with the helmsman's console alongside, separated by a curtain. The students' room has two CAAIS displays with an overhead visual control system, the Remote Control Mine Destruction System (RCMDS), a minesweeping sonar console, simulated acoustic and magnetic sweep equipment, a chart table and an echo sounder.

The system features Plessey 193M high resolution sonar simulation that includes sea bottom pictures which may be altered to meet various requirements. The computer also has digital models of both mines and mine-like objects which can be 'laid' by the instructor. The models are of sufficient fidelity to allow classification by sonar. Synthesised Kelvin Hughes Type 1006 navigation radar video signals provide displays with coastlines, sea clutter, navigation buoys and other vessels with the instructor selecting range scale, Pulse

Rate Frequency (PRF) and pulse length. Inputs from 'buoys' and 'navigation radar' can be injected into the training system.

The trainer includes a combat data system which is identical to the operational system, normal communications facilities and realistic ship control with orders from the student commander being injected into the simulation computer at the helmsman's console. Front panels of the acoustic and magnetic sweep consoles are provided with the indicators driven by simulated signals. Control of the remote-control mine disposal vehicle includes a synthesised television view and searchlight control.

'Sandown' class
The 'Sandown' class command and operator trainer (AS 1107) is currently under development. As in the 'Hunt' class simulator, major elements of the operational systems will be used to ensure a high degree of realism.

Major advances in the new simulator will include Computer-Generated Imagery (CGI) for the Remotely Operated Vehicle (ROV) camera, increased sonar modelling fidelity and more flexible exercise control facilities. The major equipments in the operations room area will be the NAUTIS command system, the Type 2093 sonar, and the Remotely Controlled Mine Disposal System (RCMDS) II. There will be a partial bridge area with a NAUTIS position and a replica ship's position control station so that bridge and operations room crew can be trained as a team.

The Type 193M mine location sonar consoles in HMS Arcturus

Status
The 'Hunt' class simulator, designated HMS *Arcturus*, is in service with the Royal Navy. It is based at HMS *Vernon*, which is now part of HMS *Dryad*'s School of Maritime Operations. The 'Sandown' class simulator contract was awarded in February 1992.

Manufacturer
Marconi International Simulation and Training, Cheadle Heath.
(Ex-Ferranti, for details see Directory of Suppliers)

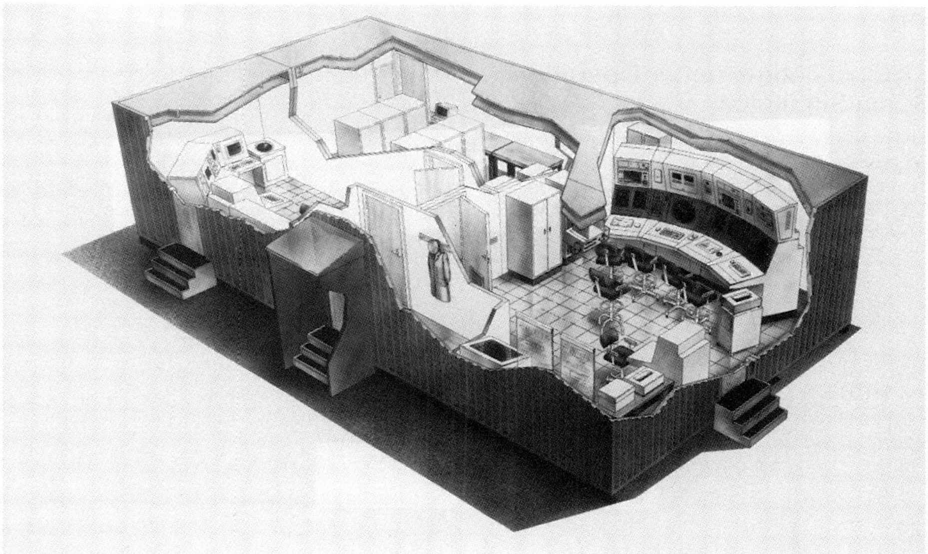

An artist's impression of the 'Sandown' command and operator trainer

Marconi Simulation Mine-Countermeasures Vessels (MCMV) Command/Operator Trainers

Marconi Simulation has used its experience in command team training to design a modular Command/Operator Trainer for Mine CounterMeasures Vessels (MCMV).

The trainer consists of a facsimile of the operations room of a specific MCMV and will provide both individual training and command team training in a realistic environment. It will be installed in transportable containers which will include consoles for the mine-hunting sonar, the command system and the Remotely Operated Vehicle (ROV) system. There will also be simulation of the navigation radar, navigation aids, the ship control system, echo sounder and communications.

The sonar simulation will be based upon a number of existing modules and will provide high resolution displays of the sea bottom in various conditions to allow both search and classification training. The command system operator will have all the required facilities at his console to plan and to execute MCM operations in conjunction with the sonar and ROV operators as well as the navigation and control systems.

The ROV operator's console includes appropriate controls and instrumentation together with a realistic simulation of the TV camera and sonar displays. Radar simulation will include digital coastline simulation techniques and it will provide realistic displays which will include sea clutter, navigation buoys and other appropriate features.

The instructor will simulate the appropriate bridge crew responses via a simulated communications system and will also monitor and control the exercise. He will have user-friendly facilities to create preprogrammed exercises which may be selected from a library. He will be able to initiate them and interact with the scenario as well as recording exercises for debriefing.

Status
Available for production.

Manufacturer
Marconi Simulation, Dunfermline.
(For details see Directory of Suppliers)

UNITED STATES OF AMERICA

AAI Pierside Combat Systems Team Trainers (Devices 20B5 and 20B4)

These systems were designed to provide shipboard combat information centre teams with realistic shipboard training in their operational environment. They can be used to provide CIC teams with refresher training as well as raising the level of individual crewmen's skills. The threat environment of the 20B4 can present up to 100 targets with up to 256 emitters.

The systems are housed in mobile containers weighing 15.88 tons which contain a computer, three student consoles and voice data communications equipment. The container is driven to the dockside and parked alongside the ship. It is then linked to the ship's sensors and data transmission systems and produces a realistic simulated environment for air, surface and anti-submarine warfare exercises.

The 20B5 system is designed to support the 'Oliver Hazard Perry' class (FFG 7) guided missile frigate. It includes simulation of the AN/SQS-56 hull-mounted sonar and the AN/SQR-19 passive towed array sonar used in half this class and the AN/SQQ-89 and LAMPS Mark III used in the remainder. In the field of countermeasures a variety of systems may be simulated including the Mark 36 SRBOC (Super Rapid-Blooming Overboard Chaff) chaff launcher, the AN/SLQ-25 Nixie torpedo countermeasures system and Prairie-Masker acoustic decoy together with the AN/SLQ-32 electronic warfare system. One system can train two ships simultaneously with provision for expansion to train up to six ships.

The US Naval Air Warfare Center Training Systems Division is upgrading Device 20B5 with new threat target data and shipboard equipment upgrade interfaces to drive operator display equipment. The 20B5 and 20B4 designs are currently undergoing a degree of miniaturisation designed to provide for future ship training to be conducted at sea, including an interface with the Distributed Interactive Simulation network to allow participation in long haul network Battle Force Tactical Training (BFTT).

Status

Two 20B5 systems were bought in 1983 for $16.1 million and two more during 1985. They are located at Maryport and San Diego. In 1986 the Royal Australian Navy purchased a system at a cost of US$10.3 million to support destroyers and frigates. The four Device 20B4 systems are currently being upgraded to provide the Mark 86 fire-control system, NATO Seasparrow missile, AN/SQQ-89 Anti-Submarine Warfare Combat System, AN/SWG-1A Harpoon missile fire-control, AN/SLQ-32A electronic warfare and Link 11 training capabilities.

A Device 20B5 simulator alongside an 'Oliver Hazard Perry' class frigate

Manufacturer

AAI Corporation, Maryland.
(For details see Directory of Suppliers)

CAE-Link Mine-hunter Operations Room Simulator

CAE-Link has delivered an operations room simulator for mine-hunters/mine-sweepers for the Belgian-Netherlands Mine Warfare School (Eguermin) in Oostende, Belgium. The facility will also be used by the navies of all NATO countries, except Iceland.

The simulator uses networked, high performance RISC (Reduced Instruction Set Computer) microprocessor-based work stations as the basis for all computing and display console nodes. It consists of nine CICs, six configured as mine-sweeper CICs and the remainder as mine-hunter CICs, an exercise control area and a debrief facility.

Each CIC contains five or seven simulated consoles, a peripheral sensors panoply, three status boards, a problem clock and a number of wall-mounted communications panels. In the mine-sweeper CIC the work stations will include radar, automatic data processing system, auto-pilot system and manoeuvring consoles. There are sweeps command consoles for simulated mechanical and influence sweeps. In the mine-hunter CIC the sweeps command console will be replaced by detector sonar, classification sonar and Remotely Operated Vehicle (ROV) consoles.

The exercise control area has eight instructors' stations to create, monitor and evaluate the exercises. They use commercial off-the-shelf equipment and each features two high resolution raster scan displays on tilt/swivel pedestals. In addition, they have a communications control panel, a qwerty keyboard and roller ball. The debriefing facility includes a large display screen and two monitors, the latter being used to display CIC sensor displays.

Eguermin can simulate 'Aggressive', 'Avenger', 'Hunt', 'Frankenthal', 'Lerici', 'Lindau', 'Oksy', 'Sandown' and 'Tripartite' class minehunters. Detection and classification sonars simulated include the Atlas Elektronik DSQS-11M/H, GEC-Marconi Types 193 and 2093, Raytheon-Thomson Sintra AN/SQQ-32, AN/SQQ-14 and AN/SQQ-14/IT as well as Thomson Sintra DUBM-21 and TSM2023, while remotely operated vehicles simulated include the Consorzio SMIN MIN, the ECA PAP 104/105, the Gaymarine PLUTO, the Honeywell SLQ-48 and STN Pinguin.

Status

The system was delivered to the Belgian-Netherlands Mine Warfare School in Oostende in September 1993.

Manufacturer

CAE-Link Corporation, New York State.
(For details see Directory of Suppliers)

Intellisys Command Tactical Trainer (Device 20F16)

This is a generic system designed to teach key members of the CIC (Combat Information Centre) and is destined for an officers training school.

It consists of a generic representation of a CIC based loosely upon that of the 'Kidd' (DDG 993) class destroyers but it also has two large screen displays which are more usually found in Aegis-equipped ships. The system is designed to train tactical action officers, weapon co-ordinators, prospective commanding officer and executive officers, as well as selected CIC personnel.

In addition to an Aegis-style command station there are four Norden 'sensor' display consoles which present Naval Tactical Data System (NTDS) symbology. There are two large display screens for the presentation of processed data and six Automated Status Boards (ASTABS) for the presentation of alphanumeric data.

The system also includes an instructor's station to supervise and monitor the exercise. The system uses a Harris Night Hawk computer, based upon Motorola 68030 microprocessors, with Ada and 'C' software language programs. By these means up to 120 platforms may be generated, each having a degree of artificial intelligence which permits one million moves per second.

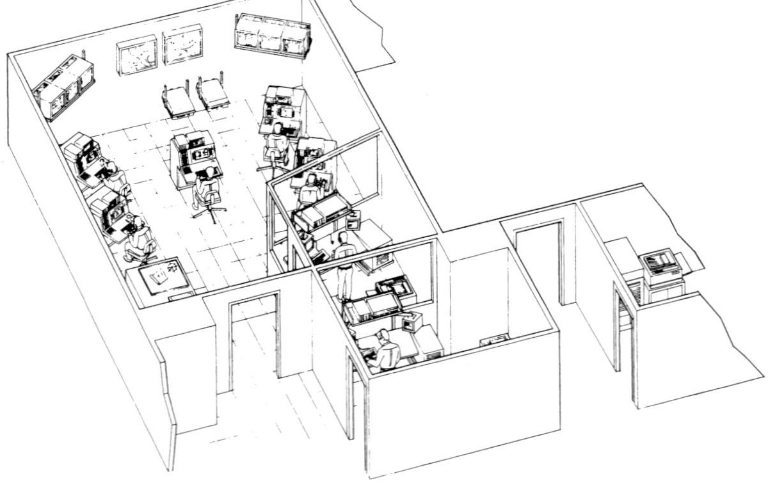

An artist's impression of the Command Tactical Trainer (Device 20F16)

Status

The system is for the Surface Warfare Officers' School in Newport, Rhode Island.

Manufacturer

Intellisys, New York State.
(For details see Directory of Suppliers)

Unisys Fleet AAW System Trainer (FAST)

The Fleet Anti-Air Warfare (AAW) System Trainer is an embedded operational training system installed in warships. It provides exercises for the whole AAW combat system team using normal operational system modes and crews.

The system consists of a TAC-3 work station and a radar simulation unit with interfaces for the AN/SPS-49 and AN/SPS-55 radars. The radar simulation unit has a single board computer based upon a Motorola 68040 microprocessor and is capable of 28 Mips. It has Ada-language software and controls in real time the exercise scenario including front-end stimulation for the radars. The system can provide 71 radar targets and 'track' 20 of them in a wide range of environmental and electronic warfare environments.

Status

In production and in service in 'Oliver Hazard Perry' (FFG 7) class frigates.

Manufacturer

Unisys Corporation, Great Neck, New York State.
(For details see Directory of Suppliers)

SUBMARINE COMMAND TEAM TRAINERS

CANADA

Atlantis Fire-Control and Tactical Team Trainer

This system is used to train command team members individually and collectively and is configured to replicate the control room of an 'Oberon' class submarine. It provides operators with a realistic series of tactical exercises. Accurate simulations of both friendly and hostile vehicles are created together with those of the Mark 37 and Mark 48 torpedoes. It is the first Atlantis training system to use the Ada software language. In 1992 it was expanded by the provision of a Marconi periscope simulator system.

Status
In service since 1990 with the Canadian armed forces at the Naval Fleet School, Halifax.

Manufacturer
Atlantis Aerospace Corporation, Ontario.
(For details see Directory of Suppliers)

FRANCE

DCN Saturne Control Room Simulator

In support of the French Navy's 'Améthyste' class submarine programme, DCN Ruelle has developed the Saturne simulator.

The system uses a distributed computer architecture with a Gould 3297 responsible for the tactical situation simulation and sound propagation model, five Bull SPS7 and Mitra computers simulating tactical data, sensors and fire-control functions. A system dedicated signal generator is used for sonar audio simulation and the system software is written in Fortran 77 and C. There are three independent data networks and data is displayed on Unigraph industrial graphic display systems.

The system is to be upgraded with a new EW system and the incorporation of a periscope simulator with photo-quality computer-generated imagery.

Status
Saturne became operational at ENSM (Ecole Navale Sous-Marine) Toulon in 1990 and was upgraded in 1992. The ENSM also includes 'Agosta' class and 'Daphné' class control room simulators.

Manufacturer
DCN, Paris.
(For details see Directory of Suppliers)

The DCN Saturne submarine control room simulator

GERMANY

Atlas Elektronik Submarine Command Team Trainer (SCTT/ AWU 206/206A)

As a major supplier of submarine electronics, Atlas Elektronik GmbH also produces the Submarine Command Team Trainer (SCTT). This is designed to provide training for all members of the command team in attacking both surface vessels and submarines, defence against such vessels as well as ASW aircraft and operations in minefields. The system, which is in a permanent structure, simulates tactical situations and trains student weapon officers and operators to respond. It can also be used to develop and analyse naval doctrines, weapon firing and the most efficient use of onboard equipment. Two have been produced for the German Navy as AWU 206 and AWU 206A to train command teams for the Type 206 and Type 206A diesel electric submarines.

The SCTT consists of an auditorium with a large screen display, an instructor's console and communications equipment, a combat information centre with real equipment and a computer room which generates the sonar, weapon handling, tactical data handling, periscope, navigation, ship control and electronic surveillance/warfare inputs for the system. The consoles in

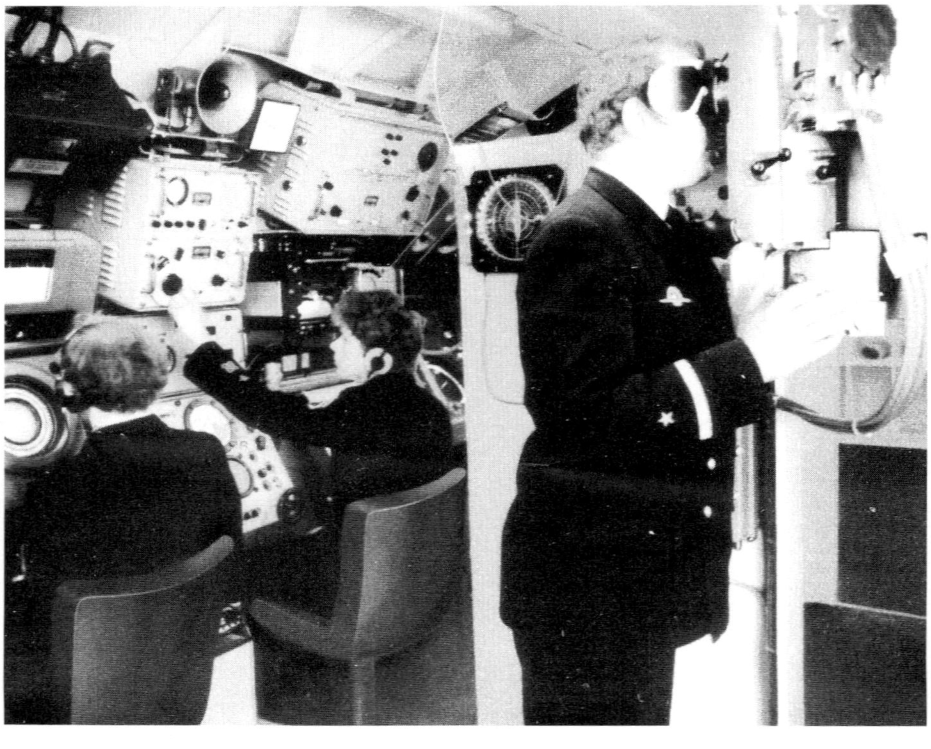

The control room in the Atlas AWU 206A simulator supplied to the German Navy

the combat information centre of the German Navy systems are for the Signaal M 8 (Type 206) or Atlas SLW 83 ISUS (Type 206A) combat system. The former is associated with the AN 410, AN 5039 and Thomson Sintra DBQS-21 and DUUX-2 sonar systems while the latter is associated with the CSU 83 sonar suite.

Status

Two systems have been sold to the German Navy: the AWU 206 was commissioned in 1978 while the AWU 206A was commissioned in 1988. A Far Eastern navy has awarded Atlas Elektronik a contract for another SCTT.

Manufacturer

Atlas Elektronik GmbH, Bremen.
(For details see Directory of Suppliers)

STN Submarine Attack Simulator

STN (Systemtechnik Nord) is offering a Submarine Attack Simulator (SAS) designed to train submarine command teams in procedures and tactical operations. It is a modular design consisting of a simulated control room, an instructor station, a briefing room and a computer system.

The control room uses genuine equipment rather than facsimiles and these will be linked to the computer system to provide accurate sensor and operational performance parameters. The periscope subsystem will be a facsimile with video-generated data. The degree of simulation will depend upon the customer requirement but the system is designed for low operating costs. It will also be capable of part-task training.

The modular concept ensures that the system may, for financial reasons, be installed at reduced cost with only the basic subsystems which may be upgraded at the customer's discretion. Fire-control consoles would be involved in the basic configuration with the addition of a periscope system, a plotting table, passive sonar classification and target assignment, a tactical data handling system and a radar.

The SAS may be mounted in a building or it may be installed on a motion system with a six degree field-of-movement.

Status

Available for production.

Manufacturer

Systemtechnik Nord GmbH, Bremen.
(For details see Directory of Suppliers)

ITALY

DATAMAT Submarine Trainer

DATAMAT has developed and produced an attack team trainer which incorporates real equipment of the MM/BSN-716(V)2 command and control system together with the A 184 torpedo fire-control system. It has a DEC mainframe computer which contains the data for tactical scenario generation while the ship simulation subsystem is based upon Motorola 68020 microprocessors.

The hardware inside the simulator includes consoles for controlling sensors and navigation equipment as well as a torpedo simulator. The instructor's station features three video terminals, three multi-function keyboards, a roller ball and communications control panel. The simulator also has a debriefing system with graphics work station and large screen projector.

Status

A simulator to train crews for the 'Toti' and 'Sauro' class submarines entered service at the Italian Navy Training Centre, Taranto, in June 1991.

Manufacturer

DATAMAT Ingegneria dei Sistemi SpA, Rome.
(For details see Directory of Suppliers)

UNITED KINGDOM

Marconi Submarine Command Team Trainers

Marconi (ex-Ferranti International Simulation and Training) has produced a number of submarine command team trainers which usually comprise an instructor's room and one or two simulated submarine control rooms with simulated sonar, radar, Electronic Support Measures (ESM), fire-control systems, plotting facilities, key ship's instruments, such as log and compass, and an attack periscope. The periscope systems initially used televised models but the latest systems incorporate computer-generated imagery. Some systems are in transportable containers while others are installed in rooms within existing training establishments.

The systems are used to train new submarine commanders, as part of operational working up, for general command team training and refresher training, as well as for training officers in specific roles such as Officer of the Watch. The earlier systems used central processing but the latest incorporate a distributed architecture. The later systems also feature an ocean model simulation which provides a real-time, multi-target environment with a range of acoustic sensor range prediction performances. Vehicle databases are based upon the latest Ministry of Defence trainer database formats which can be modified within the trainer. In addition to surface vessels, submarines and aircraft, the systems can simulate enemy torpedoes, decoys, sonobuoys, and various sonars.

In HMS *Tactician* there is a CGI system for periscope views. This enables the student to benefit from a wide variety of aircraft and ship models not only in daylight but also in simulated low-light television and using thermal imaging. Up to six target images can be generated from a library of more than 60 models.

In May 1989 a contract was awarded to build the command team trainer for the 'Vanguard' class ballistic missile submarines. It will be installed in a permanent building and will consist of two rooms, an instructor's station, viewing area and debriefing facilities. One room will consist of the 10-place command team trainer with simulated fire-control, weapons and tactical data handling system consoles. Uniquely there will be no periscope simulator because all optical/electro-optical images in the ship will be displayed on a screen at a two-place command desk. These pictures will feature the computer-generated imagery technology already used in the company's periscope simulator but with improvements including rotating features such as aircraft propellers and radar antennas.

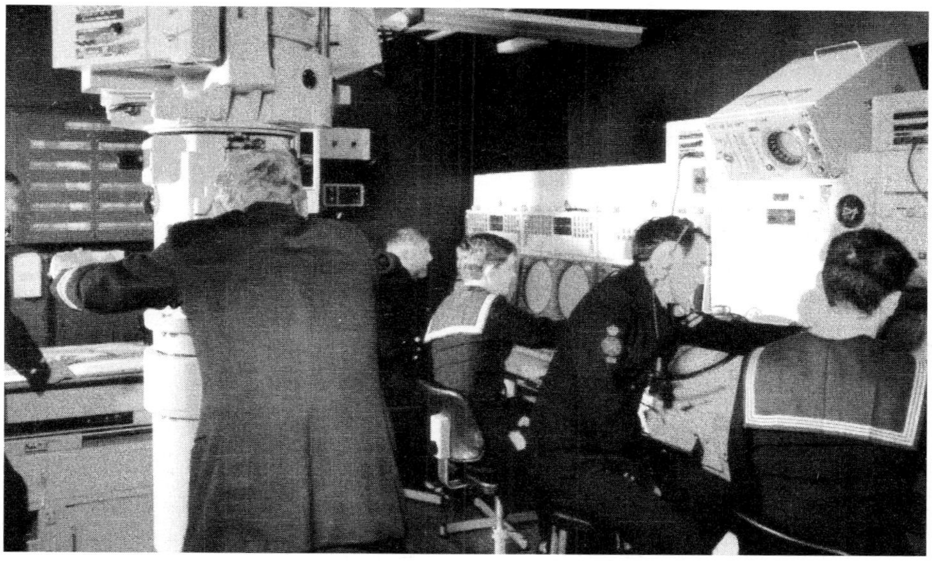

The Marconi nuclear submarine command team trainer at the Royal Navy's Faslane submarine base

Next door will be a seven-position simulated sound room where the acoustic sensor team will be trained. The Type 2054 sonar suite will be simulated to display authentic contact signatures as defined by the exercise scenario. A comprehensive software 'model' of the ocean enables the system to emulate different environmental conditions including ambient, biological and own-ship noise as affected by sea state, weather and bathythermal conditions. An additional CBT (Computer-Based Trainer) sonar trainer (the Type 2054 MMIT) is incorporated to enable operators to acquire basic classification skills with the equipment and to interpret the display presentation. The two rooms can operate autonomously or co-operatively and the whole trainer will be able to provide tactical training in association with collocated SSN trainers.

The four-position instructor's station will allow close monitoring of the student team's performance and an action replay facility will be provided for debriefing. The instructor will be able to control all aspects of the exercise scenario. Manoeuvres by the 'contact' can be injected together with various weapon and sensor emissions.

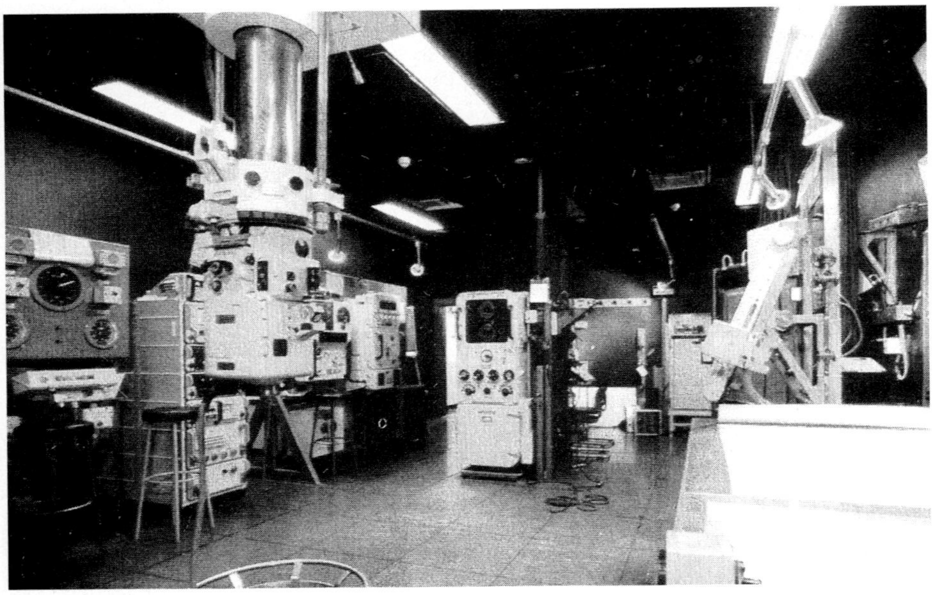

HMS Pugnacious *was one of the earliest submarine command team trainers to be supplied to the Royal Navy. The system is still in use after 21 years' service training crews of the 'Oberon' class submarines*

Status
Marconi has produced simulators for various classes of diesel electric and nuclear submarines both for the Royal Navy and for export customers including Australia, Brazil and Norway.

Manufacturer
Marconi International Simulation and Training, Cheadle Heath.
(Ex-Ferranti, for details see Directory of Suppliers)

Marconi Simulation Submarine Command Team Trainer (SCTT)

The Marconi Simulation Submarine Command Team Trainer (SCTT) provides training for submarine command team personnel in the co-ordination of attack and defensive manoeuvring tactics. The SCTT also provides both sub-team and full-team training in the operation of all related tactical functions including sonar, radar, electronic warfare, periscope, weapon, navigation and communication systems.

The training requirements for an SCTT will vary according to vessel type, equipment configuration, operational deployment and the training philosophy adopted by the designated naval authorities. These factors are taken into account during SCTT design by Marconi Simulation to ensure that operational scenarios and simulated equipment configurations are incorporated to meet customer requirements.

Comprehensive periscope simulation, based on Marconi Simulation's computer-generated graphics system, is incorporated within the SCTT. The system will enable the command teams to identify surface targets, assess target range, bearing, course and speed, and provide a comprehensive analysis of the tactical situation. For details of the graphics system refer to the entry in the *Computer-generated Image and Visual Display Systems* section.

Status
Available.

Manufacturer
Marconi Simulation, Dunfermline.
(For details see Directory of Suppliers)

TTS Submarine Command Team Trainer

Thomson Training and Simulation (TTS) (formerly Link-Miles) produces a submarine command team trainer. This trainer is designed to provide effective training in command, target detection and surveillance, weapons control and the use of combat data systems for the whole operations room team. The heart of the system is a replica of a submarine control room with Intel 386/387 microprocessor-driven facsimiles of the command, weapon control and sensor consoles, together with a periscope. All systems use Fortran language software.

Visual simulation for the periscope uses computer-generated imagery and the two systems available are Seescope, which has the widest range of facilities, and VISTA (Visual Imagery Simulation Training Aid - see entry under *Computer-generated Image and Visual Display Systems* section).

Status
In service with the Royal Norwegian Navy as part of an integrated training suite for the 'Ula' (Type 210) class at the submarine training school at Haakonsvern naval base, near Bergen.

Manufacturer
Thomson Training and Simulation, Crawley.
(For details see Directory of Suppliers)

Helmsman's station 'Ula' class technical trainer

Devonport Submarine Command Team Trainer

In July 1993 the British Ministry of Defence issued requests for tenders for three subsystems which will be used in a submarine command team trainer. The trainer will be provided in two stages and will be in portable buildings at Devonport.

The subsystems are for a periscope simulation system (for which tenders were due in October 1993), a generic sound room and the integration of a SubMarine Command System (SMCS) based upon government-furnished equipment. Tenders for the last two subsystems were required by January 1994 and by March 1994 respectively.

UNITED STATES OF AMERICA

CAE-Link Submarine Trainer

The CAE-Link Corporation is a major supplier of electronic equipment to the US Navy. There are three types of Trident training devices and four different simulators for various classes of nuclear attack submarines. They were designed to maintain proficiency and to increase crew effectiveness in a variety of tactical situations through highly advanced simulation.

The Trident command and control trainer (Device 21A42/A) simulates the Trident submarine tactical environment which consists of ocean models, contacts, weapons, sonar, periscope visual targets and countermeasures. A second device, the defensive weapons subsystem operator trainer (Device 21B67/A), is used to train personnel in countermeasures procedures and tactics. The third device known as the Trident Unique Software Support Facility (TUSSF) supports future hardware and software development for the two operator-team trainers. Currently all trainers have undergone retrofits to reflect submarine modernisation programmes. The TUSSF has been updated and reflects the newest configurations in this family of Trident submarine trainers.

Submarine combat system trainers provide fire-control team training. All four devices simulate the tactical environment which includes oceans, targets, own-ship dynamics, sonar sensors, weapons and countermeasures. Device 21A37/4 uses multiple attack centres to train crews, while Device 21A41 has a single attack centre to train one fire-control team. Devices 21A38A and 21A39A were produced as more capable upgrades to older attack and ballistic missile combat system simulators. Personnel can be trained on the Mark 113 and the Mark 117 fire-control systems using realistic simulations for the Mark 48 and the Mark 37 torpedoes, the Sub-Harpoon anti-ship missile and the Subroc anti-submarine system. Target data simulations for up to 57 targets with surface, submarine and air contacts are possible.

Status
CAE-Link supplied systems to act as the Trident Command and Control Team Trainer (CCTT) and the Defensive Weapons Subsystem Operator Trainer (DWSOT) to the US Navy's Trident Training Facility at Bangor, Washington and King's Bay, Georgia (one per site). The TUSSF is at The Naval Undersea Warfare

Center, Newport, Rhode Island. Attack and ballistic missile submarine combat system trainers have been supplied to the Fleet Ballistic Missile Training Center at Charleston, South Carolina, the Pacific Submarine Training Center at Pearl Harbor, Hawaii, the Fleet Anti-Submarine Warfare Training Centers at Pearl Harbor, Hawaii, and Norfolk, Virginia, and the Naval Submarine Base at New London, Connecticut.

CAE-Link is upgrading the Trident trainers under a $24 million contract. The work, which will be completed in 1995, involves converting the Mark 118 fire-control system to a CCS Mark II system, conversion from AN/BQQ-6 to AN/BQQ-5E sonar, conversion from AN/WLR-17 to AN/WLR-9, ESM upgrades on Device 21A42/A and DWSOT upgrades (including adding periscopes and ESM upgrades). The simulators involved are the CCTT, the DWSOT and the TUSSF.

Manufacturer
CAE-Link Corporation, Binghamton, New York State.
(For details see Directory of Suppliers)

The simulated control room in a Link Trident submarine command team trainer (Device 21A42)

Raytheon AN/BSY-2 Submarine Combat System Team Trainer (SCSTT)

The Raytheon Company won an $80.89 million contract to provide a team trainer for the AN/BSY-2 submarine combat system which was scheduled to equip the 'Seawolf' class nuclear submarines. As the AN/BSY-2 programme has now been reduced in scope there is no longer a requirement for this trainer.

Manufacturer
Raytheon Company, Submarine Signal Division, Rhode Island.
(For details see Directory of Suppliers)

TACTICAL TRAINING SIMULATORS

CANADA

Bendix Avelex Operations Team Trainer

The Bendix Avelex Operations Team Trainer (OTT) for the Canadian Navy is a flexible, fully integrated software framework featuring distributed architecture connected through a local area network. The same network can be used for building trainers and simulators of almost any level of sophistication for training users to manage discrete systems to solve three-dimensional motion problems.

The OTT facility is designed to meet a range of training requirements for effective naval operations room management. The installation consists of four simulated operations rooms, a main control room with four controller consoles and a large projection screen for monitoring and debriefing purposes. Student command teams are presented with equipment and operational scenarios similar to those normally experienced at sea.

The operations rooms are equipped with replicas of actual shipboard equipment such as a ship control unit for manoeuvring, an IFF decoder panel, a chaff launcher panel, a radar control panel and a heading marker panel. The OTT system drives this equipment to simulate the dynamic operating characteristics of the ship while representing a realistic visual and acoustic behaviour simulation of sensor and weapons data. It also provides contact association, target motion analysis, data display and weapon firing proposals together with simulation of weapons and decoys.

Instructors manipulate environmental and operational conditions during the exercise from the four-position control console. A label plan raster display, integral to each console, is used to monitor the progress of the exercise in graphic form. The content of the colour raster scan display projected onto the large screen can be defined from any controller position. This display is used for monitoring purposes during the exercise and for subsequent debriefing.

The Bendix Avelex Operations Team Trainer (OTT)

The data required for running training exercises, including land mass radar simulation, is generated on an off-line exercise preparation system. This system permits all game settings and scenarios to be tested off-line before the actual game. Parameters available to game planners include an exercise area of 1024 × 1024 nm (1896 × 1896 km) and up to 64 vehicle types, 104 manoeuvrable targets and 186 point targets per exercise. Positional accuracy is 30 m.

Status

The OTT was accepted by the Canadian Fleet Forces School at Halifax, Nova Scotia, in 1990.

Manufacturer

Allied-Signal Aerospace Canada, Montreal.
(For details see Directory of Suppliers)

GERMANY

Atlas Elektronik AGUS Anti-Submarine Warfare Trainer

As a major supplier of electronics to the German Navy, Atlas Elektronik GmbH has also supplied an anti-submarine warfare trainer known as the AGUS. AGUS was designed to train and assess command information centre teams in anti-submarine warfare. It also trains and assesses operators of differing grades in the practical use of sensors and weapon systems, provides facilities for training in convoy protection and can be used to develop and give training in anti-submarine warfare tactics. It features a distributed architecture which permits modular construction to meet customer requirements.

The system consists of a number of command information centre cubicles, an instructor's station, a debriefing auditorium with a large screen display and an electronics room. The simulator provides dynamic operating behaviour for 'ships', realistic visual and acoustic presentation of sonar radar and weapon data, together with combat situation displays. It also provides a realistic contact association, target motion analysis and data display, weapon firing proposals, and simulation of weapons and decoys.

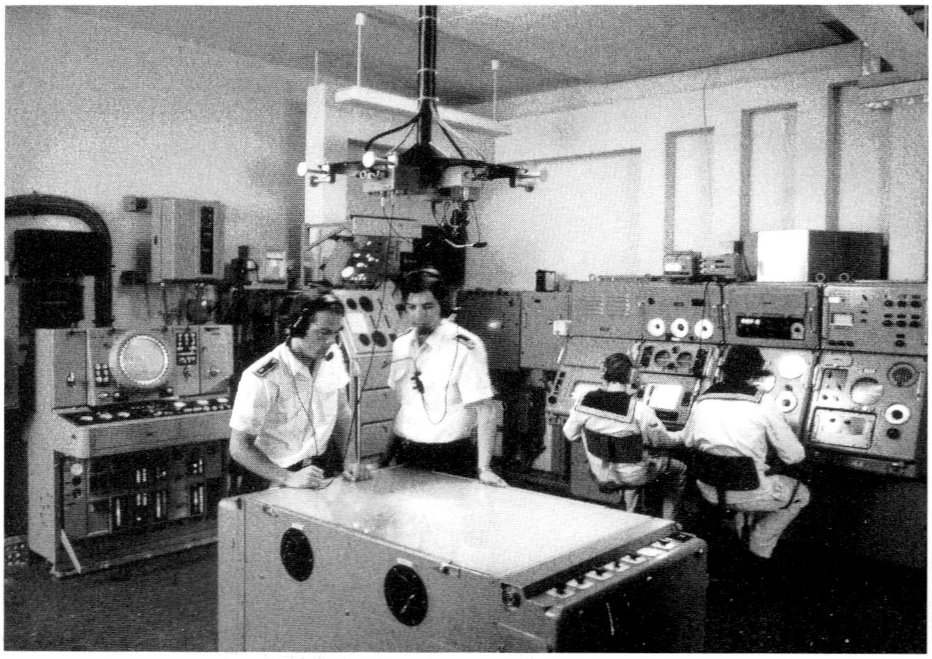

The German Navy's AGUS anti-submarine warfare trainer

Status

The German Navy commissioned an AGUS trainer in 1979 and updated it in 1986.

Manufacturer

Atlas Elektronik GmbH, Bremen.
(For details see Directory of Suppliers)

Dornier Action Speed Tactical Trainer (ASTT)

The Dornier Action Speed Tactical Trainer (ASTT) is designed to provide a simulated naval warfare environment using real-time technology in order to train command teams in tactics and operational procedures for surface engagement, anti-air, anti-submarine and electronic warfare together with amphibious operations and co-operation with aircraft.

The system features modular hardware and software based upon a distributed architecture. The ASTT is organised into cubicles representing ships, an instructors' control area and an auditorium with a large display screen for briefing and debriefing.

The cubicles provide up to five operator positions and feature a tactical display console, a helmsman's console, a chart table and a communications console. It can be configured to represent any type of ship, while for the training of larger CIC teams such as frigates, up to four cubicles may be combined. The equipment permits students to 'manoeuvre' the ship and to 'operate' both weapons and sensors.

The instructors' control area features dual-display consoles and is used to prepare, to control and to monitor exercises. It is also used to participate in the exercises. Up to 100 computer or instructor-controlled tracks of a dynamic or a fixed nature are provided with computer-controlled tracks including aircraft attack profiles, missile trajectories and helicopter manoeuvres.

Status

Available.

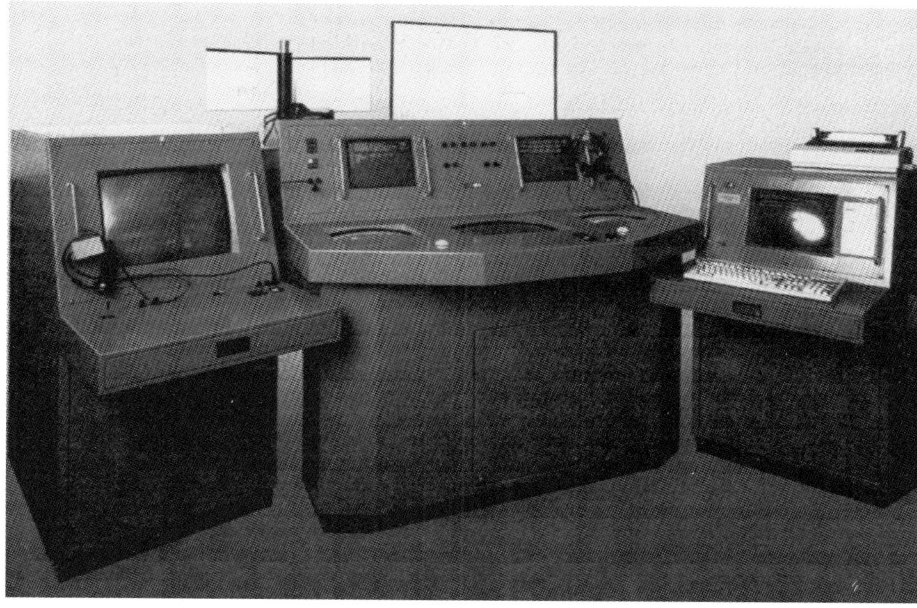

The ship cubicle hardware used in the Dornier Action Speed Tactical Trainer. On the left is the helmsman's console, in the middle is the tactical display console and on the right is the communications console

Manufacturer

Dornier Luftfahrt GmbH, Friedrichshaven.
(For details see Directory of Suppliers)

Dornier Fleetwork Trainer (FWT)

An important producer of systems, Dornier GmbH has developed and produced a Fleetwork Trainer (FWT).

The FWT system is designed to give training in fleet work and basic tactical procedures. It provides a number of cubicles, each manned by two students, representing ships' bridges and these are used to control formations of ships. The system is controlled by two instructors with their own consoles.

These cubicles are linked with a ship dynamics calculator which computes the actual position of the 'ship' in a formation based upon initial position, control inputs and ship characteristics. Helicopters can also be integrated. The cubicles can be divided into two formations with their own communications circuit.

Status

One system has been sold to the Nigerian Navy for use in its new Operator Training School. Another version is installed at the Ghana Navy Training Centre at Tema and at Dornier's training centre in Bremen.

Manufacturer

Dornier Luftfahrt GmbH, Friedrichshaven.
(For details see Directory of Suppliers)

A debriefing session at the Nigerian Navy's Fleetwork Trainer (FWT)

STN AAW Simulation Programme (ASIM/ASCAD)

The development or improvement of Anti-Air Warfare (AAW) systems has always encountered problems in determining system efficiency and survivability in various scenarios and environments. To assist the solution of this problem STN Systemtechnik Nord has developed the AAW Simulation Programme (ASIM/ASCAD) to undertake realistic and reproducible analyses of the efficiency and survivability of most AAW system configurations.

The programme is run on personal computers based upon 386/486 microprocessors and co-processors and can generate scenarios with hundreds of targets with own ship as the centre point. Each target has separate trajectories, and their manoeuvrability as well as speed may be controlled by definable real life performance parameters with the targets being capable of circling as well as horizontal and vertical weaving manoeuvres.

ASIM/ASCAD may be programmed to work with standard sensor modules such as 2D, 3D and multi-function radars. For specific systems these modules may be modified to meet the performance of the sensor. Environmental features such as multi-path propagation, 'clutter' and ECM may be considered while there is provision in the system for multiple-launch missile systems. The modular nature of the programme allows it to be adapted to meet new system configurations.

The attack and defence activities are presented on the display using colour Naval Tactical Data System (NTDS) symbology. It is possible to analyse them by various criteria such as X-Y, way-time and height-distance diagrams to calculate the efficiency of an AAW system in a given scenario, to compare the performance of different systems, to analyse the system's response in different conditions and to understand its limitations.

Status
In production since 1988.

Manufacturer
STN Systemtechnik Nord GmbH, Bremen.
(For details see Directory of Suppliers)

The STN AAW Simulation Program being run on a personal computer

INDIA

Keltron Ship Control System

The Kerala State Electronics Development Corporation (Keltron) has developed a tactical training Ship Control System. The microprocessor-based system features a number of cubicles which may represent ships, submarines or aircraft which may then conduct interactive exercises. The manufacturers state that there are panoramic display facilities for 'the complete deployment situation' but no further details are available.

Status
Believed to be in production, possibly for the Indian Navy.

Manufacturer
Keltron, Trivandrum.
(For details see Directory of Suppliers)

ISRAEL

Elbit Naval Tactical Training Centre (TTC) and Naval Tactics and Command Trainer (NTCT)

Elbit is an important supplier of electronic equipment to the Israeli Navy and has produced the Naval Tactical Training Centre (TTC) and the ASWT-2 anti-submarine warfare trainer (see following entry).

The TTC is designed to give naval commanders tactical command decision procedure training by simulating scenarios which might be encountered. It may also be used to develop and analyse combat doctrines.

A fourth generation tactical trainer, based upon a distributed architecture, was produced in 1989. It comprises 11 student stations (which may be expanded to 14) and eight instructor stations.

It features high quality colour displays, advanced MMI and inherent simulation of C^3I functions. Two exercises may be held simultaneously each with a maximum of 350 simulated objects including naval, air and shore units, unmanned aircraft, EW systems and weapons. The communications facilities include up to eight voice/datalink nets per student station as well as two internal communications nets.

The NTCT is a derivative of the TTC and uses DEC hardware. This includes VAX computers, MicroVAX II local computers and alphanumeric displays. The system also uses Genisco graphic display generators.

Status
Approximately 12 TC/NTCT systems have been sold both to the Israeli Navy and foreign customers.

Manufacturer
Elbit Computers Limited, Haifa.
(For details see Directory of Suppliers)

The instructor's console in the Elbit Tactical Training Centre system

Elbit ASWT-2 Anti-Submarine Warfare Trainer

This system was designed to train Anti-Submarine Warfare (ASW) teams and provides all aspects of operational experience including classification of sonar echoes, identifying submarines, attack procedures and the use of weapons. All vehicles involved in ASW are represented within the system which is housed in a building.

The ASWT-2 consists of a control room, a computer room, a debriefing room, two anti-submarine 'ships' each with their own bridge, an operations room and sonar room, and a number of multipurpose cubicles representing various submarine and air vehicles. Each bridge has a manoeuvring control console, a simulated bridge radar, communications and weapon release equipment. The operations room includes a plotting board, indicators to note target data, a console for simulated radar and visual detection together with an optional electronic support measures display. The sonar room has only the sonar operator's console and communications.

The exercise area covers 100 × 100 nm (185 × 185 km) and realistic maritime environments are created. Three types of torpedo, two types of depth charge and three other anti-submarine weapons can be simulated.

Status
The first generation system, ASWT-1, was delivered to the Israeli Navy in 1971 while an ASWT-2 system succeeded it in 1979.

Manufacturer
Elbit Computers Limited, Haifa.
(For details see Directory of Suppliers)

ITALY

DATAMAT Action Speed Tactical Trainer (ASTT)

An Action Speed Tactical Trainer (ASTT) has been developed by DATAMAT for training staff officers, commanding officers and operational teams of naval or airborne units in planning and executing naval operations either in one major exercise or a series of independent exercises. The system consists of five major subsystems: cubicles, control, debriefing, computer and communications.

There are ten cubicles each representing the combat information centre of a surface warship or submarine, or acting as an aircraft, and housing the equipment necessary for training command teams. Each has two or three horizontal display systems. The control room consists of five twin-operator consoles each with three display screens and three hard-copy printers. It is separated from the debriefing room by a folding screen and data may be displayed upon a large screen in the debriefing room. The communications system includes voice, data link and telegraphic communications.

The gaming area is defined as being 1000 × 1000 nm (1852 × 1852 km), 900 m deep and 27 400 m high. The vehicle characteristics, maximum speed, radar, electro-optical and optical signatures together with turn rates are faithfully simulated. For aircraft these performance parameters also include climb and dive rate together with maximum altitude, while for submarines, features such as maximum depth, surfacing and diving rates together with snorkel/periscope depth have also been included.

The system also authentically reproduces the characteristics of sensors, weapon systems and electronic warfare systems.

An instructor's console for the Datamat Action Speed Tactical Trainer

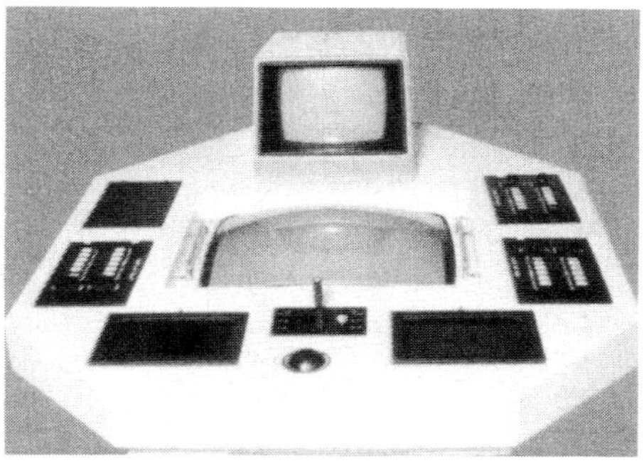

One of the Action Speed Tactical Trainer student consoles; the system will use 22 consoles in 10 cubicles

Status
In production for the Italian Navy.

Manufacturer
DATAMAT Ingegneria dei Sistemi SpA, Rome.
(For details see Directory of Suppliers)

NORWAY

MPS Action Speed Tactical Trainer (MASTT)

The MPS Action Speed Tactical Trainer (MASTT) is a flexible system for building naval tactical trainers and simulators extendable to almost any level of complexity, sophistication and training level. It is designed to fulfil the training requirements for seamen (reporting procedures, sensor and weapon operating procedures, classification training and operator training on real equipment), officers under training (decision making, team work, reporting procedures and weapon employment), and high ranking officers (tactical disposition training, tactical analysis, and exercise planning, analysis and debriefing). The system is stated to be very easy to use with students able to operate the cubicles by themselves after one or two one-hour introductory sessions.

MASTT was developed in close co-operation with the Royal Norwegian Navy and was installed at the Royal Norwegian Navy Training Establishment in Bergen in November 1986.

The installation at Bergen consists of 10 cubicle consoles, an extended control section, a standard control section, a system console, two recorder/playback computers, two weapon computers, a communication control, three sonar simulators and two anti-submarine warfare consoles. Each trainer console is software configurable and the user has full control of model and scenario generation. The distributed concept, using an Ethernet network, makes it easy to expand the trainer by adding more cubicles or to enhance it by interfacing with real equipment simulators (such as radar, sonar and weapon control).

The model preparation and storage for the complete game is performed using a commercial computer. The user interacts with the system console via a menu-driven operator dialogue. All game settings and scenarios can be prepared and tested off-line prior to the actual game, so it is possible to build up a library of preplanned games settings on disk.

A student's console with units, weapons and sensors operated via dedicated easy-to-use panels

A console in the Swedish MPS Action Speed Tactical Trainer, named Mimer

Scenarios are based on digitised map contours overlaid with various optional synthetic information. The actual exercise area is generated from a map library. Two files are made: one for the three-dimensional terrain model and the other for a two-dimensional model for presentation. These files can be stored for later use.

Each cubicle console can be configured as a surface ship, a submarine, an aircraft, a land-based coastal defence installation, or a coastal radar station. A unit configuration consists of a set of parameters describing the unit type and its dynamic behaviour. In addition, the user specifies the sensors and weapons with which the unit is equipped.

Wind, sea current, precipitation and sea temperature layers are among the data provided.

An exercise is initiated from the same console used for set-up and preparation. The system software is loaded into the different computers in each cubicle via the local area network and the trainer is started. The user then selects the game set-up to be used, downloads it and the exercise commences.

Following an exercise a playback programme provides synchronised audio and video playback with freeze and fast wind/rewind functions. During replay the screens in the cubicles and control will contain the same information as during the exercise. Recorded information can be used during a game. The playback programme can also use recordings from real exercises.

Status
The MASTT system has been operational with the Royal Norwegian Training Establishment since 1984. The Swedish Navy took delivery of a 12-cubicle MASTT system, named Mimer, at its training centre Berga Örlogsskolor (BOS) outside Stockholm in January 1991.

Manufacturer
Micro Processor Systems A/S, Kongsberg.
(For details see Directory of Suppliers)

MPS Anti-Submarine Warfare Trainer (MASWT)

The MPS Anti-Submarine Warfare Trainer (MASWT) provides realistic and efficient training in ASW and the use of various sensors and weapons. It is designed to train commanding officers as well as sensor and weapon operators in ASW vessels such as destroyers, frigates, corvettes and also helicopters.

The system provides advanced simulation of the platforms, weapons, active sonars and other sensors with facsimiles of both sensor consoles and weapon order panels. The sonar equipment includes the MPS Multibeam Sonar Simulator or MMSS (see separate entry).

The scenario is defined by a menu-driven programme. The MASWT has easy-to-use man/machine interfaces and includes full recording and playback/debriefing facilities as well as a voice communications system.

Status

In service with both the Royal Norwegian Navy and the Swedish Navy.

Manufacturer

Micro Processor Systems A/S, Kongsberg.
(For details see Directory of Suppliers)

Norwegian students under instruction in an MPS Anti-Submarine Warfare Trainer

MPS Coastal Fortress Simulator

The MPS Coastal Fortress Simulator (MCFS) provides realistic training for commanding officers and equipment operators within a coastal fortress system including artillery batteries, missiles, torpedoes, minefields and their Command, Control and Communications (C^3) elements. The training involved includes co-operation, compilation of information, analysis and decision making as well as the use of weapons, sensors and so on. Training is provided in a very realistic environment with simulated equipment being used to train personnel in the operation of the real equipment used in coastal fortresses via the use of 'look alike' Man/Machine Interfaces (MMI), digitised maps and terrain models.

The concept, which emulates the MMI of the real operator consoles, includes specially designed user-configurable consoles acting as mine, torpedo, artillery, battery and C^3 units. Scenario definition is by means of menu-driven programs. The trainer has a distributed real-time computer using a local area network (LAN - the Ethernet IEEE 803.2) and uses digitised maps and high resolution digitised terrain models. There are full recording and playback/debrief facilities and a voice communication system.

Status

The system was installed in January 1993 at the Royal Norwegian Navy training establishment, KNM Tordenskjold, and it is interfaced with the MASTT and MASWT trainers (see previous entries).

Manufacturer

Micro Processor Systems A/S, Kongsberg.
(For details see Directory of Suppliers)

SOUTH AFRICA

UEC Weapon System Training Simulator (WSTS)/Weapon System Logics Trainer (WSLT)

To support training of crews for the 'Minister' class fast attack craft of the South African Navy, UEC Projects (Pty) Ltd (UEC) has developed two simulators: the Weapon System operational Training Simulator (WSTS) and the Weapon System Logics Trainer (WSLT).

The WSTS is designed to provide a simulated operational environment for training weapon system officers and operators. It uses a mainframe computer and five personal computers together with dedicated embedded sub-processors. The students use a weapon system console, either real or facsimile, and a simulator which injects 'radar returns' into the weapon system console according to customer requirements. This allows the acquisition and tracking capabilities of the weapon system to be reproduced. Synthetic visual targets are generated for insertion into the 'optical director'. The instructor creates the simulated operational scenarios and manipulates platforms to engage the students' 'ship' and the WSTS includes a complete range of background noise.

Other elements of the WSTS include the computer, an electronic warfare facility and subsystems for operational graphics, optical director visuals, echo generation, sound generation, missile-on-launcher, ship interface and audio/visual recording. The simulator has provision for up to 70 targets and typical numbers which may be generated are 19 air, surface and shore targets, two friendly and four hostile missiles, two gun projectiles and three chaff clouds.

The instructor's console features a raster-scan display with a plan view of the engagement exercise while the MMI includes a keyboard, roller ball and touch-screen controls. It is used to prepare and to monitor the exercise with the latter function aided by a video console using up to four monitors. A view of the 'engagement' from a consort vessel may be added with an 'observer' reporting to the WSTS personnel by 'radio'.

The WSLT is based upon personal computers and is programmed to suit the training of the logics of any naval weapon system.

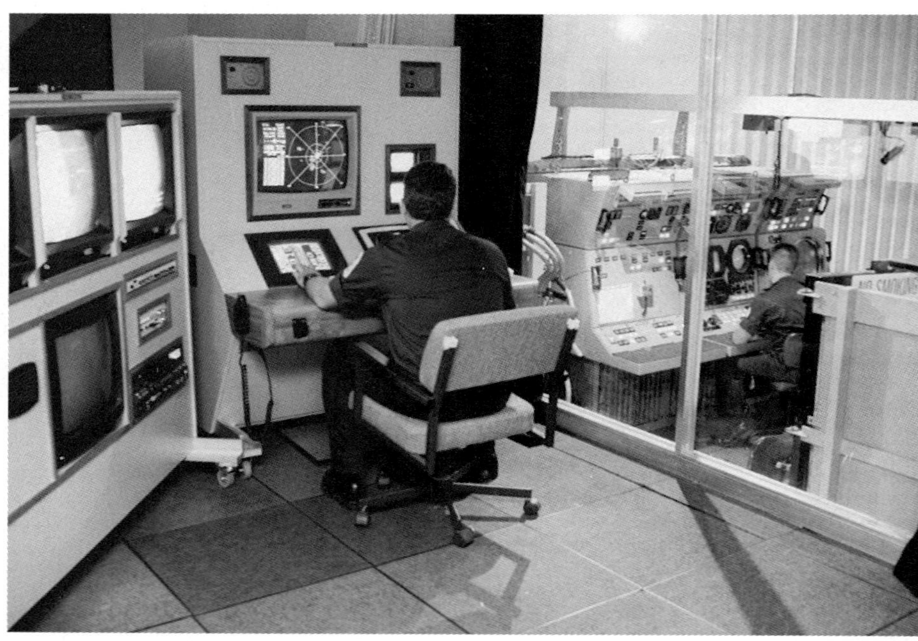

The UEC Weapon System Training Simulator's weapon system console is seen (right) from the instructor's station on the left

Status

Both are in service with the South African Navy Strike Craft Training School (SCTC). The SCTC WSTS replaces another weapon system training simulator and uses real consoles but export versions will use facsimiles and will have improved performance but the same basic software.

Manufacturer

UEC Projects (Pty) Limited, Natal.
(For details see Directory of Suppliers)

SPAIN

SAINSEL Tactical Simulators

SAINSEL produces various types of tactical simulator for naval purposes, ranging from large scale tactical simulators presenting information on a strategic scale, to local combat situations involving individual vessels. SAINSEL produces all the consoles and other hardware required as well as the supporting software and other logistic support.

Status
In production and service.

Manufacturer
SAINSEL, Madrid and Seville.
(For details see Directory of Suppliers)

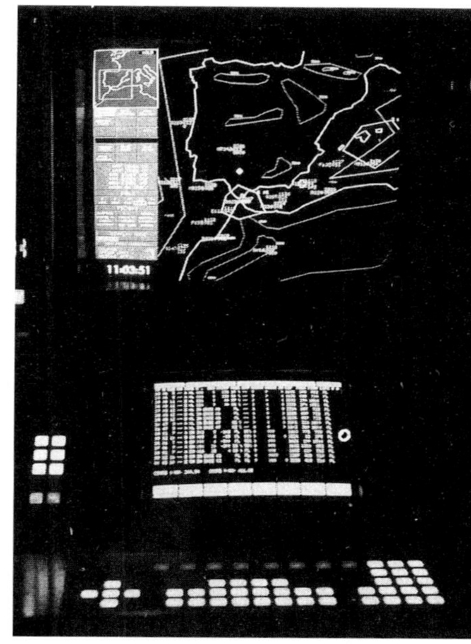

The display from a typical SAINSEL tactical simulator

UNITED KINGDOM

BAeSEMA Mine Warfare System Centre

BAeSEMA has developed an integrated support facility for mine warfare forces as the Mine Warfare System Centre (MWSC). The MWSC consists of a computer network running a modular suite of software using integrated databases, high resolution graphics and modern windowing techniques. Effective and efficient utilisation of mining and mine countermeasures assets is achieved by centralising and co-ordinating the operational, technical and training support functions in a centralised facility.

Operational support functions allow the determination of the mining threat, analysis of the risk to shipping, command and control of mine warfare operations together with the assessment of the effectiveness of vessels, equipment, operators and tactics. The technical support modules enable the integrated logistic support activities, availability and reliability assessment to be made.

The MWSC also provides the range of training aids for the review and assessment of operational data through to the development of command team tactics. With links into mine countermeasures simulators and weapon trainers, the MWSC provides crew performance evaluation and enhanced operator training through access to operational databases. Platform performance on exercise minefields and mine countermeasures ranges can also be evaluated.

Status
MWSC modules have been implemented for the Canadian, UK and US defence ministries.

Manufacturer
BAeSEMA Limited, Esher.
(For details see Directory of Suppliers)

The BAeSEMA Mine Warfare System Centre

DSD Action Speed Tactical Trainer (ASTT)

The ASTT has been produced to meet a customer's specific requirements for operational room team training in specific warships.

The system consists of six cubicles and four instruction units situated in a purpose-built training complex with an auditorium for briefing, replay and debriefing as well as on-line presentation. The system is capable of emulating the total functions of specific warships' data handling systems, weapons and sensors with one cubicle acting as the CIC of one warship.

Each CIC features generic operational panels with touch-sensitive screen displays which may be used to represent the CIC of specific warships with the system designed to emulate the operations rooms of MEKO 360 frigates, anti-submarine corvettes and fast attack craft.

Each cubicle has a tactical display console, a helm console, and a communications console and can train up to five students. Trainees must operate the systems in accordance with the capabilities of their real world equivalents. In addition to the normal tactical facilities, students have direct access to weapon and sensor control panels which reflect the procedural commands which must be generated within a warship's warfare command procedures.

Instructors are able to monitor and alter any aspect of the exercise scenario and may introduce up to 100 dynamic targets of any category and 100 elements of a non-combatant nature. The instructor consoles are designed to use touch-screens. The system also provides both internal and 'external' communications nets under full computer control.

The console of the DSD Action Speed Tactical Trainer (ASTT)

Status
A system produced for the Nigerian Navy was delivered in 1992.

Manufacturer
DSD Digital Systems and Design Limited, Eastleigh.
(For details see Directory of Suppliers)

Marconi Action Speed Tactical Trainer (ASTT)

Marconi (ex-Ferranti International Simulation and Training) has produced a number of Action Speed Tactical Trainers and this record was enhanced by the purchase of the Simulation Division of Solartron Defence Systems in the late 1980s.

The ASTT is designed to simulate the complete spectrum of naval warfare to provide comprehensive training for operational and tactical decision making. The system configuration consists of a central auditorium/debriefing suite, instructors' consoles and from 2 to 22 (or more) trainee cubicles according to the requirements of the individual customer. The vehicle, weapon and sensor characteristics are defined in a series of incorporated data libraries established by the user and are allocated to the cubicles during exercise preparation. These characteristics can be updated as required to match the introduction into service of new platforms and equipments.

The complexity of equipment contained in a single cubicle depends upon the user's requirements. The basic range consists of a single tactical console with associated communications system and peripheral equipment such as stateboards. More complex configurations include separate operational displays for radar, sonar and EW together with additional consoles to represent larger units including OTCs.

The system offers an exercise area of 2048 × 2048 nm and 200 supplementary tracks representing vehicles, weapons, decoys and sonobuoys. In addition, 100 mine tracks, 48 reference points and 100 history points for debriefing purposes may also be generated. Exercises can be run at four times action speed and can be replayed at up to 16 times action speed. These parameters are expandable according to customer requirements.

The instructor facilities provide the means of preparing, controlling, monitoring and debriefing exercises and are configured to meet specific user needs. Preset data libraries containing environmental parameters allow the instructor to define the required scenario for fast, flexible exercise construction. Advanced graphical techniques allow the instructor to concentrate on the progress of the students while flexible record, playback and storage facilities provide high quality debriefing and allow exercise material to be retained in the data library for future use.

A cubicle in a Marconi Action Speed Tactical Trainer (ASTT)

Status

ASTTs have been supplied to the Royal Navy's School of Maritime Operations (HMS *Dryad*) and to 14 overseas navies including those of Australia, Canada, Greece, Indonesia, Malaysia, Netherlands, New Zealand and Pakistan.

Manufacturer

Marconi International Simulation and Training, Cheadle Heath.
(Ex-Ferranti, for details see Directory of Suppliers)

Reflectone Helicopter Control Trainer (HCT)

The Helicopter Control Trainer (HCT) supplied by Reflectone UK Limited (formerly BAeSL) to the Royal Navy provides training for aircraft controllers in shore-to-air and ship-to-air operations. These include radar and communications procedures, the control of air traffic, weapons delivery and mission control for ASW and anti-surface warfare. The core of the HCT is the SD-Scicon Director system.

The system consists of six cubicles, game and simulation control positions and a maintenance area. It is installed in a self-contained, transportable container complex in purpose-built accommodation. Each cubicle represents a ship operating in a simulated naval scenario based upon selectable geographical locations. They may operate autonomously or interactively and each cubicle also has instructor facilities. One cubicle has extra equipment to act as a tactical command station.

Exercises are monitored and managed by a game controller assisted by six simulation control positions all fitted with simulated radio communications to the students.

Status

The system has been installed at the Royal Navy Helicopter School HMS *Osprey* at Portland, Dorset and entered service in April 1992.

Manufacturer

Reflectone UK Limited, Bristol.
(For details see Directory of Suppliers)

SD-Scicon Director Maritime Tactical Trainer (MTT)

The Director system has been produced by System Designers (SD), part of the SD-Scicon group since 1988, for many years and the manufacturer has developed a Maritime Tactical Trainer (MTT) version.

Director is designed to provide realistic action speed training in maritime tactics for naval and air personnel. It features distributed processing architecture, with each work station having its own powerful IBM-compatible computer with proven software. The work station features a raster radar and data display, a separate raster display, a custom-built keyboard and roller ball. It may be used as an instructor station, a student station, for exercise preparation or for system management. The work stations are linked by a local area network.

The exercise preparation system allows instructors to generate a wide range of exercises ranging from basic warfare skills to advanced area scenarios. The instructor inputs geographic and maritime data, platform, sensor and weapon characteristics, mission profiles and automatic events. During the exercise the instructors can introduce tactical variations, equipment failure, emergency situations, casualties and other features.

Status

The Director range was developed from 1982 and entered service in 1984. A new range of consoles was introduced in 1988. The MTT has been sold to Oman.

Manufacturer

SD-Scicon, Berners Street, London.
(For details see Directory of Suppliers)

The SD-Scicon Director Maritime Tactical Trainer (MTT) work station

Thomson C8000 Action Speed Tactical Trainers

The C8000 Action Speed Tactical Trainers are central processor architecture systems designed to teach and develop tactics for naval warfare applications. These systems can simulate an area of up to 2000 × 2000 nm with aircraft and missiles up to a height of 30 480 m and submarines to a depth of 910 m. The systems simulate the control of ships' operations rooms, as well as aircraft, other vessels and feature inputs by navigation, communications, sensor, weapons and electronic warfare equipment by means of computer-driven displays and man/machine interfaces. The central computer ensures data presented to each cubicle is up to date and that no unit exceeds its real performance. It can also inject peripheral targets from a pool of 48 additional tracks.

Various types of system are available with some representing specific ship operations rooms, while others have command cubicles to represent any kind of ship, aircraft, submarine, helicopter or surface effect vessel to create a variety of friendly and hostile forces. A typical system will have six cubicles which feature raw sensor plotting tables for manual operation or automatic systems, or combinations of both. They will usually have three 'radars' and a weapon control console with eight to ten weapons. The students can 'manoeuvre' their unit, operating sensors and weapon systems as well as communicate. All are interconnected by internal and 'external' communications systems, the latter simulating six or more HF, VHF, and UHF radio channels together with realistic noise interference, range fade and jamming effects which might be encountered at sea.

The war games are directed from a central control room where problems are posed in real time (action speed). Two types of exercise may be played; the students may fight each other or they may combine to fight the instructional staff. The games are prepared on punch paper or magnetic tape that is fed into the computer which then automatically provides the cubicle displays with their performance criteria. The course of the game is recorded for later debriefing and analysis using a large vertical screen with monochrome or colour display.

Status

Some 25 trainers of this family have been produced. Customers include Australia, Belgium, Brazil, Canada, Chile, Germany, Greece, India, Norway, Spain and the Royal Navy which purchased five systems.

Manufacturer

Thomson Training and Simulation (TTS), Crawley.
(Ex-Hughes Rediffusion, for details see Directory of Suppliers)

Westland Tactical Fleetwork Trainer

Westland System Assessment Limited (WSAL) has developed a new generation of tactical fleetwork trainer. Its purpose is to exercise individuals in the interpretation and application of tactical manoeuvring signals which are issued either by the instructor or by a selected student position. The system is also used to teach station keeping.

The system consists of a network of microcomputer work stations each representing a ship, the optimum number being eight work stations although this can be increased or decreased. Each student position is autonomous and the screen displays a scaled plan view of the 'training area' with 'own ship' and other student 'ships' shown together with targets generated by the instructor who has a similar work station. Digitised land for coastal and harbour training exercises is also shown. The student may alter the display characteristics and his ordered manoeuvring sector together with the movement and speed of his 'ship'.

The instructor position is a work station similar to that of the student's and is used to establish a training scenario, to set formation performance parameters and to provide a variety of reference points. It also provides surface ship, submarine and aircraft contact positions and their initial courses and speeds. It can control all contacts and introduce new ones. There is a replay facility and debriefing may be assisted by large screen video data projection.

The signals are transmitted through an intercom facility, and in response to the signals, the students manoeuvre their 'ships' via a computer keyboard. The manoeuvres are then displayed on the screen via a network.

Status

In production and in service with the Royal Navy in HMS *Mercury*.

The Westland System Assessment Limited Tactical Fleetwork Trainer

Manufacturer

Westland System Assessment Limited, Yeovil.
(For details see Directory of Suppliers)

UNITED STATES OF AMERICA

CAE-Link Tactics Trainers

The CAE-Link Tactics Trainers are team trainers and consist of a central auditorium and debriefing room with large screen displays and up to seven instructor consoles together with one or two communications controller consoles, a computer room and a number of individual instruction areas.

The trainers can produce up to 24 surface, submarine and air vehicles and can engage up to 280 targets of which 80 can 'fight back'. A full range of sensors and weapon systems can be simulated with voice, teletype and data link communications in a game area of 2000 × 2000 nm (3700 × 3700 km), 30 480 m in height and 1000 m in depth. The speed of the action can be accelerated 20 times.

However, the numbers vary. One trainer provides simulation of 10 combat information centres together with a land-based headquarters and has a gaming area of 1024 × 1024 nm (1894 × 1894 km). The German Action Speed Tactics Trainer simulates 14 ships of eight different types and 112 target vehicles of 10 different types. A tactics procedures and radar navigation trainer for the same navy has six cubicles and offers 57 targets, 26 radar displays and 18 plotting tables in a gaming area of 200 × 200 nm (370 × 370 km).

The instructor's station in a CAE-Link Action Speed Tactics Trainer

The student display station in a CAE-Link Action Speed Tactics Trainer

Status

CAE-Link has supplied an Action Speed Tactics Trainer and a Tactics Procedures Trainer to Germany and these are installed at Wilhelmshaven and Bremerhaven respectively. Other Action Speed Tactics Trainers have been supplied to the Italian Navy at Taranto and to both the Portuguese and Turkish navies, while the Royal Danish Navy has a Tactics Trainer at Frederikshavn. Several trainers have been supplied to a number of Asian navies. The US Navy operates a number of the company's surface, sub-surface, and airborne tactics trainers.

Manufacturer

CAE-Link Corporation, Binghamton, New York State.
(For details see Directory of Suppliers)

CAE-Link ASW Team Tactical Trainer (Device 14A12)

CAE-Link's Surface ASW Team Tactical Trainer (Device 14A12) provides Anti-Submarine Warfare (ASW) team procedural, team tactical training and evaluation for conventional ASW warships and those equipped with tactical data systems. Simulated sonars, command information centres, bridges and platforms (aircraft and ships) are used to provide trainees with the capability of exercising essential anti-submarine warfare engagement procedures in a simulated multi-threat environment.

These simulated systems contain general-purpose programmable consoles which can be reconfigured under software control to represent the anti-submarine capabilities inherent in any of the US Navy's 16 major classes of anti-submarine warfare ships. The simulated Underwater Battery Plot features reprogrammable sonar general-purpose consoles. The system's design covers six major categories of simulation: platform simulation (friendly and hostile); sensor detection; weapons employment; tactical environment; probability of interception calculations; and damage assessment. A modular, database-driven design allows for ease of future modifications and additions as new platforms, weapons and sensors are added to the fleet.

Status

Five 14A12 Devices have been delivered to the US Navy Fleet Training Centers at Norfolk, Virginia; San Diego, California; Mayport, Florida; Pearl Harbor, Hawaii; and Charleston, South Carolina.

In November 1991 CAE-Link announced an $8.25 million contract from the US Naval Training Systems Center to produce a version of Device 14A12 for the Japanese Maritime Self Defence Force as Device 14A12-J. This was due to be delivered to Yokosuka naval base, headquarters of the Escort Fleet, in the Spring of 1994, to become operational in June 1994.

Manufacturer

CAE-Link Corporation, Binghamton, New York State.
(For details see Directory of Suppliers)

A student console in a Device 14A12 system

Delex Harpoon Interactive Tactical Training System (HITTS)

This is a shore-based, multi-station naval anti-surface warfare tactical training system based upon networked commercial desktop computers and Delex-developed software. It was designed to provide tactical employment training for the Harpoon and other weapon systems.

HITTS simulates up to 40 manoeuvring targets (each with defensive capabilities) and it will simulate up to 24 Harpoon firings per scenario. It can be configured to simulate other anti-ship missiles, either ship- or air-launched, as well as guns. It also simulates active and passive sensors of all platforms in the scenario together with environmental conditions and detection horizons.

Training covers all aspects of anti-surface warfare and the use of all versions of Harpoon, air-, ship- and submarine-launched. The freeplay scenarios are run in real time and involve multiple platforms, both friendly and hostile. The instructor controls progress and may control target actions and reactions.

The students control launch platforms and can order platforms which assist in over-the-horizon targeting. They also control electronic emissions from platforms. Once decisions have been made they receive immediate feedback.

HITTS has three operating modes; the tutorial mode highlights tactically significant aspects of anti-surface warfare; the exercise generator allows the instructor to create a number of training exercises to teach specific skills; the exercise simulator provides realistic simulations and allows the instructor to control all aspects of the scenario.

Instructor and student stations of the Delex HITTS

Status

HITTS has been purchased by the US Navy and six foreign countries.

Manufacturer

Delex Systems Inc, Virginia.
(For details see Directory of Suppliers)

Delex Surface Launch Organic Tactical Training System (SLOTTS)

SLOTTS is a shipborne system designed to provide anti-surface warfare teams with training in the tactical employment of the Harpoon anti-ship missile. It helps the teams to develop tactical proficiency in the use of the Harpoon and also teaches them Combat Information Centre (CIC) procedures.

SLOTTS interfaces directly with the AN/SAG-A(V) Harpoon ship command-launch control set and simulates realistic environmental, geographic and tactical scenarios in which the team may launch up to eight missiles using the actual CIC equipment.

Status

SLOTTS has been sold to three countries.

Manufacturer

Delex Systems Inc, Virginia.
(For details see Directory of Suppliers)

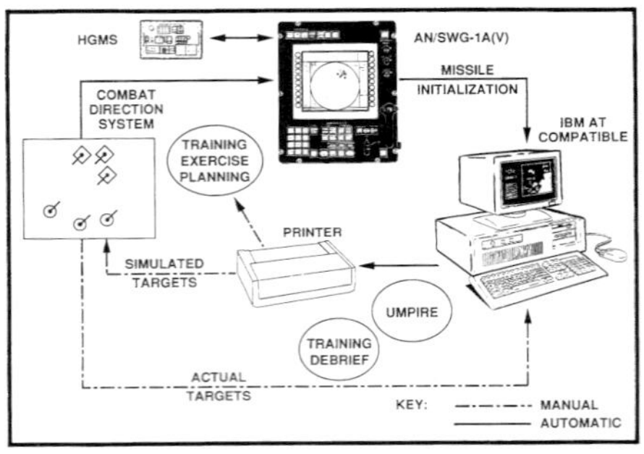

Diagram of the Delex Surface Launch Organic Tactical Training System (SLOTTS)

Hughes Surface Tactical Team Trainer (Device 20A66)

Hughes Training won a $27 million contract for a new surface tactical team trainer, Device 20A66, to replace the existing Device 14A6.

It was designed to provide Anti-Submarine Warfare (ASW) tactical team training in a multi-platform, multi-threat environment for levels from supervisor to battle group commander. There will be simulations of friendly and hostile platform dynamics, sensor systems, weapon systems and communications, while tactical threats will be presented in a realistic acoustic and electromagnetic operating environment.

The system will consist of an exercise development subsystem, an exercise control and support subsystem, a central computer complex, a digital voice switching system, a briefing/debriefing area and simulated command centres and aircraft. The exercise development subsystem will include a general-purpose instructor console and a graphic entry device while the exercise control and support subsystem will have up to nine general-purpose consoles. Both subsystems will be linked to the central computer complex by an instructor Local Area Network (LAN).

The database-driven Ada language software will

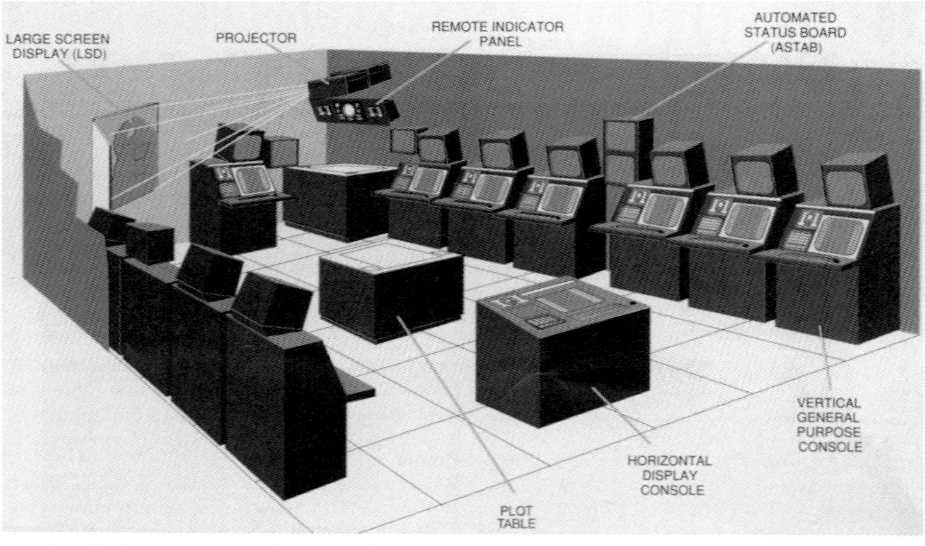

An artist's impression of the replica Aegis Combat Information Centre in Device 20A66; it was designed to reproduce platform capabilities rather than be a facsimile of a specific class Combat Information Centre

provide extensive capabilities and is designed for easy modification and update. Instructors and trainees will have the ability to manoeuvre platforms, to display and to manage sensor data, to determine target classification and identification and to perform target motion analysis. They will be able to engage targets, to 'fire' weapons, to launch chaff as well as employing both acoustic and EW countermeasures.

The simulated command centres will include a Tactical Flag Command Center (TFCC) for battle group commanders and staff, a Warfare Commander (WC) module for anti-air and anti-surface warfare commanders and their staff, an aircraft carrier ASW module for the ASW commander and his staff, reproduction Combat Information Centres (CICs) and aircraft crew stations.

The reproduction CICs will be based upon general-purpose consoles rather than facsimiles and are designed to simulate the capabilities of various command and control centres as well as platforms. There will be one aircraft carrier CIC representing 'Nimitz' (CVN 68) class ships, and five for Aegis-equipped ships of the 'Ticonderoga' (CG 47) and 'Arleigh Burke' (DDG 51) classes. Another 12 will represent either non-Aegis surface ships such as the 'Virginia' (CGN 38), 'California' (CGN 36), 'Belknap' (CG 26) and 'Leahy' (CG 16) class cruisers, 'Kidd' (DDG 993) and 'Spruance' (DD 963) class destroyers, 'Oliver Hazard Perry' (FFG 7) and 'Knox' (FF 1052) class frigates or, alternatively, attack submarines of the 'Sturgeon' (SSN 637) and 'Los Angeles' (SSN 688) classes.

There will be 11 aircraft crew stations representing fixed-wing aircraft such as the E-2C Hawkeye, the P-3C Orion and the S-3A/B Viking as well as rotary-wing aircraft such as the SH-60B/F Seahawk (LAMPS Mark III), the SH-2F Seasprite (LAMPS Mark I) and SH-3A Sea King. Communications between the instructor LAN elements and the remainder of the system are by means of a global LAN and the digital voice network. Simulated encryption and HF/UHF traffic will also be included.

Status

The first trainer was scheduled to be delivered to the Fleet ASW Training Center Pacific in San Diego, California, in 1993. The US Navy has an option on a second system which would be installed at the Fleet ASW Training Center Atlantic in Norfolk, Virginia.

Manufacturer

Hughes Training Inc, Manhattan Beach, California.
(For details see Directory of Suppliers)

Ship Analytics Naval Ship Combat Information Center

Ship Analytics offers a companion Naval Ship Combat Information Center (CIC) option to its Full Mission Shiphandling Simulator system. The CIC extends the scope of training to include co-ordinated Bridge/CIC operations in a realistic tactical environment, as required to 'fight the ship'. Modular in design and engineered for installation in a separate space adjacent to the bridge, the CIC may be procured as part of the original FMSS system or added as an upgrade at a later date. The CIC includes six subsets of equipment: radar/sonar; plotting; ship tactical status display; ship operational status display; interior ship communications; and exterior ship communications. The standard CIC is designed to provide the equipment functionality necessary to implement an interactive Bridge/CIC capability, suitable for the conduct of effective operational procedure and tactics training for modern warship bridge and CIC personnel.

Status

The Ship Analytics CIC has been delivered to the US Navy and the Turkish Navy.

Manufacturer

Ship Analytics, Connecticut.
(For details see Directory of Suppliers)

The Spruance (DD963) class Combat Information Center in US Navy Device 20B6D with air plots and simulated AN/SPA-25F navigation radar display

Ship Analytics Action Speed Tactical Trainer

The Ship Analytics Action Speed Tactical Trainer (ASTT) is a strategic and tactical combat command trainer used to enhance the decision-making and teamwork skills of junior and senior naval officers. The system provides simulated war games between two opposing forces. It has a game control suite and a number of student and command control stations representing platforms such as ships, submarines, fixed- and rotary-wing aircraft. Actual and simulated radio voice channels allow for interplatform communications under instructor control. The exercise area is played in a 2048 × 2048 nm (3795 × 3795 km) gaming area that can include land masses and minefields.

Sensor data is displayed graphically and in alphanumeric form on student stations whose man/machine interface for platform, sensor and weapon control is based upon keyboards and a roller ball. In addition to receiving processed electronic and visual data, each platform also has available unprocessed radar and sonar data using the Ship Analytics Radar Simulator. The simulated weapons systems include air-, surface- and subsurface-launched missiles, depth charges, torpedoes and guns in quantities appropriate to the platform. Platform movement through the exercise area is controlled by the students and the platforms continue to participate in the game until it is destroyed or the exercise is terminated. The instructor station creates and monitors all aspects of the exercises. The instructor monitors the data from individual platform sensor and weapon systems. The ASTT trainer allows the instructor to simulate a variety of malfunctions. These malfunctions include craft performance, weapons, sensors and communications failures. The instructor can also introduce and control simulated platforms that may be assigned to either force or be neutral.

Status

A system was produced for a major Middle Eastern navy.

Manufacturer

Ship Analytics, Connecticut.
(For details see Directory of Suppliers)

Typical ASTT cubicle and instructor equipment

Syscon Naval Tactical Game (NAVTAG - Device 16H3)

The Naval Tactical Game (NAVTAG - Device 16H3) is a war-gaming system designed to train surface warfare officers. It can be played by as few as two and as many as 31 students and it consists of three subsystems: hardware, software and courseware.

The hardware consists of three table-top computers and a printer, while the software program is designed to react to the students' inputs as well as providing an accurate simulation of ships, weapons and sensors. The game moves in a series of turns, each of which simulates one minute in real time, which allows the exercise to proceed at a pace set by the students' abilities.

Status

In service with the US Navy at shore stations, with two fleets and on warships.

Manufacturer

Syscon Corporation, Washington DC.
(For details see Directory of Suppliers)

Syscon Tactical Anti-Submarine Warfare Interim Trainer (TASWIT)

The Syscon Corporation developed the Tactical Anti-Submarine Warfare Interim Trainer (TASWIT) to enhance the existing ASW team trainer (Device 14A6) pending the arrival of Device 20A66 (see separate entry). TASWIT is based upon a Hewlett Packard HP825SRX computer supplemented by a distributed network of 64 Zenith PC AT personal computers. These provide computer-assisted individual level training, or they can provide combat information centre team training. They can also be used for multi-ship team training.

Device 14A6 has 16 simulated anti-submarine warfare ship combat information centres with Naval Tactical Data System (NTDS) displays, sonar, tactical picture and electronic warfare displays. The system can simulate 16 anti-submarine warfare ships, their individual aircraft and four submarine targets together with nine anti-submarine weapons and 64 sonobuoys.

Status

Integrated with Device 14A6 at the US Navy's Fleet Anti-Submarine Training Center, Atlantic, in Norfolk, Virginia, in August 1988.

Manufacturer

Syscon Corporation, Washington DC.
(For details see Directory of Suppliers)

Unisys Tactical Advanced Combat Direction and Electronic Warfare Trainer (TACDEW)

The Unisys, formerly Sperry, organisation produced the TACDEW (Tactical Advanced Combat Direction and Electronic Warfare) trainer (Device 20F15) which is used to train naval officers in integrated combat direction by simulating various warships and sensors in exercises with and against each other. There is one system each for the Atlantic and Pacific Fleets. Since 1982 the system has been progressively improved including the provision of 26 radar simulation-target generation systems and 22 radar display systems for each fleet system.

The US Naval Training Systems Center plans further modernisation to TACDEW during the early 1990s. These include the provision of an environmental generation control system, replacement of the radar video simulation subsystem with a digitally activated generic radar display system, an external voice communications system together with problem control and evaluation.

Status

In service with the US Navy (2).

Manufacturer

Paramax, New York State.
(For details see Directory of Suppliers)

Amphibious Warfare Tactical Trainer (Device 16C81)

Plans exist for a multipurpose, multi-echelon training system, Device 16C81, to train amphibious task force commanders, landing force commanders, their subordinates and their staffs in tactical decision making in a realistic operational environment. It will also train selected teams in previously acquired amphibious warfare-related operational skills.

In addition to training the combat elements, the trainer will be used to train the logistic command and control organisation in support of landing forces including the off-loading of the assault follow-on echelon. Refresher training will also be provided for air controllers in the Tactical Air Command Center/Helicopter Direction Center (TACC/HDC).

Status

The requirement is for two systems for the Naval Amphibious College at Coronado, California, and the Naval Amphibious College at Little Creek, Virginia.

Tactical Combat Training System (TCTS)

The US Navy has a requirement for a comprehensive Tactical Combat Training System (TCTS) capable of covering training for individual warships, for co-ordinated training involving surface ships, submarines and aircraft as well as for integrated battle groups/battle forces.

TCTS is to provide accurate, realistic and timely reports of exercise activities without artificially restraining the participants. The system is to have the least possible demands for space and support personnel and is envisaged as being based upon desktop computers and software with a data link for distributing data throughout the fleet and a satellite link.

For battle group/battle force exercises it is envisaged that four systems, one each for the Atlantic and the Pacific Fleets and two for forward deployment in the Mediterranean and the Indian Ocean, will be required. Each will be centralised through aircraft carriers attached to a battle group and will allow participation by 24 surface ships, six submarines and 100 aircraft. The software will consist of between 500 000 and 1 million lines of Ada language and TCTS will be capable of incorporating intelligence data about the weapon systems and tactics of potentially hostile nations in the regions to which the battle group is deployed.

It is anticipated that deployment of such a system will take place about the turn of the century but in October 1992 three companies or consortia were issued design concept development contracts. These call for developing system specifications, platform data, creating a technical library on the equipment and interfaces required, as well as producing documentation. The contracts were awarded to CAE-Link Training Operations, the Raytheon Company's Submarine Signal Division and Applied Data Technology Inc, as well as a consortium consisting of Loral Space and Range Systems, Grumman Aerospace Electronics' Combat Systems, United Information Systems (UIS) and the Hughes Aircraft Company's Combat Systems Engineering Center.

Status

See text.

PROPULSION/ELECTRICAL POWER GENERATION AND DISTRIBUTION SYSTEM SIMULATORS

CANADA

CAE Electronics Integrated Platform Management System

CAE supplies Integrated Platform Management Systems (IPMS) for warships to control and monitor all hull, mechanical and electrical machinery and systems. The systems are controlled and monitored through multi-functional operator consoles which are connected to the machinery via a serial databus.

To assist operator training, CAE supplies a land-based trainer and an onboard embedded trainer. Both provide full scope training for all platform systems control and monitoring functions including all modes of emergency and casualty control procedures and the use of built-in test equipment. A sophisticated instructor facility operates under programmable lesson plans which simulate all systems in real time. Embedded trainers can be used in harbour or while under way due to redundant multi-function consoles which permit immediate return to real-time operations. Land-based trainer consoles are identical to the shipborne ones.

Status

IPMS is installed in the Canadian 'City' class frigates and 'Iroquois' class destroyers, the US Navy's 'Osprey' (MHC 51) class minehunters and the Israeli 'Lahav' (Saar 5) class corvettes.

Manufacturer

CAE Electronics Limited, Québec.
(For details see Directory of Suppliers)

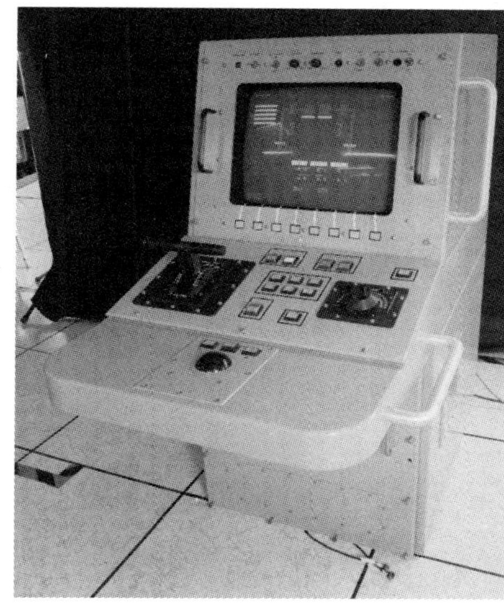

The CAE IPMS control console has an embedded trainer

Virtual Prototypes Scenario Toolkit And Generation Environment (STAGE)

Virtual Prototypes produced the MArine Systems Trainer (MAST) computer-based part-task procedures trainer for marine systems engineering training and for integrated machinery control system training. The product was substantially improved after it was unveiled and to reflect this has been renamed Scenario Toolkit And Generation Environment (STAGE).

The former MAST system consists of 27 work stations grouped in three cells of nine of which one is the instructor's station and the others are for the students. All stations have touch-sensitive screens so students can mimic the use of actual hardware systems. The system has the Virtual Applications Prototyping System (VAPS) graphic emulation software, the Virtual Prototypes scenario generator for tactical war-gaming and simulation management and other simulation software.

STAGE integrates distributed simulation systems to develop a synthetic outside world for training systems, for operational analysis and for testing Human-Machine Interfaces. Its modular nature allows it also to be used for a wide variety of tasks including naval combat systems as well as naval part-task training.

Status

The MAST system was delivered to the Canadian Department of National Defence as part of the Canadian Patrol Frigate Program.

Manufacturer

Virtual Prototypes, Montreal.
(For details see Directory of Suppliers)

The Marine Systems Trainer (MAST) is used by the Canadian Navy

FRANCE

DCN Ship Propulsion and Electrical Power Generation Machinery Control Simulators

The DCN's ECAN (Etablissement des Constructions et Armes Navales) Ruelle, F-16600 Ruelle, is a major supplier of submarine/ship propulsion and power generation machinery control simulators. It produces the Simpra, Orion and Simcor submarine propulsion machinery control simulators, the propulsion machinery control and electrical power generation/distribution simulators for the Saudi Arabian Type F2000 frigates and a reconfigurable electrical power generation/distribution simulator for French surface warships.

These systems are based upon the SEL 32/77 computer but are mounted on static platforms. They are usually configured to represent specific ship systems. Propulsion/machinery control simulators include:

(a) Simpra for 'Agosta' class submarines
(b) Orion for 'Rubis' class submarines
(c) system for 'Le Redoutable' class submarines
(d) Simcor for 'Georges Leygues' class frigates
(e) system for Type F2000 class frigates.

There is also a propulsion control simulator for the 'Daphné' class submarines, ventilation system simulators for both the 'Agosta' and 'Daphné' classes, and a 'Daphné' class trim tank simulator.

Electrical power generation/distribution machinery control simulators have been produced for the Type F2000 class frigates and a reconfigurable system has been produced for training crews of the 'Tourville' class destroyers, the 'Georges Leygues' class frigates and the 'Eridan' (Tripartite) class mine-hunters.

Status

Simpra has been produced for the French and Spanish navies. Orion, Simcor and the 'Le Redoutable' simulator are used by the French Navy, with the last used at Soument M4 in Brest. Most of the French Navy's simulators for diesel-electric submarines are at the ENSM (Ecole Navale Sous-Marine) Toulon. The Saudi Arabian Navy uses both versions of the Type F2000 class simulators. The reconfigurable power generation simulator is used by the French Navy.

Manufacturer

DCN, Paris.
(For details see Directory of Suppliers)

ITALY

Elsag Bailey SACIE Power Generation and Distribution Simulation System

Elsag Bailey, part of the IRI-Finmeccanica Group, provides the SACIE electrical power generation and distribution systems for most Italian warships. They have produced a simulator which allows students to perform all the basic operations.

The system features a facsimile control room with a Modiac distributed architecture computer system. This comprises processing nodes based upon an M3BUS parallel bus and a 16-bit DCJ11 microprocessor. The instructor station has a video terminal and colour monitors while the students operate facsimile mechanical and electrical control panels.

A subset of the ship's electronic cards are in the system together with high-speed transmission cards which simulate the input/output cards of the ship. They are controlled by a software package identical to that of the ship and this can be accessed by the instructor both to monitor the exercise and to insert malfunctions.

Status

In service with the Italian Navy at the Taranto training centre.

Manufacturer

Elsag Bailey SpA, Genoa.
(For details see Directory of Suppliers)

FIAT SEPA Propulsion System Control Simulator

SEPA, a major producer of propulsion and submarine control systems for the Italian shipbuilding industry, has supplied a propulsion system control simulator to the Italian Navy. It is a single console system representing the propulsion control unit of the 'Maestrale' and 'Lupo' class frigates and is controlled by SEPA ULP computers.

An instructor's console, near the student's console, monitors the student's performance. It is also used to alter the simulated environment and to add a wide range of malfunctions.

Status

In service with the Italian Navy training centre in Taranto since the early 1980s and believed to have been exported.

Manufacturer

FIAT, Turin.
(For details see Directory of Suppliers)

NETHERLANDS

Van Rietschoten & Houwens Propulsion System Simulators

The Dutch company Van Rietschoten & Houwens Defensiesystemen BV (R & H) is active in the fields of engineering, integration and the supply of electronic, electrical and hydraulic systems. Its main activities include data acquisition, data processing, control and monitoring, automation, communication, training and simulation for onshore and ship applications including military, naval and nautical.

Simulators and trainers for various ship systems have been produced and these include Ship Control Centres (SCC), Integrated Monitoring and Control Systems (IMCS) platform trainers for submarines, damage control trainers and hardware trainers.

The SSC simulators represent the functioning of the SCC in a surface vessel including the simulation of all propulsion and power generation functions. The computer provides all the characteristics of the engines, clutches and couplings, propellers and hull. Instructors can watch the students and can introduce about 100 simulated failures. They can also act as the 'bridge' and transmit commands by telegraph or telephone.

Status

SCC simulators for frigates and corvettes have been produced as well as an IMCS simulator.

Manufacturer

Van Rietschoten & Houwens Defensiesystemen BV, Rotterdam.
(For details see Directory of Suppliers)

Training on a Van Rietschoten & Houwens submarine Integrated Monitoring and Control Systems simulator

A corvette Ship Control Centre simulator

NORWAY

Norcontrol PPT-2000 Propulsion Plant Trainers

Norcontrol is a world leader in maritime simulation with more than 300 simulators sold to more than 40 countries. The company's products range from part-task trainers to full propulsion plant trainers.

The Norcontrol Propulsion Plant Trainers (PPT) are a series of trainers designed for the operation and understanding of ship propulsion plants. All systems are simulated realistically enabling training in resources management and efficient operation of the propulsion plant, including electrical power generation and distribution.

The PPT provides the environment for students to experience realistic surroundings. Features include vital components such as main engine remote-control, engine local control, engine-room operation panels, controllers, alarm system, switchboard and engine sound.

The PPT is delivered in three versions: replica, operational and work station. Replica trainers resemble the engine-room and engine control room of a specific vessel with simulated panels and switches together with an instructor station. Operational trainers are similar to the replica trainers except they are generic control rooms and engine-rooms. The work station-based trainers contain the same functions and features as the others, but all operation and feedback is within the work station. These trainers may be delivered as integrated elements of the other trainers.

Status

More than 80 simulators have been sold to Bulgaria, China, the CIS, the Faroe Islands, Finland, France, Greece, Hong Kong, Iceland, India, Iran, Japan, South Korea, Mexico, Norway, Poland, Portugal, Singapore, Spain, Sweden, the UAE, the USA and the former Yugoslavia. The Royal Norwegian Navy operates a medium-speed diesel engine trainer while the Republic of Singapore Navy operates a high-speed diesel engine trainer with colour graphic work stations.

Manufacturer

Norcontrol Systems A/S, Horten.
(For details see Directory of Suppliers)

A Norcontrol propulsion system simulator

UNITED KINGDOM

Haven Engine and Machinery Simulators

Haven Automation has been manufacturing simulators since 1966 and produces systems designed to train engineers. They have produced major ship machinery simulators for civil customers in the United States and the United Kingdom as well as for the Royal Navy's training school HMS *Sultan*.

Latest products are the Modeq range of modular training systems based upon IBM-compatible personal computers. These run a series of schematic interactive panels together with the instructor's station and support facilities. The existing product range consists of Modeq 100, Modeq 200 and Modeq 300.

Modeq 100 is an electrical power generation simulator designed to teach training in power generation theory and practice. It simulates a grid supply and two three-phase diesel generators rated at 1200 kVA which supply power to one of four load centres. Modeq 200 is an electrical motor and motor drive simulator while Modeq 300 is a marine diesel engine simulator.

The panel supports controls for operating equipment such as pumps and compressors. Analogue instruments monitor the 'engine's' performance and there is a command, control and telegraph section for communication with the 'bridge'. Faults can be inserted by the instructor and the student must enter 'repair or maintenance' commands.

Status

Two large simulators have been supplied to the Royal Navy together with a number of Modeq systems.

Manufacturer

Haven Automation Limited, Swansea.
(For details see Directory of Suppliers)

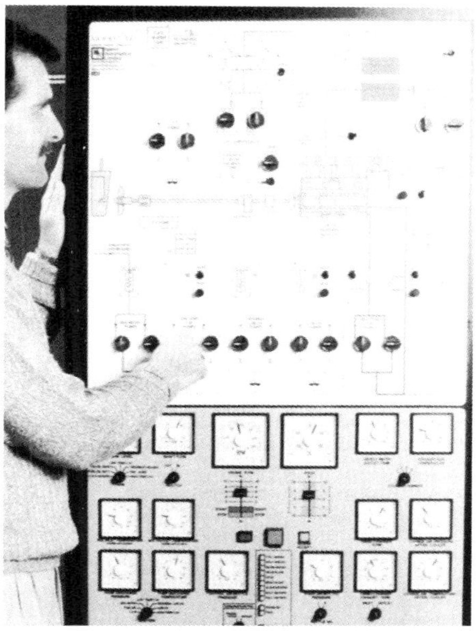

A Modeq 300 marine diesel engine simulator

Marconi Simulation Nuclear Submarine Machinery Control Room Simulators

Marconi Simulation Nuclear Submarine Machinery Control Room Simulators provide training for nuclear submarine engineering watch-keeping staff in the start up, shut-down and emergency procedures relating to nuclear submarine propulsion plant.

Within the simulated Machinery Control Room environment are installed all control, instrumentation and equipments normally visible to, and operated by, the watch-keeping staff. The appearance, location and operation of these equipments correspond exactly to those which exist in the actual submarine. Additional realism is achieved by the inclusion of simulated audio effects in response to operator actions. These audio effects include diesel generator run-up, the rod drive transmitter, circuit breaker actions and instructor-induced steam leaks.

Reactor initial conditions such as core age, power levels of previous operations, time since shut-down and desired pressuriser and heat exchange levels may be specified by the instructor as system initial conditions. The Instructor's Console has been designed to enable the training staff to overlook the interior of the simulated control room and observe trainee actions. Under the controlled lighting conditions the glazing material

A Marconi Simulation nuclear submarine machinery control room simulator

between the instructor's console and the machinery control room becomes unidirectional. This enables the instructor(s) to have full vision of trainee actions while remaining unobserved.

Full record and replay facilities are provided, operating in real or non-real time as required for debriefing purposes.

Status
Available. A total of seven of these simulators, representing four different classes of submarine, has been delivered to the Royal Navy.

Manufacturer
Marconi Simulation, Dunfermline.
(For details see Directory of Suppliers)

Thomson Training and Simulation Single Role Mine-Hunter Engineering Trainer

To assist the training of engineering personnel assigned to the 'Sandown' class Single Role Mine-hunter, Thomson Training and Simulation (formerly Hughes Rediffusion) provided a practical procedural training system which includes computer-based classroom training.

For procedural training there is a full simulator of the ship control centre and another four computer work stations of which one is used by the instructor. He may configure any work station to simulate either a local instrumentation panel or to operate a computer-based training package.

The computer-based training package includes three modules of text and full-colour animated schematics which describe the ship layout and features of several ship systems. These modules are instructor-led with the student viewing material on a 37-inch (940 mm) screen. A fourth module used for instruction on the D86 digital control and surveillance equipment is designed to be used in a self-teach mode on the work stations.

Status
In service with the Royal Navy.

Manufacturer
Thomson Training and Simulation, Crawley.
(For details see Directory of Suppliers)

Thomson Training and Simulation 'Vanguard' Class Propulsion Control System Simulator

A simulator of the control system for the nuclear propulsion of the 'Vanguard' class ballistic missile submarines was built by Thomson Training and Simulation (formerly Hughes Rediffusion). The simulator is a replica of the manoeuvring room from which the propulsion system is controlled. Plant failures and faults in the main and auxiliary power system can be simulated in the simulator which trains watch-keepers in both normal operations and emergency procedures.

Accurate real-time simulation of the instrumentation and controls related to the submarine's propulsion system is achieved using distributed processing backed up with an ultra fast reflective memory based upon flight simulation techniques. The instructor's console is equipped with the Thomson Training and Simulation Touch-Activated Simulator Control (TASC) which allows the instructor to change a training scenario instantly by selecting a function on the touch-screen. The TASC also provides the full range of control and monitoring facilities.

The simulator was delivered to HMS *Neptune* at Faslane, Scotland, in 1990. In February 1992 it was announced that the company had received a £3 million contract to supply another simulator of a nuclear submarine's propulsion control system. This will be delivered in 1995.

Status
See text.

Manufacturer
Thomson Training and Simulation Limited, Crawley.
(For details see Directory of Suppliers)

Thomson Training and Simulation Type 23 Frigate Machinery Control and Surveillance System Shore Trainer

During 1986 it was announced that an order worth over £1 million had been awarded by the United Kingdom Ministry of Defence to Hughes Rediffusion Simulation Limited (now Thomson Training and Simulation). The order was for a Machinery Control and Surveillance System Shore Trainer for Type 23 frigates.

The trainer consists of four sections: the instructor's room; the main ship control centre; a conventional classroom area; and a maintenance trainer. The instructor's room overlooks the replica ship control centre with a series of one-way windows allowing the instructor to monitor the training exercises. His console incorporates TASC (Touch-Activated Simulator Control) which combines full-colour touch-screen technology with microprocessor control.

The main ship control centre represents the engineering control console area of the Type 23. Students sit at a large console with Machinery Control And Surveillance (MCAS), together with Nuclear, Biological and Chemical Damage control (NBCD) sections. To one side are the NBCD incident board and controls for ventilation, chilled water supply and other environmental systems. To the other are the MCAS controls for the 'ship's' combined diesel-electric and gas-turbine (CODLAG) propulsion system and electrical supplies.

Also in this space are five local control consoles representing equipment in other parts of the ship. These represent the gas-turbine local control panel, the diesel local control panel and the Varister local control panel together with the forward and aft secondary electrical local control panels. Other systems such as chilled water local control panels are provided at two touch-activated colour monitors.

The classroom is allocated for briefings and debriefings. It includes a back projection screen and an active MIMIC panel linked into the main simulator for illustrated instruction. The maintenance trainer permits instruction on the maintenance of the ship's computerised control and surveillance systems either in a combined or a separate mode operation. The maintenance trainer uses Intel 280 processors with Use software.

The instructor console for the Type 23 frigate Machinery Control and Surveillance trainer with the main ship control centre visible through the one-way window

Trainee crews will be able to experience hands-on training in all procedures associated with normal, incidental and emergency procedures.

The central system computer is a Gould 32/6732 using a Hughes Rediffusion-produced Fortran 77 software language program based on the Type 23 manufacturer's data. The equipment also includes a maintenance trainer for the MCAS microprocessor-based secondary surveillance system.

Status
The trainer was opened at HMS *Sultan*, Gosport, in January 1989.

Manufacturer
Thomson Training and Simulation Limited, Crawley.
(For details see Directory of Suppliers)

Thomson Training and Simulation Submarine Machinery Control Trainer

The Thomson (formerly Link-Miles) Submarine Machinery Control Trainer system is used to train crews in an engine-room compartment. It can provide both initial and refresher training in a wide variety of conditions to give the students experience in a broad spectrum of standard and emergency operating procedures. It consists of facsimile machinery control consoles and a training management system and features a separate instructor's position with a one-way mirror. The system uses functionally distributed computing architecture with microprocessors and it can be integrated with a submarine control simulator.

Status

The company has delivered a machinery control trainer for the 'Upholder' class submarine to meet a Royal Navy requirement. This is installed at HMS *Dolphin*, Gosport, and was formally accepted in July 1989. Machinery control trainers, both stand-alone and integrated with submarine control simulators, have been sold to the Royal Navy and a number of European and Far Eastern navies. The latest contract, announced in August 1990, is to supply the Australian Submarine Corporation Pty Limited with a propulsion control simulator for the Royal Australian Navy's new 'Collins' class submarines. The system will be installed at HMAS *Stirling* near Perth, Western Australia and is part of a A$14 million contract which includes a ship control simulator.

Manufacturer

Thomson Training and Simulation, Crawley.
(For details see Directory of Suppliers)

The Thomson 'Upholder' class machinery control trainer with instructor station in the foreground

Vosper Thornycroft Surface Ship Machinery Control Simulators

Vosper Thornycroft supplies machinery control and monitoring systems for frigates, mine countermeasures vessels, and diesel-electric and nuclear-powered submarines and has produced a range of propulsion control simulators. These simulators are custom-built systems with facilities ranging from simple simulators for familiarisation with individual pieces of equipment to full-detail complete system trainers. The latest feature, with the machinery control and monitoring systems, is the D86 microprocessor-based computer system which uses the Intel 8086 and 80286 microprocessors and Pascal and Coral 66 software. All have instructor consoles which can inject emergency situations into the students' consoles. A typical console, used in the 'Hunt' class machinery controls trainer, features normal bridge controls with internal communications and a colour monitor showing MIMIC diagrams of the machinery's reactions to the student's actions.

A description of two systems indicates the product range. For the Belgian Navy there are three machinery operating consoles from the combat information centre, the bridge station and the machinery control room. There are facilities for limited maintenance at the rear of the consoles. For the Spanish Navy there are full-scale operator and maintainer machinery control consoles for corvettes, and for patrol craft a complete set of consoles including machinery space controls, electrical distribution boards, and local control panels.

Status

The company has provided machinery control simulators for the Type 22 class frigates, the 'Invincible' class aircraft carriers, and a combined one for the Type 21 frigate and Type 42 destroyer. A machinery control operator/maintainer for 'Hunt' class mine countermeasures vessels was commissioned in 1985 at the Royal Navy's Marine Engineering School at Portsmouth (HMS *Sultan*). A similar system has been ordered for the 'Sandown' class Single Role Mine-hunter.

A machinery control operator/maintainer for the Belgian Navy's 'Wielingen' class frigates was commissioned at Zeebrugge in 1981, while the Spanish Navy commissioned a machinery control and switch gear operator/maintainer trainer for 'Descubierta' class frigates and a machinery control operator for the 'Lazaga' class patrol boats at El Ferrol during 1988.

Manufacturer

Vosper Thornycroft (UK) Limited, Portsmouth.
(For details see Directory of Suppliers)

The Vosper Thornycroft engineering training simulator for 'Descubierta' class corvettes installed at El Ferrol

A complete set of consoles for 'Lazaga' class patrol craft engine room simulators. These were supplied to the Spanish Navy by Vosper Thornycroft

UNITED STATES OF AMERICA

CAE-Link Gas Turbine Propulsion Plant Trainer

The CAE-Link Gas Turbine Propulsion Plant Trainer is an expandable system designed to train engine-room personnel in the safe and efficient operation of naval gas-turbine propulsion units. The system simulates a system with two engine rooms (port and starboard) and four gas-turbines, and provides real-time simulation of both normal and emergency conditions, not only of the engines but also of ship systems and environmental effects. It comprises an instructor station, trainee stations and a computer complex.

The instructor station includes a control and monitor display terminal with an alphanumeric keyboard. The station is used to monitor and control all training problems. It features a self-contained memory, a flickerless display and an operating speed of 900 000 characters/s, while field tab pointers are positioned on the display by the computer. The student stations interact in the same way as in a real ship and they include a propulsion auxiliary control console, one propulsion local control console for each 'engine room' and an electrical plant control console.

The mini-computers in the computer complex simulate all operating aspects and a wide range of emergency conditions for the main propulsion gas-turbine, ship motion, controllable reversible propeller and key subsystems. The computer complex is capable of simulating the electrical power, reduction gear, shaft, fuel and lubrication, throttle, waste heat boiler, bleed air, combustion air, compressed air, distillation and sewage systems.

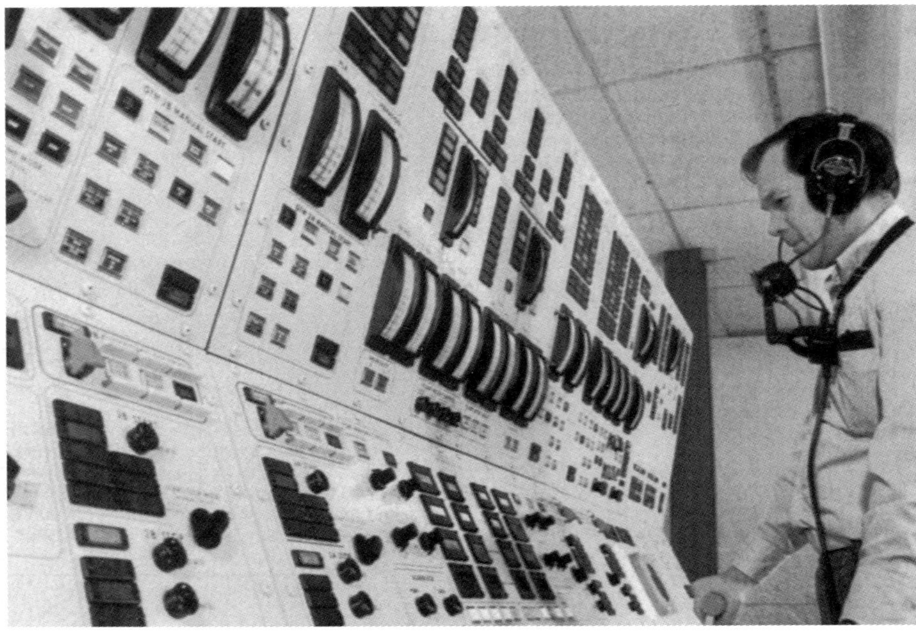

The CAE-Link gas-turbine propulsion plant trainer

Status

CAE-Link supplied three systems which simulate specific warship propulsion plants. Individual 'Spruance' (DD 963) class destroyer and 'Oliver Hazard Perry' (FFG 7) class frigate engine-room simulators have been installed at the US Navy's Great Lakes Training Center in Illinois, while a 'Perry' class simulator has been installed in Sydney for the Royal Australian Navy.

Manufacturer

CAE-Link Corporation, Binghamton, New York State.
(For details see Directory of Suppliers)

DynaLantic Gas Turbine Propulsion Plant Trainers (Devices 20H5/20H6/20H7)

DynaLantic states it is the leading supplier of gas-turbine propulsion engineering simulators in the United States and, in addition to supplying the US Navy, it has also exported simulators.

The company has supplied to the US Navy simulations and/or stimulations of the LM 2500 control system housed in trailer and configured to replicate those in specific classes of ship. They provide a full dynamic simulation of the LM 2500 gas-turbine operation from "cold iron" to "battle steaming". All of the ship's normal and casualty control procedures for the propulsion plant can be practised and reviewed by instructors.

The systems are Devices 20H5 for 'Ticonderoga' (CG 47) class cruisers, 20H5A for 'Spruance' (DD 963) class destroyers, 20H6A and 20H6B for 'Oliver Hazard Perry' (FFG 7) class frigates and 20H7B for Taiwanese 'PFG-2' class frigates.

The Device 20H7B stimulates the ship's propulsion but also includes damage control console stimulation/simulation and can be used as a maintenance trainer for electronic fault isolation and line replacement unit diagnosis.

Status

Since 1989 the company has supplied four trainers, two for 'Spruance/Ticonderoga' class systems and two 'Oliver Hazard Perry' class, to the US Navy and one to Republic of China (Taiwan) Navy for the 'PFG-2' class frigate.

Manufacturer

DynaLantic Corporation, New York State.
(For details see Directory of Suppliers)

The interior view of a Device 20H5A which provides training for 'Spruance' (DD 963) class destroyers

LM 2500 Training System (Devices 19E43/19E46)

As the manufacturer of the LM 2500 naval gas-turbine propulsion system, General Electric (which became part of Martin Marietta in 1993) produced a training system to support it. This system was designed to train engine-room crews in the workings of the LM 2500. It consists of five part-task trainers: for calibration; combustor and high/low pressure borescopes; a turbine model; and a sensor/igniter panel trainer. In the cases of the calibration and borescope trainers the appropriate elements of the propulsion unit are available in an instruction area where students may practise upon them. The sensor/igniter panel trainer is a working panel showing the operation and elements of the system.

Status

A system was purchased for the US Navy in 1984, at a cost of approximately $14 million.

Manufacturer

Martin Marietta Simulation and Automated Systems, Daytona Beach, Florida.
(For details see Directory of Suppliers)

SUBMARINE STEERING AND DEPTH CONTROL SIMULATORS

FRANCE

DCN Submarine Steering and Depth Control Simulators

The DCN's ECAN (Etablissement des Constructions et Armes Navales) Ruelle, F-16600 Ruelle, is a major supplier of submarine steering/depth control simulators including the Sienda, Sisma and Sirius systems. These systems are all based upon the SEL 32/77 dual processor computer and are mounted on platforms which can reproduce the movement of submarines in two dimensions. They are designed to reproduce the environment and characteristics of specific submarine classes. The systems are:

(a) Sienda for 'Daphné' class diesel-electric submarines
(b) Sisma for 'Agosta' class diesel-electric submarines
(c) Sirius for 'Rubis' class nuclear submarines
(d) One for 'Le Redoutable' class nuclear submarines.

Status
Sienda is used by the French (at Toulon), Spanish, Pakistani and South African navies. Sisma is used by the French (at Toulon) and Spanish navies. Sirius and the 'Le Redoutable' system are used by the French Navy, with the latter system at the Soument (Sous-Marins Entrainement) M4 at Brest.

Manufacturer
DCN, Paris.
(For details see Directory of Suppliers)

The DCN Sirius simulator for training crews of 'Rubis' class submarines

GERMANY

STN Submarine Control Simulator

STN SystemTechnik Nord GmbH (formerly MSG Marine- und Sondertechnik and previously MBB) is a major supplier of electronics and systems to the German armed forces and for 20 years has produced naval simulators including Submarine Control Simulators. These permit realistic training of crews, especially the diving officer and the hydroplane operators.

The systems are facsimile reproductions of submarine diving stations and reproduce submarine dynamics and behaviour, as well as equipment functions, through the use of mathematical models. The provision of computer-aided facilities for simple control of the training sequence and assessment of the exercise provide extremely cost-effective training.

The simulator cabin which is the students' station is installed upon a motion system which can move ±30° in the roll axis and ±45° in the pitch axis and provides realistic reproduction of submarine behaviour. The data processing subsystem has a simulation computer for the mathematical model and input/output data processing as well as a graphics computer to support the instructor's console. There are also interface computers to support the motion system and the students' controls. The mathematical model incorporates partial models of individual submarine systems together with hydrostatics and hydrodynamics of the hull. Environmental conditions such as sea states, salinity and water temperature are also taken into consideration. The hydrodynamic model is based upon standard movement equations and the related sets of coefficients as well as the results of trials.

The instructor can define environmental conditions as well as operating modes and can insert faults or system failures through his console. He can observe the students throughout the exercise.

The instructor console in a STN Submarine Control Simulator

Status
Submarine control simulators have been supplied to the navies of Germany, Greece, Turkey, Italy and Korea.

Manufacturer
STN SystemTechnik Nord GmbH, Bremen.
(For details see Directory of Suppliers)

ITALY

FIAT SEPA Sauro Submarine Control Simulator

SEPA, which produces control systems for the Italian Navy's submarines, has developed a number of submarine control simulators. The latest is being built to train crews for the 'Sauro' class submarines of the Italian Navy.

It consists of a replica control room, complete with autopilot console, trim and compensation controls and facsimile periscope, installed on a three-motion hydraulic platform. The computer has full details of the 'Sauro' submarine's dynamic performance to provide a realistic response to the facsimile controls.

An instructor's station is mounted alongside the control room to monitor the student's progress. The instructor is also able to alter the simulated environment and to add a wide range of malfunctions.

Status
The system was scheduled to be delivered to the Italian Navy training centre at Taranto in April 1993. It is also believed to have been exported.

Manufacturer
FIAT, Turin.
(For details see Directory of Suppliers)

NETHERLANDS

TNO Sea Dragon

The Sea Dragon simulator was developed by TNO-IWECO (Institute of Mechanical Engineering) to meet a Republic of China (Taiwan) Navy requirement. The simulator was designed to train crews for the new 'Sea Dragon' class submarines. It is housed in three movable containers which comprise the hydraulic system, the instructor's position and simulated engine control room, while the third contains the simulated control room. Up to six students can be given training in normal steering control, ballast-trim and engine control functions, together with emergency situations.

The instructor's position controls the activities in the student areas which are observed through one-way windows. The instructor can select about 100 control functions and has available 130 malfunctions which he can insert into the training schedule. He can also simulate the activities of other departments in the submarine which are not represented in the trainer. The control room and engine-room areas have facsimile operating and control panels. At the request of the Republic of China Navy, motion simulation is confined to pitch.

Status
The system was ordered by the submarine manufacturers, Wilton Fijenoord BV, and the system has been delivered to the Republic of China (Taiwan) Navy.

Manufacturer
TNO Physics and Electronics Laboratory, The Hague.
(For details see Directory of Suppliers)

The instructor's station of the TNO-IWECO Sea Dragon submarine handling system. The students' compartment can be seen through the one-way window on the left of the picture

UNITED KINGDOM

Thomson Training and Simulation Nuclear Submarine Control Room Simulator

The Nuclear Submarine Control Room Simulator produced by Thomson Training and Simulation (formerly Hughes Rediffusion) was originally one system based in two cabins to provide training for nuclear submarine crews, each cabin being mounted on a two-axis motion system. One cabin was configured to represent either 'Resolution' or 'Valiant' class submarine control rooms while the other represented 'Swiftsure' or 'Trafalgar' class submarines. The system was controlled by an R2000A computer with a shared linkage system and one hydraulic power unit.

The 'Swiftsure-Trafalgar' system was subsequently modified with a Gould 32/67 computer, a new linkage and advanced offboard instructors' and operators' consoles. These permit the setting up of training exercises in automatic, semi-automatic and manual modes to provide a complete range of standard and emergency operating procedures.

Status
The original system (NUSCOT) was commissioned at the Clyde Submarine Base (HMS *Neptune*), in July 1977. In the summer of 1986 the 'Swiftsure-Trafalgar' element was transferred to the Devonport base (HMS *Drake*) to become DEVNUSCOT.

Manufacturer
Thomson Training and Simulation Limited, Crawley.
(For details see Directory of Suppliers)

TTS Submarine Control Simulator (SCS)

The Thomson Training and Simulation (formerly Link-Miles) Submarine Control Simulator (SCS) incorporates a full-size replica of the forward and port side sections of a submarine control room which is mounted, with the instructor's facilities and computer, on an electrically driven, two-axis motion platform and includes an onboard training management system. The system is designed to produce as authentic an environment as possible and, as well as sea motion, these include a simulated intercom system and authentic submarine lighting. Distributed architecture with Intel 386/387 microprocessors is used in the system whose software is written in Fortran.

The instructor's console and management station is separated from the student's compartment but the instructor can observe them through a one-way mirror. Each instructor's position includes consoles for the instructor and the unit operator. The consoles can display detailed schematics which permit the instructor to adapt a wide variety of lesson plans. These facilities permit student-paced training as well as precise monitoring of performance.

The instructor's position in the 'Vanguard' system includes a control and monitoring facility based upon a high definition ship schematic display. To monitor students' progress the instructor gains access by moving the cursor to the relevant part of the ship reducing the need for displaying alphanumeric data. Faults may be injected or removed with the aid of a keyboard.

The TTS 'Vanguard' class Submarine Control Simulator demonstrating the motion system

Status
In service with the Royal Navy ('Upholder' and 'Vanguard' classes), the Indian Navy (HDW Type 1500), the Royal Norwegian Navy ('Ula' class), the Royal Netherlands Navy ('Walrus' class) and the Royal Australian Navy ('Collins' class). A simulator is in production for the Brazilian Navy ('Tupi' class).

Manufacturer
Thomson Training and Simulation, Crawley.
(For details see Directory of Suppliers)

UNITED STATES OF AMERICA

Contraves-SSI 'Ohio' Class Ship Control Team Trainer (Device 21C10A)

This system is designed to provide a range of instruction from basic instrument familiarisation and ship manoeuvring to full command and control exercises.

The system consists of a simulated ship control panel, a ballast control panel, steering and diving stations, a missile command centre digital display unit, and an officer-of-the-deck station with realistic sound and communications system. They are mounted on a manoeuvrable platform capable of providing ±45° pitch and ±25° roll motions. The instructor's console is mounted in a separate cabin on the platform and the system is operated by a Gould SEL 32/9750 which can simulate all the dynamic movements of the 'Ohio' class submarine.

Status

Gould's Systems and Simulations Division won the $6.2 million contract to build two trainers in August 1986 but in April 1987 responsibility was transferred to Systems and Simulations Inc (SSI). SSI subsequently became part of the Contraves organisation. The systems were installed at the Trident Training Facility, Bangor, Washington, in January 1989 and King's Bay, Georgia, in 1990.

Manufacturer

Contraves-SSI, Florida.
(For details see Directory of Suppliers)

DynaLantic SSN-21 Seawolf Ship Control Operator Trainer (Device 21C13)

This device is being designed and developed by DynaLantic and features a simulated ship control station with all the relevant indicators, displays, panels and controls representing the ballast, combined instrument and diving officer's panels, the helmsman/planesman stations and communications equipment. The simulator, which will house a 'crew' of 11, will be installed on a motion platform.

In addition to the ship control and emergency control training, Device 21C13 will include a high frequency sonar simulation with interactive tactical navigation using the ship control station's horizontal plotting table.

Status

Under development for the US Navy.

Manufacturer

DynaLantic Corporation, New York State.
(For details see Directory of Suppliers)

SENSOR AND ELECTRONIC WARFARE SIMULATORS AND TRAINERS

BULGARIA

Kintex Universal Radar Simulator

The Universal Radar Simulator was designed for the mass training of operators in marine radars. It consists of a number of generic radar simulators of which up to 20 may be controlled by two common 16-bit personal computers. The system is controlled by an instructor's station and several radar systems may be emulated.

The simulator can display up to 80 targets moving at different speeds and in different directions, up to 15 targets acting in groups and up to 15 navigation signs. Coastline can be added for greater realism and both active and passive electronic warfare techniques may be practised.

In stand-alone mode the system may be used for individual training. Alternatively, it may be used for group training with teams conducting different tasks.

Status
In production.

Manufacturer's Export Agency
Kintex, Sofia.
(For details see Directory of Suppliers)

The Kintex Universal radar simulator

CANADA

Computing Devices AQQ-T501 Tactical Acoustic Trainer

The AQQ-T501 Tactical Acoustic Trainer is a new modular acoustic simulator which provides the dynamic stimulation of acoustic and tactical systems in a manner which is indistinguishable from live target operations.

Realistic target signatures and operational scenarios are achieved in a multiple target environment which includes authentic oceanographic effects, shipping interference as well as fully responsive, manoeuvring targets. Targets are independently defined with each following unique parameters in position, course, speed and depth as required for the exercise. Each target reacts to programmable oceanographic conditions, sensor type and location as determined by the instructors. The result is type specific target signatures which are identical to those of actual live targets under similar conditions.

The AQQ-T501 provides authenticity through attention to detail. Both speed-dependent narrowband and non-speed-related narrowband sources are accurately portrayed. Broadband and transient sources are simulated in an equally realistic manner. Individual lobing patterns are produced for each line family in a target's signature and results in aspect dependent detection opportunities. Passive realism is further enhanced by varying the acoustic environment including sea state, ambient, shipping and biological noise, convergence zone phenomena, multipath and layering all of which can be controlled by predetermined instructions or on-line interactive manipulation. In the active environment, realism is achieved with respect to reverberation, propagation loss, vertical attenuation and Doppler.

Status
In service with the Canadian Navy.

Manufacturer
Computing Devices Canada Limited, Ottawa.
(For details see Directory of Suppliers)

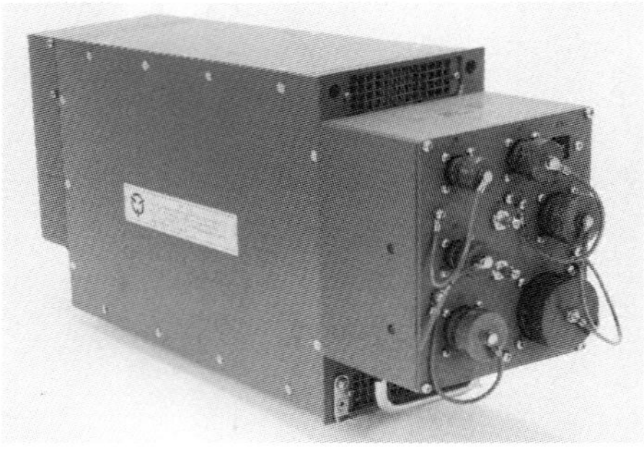

Computing Devices AQQ-T501

Westinghouse OH-5001/ SQS-505(V)6 Simulator Stimulator

This Westinghouse Canada stimulator injects noise and target signals into the front end of a sonar receiver to provide video and audio outputs simulating the signatures which operators might expect to receive in operational service.

The system can generate four realistic targets as surface ships, submarines, torpedoes or wrecks. The OH-5001 has four operational modes: at sea with the transmitter on or the transmitter off, harbour operation and test.

In the at sea mode with transmitter on, active and passive targets are simulated and the signals superimposed on the ship's noise, flow noise and reverberation. In the at sea mode with transmitter off, signals and reverberations are superimposed upon the ship's noise and flow noise from the staves. For harbour operation all signals, noise, reverberation and target data are supplied together with simulated data of the ship's course and speed.

Status
In production with 13 sets delivered for the Canadian Patrol frigate programme.

Manufacturer
Westinghouse Canada Inc, Ontario.
(For details see Directory of Suppliers)

FINLAND

Elesco FERS 800 Echo Repeater

The FERS 800 is designed for navies which wish to conduct ASW training but do not have any submarines. The echo repeater is a turnkey system for training sonar operators and assessing sonar performance. It is capable of receiving sonar transmissions and simulating their effect upon a target.

It can be rapidly adapted to a wide range of sonar equipment, from 10 to 35 kHz, as well as environmental conditions. It can include simulation of target movements even in buoy suspension modes and active depth control in towed mode. The FERS can be used in three modes: basic, telemetry and towed.

Status
Available.

Manufacturer
Elesco Oy, Espoo.
(For details see Directory of Suppliers)

FRANCE

Thomson Sintra Sygame (TSM 9001) Acoustic Training System

The Sygame is designed to provide sonar and sonobuoy operators with self-paced or instructor-guided audio training. It consists of a realistic sound generator and a work station with a set of software tools. Sygame may operate in stand-alone mode, on land or at sea, or networks of up to eight work stations may be created.

The system synthesises sounds to match the audio response from active or passive sonars while the menu-driven software enables progressive training from elementary sound recognition to target classification in a complex noise environment. It may be used to train students to recognise both a sonar pulse and noises generated by sources such as propellers, engines, hulls and transients with up to 12 pieces of machinery being mixed at any one time. The combination of these sounds can be used to create realistic signatures for submarines and surface ships.

An improved version, Sygame V1 N, which may be integrated with a network, has also been developed. It has an Internet protocol and a standard talk rate of 230 kbits/s.

Status

Sygame has been ordered for the Australian, French and Italian navies.

Manufacturer

Thomson Sintra, ASM (Activités Sous-Marines), Sophia Antipolis.
(For details see Directory of Suppliers)

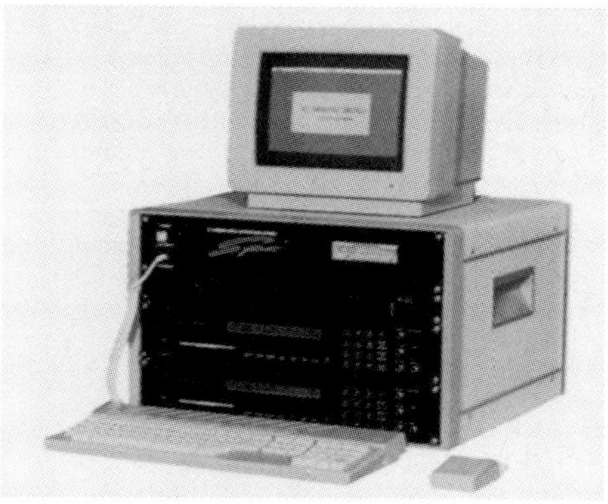

The Sygame sonar operator trainer

Thomson Sintra Symen Sonar Operator Training System

Symen is designed to provide sonar training using audio and visual references in both active and passive sonar operation. It is a modular system with an Ethernet-standard network configuration and may be used to train operators on the same kind of sonar, or on different sonars in the same ship, or on different sonars in different ships. In addition to individual training, students may be trained as members of a single ship's ASW team or as part of an ASW task group team.

The software-driven system enables each student's work station to be configured individually by the instructor to a specific sonar for the duration of the lesson and the student may then improve his skill in the use of the sonar. There is a 'mailbox' facility for communicating between the instructor and the student.

Status

Available.

Manufacturer

Thomson Sintra, ASM (Activités Sous-Marines), Sophia Antipolis.
(For details see Directory of Suppliers)

GERMANY

Atlas RASI Radar and Navigational Aid Simulator

Atlas Elektronik is a major supplier of radars and sonars to world navies and produced the RASI radar and navigational aid simulator to assist operator training.

RASI is used to train operators in any type of radar but especially the latest Automatic Radar Plotting Aid (ARPA) radars. It is also used to familiarise sea-going personnel with particular areas and harbour approaches. The system consists of students' cubicles, an instructor's station, briefing and debriefing facilities as well as a data input station.

The student's cubicle has a radar of any existing make and a slave radar, navigational instruments, chart table, a communications unit, a helmsman's unit as well as indicators to present 'own ship' and environmental data. The instructor's station has a slave radar display, colour graphic situation display, a menu display unit, X-Y plotter, communication equipment and MMI consisting of a keyboard, function keys and roller ball. The data input station is used to generate exercise areas, to update them and to generate new types of 'own ship' and traffic.

Status

A RASI system was commissioned by the German Navy in 1985, and two systems were delivered to maritime training establishments in Hamburg and Cuxhaven (both coming on-stream in 1986), while a fourth was delivered to Taiwan (Republic of China) and was commissioned in 1984.

The instructor's console of the Atlas RASI radar simulator system

Manufacturer

Atlas Elektronik GmbH, Bremen.
(For details see Directory of Suppliers)

Atlas VISIG Electronic Warfare Simulator

VISIG is an intelligent signal generator for real-time simulation of complex radar and missile guidance signals. Its prime application is in the creation of real-time electronic warfare scenarios.

The system is based upon microprocessor technology and uses a modular structure. Independent generators are used for each emitter represented, allowing their numbers to be flexible. There is multiple generator pulse and scan synchronisation as well as the generation of real video signals. The customer defines the number of antennas, radar bands and antenna characteristics.

Status

Available.

Manufacturer

Atlas Elektronik GmbH, Bremen.
(For details see Directory of Suppliers)

Dornier Electronic Warfare Trainer (EWT)

Dornier has developed the Electronic Warfare Trainer (EWT) to provide basic training to operators in evaluating signals received through ESM systems. The trainer consists of generic video and digital display units installed in cubicles with one or two students per cubicle. An instructor's station is nearby.

The students measure and evaluate the parameters of 'radar signals' received from 'shipborne ESM receivers'. They then classify and seek to identify them by means of evaluated parameters and parameter lists. Signal parameters generated include pulse amplitude, scan period, pulse repetition time, pulse repetition frequency, pulse width and carrier frequency. Manual parameter settings may be made and there is tape-controlled parameter processing and display.

Status

Available.

Manufacturer

Dornier Luftfahrt GmbH, Friedrichshaven.
(For details see Directory of Suppliers)

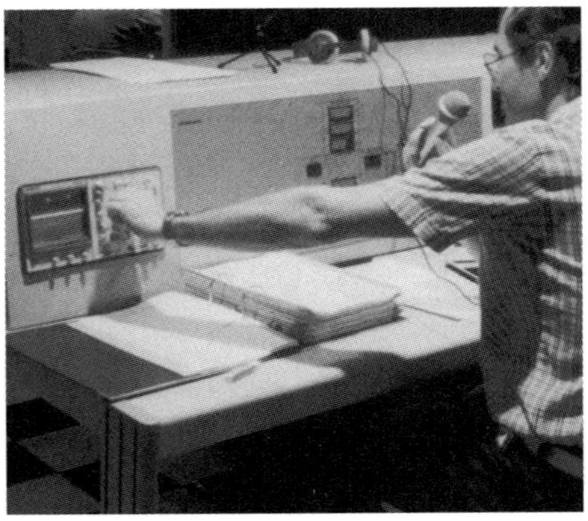

The Dornier Electronic Warfare Trainer (EWT) instructor station

INDIA

Bharat Radar Target Simulator (RTS 01)

The Bharat Radar Target Simulator (RTS) was designed for use on board frigates for operator training at sea or in harbour. It provides target identification, tracking and designation at TIU and other displays associated with the Signaal M44/45 fire-control system. It consists of the simulator console, a RUGCOM ruggedised 16-bit mini-computer based upon the Intel 80386 processor, an instructor control panel, a change-over switch with relay, and junction boxes. Within the console are a search video generator, three track video generators, course and speed synchros and a power supply.

The computer provides the target data for the search video generator once a second. The video generator controls the rotation of the PPI displays and generates target traces mixed with simulated receiver noise, sea clutter, jamming videos and coastline videos. The system can generate a maximum of 20 moving targets, both ship and aircraft. Of these 12 are preprogrammed on digital cassette tapes while one may be introduced by the instructor. Up to three M44/45 systems may be exercised simultaneously.

Status

In service with the Indian Navy.

Manufacturer

Bharat Electronics Limited, Bangalore.
(For details see Directory of Suppliers)

Keltron Radar/Sonar Simulated Display Unit

The Kerala State Electronics Development Corporation Limited (Keltron) has produced a radar/sonar simulated display unit which may be used in association with its Ship Control System. Based upon IBM personal computers it has facilities for operating in the same way as a radar or a sonar display. For the former role it has provision for simulated jamming as well as for the display of Naval Tactical Data System (NTDS) symbology.

Status

Believed to be in production.

Manufacturer

Keltron, Trivandrum.
(For details see Directory of Suppliers)

Keltron Navigation Simulator

The Keltron Navigation Simulator consists of three cubicles each containing IBM personal computers with coastline displays. It is used for training in pilotage in poor visibility or at night, based upon scenarios created by the instructor. A digital interface for coastline generation and a display upon a radar PPI screen are also available.

Status

Believed to be in production.

Manufacturer

Keltron, Trivandrum.
(For details see Directory of Suppliers)

INTERNATIONAL

Multi-Application Sonar Trainer (MAST)

Alenia-Elsag Sistemi Navali developed, in co-operation with Raytheon, a Multi-Application Sonar Trainer (MAST). As the production of MAST initially took place to meet an Italian requirement the system is described here rather than in the United States section.

MAST is used to train sonar operators to work with hull and variable depth sonars and reproduces the acoustic signals normally encountered, including passive and active signatures, together with environmental characteristics. It may also be used to create or modify new target parameter or ocean model tables.

The trainer consists of student stations which are identical to the operational displays, an instructor station, a VAX 11/750 mini-computer and a signal generator. The instructor station is used to insert data related to 'own ship' and the target's performance such as location, course and speed.

The signal generator produces both signals and noise which are injected into the front end transducer array interface. They simulate those generated by the real acoustic environment when converted from pressure to voltage by the receiving transducer array. These include all possible variations due to motion, mutual interference, propagation characteristics and environmental conditions.

The MAST sonar trainer produced under licence by Alenia-Elsag

The MAST trainer is used to train operators of the DE 1160 and DE 1164 hull variable depth sonar systems and is being expanded to include training for operators of the improved DE 1167 sonar system together with passive sonar systems.

Status

In service with the Italian Navy's Fleet Training Centre at Taranto.

Manufacturers

Alenia-Elsag Sistemi Navali, Rome.
(For details see Directory of Suppliers)

Raytheon Company, Submarine Signal Division, Rhode Island.
(For details see Directory of Suppliers)

ISRAEL

Elbit Electronic Warfare Trainer (EEWT)

As a supplier of electronic warfare equipment to the Israeli Navy, Elbit has also developed an electronic warfare trainer known as EEWT - Elbit Electronic Warfare Trainer. The EEWT is a centralised architecture system designed to improve the skills of operators by providing exercises in electronic warfare ranging from the simple to the complex, as well as teaching responses to common electronic warfare situations such as anti-missile operations and the tactical use of equipment. The system consists of an instructor's room, and a number (up to 12) of simulated ships each with an operations room, two ESM (Electronic Support Measures) operators and one ECM (Electronic Countermeasures) operator each in a separate cubicle.

The electronic warfare requirements can be met by synthetic display simulation or, alternatively, either by video or RF signal injection into genuine consoles depending upon customer preference. Elbit prefers to offer synthetic display simulation. The instructor's position has a console with situation and graphics display for monitoring the student's activities, supervising the exercise and inserting new problems. It will also feature a PPI radar display similar to that used in the operations rooms. The operations room features a conventional radar display, a 'manoeuvre' control unit and communications equipment, and ESM positions, each with two consoles: one with two IFM (Instantaneous Frequency Measurement) displays and an analysis display while the other has a goniometric display. The ECM console has a panoramic display, deception, jammer and decoy controls.

Status
Available.

Manufacturer
Elbit Limited, Haifa.
(For details see Directory of Suppliers)

Elbit Naval Electronic Warfare Simulator (NEWS)

The Naval Electronic Warfare Simulator (NEWS) is designed to train commanders, officers and crews in selecting the most effective and most efficient electronic warfare response to attacks by multiple anti-ship missiles. It may also be used to assist research teams in planning warship "soft kill" defence.

It is based upon a distributed architecture and uses work stations configured as consoles supplemented by a voice communication system. Computers are used for running and monitoring the system which features an environmental and system simulation local area network linked to a system simulation local area network. The numbers and types of processor and work station depend upon the customer requirement.

Training can be conducted in single run or random run modes. The former involves a single attack scenario in real time with a continuous display of the missile logic, the electronic warfare response and the decoy dispersal pattern. The latter involves a large number of simulation runs performed in accelerated time with each run's parameters chosen randomly.

Status
Available.

Manufacturer
Elbit Limited, Haifa.
(For details see Directory of Suppliers)

Rafael Naval EW Simulator (RNEWS)

RNEWS simulates the engagement between an anti-ship missile and a target in an Electronic Warfare (EW) environment. It was originally designed for developing missile seekers but is now also used for training purposes. The system consists of a host computer linked to a missile simulator and a missile seeker. All are linked to a target generator and an EW generator.

Real RF is transmitted to the seeker, its effectiveness being tested against real EW systems. The system can handle complex scenarios and can quickly produce data through peripherals, allowing the test and evaluation of real seekers and EW systems as well as the evaluation of new EW techniques against specific missiles.

Status
RNEWS is believed to be in service in Israel and has been sold to two foreign navies. Both export systems are believed to be in service and the manufacturer is negotiating with three other navies.

Manufacturer
Rafael, Haifa.
(For details see Directory of Suppliers)

ITALY

Sindel RS 100/B-MM Radar Simulator

The RS 100 system trains operators in operating S- and X-band radars in realistic conditions with simulated sea and rain clutter.

The system can be interfaced with any radar display upon which can be presented true motion, relative motion and stabilised relative motion. The speed, course and manoeuvring parameters of the 'ship' and the targets are variable and any kind of coastline may be injected.

The instructor has 100 exercises per disk and, because they develop automatically, he can concentrate upon monitoring the students rather than operating the simulator. He can, however, freeze the exercise to give verbal instruction. Exercises which can be carried out with the simulator include navigation, Anti-Submarine Warfare (ASW), and attacks by fast attack craft.

Status
In service with the Italian Navy at Taranto.

Manufacturer
Sindel srl, Genoa.
(For details see Directory of Suppliers)

NETHERLANDS

Radio Holland Navigation Simulators

The Radio Holland Group has been providing marine electronics for more than 70 years and it has extended this expertise into the field of simulators including satellite communications simulators which are used by several organisations in the Netherlands.

The company also produces a range of navigation simulators (Navsim) as well as two river navigation simulators (Rivsim) which are used in schools at Rotterdam and Kiev. The following description of the Rivsim at Rotterdam provides some idea of the company's simulator capability.

This system consists of four 'own ship' bridges for students, an instructor's console and a computer system. The students each have a simulated JMA606 river radar and may select one of three kinds of ship: a fast, small or an 80 m vessel which is either loaded or unloaded. Each ship has its own mathematical model which simulates its real performance. The instructor can then add a further 48 vessels, selectable from a library of 13 types, for use in the exercise. Other inputs include buoys, realistic radar and environmental performance. The students may operate autonomously or in association.

The instructor's station also has a simulated JMA606 radar display for each student cubicle, three repeater instruments for each cubicle, a simulated VHF system with six channels and an audio system for simulating real VHF traffic and weather forecasts. The processor system consists of four Z80 microprocessors and two disk drives of which one is used as an exercise database while the other is used for exercise data and for storage.

Status
The company has helped to supply a Norcontrol Navsim NMS 90 navigation simulator to the Royal Netherlands Navy for use at Den Helder.

Manufacturer
Radio Holland Group, Amsterdam.
(For details see Directory of Suppliers)

Radio Holland LCS-100 LORAN-C Signal Simulator

LORAN is an acronym for Long Range Radio Aid to Navigation and is a major maritime navigation aid. The LCS-100 dynamic dual rate LORAN-C signal simulator is designed to test and to calibrate single and dual rate LORAN-C receivers.

The simulator is controlled by an IBM PC, XT, AT, PS/2 or similar computer and all parameters of the signal are programmable. It is designed to generate high accuracy LORAN-C signals from up to two simulated chains with up to seven stations per 'chain' and, as each 'chain' has its own clock input, this makes it possible to simulate interchain drift.

Status

In production.

Manufacturer

Radio Holland Group, Amsterdam.
(For details see Directory of Suppliers)

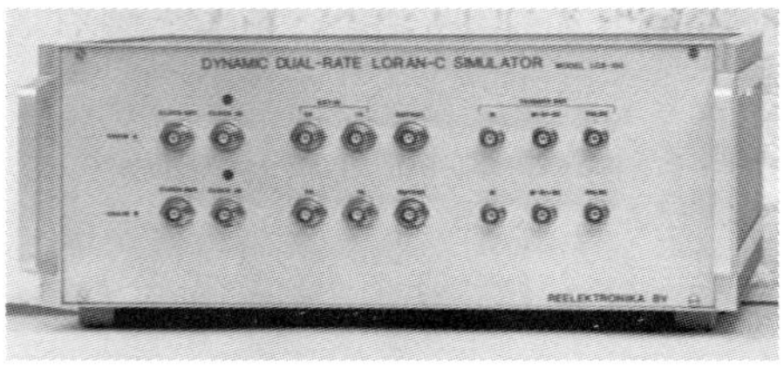

The Radio Holland LCS-100 LORAN-C signal simulator

Signaal SSE Sonar Simulator

Signaal Special Products (formerly Van der Heem Elektroniks NV) has produced a variety of shipborne simulators for training sonar operators together with ship command and weapon control teams. Their SSE sonar simulator, which generates realistic surface and submarine targets in video and audio form, can be used with any active or passive sonars. The system can be used when the ship is berthed or at sea and in the latter situation simulated targets can be inserted into the existing marine environment.

Several dedicated models are available for use with specific sonar systems including the SSS-03 with the EDO Model 610 sonar, the SSS-05 with the Westinghouse Canada AN/SQS-505 and the SSS-32 with the Signaal PHS-32. All the simulators can produce realistic simulations of two targets with accurate representations of the performance envelope in all sonar operational modes. These targets can be 'ships', 'submarines', 'wrecks', and 'false echoes'. One target can be a torpedo 'fired' from a target submarine.

Status

The system is in service with Belgium (SSS-05), Brazil (SSS-03), Canada (SSS-05), France (SSS-05), Iran (SSS-32), Italy (SSS-03) and the Netherlands (SSS-05).

Manufacturer

Signaal Special Products, The Hague.
(For details see Directory of Suppliers)

NORWAY

MPS Multibeam Sonar Simulator (MMSS)

The MPS Multibeam Sonar Simulator (MMSS) is a training system for the operators of active sonars whether hull-mounted, variable depth or dipping. Each consists of an instructor's station with system and game control consoles, a student steering console (externally identical to the game control console), a real sonar console for the student and a digitiser for preparing the 'map' and 'terrain'. The student can use all the controls on the sonar console, learn to understand the working principles of the equipment and assess the results when the control settings are altered. The system is based upon the 32-bit Multibus II real-time computer using the Intel 80386 as the main processor.

Echoes from underwater objects are produced by a digitised representation of reflected sound at different aspect angles. This produces a very accurate generation of the acoustic characteristics of the sound return which takes into account the size, geometry, location, aspect angle, and Doppler effects. The operational characteristics of the sonar, including the source level, transmission modes, scanning sectors and pulse lengths, together with the actual operator settings such as range scale and tilt, are used as the simulator model.

The operator can use either an actual sonar console or an MPS tactical console simulating the man/machine interface, presentation modes and signal processing capabilities of a specific, or a set of, real sonar set(s). In order to 'steer' and 'control' the sonar-carrying 'vessel' a manoeuvre console is connected for setting speed, course and manoeuvre models. The sensor platform and all the targets are controlled by the instructor.

Optional extras for the system are a network interface with an action speed tactical trainer such as MASTT, a plotter for analysis purposes, additional tactical displays and an interface with a video-audio recorder.

The MPS Multibeam Sonar Simulator (MMSS) supplied to the Royal Swedish Navy

Status

In service with the Royal Norwegian Navy since June 1989 at KNM Tordenskjold, Bergen, where it has been integrated with MASWT. In August 1990 a system was delivered to the Swedish Navy's training establishment at Berga Örlogsskolor. This system was designed to train commanding officers and sonar operators of 'Hugin' class fast attack craft and includes simulation of the Simrad SA950 hull-mounted and Simrad ST570 variable depth sonars. In addition to an instructor's console, it has a student steering console, a real sonar console for a student and an exercise preparation and control station.

Manufacturer

Micro Processor Systems AS, Kongsberg.
(For details see Directory of Suppliers)

Norcontrol Navsim NMS-90 Radar and Navigation Simulator

Navsim NMS 90 is a series of radar and navigation training simulators designed to train students ranging from cadets to senior officers. A number of skills may be developed on this simulator including blind pilotage, bridgemanship, team training, fleet work and communications. Several cubicles may be connected to the simulator and the cubicles may also be linked to each other for combined exercises in which bridge and operations room teams are trained together. The cubicles are usually equipped with ship-like instrumentation including steering system, engine control, radars and tactical displays. Norcontrol has an extensive library of panels/equipment which will fit almost any specific requirement without development.

The exercise area databases contain radar coastlines, navigation points, buoys, soundings, current patterns and navigation systems such as GPS (Global Positioning System), Loran-C, Decca, Omega and Transit. More than 200 exercise areas are immediately available.

The instructor station is extremely user-friendly and is based upon Intel microprocessors. It is designed in close co-operation with users and is based upon more than 20 years of experience in marine simulation.

Status

More than 150 radar simulators have been produced for Australia, Bangladesh, Bulgaria, Canada, the CIS, China, Denmark, Finland, France, Germany, Greece, Hong Kong, Iceland, South Korea, Malaysia, Mexico, the Netherlands, Norway, Peru, the Philippines, Poland, Portugal, Spain, Sri Lanka, Sweden, UAE, UK, USA and the former Yugoslavia. The navies of France, Greece, the Netherlands, Norway, Spain, Sweden as well as the Royal Navy use NMS-90 simulators.

Manufacturer

Norcontrol Systems A/S, Horten.
(For details see Directory of Suppliers)

SPAIN

SAINSEL Navigation Simulator

The SAINSEL Navigation Simulator is one of several naval-orientated operational simulators which extend to naval power-plant, submarine command simulators and operational centre simulators.

The Navigation Simulator simulates the functioning of radars such as ARPA and anti-collision, navigation equipments such as GPS, Transit, Decca and Omega and propulsion and steering systems. These facilities are combined at a student's cubicle, enabling a student to acquire the necessary training needed to carry out specific bridge watch-keeping functions.

Status
In production and service.

Manufacturer
SAINSEL, Madrid and Seville.
(For details see Directory of Suppliers)

The student's cubicle for the SAINSEL Navigation Simulator

UNITED KINGDOM

Anaren Electronic Warfare Simulators

Anaren produces a line of digital, RF and video simulators for electronic warfare training. They are designed for modularity and easy upgrade with all using personal computer-based software. The DEWS 6000 series provides real-time simulation based upon pre-compiled scenarios while the DEWS 6500 generates real-time signals using real-time scenario execution techniques.

DEWS 6000 Series
This series of digital simulators provides up to 256 simultaneous emitters and full 3D platform dynamics.

DEWS 6500 Series
This series of digital simulators uses time slice modelling which allows the customer to modify the system parameters in real time. It has a dedicated 32-bit processing unit and may be used for electronic warfare system embedding or onboard training where joystick control is required.

VTG 6000/6500 Series
The VTG range uses either a DEWS 6000 or a 6500 as a front end and adds video level outputs. A wide range of options is available including multi-port DF, multi-band frequency operation or omnidirectional outputs. The range is suitable for radar warning receiver applications as well as Electronic Counter Measures (ECM) testing.

RFES 6000 and ARTES 6500 Series
These are RF simulators which allow straightforward addition of RF sections to the DEWS 6000, the DEWS 6500 or the VTG 6000/6500 Series. A wide range of RF sources is provided from 500 MHz to 18 GHz and they can be VCO or synthesiser-based as required. Full DF modelling units are available together with low power and high power outputs. The DF modelling facilities are user-definable and up to nine ports can be individually defined in terms of antenna patterns and spatial orientation.

RSS 6000 Series
This a field-deployable, portable, fully programmable RF simulator which generates signals in the 2-18 GHz range. It can generate up to 16 signals.

RSS 6500/D
This is an RF simulator with integral Anaren Digital RF Memory (DRFM) to enable a wide range of ECM techniques to be generated.

7500 OBT
This is an electronic warfare simulator specifically designed for onboard training and can be operated in a stand-alone mode controlled by an intelligent hand-held terminal or as part of a group trainer with a personal computer or remote source control.

Chameleon
This is an ECM simulator which can be used for testing and for training both electronic warfare and radar operators as well as systems. It features a sophisticated RF processing subsystem which includes a frequency measuring receiver and DRFM. The unit is controlled with a PC-486.

Status
In production.

Manufacturer
Anaren Microwave Limited, Camberley.
(For details see Directory of Suppliers)

Aware Simulation Simulators

Aware Simulation Limited (formerly a subsidiary of Walmore Electronics Limited as Walmore Defence Systems Limited) produces customised systems for training electronic warfare operators, incorporating a number of approaches including radio frequency radiation and radio frequency injection. All simulators are reported to be user-friendly with operator-defined or preprogrammed emitters with menu techniques used for control. Frequency ranges cover from 0.5 to 18 GHz. Electronic warfare simulators have been supplied to the Royal Navy, the British Army and the Royal Air Force.

Status
In service (see above).

Manufacturer
Aware Simulation Limited, London.
(For details see Directory of Suppliers)

Caltronic Electronic Warfare Simulator

In December 1990 Caltronic Systems revealed that it had won the contract to supply an electronic warfare simulator for the Type 23 frigate shore development facility. No further details are available.

Status
In production.

Manufacturer
Caltronic Systems Limited, Portsmouth.
(For details see Directory of Suppliers)

Data Sciences Integrated Training System (ITS)

The Data Sciences Integrated Training System (ITS) provides scenario-based individual and collective skills operator training using desktop computers. A typical system consists of a network of commercial, off-the-shelf work stations running simulations of operational requirements, especially sensor training, with which the trainees interact. The system can be supplied with a fully integrated communications suite and can also be mounted into prefabricated consoles for added realism. The software for the system, which is the basis of the Electronic Warfare Training System (EWTS) and the Radar Picture Compilation System (RBOT), is written in Ada.

Status
In production.

Manufacturer
Data Sciences Limited, Farnborough.
(For details see Directory of Suppliers)

Data Sciences Electronic Warfare Training System (EWTS)

This system provides individual and collective skills training to give operators the necessary skills to operate electronic warfare equipment to the maximum effect. A version of the system is available to provide ESM, ECM and radar warning receiver training while another is under development which will add ELINT, radar and communications functions.

EWTS features an instructor's station and a number of student stations each with their own simulated electronic warfare console. The students operate their equipment independently in response to a training exercise which may be changed or altered by the instructor.

A variety of displays and scenarios is available and student progress is automatically recorded for subsequent analysis. The system permits all forms of instruction from basic electronic warfare concepts to advanced tactics and procedures.

Status

An EWTS has been delivered to a Middle Eastern country's defence forces as part of an electronic combat training system supplied by Marconi (ex-Ferranti) International. Data Sciences will also supply an EW Evaluation System (EWES) software package.

Manufacturer

Data Sciences Limited, Farnborough.
(For details see Directory of Suppliers)

The Data Sciences Electronic Warfare Training System instructor's station

Data Sciences Radar Picture Compilation System (RBOT)

This system provides training in the general principles of radar picture compilation and track management at basic and intermediate levels. It also provides familiarisation training in operational procedures.

It is based on a network of high performance work stations built into general-purpose operator consoles. Each console is equipped with a roller ball, keyboard and special purpose keypad. The work station's screen provides a high resolution radar PPI display with real-time video and synthetic track symbols. There are two instructor and 16 operator positions. Operator performance can be monitored and displayed at the instructor position.

Operators may train individually or in groups of four, where each group represents a single ship with operators co-operating to form a single shared radar picture. Track data can be passed between vessels by means of a simulated data link. A sophisticated intercom system allows students to communicate with one another and the instructor.

A complete Radar Picture Compilation System

Status

The system was delivered to the Royal Netherlands Navy in May 1992.

Manufacturer

Data Sciences Limited, Farnborough.
(For details see Directory of Suppliers)

DSD Sonar Procedure Trainers

Digital Systems and Design Limited (DSD) produces generic and dedicated Intelligent Procedure Trainers (IPT) which provide students with a full range of training facilities from basic to full tactical and weapons operations. Each student is guided automatically by verbal and visual commands in accordance with a predefined training programme. The system features high resolution animated graphic presentations, full touch-screen control operations, complete emulation of main system displays and operating control panels within the customer's requirements, together with emulation of targets and the environment.

The Mine-Hunting Sonar (MHS) system (IPT-MHS) provides a complete and comprehensive facility for training mine-hunting sonar operators at basic and advanced levels. The emulation is in real time and includes a degree of unpredictability. Each module consists of instructional lessons and exercises while upon completion of the lesson or exercise the instructor is provided with a hard copy printout for debriefing. The system may be used not only for training but also for evaluating tactics, platforms and equipment. The Anti-Submarine Warfare (ASW) system (IPT-ASW) has similar facilities and can be used for a wide range of instruction ranging from basic operator training to complex ASW exercises. This system features acoustic noise emulation, full touch-screen control and synthesised speech for commands.

Two dedicated products have been produced, the IPT-32 and the IPT-2022. The IPT-32 is an anti-submarine warfare sonar simulator for training operators of the PHS-32 hull-mounted sonar while the IPT-2022 is a mine-hunting sonar and mine disposal system trainer for training operators of TSM 2022 sonars. Both simulators have been supplied to Dornier for use with its Fleetwork Trainer.

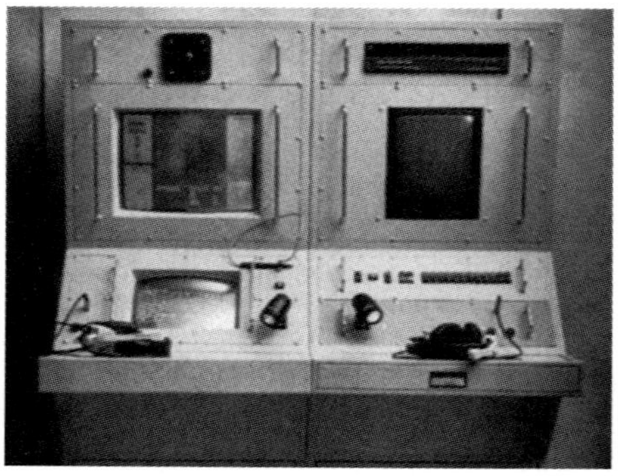

The DSD IPT-32 anti-submarine sonar trainer. The student's position is on the left and the observer's is on the right

The company has also produced a generic trainer which is software controlled and may be used for sonar or radar training. With the relevant software several IPT (Generic Trainers) may also be used as an Action Speed Tactical Trainer.

Status

An IPT-32 system was sold to Dornier for the Fleetwork Trainer supplied to Nigeria. Three IPT-2022 systems were sold to Dornier of which two have been supplied to the Nigerian Navy.

The DSD IPT-2022 mine-hunting sonar trainer

Manufacturer

DSD Digital Systems and Design Limited, Eastleigh.
(For details see Directory of Suppliers)

DSD Radar Emulator

Digital Systems and Design Limited (DSD) has developed a radar emulator which provides facilities that can be used for training in tactical displays, fire-control, tactical navigation, coastal navigation and ship handling. The presentation includes land masses, cloud, electronic and mechanical jamming and full NTDS track symbology.

The emulator can be operated in a raw state and can be programmed for use with various types of command and control systems and with various facilities including track table limits, automatic/manual extraction and tracking. Target emulation is such that students are presented with ones which are not stereotyped and which consequently assist instruction.

Status

Development complete.

Manufacturer

DSD Digital Systems and Design Limited, Eastleigh.
(For details see Directory of Suppliers)

GEC AS 2105 Sonar Simulator

A major supplier of electronics to the Royal Navy, GEC Avionics produces the AS 2105 sonar simulator. This is a compact, self-contained system designed to provide comprehensive testing and training facilities for passive sonar systems. The system, which can be installed in surface warships and submarines, can be configured to simulate towed, flank and conformal hydrophone arrays. It is a compact unit (525 × 440 × 415 mm) weighing 25 kg and capable of operating with various power supply levels.

The basic version, when used for simulation of a single sonar, can be controlled either by the integral lap computer or by a ship's computer which has been suitably programmed. The Mark II version, for integrated sonar systems, uses a central scenario computer to control several simulators, each of which is interfaced with a separate sonar.

In the test mode, calibrated signals generated in the simulator are used to measure the accuracy of a sonar's performance in processing and displaying frequencies, bearings, target movement and signal-to-noise.

As a training aid the simulator is programmed to respond to simple anti-submarine warfare scenarios, modelling 'own ship' and target information together with basic environmental effects. Exercises can be preplanned or free play. A fast time-review facility enables scenarios to be checked quickly.

A version of the Mark II is PASSIM (Passive Sonar Array Simulator) which simulates sounds received by towed sonar arrays. This is fed into the sonar system which displays the simulated data.

Status

In service with the Royal Navy. In January 1989 a contract for a PASSIM system was awarded by the United Kingdom Ministry of Defence. This will be supplied to the Type 23 Frigate Weapon System test facility at the Defence Research Agency establishment at Portsdown.

Manufacturer

GEC Avionics Limited, Sonar Systems Division, Welwyn, Hertfordshire.
(For details see Directory of Suppliers)

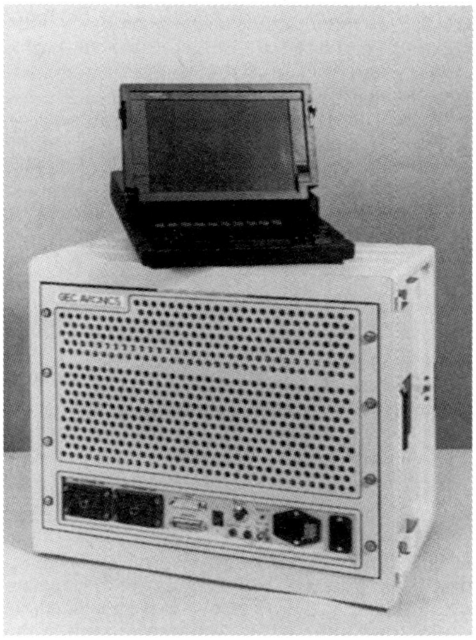

The GEC Avionics AS 2105 sonar simulator

GEC AS 2107 Sonar Simulator

The AS 2107 is a development of the AS 2105 (see previous entry) which is being remodelled and extended to provide an improved onboard training capability. It generates synthetic 'targets of opportunity' needed for opportunistic training of sonar operators, sonar subteams and command teams while on passage or on patrol. This is achieved by superimposing the training data on live data, enabling the commander to maintain surveillance and safety coverage during a training exercise.

The trainer will be available in two forms. A self-contained compact equipment will be suitable for integrating into a ship's sonar while the other version will be a portable unit.

Status

In service with the Royal Navy.

Manufacturer

GEC Avionics Limited, Welwyn, Hertfordshire.
(For details see Directory of Suppliers)

Marconi Onboard Radar Trainer

To complement its range of shore-based radar training devices, Marconi (ex Ferranti International Simulation and Training) has developed an onboard radar simulator designed to enhance shipborne combat command and anti-air warfare continuation training.

The equipment provides simultaneous simulation of any two collocated or 'own ship' surveillance radars and an associated secondary radar. Up to 80 supplementary tracks can be positioned in an exercise area of 2048 × 2048 nm. Real-time video generation techniques are used to generate target returns and IFF emissions which are mixed with live radar responses or injected into a synthetic scenario shown on the ship's own radar displays.

The system incorporates a comprehensive library of effects including clutter, jamming, environmental effects and land mass features which are introduced into the exercise. Scenarios are controlled by a handheld terminal for interactive instruction or can be programmed to allow for autonomous operation.

Status

This system has recently been supplied in some numbers to the Royal Navy. In April 1993 a £3.6 million contract was awarded to supply 15 Anti-Air Warfare On-Board Radar Simulators (AAWOBRS) for installation in four classes of Royal Navy ship from 1995.

Manufacturer

Marconi International Simulation and Training, Cheadle Heath.
(Ex-Ferranti, for details see Directory of Suppliers)

Marconi Sonar Principles Trainer

A major supplier of sonar systems to the Royal Navy, Marconi (ex Ferranti International Simulation and Training) also produces Sonar Principles Trainers.

The Sonar Principles Trainer was designed to provide comprehensive and cost-effective training of sonar operators in understanding the elements of sonar signals and the skills relating to detecting, evaluating and tracking targets in a variety of scenarios and environmental conditions. It was designed to be reconfigurable to provide training for operators of various sonar types.

It consists of an instructor's user-friendly console at which the instructor can prepare, execute, monitor and debrief exercises using an extensive library. The instructor has full control over the ocean environment 'ship movement', the sonar types involved, selection and movement of target vessels and weapon deployment. A repeat monitor and audio system allow him to follow any student's performance.

Up to 10 trainer operator consoles are available each with a multipurpose display unit, a sonar control panel, simulated audio reception and communications facilities. Each student can control his own mode of operation, sonar frequency, pulse width and beam forming. The sonar pictures and audio effects presented respond to student actions.

Status

A system with 10 student positions was supplied to the Brazilian Navy in 1989. A passive sonar analysis principles trainer has also been supplied to the Royal Navy (see separate entry - SAINT).

In June 1991 two sonar trainers were delivered to the Royal Navy. Marconi-Thomson delivered a containerised Type 2050 search and attack sonar trainer to HMS *Dryad* at Fareham, Hampshire, while Marconi delivered a stand-alone Type 2054 sonar man/machine interface trainer to HMS *Neptune* at Faslane, Scotland as part of the 'Vanguard' class submarine command team trainer (AS 1104) with eight student consoles and one instructor console.

Manufacturer

Marconi International Simulation and Training, Cheadle Heath.
(Ex-Ferranti, for details see Directory of Suppliers)

Marconi Surveillance and IFF Radar Simulation (SIRAS)/ Navigation and RRB Simulator (NARSIM)

In 1988 it was announced that Marconi had won two contracts to supply the Ministry of Defence with radar simulators to be incorporated into a Type 23 frigate shore development facility at the Defence Research Agency establishment, Portsdown, Portsmouth, Hampshire. The Surveillance and IFF Radar Simulator (SIRAS) generates a supply of representative targets as real-time outputs that are used to simulate the signal processing system of a Type 23 frigate's primary radar system along with its associated identification (IFF) system. Control of these outputs plus other relevant information is interfaced directly with the real radar equipment. SIRAS also generates video outputs to the coder/decoder of the IFF system as well as outputs for scenarios produced by the shore development facility scenario generator, although it is also capable of stand-alone operation.

The SIRAS system operates on a Digital Equipment Corporation (DEC) MicroVAX II computer. In addition to providing real-time simulation, the computer interfaces to two synchronised radar video generators. Two generators are required in SIRAS due to the complexity of the Type 996 radar. This enables the system to simulate accurately each of the radar beams and also add other effects such as IFF, weather, mutual radar interference and Electronic Warfare (EW) returns.

The other system is the Navigation and RRB Simulator (NARSIM) which simulates radar transmit and receive functions for the Type 23's Type 1007 navigation radar system. The shore development facility navigation radar equipment may be provided with input from NARSIM or from an actual navigation radar.

NARSIM produces video for the Type 1007 radar's data distribution unit and reacts to control signals from it as well as passing video to the outfit tack extractor. It generates video outputs to the RRB equipment when scenario platform transponders are triggered by the radar.

NARSIM uses the same computer as SIRAS but interfaced with one radar video generator. It has similar features to SIRAS.

Status

Both systems are in service at the Portsdown shore development facility for which Marconi received a contract to produce a data link simulator.

Manufacturer

Marconi International Simulation and Training, Cheadle Heath.
(Ex-Ferranti, for details see Directory of Suppliers)

Marconi SAINT (Sonar Analysis Initial Trainer)

The role of the Sonar Analysis Initial Trainer (SAINT) is to train sonar operators in the principles of underwater noise generation, signal display presentation and narrowband analysis from basic to advanced level. The system consists of a dual position instructor's console linked to as many as 12 student positions each for two students. As a result of this organisation each instructor can conduct independent exercises with six positions.

Each instructor position has two video monitors: one for monitoring the trainer control menu and the other to repeat any picture seen by students. There is also a joystick-controlled pointer to identify important features on any video seen, a digitiser pad and stylus for rapid repositioning of frequency lines in a lofargram or free-hand graphics 'sketching' onto video pictures. There are video record facilities and communications with each student location.

Each student position has a table top-mounted paper chart recorder with numeric display to identify the frequency window covered by the lofargram being recorded. The positions also have a video monitor and cursor generation electronics. Cursors can be superimposed upon a picture relayed by the instructor to permit the student to identify both frequencies and harmonic relationships. Numeric displays can also be incorporated.

Status

In service with the Royal Navy.

The Marconi Sonar Analysis Initial Trainer in operational mode

Manufacturer

Marconi International Simulation and Training, Cheadle Heath.
(Ex-Ferranti, for details see Directory of Suppliers)

Marconi Radar Operator Trainer

This is a versatile, multipurpose trainer which can be fitted into a building or a transportable container. It is designed to develop a wide range of skills including training operators to recognise all types of radar response, to plot and report targets, to use the controls to produce the best radar picture, to track targets in an electronic warfare environment, to monitor and control auto-tracking. The system can also be used for training navigation and blind pilotage as well as familiarisation with IFF responses. It consists of a Marconi Radar Simulator, an instructor's console and up to 10 trainee display consoles.

The instructor's console has a multi-line solid-state display and conventional keyboard with the addition of 20 rapid function keys. The display provides three separate areas for 'own ship' tote, track data and menu pages. The Radar Simulator includes a microprocessor-controlled radar video generator module which can also be installed in a host system. It provides realistic simulation of all radar effects and has a capacity of 128 tracks while the scenario generator has a 40 track capacity covering a 2048 square mile (5304 km^2) area.

Marconi radar simulators

The standard student display is a Racal Decca relative motion marine unit with pedestal or bulkhead mounting. It has a 256 mm diameter tube, electronic bearing and range markers, nine range scales and the usual display controls. However, alternative displays can be supplied to meet customer preference.

Status

A system with eight displays was supplied to the Royal Navy's School of Maritime Operations (HMS *Dryad*) in 1971. Marconi also supplies onboard trainers with radar video generators and scenario generators to customers including Brazil which uses them in the 'Niteroi' class frigates with hand-held control terminals.

Manufacturer

Marconi International Simulation and Training, Cheadle Heath.
(Ex-Ferranti, for details see Directory of Suppliers)

Marconi Onboard Sonar Trainer

Marconi (ex Ferranti International Simulation and Training) has an extensive technology base which has been applied to a wide range of onboard sonar training applications. To date, systems have been implemented either by simulation at the array interface unit or as injections into the digital data stream after the signal processing stages. The system is suited to all active and passive sonar types and provides comprehensive scenario generation and storage capabilities.

The instructor control station is an intelligent graphics terminal from which exercise scenarios are constructed and run. Scenarios consist either of synthetic target overlays on a real world background or they are completely simulated. Features include a choice of ocean models incorporating a range of thermal and acoustic conditions and realistic broadband and narrowband targets which are compatible with the MoD/DGUW database.

A Towed Array On-Board Trainer has been developed with a single interface box controlled by a laptop computer.

Status

Marconi has supplied onboard trainers for the FMS 15 sonar to the Royal New Zealand Navy and for the Type 2050 sonar which is being installed in Type 42 Batch 2 destroyers together with Type 22 and Type 23 frigates. In July 1993 Marconi, under subcontract to Marconi-Thomson Sonar Systems UK Limited, won a contract to supply simulation modules for the Towed Array On-Board Trainer.

Manufacturer

Marconi International Simulation and Training, Cheadle Heath.
(Ex-Ferranti, for details see Directory of Suppliers)

Marconi (Solartron) Radar Simulators

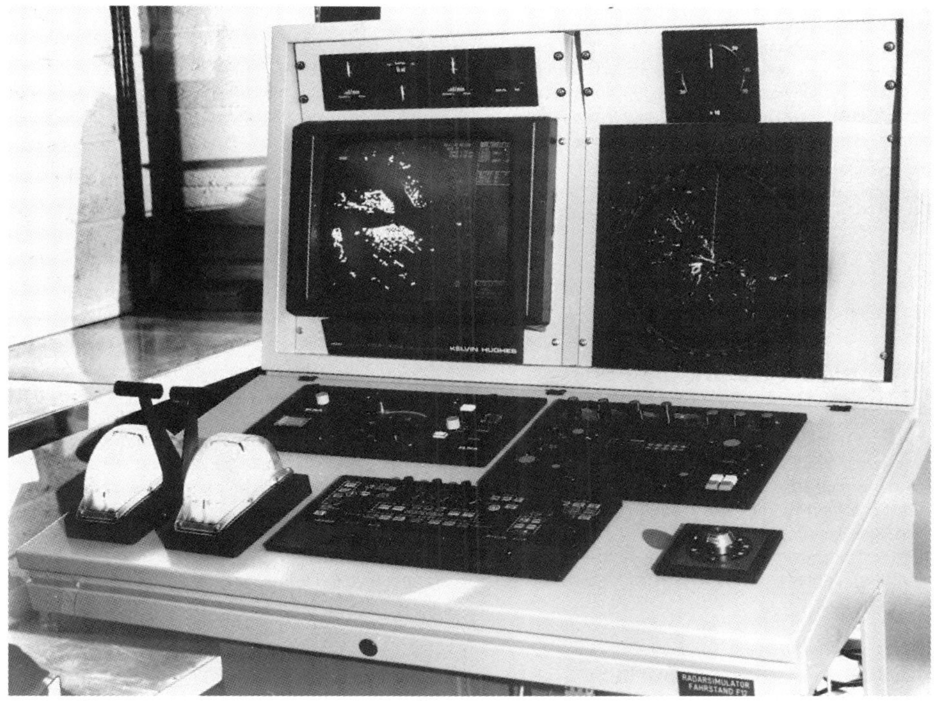

A Marconi (Solartron) radar simulator in service in Germany

During the 1960s, 1970s and 1980s Solartron produced a large number of radar and navigation aids simulators. This activity was subsequently acquired by Ferranti who were later acquired by GEC and renamed Marconi International Simulation and Training.

Typical of this range of products were the SY 2086 and SY 2094. The SY 2086 was a fully programmable simulator based upon the company's SY 2098 digital coastline generator. The latter provides highly realistic, drift-free representations of land masses and radar reflective objects for up to four 'own ships'. It generates a minimum of 24 targets, but any number could be specified, while the database could store information for about 8045 km of coastline.

The student stations represented a ship control unit with switches for 'helm' and 'engine telegraph' together with meters indicating speed, engine revolutions per minute, rudder angle and gyrocompass reading. Digital readouts of distance run and of exercise time were also provided.

The instructor's console featured the IKAT (Interactive Keyboard And Terminal) in which the instructor could monitor student progress or intervene through a menu-driven process. The progress of all ships and targets was recorded on an X-Y plotter while the entire exercise was recorded upon a floppy disk for later evaluation and debriefing.

The SY 2094 was similar but featured a more sophisticated instructor's console and 'own ship' cubicle. This included a colour graphics display which provided a detailed labelled plan of the exercise area with a roller ball for easing the positioning of vessels. A second VDU provided tote information on the location, speed and heading for all vessels in the exercise, together with details of all radar effects and weather condition set by the instructor. An optional extra was a display to show the student's radar picture.

Both systems used a variety of Solartron navigation aid simulators including units representing the Decca Navigator, satellite navigation, Omega, and LORAN-C. Radio direction finding and echo-sounding simulators were also available.

An SY 2094 was sold to the Royal Navy for use in HMS *Mercury* while radar training systems of this generation were sold to Australia, Canada, Germany, Indonesia, New Zealand, and Sweden (two). A helicopter direction trainer was also sold to the Netherlands. Digital coastline generators were incorporated into simulators for the Royal Navy and Brazil. A new development system, with enhanced capabilities, is expected to be launched shortly.

The company also developed a new onboard radar trainer, the SY 2081. This is a one 'own ship', two target system in which the ship speed and heading are set up on the trainer's panel. The exercise portrays the 'own ship' centrally deployed while one or both of the targets manoeuvre within an 80 nm (150 km) square exercise area. The target speed, course, rate of turn and range of first detection are individually controlled on the front panel. These trainers are used by New Zealand and the United Kingdom. (See Marconi Onboard Radar Trainer entry for latest systems supplied to the Royal Navy).

Status

Land-based radar trainers have been delivered to the Royal Navy as well as to Australia, Brunei, Canada, Germany, India, Indonesia, Malaysia, Morocco, the Netherlands, New Zealand, Norway, Sweden, and other countries.

Manufacturer

Marconi International Simulation and Training, Cheadle Heath.
(Ex-Ferranti, for details see Directory of Suppliers)

Marconi Submarine Periscope Simulator

The Marconi Simulation Periscope Simulator is based upon the company's low-cost, high performance graphics system and was designed to provide cost-effective training in the use of the submarine periscope during either normal or emergency operating conditions. It is designed to provide both initial and continuation training.

To ensure that the quality of the training provided by the periscope simulator meets the user's requirements, Marconi Simulation has ensured that the fidelity of its simulator encompasses not only accurate modelling of the tactical environment and equipment functionality but also the sensory interface and realism of the visual scene.

The accuracy of the physical periscope replication will depend upon the level and sophistication of the user's training requirement. The periscope replication will, therefore, range from a generic periscope station, with the visual scenes presented on a colour monitor, to a faithful replication of a specific periscope incorporating full optics to provide greater realism for the operator.

Multiple independent views from multiple periscopes and remote monitoring systems can be provided depending upon the customer's requirements. The periscope simulator may be used as an independent training device or as a fully integrated element of a command team trainer. The system simulates underwater, infra-red and low-light television views and drain-down effects.

Status

A simulator is in service with the Royal Netherlands Navy while visual software has been supplied to the Royal Canadian Navy. A full simulator has been ordered by the Royal Navy for integration with the AS1082 Attack Teacher in HMS *Dolphin*, the Royal Navy Submarine School at Gosport, Hampshire.

Manufacturer

Marconi Simulation, Dunfermline.
(For details see Directory of Suppliers)

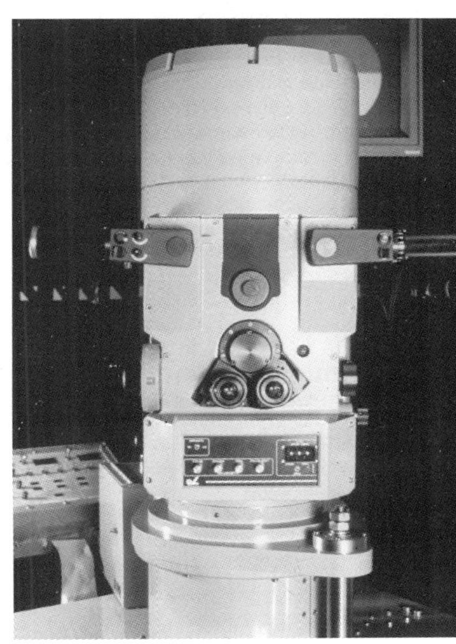

Marconi Simulation Periscope Simulator

Marconi Generic Sonar Trainer (GST)

To satisfy the continuing requirement for acoustic training, Marconi Simulation developed the Marconi Generic Sonar Trainer (GST). This system, an augmentation of the Marconi MANDARIN computer-based training system with additional hardware and software, is an extremely powerful real-time training tool.

The simulator is used for teaching sonar operators a range of skills associated with underwater acoustic analysis, providing training in such subjects as source characteristics, oceanographics, sensor characteristics and visual signal presentations such as lofargrams. By using Marconi Simulation's Programmable Audio Synthesiser (PAS) in conjunction with GST, realistic audio, synchronised with simulated passive or active sonar displays can be produced.

GST can be used for teaching the basic principles of sound generation, propagation, transmission loss, and sonar processing systems, and can provide sonar-related training requirements from the time a student enters service to his first encounter with the prime equipment.

Status

Available.

Manufacturer

Marconi Simulation, Dunfermline.
(For details see Directory of Suppliers)

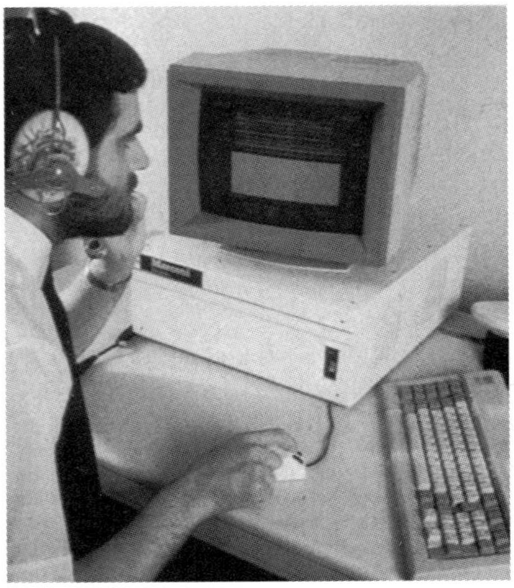

The Marconi Generic Sonar Trainer (GST) in use for lofargram training

---

Marconi Mandarin Solo For Windows Desk Top Sonar Trainer

In order to satisfy the requirement for sonar operators to classify acoustic signatures accurately and rapidly, Marconi Simulation has designed a flexible and comprehensive training tool, the Mandarin Solo For Windows Desk Top Sonar Trainer.

The trainer provides an interactive, high-fidelity training system for presentation of realistic contact displays in configurable training exercises. The system allows the student to analyse and classify real-time generated displays in a time-critical environment using standard generic facilities based upon real sonar equipment features. The system can also be provided with an audio facility and is adaptable to both active and passive sonar training.

Status

Mandarin Solo For Windows is in service with the Royal Navy and the US Navy. The 'Mandarin for Windows' software is being supplied to the US Navy as part of the maintenance trainer for AN/SQQ-89, and Mandarin will also be supplied to IBM for the US Air Force ATS programme.

Contractor

Marconi Simulation, Dunfermline.
(For details see Directory of Suppliers)

Marconi Simulation Sonar Simulation

Marconi Simulation supplies specialist equipment for the test and evaluation of passive, active and intercept sonar processing equipment. The company also provides specialist ship data system simulators providing signal sources for use in shore development facilities. The mathematical modelling expertise gained by Marconi Simulation, as a result of their association with the development of major sensor simulation for the Royal Navy, combined with extensive private venture work, has provided the company with a comprehensive and up-to-date capability for the design of sonar simulation equipment.

Sonar simulation modules are available for audio, active, intercept, passive narrowband and broadband scenarios. These modules have extensive software parametric control capabilities and may be configured for sonobuoy, towed array, and flank or bow array applications.

In addition, Marconi Simulation has supplied the Royal Navy with compact sonar simulation systems for a number of acoustic test requirements. The Sonar 2054(I) is an example of factory/setting-to-work equipment which is compact and portable for use on board ship or submarine. Dynamic high frequency signals can be supplied simultaneously to all hydrophones to allow comprehensive testing of the prime sonar equipment. In addition, as most of the parameters are under software control, the device also has an onboard training capability.

Larger versions of the equipment are available for maintainer training and shore-based development applications.

Array Signal Simulator for Sonar 2054(I)

Status

In service with the Royal Navy.

Manufacturer

Marconi Simulation, Dunfermline.
(For details see Directory of Suppliers)

Marconi Simulation Simulators for Shore Development/Integration Facilities

Marconi Simulation has been involved in the specification, design and supply of large simulation systems for testing and evaluation at shore development facilities. These systems have been designed for stress testing and proving command and weapon systems, including their data link highways.

A major contract awarded to Marconi Simulation is for the proving of the onboard combat system for 'Invincible' class aircraft carriers and the Type 42 destroyer. The company is supplying radar simulators for the Type 909, 992, 1022, 1007 and 1010/1011 radars, as well as simulators for the Type 2050 and 2016 sonars.

For the Trident shore development facility the company supplied sonar simulators to emulate the function of the acoustic transducers by direct simulation of the sonar receiver input circuitry. The generation of these signals is governed by heading/orientation of the sonar arrays with respect to the acoustic targets responding to preprogrammed or manual position control inputs.

Status

In service with the Royal Navy.

Manufacturer

Marconi Simulation, Dunfermline.
(For details see Directory of Suppliers)

THORN EMI AS 1092 Shipborne Radar and Sonar Trainer

In the 1970s THORN EMI marketed a shipborne sonar trainer, the AS 1077. An improved version of this product is the AS 1092 radar and sonar trainer. This is the latest in a series of multi-sensor onboard trainers produced by THORN EMI and designed to train command team sensor operators within their operational environment using their normal equipment. The systems are modular and compact to suit the limited space available.

The AS 1092 itself consists of a main control console, a sonar simulation cabinet and an Initial Detection and Classification Trainer (IDCT), a second control unit, weapons system and navigation radar interface units and an IDCT remote-control unit, a second control position and a hand-held helmsman's control unit. The system uses seven Motorola 68000 16-bit microprocessors and five Motorola 68020 8-bit processors. To be compatible with Marconi sonars a Z80 processor may be added to the systems.

Synthetic targets can be inserted onto screens receiving real data or an exercise may be carried out using simulated targets and simulated background with the latter covering an area of 2048 × 2048 nm (3788 × 3788 km) to heights of 80 000 ft (24 384 m) and to depths of 20 000 ft (6096 m). A ship movement simulator is included with high integrity change-over devices for switching between real and simulated data. Up to 70 vehicles (surface ships, submarines, torpedoes, aircraft) may be simulated simultaneously from a library with a capacity for 254 classes. The system can be used to provide procedures training or as part of command team training with simulated weapons firing.

The THORN EMI AS 1092 shipborne radar and sonar trainer

Status

The AS 1092 is in service with the Royal Navy in air defence and anti-submarine frigates.

Manufacturer

THORN EMI Electronics Limited, Feltham.
(For details see Directory of Suppliers)

Anti-air Warfare Onboard Radar Simulator

In February 1992 the Ministry of Defence issued a requirement for an anti-air warfare onboard radar simulator. This was to be a real-time dynamic scenario generator which would be able to provide simulated air and surface target responses from any two of a number of specified primary radars and associated secondary radars according to the class of ship simulated.

The simulation was to provide simulated radar video to the ship's operational displays and simulated tracks to radar track extraction equipments. It was also to provide environmental effects including land mass, precipitation, sea clutter and jamming.

Status

The requirement was for 15, 20 or 25 equipments for ships and shore establishments.

Towed Array Onboard Operator Trainers

In September 1992 the Ministry of Defence invited tenders for a stand-alone, onboard training facility for towed array sonar systems to maintain operator and limited ASW command team skills. The system is to be installed in both surface ships and submarines and must interface with both towed arrays and sonar processing equipment.

Status

The requirement is for about 60 systems in a two or three year period and the tenders were due in December 1992.

UNITED STATES OF AMERICA

AAI Generic Radar Display System (GRDS)

The AAI Generic Radar Display System (GRDS), also known as Device 20F15, is a microprocessor-based universal peripheral device which generates radar video simulating surface and air search radars. Communicating with the host computer through a standard Ethernet local area network, the GRDS generates radar video reflecting the tactical environment including targets, jamming, weather, clutter and IFF.

The design of the GRDS incorporates the inherent flexibility to simulate any type of search radar, since the pertinent parameters describing the features of the radar to be simulated are outputted to the GRDS from the host computer. This capability is of primary importance in providing cost-effective training.

Each GRDS video generator has the inherent capability to provide simultaneous simulation of two independent radars. The video generator provides simulated radar video, synchro antenna position and timing signals to independently driven PPI displays for two collocated radars. For team training, several GRDS systems can be employed on the same local area network to provide for joint team exercises involving a variety of ships and fleet manoeuvres.

AAI has derived a new training device from this system, known as the Digital Radar Land Mass System. This is a microprocessor-based peripheral device which can generate radar video and radars. It can also operate directly from a digitised topographical map.

A student studying with a Device 15F12

Status

The company has supplied 24 GRDS video generator systems each to the Fleet Combat Training Centers, Pacific (San Diego, California) and Atlantic (Virginia Beach, Virginia). The last units were delivered in November 1988.

Manufacturer

AAI Corporation, Maryland.
(For details see Directory of Suppliers)

AAI Device 15F12 Radar Navigation Trainer

Device 15F12 was developed by AAI as a digital land mass and target simulator for surface search radar sets. It is designed to be used for training and qualifying combat information centre teams and bridge personnel in the techniques of shipboard radar navigation and collision avoidance.

The system provides high-fidelity simulated radar video for the returns of land mass, targets, and rain squalls co-ordinated with depth indicator readings and a dead reckoning plot. These features are modified and move as the 'ship' moves on the map. All targets and rain squalls move independently.

The 'ship' helm orders, physical characteristics and response parameters are simulated during training exercises as are target characteristics. The land mass region simulated by the system is 64×64 nm (118.5×118.5 km). In addition to the 'ship', another 12 surface ship targets can be manoeuvred within a 164 nm^2 (563 km^2) area centred on the land mass region within 15 m of any map point.

AAI has recently developed a second generation Device 15F12 which has resulted in a significant reduction in size and weight as well as significant improvements and maintainability. This second generation design makes extensive use of commercial off-the-shelf VME computer components and AAI software housed in the instructor control console. One unit has been delivered to the US Navy Training Center in Dam Neck, Virginia, while a second is scheduled for installation in the US Navy Training Center in San Diego, California, in September 1994.

Status

Eight Device 15F12 systems have been delivered to various US Navy training centres. AAI is developing the next generation of shipboard radar navigation trainers which will use software-intensive simulation techniques and off-the-shelf processors.

Manufacturer

AAI Corporation, Maryland.
(For details see Directory of Suppliers)

AAI Naval Electronic Warfare Training System (NEWTS) (Device 10H1)

NEWTS (Naval Electronic Warfare Training System), officially designated Device 10H1, is a rapid, cost-effective means of training large numbers of students in electronic warfare.

The system consists of eight generic equipment suites which include superheterodyne, instantaneous frequency measurement, radar homing and warning, direction finding receivers, deception and noise jammers, chaff and automatic PRF measurement and receiver/jammer systems. There are 60 student stations which feature self-paced training through computer-aided instruction. The students learn in a dense electronic warfare environment with up to 128 emitters per problem, 32 platforms operating at the same time and with up to 64 simultaneous signals. The simulation is of high-fidelity and the threats react to student actions.

Computer-assisted instruction is a major feature of the system, prompting the student through displayed briefing pages, evaluating his performance and asking questions through an instructional cathode-ray tube display. This also assures a high student-to-instructor ratio.

Status

A 60-station system is in service at the Naval Technical Training Center, Pensacola, Florida.

Manufacturer

AAI Corporation, Maryland.
(For details see Directory of Suppliers)

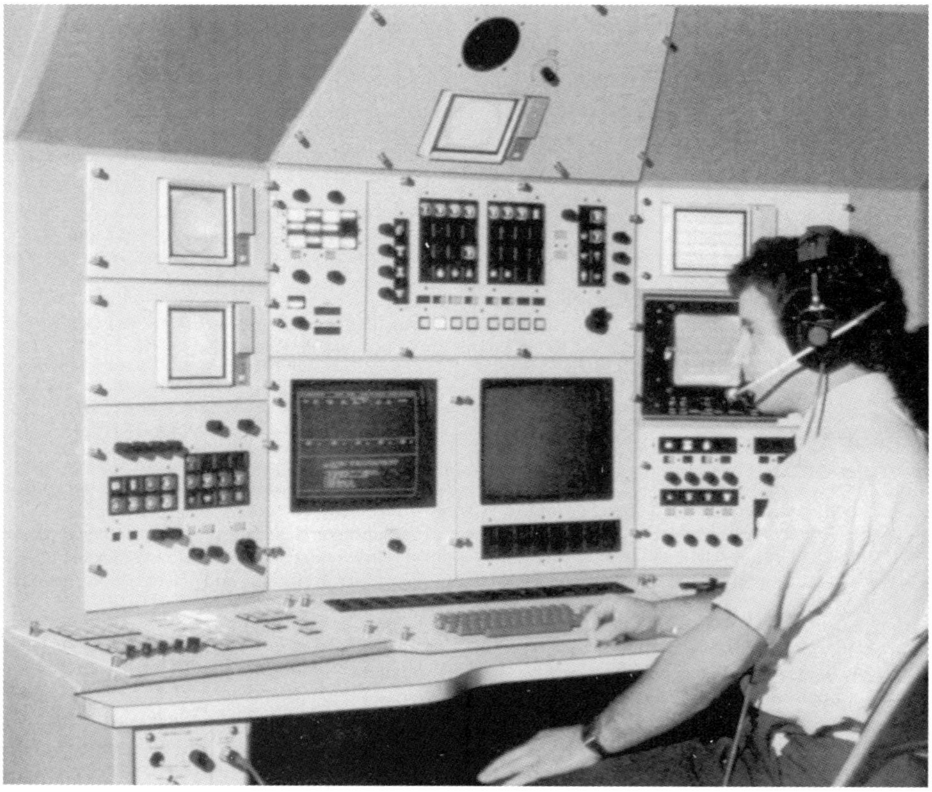

A Naval Electronic Warfare Training System (NEWTS) student's station

CAE-Link Acoustic Operator Trainer (Device 14E35)

CAE-Link has produced five acoustic operator trainers in support of the sensor system in the AN/SQQ-89 underwater combat system. These training devices have four components: the AN/SQQ-28 LAMPS Mark II sonobuoy signal processor, the AN/SQR-19 TACtical Towed Array System (TACTAS), the AN/SQS-53B or -53C hull-mounted sonar and the AN/UYQ-25 Sonar In-situ Mode Assessment System (SIMAS). The trainers allow for training in various crew configurations and individual operator training for the three sonars. The AN/SQS-53B trainer has four student stations while the AN/SQS-53C trainer has five. The trainer also provides reduced capability for training in the casualty modes of the operational AN/SQQ-89 and interface with the Mark 116 Underwater Fire-Control System.

The trainer simulates the performance of the operational systems and can also provide a variety of exercises from simple target location to complex tactical operations. Each device consists of an instructor station, four student stations, the computer system (based upon Harris Series 800 and 1200 mainframe computers with Fortran language software) and an acoustic signal generator. The acoustic signal generator is a special purpose digital signal generator using data pipelining and parallel processing techniques to produce the target composite digital time waveform which can generate up to 20 dynamic air, surface and subsurface targets to depths of 15 000 ft (4572 m) and to altitudes of 50 000 ft (15 240 m). It can also generate the signatures of diesel, diesel-electric, steam, turbine, nuclear and auxiliary machinery. Each device is configurable to

The Device 14E35 acoustic operator trainer

different ship classes such as the 'Ticonderoga' (CG 47) class cruiser, 'Spruance' (DD 963) and 'Kidd' (DDG 993) class destroyers, or 'Oliver Hazard Perry' (FFG 7) class frigates.

Status

CAE-Link received the first contract for these systems in 1982 configured to the 'Spruance' and 'Oliver Hazard Perry' classes. The first two units entered service with the Fleet Anti-Submarine Training Center at San Diego, California, late in 1987 and two more followed by 1990. A fifth was delivered to San Diego in February 1993. Two of the San Diego systems are designated 14E35C and are used to train operators of the AN/SQS-53C. Five systems have been supplied to San Diego in a programme which will cost about $107 million.

Manufacturer

CAE-Link Corporation, Binghamton, New York State.
(For details see Directory of Suppliers)

Comptek AN/SLQ-32(V) Operator Training Device (OTD)

The Operator Training Device realistically simulates the AN/SLQ-32(V) Countermeasures Set with a facsimile of the SLQ-32 Display Control Console (DCC). The OTD uses a combination of general-purpose and specific-application commercial hardware, together with Comptek emulation software, to simulate accurately all the SLQ-32 operator functions, displays and indicators together with audio output for signal-selected emitters at a fraction of the cost of using the in-service equipment.

The system may be run in a stand-alone configuration or it may be integrated with a Combat Direction System using either operator training software or fleet operational software. The operator training software may be used for predefined scenarios or it may be supplemented in this role by the Comptek AN/SLQ-32(V) Environment Generator (see next entry). The OTD is used to train operators, as a development test-bed for operational software and for use in verifying threat libraries.

Status

The OTD is in production and more than 40 units are in service with the US Navy and allies.

Manufacturer

Comptek Federal Systems Inc, West Seneca, New York State.
(For details see Directory of Suppliers)

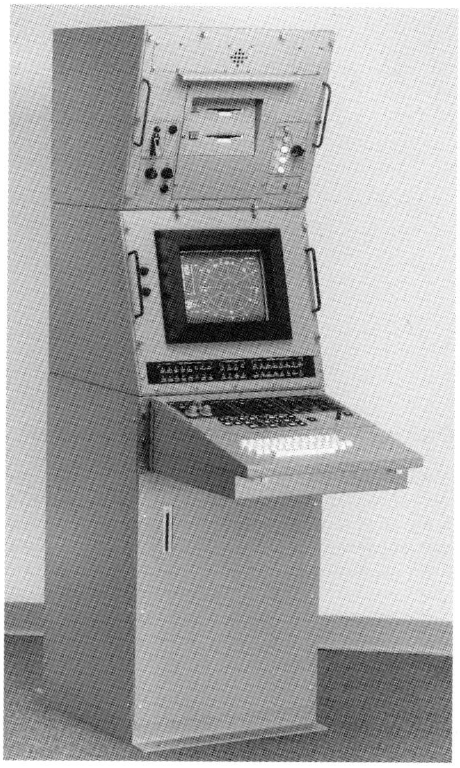

The AN/SLQ-32(V) Operator Training Device (OTD)

Comptek AN/SLQ-32(V) Environment Generator

Comptek's AN/SLQ-32(V) digital environment generator emulates the complete set of SLQ-32 ESM components for training, for threat library verification, for software development or for performance assessment. It may be linked either to an operational SLQ-32 system or to the AN/SLQ-32 (V) Operator Training Device (see previous entry). It can simulate a wide range of realistic radar environments similar to those which may be encountered in both friendly and hostile situations.

Through the use of scenarios developed and controlled using either Comptek's software or the user's specific application software, the environment generator can present a radar environment as diverse and as complex as any electronic warfare operator will encounter. The simulated RF environment may consist of up to 200 emitters configured on up to 200 dynamic air, surface and submarine platforms which can carry and launch missiles or be landed. The exercise 'area' can be 3000 nm (5500 km) in diameter and up to 100 000 ft (3050 m) in altitude with a scenario which is either scripted or controlled dynamically by operators.

The environment generator is available with options for data extraction and video generation. The former provides for the collection of operator actions, emitter data and other information using the data extraction points defined in the SLQ-32 operational software. The latter provides accurate video and audio representation of the emitter's signal selected by the SLQ-32. These can then be used for operator recognition training or by collateral pulse analysis equipment.

Status

The environment generator is in production and is in service with the US Navy and allied navies.

Manufacturer

Comptek Federal Systems Inc, West Seneca, New York State.
(For details see Directory of Suppliers)

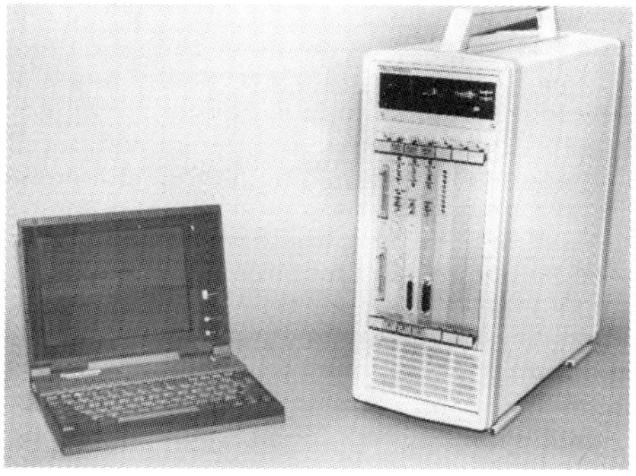

The AN/SLQ-32(V) Environment Generator

AN/SLQ-32(V) Software Support and Training System (SSTS)

The Comptek AN/SLQ-32(V) Software Support and Training System (SSTS) is a multipurpose system which may be configured to the customer's need for use in SLQ-32 training, software development, performance evaluation and threat library verification. It consists of one or more AN/SLQ-32 Operator Training Devices (OTD), AN/SLQ-32(V) environment generators and a host computer with peripheral devices.

The host computer, which is selected by the customer, provides facilities for software development and compilation, configuration management and scenario generation. The OTD provides a realistic platform for execution of the software while the environment generator defines the scenarios and/or real-time injected events provided by the host computer. Software used for SSTS applications may consist of commercial software products, Comptek proprietary software and/or customer or government furnished software.

Status

The SSTS is in service in the United States and has been exported.

Manufacturer

Comptek Federal Systems Inc, West Seneca, New York State.
(For details see Directory of Suppliers)

Computer Sciences Q-21 Emulator

The US Navy has introduced a new family of radar/sonar display systems, the Hughes AN/UYQ-21, and CSC has developed the Q-21 to assist shore-based training at reduced cost. Four versions are available to provide students with experience in Naval Tactical Data System (NTDS) consoles and each reproduces the performance of the respective console. The versions are the OJ-451(V)9 tactical data console, the OJ-452(V)10 acoustic console, and OJ-535 graphics terminal of the UYQ-21 display family, together with the OJ-194 sensor display console of the older AN/UYA-4 family.

The system comes as a basic console to which an auxiliary display is added in the OJ-451 configuration while a second display unit is added for the OJ-452. Although the Q-21 reproduces the performance of the real consoles it consumes less power for it uses the US standard 110 V 60 Hz power system and forced air-cooling.

Status

Some 100 systems have been ordered and by the Spring of 1990 some 60 had been delivered. These are used at the Aegis Educational Center in Dahlgren, Virginia and at training establishments in Orlando and Wallops Island.

Manufacturer

Computer Sciences Corporation, New Jersey.
(For details see Directory of Suppliers)

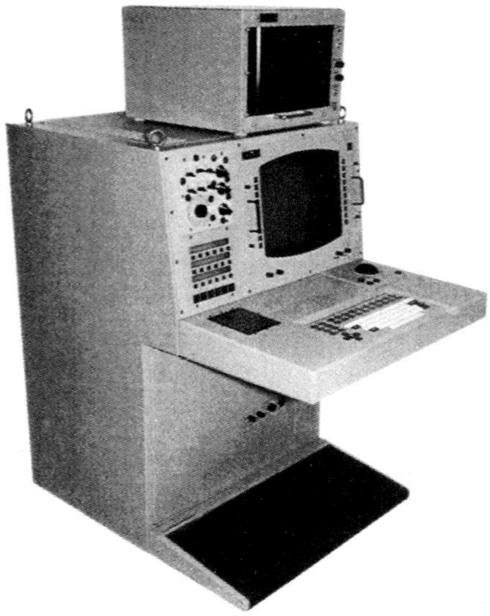

A Q-21 emulator in the OJ-451 configuration

Delex Decoy Deployment Trainer

Delex has developed a decoy deployment training system based upon a single-station personal computer. It is designed to provide decoy operators and tactical warfare officers with training in variables which affect chaff engagements, including wind direction and speed, missile type and range, ship size and speed.

The hardware and software package provides a 40-hour course at low cost. In the tutorial mode it provides a series of interactive lessons which enable the students to familiarise themselves with decoys and how they affect both radar and Infra-Red (IR) seekers in missiles. The mode also provides familiarisation with chaff and IR rounds as well as launcher systems.

In the procedures mode the student is taught launch panel operation through touch-screen menus which provide tutorial, procedure and practice modes. Finally there is an engagement module which provides the student with a full-scale scenario against 'hostile anti-ship missile attacks'. In the modules platforms manoeuvre, attack with missiles and operate in an environment prepared by the instructor who defines the number of missiles, their speed and seeker parameters, the decoy loadout and wind vector.

Status

In service with a Middle East navy.

Manufacturer

Delex Systems Inc, Virginia.
(For details see Directory of Suppliers)

Diagnostic Retrieval Systems AN/SQQ-T1 Sonar Training Set

The SQQ-T1 is a shipborne sonar training set designed to improve the efficiency of operators of the AN/SQR-17A sonar signal processing system, AN/SQR-18A sonar and the AN/SQS-26 sonar sets. It is designed to provide training in the operation of both active and passive sonars.

The system simulates realistic, high-fidelity acoustic signals through an acoustic simulator unit. It can be operated on an individual basis using up to 64 preprogrammed scenarios or it can be used on an interactive basis with an instructor who has his own control unit. The gaming area can be as much as 512 × 512 nm and include up to four acoustic targets from a library of 40 target types.

The instructor can select 16 preprogrammed ocean areas with various sea state and shipping density levels. Data from up to four simulated sonobuoys may also be reproduced in the system. The instructor control unit and acoustic simulator unit provide a core stimulation capability which may be adapted to various sonar systems.

Status

Previously installed in US Navy 'Knox' (FF 1052) class frigates. Many of these vessels are being sold to foreign navies.

Manufacturer

Diagnostic Retrieval Systems Inc, New Jersey.
(For details see Directory of Suppliers)

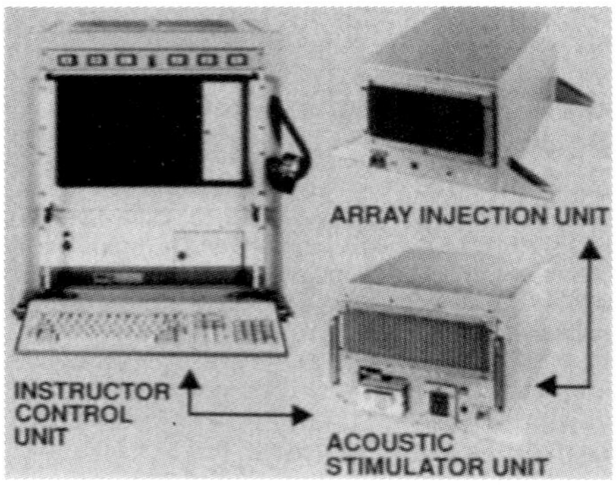

AN/SQQ-T1 sonar training set

Diagnostic Retrieval Systems AN/SQQ-T1A Sonar Training Set

The AN/SQQ-T1A is an upgrade to the AN/SQQ-T1 Sonar Training Set and adds radar simulation capability. It is designed for use with sonar signal processors and radars operating in the shallow water and littoral water environments. The system consists of an instructor console, an acoustic stimulator and a radar stimulator and it is used to optimise the efficiency of AN/SQR-17A(V)3 and AN/SPS-64 operators and can be reconfigured easily for a wide variety of sonar signal processors and radars.

The AN/SQQ-T1A is a portable training system with an instructor console hosted in a laptop computer. The system simulates realistic, high-fidelity acoustic signals through the acoustic stimulator unit. It can be operated in either an interactive mode or the instructor can select up to 128 (expandable) preprogrammed scenarios. He can also select up to four acoustic targets from a library of 128 (expandable) and 30 air targets. Correlation is provided between radar and sonar contacts. The system allows the selection of up to 36 ocean environments with various sea state and shipping levels. Instructors may also select up to four directional noise sources. Some 12 sonobuoys are available in any training scenario of which four may be active at any one time.

Status

These trainers will be provided to the US Navy in mobile vans as part of its Mobile In-shore Undersea Warfare (MIUW) programme. Five new systems are being provided to the US Navy and three systems in 'Knox' class frigates were upgraded.

Manufacturer

Diagnostic Retrieval Systems Inc, New Jersey.
(For details see Directory of Suppliers)

EMS Basic Sonar Operator/ Diagnostic Trainer (Device 14G1)

The sonar operator and diagnostic trainer known as Device 14G1 was designed to provide student sonar operators with familiarisation training with the AN/SQQ-89 (V) sonar and to provide classroom training for recruits.

It consists of four student stations each of which simulates the appearance and operations of the AN/UYK-21 console associated with the SQQ-89 sonar. The students receive from the instructor's station 6 hours of fully automated, self-paced, computer-controlled instruction. The instructor selects either individual exercises or a common training exercise for all the students. Each includes recorded narrations and visual presentations of the selected exercise.

The instructor station has a control box, a display screen for monitoring the students, an alphanumeric keyboard and a matrix printer.

Status

In service with the US Navy.

Manufacturer

EMS Development Corporation, Farmingdale, New York.
(For details see Directory of Suppliers)

A Basic Sonar Operator/Diagnostic Trainer (Device 14G1)

Hughes Combat Crew Trainer (CCT)

Hughes developed a generic signal generator, the Combat Crew Trainer (CCT), which simulates the front end of a naval radar. The system may be used to simulate combat scenarios involving up to 100 targets against a synthetic background of land masses and various meteorological and electronic warfare conditions. The CCT will interface with two- and three-dimensional radars, fire-control systems and IFF radars as well as data systems.

Status

The CCT has been used by the US Navy since 1978.

Manufacturer

Hughes Aircraft Company, Manhattan Beach, California.
(For details see Directory of Suppliers)

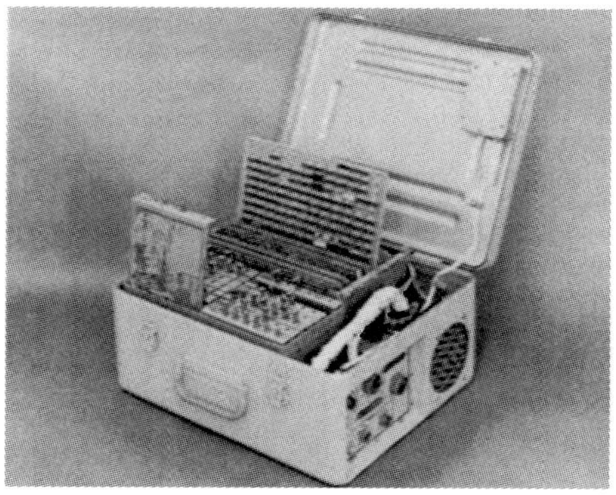

The Hughes Combat Crew Trainer for use with radar systems

Hughes Video Signal Simulator (SM-441)

The VSS is part of the Naval Tactical Data System (NTDS) AN/UYA-4 display system and is an onboard system designed to simulate tactical situations. It can also be used to test and calibrate display equipment.

The system has three modes: synthetically generated video; live radar; and a composite of both. There is manual control of 11 sensors, each of four channels with five radar beamwidths, five video durations and four video intensity levels. The basic system provides simulation only of 2D radars but options are available for a variety of radars including the AN/SPS-30, SPS-39, SPS-48 and SPS-52.

Status

In service with the US Navy on some 150 NTDS-equipped warships.

Manufacturer

Hughes Aircraft Company, Fullerton, California.
(For details see Directory of Suppliers)

Hughes Sonar Trainers

Hughes now manufactures a number of sonar operator/team trainers. Each system is dedicated to a sonar set: 14E36 for the AN/SQR-18A Tactical Towed Array Sonar (TACTAS); 14E19 for the AN/SQS-26CX bow-mounted system; 14E23 for the AN/SQS-35 variable depth sonar; 14E24 for the SQQ-23; and 14E25 for the AN/SQS-53 bow-mounted system.

Depending upon the operational system the student stations are either actual sonar consoles or high-fidelity facsimiles. In either case students are presented with sonar audio and video signals. The underwater acoustic environment is faithfully modelled in all operator trainers. Simulated sonar propagation is generated based upon ocean variables such as regional/seasonal velocity profiles, layer depth and sea state as well as bottom type, depth and slope. These simulations use complex algorithms to account for effects of propagation loss, reverberation, surface duct, time delays, convergence zones, multi-path returns and shallow water operation.

The instructor consoles are used to set up problem exercises, to monitor student performance and to modify the tactical situation in real time. The instructor also has the capability of inserting malfunctions and replaying the exercise.

Status

Hughes has some 36 sonar trainers in service with the US Navy.

Manufacturer

Hughes Training Inc, West Covina, California.
(For details see Directory of Suppliers)

Indian Head Division, Naval Surface Warfare Center, Combat System Display Console Emulators

The Indian Head Division of the Navy Surface Warfare Center (IHD/NSWC) produces a number of emulators of combat system display consoles. The AN/UYQ-21 Display Console Emulator is a facsimile of the OJ-451 and OJ-535 display consoles of the UYQ-21 display system which can be used to train crews ashore at up to half the cost of training them afloat on the real equipment.

The emulators display situation information complete with friendly, hostile and unidentified tracks via the Basic Display and Digital Display Units. The operator co-ordinates with students and updates track data, then uses the computer controlled action entry panel to transmit orders to the combat system. All the functions of the UYQ-21 display family are usually emulated except the Selective Identification Feature (SIF) and the Interrogate Friend/Foe (IFF) which are optional features.

IHD/NSWC also produces emulator products for the CV-3551 Central Data Buffer, the RD-358 Magnetic Tape Unit, the CV-3820 and DV-3893 Common Digital television generator. Emulation stimulation hardware and software are produced and the organisation is currently developing an OJ-883 Combat Display Console Emulator.

Status

These emulators are all in service with, or under development for, the US Navy.

Manufacturer

Indian Head Division, Naval Surface Warfare Center, Maryland.

(For details see Directory of Suppliers)

Indian Head Division, Naval Surface Warfare Center, Integrated Training Interface (ITI)

The Indian Head Division produces the Integrated Training Interface (ITI) which supports integrated surface ship and Light Airborne Multi-Purpose System (LAMPS) Mark I anti-submarine warfare command and control training. It links the ASW personnel in the ship at the pierside to the aircrew at a Weapons Systems Trainer (WST). The ITI is transportable.

The system is designed to provide one-way radar target data communication and two-way secure voice communications between the WST and docked 'Spruance' (DD 963) class destroyers or 'Knox' (FF 1052) class frigates by land link using existing telephone lines. The target data is displayed on ship radar consoles through a portable radar stimulator.

It is planned to modernise the ITI so it may interface with additional radar displays as well as with sonar operators, through acoustic processor simulation, for co-ordinated operations training.

Status

The ITI is in service with the US Navy.

Manufacturer

Indian Head Division, Naval Surface Warfare Center, Maryland.

(For details see Directory of Suppliers)

Kollmorgen Submarine Tactical Visual Training System (STVTS)

As the only supplier of periscopes to the US Navy, Kollmorgen also supplies a periscope training system, the Submarine Tactical Visual Training System (STVTS). This is based upon the standard Type 18 periscope and provides a training device for individual operators and submarine command teams.

Status

In production and in service with the US Navy.

Manufacturer

Kollmorgen Corporation, Massachusetts.

(For details see Directory of Suppliers)

Locus Micro-LORES Portable Radar Environment Simulator

The Locus Micro-LORES is a portable radar environment simulator designed to train EW operators and to test EW systems. The system produces a pulse-to-pulse description of the radar signals which would be seen by a receiving platform operating in the presence of live signals. The system simulates both the pulse and antenna behaviour of individual radars and the motion of the platforms which carry them. It is capable of simulating multiple moving platforms each with multiple radars.

The system consists of a digital generator and the Micro-LORES software package which runs on a standard IBM PC/XT/AT or compatible computer. The generator can produce up to 256 emitters in one exercise with up to eight being on each of 32 platforms. The gaming area is 1000 nm × 1000 nm × 100 000 ft and may involve up to 32 moving platforms.

Three optional versions are available in addition to the basic system. Option MRE100 is a direct interface to the AN/SLQ-32 electronic warfare system and provides digital and video stimuli to the SLQ-32 equipment. Option MRM101 provides an interface with the Hewlett-Packard HP-83620 microwave synthesiser while Option MRM200 provides a similar interface to the Watkins-Johnson WJ-45100 synthesiser to create the WJ-45900 simulator.

Status

In production.

Manufacturer

Locus Inc, Pennsylvania.

(For details see Directory of Suppliers)

Loral Hycor Computer-Based Training

This simulator consists of a desktop computer, based upon the Hewlett-Packard Vectra ES/12, with touch-screen and keyboard. The student has a number of software tutorial packages which provide interactive lessons. These allow weapons officers and operators to become familiar with decoy systems and their effects upon missiles. There are also lessons on the decoy launcher system hardware and how they operate.

Once the student is familiar with the system there are engagement analysis software packages which provide scenarios with appropriate responses. The instructor can define the number of launch platforms, the number and type of missile together with its performance parameters as well as countermeasures characteristics.

Status

Available for production.

Manufacturer

Loral Hycor, Woburn, Massachusetts.

(For details see Directory of Suppliers)

A student examining anti-ship missile threats on a Loral Hycor training computer and the automated best response

Norden CE535 Military Display Emulators

The Norden CE535 is a military display emulator compatible with the US Navy's AN/UYA-4 and AN/UYQ-21 family of display systems. They are based upon the OJ-194 (UYA-4) and OJ-535 (UYQ-21) terminals and are designed to provide Naval Tactical Data System (NTDS) data to support emulation of other consoles.

There is a radar video simulator which emulates the Hughes VSS and can provide programmable radar parameters compatible with any US Navy radar. The system can also generate realistic scenarios similar to those which may be encountered during the 1990s.

Status

Available for production. In addition, Norden Marine and Ground Systems has produced acoustic audio generators for the AN/BQQ-5 sonar operator trainer and the AN/BQQ-6 Trident sonar operator trainer as well as the AN/BSY-1 sonar operator trainer.

Manufacturer

Norden Service Company Inc, Maryland.

(For details see Directory of Suppliers)

The Norden OJ-194 display emulator

Raytheon DS1200 Sonar Trainers

The DS1200 family of sonar trainers is based upon common hardware and software modules to produce an expandable, programmable system. They are normally carried on warships and produce high-fidelity, electronically synthesised active and passive contacts which are manoeuvrable, three-dimensional and realistic. They feature 'front-end' target injection and sonar simulation.

The original system, DS1210, was for onboard sonar training with passive systems and was subsequently supplemented by the DS1213 which is for totally integrated sonar suites. The DS1240 is for surface warships offering passive sonar system training with the capability of providing background noise generation when used in dock. The multi-target, shore-based version is designated MAST (see entry under Italy).

Status

Some 40 DS1200 trainers have been produced for the US Navy and other navies since 1970. Of this total, 31 DS1210 are for US Navy ballistic missile submarines of which two were later modified for use in totally integrated sonar suites as DS1213. Nine DS1240s were ordered for the 'Ticonderoga' (CG 47) class cruisers.

In February 1992 Raytheon won a contract to supply hardware for the AN/SSQ-94 combat information centre trainer in Panama City, California. The hardware is part of a contract for AN/SQQ-32 variable depth minehunting sonar systems and the CIC trainer will be for the 'Avenger' class mine-countermeasures vessel.

Manufacturer

Raytheon Company, Submarine Signal Division, Portsmouth, Rhode Island.
(For details see Directory of Suppliers)

Raytheon AN/SQQ-89(V)-T Onboard Sonar Trainer

This simulator is part of the US Navy's AN/SQQ-89 anti-submarine warfare system and is used to generate manoeuvrable, three-dimensional targets. It injects synthetic data into the sonars and interfaces with the AN/SQS-53 active and passive sonars, the AN/SQR-19 towed array sonar, the AN/SQQ-28 and the LAMPS Mark III as well as the Mark 116 Mod 5, 6 and 7 anti-submarine combat system. It is available in four configurations to match those of the SQQ-89 in various ships.

It consists of a trainer control console, a scenario input device, signal generator and processor, signal injector, system status and power control, RF transmitter, RF antenna, LAMPS helicopter interface, and LAMPS helicopter navigation simulator. The simulator is supported by a printer-plotter, a status indicator and maintenance assistance modules.

Status

The development and production contract was awarded in 1985 and the first shipboard installation was made in July 1988. By April 1993 98 had been delivered to the US Navy for 'Belknap' (CG 26), 'Virginia' (CGN 38) and 'Ticonderoga' (CG 47) class cruisers, 'Kidd' (DDG 993) and 'Spruance' (DD 963) class destroyers, and 'Oliver Hazard Perry' (FFG 7) class frigates.

Manufacturer

Raytheon Company, Portsmouth, Rhode Island.
(For details see Directory of Suppliers)

A crewman monitors a control console for the AN/SQQ-89(V)-T onboard trainer

Republic Electronics Radar Environmental Simulator System (RESS)

Republic Electronics produced the RESS (Radar Environmental Simulator System) which was selected for US Navy service as the AN/USQ-93(V). The system was designed to provide onboard training, testing and performance monitoring of the New Threat Upgrade Anti-Air Warfare equipment.

The RESS will simulate accurately in real time all the radar returns which would be encountered in anti-air warfare. These include IFF, decoys, clutter and weather. The system can be controlled by the Tactical Data Systems Training Supervisor or it can come under the preprogrammed control of the ship Weapon Direction System. It is being purchased to interface with AN/SPS-48E and AN/SPS-49(V)5 surveillance radars, AN/SPS-51D and SPS-53B fire-control radars and Mark XII IFF systems.

Status

In production for cruisers and 'Kidd' (DDG 993) class destroyers in the New Threat Upgrade (NTU) and Advanced Combat Direction System (ACDS) programmes.

Manufacturer

Republic Electronics, Melville, New York.
(For details see Directory of Suppliers)

An AN/USQ-93 controller

Rockwell Sonar OnBoard Trainer

Rockwell's sonar OnBoard Trainer (OBT) is a low-cost simulator, using VME-based processors, which simulates sonar hydrophones and array geometry before beam forming. It consists of a control computer, a target signal generator and an inverse beam former. The OBT may be used to generate synthetic contacts on a real-time, real-world environment or the entire scenario may be synthetic.

Status

Believed to be in service with US Navy submarines.

Manufacturer

Rockwell International, Autonetics Electronics Systems, Anaheim, California.
(For details see Directory of Suppliers)

Rockwell Seatone Contact Signature Generator (CSG)

The Seatone Contact Signature Generator (CSG) is designed for a variety of precision, controllable, high-fidelity, signal-source applications. It is being incorporated in the company's sonar operator trainers, sonar onboard trainers, classification trainers and sonar test equipment, together with sonar analysis centre simulators and mobile acoustic targets. Its main functions are tone generation, tone parameter computation and path-sensor equalisation.

The system can generate 640 programmable tones and six contacts. Two of these are multi-path with 128 tones/contact and the remainder are direct with 96 tones/contact, including narrowband, modulated broadband and unmodulated broadband.

Status

In service with the US Navy and other navies.

Manufacturer

Rockwell International, Anaheim, California.
(For details see Directory of Suppliers)

Ship Analytics Radar Simulator

The Radar Simulator is a generic sonar and radar simulator that may be configured to represent specific displays such as the AN/SPA-25F or the Canadian Marconi LN-66. The unit duplicates all the functions of the real display. It also simulates an Automatic Radar Plotting Aid (ARPA) including the touch screen technology found in many modern shipboard ARPA radars. The Radar Simulator can be easily configured to simulate other specific radar units.

The system consists of a simulation unit based upon a personal computer, a display unit and a data acquisition subsystem with controls. The Radar Simulator may be configured as either an air search or a sea search unit and can display tactical contacts, weather, chaff, sea return, jamming and land mass features. The simulator sonar is active, operating in either a steerable sector or an omni mode. The Radar Simulator can be interfaced with any host simulator.

Status

The Radar Simulator is used in Ship Analytics Full Mission Shiphandling Simulator systems and Action Speed Tactical Trainers. Over 70 units are currently in operation throughout the world, for both military and commercial applications.

Manufacturer

Ship Analytics, Connecticut.
(For details see Directory of Suppliers)

A Radar Simulator in an ARPA configuration

TAS Radar Video Simulation (RVS)

Developed by the Technology Applications and Service Company (TAS), Radar Video Simulation (RVS) was designed to be used in training or testing applications in which a combat system requires radar video inputs. RVS provides the target, clutter, jamming and IFF/SIF video inputs with the associated trigger and timing signals. The main modes of all known two-dimensional radars can be generated while the majority of US Navy radars are instantly selectable. Three-dimensional simulation of the AN/SPS-48C and the AN/SPS-52B is also available. A mixed mode is also available in association with a live radar.

Status

In production.

Manufacturer

Technical Applications and Service Company (TAS), Maryland.
(For details see Directory of Suppliers)

Vitro OBTSIM

The OnBoard Trainer (OBT) Simulation (OBTSIM) programme is an IBM personal computer-based, graphical, real-time software program which simulates the operation of the AN/SQQ-89(V)-T onboard trainer. Using unclassified software it provides hands-on training to students who do not have access to the onboard trainer. It provides operator proficiency training and may also be used as an engineering tool for conducting various anti-submarine warfare application studies. A future OBTSIM version will be capable of creating scenario script files which will be compatible with the onboard trainer.

Status

In production.

Manufacturer

Vitro, Rockville, Maryland.
(For details see Directory of Suppliers)

NAVAL MAINTENANCE TRAINERS

GERMANY

Dornier Maintenance Rack Family

Dornier is producing a range of maintenance rack simulators designed to provide practical training without using valuable in-service equipment. There are currently four versions available for naval training.

Maintenance Rack
This is a job-tailored system which may include original equipment which is modified to meet specific training objectives. It consists of a mobile 19 in (483 mm) rack with plug-in units most of which are electrically powered. Each unit is devoted to a particular subject and can be removed from the rack for use on a work bench.

There are standard monitoring devices within the rack for 'system checks', while within the individual drawers are fault switches which are accessible only to the instructor. These are used to insert 'faults' and may be repaired if the correct procedures are followed.

Maintenance Rack (Radar)
This system is designed to provide comprehensive instruction for maintenance personnel on radar principles and applied techniques. Test points are provided and identified while signal generators are integrated for simulation of the processing circuitry and to operate the radar display. A number of fault switches is provided for training in troubleshooting routines.

Maintenance Rack (Sonar)
This system consists of a mobile 19 in (483 mm) rack with a number of plug-in units housing functional groups of the sonar system. In addition there is a real sonar display console and a water tank with transducers and a test probe. The sonar system itself is created from original components which are modified and circuited for training purposes including the provision of fault switches.

BITE Emulator
This system is for training maintenance procedures on systems which have built-in test equipment (BITE). The emulator reflects the system's BITE capabilities with fault messages appearing on a display. The student then uses the manufacturer's manuals to ascertain the cause of the fault and to recommend corrective actions.

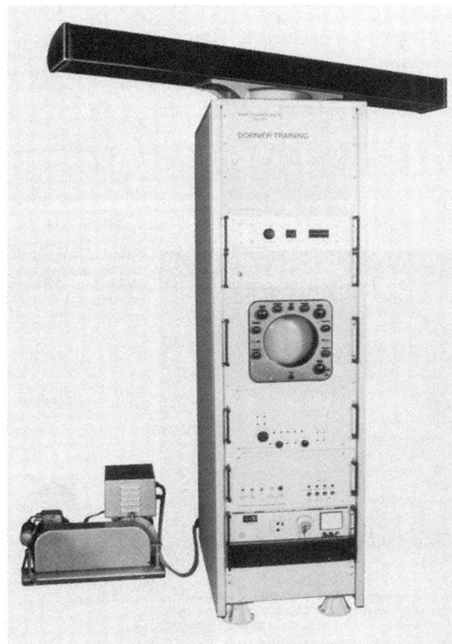

The Dornier Maintenance Rack (Radar)

Status
Available.

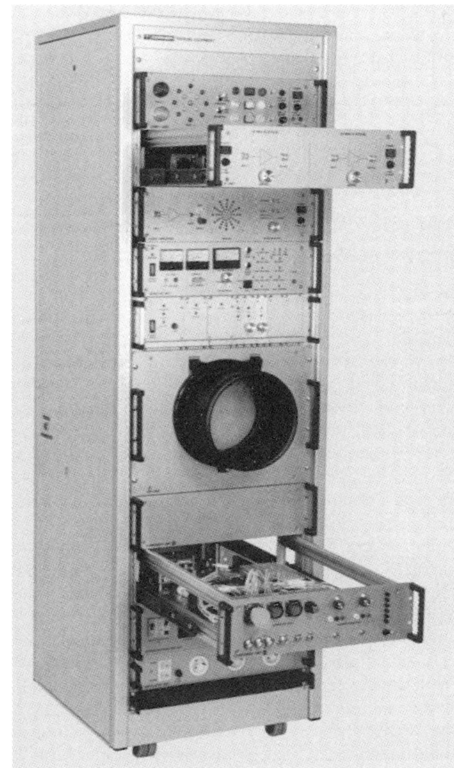

The Dornier Maintenance Rack

Manufacturer
Dornier Luftfahrt GmbH, Friedrichshaven.
(For details see Directory of Suppliers)

UNITED KINGDOM

Pennant Ship Propulsion Control System Maintainer Trainer

Pennant International has extensive experience in designing and producing maintainer trainers for ship propulsion control systems, including the 'Broadsword' (Type 22) class frigates.

The maintainer trainer is a full-scale replica of the system with a full simulation model and complete mechanical and electrical controls, indicators, components and wiring. Control of the training device is through a separate console from which the instructor can introduce the typical range of faults likely to be encountered under operational conditions. Students are expected to diagnose the fault according to procedures, using built-in and portable test equipment. They are then required to rectify the fault by the appropriate means.

Status
In service with the Royal Navy.

Manufacturer
Pennant International Limited, Cheltenham.
(For details see Directory of Suppliers)

The Pennant Ship Propulsion Control System Maintainer Trainer

VSEL Type 42/CVS Weapon System Maintainer Trainer

Vickers Shipbuilding and Engineering Ltd has won a contract to assist the maintenance of the ADAWS (Action Data Automation Weapon System) used in the Type 42 destroyers and the 'Invincible' class aircraft carriers by providing a maintainer trainer.

The system will be installed in HMS *Collingwood* at Fareham, Hampshire, and will include simulated dual redundant local area network and simulated point defence missile system, navigation radar, long-range search radar, radar track combiner, onboard radar trainer, tactical decision aid and data link processor system.

Status
See text.

Manufacturer
Vickers Shipbuilding and Engineering Limited, Barrow-in-Furness.
(For details see Directory of Suppliers)

Vosper Thornycroft Weapon System Maintainer Trainer for the Single-Role Mine-Hunter

Vosper Thornycroft has just completed the design and installation of the Weapon System Maintainer Trainer for the 'Sandown' class at HMS *Collingwood*.

A three-storey, steel-framed structure has been constructed to represent the various decks of the mine-hunter. This enables the full maintenance procedures associated with the towed body and the main sonar winch to be undertaken. A screened enclosure has been fitted out to represent the operations room, the bridge and sonar equipment instrument spaces. A third area has been configured to represent the onboard RCMDS (Remote Control Mine Disposal System) preparation and handling area.

Status

The trainer entered service in mid-1993. It will be used to train crews of the Royal Navy and the Royal Saudi Navy.

Manufacturer

Vosper Thornycroft (UK) Limited, Southampton Office.
(For details see Directory of Suppliers)

UNITED STATES OF AMERICA

AAI AN/SQQ-89(V) ASW Combat System Maintenance Trainer

AAI is a major supplier of electronics to the American Defense Department and has developed the AN/SQQ-89(V) anti-submarine warfare (ASW) combat system maintenance trainer. It consists of three maintenance training devices; 14G2 for common SQQ-89 equipment, 14G3 for the AN/SQR-19(V) towed array and AN/SQQ-28(V) LAMPS Mark III sonar signal processing systems, and 14G4 for the AN/SQS-53B(V) hull-mounted sonar ranging/detecting set. Each device has a distributed architecture based upon the Intel 80286 single board computer. It is used to provide students with practical experience in operating and maintaining SQQ-89 equipment with the focus of attention upon fault isolation, identification and the alignment procedures used to remedy faulty assemblies.

Each device consists of an instructor station, up to 14 student stations with an integral distributed computer system and a communication system. The instructor's station features two high resolution graphics displays, one of which is used to monitor all student stations, while the other can be used to intervene to give further instruction. The other display is used to control the central file server.

The student stations provide a realistic training environment using graphic representations of unit interiors and components to assist recognition skills, simulated controls and indicators, and software which simulates the real responses of test and sonar equipment. Technical manuals are provided for each station.

Each device has a rotating drum which houses up to eight subsystems on which training is performed. The drum is located between the instructional station and simulated operational equipment. Simulated test equipment is mounted on a mobile test cart beneath the rotating drum.

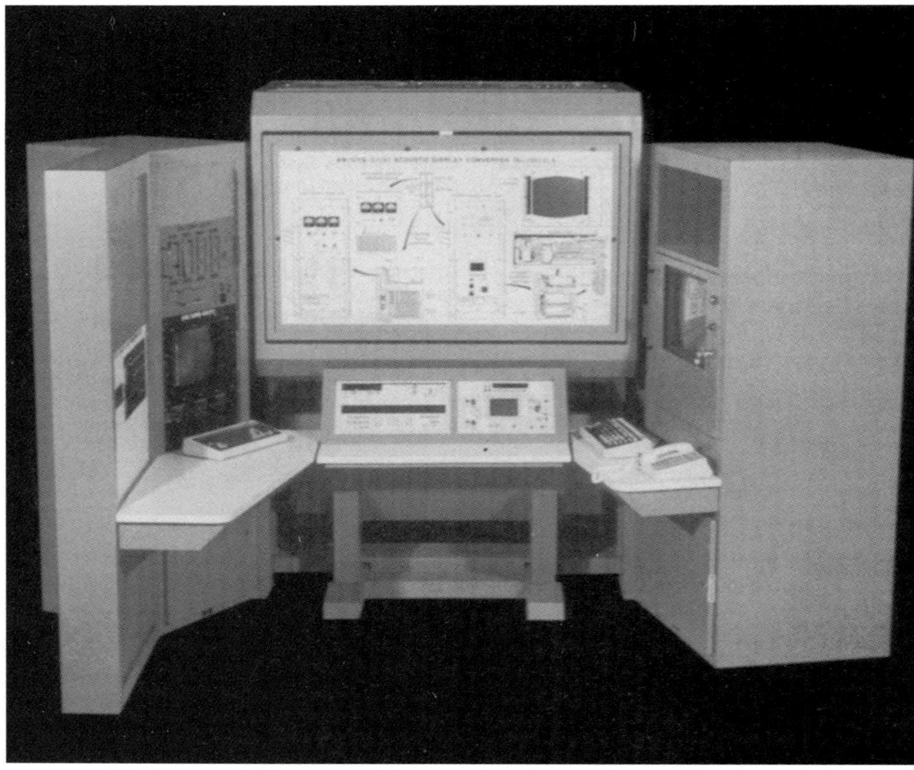
The AAI AN/SQQ-89 anti-submarine warfare combat system maintenance trainer

Status

These devices were installed and operational at the US Navy's Fleet Anti-Submarine Warfare Training Center, Pacific at San Diego, California. However, due to fleet cutbacks Devices 14G2 and 14G4 are no longer in use.

Manufacturer

AAI Corporation, Maryland.
(For details see Directory of Suppliers)

Bell Aerospace Textron AN/SPN-46 Radar Maintenance Trainer

The US Defense Department awarded Bell Aerospace Textron an $11.3 million contract to provide a maintenance trainer for the AN/SPN-46 carrier approach radar. The company will provide a trainer unit, maintenance assistance modules and two antenna pedestals. No further details are available.

Status

In service with the US Navy at the Naval Air Technical Training Center, Memphis, Tennessee.

Manufacturer

Bell Aerospace Textron, Buffalo, New York.
(For details see Directory of Suppliers)

DynaLantic Recovery Assist, Securing and Traversing (RAST) Troubleshooting Trainer (Device 11H117)

The Recovery Assist, Securing and Traversing (RAST) system is an electromechanical, hydraulically operated winch-like device which is used to land and move helicopters on destroyers and frigates. To provide training in RAST maintenance, DynaLantic developed the interactive computer-based RAST Troubleshooting Trainer.

The simulator, designated Device 11H117 by the US Navy, is a part-task trainer based upon an Intel 310 microcomputer. The student station is divided into RAST operational panel components, electrical test points and an interactive audio/video system. The operational elements are facsimiles of the actual RAST panel equipment. The electrical test points consist of full colour photographs with test points appropriately placed. The student treats the displayed 'electrical system' in the same way as a real one. In the interactive element the student selects RAST components for various actions and tests and is able to see the result.

In addition to the student's station there is an instructor's station with video monitor, keyboard and dot matrix printer. The instructor can monitor the exercises and set up new exercises.

Status

In service with the US Navy.

Manufacturer

DynaLantic Corporation, Deer Park, New York.
(For details see Directory of Suppliers)

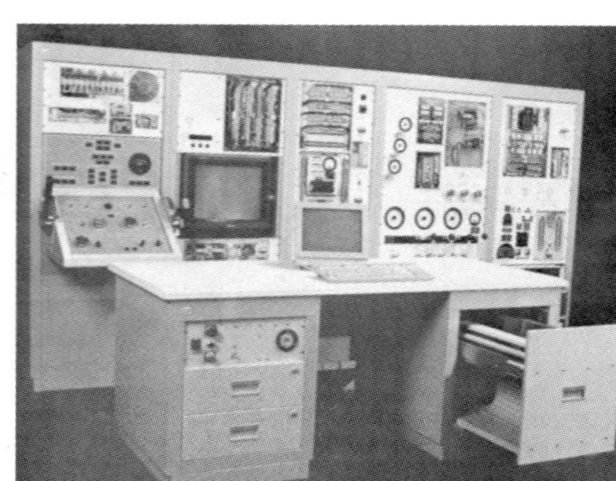
The DynaLantic Device 11H117 showing the simulated control console (far left), the simulated hydraulic test panel (right) and the instructor station in the foreground

Raytheon Trident Sonar Maintenance Trainer Front-End Simulator (TSMT FES)

The Trident Sonar Maintenance Trainer Front-End Simulator (TSMT FES) was designed to offer US Navy personnel realistic, effective maintenance training for the AN/BQQ-6 and AN/WLR-17 sonar systems. It is a follow-on from the AN/BQR-4T onboard sonar operational trainer. The system offers intermediate and advanced instruction in sonar system maintenance, repair and calibration. Realism is achieved by injecting TSMT-transmitted signals into the front end.

The system consists of 10 elements, designated Units. The contact generation and control assembly is Unit 201 and is the major electronic processing cabinet of the system. It receives input commands and processes them to develop contact signals appropriately modified to simulate propagation effects for selected ocean models. Signals are further modified to account for the effects of various arrays of the AN/BQQ-6 sonar. Signals for the spherical array are output to Unit 202 for further processing. Simulation outputs are also developed for bathythermograph, depth-sound speed recorder, and fathometer muting signals. The unit contains four major drawers each comprising a module assembly and a power supply section.

The spherical array inverse beam former is Unit 202. This receives control and the simulated acoustic signals from Unit 201 and processes them to account for the spherical array spatial effects. It combines the signals with simulated sea state noise and outputs 944 parallel channels to the AN/BQQ-6 spherical array inputs. The unit consists of four major drawers each with a module assembly and power supply.

Unit 203 is the trainer control console which programmes, controls and monitors exercises by interacting with a series of functionally dedicated plasma display panels. Unit 204 is the printer-plotter which generates hard copies from the plasma displays while Units 205, 206 and 209 provide resistive loads for AN/BQQ-6 acoustic communication and the high frequency sonar transmitters together with supplying variable loads for the power supply of the supplementary array. Unit 207 is the distress beacon dummy load while Unit 208 is the emergency communications dummy load.

The processor system is Unit 210 and consists of a powerful HP 9807 transportable, UNIX-based computer. The operating system and the application software are provided on a 3.5 in (88.9 mm) floppy disk and loaded via an RS 232 A standard interface. The trainer database consists of 16 contacts, one ocean and 40 active emitters. Up to four active and passive, manoeuvrable targets are generated at any one time.

The final elements of the system are the power control-status monitor (Unit 211) and the active emissions simulator (Unit 212). The latter provides active emissions to the AN/WLR-17 system and receives control orders from Unit 201 through an IEEE-488 interface.

Status

The US Navy has five of these systems on order.

Manufacturer

Raytheon Company, Submarine Signal Division, Rhode Island.

(For details see Directory of Suppliers)

Unisys Mark 92 Fire-Control Maintenance Trainer (Device 11G3)

Formerly the Sperry Division of the Sperry Corporation, Unisys has produced the Mark 92 Fire-Control Maintenance Trainer (Device 11G3).

The Mark 92 system is used with the Standard Missile and the Mark 75 Gun Mounting. Each computer-driven maintenance trainer consists of an instructor station and a number of student work stations. The work stations are divided into three major sections. The first has a pictorial presentation of the major circuit systems of the Mark 92 system and also includes working test jacks, major indicators and displays. The working parts are computer controlled. The second section consists of a computer display which is fully interactive with the test section. The third section contains the test equipment. In addition to working with the trainer, students also work with an operational Mark 92 system.

Status

Two systems were installed at training schools at Dam Neck, Virginia, and San Diego, California.

Manufacturer

Unisys Corporation (Naval Systems Division), Great Neck, New York.

(For details see Directory of Suppliers)

UNDERWATER PRACTICE WEAPONS AND TARGETS

FRANCE

DCN CI1 Submarine Target

One of the DCN's establishments, ECAN (Etablissement des Constructions et Armes Navales) St Tropez, F-83990 St Tropez, produces a target system, the CI1 submarine target.

The CI1 submarine target is used both to train torpedo launch crews and to evaluate new operational techniques and tactics, due to the device's accurate representation of submarine characteristics.

The electrically powered system resembles a large torpedo and carries acoustic equipment which permits it to simulate the signature of a submarine being illuminated by the active sonar of a homing torpedo. Alternatively it can act as the homing torpedo. The system has a maximum endurance, at minimum speed, of 8 hours.

A new version of this submarine target, the CI5, will have a command link system enabling the mission programme to be modified during the exercise, and it will also have a higher speed to match the speed performances of nuclear submarines.

Status

The CI1 is operational but no longer in production. The CI5 is under development.

Manufacturer

DCN, Paris.
(For details see Directory of Suppliers)

DCN CI2/3/4/7 Stationary Targets

DCN Saint Tropez has developed a series of stationary targets, CI2, CI3, CI4 and CI7, to check torpedo homing heads and fuzes. All have 24 hour endurance and all can record for up to an hour.

A typical system is the CI3 which is designed for testing the F17 torpedo warhead and fuze. It consists of a spherical body weighing 40 kg which can operate in waters up to 100 m deep. The body consists of an acoustic noise generator which simulates a target's acoustic signature, an electromagnetic transmitter which simulates the target's magnetic signature and a magnetic recorder. It can be used autonomously, with a buoy, or it may be attached to a hulk for live firing trials.

The CI2 weighs 40 kg and can be operated in waters up to 300 m deep, the CI4 weighs 2000 kg because it has a larger area to allow for self-destruction of torpedoes within its vicinity in waters down to 300 m. The CI7, which can be used in water as deep as 500 m, weighs 80 kg.

Status

In production and in service.

Manufacturer

DCN, Paris.
(For details see Directory of Suppliers)

DCN CI9 Towed Target

This towed target, produced by DCN Saint Tropez, simulates the acoustic signatures of submarines for torpedo homing heads and is towed at depths of 20 m to 400 m at speeds of 5 to 20 kts by surface ships.

The system consists of a target with 15 repeater modules on a cable 100 m long. Each repeater module consists of a receiver and a transmitter transducer with a circuit to analyse the received signals and to retransmit them and a multiplexer to transmit all incoming signals to the surface.

The target is linked to a towing system with a hydrodynamic depressor which ensures the target remains at the required depth, a 600 m towing cable and a "fish" to which the target is attached. The "fish" provides electrical supplies to the antenna modules, calculates the target's trajectory, provides signal modulation for its compass and depth finder sensors.

Specifications

Length:
(body) 2.2 m
(depressor) 1.6 m
Diameter:
(body) 320 mm
Width:
(depressor) 1 m
Weight:
(body) 150 kg
(depressor) 500 kg

Status

In production.

Manufacturer

DCN, Paris.
(For details see Directory of Suppliers)

ECA STAR Submarine Target

To replace submarines as torpedo targets in ASW warfare and for weapon evaluation, ECA and Safare Crouzet have developed the Star towed target which simulates the acoustic activity of a submarine under attack by an active torpedo.

The towed target consists of between five and 15 acoustic repeater modules, provided by Safare Crouzet, which trail behind a towed body containing the electronics, a battery, navigation and external trajectography sensors. Some 400 m ahead of the towed body is a 350 kg hydrodynamic depressor to counteract the buoyancy of the towed body and the repeater modules. They are all towed by a 600 m cable attached to the ship via a 1 tonne 15 kW hydraulic winch. Other shipborne equipment includes an electronic rack with control, monitoring and recording instruments.

Each acoustic module is checked before the target is launched and it can then be towed at speeds of between 3 and 20 kts at depths from 20 to 400 m. Sensors in the towed body provide details of its heading and depth and the target can be towed for up to eight hours at a time.

When the target antennas are partially or totally located by the torpedo's sonar, each module retransmits the signal received at the appropriate level to simulate the acoustic return from a submarine. The torpedo's signals are also retransmitted from each module to the towing ship.

Specifications

Length:
(body) 2 m
(antenna with 15 modules) 100 m
Diameter:
(body) 300 mm
(repeater module) 100 mm
Weight:
(body) 160 kg
(antenna with 15 modules) 150 kg
(depressor) 500 kg

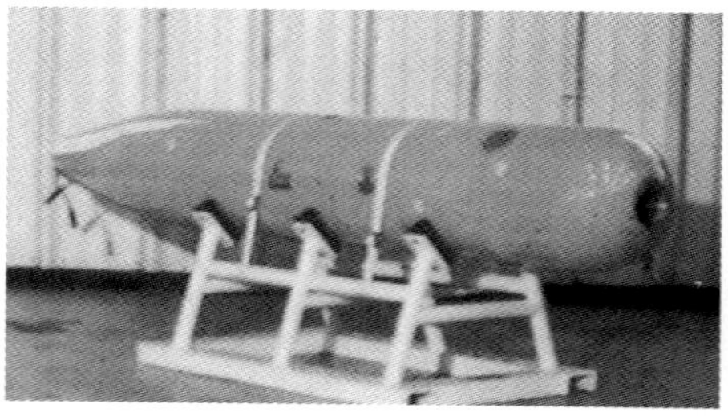

The Star towed body which contains the key 'wet-end' electronics

Status

In production.

Manufacturer

ECA, Meudon Branch.
(For details see Directory of Suppliers)

Thomson Sintra Cible Autonome Légère pour Armes Sous-marines (CALAS) Submarine Target

To provide a realistic but low-cost expendable simulation of a submarine for ASW training, Thomson Sintra and CNIM (Constructions Industrielles de la Méditerranée) have developed CALAS (Cible Autonome Légère pour Armes Sous-marines). It can be deployed easily from a surface ship to provide a simulation of submarine movements and acoustic signatures.

It is a small, torpedo-shaped vehicle, with a hydrophone which is towed by an 18 m cable, operating autonomously for up to two hours at speeds of 7, 10 or 12 kts and at depths of between 20 and 300 m. It will transmit an acoustic signature in the 300 to 800 Hz range and be capable of processing simultaneously two active sonars within the 4 to 33 kHz frequency band.

The CALAS mission parameters are programmed on a laptop computer and include mission path, speed, depth, duration and acoustic operation. The target is delivered in a container and because it is expendable there are no maintenance, retrieval or reconditioning costs.

Specifications

Length: 1.85 m
Diameter: 150 mm
Weight:
(in air) 30 kg

Status

Under development.

Manufacturer

Thomson Sintra ASM (Activités Sous-Marines), Sophia Antipolis.
(For details see Directory of Suppliers)

GERMANY

STN Torpedo Simulator

This torpedo simulator is designed to close the gap between simulated and real torpedo firings. It is a replica of up to eight torpedoes and is linked to a weapon control system instead of a live weapon.

The simulator fulfils all requirements for training by simulating the ship launching the weapons, up to eight torpedoes and up to eight targets. It may also be used for checking weapon control systems.

Status

In production and in service for simulating SUT and DM2A3 (Seehecht) torpedoes.

Manufacturer

STN Systemtechnik Nord GmbH, Bremen.
(For details see Directory of Suppliers)

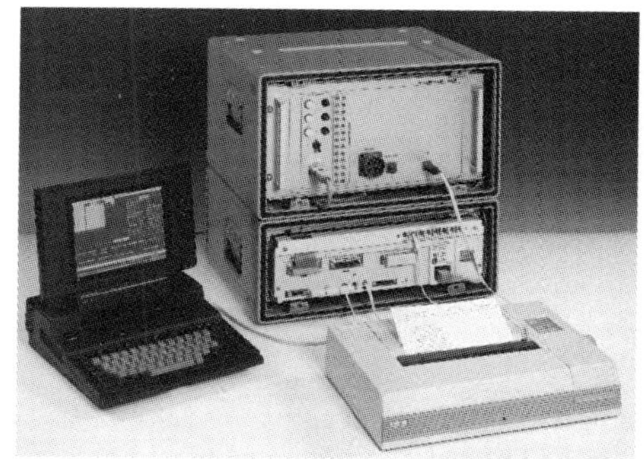

The STN torpedo simulator

ITALY

Whitehead MP-80/I Exercise Mine

Whitehead SpA (formerly MISAR SpA) is Italy's leading producer of naval mines and, for mine countermeasures training, has produced the MP-80/I exercise mine. The MP-80/I is an inert version of the operational MP-80 sea mine designed for exercises, training and trials. The system consists of a sensor and processing section and a tail section with recovery aids.

The MP-80/I exercise mine is laid by surface vessels, submarines or aircraft after the appropriate sensor simulator has been installed to cover a complete range of ground mine influences. The system can be used not only to provide operational training to mine-hunter and mine-sweeper crews with any type of influence mine but it may also be used in mine warfare research. The activation of the mine is simulated by means of a flare and at the end of an exercise a buoy is released which permits the system to be recovered.

The dimensions and weights provided below may vary according to the precise operational model of MP-80 mine being modelled.

Specifications

Length: 2.096 m
Diameter: 533 mm
Weight: 750 kg

Status

The system is in service with a number of NATO navies, many of whom also use the company's war stock naval mines. One customer was Norway which ordered a number of exercise mines in 1984. India is believed to be a non-European customer.

Manufacturer

Whitehead SpA, Livorno.
(For details see Directory of Suppliers)

Whitehead MP-80/I exercise mine

Whitehead BSS Underwater Mobile Target

The Whitehead Underwater Mobile Target (BSS) is based upon a modification kit to the company's A184 heavyweight torpedo although it may be applied to any similar type. It was designed for torpedo dynamic trials and operational evaluation.

The BSS is a self-propelled transponder whose run is preset allowing it to emulate a real underwater mobile target. The shipborne element consists of a portable control panel, an underwater telephone and junction boxes. The BSS can simulate either a submarine's reflected acoustic signature when it is struck by the sonic pulse from a homing torpedo seeker or it can reproduce the acoustic signatures of both diesel-electric and nuclear submarines.

The trajectory of the BSS can be modified during the mission by an underwater telephone and it can be tracked within a three-dimensional test range by means of a synchronised acoustic transmitter. The BSS may be launched from surface ships or submarines and with the latter may be used either with swim-out or air pulse launching systems.

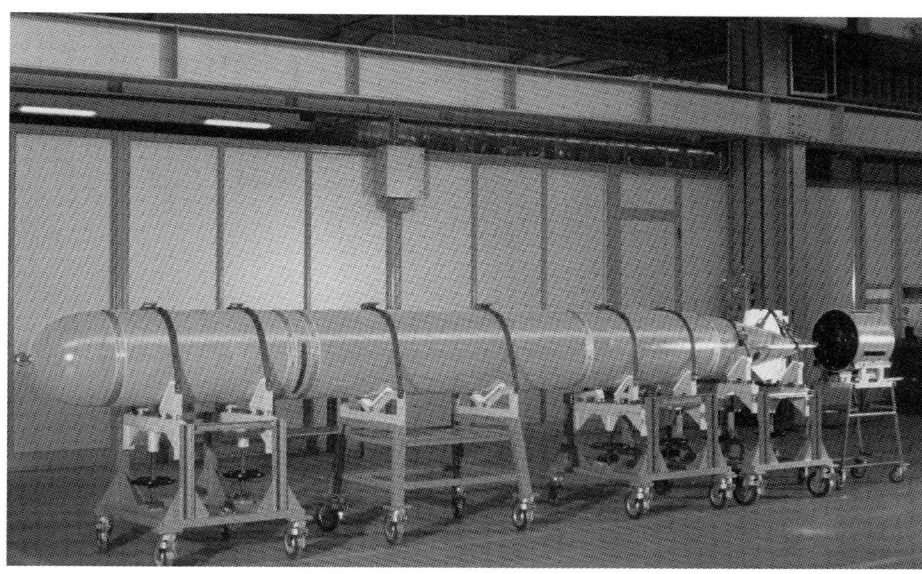

The Whitehead BSS underwater mobile target

Specifications

Length: 6.085m
Diameter: 534 mm
Weight: 1122 kg

Status

In service with the Italian Navy.

Manufacturer

Whitehead SpA, Livorno.
(For details see Directory of Suppliers)

SOUTH AFRICA

Northbend Instrumentation Matrix Practice Mines

The Matrix moored and bottom practice mines are designed to train personnel to handle explosives in assembling, testing and transporting naval mines, in carrying and laying such mines and in mine countermeasures. The mines are identical to operational weapons and retain both arming devices and detonators but the main charge is replaced by an explosive substitute.

Exercise mines are deployed at sea and are linked to a shore station by up to 4 km of lightweight cable to enable two-way communication including reprogramming a mine within a second. A data display, recorder and computer are used to monitor the exercise in real time and when a ship passes over the 'mine', data is progressively updated. When the ship's stern has passed the 'mine' an analysis of the exercise is immediately displayed to indicate whether or not the 'mine' detonated and, if not, why it did not do so.

The Matrix range also includes drill mines which are ballasted shells for practising mine laying. They may also be used to train the operators of mine-location sonars and for this role the moored drill mine can uncouple to settle at a preselected depth.

A test and simulation bench is used to play back target signature recordings into the mine electronics to simulate mine fires. It provides training in mine programming and improves target identification skills.

Status

Available.

Manufacturer

Northbend Instrumentation (Pty) Limited, Bellville.
(For details see Directory of Suppliers)

UNITED KINGDOM

BAeSEMA Versatile Exercise Mine System (VEMS) and Advanced VEMS

BAeSEMA has recently created the Advanced Versatile Exercise Mine System (VEMS) from the established VEMS. The advanced version provides greater capability in emulations, recording of data and operational aspects including a wider range of sensors. VEMS already provides a realistic target for the at-sea training of mine countermeasures forces. It may also be used for evaluating mine countermeasures tactics and equipment, to evaluate the risk to own vessels from mines and for mine/mine countermeasures technology research and development.

The VEMS simulates the actuation system of any magnetic, acoustic, pressure or combination influence naval ground mine including the latest multi-microprocessor controlled mines and those with seismic sensors, and records all data for subsequent analysis. An acoustic transponder fitted to the VEMS provides two-way communication between the mine and the exercise vessel. The transmitted data includes mine actuation signals and mine-to-ship ranges. The transponder can also be used to send commands to a VEM on the seabed and to recover recorded influence and event data from it. The VEMS can remain on the seabed for up to six months before being recovered, serviced and reprogrammed. Mine emulation programmes are available from a standard library of mine types or can be developed to meet specific customer requirements.

The system is in two sections: a heavy ballast section and a buoyant section containing the sensors, microprocessor, memory and acoustic transponder. The system is launched from a vessel either by means of a trolley, which forms part of the system, or by slipping it from a davit. It may also be laid by helicopter.

Recovery is initiated either at a preprogrammed time or by a coded acoustic signal command from the vessel. The mine recovery mechanism uses compressed air from a bottle in the ballast section to free the mine from mud and silt, then to separate the buoyant section from the ballast section. After separation the buoyant section floats to the surface tethered to the ballast section by a 200 m rope.

The Versatile Exercise Mine System (VEMS) being lowered into the water

Specifications

Length: 2.71 m
Diameter: 533 mm
Weight: 560 kg

Status

The VEMS is in service with the Royal Navy, the US and Royal Thai navies. Development plans include a buoyant version and the use of VEMS as a mobile range. The US Navy has also bought Advanced VEMS.

Manufacturer

BAeSEMA Limited, Bristol.
(For details see Directory of Suppliers)

BAeSEMA Limpet Exercise Mines

The BAeSEMA Limpet Exercise Mine family includes live mines, exercise and training mines and brings together the BAeSEMA exercise limpet mine and the M.P. Compact Energy underwater demolition unit. The two companies have now expanded their product range.

The Live Exercise Limpet Mine contains modular plastic explosive charges of between 1 and 6 kg, and is equipped with a sophisticated safety and arming device built into a modular self-contained timer unit with a firing delay of up to 24 hours. Slave explosive units, attached by special couplers, are available to increase explosive power. The magnetic and strap attachment methods are standard as is built-in positive buoyancy.

The exercise and training variants simulate all the BAeSEMA live and other mines with a high intensity, flashing, light simulating detonation. The training version of the BAeSEMA mine simulates the live mine to component level and can be used for training through all operational stages.

Status

In production and in service with a number of navies worldwide.

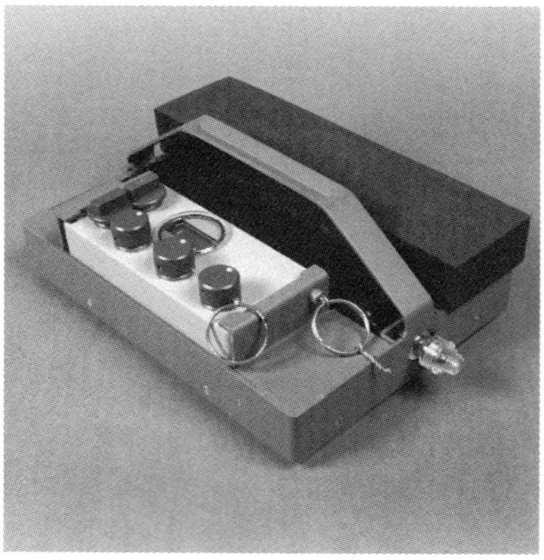

A BAeSEMA Live Exercise Limpet Mine

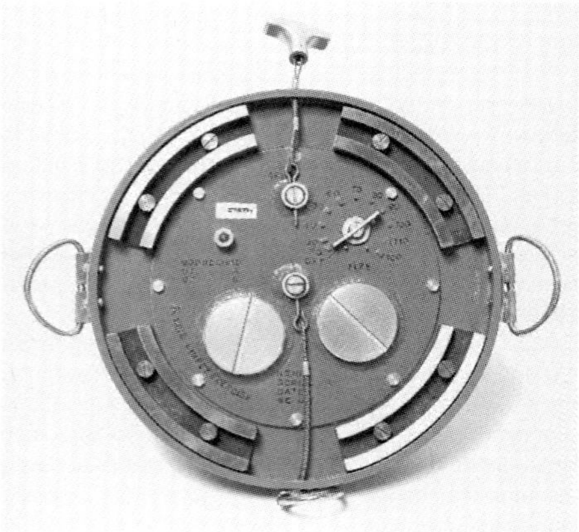

The under surface of a BAeSEMA Limpet Exercise Mine

Manufacturer

BAeSEMA Limited, Bristol.
(For details see Directory of Suppliers)

GEC-Marconi Torpedo Exercise Target (TET)

The Torpedo Exercise Target is designed to provide a full assessment of the performance of advanced homing torpedoes under realistic firing conditions but without using ships as targets.

The system consists of a small underwater unit which houses transducers, attitude and depth sensors. It is suspended from a multi-core strain cable which links it to the main processing unit in the target support boat. The TET features onboard data recording for post-run analysis, automatic real-time selection of either broadband or narrowband operation and a remote-control option through a UHF link.

Status

In production.

Manufacturer

GEC-Marconi Avionics Limited, Basildon.
(For details see Directory of Suppliers)

The Torpedo Exercise Target shown suspended from a target support boat

GEC-Marconi Practice Delivery Torpedo (PDT)

The Practice Delivery Torpedo (PDT) is a low-cost training system which is a facsimile of the American Mark 46 lightweight torpedo. It is designed only for launching from aircraft, helicopter and shipborne torpedo tubes and once it enters the water it floats to the surface. The system can be recovered, checked and replaced in the launcher system within an hour.

The PDT can be launched a minimum of 10 times before it requires a major overhaul and it can also withstand up to 150 air drops. For launch record purposes it includes a readout subsystem while a three-mode recovery air system assists retrieval.

Status

In production and in service with the Royal Navy together with those of Brazil, France and Pakistan.

Manufacturer

GEC-Marconi Underwater Weapons Division, Waterlooville.
(For details see Directory of Suppliers)

Graseby Dynamics G 733 SLUTT (Ship-Launched Underwater Transponder Target)

Graseby Dynamics Limited has considerable experience in producing not only sonar systems but also acoustic decoy and countermeasure systems. It has also produced the G 733 SLUTT (Ship-Launched Underwater Transponder Target).

The G 733 is a small towed transponder system designed to simulate a variety of underwater targets to permit functional training of sonar operators as well as to assist calibration and alignment tests. Some 24 versions have been produced to meet various requirements.

Status

In service with the Royal Navy and a number of other navies around the world.

Manufacturer

Graseby Dynamics Limited, Bushey.
(For details see Directory of Suppliers)

Graseby Dynamics Programmable Acoustic Target (PAT)

PAT is a user-programmable version of the SLUTT (see previous entry). It provides a range of simulated outputs for a realistic anti-submarine warfare exercise to be carried out using modern sonars. PAT also produces a wideband modulated noise to mimic the propeller beat-rate of a surface ship, a submarine or a torpedo.

Status

In service with the Swedish Navy.

Manufacturer

Graseby Dynamics Limited, Bushey.
(For details see Directory of Suppliers)

Graseby Dynamics Underwater Torpedo Target (UTT)

UTT is a self-programmable acoustic target for use in testing passive homing torpedoes. It produces broadband noise which is amplitude-modulated to provide a good imitation of the acoustic signature of a surface ship or submarine propeller. To simulate a target which is getting nearer or further away, the noise output can be programmed to increase or to decrease gradually.

Status

A version of UTT is in service with the Royal Navy.

Manufacturer

Graseby Dynamics Limited, Bushey.
(For details see Directory of Suppliers)

MSI Fleet Exercise Minelaying System (FEMS)

To meet a British Ministry of Defence requirement, MSI-Defence Systems has designed a system which allows exercise mines to be laid or recovered in order to assist the training of Mine Countermeasures Forces. The system is the Fleet Exercise Minelaying System (FEMS).

The FEMS is portable and may be fitted to a variety of vessels, including offshore supply ships, within 24 hours and may be removed within eight hours. It consists of two laying rails and a stowage rail section which are bolted to pre-drilled deck pads welded to the ship's after deck, together with two electrohydraulic winches (one for each laying rail) operated from a centrally located remote-control unit. FEMS weighs 5.5 tonnes empty and 35 tonnes fully loaded.

The Fleet Exercise Minelaying System loaded with bottom and moored exercise mines

The exercise mines are moved to the laying rails by the winches and are released from the rails upon command by the minetrap in conditions up to Sea State 4. A sponson ensures the mine enters the water clear of the ship's stern. The centrally located mine storage rail can carry extra mines or it may be used to stow defective mines.

Status
In service with the Royal Navy and available for export.

Manufacturer
MSI-Defence Systems Limited, Weymouth.
(For details see Directory of Suppliers)

THORN EMI SoundTrak (Type 2058) and SoundTrak II Target Simulator

The SoundTrak family of acoustic target simulators is designed to be used with surface ships, submarines and aircraft in anti-submarine warfare training, in system performance assessment and other operational applications which require a versatile, reliable underwater acoustic source.

The system consists of a towed body containing electro-acoustic transducers which is towed behind a surface ship and is connected to an inboard electronics rack containing signal generation and drive circuits. SoundTrak produces realistic acoustic target signatures, transmitting simultaneously a mixture of narrowband tones and broadband noise. The simulators can be deployed from a variety of vessels using standard winch systems.

The nature of the output in terms of narrowband frequencies, noise bands, signal amplitudes and modulation can be programmed manually and the system can store up to eight preprogrammed pages of data. Additionally it can retransmit acoustic signatures which have been recorded previously or which have been generated using large scale computer-generated simulations.

SoundTrak has exceptionally low frequency performance in relation to its very small physical size and it has been designed to operate with passive sonar systems. SoundTrak II also includes both narrowband and broadband high frequency signals and is able to echo-repeat signals received from active sonars.

Status
SoundTrak is in service with both the French and Royal navies; the latter designates the system Type 2058. SoundTrak II is in service with an unidentified Far Eastern customer.

Manufacturer
THORN Automation, Rugeley.
(For details see Directory of Suppliers)

A SoundTrak target simulator system being recovered during trials

UNITED STATES OF AMERICA

Alliant Techsystems HOTTorp

In 1988 Alliant Techsystems began to address a Royal Australian Navy requirement for a Mark 46 practice torpedo. This is now available as HOTTorp.

HOTTorp is a low-cost operational training torpedo which can be used for handling, loading, preparing, presetting and launching, training and recovering. It has movable ballast weights to adjust the centre of gravity, stationary fins and windmilling propellers. At the front are six drop weights and behind them is a buoyancy chamber. A simulated fire-control system records and displays all preset information.

The HOTTorp is loaded and launched in the same way as a war-shot torpedo. Upon entering the water an actuator in the nose releases the weights and the buoyancy chamber helps bring the system to the surface where a seawater-activated dye container helps

An Alliant Techsystems HOTTorp

recovery. The maximum plunge depth from a surface ship is 23 m and from a helicopter it is 37 m. The system can be recovered and used again within 60 minutes. The manufacturer states it has a minimum operating life of 150 launches.

Specifications

Length: 2.591 m
Diameter: 325 mm
Weight:
(in air) 230 kg
(in water) 52.36 kg

Status

Available for production. A similar system is being offered to meet a US Navy requirement.

Manufacturer

Alliant Techsystems Inc, Hopkins, Minnesota.
(For details see Directory of Suppliers)

Loral Mark 30 Mod 1 ASW Mobile Target

The Mark 30 Mod 1 anti-submarine warfare (ASW) mobile target was developed by the US Navy Sea Systems Command with technical support from the Naval Underwater Systems Center and since 1978 Loral Defense Systems-Akron has been the sole source contractor.

The Mark 30 is a self-propelled, multi-speed, deep-running, active and/or passive acoustic target which simulates the acoustic, dynamic and magnetic characteristics of a submarine. It enables submarine, ship and aircraft personnel to participate in ASW exercises in a realistic environment without using a submarine. It can be launched from a ship or a helicopter.

Mark 30 Mod 1 mobile targets being inspected before delivery

The target is a torpedo-shaped vehicle with an acoustic generator and signal processors in the nose section. The generator can produce 12 discrete low frequency lines simulating periodic energy emitted by rotating machinery. The battery section has eight silver-zinc modules producing 300 V DC and behind it is the guidance and control section. In the rear of the vehicle are the transducer section and the tail section which includes a 107 hp contra-rotating propulsion motor, the steering mechanism and the two propellers. A towed sonar array may be attached to the tail section.

The Mark 30 is capable of speeds between 7 and 30 kts and can operate in depths from 7.5 to 610 m. Endurance at 7 kts is 5 hours and at 30 kts it is 30 minutes.

Specifications

Length: 6.22 m
Diameter: 533 mm
Weight: 1225 kg

Status

In service with the US Navy since 1978 mostly at the Barking Sands Tactical Underwater Range in Hawaii, the Atlantic Fleet Weapons Training Facility in the Virgin Islands, the Atlantic Undersea Test and Evaluation Center in the Bahamas and the Naval Undersea Warfare Engineering Station at Keyport, Washington.

Manufacturer

Loral Defense Systems-Akron, Ohio.
(For details see Directory of Suppliers)

Sippican Mark 39 Expendable Mobile ASW Training Target (EMATT)

The Target Mark 39 or Expendable Mobile ASW Training Target (EMATT) is an expendable self-propelled target which may be deployed from surface ships or anti-submarine aircraft. It has a maximum speed of 8 kts, or 7 kts with MAD (Magnetic Anomaly Detection) device attached. The operating depth is between 23 and 183 m and air launches may take place from altitudes between 152 and 3048 m at speeds of 40 to 200 kts while surface launches may take place from less than 6 m above the surface at speeds up to 10 kts in conditions up to Sea State 4.

The EMATT may be launched down an aircraft sonobuoy chute or thrown over a ship's side. Before launch one of three preprogrammed operational patterns, with and without MAD capability, is selected. When launched from the air the Mark 39 separates from its air-launch assembly as it enters the water.

After entering the water the target will run for up to three hours generating acoustic signatures over four discrete frequencies. It has a unique echo-repeat system which receives, stores, then transmits active sonar signals enhanced to simulate the echo from a submarine. A transponder is also included for use with the Mark 46 torpedo to provide acquisition and homing responses.

For use with MAD-equipped aircraft the Mark 39 EMATT deploys a 30.5 m wire through which DC power is pulsed to create a magnetic field. This unique signature provides excellent tactical training. The MAD may be selected before launch or it may be acoustically commanded during the target's run.

The Mark 39 Expendable Mobile ASW Training Target (EMATT) target system shown attached to a US Navy helicopter

Specifications

Length: 90 cm
Diameter: 12.2 cm
Weight: 10 kg

Status

In production and in service with the US Navy which has awarded a contract for 3200 units for fleet use. Another 300 specially configured units will be supplied to US Navy target ranges.

Manufacturer

Sippican Inc, Marion, Massachusetts.
(For details see Directory of Suppliers)

Westinghouse Self-propelled Acoustic Target (SPAT) Mod 1

The Self-Propelled Acoustic Target (SPAT) was designed to assist the training of sonar crews in the detection, classification and tracking of diesel-electric submarines. SPAT Mod 1 adds new capabilities, higher reliability and simplified maintenance. It uses noise-generating circuitry to create a submarine's unique acoustic signature in the 350 Hz to 85 kHz range. SPAT Mod 1 also trails a hydrophone to pick up an acoustic pulse which is then elongated by onboard electronics and transmitted.

The trainer-programmable electronics enable SPAT Mod 1 to take up to 10 different headings during an exercise which can last from 1 to 180 minutes. The target's speed may be programmed between 7 and 12 kts. Run time at 7 kts is 180 minutes. SPAT Mod 1 can dive to a maximum of 250 m.

Preparations to launch a Self-Propelled Acoustic Target (SPAT) from a small boat

The SPAT Mod 1 modifications include the ability to operate by day or by night, extended run time, programmable training levels, extended vehicle life, an improved vehicle locator and an automated launcher.

Specifications

Length: 3.28 m
Diameter: 25.4 cm
Weight: 150 kg

Status

In production.

Manufacturer

Westinghouse Electric Corporation (Naval Systems Division), Cleveland, Ohio.
(For details see Directory of Suppliers)

MISCELLANEOUS NAVAL SIMULATORS AND TRAINERS

GERMANY

Dornier Communication Equipment Operator Trainers (CEOT)

The Communication Equipment Operator Trainer (CEOT) can be tailored to meet a specific customer ship-fit communications requirement and is used to train operators in voice, Morse and teleprinter communications as well as equipment maintenance. Individual operator positions for all communications modes are provided as they would be on board a ship. There are separate radio and audio frequency sections, a selection of different types of antennas or dummy loads, as well as a modular layout to meet customer specifications.

Status
The CEOT has been sold to Nigeria.

Manufacturer
Dornier Luftfahrt GmbH, Friedrichshaven.
(For details see Directory of Suppliers)

A Communication Equipment Operator Trainer (CEOT) in use with the Nigerian Navy

---

INTERNATIONAL

Fire-fighting and Damage Control Simulators

Some navies have created their own fire-fighting and/or damage control simulators. These include the following:

Finland
A fire-fighting simulator constructed by the Finnish Defence Ministry is operated by the Finnish Navy and Merchant Marine at Helsinki naval base. Damage control training is also given in the simulator.

Sweden
A damage control simulator built by the Swedish Navy is located at Berga Naval School.

Venezuela
The Venezuelan Navy has built its own damage control simulator at the Centro de Adistramiento Naval C. N. Felipe Santiago Estevez in Catia La Mar.

Status
See text.

RDM Damage Repair Instruction Unit (DRIU)

The Dutch company de Rotterdamsche Droogdok Maatschappij bv (RDM) has developed a simulator for training personnel in damage control and repairs.

The Damage Repair Instruction Unit (DRIU) is a 10 × 5 × 6 m two-deck simulated ship section on a hydraulic platform which can tilt to 15° on either side. It includes watertight doors, hatches, pipe runs, stairways and lockers as well as a changing room and a classroom. There is a low pressure water supply system providing 'leakage' at 150 m^3/h as well as a high pressure water supply system for fire-fighting, both of which systems are recyclable. There is also a smoke-injection system.

Status
A system representing both the 'Kortenaer' class frigates and 'Alkmaar' class mine-hunters has been ordered by the Royal Netherlands Navy for installation at Den Helder.

Manufacturer
de Rotterdamsche Droogdok Maatschappij bv, Rotterdam.
(For details see Directory of Suppliers)

SINGAPORE

Vosper Naval Damage Control and Fire-fighting Training Simulator

This simulator consists of two training areas each of a replica ship's compartments on two decks. The first training area is for fire-fighting and contains portable equipment including a fire pump as well as systems for creating artificial fires. Here students are trained not only to fight a wide variety of fires but also in search and rescue techniques as well as the use of breathing apparatus.

The second training area is for teaching basic and advanced damage control and consists of a floodable compartment in which leaks can be simulated. The water comes from an integral tank.

Status
The system is in production and has been bought by the Brunei Navy.

Manufacturer
Vosper Naval Systems Pte Limited, Singapore Branch.
(For details see Directory of Suppliers)

UNITED KINGDOM

Marconi Simulation Nuclear, Biological and Chemical Defence Command, Control and Communications (NBCDC³) Training Systems

The Marconi Nuclear, Biological and Chemical Defence Command, Control and Communications (NBCDC³) Training System has been designed to ensure personnel are trained to fulfil their appropriate NBCD operational role. The training is achieved through a layered system of shore-based training.

The system has two teaching areas. There is a desktop computer-based trainer which provides self-paced learning in basic NBCD and specific ship NBCD procedures, in ship stability and in communications teamwork. This trainer has 16 student work stations linked for team communications training, and two instructor stations which monitor student performance.

The full-scale simulator is a replica of specific ship control (SCC) centres using a reconfigurable 'touchscreen' display for status monitoring and operator interaction. These SCCs represent those of the Type 22 or Type 23 frigates functionally, visually and spatially. Comprehensive instructor station facilities enable the instructor to create highly realistic and stressful scenarios for the NBCD command teams.

Status
Under development to form part of the relocated HMS *Phoenix* NBCD and fire-fighting school at Whale Island, Portsmouth.

Manufacturer
Marconi Simulation, Dunfermline.
(For details see Directory of Suppliers)

Marconi Simulation NBC Protection Training Unit

The Marconi Simulation NBC Protection Training Unit (PTU) provides realistic training to NBC protection, cleansing, monitoring and decontamination teams on board naval vessels. The teams are trained to carry out their tasks by following set procedures encountered in a contaminated environment following a NBC attack.

The PTU consists of realistic practical training areas which include a large, waterproofed floor area which represents a warship's upper deck with equipment including bollards, lockers, instrument binnacles, guns and hatches. Other training areas represent a ship citadel airlock, NBC cleansing stations, a ship magazine, a chemical training balcony and various out stations. The PTU also includes an instructor's control room.

Status

In service with the Royal Navy.

Manufacturer

Marconi Simulation, Dunfermline.
(For details see Directory of Suppliers)

MSI Defence Systems TN4 Plotter Navigator Training System

MSI Defence Systems Limited manufactures a wide range of naval and maritime products, including naval plotting tables, and, to assist training, it has developed the TN4 Plotter Navigator Trainer.

This system is designed to give navigators experience in the procedures and techniques for plotting a ship's course and contacts within the ship sensor range. The system consists of an instructor's console with a ship-contact simulator, communications and recording facilities and up to four DESL Type TN4 or SPL plotting tables each of which can accommodate up to three students.

The microprocessor in the instructor's console simulates the ship course and speed based upon predetermined ship dynamic characteristics and transmits simulated target track data to the plotting tables either automatically or orally. The system can simulate any time of the day or night and features a 'freeze' facility.

Status

Has been sold both in the United Kingdom and abroad.

Manufacturer

MSI Defence Systems Limited, Norwich Branch.
(For details see Directory of Suppliers)

A TN4 naval plotting trainer system in use

Vosper Thornycroft Damage Repair Instruction Unit

As a major shipyard and naval systems supplier Vosper Thornycroft has produced damage repair instruction units. Static models are in service with the Sultan of Oman's Navy and the Kenyan Navy but the most comprehensive unit is installed at the Royal Navy's shore training establishment at HMS *Raleigh*, Plymouth. This computer-controlled system can simulate flooding from splinter holes, distorted hatchways, holed bulkheads or distorted pipework and there can be 'electrical damage', all with realistic battle sounds. Students can be trained to repair battle damage, to handle casualties and practise search and rescue techniques in an environment which can be flooded with up to 95 tons of water or into which simulated smoke may be fed.

The system consists of a training unit measuring 14.6 × 6.7 × 7.77 m. It is three decks high and is outfitted to represent the interior of a typical frigate. The training unit is attached to a steel frame which can tilt it 15° either side of vertical at varying speeds. There is a control unit and also a pump house to flood the training unit.

Status

In service with the Royal Navy at HMS *Raleigh*, Plymouth. Static models are in operation with the Sultan of Oman's Navy and the Kenyan Navy.

Manufacturer

Vosper Thornycroft (UK) Limited, Southampton Branch.
(For details see Directory of Suppliers)

The Vosper Thornycroft Damage Repair Instruction Unit installed at HMS Raleigh, *Plymouth*

UNITED STATES OF AMERICA

AAI Fire-fighting Trainer

This training system consists of representations of various ship compartments in which there are metal reproductions of bunks, storage lockers, wire, boiler fronts, electrical panels, bombs, stoves and clothes dryers.

Beneath or behind each 'compartment' is a computer-controlled propane burner system which simulates fire growth. There are also sensors underneath which detect the various fire-fighting agents such as water or powder to ensure the right ones are used by the students. If the correct techniques are applied the computer extinguishes the fire but, if not, the computer will cause the 'fire' to flare. Instructors can also introduce non-toxic smoke.

Status

Devices 19F3B, 19F1B and 19F3C are in production while Devices 19F5A and 21C12A have entered service with the US Navy.

Manufacturer

AAI, Hunt Valley, Maryland.
(For details see Directory of Suppliers)

Symtron Systems Fire-fighter Trainers (19F1A/19F3/19F3A/19F4/19F5/21C12)

These trainers, reportedly the most advanced fire-fighting training devices in the world, have been developed in response to growing environmental concerns associated with traditional training methods. They use clean-burning Propane as a fuel source, non-toxic synthetic smoke generators for obscuration and computer control for flexibility and safety.

Mock-ups of prop-shaft bilge, oil spray bilge, wire bundle, electrical panel, motor generator, bunk, storage compartment, radar console, trash can, deep fat fryer/stack hood, submarine hull insulation and aircraft carrier deck Class A, B and C fires are available. The systems use industrial-type programmable controllers to monitor and control all aspects of the training scenarios. The system monitors the training compartment temperature and initiates ventilation if the preset limits are exceeded. Emergency stops placed at the entrance to each compartment and at the instructor's station allow the system to be shut down immediately if an emergency arises.

The basic trainers, Devices 19F5 and 19F3A, provide for basic instructional training responses to each class of fire. The 19F1A and 19F3A devices provide for advanced team training and have the capability of running as many as eight fires at one time from a selection of 13. The 21C12 device replicates the engineering compartment of a submarine and the 19F4 device provides for aircraft and helicopter deck fires.

One of the simulated fires in a Symtron Systems fire-fighter trainer

Status

Devices 19F1A and 19F3 are operational at Mayport, Florida, with one of each scheduled to be completed at Treasure Island during 1992. Device 19F5 is operational at Great Lakes, Device 21C12 is operational at New London and San Diego and Device 19F3A is at Newport, Rhode Island.

Manufacturer

Symtron Systems Inc, New Jersey.
(For details see Directory of Suppliers)

Systems and Simulation Landing Signal Officer Trainer (LSOT)

The Landing Signal Officer Trainer (LSOT) has been developed to provide an interactive instructional medium within a controlled classroom environment for training aircraft carrier Landing Signal Officers (LSOs).

The system, based upon a Gould 9705 computer, features an instructor operator station with a 19 in (482.6 mm) touch-sensitive, multifunction cathode-ray tube screen. This provides the instructor with a repeater to the LSO head-up display, or PLAT (Pilot Landing Aid Television) and allows him to monitor and to control instruction. He can control the visual displays, scenario selection, malfunction insertion and voice communications.

The student station, which can be used for self-training or under the supervision of the instructor, consists of a head-up display console with a 9 in (228.6 mm) cathode-ray tube that presents the simulator PLAT. There is a base station, a manually operated visual landing aid system controller and a landing area status light.

A computer-generated visual system provides visual simulation of up to six different moving aircraft types including the S-3A Viking, A-6E Intruder, the EA-6B Prowler, the E-2C Hawkeye, the F-14A Tomcat and the F/A-18 Hornet. The background scene provides varying seascape, light and environmental conditions and there is a 270° display system with a 6 m radius. The voice generation system provides deck noise, loudspeaker calls, communications on the same radio frequency and wind sound, together with authentic engine sounds.

Status

In production for the US Navy.

Manufacturer

Systems and Simulation Inc, Tampa.
(For details see Directory of Suppliers)

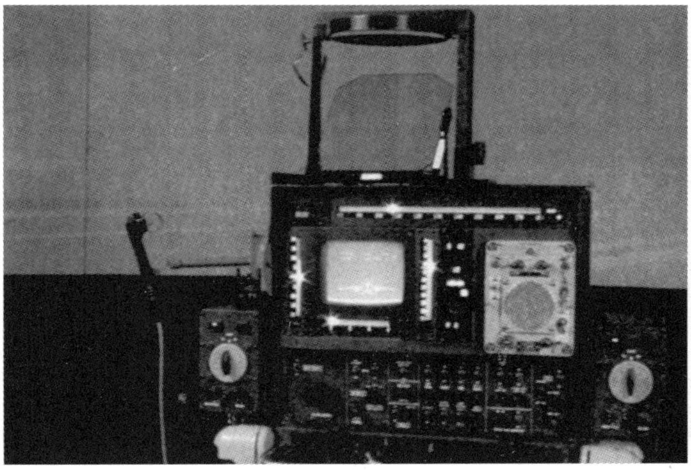

The Pilot Landing Aid Television (PLAT) repeater, incorporating a touch-screen, is part of the Landing Signal Officer Trainer (LSOT) student's station

The Landing Signal Officer Trainer (LSOT) student station with a computer-generated image view of an 'aircraft carrier' deck

AVIATION SYSTEMS

Flight Simulators and Flight Training Devices
Trainers for Non-Pilot Aircrew
Airborne Trainers - Non-EW
EW Trainers for Air Systems
Aircraft Maintenance Trainers
Air Traffic and Fighter Control Simulators

FLIGHT SIMULATORS AND FLIGHT TRAINING DEVICES

INTRODUCTION

This section is essentially about ground-based simulators and trainers for pilots. Where other aircrew are involved with the pilot or pilots in a simulator or training device (such as navigators, weapon system officers, EW officers, air observers, air engineers, loadmasters and so forth), their training stations in the appropriate simulator will be described in this section. Otherwise other aircrew stations are covered in the next section which is titled 'Trainers for Non-Pilot Aircrew'. Generally the pilot training devices in this section have either a full flight programme or at least some elements of one so that they can be 'flown' like the aircraft. A few devices such as CBT systems, disorientation trainers, aircrew selection trainers and some specialised weapons trainers, may not have a flight model and cockpit controls in the orthodox sense. Also there are some entries for Russia and the USA for Spacecraft Simulators. All of these devices are included in this section because they are designed primarily for the training of pilots and there are too few of them to create other sections of the book specifically for them.

Regulatory Levels. Before hardware is described, this introduction gives some tables describing the various levels of Flight Simulator (FS) and Flight Training Device (FTD) according to regulatory bodies such as the US Federal Aviation Administration (FAA) and the UK Civil Aviation Authority (UK CAA). No apology is made for quoting civil criteria in a book about military training. In many cases, military simulators are specified in terms of the civil criteria, and in any case many military types such as transport, VIP and tanker aircraft are based on standard civil aircraft. Also, manufacturers of flight simulators and flight training devices often quote the civil regulatory approval levels in their equipment specifications, and in the section that follows the different approval levels will frequently be mentioned.

Terminology Pitfalls. Care should be taken in differentiating between the same numerical categories for Flight Simulators (FS) which are used by the UK CAA and were used by the FAA (Called 'Phases') before 1991 on the one hand, and the lower-category FAA Flight Training Devices (FTD) on the other. Normally the level will be prefixed appropriately such as 'CAA FS-3' or 'FTD-3' for a UK CAA Level 3 Flight Simulator and an FAA Level 3 Flight Training Device, respectively. The old FAA Phase 3 category is now Level D, see the following table. There are also JARSIM levels. Terminology is somewhat of a minefield, hence these notes and the tables that follow.

Tables. Five tables follow:

Table 1 is a summary of the US FAA regulatory levels for Flight Simulators (FS).

Table 2 is a summary of the FAA training credits allowed for the various FS levels. Although designed for the civil market, similar systems for training credits are in existence for training on large, multi-engined military aircraft, at the discretion of the military aviation authorities in the nation concerned, who in general are their own national military aviation regulatory authority. It is not believed that any military organisation is carrying out 'Zero Flight Time' conversions because of the need to practise types of flying not allowed in civil passenger-carrying operations, such as tactical flying and activities such as Air to Air refuelling in large military aircraft.

Table 3 is a summary of the FAA regulatory levels for Flight Training Devices (FTDs).

Table 4 is a summary of the FAA training credits allowed for the various FTD levels.

Table 5 is a summary of the UK regulatory levels for Flight Simulators (FS).

International Simulator Regulatory Procedures. A note then follows on regulatory procedures in general, mentioning the role of the Royal Aeronautical Society (RAeS) in London whose Flight Simulation Group (FSG) commitee has co-ordinated and led an International Working Goup (IWG) which drafted and then defined an international standard for the acceptance of flight simulators. The RAeS standard has been accepted by the International Civil Aviation Organisation (ICAO) and by regulatory authorities such as the US FAA and the UK CAA.

JARSIM. A note follows on the European Joint Aviation Authorities (JAA) because flight simulator requirements (JARSIM) are being developed. A JARSIM summary is at Table 6. Prior to this, in Europe some flight simulators were being validated by up to five different national regulatory bodies. The US FAA is an observer to the JAA. A long-term intention has been declared of producing a world standard for both airworthiness and flight simulation criteria.

Relationships Between FAA, CAA, JAA and RAeS/ICAO Simulator Levels. Finally, Table 7 gives a comparison between the various simulator levels dealt with in more detail in the earlier individual tables.

Table 1 - Flight Simulators (FS) - FAA Levels
Regulatory Document: FAA AC 120-40B dated 29 Jul 91
Blank = not required

FAA FTD Level under AC 120-40B	A	B	C	D
Old FAA Phase under AC 120-40A	Vis	1	2	3
Motion System				
Motion System minimum DoF	3	3	6	6
Motion System special effects; ground rumble, buffet from thrust reverse & spoilers, bumps simulating gear travel		Yes	Yes	Yes
Motion System - Buffet from Mach effects, gear, flaps, stall, scuffing nosewheel				Yes
Visual System				
Visual System minimum FoV (Horizontal × Vertical) at each pilot's seat	45 × 30 per pilot, simultaneously		Collimated, 75 × 30 per pilot, 150° horizontal in total	
Visual cues to assess sink rate for landing and depth perception for takeoff (eg texture for judgement of height on landing, perspective & other cues for takeoff)		Yes	Yes	Yes
Quick Tests to confirm Visual System Performance			Yes	Yes
Dusk Scene to give horizon, fields, roads, lakes, etc			Yes	Yes
Minimum of 10 levels of occulting (correct lines of sight with respect to terrain contours and 3-D objects			Yes	Yes
Daylight Visual Scenes, minimum 400 edges or 1000 surfaces (polygons), minimum brightness 6 foot-lamberts.				Yes
Night Scenes minimum 4000 light points				Yes
Maximum Response Times (Latencies)				
Maximum for motion, visual & flight instrument response (millisec)	300	300	150	150
Modelling and Operational Realism				
Taxying & Ground Operations	Generic	All	All	All
Ground-effect for Takeoff and Landing		Yes	Yes	Yes
Reverse thrust aerodynamic and ground reaction model		Yes	Yes	Yes
Significant Sounds			Yes	Yes
Windshear Model			Yes	Yes
Realistic Landing Decell. Modelling runway, brakes & tyres			Yes	Yes
Quick test of programming & hardware (eg ATE)			Yes	Yes
Control Feel Dynamics to replicate aircraft			Yes	Yes
Realistic Sounds; cockpit noises, precipitation, wipers etc				Yes
Aerodynamic modelling inc Mach effects, aeroelasticity, icing, non-linearities in sideslip				Yes
Self-test for nominated performance tests; plus diagnostic analysis print-outs				Yes

Table 2 - FAA Flight Simulator Training Credits - Summary

Sim Level	Training Credits (FAR-121 Appendix H)
A	Crew procedures training Instrument flight basic and procedural training Limited transition, conversion, recurrent training, testing and checking except for takeoff and landing (TOL)
B	As Level A plus: Recency of experience Transition training for TOL Transition testing and checking except for TOL
C	As Level B plus: Transition, conversion, testing and checking of suitably experienced crews including TOL
D	As Level C plus: Zero Flight Time (ZFT) conversions of suitably experienced crews from a similar aircraft type

Evaluations of Individual Simulators: On initial validation and then annually, tests are made to an Acceptance Test Guide (ATG), generally over 2 days, together with other objective and subjective tests. Such tests are made by CAA pilots and a suitable technical inspector; the pilots are normally a Qualified Test Pilot (TP) plus the flight operations inspector for the type concerned. Between 5 and 8 hours of simulator time is taken for these tests.

Flying Conditions for ZFT Conversions

A general indication of the minimum conditions before a pilot may carry out a Zero Flight Time (ZFT) conversion to a new type are given on the next page and are taken from the UK CAA's CAP 453. At the international level it is not possible to be absolutely precise because the conditions vary between regulatory authorities (FAA (US), CAA (UK), LBA (Germany) etc). ZFT conversion

Table 3 - Flight Training Devices (FTDs) - FAA Levels
Source: FAA AC 120-45A

All FTDs require a replica cockpit, none essentially require either Motion or Visual Systems but a customer is at liberty to fit systems in excess of these minimum requirements
Blank = not required

FAA FTD Level	1	2	3	4	5	6	7
Type of Trainer		Generic (Rep = Representative of aircraft class)			Aircraft specific		
Daily Pre-flight documentation Light & C-Bs suitable for training task Instructor controls for activation of normal, abnormal & emergency conditions, for automatic action of FDT response without further Instructor input Simple seats for Instructor & FAA inspector with adequate view of crew members activities Computing including system evaluation & test Timely modifications to match aircraft mods		Yes	Yes	Yes	Yes	Yes, as aircraft where relevant	Yes, as aircraft where relevant
Aircraft System(s) installed and modelled to include normal, abnormal & emergency operation of the system(s) included, includes controls & switches Levels 2 & 5 require aerodynamics & controls/forces to fly a manual approach		Flt & nav system	Flt & nav systems for IFR ops	≥1 system	Flt & nav systems	All systems	All systems
Aerodynamics for forces on the aircraft & atmospheric variables		Rep	Yes, inc C of G effects		Yes	Yes, inc C of G effects	Yes, inc C of G effects
Instruments & nav equipment with automatic responses		Yes	Yes		Yes	Yes	Yes
Correct eye-point at crew seats			Rep		Yes	Yes	Yes
Simulator Systems Test Capability			Yes			Yes	Yes
Cockpit sounds resulting from pilot actions Intercomm & simulated R/T, Oxygen masks where appropriate to the aircraft							
Cockpit controls/switches replicate aircraft			Rep			Yes	Yes
Aerodynamic modelling inc Mach effects, aeroelasticity, icing, non-linearities in sideslip. Reverse thrust; modelling of aerodynamics and ground reaction. Control Feel Dynamics to replicate aircraft. Sounds of cockpit noises, precipitation, wipers etc in normal abnormal & emergency ops							Yes
Visual & Motion Systems (not compulsory for FTDs, but if fitted, to comply with AC 120-40 for Level A Flight Simulators)							

Table 4 - FAA Flight Training Devices Training Credits - Summary

FTD Level	Training Credits
1	Reserved
2	Crew procedures training, generic *ab initio* training, limited training for Instrument Flying (IF) and non-precision instrument approaches
3	As Level 2 plus: generic *ab initio* training for a particular aircraft family
4	Crew and systems procedures training, aircraft performance limitations training
5	As Level 2 but aircraft specific
6	Aircraft-specific training: pushbacks, engine starting, pre-flight checks, regulated takeoff, climbs and descents, steep turns, in flight engine shutdown and restart, NDB, VOR, RNAV, LORAN & other Nav procedures, back-beam approaches, training and testing IR non-precision approaches on all engines
7	All training equivalent to Full Flight Simulator Level C less visual and motion aspects

Table 5 - UK CAA - Regulatory Document CAP 453 (Sep 89)

CAA Level	Description of Device	Motion Platform	Visual System	Training Credits
1	Synthetic Flight Trainer, not type-specific	No	No	Intended for use as an *ab initio* training aid Some flying training (instrument) credits
2A	Non-visual Flight Simulator, type-specific	3 DoF	No	Type-specific conversion training and skill testing Recurrent training and checking, alternating with the aircraft
2B	Visual Flight Simulator	3 DoF	90° × 30° (each pilot 45° × 30°)	As for level 2A but without the need to alternate with the aircraft
3	Visual Flight Simulator	4 DoF	150° × 30° (each pilot 75° × 30°)	As for level 2B but greater credits are given towards training and checking, in that the night flying portion of type conversion and airline recency of takeoff and landing may be accomplished without recourse to the aircraft
4	Visual Flight Simulator, suitable for Zero Flight Time conversions	6 DoF	150° × 30° (each pilot 75° × 30°)	As for level 3 plus Zero Flight Time (ZFT) conversions of experienced crews from a similar type

conditions also vary within a regulatory authority depending on the training schedule and standards of training aids and flight simulators used by a particular airline, and the appropriate approvals given by the National Regulatory Authority.

Minimum Flight Times and Route Sectors Flown

Aircraft Commanders converting to Commander of the new type at least 2000 hours or 500 sectors as Commander, which must include at least 1500 hours or 400 sectors on similar types.

Copilots converting to Copilot of the new type: at least 1500 hours or 400 sectors, which must include at least 1000 hours or 300 sectors on similar types.

Copilots upgrading to Commander of the same type: at least 3000 hours or 800 sectors, which must include at least 1500 hours or 400 sectors on similar types and 500 hours or 200 sectors on the same type.

Closeness of Aircraft. Closeness of aircraft already qualified on, compared to the intended new type: Large jet to large jet, not prop to jet and vice-versa.

Supervision after Conversion. Supervision by a Training Captain after conversion before flying in the role to which the pilot has been converted:

Route Legs: not less than 12 of which 6 are supervised by a Base Training Captain and the rest by a Line Training Captain.

National and International Regulation of Flight Simulators

This ia an extract from a Position Paper on Flight Simulation produced by the Flight Simulation Group (FSG) of the Royal Aeronautical Society (RAeS) in London and published by the RAeS in 1994. The RAeS have kindly agreed to this extract being published in JMTSS. The full position paper may be obtained form the RAeS; their address is given in the Directory of Suppliers.

Civil Regulatory Procedures. In most of the leading aviation nations, the standard of flight simulators used for the higher levels of civil training is subject to close control by the national aviation regulatory body. In the USA, this body is the Federal Aviation Adminstration (FAA), in the UK the Civil Aviation Authority (CAA), in Germany, the Luftfahrt Bundesampt (LBA), and so forth. In the general training of individual pilots, appropriate training exercises and flight checks or tests may be carried out on approved simulators instead of in the aircraft itself, depending on the nature of the exercise, the experience of the pilot, and the type of simulator or trainer concerned. Such exercises conducted under these highly controlled conditions are said to gain 'credits' towards the appropriate qualification, in lieu of aircraft flight time. The highest level currently approved for this simulator training is the so called 'zero flight time' conversion where it is possible for an already experienced airline captain to convert to a similar aircraft type entirely on a high quality simulator, flying the aircraft itself for the first time on a revenue flight but under the supervision of a captain already fully type qualified. Simulators approved for 'zero flight time' training not only have to be built to certain standards, but they also have to have the appropriate level of motion and visual systems; they are not only independently checked by the regulatory body after manufacture, but are regularly re-checked to ensure that they continue to match the real aircraft characteristics in the critical areas for training. Lower conditions apply to simulators and other training devices which are used together with the aircraft itself. The regulations for simulator training 'credits' have been developed over many years and have kept pace with developments in flight simulation such as improved visual systems and rapid-reacting large throw hydraulic motion platforms; they have therefore been generally uncontroversial.

International Regulations - the Role of the RAeS. In 1989 the Royal Aeronautical Society (RAeS) established an international working group (IWG) chaired by a member of the RAeS Simulation Group Committee, with representation from industry, airline operations, and national regulatory bodies. The intention was to draft an international standard for the acceptance of flight simulators, using the RAeS connection in order to establish a neutral, apolitical environment, more likely to achieve acceptance than from other established bodies known to have special interests. After meetings in the USA, Europe, and at the RAeS Headquarters in London, a document was drafted and agreed unanimously at the RAeS in early 1992. This document covered proposed international standards for objective and subjective testing of flight simulators. The success of this process may be measured by the fact that the document has been accepted without alteration by the International Civil Aviation Organisation (ICAO), and has greatly facilitated the writing of national regulatory rules on the subject. As an example, the European Joint Aviation Authorities (JAA), is in the process of incorporating the RAeS document into its Joint Aviation Requirements for Simulators (JARSIM). The IWG has greatly facilitated the building of trust between various national regulatory bodies, and a number of joint simulator evaluations have already taken place between the regulatory bodies in the UK, USA, Canada, Germany, and Australia. A number of re-evaluations of individual flight simulators have been accepted by these five authorities when completed to the RAeS standard by one authority.

International Civil Aviation Organisation (ICAO). ICAO are producing an agreed standard, using the RAeS document almost unaltered. The US and UK civil regulatory authorities (FAA, CAA) have accepted the RAeS document as an alternative means of compliance.

JAA Requirements for Flight Simulators (JARSIM). These are common European requirements, under development by a working group on which all JAA nations are represented, and chaired by Capt Paddy Carver of the UK CAA. The managers of the FAA National Simulation Programme (Ed Cook and Paul Ray) are represented on the JARSIM working group. JARSIM may come into force in the period 1995-96. The long term intention is to achieve common criteria between the FAA and JAA which could then become an accepted world standard. Current proposals are: Annual evaluation of individual simulators; Qualification process with added approval for individual companies; Requirement for maintenance standards; Training Credits for four levels of simulation; RAeS document, modified, to form Advisory Circular, Joint (ACJ); Older simulators to be accepted as before (so-called 'Grandfather Rights').

JAA. The Joint Aviation Authorities (JAA) have a headquarters in Hoofddorp, the Netherlands (see under JAA in the Directory of Suppliers) headed by a Director General. The 23 nations which currently constitute the JAA are all the European Community (EC) nations plus others, the current nations are: Austria, Belgium, Cyprus, Denmark, Finland, France, Germany, Greece, Iceland, Ireland, Italy, Luxembourg, Malta, Monaco, Netherlands, Norway, Poland, Portugal, Slovenia, Spain,

Table 6 - JARSIM Levels - Summary

JAR Sim Level	Technical Standard	JARSIM Training Credits
A	FAA Level A	Type conversion training, testing and checking credits excluding takeoff and landing (TOL) Recurrent training, testing and checking except for TOL
B	FAA Level B	As Level A plus: Recurrent testing and checking extended to TOL
C	RAeS/ICAO Level 1	As Level B plus: TOL credits for pilots with certain minimum experience (Zero Flight Time, ZFT) TOL recency permitted
D	RAeS/ICAO Level 2	As Level C plus: Zero Flight Time (ZFT) may not require a minimum experience requirement for the pilot

Sweden, Switzerland, UK. The US FAA has observer status. Joint certifications, accepted by all of the JAA nations, have been achieved for aircraft from Airbus, Canada, Russia, Sweden, UK and the USA; and for engines from France, the UK and USA.

JARSIM. The current JARSIM requirements use the technical standards of the US FAA Levels A and B (see table 1) for JARSIM Levels A and B; Levels C and D use the technical standards from the RAeS/ICAO standard mentioned above.

Table 7 - Relationships Between Flight Simulator Regulatory Levels

Note: The levels increase in standard down the table, the highest level being at the bottom (Level D etc)

FAA (US) Level	JAA (European) Technical Standard	CAA (UK) Level	RAeS/ICAO (International) Level
A	FAA FS-A	2B	–
B	FAA FS-B	3	–
C	RAeS/ICAO Level 1	4	1
D	RAeS/ICAO Level 2	–	2

AUSTRALIA

HDH Hawker Pilot Trainer (HPT)

Hawker de Havilland (HDH) Victoria Limited manufactures the Hawker Pilot Trainer (HPT) which was developed by HDH together with the Royal Melbourne Institute of Technology. The HPT is a two-seat fixed base low-cost generic FTD with eight cockpit TV display screens. Two screens, one for each pilot, display the outside world, and the others are used for display of cockpit instruments and displays. Since the displays are computer-generated, they are inherently reconfigurable, and the trainer software currently allows the following options:

Single-engined, fixed prop, fixed gear, 2000-2500 lb
Single-engined, CS prop, retractable gear, 2000-2500 lb
Twin-engined, CS prop, retractable gear, 5000-5500 lb
Twin-turboprop, 9000-11 000 lb
Twin-jet, 10 000-12 000 lb

Exact layouts are configured to customer requirements.

Additional Graphics for Approach Control. One feature of the outside-world software is additional graphics which show the trainee indications of the ideal approach path. These additional symbols are in the form of T-shaped stalks which originate from the ground at frequent intervals on the extended approach, the height of the bar of the T being the ideal approach path for the range concerned. The symbols are then withdrawn when the student has made the appropriate progress in judging approach paths.

Display of In-Flight Forces. For basic tuition, a display is available of aircraft in-flight forces such as lift, thrust and drag; this display changes in accordance with flight conditions and control inputs and can be viewed in real time simultaneously with the outside world and other cockpit displays.

The HDH Hawker Pilot Trainer

Status
Available

Manufacturer
Hawker de Havilland Victoria Limited, Port Melbourne. (For details see Directory of Suppliers)

BRAZIL

ABC Dados Flight Simulator for Tucano

ABC's first simulator was of the EMB-312 Tucano turboprop trainer and followed the prototype built for CTA, the Brazilian Air Force Technology Centre. The work was carried out in association with the American company Intersim, and took 30 months to develop and manufacture.

Visual Image Generation (IG). CGI with variable visibility, runways, clouds and terrain.

Visual Display System (DS). Collimated monitor.

Motion Platform. Originally 2 DoF, upgraded to 4 DoF.

Main Computer. Originally a Brazilian microcomputer, later upgraded to a Gould Concept 32/27.

Instructor Operating Station (IOS). Mounted off board, the IOS can have either a closed circuit monitor of the cockpit or a duplicate set of the main cockpit instruments. A dedicated keyboard and colour monitor used for exercise control including changes in environmental conditions, the aircraft configuration and the choice of about 100 malfunctions.

The first of the ABC Tucano simulators

Cockpit. Uses real instruments linked to the computer system and to the instructor station.
Other Features. A sound generation system is fitted.

Status
The Tucano simulator for the Brazilian Air Force was produced in 1984 and series production began in 1986. Since then, five have been produced for the Iraqi and Egyptian Air Forces and two for the Brazilian Air Force Academy. The latter two entered service between June 1991 and January 1992.

Manufacturer
ABC Dados Informática SA, São José dos Campos.
(For details see Directory of Suppliers)

ABC Dados Flight Simulator for Brasilia

ABC Dados, in association with the American company Intersim, developed a flight simulator for the Embraer EMB-120 Brasilia range of light transport aircraft.
Visual Image Generation (IG). 2 or 4 channels of Vital IV CGI by FlightSafety International (ex McDonnell Douglas). Details in Image Generation section.
Visual Display System (DS). 2 or 4 collimated monitors.
Motion Platform. 6 DoF.
Main Computer. Gould Concept 32/27.
Certification. FAA Phase 2.
Instructor Operating Station (IOS). On board, with a colour video terminal and keyboard.
Other Features. Sound system.

Status
The system began training in 1988 in the USA and in 1990 in Brazil.

Manufacturer
ABC Dados Informática SA, São José dos Campos.
(For details see Directory of Suppliers)

ABC Dados simulator for the EMB 120 Brasilia aircraft

CANADA

Atlantis Cockpit Procedures Trainers (CPT)

These CPTs include a facsimile cockpit and an instructor operating station (IOS). The IOS is not only capable of setting up lessons for students but also of inserting faults during the exercise. The trainers provide simulation of engine aspects, and sound effects are included.
Customer Options. The customer may specify any number of features including open or closed cockpits, real or facsimile instruments and monochrome or full colour displays.
Aircraft Simulated. Aircraft simulated include the F/A-18 fighter, the OH-58 helicopter, the Canadair Challenger and the Lear-35 business jets.

Status
Available.

Manufacturer
Atlantis Aerospace Corporation, Woodbridge, Ontario.
(For details see Directory of Suppliers)

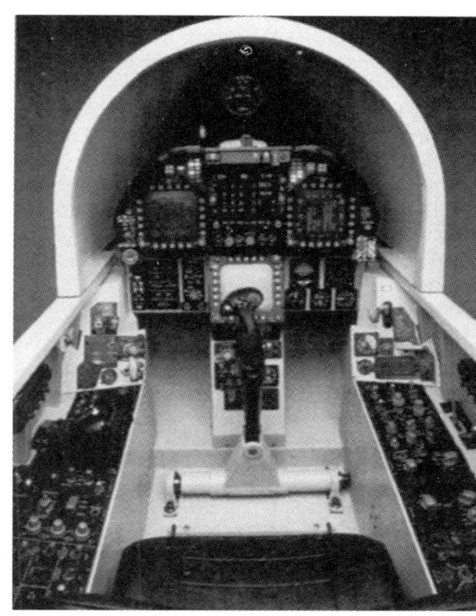

An F/A-18 cockpit procedures trainer produced by Atlantis for the RAAF

Atlantis Flight Training Device (FTD)

Atlantis has delivered a Level 5/6 King Air C-90A Flight Training Device (FTD) to the Canadian Department of National Defence for the Contracted Flight Training and Support Program. The trainer features an enclosed replica cockpit, an instructor station and student evaluation facilities. No motion or outside-world visual systems are fitted.

Digital electrical control loading is employed. Some 40 malfunctions specified by the customer can be simulated. A sound system not only includes communications, navigation and intercom facilities but also an audible feedback of engine, airframe and system sounds.

Status

In service with the Canadian forces as the Multi-Engine Operations and Procedures Trainer (MEOPT).

Manufacturer

Atlantis Aerospace Corporation, Woodbridge, Ontario. (For details see Directory of Suppliers)

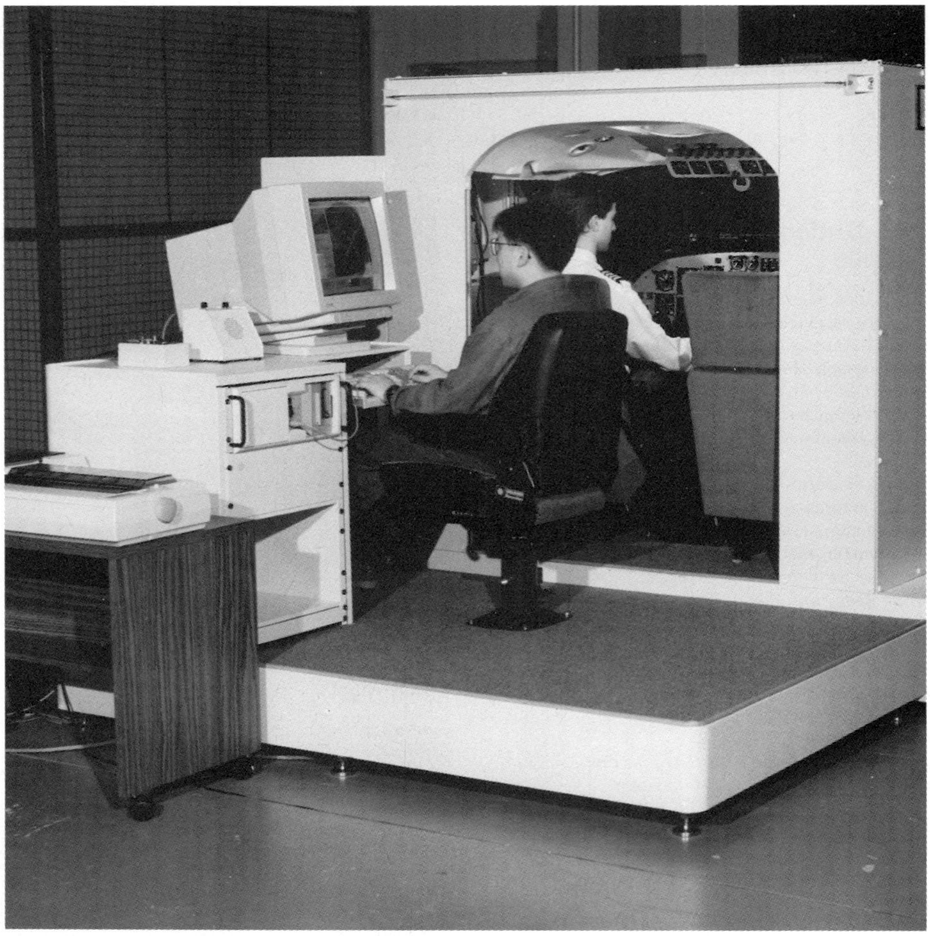

The Atlantis Multi-Engine Operations and Procedures Trainer

Atlantis Helicopter Instrument and Operational Procedures Simulator (HELIOS)

This is a procedures trainer for the OH-58/Bell 206 JetRanger III. It is designed to provide basic instrument flight training together with training in operational and emergency procedures. It has no motion or outside-world visual systems.

Components. It consists of a simulated cockpit with real or facsimile instruments, and an instructor station capable of fault insertion.

Features. The trainer simulates the appropriate helicopter systems including audio simulation of engine, rotor and warning systems. Synthesised voice instructions may be used, and simulated map areas may be presented to represent real maps for navigation purposes.

Status

Available. In service with the Canadian armed forces.

Manufacturer

Atlantis Aerospace Corporation, Woodbridge, Ontario. (For details see Directory of Suppliers)

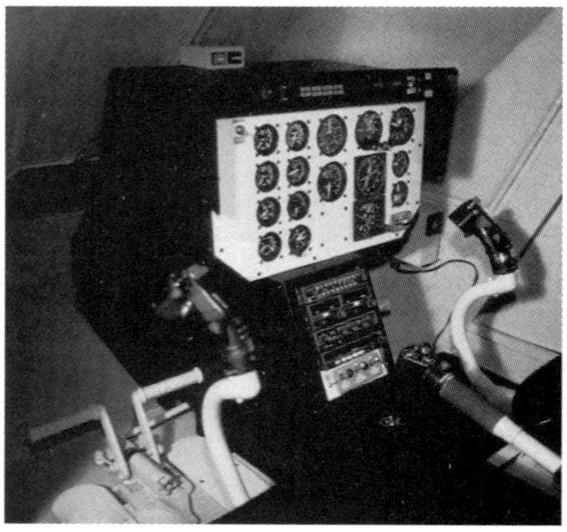

Atlantis Instrument Trainer (HELIOS) for Bell 206 Helicopter

CAE Flight Simulation

CAE Electronics of Montreal is a member of the Aerospace and Electronics Group of CAE Inc which also includes CAE-Link in the USA. CAE Electronics (subsequently abbreviated as 'CAE') produces both civil and military simulators and currently has about a 50% market share of the world production of civil Full Flight Simulators (FFS). As well as FFS, CAE produces Flight Training Devices (FTDs), computer-assisted training systems and Computer Based Training (CBT).

Visual Systems. CAE produces the MAXVUE™ visual image generation and display system (IG & DS), and the Fibre Optic Helmet-Mounted Display (FOHMD) DS. CAE will also fit other visual systems to a CAE simulator base, both IG and DS.

Motion Platforms. CAE produces a range of 6-DoF synergistic motion platforms, see the motion section later.

Instructor Operating Station (IOS). A full range of IOS facilities is produced, mounted on or off board at customer choice.

Certification. All certification categories are covered, including those of the FAA and European JAA.

Civil Fixed-Wing Production. CAE has built simulators for the Airbus A300, A310, A320, A330 and A340, Boeing 727, 737, 747, 757, 767 and 777, McDonnell Douglas DC-8, -9, -10, MD-80, MD-81, MD-82, MD-87, MD-88, MD-90 and MD-11 series, Lockheed L-1011, Fokker F 28, Fokker 50 and Fokker 100, Alenia/Aerospatiale ATR 42, Cessna Citation 500 and 550, Dash 8, Saab 340, Piper Cheyenne IIIA, Canadair Challenger CL-600 and 601-3R, and Canadair Regional Jet. Boeing awarded CAE a contract to develop and build the first full-flight and fixed-base simulators for the 777. CAE is also building the first MD-90 simulator.

FTDs. CAE's modular Flight Training Devices (FTDs) are custom-built to operate in a classroom environment. CAE offers a full range of FTDs which meet Federal Aviation Administration (FAA) levels and their international equivalents such as the European JARs and others. FTDs available include stand-alone designs which are suitable for commuter and general aviation aircraft and CBT devices.

Military Fixed-Wing Production. CAE has built simulators for the Boeing E-3A AWACS, Lockheed C-5B Galaxy and C-130 Hercules and P-3C Orion, Panavia Tornado, McDonnell Douglas CF-18 and A-4S Skyhawk and the Northrop F-5E Tiger II.

Canadian Forces. The CAF has a CAE C-130 Hercules flight simulator at Canadian Forces Base (CFB) Trenton. CAE designed and developed the hardware and software for the Weapons Systems Software Unit (WSSU) located at CFB Cold Lake, Alberta. WSSU includes the avionics of the CF-18 Hornet and a CAE CF-18 weapons simulator. It is used to assess software problems and then develop solutions for the aircraft, and to evaluate proposed software changes to the CF-18 and to carry out normal software maintenance and configuration management. In May 1990 CAE was awarded contracts to develop and produce CF-18 radar software test stations and to upgrade the CF-18 flight and mission simulator computers.

German Air Force. The GAF operates six CAE Tornado flight simulators which were built at CAE's facility in Stolberg, Germany. Originally fixed base, an enhancement programme exists to fit motion platforms and a wide angle visual system using the CAE Fibre Optic Helmet-Mounted Display (FOHMD) visual display system.

USAF. The USAF has six CAE C-5B Galaxy System

Trainers at Altus Air Force Base (AFB), Oklahoma, Travis AFB, Colorado and Dover AFB, Delaware. A seventh was installed at Westover AFB, Massachusetts in 1992.

Helicopter Simulation. Helicopter simulators have been built for the Agusta AB-205 and AB-212, Bell UH-1D, Boeing-Vertol CH-47, Sikorsky CH-53, the Westland Sea Lynx and the Sea King Mark 41. The Lynx ASW helicopter simulator for the Royal Netherlands Navy is also used by the Danish, German and Norwegian navies. CAE is building an Agusta A109 helicopter simulator for Siemens (Belgium) and this will be delivered to the Belgian Army in 1994. CAE is also building a Utility Tactical Transport Helicopter (UTTH) simulator for Bell Helicopter as part of the CF UTTH programme which will feature an eight-channel CAE MAXVUE™ visual system.

Blade-Element Modelling. A mathematical rotor blade-element model has been developed. The model treats the helicopter, the main rotor hub and hinges, together with the rotor blades themselves as separate elements. Blade flapping and lagging are modelled. Velocities are calculated at five points along each blade to allow for the varying Mach and Reynolds number effects. The use of a high-speed computer (six to ten times faster than normally used in FW simulators) enables a typical rotor blade to be simulated at every 7° during its rotation, thus enhancing the realism of the simulation.

Improved Rotorcraft Simulation Programme. CAE is presently involved with the Simulation and Improved Rotorcraft Integration Technology (SPIRIT) programme. This is to develop, demonstrate and deliver advanced rotorcraft simulation technology systems, airborne helmet displays and hand controllers. SPIRIT will feature a two-channel CAE MAXVUE™ image generator using the CAE FOHMD visual display system.

Status

All designs of full flight simulators, FTDs and CBT are available for civil and military fixed- and rotary-wing aircraft, as is the MAXVUE™ visual system and the FOHMD ultra-wide angle display system.

Manufacturer

CAE Electronics Ltd, Montreal.
(For details see Directory of Suppliers)

CAE Full Flight Simulator with 6-DoF motion platform and CAE MAXVUE™ Visual System

CAE Crew Station Research and Development Facility (CSRDF)

The Crew Station Research and Development Facility (CSRDF) was developed for NASA and is based at the Ames Research Center at Moffett Field near San Francisco (for details of NASA Ames see the Directory of Suppliers)

General. CSRDF is a generic model which accommodates many helicopter configurations and provides comparative evaluation of the effectiveness of crew performance as well as associated MMI considerations of one- and/or two-crew cockpit configurations while employed in combat mission tasks.

Visual Image Generation (IG). 3 channel CGI.

Visual Display System (DS). CAE FOHMD, giving essentially an unlimited FoV. Details in Visual Display Systems section.

Motion Platform. None, fixed-base.

Main Computer. Gould Concept 32/27.

Instructor Operating Station (IOS). An experimenter/operator console is designed for flexible control of sorties.

Cockpit. Two helicopter cabins for one- and two-man configurations. A multi-axis single-hand controller is fitted, which can be used to fly the aircraft with one hand, without the need for separate cyclic and collective controls.

Other Features. Three blue/red team stations, a mission communications support station.

Tasks. CSRDF has been used to help determine the cockpit configuration of the future US Army Light Helicopter (LH). It was also instrumental in the development of the Boeing-Sikorsky RAH-66 Comanche helicopter prototype for the US Army. The CSRDF has been used to simulate five classes of mission tasks: anti-armour, armed reconnaissance, anti-personnel/anti-material attack, security operations and command/control operations. The combat mission software package has simulated the tactical environment in terms of own ship and blue/red players' weapons, acquisition systems and countermeasures. It also simulates own ship and blue/red player target acquisition and designation equipment as well as RWR equipment.

Status

In service since 1986.

Manufacturer

CAE Electronics Limited, Montreal.
(For details see Directory of Suppliers)

CAE Simulator Complexity Test-Bed (SCTB)

CAE has developed a Simulator Complexity Test-Bed (SCTB) for the US Army Research Institute at Fort Rucker, Alabama. This is used to investigate different levels of simulator sophistication as they impact on transfer-of-training. CAE developed this simulator using a modular hardware and software architecture which can be reconfigured rapidly to represent different levels of simulator complexity up to that of the AH-64 Combat Mission Simulator (CMS). The SCTB uses CAE's FOHMD wide-angle display system together with CGI image generation. Issues of 'virtual reality' are also being addressed. SCTB may also be employed in a portable system which can be used for full mission training in a reduced area such as a field trailer.

Status
See text.

Manufacturer
CAE Electronics Limited, Montreal.
(For details see Directory of Suppliers)

The pilot station in the SCTB with the pilot wearing the Fibre Optic Helmet-Mounted Display (FOHMD)

The CAE Lynx helicopter simulator supplied to the Royal Netherlands Navy

FRANCE

CGA Alcatel Training Devices

CGA Alcatel offers a range of fixed base Flight Training Devices (FTDs) for single and twin turboprop aircraft. They have a central processing unit, simulated instruments, provision for IFR and radio navigation, instructor station, control loading, digital sound systems and a visual system, all to customer specifications.

Status
Available.

Manufacturer
CGA Alcatel Thomson, Paris.
(For details see Directory of Suppliers)

Sogitec Simulation

Sogitec is a subsidiary of Dassault Aviation and, amongst other products in the electronics and communications fields, manufactures flight simulators and trainers ranging from full mission devices to CPTs. It also produces the Apogée, GI 510 and GI 10 000 CGI systems and the SINDBAD database creation tool (The acronym SINDBAD is derived from Synthetic Image Numeric Data Base Aided Design).

Sogitec Full Mission Simulators

Sogitec manufactures mission simulators for the Mirage and other aircraft with dome-projected visuals.
Visual Image Generation (IG). Sogitec GI 10 000 CGI system for Air-to-Ground scenes. (Details in Image Generation (IG) Systems section). For Air-to-Air combat, two targets are generated from a choice of four types of aircraft.
Visual Display System (DS). Projection on inside of 8 m diameter dome. A horizon projector and two air target projectors are used for air combat exercises.
Motion Cueing. No motion platform is fitted, but pilots use their own anti-G suits in the simulator which allows realistic cueing for simulated G as the simulator inflates the suits appropriately. Also, a Sogitec design of simulator-specific G-seat is fitted; this is described in the section Motion Systems under G-Seats.
Control Loading. Digital.
Instructor Operating Station (IOS). The IOS can be used by one or two people. A television screen displays

Sogitec Mirage 2000 combat simulator dome showing the Sogitec GI 10 000 CGI imagery

the computerised ground image with superimposed HUD symbology as seen by the pilot. The instructor can 'fly' one of the target aircraft against the trainee. He can also introduce malfunctions.
Other Features. The simulated tactical area is 2000 × 1400 km.

Status
The system has been exported to several countries, including Peru. It has been configured for the Alpha Jet, Mirage F1 and Mirage 2000.

Manufacturer
Sogitec Industries SA, Suresnes.
(For details see Directory of Suppliers)

Sogitec Combat Tactics Trainers (CTT)

Sogitec has developed the Combat Tactics Trainer (CTT) as a low-cost aid to training in air-to-air and air-to-ground weapons systems operations. They are deployed at squadron level and provide real-time training in tactical scenarios.

Status
Three Mirage 2000 CTTs are in service with the French Air Force.

Manufacturer
Sogitec Industries SA, Suresnes.
(For details see Directory of Suppliers)

Sogitec Puma and Super Puma FMS

Sogitec has produced an FMS for the French Army's light aviation (ALAT) for their Puma and Super Puma Helicopters.
Visual Image Generation (IG). Sogitec three-channel CGI with imagery for landings in small areas or on ships, in severe weather, and for NVGs.
Visual Display System (DS). Projection onto inside of a Sogitec-manufactured dome of 8 m diameter inside which the cockpit and motion platform are mounted. The three-channel visual display has a 200° horizontal and 100° vertical FoV.
Motion Platform. 6 DoF by Hyraudyne with throws of ±25° pitch, ±25° roll, ±32° yaw, ±0.88 m heave, ±0.55 m sway and ±0.6 m surge. Real-time BITE is provided.
Instructor Operating Station (IOS). Mounted off board.
Cockpit. May be configured to Puma or Super Puma.
Other Features. Sogitec's SINDBAD interactive database creation tool is provided which will create scenarios.

Status
In service at the ALAT training centre at Le Cannet des Maures in the South of France.

Manufacturer
Sogitec Industries, Suresnes.
(For details see Directory of Suppliers)

Cockpit view in the Sogitec Super Puma FMS

Sogitec Super Puma Full Mission Simulator for the French Army showing projection dome surrounding the motion platform and cockpit

Sogitec Cockpit Procedures Trainers (CPTs)

Sogitec has produced a number of trainers which are designed to provide familiarisation with the cockpit, training on normal and emergency procedures, instrument flying, and radio navigation system operations.

Status
Alphajet CPTs are in service with the French Air Force and with foreign customers.

Manufacturer
Sogitec Industries, Suresnes.
(For details see Directory of Suppliers)

Thomson Training and Simulation (TTS)

Thomson Training and Simulation (TTS) Air Combat Simulators
The Air Combat Simulator (ACS) developed and built by Thomson entered service at the French Air Force base at Mont de Marsan in 1985. This multi-dome system consists of a hall containing three 8 m diameter domes. The simulator provides combat training for two-against-two Mirage F1 or Mirage 2000 aircrew.
Visual Image Generation (IG). Target aircraft images are generated by a Visa CGI system. The dimensions and attitude of the target are calculated for ranges from 60 to 6000 m. Details on Visa in the Image Generation (IG) Systems section.
Visual Display System (DS). Images of the sky, ground, enemy aircraft and missiles fired are projected onto the inner surface of the domes by a horizon projector and a target projector. The horizon projector gives a view of the ground varying with altitude and provides roll and pitch angle references. One of the three domes is also equipped with the MARS (Multi Aircraft Raid Simulation) laser-projection system; up to four target aircraft and two missile images may be projected at any one time.
Motion Cues. The simulators are fixed-base but G cues are produced by a G-seat and by the use of the pilot's own anti-G suit which is inflated appropriately by the G-seat pneumatic system in response to computed G.
Control Loading. Digital.
Instructor Operating Station (IOS). Mounted off board. The flight director's station has three colour graphic screens and four monochrome screens giving each pilot's radar screen and head-up display and continuous information and views of the combat sequences. Eye-points include perspective, cockpit and combined views. Utilisation and initialisation pages are also available.

Instructor-Flown Target. A target can be manually flown from the flight director's console with a joystick and a throttle.

Record/Replay is fitted.
Cockpits. The cockpits contain only those systems needed during the combat phase. This includes flight controls and air-to-air weapons system functions.
Deployment. In service at Mont de Marsin and with the Mirage 2000C training centre at Orange.

Thomson Mirage 2000C Simulator
The Mirage 2000C simulator is to train pilots in all missions, and in particular against low altitude penetration raids. The 'look-down, shoot-down' RDI radar and fire-control system is simulated and is correlated with the visual imagery.

Visual Image Generation (IG). Thomson Visa 4 CGI is used for ground imagery, details in Image Generation (IG) Systems section. A separate system generates

Thomson Air Combat Simulator IOS

Thomson Air Combat Simulator 8 metre domes at the French Air Force base at Mont de Marsan

images of target aircraft and missile imagery for projection by the MARS laser target projectors.

Visual Display System (DS). Projection on the inside of an 8 m diameter dome gives almost a full 360° FoV.

The MARS (Multi-Aircraft Raid Simulation) projection system uses laser scanning. The laser projectors are located on either side of the simulated cockpit outside the pilot's field-of-view and, due to the essentially parallel laser beam, produce high resolution images. Four target aircraft and two missile images may be projected at any one time.

The Janus image projection system includes two fish-eye lenses, one in front of the cockpit and one behind it. This system projects the CGI images of ground, horizon and sky background over the whole dome.

Motion Cueing. Fixed-base but with G-seat and use of the pilot's anti-G suit.

Control Loading. Digital.

Instructor Operating Station (IOS). Mounted off board.

Other Features. Simulation of the secondary ground attack role is also included.

Deployment. Ready-for-Training (RFT) dates were March 1990 at Orange Air Base and July 1992 at Cambrai Air Base. Two more are to be delivered to the French Air Force and three are on order for export.

Thomson Mirage 2000 simulator can be used for air refuelling training

Mirage 2000D Flight Simulators

Thomson has developed, in co-operation with Sogitec Industries SA (a subsidiary of Dassault), a flight simulator for the Dassault Mirage 2000D optimised for the ground attack role.

Visual Image Generation (IG). 5 channel CGI.

Visual Display System (DS). Projection onto inner surface of 8m wide spherical dome. FoV is ±130° horizontal and +90°/−35° vertical, total 260° horizontal and 125° vertical. There are six projectors, three light-valve type for the lower resolution side windows and a centre window, and three tri-tube type for the high resolution centre windows.

Motion Platform. 2 DoF (pitch and roll), electrically actuated.

Cockpit. Two seats in tandem.

Other Features. Simulation of the following: Antilope radar including Terrain Following (TF), Uliss 52 Navigation system, GPS, Thermal Imaging, Laser Guided Weapons, Laser Target Designation, and ECM systems.

Deliveries. Eight Mirage 2000D simulators have been ordered, five for the French Air Force for their base at Nancy/Ochey, and three for overseas nations. The FAF simulators are to be Ready-for-Training (RFT) in the first quarter of 1996.

Mirage 2000N Flight Simulator

Visual Image Generation (IG). 5 channels of Thomson's Visa 4 CGI. Details in Image Generation (IG) Systems section.

Visual Display System (DS). 5 projectors provide a FoV of 105° vertical by 160° horizontal.

Other Features. The radar and visual databases are correlated.

This fixed-base simulator is capable of training for a hostile environment at aircraft speeds up to 250 m/s (486 knots) and altitudes of less than 30 m (98 ft).

Transall C160 Flight Simulators

Luftwaffe

In April 1992 Thomson was awarded a contract by the German Defence Ministry for a simulator for the Transall C160 transport aircraft. This was the first Luftwaffe simulator contract won by Thomson and it will be delivered to the Wunsdorf Luftwaffe base in 1996. It will be used to train crews for transport, logistic and tactical low level flights.

French Air Force (FAF). The French Defence Ministry (DGA/DCAe) has ordered two C160R simulators for the FAF, to be delivered in 1995 to the Evreux transport base and in 1996 to the Toulouse transport training centre.

Visual Image Generation (IG). 5 channels of E&S ESIG 3000 CGI. Details in Image Generation (IG) Systems section.

Visual Display System (DS). WIDE with 220° horizontal and 50° vertical FoV. Details in Visual Display Systems section.

Motion Platform. Thomson 6 DoF.

Instructor Station (IOS). The IOS features a touch-screen, 30 visual techniques, automatic exercise control, and intelligent management of tactical scenarios.

Other features. Provision for NVG, radar, ECM, and AAR simulations. Also for assault landings and low-level delivery (skim release) of stores.

Helicopter Mission Simulators

Thomson has delivered a Super Puma simulator to the Swiss Air Force, and is producing for overseas customers 2 further Super Puma simulators and a Eurocopter Dauphin simulator for delivery in 1994 and 1995.

Visual Image Generation (IG). Thomson Visa 4 CGI for visual and FLIR imagery. Details in Image Generation (IG) Systems section.

Visual Display System (DS). Partial dome, FoV 150°

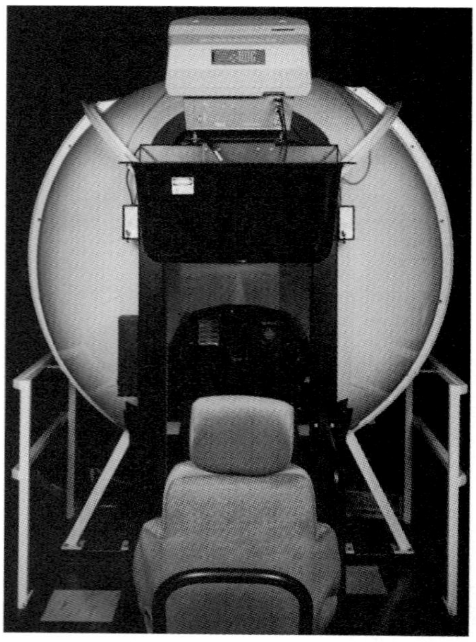

TTS COSMOS mini-sphere Air Combat Trainer

The Mirage 2000N Simulator

Thomson Super Puma simulator for the Swiss Air Force

IOS for the Thomson Mirage 2000N FMS

Cockpit of the Thomson Atlantique 2 flight simulator

horizontal and 70° vertical, Phebus projectors. Chin windows are included.
Motion Platform. Thomson 6 DoF.
Instructor Station (IOS). The IOS features a touch-screen, 30 visual techniques, automatic exercise control, and intelligent management of tactical scenarios.
Other Features. Blade-element mathematical modelling.

These simulators are designed for all aspects of helicopter activities including taxying, hover-taxi, take off, hovering, approach and landing.

Atlantique 2 Flight Simulators

Two simulators have been supplied to the French Navy for the Atlantique 2 (ATL2) Maritime Patrol Aircraft (MPA).
Visual Image Generation (IG). Thomson Visa 4 CGI. Details in Image Generation (IG) Systems section.
Visual Display System (DS). Monitors.
Motion Platform. Thomson 6 DoF.
Control Loading. Digital, 1000 Hz.
Other Features. A General On-Line Diagnostic (GOLD) system is fitted, providing rapid fault detection and is designed for 99 per cent simulator availability.

COSMOS Mini-Sphere Air Combat Trainer

COSMOS is a fixed-base very wide FoV trainer which has been developed in co-operation with the French Armed Forces Services, primarily for air combat simulation. It also has other applications where wide FoV is required at reasonable cost and collimation or long focal lengths are not critical factors. It could also be useful as research equipment to investigate wide FoV at lower cost than other systems.
Visual Image Generation (IG). CGI for background scene, laser projectors overlay up to four independent targets per COSMOS system with a resolution of 0.7 arcmin.
Visual Display System (DS). Projection on the inner surface of a dome of 2 m diameter. FoV ±180°.
Coupling and Networking. Can be coupled to existing simulators, used as a stand-alone device, or networked with other COSMOS trainers or other systems.
Transportability. Transportable due to its small size.
Deliveries. Two delivered to the French Armed Forces Services (DGA/DCAe) in 1993, for study of AD and GA for the Rafale and Mirage 2000 programmes.

Status

Available, in service and on order.

Manufacturer

Thomson Training and Simulation (TTS), Cergy-Pontoise and Trappes.
(For details see Directory of Suppliers)

GERMANY

CAE Flight Simulators

CAE Electronics GmbH at Stolberg near Aachen is a member of the Aerospace and Electronics Group of CAE Inc. The company maintains and builds simulators for, amongst others, the German military. They built the original 7 Tornado simulators for the German military, and have recently built an advanced Tornado Simulator Test-bed. The Stolberg facility also manufactures Aircrew Selection Systems and PTTs.

Tornado Simulators. CAE Stolberg built 5 Tornado simulators for the Luftwaffe and 2 for the Bundesmarine. They were delivered between 1981 and 1988. They are fixed-base devices with a 3-channel CGI visual system displayed on three collimated monitors. The image generator was developed by MBB from a basic Compuscene II system and incorporates real-world DLMS-based imagery. The CAE (Montreal) G-Seat is fitted which gives a number of motion cues and is described under G-Seats in the Motion Cueing section of this book.

Tornado Simulator Test-bed. The contract for a test-bed for a more advanced Tornado simulator was issued in June 89. This simulator has a modern 6 DoF platform and a visual system with essentially unlimited FoV; details are given below:

Visual Image Generation (IG). Evans and Sutherland ESIG 1000 with an eye-tracked centre (AOI) channel and a background channel. Details of ESIG 1000 in the Image Generation (IG) Systems section.

Visual Display System (DS). CAE FOHMD display system with a centre (AOI) display and a lower resolution background display. Details in Visual Display Systems section.

Motion Cueing. CAE (Montreal) Series 600 6 DoF motion platform; details in the Motion Platform section. The CAE (Montreal) G-seat is also fitted, which has 4 pneumatic seat pads and 4 pneumatic back pads, and lap belt tensioning. The G-Seat also incorporates pneumatic pressure to drive the pilot's normal aircraft anti-G suit; details in the G-Seat section of the Motion Cueing section of this book.

Control Loading. Digital, servo loops running at 2000 Hz.

Instructor Operating Station (IOS). Mounted off board.

Other Features. Briefing/Debriefing station is included.

Deployment. In use by the Luftwaffe and the Bundesmarine at CAE's plant at Stolberg.

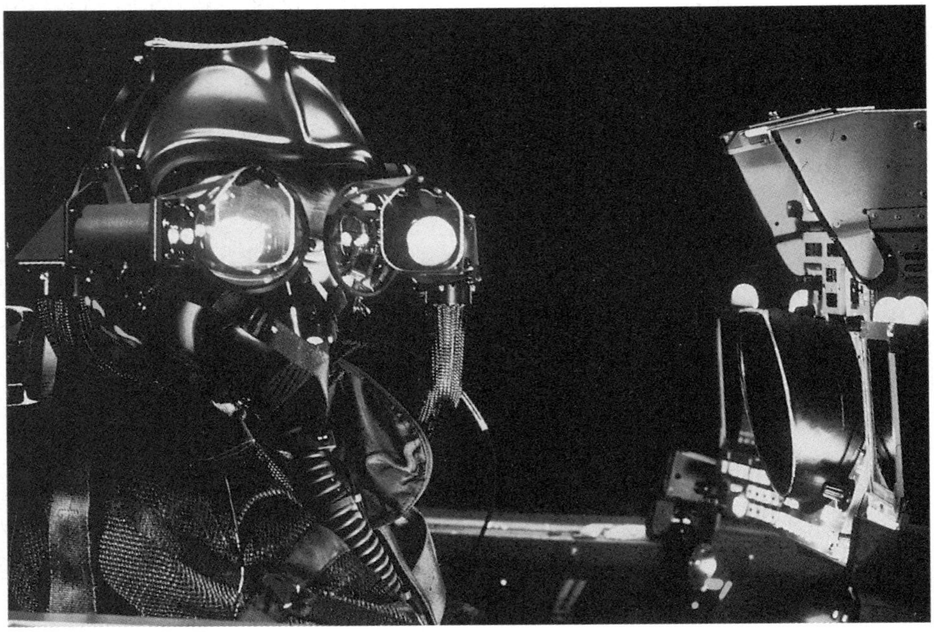

CAE Fibre Optic HMD on the Tornado Simulator

Aircrew Selection System. CAE Stolberg has developed a range of aircrew selection systems for both civil and military applications. They are designed to assess candidates for their abilities in information processing, problem solving, spatial abilities and psychomotor co-ordination.

Components. The main components are a Test Cockpit (TC) and a Test Attendant Station (TAS) which is essentially an IOS. The TC replicates basic aircraft controls and out-of-window views and essential instruments by colour monitors. Customer options include choice of controls, displays and visual channels. The TAS can control several TCs at once.

Tasks. The candidate is presented with tasks varying from the routine to the intensively dynamic, all designed to be relevant to the aircrew role for which the candidate is being considered.

Additional Measurements. Additional measurements made include stress tolerance and the willingness (or otherwise) to take risks.

Data Analysis. Data is collected not only on performance but on the underlying psychological processes. It is possible to differentiate between candidates who will always have problems against those who may, for instance, suffer test anxiety and so be likely to be trainable.

Deployment. In service with the Flugmedizinisches Institut der Luftwaffe at Fürstenfeldbruck, Germany, and the Nationale Luchtvaart School (NLS) at Maastricht, the Netherlands. A further two systems have been ordered by Chinese Airlines, one of which will go into service in 1994 and one in 1995.

Status

As indicated in text.

Manufacturer

CAE Electronics GmbH, Stolberg.
(For details see Directory of Suppliers)

CAE Tornado Simulator on 6-DoF platform

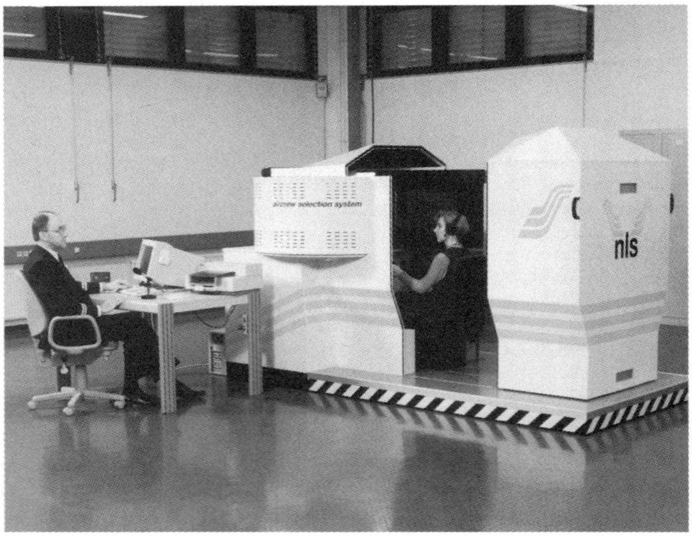

CAE (Stolberg) Aircrew Selection System equipment

Dornier Alphajet Air Mission Simulator

This simulator for the Alphajet is to train for the flight, navigation and weapons procedures of the Luftwaffe Tactical Combat Training Programme 80.
Visual Image Generation (IG). Three channel CGI.
Visual Display System (DS). Three collimated monitors, with FoV 140 × 32°.
Instructor Operating Station (IOS). Mounted off board.
Cockpit. Contains facsimiles of flight and other controls and the head-down displays.
Computers. Two digital computers are used to establish mathematic models for the weapon systems and also for data recording and evaluation.
Other Features. A sound generation system is fitted.

Status

In service with the German Air Force.

Manufacturer

Dornier GmbH, Friedrichshafen.
(For details see Directory of Suppliers)

Dornier Alphajet simulator

Simtec Flight Simulators

Simtec Simulation Technology of Braunschweig (Brunswick) manufactures a range of flight simulator components and also its own product line of procedure trainers and flight simulators. They have made a full flight simulator (FFS) for the Dornier 228 STOL twin engined transport, and a Fairchild Metro III simulator. Some of the products are listed below:
Modular Aircraft Simulation System (MASS). MASS is a system of modules and computer architecture designed to build up training aids for a customer starting with a basic cockpit procedure trainer (CPT). Then by adding items such as a sound system and Instructor Station (IOS) the basic CPT can be made into a Cockpit Emergency Procedures Trainer (CEPT). The addition of more type-specific data and more instrumentation would convert the CEPT into one of the higher levels of Flight Training Device (FTD) suitable for FAA and other approvals (see the table of FAA FTD levels at the beginning of this section). A customer might also wish to have a visual system fitted, although under the FAA FTD rules this is not essential for certification as an FTD. Finally, with the addition of a motion platform and a visual system with at least 150° × 40° FoV, a Full Flight Simulator can be produced which can be certificated to FAA Level C or D.
Instructor Operating Station IOS. The Simtec IOS features its Modular Instructor Panel System (MIPS) with standard modules including LCD display and software libraries.
Motion Platforms. Simtec makes a 3-DoF and two types of 6-DoF synergistic platform; details are tabulated in the Motion Platforms section of this book.
Wind Shear Modelling. Simtec has a MOdular Simulation of Environment Situation (MOSES) model which provides real-time 3-D prediction of wind, gust and turbulence, including microburst conditions. It meets US FAA requirements for windshear modelling.
Main Computer. VAX-VMS, PDP-RT11 and PC systems may be used. Flight dynamics calculated at 59 Hz.
Control Loading. Digital, 1000 Hz.
Sound Simulation. A Synthesised Aircraft Sound System (SASS) is available, based on measured sound characteristics from the real aircraft cockpit or the operating station for applications such as cars and locomotives.
Avionics Simulation. Avionics interface is compatible with ARINC 429 and MIL-1553-B formats.
Modular Instrument Interface System (MIIS). The Simtec MIIS fits to any host peripheral databus and interfaces with digital (including EFIS), analogue and converted instruments and indicators.
Certification. Up to FAA Level C, depending on customer requirements.

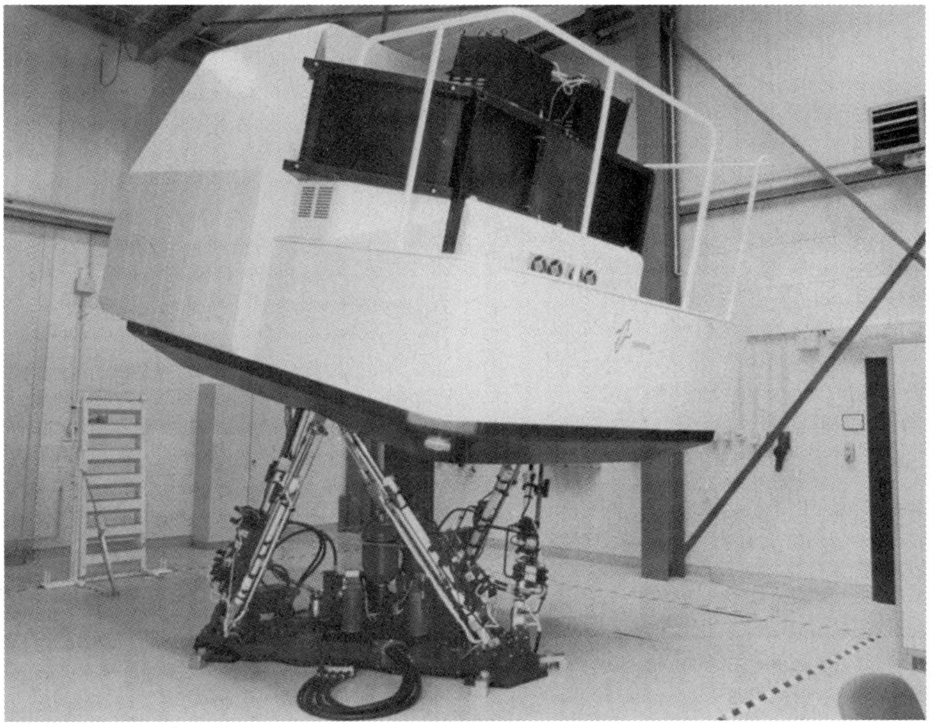

Simtec flight simulator for the Dornier 228 STOL twin turboprop. Note the Simtec compact 6-DoF synergistic platform

Status

Available.

Manufacturer

Simtec Simulation Technology GmbH, Braunschweig (Brunswick).
(For details see Directory of Suppliers)

INDIA

ADE Kiran and Ajeet Flight Simulators

The Aeronautical Development Establishment (ADE) of the Defence R&D Organisation (DRDO) of the Indian Ministry of Defence designed and developed simulators for the HJT-16 Kiran trainer and the Ajeet fighter. These simulators are in service with the Indian Air Force.
Visual Image Generation (IG). CGI with 6-12 databases for airfields, air to air and air to ground weapon training.
Visual Display System (DS). Single collimated monitor with FoV 48° horizontal and 36° vertical.
Motion Platform. 6 DoF, with throws of ±25° in Pitch; ±20° Roll; ±20° Yaw; ±2.5 ft Heave; ±3.33 ft Sway; ±3.33 ft Surge, with a 0.9 g capability in heave at an onset rate of 3 g/sec.
Main Computer. VAX11/780, update rate 20 Hz.
Control Feel. Hydraulic.
Instructor Operating Station (IOS). Graphics based with 45 minutes playback facility.
Cockpit. Based on an original aircraft shell.
Other Features. Sound system.

Status

In service. By early 1994 the Kiran simulator had been operated for over 3000 hours by the Indian Air Force.

Manufacturer

Aeronautical Development Establishment (ADE), Bangalore.
(For details see Directory of Suppliers)

ISRAEL

BVR HOTSHOT Air-to-Air Combat Tactics Trainer

The HOTSHOT air-to-air combat tactics trainer (CTT) made by BVR is a compact system intended for use at squadron level. It can be used in a stand-alone training mode, or with an instructor.

Cockpit and Displays. A typical trainer for the F-16A has three displays; a radar screen, a threat warning and an outside world display with a head-up display superimposed. The type of outside world display is a customer choice and can be a single monitor or a three channel CGI system projected on screens.

Engagements. The pilot can fly various engagement scenarios against a maximum of 20 enemy aircraft. Target aircraft will fly in response to the trainee pilot's actions, and try to intercept him and shoot him down.

Computers. An F-16A HOTSHOT trainer system incorporates five CPUs from the Motorola 68000 family. The main CPU performs flight calculations, another is used for the user interface and each display is supported by its own CPU.

Debrief. Debriefing facilities include recording for replay, storing for comparison and changing to the view from the other aircraft's cockpit.

Networking. The trainer can be networked, enabling fellow pilots to team up against multiple targets or to fly practice missions against each other.

Flexibility. Aircraft and weapon parameters can be changed at user level as can the type of scenario and the level of intelligence in the opposing aircraft's reactions. The trainer may be tailored for air-to-ground missions as well as air-to-air combat. It can be adapted to any existing or future combat aircraft.

Cost. A price of about $400 000 per unit has been quoted, but will depend on the visual system used.

The first F-15 system entered service with the Israeli Air Force in August 1987. All Israeli F-16A, F-16C and F-15 squadrons have HOTSHOT trainers, and the system has been demonstrated to several NATO air forces.

Marketing in the USA. HOTSHOT trainers are marketed in the USA by Loral Defense Systems Division, of Akron, Ohio.

Status

In production. In service with the Israeli Air Force and the Royal New Zealand Air Force (three systems for Skyhawk).

Manufacturer

BVR Technologies Limited, Gyvataim.
(For details see Directory of Suppliers)

BVR HOTSHOT trainer configured for F-16C

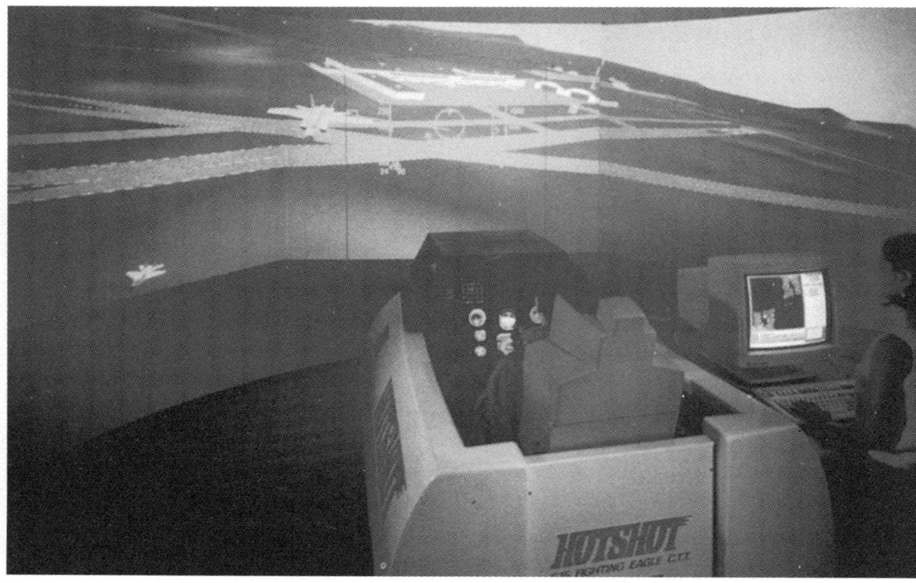

BVR HOTSHOT trainer with three channel projected visual system

ITALY

Agusta Flight Simulators and Trainers

Agusta Sistemi has been involved in a number of simulators and trainers. It is prime contractor for the A129 Mangusta Attack Helicopter Combat Mission Simulator (CMS) suite described below. Agusta Sistemi has also manufactured a Maintenance Simulator for the Italian Navy AB212 helicopter, CPT for the SIAI-Marchetti S.211 for the Philippine Air Force, and CBT for Alitalia and other airlines, for training in aerodynamics, meteorology, and general pilot procedures.

Simulators for the A129 Mangusta Attack Helicopter. Agusta is the prime contractor to the Italian Army for a Combat Mission Simulator (CMS) suite for the Agusta A129 Mangusta attack helicopter. The suite includes a Link-Miles (now part of Thomson Training and Simulation (TTS)) flight simulator, and a Link-Miles (now TTS) maintenance trainer and part-task trainers (PTTs). Agusta supplies the data package, the aerodynamic database, the systems simulation and the interface between the simulation and onboard computers.

Flight Simulator Two-Cockpit Layout. The TTS (ex-Link-Miles) A129 Flight Simulator has two replica cockpits for the pilot and co-pilot/gunner, each with its own dome-projected visual display system and mounted on its own 6-DoF motion platform. They are housed together and share common services, computing and image generation facilities. The layout is similar to that of the Apache CMS produced for the US Army by CAE-Link (qv in US section). The image generator is the Martin Marietta (previously GE) Compu-Scene IV. The two cockpits can be operated in independent or combined modes.

Status

In service.

Manufacturers

Agusta Sistemi Srl, Tradate and Link-Miles (now part of Thomson Training and Simulation (TTS), Crawley, UK).
(For details see Directory of Suppliers)

Alenia Flight Simulators

Alenia Defence Systems produces simulators and trainers ranging from Full Mission Simulators (FMS) on motion platforms to fixed-base basic FTDs. A Tornado FMS has been produced with a 6-DoF platform and has recently had improvements to the visual system and the addition of an EW module. Alenia also produces a CPT for the SIAI-Marchetti S.211 and a generic twin turboprop flight simulator with a two channel visual system. Other PTTs and CBT systems are produced for a variety of customers.

FMS Proposals. Alenia has proposed FMS designs for the Aeritalia AMX, the European Fighter Aircraft (EFA), the McDonnell Douglas AV-8B Harrier II and S.211.

Status

Available. In service and proposed, see text.

Manufacturer

Alenia Defence Systems, Simulators and RPVs Division, Goriza.
(For details see Directory of Suppliers)

The Alenia Tornado FMS on a 6-DoF motion platform

JAPAN

Mitsubishi Spatial Disorientation Pilot Training Device

The Mitsubishi Precision Company Limited of Tokyo produced a vertigo simulator for Japanese Air Self-Defence Force pilots. Trainee pilots using the simulator can experience realistic spatial disorientation conditions and learn how to recover from them.

Components. The simulator consists of a simulated cockpit station with visual display units mounted on a two-degrees-of-movement motion system plus signal converter electronics cabinets, all of which are located on a rotating platform. An instructor's console is also included, together with interface, visual, computer and power cabinets and a hydraulic power unit.

Cockpit. The simulated cockpit is equipped with instruments and displays which allow simulated VFR and IFR flight. The visual system can provide various images including a lead aircraft, lights on the ground, stars, a sun or moon, sun or moon rays, clouds, fog, and the horizon and landscape necessary to create conditions leading to spatial disorientation.

IOS. The Instructor Operating Station (IOS) is provided with the controls, indicators and displays necessary to set training scenarios, control training programmes, manage and teach a trainee pilot, and manipulate the leader aircraft. The simulator has been provided with 56 different training scenarios which can be used effectively in a variety of simulated flight conditions.

The conditions of spatial disorientation that can be experienced in this simulator can be described under the following headings:

Visual Autokinesis. In the darkened cockpit of the simulator a trainee pilot will be asked to look at the wingtip lights of a lead aircraft which is shown in the visual display lights or simulated cockpit windows. The trainee's aircraft and the leader aircraft fly in formation and maintain their relative positions constantly. Within a relatively short period the trainee will observe that the wingtip light of the leader aircraft appears to move up or down as the trainee or wingman trails the leader aircraft, despite the stationary geometry between the two aircraft. This illusion is caused by the visual autokinesis effect.

Relative Motion Illusion. An adjacent aircraft overtaking can give the illusion that the trainee's own aircraft is decelerating.

False Horizon Illusion. A long cloud formation which is tilted at an angle to the true horizon is displayed. The trainee may be subjected to vibration induced by the motion system and will soon assume that the cloud gives a true horizon unless reference is made to cockpit indicators.

Light Point Illusion. A trainee pilot will be asked to fly his aircraft in a night environment without a visible horizon. He will observe simulated stars and ground lights and sooner or later will not be able to distinguish between them. As a result he may start to feel he is in inverted flight.

False Vertical from Sun or Moon Beams. A trainee pilot can be shown sun or moon beams coming through or round clouds and may assume that the direction of the sunlight is the true vertical direction.

The 'Leans'. This illusion is caused by the semicircular ear canals which cannot perceive angular motion below a certain threshold. A trainee can be exposed to a long and slow turn in simulated cloud and then be asked to roll out suddenly. He will then feel as though he is turning in the opposite direction. As a slight variation the trainee is subjected to a long and slow turn induced by the motion system. The trainee will be told to make a roll out which will produce an erroneous feeling in which the trainee will feel he is tilted in the opposite direction.

Somatogravic Illusion. In a low visibility environment a pilot may assume the direction of a combined force, a result of the gravity force and a linear acceleration force, as the true direction of the gravity force or the local vertical direction. He will then misjudge the true attitude of his aircraft. This is known as a somatogravic illusion. On the Vertigo Simulator the motion system will induce a nose-up or nose-down feeling during a sudden acceleration or deceleration environment so that a trainee will experience an erroneous attitude feeling. On acceleration after take off or on a missed approach, the illusion is of a greater pitch angle than in fact is present, and there is a tendency to compensate by lowering the aircraft's nose, with the danger of striking any high ground that is close to the climb-out path.

Coriolis Effect. Coriolis is another term for effects due to the apparent centrifugal force (outwards) and the true centripetal force (inwards) in an aircraft turn. If a pilot moves his head out of the plane of rotation during a turn or spin the Coriolis effect can cause an overwhelming illusion of a change in aircraft attitude. For example, if the pilot moves his head to the right during a left turn he feels the aircraft is going up. If he turns his head to the left in the same turn he will feel the aircraft is going down. On the Vertigo Simulator the turning platform will be rotated continuously and a trainee will be asked to move his head away from the plane of the rotation, that is, he will be instructed to operate a control lever on the side of the simulator cockpit to feel the Coriolis effect. This will teach the trainee not to move his head unnecessarily during a turn.

The Mitsubishi spatial disorientation trainer

Aim of Training in Spatial Disorientation. This training is designed to alert a trainee pilot to the possibility of an accident caused by erroneous sensations of aircraft attitude, particularly in critical flight phases such as during an approach to land in low visibility. During the simulated flight the trainee will be shown 'tilted' cloud formations, sun rays off the vertical, and will be subjected to various motion cues and will begin to lose his sense of true verticality. The flight will continue into a limited visibility approach to land. A missed approach procedure will be carried out and the trainee will experience the erroneous pitch up feeling due to acceleration. The turning platform will be rotated so that the trainee will be subject to the Coriolis effect, motion below sensory threshold, and other types of illusory perceptions. The trainee, having experienced these conditions, will be prepared for them in the real aircraft.

Status

Available. In service with the Japanese Air Self-Defence Force.

Manufacturer

Mitsubishi Precision Company Limited, Tokyo.
(For details see Directory of Suppliers)

NETHERLANDS

NLR, the Nationaal Lucht- en Ruimtevaartlaboratorium of the Netherlands

Nationaal Lucht- en Ruimtevaartlaboratorium stands for the National Aerospace Laboratory (NLR) of the Netherlands. NLR carries out theoretical and experimental research for Dutch and foreign aircraft industries, civil and military aircraft operators and organisations concerned with aviation and space flight. It obtains 70 per cent of its funding from commercial contracts for research and the rest from government for NLR's basic programme of research. NLR operates several wind tunnels, two laboratory aircraft, research flight simulators, a computer network, and has equipment for research in the areas of aircraft noise, Air Traffic Control, structures and materials, space flight, and electronics.

Flight Simulation Department

The Flight Simulation Department is responsible for the basic technology of flight simulation, its application in training research, performing experiments for customers and for use of flight simulation techniques in the integration of systems in flight. The facilities may be hired at negotiated rates for specific projects, subject to being compatible with the planned programme of research and trials, and appropriate security and other clearances. For contact addresses and telephone/fax numbers, see under NLR in the Directory of Suppliers before the indexes.

Modular Systems

NLR's research flight simulation facility is essentially modular for flexibility of research. It includes a number of cockpits, and systems such as visual, motion, computers, and a large amount of simulation software and tools.

Cockpits. The following cockpits are available:

A reconfigurable single-seat fighter cockpit with interchangeable pilot controls.

A side-by-side 'glass' transport cockpit, equipped with six and three CDUs, all fully programmable CRTs in the observer's seat are interchangeable with conventional instruments or real EFIS displays through an ARINC-429 bus.

An instrumented F-16 cockpit in Mid-Life Update (MLU) configuration (comparable with F-16 C/D block 50).

A basic F-16 mock-up equipped with CRTs and Touch Screens.

All cockpits, except the F-16 mock-up, can be placed either on the 6-DoF or 4-DoF motion platforms. The single-seat fighter and transport cockpits are equipped with electrohydraulic control loading systems with various controls such as wheel/column, centre stick or sidestick. Pilot primary flying controls, throttle, flap and trim handle are servo-driven.

Image Generation

Two channels of Evans and Sutherland ESIG-3000 are available.

Database Modelling. An Evans & Sutherland 'EaSIEST' database modelling system is available. Several geo-specific large area databases have been created including the Hunter-Liggett area of the USA. Customer-specific database development can be carried out, and the system can import DTED/DFAD and

A Hydraudyne 6-DoF motion platform with projection dome at NLR Amsterdam

MIL STD 1821 (ex-Project 2851) data in combination with photos for geo-specific terrain modelling.

TV Model Board. A 1:2000 scale model board is available for tasks not requiring CGI. Its gaming area is 24.5 × 9 km and it includes an airfield with three runways and Cat II lighting. This is normally displayed on a collimated monitor with a FoV of 48°H × 36°V. The image from the model board can be enhanced by graphical workstation techniques and, day, dusk and night images can be created.

Display System

A Vista View head-tracked projection system is inside a dome of 8.5 ft radius which is mounted on a 6-DoF motion platform. This gives a total possible Field-of-Regard (FoR) of nearly the complete 360°. The head-slaved image consists of an oval background of FoV 120° horizontal by 90°vertical, in the middle of which is a high resolution AOI inset with a FoV of 40°H × 30°V.

Motion Platforms

A 6-DoF synergistic motion platform specially made for NLR by Hydraudyne has movements of ±29° in pitch, ±30° in roll, ±41° in yaw, ±1.1m heave, ±1.4 m sway and +1.7/−1.3 m surge. Other characteristics of this platform are in the table in the Motion Platform section of this book.

Alternatively, a 4-DoF platform made by the Delft University of Technology may be used. This is also described in the table in the Motion Platform section, and has movements of +40°/−20° in pitch, ±19° roll, ±29° yaw and ±0.28 m in heave.

Sound Generation

A digital CHeSS/Paradigm Emax II sound simulation system is available. This creates cockpit-sampled sounds such as aerodynamic, engine, hydraulics, missile and gun-firing effects and so forth. A development station is available to modify or create new sound libraries and the system can create three-dimensional sounds.

Human Factors Measurements

Instrumentation and fittings in the cockpits allow measurements of variables such as heartbeat, respiration rate, blood pressure, EEG, eye-pointing and so forth.

Computing

Silicon Graphics Challenge L for real-time operation:

Four R4400 Mips processors (total 180 MWhetstones), expandable to eight processors
64 Mbyte main memory
Two disk units, total 2 GByte storage capacity
One Ethernet adapter
One SCRAMNet interface.

Another Silicon Graphics Challenge L is used for software development purposes, with the same characteristics as the above and also has a CD-ROM drive, 4 mm DAT recorder and 10 19 inch X-terminals.

Concurrent Computer Corporation Micro Five MultiProcessor System for real-time operations:

Two processors, expandable to five processors
16 MByte memory
Three disk units with a total of 900 MBytes
One Ethernet adaptor.

A dedicated real-time network can be used during simulator experiments next to a non-real-time computer network for data exchange between development computer, X-terminals, and graphics computers. A fibre optic-based Flight Simulator Interface System (FSIS) connects computer and hardware systems in a ring via SCRAMNet. Integration of actual flight hardware can be achieved through the FSIS Cockpit Avionic Node capable of connecting systems with the military bus system MIL-STD-1553B or the civil bus system ARINC-429.

Graphics Work Stations

Silicon Graphics Integrated Raster Imaging System (IRIS) work stations as follows: 4D/70 GT, 4D/85 GT, 4D/210 VGX, 4D/310 VGX, three 4D/310 GTXB, Indy "Eurocad", Indigo 2 Extreme, and several other Indigos in various configurations. These graphics work stations are used for cockpit display generation, out-of-the-window scene generation (limited FoV), and F-16 mock-up interface purposes.

Software

Aircraft models are available for simulation of: Fokker F28, Fokker 100, Fokker 50, Boeing 747-200/400, Cessna Citation, F-16, Metro II and also for a generic helicopter. All aircraft systems are simulated by a software counterpart, enabling easy modification of systems for research purposes. For instance, an FMS has been extended with data link functions and uses a touch-screen input device instead of the CDUs. The modular software allows the mixing of systems with any aircraft model: for example the Boeing 747 may be simulated with Fokker 100 EFIS or the B747-FMS may be utilised by the Fokker 100, and the F-16 HUD may be used in the Citation. All simulation models are written in FORTRAN-77; the 'C' language is used only for graphical models and interface software.

Software development tools. A library of mathematical models is continuously being updated or new ones are being added. NLR has at its disposal a wide range of software development tools, some tailored for the real-time flight Simulation application. A structured approach in software design is used for low-risk development.

Status

Facilities described in the narrative are available for contract work.

Supplier

NLR, National Aerospace Laboratory of the Netherlands.
(For details see Directory of Suppliers)

PORTUGAL

OGMA Gyromag 87

Gyromag is a spatial disorientation trainer designed and manufactured by the Portuguese Air Force's Industrial Department.

Components. It consists of a control station and an enclosed cockpit mounted on an arm some distance from a central pivot.

Performance. The cockpit can be rotated to a speed of 19 rpm while the arm on which it is mounted may be rotated at up to 25 rpm. Radial acceleration is up to 1.5 *g* at the cockpit mounting while pitch and roll movement of the cockpit is a total of 40° at a maximum rate of 0.5 rad/s.

Deployment. In service with the Portuguese Air Force at the Physiological Training Unit in Lisbon.

Status

Available. In service with the Portuguese Air Force.

Manufacturer

OGMA, Alverca.
(For details see Directory of Suppliers)

The Gyromag 87 spatial disorientation trainer

ROMANIA

SIMULTEC Flight Simulators

SIMULTEC of Bucharest makes a number of flight simulators and training devices which are described below. It also makes the TCGI-1000 and GEN-500 CGI systems, and a 6 DoF motion platform whose brief specification is given in the first entry below.

Fixed-Wing Flight Simulators. Simulators have been produced for the IAR-99 Hawk and the IAR-93 Avioane (ex-Yurom) aircraft.

Visual Image Generation (IG). 3 channels of SIMULTEC TCGI-1000 CGI system. Details in Image Generation (IG) Systems Section.

Visual Display System (DS). 3 projectors.

Motion Platform. SIMULTEC 6 DoF, unit cost about US$200 000. Throws +33°/-36° in pitch; ± 32° roll; ±39° yaw; 2 m heave (total); ±1.1 m sway and ±1.15 m surge. Heave acceleration ±0.8 Gz at an onset of 3 g/sec, at a payload of 2500 kg; max payload 4000 kg. Response time to step input (latency) 120 msec. FAA Phase B compatible.

Main Computer. Either VAX 3500 or 486/66 PC, using DOS (PC version) and FORTRAN, C.

Instructor Operating Station (IOS). Mounted off board, 16 system failures and emergencies can be selected by a keyboard.

Control Loading. Electromechanical.

Other Features. Sound system.

Cost. Unit cost about US$1 500 000.

SIMULTEC Helicopter Simulators

Simulators have been made for the Alouette and IAR-330 Puma. Their characteristics are such that difficult training manoeuvres such as autorotations and Engine-Off Landings (EOLs) may be carried out.

Visual Image Generation (IG). 3 channels of SIMULTEC textured GEN-500 CGI system. Details in Image Generation (IG) Systems Section.

Visual Display System (DS). 3 projectors.

Motion Platform. Digitally controlled SIMULTEC 6-DoF platform, brief specification as for item above on IAR 93 and 99 aircraft.
Main Computer. VAX 3500, IRIS Indigo or 486/66 PC, operating systems VMS, DOS or UNIX. Language (HLL) FORTRAN, C.
Instructor Operating Station (IOS). Mounted off board, IRIS Indigo graphic station working under UNIX on Ethernet.
Control Loading. Digital, hydraulic.
Other Features. Sound system.
Cost. Unit cost about US$2 000 000.

SIMULTEC EFIS/FMS Simulators

These are for the Electronic Flight Instrument Systems (EFIS) and Flight Management Systems (FMS) of aircraft such as the Airbus 310.
Components. They use two 486/66 PCs and the C language operating under DOS. Stylised flight controls, throttles, a runway visual image, and a sound generation system are available.
Cost. Unit cost about US$20 000.

Status

In service and available for export.

Manufacturer

SIMULTEC SA, Bucharest.
(For details see Directory of Suppliers)

IOS for the SIMULTEC simulator for the IAR-99 aircraft

RUSSIA

TsAGI Flight Simulators

The Tsentral'ny Aerogidrodinamicheski Institut (TsAGI) at Zhukovsky, Moscow Region, is the major centre for aeronautical science in Russia. As part of its activities, it produces simulators for Russian military and civil aircraft and spacecraft. These simulators are to assess flight characteristics (research role), and also used for crew training.
Su-27 and MiG 29 Simulators. Simulators for the Su-27 and the MiG 29 combat aircraft have been produced. A 5 DoF motion platform was provided. Computers were based upon IBM PC 286/287 and 386/387. These systems were used to train 70 test pilots in the Su-27 and assisted development of the 'Pugachov Cobra' manoeuvre. A similar simulator was used for deck-landing training for the Su-27K.
Air Combat Simulation TsAGI has also produced an air-to-air combat simulator in a 12 m diameter dome. Two targets were provided, one projected by laser and the other by optical projection. This system may be networked with others so that combats involving up to three pilots may be conducted.
Helicopter Simulation. A helicopter simulator with a 3 DoF platform has been produced.
Buran Spacecraft Simulator. The Buran spacecraft has been simulated.
CBT/FTDs. Flight-training systems based upon PCs have also been developed by TsAGI and 100 have been supplied to the Russian Air Force.
Froggy Low-Cost Simulator. A low-cost PC-based generic flight simulator named Froggy is under development. TsAGI is proposing a collaborative venture with Western manufacturers. The system will be on a motion platform but will be transportable.

Status

See text.

Manufacturer

TsAGI, Zhukovsky.
(For details see Directory of Suppliers)

SPAIN

CESELSA A-7P Mission Simulator (MS)

CESELSA has produced a fixed-base Mission Simulator (MS) for the A-7P Corsair II for the Portuguese Air Force.
Components. The MS consists of six major systems; the cockpit, the IOS, the host computer, the visual system, the real-time I/O system and the radar (DRLMS) simulator.
Training. The MS can be used for all phases of training including cockpit familiarisation and procedures, operational flight and manoeuvring together with tactical training and tactical missions. The simulator has four main operating modes; flight, navigation, air-to-air and air-to-ground. In the air-to-air mode the MS can provide single or formation targets under either instructor or automatic control.

Status

In service with the Portuguese Air Force.

Manufacturer

CESELSA, Aeronautical Applications Branch, San Fernando de Henares.
(For details see Directory of Suppliers)

The CESELSA A-7P Mission Simulator IOS

CESELSA AV-8B Mission Simulator (MS)

CESELSA built the AV-8B MS combining basic and operational flight training together with weapons and tactics training in one simulator.
Components. The system consists of a cockpit, image generation and display systems, an Instructor Operating Station (IOS), computers, a real-time I/O system and a motion system.
Visual Image Generation (IG). Evans and Sutherland CT5A.
Visual Display System (DS). A 6 m diameter dome surrounds the cockpit section and images are projected on the inside of the dome.
Motion Platform. Pneumatic-hydraulic.
Cockpit. Facsimile.
IOS. The IOS has two interchangeable touch-screen graphics displays. The instructor can select exercises and problems during the exercise as well as providing a variety of other features including a simulated ground-control station.
Computing. Gould 32/87, supplemented by CESELSA SES microprocessors and an AN/AYK-14 aircraft computer.

Status

In service at the Rota naval air base since 1988.

Manufacturer

CESELSA, San Fernando de Henares.
(For details see Directory of Suppliers)

The cockpit of the CESELSA AV-8B Mission Simulator showing the visual scene generated by the E&S CT5A system

CESELSA Operational Flight Trainer (OFT) for Casa C-101

The OFT was designed to familiarise student pilots with the Casa C-101 two-seat trainer. It does not incorporate a visual system, but is compatible with the retrofit of a visual system. It can simulate normal and emergency procedures, take off, approach and landing, instrument flight and navigation, and tactical missions. Emphasis has been given to the accurate simulation of control forces, and the feel system is provided by a high performance hydrostatic control loading system. The central processing system is based on a 32-bit VAX 11/780 computer, with 1.5 Mbytes of main memory and 56 Mbytes of disk storage.

Status
Two OFTs have been used by the Spanish Air Force Academy at San Javier, near Murcia. Two more advanced OFTs were delivered to the Spanish Air Force in 1989.

Manufacturer
CESELSA, San Fernando de Henares.
(For details see Directory of Suppliers).

CESELSA C-101 OFT with instructor station

CESELSA C-101/CC Operational Flight and Tactics Trainer (OFTT)

The C-101 Operational Flight and Tactics Trainer (OFTT, or SOTV in Spanish) produced by CESELSA provides training for the C-101/CC Aviojet light attack aircraft.

General Layout. The OFTT/SOTV is a fixed-base device with one collimated monitor displaying images generated by the FlightSafety International Vital V (ex-McDonnell Douglas) CGI system.

Training Roles. The OFTT/SOTV can be used for training in start up, taxying, take off, normal flight, normal and abnormal manoeuvres, radio navigation, visual and instrument flying, approach and landing procedures, and emergency conditions. The simulator also provides training in weapons delivery, both air-to-air and air-to-ground.

Status
In service with the Royal Jordanian Air Force.

Manufacturer
CESELSA, San Fernando de Henares.
(For details see Directory of Suppliers)

IOS of the CESELSA OFTT for the C-101/CC aircraft used by Royal Jordanian Air Force

CESELSA F-16A Mission Simulator (MS)

A Mission Simulator (MS) is being developed by CESELSA for the F-16A for the Portuguese Air Force.

This MS is being designed for commonality of technology and resources with CESELSA's A-7P MS to allow flexible use of the two simulators. Five of the F-16A MS subsystems will be used jointly by both simulators; these are the instructor station, the host computer, the visual system, the real-time input/output system and the DRLMS simulator. An interface console and the cockpit, with half dome projection system, will be specifically for the F-16A.

Computing. The system will use the latest generation of RSX Encore computer linked to the visual subsystem, digital radar land mass simulator and instructor station via an Ethernet-standard local area network.

Training. Cockpit familiarisation, operational flight and manoeuvring, emergency procedures and navigation.

Visual Image Generation. The system will have a FlightSafety International (ex-McDonnell Douglas) VITAL VII CGI system. The database includes fixed and moving targets on land, sea and air.

Status
Under development for the Portuguese Air Force.

Manufacturer
CESELSA, San Fernando de Henares.
(For details see Directory of Suppliers)

CESELSA F/A-18 Operational Flight and Tactics Trainer (OFTT)

CESELSA produces a fixed-base Operational Flight and Tactics Trainer (OFTT, SOTV in Spanish) for the F/A-18 Hornet.

Components. The F/A-18 OFTT consists of five major subsystems; the cockpit, the instructor station, computer, instructor's presentation subsystem, and the visual system.

Visual Image Generation (IG). 3 channels of FlightSafety International (ex-McDonnell Douglas) Vital IV. Details in Image Generation (IG) Systems section.

Visual Display System (DS). Three collimated monitors.

Training. The OFTT can be used for all phases of training including cockpit familiarisation and procedures, operational flight and manoeuvring, plus tactical training and tactical missions. Training may be carried out in two ways, either under complete instructor control (free flight mode), or in the automatic or semi-automatic mode (mission mode). The simulation includes the three main operation modes of the F/A-18; Navigation, Air-to-Air and Air-to-Ground. This includes flight refuelling, formation flying and landing in bad weather.

Air-to-Air Combat. In the Air-to-Air mode the trainee is able to perform all the air combat phases such as detection and avoidance of detection, closing, attack manoeuvres and disengagement. The simulator provides up to 10 targets, either as single units or in formation, under either instructor or automatic control.

Air-To-Ground Combat. The trainee can practise target acquisition using the various active and passive sensors available (for example radar, FLIR, and laser spot tracker/strike camera (LST/CAM).

Status
In service with the Spanish Air Force. Two systems have been ordered.

Manufacturer
CESELSA, San Fernando de Henares.
(For details see Directory of Suppliers)

IOS for F/A-18 OFTT

CESELSA SH-60B LAMPS Mark III Mission Simulator (MS)

To meet a Spanish Navy requirement CESELSA is developing a Mission Simulator (MS) for the Sikorsky SH-60B LAMPS Mark III ASW helicopter.
Visual Image Generation (IG). 4 channels of CGI.
Visual Display System (DS). 5 windows.
Motion Platform. The cockpit is mounted on a 6-DoF motion platform whereas the sensor operator's station is fixed base.
Main Computer. One of the helicopter's two AN/AYC-14 computers is used while the second, which controls ESM, is emulated by software.
Instructor Operating Station (IOS). The instructor station has two real hardware displays which relay data from the cockpit. There are two 19 in high resolution touch-screen displays to act as the main interfaces between the instructor and the students. The instructor also has a display which repeats the images shown on any of the four cockpit displays.
Sensor Operator Station. The sensor operator station has synthetic and/or recorded sounds in order to train the sonar operators. It also has an instructor's station for overall control of the simulator while it is in use.
Other Features. The pilot and sensor operator stations may be used for other separate or integrated crew training.

Status

Under development for the Spanish Navy.

Manufacturer

CESELSA, San Fernando de Henares.
(For details see Directory of Suppliers)

SAINSEL Flight Simulators

SAINSEL produces both fixed base and full flight simulators for aircraft such as the Airbus A-320 and the CASA CN-235. SAINSEL produces both the hardware and the supporting software.

Status

In production and service.

Manufacturer

SAINSEL, Seville and Madrid.
(For details see Directory of Suppliers)

SWEDEN

Saab Simulator for JAS 39 Gripen

Saab-Scania has developed several flight simulators for the JAS 39 Gripen, using units such as Ericsson computers and cockpit displays. Saab-Scania previously developed two flight simulators for the Saab Viggen programme.

Status

In service in the JAS 39 programme.

Manufacturer

Saab-Scania AB, Linköping.
(For details see Directory of Suppliers)

SWITZERLAND

Initiative Computing (IC) ELITE software and cockpit controls

IC's ELITE flight training software is designed to work on PCs of 386 standard and above (with maths co-pro) or on Apple Mac SE/30 and the MAC II family. It produces a flight instrument display and an outside-world window, similar to the Microsoft Flight Simulator series but being specifically designed for the aircraft training role and not the computer-games market. Although not specifically designed for the military market it has equal application for basic pilot training in both the civil and military field.
Components The system consists of a PC or Apple Mac computer with a colour screen, and hardware modules such as a pilot's control column, rudder pedals, throttle box and avionics panel. It can be free-standing but is more effective if mounted in a small, darkened cabin as in the photo.
Software Modules. Software currently exists for the following aircraft: SOCATA Tobago, SOCATA Trinidad, Mooney 200 hp, Cessna 172, Arrow, Basic Piston, Basic Jet, MD 81. Other software can be developed to customer choice.
Hardware Modules. Hardware is available as follows: Control Column (single stick and yoke types), Three-lever box which can be configured for throttle(s), mixture, gear, flaps, rudder pedals, and an avionics and switch panel for control of VOR, DME, NDB, Autopilot, Flight Director, ADI, HSI and so on.
Flight Analysis. Freeze, repositioning, restart, replay and elevation and azimuth analysis are available (the latter for approaches).
Navigation Databases. 17 databases of navigation aids are available for various regions of Europe; these include about 550 different regularly used airfields. 29 databases are available for the USA.

Status

Available, in service with Swissair amongst others.

Manufacturer

Initiative Computing AG, Winkel-Rütl.
(For details see Directory of Suppliers)
Supplied also by Aviation Software, Geneva.
(For details see Directory of Suppliers)

ELITE PC-based CBT in a small cubicle. Note the switch/avionics panel on the right, pilot's control yoke, and throttle box. The rudder pedals are out of sight

UNITED KINGDOM

BAe Air Combat Simulator

The BAe Air Combat Simulator (ACS) consists of two domes each with a fixed-base cockpit. The twin-domed configuration of the ACS allows interactive combat between two pilots. Combat can also be against a computer-driven opponent or against an instructor-flown opponent. Multi-player scenarios are increasingly used to allow realistic combinations of threat aircraft and weapons to be evaluated.
Visual Image Generation (IG). CGI is used for target images. The horizon and ground scene can be through CGI channels or more simple ground/horizon projection slides, at customer choice.
Visual Display System (DS). The domes are of 30 ft diameter and are made of flexible material which is inflated and kept under slight positive air pressure through an air pump. This gives a very accurate shape of screen surface with no discontinuities or distortions. Targets are projected by scanning laser projectors giving high resolution for visual identification.
Motion Cueing. The cockpits are fixed-base but pneumatic pressure is provided for the pilot's anti-G suit, programmed in accordance with the computed G of the manoeuvre demanded. A seat buffet mechanism is also provided to simulate high-G and other buffet and vibration conditions.
Instructor Operating Station (IOS). Mounted off the domes.
Cockpits. The cockpits are stylised and are only fitted with the instruments and controls necessary for air combat. They are modular and can be reconfigured to represent any aircraft, missile or weapon system.
BVR Engagements. Beyond-visual-range (BVR) combat may be included at customer choice.
Other Features. A sound generation system is fitted.

Status
In service and production.

Manufacturer
British Aerospace Defence Limited, Warton.
(For details see Directory of Suppliers)

BAe Air Combat Simulator cockpit view

BAe Fox 1 Beyond-Visual-Range (BVR) Trainer

Fox 1 is a desktop trainer for use in training, tactics development and evaluation of scenarios. It is designed to provide experience in the transition between beyond-visual-range and in-visual-range (BVR/IVR) either in stand-alone exercises or under instructor control.
Crew Stations. Fox 1 consists of several networked crew stations (single or twin seat) allowing engagements against up to twenty opponents. Each trainer has the primary flight and systems instruments, radar, radar warning receiver, weapon and Head-Up Display symbology together with the outside world, visual targets and friendly aircraft. A desk and VDU layout is provided with a stick and throttle as well as all HOTAS (Hands-On-Throttle-And-Stick) switches.
IOS. The IOS allows scenario setting, operational analysis and record/replay. Different aircraft, weapons and avionics can be simulated, while a database of development programs allows the instructor to edit models or to create new datasets for use in the simulator.

Status
Available.

Manufacturer
British Aerospace Defence Limited, Warton.
(For details see Directory of Suppliers)

BAe Fox 1 BVR trainer

BAe Avionics Part-Task Trainer (APTT)

The APTT enables aircrew to practise avionics systems management and to become familiar with instrument display symbology within the classroom environment. The APTT is also designed to ease the transition from cockpits with conventional analogue instrumentation to EFIS displays (glass cockpits). Typical instrument configurations for this device include Head-Up Displays (HUDs), radar and HOTAS (Hands-On-Throttle-And-Stick) controls. The device is open plan with a separate instructor console to allow monitoring of student performance, direct control by the instructor and for demonstration purposes.

Status
Available, in production for several air forces.

Manufacturer
British Aerospace Defence Limited, Warton.
(For details see Directory of Suppliers)

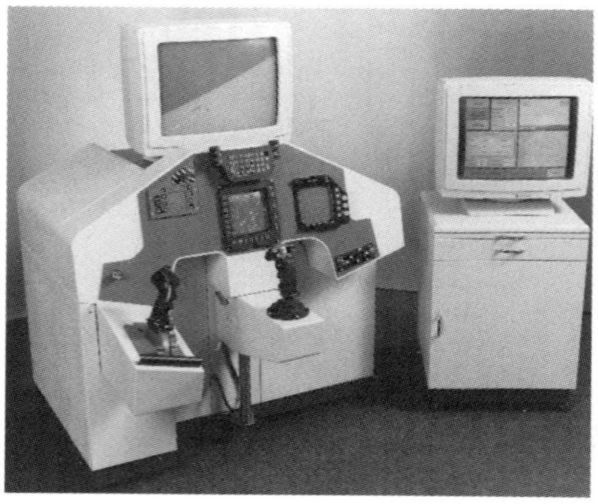

BAe APTT with HUD and HOTAS controls

BAe Cockpit Procedures Trainer (CPT)

The CPT is designed to provide aircrew with cockpit familiarisation and procedures training and to provide ground crew with cockpit-related maintenance training. Each device is designed to meet specific training objectives but a typical CPT consists of a full-scale cockpit with fully functioning instruments and controls, or realistic facsimiles, as well as a separate mobile instructor station with the ability to inject faults.

Status
Available and in service with several air forces.

Manufacturer
British Aerospace Defence Limited, Warton.
(For details see Directory of Suppliers)

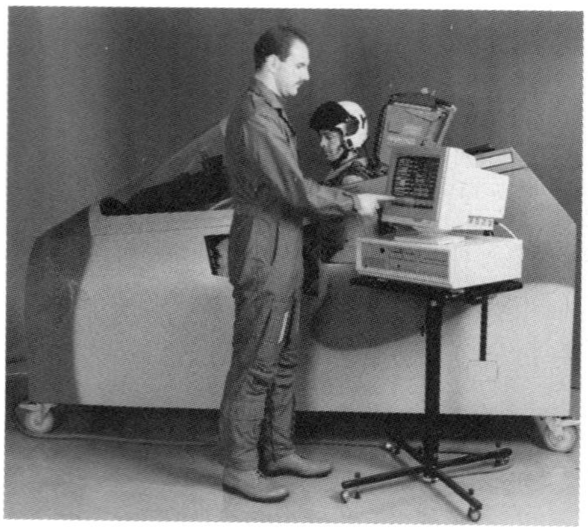

A BAe CPT with offboard PC-based IOS

BAe Computer Based Training (CBT)

BAe CBT uses the electronic display of graphics, video stills and sequences, text, audio commentary and sound effects for training through a keyboard and monitor screen. Lesson planning and progress is controlled through the instructor console. Typically the system consists of a PC-based instructor work station linked to a student work station. CBT provides self-paced tuition in all aspects of airframe, engine and avionic systems using a network of individual student work stations hosted by a single master workstation. Networked workstations can be brought under instructor control to present students with identical displays if required. Students have a mouse or touch-screen and are provided with headphones for audio instruction.

Status

In service with a number of air forces.

Manufacturer

British Aerospace Defence Limited, Warton.
(For details see Directory of Suppliers)

A Hawk UHF communications CBT student work station

Defence Research Agency (DRA) Aircraft Simulators and Trainers

The Defence Research Agency (DRA) of the UK MoD is concerned with a number of training devices and simulators, some of which are described below.

DRA Advanced Flight Simulator Complex (AFS)

The DRA Advanced Flight Simulation (AFS) complex is situated at the DRA site at Bedford. It has a very large throw motion platform, and a variety of visual systems varying from model boards to CGI. Simpler motion platforms are also available, as are modular generic cockpits for various classes of aircraft including helicopters. A G-seat is also available for investigation of off-platform cueing for motion and vibration, and for simulating high Gz by buttock pressure, seat-pan lowering and strap tightening and loosening.

Motion Platform. A large-throw electro-hydromechanical cable-operated system was made specially for the DRA by Vickers Engineering Ltd and was commissioned in 1990. It is believed to be second only in size of throw to that at NASA Ames but is of more recent construction. It was designed not for the training industry but for the use of simulation in flight research, and is capable of producing motion cues which match the aircraft for most flight regimes except sustained high Gz. A table of characteristics follows:

DRA Bedford Advanced Flight Simulator Large Motion System

Motion	Displacement ±	Velocity ±	Acceleration ±
Pitch	28°	28°/s	115°/s²
Roll	(56° Total)	57°/s	172°/s²
Yaw		28°/s	86°/s²
Sway or Surge (1)	4 m (8 m total)	2.5 m/s	5 m/s² (0.5 g)
Heave	5 m (10 m total)	3 m/s	10 m/s² (1.0 g)

Notes: (1) Either Sway or Surge depending on which way the cockpit is mounted.
(2) The motions above may be obtained both individually and simultaneously with any other motion.

The Large Motion System at DRA Bedford with 10 metre heave travel

Visual Image Generation. Link-Miles (now Thomson Training) Image 600PT, also various model boards used for high resolution low flying simulations.

Visual Display System. Collimated monitors giving a maximum simultaneous FoV of ±63° x 24° which can be positioned in the pilot's visual field at positions appropriate to the trial in hand.

DRA Expertise. The DRA lists expertise in the following areas:

Flying qualities requirements and assessments.
Cockpit displays, HUD, HDD, HMD and symbology.
Helicopter mathematical modelling, control system design and evaluation.
Systems Integration.
Development of aids and flying techniques for launch and recovery to ships.
Modelling of threat and atmospheric environments.
Simulation technology and validation of simulation techniques.
Project management and trials planning.

DRA HOVERS Simulator

HOVERS stands for Helicopter Operation Visual Engagement Real Time Simulation. Based at the DRA's Farnborough site, this is a fixed-base programmable helicopter simulator designed for research into operations in a nap-of-the-earth environment. It is not enclosed by a cockpit structure but the crew stations and displays are ergonomically and spatially correct. It can be configured for either one or two-crew operation, and for any combination of flight characteristics. Six visual channels are available, and can present outside-world scenes, sensor information, and special flight instrument displays.

DRA Farnborough Helicopter Simulator

This simulator has five visual channels and a Rediffusion 3 DoF platform giving pitch, roll and heave. It also has multifunction displays and a Helmet-Mounted Display capability. Images may be derived from high resolution model boards designed for low flying fidelity, or from an extensive library of CGI visuals which also include FLIR imagery.

Thermal Image of helicopter from HOVERS simulator at DRA Farnborough

Air Combat Dome

Situated at DRA Farnborough, exercises and trials involving Air Combat may be carried out using this facility.

Advanced Reconfigurable Mission System (ARMS)

The ARMS is situated at DRA Farnborough and has a FoV of 150° horizontal and 90° vertical. A range of adaptable cockpits allows rapid changes between aircraft categories and types. A complete range of weapons and sensors can be modelled, and there is an extensive library of CGI, EW effects and performance matrices.

JOUST

JOUST is an Air Defence Beyond-Visual-Range (BVR) simulation and allows fixed-wing pilots to participate in man-in-the-loop two versus two combats in a scenario involving other enemy aircraft. Cockpit stations are generic based on networked Silicon Graphics engines. Computation includes flight models, sensors, weapons and visual scene generation.

Farnborough Computing Complex

This consists of over 60 computers and work stations whose total capacity is in excess of 2000 Mips. These machines have a built-in graphics capability and communications, support the UNIX system and a variety of

Example of generic helicopter simulation at DRA Bedford

compilers, providing a productive programming environment. This enables rapid prototyping of systems to customers' requirements.

DRA Parallel Processing and Military Simulation Section

This section is based at the UK DRA facility at Malvern, the former historic Telecommunications Research Establishment which, during World War II, played a crucial role in the development of radar. The section studies the application of parallel and networked systems in simulation.

Facilities. Shared memory SparcCenter 1000 with eight 50 MHz SPARC processors; distributed memory Meiko CS-2 with 22 SPARC processors and a high-speed network comms backbone; network of UNIX work stations linked by Ethernet and FDDI connections.

Developments. Tools for parallel simulation programming have been developed for: C++ simulation libraries, graphical display libraries, and new simulation language compilers. Tools and libraries also exist for DIS programming and connectivity via ISDN and Internet for linking into confederations of distributed simulations. Some aims of the section are listed below:

Parallel Architectures. To exploit parallel architectures to speed up simulators and simulations for all military purposes. This includes simulation for training, OA, mission planning and procurement simulations.

Programming. To develop new simulation programming paradigms and languages.

Distributed Interactive Simulation (DIS). To advise on DIS applications to connecting simulations and to study standards for networking.

Artificial Intelligence (AI). To apply AI to simulators in order to include intelligent entities in the simulation.

Status

These equipments and facilities are in service with the DRA of the UK MoD, and are available for hire or supply subject to appropriate security clearances and to other DRA customer priorities.

Supplier

The DRA of the UK MoD.

(For details see under DRA in the Directory of Suppliers; note the different locations at Farnborough, Bedford and Malvern)

EASAMS Tornado Air Intercept Trainer (TAIT)

EASAMS produced two Tornado Air Intercept Trainers (TAIT) which were delivered to the RAF in 1985 and 1986. The trainer is used by aircrew at No 229 OCU at RAF Coningsby. 229 OCU converts aircrew to the Air Defence Variant (ADV) of the Tornado, the Tornado F3.

Modelling. The TAIT includes mathematical modelling of the aircraft Air Intercept (AI) radar, avionics and weapons systems, its flight characteristics, and its operational scenarios.

Trainee Stations. The pilot and navigator under training sit in a full-size replica of the Tornado's tandem cockpit; there is no cockpit canopy and so direct verbal instruction can be given and training can be in a relaxed 'shirtsleeve' environment if required.

IOS. The IOS has a colour touch-screen scenario display and other facilities for exercise control and monitoring.

Computer. DEC VAX 400. Software is compiled in VAX Fortran 77.

Updates. Facilities for Link 16 JTIDS (Joint Tactical Information Distribution System) have recently been fitted.

Status

In service with the Royal Air Force.

Manufacturer

EASAMS Limited, Camberley.

(For details see Directory of Suppliers)

Marconi Tucano Flight Simulator

Marconi (ex-Ferranti International Simulation and Training) manufactured the flight simulators for the Tucano T Mk 1 basic training aircraft for the RAF.

Visual Image Generation (IG). 3 channels of IVEX VDS-1000 CGI. Details in Image Generation (IG) Systems section.

Visual Display System (DS). SEOS PRODAS 300 direct projection system with 3 projectors. FoV 140° horizontal x 35° vertical, giving 525 lines at 30 Hz interlaced. Details in Visual Display Systems section.

Motion Platform. FlytSim 4-DoF non-synergistic, in a version modified by Marconi. Pitch, Roll, Heave and Yaw motions.

Instructor Operating Station (IOS). The offboard IOS houses the usual simulator control equipment and a simulator operator's seat. PCs are used together with touch-screen control and Marconi's proprietary Control And Display System (CADS) and Record/Replay facilities are fitted. A separate station for a Qualified Flying Instructor (QFI) is also part of the IOS and is described below.

The QFI Station. An unusual feature is the separate offboard station for a Qualified Flying Instructor (QFI) which is part of the IOS. The QFI station is provided with full controls, basic flight instruments, and monitor screens of the visual system. This is not only for the purpose of monitoring the student but also to enable the instructor to carry out demonstrations of flying exercises to the student in the simulator cockpit, using the flying controls provided (Throttle, Control Column, Rudder Bar, Flap, Gear, Trim and so forth). The QFI station is not enclosed in a cockpit model but is highly stylised in order to achieve the purpose of making flying demonstrations at minimum extra cost. This layout was chosen rather than providing the QFI with a tandem on-motion cockpit not only for reasons of cost but also because each QFI deals with several students and cannot be available all the time to occupy a rear simulator cockpit for a whole student simulator sortie.

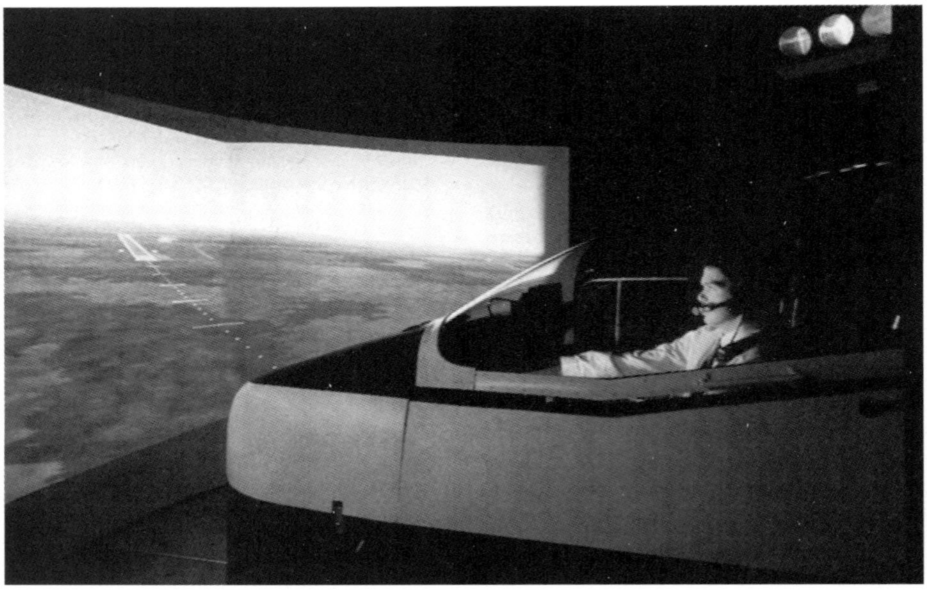

Marconi Tucano flight simulator mounted on a 4-DoF platform and showing the SEOS PRODAS display of three channels of the IVEX VDS-1000 CGI

Cockpit. Only the front cockpit of the tandem Tucano aircraft is modelled in detail and placed on the motion platform.

Control Loading. Digital.

Other Features. Sound system.

Status

One simulator was delivered to RAF Church Fenton and the others are at RAF Cranwell and RAF Linton-on-Ouse (two each). A simulator together with a Cockpit Emergency Procedures Trainer (CEPT) for the Tucano has also been delivered to Kuwait.

Manufacturer

Marconi Simulation and Training, Cheadle Heath, Stockport.

(Ex-Ferranti, for details see Directory of Suppliers)

Marconi Simulation Avionics Systems Trainers (AST)

The Marconi Simulation Avionics Systems Trainers (AST) can provide relatively low-cost training in the operation of avionics systems.

Components. A typical AST consists of a cockpit incorporating electronic head-up and head-down displays and the associated controls for weapons and avionics systems.

Visual Scenes. A simplified visual scene is provided to enable a student pilot to practise target location and attack configurations using HUD and HOTAS, where fitted. Visual air and ground targets are derived from Marconi Simulation CGI systems. Various avionics systems are simulated together with their controls while non-functional control and other panels are represented by two-dimensional screens.

Application. The cockpit may be used in conjunction with two identical class display consoles for use by student observers, an instructor's station, and a computer cabinet. The six class consoles each contain three colour monitors. One monitor reproduces the picture displayed by the cockpit multipurpose display (MPD) while a second monitor shows the current head-up display picture. The third monitor reproduces pictures of the essential cockpit control panels with buttons lighting up and switches moving as they are operated by the student pilot or by the computer in demonstration or replay modes.

Scenarios. Preprogrammed mission scenarios may be introduced together with target attack options.

Assessment. Evaluation of student performance is available using record replay facilities for exercise debriefing purposes.

Maintainer Training. The system is also applicable to

maintainer training using the built-in test (BIT) functions.
Customers. A Harrier Avionics Systems Trainer (HAST) is in service with the RAF at Wittering. It provides training in the operation and interpretation of the Harrier GR5/GR7 cockpit controls and display systems, and complements the TTS (ex Link-Miles) Harrier Full Mission Simulators.

Status
Available. In service with the Royal Air Force.

Manufacturer
Marconi Simulation, Dunfermline.
(For details see Directory of Suppliers)

Marconi Simulation Harrier Avionics Systems Trainer (HAST) in use showing student and instructor stations. Note the single channel visual scene at the top of the student station, and the student's HOTAS controls

Marconi Simulation Flight Trainers and Simulators

Marconi Simulation designs and manufactures cockpit procedure trainers (CPT), part-task trainers (PTT), operational flight trainers (OFT) and flight simulators for training both commercial and military aircrew.

These training systems are designed to provide training in instrument flying, radio navigation system operation, and flight systems management.

The trainers may be installed either in a trailer or at a fixed site. Full flight simulation systems can also be provided with motion platforms, digital control loading and computer-generated image displays representing airfield and air-ground attack configurations.

Status
An A36 Beechcraft flight simulator has been supplied to the Netherlands Government Aviation Flying School. CBT equipment has been supplied to the Oxford Air Training School for training in aerodynamics and gas-turbine engine operation.

Manufacturer
Marconi Simulation, Dunfermline.
(For details see Directory of Suppliers)

A Marconi Simulation Flight Trainer

MoD UK Simulator and Trainer Requirements

MoD UK has notified requirements for the following:
Sea King simulator and radar trainer. A cockpit dynamic simulator and a stand-alone radar crew trainer are required for the Sea King HAR3A which is the SAR version of the Sea King operated by the RAF. The dynamic simulator has been specified to have a 6-DoF motion platform, day colour CGI and display system with a FoV of 210° horizontally and 40° vertically. The radar crew trainer will feature a replica of the radar operator's console with simple aircraft noise effects. It will simulate all of the radar's operating modes with effects which include sea state, precipitation and target cross-section. Both systems will fit into an existing building which used to house a previous simulator.
Tristar Tanker Simulator. This has been specified to have a 6-DoF motion platform, day colour CGI and a display system with a FoV of 150° horizontally and 40° vertically. An onboard instructor station is required with two places. The simulator is to fit into an existing building.

Status
See text.

Sponsor
MoD (UK). Notification is through the MoD contracts bulletin which is freely available from Contracts branches of the MoD UK Procurement Executive (MoD(PE)).

Pennant Tucano CPTs

Pennant International has supplied UK MoD with 7 Tucano T1 Cockpit Procedures Trainers (CPTs). Each consists of a full-scale representation of a Tucano front cockpit. An instructor's control panel may be installed on either side of the cockpit to the rear of the seat.
Sound Generation System. There is an audio system to provide engine noise and the sound of propeller wash, audible warnings and intercommunications facilities.
Ejection Seat Procedures. A full facsimile seat allows for preflight checks and strapping-in procedures to be carried out.
Instructor Aspects. The instructor can set up a training scenario under normal or fault conditions using his own control panel and a VDU mounted beside the cockpit where it may be seen by him and/or the student. In the self-instruction mode, students can operate the CPT and conduct a training sequence using a hand-held key unit and prompts from the VDU.
FRCs. The VDU is used to display Flight Reference Card pages and Emergency Procedures.
Support Role. In a support role it is used to monitor and to sequence BITE checks and to store maintenance data including spares listings.
Rear Cockpit Layout for Navigator Training. In 1992, in response to a requirement for the provision of specialised navigator training, a CPT was reconfigured to represent a rear cockpit.

Pennant CPT for the RAF Tucano T1

Status
Seven Tucano T1 CPTs have been produced for the Royal Air Force.

Manufacturer
Pennant International Limited, Cheltenham.
(For details see Directory of Suppliers)

Reflectone UK (RUKL) Trainers and Simulators

Reflectone UK Limited (RUKL, ex-BAeSL) manufactures a range of Flight Trainers based on low-cost visual systems and user-friendly instructor operator stations.
PC-9 Simulators. RUKL, in conjunction with BAe Military Aircraft Division, constructed and delivered two PC-9 operational flight simulators for the Royal Saudi Air Force. Each device incorporates the two-cockpit configuration of the aircraft, allowing flying instruction to be given from the rear seat. A McFadden short stroke six DoF motion platform is fitted. Visual display is from an offboard projected system using images derived for the Image IIT CGI system from Link-Miles (now part of Thomson Training Systems (TTS)).

Status
Available, delivered and in production.

Manufacturer
Reflectone UK Ltd (RUKL), Bristol.
(For details see Directory of Suppliers)

PC-9 flight simulator

Thomson Training and Simulation (TTS) Concept 90 Full Flight Simulators

The Concept 90 flight simulators were introduced by the then Hughes Rediffusion company, subsequently taken over in 1993 by Thomson-CSF and now part of the Thomson Training and Simulation (TTS) organisation.
General Design. Concept 90 was designed in conjunction with MGA Developments of Coventry, and makes extensive use of composite moulded panels for functional as well as aesthetic reasons. The design is also intended to speed up manufacture.
Maintainability. Access and maintainability is enhanced by hinged doors underneath and around the simulator. Cabling is standardised with improved access to connectors. Air-conditioning, smoke detection and fire suppression systems are integrated into the aft simulator cabin.
Customers. 30 Concept 90 simulators had been ordered by 1994 by customers in Ethiopia, Japan, Switzerland, UK and the USA. Aircraft simulated include the Boeing 747-400, 767-300, and 777, and Airbus models including the A321.
Visual Image Generation (IG). The standard IG offered was the Evans and Sutherland (E&S) SP-X. Subsequently the newly developed TTS SPACE IG will be available, but TTS chairman Louis Le Portz has stated that, although E&S is now independent of TTS, its systems will continue to be available and supported as a customer choice. Whichever IG is chosen, either three or five channels can be fitted depending on the horizontal FoV required by the customer. Three channels give about 150° horizontal FoV and five channels, 200°. Details in Image Generation (IG) Systems section.
Visual Display System (DS). WIDE™ wide angle collimated display system with either three or five projectors to match the number of IG channels ordered. Details in Visual Display Systems section.
Motion Platform. Rediffusion (now TTS) standard 6-DoF platform.
Main Computer. Motorola 88000 single-board RISC processors.
Instructor Operating Station (IOS). The IOS console is mounted on board the motion platform and is known as ACCESS (previously the Rediffusion TASC II IOS). It is provided with touch-screen controls and computer graphics. The instructor seat is a fully adjustable aircraft pilot seat. This allows the instructor to select the best position to monitor the actions of the trainees. Once this position has been selected the console can be moved towards the instructor along a floor rail.
Control Loading. Digital.
Other Features. Digital stereo sound system.
Certification. Compliance has been demonstrated for all Phases or Levels set by the world regulatory bodies (FAA/CAA and so on) and zero flight time (ZFT) training is possible on these simulators.

Status
Available and in service.

Manufacturer
Thomson Training and Simulation, Crawley.
(For details see Directory of Suppliers)

Exterior of the Concept 90 simulator

TTS Concept 90 ACCESS IOS

TTS Vantage FTDs

Derived from the Concept 90 flight simulator programme, Vantage is a modular range of flight training devices designed to meet the latest FAA requirements and the anticipated requirements of regulatory bodies worldwide. Vantage takes account of open systems so that commonality of processor with other flight training equipment can be achieved.
Components. Construction is on a rigid base frame incorporating control loading and the main services. Polycarbonate materials are used with the intention of improving build quality and durability. The 400 and 500 variants incorporate the computer, power supplies and interface electronics so that the entire system is self-contained and free-standing. On the higher specification variant, the electronics bays can be integrated either with the main device or located remotely.
IOS. IOS options available are either a twin screen PC-based system, or the ACCESS touch-screen graphics-based IOS as fitted to the Concept 90 Full Flight Simulators (see above).

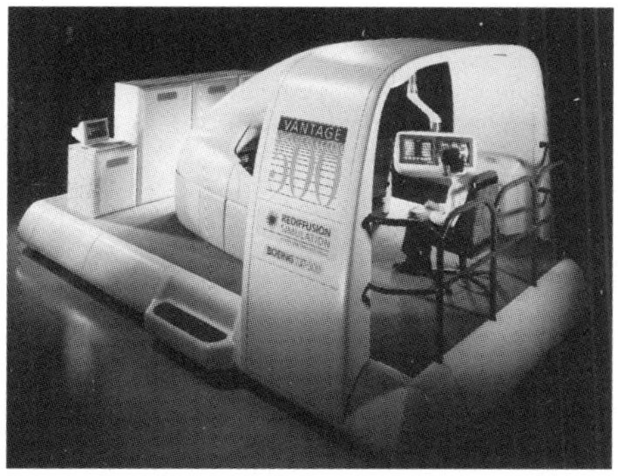

TTS Vantage 500 FTD configured for the Boeing 737-300

Vantage 400. The Vantage 400 meets FAA level 4 requirements and provides cockpit procedures training in an open cockpit format. The design includes a movable, cantilevered console.
Vantage 500. The Vantage 500 extends the training capability to meet FAA level 5 requirements and can be configured to provide full aircraft systems and flight management systems training. Whilst retaining the open cockpit of the Vantage 400, the Vantage 500 includes full aircraft controls and instruments with a flight programme which allows the device to be flown in manual or automatic modes. The IOS has touch-screen control and includes set-up pages for up to 100 malfunctions.
Vantage 600. The Vantage 600 was designed to meet the FAA's proposal for level 6 devices and features a closed full replica cockpit with digital control loading and full systems simulation. The ACCESS IOS is offered as an option.

Status
Available and in service.

Manufacturer
Thomson Training and Simulation, Crawley.
(For details see Directory of Suppliers)

TTS Military Flight Simulators

TTS Hawk 200 Combat Mission Simulator (CMS)
Thomson Training and Simulation (ex-Rediffusion element) is supplying a Combat Mission Simulator (CMS) for the BAe Hawk 200 light attack aircraft to the Malaysian Government. At a total value of £18 million (US$32 million), the contract includes an avionics and weapons system trainer (AWST) and a CBT system for both flight and ground maintenance crews. The CMS is described below and is a development of the design for the Hawk simulator previously delivered to the Swiss Air Force.
Visual Image Generation (IG). Martin Marietta Compu-Scene VI. Details in Image Generation (IG) Systems section.
Visual Display System (DS). Direct projection on the inside of an 8 m diameter partial dome.
Motion Cueing. The simulator is fixed base, but is provided with a vibration system to simulate aerodynamic buffet, pilot's anti-G suit and a G-seat which gives seat pressure under simulated G, seat pan lowering under G, and strap tightening.
Control Loading. Digital.
Other Features. Digital stereo sound system.
Customers. The Malaysian order is scheduled for installation late 1994, one ordered in 1993 by Indonesian Air Force.

Swiss Hawk
TTS (ex-Rediffusion element) produced a Hawk Mark 66 simulator for the Swiss Air Force. It was installed at Emmen air base in 1990.
Visual Image Generation (IG). Five channels of Evans and Sutherland SP-X500. Databases were created to include mountainous terrain, lakes, valleys and airfields. Other details in Image Generation (IG) Systems section.
Visual Display System (DS). Direct projection on inside of 8 m diameter partial dome, five projectors. FoV 200° horizontal x 60° vertical.
Motion Cueing. The simulator is fixed-base, but is provided with a vibration system to simulate aerodynamic buffet, pilot's anti-G suit equipment and a G-seat which gives seat pressure under simulated G, seat pan lowering under G, and strap tightening.
Control Loading. Digital.
Other Features. Digital stereo sound system.

Simulators for BAe Hawk - TTS (Ex-Link-Miles Element)
TTS (ex-Link-Miles element) produced several simulators for the BAe Hawk. The principle of Functionally Distributed Simulation (FDS) was first used on a Hawk simulator for the Finnish Air Force. Simulators for the Hawk were then ordered for the Kuwaiti and United Arab Emirates air forces, two similar systems were supplied to the Royal Saudi Air Force. Finland ordered a more advanced Hawk simulator, and the Korean Air Force has ordered a simulator for the Hawk T67. These simulators incorporate microprocessor simulation technology (MST), the IMAGE III-T CGI system is displayed through three collimated monitors, and the ex-Link-Miles standard 6-DoF motion platform is used.

Nimrod MR2
TTS (ex-Rediffusion element) produced simulators for the Nimrod MR Mark 2 MPA for RAF St Mawgan and RAF Kinloss. These simulators replaced earlier Redifon Nimrod simulators as part of a £15 million contract awarded in 1988 for three devices, subsequently reduced by MoD UK to two.
Visual Image Generation (IG). Three channels of Evans and Sutherland SP-X500HT. Special visual databases were developed representing a number of airfields, 20 different NATO and Eastern Bloc ships, and

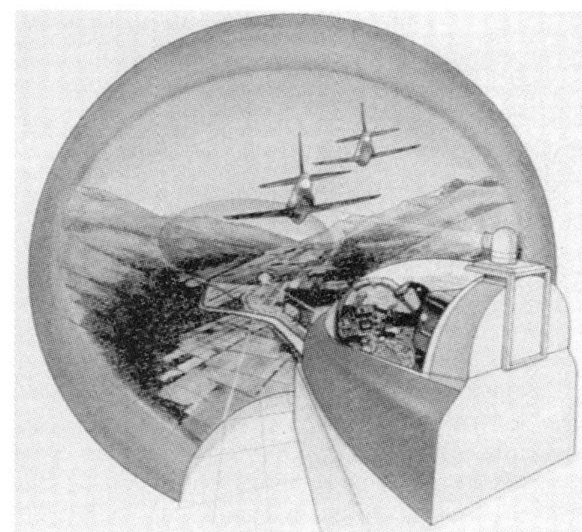

Drawing of the TTS Hawk 200 CMS

RAF VC10 and TriStar tanker aircraft. Other details in Image Generation (IG) Systems section.
Visual Display System (DS). TTS WIDE™ with three projectors; FoV 150° horizontal x 40° vertical. Details in Visual Display Systems section.
Motion Platform. TTS (ex-Rediffusion design) 6-DoF.
Instructor Operating Station (IOS). TASC system with touch-screen control, mounted on board.
Control Loading. Digital.
Other Features. Air-to-air refuelling is simulated using the probe and drogue method.

Sentry AEW Mark 1
TTS (ex-Rediffusion element) produced a simulator for the RAF Sentry AEW Mark 1 (Boeing AWACS) airborne early warning aircraft, contracted through the Boeing Military Aircraft Company. It is in service at RAF Waddington.

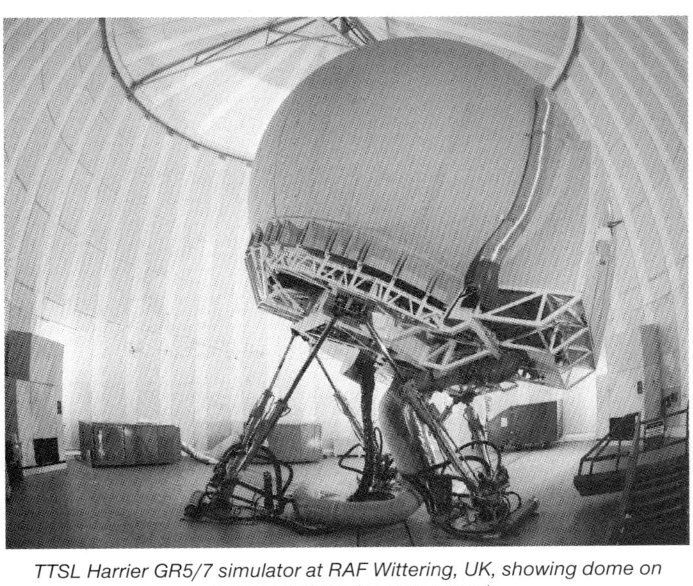

TTSL Harrier GR5/7 simulator at RAF Wittering, UK, showing dome on motion platform (see next page)

TTSL Harrier GR5/7 simulator with CAE-Link ESPRIT display – note the high resolution eye-slaved inset and lower resolution background giving nearly 360° FoV

Visual Image Generation (IG). Two channels of SP3T. Details in Image Generation (IG) Systems section.
Visual Display System (DS). WIDE™ with three projectors; FoV 150° horizontal x 40° vertical.
Motion Platform. TTS (ex-Rediffusion design) 6-DoF.
Instructor Operating Station (IOS). TASC system with touch-screen control, mounted on board.
Control Loading. Digital.
Other Features. Air-to-air refuelling is simulated using the probe and drogue method.

Aermacchi MB.339

TTS (ex-Rediffusion element) produced four MB.339 simulators which were delivered to the Italian Air Force at Lecce in southern Italy in 1990 and 1991. The design is fixed base and the cockpit section was supplied by Aermacchi. A similar simulator was ordered by the Royal New Zealand Air Force and was installed at Ohakea Air Force Base by 1993.

Advanced Panoramic Helmet Interface Demonstrator System (APHIDS)

TTS (ex-Rediffusion element) started a design study in July 1991 for an integrated helmet display system technology demonstrator. The contract is to establish the specifications for an engineering simulator for the DRA Farnborough Advanced Panoramic Helmet Interface Demonstrator System (APHIDS). It is hoped that APHIDS will eventually lead to a helmet display system for future combat aircraft. Hardware options and human factors will be studied.

Sea King 5 Mission Simulator

TTS (ex-Rediffusion element) produced a full mission simulator for the Sea King HAS Mark 5 which is installed at RN Air Station Culdrose in Cornwall. The pilot simulator may be integrated with one of three ASW trainer cabins for the Sea King Observer crew which were produced by the then Ferranti Simulation company (now Marconi Simulation and Training) at Cheadle Heath, Stockport. The simulator may be operated in the ASW or SAR roles.
Visual Image Generation (IG). Five channels of E&S SP3T, 17 visual databases fitted. Other details in Image Generation (IG) Systems section.
Visual Display System (DS). WIDE™II with five projectors. FoV 200° horizontal x 40° vertical. Details in Visual Display Systems section.
Motion Platform. TTS (ex-Rediffusion design) 6-DoF.
Control Loading. Digital.

Tornado Simulators and Trainers

Ten Tornado simulators were produced for the RAF in a work-share programme between Rediffusion and the then Link-Miles (both now part of TTS). Six were for the GR1 Interdiction-Strike (IDS) version and used the Rediffusion design (now obsolete) 3-DoF motion platform. The four F3 Air Defence Variant (ADV) simulators were fixed base. None of these simulators was fitted with visual systems, although provision was made in the design for the GR1 simulators for a future system.

Link-Miles was prime contractor for two IDS and one ADV simulator for Saudi Arabia which was also a work-share with Rediffusion and had a substantial number of improvements over the RAF versions. These included a 6-DoF motion platform, improved computing, and a three channel Image 3T CGI displayed by three collimated monitors.
The then Link-Miles company also made a number of PTTs for Saudi Arabian Tornados. These include cockpit procedures trainers (CPTs) for IDS and ADV, navigation attack systems trainer for the IDS.

Other Simulators and Trainers - TTS (ex-Rediffusion Element)

Flight Simulators. TTS (ex-Rediffusion element) produced a wide range of simulators for aircraft types including the Lynx for the UK Army Air Corps, the VC10 in its transport and tanker roles, and five Hawk simulators for the RAF with 3-DoF motion but no visuals. Simulators have also been supplied to the air forces of Austria (AB 204), Denmark (F-35 Draken), Germany (C160), Netherlands (RF-5B), Norway (four F-5A), South Africa (four Impala, one Impala-Macchi), and Tunisia (MB 326).
Trainers. Flight and procedures trainers were produced for the Brazilian and Egyptian air forces. A Tornado F Mark 3 Cockpit Emergency and Procedures Trainer (CEPT) was produced for the RAF.
USAF B-1B. Equipment has been supplied to the Boeing Military Aircraft Company (BMAC) for training for the US Air Force B-1B bomber. The equipment includes two full flight stations, two WIDE display systems and two offensive/defensive operator's stations, for incorporation in two full Weapon System Trainers (WST).
PTTs. A low-cost station-keeping equipment (SKE) part-task trainer was supplied to the RAF for training Hercules aircrew.

Eurofighter Training Needs Analysis (TNA)

TTS (ex-Rediffusion element) was awarded a contract by the UK MoD in February 1989 for a TNA for the Eurofighter 2000.

Agreement with MiG

TTS (ex-Rediffusion element) made an agreement at the Farnborough Air Show in 1992 with the Russian aircraft manufacturer Mikoyan and the avionics integration organisation GosNIIAS to explore training systems development associated with Mikoyan military and training aircraft including the MiG 29 and MiG 31.

Harrier GR5/7 Full Mission Simulators (FMS)

TTS (ex-Link-Miles element) has supplied two FMS for the RAF Harrier GR5/7, one of which was to be based at RAF Gutersloh in Germany and the other at RAF Wittering in the United Kingdom. With recent withdrawals from Germany it is probable that both simulators will now be installed at RAF Wittering. The total contract value has been estimated at around £35 million, and at the time of contract were the most sophisticated simulators ordered for a fast-jet for the RAF. The total system comprises a simulator room, computer area, visual database work station, instructor's operating room and remote debrief facility.
Visual Image Generation (IG). Two channels of MOD DIG CGI. FLIR imagery is included. Real world scenes are generated from DTED and DFAD files using the DLMS and DoD Project 2851 protocols; manual editing then enhances the basic scenes. A work station for customer generation of scenarios is included.
Visual Display System (DS). Projection on the inside of a 7.35 m dome mounted on the motion platform. The CAE-Link ESPRIT (Eye Slaved Projected Raster InseT) head and eye-slaved system is used giving a FoV of 240° horizontal x 130° vertical. A high resolution (HR) projector is head- and eye-slaved so that the small Area of Interest (AOI, about a 20° cone angle) area of HR imagery is always in the centre of the pilot's vision. A lower resolution wide-angle background projector produces peripheral visual cues. Details in Visual Display Systems section.
Motion Cueing. TTF (Link-Miles design) 6-DoF motion platform, plus pilot anti-G suit and the Cranfield design of G-seat giving seat pressure under simulated G, seat pan lowering under high G, and strap tightening.
Instructor Operating Station (IOS). Mounted off board. This is a comprehensively equipped IOS with monitor screens of the pilot's visual system, graphical presentation of the tactical situation including threat systems.

RAF Tornado simulator

TTS simulator for Swiss Air Force Hawk 66 installed at Emmen

Debrief Console. Playback and sortie analysis takes place at a separate console. Computer graphics are used to view the sortie from a number of eye points as well as from the pilot's cockpit.
Control Loading. Digital.
Computers. Three Concurrent Computers Corporation 3280MPS parallel processors are used each with a processing power of 33 Mips.
Other Features. Aircraft systems simulated include the Angle Rate Bombing Set (ARBS), the forward-looking infra-red (FLIR), and the wide-angle head-up display (HUD). Weapon effects include probability programs to predict hits, kills, duds, misfires, near misses and collateral damage, both from 'own ship' weapons and threat systems.

Sea Harrier Dynamic Simulator

Thomson Training and Simulation (ex-Link-Miles element) supplied a Sea Harrier dynamic simulator to RNAS Yeovilton.
Visual Image Generation (IG). Three channels of Image 2T CGI. Details in Image Generation (IG) Systems section.
Visual Display System (DS). Three collimated monitors.
Motion Platform. TTS (ex-Link-Miles) standard 6-DoF platform, with Cranfield design of G-seat and use of the pilot's own anti-G suit.
Control Loading. Digital.
Instructor Operating Station (IOS). Mounted off board.

TTS Sea Harrier Operational and Radar Trainer (SHORT)

Thomson Training and Simulation (ex-Link-Miles element) supplied two SHORT trainers to the RN. The system consists of fixed-base student stations and an IOS in a relocatable container. The IOS has two Vistagraphic 4500 colour screens, and controls are claimed to be user-friendly controls and the layout to be designed on ergonomic principles. Each student station has facsimile instruments and displays. There is a single-channel VISTA image generator which provides day/dusk/night scenes projected through the head-up display (HUD).

Two systems, one configured to Sea Harrier FRS1 and the other to FRS2 standards, have been delivered to the RN Air Station at Yeovilton. They will relieve the Sea Harrier main flight simulator of some radar training. The FRS1 system can be upgraded to FRS2 standards.

Sea Harrier Operational and Radar Trainer (SHORT)

TTS Microflite procedures and part-task trainer in the final stages of assembly

Simulators for Agusta A129

TTS (ex-Link-Miles element) supplied an Agusta A129 Mangusta helicopter simulator to Agusta Sistemi, which was prime contractor for a Combat Mission Simulator (CMS) suite for the Italian Army.
A129 Flight Simulator Two-Cockpit Layout. The TTS (ex-Link-Miles) A129 Flight Simulator has two replica cockpits for the pilot and co-pilot/gunner, each with its own dome-projected visual display system and mounted on its own 6-DoF motion platform. They are housed together and share common services, computing and image generation facilities. The layout is similar to that of the Apache CMS produced for the US Army by CAE-Link (qv in US section). The image generator is the Martin Marietta (previously GE) Compu-Scene IV. The two cockpits can be operated in independent or combined modes.

TTS (ex-Link-Miles element) also produced a maintenance trainer for the A129.

Helicopter Simulators (ex-Link-Miles Element)

Lynx. The Westland Lynx helicopter simulator supplied to the British Army Air Corps (AAC) in Germany is mounted on a 6-DoF platform and originally employed three channels of the IMAGE IIT CGI system displayed by three collimated monitors. It is understood that this has been upgraded to a colour IG.

EH101. Westland's development programme for the EH 101 includes a Link-Miles simulator at Yeovil operated in the R&D role. This has a 6-DoF motion platform and 4 channels of IMAGE IIT Textured CGI, displayed through four collimated monitors. The monitors can be reconfigured for the trial in hand, and can be set up to include the chin window for hovering. Software is designed to make it easy to alter gains and control laws for aircraft development purposes and for assessing handling qualities. It was this simulator of which the Chief Test Pilot of Westlands, Colin Hague, spoke after the first flight of the EH101 helicopter when he was quoted as saying "the aircraft felt like an old friend, and flew just like the simulator".

Sea King. Sea King flight simulators have been supplied by Link-Miles to the Royal Navy, the Royal Australian Navy, Egypt (Mk 47 Commando), India, and Qatar.

Blade Element Modelling. Recent Link-Miles helicopter flight simulators employ computing which involves blade-element theory. This mathematically treats each rotor blade as an independent aerofoil. Calculations are made at several places along the length of the blade from root to tip. The aerodynamic and inertia loads are computed over each 7° of blade rotation, at iteration rates up to 180 Hz.

Other Simulators (ex-Link-Miles Element)

TTS (ex-Link-Miles element) has produced simulators for the BAe Buccaneer, Lockheed C-130 Hercules (seven to the Royal Air Force), Sepecat Jaguar GR1 (six, of which three remain in service) and McDonnell Douglas KC-10 (three). Twelve J21 flight simulators were also supplied to the then Yugoslavia. Link-Miles also produced the Microflite range of PTTs (see photo).

Status

Simulators, Trainers, SPACE IG and WIDE and AWARDS Display Systems all available, otherwise see text.

Manufacturer

Thomson Training and Simulation, Crawley.
(For details see Directory of Suppliers)

Westland CBT for the Lynx Helicopter

Westland, with Aerospatiale of France the makers of the Lynx helicopter, offers a supporting package of CBT which students can use on PC-based work stations. The low-cost packages combine computer graphics with moving video, still images and digital sound. The packages are based upon 18 software modules. These include an introduction to the aircraft and its roles and briefings on aircraft parts and systems such as its airframe, electrical systems, fuel system, transmission, flying controls, turnround servicing and ground handling.

Status

In service.

Manufacturer

Westland System Assessment Limited, Yeovil.
(For details see Directory of Suppliers)

C R Wood Low Cost Aircraft Trainers

Consulting Engineers C R Wood have taken over the facilities of the former Quality Training Equipment Ltd and continue to make low-cost aircraft flight training devices and upgrades to existing devices. The company also supplies cockpit hardware and simulated instruments to other manufacturers and software houses.

Status
Available and in service with 25 operators in Belgium, Holland, Luxembourg, Norway, and the UK.

Manufacturer
C R Wood, Brize Norton.
(For details see Directory of Suppliers)

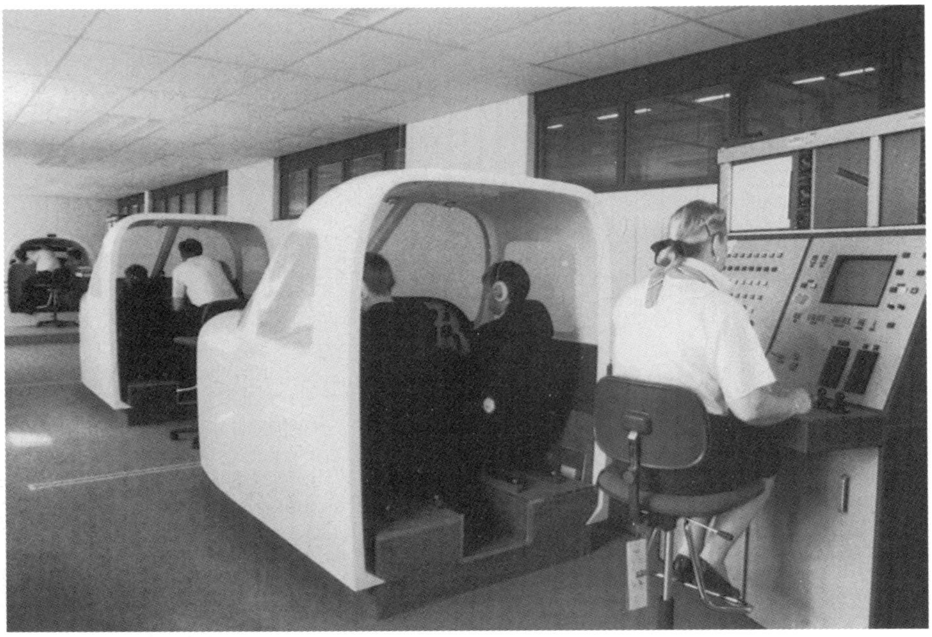

C R Wood (Ex QTE) Beechcraft 200 trainer and two basic trainers in use at the National Luchtvaart School at Maastricht, Holland

UNITED STATES OF AMERICA

AAI EF-111A Operational Flight Trainer

The AAI Corporation developed the EF-111A Operational Flight Trainer (OFT) to meet the US Air Force's requirement for crew co-ordination training in a realistic electromagnetic combat environment.

Radars Simulated. Pilot radar navigation and terrain-following simulation is accomplished through simulation of the AN/APQ-160 navigation radar indicator and the AN/APQ-110 terrain-following radar display.

Radar Database. The OFT radar database contains more than 500 000 nm^2 (1.71 million km^2) of digitised radar land mass (DRLMS) data. Radar data is predicted from topographical data from the Digital Land Mass System (DLMS) which is sponsored and co-ordinated by the Defense Mapping Agency (DMA) and contains both terrain (DTED) and cultural feature (DFAD) files. The DLMS database is used to ensure correct sight lines (occultation) of simulated threat signals and targets, with respect to terrain and propagation conditions.

IOS. The EF-111A OFT includes an IOS for controlling and monitoring the training mission. The IOS includes three instructor positions and incorporates repeaters for flight instruments together with radar and electronic warfare displays.

Status
In service with the US Air Force.

Manufacturer
AAI Corporation, Hunt Valley, Maryland.
(For details see Directory of Suppliers)

AAI OFT for the EF-111A assigned to the 430th Electronic Combat Squadron, USAF. Note the three-position IOS in the foreground

ARI Trainers and Simulators

Aviation Resources Inc makes a range of trainers and simulators for the aviation industry ranging from Full Flight Simulators (FFS) mounted on the ARI 6-DoF platform, through fixed-base Flight Training Devices (FTDs), to CBT and other trainers such as cabin trainers, door and exit trainers and so forth. See the information given in the Directory of Suppliers.

Customers. Customers include Aero California, Air Canada, Air New Zealand, Airborne Express, ALG Leasing, American Airlines, American Airlines Training Corporation, Ansett Airlines, British Airways, Canadian Airlines, Continental Airlines, Dalfort Aviation, Delta Air Lines, Eastern Airlines, Egyptair, Embraer, First Air, Finnair, Garuda, Johnson Engineering, Kalitta Air Services Inc, Libyan Arab Airlines, NATCO, Northwest Airlines, Ozark Airlines, Pakistan International Airlines, Qantas Airways, Saudia, Scandinavian Airlines System, Swissair, TAESA, UPS, TWA, Transbrasil, USAir, United Airlines and the USAF.

Aircraft Covered. Flight simulators, upgrades, or moves have been accomplished for the Airbus A300, ATR42, Boeing 727, 737, Citation, DC-8, DC-9, DC-10, Embraer Brasilia, Fokker F-27, Learjet, Saab SF 340, and Tristar.

Recent Work. ARI supplied to the Johnson Space Centre at Houston a Space Shuttle Orbiter Neutral Buoyancy Trainer (ONBT), and in February 94 was contracted to supply extra equipment for this trainer. In January 94 a Fokker F-27 flight simulator was

ARI's 6-DoF platform on a load test

completed for Merpati at Surabaya, Indonesia. In November 1993 an upgrade of an Airbus A300 flight simulator was completed for Pakistan International Airlines, and in July 1993 the first of two L-1011 Tristar flight simulators was upgraded and certified to FAA Level C standard for Delta Air Lines.

Control Loading. For Full Flight Simulators, ARI uses the Fokker Microprocessor-based Control Loading (MiCol). Iteration rates are 5000 Hz for the inner loop and 2000 Hz for the outer loop.

Visual Systems. ARI has fitted a number of different CGI systems and display systems including projection and cross-cockpit viewing systems (WIDE™ and equivalents).

IOS. ARI uses touch-screens and its IOS has automatic lesson plan capability with manual override, and an intelligent Portable Control Unit (PCU) is available.

FTDs. ARI's Flight Training Devices (FTD) are modular in design and can be enhanced with minimum disruption to training.

Status
See narrative above.

Manufacturer
Aviation Resources Inc (ARI), Broken Arrow (Tulsa), Oklahoma.
(For details see Directory of Suppliers)

ARI open-plan FTD

ATC-212H IFR Flight Trainer for Helicopters

The ATC-212H is an IFR flight trainer representing the general characteristics of light to medium turbine-powered helicopters. Rotor torque can be switched to represent clockwise or anti-clockwise rotation. Realistic response to control inputs is provided by hybrid computer circuitry. Cyclic and collective-pitch controls can be adjusted to provide the desired control stabilisation. The computer is programmed for a speed range of 40 to 160 knots. Winds of up to 50 knots can be introduced from 12 directions, and five levels of turbulence can be set up to affect roll and pitch attitude.

Accessories. Accessories include an X-Y flight plotter, a monitor communications system, a headset, and an enclosure for IFR training in accordance with FAA FAR Part 141. The flight plotter can be used with standard low altitude charts to provide records for post-exercise study.

FAA Credits. The FAA allows 5 hours of the time a student has spent on the ATC-212H towards a rotary-wing flight training course.

Dimensions:
(simulator) 864 × 1168 × 1626 mm
(flight plotter) 483 × 152 × 356 mm

Status
In production.

Manufacturer
ATC Flight Simulator Company, Santa Monica, California.
(For details see Directory of Suppliers)

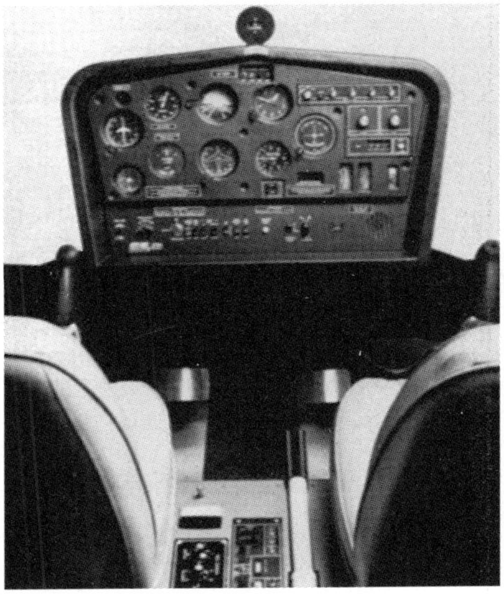

ATC-212H helicopter IFR trainer

Boeing Flight and Mission Simulation for B-1B

Boeing's Defense & Space Group has produced a flight and mission simulator system for the Rockwell B-1B bomber under contracts which included provisions for five weapons system trainers (WSTs), two mission trainers, one software support centre, and six cockpit procedures trainers (CPTs).

WST. Each WST consists of a flight station for pilot and co-pilot, and a compartment for the operators of the offensive and defensive systems.

Mission Trainer. The mission trainer is identical to the offensive/defensive station of the WST, and trains operators before they are integrated with flight crews in the WST.

Rediffusion-supplied Equipment. Rediffusion UK (now part of Thomson Training) was a major subcontractor and supplied two fully integrated flight decks, two motion platforms, two Total Sound Environment systems, instructor-station equipment and a WIDE (wide-angle infinity display equipment) visual display system, as well as linkage and control loading.

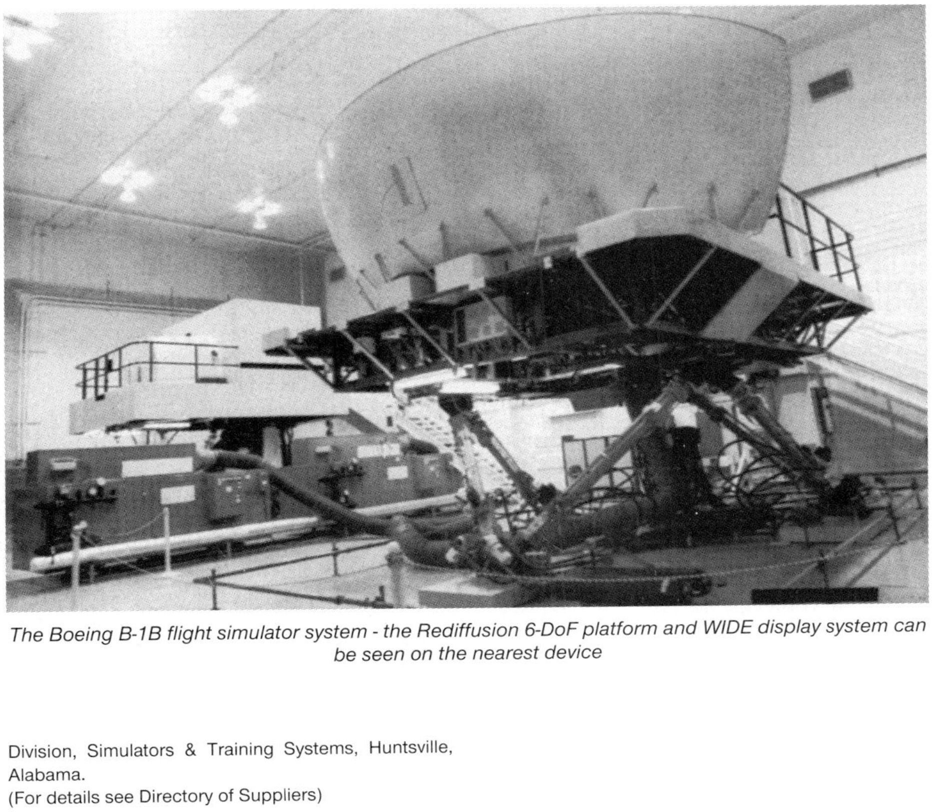

The Boeing B-1B flight simulator system - the Rediffusion 6-DoF platform and WIDE display system can be seen on the nearest device

Status
The first simulator was delivered in December 1988 and B-1B simulators have been installed at Ellsworth AFB (Air Force Base), South Dakota, Dyess AFB, Texas, Grand Forks AFB, North Dakota, and McConnell AFB, Kansas. The systems are upgraded to maintain concurrency with the aircraft.

Manufacturer
Boeing Defense & Space Group, Missiles & Space Division, Simulators & Training Systems, Huntsville, Alabama.
(For details see Directory of Suppliers)

Boeing E-3A Training Package

The E-3A Airborne Warning and Control System (AWACS) training package, was produced by Boeing's Defense & Space Group under a contract with the US Air Force.

Components. The training package includes academics, training requirement analysis, courseware development, aircrew training devices, ground training, flight training support, in-flight trainers, a Boeing 707 aircraft trainer and logistics support.

Licence Agreement with Rediffusion. The full flight simulator which trains military flight crews on the AWACS was produced by Boeing under a licensing agreement with Rediffusion Simulation Limited, UK (now part of Thomson Training).

Savings. This total approach to training helps to reduce the time required to train aircrew from 103 to 59 days, and requires a third of the personnel of previous training systems.

Status

Delivery of the training package elements to the US Air Force was completed in 1986. Two E-3A training packages are in service at Tinker Air Force Base in Oklahoma and another is in service with the Royal Saudi Air Force in Riyadh.

Manufacturer

Boeing Defense & Space Group, Training Systems, Huntsville, Alabama.
(for details see Directory of Suppliers)

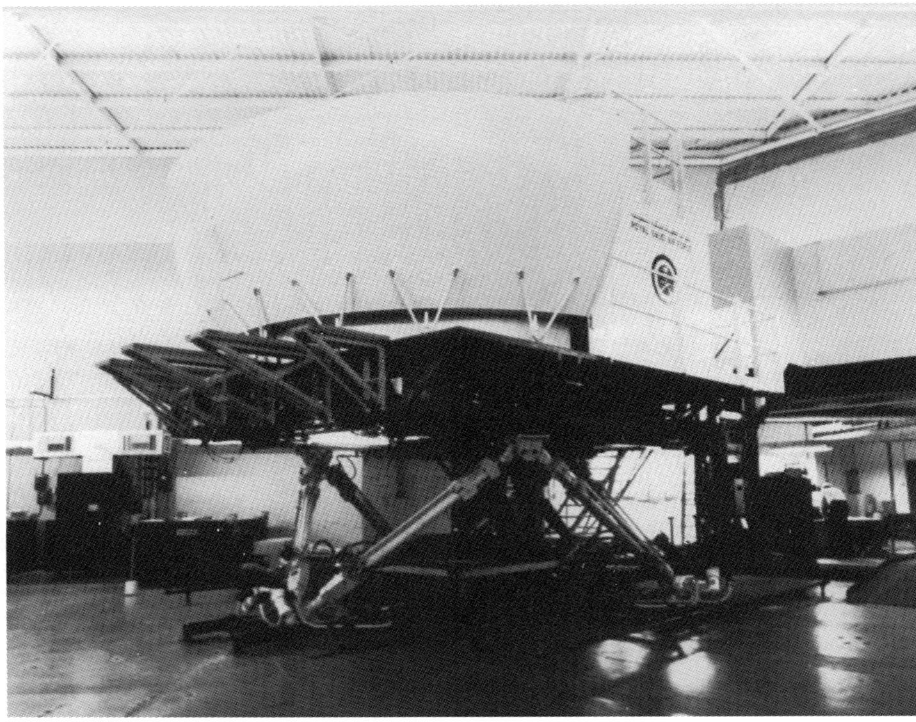

An E-3A simulator of the Royal Saudi Air Force based at Riyadh

Boeing ModSim

Boeing Defense & Space Group has established criteria for the design of reconfigurable simulators as modular simulator (ModSim) architecture. The first application was in 1991 when Loral Western Development Labs awarded the company a contract to build a series of reconfigurable rotary-wing aircraft simulators for the US Army's Fort Rucker and Fort Knox simulation centres.

Under the plan, multiple simulators will be able to operate as a team on an electronic battlefield along with a wide range of other simulators for ground forces, air defence and fixed-wing aircraft. The rotary-wing simulators will be reconfigurable into OH-58D, AH-64A/B/C and RAH-66 aircraft. The architecture, developed under a four-phase ModSim contract supported by the US Air Force, Army and Navy, will be used as the basis for the design. Boeing will develop major portions of the simulation software, and will lead the integration, delivery and installation of these simulators for Loral.

Status

Funding has been authorised for the development of the RAH-66 simulator as well as for the testing and evaluation of system avionics.

Manufacturer

Boeing Defense & Space Group, Huntsville, Alabama.
(For details see Directory of Suppliers)

Boeing Simulator Package for F-22

The F-22 air superiority fighter development team includes Lockheed, Boeing, and General Dynamics. Part of Boeing's role is to perform the work package in support of the design and development of an integrated training system. Boeing's Defense & Space Group will be responsible for the training system while Boeing's Missiles & Space Division in Huntsville, Alabama, will develop the maintenance training system, training management system, the training system support centre, computer-based training system, courseware and components of the pilot training system.

Status

Design of the training system began in 1991 after the US Air Force awarded the Lockheed-Boeing-General Dynamics team a $9.55 billion contract for engineering and manufacturing development of the F-22. This work will continue over an eight-year period.

Manufacturer

Boeing Defense & Space Group, Huntsville, Alabama.
(For details see Directory of Suppliers)

Boeing Simulator Upgrade Programme for KC-135

Boeing modified the KC-135A weapon system trainers (WST) to KC-135R standard operational flight trainers (OFT) as the US Air Force re-engined the KC-135. In March 1986 the Boeing Defense & Space Group received a contract to develop KC-135A and KC-135R OFTs and a software support centre to replace the ageing MB26 (KC-135A) cockpit procedures trainer (CPT).

OFT Motion and Visual Systems. The OFTs have the Evans and Sutherland SP1 night/dusk CGI for the pilot and co-pilot and an onboard IOS; some have Rediffusion (UK) 6-DoF motion platforms and an upgrade is planned in order to fit all with 6-DoF motion.

Numbers. Four production options have been awarded for a total of 19 OFTs. Contractor logistics support is also provided.

Status

Completed and in service with a total of 12 KC-135R and seven KC-135A flight simulators (OFTs) at one Air National Guard and 15 US Air Force locations.

Manufacturer

Boeing Defense & Space Group, Huntsville, Alabama.
(For details see Directory of Suppliers)

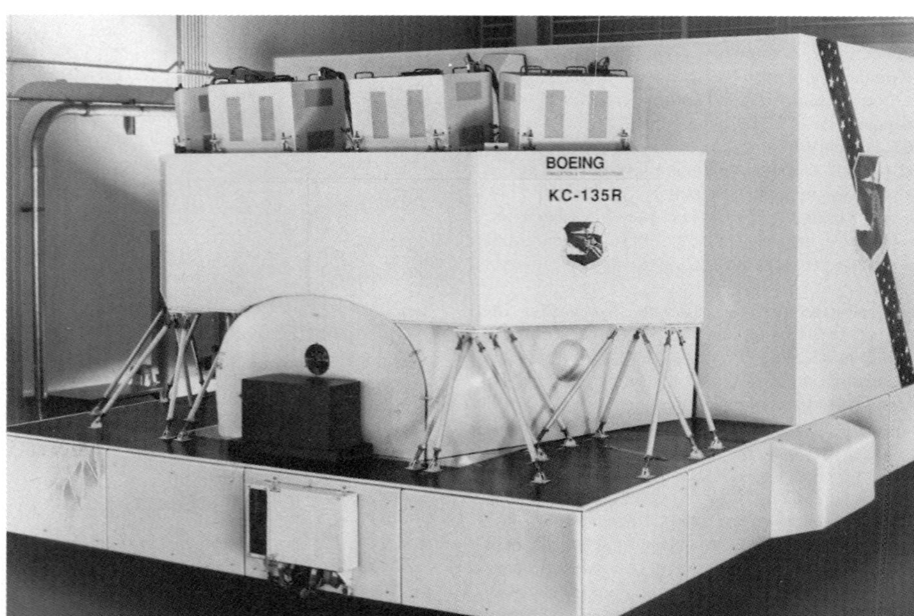

One of 19 KC-135 OFTs used by the US Air Force - this one is fixed base but some have 6-DoF motion and others are being fitted

Burtek Flight Simulation (Thomson Training and Simulation)

Founded in 1939, Burtek Inc was purchased by Thomson-CSF in 1979 and is part of the newly titled Thomson Training and Simulation (TTS) division of Thomson-CSF. TTS, as well as Burtek, includes the previous Thomson-CSF simulation facilities in France, and also the facilities of the ex-Link-Miles and Rediffusion companies in the UK.

Products. Burtek designs and manufactures flight training devices and cabin simulators for commercial airlines as well as a wide range of military flight and weapons system simulators (WST).

Deliveries - Flight Simulators. Burtek's military simulators include seven C-141B StarLifter cockpit procedures trainers (CPT), eight C-141B full mission simulators (FMS), five F-5E operational flight trainers (OFT) and an F/A-18 HOTAS (Hands On Throttle And Stick) training aid.

Maintenance Trainers. Maintenance trainers have been built for the F/A-18, CF-18, UH-60 and B-1B.

Status

See text.

Burtek CPT for the US Air Force/Lockheed C-141 StarLifter transport

Manufacturer

Burtek Inc (part of Thomson Training and Simulation), Tulsa, Oklahoma.

(For details see under Thomson Training in the Directory of Suppliers)

CAE-Link Training Systems

As well as the manufacture of simulator and training devices, CAE-Link has been involved in complete training systems for many years. The total training system (TTS) approach calls for one contractor source to assume responsibility for the entire training process from design of the training programme to the production of the required courseware, hardware and software and the provision of facilities, management, instruction and maintenance. This differs from previous practice where the customer acquired and managed the individual components of a training programme. The TTS approach is becoming progressively more prevalent in DoD requirements. Some CAE-Link training systems are described below:

CAE-Link C-130 Aircrew Training System

In April 1987 the US Air Force selected CAE-Link to provide the C-130 Aircrew Training System (ATS), the first system to include tactical mission training for USAF transport aircraft crews. On 31 March 1988, the Military Airlift Command (MAC) transferred responsibility to CAE-Link for C-130 refresher training, logistics support and simulator operations. This is claimed to be the largest contractor-managed military aircrew training system in the world. Crew members trained include the pilot, co-pilot, navigator, flight engineer, loadmaster and maintenance engine-run personnel.

Contractor Personnel. The system uses contractor personnel to conduct ground-based training and to maintain the associated training aircraft.

USAF/MAC Responsibilities. MAC will continue to provide instruction for all in-flight training and evaluation.

Locations. Training by CAE-Link takes place at eight US Air Force bases; Little Rock Air Force Base (AFB), Arkansas (C-130E and two CPTs); Pope AFB, North Carolina (C-130E All Weather Delivery System (AWADS) configuration); Kirtland AFB, New Mexico (HC-130P in the Special Operations Forces (SOF) configuration); McChord AFB, Washington (C-130E); Dyess AFB, Texas (C-130H); Elmendorf AFB, Alaska; Rhein-Main AB, Germany; and Yokota AB, Japan. There are currently eight flight simulators, all originally built by CAE-Link, at five bases; Little Rock (four), Pope, Kirtland, McChord and Dyess. A ninth simulator (C-130E), originally at Clark Air Base, Philippines, has been placed in service with the Minnesota Air National Guard at Minneapolis.

System Components. Each simulator consists of an Instrument Flight Simulator (IFS) and a Satellite Navigation System (SNS). Each IFS has a wide-FoV visual system. The IFS and SNS can either operate in stand-alone modes or be integrated, whereby the functions of the IFS navigation station are replaced by the SNS. All simulators are progressively modified to be current with changing aircraft systems. The present C-130 training programme includes interactive, self-paced computer-based training (CBT) and an automated logistics management capability to analyse cost, reliability and maintainability.

CAE-Link Training Management System (TMS). A TMS was developed for the C-130 ATS. It is a multi-module, comprehensive relational database about students, instructors, media, equipment and facilities. It is designed to make more effective use of resources. New courseware, based on a detailed Instructional Systems Development (ISD) analysis and using modern instructional strategies and technologies, was implemented incrementally and was issued to all sites.

Scope of Training. The ATS includes all aspects of MAC tactical airlift training. This ranges from initial classroom training to special mission qualification and operational mission proficiency readiness. Types of training include initial qualification, upgrade, refresher, instructor and special qualifications. The programme is based on measurable proficiency standards which are agreed with the USAF and form a quality guarantee.

Mission-Orientated Simulator Training (MOST). MOST scenarios include mission preparation, detailed preflight briefings and an extensive simulator mission involving all aircrew functions and incorporating real-world operational problems, crew co-ordination requirements and critical mission decision points. The training concludes with an extensive mission debrief conducted by CAE-Link instructors.

Contract term. The Link C-130 ATS contract will last for 12 years, a three-year initial development phase ended in 1991 and overlapped the first year of 10 one-year options and maintenance options, extending currently to the year 2000.

KC-10 Aircrew Training System

In September 1991 CAE-Link announced it had won a $13 million contract to manage the KC-10 Aircrew Training System (ATS) for the Strategic Air Command (SAC). The previous contractor was American Airlines Training Corporation, which was purchased by McDonnell Douglas in 1986.

Training Provision. CAE-Link will provide initial,

CAE-Link C-130 CPT

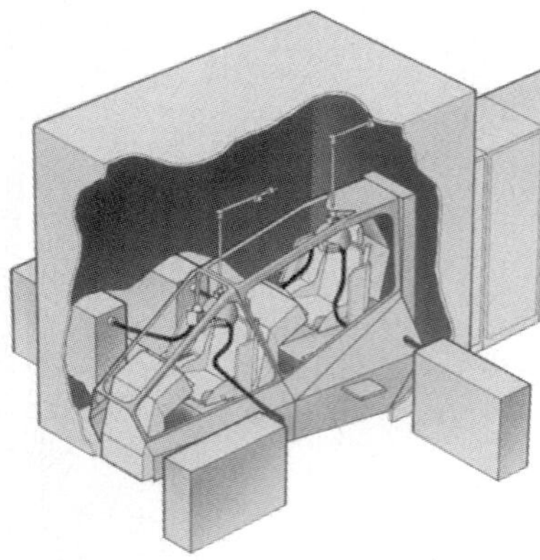

Drawing of the Team/Combined Arms Trainer (T/CAT) for the RAH-66 Comanche

refresher, upgrade and instructor training for all aircrew of the KC-10A Extender with academic and simulator training at Seymour Johnson AFB, North Carolina, Barksdale AFB, Louisiana and March AFB, California.

RAH-66 Comanche Integrated Training System

The Integrated Training System (ITS) devices for the new US Army RAH-66 Comanche light helicopter (formerly LH) will be developed by CAE-Link, a member of the Boeing-Sikorsky 'Team Comanche'.

Training Devices. The RAH-66 training design will encompass a variety of tutorial and hands-on media, among them computer-aided instruction, interactive video-disk, task trainers, operational flight simulators, weapon system and combat mission simulators, networked combined arms trainers as well as aircraft-embedded training, together with courseware, instructors, automated management tools and other support resources.

Aircrew Training. An Operator Training Device System (OTDS). OTDS will consist of Cockpit Procedures Trainers (CPT) based upon simple, single-screen visual displays, Combat Mission Simulators (CMS) and Team/Combined Arms Trainers (T/CAT). The last two systems will be static or mobile with the two CMS at Fort Rucker, Alabama, having a motion system. The two seats of the T/CAT may be operated independently.

Use of CAE FOHMD Display System. Compact, mobile visual display systems will use the Fibre Optic Helmet-Mounted Display (FOHMD) developed by CAE Electronics of Montreal, with CAE-Link a fellow member of the CAE Aerospace and Electronics group of CAE Inc. The use of the FOHMD is claimed to allow pilots to detect and identify targets at greater distances than in traditional dome projection systems; it is also much more compact. The display will feature a large instantaneous field-of-view (IFoV) and an unlimited field-of-regard (a properly orientated display will be produced for any head position including looking to the rear). Both CMS and T/CAT will replicate the tandem cockpit of the aircraft and will incorporate the CAE FOHMD display system (see also the Display Systems section of this book). The CMS will use overhead sensor tracking for the FOHMD while the T/CAT will use a mechanical head-tracker. The T/CAT will feature two-channel, medium resolution, image generators which will supply a 100 × 45° instantaneous field-of-view.

Maintainer Training Device System (MTDS). The MTDS will consist of six devices; an integrated composite maintenance trainer and five dedicated modules. The modules will be for: 1. cockpit, sensor, turret/gun; 2. rotor, transmission; 3. weapon system; 4. engine interface and mission equipment package; 5. fantail and drive train; 6. landing gear and hydraulics.

Fielding - IOC. The RAH-66 ITS will be produced and fielded simultaneously with aircraft Initial Operating Capability (IOC).

Force Level Simulation

Force Level Simulation (FLS) is a real-time, tactical force-on-force (blue on red) war game simulation. It can also be operated in non-real time and replay modes. Each operational mode has its own graphical display and control capabilities, and a scenario can be viewed from any eye point.

Real-Time Mode. The real-time mode supports one or more simulators with computer-generated forces (friendly and threat) with multi-echelon modelling of C^2 and C^3 with reactive movement and weapons tactics. Manned simulators may be connected to FLS through device-specific interface protocols or DIS protocols (Distributer Interactive Simulation).

Non-Real-Time Mode. The non-real-time mode allows analytical studies for weapon effectiveness, capability analysis, design modification analysis, force correlation and tactics development.

Replay Mode. The scenario replay mode plays stored real-time or non-real-time modes on graphics displays at any replay speed both forwards in time and also backwards if required.

Modelling. FLS models distinct command echelons, detailed simulations of the control processes at each command position for force allocation, and communication networks which each C^2 organisation will have. All units, platforms and their associated weapon systems are modelled not only with respect to system performance, but for tactical behaviour. Units in the scenario will be affected (constrained at times) by rules of engagement which are applicable to the nature of the conflict being examined.

Results. The result is a dynamic mission simulation which portrays the interaction of multiple participants, be they simulated or man-in-the-loop, as they manoeuvre with respect to each other and to the threat presented by the opposing force or forces. The flexibility of FLS allows the creation of large scale war-games ranging from mission environments expected during multi-sided, low intensity regional conflicts to a two-sided global war-type scenario.

Architecture - Other applications. Although FLS architecture was designed initially for aircrew training, it also represents an analytical tool which can be used for examining various aspects of weapon system performance and those other systems tasked to perform the tactical mission environment. As an example, FLS was demonstrated in 1990 by networking two US Army advanced helicopter simulators; further work is described below.

Networking - the STAARS programme. In 1991 CAE-Link completed the first field application, Project Desert STAARS (Sustainment Training for Army Aviation Readiness through Simulation). This included correlation with a real-world database using Defense Mapping Agency (DMA) data and photographic information. This was used for theatre-oriented training of AH-64 helicopter crews against real-world targets. FLS currently features up to 30 interactive players, some man-in-the-loop and some computer-generated, in a dynamic Red v Blue scenario with multi-echelon command, control and communications.

Crew Resource Management (CRM) Training

CAE-Link has developed Crew Resource Management (CRM) Training which provides specific tools to strengthen skills in the areas critical to recognising and dealing with crew error. These include planning, command authority, crew and mission assessment, communication, conflict management, risk assessment and management, judgement, decision making, situational awareness and workload management.

CRM Courses. The basic CRM course will be adapted to the weapon systems, mission and crew experience. Normally the course is in one- or two-day formats, but can be modified to suit the customer. A CRM 'facilitator' course prepares for leading and presenting the basic CRM programme. Other applications include Air Traffic Control (ATC), aircraft maintenance and AAA teams.

Status

See text.

Manufacturer

CAE-Link Corporation, Binghamton, New York.

(For details see Directory of Suppliers)

CAE-Link B-2 Aircrew Training Devices

CAE-Link has produced aircrew training devices for the Northrop B-2 Advanced Technology Bomber (ATB), each simulator having 20 Concurrent Computer Corporation Model 3280MPS super mini-computers. These systems include three Weapon Systems Trainers (WST) and two part-task Mission Trainers (MT) for initial qualification of the mission commanders.

WSTs. The WST is to include a simulated operational area of 20 million miles². Initially, 5 million miles² of radar and 500 000 miles² of visual data were stored. The threat database or Electronic Combat Environment will include 12 000 dynamic threats or emitters of which 2000 may be active within the B-2's visual and radar range. Each WST uses 1.7 million lines of Ada-language software and is capable of 75 Mips, eight times greater than the B-52 weapon system trainer. The out-of-window visual system uses the Martin Marietta (ex-GE) CompuScene PT 2000 image generator and the AWARDS collimated display system which was developed by CAE-Link's UK associate Link-Miles (now part of the Thomson Training group). More details are in the Visual Systems section of this book.

The cockpit of the CAE-Link B-2 Weapon Systems Trainer

Mission Trainers. There are also two part-task Mission Trainers (MT) for initial qualification of the mission commanders. An MT consists of the right seat of a WST. As with the F-117A training system, simulated scenarios are created using actual operational flight plan tapes from the aircraft supplemented by simulated visual, radar and threats. The system can be 'flown' automatically through the software or manually by the instructor.

Status

Delivered in the Summer of 1993.

Manufacturer

CAE-Link Corporation, Binghamton, New York.
(For details see Directory od Suppliers)

CAE-Link F-117A Pilot Training System

The contract to build the simulator for the F-117A Stealth Fighter was awarded in August 1982 and the complete CAE-Link Pilot Training System entered service in January 1986 at Nellis Air Force Base, Nevada.

Integrated Support Facility (ISF). The F-117A programme incorporates an Integrated Support Facility (ISF) which is a near duplication of the Weapon Systems Trainer (WST), lacking only the canopy and two of the visual system windows. The ISF provides an engineering tool to test and to prove all aircraft changes before their incorporation into the WST. The ISF approach provides the maximum use of the flight simulator for pilot training. The training system also uses actual operational flight plan tapes for mission scenario generation.

CAE-Link Responsibilities. In addition to designing, manufacturing and updating the flight simulator, CAE-Link also maintains both the WST and the ISF. It is also responsible for courseware development as well as academic and simulator instruction under the pilot training system contract.

Currency with Aircraft Fit. Ensuring that the build-standard of hardware and software in both training devices and the aircraft itself is kept in step, is achieved through a specifically planned interface between Lockheed, CAE-Link, the USAF and the Systems Program Office.

Status

In service.

Manufacturer

CAE-Link Corporation, Binghamton, New York.
(For details see Directory of Suppliers)

CAE-Link Military Aircraft Simulators - Fixed-Wing

B-52 WST

The B-52 weapon system trainer (WST), designed and built by CAE-Link, is a sophisticated fully integrated strategic mission simulator. It allows the six-man crew to carry out combat mission training under realistic conditions, including hostile environments. Various modifications have been carried out so the simulated equipment matches improvements made to the aircraft fit.

F-4 Phantom Simulators and Trainers

CAE-Link has produced 71 F-4 simulators for a number of customers and is currently installing refurbished F-4E trainers in Greece and Turkey.

USAF F-4 Trainers. The USAF has reduced its inventory of F-4 trainers to four, an F-4E, two F-4Gs, and an RF4-C. Recent modifications have been software changes and have included an update to the Threat Libraries of all the remaining trainers.

F-4G OFP Development. An F-4G trainer is at CAE-Link's main base at Binghamton NY and is used for the development of Operational Flight Programmes (OFPs) for the other simulators.

F-4E. An F-4E trainer is at Holloman AFB and is also used for German Air Force training.

F-4G. An F-4G trainer is with the Air National Guard at Boise, Idaho. Its APR-47 Radar Warning System has been updated to OFP 9000 standard and was fielded in October 94, ahead of the aircraft OFP update. It is claimed that the fit of F-4G trainers has been continuously updated together with aircraft modifications and that they have always matched the aircraft fit.

RF4-C. An RF4-C trainer is with the Air National Guard at Reno, Nevada. OFP 3.1 and the SAM flyout update to the visual system have been delivered.

F-16 Training Programme

The CAE-Link F-16 training and support programme includes the provision of tactical flight simulators, simulation of specialised sensor systems such as LANTIRN, part-task devices for air intercept and air-to-ground training, academic courseware, classroom and simulator instruction. It also includes simulator maintenance and logistic support.

Training. The F-16 training curriculum includes task listings, objective hierarchies, training device analysis, lesson format guides, a training development and support plan, a quality control plan as well as student materials in a variety of media for basic, transition, conversion, instructor and international operator courses. They cover the F-16A/B/C/D used by the US Air Force and foreign air forces as well as the F-16N used by the US Navy for adversary pilot training. In addition there are more than 70 computer-based training interactive video disk systems available for F-16 conversion, air intercept and ground attack training.

Simulators ordered. CAE-Link has delivered, or has received orders for, 46 F-16 Tactical Flight Simulators (TFS) with users including the US Air Force, the US Air Force Reserve, the US Air National Guard, Belgium, Denmark, Egypt, Greece, the Netherlands, Norway, Turkey and Venezuela.

The CAE-Link F-16 Tactical Fighter Simulator (TFS) with the new instructor operator station

F-16 TFS. The simulator consists of cockpit/avionics, instructor station, digital computer, interface equipment, and anti-G suit/G-seat motion cues. It is used to provide realistic tactical training environments including moving targets in various atmospheric conditions. Also simulated is air refuelling with the KC-135 tanker, including slipstream buffet. The TFS are co-produced multinationally with Danish and Norwegian companies assisting CAE-Link.

F-16 Air Defence Simulator. The F-16 Air Defence Fighter Simulator has been ordered by the US Air National Guard. This modified F-16A simulator includes radar improvements and the addition of an AIM-7 Sparrow MRAAM capability.

LANTIRN Simulator. CAE-Link has developed an F-16 LANTIRN (Low Altitude Navigation and Targeting Infra-Red system for Night) simulator. It provides complete simulation of the system including the navigation pod's forward-looking infra-red (FLIR), the targeting pod's narrow and wide field-of-view FLIR, the AGM-65D Maverick missile infra-red, laser ranging and designation, the terrain-following radar (TFR) and the radar altimeter. The student views the sensor imagery through the wide area Raster HUD (RHUD) or on the cockpit multifunction displays.

Turkish F16 Order. In October 1992 CAE-Link announced that it had won a contract to supply an F-16C Block 40 weapon system trainer to the Turkish Air Force under the 'Peace Onyx III' programme. This will include electronic warfare and LANTIRN simulation, the Interactive Tactical Environment Management System (ITEMS) and programmable Digital Radar Land Mass Simulation (pDRLMS). It will also feature an Evans & Sutherland ESIG-3000 two-channel image generator and a 24 ft diameter fibreglass dome.

F16 PTTs. Part-Task Trainers (PTTs) are also produced by CAE-Link to provide a low-cost approach for air combat training (ACT). They give training in interpreting and employing the F-16 radar, the Head-Up Display, weapons and controls during air-to-air missions. The expandable design will also permit cover of radar warning and self-protection, engine emergency procedures, air-to-ground avionics and other unit training requirements. A library of more than 88 preprogrammed scenarios is available, including 190 air targets.

Touch-Screen IOS. CAE-Link is introducing a full colour, touch-screen instructor operating station. The microprocessor-based station is being installed on all new F-16 simulators in production and retrofitted to existing simulators.

CAE-Link Contractor Training. The company also provides contractor training/academic training for F-16 pilots at Luke, MacDill, Hill, Homestead and Shaw AFBs.

Simulator Modifications for F-111

CAE-Link is carrying out an upgrade of the US Air Force's F-111 simulators. Improvements include DRLMS (Digital Radar Land Mass Simulation), new computational systems, re-hosting the computer assembly language, built-in diagnostics, and modern instructor/operator stations. Also being embodied is the introduction of simulation of the Pave Tack laser designator/ranger. The programme also includes the provision of two software support centres to manage the simulators' software and documentation, including modifications to radar databases and configuration control to achieve and maintain concurrency with modifications to the aircraft.

Initial Contract. The US Air Force contract with CAE-Link was for five simulators: 1. A FB-111A bomb/navigation simulator, delivered to Plattsburg AFB in January 1987; 2. A F-111D mission simulator (MS) to Cannon AFB in July 1987; 3. A FB-111A MS to Plattsburg AFB in March 90; 4. A F-111F MS to Lakenheath in the UK in June 1990; 5. A FB-111A to Cannon AFB in September 1990. Two software support centres were delivered, to

Cannon AFB in July 1987 and Plattsburg in March 1990. There was an option for five further simulators; for the FB-111A and the F-111A, D, E, and F versions. There was also an option for modification for the two EF-111A flight simulators.
Current Situation. Force restructuring has caused the retirement of all except the F-111E and F-111F MS, the two EF-111A simulators and the software support centre at Cannon AFB. An EF-111A simulator has been relocated to Cannon AFB and the second EF-111A will also be relocated to Cannon in 1994. The F-111E and F simulators have already been relocated to Cannon AFB.

KC-130R Operational Flight Trainer (OFT) Upgrade

Link has completed a host-computer change and a software conversion programme for the US Marine Corps KC-130R OFT (Device 2F107) at MCAS El Toro, California. This incorporates a four-channel digital control loading system and an Avionics System Improvement Program (ASIP). The ASIP will update the simulator with a new FCS-105 flight director/autopilot system, a Mark II Ground Proximity Warning System (GPWS), an Altitude Alerter/Preselect System (AAPS) and a Combined Altitude Radar Altimeter (CARA).

Status

See text.

Manufacturers

CAE-Link Corporation, Binghamton, New York.
(For details see Directory of Suppliers)

CAE-Link Military Aircraft Simulators - Rotary-Wing

Synthetic Flight Training System (SFTS) for US Army

SFTS is the result of R&D by CAE-Link into training methods, computer technology and simulation techniques, and was specifically developed for the US Army's flight training programme. SFTS includes a simulation system for each of the helicopters expected to be in the US Army's inventory to the end of the century.
Simulators Involved. 131 Simulators (Cockpits) are in service or on order from CAE-Link. These include the UH-1H (22x4-simulator systems, 88 cockpits), CH-47D Chinook (six, of which four were converted from CH-47C), AH-1S Cobra (nine), UH-60A Black Hawk (18), and AH-64A Apache Combat Mission Simulator (eight) while MH-47E and MH-60K simulators have been provided for Special Operations.
Estimated Savings. According to US Army estimates, when fully fielded the SFTS simulators will save $608.6 million per year or $4.65 million per simulator (that is, per independent simulated Cockpit unit).

Special Operations Aviation Combat Mission Simulators (SOACMS)

CAE-Link has delivered two Special Operations Aviation Combat Mission Simulators (SOACMS); an MH-47E Chinook derivative (Device 2B46) and a MH-60K Black Hawk derivative (Device 2B47A). They were being provided under a $60 million contract for the US Army Special Forces (SF).
Training. The simulators provide a range of aircrew training from conversion-to-type, up to full task loading in a simulated high threat environment. They also provide training in the use of the AN/APQ-174 multi-mode radar. They provide training for covert, deep penetration missions which will be carried out in all types of weather, day or night, over every kind of terrain. They will be used for the rehearsal of actual SF missions.
Components. Each CMS consists of a single cockpit with pilot and co-pilot stations and an onboard IOS. The cockpit is mounted on CAE-Link's standard 6-DoF synergistic motion platform. The pilot stations are replicas of the 'glass cockpit' for the Chinook MH-47E and the Black Hawk MH-60K, based upon the IBM Model 360 avionics management system. Image generation is by 8 channels of the enhanced ATACDIG CGI system which includes models for air-to-air refuelling and shipboard landings. The exercise area covers 80×100 km.

Pilot and co-pilot training in the UH-60A Black Hawk simulator

Simulator for AH-1S Cobra

Seven AH-1S Cobra simulators are in service with the US Army at Fort Hood, Texas; Fort Campbell, Kentucky; Fort Lewis, Washington State; Illesheim, Germany; Fort Rucker, Alabama and at two National Guard establishments in Pennsylvania and Arizona.
Visual Image Generation (IG). LIG system. Four out of the seven systems feature a full colour laser image generation (LIG) system. LIG provides a good nap-of-the-earth (NOE) model for flight and tactical training across a terrain model board representing a 140 km^2 gaming area. LIG was produced as a result of a US Army R&D contract with CAE- Link. It uses an optical probe with the three rotational freedoms (pitch, roll, yaw) mounted on a gantry which provides horizontal and vertical movement relative to the model board and so simulated speed and altitude. The colours of the laser beams are reflected from the model board back to banks of photomultipliers. Due to the narrow laser beams, a high resolution is achieved and the result is a very high-fidelity visual scene of the low level environment. This is a related method to that developed by the UK DRA in the late 1980s at their R&D establishment at Farnborough, UK, similarly for helicopter NOE scenarios. Details in Image Generation (IG) Systems section.
ATACDIG. The other three simulators have CAE-Link's ATACDIG CGI system with realistic moving targets and improved TSU (see ATACDIG entry in the CGI section of this book).
Visual Display System (DS). Three CAE-Link wide-angle collimated (WAC) monitors per cockpit.
Motion Platform. Each Cobra simulator has two CAE-Link standard 6-DoF platforms, one for each of two tandem cockpits.
Main Computer. Five DEC PDP 11/55 computers are fitted.
Instructor Operating Stations (IOS). Mounted on board each of the two cockpits.
Cockpits. Two separate cockpits for pilot and navigator/gunner each mounted on its own motion platform with its own visual display system. Pilots and gunners may train individually, or the two cockpit modules may be integrated for crew training.
Control Loading. Digital.
Other Features. Systems simulated include Doppler navigation, fire-control computer, laser rangefinder, and head-up display (HUD).
Weapons. The pilot can simulate firing all weapons except the TOW missile. The gunner, who also serves as navigator, can fire all weapons, including the wire-guided TOW, 20 or 30 mm cannons, and 2.75 in (70 mm) free-flight rockets - all through a telescopic sight unit (TSU). An improved TSU is incorporated on the AH-1S simulators already delivered which uses the Link CRISP laser disk visual system.

Weapons System Trainers for AH-1W SuperCobra

The US Marine Corps Air Stations at Camp Pendleton, California, and New River, North Carolina, are the sites for the first SuperCobra AH-1W Weapons System Trainers (WST), designated Device 2F136.
Visual Image Generation (IG). Evans and Sutherland ESIG 3000. Details in Image Generation (IG) Systems section.
Visual Display System (DS). Projection on inside of domes mounted one on each motion platform.
Motion Platform. Each simulator has two CAE-Link standard 6-DoF platforms, one for each of two tandem cockpits.
Instructor Operating Stations (IOS). Mounted on board each of the two cockpits. Features include auto-fly, store/reset, freeze, hard copy for record keeping and syllabus development, and record/playback - enabling a trainee to observe a just-completed sequence, evaluate it with his instructor, and then repeat the mission correctly.
Other Features. The TOW & Sidewinder missiles are modelled. Flight modes simulated include hover, transition, cruise, and auto-rotation. Atmospheric effects are also present: pressure variation, non-standard temperatures, wind velocity and direction, and icing conditions. The simulation includes the tactical systems, such as the HMS, RWR, the telescopic sight unit (TSU) controls and eye-piece, and the armament control and delivery systems panel.
Matching Changes in the Aircraft - Configuration Control. The use of actual aircraft equipment and displays combined with modular simulator software facilitates updating the simulator to match future aircraft changes.

Combat Mission Simulator (CMS) for AH-64 Apache

AH-64 Apache CMSs (Device 2B40) for the US Army are installed at Fort Bragg, North Carolina; Fort Campbell, Kentucky; Fort Hood, Texas (two), and Fort Rucker, Alabama (two) as well as at Hanau and Illesheim, Germany. It is planned to deploy others at Camp Humphreys, Korea, and at a National Guard establishment at Marana, Arizona. CAE-Link has installed a CMS for the Egyptian Air Force under a $25.5 million contract announced in March 1991; the first export contract for the Apache CMS. In February 1993 a further $15 million contract was awarded for a CMS for the US Army.
Visual Image Generation (IG). Six channels of ATACDIG CGI (3 per cockpit). Details in Image Generation (IG) Systems section.
Visual Display System (DS). Three CAE-Link wide-angle collimated (WAC) monitors per cockpit.
Main Computer. Seven Concurrent 3250 and two 3280 32-bit digital computers and 29 auxiliary processors integrated into a parallel architecture.
Instructor Operating Stations (IOS). Mounted on board each of the two cockpits. Features include

auto-fly, store/reset, freeze, hard copy for record keeping and syllabus development, and record/playback - enabling a trainee to observe a just-completed sequence, evaluate it with his instructor, and then repeat the mission correctly.
Cockpits. Two separate cockpits for pilot and navigator/gunner each mounted on its own motion platform with its own visual display system. Pilots and gunners may train individually, or the two cockpit modules may be integrated for crew training.
Control Loading. Digital.
Sensors. All onboard sensors are also simulated including day television, direct view optics, laser tracker, laser rangefinder and designator, image autotracker and forward-looking infra-red (FLIR). These systems are linked to the pilot's cockpit for navigation and the gunner's cockpit for target acquisition and weapon firing, as appropriate.
Other Features. The CMS teaches Apache crews the skills required in a high technology battlefield. The database includes a wide variety of threats programmed to engage and destroy the Apache. The simulator integrates tactics, terrain, weather effects, aircraft sensors, crew communications, enemy threats, and weapons delivery.
Cost. Unit cost about $15 million.

Apache CMS Update

Three Block Upgrade Configuration (BUC) programmes have been developed for the Apache CMS.

BUC I was completed in September 1990 and included updating the fire-control computer and expanding the threat library.

BUC II, begun in March 1990, included adding the Terrain Information System and expanding the database to provide an 80 × 100 km gaming area.

Future updates planned include adding the Airborne Target Handover System, air-to-air Stinger, GPS and Longbow millimetre wave radar.

Simulators for CH-47D Chinook

CH-47D Chinook helicopter simulators (Device 2B31) have been delivered to the US Army at Fort Lewis, Washington and Fort Rucker, Alabama, in addition to four CH-47C systems which have been upgraded to 'D' configuration at Fort Hood and Fort Campbell and Mannheim, Germany, and Camp Humphreys, Korea.
Visual Image Generation (IG). Three channels of CAE-Link ATACDIG CGI. Details in Image Generation (IG) Systems section.
Visual Display System (DS). Four CAE-Link wide-angle collimated (WAC) monitors, the two front monitors presenting the same scene to each of the two pilots. Switchable eye point between the two pilots depending which is flying the aircraft.
Motion Platform. CAE-Link standard 6-DoF.
Main Computer. Each simulator originally had three DEC PDP 11/55 processors but in 1993 CAE-Link completed an upgrade to one Harris Nighthawk microprocessing system with Ada software.
Instructor Operating Station (IOS). Mounted on/off board.
Cockpit. Single cockpit with side by side seating.
Control Loading. Digital, 1000 Hz.
Other Features. Basic training tasks supported include ground operations, take off, hovering, basic flight, auto-rotation, navigation, approach and landing, advanced operations such as those in confined areas or with slung loads, advanced flight training, formation flight manoeuvres, instrument, flight and emergency procedures. Special training features are the fore and aft rotors, landing gear, battlefield positioning of artillery and other equipment, nap-of-the-earth (NOE) flight and night vision navigation including Automatic Flight Control System (AFCS) and Doppler.

Operational Flight Trainers for SH-60B LAMPS

Mark III Sea Hawk Helicopter

CAE-Link delivered four simulators for the US Navy's LAMPS Mark III (Light Airborne MultiPurpose System) Sea Hawk helicopter. The first two, originally configured as Operational Flight Trainers (Device 2F106) and subsequently integrated as Weapon Systems Trainers (Device 2F146), are at North Island NAS, California, and Mayport NAS, Florida.

Simulators numbers 3 and 4 are configured as Weapon Systems Trainers (WSTs) and were delivered to North Island and Mayport during 1986 and 1987. All four simulators are mounted on 6-DoF motion platforms and feature six-window dusk/night visual systems. All normal and emergency flight procedures and anti-submarine missions can be simulated, including take offs and landings using a small deck pitching up to 20° in all axes.

In August 1991 AAI-SMI received a contract to conduct visual system upgrades on a number of simulators including the two SH-60B OFTs which were fitted with dome display systems and the Star Technologies G 2000 CGI system.

Simulators for UH-1H Huey

The UH-1H Huey simulators were the first to be delivered to the US Army's SFTS programme, beginning in 1971. CAE-Link supplied 22 systems each with four cockpits (88 training cockpits in total), which are installed at Fort Campbell, Kentucky, Fort Belvoir, Virginia; Fort Bragg, North Carolina; Fort Rucker, Alabama (eight systems, 32 training cockpits); Fort Sill, Oklahoma; Fort Knox, Kentucky; Fort Riley, Kansas; Fort Hood, Texas; Fort Lewis, Washington and Fort Ord, California; Fort Shafter, Hawaii as well as at Indiantown Gap, Pennsylvania; Hunter AAF, Georgia; Hanau, Germany; and Seoul, South Korea.
Components. Each simulation unit includes four cockpit trainee stations each mounted on 5-DoF non-synergistic motion systems, and an IOS with two Honeywell 790s and an ADD 900 digital computer. The motion platform is pre-synergistic and is mounted on short laterally orientated rails; jacks are used in the direction of the rails in order to produce sway. Yaw is produced by jacks pushing on a turret ring, and pitch, roll and heave through other jacks in the conventional way.
Simulations Covered. The simulators model flight characteristics, the engine and its systems, radio communications and navigation. A total of 93 radio communications and navigation facilities is provided within the simulator's 663 km^2 gaming area.
Visual Systems. Originally no visual systems were fitted. Subsequently a number of experiments has been made with a variety of low-cost CGI visuals displayed by monitors.

Simulators for UH-60A Black Hawk

CAE-Link has delivered 18 simulators (Device 2B38) of the US Army's UH-60A Black Hawk helicopter. They have been installed at Fort Campbell, Kentucky (two); Fort Bragg, North Carolina; Fort Rucker, Alabama (two); Fort Hood, Texas and Fort Lewis, Washington as well as at Seoul, South Korea, and Hanau and Illesheim in Germany. Others are installed at Fort Stewart, Georgia; Fort Riley, Kansas; Fort Carson, Colorado; Fort Ord, California; Fort Richardson, Alaska and Fort Shafter, Hawaii, as well as at the Aeromedical Research Center in Fort Rucker.
Visual Image Generation (IG). Three channels of CAE-Link ATACDIG CGI. Details in Image Generation (IG) Systems section.
Visual Display System (DS). Four CAE-Link wide-angle collimated (WAC) monitors, the two in front of each pilot displaying the same CGI channel but having the facility of an eye point switchable between the pilots depending on who is flying the aircraft.
Motion Platform. Standard CAE-Link 6-DoF.
Main Computer. Concurrent 3252 32-bit super mini-computer.
Instructor Operating Station (IOS). A windows-based ADA IOS is mounted on board. This exploits colour graphics display capabilities and can be used to display cockpit instruments and non-safety hard switches. The system can be added to other simulators and is already used in other CAE-Link simulators.
Cockpit. One cockpit with side by side seating mounted on the motion platform.
Control Loading. Digital.
Other Features. Pilots and co-pilots will be able to train in all types of flying, including NOE; emergency procedures can also be simulated. One simulator at Fort Rucker is configured for aero-medical research and includes an environmentally controlled cockpit for research into heat and cold, bio-medical sensors, signal conditioning equipment and specialised software.

Black Hawk Update

As part of a $7 million block update (BUD), enhancements have been made to the 18 UH-60A Black Hawk flight simulators. Included are out-the-window display improvements, NVG-compatible lighting and the addition of several new malfunctions.
CGI Upgrade. The upgrade to the ATACDIG CGI incorporates the blowing sand/water/dust created by rotor wash (as a result of experience in the Gulf War), capabilities for landing aboard assault carriers and frigates, the addition of several friendly and threat vehicles to the database, and the ability for air targets to fire upon the simulated Black Hawk.
Future GPS Fit. Plans exist to retrofit a Global Positioning System (GPS) capability into UH-60 simulators. The last UH-60A simulator, scheduled for installation in Fort Richardson, Alaska, has been retained for integration and testing of such a system.

Desert Hawk

In 1989 CAE-Link obtained a $20.4 million contract to supply a UH-60 Desert Hawk flight simulator (Device 2B38A) to Saudi Arabia. This was delivered in 1993 and provides aircrew with basic, transition, refresher and advanced training. It is similar to the US Army UH-60A simulator but includes features unique to the Saudi Arabian aircraft. It has an enhanced ATACDIG image

AH-64A Apache Combat Mission Simulator - note the separate cockpits for pilot and gunner, each mounted on its own 6-DoF motion platform and with its own IOS and visual display system

generator with 256 texture patterns and a real-world visual database 80 x 100 km in area.

S-70B-2 Seahawk Mission Simulator for Royal Australian Navy

This system for an ASW mission simulator was ordered by the Australian DoD for the Royal Australian Navy (RAN). It was produced by CAE-Link in conjunction with Qantas Airways Limited and Computer Sciences of Australia Pty Limited (CSA) and delivered to the Nowra NAS in 1990. The simulator provides individual and integrated crew flight and mission training.

Visual Image Generation (IG). Five channel Link-Miles (now part of Thomson Training) IMAGE II-T dusk/night textured IG. Details in Image Generation (IG) Systems section.

Visual Display System (DS). Five collimated monitors.

Motion Platform. CAE-Link standard 6-DoF.

Instructor Operating Station (IOS). The onboard instructor station permits individual training while an offboard station permits integrated crew mission training.

Cockpit. For three crew members.

Control Loading. Digital.

Sensor Simulation. Includes radar, acoustic/sonobuoys, MAD.

Other Features. Simulation is provided for ship aerodynamic wake and ship deck motion for practising landings in all sea states. Exercises are conducted with simulated ships, aircraft and submarines.

Companies Involved. CAE-Link developed the mission simulator, with CSA responsible for the weapon systems element and Marconi (ex-Ferranti) Computer Systems (in both Australia and the United Kingdom) acting as subcontractor for the acoustics and radar simulation. Final assembly, integration, test equipment and support was carried out by Qantas with supporting documentation by Publication Perspectives Pty Limited.

Cost. Contract cost A$48 million.

Simulator Networking - Multisim

CAE-Link has demonstrated the networking of UH-60 Black Hawk and AH-1S Cobra simulators with an AH-64 Apache CMS to allow US Army helicopter crews to train together as a team. The networking also demonstrated the capability of the Apache CMS to adapt to air-to-air combat training. The research follows increasing US Army interest in both the networking of training devices and in developing air-to-air training combat capabilities for helicopters.

Threat Simulation. The Black Hawk simulator was flown by an instructor to portray the threat and provide a realistic, unpredictable foe for the crew of the Apache whose crew saw the threat portrayed as a Hind helicopter, modelled by the ATACDIG computer-generated image visual system. Quasi-intelligent, reactive threats from ground and air are already part of the Apache CMS but none involving threat helicopters was specified for the original modelling.

Results. The research has demonstrated that the Multisim networking of flight and weapons simulators is less complex than anticipated and is feasible for simulators located either at a single site or at different sites.

Status

See individual entries.

Manufacturer

CAE-Link Corporation, Binghamton, New York.
(For details see Directory of Suppliers)

CAE-Link NASA Training Systems Contract (TSC)

In 1989 the National Aeronautics and Space Administration (NASA) awarded CAE-Link a Training Systems Contract (TSC) to perform systems engineering and integration activities for the development of crew training and simulation systems. The contract involves the training requirements for specific missions and also incorporating technologies into NASA's training suite in a cost-effective way which prepares for future training needs. CAE-Link is the prime contractor supported by IBM, Booz-Allen & Hamilton and Eagle Aerospace.

Project Scenario. The TSC project scenario was derived from the planned human exploration initiative which involves many parallel, interdependent training and simulation projects. The Shuttle Mission Simulator (SMS), for example, must train for the operation of the orbiter as well as for assembling the space station using the robotic manipulator arm and for docking with the space station.

The Shuttle Mission Training Facility (SMTF)

The SMTF is to train astronauts and mission control ground crew to control the space shuttle through all aspects of the mission from lift-off to landing. It includes four trainers; a motion-based trainer which simulates the cockpit fore and aft decks with 6-DoF motion; a fixed base trainer which simulates the fore and aft decks and the crew cabin below; a fixed base guidance and navigation simulator which simulates only the shuttle aft crew station; a Spacelab trainer.

The SMTF was built in 1978 and is used both to provide primary full systems training for shuttle missions and to test flight software in a realistic environment using flight-equivalent onboard computers. It is currently receiving Step 2-3 upgrades to provide a modern CGI capability, to replace the Spacelab computer and NASCOM Network simulation components and to upgrade both the voice system and the guidance-navigation trainer.

Step 4 will upgrade the SMTF computer hardware and software architecture so it will be compatible to, and interoperable with, the Space Station Training Facility (SSTF). Step 5 will integrate expert systems to aid instructors and to reduce the complexity of trainer operations by increasing automation.

Space Station Training and Verification Facility (SSTVF)

The SSTVF will be the primary system used to train in space station operation when it is unmanned, man-tended and permanently manned. It will simulate the space station elements such as the laboratory module and the nodes as well as the distributed systems such as electrical power, propulsion and data management. It will also simulate station communication network functions which are visible to, and controllable by, flight crew members and ground controllers. Long simulations with ground control centres, often exceeding 48 hours, will be supported by SSTVF to prepare specific teams for the time-critical handover procedures which occur when the shuttle is docked at the space station. It is estimated that more than 2 million lines of Ada-language software code will be required for SSTVF.

The SSTVF will consist of a set of workstation-based PTTs and a network-based training complex. The trainers will provide low-cost, early training on single space station systems. The complex will provide up to two simultaneous whole station training sessions on the complete interaction of space station systems. The complex will be able to assign any combination of training devices such as visual display systems, robotics processors, telemetry simulation equipment and astronaut crew stations as well as instructor stations to a training session.

The SSTVF programme is scheduled for completion in 1999. The PTTs were delivered in 1994 and the first full system training capability in mid-1995.

Assured Crew Return Vehicle (ACRV)

The ACRV is a space 'lifeboat' planned to give personnel working in space the opportunity to return to earth in an emergency. Through the TSC, CAE-Link is supporting the ACRV Project Office by reviewing ACRV operations concepts.

Heavy Lift Launch Vehicle

A special NASA/Defense Department force is to assess development of an evolutionary heavy lift vehicle to carry equipment which is either too large or too heavy for the space shuttle cargo bay. TSC has assisted NASA's conceptual design and budget estimate for the Shuttle C, a shuttle-derived unmanned heavy lift launch vehicle.

Lunar/Mars Exploration

TSC has reviewed the current studies on Lunar and Mars exploration. These studies identify crew training tasks such as surface excavation, large equipment assembly, rendezvous and docking, Lunar/Mars control of unmanned vehicles and surface communications with space vehicles and earth. The studies also identify mission operation tasks such as planning and analysis, ground-control of manned/man-tended space vehicles.

Status

See individual entries.

Manufacturer

CAE-Link Corporation, Houston, Texas.
(For details see Directory of Suppliers)

CAE-Link Flight Training Devices (FTDs)

US Federal Aviation Administration (FAA) Advisory Circular 120-45A includes training devices not categorised as Flight Simulators (FS) but as Flight Training Devices (FTDs). Level 2 and 3 FTDs are generic and represent a broad category of aircraft whereas Level 4 and 5 FTDs are aircraft-specific. CAE-Link believes that user response has shown that a Level 3 FTD meets many requirements for flight training schools, and also for initial training in airlines and military organisations for multi-crew, multi-type training. Students can log more than 100 hours in the FTD before transferring to training on a Flight Simulator (FS) specific to an aircraft type. Under the FAA Advanced Qualification Program (AQP), those training programmes subject to Parts 121 and 135 can develop new programmes incorporating devices such as computer base training (CBT), cockpit procedures trainers (CPTs), part-task trainers (PTTs) and FTDs, as well as conventional Flight Simulators. CAE-Link believes that the FTD 3 is a cost-effective way of bridging the gap between very low cost, lower fidelity training devices, aircraft-specific Level 4-6 FTDs, and the Flight Simulator (FS) for a type.

CAE-Link Level 3 Flight Training Devices (FTD-3)

An example of a CAE-Link FTD-3 is the FAA-approved Level 3 device for twin-turboprop aircraft shown in the photograph. CAE-Link FTD-3 devices are designed to the following criteria:

To satisfy FAA certification requirements;

To use commercial off-the-shelf components (COTS) for low cost;

Do not require skilled technicians to operate and maintain;

Use a modular design approach for ease of expansion;

Operate in a classroom environment.

Generic software operates through an aero/engines package suitable for a wide range of aircraft and complying with FTD-3 requirements. These data packages are based on a selection of data from a variety of aircraft in the class.

CAE-Link Reconfigurable General Aviation Trainer (RGAT)

The CAE-Link Reconfigurable General Aviation (GA) Trainer is an FTD-3 device that can provide *ab initio* and primary training on two or more different types of aircraft. The RGAT design is similar to CAE-Link's other FTD-3 devices. It occupies only a small amount of space and can be placed in an office or classroom environment. Little maintenance is required on the device and it is designed for ease of making changes including reconfiguration to new aircraft types.

The RGAT can also be integrated with CAE-Link's Virtual Controller™ and Virtual Tower™ Air Traffic Control simulators.

Status

Delivery of a Twin Turbo FTD was completed to Alenia

in 1992. Delivery of three Reconfigurable General Aviation Trainers to the University of North Dakota will be completed in August 1994.

Manufacturer
CAE-Link Corporation, Binghamton, New York.
(For details see Directory of Suppliers)

CAE-Link Level 3 FTD for a twin turboprop operated by Alenia

Contraves-SSI Training Systems

Gould's Systems and Simulation Division was subject to a management buy-out in April 1987 and renamed Systems and Simulations Inc (SSI). Subsequently it has become Contraves-SSI. Facilities are located in Tampa, Florida, and Melville, New York. The organisation has experience in the design, production, integration and support of aviation, electronic, marine and land-based simulation and training systems.

Aircraft Trainers Produced. Aircraft training experience includes full flight simulators (FFS), operational flight trainers (OFT), cockpit procedures trainers (CPT) and part-task trainers (PTT) for aircraft including the McDonnell Douglas F-4 and F/A-18, Fairchild A-10, Grumman E-2C, Lockheed C-130 and C-5, and Beechcraft T-34 and T-44. Current programmes include the following:

Aircrew Trainers for T-34.
The US Navy was provided with a total of four T-34 cockpit procedures trainers (CPTs), and 25 T-34 instrument flight trainers (IFTs). The CPTs (Device 2C42), located at NAS Milton, Florida, provide basic training for pilots in normal and emergency aircraft procedures. The T-34C IFTs (Device 2B37) facilities include ground handling, take off, climb, cruise, descent, landing, navigation and communications, as well as procedures training. The IFTs are located at NAS Pensacola, Whiting and Corpus Christi.

Aircrew Training System for C-5B.
Teamed with United Airlines and CAE, Contraves-SSI is responsible for the development of one new C-5B cockpit procedures trainer (CPT) and refurbishment of two existing systems previously delivered to the US Air Force by Gould.

E-2C OFTs. The fixed-base E-2C OFTs use a Harris H800 computer with a 300 Mb hard disk and have the Fokker MiCoL digital control loading system. Their visual system consists of three channels of the E&S SP1T dusk/night image generator with a texture capability. The IG data is displayed through 4 monitors, the monitors immediately ahead of the two pilots displaying the same IG channel. In December 1992 Contraves-SSI received an $11.17 million contract to modify four E-2C Operational Flight Trainers into three E-2C Flight Trainers and a C-2A Flight Trainer at NAS Norfolk, Virginia, and NAS Miramar, California by December 1997.

Contraves-SSI PTT for McDonnell Douglas F/A-18

Status
See text.

Manufacturer
Contraves-SSI, Tampa, Florida.
(For details see Directory of Suppliers)

Contraves-SSI Twin Dome Combat Mission Simulator

Contraves-SSI has developed a twin dome flight simulator for a foreign customer, for air-to-air and air-to-ground training.

Visual Image Generation (IG). CGI.

Visual Display System (DS). Two 40 ft diameter domes, each with three 1024-line optical projectors which can produce up to two targets per dome.

Motion Cueing. Fixed-base cockpits but with pneumatic pressure for pilot's anti-G suits.

Computers. Three Gould 32/9780 computers of which one drives the visual system.

Control Loading. Digital.

Instructor Operating Station (IOS). Mounted off board.

Database. A simulated area of 1156 × 1156 km is modelled.

Training Scenarios. Training scenarios will include

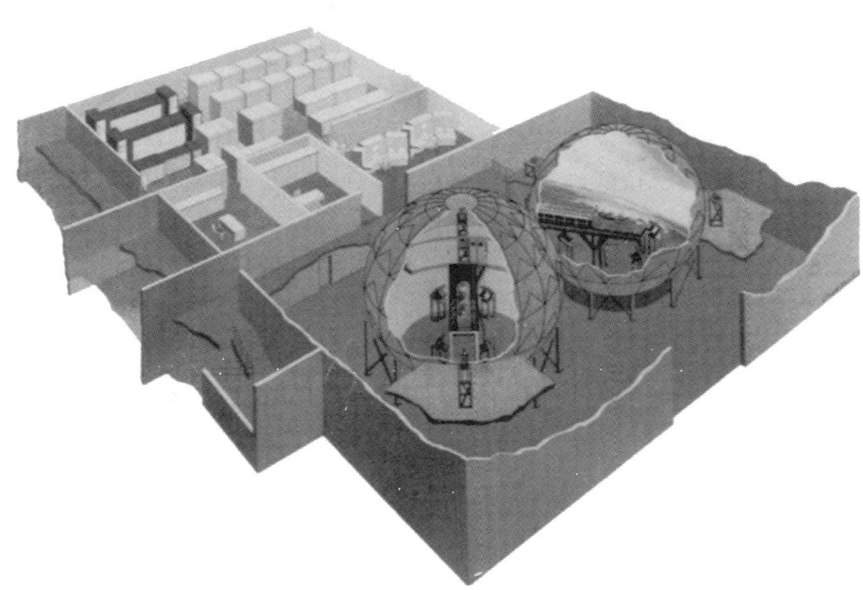

Artist's impression of the SSI twin dome combat mission simulator

independent one-on-one against the instructor operator station or Adaptive Manoeuvring Logic (AML) driven computer-controlled targets; independent one-on-two and integrated one-on-one, one-on-two.
Other Features. The system will be used to perform flying qualities refinement and provide the capability for avionics and weapon system development testing and tactical analysis. In addition the simulator will provide training in all phases of flight, including preflight, take off and landing, malfunctions, as well as aircraft control for handling qualities evaluation.

Status
Delivered.

Manufacturer
Contraves-SSI, Tampa, Florida.
(For details see Directory of Suppliers)

CTA Part-Task Trainers

The Simulations Systems Division (SSD) of CTA Inc, at Englewood, Colorado, has made a number of trainers for the A-6E, F-15, F-16, F-18, KC-135 (Navigator RV Trainer), T-45 (Navigator Trainer), a UAV trainer, and the UH-1 helicopter. It has also made a pilot selection trainer for the USAF, and has demonstrated compliancy with DoD MIL-Q-9858 and MIL-I-45208. On the civil side it makes FTDs and has delivered a Boeing 747 trainer to the FAA which had a reconfigurable cockpit.

F-15A Part-Task Trainers
The CTA F-15A PTTs are deployable devices for pilot training and can be used in a stand-alone training mode or under instructor control. They are designed to train for situations varying from take off and landing to weapons training with multiple threats. A range of cockpit fidelities is available depending on customer requirements. The design is modular and has a small footprint; it is suitable for installation in the field or areas not normally used for training devices. It is constructed from COTS components and uses the CTA Mission Simulator System software which is used on a number of other devices and allows networking with other systems if required.

F-16 Part-Task Trainers
The CTA F-16 PTTs are low-cost deployable devices for pilot training and can be used in a stand-alone training mode or under instructor control. Training includes standard and emergency conditions. Threats can be preset, autonomous or instructor controlled. A range of cockpit fidelities is available depending on customer requirements, and a number of different visual systems can be fitted. The design is modular and has a small footprint; it is suitable for installation in the field or areas not normally used for training devices. It is constructed from COTS components and uses the CTA Mission Simulator System software which is used on a number of other devices and allows networking with other systems if required. The CTA F-16 family includes 'A' and 'C' model configurations; the associated aircraft Operational Flight Programs (OFP) are included.

CTA FTD with monitor-based visual display

F/A-18 Deployable Combat Trainer
The F/A-18 Deployable Combat Trainer (DCT) is for pilot training in weapon systems and multiple threats. It can be used in a stand-alone training mode, for preset exercises, or under instructor control. A range of cockpit fidelities is available depending on customer requirements. The design is modular and has a small footprint; it is suitable for installation in the field or on ships. It breaks down into modules for transportation. It is constructed from COTS components and uses the CTA Mission Simulator System software which is used on a number of other devices and allows networking with other systems if required.

Status
F-15A PTT; units fielded and in production. F-16 PTTs; units fielded and in production. Additional aircraft configurations available. F/A-18 DCT; available.

Manufacturer
CTA Inc, Englewood, Colorado.
(For details see Directory of Suppliers)

CTA F/A-18 Deployable Combat Trainer with monitor-based visual display

CTA F-15A PTT with three-screen projected visual system

Delex Tactical and Procedural Part-Task Trainer (PTT)

Delex has produced a tactical and procedural PTT for the crews of US Navy F/A-18 and A-6E aircraft. It can be used both ashore and at sea and is compact and easily transported. The A-6E Systems Weapons Integration Program (SWIP) PTT provides training in the use of the Harpoon Block IC anti-ship missile, the High-Speed Anti-Radiation Missile (HARM), the Maverick Missile and the Stand-off Land Attack Missile (SLAM).

The PTT is known as Trainer Device Number T0402 and it simulates the cockpit of either the A-6 or the F/A-18 (T0401) using changeable facsimile panels and software. Cockpit displays are represented by two 19 in (482 mm) colour touch-screen monitors. The system is based upon Motorola 68030 and Intel 80486 microprocessors and latest versions are one-third smaller than previous models.

Status
24 are in service with the US Navy.

Manufacturer
Delex Systems Inc, Vienna, Virginia.
(For details see Directory of Suppliers)

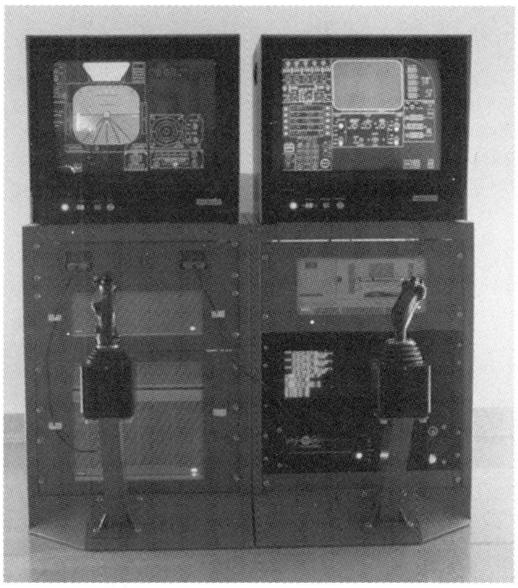

The Delex PTT for F/A-18 or A-6

ECC Combat Mission Trainer

The ECC International Corp produces a Combat Mission Trainer (CMT) which is a low-cost squadron-level trainer for the air-to-air and air-to-ground roles.

Components. Cockpits are fixed base and open-plan without canopies. Seats are not aircraft replicas and are without straps. A typical system consists of up to six cockpits and an IOS which is known as the Air Battle Instructor/Manager Control Station (ABI/MCS). Each cockpit may be fitted with a three channel CGI system, displayed by direct projection on screens in front of the pilot.

Training Scenarios. Aspects which can be included in the simulation include Beyond Visual Range (BVR) radar and missile training, Within Visual Range (WVR) combat, and cockpit aspects such as HUD, HOTAS and LANTIRN. The six cockpits can be networked to enable pilots to fly against each other, or against simulated adversaries.

Status
Available

Manufacturer
ECC International Corp, Wayne, Pennsylvania.
(For details see Directory of Suppliers)

ECC squadron-level combat trainer

ECC Flight Academy CBT

ECC International Corp has developed and installed a CBT system for the air force academy of a Middle East air force. It involves computer-aided instruction (CAI) and interactive video disk (IVD) hardware and software. Students are taught using the ECC CAI/IVD courses while pre-flight training in the PC-7 is given in an ECC cockpit procedures trainer.

Student Terminals. The CAI/IVD system includes terminals with two screens, graphic and test overlays and computer-managed instruction and testing. The students have self-paced courses but large screen courses led by instructors are also provided.

Courseware. Course material is provided in the work stations and provides each student with a uniform programme of instruction which is paced to their abilities. The courseware can be updated and modified to reflect revised training objectives and to ensure that the technical material keeps pace with aircraft modifications.

Status
In service.

Manufacturer
ECC International Corp, Wayne, Pennsylvania.
(For details see Directory of Suppliers)

Flight academy classroom CBT equipment supplied by ECC to a Middle East air force. Note the large screen at the end of the room which is used by the instructor for class tuition

ECC Part-Task Trainers (PTTs)

ECC has produced a wide variety of PTTs. These include an F-16 engine trainer for the US Air Training Corps' technical training centres, a C-5B aerial refuelling receptacle training system, and LANTIRN part-task trainers for both the F-16C and the F-15E.

LANTIRN PTT. LANTIRN (Low Altitude Navigation and Targeting Infra-Red at Night) part-task trainer features a model of the aircraft cockpit. In the F-15E trainer at Luke Air Force Base, Arizona, the rear cockpit of the trainer is used by student weapons system officers (WSOs). In the F-16C configuration a single cockpit is used.

CBT - BATs. The company has also produced a range of 15 CBT devices known as Basic Aviation Trainers (BAT), each dealing with a subject to be taught. Each is a self-contained, bench-top unit designed for training an individual or a small group. Each trainer gives a graphic display with a schematic diagram of the system being considered; the instruments and equipment needed to test and analyse each system are included. A similar approach is adopted with the airborne radar maintenance simulator and the flight control surfaces trainer for the US Navy.

Status

In service.

Manufacturer

ECC International Corp, Wayne, Pennsylvania.
(For details see Directory of Suppliers)

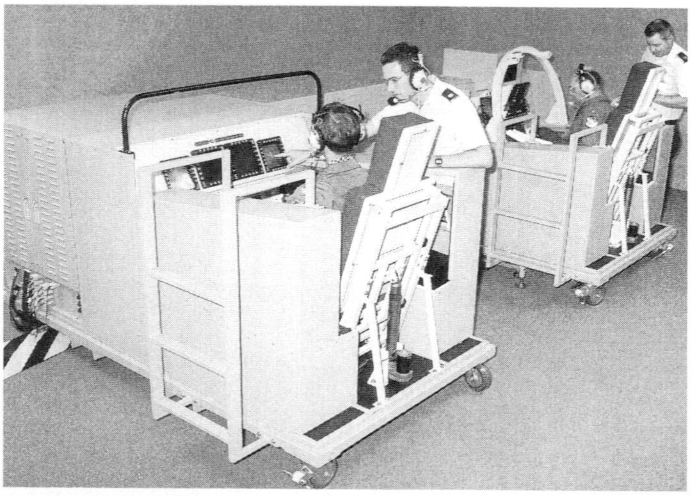

The LANTIRN Part-Task Trainer for F-16C pilots. Note the simpler device on the left and the more detailed cockpit on the right

Detail of the cockpit of the LANTIRN Part-Task Trainer for the F-16C

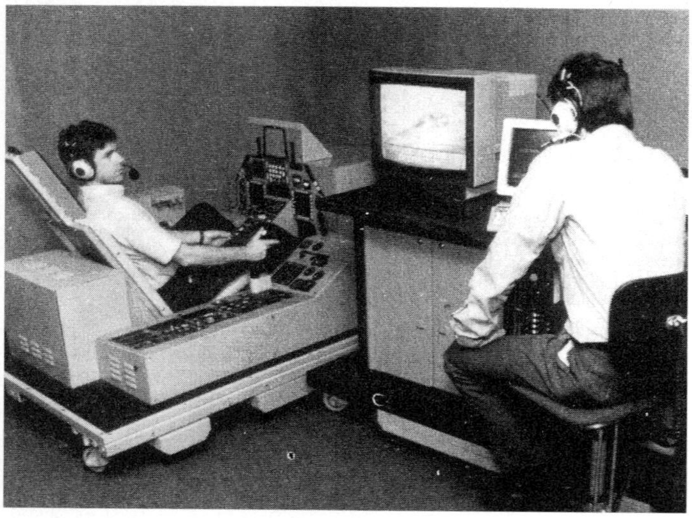

F-16 trainer for engine handling

ECC Pilatus PC-7 Instrument Flight Trainer

ECC International Corp makes a variety of aircraft trainers. Over 200 maintenance trainers have been produced. Examples include the A-4, A-6, AV-8, C-5B, C-130, EC-130, F-5E/F, Sea Harrier, T-34C and T-44 fixed-wing aircraft, together with the AH-1S/T, Sea King, Puma, Gazelle, CH-47, Bell 205, Bell 206, and Bell 214 helicopters. It has also produced flight trainers for the Pilatus PC-7, the T-44, the Bell 205 (UH-1) and Bell 206. ECC also produces various training aids for ground-based equipments.

PC-7 Instrument Trainer. The Pilatus PC-7 instrument flight (IF) trainer uses a facsimile PC-7 cockpit. An EC-3X computer is used for the flight model and to run exercises. An instructor's station has two monitor screens. Pilots learn procedures including pre- and post-flight checks, engine start, taxi, take off, instrument flight, approach and landing.

One Middle East customer complements the system with the EC-10 computer-based instructional system which presents lessons to individual student video display terminals for interactive training. The programmes relate to basic pilot skills, the PC-7 and the British Aerospace Hawk.

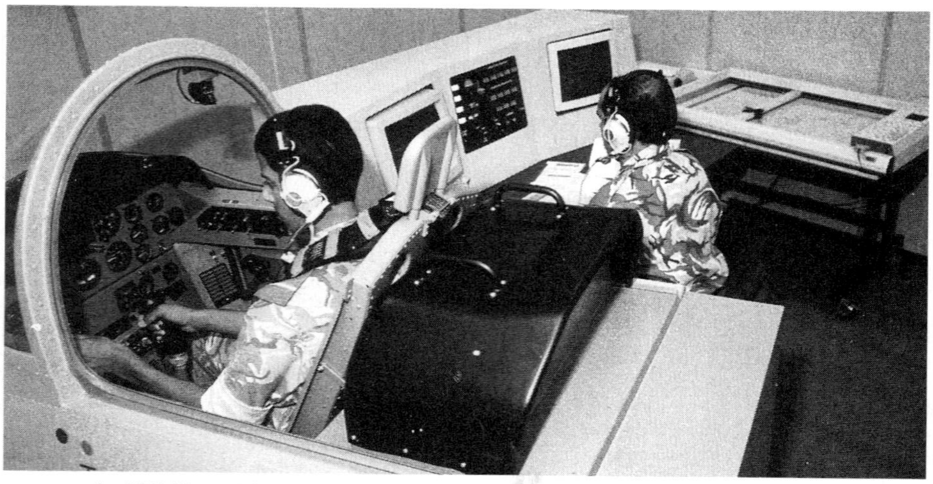

An ECC Pilatus PC-7 instrument flight trainer used by a customer in the Middle East

Status

In service.

Manufacturer

ECC International Corp, Wayne, Pennsylvania.
(For details see Directory of Suppliers)

ECC T-34C Cockpit Procedures Trainer (CPT)

This is a low-cost CPT designed for the US Naval Undergraduate Training Program. It is designed for use in the classroom and essentially consists of a flat panel on which oversize instruments and controls are displayed. It is not, however, a simple photo-montage because audio cues and interactive video disks are used to present the appropriate indications on the panel. Two more interactive backlit panels illustrate engine performance and the aircraft conditioning and electrical systems; these may be operated independently or together with the cockpit controls and instruments.

Status
In service with the US Navy. One system was supplied to the Republic of China Air Force Academy near Kaohsiung, Taiwan.

Manufacturer
ECC International Corp, Wayne, Pennsylvania.
(For details see Directory of Suppliers)

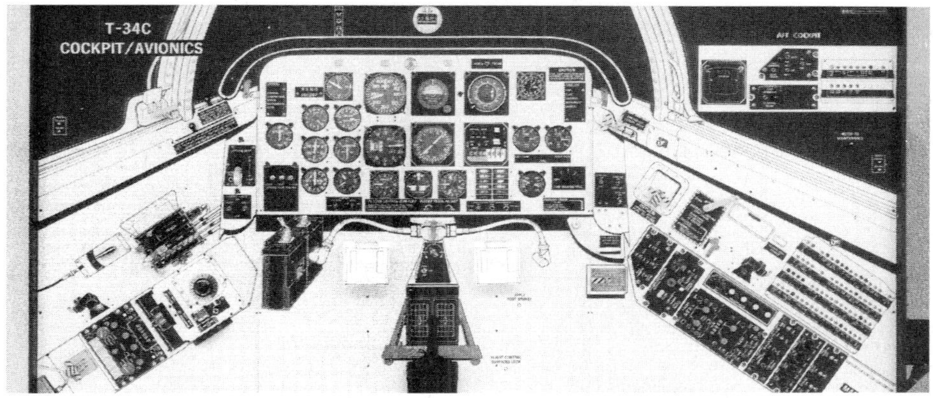

The T-34C cockpit procedures trainer which uses interactive video disc and audio cues, showing what can be achieved with a simple panel-based layout for classroom tuition

EMS TH-57B Cockpit Procedures Trainer

The EMS Development Corporation produces a low-cost Cockpit Procedures Trainer (CPT) for the TH-57B helicopter of the US Navy.
Components. The system consists of a replica of a Bell 206 Jet Ranger cockpit, an instructor station (IOS), and an equipment cabinet.
IOS. The instructor station consists of a desk, a terminal with a keyboard and an intercom panel. The instructor can insert any one of 48 malfunctions which will give corresponding instrument responses. The system can arrange three instructor-selected malfunctions in succession in the engine, electrical system, transmission, fuel system and hydraulic system.

Status
Three systems are in service with the US Navy and are designated device 2C67.

Manufacturer
EMS Development Corporation, Farmingdale, New York.
(For details see Directory of Suppliers)

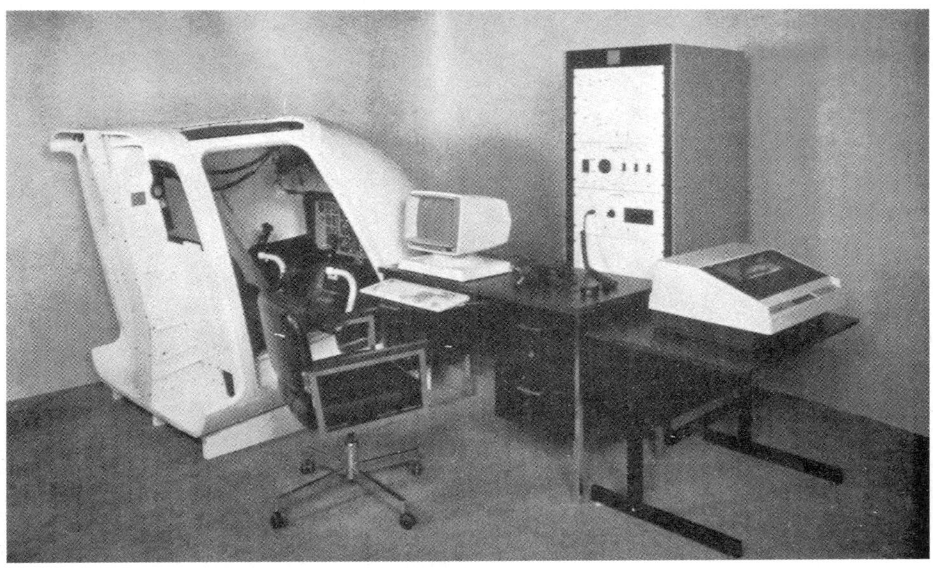

EMS TH-57B helicopter CPT

Frasca Flight Simulators and Flight Training Devices

Frasca International Inc (FII) was founded in 1968 and designs and manufactures a range of low-cost fixed-base FTDs for helicopters and fixed-wing aircraft. Since 1990 they have entered the field of Full Flight Simulators and have produced models with 3-DoF and 6-DoF motion platforms. They also manufacture CPTs and have their own product line of low-cost CGI systems. Frasca has sold over 1200 training devices in 60 countries. Some of their current product lines are described below. The CGI systems are described in the Image Generation section later in the book. Frasca also offer upgrades, refurbishment, relocation and engineering services.

Frasca IOS with plotter - a PC-based version ('GISt') is also available using a mouse, icons and menus to replace the plotter

Certified Devices Group
A Certified Devices Group (CDG) specialises in ensuring that devices and simulators meet the appropriate FAA criteria, levels A-D for Flight Simulators and Levels 2-7 for Flight Training Devices.

Frasca Full Flight Simulators (FFS)
Frasca FSS generally employ the following systems:
Aircraft Model. Specific.
CGI System. Customer choice. Frasca systems available.
Display System. Customer choice. Collimated Monitor or screen projection systems available.
Computer. Motorola 88000 RISC, mounted on board with appropriate I/O systems.
IOS. One or two touch-screens.
Motion Platform. Digital.
Control Loading. Digital.
Sound Generation. Digital.

Frasca FTDs
Frasca FTDs generally employ the following systems:
Aircraft Model. Generic or specific, depending on the FAA Level of the device and customer choice.
Host Computer. Motorola 68000.
IOS. One or two touch-screens.
Control Loading. Electric, hydraulic or spring, depending on the FAA Level of the device and customer choice.
Sound Generation. Digital.

Frasca Model 131 Trainer for Single-engined Aircraft
This is Frasca's simplest model and is mounted in a small cabin which is open at the back for ease of access and for 'over-the-shoulder' observation and instruction. A single-channel visual system has one CGI channel and a directly-viewed TV monitor screen. The cabin is deliberately narrow so that it can be installed in one piece without removing doors.

Frasca Model 132 Trainer for Twin-engined Aircraft
Similar to the Model 131 but configured for two engines.

Frasca Model 135 Trainer for Helicopters
The basic model is similar to the other 100-series trainers but is configured for light helicopters with a collective and cyclic stick, manual throttle on the collective hand grip, and appropriate helicopter instruments.

A version for the Robinson R22 is a Helicopter Instrument-Flying Procedures Trainer (HIFPT) and is mounted in a double-width cabin in order to accommodate the cyclic control column of the R22 which spans both seats from a mounting between the seats.

Frasca Model 140-series Trainers for Single- and Twin-engined Aircraft
The 140-series, like the earlier 100-models, are compact cabin-mounted trainers with an open back to the cabin. There are a number of improvements over the earlier models, one being a flexible navigational database with over 200 programmable ground stations. The general fidelity of the simulation is also improved by features such as response to control movements varying

Frasca Level C simulator for Bell 412 helicopter. In the foreground is the company's founder, Rudy Frasca, and on the right is the company's first trainer

Frasca Model 342 helicopter trainer configured to be similar to a Bell 206L

with airspeed, allowance for stability factors and the low frequency longitudinal oscillation (phugoid), torque and (for twin-engine simuations) asymmetric thrust effects, stall speed variation with configuration and power, weight variation with fuel burn, and so forth.

Frasca Model 146 GAT-1 Upgrade Conversion

The General Aviation Trainer (GAT-1) can be refurbished with new instrument panels and virtually any option from the current 100 series devices including the IOS. The bellows-type motion system of the GAT-1 can also be renovated.

Frasca Model 242 Trainer for Twin-engined Aircraft

The Model 242 trainer includes two pilot seats and two sets of flight controls, and has dynamic control loading. The standard fit has only one set of flight instruments but the co-pilot can be provided with flight instruments if required.

Model 242T. The T model is for light turboprop twins.

Frasca Model 342 Helicopter Trainer

The Model 342 is a generic trainer typical of a light single turbine-powered helicopter. The standard fit is described below, various enhancements and different fits are available at customer choice.

Engine Modelling. Normal operation is provided, and start and shutdown is provided as a option. Engine failure can be achieved through the fuel valve or at the IOS. Fuel quantity depletes or can be set by the instructor.

Flight Instruments. The standard package includes the basic six flight instruments; ASI, T&S, Attitude Indicator, Directional Gyro, Pressure Altimeter, and VSI.

Sound. Turbine sounds and rotor chop are simulated, and air noise increases with airspeed.

Avionics. Standard fit includes full IFR avionics.

Frasca Hawker 800 simulator on 3-DoF motion platform

Frasca Hughes/Schweizer 300 trainer with single channel of IVEX CGI and back-projection screen

Image Generation. One channel of Frasca FVS2000 CGI system.

Display System. Screen projection.

Controls. One set provided, second set optional. A force-trim system is provided.

IOS. Compact monitor, keyboard and X/Y plotter. PC option available with mouse and Windows-based screen with menus.

Frasca Aircraft-Specific Trainers

Frasca have well-proven fixed-base designs for the following aircraft:

Fixed-Wing. Cessna Citation, Pilatus PC-7.

Helicopters. Bell 205, Bell 212, Hughes/Schweizer 300, and the MBB Bo-105.

Frasca Reconfigurable Simulator for Engineering Education (R-SEE)

A reconfigurable simulator for engineering education allows students to test their own aircraft designs and flight laws. The simulator has a cockpit with a control stick, rudder pedals, throttle, and controls for gear, flaps and trim. A Frasca FVS-100HR image generator provides an outside world view, displayed through a monitor screen. Customers include the Department of Aeronautical Engineering at the University of Illinois.

Frasca Graphical Instructor's Station (GISt)

The Frasca GISt is a PC-based IOS working through a mouse and a Windows-based screen presentation with

Frasca 135 trainer showing compact size for ease of installation

icons and menus. It is available on all 100, 130, 140 and 240 series trainers, and as an upgrade if required. It replaces the plotter, menu terminal and control panel in earlier IOS designs.

Frasca Pilot Automated Selection System (PASS)

The Frasca Pilot Automated Selection System (PASS) is designed to assess a pilot applicant's potential. The applicant flies a model 141 single-engine trainer through a prescribed pattern and so the co-ordination skills of a range of applicants can be compared. A more elaborate IOS is provided which has four screens and as well as the exercise monitoring function, is used to create training patterns for use in assessment and to output applicant's performance for evaluation and comparison purposes.

Bendix/King Avionic Option

In addition to the standard Frasca avionics, in a number of trainers an option for Bendix/King avionics can also be offered.

Frasca 140-series trainer showing IOS on the right

Visual Systems. Frasca can fit a number of visual systems as customer options. In the past these have included the IVEX VS 1000 image generator, displayed through a collimated monitor, and Frasca's own range of low cost CGI systems. Frasca CGI systems are priced at between $10 000 and $35 000 per channel and include the FVS100, FVS100HR and FVS200HR models; for details see Frasca entry in the Image Generation section.

Status

See text. Frasca-built equipment in military use include a helicopter instrument procedures trainer (HIFPT) for the Gazelle AH Mark 1 for the British Army Air Corps at ACC Middle Wallop supplied in 1990. Two HIFPT for the Bell 205/UH-1 are operated by the Spanish Air Force while a Bell 205 trainer is operated by the Royal New Zealand Air Force. Other equipment is operated by the Columbian Air Force, Mexican Navy, and the US Navy operates 9 Frasca CPTs.

Manufacturer

Frasca International Inc, Urbana, Illinois.
(For details see Directory of Suppliers)

FlightSafety International (FSI) Simulators

FlightSafety International (FSI) is a company entirely devoted to training. It owns 38 training centres and academies for pilots and maintainers, located in the USA, Canada and France. It also is involved in maritime training through its subsidiary MarineSafety, and in the power industry through its associate (with Babcock and Wilcox) PowerSafety. It owns some 135 FAA-certified flight simulators, claimed to be the world's largest fleet of flight simulators owned by one organisation. Most of these flight simulators are made by FSI itself.

The company headquarters is at La Guardia Airport, New York, and there are three company divisions: the Instructional Systems Division at Hurst, Texas; Simulation Systems Division (SSD) at Broken Arrow near Tulsa, Oklahoma; and Visual Simulation Systems Division at St Charles, Missouri. The latter produces the VITAL product line of CGI systems, purchased from McDonnell Douglas Electronic Systems in 1993 (details in the Visual Systems section of this book). FSI SSD manufactures two types of 6-DoF synergistic motion platforms with different throws, known as the 'big base' and 'small base'; the small base has a maximum payload of 16 000 lb and the big base, 26 000 lb; details are in the Motion Platform section of this book. Recent flight simulators made by SSD have been approved by the FAA to level D (FS-D) standards. FSI SSD also designs and manufactures Part-Task-Trainers (PTTs) and Flight Training Devices (FTDs) for both the civil and military markets.

USAF C-17 Training

FSI provides flight simulators for the C-17 Globemaster, through MDTS as the prime contractor. This involves 12 C-17 flight simulators to FAA level FS-C standard (ex-Phase 2), and a cockpit systems trainer. The 12 FS-C devices have the following characteristics:

Visual Image Generation (IG). Five channels of Martin Marietta (ex-GE) PT2000 CGI. Details in Image Generation (IG) Systems section.

Visual Display System (DS). FlightSafety (ex-McDonnell Douglas) Multiview collimated cross-cockpit display system with a 225° horizontal FoV. Details in Visual Display Systems section.

Motion Platform. FSI 6-DoF.

Certification. To FAA Level FS-C standard.

Special Features. Six of the simulators are coupled to a separate device which simulates the lower-deck loadmaster station. The loadmaster device is equipped with a visual system portraying action in the cargo compartment during air-drop operations, and can be used independently of the pilot simulator for stand-alone loadmaster training.

USAF C-5 Galaxy Training

FSI provides all pilot training for the US Air Force C-5 Galaxy programme. The company designed six C-5 aircrew training systems at Altus AFB, Oklahoma; Dover

FSI Simulation Systems Division plant at Broken Arrow (Tulsa) with three simulators in the final stages of build

FSI SSD simulator for the USAF C-17 which also can include a lower-deck loadmaster trainer

FSI flight simulator for USAF C-5B Galaxy with 6-window visual system

FSI CBT for the USAF C-5B Galaxy. Note the complex systems panels in the background

AFB, Delaware; Kelly AFB, Texas; Stewart International Airport, New York; Travis AFB, California; and Westover AFB, Massachusetts. The C-5 Flight Simulators have the following characteristics:
Visual Image Generation (IG). Six channels of the Evans and Sutherland SP1 Dusk/Night CGI system. Details in Image Generation (IG) Systems section.
Visual Display System (DS). Six collimated monitors.
Motion Platform. FSI 6-DoF.

USAF C12J Training

In January 1990 the US Air Force awarded the company a contract to train C-12J (Beech 1900) pilots at La Guardia, New York.

USAF KC-135 Training

In 1992, FSI was awarded a contract to manage KC-135 aircrew training for USAF Materiel Command. This involves conversion and refresher training for some 2700 pilots, navigators and boom operators (boomers).

USN Training

FSI trains crews for the C-9B, the naval version of the DC-9. This contract includes the use of the two DC-9-30 simulators at FSI's centre in St Louis, Missouri. Another contract with the USN is for training Sabreliner and Gulfstream pilots and Sabreliner ground crews. FSI also trains pilots for the USN's Cessna Caravan (U-27A) and Cessna 441; this training is conducted at the FSI's Wichita and Vero Beach facilities.

US Army Training

In January 1990, FSI received a contract from the US Army to train helicopter pilots to fly the U-21 (King Air 90) fixed-wing aircraft. This is carried out at FSI's Daleville and Dothan Learning Centres in Alabama.
Helicopter Simulation for the Military. The FSI Bell 412SP/212 simulator at Fort Worth, Texas, has been used to train pilots of the Royal Norwegian Air Force since 1991, while the FSI Sikorsky S-76 simulator at West Palm Beach, Florida, has been used to train pilots from Brunei, Hong Kong, Jordan and the Philippines.

Status

See text.

Manufacturer

FlightSafety International Simulation Systems Division (FSI SSD), Broken Arrow (Tulsa), Oklahoma.
(For details see Directory of Suppliers)

Hughes Combat Simulator for F/A-18

Under contract to the US Navy, Hughes has produced five combat simulators (designated Device 2E7) to train US Navy and Marine Corps pilots for the McDonnell Douglas F/A-18 Hornet. Each device has two domes each having a replica cockpit.
Visual Image Generation (IG). CGI. Aircraft carriers are modelled so that flight-deck operations can be carried out by day and night. Air-to-air targets are produced with details such as afterburner and control surface positions. During air-to-air engagements, missile smoke trails, explosions and gunfire are portrayed.
Visual Display System (DS). Two 40 ft diameter domes are used. An earth-sky background is projected on the inner surface of the dome by fixed light-valve projectors, while target imagery is projected by four high resolution rotatable projectors mounted around the outside of the cockpit.
Motion Cueing. The cockpits are fixed base but the pilot uses his own anti-G suit in the simulator. In addition, a G-seat is fitted with inflatable bladders in the seat and variations in shoulder harness tension which are set to respond to computed G. A seat vibration system simulates aircraft buffet, stalling, gunfire and speedbrake extension.
Cockpits. Each cockpit is an exact replica of a McDonnell Douglas F/A-18 cockpit. Canopy, windscreen and leading edge extensions are reproduced to give the same exterior field-of-view as that of the actual F/A-18. A head-up display (HUD) is fitted, together with the full complement of aircraft controls, displays and instruments.

In addition to giving pilots experience in combat tactics and aircraft management, the system also familiarises them with the variety of modes available with the Hughes Aircraft AN/APG-65 radar on the F/A-18, and the display of its functions and commands on the head-up display.
Control Loading. Digital.
Instructor Operating Station (IOS). Mounted off board. The central control room has an instructor's station with displays which mirror the cockpit indications. Additionally, he can introduce a range of malfunctions of increasing severity which are realistically reproduced in the cockpit, with caution lights, audio alerts and voice alert. There is a freeze and playback facility so that the trainee can see the effects of his actions. The instructor's position has a stick and throttle to enable him to manoeuvre the target aircraft projected onto the dome. His display at the console can be selected to give an ACMR presentation. Multiple targets can be presented on his screen and can be viewed from any angle with the chosen viewpoint selected by rotating a rolling ball inset into the console. The target can be an instructor-flown adversary target or the instructor can select one of five levels of difficulty with the computer flying the target.
Operating Modes. The simulator can be operated in an independent or an integrated mode. In the first, one or two pilots, each in cockpits in his own dome, can fly against 'targets' that are instructor-flown or flown by an intelligent adversary computer program called Adaptive Manoeuvring Logic (AML). The AML can fly the targets in optimal offensive or defensive manoeuvres. In the integrated mode, the two pilots can engage in combat against one another or against IFT or AML targets.
Database Gaming Area. A large gaming area, over 1110 × 1110 km, is provided to create the scene from the cockpit and air-to-ground sensor imagery. The gaming area includes a specifically modelled area depicting Western Arizona and Southern California as well as the Fallon, Nevada, bombing range. The specific area includes detailed models of several military airfields and target areas. In addition detailed target images can be inserted onto the gaming area by the instructor. These target images can be used for air-to-ground weapon delivery training and scoring.
Air Targets. The computer program driving the target aircraft has total knowledge of the combat situation and selects the optimum manoeuvre relative to the trainee's actions. Level 1 is a non-manoeuvring target and the levels increase in difficulty until Level 5, the most difficult, which is virtually unbeatable. Targets programmed into the simulator are the Northrop F-5E, General Dynamics F-16, McDonnnell Douglas F-15 and F/A-18, and two foreign aircraft, reportedly the MiG-21 and MiG-23. Each has the appropriate flight dynamics for that particular adversary aircraft.

Hughes Aircraft combat simulator for US Navy/McDonnell Douglas F/A-18 with manoeuvring target

Status

In production. Simulators have been delivered to the US Naval Air Stations at Lemoore and Cecil Field and to the US Marine Corps at El Toro. The fourth simulator is destined for NAS Lemoore, and the fifth for Cecil Field.

Manufacturer

Hughes Aircraft Company, Training and Support Group, Manhattan Beach, California.
(For details see Directory of Suppliers)

Hughes F-14A Mission Trainer

The Hughes Training Inc (HTI) F-14A Tomcat mission trainer (Device 15C9A) was designed to train US Navy aircrews to carry out air-to-air interceptions. It uses actual aircraft mission tapes and an onboard computer to create a realistic combat scenario.

The system simulates the AN/AWG-9 weapon control system including the radar and the long-range air-to-air Phoenix missile. The trainer also provides full crew training for the radar intercept officer (RIO) in communications, navigation and ECM.

Status

In production and in service with the US Navy at NAS Miramar and NAS Oceana.

Manufacturer

Hughes Aircraft Company, Training and Support Group, Manhattan Beach, California.
(For details see Directory of Suppliers)

HTI F-14A mission trainer

Hughes F/A-18 OFT

The Hughes Training Inc (HTI) OFT for the McDonnell Douglas F/A-18 provides not only general flying training on the aircraft but also various aspects of carrier operations including approach, arrested landing, wave-off, bolter, touch-and-go, barrier arrestment, deck taxi, and catapult tensioning and launch.

Visual Image Generation (IG). Three channels of Vital IV CGI generate scenes for both shore- and carrier-based operations.

Visual Display System (DS). Three Collimated Monitors.

Motion Cueing. No motion platform is fitted, but pilots use their own anti-G suits in the simulator which allows realistic cueing for simulated G as the simulator inflates the suits appropriately. Also, a simulator-specific G-seat is fitted which gives seat pressure with simulated G, seat pan lowering under high G (to model eye point lowering due to high G forces), and strap tightening and loosening in response to simulated G (modelling body pressures against the straps as G varies in the aircraft).

The HTI F/A-18 operational flight trainer

Computer. A Gould/SEL 32/8750 processor with 256 Mbytes of memory is supplemented by two AN/AYK-14 computers identical to the aircraft's mission computers.

Instructor Operating Station (IOS). The IOS can be operated individually or jointly by an instructor and a WST operator, and can also accommodate an observer. Faults can be inserted or programmed from the display keyboard. Two colour CRTs are included; one relays information being presented on the pilot's instrument panel and the other shows the pilot's view through his canopy. A computer-aided performance measurement system compares the trainee's performance with preset criteria; a hard copy printout is provided at the end of the lesson.

Australian Deliveries. The Australian Government contracted with Hughes to produce two F/A-18 operational flight simulators at a cost of $22.5 million, along with supporting equipment and services, to support the fleet of 75 aircraft ordered by the RAAF. The first simulator was delivered in 1985 and the order was expanded to seven further WSTs. A factor in the choice was a significant work-sharing programme with Australian industry. Major partners were C-3 Limited, a Canberra-based software development and maintenance company, and THORN EMI (Australia).

Head-Tracking Enhancement. In February 1992 the Gaertner Research Division of GEC Ferranti obtained a contract from HTI to supply a new optical head-tracking system for two dual pilot and one single pilot system. The tracker optically monitors a pilot's head position and this allows the mission computer to know exactly where the pilot is looking in order to slave steerable missiles or sensors.

Status

Some 22 in service with the US Navy, eight with the RAAF.

Manufacturer

Hughes Training Inc, Herndon, Virginia.
(For details see Directory of Suppliers)

Hughes C-141 Aircrew Training System (ATS)

Hughes Training Inc (HTI) has the current contract for the USAF C-141 Starlifter Aircrew Training System (ATS). The contract was awarded in 1989 and over its 15-year period, HTI will provide C-141 aircrew training and contractor logistics support at several Military Airlift Command, US Air Force Reserve and Air National Guard bases in the United States. The seven C-141 weapon system trainers (WSTs) will be refurbished and new courseware will be developed. In addition, Hughes will provide systems definition, aircrew training system operation and management, system integration and a training management system.

Site. Altus AFB will be the site for the refurbished C-141 weapon systems trainers and the computer-based system. C-141 crew members will begin their course on the self-paced, computer-based system. Depending upon their aircrew speciality (pilot or loadmaster) they will then be given further training on the cockpit procedures, the loadmaster or the weapon system trainers.

Altus AFB will also be the site for the C-141 Training System Support Center (TSSC). This will provide support to modify or update the C-141 ATS in order to preserve concurrency with the aircraft fit.

Refurbished C-141 Flight Simulators

Visual Image Generation (IG). Five channels of

Evans and Sutherland SP-X500. Details in Image Generation (IG) Systems section.
Visual Display System (DS). Thomson Training (ex-Rediffusion) WIDE II display systems with 200° horizontal and 40° vertical FoV. Details in Visual Display Systems section.
Motion Platform. The original 6-DoF platform will be retained.

Status
In service and being delivered.

Manufacturer
Hughes Training Inc, Arlington, Texas.
(For details see Directory of Suppliers)

HTI C-141 WST

Hughes CH-53 OFT

Hughes Training Inc (HTI) has produced training equipment for the Sikorsky CH-53, HH-3F, HH-53C and HH-52A, Bell UH-1E and TH-1L, and Boeing Vertol CH-47 helicopters. HTI mathematical models include buffet and rotor stall effects, auto-rotation, power settling, ground effect, ground resonance, and variable turbulence. The simulators for the US Marine Corps CH-53D and CH-53E assault helicopters (Devices 2F120/2F121) have the following characteristics:
Visual Image Generation (IG). Six channels of Evans and Sutherland's CT5A CGI system. Details in Image Generation (IG) Systems section.
Visual Display System (DS). Six collimated monitors with 190° horizontal and 60° vertical FoV.
Motion Platform. 6-DoF.
Main Computers. Two Gould/SEL 32/77 computers.
Instructor Operating Station (IOS). The IOS has two cathode-ray tubes and a digital computation system.

Status
Three CH-53 simulators are operated at MCAS (Marine Corps Air Station) New River and Tustin.

Manufacturer
Hughes Training Inc, Herndon, Virginia.
(For details see Directory of Suppliers)

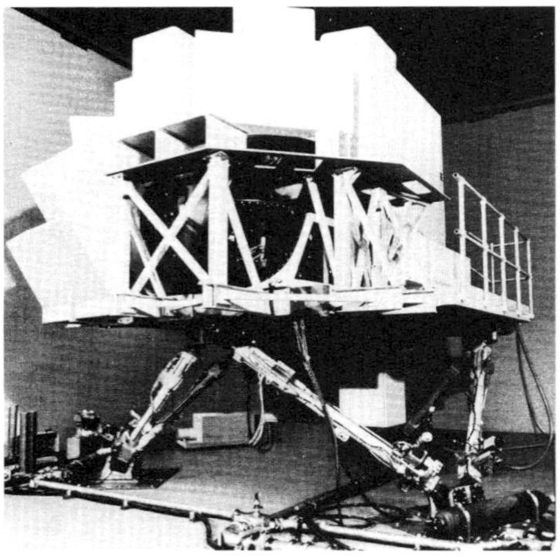

Full flight simulator for US Marine Corps/Sikorsky CH-53 helicopter

Hughes EA-6B WST

The Hughes Training Inc (HTI) Grumman EA-6B Prowler WST (Device 2F156) is unusual in that it has a cockpit section for four aircrew members mounted on an industry-standard size of motion platform. This is believed to be the largest number of aircrew trained simultaneously on a large-throw 6-DoF motion platform at any one time. The WST has the following characteristics:

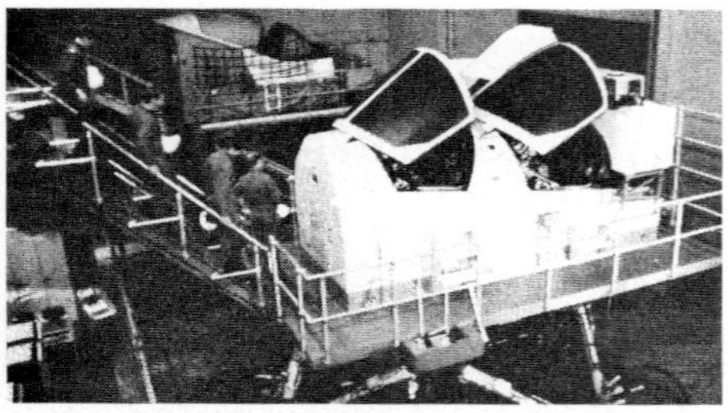

HTI weapons system trainer for US Navy/Grumman EA-6B

Visual Image Generation (IG). CGI generates outside-world scenes for both shore- and carrier-based operations.
Visual Display System (DS). Collimated Monitor, displayed to the two pilots only.
Motion Platform. 6-DoF.
Main Computer. Gould Concept 32/27.
Instructor Operating Station (IOS). Mounted off board. Up to three instructors can control a mission, integrating the operation of the aircraft with the tactical environment. The IOS has five CRT graphics displays with interactive terminals, and repeat displays of the crew's out-of-window scenes, radar and other displays.
Cockpit. The four-seat cockpit is fully representative of the EA-6B.
Databases. Simulation includes DRLMS for the search-radar returns, with a storage capacity of 1.6 million square miles. Terrain and other features that can be stored for display include shadows, refraction and earth curvature, far-shore brightening, range attenuation, moving targets, and emitter occulting.
EW. Electronic countermeasures simulation covers tactical and defensive jamming, communications jamming, and the effects of chaff.

Status
In service with the US Navy. They are installed at NAS Oceana (2), NAS Whidbey and the Marine Corps Air Station El Toro.

Manufacturer
Hughes Training Inc, Herndon, Virginia.
(For details see Directory of Suppliers)

Hughes MH-53E OFT

Hughes Training Inc (HTI) received a contract for two OFTs (Device 2F141) for the US Navy's MH-53E Sea Dragon mine-sweeping helicopters. They are capable of visual and instrument flight training in both day and night environments as well as mission-orientated pilot training in Airborne Mine CounterMeasures (AMCM) with various towed mine-sweeping devices. Additional capabilities include air-to-air refuelling, hover in-flight refuelling, search and rescue, and external cargo transport operations in land and seaborne environments. Characteristics feature the following:
Visual Image Generation (IG). Eight channels of FlightSafety (ex-McDonnell Douglas) Vital VII CGI. Details in Image Generation (IG) Systems section.
Visual Display System (DS). Four IG channels supply a FlightSafety (ex-McDonnell Douglas) Multiview collimated cross-cockpit display system which has a 180° horizontal and 40° vertical FoV. Details of Multiview are in the Visual Display Systems section. Two IG channels are used for collimated monitors which cover the FoV of the lower cockpit chin windows (particularly used when hovering) and the final two IG channels are used to simulate rear-facing MCM mirrors unique to mine-sweeping helicopters.
Motion Cueing. A 6-DoF motion platform is fitted, and extra vibration cues are provided by a seat vibration system.
Main Computer. Gould/SEL 32/9780 multi-tasking computer system.
Instructor Operating Station (IOS). Mounted on board, the IOS has two high resolution colour graphic CRTs and a hand-held keypad. The IOS permits over-the-shoulder supervision and instruction.
Other Features. Digital Stereo Sound system.

Status
Installed at NAS Norfolk.

Manufacturer
Hughes Training Inc, Herndon, Virginia.
(For details see Directory of Suppliers)

Hughes OFTs and IFTs for the T-45 Goshawk

Hughes Training Inc (HTI) is supplying 22 operational flight trainers (OFTs, Device 2F138) and 10 instrument flight trainers (IFTs, Device 2F137) for the USN T-45 Goshawk. The OFT design is generally similar to that of the Rediffusion (ex-Hughes Rediffusion, now TTS and described under Thomson Training UK) Hawk simulators built for the Swiss Air Force and Malaysia. The T-45 OFT has the following characteristics:
Visual Image Generation (IG). Three channels of Evans and Sutherland SP-X500 CGI. Special modelling includes carrier landings and formation flying, and all manner of maritime weather conditions. Details of SP-X in Image Generation (IG) Systems section.
Visual Display System (DS). Projection on inner surface of a partial dome, FoV 160° horizontal by 60° vertical.
Motion Cueing. No motion platform is fitted, but pilots use their own anti-G suits in the simulator which allows realistic cueing for simulated G as the simulator inflates the suits appropriately. Also, a simulator-specific G-seat is fitted which gives seat pressure with simulated G, seat pan lowering under high G (to model eye-point lowering due to high G forces), and strap tightening and loosening in response to simulated G (modelling body pressures against the straps as G varies in the aircraft).
Main Computer. Gould/SEL 32/8750 computer with 256 Mbytes of memory.
Instructor Operating Station (IOS). The IOS is immediately behind the trainee pilot, allowing 'over-the-shoulder' instruction. Cockpit instrument information is repeated for the instructor, who can also control such facilities as intercom and the G-seat. The trainee's performance can be recorded for future analysis.

Status
In production.

HTI OFT for USN T-45A Goshawk - the E&S SP-X500 image generator showing a carrier approach

Manufacturer
Hughes Training Inc, Herndon, Virginia.
(For details see Directory of Suppliers)

Hughes Osprey MV-22A OFT and AST

Hughes Training Inc (HTI) produces the OFT for the MV-22A Osprey tilt-rotor, twin engine VSTOL aircraft. An Aircraft Systems Trainer (AST) is also produced. If the Osprey programme goes ahead, there are planned to be eight OFTs and seven ASTs. The OFT has the following characteristics:
Visual Image Generation (IG). Seven channels of Martin Marietta (ex-GE) Compu-Scene PT2000 CGI. Details in Image Generation (IG) Systems section.
Visual Display System (DS). Projection on the inside of a 24 ft diameter dome mounted on the motion platform.
Motion Cueing. A 6-DoF motion platform, extra vibration cues will be provided by a separate seat vibration system.
Main Computer. Seven Multibus II/80386-based computers and a UNIX operating system. It will contain 70 Mbytes of memory and there will be two 1553B databus interfaces.

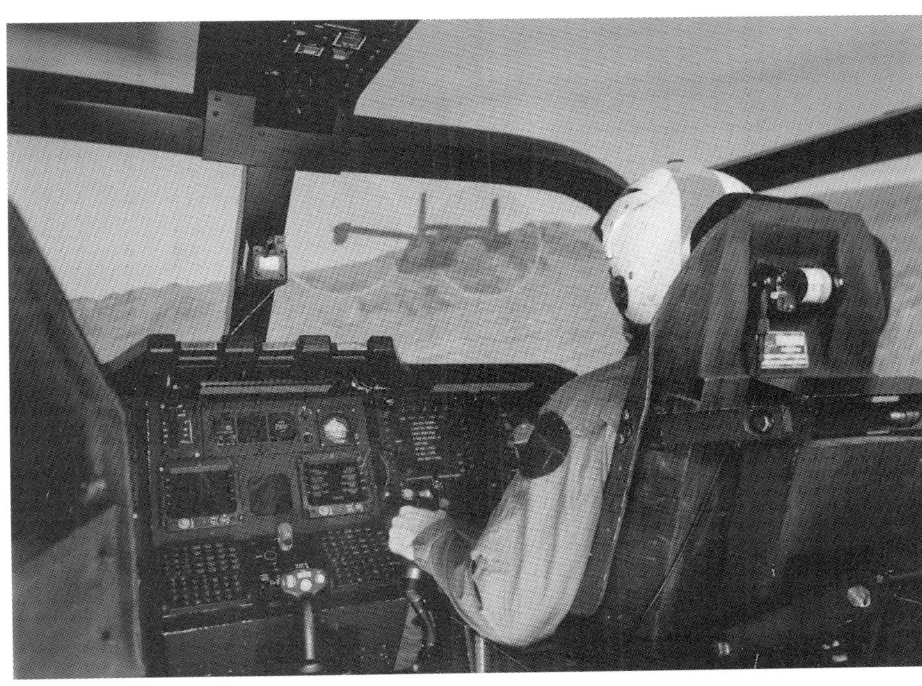

HTI OFT for the MV-22A Osprey, with a 6-DoF platform and projection of seven channels of Compu-Scene PT2000 CGI on a dome mounted on the platform

Instructor Operating Station (IOS). Mounted on board, with two flat-panel graphic displays and bezel-mounted function switches.

Status

Undergoing integration and test.

Manufacturer

Hughes Training Inc, Herndon, Virginia.
(For details see Directory of Suppliers)

Hughes Simulators for the US Coast Guard

Hughes Training Inc (HTI) has supplied the US Coast Guard with flight simulators for the Dassault HU-25A Falcon tri-jet and the Aérospatiale HH-65A Dauphin 2 helicopters. The simulators are installed at the US Coast Guard Aviation Training Center at Mobile, Alabama.
Visual Image Generation (IG). Evans and Sutherland SP-1 dusk/night CGI system. Details in Image Generation (IG) Systems section.
Visual Display System (DS). Details in Visual Display Systems section.
Motion Platform. 6-DoF.
Main Computer. Gould 32/77.
Instructor Operating Station (IOS). Mounted on board. The HU-25A simulator IOS allows the instructor to sit either at a console, or between the two pilots, controlling the exercise via a hand-held keyboard.

Status

See text.

Manufacturer

Hughes Training Inc, Herndon, Virginia.
(For details see Directory of Suppliers)

HTI simulators for Dassault HU-25A Falcon (foreground) and Aérospatiale HH-65A Dauphin

Lockheed (ex-General Dynamics) F-16 Simulators

Lockheed Fort Worth Company (ex-General Dynamics) produces the F-16 fighter and also has produced a number of simulators to support the aircraft.
F-16 Weapon System Trainer (WST). The F-16 WST has an open plan cockpit and is housed in a dome in which is mounted a visual system. Realistic terrain, weather, ECM and threat models are generated. An instructor station to one side of the cockpit controls the scenario and allows over-the-shoulder instruction. The WST makes extensive use of F-16 avionics and software to ensure concurrency with the aircraft.
The F-16 Part-Task Trainer (PTT). The F-16 PTT is for squadron-level training and allows pilots to practise the use of aircraft avionics and weapons. It consists of a cockpit with facsimile controls, a display unit for the scenario and for debriefing, and an instructor station.

Status

In service.

Manufacturer

Lockheed Fort Worth Company (ex-General Dynamics Tactical Military Aircraft Division), Fort Worth, Texas.
(For details see Directory of Suppliers)

Loral F-15 OFT

Loral Defense Systems-Akron designed and produced the F-15 Operational Flight Trainer (OFT) in the 1970s. The OFT provides training from TOL to environments which may involve up to 79 different emitter targets of which 15 may be active at any one time.
Motion Cueing. The first seven F-15 OFTs were built with a 6-DoF motion system. However, the early 6-DoF platforms of the mid and late 1970s did not live up to their promise and had high latencies and few monitoring and diagnostic systems, sometimes in service becoming out of synchronisation with other simulator systems without this being detected. Consequently, in the seventh and subsequent simulators, motion cueing was confined to the use of the pilot's anti-G suit and a simulator-specific G-seat.
Computing. The OFTs use Encore 32/8780 computer systems and Fortran software language.
Enhancements. The OFTs were updated by 1982 to include the F-15 Multi-Stage Improvement Program (MSIP).
Version for Japan. A version was produced for the Japanese Air Self-Defense Forces by Loral and the Mitsubishi Precision Company. These are at Chitose, Hyakuri, Nyutabaru, Komatsu and Tsuiki, and have been upgraded to match aircraft upgrades.

Status

Some 14 F-15 OFTs in service with the USAF, three with the Royal Saudi Air Force, five with the Japanese Air Self-Defense Forces.

Manufacturer

Loral Defense Systems-Akron, Akron, Ohio.
(For details see Directory of Suppliers)

Loral F-15E WST

Loral Defense Systems-Akron produced the F-15E Weapon System Trainers (WST) for the USAF.
Sensor Simulation. The sensor simulation includes both real-beam mapping and synthetic aperture radar as well as the infra-red and electro-optical sensors of the F-15E including LANTIRN simulation.
Visual Image Generation. The WST uses an Evans & Sutherland CT6 image generation system. Databases are derived from Defense Mapping Agency (DMA) data to provide fully correlated radar, sensor, and weapons simulation.
Computing. Encore 32/8780 computers using Fortran are used, with C language for the graphical display system.

Status

WSTs have been delivered to Luke AFB, Arizona, and Seymour Johnson AFB, North Carolina.

Manufacturer

Loral Defense Systems-Akron, Akron, Ohio.
(For details see Directory of Suppliers)

Loral JAS 39 FMS

Loral has won a contract to supply four Full Mission Simulators (FMS) to the Swedish Air Force for the JAS 39 Gripen. Two will have two channels of E&S ESIG-3000 CGI displayed by direct projection on the inside of a 24 ft diameter dome through the E&S head-tracked Area-Of-Interest (AOI) system; the two channels are the high resolution AOI and the low resolution background (details in the Visual Systems section of this book). Radar land mass simulation will be based upon the Sun SPARC 1000. The system will also include two classrooms for specialist training. The other two simulators are designed for squadron-level training and will have the ESIG 2000 CGI system.

Status

In production.

Manufacturer

Loral Defense Systems-Akron, Akron, Ohio.
(For details see Directory of Suppliers)

Loral Special Operations Forces Aircrew Training System (SOF-ATS)

In July 1990 Loral was awarded prime contractorship for the design, development, operation and support of a Special Operations Forces Aircrew Training System (SOF-ATS) for the USAF. The SOF-ATS is a total training system (TTS) designed to provide aircrew with realistic Mission Rehearsal Training (MRT) for special operations missions anywhere in the world and at short notice. A key objective is to provide multi-aircraft mission rehearsal capabilities within 48 hours.
C-130 Hercules Variants. Training systems are being produced for five Hercules variants: MC-130E Combat Talon I; MC-130H Combat Talon II; AC-130U and AC-130H gunships; and the HC-130N/P tanker.
Helicopters. Helicopters in the programme include the Sikorsky heavy lift MH-53J Pave Low (Stallion) and the medium lift Sikorsky MH-60G Pave Hawk (the USAF version of the Army Black Hawk).

High resolution real-world ESIG-4000 CGI developed by Evans and Sutherland for mission rehearsal as part of the Special Operations Force Aircrew Training System

Training Aids and Devices. The system will employ Computer-Based Training (CBT), stand-alone Crew Station Trainers (CSTs), Operational Flight Trainers (OFTs) and Weapon System Trainers (WSTs). The programme requires contractor-operated training systems for all seven SOF aircraft and this will eventually involve about 48 crew station training positions. Loral subcontracts the production of some devices to other companies with Flight Simulation expertise, such as Reflectone, also to suppliers of CGI systems, such as Evans and Sutherland and Martin Marietta, the latter now owning the Compu-Scene product line of IGs previously owned by GE.

Visual Image Generation. The exact IG used depends on the particular training aid, and CGI systems used include the Evans & Sutherland ESIG-4000 and Martin Marietta (ex-GE) Compu-Scene V, PT2000 and SE 2000. See the Image Generation Systems section in this book for more details of these CGI systems.

Database Generation System (DBGS). System architecture and work stations are specifically designed for rapid high-fidelity modelling of the intended special operations scenario, based on all available data. The basic data format is DMLS (DTED & DFAD) and data protocols of the DoD Standard Database Facility (SDBF, formerly known as Project 2851) are used. Data is derived from mapping, satellite, radar, photographic and intelligence sources. The DBGS uses an integrated network of Sun and Silicon Graphics computers and work stations; operating personnel include both contractor database modellers and military intelligence specialists. The final product is a correlated database for the visual, IR and radar parts of the spectrum. The DBGS is sited at Kirtland AFB and is collocated with the DoD STBF (Project 2851) programme, creating a centre of excellence for database modelling.

Software. Software for the mission rehearsal display is being produced by Celeris Inc.

Details of Simulators. Some brief details of some of the simulators involved are as follows:

The MH-60G WST is on a 6-DoF motion platform and its visual system is based on 8 channels of the Compu-Scene V IG producing both visual and IR imagery. DRLMS simulation is also included.

The MH-60G OFT is a fixed-base device but has seat shakers to simulate some motion cues. Its visual system is based on 6 Compu-Scene PT2000 channels, displayed through a wide-angle collimated cross-cockpit display system.

The TH-53 OFT was produced by an extensive refurbishment of a previous simulator. The entire motion platform was replaced with a new 6-DoF platform, and the visual system is based on the Compu-Scene V image generator.

A unique Aerial Gunner and Scanner Simulator (AGSS) is being produced for training gunners and para-rescue crew in NVG scanning and gunnery. This simulator is on a simple motion platform and when operating in a stand-alone training mode, will employ 2 channels of Compu-Scene SE2000 CGI. When networked to one of the other devices it will utilise the Compu-Scene V CGI of the other device. Display is through head-tracked Helmet-Mounted Displays.

Hercules Simulators: The HC-130P WST has been upgraded with Compu-Scene V image generation and DRLMS, and the new-build MC-130H WST will use the E&S ESIG-4000 IG and is forecast to be RFT during 1995.

Contracts. The initial contract was for $71.5 million. Contracts for the full complement of SOF-ATS systems are valued at $650 million to the year 2004.

Further Details. A public domain paper was presented at the Rotorcraft Simulation conference held in London in May 1994 at the Royal Aeronautical Society by Lt Col Edward T Reed and Capt Kyle P Cunningham entitled "Advanced Mission Training and Rehearsal for the US Air Force Helicopter force using high-fidelity simulation". Copies available from the RAeS.

Status

The TH-53A OFT was RFT in October 1992, other deliveries began in the Autumn of 1993. The helicopter WSTs are being supplied to the 58th Special Operations Wing (SOW) at Kirtland AFB, New Mexico, and other devices are delivered to Hurlburt Field, Florida.

Prime Contractor

Loral Defense Systems-Akron, Akron, Ohio.
(For details see Directory of Suppliers)

Loral F-22 Training Assessment

Loral Defense Systems-Akron was selected to provide the Pilot Training System Device requirements assessment for the Lockheed F-22 Advanced Tactical Fighter (ATF). Loral is tasked with ensuring concurrent development of an interactive training device for the aircraft.

Status

System definition is in progress.

Manufacturer

Loral Defense Systems-Akron, Akron, Ohio.
(For details see Directory of Suppliers)

Martin Marietta Flight Simulators

GE Aerospace's Simulation and Control Systems Division at Daytona Beach was acquired by Martin Marietta in April 1993. One of the plant's principal products is the Compu-Scene series of high resolution image generators which are fitted to many simulators described in this book. Some other products have included the following:

F-5 Flight Simulators. A simulator for the Northrop F-5 was produced in either single or dual cockpit configurations.

Components. The single cockpit system comprises the pilot's cockpit, instructor station, and three-channel Compu-Scene image generator. The two-cockpit configuration permits simultaneous use of the image generator channels by two pilots, in order to practise formation flying, one-versus-one combat, or even two-versus-one if a computer-controlled model is programmed into the scene.

IOS. The IOS is directly behind the student, permitting full communication and also allowing the Instructor to see the pilot's visual displays.

CGI Database. The Compu-Scene CGI system depicts ground scenes and sky equivalent to an area of 340 × 340 nm (630 × 630 km) and extending up to 60 000 ft (18 288 m).

Operating Costs. Operating costs of the mission simulator system, according to GE, are estimated at between 10 and 15 per cent of the cost involved in operating an F-5.

Customers. No longer produced but in service with several countries including Brazil, Taiwan and Thailand.

Manufacturer

Martin Marietta, Simulation and Automated Systems Division, Daytona Beach, Florida.
(For details see Directory of Suppliers)

McDonnell Douglas Training Systems (MDTS)

McDonnell Douglas Training Systems (MDTS) was formerly the American Airlines Training Corporation which was purchased by McDonnell Douglas in August 1986. MDTS develops flight simulators for the McDonnell Aircraft (McAir) Company and other customers. Some projects are described below:

AV-8B Harrier Simulators

MDTS supplied the flight dynamics software for the US Marine Corps' AV-8B Harrier II training facility at Cherry Point Air Station.

OFT. Two fixed-base operational flight trainers (Device 2F133) use the Evans & Sutherland CT5 CGI system displayed by projection in a dome.

WST. Two fixed-base weapons systems trainers (Device 2F134) permit exercises in an EW environment.

NAWST. MDTS also produced two Night Attack Weapons Systems Trainers (NAWST) for Cherry Point and MCAS Yuma, Arizona. This simulator (Device 2F150) is designed to provide night attack training for pilots already familiar with the AV-8B. Its visual system uses the Compu-Scene IV image generator displayed through a head-slaved Area-of-Interest (AoI) system inside a dome. Night imagery including FLIR and NVG simulation is included.

Simulators for A-6E

Device 2F156 is a fixed-base WST. The visual system uses the E&S ESIG 600 CGI system and display is by projection in a 30 ft diameter dome. A radar simulation system, FLIR simulation, and G-cueing by use of the pilot's anti-G suit are provided. Aircrew are trained in take off and landings, combat skills, ground attack, air-to-air missile firing, radar navigation, and EW. Four simulators are installed at NAS Oceana (two), NAS Whidbey and MCAS El Toro. The project was undertaken by MDTC in conjunction with the AAI Corporation.

Simulators for F-14 Tomcat

The fixed-base F-14 WST (Device 2F153) is similar to the A-6 WST described above and is used for day and night carrier landing training as well as low level training for the TARPS mission. The WST can be used alone or linked with others in the Target Environmental System. The US Navy has six systems, three each at NAS Miramar and Oceana, the first being delivered to Miramar in February 1992. Some of these systems have a smaller dome. The project was undertaken by MDTC in conjunction with the AAI Corporation, on a $300 million contract which also included the A-6E simulators described above.

Tomcat Air Combat Simulator. On a separate $9.2 million contract, MDTS updated the twin dome F-14 Tomcat Air Combat Simulator at NAS Oceana (Device 2E6). Part of this update included the fit of the Compu-Scene III CGI system.

Modular Aircrew Simulation System (MASS) ReConfigurable Cockpit (RCC)

MASS/RCC is a lightweight, transportable system giving simulated cockpit layouts using touch-screens

MASS-Attacker is an Air Combat version of the MDTS ReConfigurable Cockpit (RCC) trainer

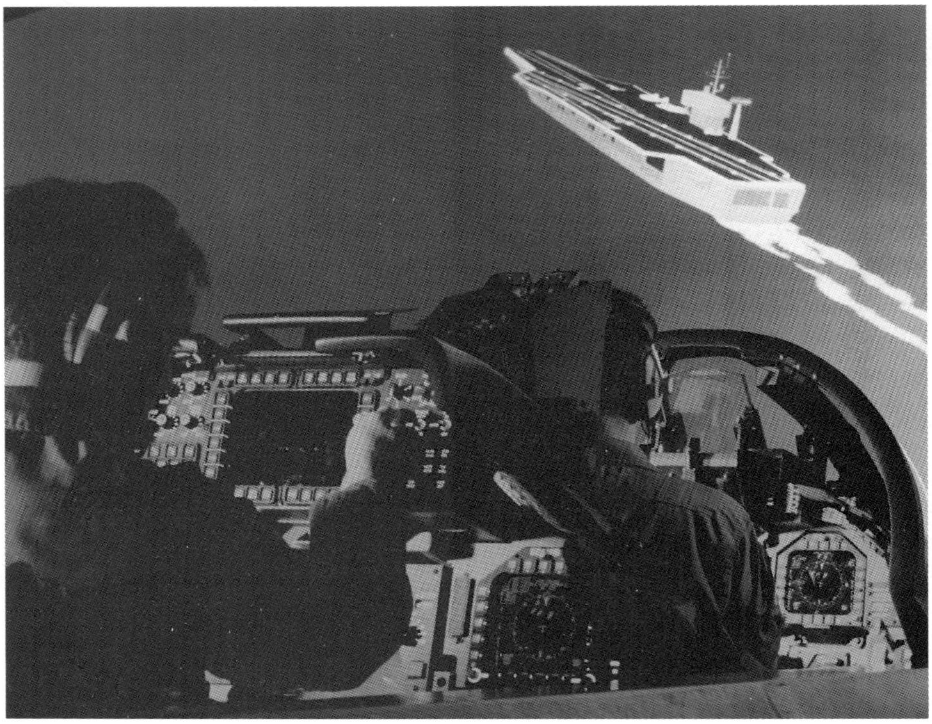

MDTS F-14D trainer with image projected on a 40 ft diameter dome

McDonnell Douglas air combat simulator for two-seat F-15

instead of replicated switches and instruments. The system fits on a standard pallet and uses a 110 V 60 Hz power supply, taking about 2.2 kW of power. The system uses commercially available RISC computing and a choice of software languages ranging from Ada to C.

Out-of-window visuals can be included at customer choice, and can consist of one or more television monitors (collimated or otherwise), or one of the many projection systems on the market (see the Visual Display Systems section of this book).

MASS/RCC has been demonstrated for the F/A-18 strike aircraft, the F-16 fighter and the USN T-45A Goshawk trainer. A derivative, MASS Attacker, can be used for air combat training and is described below.

Modular Aircrew Simulation System (MASS) Attacker Air Combat Training System

MASS Attacker is a derivative of the MDTS MASS/RCC (ReConfigurable Cockpit) system described above. The cockpit of MASS Attacker uses touch-screens instead of replicated switches and instruments.

In the Air Combat role, a wide FoV visual system is fitted, normally employing multiple projectors and a dome or partial dome. Since the cockpit is inherently reconfigurable, MASS Attacker can be used for a variety of different aircraft types.

F-15 Weapons and Tactics Trainers (WTT)

Some 18 F-15 WTTs have been ordered by the USAF for squadron-level training. It is based on the MDTS MASS/RCC system described above, configured for the F-15.

The WTT is equipped with an F-15 throttle and stick, right and left panel control switches, a 27 in colour touch-screen display, a 19 in monitor-based simulation of the aircraft HUD and the out-of-window scene and a 19 in auxiliary colour display. The trainer features simulated avionics, air-to-air, radar, weapons and defence systems and pilots can practise on a library of up to 30 mission scenarios, which can engage up to 16 enemy aircraft. The system entered service in March 1992.

F-15 Courseware

MDTS also has a $6.08 million contract for F-15E aircrew training and training courseware.

MDTS Dome Simulators at St Louis

The simulation facility at St Louis, has a total of 11 domed simulators of diameter 12.2 m and 6.1 m. The facility was originally designed to assist in the design and development of the F-15 fighter aircraft. It has grown to include simulations of the US Navy/McDonnell Douglas F/A-18 fighter and AV-8B V/STOL aircraft and for more general development and investigation of advanced fighter tactics.

F-15 Dome Simulator

One of the 12.2 m diameter projection domes at St Louis houses a cockpit for the McDonnell Douglas F-15E Strike Eagle. This has been operating with US Air Force pilots since March 1983. It includes simulations of the improved Hughes APG-63 radar, terrain-following system, forward-looking infra-red (FLIR) sensors and navigation equipment.

Visual Image Generation (IG). CGI. The detailed

ground scenes are provided by a computer-generated image visual system and multiple target projectors.
Visual Display System (DS). Projected ground scenes and additional target projectors for air-to-air targets.
Motion Cueing. Fixed base but pilot's own anti-G suits are used.
Computer. Gould 9780.
Control Loading. Digital.
Instructor Operating Station (IOS). Mounted off board.

US Navy ACMS

A similar concept to the dome F-15 simulator was developed for the US Navy as the Air Combat Manoeuvring Simulator (ACMS), designated Device 2E6, which is located at NAS Oceana, Virginia. The Device 2E6 is used for the visual air-to-air combat training of F-4 and F-14 crew members.

USAF Tanker and Transport Training System (TTTS)

In February 1990 a team led by MDTS received an $8.89 million contract from the US Air Force for the initial phase of the Tanker Transport Training System (TTTS) programme.

The TTTS uses the T-1A Beech Jayhawk aircraft (a derivative of the Beechjet 400A) in the training of potential tanker and transport pilots and is part of the USAF Specialised Undergraduate Pilot Training (SUPT) system. After primary flight training crews scheduled for multi-engined aircraft will be trained on the T-1A Jayhawk and those scheduled for fighter aircraft will fly the T-38 Talon.

MDTS is the prime contractor for the TTTS programme. Beech Aircraft Corporation is delivering up to 211 aircraft, the first in November 1991 and the last is scheduled for delivery in 1997.

The Quintron Corporation of Chantilly, Virginia, will supply 11 OFTs which will be to FAA Level C standard.

The T-1A system will be installed at the following Air Training Command bases: Randolph AFB, Texas; Reese AFB, Texas; Vance AFB, Oklahoma; Columbus AFB, Mississippi; Laughlin AFB, Texas and Williams AFB, Arizona. Reese AFB was the first base equipped with the system.

C-17 Training Systems

MDTC was awarded a $420.99 million contract in October 1988 to design, build and operate an aircrew training system for the C-17 Globemaster transport.

The system will include classroom instruction and self-paced CBT on 86 interactive computer terminals equipped with video disks. The bulk of the instruction will be given on 12 Full Flight Simulators which are subcontracted to FlightSafety International (FSI) and are described in the FSI entry in this book. A cockpit procedures trainer (CBT) will be similar to the Full Flight Simulators but will have neither the visual nor the motion systems.

A full-scale cargo compartment trainer will allow training of loadmasters on the various C-17 cargo compartment features without having to use the aircraft. It will also allow practice loading of cargo or vehicles and practice in the reconfiguration of the cargo compartment. The cargo load model will allow cost-effective training in load planning for weight and balance. It is similar to the device used for training KC-10 loadmasters-boom operators and enables students to visualise critical clearances within the cargo compartment.

The training systems will be installed at five US Air Force bases, two Air Force Reserve locations and two Air National Guard bases. Altus AFB, Oklahoma, will serve as the central training base with five weapon system trainers, the cockpit systems simulator, the cargo compartment trainer and the cargo load model, while Charleston AFB, South Carolina, will be the first base to have operational C-17s. The first simulators were installed at Charleston in 1991 and training began in August 1992. Simulators will be delivered to Altus from 1993 and deliveries of all the simulators will be complete in 2003.

Status

See individual entries.

Manufacturer

McDonnell Douglas Training Systems (MDTS), St Louis, Missouri.
(For details see Directory of Suppliers)

Perceptronics Avionics Situational Awareness Trainer (ASAT)

The Avionics Situational Awareness Trainer (ASAT) is a modular family of trainers designed to make pilots familiar with the avionics systems in their cockpits. All the systems are real-time and interactive with easily generated scenarios through the use of simple-to-use scripting tools. Pilots can choose between preset scenarios or those generated by a script. The systems can operate in a stand-alone or in networked modes.

The Basic Table Top (BTT) model is designed to provide practice weapons experience using head-up display (HUD) and instruments interpretation, radar sorting, targeting, weapons delivery, defensive countermeasures. It has a 14 in monitor upon which is displayed F-16 instruments, a side-stick controller and throttle.

The Enhanced Table Top (ETT) model has a 1024 line monitor to improve the fidelity of both the HUD and the beyond-the-canopy view. There is also a graphics co-processor card to enable a screen refresh rate of approximately 30 frames/second.

The Beyond Visual Range (BVR) model is driven by three microcomputers. High-fidelity speakers, subwoofers and vibrators within the inclined seat provide realistic cueing for the onset of G, weapons launch, ambient flight noise and weapons hit effects. There is an after-action review facility. A fourth microcomputer and a stores control panel may be added; the system is then designated the Model AG for air-to-ground training.

There are two CPTs based upon the Model AG; with the addition of flight and navigation instruments the AG becomes the Instrument Procedures Training (IPT) Model, and the other CPT is the Emergency Procedures Trainer (EPT) model which includes switches, handles and panels.

Versions of the ASAT systems mounted in domes for wide angle visual projection are Model BVR/D, AG/D, IPT/D, EPT/D. These versions use a 42 ft dome with a field-of-view of between 180 and 240°. For integration into a network there is the Model IOS instructor station which provides full control of the simulation. This provides viewpoints including those from the cockpit, tower, ground and even from missile seekers.

Status

In production.

Manufacturer

Perceptronics Inc, Training & Simulation Division, Woodland Hills, California.
(For details see Directory of Suppliers)

Reflectone Helicopter Simulators

Reflectone MH-60G WST

In September 1990 Reflectone was awarded a $7 million contract by Loral under the USAF SOF-ATS programme (see under Loral entry) to supply a WST for the MH-60G, the US Air Force Special Operations Forces (SOF) version of the UH-60 Black Hawk.
Visual Image Generation (IG). Eight channels of Martin Marietta (ex-GE) Compu-Scene V CGI. Details in Image Generation (IG) Systems section.
Visual Display System (DS). Eight Collimated Monitors including chin windows.
Motion Platform. Reflectone 6-DoF.
Control Loading. Digital.
Instructor Operating Station (IOS). Mounted on/off board.
Other Features. DRLMS simulation.
Deployment. The WST will be installed at Kirtland Air Force Base, New Mexico.

Reflectone CH-46 Sea Knight OFT

Produced for the US Navy and Marine Corps as Device 2F117A/B, one is for the Boeing CH-46D and two for the CH-46E.
Visual Image Generation (IG). E&S CT 5A. Details in Image Generation (IG) Systems section.
Visual Display System (DS). Collimated Monitors.
Motion Platform. Reflectone 6-DoF.
Computers. Harris Dash-4 computer with 96 k words of 24-bit memory and more than 40 Mbytes of disk storage.
Control Loading. Digital.
Instructor Operating Station (IOS). Mounted on board. A remote trainer control panel at the centre console of the flight-deck, permits the instructor to fly as a pilot or co-pilot and so exercise limited control of the training session. In addition it allows the student to conduct self-training, without instructor assistance, and enables playback of the most recent 5 minutes of flight.
Other Features. The system permits training in TOL in confined areas, operations aboard LST-class ships, IFR and VFR flying in poor weather, flight with a slung load, and formation flying.
Deployment. A CH-46D trainer was installed at NAS North Island, California, while two CH-46E trainers were installed at MCAS Cherry Point, North Carolina and El Toro, California.

Reflectone SH-3H Sea King WST

Designated Device 2F64C for the USN is designed to train in all aspects of flight, including the detection, classification, tracking and attacking of hostile submarines.
Visual Image Generation (IG). Seven channels of FlightSafety (ex-MDTS) Vital IV CGI system. Details in Image Generation (IG) Systems section.
Visual Display System (DS). Seven Reflectone Collimated Monitors including chin windows.
Motion Platform. Reflectone 6-DoF.
Computers. Two Harris H1000 computers, each with 64 k words of memory and 40 Mbytes of disk storage.
Control Loading. Digital.
Instructor Operating Station (IOS). Mounted on/off board.
Other Features. A separate station is provided for the sensor operator and his sonar and MAD equipment. It can be operated in a stand-alone mode or with the rest of the WST. An acoustic generator was incorporated during 1984 to provide realistic target acoustic signatures for the sonar data computer.
Deployment. Two trainers were built (the last in January 1983) for the US Navy in NAS North Island, California, and NAS Jacksonville, Florida.

Modifications to SH-2F/SH-3H WSTs

During 1990, helicopter WSTs at NAS Norfolk, NAS Jacksonville, and NAS San Diego (two) were modified with improved computers and with enhancements to radar, acoustic, ESM and tactical navigation simulation. In addition, upgraded avionics were fitted and synchronisation of the visual system was improved.

Flight Instrument Trainer for TH-57C

Device 2B42 provides full flight training for the US Navy's TH-57C (Jet Ranger) pilots.
Visual Image Generation (IG). Three channels of E&S SP-1 image generator. Details in Image Generation (IG) Systems section.
Visual Display System (DS). Four Reflectone Collimated Monitors, the centre two displaying the same IG channel.
Motion Platform. Reflectone 6-DoF
Deployment. The US Navy purchased nine TH-57C trainers, six full-flight (valued at $21.6 million), the others part-task cockpit procedures trainers and all were delivered by early 1988. Two of the full-flight trainers have full-colour computer-generated image visual system. All the systems are at NAS Whiting.

Weapons System/Acoustic Trainers for SH-60F Sea Hawk

Reflectone produced simulators for the SH-60F Sea Hawk helicopter for the USN (the naval version of the Army Black Hawk medium lift helicopter). The Norden company provided the sensor suite simulation. The

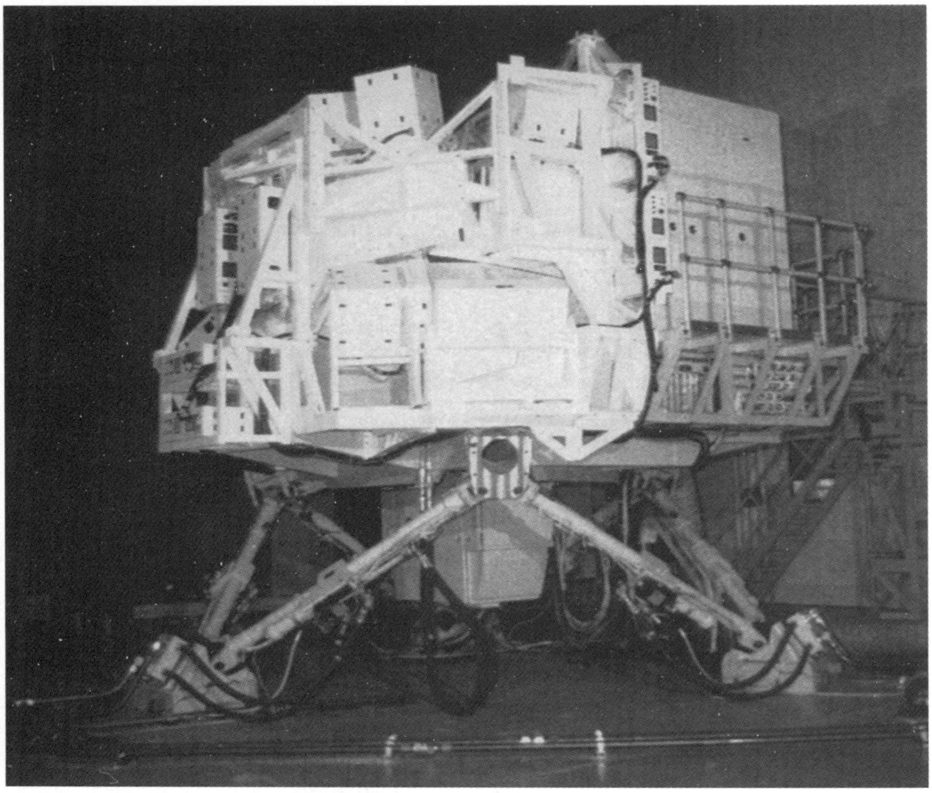

Reflectone SH-60F flight simulator

equipment forms two adjacent training suites each containing two Operational Flight Trainers (Device 2F146), one acoustics trainer with two sonar student stations (Device 14D3), and a maintenance trainer.

WST. The Weapons System Trainer (WST) consists of the Reflectone OFT and Norden's sensor operator trainer each of which can operate independently or in various integrated modes.

Visual Image Generation (IG). Six channels of FlightSafety (ex-MDTS) Vital V CGI system. Details in Image Generation (IG) Systems section.

Visual Display System (DS). Eight Reflectone Collimated Monitors, the two centre monitors displaying the same IG channel.

Motion Platform. Reflectone 6-DoF.

Control Loading. Digital.

Instructor Operating Station (IOS). Mounted on board.

Other Features. The sensor operator trainer has two stations and both are supported by the Reflectone IOS.

The WSTs may be networked to form two mission system trainers using an Ethernet software protocol developed by Reflectone.

Deployment. The SH-60F trainers are at NAS North Island, California, and NAS Jacksonville, Florida. Deliveries were completed in October 1991.

Upgrade. In August 1991 AAI-SMI won a contract for visual upgrades.

Status

See individual entries.

Manufacturer

Reflectone Inc, Tampa, Florida.
(For details see Directory of Suppliers)

Reflectone Military Aviation Simulation - Fixed-Wing

Reflectone A-10 OFT

The OFT trains for flight from preflight and start up through navigation and combat, visual and GCA approach, to landing and post-flight debriefing.

Visual Image Generation (IG). One channel of FlightSafety International (ex-McDonnell Douglas) Vital V. Details in Image Generation (IG) Systems section.

Visual Display System (DS). One collimated monitor.

Motion Cueing. Mounted on a fixed base, a G-seat is used with pneumatic cushions to provide 'seat-of-the-pants' cues.

Computers. Three SEL 32/35 computers each with 128 k words of core memory and three 40 Mbyte disks. One computer is largely dedicated to EW simulation.

Control Loading. Digital.

Instructor Operating Station (IOS). Separate EW training is provided through a dedicated EW IOS station.

Other Features. Weapons simulation includes trajectories and guidance of 'smart' bombs. Scoring algorithms for the 30 mm cannon on this aircraft, and its bombs and rockets, are also included.

Deliveries. The production run of 14 systems was completed in 1981, and the OFTs were subsequently updated by Burtek (now part of Thomson Training).

Reflectone EA-6B OFT

Two OFT and Navigation Trainers full-mission training systems (Device 2F143) were produced by Reflectone. They permit simulation of all flight procedures, including emergencies, carrier operations, in-flight refuelling, formation flying and stall/spin recovery.

Visual Image Generation (IG). Three channels of E&S SP-X 500. Details in Image Generation (IG) Systems section.

Visual Display System (DS). Rediffusion (now Thomson Training) WIDE™ giving a FoV of 180° × 40°. Details of WIDE™ in Visual Display Systems section.

Motion Platform. Reflectone 6-DoF.

Instructor Operating Station (IOS). Mounted on board.

Deliveries. The first OFT was delivered to NAS Whidbey Island, Washington, and the second was delivered to MCAS Cherry Point, North Carolina, in November 1990.

Device 2F119 Upgrade. Reflectone also upgraded the EA-6B flight trainer (Device 2F119) at NAS Whidbey Island. The flying and motion qualities were improved while texture was added to the original SP-1 visual system, making it an SP-1T version. Reflectone also created a Team Tactics Trainer from tactical work stations in the WST and made it a unit which operates autonomously or integrated with the remainder of the flight trainer.

Aerial Refuelling Part-Task Trainer for C-5A and C-141

This PTT is for Air-to-Air Refuelling (AAR) receiving

IOS for the Reflectone EA-6B OFT

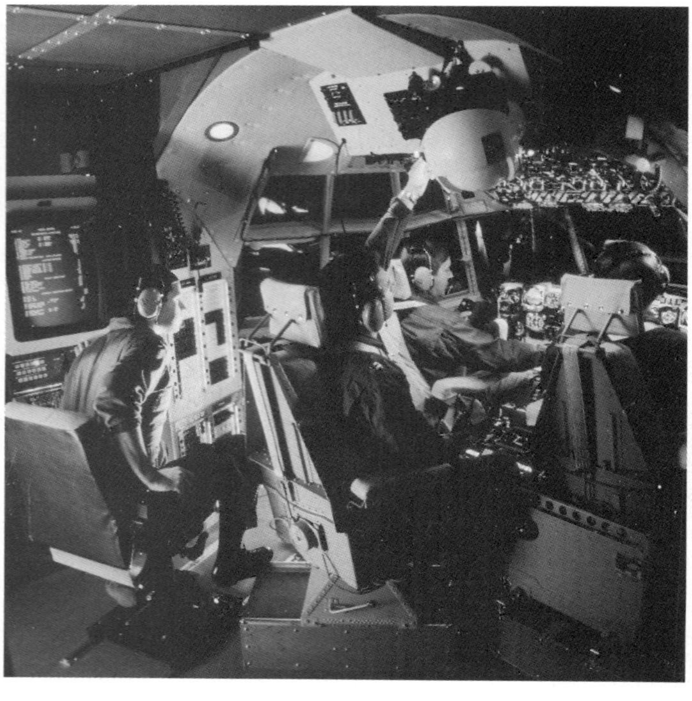

Reflectone C-130 OFT

techniques for the crews of the C-5A Galaxy and C-141 Starlifter aircraft. Tanker aircraft modelled include the KC-10 and KC-135.

Modelling. The simulation features tanker downwash and receiver upwash models, tanker tail aerodynamics and receiver bow wave, engine efflux and tanker-boom dynamics. Digital control loading is fitted.

Visual System. In a unique Direct-View-With-Depth visual system, scale models of tanker aircraft are viewed through the optics, giving realistic depth perception.

Deployment. The system entered service at Altus AFB, Oklahoma in March 1989 and is now also used at Travis AFB, California; McChord AFB, Washington; Dover AFB, Delaware; McGuire AFB, New Jersey; Norton AFB, California and Charleston AFB, South Carolina.

Reflectone KC-130F/R/T OFT

This OFT is designated Device 2F107 and provides full flight training for the three-man crew of the US Marines' C-130F/R/T tanker. All normal and emergency flight procedures can be replicated, as well as all aspects of in-flight refuelling. Landing on soft ground and ditching can also be practised.

Visual Image Generation (IG). Five channels of Martin Marietta (ex-GE) Compu-Scene II CGI. Details in Image Generation (IG) Systems section.

Visual Display System (DS). Six Collimated Monitors, the two monitors in front of the pilots displaying the same front IG channel.

Motion Platform. Reflectone 6-DoF.

Instructor Operating Station (IOS). Mounted on/off board.

Deployment. Installed at MCAS Cherry Point, North Carolina, in 1988.

Reflectone C-130H OFTs

In December 1990, Reflectone delivered a C-130H Operational Flight Trainer (OFT) to the European Aircraft Training Centre (EATC), Brussels Airport, Belgium. A similar OFT was delivered to the Singapore Regional Training Centre of Asia-Pacific Training and Simulation and was RFT in February 1992.

Visual Image Generation (IG). The Brussels device has three channels of FlightSafety International

Reflectone Air-to-Air Refuelling PTT for C-5A and C-141

(ex-MDTS) Vital V CGI. The Singapore model has three channels of the E&S SP-X250 CGI system. Details in Image Generation (IG) Systems section.

Visual Display System (DS). Four Collimated Monitors, the centre two presenting the same IG channel.

Motion Platform. Reflectone 6-DoF.

Control Loading. Digital.

Computer. The Singapore device system has a Multi-SEL computer system.

Instructor Operating Station (IOS). Mounted on board.

Other Features. More than 500 simulated malfunctions are available.

USAF C-130H Aircrew Training

In August 1991, Reflectone won a five-year US Air Force Training Command contract worth $3.3 million to provide C-130H pilot and aircrew refresher training to the US Air Force Reserve and the US Air National Guard.

Flight Crew Station Trainer for MC-130H

In January 1991, Reflectone was awarded a contract by Loral Defense Systems to produce a full flight simulator for the MC-130H Combat Talon II Weapon System Trainer to be part of the Special Operations Forces Aircrew Training System (SOF-ATS).

Visual Image Generation (IG). Five channels of Evans & Sutherland ESIG-4000 CGI. Details in Image Generation (IG) Systems section.

Visual Display System (DS). Six Collimated Monitors with the centre two displaying the same IG channel.

Motion Platform. Reflectone 6-DoF.

Control Loading. Digital, electric actuators.

Instructor Operating Station (IOS). Mounted on/off board.

Other Features. Ada-language software and a micro-digital signal processor communication system.

Deployment. RFT late 1993.

Reconfigurable Mission Rehearsal Device (RMRD)

Under the sub-contract from Loral for SOF-ATS, Reflectone will produce electric control loading systems, digital audio recording and playback as well as digital communications systems for the reconfigurable Helicopter/Fixed-Wing mission rehearsal device (RMRD).

Technology Demonstrator C-130H OFT

A C-130H OFT demonstrator has been available from Reflectone's Tampa, Florida, facility. The device features a microprocessor-based instructor station with two colour monitors and touch-screen windows technology. It also features a new 6-DoF hydrostatic motion system, digital control loading with electric actuators, micro-digital signal processor communications system, digital sampling aural cueing with CD fidelity as well as a new wiring philosophy with pre-installed wiring for future expansion. The computer system is the UNIX-based Concurrent two CPU Model 8400.

Status

See individual entries.

Manufacturer

Reflectone Inc, Tampa, Florida.
(For details see Directory of Suppliers)

SimCom Flight Trainers

SimCom International Inc of Orlando makes a number of Flight Training Devices (FTDs) for light aircraft up to turboprop twins. These are used in various SimCom flight training centres.

Aircraft Simulated. The following aircraft are currently simulated: Beech Baron 58P, Beech Bonanza, Cessna 340 and 400 series, Cessna Centurion, Piper Cheyenne, Piper Navajo Chieftain, King Air C90.

Design Concept. The SimCom concept is not to use motion platforms on the grounds of expense, but to otherwise enhance realism by using a real aircraft cockpit section for the simulator cockpit, combined with a wide-angle projected outside-world display. The real cockpits are normally purchased at low prices and are from time-expired or damaged aircraft.

Visual System. The Visual System was developed under licence from Telludrian, uses four CGI channels and is displayed by direct screen projection with a FoV of 170° x 40°. SimCom has a database modelling tool and uses real-world databases for the Orlando, Florida, and Denver, Colorado areas.

Status

In service at SimCom flight training centres.

Manufacturer

SimCom International Inc, Orlando.
(For details see Directory of Suppliers)

SimCom Cessna FDT showing real aircraft front end and the 170° wide visual system

SimCom's use of real aircraft front ends is shown here with (left to right) a King Air C90, Super King Air B200 and Piper Cheyenne 2 being converted to flight simulators

SimuFlite Training

SimuFlite is involved in various training programmes. It provides flight-deck management courses lasting three days, focusing on cockpit crew co-ordination, communication and other human factors. It also has an international procedures course lasting three days, which provides instruction in preparing flight plans and itineraries, obtaining overflight and landing permits, customs procedures, and navigation techniques. The following programmes also exist:

SimuFlite F-16 training

In March 1988 the company was awarded a $3.14 million contract by the DoD for training F-16 pilots of Tactical Air Command. The work, which includes academic and simulator training, was at Luke AFB, Arizona; MacDill AFB, Florida; Hill AFB, Utah; Homestead AFB, Florida, and Shaw AFB, South Carolina.

C-21A Learjet training

In 1988 SimuFlite won a $2.6 million contract for C-21A Learjet simulator and academic work for the US Air Force's Military Airlift Command (MAC). The C-21A Learjet is used for operational support as well as providing new MAC pilots with both experience and flight time.

Visual Image Generation (IG). Three channels of Link-Miles (now Thomson Training) IMAGE III CGI system. Details in Image Generation (IG) Systems section.
Visual Display System (DS). Four Collimated Monitors, the centre two displaying the same IG channel.
Motion Platform. Reflectone 6-DoF.
Control Loading. Digital.
Instructor Operating Station (IOS). Mounted on board.
FAA Certification. Level C.
Deployment Based at Dallas/Fort Worth Airport, Texas.

C-130/L-100 training

SimuFlite has operational responsibility for the Hercules Flight Training Center in Marietta, Georgia, which houses an FAA-approved C-130/L-100 simulator. SimuFlite also is responsible for the Southern Air Transport/Reflectone Inc partnership for C-130/L-100 simulator training in Tampa, Florida.

Status

See text.

Manufacturer

SimuFlite Training, Dallas, Texas.
(For details see Directory of Suppliers)

SMI HH-60J OFT

The HH-60J Sea Hawk simulator for the US Coast Guard has the following characteristics:
Visual Image Generation (IG). Five channels of E&S ESIG-600 CGI. Details in Image Generation (IG) Systems section.
Visual Display System (DS). Projected on the inside of a dome.
Motion Platform. Reflectone 6-DoF.
Computers. Harris Night Hawk.
Control Loading. Digital.
Instructor Operating Station (IOS). Mounted on board. An observer station which permits over-the-shoulder instruction during training exercises.

Status

Installation at the US Coastguard training centre at Mobile, Alabama.

Manufacturer

Systems Management Incorporated (SMI), Hunt Valley, Maryland.
(For details see Directory of Suppliers)

AAI/SMI HH-60J Sea Hawk OFT

Tracor Tactical Combat Simulation

Tracor has developed a range of tactical combat simulation packages designed to analyse, develop, test and operate weapon systems. The work involves engineering studies and analyses, producing software and hardware, then integrating them with simulators, avionics laboratories and RF emitters.

Tracor TACSIM

Tracor Flight Systems Tactical Simulation (TACSIM) software exploits experience gained with the F-15B, EA-6B and B-1B. It simulates realistic, multiple-platform, air-land-sea combat environments and includes both fixed- and rotary-wing aircraft and land-based systems.

Tracor TES

The Tactical Environment Simulation (TES) software developed by Tracor is designed to simulate many-on-many mission scenarios in real time. Interfacing with manned simulators, TES allows the simulator pilot to acquire, attack, kill, jam or evade air and ground threats which are attempting to acquire, engage and destroy the simulated aircraft. TES can be used with part-task mission simulators which include sufficient displays and controls for TES interaction.

Status

In service.

Manufacturer

Tracor Flight Systems Inc, Santa Ana, California.
(For details see Directory of Suppliers)

Veda Genesis 2000 Aerodynamic Simulation System

The Genesis 2000 system is a real-time engineering flight simulator for rapid prototyping and modification of aerodynamic models through an interactive menu system. It has a library of external visual scenes, a programmable target aircraft for conducting air-to-air tracking tasks, and a data recording system emulating typical flight test instrumentation for analysis of open and closed loop control systems and their associated handling qualities.
Components. Genesis 2000 consists of a generic cockpit with two displays; a 19 in (483 mm) real-world view and a 14 in (355 mm) synthetic head-down flight instrument display. Flight controls include a force-transducer side-stick (as in the original F-16 aircraft until later modified with a semi-compliant side-stick), twin

Veda Genesis 2000 aerodynamic simulation system

throttles and rudder pedals. A full range of secondary flight controls is provided including landing gear, flaps, speedbrakes and three-channel auto- stabilisation.
Features. The system features distributed parallel bus architecture, using Motorola 68020/68881 32-bit processors. Other options are available including centre stick or wheel with or without feel forces (control loading), a programmable HUD, high-speed 32-bit SPARC architecture processors, a controller with programmable touch-sensitive panels, and a FBW control system-modelling package.

Status
Available.

Manufacturer
Veda Incorporated, California, Maryland.
(For details see Directory of Suppliers)

USAF Materiel Command R&D Mission Simulator-1 (MS-1)

The US Air Force's Wright Research and Development Center has created a flight simulator to provide a full-mission combat environment for the analysis and evaluation of advanced aircraft. The system is known as Mission Simulator-1 or MS-1. It is used for a number of research projects ranging from aircraft handling qualities and safety-of-flight evaluation to integrated system performance as part of a multi-aircraft engagement.
Components. The simulator consists of a generic reconfigurable cockpit housed in a 40-foot (12.2 m) diameter dome. Up to four laser-projected images are projected on the inside of the dome to represent friendly or hostile aircraft and missiles. There are separate conventional colour projectors which provide images such as aircraft targets or wingmen. A high resolution projected inset provides additional images, such as ground targets.
Sound Generation System. A digital stereo audio-generation system is used to create aircraft sounds and cockpit audio tones, such as from the RWR and missile seekers.
Computing and Networking. The MS-1 is controlled by a real-time multi-processor computer network and the simulator can interact with other simulators.

Status
In service at Wright-Patterson AFB, Ohio, since September 1990.

Sponsor
US Air Force, Air Force Materiel Command, Aeronautical Systems Center, Wright-Patterson Air Force Base.
(For details see Directory of Suppliers)

YUGOSLAVIA (SERBIA AND MONTENEGRO)

SL-G4 Flight Simulator

The SL-G4 is a full flight simulator for the SOKO G-4 Super Galeb light combat aircraft.
Components. The cockpit section is mounted on a motion platform and the simulator has an instructor's station, computer and video terminal.
Motion Platform. 3 DoF. The two main suspension beams when moved together give heave and when moved differentially give pitch. A separate small jack, laterally displaced from the centreline, gives roll.
Activities Simulated. The SL-G4 provides full simulation of instrument take offs and landings, engine handling, communications and navigation equipment. The system has partial simulation of the hydraulic, fuel supply, electrical, cockpit and instrument lighting systems.
IOS. The instructor's station has duplicate instruments and can introduce simulated malfunctions.
Visual System. A visual system can be fitted which uses CGI and displays through a collimated monitor.

Status
Was in production for the former Yugoslav Air Force. In view of the current situation in Serbia, its status is currently not known.

Contractor
Holding Company YUGOIMPORT-SDPR Ltd, Belgrade.
(For details see Directory of Suppliers)

Galeb Flight Simulator

Galeb Flight Simulator Motion Platform - the two main beams produce heave and pitch and the separate small jack produces roll motion

TRAINERS FOR NON-PILOT AIRCREW

These trainers are differentiated from those in the previous section (Flight Simulators and Flight Training Devices) in that they do not have aerodynamic flight and/or aircraft handling programmes designed for handling fidelity of the pilot's flight controls. Most are clearly for the training of non-pilot aircrew such as navigators, weapon system officers (WSOs), EW officers, air observers, air engineers, loadmasters and so forth. Where there is a certain amount of ambiguity such as in trainers for E-O weapons which could be operated by either pilots or by WSOs or equivalents, the trainer has been placed in this section because it is essentially a system trainer rather than a simulator with a dynamic flight model for training in aircraft handling.

FRANCE

DCN Estocade Maritime Aircraft Acoustic Trainer

The DCN (Direction des Constructions Navales) Estocade acoustic training simulator provides Maritime Patrol Aircraft (MPA) tactical crews with in-flight experience of synthetic acoustic signatures. The system is coupled to sensor receivers on the aircraft and its signals simulate a target and weapons effects; it is available in five languages.

Components. The system consists of two units, each carried in 25 kg cases, one with the computer and the other with the simulator. The simulator contains a small display screen and a plug-in miniature keyboard.

Operation. A simulated submarine is controlled by the operator and the dropping of up to four active sonobuoys is modelled. The aircraft's real sensors receive 'data' from these simulated buoys and when the 'target' has been located, simulated torpedoes may be dropped either manually or automatically through Digibus coupling in the Atlantique 2 simulator (see separate entry under Thomson-CSF in this section). A developed version will simulate inputs from eight passive sonobuoys.

Aircraft Interface. A 115 V 400 Hz AC power supply is required, and a single coaxial cable to couple to the VHF aerial wiring which would receive signals from real sonobuoys.

Status
In production for the French Navy.

Manufacturer
DCN, Paris.
(For details see Directory of Suppliers)

ISRAEL

Rafael Electro-Optical Trainer (EOT)

The Electronics Systems Division of Rafael makes an Electro-Optical Trainer (EOT) for training aircraft operators of E-O systems.

CGI System. A CGI system is used based on GP off-the-shelf hardware and software. Images are derived from real photographic sources and a precise model of terrain elevation. Typical terrain cover is 300 x 50 km.

Instructor Station. The Instructor Operating Station (IOS) has a display of the trainee's controls and display. Debriefing may be on- or off-line, and automatic performance evaluation is provided.

Status
Available. In service with the Israeli Air Force.

Manufacturer
Rafael, Electronics Systems Division, Haifa.
(For details see Directory of Suppliers)

Rafael Electro-Optical Trainer - Operator's Display

ITALY

DATAMAT Radar Operator Training System (ROST)

The DATAMAT ROST (Radar Operator Training System) was designed to train radar operators in helicopters and can train two students simultaneously under the control of an instructor.

Components. The ROST consists of a computer system, a video signal generator, one instructor station and two students' stations.

Trainee Station. The student's station consists of a real radar display with associated controls, a communication link with the instructor and an altitude meter with simulated altitude.

Instructor Station. The instructor's station features a computer terminal with raster and graphic displays.

Scenarios. The operational scenario covers a radius of 400 nm (740 km). The scenario covers a minimum of 20 static and 30 moving targets together with a coastline. The aircraft altitudes simulated are 30, 91, 152 and 457 metres (about 100, 300, 500 and 1500 ft) and its maximum simulated speed is 150 knots. The system can generate sea and ground clutter and EW environments.

Status
In production.

Manufacturer
DATAMAT Ingegneria dei Sistemi SpA, Rome.
(For details see Directory of Suppliers)

SPAIN

CESELSA ASW Trainer for Airborne Sonar Systems

This system is designed by CESELSA to provide sonar operators in ASW helicopters with training in the use of dipping sonar equipment. The student uses real sonar equipment with a simulated acoustic and tactical environment.

Components. The system consists of a student console with 'dry-end' sonar components, supplied by the user, and the layout of a typical helicopter sonar operator station.

IOS. The instructor's console features a man/machine interface based upon touch-screens and keyboards and a high resolution CRT display. The instructor can select or create the scenario then select environmental, target and acoustic conditions.

Software and Processing. The software is reconfigurable to the Bendix AN/AQS-13B, with and without adaptive processing sonar (APS), the Bendix AN/AQS-13E and the Bendix AN/AQS-18.

Status
Available for production.

Manufacturer
CESELSA, S Fernando de Henares.
(For details see Directory of Suppliers)
US Enquiries to: AlliedSignal Ocean Systems, Sylmar, California.
(For details see Directory of Suppliers)

The CESELSA Anti-Submarine Warfare System Trainer

UNITED KINGDOM

Marconi Airborne Warning and Control System (AWACS) Mission Simulator

In January 1990, Marconi (ex-Ferranti International Simulation and Training) won a £1.7 million contract to provide a mission simulator for Royal Air Force Boeing AEW 1 Sentry AWACS (Airborne Warning and Control System) aircraft. The simulator is a facsimile of the aircraft's rear cabin and is used to train aircrew in the operation of radar, communications and control systems.

Status

The system was delivered to RAF Waddington in 1990.

Manufacturer

Marconi Simulation and Training, Stockport.
(For details see Directory of Suppliers)

Marconi Air Navigation Trainer (ANT)

In 1990 Marconi (ex-Ferranti International Simulation and Training) was awarded a contract to produce an Air Navigation Trainer (ANT) which is designed to train navigators in basic and advanced radar and navigation skills. It will also be used to provide Air Interception (AI) training.

Components. Some 18 student work stations are linked to nine IOS control units. The student stations are based on touch-screen graphics work stations rather than analogues of cockpit instruments and controls, and so are inherently reconfigurable to different layouts, cockpits and aircraft.

Radar Simulation - DRLMS. The system uses digital emulation for radar simulation and has a software development of the CREST system for DRLMS data.

Status

In service at RAF Finningley since 1992, design available for other customers.

Manufacturer

Marconi Simulation and Training, Stockport.
(Ex-Ferranti, for details see Directory of Suppliers)

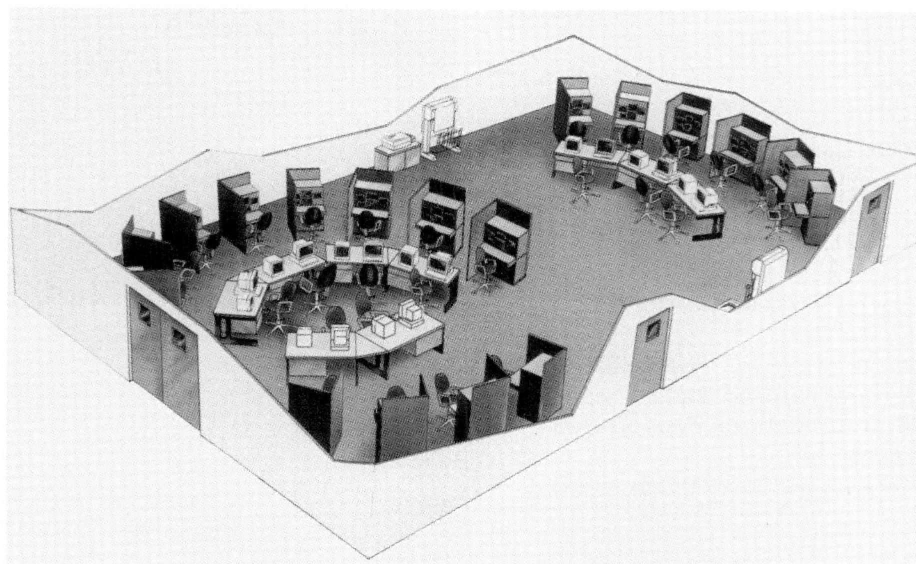

The Marconi Air Navigation Trainer (ANT)

Marconi Jetstream Radar Procedures Ground Trainer

The Radar Procedures Ground Trainer at the Fleet Air Arm Observer School has ten student positions. Each is a facsimile of the student observer's position in the Jetstream aircraft with simulated instruments, radar, air navigation system and radio. The system is used to give basic Observer training in air navigation and radar operation.

Status

In service with the Royal Navy at RN Air Station Culdrose, Cornwall.

Manufacturer

Marconi Simulation and Training, Stockport.
(Ex-Ferranti, for details see Directory of Suppliers)

IOS and student positions in the Jetstream radar procedures ground trainer

Marconi Lynx Observer Procedure Trainer (LOPT)

The Lynx helicopter Observer Procedure Trainer (LOPT) at HMS *Osprey,* Portland, comprises two independent cubicles, each simulating a Lynx observer cockpit.

The trainer is based upon Argus 700 computers linked in a multiprocessor configuration. Radar and EW simulation is included.

The simulator is easily converted to other aircraft types and includes both library and scenario generation capabilities. Classified information is designed to be implemented easily by the customer.

Status
Operational at HMS *Osprey*, Portland, Dorset.

Manufacturer
Marconi Simulation and Training, Stockport.
(Ex-Ferranti, for details see Directory of Suppliers)

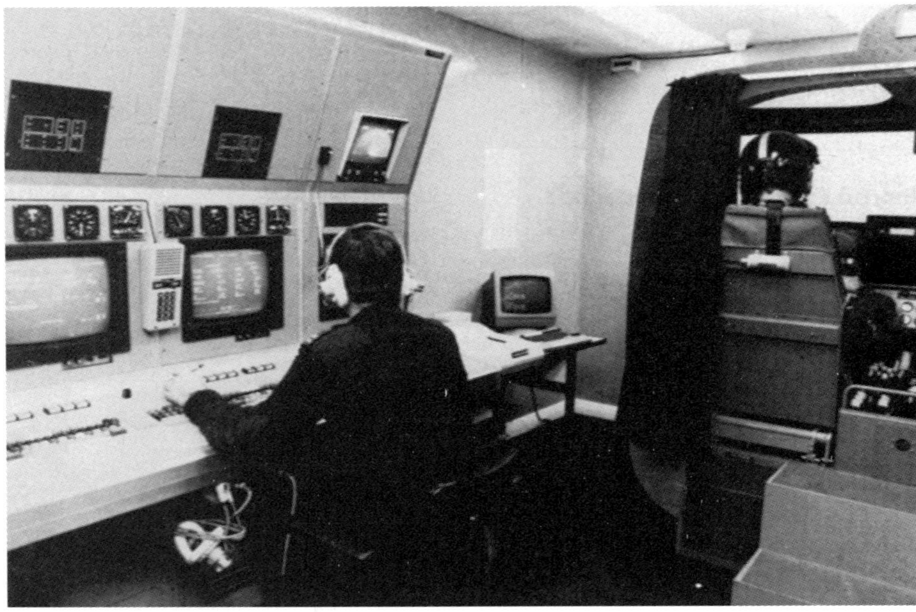

IOS of Lynx observer procedure trainer

Marconi Nimrod Basic Communications Procedures Trainer

This trainer is used to train up to eight students in communications procedures. Each sits in his own booth which simulates the communications suite of the Nimrod MR Mark 2 maritime reconnaissance aircraft. Each student has an instructor and up to six 'ground stations', simulated by other instructors, to train in anti-submarine warfare, maritime reconnaissance and search and rescue.

Status
In service since early 1990 at RAF Finningley, Yorkshire.

Manufacturer
Marconi Simulation and Training, Stockport.
(Ex-Ferranti, for details see Directory of Suppliers)

Marconi Sea King Mark 6 Rear Crew Trainer

The Marconi (ex-Ferranti) Sea King Mark 6 rear crew trainer (RCT) is used to train non-pilot aircrew in anti-submarine warfare (ASW) operations.

Components. It comprises three rear crew trainer cabins (RCTCs) and three additional cabins designated as procedure trainers (PTs). Each RCTC consists of a replica of a Sea King rear cabin, and the other cabins contain the computer, instructor's console and simulation equipment.

Conditions Included. Environmental factors such as sea state, tidal and bathythermal conditions and bottom topography are all carefully modelled. Similarly, flight profiles, weapon trajectories and ship and submarine behaviour characteristics are all accurately generated according to type.

Control of Training. Training sessions are controlled by the instructors either from VDU-based consoles or from a portable local control panel. Each exercise is set up from the instructor's console which has facilities for monitoring the complete tactical situation and injecting faults into the system. The local control panel may be used within the rear cabin for more direct control and closer monitoring of trainee performance.

Networking with Dynamic Simulator. Each RCTC is capable of linking up to a dynamic cockpit simulator to form an integrated training system. This will enable three teams to operate independently, allowing individual levels of training, or jointly to permit full mission training.

Other Platforms. Although the system has been designed specifically to train Sea King crews, similar designs may be used to train crews on any airborne ASW platform and they have already been delivered for training P-3C Orion, Lynx, CH-124, SH-76 and Sea Hawk crews.

The Marconi Sea King rear crew trainer upgraded to Mark 6 standards

Status
In service with the Royal Navy at RNAS Culdrose. Similar equipment is also in service with the armed forces of Canada, Egypt, India and Qatar.

Manufacturer
Marconi Simulation and Training, Stockport.
(For details see Directory of Suppliers)

Marconi Simulation Airborne Early Warning Helicopter Team Trainer

The Marconi Simulation AEW helicopter team trainer provides training for the observers of Searchwater radar-equipped Sea King AEW Mark 2A helicopters. It is housed in a self-contained, air-conditioned, transportable container.

Training. The trainer provides operators with a comprehensive radar, ESM, navigation and communications environment.

Components. In addition to the trainee crew facility the container also incorporates an instructor's console area, an office and a debrief facility.

Status

Available to customer specification. In service at RNAS Culdrose, Cornwall.

Manufacturer

Marconi Simulation, Dunfermline.
(For details see Directory of Suppliers)

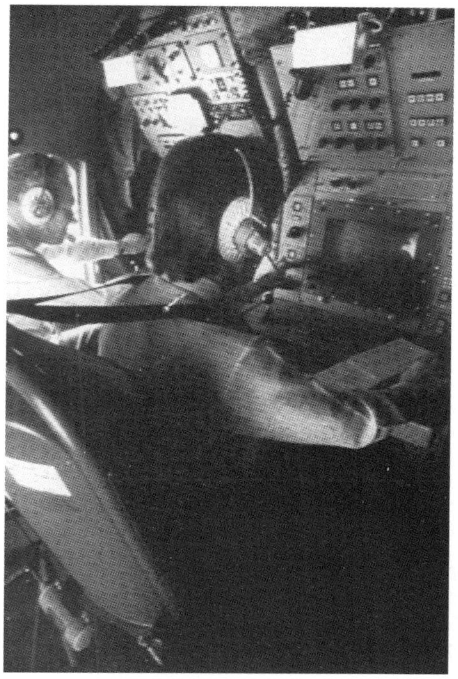

Interior of Marconi Simulation AEW helicopter team trainer

Marconi Simulation Maritime Crew Trainers (MCT)

Marconi Simulation produced the Nimrod MR Mark 2 Maritime Crew Trainer (MCT) which, when coupled by an Ethernet link to the TTS (ex-Rediffusion) Flight Simulator was the first fully integrated Full Mission Simulator in the Royal Air Force. The MCT trains the 'tactical crew' (that is the crew concerned directly with ASW operations except the pilots) simultaneously in a fully interactive operational environment.

Training. The MCT has accurate simulations of the acoustics, radar, electronic warfare, communications, navigation and weapon systems in real-time training scenarios of various levels of complication.

IOS. A comprehensive instructional, record/replay and debriefing facility is provided. Tactical scenarios can be developed for any large scale geographical exercise area under varying weather, sea state and oceanographic conditions. A wide range of airborne, surface or subsurface targets may be introduced which are controllable to provide trainees with realistic tactical scenarios.

Status

In service with the Royal Air Force.

Manufacturer

Marconi Simulation, Dunfermline.
(For details see Directory of Suppliers)

Part of a Nimrod MR Mark 2 Maritime Crew Trainer (MCT)

Pennant Air-to-Air Refuelling Simulator

Pennant International produces simulators for Air-to-Air-Refuelling (AAR) tanker operations. The simulators replicate the tanker AAR fuel dispensing operator stations.

Simulator of Rearward-Looking TV. Pennant has installed a TV simulator on the RAF Tri-star and VC10 AAR simulators. This is an interactive replica closed-circuit television (CCTV) system emulating the aircraft system which shows the refuelling operator the aircraft being refuelled.

IOS. There is also an instructor's station for creating scenarios, introducing faults and emergencies and monitoring the students. The exercise scenarios provide an interactive and authentic representation of the tanker and the AAR system operation throughout all flight and ground conditions.

Status

In service in RAF Tristar and VC10 tanker aircraft. An AAR training needs analysis (TNA) has been completed for another air force.

Manufacturer

Pennant International Limited, Cheltenham.
(For details see Directory of Suppliers)

The Pennant AAR Simulator showing the refuelling operator's switch panels

UNITED STATES OF AMERICA

AAI Undergraduate Naval Flight Officer (UNFO) Radar Training System

To provide the US Navy with air-to-air and air-to-ground training with the AN/APG-66 NT radar the AAI Corporation has provided, under contract to Saberliner, the Undergraduate Naval Flight Officer (UNFO) radar training system.

Databases. DMA DLMS Level I and Level II land mass (DTED) and feature (DFAD) data is used to make predictions of real-world radar returns. In turn these predictions are used to provide simulated radar video which also takes into account weather, moving targets and EW effects. Coverage includes more than 1 million nm² of land mass covering the eastern part of the United States.

Design. UNFO incorporates a software-intensive simulation approach to digital radar simulation. It consists of VME-based commercial off-the-shelf hardware and AAI software. A colour graphic high resolution monitor provides instructor interfaces. UNFO has separate air-to-air and air-to-ground configurations.

Status
In service.

Manufacturer
AAI Corporation, Hunt Valley.
(For details see Directory of Suppliers)

AAI Undergraduate Naval Flight Officer (UNFO) Radar Training System

CAE-Link Anti-Submarine Warfare Trainers

CAE-Link provides anti-submarine warfare training systems for the US Navy and also for the Canadian armed forces. The systems include acoustic and non-acoustic sensor operator trainers for individual training together with team trainers. These programmes include the following:

P-3C Tactical Operational Readiness Trainers (TORT)
Designated Device 2F140(T) these systems are designed to provide integrated team training for the tactical crews of the Lockheed P-3C Orion. Training is provided in a high density acoustic/RF environment.

The trainer is a facsimile of the anti-submarine warfare compartment and uses actual hardware. There is also an instructor's station divided into non-acoustic, acoustic, tactical/navigation and pseudo-pilot areas. The simulated operational area is 1326 km² and up to 22 860 m deep. Up to 35 targets, including eight acoustic, are available at any one time while up to 31 possible radar contacts are available together with up to 12 radar jammers.

CAE-Link has provided four TORTs to the USN, and is providing an acoustic generator to Mitsubishi Precision Company of Japan for that company's P-3C Operational Flight Tactical Trainer (OFTT) which is based upon the P-3C Update III TORT.

S-3 Position Trainer
Designated Device 14B49A, this is a two station, six position trainer for the Lockheed S-3A Viking. The stations simulated are those of the co-pilot, the tactical co-ordinator and the sensor operator each of which is triplicated for multiple training. The system is designed to provide each crew member with individual skill levels. Recent enhancements include increased acoustic and electronic support measures (ESM) capabilities, and the addition of a Harpoon anti-ship missile system capability.

A sensor station in a Device 2F140(T) TORT

S-3B WST Trainer
Designated Device 2F92A, consists of a tactics and flight trainer which simulates the pilot, co-pilot, tactical co-ordinator and the sensor operator stations of the S3B aircraft. The system is designed to give team training in systems and mission aspects to the whole crew.

P-3C ASW Tactics Trainer
This system is designated Device 2F87(T) by the US Navy. The system simulates the search radar, acoustics, ESM, magnetic anomaly detection, navigation and communications systems. It is designed to provide crew members of the Lockheed P-3C Orion with complete tactical training in the anti-submarine warfare (ASW) mission.

CP-140 Operational Mission Simulator
This system is similar to the tactical trainers used to train US Navy P-3 and S-3 aircrew but has been adapted to incorporate specific Canadian systems as used in the CP-140.

P-3 Acoustic Target Generator
CAE-Link provided an acoustic target generator which was integrated into the third P-3 Operational Flight Tactical Trainer being built by Mitsubishi for the Japan Maritime Self-Defense Agency.

The generator provides training in a high density acoustic/RF environment for P-3 tactical crew members. It simulates the real-world acoustic signals from sonobuoys used to detect, identify and track submarines. The Japanese trainer is based upon the P-3C Update III TORT. The system was delivered in May 1992.

Status
Device 2F140(T) is operational at NAS (Naval Air Station) Barbers Point, Hawaii; Brunswick, Maine; Jacksonville, Florida; and Moffet Field, California.

Devices 14B49 and 2F92A are operational at NAS North Island, California and Cecil Field, Florida.

Manufacturer
CAE-Link Corporation, Binghamton, New York.
(For details see Directory of Suppliers)

ESCO AN/APQ-167 Airborne Radar Training System

The AN/APQ-167 airborne radar training system is installed in a Cessna T-47 Citation to train undergraduate Naval Flight Officers. A 100 kW J-band radar is used which includes signals which simulate air-to-air or air-to-ground engagements. The system weighs 95.25 kg.

Aircraft Components. The aircraft has two student and one instructor stations. An antenna change allows a change from air-to-air to air-to-ground training (or vice versa) within 30 minutes.

Air-to-Air. For the air-to-air mission the system has a simulated detection range of 74 km and a lock-on range of 55 km. AIM-9 Sidewinder short range and AIM-7 medium range missile engagements can be simulated together with gun engagements.

Air-to-Ground. In the air-to-ground role the system has a simulated 200 ft (61 m) resolution. The system has a ±45° search/map mode and a ±20° sector scan/sector map mode.

Status
In service at the Naval Air Station Pensacola since 1983.

Manufacturer
Electronics and Space Corporation (ESCO), St Louis, Missouri.
(For details see Directory of Suppliers)

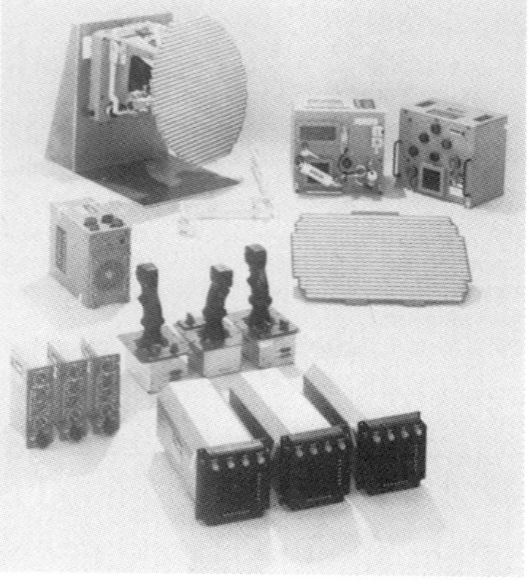

The elements of the AN/APQ-167 radar training system

Hughes GBU-15/AGM-130 PTT

This PTT is designed to give airborne weapon systems officers (WSOs) practice in operating the GBU-15 and AGM-130 stand-off weapon systems. The student station is a replica of the rear seat of an F-4E and contains all the relevant controls and instruments. Simulated video from the weapon is presented on a small CRT and a simulated integrated hand controller is used for guidance. Visual presentations are generated by a PhotoVision 3 CGI photo-based system. The database and processor are both expandable.

Status

In service with the US Air Force.

Manufacturer

Hughes Training Inc, West Covina, California.
(For details see Directory of Suppliers)

The HTI GBU-15/AGM-130 part-task trainer

Martin Marietta Target Acquisition Designation Sight (TADS) Selected Task Trainer (TSTT)

The TADS Selected Task Trainer is designed to help train student co-pilot/gunners in AH-64 Apache attack helicopters in the use of the Target Acquisition Designation Sight (TADS). Multi-Target Detector and Automatic Tracker (MTDAT) software is included.

The TSTT consists of a replica Apache co-pilot/gunner station. It supports training in all phases of TADS operation from power-up to system shut down.

Status

In service with the US Army.

Manufacturer

Martin Marietta Electronic Systems, Florida.
(For details see Directory of Suppliers)

Trainer for the Target Acquisition Designation Sight

McDonnell Douglas Training Systems C-17 Globemaster Loadmaster Training

MDTS was awarded a $420.99 million contract in October 1988 to design, to build and to operate an aircrew training system for the C-17 Globemaster transport.

The system will include classroom instruction and self-paced CBT on 86 interactive computer terminals equipped with video disks. The bulk of the instruction will be given on 12 Full Flight Simulators which are subcontracted to FlightSafety International (FSI) and are described in the FSI entry in the Flight Simulator section of this book.

A full-scale cargo compartment trainer will allow training of loadmasters on the various C-17 cargo compartment features without having to use the aircraft. It will also allow practice loading of cargo or vehicles and practice in the reconfiguration of the cargo compartment. The cargo load model will allow cost-effective training in load planning for weight and balance. It is similar to the device used for training KC-10 loadmasters and boom operators and enables students to visualise critical clearances within the cargo compartment.

The training systems will be installed at five US Air Force bases, two Air Force Reserve locations and two Air National Guard bases. Altus AFB, Oklahoma, will serve as the central training base with five weapon system trainers, the cockpit systems simulator, the cargo compartment trainer and the cargo load model while Charleston AFB, South Carolina, will be the first base to have operational C-17s. The first simulators were installed at Charleston in 1991 and training began in August 1992. Simulators will be delivered to Altus from 1993 and deliveries of all the simulators will be complete in 2003.

Status

See individual entries.

Manufacturer

McDonnell Douglas Training Systems (MDTS), St Louis, Missouri.
(For details see Directory of Suppliers)

Merit Radar Simulators

Merit Technology produces a number of radar simulators for use in US Air Force and US Navy aircraft.

AN/APQ-174 Radar Simulator.
This simulator is used in the CAE-Link Special Operations Aviation Combat Mission Simulators (SOACMS) to train crews in covert, long-range missions using terrain-following and terrain-avoidance techniques with the AN/APQ-174 radar. It consists of a single-bay rack and a work station and simulates the performance of the radar. Training options include a variety of simulated radar failures and malfunctions which may be inserted. The visual 80 × 100 km tactical database has been converted to a 12.5 × 12.5 km gridded radar database to ensure correlation between the visual scene and the radar simulator.

P-3 Simulators.
Merit produces a number of modules for P-3 training. Their weather/land mass radar trainer is used in Devices 14B40(A) and 2F140(T). The system simulates the performance of the AN/APS-115 ocean surveillance radar and provides realistic training in weather avoidance and land mass navigation as well as accurate target/IFF modelling of both surface and airborne targets.

The trainer consists of work stations each with a 19 in (483 mm) cabinet which contains the real-time radar simulation. Land mass and weather simulation is accomplished by using commercial off-the-shelf hardware and a VME-bus architecture. The land mass database is derived from Defense Mapping Agency (DMA) DLMS terrain and feature files (DTED and DFAD). Weather cell editing features allow the user to generate realistic weather databases for use by the dynamic weather environment trainer.

In December 1990 Merit was awarded two contracts worth a total of $9.7 million by the US Navy to provide radar and Forward-Looking Infra-Red (FLIR) simulators for incorporation in the CAE-Link P-3 Update IV simulator. The radar simulator will emulate all operational modes of the AN/APS-137 radar including land mass and inverse synthetic aperture radar capabilities. The FLIR simulation is to train operators in the use of the AN/AAS-36 IR detection system.

Status

In service and in production.

Manufacturer

Merit Technology Inc, Plano, Texas.
(For details see Directory of Suppliers)

AIRBORNE TRAINERS - NON-EW

This type of training is sometimes known as embedded training. Often the principle involved will be that of STimulation or triggering of an aircraft's actual sensor or other systems in flight, rather than a computer-based Emulation of systems as is done in most ground-based flight simulators and training devices.

AUSTRALIA

HDH Harpoon Captive Carriage Weapons Simulator (CCWS) and Stand-off Weapons Active Training Simulator (SWATS)

The Harpoon Captive Carriage Weapons Simulator (CCWS) is an inert training aid which simulates the AGM-84 Harpoon anti-ship missile for practising maritime strike operations. It is carried by the F-111C, P-3 and the F/A-18 and simulates the Harpoon both aerodynamically and electronically so that it may be used for both ground crew and pilot training.

The Stand-off Weapons Active Training Simulator (SWATS) is being developed as an advanced derivative of CCWS. It will incorporate an active microwave seeker head and solid-state recording system.

Status
CCWS in service with the RAAF.
SWATS under development.

Manufacturer
Hawker De Havilland Limited, Salisbury.
(For details see Directory of Suppliers)

HDH Harpoon Captive Carriage Weapons Simulator and Stand-off Weapons Active Training Simulator

UNITED KINGDOM

GEC Airborne Sonar Crew Trainer (ASCT)

The Airborne Sonar Crew Trainers (ASCT) were designed by GEC Avionics to produce synthetic acoustic signals as if from sonobuoys and dipping sonar. The synthetic signals are used to stimulate the equipment in a maritime aircraft itself in order to give 'embedded training' in the air. Synthetic target and environmental data is processed by the aircraft systems and aircrew react to the synthetic environment by employing normal procedures and tactics. Airborne training is carried out without the need to deploy real sonobuoys or dipping sonar. There are two versions:

ASCT 1. ASCT 1 is a software-based system developed specifically for the GEC Avionics AQS 901 acoustic processing system fitted into the Nimrod MR2 aircraft of the RAF and the P-3C Orion aircraft of the RAAF. It comprises an Exercise Control Unit (ECU) and a computer program which is loaded into the AQS 901 from a magnetic tape unit. The simulated target position, course, speed and depth are keyed in to the ECU which can also control subsequent target manoeuvres and ocean conditions. The system then runs automatically during flight.

ASCT 2. ASCT 2 is a hardware-based system consisting of an ECU and an ASCT simulation unit (ASU) which provides ASW scenarios and generates acoustic data from simulated sonobuoys or dipping sonars. A typical ASU weighs 12 to 16 kg for an 8- and 16-channel system. There are two types of ECU; a small hand-held unit weighing 3 kg and a laptop weighing 6 kg. It is supported by the exercise support system (ESS), a PC-based software program; a printer is used to record the exercise scenarios and progress.

Status
ASCT 1 is in service with the Royal Air Force and the Royal Australian Air Force. An Acoustic Simulation System (ASS), based upon ASCT, was delivered to Boeing Defense & Aerospace for the P-3 Update IV programme at the US Naval Air Warfare Center, Patuxent River, Maryland.

Manufacturer
GEC Avionics Limited, Maritime Aircraft Systems Division, Rochester.
(For details see Directory of Suppliers)

UNITED STATES OF AMERICA

Cubic AN/USQ-T2(V) ACMI System

Cubic produces TACTS (Tactical Aircrew Combat Training System) or ACMI (Air Combat Manoeuvring Instrumentation).

Components. The system consists of the Aircraft Instrumentation Subsystem (AIS) pod, the Tracking Instrumentation Subsystem (TIS), the Control and Computation Subsystem (CCS) and the Display and Debriefing Subsystem.

System Operation. The AIS is attached to the aircraft, records flight and weapons data and transmits it to the ground. This data is received by the TIS which relays it to the CCS where a master computer compiles, computes and records it. The data is then presented on multicolour video displays by the Display and Debriefing Subsystem while the monitored flight is in progress and later will be replayed during debriefing.

36 Aircraft Capability. TACTS/ACMI can track up to 36 aircraft simultaneously, recording their location throughout the exercise. Consequently mock dogfights and bombing missions can be flown without using live ammunition.

The Cubic AIS (Aircraft Instrumentation Subsystem) ACMI pod fitted on an F-15. Since 1987 Kollsman has been producing similar pods as the P-4A

Production Status

Cubic has produced 21 systems with the first becoming operational in December 1973. TACTS is in service with the US Navy and US Marine Corps at the Oceana and Yuma ranges as well as NAS Fallon. The ACMI serves with the US Air Force and Air National Guard at Eglin, Holloman, Luke, Nellis and Tyndall Air Force Bases, with US Air Forces Europe (USAFE) at Decimomannu Air Base, Sardinia, and with Pacific Air Forces Command at Kusan Air Base, Korea (shared with the South Korean Air Force) and Okinawa.

In August 1987 Cubic was awarded a $74 million contract to supply four ranges which will be installed at the US Marine Corps Air Stations at Cherry Point, North Carolina, and Beaufort, South Carolina, Homestead AFB, Florida, and the Air National Guard Field Training Site at Gulfport, Mississippi. These systems will use Concurrent Computer Corporation Models 3250 and 3260MPS computers. Cubic has supplied the Combat Maneuver Training Center (CMTC) to the US Army at Hohenfelds, Germany, and is installing the Air Warrior System at the National Training Center, Fort Irwin, California. This is the first to integrate air combat/close air support and ground training and will become operational late in 1993. Applied Data Technology is expanding the Beaufort TACTS which will be integrated with the US Navy's Pinecastle Range in Aston, Florida.

Export Orders. The system was exported to Canada in 1981 (CFB Cold Lake) and to the Royal Thai Air Force. In October 1988 British Aerospace placed a $50 million contract with Cubic for TACTS, using Model 3200 computers, for the BAe air combat range which it operates in the North Sea between England and the continent of Europe and is used by the Royal Air Force and several other NATO air forces.

Manufacturer

Cubic Defense Systems, San Diego, California.
(For details see Directory of Suppliers)

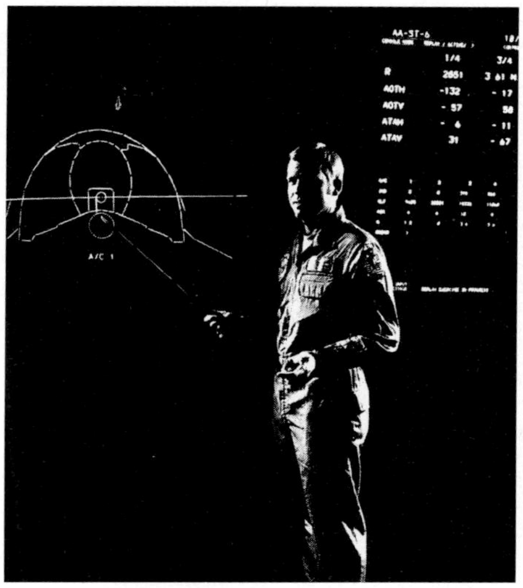

An instructor reviews data from Cubic's ACMI Air Combat Training System

Loral Laser-Guided Training Round (LGTR)

The Laser-Guided Training Round (LGTR) was designed to complement training with live rounds. It consists of a lightweight, low-cost, inert round which uses semi-active laser and pneumatic actuation. It is 1.9 m long, 101.6 mm in diameter and weighs 40.37 kg.

The LGTR is used instead of the live round for loading, airborne release and designation training and may be used to enhance designation timing and accuracy. A smoke/flame marker simulates the impact of a live missile on the target.

Status

In production with 1800 units being ordered by the US Navy for training A-6 and F/A-18 crews.

Manufacturer

Loral Control Systems, Archibald, Pennsylvania.
(For details see Directory of Suppliers)

NSWC Airborne Training Missiles

The Indian Head Division of the Naval Surface Warfare Center (IHD/NSWC) produces a number of airborne training missiles which represent AGM-84 Harpoon, Stand-Off Land-Attack Missile (SLAM), and AGM-119 Penguin.

Captive Air Training Missile (CATM) versions of Harpoon (CATM-84D) and Penguin are produced, the former in instrumented and uninstrumented versions. For Harpoon and for SLAM there are dummy air training missiles and jettison test vehicles.

Air training missiles are inert training shapes which replicate the external appearance, form, fit, weight and centre of gravity of the real missiles. They are carried by aircraft and used in both classroom and fleet training exercises to train air crews carrying the missile in tactical missions. They are also used to train ground crews in loading, unloading and handling the missile. The breakout, assembly, strike up, strike down, loading, unloading, stowage and organisational maintenance of the training missile is identical to that of the real missile.

Status

These airborne training missiles are produced for the US Navy.

Manufacturer

Indian Head Division, Naval Surface Warfare Center, Indian Head, Maryland.
(For details see Directory of Suppliers)

EW TRAINERS FOR AIR SYSTEMS

CANADA

Innotech Electronic Support Training

Innotech teamed with Canadian Marconi and Litton Systems Canada in the Canadian Airborne Electronic Support and Training (EST) programme. This is based on Canadair CC-144 Challenger aircraft and entered service with the CAF during 1988. The CC-144 is equipped with ELINT, COMINT, active ECM, and chaff dispensers in order to train operators of ground, air and shipborne systems.

Status
In service with the Canadian armed forces.

Manufacturer
Innotech Aviation, Dorval, Québec.
(For details see Directory of Suppliers)

One of seven CC-144 Challengers used in the CAF Electronic Support and Training programme

FRANCE

Dassault Electronique MESA EW Trainers

MESA stands for *Modelisation de l'Environnement et du Système d'Autoprotection* and is a family of interactive EW simulators for modelling the electronic characteristics of weapon systems emissions and onboard self-protection systems either in an all-digital or hybrid (digital and RF) context.

First an unclassified generic model is produced of the appropriate weapon system, ship or aircraft self-protection systems, together with operational scenarios and appropriate terrain factors. The basic model is then enhanced from a more refined database containing classified details of the systems in question.

The device is a stand-alone system using a PC running under the UNIX operating system.

Up to 100 simultaneous threats and up to 10 ESM/ECM platforms can be simulated. For qualification, tests or equipment acceptance, hybrid configurations can also be produced.

Status
Available.

Manufacturer
Dassault Electronique, Saint-Cloud.
(For details see Directory of Suppliers)

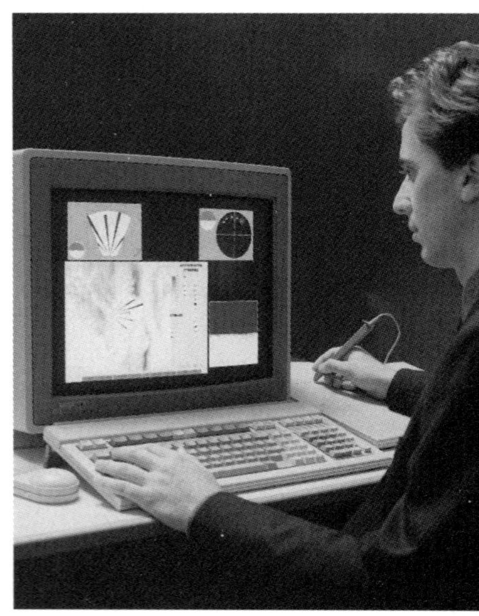

Dassault MESA EW system being used to evaluate a self-protection system

ISRAEL

Elisra EW Systems

EWT-100
The EWT-100 is a computer-controlled EW trainer. The student pilot operates the trainer on a simulated combat mission. The trainer may be used for teaching theory, tutorials, system operation training and responses, as well as mission training.

Elisra NS-9002/NS-9002R
The NS-9002 is a vehicle-mounted self-contained EW simulation system. It can be employed for system evaluation, maintenance and operator training. It is capable of generating any known electromagnetic environment both at digital and RF levels.

Emitters. The system generates up to 256 independent emitters in the 0.5 to 18 GHz range (C to J-bands), with an extended frequency range available. All emitter parameters are stored in a library which is easy to access. The NS-9002 features multi-emitter generation up to 2 million pulses/sec, RF transmissions being made through antennas or direct coupling to the system under test. Dynamic variation of emitter parameters is possible and software is designed to be easy to use.

NS-9002R. The NS-9002R is used for the evaluation and maintenance of radar systems as well as for operator training. It includes a receiver/transmitter and a processor which enables reception and synchronisation of the radar signal and transmission of targets as well as ECM jamming and deception signals. The simulator can work with different radar types and modes.

Elisra REAS (Realistic Electromagnetic Arena Simulator)
REAS is a tool for testing the reception and identification capabilities of aircraft Radar Warning Receivers (RWR). The system simulates CW or pulsed RF signals between 0.7 to 18 GHz as would be received by the antenna of an aircraft RWR. The system provides a three-dimensional simulation of the aircraft flight path which includes radiation patterns of transmitting and receiving antennas, antenna scan patterns and various modes of operation of scan and control radars. Up to 128 emitters may be simulated in any one scenario.

REAS can be programmed to simulate a number of emitters by determining their location relative to the position of the imaginary aircraft and their parameters. The parameters of one or more emitters may be altered to enable the customer to examine the identification capability of the system under test, to correct the definition of the threat target or to check target separation.

Status
In production and service.

Manufacturer
Elisra Electronic Systems Limited, Bene Beraq.
(For details see Directory of Suppliers)

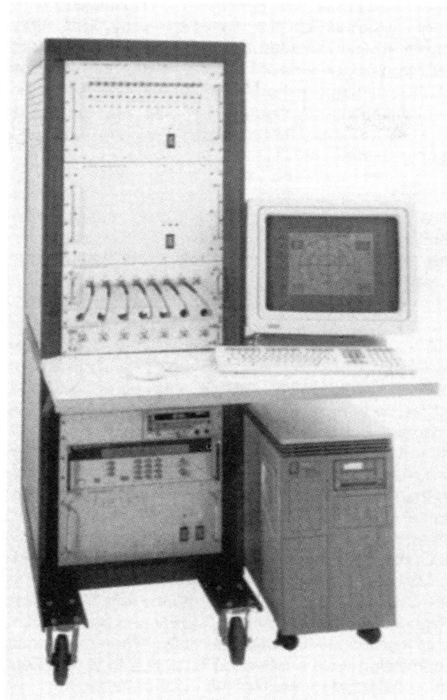

The Elisra Realistic Electromagnetic Arena Simulator (REAS) for use with aircraft RWRs

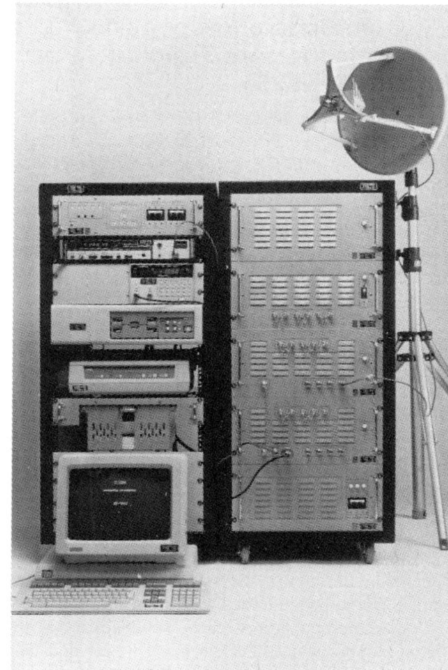

Basic Modules of the Elisra NS-9002 EW System

Rafael Electromagnetic Pulse (EMP) Devices

EM-101. The EM-101 transient generator system is used for testing subsystems including various current and voltage waveforms. Typically they are subjected to damped sine wave signals at frequencies which range from 10 kHz to 100 MHz.
EM-104. The EM-104 is used for testing long cables which are used for communications and power. Currents of thousands of amperes can be generated and this system can be used for testing on complex systems such as those in bunkers and communications centres.
Testing Complete Equipment Systems. Various systems are available for testing complete systems such as aircraft. The bounded wave simulator generates a 100 kV/m electric field and 5 ns rise time for the pulse. The simulators have high voltage generators which are rated up to 2 million volts.

Status
Available.

Manufacturer
Rafael, Haifa.
(For details see Directory of Suppliers)

SWEDEN

Ericsson Erijammer A100 and Erijammer B100

Erijammer A100
The Ericsson Erijammer A100 is a high band jammer pod or internally mounted system used for training fighter interceptor crews and ground-based air defence forces. It may be either manually or automatically controlled. The system consists of two units, a control panel together with a pod featuring built-in antennas and electronics.
Pod. The pod is self-contained, requiring only power from the aircraft, with single or multi-threat capabilities derived from a frequency-memory loop, a set-on receiver and selectable bandwidths. It also features a built-in radar warning receiver, smart noise and 45 deception modes which vary in range, angle and velocity.
Operation. The ECM operator controls the system by selecting one of three 120° antenna lobes. The analysis and subsequent jamming of incoming signals is initiated either manually or automatically. A frontal antenna with a higher gain and a narrower beam can be chosen for specific missions such as escort jamming and radar simulation.

Specifications
Frequency coverage: G, H, I and part of J-bands
Output power: 350 W
Noise bandwidth: 10-1000 MHz (wideband), 2-100 kHz (Doppler)
Antenna coverage: (horizontal) 360°; (vertical) ±20°
Polarisation: circular (or front linear)
Cooling: ram-air
Length: 3.23 m
Diameter: 426 mm
Pod weight: 210 kg
Deployment. In service in Canada and Switzerland.

Ericsson Erijammer A100 training pod on a Pilatus PC-9

Erijammer B100
The Erijammer B100 is a training low band jammer system developed by Ericsson and Rodale Electronics Inc for ECCM training of search radars in the D to G bands. It is designed to train radar operators in surface-to-air weapons or air-to-air weapons elements of air defence.

The system will be housed in an A100 pod shell, or internally, and will feature Doppler, smart and spot noise together with false targets. There will be 360° full antenna coverage, with selectable forward high and low gain antennas, and set-on receivers.

Specifications
Frequency coverage: D, E, F and G-bands
Output power: 350 W
Noise bandwidth: 5-500 MHz (wideband), 0-100 kHz (Doppler)
Antenna coverage: (horizontal) 300°; (vertical) ±70°
Polarisation: circular (or front linear)
Cooling: ram-air
Length: 3.23 m
Diameter: 426 mm
Pod weight: 200 kg

Manufacturers
Ericsson Radar Electronics AB, Stockholm.
(For details see Directory of Suppliers)
The B-100 is made in conjunction with Rodale Electronics Inc. of Garden City, New York State, USA.
(For details see Directory of Suppliers)

Ericsson/Rodale Responsive Electronic Warfare Training System (REWTS)

Ericsson and Rodale Electronics Inc of New York State, USA, have developed a Responsive Electronic Warfare Training System (REWTS) for training electronic warfare officers (EWOs).

The system is designed to provide realistic scenarios so the EWO and/or the pilot can learn to use his radar and electronic warfare equipment effectively. A built-in RWR and set-on receiver are provided. The instructor knows the location of the aircraft, the student's reaction to jamming and his use of the radar's ECCM capabilities. The system is offered with Ericsson's Erijammer A100 and B100 as well as Rodale's 8102 Smart Crow or DARTS.

Status

Available.

Manufacturers

Ericsson Radar Electronics AB, Stockholm; and Rodale Electronics Inc, Garden City, New York, USA.

(For details see Directory of Suppliers)

UNITED KINGDOM

Aware Simulation Portable Emitter Simulators

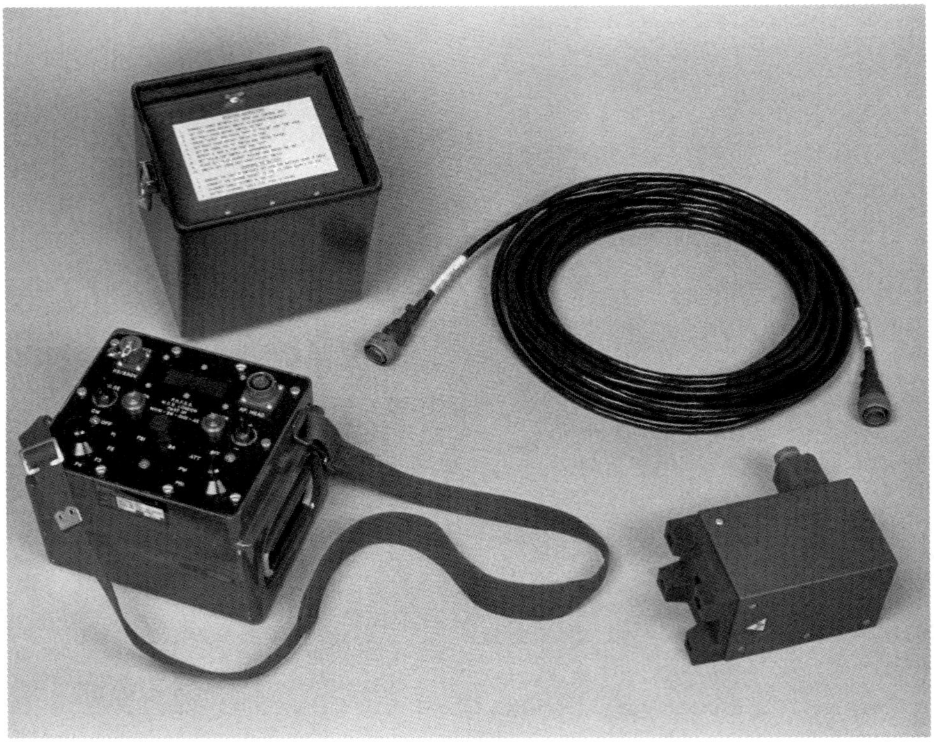

Aware Simulation Limited P200 emitter simulator

P200

The Aware Simulation Limited P200 Portable Emitter Simulator is a manportable, battery-operated, wideband emitter simulator designed for first-line non-interference testing of aircraft electronic warfare systems.

Four frequencies may be selected while the pulse width, pulse repetition frequency, mode and level may be programmed by the operator. The unit includes an integral charger and BITE. The test antenna and the control-processor unit may be separated by distances of up to 20 m, allowing operation of both the P200 and the electronic warfare system by a technician working in the cockpit.

PM200

The PM200 is a manportable power meter designed for use with the P200 Portable Emitter Simulator but can be used whenever microwave power measurement is required. The power sensor and processor may be up to 20 m apart, thus allowing operation of both the PM200 and the aircraft EW system by a maintainer working in the cockpit.

Status

In service with the Royal Air Force.

Manufacturer

Aware Simulation Limited, London; marketing and support is by Reflectone UK Limited.

(For details see Directory of Suppliers)

Data Sciences Evaluation Models for Training

Data Sciences Limited produces three software evaluation models which can be used within a classroom environment by instructors to assist in the demonstration of concepts and/or tactics to students. All run on off-the-shelf work stations and have realistic threat scenarios, easy instructor interfaces and graphic representation for ease of understanding by the student.

Flexible Attrition Model (FAM)

FAM is a generic model to simulate the land/air battle. It simulates the interaction between airborne systems and ground-based air defence systems to help assess the survivability of aircraft in hostile environments. Digital terrain maps can be read in the system to provide accurate engagement assessments.

Electronic Combat Modelling and Evaluation System (ECMES)

ECMES is a tactical level model which simulates all elements of modern electronic combat at a pulse-by-pulse level of detail. It can fully simulate ESM systems, surveillance and tracking radars, airborne intercept radars, missile seekers, electronic warfare control processors, ECM systems, decoys and weapons.

Electronic Warfare Evaluation System (EWES)

This is an engineering level model which provides a totally integrated ESM evaluation system. It enables the instructor to create and record three-dimensional scenarios in which he specifies platforms and their emitters so he can analyse ESM system performance against them.

Status

In service with the UK MoD and overseas authorities.

Manufacturer

Data Sciences Limited, Farnborough, Hampshire.

(For details see Directory of Suppliers)

Marconi Radar Environment Simulator

The Marconi Radar Environment Simulator was designed to provide a controlled, dynamic environment for training Electronic Counter-CounterMeasures (ECCM) operators. It provides them with a realistic ECM training environment in the form of IF signals to the radar sensor.

The environment is controlled by the user so that exercises may be repeated. However, it may also be changed by preplanning or through interactive use with the student.

Status

In service with the Royal Air Force and other NATO air forces since 1986.

Manufacturer

Marconi Radar and Control Systems Limited, Chelmsford, Essex.

(For details see Directory of Suppliers)

Marconi Simulation 8502 Electronic Warfare Scenario Generator

The Marconi Simulation 8502 Electronic Warfare Scenario Generator is a combination of RF hardware, computer processing and simulation software, all brought together to provide an EW validation facility.

System Testing. For system testing, RF and microwave signals from the 8502 provide dynamic testing. Combinations of scanned, pulse-modulated or continuous wave signals can be downloaded from a library of threats into the system under test, and the responses analysed to determine the serviceability or accuracy of EW systems.

Scenario Generation. The 8502 EW scenario generator also provides the means to validate new mission software programs by subjecting a fully representative EW system to an electromagnetic environment based on the postulated scenario.

Dynamic Simulation. Dynamic simulation is performed using fly-through, three-dimensional scenarios in real time. The response analysis capability of the simulator enables the operator to determine the level of functionality available with the system involved.

Operator Training. For operator training the 8502 EW scenario generator provides the user with the facility for operator and tactical analysis training. The EW prime equipment being simulated can be subject to scenarios representing the dense and constantly changing environment normally experienced during times of tension, providing the user with the opportunity to create a battle plan in real time which may be re-enacted over periods of hours or days, as required. In addition the scenario generator may be used to simulate the operational equipment, thereby exercising the electronic counter-countermeasures (ECCM) circuitry and providing the operator with the opportunity to become experienced in the techniques of signal detecting, analysis, jamming and deception, using the prime equipment controls.

Status

In production. In service with the United Kingdom armed forces. RF generators have been supplied for the development trials of the Tornado Radar Homing and Warning Receiver (RHWR), and the Harrier GR5 Zeus electronic countermeasures (ECM) system.

Marconi Simulation 8502 electronic warfare scenario generator

Manufacturer

Marconi Simulation, Dunfermline.
(For details see Directory of Suppliers)

Marconi Simulation Electronics Analysis Systems Trainer

The Electronics Analysis Systems Trainer developed by Marconi Simulation for the Royal Air Force provides operators with instruction and training in the skills needed to operate support systems. The trainer is a portable self-contained unit and is capable of simulating an EW scenario as defined by an instructor. Controls and displays are provided to enable a student operator to analyse the scenario displayed with reference to data stored in the system library.

Operator Training. The training system develops the operator's ability to identify platform emitters used in airborne, shipborne and land-based systems and correlates this information with known emitter platforms and platform threat data. It gives cost-effective signal analysis training, and provides ESM operators with training prior to their introduction to a prime equipment.

Status

Available. In service with the Royal Air Force.

Manufacturer

Marconi Simulation, Dunfermline.
(For details see Directory of Suppliers)

The basic Marconi Simulation Electronics Analysis Systems Trainer

Reflectone UK Talisman II Electronic Warfare Training System

The Reflectone UK Limited Talisman II Series of EW Training Simulators has been designed for the training of *ab initio* students in EW principles.

Talisman E. Talisman E is for non-communications EW, such as radar warning receivers (RWRs), ESM and ELINT training.

Talisman C. Talisman C is for communications EW, such as communications ESM and COMINT training. For advanced ESM/ELINT operator training the core system may be used to drive an RF signal generation extension with cable-coupled connection to stimulate operational in-service electronic warfare equipment.

Components. Each Talisman classroom system consists of an instructor station, master fileserver/network controller and up to 10 PC-based student consoles based upon Intel 80386/80486 microprocessors. The main fileserver is VAX work station-based with all software written in VAX Ada while student console software is written in 'C'. The consoles provide representative displays and controls of modern EW systems and may be customised to match in-service equipment where necessary. High definition colour graphics are used with signal parameter measurement on real oscilloscope and/or spectrum analysers.

Signal Generation. Signal generation within the classroom is achieved at the video signal level for radar training and at full RF for communications training.

Training Scenarios. Although all training is conducted under instructor control the maximum independent student participation takes place and students with different skill and experience levels may train simultaneously and at their own pace. Multiple scenarios may be generated and held on hard disk, floppy disk or cartridge tape units which are reusable and may be edited at any time.

Enhanced Talisman E. An enhanced Talisman E has been produced for RF stimulation of mobile ESM systems for the British Army. It is housed in a transportable RF-screened and air-conditioned ISO container and drives 2 ESM systems simultaneously.

Status

In service and production.

Manufacturer

Reflectone UK Limited, Bristol.
(For details see Directory of Suppliers)

A Talisman E student console. Not shown is the audio generator

The elements of the Talisman E trainer

UNITED STATES OF AMERICA

AAI Onboard Electronic Warfare Simulator (OBEWS)

AAI produces EW trainers for the A-10, F-16 and EA-6B.
EW Trainer Pod. The OnBoard Electronic Warfare Simulator (OBEWS) provides in-flight EW training for the aircraft to which it is fitted. It simulates the cockpit RWR with up to 30 accurate threat signatures. The pilot can respond using electronic countermeasures, chaff/flare launchers and terrain screening causing the threats realistically to change mode, break radar lock-on or cease operation. The system records the situation for debriefing.
Components. The system consists of a ground support subsystem and an onboard subsystem which fits into the shell of the AIM-120A AMRAAM (Advanced Medium Range Air-to-Air Missile) and it is usually used on F-16s.
Ground System. The ground support subsystem is used to define the training area boundaries within a 409.6 km^2 area in which the electronic order of battle of up to 30 threats is located, with any eight active at one time. The data is then downloaded into an EPROM module which is inserted in the onboard subsystem. The latter not only generates the threats and records the pilot's reaction but also monitors the aircraft position.

Status
Available and in service.

Manufacturer
AAI Corporation, Hunt Valley, Maryland.
(For details see Directory of Suppliers)

Loading the AAI OBEWS on an F-16 - The pod used is the outer shell of an AMRAAM missile

AEL AN/FSQ-T22 Electronic Combat Simulator

The AEL Defense Corporation has developed the AN/FSQ-T22 electronic combat simulator. It consists of five subsystems; monitor and control, signal generation, antenna, evaluation and surveillance. It may be used for tactical scenarios with a variety of radars from D to J-band (1-18 GHz) with high pulse densities, 15 simultaneously per pulse band and up to 3 million pulses per second, and complex waveforms as well as scan patterns. The system has a library of 5000 emitters and can use 300 in each scenario with up to seven pulsed and three continuous wave bands.

Status
In service.

Manufacturer
AEL Defense Corporation, Lansdale, Pennsylvania.
(For details see Directory of Suppliers)

The monitor and control subsystem of the AN/FSQ-T22

Aydin AN/MST-T1A Multiple Threat Emitter System (MUTES)

The Aydin AN/MST-T1A Multiple Threat Emitter System (MUTES) is a transportable EW training centre designed to emit multiple signals under computer control. It is capable of simulating threats over the specified frequency spectrum with sufficient accuracy to activate radar warning receivers and ECM systems.
Components. The system consists of eight subsystems; tracking, antenna, transmitter/modulator, system control, control console, van/shelter, communications equipment and remote-control.
Frequencies. A variety of frequencies is covered by the transmitters from 0.67 to 15.4 GHz.
Antennas. The antenna group consists of dish antennas and feeds mounted on the transportable complex.
Training. MUTES models a variety of radars including early warning, height finding, ground control intercept (GCI), aircraft intercept, SAM acquisition/tracking, missile guidance and AAA radars. Computer control of the signal types and characteristics allows the operator to switch rapidly between radiated sources to enable a wide range of multiple threats to be simulated during a single exercise.

Status
In service.

Manufacturer
Aydin Corporation, San Jose, California.
(For details see Directory of Suppliers)

Cincinnati Electronic Combat Trainer (Device 15E34A)

The Cincinnati Electronic Combat Trainer (ECT) enables the crews of the EA-6B aircraft to fly in simulated hostile environments and to practise ECM techniques which the system computer automatically monitors.
Components. The ECT has six subsystems; signal generation, antenna, evaluation, monitor and control, built-in test, tracking and surveillance. It simulates 75 independent multiplexed pulse signals between the frequencies 500 MHz and 12 GHz with 15 signals in any band.
Training Scenarios. Five training scenarios, each with up to 300 emitters, are provided with the trainer. The operator can modify them or write new ones.
IOS. The instructor station monitors and controls the exercise in real time. It is used also to generate the mission scenario and interacts with it to modify emitter parameters and timed sequence of events.

Status
In service with the US Navy.

Manufacturer
Cincinnati Electronics, Ohio.
(For details see Directory of Suppliers)

Cross Systems 0-1843(V)/ ULQ-21(V) Waveform Controller Oscillator

This Cross Systems Inc system is designed to produce low power RF jamming signals for training and laboratory testing of radar systems. It can be pod-mounted or internally installed in both aircraft and unmanned aerial vehicles. A high power amplifier and antenna can be added to make up an ECM system.

System Control. The system is controlled by a microprocessor and can store up to 166 technique/parameter sets for a preplanned response or it may be controlled through an RS-232 standard interface.

Types of Transmission. It can generate various types of noise and deception techniques including continuous noise, blinking noise, noise with swept amplitude modulation, swept noise, repeater with swept amplitude, modulation and range gate stealer, range gate stealer with swept amplitude modulation and range gate stealer with swept amplitude hold. It may also be combined with a velocity deception techniques generator and set-on receiver to provide Doppler and fast set-on techniques.

Status

In production and service with the US Navy and NATO countries.

Manufacturer

Cross Systems Inc, Atlanta, Georgia.
(For details see Directory of Suppliers)

Cross Sytems Radar Return Environment Simulators (RRES)

The Cross Systems Inc RRES is a series of RF generators which provides a simulated real-world environment generating the radar return signals which a radar would encounter during operation. The systems provide realistic target returns, a complete array of programmable countermeasures and computer-controlled clutter generation.

Status

In production and in service.

Manufacturer

Cross Systems Inc, Atlanta, Georgia.
(For details see Directory of Suppliers)

Ford/Datron Threat Radar Emitter Simulator (TRES)

The Ford/Datron Threat Radar Emitter Simulator (TRES) is developed by Ford Aerospace and Datron Systems for training, test and evaluation applications.

Components. The system consists of unmanned field-locatable terminals which include antenna-mounted RF transmitter modules and power generation units. A central processing subsystem exercises command and control functions.

Each TRES configuration consists of control and display, RF emission, antenna beam pointing, diesel generator and equipment enclosure subsystems.

Operation. The system is designed to simulate multiple target track and missile guidance threat radars.

Status

In service at the US Naval Weapons Center, China Lake.

Manufacturers

Ford Aerospace, Sunnyvale, California, and Datron Systems, Simi Valley, California.
(For details see Directory of Suppliers)

Harris AN/MST-T1V Mini-MUTES Radar Simulator

This Multiple Threat Emitter System (MUTES) consists of a master control group and from one to five remote emitter units. It is used to simulate radar emissions which might be encountered in an integrated defence system.

Components. The remote emitter units can generate up to 10 realistic threat signals and each is able to acquire and track a co-operative aircraft and to direct emitter signals. Each is able to permit multiple IFF acquisition and tracking and can also record ECM and student response. They can process detailed electronic warfare data and then transmit it to the master control group.

Operation. The system is manned by a single operator who sets up the operational parameters and monitors the exercise and the status of the emitter units which may be up to 80 km from the master control group. They may be linked through a telephone system.

Status

In service with the US Air Force.

Manufacturer

Harris Corporation, Melbourne, Florida.
(For details see Directory of Suppliers)

Hercules Transportable Radar and Communications Simulators (TRACS) Vans

The TRACS is an automatic tracking, computer-controlled, multi-emitter radar and communications simulator and jammer system housed in a van-mounted transportable shelter. The self-contained system is composed of five linked subsystems to train radar, communications and combat systems operators. The system can operate on land and on naval vessels and is controlled by two operators either in manual or computer mode. One operator controls the communications/communications jamming subsystem while the other controls the radar simulators and jammers as well as receiver/DF subsystem.

Components. The subsystems are the radar simulator, the radar jammer, communications/communications ECM, receiver and direction finding (DF) and computer control and monitoring.

Operation. The radar simulator generates four independently controlled RF signals whose characteristics may be either manually selected or software controlled. The radar jammer generates two independently controlled countermeasures signals whose parameters may be selected automatically or manually. The communications/jamming subsystem is capable of voice and encrypted voice communications and covers the HF, VHF and UHF spectrums. The receiver/DF subsystem provides surveillance, analysis and DF for ESM and jammer alignment.

Status

Three systems have been operational since the mid-1980s with the NATO Multination Electronic Warfare Support Group (MEWSG).

Manufacturer

Hercules Defense Electronics Systems Inc (HDES), Clearwater, Florida.
(For details see Directory of Suppliers)

Lockheed Modular Adaptable Radar Simulator (MARS)

The Lockheed Fort Worth Modular Adaptable Radar Simulator (MARS) employs real RF signals to simulate target radars. The system accommodates aircraft flight profiles and threat weapon simulation with links to real or simulated ECM equipment.

Components. Multifunction threat radar consoles are provided to train radar operators. MARS includes radar receivers and tracking loops which are adaptable through software. The system can display radar tracking errors and missile miss distances.

Operation. Detailed parameters of the radars, missiles and target aircraft are all under software control and may be easily adjusted. There are two receiver types to accommodate the signal processing requirements. The MARS digital receiver emulates various radars which are in common use while the receiver provides a conventional pulse radar type capability.

Status

Four systems are in service.

Manufacturer

Lockheed Fort Worth Company, Fort Worth, Texas.
(For details see Directory of Suppliers)

Lockheed AN/MST-1(V) Multi-Threat Training System

The Lockheed Fort Worth (previously General Dynamics) AN/MST-1(V) creates a multiple threat training signal environment characteristic of an integrated air defence system.

Components. The centrally controlled remote emitters are dispersed, with each transmitting the radar signals associated with a specific type of fire-control radar or search radar.

Operation. Receivers at each unit intercept the jammer response to the threat signals to assess the noise jamming and range gate or velocity gate pull-off jamming. This is then analysed for debriefing purposes.

Status

In service.

Manufacturer

Lockheed Fort Worth Company, Fort Worth, Texas.
(For details see Directory of Suppliers)

LFW AN/MST-1(V) remote emitter

Lockheed Sanders Onboard Electronic Warfare Simulator (OBEWS)

In November 1992 Lockheed Sanders was awarded a $4.6 million engineering and manufacturing development contract for an F-16 Onboard Electronic Warfare Simulator (OBEWS).

Components. The OBEWS is installed in an existing cartridge slightly larger than a VHS cassette.

Operation. The OBEWS will allow F-16 Blocks C and D to be flown as electronic warfare threat simulator trainers from a base equipped with the US Air Force Mission Support System (AFMSS). The system will activate the aircraft's radar warning receiver as if a real engagement was taking place. The pilot will then undertake self-protection electronic countermeasures. The system will also be capable of recording the exercise for debriefing purposes.

Status

The contract will be completed by 1996.

Manufacturers

Lockheed Sanders Inc, Nashua, New Hampshire.
(For details see Directory of Suppliers)
A pod-mounted OBEWS is available from AAI.

LTV AN/TPT-T1(V) Threat Emitter Simulator

The LTV AN/TPT-T1(V) is an unmanned threat emitter which simulates the radars of various surface-to-air and anti-aircraft artillery systems. Each system has a master controller and up to five of the emitters are under a master controller. The system is configured to interface with a rangemaster computer which allows the whole mission to be controlled.

Status

In June 1988 LTV received a contract to build 40 AN/TPT-T1(V) systems for ranges at Nellis Air Force Base, Crow Valley, Avon Park, Spadeadam, UK, the Alaskan Air Command and Pil Sung.

Manufacturer

LTV Missiles and Electronics Group, Buffalo, New York.
(For details see Directory of Suppliers)

Martin Marietta M AN/MSR-T4 Countermeasures Training Equipment

The Martin Marietta AN/MSR-T4 is designed to evaluate the performance of threat emitters and airborne ECM.

Components. It uses microprocessor-controlled receivers and table-driven software algorithms to provide a detailed analysis and evaluation of ground-based threat emissions and aircraft ECM responses.

Operation. The range control centre and the aircraft parent unit use the real-time analysis to evaluate equipment readiness and crew proficiency in airborne ECM. Using a multiple antenna and automatic receiver system the MSR-4T intercepts, measures and analyses signals in the 30 MHz to 18 GHz range. Aircraft tracking data is provided by other equipment or by a separately housed IFF tracking subsystem.

Status

In service with the US Air Force.

Manufacturer

Martin Marietta Corporation, Denver, Colorado.
(For details see Directory of Suppliers)

Metric Radar Simulators

The Metric Systems Corporation makes a number of tracking radars and radar/threat simulators in all radar bands up to 35 GHz. The radar simulator emissions are programmable in order to cover all training situations. Most are ground based, but an airborne pod is also available. Details of some equipment follow:

Models 3131 and 8877 Modular Threat Simulators

The Model 3131 and Model 8877 are low-cost training devices.

Model 3131. This consists of a master control and up to five remote emitters which are controlled by UHF radio, fibre-optic or telephone links. The operator controls the antenna mode, RF pulse characteristics and frequency. Each emitter is capable of bands up to 35 GHz with programmable PRF and PRI. The emitters feature various configurations of single or dual axis pedestals, parabolic and shaped antennas, track or guidance transmitters and manual or slaved tracking. Optional extras include an electronic warfare response monitor, IFF tracking system and an optical television tracking system.

Model 8877. This is a portable, modular threat radar simulation system with RF transmitters available from 850 MHz to 17.5 GHz frequencies with either parabolic or cosecant-squared antennas. Each system has an antenna, control, high voltage and RF units.

MBE System

The Multi-Band Emitter (MBE) system consists of 3 RF emitters operated through a remote-control and display unit. The emitters operate in the L, S and X bands at 0.85-0.94, 2.1-2.4 and 9.9-10.2 GHz respectively, at peak powers up to 300 kW (in L band) and at agile PRFs up to 100 kHz (in X Band). Horizontal, vertical and circular polarisation is available.

Metric AN/UPT-2 Airborne Radar Transmitter Simulator

The Metric AN/UPT-2 radar transmitter simulator models the emissions of several types of radar transmitters. It is designed for installation in manned or unmanned aircraft. It consists of a magnetron mounted to a solid-state high power modulator, a high voltage power supply and associated circuitry. One of seven magnetrons operating in four power levels in H-, I- or J-bands can be employed. It is capable of providing signals over a wide range of preselected output frequencies, pulse repetition rates and pulse widths.

Metric Airborne Emitter

The Metric airborne emitter is pod-mounted and is a computer-controlled system providing up to 8 programmes which may be remotely selected. Frequencies from 7.8 to 18 GHz are provided with PRFs up to 5000 (including pulse Doppler modes) and antenna movement up to 70°/s. High- or low-power pods may be used.

AN/MPS-T9 Early Warning Radar Simulator

This radar simulator operates in E and F bands. It is housed in two mobile containers (shelters) consisting of an operations van and an antenna van. The operations van contains systems displays, controls and communications circuits. The antenna van contains the power distribution, digital Moving Target Indicator

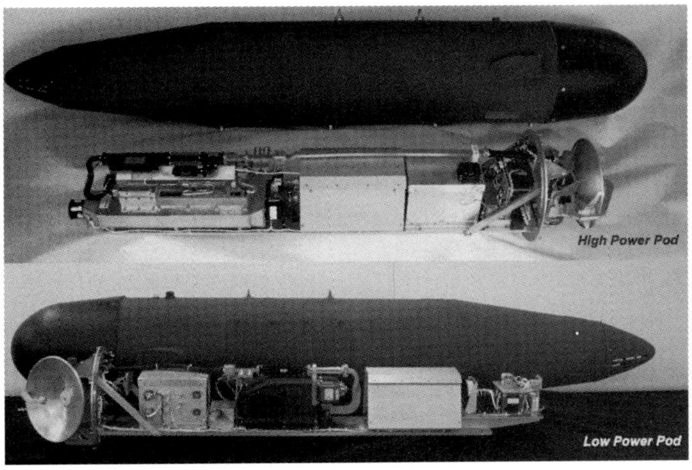

Metric Airborne Emitters

Metric range of ground radar threat simulators

(MTI), IFF interrogator and six transmitter/receiver subsystems which transmit 1 MW of peak power and cover the frequency band from 2.7 to 3.1 GHz.

AN/MPQ-T3 Multiple AAA/SAM Threat Radar Simulators

The basic model consists of three search and track radars mounted on one 12 m transportable trailer. It operates in the E, I and J bands and has its own aircraft tracking capability. Scan patterns, PRF and pulse width are under computer control.

The AN/MPQ-T3A has automatic track capabilities for each radar transmitter and can simulate any H/I-band threat and any I- or J-band threat simultaneously using up to four of its six transmitters at the same time. Its instrumentation includes provision for digital data recording and provisions for radar video recording and playback.

The AN/MPQ-T3S can accept and/or provide slaving data, and perform parallax correction on in-coming and outgoing data.

Deployment

Metric radar simulators are deployed at Nellis AFB, Nevada; NWC China Lake, California; Fallon Test Range, California; Missile Test Centre, Point Magu, California; Pinecastle Range, Florida; Eileson AFB, Alaska; Spadeadam EW range, UK; Polygon Test Range, France/Germany; Mountain Home AFB, Idaho; and Eglin AFB, Florida.

Status

Available.

Manufacturer

Metric Systems Corporation, Walton Beach, Florida.
(For details see Directory of Suppliers)

Republic RES-207 Radar Environment Simulator

The Republic RES-207 is an air defence radar simulator which can operate with 2D or 3D systems. It injects coherent and non-coherent RF signals at the front end of a real radar to produce a simulated environment which includes multiple aircraft and missile targets, and ECM features. A scenario generator is provided to prepare scenarios in advance and a data tagging unit monitors the target reports from the radar to the command post.

Status

In service with the Hughes Air Defence Radars of a NATO country.

Manufacturer

Republic Electronics Company, Hauppauge, New York.
(For details see Directory of Suppliers)

Republic Radar Environment Simulators

Republic produces a number of radar environment simulators for use in aviation training.

REES-203

The Radar Electromagnetic Environment Simulator (REES-203) is designed to provide realistic interceptor controller training. It is adaptable to 2D or 3D radars and consists of an exercise control console, up to four pilot position consoles, simulated ground-to-air and air-to-ground communications, a computer cabinet and a radar cabinet.

The central processing unit is based upon Intel 8086 microprocessors and the software is in PL/M-86 HLL. The system is capable of generating 36 manoeuvrable and six chaff targets and six IFF targets in a circle up to 300 miles from the site and to an altitude of 70 000 ft (21 335 m).

RES-205

The Radar Environment Simulator (RES-205) is designed to provide operational training, surveillance radar testing and performance monitoring. Like REES-203 it is adaptable to both 2D and 3D radars. It consists of a controller unit with central processor and an RFP-3000 unit with power supplies, real-time processor and RF section. Processing hardware and software are identical to REES-203.

RES-205 simulates up to 63 dynamic, controllable targets of selectable sizes. Up to six simulated aircraft can be designated as ECM platforms providing barrage/wideband, synchronous repeater, CW, spot and click jamming. The modelled exercise area covers a 256 mile (412 km) circle around the site to a height of 100 000 ft (30 480 m).

Status

REES-203 is in service in Norway while RES-205 is in service with Greece, Italy, Portugal and Turkey.

Manufacturer

Republic Electronics Company, Melville, New York.
(For details see Directory of Suppliers)

Rodale Darts

Darts is a magnetron-based radar simulator which can simulate a wide variety of airborne and ground-based search and track, missile guidance and fire-control radars (FCR). It may be used to train ESM operators, to test and to evaluate radar warning receivers and for electronic warfare training.

Housed in an AN/ALQ-167 pod shell or internally mounted it is microprocessor-controlled and parameters which may be selected in flight include emitter frequency (from 7.8 to 9.6 GHz and from 14 to 15.2 GHz), pulse repetition frequency (PRF), pulse width, PRF mode, antenna scan mode and parameter. These adjustable parameters include pulse widths of 0.1 to 2.0 μs, PRFs of 3 to 6000 Hz and scan rates/widths of 0.1-30 s/0.1-15 s.

Rodale AN/ALQ-167 Darts on the right and Ericsson Erijammer on the left

Specifications

Frequency coverage: H, I and part of J-bands
Typical output power: 12.5 MW
Noise bandwidth: 7.8-15.2 GHz
Length: 3.12 m
Diameter: 250 mm
Weight: 147 kg

Status

Available.

Manufacturer

Rodale Electronics Inc, Garden City, New York.
(For details see Directory of Suppliers)

Rodale 8102 Smart Crow

Smart Crow is a pod or internally mounted ECM system for broadband ECM training and is used principally in single-seat aircraft or business-jet aircraft modified for the ESM/ECM role where the Electronic Warfare Officer (EWO) has other means of determining mode selection and timing. The pod-mounted version uses the AN/ALQ-167 shell. The system is band reconfigurable with the technique generation, control, monitoring and power-handling assemblies identical for each band but with RF sources, power amplifiers and antennas selected for the appropriate band.

Specifications

Frequency coverage: D, E, F, G, H, I and part of J-bands
Noise bandwidth: 5-500 MHz (wideband); 0-100 kHz (Doppler)
Antenna coverage: (horizontal) 60°; (vertical) 60°
Cooling: ram-air
Length: 3.47 m
Diameter: 254 mm

Status

In service with the US Navy at the Pacific Missile Test Center, Point Magu, California and also in NATO's MEWSG programme.

Manufacturer

Rodale Electronics Inc, Garden City, New York.
(For details see Directory of Suppliers)

Rodale Responsive Electronic Warfare Training System (REWTS)

Rodale Electronics Inc is teamed with Ericsson Radar Electronics AB of Sweden in the REWTS system, which is described under the Ericsson entry in the Swedish section.

Sierra Model 8819 Aircraft Survivability Equipment Trainer

The Sierra Research Division has developed the Model 8819 Aircraft Survivability Equipment Trainer which can be installed in vehicles to simulate air defence batteries in order to train fixed- and rotary-wing aircrew how to penetrate and to neutralise air defences.

Threat Signatures. The simulators are designed to provide electronic, visual and thermal signatures of threats. They simulate radar warning receivers and can provide the threat operator/system indications of jammer activity for realistic application of software-activated 'break lock' conditions to produce realistic training. All Sierra systems incorporate time-coded optical engagement recording for accurate visual debriefing.

Status

Available.

Manufacturer

Sierra Technologies Inc., Buffalo, New York.
(For details see Directory of Suppliers)

Teledyne Airborne Electronic Warfare Receiver Training System (ALERTS II)

ALERTS was designed for integrated training for the US Army Aviation electronic warfare equipment on helicopters although it may also be used on fixed-wing aircraft with minor modifications. The system, which was updated in 1989, provides exact threat visual and audio simulation including search, acquisition, tracking and/or launching and the aircrew must respond to these simulated threats in the same way they would respond to real threats, that is, by masking, electronic countermeasures and so on.

Components. The system has four units; the Airborne Instrumentation Unit (AIU), the Threat Transmitter, the Ground Support Unit and the Ground Training Subsystem.

Operation. The AIU is installed in the helicopter and controls inputs into the AN/APR-39 radar warning receiver using data from the Threat Transmitter which is a low power, 2 W, omnidirectional 2.2065 GHz coded transmitter placed at selected locations and used to determine intervisibility from the threat location to the aircraft. Up to 15 Threat Transmitters may be deployed at any one time to simulate a high intensity conflict. Programming of the system and post-flight debrief is carried out on the Ground Training Subsystem which consists of two displays, one presents the course of the aircraft while the other provides data in tabular form.

Status

In service with the US Army.

Manufacturer

Teledyne Brown Engineering, Huntsville, Alabama.
(For details see Directory of Suppliers)

Tracor Radar Emitter Simulators (RES)

The Tracor Radar Emitter Simulators (RES) are housed in a modified BLU-27 canister with radome and include high power and pulse Doppler/Continuous Wave (CW) threat radar replicas and CW/ECM pods.

Components. The high power RES uses a modified AN/DPT-1 type modulator with a separate magnetron for each frequency type with output power between 70 kW and more than 200 kW. The CW RES uses a standard travelling-wave tube (TWT) transmitter which generates at least 200 W.

Operation. The RES use a digital control unit to control the pulse characteristics and scan generation. The user defines the threat by storing a program in a laptop computer with each program consisting of up to eight bars and 40 different nodes.

Status

In production and in service.

Manufacturer

Tracor Flight Systems Inc, Santa Ana, California.
(For details see Directory of Suppliers)

Tracor Real-Time Threat Environment Simulator

Tracor has developed a real-time RF threat simulator which enables the user to pass through an area in which there are simulated radars, surface-to-air missile systems and anti-aircraft guns which generate the RF and audio signals the pilot would experience.

Components. A sensor display shows the radar, infra-red and RWR contacts whilst a tactical situation display shows position in the gaming area as well as the location of both air and ground threats. An out-the-window display is provided for visual navigation and target location.

Operation. Two players can fly a simulated strike against a defended ground target complex. The simulated aircraft will fly with the characteristics of the selected platform or may be programmed by the user. As the target is approached, the simulation assigns one or more of the threats to engage the aircraft. As the assigned threat switches from search to acquisition, to track, to fire, the symbols on the tactical situation and sensor displays change colour to indicate the mode. The mode is passed to the threat generator control box where the appropriate waveform is created and an audio tone is generated.

Outputs may be fed to amplifiers generating RF signals to illuminate the antennas or other sensors of the system to be simulated or they may be fed directly to the detector-antenna terminals. Optional subsystems under development will generate digital signals which may be fed to the avionics system bus. The player stations may both be configured as 'blue' aircraft or one may engage the other.

Status

Under development.

Manufacturer

Tracor Flight Systems Inc, Santa Ana, California.
(For details see Directory of Suppliers)

Vitro Radar Simulators

The Vitro Services Corporation produces a number of radar simulators. In 1957 Vitro began work on the USAF Eglin EW Range Complex in NW Florida and by 1959, nine threat radar simulators were in use. Currently the Eglin EW range has about 25 Vitro threat radar simulators. Vitro currently supports more than 175 surveillance, tracking and threat simulation radars at 37 ranges in Argentina, Canada, Germany, Italy (Sardinia), Korea, Kwajelin Island (Pacific), Puerto Rico, Spain, Sweden, Taiwan, UK, and the USA. Some 300 threat radar simulators have been delivered. The radar simulator units are cased or racked modules which then have to be connected to an antenna unit either singly or in multiple arrangements such as where a pulsed simulator is connected to an antenna together with a CW simulator to model a particular threat radar. Some Vitro radar simulators are described below:

Vitro threat radar simulator - dual transmitter model RS 573

RS 561M(V)

This is a solid-state radar simulator designed for applications which do not require the high power of the RS 565M(V) series (see later). The solid-state modulator and constant voltage/constant current power supply permit the use of a variety of magnetrons up to 8 kW peak power. It is housed in two cases of which one contains the transmitter and the other the high voltage power supply/converter. Transmitter characteristics include four selectable PRFs which can be set in the field, two pulse groups with up to four pulses in each group, adjustable pulse width for each group and adjustable spacing between pulses.

RS 565V Series of Variable Energy Radar Simulators

The RS 565V is a modular variable energy radar simulator which can duplicate the RF characteristics of a wide range of threat radars. It has a hard tube modulator which gives pulse flexibility. PRFs between 100 and 10 000 PPS can be set, and unlimited combinations of pulse width, spacing, and pulse groups are possible. Some 11 models are available covering the frequency range 2.7 to 17.5 GHz at peak powers up to 600 kW. It consists of a control unit and a transmitter unit and may be extended by adding transmitter units with different power and frequency outputs. It is transportable by two personnel.

RS 565M(CW)

This simulator emulates Continuous Wave (CW) signatures. It can provide more than 2 kW of CW power at any frequency either in stand-alone configuration or with a Vitro pulse emitter such as the RS 565V. Options include a remote-control unit, pedestals and antennas as well as interface units to process antenna steering data from Tactical Aircrew Combat Training Systems/Air Combat Manoeuvring Instrumentation (TACTS/ACMI) ranges, IFF trackers, full-scale threat emitters or any tracking radar.

RS 572 Tactical Steerable Emitter Threat Simulator (TASETS)

Consists of the RS 565 threat emitter, an ACMI interface unit, and a Model 910 pedestal and antenna.

RS 573 Dual Transmitter

Consists of the RS 565M2 pulse unit, the RS 565M3 CW unit, with the Model 910 pedestal and antenna.

RS 575 Line-Type Radar Simulator

This is a low-cost, line-type modulator transmitter for applications that do not need the flexibility of pulse given by the RS 565 series. It weighs 63 kg. It can have either one fixed pulse width or two selectable pulse widths. PRFs from 100 to 5500 are available at frequencies from 2.7 to 18 GHz at peak powers up to 500 kW. Amongst other applications, the RS 575 can be used as a target radar for live anti-radiation missile (ARM) firings. Vitro claims that the use of RS 575 devices can transform an air weapons range or training area into a multi-threat electronic combat environment at very low cost compared to the use of more complex radar simulators.

Status

Available for export. In production and in service.

Manufacturer

Vitro Services Corporation, Fort Walton Beach, Florida.
(For details see Directory of Suppliers)

Whittaker Radar Simulators

Whittaker Electronic Systems of Simi Valley, California, makes a number of radar simulators, some of which are described below:

AN/ALQ-167 Type Airborne EW Jammer Pods

These versatile electronic warfare jammers are used in training, testing, evaluation and tactical applications. The self-contained ram-air cooled pods, complete with transmit/receive antennas, are designed for carriage on rotary or fixed-wing aircraft. Numerous jamming scenarios, including noise and deception types, are fully programmable via onboard or ground-based computer terminals. Dual- and single-band frequencies from C-Band (850 MHz) to K-Band (36 GHz) are provided. BITE is also featured.

AN/MPS-T10 Missile Fire-Control Radar Simulator

This is a self-contained replica of a surface-to-air missile (SAM) fire-control radar (FCR). It simulates the target tracking and missile uplink systems and the target function is supplemented with a dual television camera optical subsystem. The system is housed in a trainer and requires only a 60 kW generator to become operational.

AN/MSQ-T43 Threat Radar System Simulator

The AN/MSQ-T43(V) is a modular threat emitter designed to simulate fire-control radars of surface-to-air missile or anti-aircraft artillery systems.

The system is a computer-embedded in-band monopulse tracking radar consisting of a pallet-mounted antenna/transmitter/pedestal group and an operations group in a shelter. The latter includes a computer processor which controls the system. It is in service with the US Air Force and US Navy.

AN/MSQ-T4/T6 Emitter Tracker Threat Simulator

This system is an alternative means of positioning the threat signal of the training aircraft. The system locks onto, and automatically tracks, an aircraft's IFF return. The system provides realistic RF threat signatures which can be programmed to represent track-while-scan, conical, monopulse and any other type of scan applicable to the threat as well as missile uplink, interrogation and guidance signals.

AN/MSQ-T32 Radar Signal Simulator

This is a universal radar simulation system which is palletised and uses a standard S-280G operations shelter. The palletised element consists of the pedestal assembly, three antennas and transmitters, a local control assembly and a TV camera assembly. The shelter houses the operator's console and a digital electro-optical tracker assembly.

The system selectively simulates two different threat radars. One threat can operate simultaneously in H-band and two J-band frequencies while the other operates only in J-band.

AN/AST-7 Multiple Signal Simulation Pod

The Multiple Signal Simulation (MSS) pod is an airborne emitter designed to provide simulation of airborne search radars, targeting radars, terrain-following radars and missile seekers.

The pod-mounted system has factory preset and user selected emitter frequency, pulse repetition frequency and pulse width parameters. When operated with the SG-1189(V) PRF/SCAN generator cockpit control box, in-flight adjustment of PRF is available together with PRF mode selection. As an optional extra the emitter frequency can be tuned in flight.

The transmitter is an SCR-magnetic modulator specifically designed for the mission environment and consists of a power supply and a modulator. The standard antennas have beamwidths which cover ±9.5° off pod boresight but special antennas can be provided. The models are MSS-01 (H-band), MSS-02 (I-band, 8.5-9.6 GHz), MSS-03 (I-band, 9.7-10 GHz), MSS-04 (J-band, 12-13 GHz), MSS-05 (J-band, 14-15.2 GHz), MSS-06 (J-band, 15.5-17 GHz) and MSS-07 (K-band).

Status

In production and in service.

Manufacturer

Whittaker Electronic Systems, Simi Valley, California.
(For details see Directory of Suppliers)

AIRCRAFT MAINTENANCE TRAINERS

CANADA

Atlantis Maintenance Trainers

The Atlantis trainers have full free-play simulation with interactive video disk. Instructor and student stations are provided with the former capable of setting up exercises, inserting faults and monitoring student performance.

Role. The trainers are designed to teach students the theory of system operation and/or organisational level system maintenance and troubleshooting.

Components. The trainers make the maximum use of commercial hardware, although real or simulated equipment can be used, and optional features include audio simulation and voice synthesis.

Deployment. Systems have been produced for the Canadair CC-144 Challenger, the US Navy's Boeing-Vertol CH-46 Sea Knight helicopter (conceptual maintenance trainer), the McDonnell Douglas F/A-18 Hornet for the Canadian armed forces (CF-18) and Australian Air Force (maintenance and part-task trainers). Maintenance trainers have also been built for the US Navy for the Boeing E-6A, and 8 for the US Air Force for the systems aspects of the McDonnell Douglas F-15E Strike Eagle. The latter consists of simulated cockpits, two vertical simulation panels, an interactive control console with a remote monitor and printer, as well as a control and processing system. The Australian Army has been supplied with a flight control and hydraulic maintenance trainer and an avionics and external stores maintenance trainer for the Sikorsky S-70A-9 Black Hawk. The Australian Navy has been supplied with a maintenance trainer for the S-70B-2 Seahawk helicopter.

Status
See text under Deployment.

Atlantis F-15E maintenance trainers supplied to the US Air Force

Manufacturer
Atlantis Aerospace Corporation, Brampton, Ontario.
(For details see Directory of Suppliers)

FRANCE

CGA Maintenance Trainer for Mirage 2000

The CGA maintenance and armament trainer can simulate over 200 different systems failures selected by the instructor for the trainees to rectify. The instructors have a series of interactive displays, each controlling an aircraft system.

The first model of this trainer was delivered to the French Air Force at Mont de Marsan in July 1984 and subsequently orders were received from India, Abu Dhabi and Egypt.

Status
In production. In service with the French Air Force.

Manufacturer
CGA Cegelec, Brétigny-sur-Orge.
(For details see Directory of Suppliers)

Sogitec Maintenance Trainers

Sogitec produces maintenance trainers for aircraft or weapons systems. The trainers operate in three modes; illustration for theoretical training, correct operation for the study of scheduled maintenance procedures, and degraded operation for failure diagnosis and correction training.

SEMA Aircraft System Trainers
The SEMA (Simulateur d'Entraînement à la Maintenance système Avion) aircraft maintenance trainers simulate electrical, fuel and hydraulics systems together with flight controls and other aspects depending upon customer requirements. They have a modular configuration and secondary systems may be plugged into the main one. The main system, which incorporates the prime processor, features a diagram of a cockpit incorporating active controls of the simulated systems. The secondary systems consist of teaching aids with animated panels and a set of plug-in units on which are illustrated the front panels of test sets and other maintenance equipment.

SEMSA Weapon System Trainers
The SEMSA (Simulateur d'Entraînement à la Maintenance des Systèmes d'Armes) weapons system simulator features an instructor's console, cockpit and aircraft racks. The cockpit rack is a two-dimensional

Sogitec maintenance trainer for Mirage 2000

representation of the cockpit on a panel and features all items of equipment and functional controls associated with the weapons system including the Head-Up Display (HUD), radar and electronic warfare suite controls. The aircraft rack is a similar representation of the controls which permit 'repair' and 'maintenance' of the system with the rack being divided into stores and weapons, external test equipment together with electrical distribution areas.

Status
Sogitec has delivered 26 aircraft and weapons system maintenance trainers to export customers for aircraft which include the Alphajet, the Mirage 50, the Mirage F1, the Super Etendard, Mirage 2000 (three customers) and Aermacchi MB-339A. Customers include Greece (Mirage 2000) and Italy (Aermacchi MB-339A at Lecce).

Manufacturer
Sogitec Industries SA, Suresnes.
(For details see Directory of Suppliers)

GERMANY

Atlas Elektronik Electronic Radar Maintenance Trainer (EMT)

The Atlas Electronic Radar Maintenance Trainer (EMT) is designed to provide low-cost training in maintaining radars. It is a modular system which can be adapted to meet specific customer requirements. A typical system, in service with the German Air Force, is described below.

Components. The system consists of eight students' stations, an instructor station and an editor station where exercises are prepared and evaluated. Each station features a high resolution 19 in display, a text terminal, a keyboard and an electronic 'mouse'. The students use their stations to find 'faults' injected through the instructor station.

Student Activities. The student stations receive on the screen diagrams which will be of use in diagnosing the 'fault' which they then identify by connecting instruments such as oscilloscopes. The system provides the appropriate responses in real time. The student then learns to correct the 'faults' as well as practising basic operating procedures and learning the characteristics of the system and its subsystems.

Status
In service. Users include the German Air Force.

Manufacturer
Atlas Elektronik GmbH, Bremen.
(For details see Directory of Suppliers)

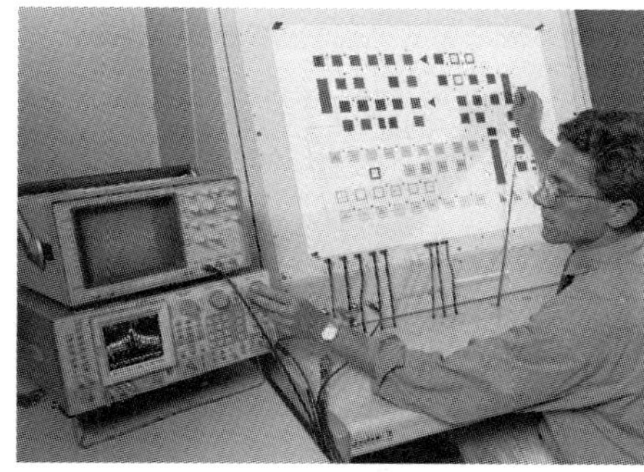

Atlas Radar Maintenance Trainer

Dornier Electronic Flight Instrument Systems (EFIS) Maintenance Trainer

The Electronic Flight Instrument System (EFIS) Trainer is designed to provide training in the maintenance of 'glass cockpit' displays, digital avionics, bus systems as well as combined indicators and controls.

Components. The EFIS Trainer consists of two 19 in racks with display and control units for attitude, location, navigational data and weather radar subsystems together with simulation units providing all the necessary interfaces and signals.

Status
In production.

Manufacturer
Dornier Luftfahrt GmbH, Training and Special Projects Division, Bremen.
(For details see Directory of Suppliers)

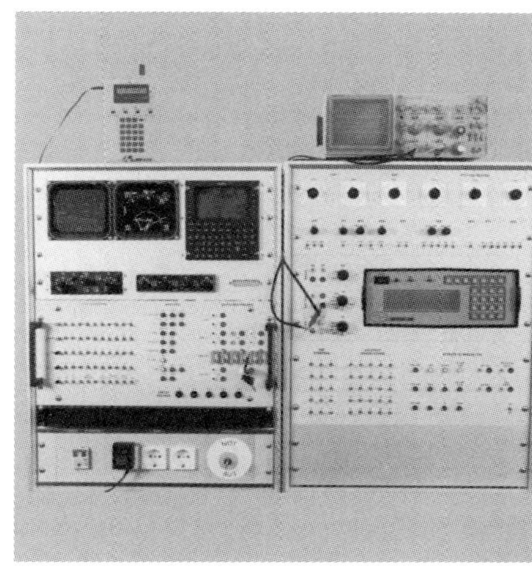

Dornier EFIS Maintenance Trainer

Dornier NARCOS Avionics Maintenance Trainer

The Dornier NAvigation, Radar, COmmunications Simulator (NARCOS) consists of complete aircraft avionics and appropriate test equipment integrated in a mobile four-rack arrangement with plug-in units or drawers housing the functional groups. Each unit may be removed from its rack and operated independently of the work bench. It may be used for training at flight line and workshop level.

Status
In service.

Manufacturer
Dornier Luftfahrt GmbH, Training and Special Projects Division, Bremen.
(For details see Directory of Suppliers)

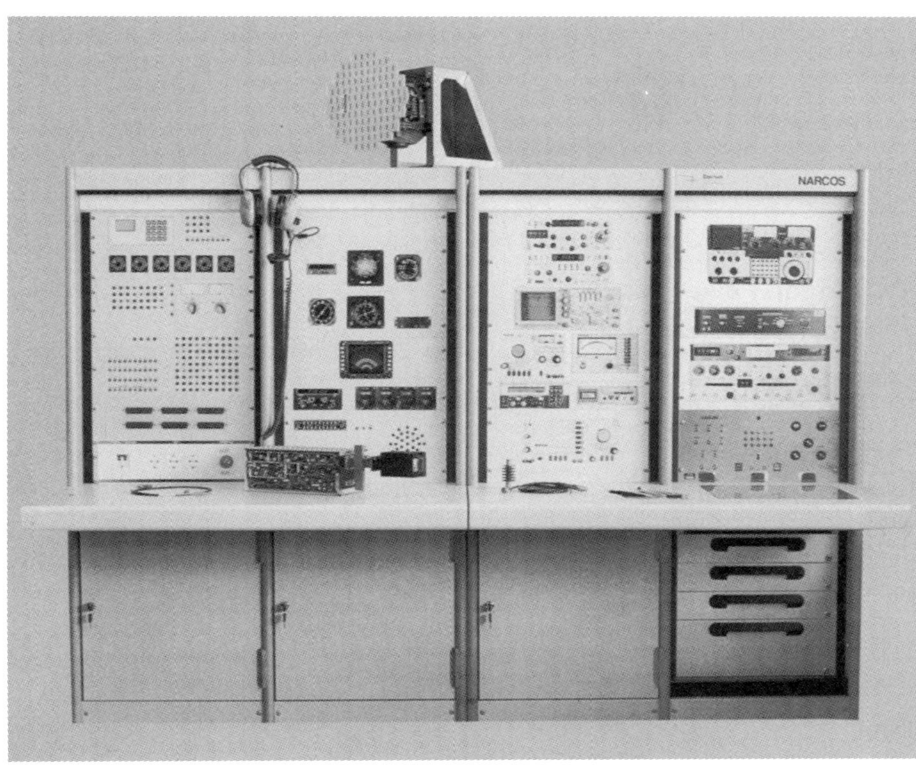

Dornier NARCOS Avionics Maintenance Trainer

ITALY

Gajon System 2000 Maintenance Trainer

For training of ground crews in maintenance the Gajon Institute of Technology has developed a computerised system which uses interactive video disk and is known as the System 2000.

Components. The system includes an integrated network of student stations, instructor stations, processors, laser video disks, two-directional audio, video and computer lines using fibre optics and local area networks.

Training Use. The 2000 systems are used in classrooms and in the field. They are used for basic and advanced training in all technical subjects and also provide real maintenance stations for radars, fire-control equipment, communications, electronic warfare, propulsion and power generation plants.

Status

In service with the Italian Air Force and the other Italian services as well as export customers who have received Italian weapon systems.

Manufacturer

Gajon Institute of Technology, Genoa.
(For details see Directory of Suppliers)

Gajon System 2000 at the Italian Air Force technical training centre

UNITED KINGDOM

BAe Aircraft Systems Maintenance Trainer

The British Aerospace Aircraft Systems Maintenance Trainer is used to assist in the classroom teaching of aircraft systems maintenance.

Components. Typically such a device consists of a cockpit module, aircraft systems module and instructor console and uses simulation software, video integration techniques and realistic system simulation. System line-replaceable units are contained within interchangeable system panels fitted with associated connectors, test sockets and adjustment devices.

Operation. The student is provided with 'hands on' system maintenance training. Video sequences illustrate the more complex maintenance procedures and provide automatic feedback if the system detects a dangerous or non-standard procedure. The instructor console is used to configure system operation and allows for extensive 'fault' injection.

Status

In service.

Manufacturer

British Aerospace Defence Limited, Military Aircraft Division, Warton.
(For details see Directory of Suppliers)

BAe Aircraft Systems Maintenance Trainer

ECC Aircraft Power Supply Trainers (APST)

The ECC Aircraft Power Supply Trainers (APST) consist of simulated aircraft components and an animated electrical system schematic, both panel-mounted, simulated first line test equipment and a separate instructor's station.

Role. They show students the principles of operation of paralleling/load-sharing and non-paralleling/load-takeover with aircraft primary electrical generation systems and secondary power supplies in normal and malfunctioning conditions.

Facilities. Facilities are included to enable instruction of safety procedures, the use of typical first line test equipment and fault diagnosis. Additionally students are able to use the trainers for experience of fault diagnostics and for examinations.

Status

In service with the Royal Navy and the Royal Air Force.

Manufacturer

ECC Simulation Limited, Shoreham-by-Sea.
(For details see Directory of Suppliers)

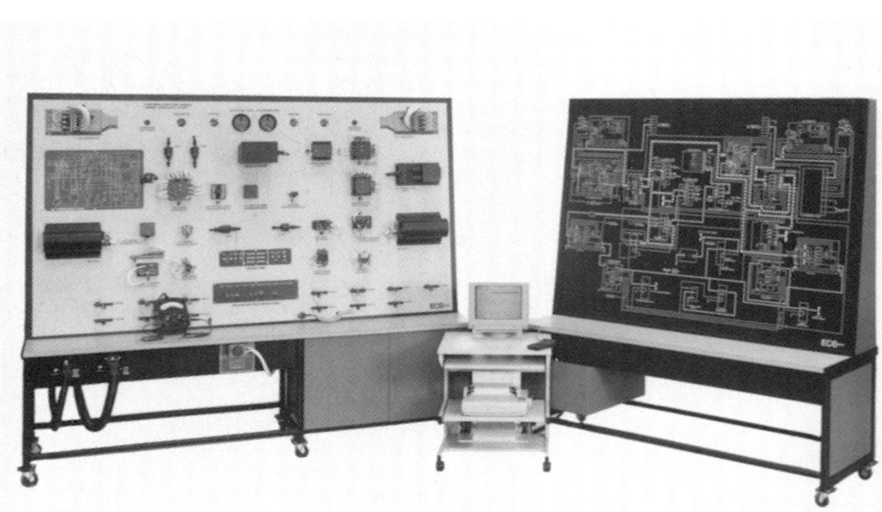

ECC Aircraft Power Supply Trainer (APST)

Pennant Fuel Management Training System (FMTS)

The Pennant Fuel Management Training System (FMTS) has been designed to provide training in the theory, skills and procedures of bulk fuel management. It covers all aspects from the receipt of fuel in the storage area to its transfer via pipelines to facilities such as airfields into local storage then distribution to refuelling positions for aircraft.

Components. The FMTS consists of an instructor station with large screen display, student work stations, and a full-scale working simulation of a bulk fuel installation panel. Students can carry out self-paced tutorials or the instructor can lead lessons; student work stations can be networked. The system entered service in October 1989. It has since been upgraded to incorporate five new emergency bulk fuel systems, the addition of an Air Staff Management Aid, VGA graphics and Ocean Tanker Accounting.

Status
In service with the Royal Air Force.

Manufacturer
Pennant International Limited, Cheltenham.
(For details see Directory of Suppliers)

Pennant Fuel Management Training System (FMTS)

Pennant Weapons System Demonstration Rig (WSDR)

In March 1993 Pennant International received a contract from MoD UK for a Weapons System Demonstration Rig (WSDR). It will be used to introduce weapons loading teams to aircraft and weapon system operations before training with real aircraft.

Sea Harrier and Lynx Configurations. The WSDR will be configured to represent on one side a Sea Harrier and on the other a Lynx helicopter. On the representative wing it will be possible to attach weapon pylons, release units and typical loads of inert bombs and missiles. On the helicopter side it will be possible to attach inert missiles and torpedoes to the weapons carrier.

IOS. At the front of the trainer will be the instructor's control console which will create training and exercise scenarios in one of four operational modes (fixed-wing bomb, fixed-wing missile, rotary-wing anti-surface vessel and rotary-wing anti-submarine) and then monitor the students.

Control Boards. Essential aircraft circuits and controls for loading, arming and releasing weapons will be collected onto control boards for operation by the team. Four such boards will be available, one for each operational mode.

Status
In service with the Royal Navy.

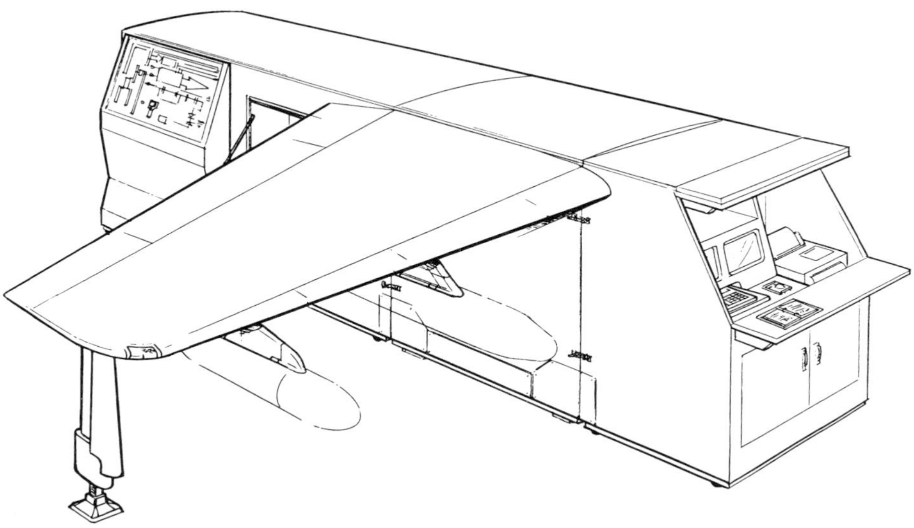

Pennant Weapons System Demonstration Rig seen from the Sea Harrier side

Manufacturer
Pennant International Limited, Cheltenham.
(For details see Directory of Suppliers)

TTSL Aircraft Maintenance Trainers

Procedures and Part-Task Trainer
A Propulsion Systems Trainer (PST) for the RAF Jaguar was made by the ex-Link-Miles Element of Thomson Training and Simulation Limited (TTSL). It consists of a mock-up of the appropriate parts of the aircraft with engine instruments 2.5 times normal size on an enlarged panel. It is intended for classroom instruction.

A Tornado Weapon System Training Rig (WSTR) was also made by Link-Miles and consists of a modified Tornado shell with relevant simulated instruments and equipment. Live weapon loading takes place under the supervision and management of an instructor using an adapted PC with a touch-sensitive control panel.

Status
In service.

Manufacturer
Thomson Training and Simulation Limited, Crawley.
(For details see Directory of Suppliers)

UNITED STATES OF AMERICA

AAI Advanced First Term Avionics (AFTA) Maintenance Trainer

The Advanced First Term Avionics (AFTA) maintenance trainer consists of one instructor station and 15 student stations.

Operation. It simulates a variety of test equipment from a simple multimeter to complex equipment such as an oscilloscope. Students use 14 items of test equipment to locate simulated faults in composite integrated weapon systems. An interactive video disc presents the simulated avionics equipment down to the component level. Assemblies or components to be tested are 'activated' by a touch-screen. The simulated test equipment is connected via a test point panel.

Flexibility. The type and number of the stations may be changed to meet customer requirements as may the test equipment.

Instructor Aspects. The scenarios are generated by an instructional training system which may be used to alter the scenarios to meet the needs of fixed- and rotary-wing aircraft as well as ships and armoured vehicles.

Status

In service with the US Navy at the Naval Aviation Training Center, Millington, Tennessee.

Manufacturer

AAI Corporation, Hunt Valley, Maryland.
(For details see Directory of Suppliers)

AAI Advanced First Term Avionics (AFTA) maintenance trainer

Boeing E-6A Maintenance Trainers

Boeing Defense & Space Group makes a number of maintenance trainers for the US Navy for the Boeing E-6A TACAMO (Take Action Charge And Move Out) aircraft.

Avionic Trainer. An Integrated Avionic Trainer (IAT) provides advanced aircraft maintenance by providing students with 300 simulated aircraft system faults. The students learn how to isolate the faults and conduct troubleshooting and component removal and/or replacement procedures.

Other Trainers. Boeing also provides trainers which use actual aircraft sections which have hardware upon which the students may practise maintenance skills for the following:

- the fuel system
- the environmental control system
- the auxiliary power unit
- active landing gear trainer
- flight control trainers.

Logistic Support. Boeing Defense & Space Group also provides contractor logistics support to the US Navy for the E-6 maintenance training programme.

Status

Delivery of the maintenance training equipment was completed in May 1990. The E-6A Training Facility is at Tinker Air Force Base, Oklahoma.

Manufacturer

Boeing Simulators & Training Systems, Huntsville, Alabama.
(For details see Directory of Suppliers)

The Boeing E-6A Landing Gear Maintenance Trainer

DynaLantic Maintenance Trainers

The DynaLantic Corp make a number of maintenance trainers for aircraft which are described below:

DynaLantic Aviation Ordnance Trainer (AOT)

The DynaLantic Aviation Ordnance Trainer (Device 3B64) is computer-based full-scale maintenace trainer designed to provide fault isolation and hands-on maintenance training in the aircraft ordnance systems of the F/A-18 Hornet, the A-6 Intruder and the A-7 Corsair II aircraft. It could be modified for use with the other aircraft such as the F-15, F-16, Mirage series, or Tornado.

Components. Each trainer has a simulated cockpit with appropriate ordnance panels, a full-scale representation of an aircraft fuselage with ordnance side-panels, facsimile wing pylons, facsimile gun station of an F/A-18 and missile stations for both the Sparrow and Sidewinder missiles.

Operation. Flight line ordnance personnel can perform 'pre-flight checks' and conduct maintenance exercises with simulated and computer-controlled 'faults' inserted from the instructor console. Real automatic test and ground support equipment together with simulated support equipment may be used with the trainer.

Deployment. In service with the US Navy which received 12 Device 3B64 at NAS Memphis between 1988 and 1990.

DynaLantic Air Cycle Trainer

The DynaLantic Aircraft Generic Air Cycle System (AGACS) Instructional Trainer simulates air conditioning and pressurisation systems (the air cycle system) and consists of two parts. These are a classroom automated system trainer (Device 11F20) and a hands-on trainer with one IOS and 15 trainee stations (Device 11F19). The classroom trainer consists of a large model of the air cycle system which is automated with lights to illustrate flows of air, valve position, and so forth within the system. The trainee stations of the hands-on trainer have the following equipment:

- Air cycle system
- Computer

Helicopter being winched down by the RAST system

DynaLantic RAST Maintenace Trainer - IOS on the left, trainee station on the right

I/O electronics
Video disk player and TV monitor
Aircraft controls and instruments relevant to the system
Key pad and terminal
Vortex tube equipment which control the air cycle system.

DynaLantic Ship Helicopter Recovery Trainer

Helicopters often have to land on ships' decks which are rolling and pitching. To aid this process a system known as RAST (Recovery Assist, Securing and Traversing) enables a helicopter in the final stages of deck landing to be winched into the deck by means of a strong cable attached to the underside of the fuselage. Device 11H117 is a maintenance trainer for the RAST equipment and consists of an IOS and a student console. The trainer simulates over 40 electrical faults and over 35 hydraulic and mechanical faults.

Status

See text.

Manufacturer

DynaLantic Corporation, Deer Park, New York.
(For details see Directory of Suppliers)

DynaLantic Aviation Ordnance Trainer - As well as the wing hard points, the fuselage distribution panels are also included

ECC Maintenance Trainers

ECC produces a range of maintenance and avionics systems trainers for a number of projects.

C-17A Maintenance Trainers

The US Air Force awarded ECC a contract to produce 12 maintenance trainers for the C-17A transport. The maintenance training devices include the following:
Tack Procedures Interactive Training System (TPITS)
Simulator for training evaluation/performance
Flight controls system trainer
Engine cowling maintenance procedures trainer
Landing gear maintenance trainer
Air-conditioning maintenance procedures trainer
Cargo doors/rails system trainer
Universal aerial refuelling receptacle slipway installation maintenance procedures trainer
Organisational level engine maintenance trainer
Auxiliary power unit organisational maintenance trainer
Consolidated maintenance trainer
Fuel tank maintenance trainer.
The contract is the largest maintenance training device contract awarded by the US Government. There will be several training systems at each C-17A base.

Flight Surface Control Trainer

These trainers provide students with the system knowledge and the skills to perform organisational-level maintenance on flight surface controls. The student learns how they operate normally and, when they malfunction, how to maintain subsystems and isolate faults. The system is in service with the US Navy.

F/A-18 Avionics Maintenance Simulators

ECC has produced five maintenance simulators for the F/A-18 Hornet for avionics, flight controls, fuel, hydraulics and armament systems.

These systems are designed for 'free play' interaction and use video discs, graphics displays, a high-fidelity cockpit, simulated test equipment and synthesised audio responses. They are designed for the 'memory inspect' maintenance procedure associated with the Hornet. This requires the technician to understand all the avionics subsystems and to be able to interpret computer-generated alphanumeric codes obtained from the cockpit displays to isolate faults in specific subsystems. These systems are in service with the US Navy.

F-16 Trainers. The company also produces F-16 maintenance simulators for the US Air Force.

Sea King Maintenance Trainer

These trainers provide maintenance training for electrical, hydraulic, weapons control, flight control, engine and tactical missions systems. Westland procured 10 for its Product Support Centre in Sherborne, Dorset, and the Indian Navy also purchased a number.

Generic Airborne Radar (GAR) Trainer

This system is used to teach basic radar skills, understanding of radar system theory together with operation, troubleshooting and repair techniques for first and second level maintenance.

Components. GAR provides a facsimile of a typical avionics repair shop test bench. It includes simulated radar assemblies antennas, a plan position indicator, synchroniser, transmitter and receiver.

Operation. The computer-controlled system is interactive and operates on a 'free play' mode permitting the student to make decisions and to follow on procedures as they would in a real avionics workshop. A wide variety of simulated test equipment is available. Students have at their disposal both built-in test (BIT) and built-in test equipment (BITE) to isolate faults of varying complexity.

Radars Modelled. The system can act as an air-to-air, air-to-ground, ground mapping, general search or IFF radar system.

ECC consolidated maintenance trainer for the C-17. A full size fuselage mock-up including the cockpit and the underfloor maintenance tunnel

ECC F-16 Maintenance trainer - cockpit module

Malfunctions. As many as 350 different system malfunctions can be inserted at the touch of a button including combinations with more than one concurrent malfunction.
Deployment. The system is used by the US Navy Avionics Technician School in Memphis, Tennessee.

Status
See text.

Manufacturer
ECC International Corporation, Wayne, Pennsylvania.
(For details see Directory of Suppliers)

Hughes Training Inc Maintenance Trainers

Hughes Training Inc (HTI) produces a wide variety of maintenance simulators for air forces. These include the following:

F-15 Aircraft Maintenance Trainers
These consist of facsimile cockpits, simulation panels and master simulation control consoles (MSCC). Training is provided on the integrated avionics suite, electrical power and lighting systems and the twin Pratt & Whitney engines.

Each integrated avionics student station consists of five panel consoles and a facsimile cockpit. The electrical power and lighting trainer consists of three panels and a facsimile cockpit while the engine diagnostics trainer has only panels. The MSCC houses the system computer, visual imagery system, and touch-screen monitoring system. The MSCC can be used by the instructors to create malfunctions and to monitor the students, and by the students for lesson presentation and tutorial videos.

F-16 Avionics/Armament Maintenance Trainers
Each trainer consists of a facsimile cockpit, a flat panel with two- and three-dimensional models and an interactive video station. Some also have replica flight line test sets.

The lesson is set up on the video station which the student then uses to localise and manipulate aircraft systems and components.

E-3A Radar Maintenance Trainer
To assist maintenance of the AN/APY-1 radar used in the Boeing E-3A Sentry AWACS (Airborne Warning and Control System) the US Air Force has a radar maintenance trainer. This consists of 10 student stations, an instructor station and a variety of facsimile equipments.
Components. Each student station contains a radar maintenance technician's console, simulated standard and unique test equipment, together with simulated radar cable connectors and jacks.
Operation. The student uses this equipment and a computer-aided instruction terminal to exercise with the simulated radar. Training follows a procedural programme and the facsimiles permit accurate responses. The instructor monitors the student from his station. All lessons are implemented by using computer-controlled simulation of the radar and test equipment functional characteristics.

Status
See text.

Manufacturer
Hughes Training Inc, Herndon, Virginia.
(For details see Directory of Suppliers)

Hughes F-15 maintenance trainer

McDonnell Douglas (MDTS) Systems Trainers

McDonnell Douglas Training Systems (MDTS) produces a number of maintenance trainers which are described below:

Classroom Trainers
These trainers are used in a classroom environment to teach system operation to aircrew and maintainers, and also to teach maintenance techniques. They consist of a suite of 13 devices which demonstrate the operation of various aircraft systems including hydraulics, electrics and weapons. There are simulated cockpit controls, management panels and animated system schematic display panels together with hand-held controllers. A screen is used to display enlarged pictures of specific systems.

Aircraft Equipment Trainers
A variety of training systems has also been produced. These usually consist of full-size replicas using aircraft components wherever possible. These include integrated avionics, flight controls/pneumatics and hydraulics, and weapons trainers. A full-size replica of an aircraft such as the AH-64A Apache is called a Composite Trainer. This latter is used to train maintenance personnel and has remove-and-replace features. Some 120 faults may be inserted.

Status
In production and in widespread service.

Manufacturer
McDonnell Douglas Training Systems (MDTS), St Louis, Missouri.
(For details see Directory of Suppliers)

Reflectone Maintenance Trainers for AV-8B

Reflectone provided the US Marine Corps with eight different maintenance trainers covering various aspects of AV-8B systems. The equipment provides training in such areas as engine, avionics, electrical, crew escape, environmental and fuel systems. A wide range of simulated faults can be injected into each system and new maintenance methods can be examined and taught before new equipment and procedures enter the squadrons.

Electrical Systems Trainer. In August 1991 Reflectone completed an upgrade for the electrical systems trainer (Device 11H95) for the night-attack version of the AV-8B. These changes include provision for a FLIR, a cockpit management system, a digital map set, night vision lighting, the angle rate bombing system and a data storage set. The updated system was RFT in September 1991.

Status

In service with the US Marine Corps at MCAS Cherry Point, North Carolina.

Manufacturer

Reflectone Inc, Tampa, Florida.
(For details see Directory of Suppliers)

Reflectone AV-8B electrical systems maintenance trainer

AIR TRAFFIC AND FIGHTER CONTROL SIMULATORS

Some Relevant Abbreviations

ATC	Air Traffic Control
ATCC	Air Traffic Control Centre
ATCO	Air Traffic Control Officer
ATCRU	Air Traffic Control Radar Unit

CANADA

ATS ATC Trainers

ATS Aerospace makes a number of products related to ATC training. These include integrated ATC Visual Tower and Radar Simulators, ATC Radar Simulators, the FlexComm Digital Voice Communications system, an Air Defence Simulator, an ATC Scenario Editor (ATCSED), and the Vizion 2k x 2k Graphics Controller.

Integrated ATC Simulators

A range of integrated ATC Tower, Radar and Procedural simulators is available. These products cover desktop skills trainers through to full environmental total immersion ATC simulators.

ATS ST/80 Radar Skills Trainer

The ST/80 uses a single PC to provide an ATCO position and two pseudo-pilots who 'fly' the aircraft in the ATC simulation. Several ST/80s may be networked together, and with the FlexComm voice communications system form a multi-sector system for ground, approach, precision approach and area control training.

Number of Aircraft and Radars Simulated. Up to 80 aircraft may be simulated per computer, with a total of 200 aircraft and 8 radars per exercise.

Training Features. The ST/80 includes many training features such as student assessment, voice recognition, record and replay and self-training mode.

Scenario Editor

The ATC Scenario Editor (ATCSED) is used in all ATS simulators for creating scenarios and exercises. It is menu-driven, and incorporates an on-line context-sensitive help facility.

ATS Full Environment Radar Simulator

This UNIX-based simulator can emulate any modern operational ATC system by providing realistic controller positions and powerful, flexible pseudo-pilot positions. The ATS Vizion 2 k × 2 k controller is used with Sony displays. Aircraft, weather and radar simulation are included.

Targets. The simulator generates 400 active targets.

Interfaces. The radar simulator interfaces to several different operational systems to provide approach and area control training in the operational environment.

ATS Vega/32 Visual Tower Simulator

This desktop skills trainer can be configured with one or more visual windows showing the view from any ATC tower. Image generation can be by CGI or projected from photographs. The CGI option allows the trainee to switch eye-points and where necessary for training and debrief purposes, view the scene from the pilot's viewpoint.

ATS Full Environment Visual Tower Simulator

A 360° Visual Tower simulator has been delivered to Vienna, Austria. It is integrated with the Radar Simulator to provide a complete ATC training system.

CGI System. Imagery includes runways, taxiways, lighting systems, buildings and other airport obstacles, and the surrounding terrain. The visual system uses photo-texturing, anti-aliasing and other features typical of a modern CGI system.

Radar Displays. The system can include ground movement radar displays, surveillance radar displays, airport information and weather displays, D/F, distance from touchdown indicators, signal light guns, airfield lighting controls, NavAid status and control, flight strip printing, and digital voice communications.

ATS Air Defence Simulator

This is a simulator for the training of fighter controllers. The system can either stimulate the real operational display system, or can provide emulations of radar displays. It provides a controllable outside world scenario, feeding radar data including weather, clutter and jamming, and voice data into the operational system. Up to 20 fighter controllers may be trained simultaneously. At the simulator positions, the instructors and target pilots have a graphical representation of the current air situation, from which they can control up to 250 flight plans from a total of 10 000 available. Fighter pilot positions and a system operator are also included. Several exercises may be played, recorded and replayed independently and simultaneously.

FlexComm Voice Communications

FlexComm is the ATS Digital Voice Communications System Simulator. It replicates the functions of the communications systems in standard ATC and air defence environments. It provides multi-channel communications switching for intercom, telephone, and simulated ground-to-air and ground-to-ground radio.

Computer Aspects. FlexComm is digital. The software driving the system is built around a 100 Mbits per second network, capable of interconnecting over 300 distributed positions. No central hardware is required.

User Control. Users control their stations through simple touch-screen input commands. Each user may have a separate touch screen layout which is readily modifiable. Traditional switch panels are also available.

Programming and Editing. The system is easily programmable and adaptable. A communications editor is offered to allow system reconfiguration.

IOS. The instructor has the flexibility to monitor any position for communications activities, and to control such features as noise, signal level, channel failure, and digital voice recording and replay.

Interfaces. FlexComm may be interfaced to any computer-based simulation system via Ethernet thus allowing dynamic real-time control of radio coverage, aircraft background noise, active jamming, and synchronised voice and exercise replay. Interfaces to operational telecommunications systems are available via special gateways.

Vizion 2 k × 2 k Graphics Controller

Vizion is a high resolution graphics adapter designed for the IBM Microchannel Architecture. Vizion displays up to 2048 lines of 2048 pixels, making it ideal for demanding applications. Vizion can be utilised for Air Traffic Control (ATG), Command and Control, medical, Geographical Information Systems, and high end Computer Aided Design applications. It takes up a single slot in an IBM RS/6000, so minimising footprint and hardware. No external boxes, power supplies or interfaces are required.

Status

In production. Nearly 100 systems are in service in Austria, Canada, China, Egypt, El Salvador, India, Libya, Malaysia, Norway, Qatar, Sudan, Sweden, Vietnam, and the USA.

Manufacturer

ATS Aerospace Inc, St-Bruno, Quebec.
(For details see Directory of Suppliers)

ATS Vega/32 visual tower simulator showing four visual windows out of a possible eight which can fill the whole 360°

ATS Vega 32 visual tower simulator

CHINA, PEOPLE'S REPUBLIC

DRS-90 Air Traffic Control Radar Simulator

The DRS-90 Air Traffic Control Radar Simulator was developed at Sichuan University and is used at the CAAC (China Airlines) Flight College, also situated in Sichuan. It was designed to simulate all aspects of ATC radar training, based on the use of a 26 in/660 mm screen, colour display and distributed information processing techniques. The DRS-90 can be used to simulate both first and secondary air traffic control radar systems.

Components. The DRS-90 system comprises a trainee station, a pilot/instructor station and an overall supervisor station. The stations are linked by a microcomputer-controlled network and analogue voice communication channels. Several trainee and pilot stations can be combined to simulate a medium-scale ATCC. Training can be conducted from basic up to refresher training for trained controllers.

Operation. The DRS-90 can present all radar displays and procedures likely to be encountered by a radar controller and the instructor can control all standard aircraft manoeuvres, along with emergency and communication problems. Exercises can be stored or prepared to meet particular requirements and exercises may be 'held', recorded and replayed as needed.

Status

In service.

Manufacturer

Sichuan University, Sichuan.
(For details see Directory of Suppliers)

GERMANY

DASA Tower Control Simulator

The simulation department of the Military Aircraft Division of Deutsche Aerospace (DASA) is involved in flight simulation, ATC and apron control simulators, aural cueing simulators, data base generation systems, visual systems, and in the simulation of sensor and EW systems. The following description is of their Tower Control Simulator (TOWSIM).

Components. The Tower Control Simulator consists of the following:

tower controller stations, fitted with ground and approach radar, control and information consoles, and with a clear view of the outside-world visual display system
one instructor station
pseudo-pilot stations
one system controller station.

Visual Image Generation. Three performance levels of day/dusk/night CGI systems are available with the appropriate number of channels to obtain the customer's choice of performance and FoV. Common features include airport, aircraft and vehicle lighting, weather variations, and controller-positioned monocular distance viewing. The time of day changes in real time and includes sun angle variation. The scene can show up to 60 aircraft and vehicles, and up to 20 vehicles and 10 aircraft moving simultaneously. Phototexture is provided.

Database. DASA can provide customer-specific databases. The customer can also build his database by using a Database Generation software tool utilising either the TOWSIM CGI system, or a separate workstation.

Visual Display System. Front or rear projection or large monitors can be provided. FoV is at customer choice and up to 360° can be provided with the appropriate number of CGI channels and display projectors.

Approach Radar Display. The hardware and software of this system and of the pseudo-pilot station is basically the same. It allows the selection of different backgrounds such as contour, blanking, shadow or image with different scaling and offset factors. Weather fronts also can be displayed.

Ground Radar Display. This system displays moving aircraft and ground traffic.

Control and Information Consoles. The control and information consoles display the information needed to control the airport traffic. Menu-driven commands are used by the tower controller to control the flight information system, runway and taxiway lighting and communication, and to receive information on these systems and on weather conditions.

IOS. The instructor console consists of a pseudo-pilot station with the following instructor facilities:

Exercise control by start, stop, freeze/reset and continuation of exercises, exercise recording and replay, weather control, malfunction control
Direct control of aircraft and ground vehicles
Aircraft flight dynamics modelling and control
System test and configuration
Exercise generation using a flight plan editor, exercise editor, map editor, symbol editor, take off/approach editor for standard procedures, weather editor.

Pseudo-pilot Stations. The pseudo-pilot stations (PPS) are used for real-time control of aircraft and ground vehicles in the overall simulation. Typically, a PPS controls five moving objects, and a maximum of 10 objects is possible. A maximum of 14 PPS can be fitted. The multifunction capability of the PPS can also be used by the instructor to generate and control exercises. Some system characteristics include the following:

LAPIS (TM) design of user interface; menu-driven aircraft control using macros; 1280 x 1024 pixel display; trackball or mouse inputs; up to 10 selectable navigation and position maps; simulation environment size from 0.5 nm to 400 nm radius.

System Control Console. This console allows system access authorisation in order to perform tests and to maintain the system.

Communication System. The communication system provides telephone, radio and intercom and is controllable at the tower controller and at the pseudo-pilot stations.

Status

Available.

Manufacturer

Deutsche Aerospace (DASA), Simulation Department of the Military Aircraft Division.
(For details see Directory of Suppliers)

DASA Visual Tower Simulator - Visual Imagery for Munich Airport in reduced visibility

DASA Visual Tower Simulator - Visual Imagery for Munich Airport in good visibility

ISRAEL

Elbit Flight Controller Training Simulator (FCTS)

The Elbit Flight Controller Training Simulator (FCTS) gives flight controller training from basic to refresher training in emergency situations or training in new procedures. Traffic loads, density and communication loads can be adapted to a trainee's capabilities and learning level.

Components. The FCTS is based on three main elements:

Commercial computers and peripherals
Work stations with graphic and alphanumeric displays and communication controls. Systems can vary from a few work stations to those involving dozens of work stations
Software developed by Elbit for the purpose.

Operation. The FCTS can be used to simulate various aircraft targets, with simulated aircraft flying characteristics, various meteorological conditions, flight plans and flight information, geographical data using maps and grids, various sensors such as primary and secondary radars, navigational aids and procedures.

Students are presented with a colour graphic display of a dynamic air situation picture in a designated airspace. There is also an alphanumeric display for flight plan data and aircraft flight data. Track details can be added, as can various background maps. The level of detail, scale, and geographic area can be selected as required.

Instructor Aspects. FCTS permits fast and simple preparation of exercises adapted to suit trainee needs and data can be changed during an exercise. Freeze, restart, and record/replay are fitted.

Future Airspace Planning. The FCTS can also be used to plan future airspace structures and compare alternative solutions, to plan future planning procedures, and to plan the configuration and internal layout and control centres. The FCTS can be used to examine traffic loads and flow, run and analyse tactical situations and operational plans, and examine and compare various procedures.

Status
In production.

Manufacturer
Elbit Computers Limited, Haifa.
(For details see Directory of Suppliers)

Elbit Flight Controller Trainer (FCTS) work station

SWEDEN

CelsiusTech ASIM Air Traffic Control Simulator

The main contractor for the CelsiusTech ASIM Air Traffic Control Simulator was Ericsson Radar Electronics AB, now part of CelsiusTech Systems AB. The system allows a number of ATC exercise scenarios from simple initial radar training with trainees working independently of each other, to complex system exercises where a number of working positions co-operate to simulate a small ATCC or part of a larger unit.

Operation. The trainees interact with 12 pseudo-pilots, each controlling up to 10 aircraft. A comprehensive traffic procedure library is included.

Components. An ASIM simulator comprises the following:

Eight training suites for radar and co-ordination controllers equipped with 23 in (584 mm) high performance synthetic radar displays, 2 keyboards and 2 rolling balls

Twelve pilot positions with alphanumeric colour displays; 8 positions are also equipped with 23 in (584 mm) radar displays

Two dummy positions equipped as the pilot display and sharing a radar display, both representing the operational environment such as adjacent ATC units, search and rescue units and such like

Two supervisor positions for operational and technical supervision, equipped as the radar positions plus alphanumeric displays

One position for exercise preparation equipped with an alphanumeric display and a plotter

A distributed and decentralised system concept with a large number of nodes connected to a local area network

A computerised switching system for voice communication

A computer-controlled TV system with 64 channels

Voice recording and synchronised playback of voice and air situation display

Capabilities for the simultaneous simulation of 180 targets in up to 8 parallel exercises - 1000 flights may be contained in one exercise

Storage capacity for 10 000 simulated flight plans and 100 exercise plans

A complete simulation of operational radar data and flight plan processing systems

A user-defined set of basic data parameters including aircraft performance data for 150 aircraft types, 100 reporting points (navaids, fixes, geographical points), 300 ATS routes, 2000 SIDs, 1000 standard arrival routes, 600 airports, 200 operational sectors, 50 exercise areas, 100 upper wind definition lists, 100 met reports, and 100 local weather reports.

There are also on-line computerised aids for exercise creation and evaluation.

ASIM is based on the use of DS86 radar displays with a graphical software support package; APN 167-11 computer nodes, including the display computer in the DS86 radar displays, connected to a high-speed local area network; EriPascal real-time programming HLL with inter-computer process-to-process communication over the LAN; and built-in technical monitoring and control system with status reporting, diagnostics and provision for reconfiguration of hardware and software.

Status
Two systems are located at the German National Air Traffic System (NATS) training academy in Langen. One system is used by the German Air Force at Kaufbeuren.

Manufacturer
CelsiusTech Systems AB, Järfälla.
(For details see Directory of Suppliers)

CelsiusTech ASIM ATC simulator for BFS, Germany

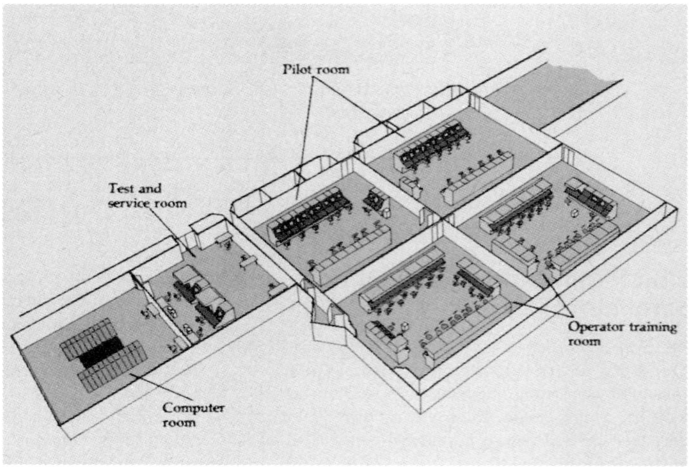

CelsiusTech ASIM air traffic control simulator at Langen, Germany

UNITED KINGDOM

Hughes UK - Air Traffic Control Trainers

Hughes UK Limited (HUKL) ATC Simulation and Training, make a series of ATC trainers including tower visual simulators, area and approach radar simulators, ramp and apron control trainers, procedural control trainers, and voice recognition systems. HUKL ATC Simulation and Training used to be the ATC division of the ex-Hughes Rediffusion company (HRSL), and this division was retained by Hughes when the rest of HRSL was sold to Thomson-CSF.

HUKL ATC Radar Trainers

Hughes UK ATC radar trainers are based on the FIRST (Flexible Independent Radar Skills Trainer) concept. The FIRST trainer simulates an airport, its environment, aircraft and ATC equipment. It is modular in design, in its basic mode is compact for ease of installation, but can be expanded into a complete tower simulator with a visual system.

Components. Two radar displays are simulated, one for a trainee ATCO and one for a pseudo-pilot. A third display simulates flight progress strips, provides on-line help, and gives second pseudo-pilot controls for running large exercises.

Operation. The trainee ATCO and pilot(s) communicate over a simulated radio link. Facilities are provided for instructors to participate according to their training requirements. In a typical exercise, the manoeuvres of all aircraft are according to their flight plans and performance. When the controller wishes to call an aircraft and direct its progress, he calls the role-playing pilot and communicates with him for the duration of the transaction. The computer then takes over again.

Instructor Aspects. The instructor can manage an exercise using commands such as record, replay, speed-up and freeze. He can change weather effects and introduce special problems during exercises. Automatic reports are provided on student performance to assist teaching reviews. The simulation can be adapted easily to different airports, aircraft and equipment.

Future Facilities. FIRST can incorporate some facilities, such as position prediction and ground speed display, which are expected to be introduced in ATC during the next few years.

Status

Available.

Manufacturer

Hughes UK Limited, ATC Simulation and Training, Crawley.

(for details see Directory of Suppliers)

Hughes Flexible Independent Radar Skills Trainer (FIRST) for ATCOs

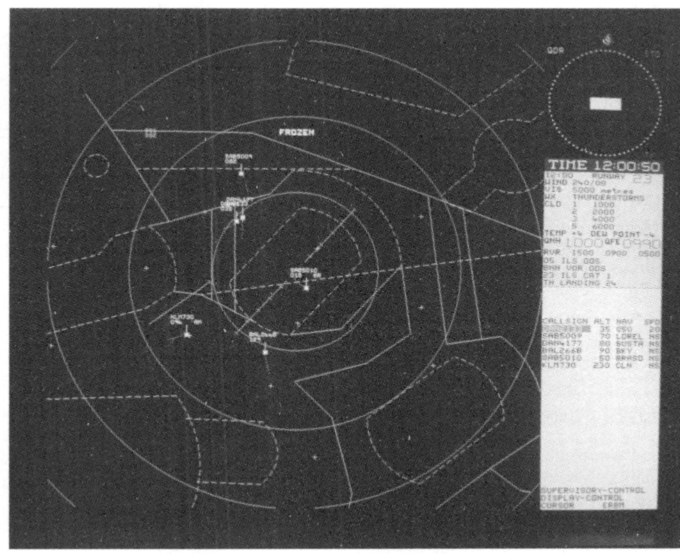

Hughes FIRST screen display for ATCO training

Invertron VISRAD Air Traffic Control Radar and Tower Visual Simulator

The Invertron VISRAD (VISual and RADar) simulator trains for all aspects of ATC, including visual and radar control, and control of ground movements. It can be used for air defence and fighter control training. VISRAD provides graphics generation of aircraft and vehicles, digitised airport background scenes and conventional model consoles and instruments at control positions. It can also be configured to provide apron control training.

Operation. Seated in a simulated control tower, trainee ATCOs have a panoramic view of runway(s) and the surrounding area, allowing observation and control of aircraft during landing and take-off, together with ground movements of aircraft and vehicles. Aircraft and other vehicle control is computer-driven under pilot or supervisor control, utilising touch-screen or other input devices.

Visual Display System. Movements within visual range are presented on a panoramic screen made up of panels each measuring 1.8 × 2.4 m and providing a visual coverage of from 60 to 360°.

Computing and Networking. VISRAD is based on an Ethernet networking system from which the various work station and visual display projectors are driven using their own computers. The architecture is based on distributed DEC VAX work stations although other computers may be used.

Instructor Aspects. Exercises may be recorded and replayed for briefing and debriefing and for assessment of student performance. The GENIF Instructor Facilities Software package enables users to modify over 40 000 software variables in the 'modelled environment' without programming skills. The creation of different airport views is accomplished by the user using the GENIF DRAW Package. Each visual display panel can show up to 16 aircraft and/or ground vehicles and the user has a choice of 256 colours at his disposal. Further work positions may be added without the need to rewrite software or replace existing hardware.

Overall layout of the Invertron VISRAD air traffic control simulator at RAF Shawbury in the UK

Radar Simulation. The radar display can accommodate any major proprietary radar system. As an alternative a radar emulation package can produce high quality graphic representations of 'real' radar devices. The system can accommodate radar approach, departure and *en route* displays plus procedural control.

Other Simulations. All communication facilities (R/T, intercom and land-line) are provided. Auxiliary control information, such as weather, messages, direction finding and time can be presented on high definition colour displays or conventional instrumentation.

Scenarios. Multiple exercise scenarios are possible. Exercises can include single or individual independent exercises, each with a full complement of aircraft types, performance groups, flight plans, airways, beacons, procedures and holding patterns, and digital recording.

Features of VISRAD include the following:

Primary radar: raw, digitised, plot processed area, approach, terminal, precision approach, Airfield Surface Movement Indicator (ASMI/ASDE); DFTI in monochrome or colour.

Secondary radar: code call sign conversion mode, code, label control, squawk/height filters; coast/suspend, jamming, sector radar, arrival and departure.

Aircraft: automatic and manual performance; full manoeuvre modelling for *en route*, circuit and altitude performance.

Environment: full atmospheric modelling of pressure, indicated air speed/true air speed (IAS/TAS)/Mach number, wind conditions, outside air temperature (OAT) and Runway Visual Range (RVR).

Navaids: radar, beacons, D/F, instrument landing system (ILS), microwave landing system (MLS), distance measuring equipment (DME).

Communications: R/T, intercom, land-lines, audio recording.

Exercise area: vertical - 5000 ft (1524 m) to Flight Level 999; horizontal - unlimited.

Graphics: menu-driven graphics packages and software tools for production of display and 'out of the window' presentations including maps, permanent echoes (PEs), weather, aircraft and vehicle images, visual backgrounds, colour selection, and occulting.

Displays: radar; programmable wind, pressure settings (RQNH, AQNH, QFE, OAT), time; airfield weather; multi-mode analogue/digital D/F and message area; ASMI/ASDE.

Status

In production. In service with Egyptian CAA and installed at the UK Central Air Traffic Training School at RAF Shawbury.

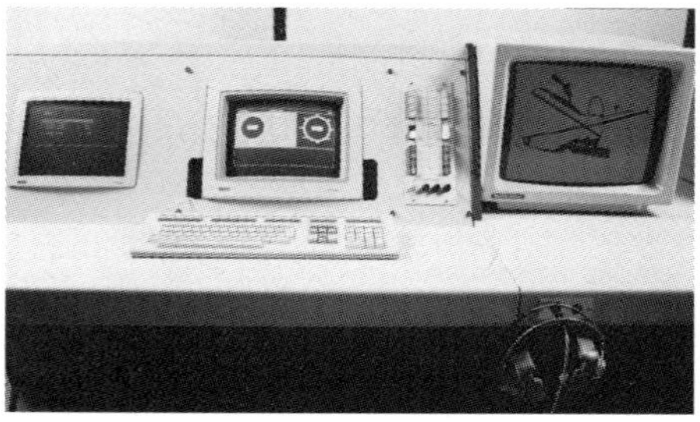

Invertron VISRAD ASMI/ASDE (Airfield Surface Movement Indicator) work station

Manufacturer

Invertron Simulated Systems Limited, Burgess Hill, West Sussex.
(For details see Directory of Suppliers)

Marconi Air Traffic Control Simulation

The Marconi Simulation ATC simulator provides training in the use of both primary and secondary radar and associated sensors in an ATCC environment.

Operation. Programmed training exercise scenarios are used to ensure that the student controllers and exercise supervisors are trained in the operation of an ATC environment under both normal and emergency conditions. Up to 75 aircraft and 40 different types can be displayed at any one time operating in an area of 100 nm (185 km) radius from the simulated airfield and from ground level up to 60 000 ft (18 288 m).

Evaluation of Performance. Record/replay facilities are provided for use in an exercise debrief facility.

Status

In service with the Royal Air Force at the Central Air Traffic School at RAF Shawbury.

Manufacturer

Marconi Simulation, Dunfermline.
(For details see Directory of Suppliers)

MST Air Defence and Air Traffic Control Simulators

Marconi Simulation and Training (MST, ex-Ferranti, Cwmbran, Wales) produces a number of facilities and trainers in the ATC and Air Defence training area.

CBT for ATCOs and AD Controllers

This activity is based at Dundridge College, Totnes, Devon, UK (for details see Directory of Suppliers) and includes a range of CBT and PTTs. Military applications include a range of instructor-led air defence, EW and ATC CBT modules running on PC-based equipment. Dundridge College has trained over 600 military and civil personnel from more than 25 countries.

Air Defence and Air Traffic Control Simulators

The range of MST ATC trainers includes stand-alone devices or systems emulating full airspace management. Each system is supported by familiarisation training for maintainers and operators. This can also include instructor training packages incorporating computer-based courseware modules.

Air Defence Simulators

These can be used for candidate aptitude evaluation, and controller training at all levels. They can cater for large battle scenarios and feature long-range surveillance radars, battle management tote displays, fighter intercept profiles and an EW environment, together with simulated environmental conditions.

Operation. Combinations of background tracks may be used in individual console exercises, or consoles may be combined in groups for simultaneous exercises or configured for a single large scale air defence exercise. Aircraft performance profiles are preprogrammed for each aircraft type and are realistically affected by altitude, weapons and fuel loading and consumption. Screens show the status of attacks, when targets are acquired, when weapons are fired, and attack results. BVR front-hemisphere attacks by long-range missiles may be simulated, followed by closer range engagements including frontal attacks followed by rear re-attack by gun or missiles. Visual identification (VISIDENT) conditions can also be simulated.

Military Air Traffic Control Simulators

MST supplies ATC simulators for instruction in area and approach radar control, procedure control and Precision Approach Radar (PAR) control. Fully integrated operation with Tower Visual Simulators offering state-of-the-art visual simulation can be provided.

Components. The Marconi ISCA 2000 Air Space Management Simulator family uses IBM PCs, Ethernet, LANs and Lynx Real-Time UNIX 'open system' standards. Functions are programmed in Ada (for scenario generation) and C (for X-window display MMI).

Operation. Use of Microsoft Windows permits convenient viewing of data, and a range of standard manoeuvres is provided. The latter include Standard Instrument Departures (SID) and STandard ARrivals (STAR) with speed or altitude restrictions; turns, intercept and fly radials and arcs, perform orbits, holding patterns, and flying formations. Cue points may be inserted where automatic changes of altitude, course and speed may take place. Approach manoeuvres include instrument landings (ILS), and visual approach and landings where aircraft follow segments on each leg with speed adjusting automatically. Specific military air traffic control procedures (eg spiral descent with simulated or actual engine failure) are incorporated. Any approach pattern can be applied to any runway.

Aircraft Control. During automatic manoeuvres the aircraft 'pilot' monitors progress and can take over

The MST Air Traffic Control Area Radar simulator at the RAF Central Air Traffic Control School, RAF Shawbury, UK

Part of the MST 36-position Air Defence simulator at the RAF School of Fighter Control at RAF Boulmer, UK

manually if required. The 'pilot' role may be undertaken by ATCO students both in order to reduce running costs and to allow the student to gain experience of the 'pilot's view'. Preprogrammed tracks are used, whilst extra aircraft may be introduced by the Exercise Controller.

Networking. ISCA 2000 simulators may be configured to embrace up to 50 work stations linked by Ethernet LANs to a file server, with fibre optic communication. Smaller systems can also be configured. Similarly, ISCA 2000 system software suites can be ported to alternative open systems hardware (such as DEC and other RISC processor-based work stations) suitable for inclusion as training stimulators for real ATC work stations within live military ATC systems.

Status

During 1991 Marconi (then Ferranti) supplied a military ATC simulator to RAF Shawbury and Phase 1 of an Air Defence simulator to the School of Fighter Control, RAF Boulmer. In 1992 the latter was expanded to Phase 2 specification, involving 36 positions. A variation of this system was delivered to the Royal Jordanian Air Force in mid-1993.

An ISCA 2000 air traffic management simulator was chosen by the Swedish Civil Aviation Authority for the Swedish Air Traffic Services Academy (SATSA) training facilities at Malmö-Sturup for completion during 1994.

Manufacturer

Marconi Simulation and Training, Cwmbran.
(For details see Directory of Suppliers)

RUKL Air Traffic Control Tower Simulator

The Reflectone UK Limited Air Traffic Control Tower Simulator is designed for *ab initio* controller training, continuation and refresher training, and procedures testing.

Components. The simulated ATCO console may include a number of radar and direction-finding displays, airfield and environmental data displays, airfield system control panels, flight plan data displays, flight strip facilities and communications systems.

Visual System. The trainer can use digitised photographs of real-world airfields to produce the static views from the simulated towers, overlaid with computer-generated dynamic three-dimensional images of aircraft and vehicles. The visual scene may be presented on monitors for classroom use, linked to a table-top trainee work station, or may be projected onto screens arranged to give the appearance of control tower windows. Up to 360° of outside world scene may be reproduced, depending on the number of image generation channels and display windows chosen by the customer.

IOS. Instructor and pseudo-pilot facilities permit full exercise, aircraft and communications monitoring and control. Traffic manoeuvres may be performed through preprogramming, via pilot or instructor intervention, or using a combination of automatic and manual input. A voice recognition and response system can replace or augment manned pseudo-pilot functions. Users can create or amend exercises and aspects of airfield, tower systems and aircraft databases, including panoramic scenes and CGI models.

RUKL ATC tower simulator - Visual Scene

Status

In production. One 300° FoV system delivered to Switzerland and one on order for the Belgian Air Force.

Manufacturer

Reflectone UK Limited, Bristol.
(For details see Directory of Suppliers)

RUKL Aircraft Control and Air Defence Simulators

Reflectone UK Limited (RUKL) offers a range of aircraft control and air defence simulators which provide training for civil and military ATCOs and air defence controllers. Facilities include basic radar skills, *en route* and area control, surveillance and approach radar. For military users facilities can include air intercept, ECM, target allocation and tactical control. Systems for air defence can be linked to computer-based digital mapping and can have large-screen tactical displays.

Components. The simulators comprise a network of student and supervisor consoles, each of which can operate both stand-alone for individual training or linked together for joint exercises. Each console has its own microprocessor-based computing system, a central high resolution display showing combinations of primary raw and processed radar returns, secondary labels, video maps, and a separate tote display. A communications system provides simulated radio channels and intercom between the students, instructors and pilot operators.

Instructor Aspects. Exercises can be prepared, developed and edited at any console, and during an exercise the instructor can amend parameters, freeze/rerun and record for later debriefing. Tracks of simulated aircraft and ships can be automatically preprogrammed and interactive in response to student inputs, or controlled from pilot work stations.

Status

In production. A helicopter control trainer entered service with the Royal Navy in 1992.

Manufacturer

Reflectone UK Limited, Bristol.
(For details see Directory of Suppliers)

THORN EMI Air Traffic Control Simulator

THORN EMI produced the Initial Radar Training Simulator (IRTS) for the Royal Air Force which was RFT at RAF Shawbury, UK, in October 1987. The contract cost was about £1 million. The system is used to train ATCOs in terminal area control techniques. It may also be used for training fighter control and air defence controllers.

Components. The system consists of three identical autonomous radar simulators with PAR displays and simulated communications. The facility includes a software development computer which can be used to generate new training exercises, different outline video maps and radar types.

Operation. The student is taught a full range of terminal control skills including in-flight emergency procedures and talk-down techniques supervised by an instructor who has a separate IOS.

Status

In service with the Royal Air Force.

Manufacturer

THORN EMI Electronics Limited, Wells, Somerset.
(For details see Directory of Suppliers)

UNITED STATES OF AMERICA

CAE-Link ATC Simulation

The CAE-Link Virtual Controller™ is an ATC radar simulator, and the Virtual Tower™ is an ATC Control Tower Visual Simulator.

CAE-Link Virtual Controller™ ATC radar simulator

The Virtual Controller ATC Simulator is a complete training system offered to customers on turn-key terms. It was originally developed by the University of North Dakota for the FAA-approved Airway Science Program. The system uses the Microsoft Windows environment with easy to use menus.

Components. Components used for real-time simulation include a real-time server, communication control unit, controllers, and pseudo-pilot and instructor stations. Off-line development components include an airspace editor, scenario builder, performance editor, and a strip writer.

Communications System. Radios, headsets, direct and indirect land lines connect all training positions.

Computing and Networking. The system utilises PCs as the interface with the computational subsystem and work station computers. The work stations are connected by an Ethernet network. A Real-Time Server is employed as the central processor. The system can be configured as a stand-alone simulator or multiple units can be networked together for large class training situations. Networking facilities allow the running of sector hand-off or ATC facility hand-off training exercises.

CAE-Link Virtual Tower™ Control Tower Visual Simulator

The tower visual simulator has stations for student controllers and pseudo-pilots, and modular design is employed.

Communications. A CAE-designed digital voice communication system is incorporated within each station. The radio models for HF and VHF communication consider distance, weather, and mode of operation.

Visual Display System. A Tower Visual system is provided for the local, ground, and flight data and assistant controllers. The visual scene can be displayed on desktop CRTs using 37in or 60in displays or provide up to 360° FoV using projectors.

Visual Image Generation. A CGI system generates a local airport scene complete with surrounding terrain and animated mobiles. Special effects such as weather changes and time of day are also portrayed.

Integration and Networking. The tower simulator is integrated with all of its subsystems via Ethernet. Each controller station can be used as a stand-alone device or connected to any of the other stations.

Status

Available. The first ATC products upon which the designs are based were developed by and sold through the University of North Dakota (UND). Ten of these devices are installed at UND, one at the Florida Institute of Technology, and one at the Florida Memorial College.

In 1993 Virtual Controller ATC radar trainers were delivered to Auburn University and to the Pan Am International Flight Academy in Miami, Florida. The Auburn devices are used under the FAA's AWS training programme.

CAE-Link provided a 180° FoV Virtual Tower simulator to the Pan Am International Flight Academy in 1994.

One 360° FoV Virtual Tower simulator and eight Virtual Controller radar simulators are due for delivery to the Finland CAA in early 1995.

Three Radar Trainers were delivered to Aeroportos e Navegacao Aerea (ANA) of Portugal in 1994.

Manufacturer

CAE-Link, Binghamton, New York.
(For details see Directory of Suppliers)

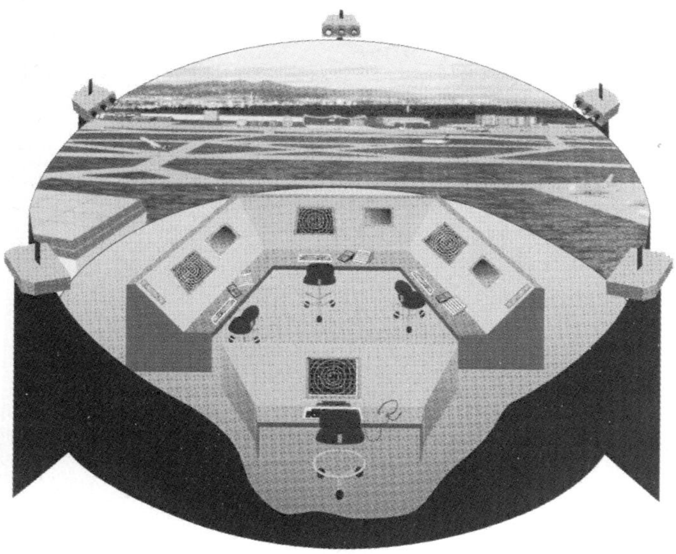

CAE-Link Virtual Tower™ ATC Visual Tower Simulator with five-projector visual display system

CAE-Link Virtual Controller™ ATC Radar Simulator

Contraves Air Traffic Control Tower Training Simulators

The Contraves ATC Tower Training Simulator (CATTS) simulates the interior of an ATC tower, with a simulated external view projected on screens.

Components. CATTS employs Motorola Delta series UNIX work stations, Silicon Graphics Iris image generators, CRT projectors, and a Contraves modular communications system. Visual image generation is by a CGI system and display is on a patented Ellipsoid high-reflectivity projection screen up to a FoV of 360° depending on the CGI channels and number of display projectors employed. There are pseudo-pilot stations for the control of up to 15 active aircraft and ground vehicles. Simulated radio, intercom and land-line communications are provided.

Instructor Aspects. Record/replay is fitted including synchronised voice traffic. A central IOS authorises and controls all exercises, while optional voice recognition can automate the pseudo-pilot task.

Radar Training. Optional radar facility consoles allow combined radar and tower training.

Contraves ATC Tower Simulator (CATTS) drawing showing four projectors

STAR ATC Simulator

Contraves also produces a training simulator for all air traffic control positions known as STAR. STAR has automated simulated aircraft responding to trainee commands. Up to ten multiple air traffic control positions can be networked.

Status

Available.

Manufacturer

Contraves-SSI, Tampa, Florida.
(For details see Directory of Suppliers)

Hughes Air Traffic Control Tower Simulation

Hughes Training Inc (HTI) upgraded the US Air Force's AN/GSN-T3 Control Tower Simulator with an improved visual system. This simulator is sited at Keesler AFB and is used to train ATCOs and associated personnel in VFR operations.

Visual Image Generation. Hughes MicroPOLY 1 CGI replaced the simulator's original large format slide and film-based system. MicroPOLY 1 generates interactive scenarios with visual background, moving aircraft and airfield vehicles. The MicroPOLY 1 image generator upgrade consists of an integrated set of 4 background channels and 12 target generation units that operate under control of a computer. Airfield and sky image generation includes images of runways, taxiways, hangars, buildings, parking ramps, navigation facilities, trees and clouds.

Target Models. Target model data, stored on the visual computer's hard disk, is transferred to individual MicroPOLY 1 target generators during system initialisation. During subsequent processing each target generator produces an individual real-time colour image that responds to proper target motion. Up to 12 simultaneous moving objects are provided from a library of 25 air and ground vehicles for presentation to student operators. The targets, consisting of 20 modelled aircraft and 5 ground vehicles, operate independently of each other.

Visual Display System. Airfield and sky images as well as the moving targets are displayed by 12 projectors over a 210° horizontal by 46° vertical FoV. The resolution and image sizes allows students to detect and identify large aircraft at 10 miles (16 km), smaller aircraft at 5 miles (8 km), landing gear down at 1.75 miles (2.8 km), and ground vehicles at 1.25 miles (2 km).

Status

Delivered during 1991.

Manufacturer

Hughes Training Inc (HTI), Arlington, Texas.
(For details see Directory of Suppliers)

Logicon Air Traffic Control Training Systems for the USN

Logicon is producing for the USN six ATC radar training systems for the Advanced Shipboard Air Traffic Control Training Systems (ASATS) Programme (Device 15G30). In addition it is producing the Shore-Based Air Traffic Control Training System (SATS), Device 15G31.

Shipboard ATC Training Systems. ASATS Device 15G30 includes Carrier Air Traffic Control Center (CATCC) and Helicopter Direction Center (HDC) trainers. The CATCC trainer is for close control flight operations for radar controlled aircraft. The HDC trainer is for control of helicopters and fixed-wing aircraft which support amphibious operations.

Shore-Based ATC Training. SATS Device 15G31 includes four trainers for shore-based training. Three are for basic air traffic control skills such as aircraft identification, radar vectoring using ATC phraseology, use of radar display equipment, use of air surveillance radar, air traffic control procedures and co-ordination, precision approach radar landing control, radar arrival control, and inter-facility co-ordination. The fourth trainer is for Radar Air Traffic Control tasks.

Common Features. Features of the two systems include: automated voice recognition of student communications to enable direct student interaction with simulated pilots; 79 high-fidelity replications of several different types of surveillance and precision approach radar consoles; distributed architecture of microcomputers interconnected via a local network.

Status

In production for the US Navy.

Manufacturer

Logicon Tactical and Training Systems Division, San Diego, California.
(For details see Directory of Suppliers)

Logicon Tower Operator Training System (TOTS)

Logicon produces the Tower Operator Training System (TOTS), Device 15G32 which is used to train US Navy and Federal Aviation Administration (FAA) ATCOs.

Components. A 210° FoV visual display simulates an airfield with moving aircraft on the ground and in the air. Students work at consoles that replicate standard tower hardware. The inputs are transmitted to the computers via a voice recognition system.

Instructor Aspects. Instructors can provide students with immediate feedback through a voice system or through student console displays. The voice recognition and voice generation systems of TOTS enable students to communicate directly with simulated aircraft, eliminating the need for pseudo-pilots.

Software HLL. System software is written in the Ada HLL.

Status

In production. In service with the US Navy's Air Traffic Control School in Millington, Tennessee, and the FAA Academy at Oklahoma City, Oklahoma.

Manufacturer

Logicon Tactical and Training Systems Division, San Diego, California.
(For details see Directory of Suppliers)

IMAGE GENERATION (IG) SYSTEMS

IMAGE GENERATION (IG) SYSTEMS

This section includes systems concerned with Image Generation (IG) such as Computer-Generated Image (CGI) systems using computer or video disks for data storage, systems synthesising imagery from digitised photographs of the real world (Computer Synthesised Imagery or CSI), and systems generating databases in the visual, IR and radar spectrums for use with CGI/CSI systems and their associated displays.

Relevant Terms and Abbreviations. The following terms, used in Image Generation technology and frequently referred to in this section, are defined or explained in the Glossary of Terms section at the beginning of this book.

- Aliasing, Anti-aliasing
- AOI
- CGI
- CSI
- Channel, Visual Channel
- DLMS/DFAD/DTED
- Dusk/Night CGI System
- Edges
- Face, Faces; as Polygon
- Frame Rate (same as Refresh Rate)
- IG
- Interlaced Scan
- LoD
- Mflops
- MIL STD 1821/Project 2851
- MIP-map
- Mips
- Occultation
- Pixel
- Photographic Texture
- Polygon; for specification purposes generally assumed to be the simplest shape, ie, a triangle.
- Processing Power, see Resolution.
- Project 2851/MIL STD 1821
- Refresh Rate
- Resolution and Processing Power
- Surface, as Polygon
- Texel/Textel
- Texture
- Update Rate (of data)

Real-World Imagery for use in CGI Systems

General. Many Computer Generated Imagery (CGI) systems are now capable of accepting real-world (geo-specific) data in a standard form. The fact that real-world data is available for any area in the world without locally-produced survey or intelligence information is dramatically demonstrated by the quality of commerically available satellite photography. Some examples are given in the pictures below, taken from Russian satellites. For military simulators with access to classified higher-resolution imagery, the quality of data is correspondingly higher. Further details of satellite aspects are in *Jane's Space Directory,* published annually.

Simulation Applications. In terms of simulation applications a series of such photographs from a satellite transiting an area of interest, can be transformed by existing software programmes into 3-D real-world imagery. This imagery can then be formatted using protocols such as DFAD/DTED and MIL STD 182 (ex-Project 2851), so that the real-world information can be used by any CGI system capable of utilising such a database format. (See the entry under Defense Mapping Agency (DMA).) Military mission rehearsal using realistic visual imagery of terrain and cultural features is therefore feasible for any of the world, irrespective of its accessibility by ground or by aircraft.

Satellite photography image from a KFA-3000 camera of the Vienna area at a resolution of 2 metres

Satellite photography image from a KFA-1000 camera of the Sarajevo (Bosnia) area at a resolution of 5 metres

Satellite photography image of the Farnborough (UK) area provided to the customer in digitised form at a resolution of 2 metres. Known as DD-5 data (DD = Digital Data) which was originally to a higher resolution from a military intelligence satellite and degraded for commerical use to a 2 m resolution

CANADA

CAE MAXVUE™ Visual System - Image Generation Aspects

The CAE MAXVUE™ visual system was introduced in early 1992 and is CAE's first entry into the visual system market. It is designed to exceed FAA Level D standards while also meeting military requirements. MAXVUE consists of both Image Generation and Display System elements. The IG element is covered here and the Display System is addressed in the next section.

General. The MAXVUE Image Generator employs CGI techniques and uses integrated circuit (IC) technology. The system employs a flexible modular architecture which enables the CGI system to be upgraded in the future. Channels can be linked together via an industry-standard Ethernet bus.

Compactness. Each channel is installed in a single industry-standard chassis. The use of multi-layer boards, CMOS technology and Application Specific Integrated Circuits (ASICs) has resulted in an architecture with few circuit boards. This is designed for high reliability and low spares holdings.

Diagnostics. Centralised diagnostics allow on-line monitoring together with chassis, board level and component level failure identification.

Sensor Integration. The CGI system may be integrated with weather radar, traffic and collision avoidance systems and moving models. Military features include a programmable FLIR post-processor as well as a powerful ancillary processor for model control, line-of-sight determination, and so on. The CGI can be integrated with CAE's Digital Radar Land Mass Simulation (DRLMS) ensuring correlation of database features.

Databases. Databases can be constructed using conventional contour maps and local photographs. They also may be derived from satellite stereo-pair photography and can be supplemented with aerial reconnaissance photographs. Topographical features such as buildings, roads, vehicles, runways, rolling terrain and surrounding areas can be accurately represented. The database may be enhanced using a library of scale models, generic structures and full-colour photographic textures.

Texture and Special Effects. Texture patterns may be scaled, rotated and translated in real time to create dynamic effects such as moving cloud layers, squall lines, ocean waves, smoke, rotating wheels and propellers, rotor down wash, and blowing sand, snow or dust.

Weather Effects. A full range of weather conditions and related visual and directional effects is included.

FoV. Calculations for the following uninterrupted FoV are available: 40 or 45° vertically by 150, 180, or 210° horizontally, employing three to five CGI channels depending on FoV and resolution required.

Display Systems. A number of display systems is available including the MAXVUE back-projected collimated continuous cross-cockpit display system, CAE's Fibre Optic Helmet-Mounted Display (FOHMD) and conventional CRT displays. See the next section on Display Systems for more details.

Display Modes. Raster and calligraphic.

Channels. Up to 8 sharing a database subsystem

Detailed Specifications

Scan Patterns: raster scan/calligraphic
Day/Night Modes: day/dawn/dusk/night
Polygons: up to 5000 per channel. Polygons can be self-textured, smoothly shaded, transparent and self-luminous
Light points: up to 1500 in day scenes, 7500 in night scenes, per channel
Lights: any colour, directional, flashing, rotating beacons and strobes. Landing light lobes steerable or fixed

Note that the total polygon/light point system capacity is dependent upon the specific configuration

Pixels per frame: up to 1.4 million (1180 x 1180)
Data update rate: 50 or 60 Hz for day scenes, 30 Hz for night scenes in order to generate the calligraphic light points
Latency: 60 msec at 50/60 Hz, 100 msec at 30 Hz
Moving models: up to 32 platforms
Anti-aliasing: sub-pixel level
Texture: all colours, 1500 patterns on line (128 x 128), 24 Mtexels in macro texture, 1.4 Mtexels in micro texture, max size 1024 x 1024, dynamic texture included

10 levels of MIP-Maps for smooth transitions at different ranges

LoD Management: smooth LoD management to prevent image popping. In synchronised modes, LoD control/range rings, and unsynchronised LoD control, range rings and/or frame stretch
Collision detection: included
Height above terrain: included

Status

Available. Up to mid-1994, CAE had sold 17 MAXVUE™ visual systems worldwide, including simulators for commercial transports, military helicopters and a tactical fighter using CAE's FOHMD display (qv in Display System section).

Manufacturer

CAE Electronics Ltd, Saint-Laurent, Québec.
(For details see Directory of Suppliers)

CAE MAXVUE™ image with fighter aircraft

CAE MAXVUE™ runway and airport

CAE MAXVUE™ airport apron at Montreal Dorval with buildings and vehicles

CAE MAXVUE™ runway at Ottawa showing dusk scene with calligraphic light points

FRANCE

Sogitec CGI Systems

Sogitec GI 10 000 CGI System

The Sogitec GI 10 000 CGI system has been in service since 1986 when it was delivered to the Centre Electronique de l'Armement (CELAR). It is designed for aircraft and tank simulation, including helicopter applications.

General Characteristics. Day, Dusk, Night system using full colour, and having a texture capability within polygons. Modular architecture is employed.

Update Rates. The update rate is adjustable from 25 to 60 Hz depending on the customer's application and the polygons and resolution required.

Polygons per Channel. 1000 to 10 000 polygons per channel at an update rate of 25 Hz, extendable to 20 000 polygons for special applications but at a lower update rate.

Texture. Textures of various kinds are available within polygons.

Resolution. Computing is for display resolutions from 512 × 512 to 1024 × 1024 (262 k to 1,049 k computed pixels per scan).

Eye Points. Multiple eye points are included and also the ability to simultaneously generate and display separate images from these eye points, such as ownship, adversary, SAM site or 'God's eye view'.

Other Imagery. Infra-red and low level light modes are included.

Enhancements in the 'New Generation' (NG) version. Latest enhancements to GI 10 000 in its so-called 'New Generation' (NG) version include improvements to processing power, reduced dimensions, and the use of generalised phototextures. Processing power is adjustable to user requirements. As an example, at 60 Hz and a resolution of 1024 x 1024 pixels per channel (1049 kpixels), 10 000 polygons per channel can be provided.

Cost. From US$156 000 per channel, although now replaced by the Apogée system described below.

Apogée CGI System

Sogitec has developed a new Image Generator known as Apogée. Some demonstration images were shown at ITEC at The Hague in April 1994 and the development programme is scheduled to be complete by late 1994 for sales in 1995.

Main Features. The main intention has been to improve the richness of scene content with large scale phototexturing, transparencies, shadow and lighting effects, sensors, and visibility effects with range.

Level-of-Detail Management. LoD management techniques are employed to be able to move smoothly from scene to scene, for instance from long-range views derived principally from processed satellite imagery, to close battlefield scenes such as bushes with camouflaged troops.

Architecture and Processing Power. Processing power of Apogée is quoted as 1.6 Gflops. Parallel architectures are used, combined with Digital Signal Processors (DSPs). This has led to greater power in a smaller rack space of office filing cabinet size. As an example, only four processors are used, each on one board in the basic system. For Image Generation, Sogitec has developed four Application Specific Integrated Circuits (ASIC) in 0.7 micron Complementary Metal Oxide Semiconductor (CMOS) technology, each ASIC having 120 000 gates.

Special Effects. Shadows, smoke, dust, rotor wash, patchy mist, breaks in cloud cover and multiple cloud layers. Real-time rendering of lights, for instance lobe shapes and range variation for aircraft and vehicle lights. Shell and bomb impacts will automatically create appropriate craters in the terrain, and destruction of objects such as walls or buildings will automatically reveal the scene behind. Gun muzzle flash and missile plumes.

Animation Effects. Some effects of movement such as a rotating radar antenna, can be automated locally on a geometry processor without increasing demand on the main simulation computer. A helicopter rotor in motion or a soldier running across the battlefield is achieved by first defining initial parameters for the movement required. The appropriate movement kinematics are then stored in memory. These are then called up during real-time play by simply activating the animation sequence when it is required.

Specification Details

Occultation and Z-buffer: Sub-pixel using AZtec™ Z-buffer process

Scan Patterns: raster scan/calligraphic capability

Modes: full day/dusk/night, time of day illumination, horizon glow; IR, FLIR, LLTV, NVG

Channels: up to 12

Imagery from Sogitec GI 10 000 CGI system showing air-to-air refuelling

Imagery from Sogitec GI 10 000 CGI system showing helicopters attacking AFVs on the far side of a village

Imagery from Sogitec Apogée CGI system showing aircraft carrier scene and Rafale aircraft

Imagery from Sogitec Apogée CGI system showing view from ground level of electricity cables and helicopters. Notice the fine detail of the wires and the pylons/gantries/power poles supporting them

Polygons: up to 16 000 per channel at 30 Hz. Three visual attributes per polygon, giving the potential for two non-visual characteristics such as for LLTV/NVGs and TI/FLIR as well as the basic colour visual information. Flat or Gouraud shading
Light points: up to 15 000 per channel, raster and calligraphic processing

Note that the total polygon/light point system capacity is dependent upon the specific configuration
Active display lines: 525 to 1023 lines
Pixels per frame: up to 2 million
Data update rate: variable from less than 25 Hz to more than 70 Hz
Latency: 50 msec at 60 Hz
Moving models: up to 256 with up to 16 articulated parts
Colours: full RGB plus transparency; 4 real-time loadable colour translation tables for sensor simulation
Anti-aliasing: AZtec™ process, 8 level blending on 4 x 4 sub-pixel masks each with Z, area, colour and attributes
Texture: full colour RGB, alpha MIP-map phototexture with tri-linear interpolation. Modulation, transparency and contour texture. Moving texture for sea/cloud animation and other effects. Multiple texture maps per polygon. Microtexture included. Memory of 8 - 32 Mtexels. Management of 4 x 4096 texture maps of 32 x 32 (1024) to 1024 x 1024 (1.05M) texels
MIP-map: included
Height above terrain: included
Collision detection: with ground and with fixed and moving models
Rangefinding: multiple point
Distortion Corrections. Linear and non-linear, static and dynamic, off-axis and asymmetric FoV.

Compatibility with display systems: Compatible with projected and monitor-based displays, HMDs and head/eye-slaved AOI methods.

Databases. Large geographical databases are created using the SINDBAD database creation software (see below) from reconnaissance data and digitised terrain such as from the NATO DLMS system (DTED/DFAD files). Three visual attributes per polygon are stored. The local data base contains more than 650 000 polygons (irregular mesh), more than two million polygons using regular mesh, and more than 300 000 feature instances. The library contains descriptions for more than 4000 features.

Imagery from Sogitec Apogée CGI system showing soldiers getting out of a truck. The men are modelled as 3-D animations

Sogitec Database Tools.
Sogitec has developed two user tools, SINDBAD and CACTUS which allow customers to create and modify their own databases. The acronym SINDBAD is derived from Synthetic Image Numeric DataBase Aided Design. It produces outside-world imagery in the visual, near and far IR and radar spectrums, collision detection, terrain-following altimetry and intervisibility (LoS) calculations. SINDBAD accepts the following source data:

- DMA files (DTED/DFAD data)
- Geographic information systems
- US Project 2851 Database formats
- CAD/CAM models, such as for specific buildings or targets
- Satellite Imagery
- Reconnaissance and other imagery
- Paper maps
- Ground level photographs.

Status
Apogée in development for sales in 1995. GI 10 000 in production. More than 20 Sogitec New Generation GI 10 000 CGI systems have been delivered for simulators operated by the French Air Force or delivered to overseas users by Dassault Aviation and Thomson Training Systems (TTS).

Manufacturer
Sogitec SA, Suresnes.
(For details see Directory of Suppliers)

Thomson Visa CGI Systems

The Visa CGI system was in production from 1982 to 1993 and has had several updates, the latest being Visa IV. It has now been superseded by the Thomson Training and Simulation (TTS) SPACE CGI system (see under International in this section) but over 70 Visa systems are in service totalling over 90 channels. Visa was used in flight simulators for aircraft and helicopters, air combat simulators, armoured vehicle driving and gunnery and ship simulators.

Visa IV. Visa IV incorporated features such as phototexture, detailed three-dimensional surface definition, and a special occultation algorithm (patented) to allow freedom of penetration of moving vehicles and targets in a dense landscape. It is suited to nap-of-the-earth and ground training operations in hilly and wooded country and can be used in low level strike, helicopter flight and tank driving and gunnery simulators. Visa IV can generate images from E-O devices such as intensifiers (NVGs and low-light TV) and IR sensors. Visa IV is also DLMS-compatible and can accept digital terrain data files (DTED data). It was chosen by the French Army for the Leclerc MBT driving and full crew simulators.

Specification of Visa IV
Polygons: 1 million per second or 20,000 at a 50 Hz data update rate
Refresh rates: 25, 30, 50 and 60 Hz at customer choice
Resolution: 2 million pixels (1414 x 1414)
Viewpoints: 5 independent
Simultaneous Windows: 10
Moving objects: 30
Meteorological effects: mist, fog, clouds, cloud layers, rain, snow
Special Effects: interpenetration of targets, dust clouds, weapon firing effects, articulation of moving objects
LoD on Terrain: 4
LoD on Objects: 8
Deployment. Some 90 channels of Visa have been delivered. Applications of the various models of Visa systems have included the following:

- 24 systems for tank driving simulators for the German Army.
- In service at the French Air Force Test Flight Centre at Istres, used for simulation research into future tactical and naval fighters, the so-called ACT and ACM.
- The Swiss Air Force Super Puma helicopter simulator.
- Two systems for Atlantique 2 flight simulators.
- A ship handling simulator for the French Merchant Navy.
- The three-dome air combat simulator at Mont de Marsan for Mirage F1 and Mirage 2000.
- The Mirage 2000C simulator at Dijon.
- The Mirage F1CR simulator at Strasbourg.

Status
In service (see text), but now superseded by the TTS SPACE CGI system.

Manufacturer
Thomson Training and Simulation, Cergy-Pontoise.
(For details see Directory of Suppliers)

Imagery from Thomson Visa IV CGI with phototexture

Typical tank driving simulator image from Visa IV CGI

GERMANY

Atlas DISI4 Computer-generated Image System

The Atlas DISI4 is a third generation CGI system producing real-time anti-aliased images together with effects such as day/night, fog and haze with a high degree of three-dimensional realism.

Features. Features of DISI4 include the following: thermal imaging, FLIR and LLTV images; continuous visibility and lighting conditions; variable fields-of-view; anti-aliasing; transparency; three-dimensional texture; pixel-by-pixel Z-interpolation; sun shading; level of detail and overload management; a built-in test and diagnostic system; a large screen projection capability; and a laser rangefinding capability.

Database Generation. Compatible with the Atlas Elektronik DataBase Generation System (DBGS), operable by the user.

Applications. DISI4 has applications in the fields of flight and combat simulation, armoured combat and gunnery simulators, ship handling and other naval simulators, and driving simulators. It is also used on tank gunnery training systems.

Status

In production. In service for a number of applications.

Manufacturer

Atlas Elektronik GmbH, Bremen.
(For details see Directory of Suppliers)

Imagery for low flying from the Atlas DISI4 CGI system

Imagery from ground level from the Atlas DISI4 CGI system

Dornier Visual Sight Simulation System

This system is intended to simulate the view through weapon and other sights. It consists of two subsystems: a Landscape Generator LSG 1000 produced by Dornier GmbH; and a Target Generator ZG 1000 produced by Deutsche Systemtechnik GmbH. Development commenced in 1985/86 and the system is now in production.

Applications. This modular sighting system has simulator applications for surface-to-surface as well as surface-to-air weapon training systems. It presents air and ground targets moving in a natural environment.

Instructor Aspects. Targets, lighting conditions and special effects can be controlled on-line by an instructor.

Background Scene. The scenery is generated by the LSG 1000 system through digital processing of photographed landscapes. Landscape reproduction in full colour is provided, from a number of eye points as required.

Target Images. The ZG 1000 system creates three-dimensional air and ground targets using CGI techniques.

Databases. Generation of user-specific databases is included.

Resolution. Resolution is 1024 × 1024 pixels at an update rate of 50 Hz (52.4 Mpixels/sec).

Status

In production.

Imagery from the Dornier Visual Sight Simulation System which uses digitised photographs for the background scene and CGI for overlaid targets

Manufacturer

Dornier GmbH, Friedrichshafen.
(For details see Directory of Suppliers)

MBB Alphavision CGI Systems

The MBB Alphavision CGI system is produced in two versions called the 1000 and 2000.

Alphavision 1000

Alphavision 1000 is the more compact of the two versions and is based on a VME-bus system. The system generates flat shaded faces using binary space partitioning technology for hidden surface solution. The number of faces can be increased or decreased by changing the system update rate.

Components. The main elements of Alphavision 1000 are a preprocessor, a geometry processor and a display processor.

The preprocessor consists of an industrial computer configuration including hard and floppy disks for data storage and programs. The software applicable to the host processor contains an operating system, real-time software and system test capability. The geometry processor is a high-speed processor which selects, orders and projects items from a memory resident 3D database into a 2D screen polygon list according to viewer position. The conversion of the 2D display list is performed by the display processor and colours and patterns are added. An integrated digital/analogue converter performs the video output.

Configuration of Channels. The standard configuration consists of a one-channel system installed in a 19 in (483 mm) rack which can take up to three channels if required. A maximum of nine channels can be installed in a 19 in (483 mm) cabinet.

Resolution and Data Update Rate. The standard display output is 800 000 pixel/channel (about 894 x 894) in a 60 Hz non-interlaced format.

Special Effects. Special features are time of day, fog simulation and up to 200 different dynamic controllable lights.

Database Generation. Database generation for Alphavision 1000 can be carried out using a database software package running on a personal computer suite including a mouse or optional digitiser, model/object library modules, scene presentation on a high resolution colour monitor, and an Alphavision module for on-line verification of the generated scene.

Alphavision 2000

Alphavision 2000 was designed using three different

eight layer printed-circuit board types complemented with VLSI/CMOS technology, digital signal processors and MBB-developed Application Specific Integrated Circuits which together provide approximately 10 Gflop in computing power for a single-channel configuration.

Components. Alphavision 2000 has a modular reconfigurable architecture based on three board types; a database manager, a geometry processor and a raster processor.

The database manager stores the on-line database, computes the potential visible units, determines the appropriate level of detail and initiates the transfer of face and vertex data to the geometry processor. This performs the transformation of vertex data (3D into 2D), polygon clipping at channel boundaries, and visibility and illumination calculations. In addition it computes the coefficients for shading, translucency, texturing and initiates the transfer of data to the raster processor array. The raster processor clips those polygons which contribute to its particular window, generates pixel entries based upon a Z-buffer algorithm, texture maps and finally outputs the resulting colour to the video generator.

Channels. Up to eight channels can be employed per system. Within a single channel a number of hardware configurations is available on anti-aliasing and channel resolution, at customer choice.

Anti-Aliasing. The level of anti-aliasing can be selected from the combinations of 1 × 1, 2 × 2 or 4 × 4 sub-pixels. Aliasing effects are reduced by spatially filtering the rendered image using up to 16 sub-pixels for each displayed pixel. Calculations are carried out to a level of a sub-pixel for colour, texture, shading, fading and translucency.

Resolution. Channel resolution is selectable up to 24 million pixels/second (eg 1000 x 1000 at 24 Hz).

Z-Buffer. Alphavision 2000 utilises the Z-buffer algorithm to resolve priority problems by using the true depth from the eye point to the sub-pixel. The advantage of this approach over a polygon or object-orientated approach is that priority issues are independent of polygon aspect ratios, complexity of object architecture and movement of models within a complex scenario.

Photo Texture. Alphavision 2000 supports the application of photo-derived colour texture to any type of polygon. All texture-specific calculations are done on a sub-pixel basis resulting in sharper textures even at short range.

Imagery from the MBB Alphavision 2000 CGI system

Shading. The Alphavision 2000 system supports two shading capabilities which can be applied simultaneously to any textured polygon. Curved surface shading minimises the number of polygons required in the representation of smooth three-dimensional objects. Sun angle shading improves the three-dimensional perception by accounting for sun illumination in a complex scenario.

Visibility Calculations. The system uses the depth values stored in the Z-buffer algorithm to perform visibility calculations on a sub-pixel basis. Visibility within each of the 256 fog layers is selectable, thus providing a more realistic scenario even when mountains protrude through clouds or fog banks.

Database Generation. The configurable MBB database generation system (Alpha-Gen) comprises hardware and software tools which allow a user to create, modify and verify databases rapidly for Alphavision 2000.

Status

In production.

Manufacturer

Messerschmidt-Bölkow-Blohm GmbH (MBB) Training Systems, Munich.

(For details see Directory of Suppliers)

INTERNATIONAL

Thomson Training and Simulation (TTS) SPACE CGI System

Thomson Training and Simulation (TTS) has developed the SPACE CGI system after a 2 year development programme involving teams from TTS units in France and the UK . This is why this system is placed under an 'International' category rather than being listed under any one country. After Thomson acquired Link-Miles (UK) in 1990 with its IMAGE family of CGI systems, some rationalisation with the Thomson-CSF Visa CGIs has been likely. SPACE is the result of that process.

Acquisition of Rediffusion from Hughes. The acquisition of Rediffusion (UK) from Hughes in late 1993 did not alter the situation as far as CGI was concerned since Rediffusion did not make CGI systems themselves but had a business agreement with Evans & Sutherland of Salt Lake City, Utah for their SP, CT and ESIG range of CGI systems, marketed by Rediffusion under the title Novoview.

Display Systems. Display systems are another matter and are covered in more detail in the next section. TTS now owns the ex-Rediffusion WIDE™ collimated cross-cockpit system and also the ex-Link-Miles AWARDS system, both achieving the same object, in the case of WIDE by flexible mirrors held in place by suction, with AWARDS by glass mirrors.

Details of SPACE CGI

Details of the TTS SPACE CGI system are as follows:

Main Characteristics of the SPACE System. High-Definition TV (HDTV) and RISC processing techniques are used. The use of HDTV architecture means that no interlaced techniques are needed in image generation and display interfaces and data update rates are the same as display refresh rates.

TTS SPACE CGI system showing airfield and surrounding terrain

TTS SPACE CGI low-level battle scene with buildings, explosion and helicopters

TTS SPACE imagery for driver training

TTS SPACE scene with tanks

Fine Grain Features. High resolution inputs into scene content are provided by MicroTerrain, MicroFeatures and MicroTexture.

Texture. Sophisticated texture capabilities are provided in full colour including micro-texture, photo-texture and translucencies. See the more detailed specification below.

Use of PCs. PCs are used for set up, software loading, disk management, diagnostics and so forth. The Windows interface is used with the PC.

VCR Compatibility. SPACE is compatible with a standard TV VHS Video Recorder (VCR).

Image Storage. Images may be stored on a PC in TIF format.

Channels. Channels are configured as graphic pipelines for input into a number of different display systems in multi-channel layouts in order to achieve the customer's required FoV and resolution.

SPACE Specifications

Scan Patterns: Compatible with raster and raster/calligraphic systems, and Helmet-Mounted Displays including commercial 'Virtual Reality' interfaces

Day/night modes: day/dusk/night with continuous time of day or night

Time of day control: sun/moon.

Horizon glow

Shadows

Directional light sources

Channels: up to 8 per system, each channel up to 4 display windows, giving a total of possible display windows of 32

Viewports: multiple

Polygons: up to 2000 per channel at 50 Hz

Light points: up to 6000 per channel at 50 Hz, traded off 3:1 with polygons. Calligraphic and raster light points provided. Up to 2 million texel (texture-pixel) light points

Note that the total polygon/light point system capacity is dependent upon the specific configuration

Active display lines: up to 2000 x 2000 per channel, either interlaced or non-interlaced

Weather effects: multi-layer broken cloud and scud

Fog with RVR control and patchiness

Visibility blending with terrain including fog and/or cloud breakthrough

Snow with blowing, drifting and build-up

Rain, thunderstorms, lightning

Special weather processor for precipitation effects

Pixels per frame: up to 4 million

Data update rate: 50 Hz, other rates available including 60 Hz

Latency: 80 msec at 50 Hz

Moving models: up to 16 platforms all with 6 DoF, processed simultaneously. TCAS supported

Colours: all colours available (nominally 16M)

Anti-aliasing: sub-pixel

Transparency: 32 translucency levels

Texture: full colour, two-colour, monochrome and translucent photo texture;

up to 4 texture maps can be combined per polygon to create texture modulation for contours and special effects;

programmable texture map resolutions from 32 x 32 to 2048 x 2048 texels (1024 to 4.19 Mtexels);

Gouraud shading and texturing simultaneously;

Infinity Texture provides complete scene cover with geospecific photo texture and is continuously refreshed;

geospecific phototexture derived from satellite, aircraft or ground sources;

up to 12 MIP-map levels of detail for smooth transition at all ranges;

up to 5 Mtexels per second dynamic update with image rendering maintained (100 000 texels at 50 Hz update rate);

up to 20 Mtexels in active memory (both generic and geospecific);

Distortion Correction: Non-linear distortion correction included, including keystone with dynamic option.

Scene Management. Continuous Level of Detail (LoD) control including Adaptive Terrain Polygonisation (ATP) in order to maintain critical terrain features at longer ranges. MicroTerrain and MicroFeatures use real-time algorithms to enhance foreground detail. Up to 8 LoD for objects, continuously blended between levels.

Occultation and Z-buffer. SPACE includes the MSA facility which is an occultation algorithm which is claimed to remove the constraints of priority modelling of earlier systems. It includes a 1/16 pixel Z-buffer with a precision of 64 bits in IEEE floating point format. This provides precision occulting, interpenetrating surface capability and freedom for fixed and moving objects.

Collision Detection: included.

Height Above Terrain: multi point calculations included.

Database. SPACE features InstantScene (SCENEMAKER), a database modelling tool which allows users to easily model limited databases by inserting library objects into the visual scene. In addition, EventProcessing gives the ability to modify the database as a result of an event such as an explosion, impact of rounds, and so forth. Database tools are based on the Multigen interface. Some 240 000 polygons can be in the active database including MicroTerrain and MicroFeatures.

Computing: VME 88000/i860 with Ethernet, HSD, SCSI and other interfaces.

Diagnostics: Fuzzy logic and JTAG provide test coverage and on-line monitoring.

Status

Available.

Manufacturer

Thomson Training and Simulation (TTS) France, UK and USA.

(For details see Directory of Suppliers)

TTS SPACE scene on approach to runway

TTS SPACE airport scene for aircraft taxying towards passenger gangway

ITALY

Alenia CGI Systems

Alenia is involved with a number of CGI systems. Some high-end systems are manufactured under licence but Alenia has its own low-cost system, developed for Computer-Aided Training. Some characteristics include the following:

General. The Alenia Low-Cost CGI System (ACS) depicts real-world scenes with 3D terrain and has correlation with radar and other sensor information.

Database. Use of DMA/DMLS digitised information in DTED and DFAD file formats. Ability to process satellite imagery. 3D model creation.

Texture. Photographic cell texturing. Microtexture.

Height above Terrain. Included.

Collision Detection. Included.

Data Update Rate. 30 Hz to 60 Hz.

Polygons. up to 80 000 on-line.

Latency. 68 msec.

Models. Multiple 6 DoF moving models.

Display Compatibility. Raster, calligraphic, monitors, collimated monitors, light valves, video projectors, back-projected collimated cross-cockpit systems, direct projection on screens of domes.

Status

Available.

Manufacturer

Alenia, Goriza.

(For details see Directory of Suppliers)

Alenia Low-Cost CGI System with Tornado low flying in rugged terrain

Alenia CGI runway approach scene

NORWAY

Norcontrol NorView Visual Systems

Norcontrol has delivered more than 300 simulators to more than 40 countries. Products range from PTTs to full mission ship handling simulators and propulsion trainers. Norcontrol NorView visual systems are designed principally for ship's bridge and ship handling simulators. They can be configured to a horizontal field-of-view of 360° if required by using an appropriate number of CGI channels and associated display windows. NorView versions in increasing order of capability are numbered 1000, 2000, 3000 and 4000. The complete visual system consists of a CGI element and projected displays. The CGI element is described in this section.

Basic Characteristics. The Norcontrol CGI is a full colour day/dusk/night system and includes features such as shading, light anti-aliasing, fog, flashing/occulting/sectored lights, range dependent light intensity, wind dependent sea texture, target ship bow wave and stern wake, urban textures, bow image, and others.

Update Rates. NorView-1000 updates images at 15 Hz and NorView-4000 at 30 Hz with the 2000 and 3000 systems at intermediate values.

Resolution. Display resolutions range from 768 × 512 (393 kpixels) to 1280 × 1024 (1311 kpixels).

Polygons. Polygons per channel per second are specified as varying from 12 000 to 60 000; at 15 Hz this is 800 - 4000 polygons per channel and at 30 Hz, 400 - 2000 polygons per channel. Polygons are Gouraud shaded and anti-aliased in systems except the basic NorView-1000.

Targets. The maximum number of moving targets per channel ranges from 10 to 60.

Lights. The maximum number of lights per channel ranges from 800 to 3000.

2000 - 4000 Versions. NorView-2000/3000/4000 versions include features such as surface anti-aliasing, and variable sun angles.

3000 and 4000 Versions. NorView-3000/4000 features include cloud and rain, target vessel texture and spherical earth effects.

Computers. The hardware is based on the 80860 RISC processor from Intel.

Hardware. Norcontrol visual systems use custom-built hardware. The hardware is flexible in design, and can be configured to meet customer specifications.

Imagery from Norcontrol NorView CGI system

Status

NorView-1000/2000/3000 is in volume production.

NorView-4000 has been sold in Canada and the United States.

More than 60 visual systems have been sold to customers in Australia, Bulgaria, Canada, Finland, Iran, Norway, Russia, United Kingdom and the United States, totalling more than 200 visual channels.

Manufacturer

Norcontrol Systems AS, Horten.

(For details see Directory of Suppliers)

ROMANIA

Simultec CGI Systems

SIMULTEC manufactures the TCGI-1000 and the GEN-500 CGI systems.

TCGI-1000 CGI System

The Transputer-based Computer Generated Image (TCGI) -1000 is a low-cost real-time 3D graphics generator for visual simulation applications. One channel of TCGI-1000 corresponds to one display unit which can be a monitor-based system or a projector

Components A PC 286, 386 or 486 is used as a host, and all software and visual databases are held on the PC hard disk, downloaded as required to the TCGI-1000 module. This module is made up of 3 types of boards: The MS-08 board is a PC AT motherboard for Transputer Modules (TRAMS) which installs up to 10 TRAMS in a single 8-bit PC ISA slot. The MS-26 board is a T805 20 MHz unit with 4 Mb DRAM and one TRAM. The MS-100 board is a graphic module with T800 20 MHz IMS G364 video controller with 2 Mb of video RAM.

Performance. Listed performance with a network of 16 transputers:

Resolution. 640 x 480 (307 kpixels) to 1024 x 1024 (1049 kpixels), non-interlaced.

Data Update rate. 60 Hz.

Modes. Day, dusk, night.

Screen Refresh Rate. 30 Hz.

Latency. 83 msec.

Channels. up to 4.

Independent View Points. 1 per channel.

Polygons. 1000 per channel.

Lights. 500 per frame per channel.

Weather. Models for fog and haze.

Database. 3D objects using binary space partition technology for hidden surface solution. Objects modelled with eight levels of detail. Database generation from a PC-based software package. On-line database capacity is 10 000 polygons.

GEN-500 CGI System

The GEN-500 CGI system is a real-time 3D graphics image generator designed for simulation applications. It is based on parallel architecture, RISC processors, and Intel I860 and T805 transputers for the geometry engine and specialised hardware for graphic rendering.

Components. Four boards comprise the GEN-500 system: An MS-08 motherboard for connection with a PC 386 or 486 host computer. MS-860 processor board with Intel I860, T805 transputer, 16M DRAM, four modules per channel. MS-ARM anti-aliasing render module, four modules per channel. MS-TM texturing module, four modules per channel.

Performance. Listed performance is as follows:

Display. Raster.

Resolution. 640 x 480 pixels (307 kpixels), 24 bit RGB, double buffer.

Data Update Rate. 30 Hz.

Latency. 83 msec.

Polygons. 500 per channel at 30 Hz, textured, anti-aliased, displayed.

Anti-aliasing. 4 x 4 sub-pixel mask.

Texture. 12 photo-derived, MIP-mapped, colour (R, G, B, A).

Weather. Haze, fog, time of day.

Scene Management. Frame synchronisation, variable visibility range.

Status

Available.

Manufacturer

SIMULTEC SA, Bucharest.

(For details see Directory of Suppliers)

UNITED KINGDOM

TTS (ex-Link-Miles) Image Generation

Although the IMAGE series has been superseded by the TTS SPACE visual system, many IMAGE systems are in use and rather than remove this entry it was felt that it would be useful to continue to describe the IMAGE CGI systems still in use.

IMAGE II

IMAGE II provides dusk/night visual scenes in raster and calligraphic light point outputs and complies with FAA Level C standards. IMAGE IIT has simple texture on horizontal surfaces and was the equivalent of the Rediffusion/E&S SP 1 CGI system.

IMAGE III

IMAGE III is a full colour day/dusk/night visual system complying with FAA Level D requirements. IMAGE IIIT has texture on horizontal surfaces. Upgrade from dusk/night IMAGE II to full day colour IMAGE III, from a computing point of view, involves nothing more than the addition of the required circuit boards. Similarly texture can be added to either II or III by adding the necessary printed circuit boards. Parallel processing allows the addition of further displays without the need to purchase the next more powerful mini-computer. IMAGE III was the equivalent of the Rediffusion/E&S SP 3 series of CGI systems.

IMAGE IV

IMAGE IV includes: full planar texture (that is on all 3D surfaces), greater scene content, more weather effects,

TTS Seescope target display (see next page)

IMAGE IV showing textures

IMAGE IV showing photo texture on blocks of apartments

translucency, multiple targets, in-sun and down-sun visibility effects, and multiple viewpoints. Sensor simulation such as forward-looking infra-red (FLIR) and low-light television can also be included.

Resolution is 756 000 pixels/channel (869 x 869) and multi-sided polygons, surface overlay techniques and texture, realistic shadows and variation of light source, lightning, thunderstorms and rain, together with the effects of drifting snow, moving clouds, fog and sea surface. IMAGE IV may be specified as a raster and calligraphic system or raster alone. It comes in -250, -500 and -600 versions corresponding to the number of polygons required by the customer.

Sales of IMAGE Systems

Link-Miles sold 98 IMAGE systems including 20 IMAGE IVs of which 16 feature the company's AWARDS display system.

Seescope Periscope Visual System

The Seescope submarine periscope display has a 360° background scene and surface contact/target imagery. There are day, dusk, night, mist and thermal imaging modes. Seescope has a textured sea surface, dynamic bow waves and stern wakes, own mast, and wash-over and green water effects. The user has freedom in the type of contact/target to be displayed, including surface and air contacts. Vessels are presented in the correct orientation, size and speed for the selected scenario relative to own vessel. Simulation of rangefinding using either optical or laser techniques, plus weapon effects, can be provided. The ocular box can be supplied by the user or a simulated ocular box is available which provides all the functions necessary for simulation.

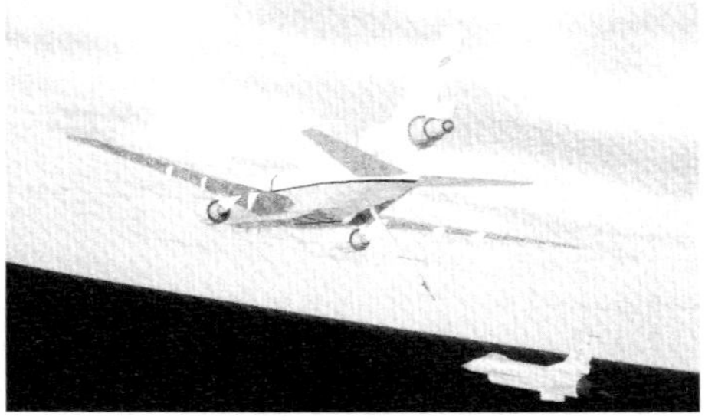

IMAGE IV database showing KC-10 flight refuelling scene

Interactive Database Editor and Linker (IDEAL)

IDEAL is a tool for the preparation of CGI databases by the user. Manual, off-line coding methods are replaced with on-line interaction in real time using a work station; database modellers can modify any part of an existing database or produce new scenarios. Once prepared, databases can be transferred to and from IMAGE systems and from VISTA (see below).

Visual Imagery Simulation Training Aid (VISTA)

VISTA is a low-cost CGI system. Potential uses include basic flying training, aircraft recognition and basic weapon aiming and gunnery procedures. A full daylight, dusk or night full-colour image can be viewed either as a projected image or on a screen. VISTA has been selected by the Royal Navy as part of the SHORT training package (qv in the Flight Simulator and FTD section under TTS).

Status

The IMAGE series of CGI systems has been superseded by the TTS SPACE system, but continues to be supported in service.

Manufacturer

Thomson Training and Simulation Limited (TTSL), Crawley.
(For details see Directory of Suppliers)

Tector Vistar CGI Systems

Tector Visual Systems has produced low-cost CGI systems and Collimated Monitor displays since the company was formed in 1974. The Tector OPDIS CGI system was probably the first low-cost real-time CGI system in the world, and was based on a flat world made up of an infinite series of rectangular field patterns with overlaid airfield and other features. OPDIS has now been superseded by the Vistar series described below.

Tector Vistar 1

The Vistar 1 low-cost CGI system is all-digital and generates a full-colour, day-through-night image. It has modular design and construction and a high expansion capability. The system is available singly or in multichannel layouts and the image may be displayed via Tector's collimated or projected display systems.

Main Features. Day-through-night; infinitely textured full-colour terrain area; flexible high, medium and low texture; single or multichannel operation; four airfields modelled from real-world data with two on-line at any one time; lights; cloud (variable cloud base and top), scud and fog; runway visual range; angle of approach indicators; unlimited two-dimensional surfaces.

Resolution and Scan Rates. Vistar 1 has an 875 line × 1024 pixel (896 kpixels) scan with a constant field rate of 50 Hz. The scan system is a 2:1 interlaced raster image and so update rate for complete replacement of data is 25 Hz and 22.4 Mpixels/s. If required 525 or 625 lines scan can be provided and the field rate can be provided at 60 Hz.

Texture Maps. The system provides unlimited two-dimensional surfaces which are enhanced by six full-colour multi-layered texture maps. These maps provide field control and pattern, medium and fine texture, runway control and markings.

Airfield Models. Within the scene up to four airfield models are provided. These are modelled from real-world data and contain the normal runways, taxiways, aprons, lights, glide path indicators and building areas.

Terrain Patterns. Field patterns are all irregular in construction and use 256 colours. Urban areas are depicted using special colourings and light clusters whilst forest areas and lakes are shown via colour and texture.

Anti-aliasing. All lights and markings are anti-aliased to one-sixteenth of a pixel and texture maps utilise a separate anti-aliasing algorithm.

Lights. Vistar 1 can instantly display 2000 raster lights with an option for 4000. These are drawn from a larger database and use 256 colours. Groups of lights utilise up to 16 intensity levels. Outputs for calligraphic light control can also be provided.

Polygons. Because of the system architecture a polygon count in the normal way cannot be quoted. However, each field pattern map contains the equivalent of between 50 and 250 polygons and each airfield model between 3000 and 75 000 polygons although not all of these will be on the screen at any one time.

Weather. Vistar 1 provides full meteorological effects as part of the standard package. The cloud top and base are variable on line in increments of 1 m, allowing the trainee pilot to fly below, through and above cloud. The instructor may select a random variation in cloud base and top to present a trainee with scud.

Visibility and runway visual range (RVR) is adjustable on-line in both slant range and RVR. Visibility is adjustable from zero to maximum in small increments. The effect due to RVR is displayed below a height of approximately 30 m. The visual effect due to slant range is displayed above this height.

Interfaces. All facilities are controlled from the simulator host computer. Interfacing is via a 16-bit parallel link operating at 50 Hz. The interface contains a simple handshake facility to synchronise the host computer to the visual system or generate any requests or interrupts. The interface between the image generator and display system is by means of standard RS170/RS343 video cables.

Latency. The maximum transport delay is 80 ms.

Vistar 1A

Vistar 1A is an enhancement to Vistar 1 to meet FAA Level C (Phase II) requirements but with the emphasis remaining on low cost.

3D Objects. In addition to the Vistar 1 circuitry, Vistar 1A contains a 3D image generator to provide a number of ground-based and/or dynamic 3D objects representing, for instance, buildings, hills or conflicting traffic.

Polygons. The standard Vistar 1A is currently providing about 400 polygons for 3D features, with the option for 4096. The objects are clipped by cloud base and cloud top and their face colours are processed to provide the appropriate atmospheric and sun illumination effects.

Database Package

Both Vistar 1 and Vistar 1A include as standard a full database preparation package allowing the user to create or modify databases with no additional equipment.

Cost

Vistar 1 is quoted at about £55 000 per channel and Vistar 1A at about £70 000 per channel.

Vistar 2

Vistar 2 is a full 3D high resolution, high performance image generator designed for FAA Level D (ex-Phase III). Design of this system is continuing and it is forecast to be available towards the middle of 1995.

Status

Vistar 1 and Vistar 1A available. Vistar 2 available 1995.

Manufacturer

Tector Visual Systems, Newark, Nottingham.
(For details see Directory of Suppliers)

Tector Vistar 1A imagery showing buildings on an airfield

Tector Vistar 1A imagery showing coastline and texture features

Tector Tecstar CGI System

Formerly known as ATIGS (Aerial Target Image-Generation System), the Tector Tecstar system computes three-dimensional, fully manoeuvrable images representing targets or objects of interest on ground, sea or in the air. These images can be keyed in to the OPDIS or Vistar 1 computer-generated ground/sky background, or used in isolation as targets projected onto the domes of air combat simulators. A Tecstar 2 version, with 256 surfaces, was announced in 1986.

Tecstar has been supplied to British Aerospace at Warton and the Royal Aircraft Establishment at Farnborough for air combat domes. The BAe system was integrated into a twin-dome simulator in late 1985. Tecstar was supplied to the Austrian Army as part of the Invertron Skyfire System A anti-aircraft gunnery trainer.

Status

Available. In service in Austria, the United Kingdom and elsewhere.

Manufacturer

Tector Visual Systems, Newark, Nottingham.
(For details see Directory of Suppliers)

A view through the sight of an air defence part-task trainer showing a Tecstar-generated 'Fitter' target

Real World Simulation CGI Systems

Real World Simulation (RWS) used to be called Real World Graphics and during 1994 moved from Hertford to a new site at St Albans.

RWS Reality Low-Cost CGI System

The RWS Reality CGI system uses parallel microprocessor technology and is a modular and expandable real-time 3D graphics computer system designed for visual simulation applications. Based on the industry-standard VME bus format, the system may be configured as a single channel with an unlimited number of channels providing an associated display system with wide panoramic views. The system has been used for ship bridge simulators produced by Norcontrol of Horten, Norway (qv under Norway entry). Since this entry was prepared more details have become available of the latest version of Reality and are given in the Addenda section.

RWS Super Reality

Announced in 1991, the RWS Super Reality CGI system is based on the VME 6U board format and can be integrated with all work stations and industrial computer systems using this bus.

Resolution. Resolutions are programmable from the US TV NSTC format up to 1024 × 768 non-interlaced (786 kpixels), all with 24 bit, true colour.

Polygons. A fully configured system will have a performance of 500 000 textured or Gouraud-shaded and anti-aliased polygons/sec.

Data Update Rate. 60 Hz maximum update rate.

Features. Depth cueing, LoD control by culling, height-above-terrain and collision detection are included in the system's Reality Graphics Environment (RGE). RGE is a real-time, multi-tasking, parallel processing, graphics-specific operating system that runs on the Super Reality system and provides features needed for integration with other simulation systems.

RWS Super Reality CGI imagery with runway and aircraft

Compatibility. Super Reality is software-compatible with Reality PC. It is based on an array of up to four Intel 1860 RISC microprocessors. Reality PC is a single PC/AT-compatible board product.

Status

Reality in service in Bulgaria, Canada, Norway, Russia and USA. Super Reality available.

Manufacturer

Real World Simulation (ex Real World Graphics), St Albans.
(For details see Directory of Suppliers)

Primary Image CGI Systems

Primary Image (formerly Akebia) has manufactured a number of CGI systems since the early 1980s.

Advanced Graphics Engine (AGE)

The Advanced Graphics Engine (AGE) family of CGI products is no longer manufactured. Details of in-service AGE systems can be found in previous editions of this book.

STRIDE PC-based CGI System

The Primary Image STRIDE is a CGI system capable of generating high resolution Gouraud shaded visual images, and was launched in 1991. STRIDE comprises a single PC base card and a number of optional plug-in modules and cards. It is claimed that it can meet the requirements of many training systems which have previously used more expensive computing.

Typical Systems. Typical STRIDE systems comprise from one to three cards although more complex systems are possible. Modular design ensures that the system can be closely tailored to a required functionality. The basic STRIDE can provide generation of targets, targets on a digitised background, or a completely computer-generated display which may be a highly realistic image or a schematic. The addition of options to the basic card, or additional special purpose cards, allows both the speed and functionality to be enhanced. The number of targets, the target complexity and the sophistication of the image generation can be determined according to the application requirement in both software and hardware.

Roles. STRIDE is suitable for target generators, visual systems for part-task trainers, instrument displays, simulated radar displays, control tower visuals and similar applications, viewing systems such as periscope visuals, and schematic displays.

Components. STRIDE comprises a number of processors and co-processors selected to provide the best overall drawing and rendering capacity. This enables bottlenecks caused by interface speeds or limited capability in any one area to be avoided. The base card is fully PC-AT bus-compatible but avoids bus bandwidth limitations by keeping all the I/O intensive elements on board. Separate internal pixel and polygon interfaces

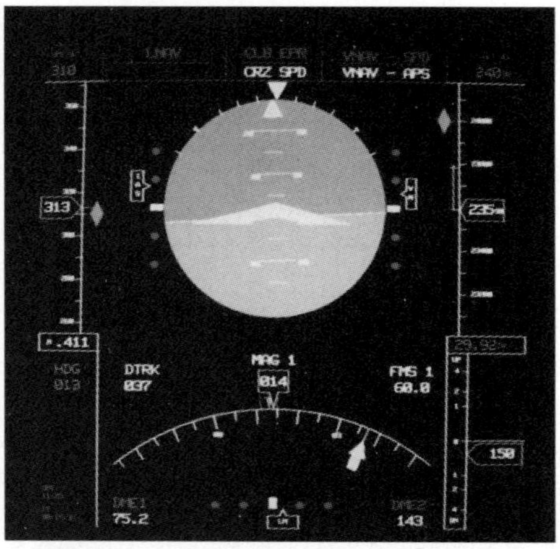

Primary Image Instrumentation Graphics Environment (IGE) instrument display

Primary Image STRIDE imagery

are provided for the optional rendering cards. This approach also ensures that minimal load is placed on the PC's own processor(s), allowing the PC to manage the training while STRIDE creates the visuals.

Video Formats. A fully programmable pixel clock ensures that all standard interlaced and non-interlaced video formats can be generated.

Resolution. A resolution of 1280 × 1024 non-interlaced (1311 kpixels) is supported, with resolution being increased if an interlaced video format is appropriate.

Polygons. The current rendering engine is capable of up to 35 000 polygons/sec (1400 at 25 Hz update), double with a second rendering engine. The polygons are 24-bit Gouraud shaded, on-screen Z-buffered and anti-aliased.

Flexibility. The multi-processor hardware concept and internal bus structure is designed for easy expansion and upgrade. Similarly, the software is object-oriented and allows a relatively high degree of independence from hardware. For example, all databases are initially created in a hardware-independent form. The database concept is itself extensible. This allows development of STRIDE in a phased and controlled manner, incorporating new processor technology whenever required.

Software Support. Software support is available for the types of application described or Primary Image will supply an image generation solution with one or more STRIDE channels.

STRIDE XMEM. In April 1993 the STRIDE XMEM was introduced and, as its name implies, it provides an extension of image memory. It can be used in standalone applications or with other STRIDE configurations. It is particularly suited to applications where CGI targets are used with a digitised photographic background image, a combination sometimes known as Computer Synthesised Imagery (CSI). CSI provides the facility for real-time panning over large area, high-fidelity images.

STRIDE-PLUS. STRIDE-PLUS is a special configuration with no SCSI support and less memory, coupled with one or more STRIDE basecards. This provides scalable power in applications such as the emulation of electronic displays such as EFIS.

Costs. Cost per channel varies from £3000 to £19 000 depending on the precise configuration.

Instrumentation Graphics Environment (IGE)

IGE is a system designed for the prototyping and implementation of real-time dynamic graphics systems.

Applications. Applications include:

electronic instrumentation systems, especially for aircraft trainers (such as the Electronic Flight Instrumentation System - EFIS),

mimic display systems, both for trainers and actual systems,

real-time graphical user interface (GUI), both for trainers and actual implementations of complex systems requiring GUI.

Graphics Design. The prototyping or display design is performed using a high level application graphics design system originated by the SL Corporation and called SL-GMS. This can be used on almost any standard work station and is object-orientated and flexible.

Real-Time System. The real-time system is based on the Primary Image STRIDE system (see above). The software is able to interpret the applications graphics designed using SL-GMS and animates them in real time using data from the application computer system. Various interfaces can be provided according to system requirements. Primary Image can design the application graphics if required or supply the real-time system.

IGE Concept. The IGE concept allows the application graphics design to be separated from programming and software design, and the design changes can be implemented with ease and speed using the power and user-friendly functionality of SL-GMS on a work station. The system imposes the minimum requirements on the application computer. For instance, a GUI can be implemented using IGE when the application computer has neither the spare capacity to run a real-time GUI, nor the high level software tools to facilitate the development.

Status

In service and in production. Systems sold to France (3 customers), Japan (2), Switzerland (4), UK (19) and the USA (4),

Manufacturer

Primary Image Limited, Kingston-upon-Thames.
(For details see Directory of Suppliers)

MST CGI Systems

MST CGI with eight target images

Marconi (ex-Ferranti) Simulation and Training (MST) has been involved in the simulation of narrow angle/high resolution visual sensor systems since the late 1960s and has produced systems for submarine periscopes, weapon sights and missile guidance applications. Early systems were based on closed-circuit television although these have now been replaced by CGI systems.

General Characteristics. The CGI system generates images in full colour and target occultation calculations are included.

Resolution. Calculations are made to a 1024 line standard (about 1 Mpixels).

Database. Operating from a digital database the system has a built-in database generator with modification facilities.

Other Sensors. Low-light level television and thermal imagery simulation can be incorporated.

Targets. There is a multi-target capability; the targets are detailed images with smooth shading. The targets can feature bow waves and 'heel under wheel', and can be selected from a library. The target can be located in a variable environment that includes textured sea and sky, variable sea states and visibility, plus time of day effects. Other effects include periscope lens washover.

Latency. Response time is 20 ms.

Other applications. The system can be used in other narrow angle applications where high resolution and rapid response are important, for example with certain weapon training programmes.

Submarine Periscope-View Simulation

A multi-target scenario can be provided with colour, sea state and visibility. The image is displayed on a high resolution monitor incorporated into a replica periscope built to a customer's submarine fit. Range, bearing and elevation estimation systems are included. The periscope structure is floor-mounted on a column and does not require a load-bearing ceiling. The latest of these systems was developed to meet a requirement for periscope view simulation for submarine command team trainers.

Status

Available. In service with the Royal Australian Navy, the Royal Navy and the US Navy.

Manufacturer

Marconi Simulation and Training (MST), Cheadle Heath, Stockport.
(For details see Directory of Suppliers)

MRCS MILSIM CGI System

MILSIM is a low-cost, high performance CGI system that can be supplied as a complete training system or for integration into other training equipments.

MILSIM provides computer-generated targets, terrain scenes and weapon effects that are injected into a weapon system operator's sight to provide training using the operational weapon system.

R & M. MILSIM is ruggedised for effective training in the field or classroom. Qualified components are used to ensure high reliability and long-term supportability.

Instructor Aspects. An instructor can set up a training exercise from a library of attack profiles and target types. The illumination and visibility range can be modified to allow training under adverse environmental conditions. Objective performance assessment is provided for each training exercise, allowing precise evaluation of each trainee's performance. Results are immediately available to both the trainee and instructor.

Roles. The MILSIM system can be used with many types of weapon system, from armoured vehicles to air defence weapons and is in production for the integrated simulator for the Bofors RBS 90 surface-to-air missile system. Other Marconi visual training systems are in service with the Royal Navy, Sweden and other nations.

MRCS MILSIM image as used by a weapon operator

Specifications

Targets displayed: 1, 2 or 4 with up to 128 model points
Target types: 16 stored concurrently, unlimited by interchangeable module
Target tracks: 40 stored concurrently, unlimited by interchangeable module
Tracers: 16 simultaneously visible
Missile model: supplied in lieu of tracer
Background: high-fidelity, real-world terrain, with multiple range planes
Target speed: 0-600 m/s
Target range: 100 m to 15 km
Reticle: supplied as required
Orientation update rate of targets: 50 Hz
Position update rate of targets: 50 Hz
Position update of sights: 50 Hz
Video output: composite video CCIR 624B - other standards available as an option
Visual spectra: daylight and thermal simulation
Field-of-view: selectable with 360° slew capability
Weight in a typical configuration: 45 kg
Dimensions in a typical configuration: 600 × 400 × 400 mm

Status

In production. In service with the Royal Navy, Sweden (RBS 90) and other nations.

Manufacturer

Marconi Radar and Control Systems Limited (MRCS), Leicester.
(For details see Directory of Suppliers)

Marconi Simulation and Training (MST) Visual System

The (MTS) full-colour low-cost day/dusk/night CGI system can be supplied with a range of display options to meet specific customer requirements. The CGI aspects are covered here.

Applications. The MST CGI can support multi-channel applications and provide video in NTSC, PAL and high resolution (1290 × 1024 pixel) formats.

Main Characteristics. Characteristics include full texturing of surfaces and three dimensional modelling of objects. Transparency and blending effects are utilised to provide reduced visibility for the generation of haze, fog, scud, cloud and for localised weather effects such as rain squalls. Continuously variable ambient and directional lighting is provided allowing for time of day and vehicle orientation.

Database Generation. Database Generation Facilities (DBGF) are available as an optional extra to the visual system. The DBGF allows users to generate their own databases and models without specialist training. This is achieved by the use of MMI techniques and a structured set of models and database building tools. DBGF options include stand-alone facilities or 'software only' packages to run on the IG if the operational role of the application permits.

Roles. The system has applications in flight simulation, weapon trainers such as gun and missile aiming training, ship's bridge simulation and periscope simulation, in addition to the simulation of TV-based weapon aiming systems. The system can also provide simulation of image intensification or infra-red systems.

Marconi CGI system imagery in a periscope trainer

Status

In service with the Royal Netherlands Navy, the KLM Flying Training School, the Royal Canadian Navy and the Defence Research Agency. Ordered by the Royal Navy, Vickers Defence Systems and GEC-Marconi Avionics.

Manufacturer

Marconi Simulation and Training (MST), Dunfermline.
(For details see Directory of Suppliers)

TTSL (ex-Rediffusion element) and Evans and Sutherland Novoview Visual Systems

Current Situation

In 1994 it was announced that CGI manufacturer Evans and Sutherland (E&S) of Salt Lake City, Utah, USA, would no longer have a special marketing agreement with the ex-Rediffusion element (Crawley) of Thomson Training and Simulation Limited (TTSL). TTSL now markets its own SPACE CGI system, but has stated that it will fit other CGI systems, including those of E&S, at customer request. Since several hundred of the Rediffusion/E&S Novoview systems are in service, it was felt that rather than deleting this entry, it should continue to be included in order to give details of the many systems still in use.

Novoview History

Under the terms of a joint collaboration agreement only Rediffusion had the marketing rights for Evans & Sutherland ESIG image generators in the worldwide civil flight simulator market and the UK military simulator market. Novoview was the brand name under which Rediffusion supplied complete visual simulation systems which included the E&S CGI system, the display system (normally the Rediffusion WIDE series), visual

TTSL Hawk Simulator used by the Swiss Air Force at Emmen showing imagery from five channels of the SP-X 500 CGI System

Novoview System 3000

databases, and integration of the visual system with the rest of the simulator. Simulators for which Novoview systems were supplied were not confined to Rediffusion-built devices, and included many CAE (Montreal) simulators before 1992 when CAE commenced sales of its own MAXVUE™ CGI and Display system.

Novoview SP Visual Systems

More than 420 Novoview SP CGI systems were supplied by Rediffusion/E&S for Rediffusion and other manufacturers' simulators. SP stood for 'Special Product'.

SP 1 and SP 1T

SP 1 is no longer marketed, but is still supported on simulators using it. It provided night and dusk scenes only (not full day colour). When texture is added, SP 1 becomes SP 1T. Further details are in the Evans and Sutherland entry in the USA section which follows.

SP 3 and SP 3T

SP 3 is no longer marketed, but is still supported on simulators using it. It produced night, dusk and full daylight conditions. When texture is added SP 3 becomes SP 3T. Further details are in the Evans and Sutherland entry in the USA section which follows.

SP-X.

Rediffusion/E&S announced the SP-X CGI system in 1985 and it was marketed until being superseded by ESIG 3000 in 1993. It is still supported on simulators using it. It was a substantial advance on the earlier SP 3 system. SP-X was a day/dusk/night system, texture being applied to all surfaces irrespective of orientation. Further details are in the Evans and Sutherland entry in the USA section which follows.

System 3000

Novoview System 3000 was launched in May 1993 to replace the SP-X system, but since the take-over of Rediffusion by Thomson it is only marketed in its two separate elements. The system was claimed to have a cost/performance ratio some five times that of the earlier SP-X systems.

Novoview System 3000 was based on the E&S ESIG 3000 CGI system but was marketed with an improved WIDE display system which featured new projectors, auto-picture alignment, new mirror image technology and different FoV at customer choice. The CGI retains the most successful aspects of the SP-X system, including database compatibility, phototexture and calligraphic light points.

Orders. By 1994, 19 Systems 3000 had been ordered.

Status

SP-series now obsolete but still supported. ESIG 3000 available from Evans and Sutherland (qv).

Manufacturers

Thomson Training and Simulation Limited (TTSL), ex-Rediffusion element, Crawley, for enquiries about systems still in service.

(For details see Directory of Suppliers)

For CGI aspects see Evans and Sutherland, Utah, in the USA section.

UNITED STATES OF AMERICA

ATC SignalVision 5000 CGI System

SignalVision 5000 is a low-cost image generator made by the ATC Flight Simulator Company of Santa Monica. No further details at this time.

Manufacturer

ATC Flight Simulator Company, Santa Monica, California.

(For details see Directory of Suppliers)

Ball 944 CGI System

The Ball 944 CGI system is an off-the-shelf, relatively low-cost computer-based IG. It can be configured for many applications ranging from desktop procedure trainers up to full mission, multichannel trainers and 'force on force' networked crew training.

Main Characteristics. The Ball 944 combines high polygon processing performance, 24-bit full-colour texturing, weather effects such as fog/haze, and real-time display list paging.

Detailed Specifications

Scan Patterns: raster; NTSC, PAL, VGA/SVGA

Polygons: Shaded, anti-aliased triangles, per channel; up to 16 000 at 30 Hz, 8000 at 60 Hz, 4000 at 120 Hz

Light points: up to 1 million transformed per second per channel; 16 667 at 60 Hz, 8333 at 120 Hz

Note that the total polygon/light point system capacity is dependent upon the specific configuration

Active display lines: 512 to 1024 (262 to 1049 kpixels)

Pixels processed per frame: up to 1.05 million

Data update rate: Up to 120 Hz depending on configuration

Screen refresh (frame) rate: All output formats except NTSC are non-interlaced and so data update rate is the same as screen refresh rate; for NTSC the update rate is half the screen refresh (frame) rate (60 and 30 Hz respectively)

Latency: 83 msec at 1024 lines at 30 Hz data update, 50 msec at 60 Hz. Can be reduced to 25 msec at 512 lines at 120 Hz update

Moving models: up to 370 platforms. The number of simultaneous moving objects is limited only by the system processing capacity

Visibility: 256 visibility levels processed at pixel level

Colours: full (16.7 M nominal)

Anti-aliasing: Real-time sub-pixel colour blending via FLASH™ algorithm for all primitives

Texture: Photo-based pattern mapping, colour, pattern sizes from 16 x 16 to 512 x 512, texture memory 2 or 8 Mbytes. Tri-linear interpolation to eliminate texture-crawl, inverse transformation per pixel for real-time perspective correction, 7-bit translucency, contour modulation, 8-bit intensity modulation per RGB component for all texture patterns

Occultation: Resolution with 32-bit FP accuracy at the pixel level by XPAN™ algorithm. This removes hidden surfaces and resolves any level of occultation within a visual database at the pixel level, and also eliminates the necessity to pre-sort any database elements or moving objects

Database: Hierarchical object-orientated, modelling in 3D co-ordinate space, 32-bit IEEE floating point format, 16 Mbyte display list memory, LoD support with object-level culling algorithm, MultiGen flight format capability.

Processing Capacity: Up to 140 Mips and 2.3 Gflops

Computing: Either the Encore 91 Series or the SUN 4/330 computer operating under SunOS UNIX. The open bus architecture of the 944 allows the integration of virtually any single board computer into its chassis. Expansion of memory and the integration of high-speed interfaces is simplified by the VME-based structure

Software: RSR software library written in the C lan-

Ball 944 CGI system showing street scene produced for a military driver training simulator

Ball 944 CGI system showing bombing of ground targets

guage; CARRERA package for rapid prototyping and integral window manager; DIGITIZE! 3D object modelling programme for moving models; MultiGen, industry-standard database generation software with optional import functions for DMA data (DLMS, MIL STD 1821 (Project 2851) and so on), model library for use with DLMS DFAD and equivalents

Status
In production and widespread use.

Manufacturers
Ball Imaging Products, San Diego, California.
(For details see Directory of Suppliers)
Ball Systems Technology Limited, Coventry, UK.
(For details see Directory of Suppliers)

CAE-Link ATACDIG Image Generator

The ATACDIG (Army Tactical Digital Image Generator) system was first developed for the AH-64 Apache Combat Mission Simulator (CMS). At that time it was the latest in a series of Link Digital Image Generator (DIG) systems which previously were numbered (for example DIG 1, 2, 3 and so on). ATACDIG is also fitted to simulators for the Black Hawk, Chinook, and Cobra helicopters and the Special Operations MH-47E and MH-60K helicopters.

Characteristics. Full-colour day/dusk/night images include textured terrain, environmental effects, horizon effects, airport lighting effects and confined landing sites such as a wood or hilltop.

Gaming Area. The tactical gaming area includes friendly and hostile sites as well as stationary and moving ground vehicles and aircraft. It can be configured to real-world tactical areas using digitised terrain and feature information from the NATO DLMS system (DTED/DFAD data).

Status
Out of production; delivery of last ordered unit scheduled for early 1995. More than 40 ATACDIG systems are in use.

Manufacturer
CAE-Link Corporation, Binghamton, New York.
(For details see Directory of Suppliers)

CAE-Link ATACDIG CGI system showing an Mi-24 Hind seen from the ownship cockpit

CAE-Link CRISP Laser Disk Imaging System

To provide the type of high scene detail important for target detection and recognition and to provide occultation which properly represents the real-world difficulty in detecting partially obscured, camouflaged threats, CAE-Link developed the Computer Reconstructed Images from Scene Photographs (CRISP) technique. This uses laser disk technology to store, digitise, access and transfer an array of still frame images, including terrain, targets, fog/smoke/dust, and weapons effects, as well as associated range data, which can be assembled in an image processing pattern to represent a full 360° panorama.

Scene images are digitised from photographic colour transparencies of actual terrain and target images are also stored on the disk. Special effects are handled as target images or as inserted texture images; the simulation of forward-looking infra-red (FLIR) images is handled by a separate database containing thermal radiation signatures and conduction and convection images. Occultation is resolved as a function of 256 range levels for each pixel.

The CRISP system uses modular, off-the-shelf image processing components and includes a personal computer as the operator station. The system has many applications and was selected by the US Army to provide training in the use of the telescopic sight units (TSU) for the gunners on AH-1S Cobra helicopters. CAE-Link has upgraded the TSUs on six AH-1S simulators. Low and high magnification of the TSU is simulated with separate image databases, enabling the gunner to complete the tactical training requirements of target recognition, acquisition and engagement. The student gunner receives a score based on his actions and control of the simulated TOW anti-tank missile.

Status
In service.

Manufacturer
CAE-Link Corporation, Binghamton, New York.
(For details see Directory of Suppliers)

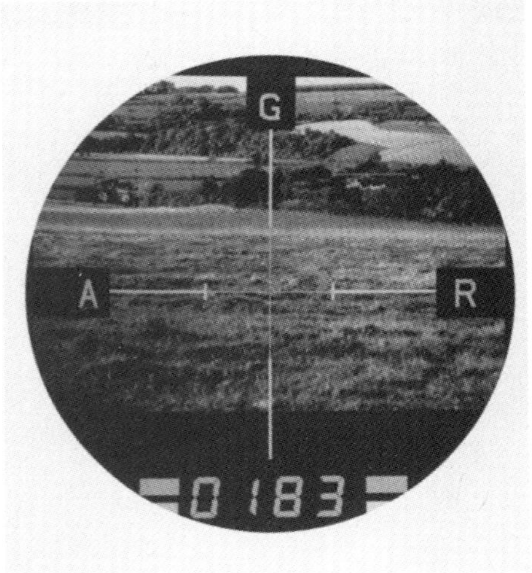

Typical gunsight view produced by the CAE-Link CRISP laser disk visual system

CAE-Link Programmable Digital Radar Land Mass Simulation (pDRLMS)

The pDRLMS is a new generation in radar simulation systems which is made possible by developments in commercial off-the-shelf Reduced Instruction Set Computing (RISC) microprocessor technology.

Computing Aspects. The computer is essentially a powerful GP parallel processor and is not limited just to radar simulation. HLL software is used. The algorithms employed to generate radar images are enhanced versions of those which have proved successful in the past. All features of the environment, except for very low resolution requirements, are treated individually to produce a faithful replication of real performance under all conditions and at all ranges.

Interfaces. The pDRLMS may be interfaced with any simulation computer or display set. Ethernet and VME interfaces are offered as standard features as are most common display formats. Special interfaces can be accommodated easily.

Programmability. The system is programmable to allow modifications and enhancements without hardware changes. Only the essential hardware to accomplish specific tasks is needed. Radar systems training supported by pDRLMS can be tailored to the needs of the user and can be modified to meet future needs. Real-beam, synthetic aperture radar (SAR), TFR and IFF modes are supported.

Diagnostics and Product Support. The pDRLMS is supported by diagnostics which can easily test the radar processing hardware while the system consists of commercially available devices with a great deal of product support.

Deliveries. pDRLMS has been delivered with software for both E-2C and HC-130P simulators, and is being developed for the Turkish F-16 simulator.

The CAE-Link programmable Digital Radar Land Mass Simulation (pDRLMS) system hardware

Status
Available.

Manufacturer
CAE-Link Corporation, Binghamton, New York.
(For details see Directory of Suppliers)

Defense Mapping Agency

The US Defense Mapping Agency (DMA) is the national military mapping authority for the USA; similar organisations in other nations include DG Military Survey for the UK.

DLMS Programme. The US DMA is involved in the multinational DLMS programme which involves Digitised Terrain Elevation Data (DTED) and Digitised Feature Analysis Data (DFAD) to a number of levels of fidelity. 3D simulator databases may be constructed from this DTED and DFAD data (see photos) and then refined by adding the texture and other cultural features of the proprietary Image Generation system.

DLMS Levels. DLMS data is collated at two levels of density:

DTED. Level 1 terrain data is recorded in the form of a grid with data points at 3 arc second (about 300 ft) intervals, and Level 2 at 1 arc second (about 100 ft) intervals.

DFAD. DFAD Level 1 feature (cultural) data is approximately equivalent to the features on a 1:250 000 topographical map; Level 2 data is approximately equivalent to the features on a 1:50 000 topographical map.

3D terrain derived from DTED and DFAD data in raw form before enhancement by a Database editing system specific to the particular CGI system

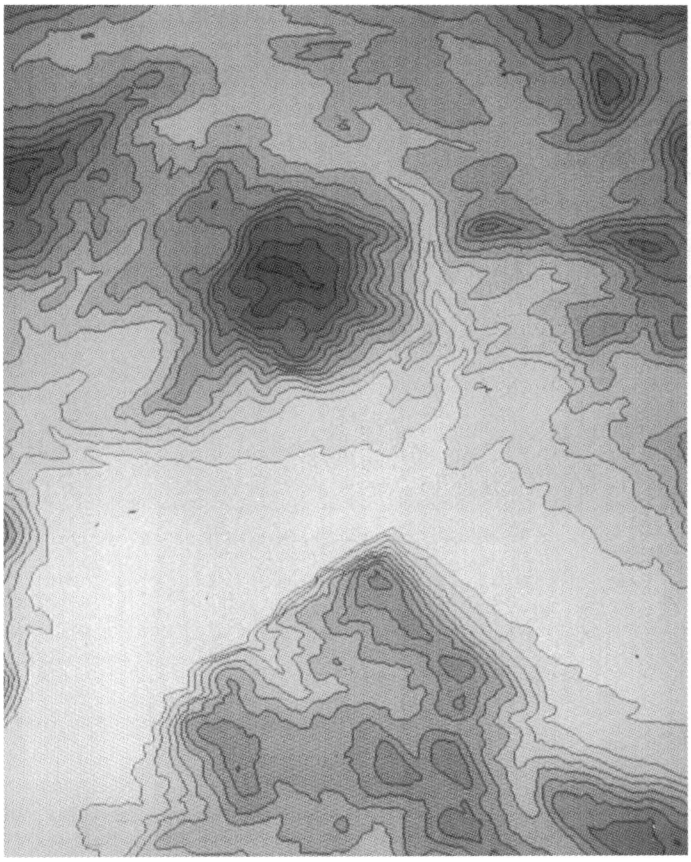

DLMS DTED-Generated Contours

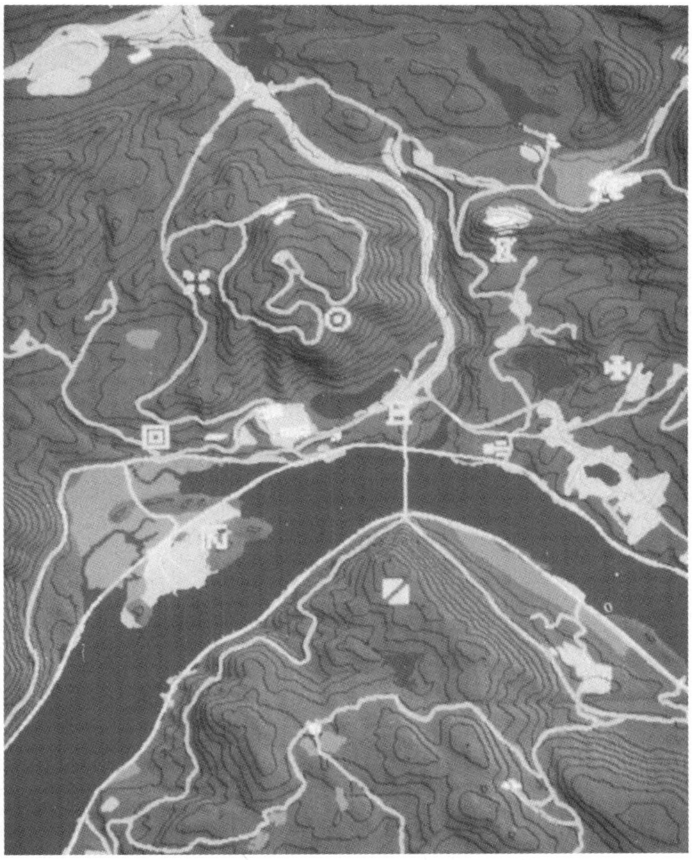

DLMS DTED-Generated Contours with overlay of DFAD data in plan view

MIL STD 1821 (Project 2851) Mapping Data. The DMA is also involved in the successor programme to DMLS known as MIL STD 1821 (Project 2851) which has a more sophisticated architecture and potentially a much greater data density.

Function of Digitisation Programmes. These programmes involved in the digitisation of real-world data are not only for the construction of databases for simulators and training aids, but also for the long-term replacement of paper maps with digital data, and with changes of mapping and navigational datum points to worldwide geographic reference systems such as the WGS 84 Geodetic System, and with associated navigational aids such as the satellite-based US GPS which gives accurate ground positions from simple receivers anywhere in the world.

Status

Some digitised real-world data is releasable to simulation manufacturers and some to the public domain, but some is classified for military use only. Most of world data to generalised accuracies is available since commercial satellite data are now available worldwide, from which the same terrain and feature data can be derived.

Co-ordinator and Data Supplier

Defense Mapping Agency (DMA), Washington DC, and Reston, Virginia.
(For details see Directory of Suppliers)

Evans and Sutherland CGI Systems

Evans and Sutherland has supplied CGI systems since the 1970s in product lines prefixed SP (Special Product), CT (Continuous Tone) and ESIG (Evans and Sutherland Image Generator).

Marketing with Rediffusion Display Systems

For civil simulators and simulators for the UK military, E&S had an arrangement with the Rediffusion company of Crawley, UK, where E&S CGI systems were marketed by Rediffusion as part of complete visual systems under the name Novoview. After the acquisition of Rediffusion by the Thomson group in December 1993 and the development of Thomson's new SPACE CGI system, in July 1994 it was announced that the special marketing agreement with the ex-Rediffusion element (Crawley) of Thomson Training and Simulation Limited (TTSL) would be dropped. TTSL now markets own SPACE CGI system (qv), but has stated that it will fit other systems, including those of E&S, at customer request. Since several hundred of the Rediffusion/E&S Novoview systems are in service, it was felt that rather than deleting this entry, it should continue to be included in order to give details of the many systems still in use.

Novoview History

Novoview was the brand name under which Rediffusion supplied complete visual simulation systems which included not only the E&S CGI system, but also the display system (normally the Rediffusion WIDE series),

ESIG-2000 CGI system imagery - land battle scene

ESIG-2000 CGI imagery on a ship handling simulator

ESIG-3000 imagery showing ship and helicopter

ESIG-4000 imagery showing terrain and aircraft

visual databases, and integration of the visual system with the rest of the simulator. Simulators for which Novoview systems were supplied were not confined to Rediffusion-built devices, and included many CAE (Montreal) simulators before 1992 when CAE commenced sales of its own MAXVUE ™ CGI and Display system.

Novoview SP Visual Systems

More than 420 Novoview SP CGI systems were supplied by Rediffusion/E&S for Rediffusion and other manufacturers' simulators. SP stood for 'Special Product'.

SP 1 and SP 1T

SP 1 is no longer marketed, but is still supported on simulators using it. It provided night and dusk scenes only (not full day colour). SP 1 has 6000 light points, with five light colours, 200 to 400 surfaces, 64 grey shades, 256 edges, and an update rate of 30 Hz. When texture is added SP 1 becomes SP 1T, but texture can only be applied on horizontal surfaces (so-called 2D Texture).

SP 3 and SP 3T

SP 3 is no longer marketed, but is still supported on simulators using it. It produces night, dusk and full daylight conditions. Under daylight conditions the update rate is increased to 50 Hz, but for night scenes this is reduced to 30 Hz in order to deal with light points by calligraphic beam pointing. When texture is added SP 3 becomes SP 3T, first marketed in 1977. Texture can only be applied on horizontal surfaces (so-called 2D Texture).

Deployments of SP 3T. Some known simulators using SP 3T include: 2 BAe 146, RAF AWACS, RN Sea King Mk 5, 35 USAF simulators, USN CH-46B and EA-6B.

SP-X

Rediffusion/E&S announced the SP-X CGI system in 1985 and it was marketed until being superseded by ESIG-3000 in 1993. It is still supported on simulators using it. It was a substantial advance on the earlier SP 3 system. SP-X was a day/dusk/night system, texture being applied to all surfaces irrespective of orientation.

Features. Other features include transparencies, luminous polygons, pixel fog, fade level of detail management, and polygon formatting for ease of modelling.

Light Points. A combination of calligraphic light points and raster lights is used.

Texture. Hardware texture is also available. SP-X texture modulates from colour to colour.

Aliasing. An anti-aliasing clamping feature is included.

Lights on Objects. Steady, flashing, strobe or 'own ship' lights.

Moving Objects. Up to 13 moving objects in the scene.

Weather. All manner of weather can be replicated, including flying below, in and above cloud.

Channels. There can be up to eight channels.

Polygons and Light Points. 500 polygons and light strings and 5000 light points/channel.

Resolution. 720 000 pixels/channel (about 850 x 850).

Update Rate. 30 Hz by night and 50 Hz by day.

SP-X Range. The SP-X range includes three basic systems:

SP-X 200. A basic dusk/night system without day capability.

SP-X 500. Full day colour with the potential to use digitised photography for a more realistic scene. Can be upgraded to the SP-X 550 standard.

SP-X 550. As the 500 version but includes a more advanced phototexture capability. Contour texturing allows the definition of complex shapes such as trees, clouds and lakes, providing fine control of the edges. Four levels of bi-linear blending support a variety of new texture effects. High levels of blending give a smooth/soft effect for applications such as cloud texture or ground gradient effect.

E&S CT Series of CGI systems

E&S developed the CT (Continuous Tone) series of CGI systems principally for the military market. They employed raster scanning outputs only (hence the 'continuous' in the title) since the scan did not have to embrace calligraphic techniques more appropriate to creating light points for airfield night scenes for civil training. Systems were numbered from the early CT-1, 2, 3 and 4 and were operated by Case-Western University, Lufthansa, NASA, and the US Maritime Administration.

E&S CT-5 CGI System

CT-5 was introduced in 1979 for the military market and had an area architecture, good polygon capability (for the time), transparency and shading within polygons but it had no texture. CT-5A was introduced in 1983 and had improved level of detail (LoD) fade control. However, the lack of texture was noticeable when involved in military low flying tasks and some operators preferred less capable systems with texture to CT-5 or CT-5A without it. CT-5A was superseded by the fully textured CT-6 system which is described next.

Sales: Operators of CT-5 and CT-5A systems include Boeing, Daimler-Benz, Douglas Aircraft, GD Corp, NASA, Northrop, Rediffusion, the Spanish Navy, US Army, US Marine Corps and US Navy.

E&S CT-6 CGI System

CT-6 was introduced in 1985 and featured a number of improvements on CT-5A including a variety of textures. These included contour, smooth-shaded, and dynamic texture with some 300 texture maps on-line. Dynamic texture featured the texture map moving relative to the polygon for effects such as rotor wash, running water, and so on. Geo-specific phototexture was added in about 1988. Other specifications included the following:

Resolution: One million processed pixels per channel (1000 x 1000).

Special Effects: Explosions, tracers, muzzle flash, smoke, missile launch, weapons scoring.

Lights: Strobe, flashing, rotating, VASI, PAPI, FLOLS, REILS, steerable, eight lobe patterns/channel.

Weather: Cloud layers, fog, haze, horizon glow, thunderstorms.

Sensors: NVGs, FLIR.

Models: Up to 16 moving models.

Occulting and Anti-aliasing: Performed at sub-pixel level.

Collision Detection: Included.

Height Above Terrain: Included.

Sales: Operators of CT-6 include Boeing, Bell, Fuji, GD Corp, Goodyear, Loral, Grumman, McDonnell Douglas, NASA, Science Applications Inc, US Air Force and US Navy.

ESIG Nomenclature

In 1990 E&S changed their nomenclature from SP to ESIG (Evans and Sutherland Image Generator). The low-cost version of SP-X 200, became ESIG-200, and the other versions of SP-X were retitled ESIG-500 and ESIG-550.

ESIG-2000

The Evans & Sutherland ESIG-2000 CGI system was announced in 1991 and incorporates the technology and architecture of the ESIG-3000 (see following entry) streamlined into a compact configurable system. Standard features include full-colour texture, a large area database and advanced scene management, atmospheric effects, sensor simulation, anti-aliasing, R-buffer and image mapping.

The ESIG-2000 uses very large scale integration (VLSI) circuitry and custom application specific integrated circuitry (ASIC) technology designed specifically for the visual system market. Features of ESIG-2000 include the following:

Configurable design. The system allows tailoring of update and refresh rates, surface and texture capacity, video output resolution and format, and supporting peripherals. Modular design allows easy expansion for future growth.

Multiple viewports. Multiple high resolution displays from a single system, each with separate host-controlled viewing direction and field-of-view.

Choice of display systems. Fully programmable video buffers enable the use of cost-effective commercial displays requiring non-linear image mapping.

Texture. The system provides crisp, stable texture on any surface in the scene regardless of altitude or viewing distance.

Realistic sensor simulation. The system has built-in circuitry for night vision goggle and infra-red simulation.

Compact packaging. The system may be mounted in a 25 in (635 mm) high floor cabinet or a standard 19 in (483 mm) rack mount.

Applications for ESIG-2000 include networked air and ground tactical trainers, low-cost flight simulators, part-task trainers, maritime trainers (including ship handling and periscope trainers), and driving simulators.

In November 1992 ESIG-2000 was selected by the US Army as the visual system for the Close Combat Tank Trainer (CCTT) programme. ESIG-2000 was also selected for the German Army's ASPG combat tactical trainer and for the British Army's Challenger MBT training programme.

ESIG-3000

ESIG-3000 was developed from CT-6 and was introduced in June 1990. It has an area-based rendering architecture, a powerful database management system, robust texture, and universal real-time software. Up to 3000 potentially viewable, textured and smooth-shaded polygons can be processed in each of up to eight channel processors at a 60 Hz update rate, with all polygons contributing effectively to the scene. Additional hardware can be added in a modular fashion in each channel processor to produce up to 1.5 million pixel resolution at a 60 Hz update rate.

Novoview System 3000

Novoview System 3000 was launched in May 1993 to replace the ESIG-200/500/550 systems (ex-SP-X), but since the take-over of Rediffusion by Thomson it is only marketed in its two separate elements. The system was claimed to have a cost/performance ratio some five times that of the earlier systems.

Novoview System 3000 was based on the E&S ESIG 3000 CGI system but was marketed with an improved WIDE display system which featured new projectors, auto-picture alignment, new mirror image technology and different FoV at customer choice. The CGI retains the most successful aspects of the SP-X systems, including database compatibility, phototexture and calligraphic light points.

Orders. By mid-1994, 19 Systems 3000 had been ordered.

ESIG-4000

The ESIG-4000 was introduced in June 1990 at the same time as ESIG-3000 and is designed primarily to reduce database generation time and costs. By taking separately processed terrain and feature databases and then combining them in its hardware, the ESIG-4000 eliminates one of the most time-consuming aspects of image generation. Known as 'parallel database development', this permits the independent creation of terrain and cultural feature databases without the need for hand modelling by database engineers. During the terrain and feature merging stages a process known as 'conforming' enables features to conform accurately with slopes and other variations in the terrain. This feature also permits late-arriving intelligence updates to be added to a database. ESIG-4000 also incorporates other features such as full-colour phototextured terrain, range-buffered visual priority, and 'betweening' for a smooth and continuous evolution of terrain and features.

Status

ESIG-2000, 3000 and 4000 available, Novoview System 3000 is no longer available (see text).

Manufacturer

Evans & Sutherland, Salt Lake City, Utah.
(For details see Directory of Suppliers)

Evans & Sutherland Radar Image Generator (ESRIG)

In November 1990 Evans & Sutherland (E&S) was awarded a contract worth $3.5 million by the German Defence Ministry for a radar simulator which will replace the GE digital land mass radar systems (DRLMS) simulation in German Navy and Air Force Tornado simulators.

ESRIG is part of the family of ESIG image generators and it includes two in-cockpit displays, one for ground-mapping and the other for terrain-following. This was the first sale of the ESRIG system which will work with the Tornado ESIG-1000 image generation system in the Tornados. E&S is teamed with Telefunken Systemtechnik GmbH of Ulm who will act as prime contractor.

Status

In service.

Manufacturer

Evans & Sutherland, Salt Lake City, Utah.
(For details see Directory of Suppliers)

Hughes Mission Rehearsal Work Station

The Hughes Mission Rehearsal work station is designed to help aircrew plan and rehearse a mission. This system allows detailed mission rehearsal planning or the evaluation of previous missions.

Work Station. The work station uses 12 VME boards and three processing units with Motorola 68030 microprocessors. Storage is on a 1.2 Gbyte disk and 200 Mips of programmable graphics processing power and 1300 Mips of image processing power are provided.

Database. Database is an 18 000 square mile area with 10 m satellite imagery sources or 320 square miles with 4-foot (1.22 m) reconnaissance imagery sources. The system is capable of showing specific topography of large mission areas, it can provide high resolution terrain, cultural and target features at appropriate mission altitudes and speeds and it can display them in varying illumination and visibility levels. The system can provide views from a variety of sensors and can adjust light, environmental and sensor conditions to generate a true perspective image from the computed vantage point. The database is rapidly accessed and uses a photo-based image generation system which accepts data from a wide variety of sources including aerial reconnaissance, satellite images and oblique photographs.

Status

Available.

Manufacturer

Hughes Training Inc, Herndon, Virginia.
(For details see Directory of Suppliers)

Frasca CGI Systems

Frasca make a range of low cost CGI systems prefixed FVS (Frasca Visual System). The original FVS 100 is no longer made as new-build but a limited number of reconditioned systems may be available. Main system characteristics are tabulated below. It should be noted that the 2000 series has substantially greater capacity and features compared to the earlier versions, in addition to its main characteristics which appear in the table:

Frasca FVS 200HR CGI imagery showing airfield at dusk with two lit runways

Frasca FVS 2000 CGI imagery showing an aircraft taxying on to the runway and the more complex runway lighting available compared to earlier Frasca systems

Status
FVS 2000, 200HR, 100HR available. FVS 100 refurbished channels only in limited numbers.

Manufacturer
Frasca International Inc, Urbana, Illinois.
(For details see Directory of Suppliers)

Feature	FVS 100	FVS 100HR	FVS 200HR	FVS 2000
Cost per channel	$10 000	$25 000	$35 000	Price depends on display system required
Resolution	640 × 480	1024 × 768	1024 × 768	1024 × 768
Time of Day				
day	Yes	Yes	Yes	Yes
dusk	No	No	Yes	Yes
night	Yes	Yes	Yes	Yes
Colours				
available	32	16.7 Mill	16.7 Mill	16.7 Mill
simultaneous	16	256	256	16.7 Mill
Number of Channels	1	1	1	up to 6
Display	19 inch Monitor	17 inch Monitor	19 inch Monitor	Up to 7 windows 19 inch, collimated or projected
VASI Models	4	4	6	6
PAPI Models	0	3	3	3
Approach lighting systems	6	6	12	12
REILS (Runway End Lights)	No	Yes	Yes	Yes
Centreline and Taxiway lights	No	Yes	Yes	Yes
Tower				
beacon	No	1	8	Yes
light gun signals	No	No	6	No
Cloud layer and visibility	Yes	Yes	Yes	Yes
Ragged Cloudbase	No	No	Yes	Yes
Ground Scud	No	No	Yes	Yes
Ground Fog	No	No	No	Yes
Special effects: Wind sock, aircraft shadow, landing light lobes, ground traffic, runway numbers	No	No	Yes	Yes
Stars at night, city lights	No	Yes	Yes	Yes
Automatic runway configuration	Yes	Yes	Yes	Yes
Geo-specific database	No	Yes	Yes	Yes
Data Update Rate Hz	15	20	30	30
Screen Frame Rate Hz	60 Hz	60 Hz	60 Hz	60/30 Hz
Interlaced 2:1	Yes	No	No	Yes

FSI Vital Visual Simulation Systems

FlightSafety International (FSI) acquired the Visual Simulation Systems (VSS) business unit of the McDonnell Douglas corporation in February 1993. This included the St Charles, Missouri, facility and the Vital product line of simulation systems. This includes CGI systems and display systems such as MultiView (qv in Display Systems section). Although the latest CGI product is Vital VIII, many earlier systems remain in service. A list of the Vital range of CGI systems is given below in chronological order:

Vital I
Vital I was a laboratory prototype produced in 1969 in an effort to replace the model-board and travelling TV camera systems which were then in use for simulator visual systems.

Vital II
In 1972 the Vital II night-only system became the first such CGI system to be approved by the FAA. In 1974, the Vital II/74 version gave an extended capability of up to 2400 lights.

Vital III
Vital III was introduced in 1975 and offered improvements on Vital II in the representation of runway surfaces, together with up to 6000 light points.

Vital IV
Vital IV was introduced in 1977 and had up to 8000 light points, 300 multicoloured surfaces, a wider range of low light conditions, realistic landing light lobes and moving air and ground hazards. It was granted FAA approval to the then Phase II (now Level C) standards in 1982.

Vital V
Vital V was introduced during 1982 and extended the range of colours to include cyan to red, using a high resolution beam penetration display. It used parallel distributive processing and incorporated individual surface texturing for a realistic representation of blowing sand and snow, cloud tops and water.

Vital VI
Vital VI was launched in 1982 and is a full-colour daylight system designed to comply with FAA Phase III requirements (now Level D). It incorporates all the features of the previous systems.

Vital VII
Vital VII was introduced in 1986 and uses full-colour raster-based technology. It is a modular design that offers a wide variety of configurations and has been approved to FAA Level D (Phase III) standards.
Texture. The Vital VII texture provides stronger height and speed cues with colour modulation, texture tiling, transparency and three-dimensional texturing. Photographic and matt texturing capability is included. Smooth surface shading is provided by intensity interpolation techniques. Surface transparency provides visual cues for window canopy, smoke and shadow simulation.
Database Management. Database management is a standard feature which permits environment transitioning, partitioning and various levels of scene modelling detail.

FSI Vital VII imagery - the approach to the USS Nimitz

Cobra helicopter flying over the sea, simulated by Vital VII

FSI Vital VIII coastal scene showing combination of digital terrain and aerial photography in database construction

FSI Vital VIII airfield scene with cloud effects above

1990 Vital VII Upgrades. In 1990, Vital VII was upgraded to Vital VII(e), standard resolution; and Vital VII(i), high resolution. New RISC technology was incorporated, increasing scene content by 50%. The new technology also allowed costs to be reduced.
Display Methods. Imagery can be displayed on a variety of collimated monitors or on projector-based collimated cross-cockpit systems such as the FSI Multiview (see the Display Systems section following).

Vital VII Specifications

Scan Patterns: raster scan/calligraphic
Modes: day/night/dusk
Channels: up to 8
Polygons: up to 1000 per channel
Light points: up to 8000 per channel
Note that the total polygon/light point system capacity is dependent upon the specific configuration
Active display lines:
Standard resolution: up to 622
High resolution: up to 1022
Pixels per frame:
Standard resolution: 500 000
High resolution: 1 million
Display update rate: 50 Hz
Latency: 80 ms
Moving models: up to 8
Colours: 256
Anti-aliasing: standard
Transparency: standard
Texture: surface, photographic and matte
Dynamic load control: standard
Database management: standard
Occulting: unlimited
Database size: 2761 nm (5115 km)

Vital VIII

Vital VIII was introduced in 1993 for FAA Level C or D training and appropriate military applications. It uses Application Specific Integrated Circuit (ASIC) technology and architectural enhancements to improve performance and flexibility at reduced cost. All Nippon airlines was the launch customer.
Texture. Over 2000 texture patterns can be on-line simultaneously. The effective texture capacity is improved by real-time paging from disk.
MultiVis Atmospheric Simulation. It incorporates the MultiVis atmospheric simulation which, amongst other things, allows textured fog and realistic penetration of objects through the fog.
Database Aspects. Quick Build methodology allows quick reaction database changes and is compatible with DoD Project 2851 DIF/HDI so that a new Vital VIII database can be RFT in 72 hours from starting to assemble it. A feature extraction capability uses computer graphics techniques to derive 3D models and colour texture from source photographic data.
Display Methods. Imagery can be displayed on a variety of collimated monitors or on projector-based domes or collimated cross-cockpit systems such as the FSI Multiview (see the Display Systems section following).

Vital VIII Specifications

Scan Patterns: raster scan/calligraphic
Modes: continuous time of day or night
Channels: up to 8
Viewports: up to 32
Polygons: up to 3000 per channel
Light points: up to 16 000 per channel
Note that the total polygon/light point system capacity is dependent upon the specific configuration
Active display lines: 1280 x 1024 (1311 kpixels)
Pixels per frame: up to 1.3 million
Display update rate: 50 Hz
Latency: 50 ms
Moving models: up to 64 platforms
Colours: 24 bit RGB, 256 intensities
Anti-aliasing: sub-pixel
Transparency: 4096 levels
Texture: Full RGB, photographic, matte, sharp contour, fine, mirror and illumination.
High level MIP-map.
Dynamic load control: standard, pixel stretch
Database management: dynamic load, multi-fade LoD.
Occulting: unlimited
Database size: dynamic loading

Status

Vital systems are in production and are in widespread service with both military and commercial users (over 370 systems in more than 42 countries). Known military users or military applications include the following: Algerian Air Force; C-130, Vital IV; Argentine Air Force; Mirage/Dagger, Vital IV; Royal Australian Air Force; F-18 and Boeing 707, Vital IV and V; Belgian Air Force; F-16, Vital IV; British Aerospace; Harrier GR5, Vital IV Canadian armed forces; CP-140 and CF-18, Vital IV; Royal Danish Air Force; F-16, Vital IV; Egyptian Air Force; F-16, Vital IV; Finnish Air Force; Hawk, Vital VII; Indian Air Force; Jaguar, Vital IV; Indian Government; various, Vital IV; Israeli Air Force; F-16, Vital IV; Israeli Military Industries; Kfir (IV), Kfir (V), Lavi (IV); Italian Army; AB-205, Vital IV; Japanese Self-Defense Force; T-2B, Vital IV; Royal Jordanian Air Force; C-101, Vital V; Royal Moroccan Air Force; C-130, Vital III; Royal Netherlands Air Force; F-16, Vital IV; Royal Netherlands Navy; P-3C, Vital V-A; Royal Norwegian Air Force; F-16, Vital IV; Portuguese Air Force; A7P, Vital VII; Saudi Arabian Air Force; C-130, AB-212, Vital IV; Singapore Air Force; F-5E and A-4S, Vital IV; Roller Helo, Vital VII, MultiView; Spanish Air Force; EF-18, Vital V; Royal Swedish Air Force; JA-37 and C-130H, Vital IV; Venezuelan Air Force; F-16, Vital IV;

US Air Force: A-10, Vital IV; A-7D, Vital IV; F-4E, Vital IV; F-4G, Vital IV; T-37, Vital IV; T-38, Vital IV; T-37/T-38, Vital IV; C-17, MultiView; C-5B, MultiView;

US Marine Corps: A-4M, Vital IV; A-6E, Vital IV; F-4J, Vital III;

US Navy: F-14A, Vital IV; F-18A, Vital IV; S-3A, Vital III-6000; SH-3H, Vital IV; SH-60B, Vital IV; P-3C, Vital IV; P-3C, Vital VI, MultiView; SH-2F, Vital III-6000; MH-53, Vital VII, MultiView; SH-60F, Vital V.

Manufacturer

FSI Visual Simulation Systems Division (Ex-McDonnell Douglas Visual Simulation Systems), St Charles, Missouri.
(For details see Directory of Suppliers)

IVEX CGI Systems

IVEX has developed an original principle where the CGI database is stored on high-capacity optical disks and is then retrieved for processing into real-time imagery. The advantage of this system compared with hard disk-based CGI systems is that storage density of optical disks is high and only a simple interface with a host computer is required. Because the scenes are stored on the optical disk with all their detail already included, this eliminates the polygon-building limitations of traditional real-time CGI systems. A typical IVEX-generated scene is claimed to contain the equivalent of one million polygons in terms of scene detail. Runway markings and landscape texturing are built into the database. It is claimed that IVEX systems give the richest scene detail at the lowest cost. The principal limitation is that the highly textured world is flat (2D surface) although three-dimensional objects such as hangars and control towers can be added to the scene. Multiple optical disks can be connected in a modular fashion so that independently controlled moving objects can be inserted into a scene.

IVEX VDS-1000 CGI System

The IVEX Visual Display System (VDS)-1000 system has a standard daylight continuous database of 50 × 50 miles (80.5 × 80.5 km) which includes a terrain model depicting rivers, lakes, beaches, farm land and several types of vegetation; 6 DoF ownship motion; user-selectable range; and a 48 × 36° FoV per channel. Optional features include natural-looking cloud base permitting pilots to fly through and above clouds with blue sky overhead; realistic time-of-day lighting from dawn/dusk to night, with runway and taxi-way lights and landing aids; and interactive three-dimensional moving objects (air and ground) with 6 DoF. Three channels are standard in Full Flight Simulators (FFS) and give a FoV of about 150° x 36° depending on the display system used which can be collimated or direct projection.
IG Types. The IVEX 1000 product line has three levels depending on customer requirements. All are day/dusk/night systems with real-world databases and variable cloud cover features.

The 1000S is the low-cost, entry-level, single channel system designed to meet a range of basic training needs such as FTDs. A generic textured terrain database is included.

IVEX VDS-2000 CGI system showing runway threshold, hills in the distance, and some 3D objects on the otherwise flat terrain

The 1000ST is designed for the FAA 'visual simulator' category (Level A/B). A light point modelling capability is included. Single or multiple channel configurations are available, and typically a three channel system would be used. The 1000ST can be upgraded to the 1000STS.

The 1000STS includes a 3D object database in addition to the light points and multichannel capabilities of the 1000ST. The light points and polygon databases can be tailored to the needs of the customer.

Features. As well as the Video-Disk-produced richly textured terrain, other features include: selectable weather such as cloud and visibility conditions; 24-bit colour resolution; easily adjustable and movable light points; and polygons with 450 edges (for example 150 triangles or 112 rectangles). HUD imagery is also available.

Image Processing ICs. To ease the task of producing high-speed, low-cost image processing systems, IVEX has designed two image processing Integrated Circuits (ICs), the RXC-01 and the PWA-01. Both provide image and video processing building blocks for performing image transformations and filtering with a minimum of external components. The ICs, when used together in an integral system, provide the ability to transform geometrically and filter a video image to produce high quality imagery in real time.

IVEX VDS-2000 CGI system showing airport apron with a number of 3D objects constructed from the system polygon generator

IVEX VDS-2000 CGI System

The VDS-2000 used the same video disk principles as the earlier 1000 systems but has a number of enhancements. Some specification features are as follows:

Channels: Up to 4

Scan Patterns: raster scan with calligraphic option.

Polygons: up to 250 per channel, used for placing objects such as hangars, control towers, other aircraft and so on, in the 2D world scene, Gouraud shaded and anti-aliased.

Light points: up to 5000 raster lights per channel, with option for 1000 calligraphic light points; 256 intensity levels. Runway, city, VASI, PAPI, Mobile and other special types provided.

Resolution: 736 horizontal x 648 (477 kpixels) at 56 Hz (26.7 Mpix/sec).

Data update rate: Up to 60 Hz, non-interlaced scanning.

Latency: not quoted.

Moving models: up to 10 platforms.

Weather Effects: Visibility 0 - 50 miles, fog, rain.

Colours: Full, nominally 16.7 million.

Anti-aliasing: Weighted average sampling for terrain, alpha blending for light points, 15 samples per pixel for 3D objects.

Texture: Rich textures using recursive fractal techniques pre-stored on video disk.

Shading: With illumination, sun angle/time of day.

Collision Detection: Included.

Height Above Terrain: Included.

Database: Extensive library of airfield scenarios and local areas. Modelling work stations available for customer modification of databases.

Computing: Host interface BIT-3, DR11W, Ethernet, synchronous or asynchronous.

Display Options: Direct Projection, collimated monitor and collimated back projection (cross-cockpit) systems allowed for.

Status

Available and in widespread service in Australia, China, Germany, Japan, Korea, Kuwait, Mexico, Norway, Singapore, Switzerland, Thailand, UK, and the USA. Military customers include the USAF, Kuwait Air Force, Royal Thai Air Force and the UK Royal Air Force who operate three-channel VDS-1000STS systems with SEOS screens on five Tucano simulators. Latest orders for the VDS-2000 include 3 systems for the Japan Civil Aviation College and two for Reflectone Jetstream 41 simulators in the UK and USA. The systems for Japan will be displayed by direct projection (SEOS PRODAS) and the Jetstream units by collimated back projection (cross-cockpit) systems (SEOS Panorama).

Manufacturer

IVEX Corporation, Norcross, Georgia.
(For details see Directory of Suppliers)

IVEX VDS-1000 CGI system showing visibility effects at the runway threshold and the rich texture possible on this video disk-based system

Loral CGI Systems

Loral produces the GT 100 and GT 200 CGI systems. More than 350 of the low-cost GT 100 systems have been delivered and the GT 200 is described below. Loral acquired the simulation interests of Bolt, Baranek and Newman in 1993 which included the Delta Graphics product line of CGI systems used in the US Army's SIMNET programme.

Loral GT 200 CGI System

The GT 200 system uses a combination of GP embedded processors and special bus structures between polygon and pixel processors in order to distribute the image processing without the conflicts that sometimes occur in parallel processing systems. Some specification details are as follows:

Moving Objects: Up to 1024.

Viewing Channels: Up to 12 per graphics subsystem.

Data Update Rate: Up to 60 Hz, fixed or variable.

Computed Resolution: Up to 1024 lines.

FoV: Up to 75° per channel, adjustable in real time if required.

Polygons: 480 polygons/s per graphics subsystem (19 polygons at 25 Hz).

Light Points: Included.
Texture: Up to 1024 on-line texture maps consisting of between 16 x 16 and 256 x 256 texels (256 - 65 536). Full blending, anti-aliased. Texture mapping includes generic and photographic in full colour.
Database: Real-time retrieval from 1.1 Gbyte disk. Library includes models, targets, textures, gaming areas.
Graphics Subsystems: Up to 10 graphics subsystems may be combined.
Occulting: Mask filter and Z-buffer.
Special Effects: 3D smoke simulation (patent pending), haze, fog, rain, snow, horizon effects, flat and smooth shading.
Display Formats: Non-Interlaced, 2:1 Interlaced, and standard video formats are available.
Host Interfaces: Ethernet, CR11W, VME/shared memory.
Sensors: TI/FLIR and NVGs with real-time control of gain, contrast, polarity (for IR) and noise.
Collision Detection: Included.
Diagnostics: BITE.

Status

Available.

Manufacturer

Loral, Cambridge, Massachusetts.
(For details see Directory of Suppliers)

Loral GT 200 CGI System showing a tank, burning vehicle and troops

Loral Vought (LV) Mission Rehearsal System

Loral Vought (LV) has been involved in the development of mission rehearsal systems which present realistic scenes of a potential combat area. One such system is a DoD programme designated Tactical Operational Preview Scene (TOPSCENE).
Operation. TOPSCENE uses high-resolution satellite imagery to provide personnel with detailed scenes in an otherwise unfamiliar area. It converts snapshots of 2D data taken at different times and angles into a real-world 3D simulation of an area. Once developed, the 3D model can be used either in slow time or real time to "fly, walk or manoeuvre through" an area with a view to exploring the area and then planning and practising a mission. Operations and combat personnel can repeatedly view selected routes, and practise taking advantage of visually significant features and terrain masking.
Components. The TOPSCENE system consists of a theatre level system (parent) and unit-level deployable stations. The theatre-level system produces databases from photographic imagery for use on the unit-level stations.
Database Development. An image mosaic/orthomap construction provides the capability to generate mosaic imagery from discrete, digital or photographic images up to 18 kpixels x 18 kpixels in size (324 Mpixels) using an interactive image-processing software package. An image mapping system then provides the end user with the capability to generate images from different perspectives such as low-flying profiles, oblique or vertical views, and so forth. The database system contains a minimum of 20 Gbytes of on-line storage with a data transfer capability of 10 Mbytes/s and computing power of 300 Mflops.
Deployable Unit-Level Stations. The deployable stations are held at unit level and provide aircrews, planners and unit commanders with visual and sensor scenes of contingency areas. Mission preview and rehearsal are performed at the operator station using a throttle, control stick and high resolution monitor. The terrain and cultural features are all correctly registered to geographical co-ordinates so that aircrew can gain familiarity with appropriate navigation cues for contingency areas while in a training environment. The station interfaces with a planning system that enables aircrews to plan routes and missions.
Mission Planning System. The planning system defines a specific mission profile consisting of mission legs or route segments, logistics, weather and threat information. Specific identification points, navigation updates and targets are assigned their associated latitudes and longitudes.

Status

Operational systems are being delivered.

Manufacturer

Loral Vought (LV), Dallas, Texas.
(For details see Directory of Suppliers)

Loral Vought TOPSCENE Mission Rehearsal Station

Martin Marietta Compu-Scene CGI Systems

Martin Marietta acquired the General Electric (GE) Simulation and Automated Systems Division's Daytona Beach facility in April 1993, with its Compu-Scene product line of CGI systems. The history of Compu-Scene originated from a CGI system supplied in 1958 to the US Army and Navy for a joint programme called JANIP (Joint Army Navy Instrumentation Program). In 1962 NASA received a CGI system to train astronauts in space rendezvous and docking manoeuvres. This was followed in 1972 and 1974 when Compu-Scene I CGI systems were delivered to the US Navy and US Air Force for use in flight simulators. In 1975 Compu-Scene II full day/night CGI systems were produced for commercial flight training, and in 1982 the more capable Compu-Scene III became available; this is described below, followed by subsequent systems:

Compu-Scene III

C III was introduced in 1982 and is a full colour day/dusk/night system which can be used in a three-channel layout. It processes 4000 edges which are the equivalent of 1333 triangular polygons per channel. Texture is an additional option, and for military applications, features such as weapon effects, 6-DoF moving models and aiming reticles are included.

Compu-Scene IV

C IV was introduced in 1984 as a state-of-the-art high-range system with, for the time, many innovations, and is still available for sale. It has photographic cell texture with on-line 'cell maps' to store photographic real-world detail; accurately correlated radar, FLIR, NVG, LLTV; circular feature generation and curved surface shading. Texture algorithms produce modulated patterns to represent sea, clouds, crops or various types of terrain.
Circular Feature Generation. Circular feature generation software creates circles, ellipses, spheres and ellipsoids of selectable colours to represent clusters of circular objects such as trees, clouds, storage tanks, silos and similar cultural objects. These features can be generated with curves instead of straight lines.
Resolution. Choice of 625, 763, 875 and 1023 lines.
Sales. Customers include Aeritalia, Aermacchi, Boeing Military Airplane Company, CAE Electronics, Hughes Aircraft, Kawasaki, Lockheed, McDonnell Douglas, MBB, NASA, Sikorsky Aircraft, US Air Force, US Army, US Navy and US Marine Corps.

PT2000

The Compu-Scene PT2000 image generator was introduced in mid-1988 as a medium-cost CGI system with many of the features of Compu-Scene IV but at less cost. PT2000 employs very large scale integration (VLSI) technology and phototexturing. Some specification features include the following:

MM Compu-Scene VI imagery showing airfield lighting, helicopter and searchlight pattern on ground

MM Compu-Scene VI imagery showing a B-2A bomber flying over Southern California

MM Compu-Scene V imagery of airfield installations

MM PT2000 imagery for platoon gunnery trainer

MM PT2000 imagery for the MV-22 Osprey OFT with aircraft carrier

Modes: Day/Dusk/Night/NVG.
Channels: 32 channel capability.
Texture: Up to 48 texture maps from 32 x 32 to 256 x 256 (1024 to 65 536 texels) with up to nine levels of detail. Two patterns per polygon with dynamic effects for effects such as smoke, rotor wash and so on. Full colour, phototexture, microtexture, fractal terrain textures. Simultaneous shading with texture.
Polygons: 4000 at 60 Hz, total 32 000 per channel.
Light Points: 4000 at 60 Hz with zero polygons; total 32 000 per channel, traded off against polygons.
Lights: 6 lobe patterns. Flares and steerable searchlight facilities available. Directional, strobe, PAPI, VASI, rotation, FLOLS, flashing, REILS.
Moving Models: up to 256 6-DoF moving models. 8 levels of detail.
Scan Format, Refresh and Update Rates: Standard is raster, 2:1 interlaced with screen refresh (frame) rate 60 Hz and data update (2 scans) 30 Hz; higher rates available with trade offs in other capabilities. Non-interlaced configurations also available.
Latency: 67 msec at 60 Hz.
Resolution: up to 2 Mpixels per channel (1414 x 1414).
Eye Points: 2, more available for special applications.
Colours: 8 colour tables for polygons and 8 colour tables for lights to support normal visual imagery, NVG and other sensor applications. Colours updated in real time for diurnal variations.
Shading: Curved surface shading provided.
Shadows: Shadows respond to sun angle.
Occulting: PT2000 employs a depth-buffer system which is claimed to make savings of up to 35% in modelling effort because separation planes are not needed. Intersecting planes are simulated where one polygon has visual priority over, but does not obscure or displace the other. For every polygon, range within a pixel is calculated to 4 sub-pixels to a floating point accuracy of 32 bits. Edges formed by intersecting polygons are anti-aliased.
LoD and Scene Management: Objects and polygons are assigned one of 64 feature groups in order to establish a load management hierarchy. An automatic overload protection feature changes the update rate in order to prevent obtrusive effects such as feature popping or system crashes.
Collision Detection: Included, with weapon impact detection.
Height Above Terrain: Included.
Weather Effects: Layered clouds, scud, patchy fog, haze, fog, rain, glare, ground fog, thunderstorm cells, visibility.
Database Management on-line: An active database is maintained for 360° round the eye point in order to cope with high slew rates.
Database Generation: A SUN-based system is used for new databases or models. There is also an extensive model library available. Source data of any sort may be included and the system is configured for ease of accepting formats such as DTED, DFAD and MIL STD 1821(Project 2851) data.
Sales. Amongst other applications, PT2000 is used for the US Air Force's C-17 WST and is also used with the US Army's Conduct of Fire Trainers (COFT).

Compu-Scene V

Announced in April 1989, Compu-Scene V retains all the features of Compu-Scene IV but adds a capability for processing photographic detail in real time. Texture maps with photographic detail are applied in real time to terrain, moving models and objects such as airfields and buildings.
Source Data. Source data ranges from satellite photography, to aerial and other photographs.
Database Generation. Database generation tools provide rapid photo image and geometry processing so that visual databases can be created and updated in the shortest possible time. The system can utilise computerised Defense Mapping Agency data such as DTED and DFAD files, and is compatible with the DoD Project 2851 formats.
Microtexture. A microtexture feature automatically blends into the visual scene at close-in distances to add surface detail. Microtexture patterns can be photographic or synthetic in nature.
Sensors. Sensors are simulated FLIR, NVGs, LLTV and radar, correlated with the visual scene.
Sales. In July 1989 it was announced that the US Air Force Military Airlift Command and Special Operations Forces had selected Compu-Scene V as part of a programme to upgrade a CH-3E helicopter simulator into an MH-53J Helicopter Weapon Systems Trainer.

Existing Compu-Scene IV systems can be upgraded to Compu-Scene V standard in the field and existing databases can continue to be used.

Compu-Scene VI

In November 1990 Compu-Scene VI was added to the Compu-Scene range by combining the architecture of PT2000 with the mission rehearsal capability of Compu-Scene V. Compu-Scene VI is a modular system that can be configured in small increments from moderate performance flight trainers to powerful mission rehearsal systems.

Status

Compu-Scene IV, V and VI and PT2000 in production. Other versions of earlier Compu-Scene CGI systems are in use in military and commercial aircraft simulators.

Manufacturer

Martin Marietta Corporation, Simulation and Automated Systems, Daytona Beach and Orlando, Florida.
(For details see Directory of Suppliers)

Martin Marietta Compu-Scene SE Series CGI Systems

In November 1992 the SE (Synthetic Environment) family of CGI systems was launched, evolved from the PT2000 system (see above). The SE series uses common firmware and shares the same 'Smart Buffer' architecture derived from the real-time depth buffer used with PT2000. This speeds processing by automatically eliminating hidden polygons and incorporates dynamic scene management which smoothly allocates polygons to image areas of highest density.

Compu-Scene SE1000

The Compu-Scene SE1000 CGI system is packaged in a small footprint chassis. It is intended for low-cost applications such as part-task trainers, virtual reality-based systems and driver trainers. It claims to deliver simulator performance at graphics-engine prices. Some specification features for a low-cost variant are as follows, the figures for growth potential refer to higher-cost configurations:

Gaming Area: Unrestricted, paged from disk.
Environment: Weather, lightning, diurnal cues.
Feedback: Collision detect, height above terrain, weapons impact.
Polygons: 2600 at 30 Hz with potential to 9000. Immediate database contains 32 000 polygons with potential for 100 000.
Resolution: 1 million pixels displayed (1000 x 1000), interlaced.
Channels: 1, potential for 4.
Light Points: Raster, 7800 at 30 Hz with no polygons, 3:1 trade off with polygons.
Moving Models: 24 6-DoF models; potential for 128.

MM SE1000 low-cost CGI system showing monitor and projected displays and the small cabinet size

Texture: Generic and phototexture, microtexture, anti-aliased.
Translucency: Multi-level.
Size: 24 W × 32 D × 38 H in.

Compu-Scene SE2000

Compu-Scene SE2000 compared to SE1000 is a larger, more versatile CGI system designed for complex tactical simulation scenarios. Some specification features are as follows:

Gaming Area: Unrestricted, paged from disk.
Database: From DTED/DFAD, MIL STD 1821 (Project 2851), Multigen interface, WGS 84 model.
Diurnal Effects: Day/Dusk/Night, moon, horizon glow, shadows, thermal characteristics.
Weather: Layer cloud, scud, haze, fog, patchy fog, rain, lightning, glare, wet runway.
Lighting: Directional, strobe, rotating, VASI, PAPI, FLOLS, REILS.
Feedback: Collision detect, height above terrain, terrain-following, LoS ranging, weapons impact.
Polygons: 24 000 per system.
Resolution: Line rates 1000, 675 or 500. Processing of up to 8 million display pixels (2828 x 2828).
Channels: Up to 8.
Eye Points: Up to 4.
Data Update Rate: 30 Hz, 60 Hz available with scene trade offs.
Latency: 117 msec, 67 msec at 60 Hz.
Models: 255 models in scene, 24 simultaneously moving 6-DoF models with 8 levels of articulation for components such as booms and turrets.
Texture: Generic and phototexture, microtexture, anti-aliased. Up to 1024 texture maps, 3 Mtexels (1732 x 1732).
Translucency: Multi-level.
Sensors: FLIR, NCG, LLTV.

Status

Available.

Manufacturer

Martin Marietta Corporation, Daytona Beach and Orlando, Florida.
(For details see Directory of Suppliers)

Martin Marietta Database Systems

Martin Marietta products include a number of database generation devices, matched to the particular CGI system used (PT2000, Compu-Scene V, and so forth). A new system now available is the Compu-Scene 'TARGET' database toolkit which is described below, as is a DRMLS database system for radar imagery:

Martin Marietta Compu-Scene 'TARGET' Database Toolkit

The TARGET is independent of individual CGI systems and provides a range of tools to capture detail from data such as maps, photographs, DTED, DFAD and Project 2851 data. Raw data has to be processed and merged ('fused') into a common, consistent database for use in a real-world simulation; TARGET is capable of fusing different source data into a consistent world model. As an example, a shoreline derived from photography or mapping is smoothly combined with a suitable shore image extracted from a library of models. Some specification features are as follows:

Woodland: Woodland and forest boundaries are automatically raised as a 3D extrusion in the visual scene.
Satellite Imagery: SPOT and LANDSAT compatible.
Universal Feature Scattering: If required, a random scattering of features such as trees can be selected for a scene.
IR Model: A mathematical model calculates IR reflectivity of objects automatically so that a TI image of a scene or of an object in the scene is available whenever required.
Automatic Database Processing: Resource control algorithms are included in order to automate the process as much as possible.
Components: A Sun Sparcstation 10 is used with a graphics accelerator. It can be configured as a network in order to share resources if required.
Peripherals: 2 Gbyte SCSI disk, digitising tablet, image scanner, tape drives, 644 Mbyte CD drive, and printer.

Martin Marietta Digital Radar Land Mass System (DRLMS)

The DRLMS simulation programme was begun in 1971 with an experimental radar prediction device. Several families of radar-image simulation equipment are now produced for all three US armed services and overseas customers.
Radar Database. Radar predictions are made from a digital database which is correlated with the visual system scenes. The system then generates a real-time simulated radar image including the scan pattern and other radar system characteristics.
Features. Features of real radars are included such as Doppler beam sharpening, terrain-following and terrain-avoidance, as well as special effects such as chaff dispensing and jamming, weather, beacons, and radar faults.
IOS. Instructors can monitor the radar image while the simulator is running a mission, and control the DRLMS for maintenance and test purposes. The data displays are 360° plan-position indicator-style representations of the information held in the system's memory about the region being displayed.
Deliveries. DRLMS simulation equipment has been

MM TARGET - Database Toolkit

delivered to the US Navy for use with Grumman/A-6E Prowler simulators and to the US Air Force for Boeing B-52 simulators. Nine systems have been supplied to MBB in Germany for use with Panavia Tornado simulators. Other systems have been built to train operators on naval ships. MM is also the contractor for the DRLMS systems used in simulators for the F-16 fighter.

Status

Available.

Manufacturer

Martin Marietta Corporation, Daytona Beach and Orlando, Florida.
(For details see Directory of Suppliers)

Star Graphicon CGI Systems

The Star Technologies Graphicon is a family of low-cost CGI systems. It includes the Graphicon G2000/STX and 2000/PTX; the Host Integrated Graphicon 2000 series; and the Graphicon 1700 simulator for outside-world scenes involving less demanding flat and shaded applications.

Star G2000/STX CGI System

The G2000/STX CGI system creates textured, smooth-shaded, anti-aliased scenes.
Components. The system utilises three types of board:

an Object Processor incorporating a 17-Mips RISC processor and a 25-Mflops DSP to handle communication with the host, builds, updates and traverses the database residing in the display list memory;

an Arithmetic Processor which utilises multiple DSPs to provide more than 200 Mflops for manipulation of geometric objects and transformation from three-dimensional polygon data to two-dimensional screen display co-ordinates;

an Image Processor which performs the polygon shading, anti-aliasing and hidden surface removal while a daughter-board provides the texturing (each Image Processor delivers more than 2000 Mips of dedicated processing power).

Star Graphicon 2000/STX imagery in an air-sea rescue flight simulation application

Resolution: 1290 × 1024 (1320 kpixels).
Polygons. It manipulates and renders up to 8000 polygons per channel at 30 Hz, or 4000 polygons at a 60 Hz data update rate.
Texture. Textured, anti-aliased, smooth-shaded. Phototexture is available.
Integral Host. Available with an integrated host computer and peripherals under the designation HG2000/STX.
Interfaces. The system can interface with the user through the Visual Interface Protocol (VIP) or Real Time PHIGS (PHotographic Image Generation System). VIP offers a standardised interface method which simplifies host integration and supports many IG features. Real Time PHIGS was designed primarily for a research and development environment.
Applications. Applications include flight simulation, tank and gunnery training, naval (ship pilot) training, land vehicle driver training and command, communication, control and intelligence use.

Star Graphicon 2000/STX imagery showing tanks

Star Graphicon 2000/STX imagery showing infantry

Status

In production. More than 825 installations in service in various applications.

Manufacturer

Star Technologies Inc, North Carolina.
(For details see Directory of Suppliers)

VISUAL DISPLAY SYSTEMS

Introduction. This section includes systems concerned with visual display to trainees, crew members or operators of training devices and simulators. The display may be of the outside world visual scene, of sensor imagery, or of monitor-based crew displays which can also be touch-screen controlled. The outside world picture is often termed the OTW display (Out-of-The-Window) and is assumed to be viewed directly by eye in the normal visual band at wavelengths of 0.4-0.7 microns. Sensors include intensifiers such as NVGs and LLTV operating in the near IR between 0.7 and 1 micron, Thermal Imagers (including FLIR) operating in the far IR at 10 microns, or radar displays operating at millimetric wavelengths and above. Imagery for all of these displays will originate from an Image Generator (see previous section), generally a Computer Generated Image (CGI) system. Monitor-based OTW displays are generally referred to as 'windows', each driven by its own CGI channel with the exception of side-by-side crew seating where each crew member may have a display window directly ahead using a common CGI source.

Collimation. Many OTW display systems use curved mirrors or Fresnel lenses to present the image at a distant focus, typically at distances between 100 and 400 ft; these are known as collimated or infinity displays. This is not only more realistic in most applications but has the advantage that the viewer has a range of head and eye movement over which the image does not change its perspective, whereas in direct projection systems as the head is moved from a calculated eye point, perspective errors occur which become progressively greater with distance from the calculated eye point.

Side-by-Side Crew Seating. These perspective errors in non-collimated systems can be critical in simulations where two crew members are seated side-by-side, and where two widely separated eye points have to be allowed for simultaneously. For this reason, and also the enhanced realism of the distant focus, collimation is a requirement for certification of a flight simulator to the higher civil regulatory levels such as the US FAA levels C and D. Some relevant terms and diagrams of commonly used layouts follow.

Relevant Terms and Abbreviations.

The following terms, used in Visual Display technology and frequently referred to in this section, are defined or explained in the Glossary of Terms section at the beginning of this book.

- AOI
- BPC
- CGI
- CSI
- Channel, Visual Channel
- Collimation, Collimated Display
- Frame Rate (same as Refresh Rate)
- IG
- Interlaced Scan
- Pixel
- Refresh Rate
- Resolution
- Update Rate (of data)

Domes and Partial Domes

This section contains at the end a small sub-section on domes and partial domes.

SOME PRINCIPLES OF VISUAL DISPLAY SYSTEMS

The following diagrams showing the general principles of some commonly used visual display systems were produced by and are reproduced with the kind permission of SEOS Displays Limited. The accompanying narratives are the editor's.

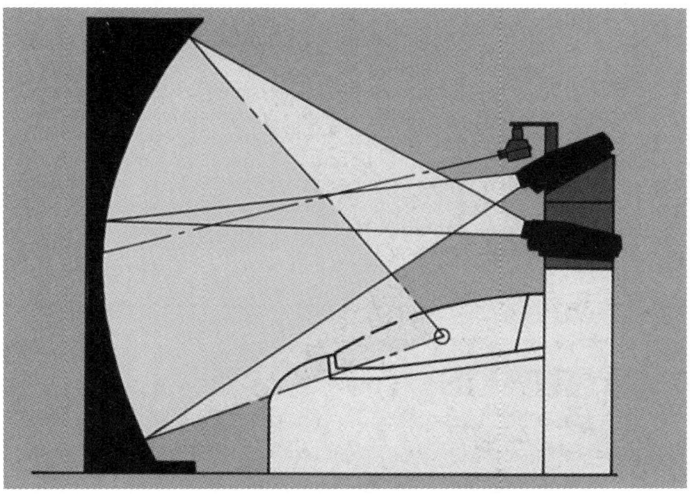

Direct Projection Display System Viewed from the Side - On the right are two projectors for the Outside-World Image (in this example one above the other for good vertical FoV), and above them is a separate projector which could be used for targets or for calibration purposes

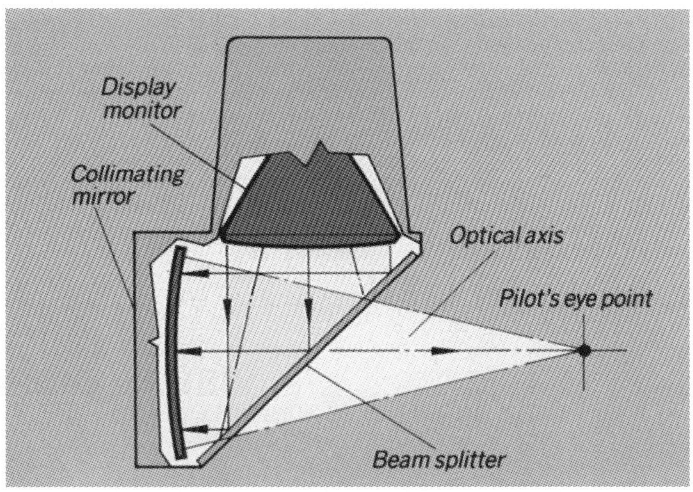

Collimated Monitor - Optical Principles of Distant Focus

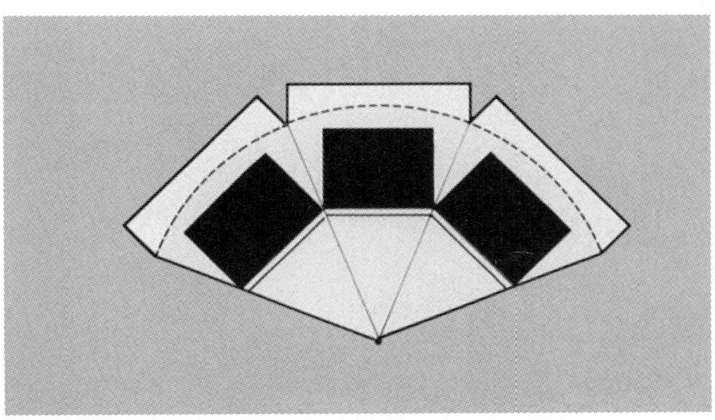

Display with 3 Collimated Monitors and a single subject's eye point - Three visual windows driven by three Image Generator channels

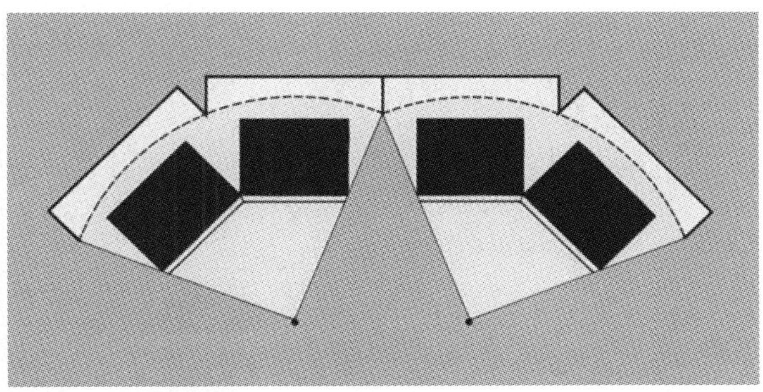

Display with 4 Collimated Monitors and two subjects' eye points side-by-side - Four visual windows are driven by three Image Generator channels, the centre two windows sharing the same IG channel and presenting identical information

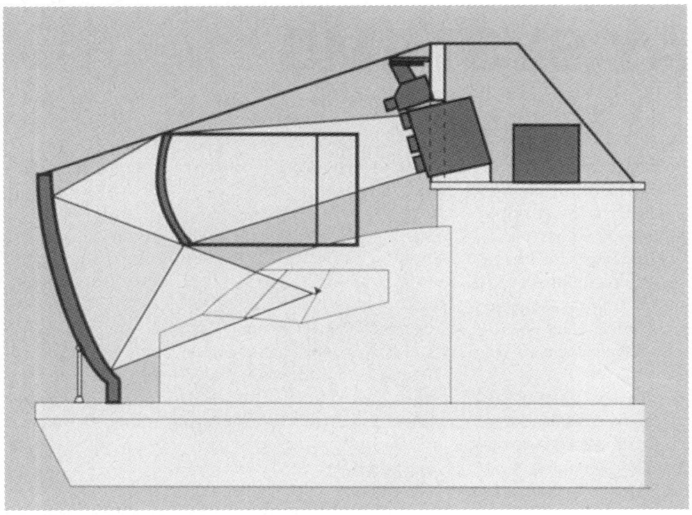

Back-Projected Collimated (BPC) Display with Cross-Cockpit Capability - Side view of optical principles. The projectors above the simulator cabin back-project images onto an intermediate screen above the crew compartment. The crew members do not see this screen directly but view its image in a curved mirror positioned ahead of the crew eye points. The mirror's vertical curvature gives the distant focus implicit in collimated systems and the horizontal mirror curvature gives a continuous, identical, horizontal view without discontinuities from both crew eye points (or from elsewhere in the cockpit). Due to collimation, scene perspective is correct from both eye points simultaneously, and also from other points in the cabin. Examples of BPC display systems include the original BPC product, the Rediffusion (now TTSL) WIDE series, the Link-Miles (now TTSL) AWARDS, the FSI MultiView, the CAE (Montreal) MAXVUE™, and the SEOS PANORAMA display systems.

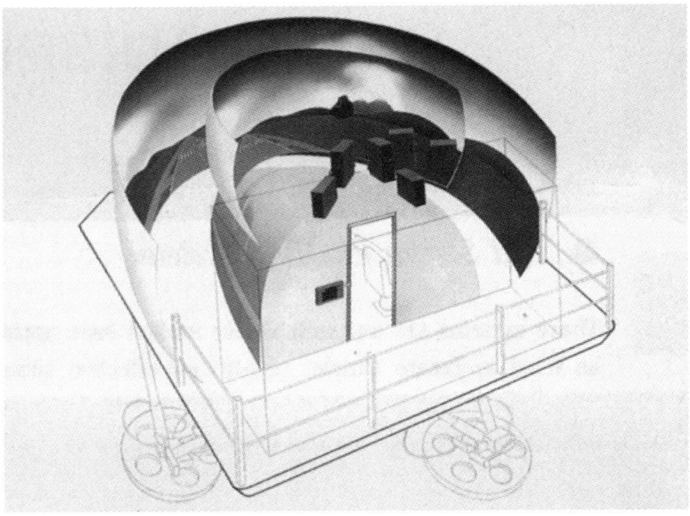

Back-Projected Collimated (BPC) Display, Artists Impression - Several projectors (Dark rectangles) may be seen above the door to the simulator crew compartment. They back-project imagery on the intermediate screen. The images on this screen are then seen by the crew as reflections in the curved collimating mirror positioned below the screen and directly ahead of the simulator cabin. The images give the illusion of being directly ahead and, due to the mirror's vertical curvature, at a distant focus.

Comparison of BPC displays with other systems

Compare a BPC display with a multiple collimated-monitor arrangement giving a similar FoV; each monitor has a restricted viewing area (generally about six inches by three) within which its full image can be seen. Since monitors in simulators for two crew seated side-by-side have to take into account two laterally separated eye points, the port side monitors will be set for the left seat crew member's eye point and the starboard monitors for the right seat occupant. It is likely that little, if any, cross-cockpit view of images on monitors on the other side of the cockpit will be seen and cockpit windows on the side away from the pilot's seat will often appear blank. This was one of several reasons why the BPC displays were developed.

Limitations of BPC Displays. BPC displays have good horizontal FoV, limited only by the number of projectors, the mirror size, and cost. Their disadvantage is that vertical FoV has so far been limited to about 40° which is adequate for airliner simulators but not for aircraft which themselves have a large vertical FoV (depending on the training task of the simulator concerned - for some tasks, having a vertical FoV substantially less than that of the aircraft, may be acceptable).

Direct Screen Projection. For simulators for which the customer requires large vertical FoV, direct projection onto curved screens or on the inside of complete domes is generally employed. With direct projection, the display eye point (from which the scene perspective will be correct) can only be calculated for one point at a time, although it can be switchable in multi-crew simulators or set at a mean point between the crew.

Helicopters. For helicopters, a BPC main display can be combined with monitors mounted below it which provide the chin-window cues for hovering and hover-taxi. The upward view needed for realistic training for auto-rotations and engine-offs is, as yet, difficult to achieve with collimated systems although it is straightforward if direct projection is employed.

VISUAL DISPLAY SYSTEMS

CANADA

ATS Wide FoV Display Screen

The ATS Vega/32 Visual Simulator may be configured with fields-of-view from 40° (a single window) to 360° (seven or eight windows). Its original purpose was ATC Tower simulation but the visual system could be used for other roles if required.

Manufacturer

ATS Aerospace Inc, St-Bruno, Québec.
(For details see Directory of Suppliers)

ATS Vega wide-angle projection system originally designed for ATC tower simulators

CAE MAXVUE™ Visual System

Introduced in early 1992 as CAE's first entry into the visual system market, the CAE MAXVUE™ visual system was designed to comply with FAA FFS Level D standards, while also being suitable for military requirements. It comprises both Image Generator and Display Systems. The IG element was covered in the previous section and the Display System is addressed here.

MAXVUE Display System

The CAE MAXVUE™ display is a Back-Projected Collimated (BPC) simulator display system providing imagery over a continuous FoV. See the definition of BPC in the glossary and also the diagrams of BPC systems at the beginning of this section on display systems.

Mirrors. Glass or Mylar mirrors are available at customer choice.

Projectors. Full colour projectors capable of both raster and calligraphic scans are located above the cockpit.

Setup. Set up is microprocessor-controlled with distortion correction, convergence, shading, contrast and colour matching. A portable remote-control unit is included for set up and adjustment.

FoV. The following uninterrupted FoV are available: 40 or 45° vertically by 150, 180, or 210° horizontally, employing three to five projectors depending on FoV and resolution required.

Status

Available. By mid-1994, CAE had sold 17 MAXVUE™ visual systems worldwide.

Manufacturer

CAE Electronics Ltd, Saint-Laurent, Québec.
(For details see Directory of Suppliers)

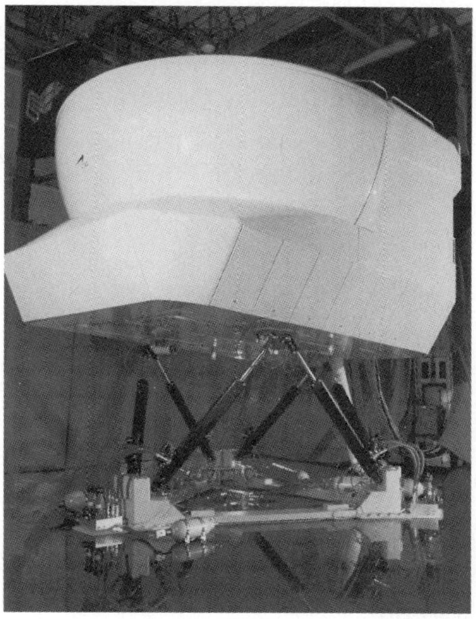

CAE MAXVUE™ Back-Projected Collimated (BPC) display system showing the characteristic smooth external shape of the double-curvature collimating mirror unit at the front of the cockpit section

CAE Fibre Optic Helmet-Mounted Display (FOHMD)

The CAE Fibre Optic Helmet-Mounted Display (FOHMD) consists of a helmet assembly and does not need projection domes, screens or collimated monitors. It is therefore much lighter and smaller than conventional display systems and yet can provide a substantially unlimited FoV both vertically and horizontally. The FOHMD displays the image in two windows directly in front of the subject's eyes.

General Characteristics. The FOHMD provides separate images to the left and right eyes and so stereoscopic images are possible, unlike most other display systems. Each eye sees a large-area, low resolution background with a smaller, high resolution inset where the eye is actually looking. CGI data is relayed to the helmet optics through two fibre optic cable bundles.

Operation. The FOHMD consists of a helmet on which are mounted two Pancake Window displays. Imagery is transferred to the helmet from four high brightness light valve projectors. An IR-based helmet tracking system provides feedback of head position information while an angular rate sensor package (mounted on the helmet optics) provides data for accurate lead predictions to compensate for computer image generation transport delays. An eye-tracking system is incorporated on some versions and is used to position the high resolution inset where the user is looking within the lower resolution wide FoV background visual channel.

Helmet Size and Fit. The FOHMD is based on a custom helmet made in two sizes with moulded internal liners to suit each individual. Comfort is claimed to be the same as a standard aircrew helmet.

Optics. The display optics consist of three basic items; relay combining optics, fibre optic cables, and the helmet display. The relay combining optics serve to transmit the images produced by the light valve to the input of the fibre optic cables. They provide correction for size and scale of the image, mapping to match the helmet display optics, and combining the high and low resolution images for each eye.

Fibre Optic Bundles. The fibre optic cables are a large number of fibres formed into a large diameter coherent optical cable. Standard cables are 1.83 m in length and are sheathed with a light, flexible protective covering.

Pancake Windows. The Pancake Windows mounted in front of the eyes are of 76.2 mm diameter. They are compact, in-line, wide-field reflective distant-focused displays which provide overlapping, full-colour, high brightness images to each eye. The windows are semi-transparent and permit cockpit controls, head-up displays and indicators to be viewed normally; a cockpit mask blanks out any unwanted CGI images where the cockpit is to be viewed. The exit pupil is large and the eye relief is sufficient to allow the use of aviator's spectacles. The overlapping display provides an instantaneous field-of-view of 65° vertically by 125° horizontally. High to low resolution boundaries are masked with a soft edge using video blending techniques.

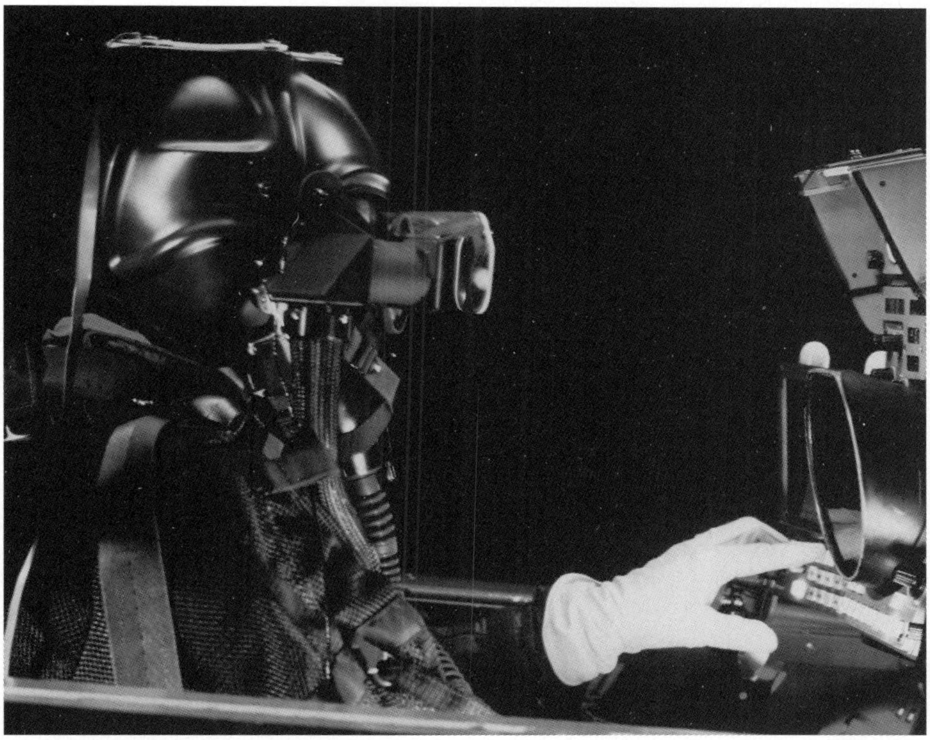

The CAE Fibre Optic Helmet-Mounted Display (FOHMD) as used for the German Tornado Simulator Upgrade Programme

Light Valves. The video input to the system is provided by four light valve large screen television projectors, two for each eye. The projectors provide a high resolution, high brightness, full-colour image.

CGI Channel Requirements. The FOHMD requires three IG channels. High resolution is needed only on the centre AOI channel.

Brightness, Contrast and Distortion. The display provides a minimum of 30 foot-lamberts at a contrast ratio greater than 50:1. Geometric distortion is claimed to be less than 1.5 per cent.

Status

In production. Applications include the US Army's Crew Station Research and Development Facility at NASA Ames, the US Army Research Institute's Simulator Complexity Test-Bed at Fort Rucker, Alabama, Rockwell International, the US Air Force Human Resources Laboratory at Williams AFB, Arizona, the Tornado Low Level Test-Bed for the German Air Force, Germany's F-4F air combat simulator upgrade, the US Army Simulation Program for Improved Rotorcraft Integration Technology, and the US Navy.

Manufacturer

CAE Electronics Limited, Saint-Laurent, Québec.
(For details see Directory of Suppliers)

FRANCE

TTS Visual Display Systems

Thomson Training and Simulation (TTS) has developed a range of visual display systems which is described below:

Janus Background Projector System. Janus uses two fish-eye wide-angle panoramic lenses, one behind and one in front of the simulator crew station. The FoV cover is virtually 360°, and it is typically fitted in a projection dome. Images of the sky and ground are derived from CGI systems such as Thomson's Visa. It is fitted to a number of simulators including those for the Mirage 2000.

Mars Laser Target Projector. Mars uses laser beams to project up to four targets, typically on the inside of a projection dome. Mars is used on a number of simulators including those for the Mirage 2000.

Phebus Projectors. The Phebus multiple projector system is designed for accurate edge-matching of images in order to build up a large FoV by using multiple projectors. For instance, five projectors can give a FoV of 105° vertical by 160°. A simulator for the Mirage 2000N uses such a five-projector system, images being generated by one channel of Visa IV CGI per projector.

Collimated Monitors. SMBC shadow mask monitors are also marketed by TTS and have a raster scan and calligraphic display capability.

High Brightness Projector. TTS has developed a full-colour high brilliance projector with correction capabilities for multi-window image projection.

Status

Available. In widespread service (see text).

Manufacturer

Thomson Training and Simulation, Cergy-Pontoise.
(For details see Directory of Suppliers)

UNITED KINGDOM

Invertron Panoramic Display System

The Invertron Panoramic Display System is presented on a screen made up of panels each measuring 1.8 × 2.4 m and providing a visual coverage of from 60 to 360°.

Manufacturer

Invertron Simulated Systems Limited, Burgess Hill, West Sussex.
(For details see Directory of Suppliers)

Invertron VISRAD with three screens out of a possible eight

Marconi Panoramic Screen Display

The Marconi Panoramic Screen Display is a large wrap-around curved screen of 10 m radius, 320° horizontal FoV and ±20° vertical FoV. It is used in Marconi's Ship Bridge Simulator.

Manufacturer

Marconi Simulation, Dunfermline.
(For details see Directory of Suppliers)

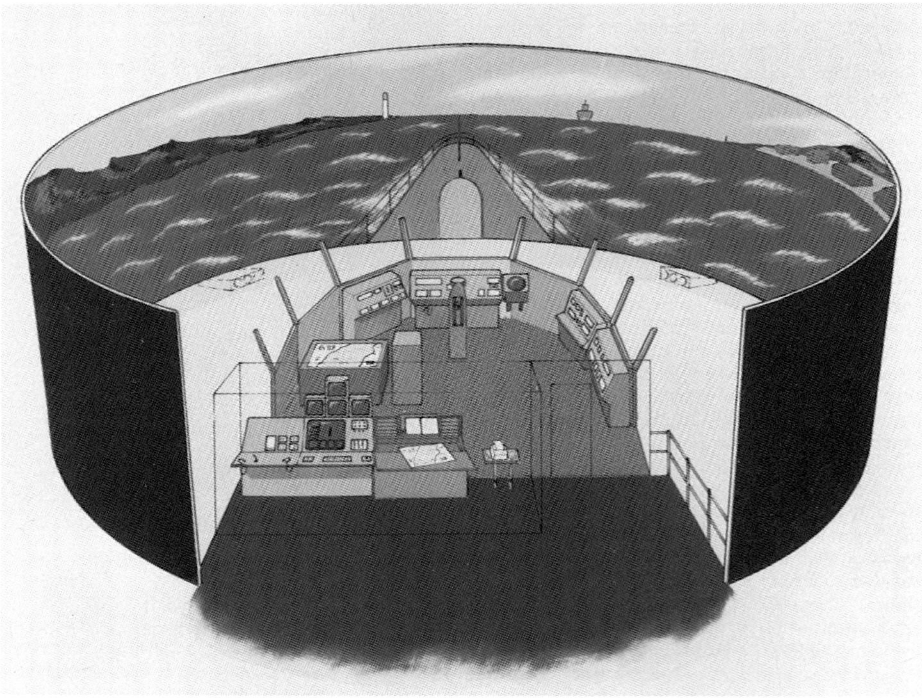

Marconi Panoramic Screen Display used in a Ship's Bridge Simulator for the Royal Navy

RUKL 360° Display System

Reflectone UK Limited (RUKL) manufactures a display system giving up to 360° of cover, depending on the number of Image Generation channels and Display Windows chosen by the customer. It is offered as an option with the RUKL Air Traffic Control Tower Simulator but could be used for other purposes.

Manufacturer

Reflectone UK Limited (RUKL), Bristol.
(For details see Directory of Suppliers)

SEOS Display Systems

SEOS Displays Limited manufactures a number of visual display systems including collimated monitors, direct screen projection systems of various FoV, a back-projected collimated cross-cockpit display, and various simulated weapon and observation sights for gunners and AFV commanders.

PRODAS Direct Projected Displays

PRODAS™ is derived from PROjected DisplAy System. PRODAS displays are direct screen projection systems for use with one or more projectors depending on the customer's requirement for FoV. A variety of screens is available as follows:

Standard PRODAS Screens. The standard screens have double curvature to enable sharp focus, and are compatible with motion platforms. They are produced in single, three and five-projector configurations designated S100, S300 and S500 respectively. FoV depends on customer requirements and may be between 35 to 50° vertically and 45 to 60° horizontally per projected channel, giving between 135 and 180° for three channels and 225 to 300° for five channels, the trade-off at the larger figures being with resolution.

Hi-View Screen. The Hi-View screen has an extended vertical area compared to the standard screen, and is also motion platform-compatible. A typical six projector layout is with two banks of three projectors and gives a FoV of 140° horizontally and −20 to +70° vertically. The vertical split can be altered to suit customer requirements and special configurations are available if larger FoV is required.

Cylindrical/Conical Screens. These screens are not

SEOS PRODAS C300 direct projection display system - MBB Alphavision CGI imagery used in an MBB ATC simulator

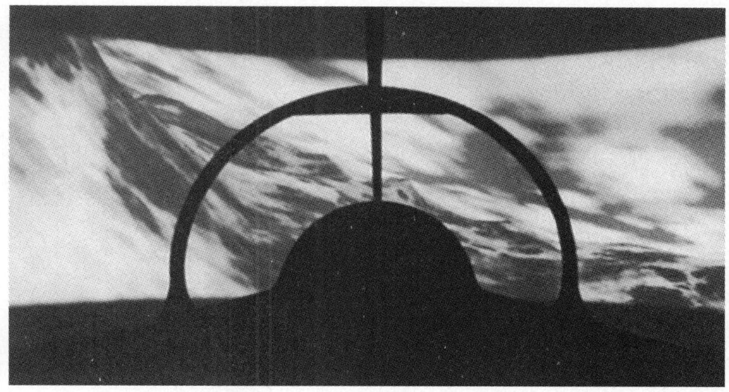

SEOS PRODAS C300 direct projection display system - E&S ESIG-2000 CGI system imagery

generally compatible with motion platforms or large vertical FoV. They provide a cost-effective alternative to the standard screen curvatures, particularly where the application requires a screen of non-standard radius.
Flat Screens. These screens are not compatible with motion platforms but can be used for low-cost display to many users at once. Large horizontal FoV can be provided but without the benefits of edge-blending between projectors.
Edge Blending. Soft edge blending between channels and colour match capability is provided by means of a Digital Display Interface Unit which is available in either analogue or digital format. The unit provides a generally improved performance including full microprocessor control. This allows all adjustments to be made via a handset which can be fitted to all PRODAS systems to make the task of aligning a multi-channel display as fast and straightforward as possible.
Colour Adjustments. IRIS Auto Convergence is also offered as an option. This allows the red and blue adjustments to be carried out automatically, although the green geometry still has to be aligned manually using the remote control handset and spherical reference grid projector. The latter provides a projected grid which is correct from the design eye point.
Sales. PRODAS displays are used for a variety of simulation applications such as military and commercial pilot training, ship's bridge simulation, air traffic control, ground-based weapons simulation, vehicle driving simulation and research systems. By 1994, 47 PRODAS displays had been supplied to customers in Australia, Germany, Italy, Netherlands, Switzerland, UK and USA.

Status

In production and available.

SEOS MIDAS Collimated Monitors

MIDAS is derived from Monitor-based Infinity DisplAy System and is a product range of collimated monitors. It is compatible with motion platforms and conforms to UK CAA and US FAA standards.
Components. Each MIDAS unit consists of a vertically mounted TV monitor, a curved collimating mirror, and a semi-transparent glass plate (the beam-splitter). The layout is shown in the diagram at the beginning of this section on display systems. The components are all mounted within a secure casing. All units have a highlight brightness of 6 Ft Lamberts, contrast ratio 15:1, and limiting resolution 3 arc minutes.
Versions. There are two versions;

MIDAS 20 with a 20 in (508 mm) monitor and a Design Eye-Point (DEP) 28 inches from the beam-splitter. FoV 43 x 35° total, 44 x 27° instantaneous FoV, weight 165 lb, minimum collimation (focusing) distance 230 ft (centre field) and 33 ft (corners), and a viewing area with no image loss of 4.5 x 2.2 inches.

MIDAS 26 with a 26 in (660 mm) monitor, DEP 37.3 inches from the beam-splitter. FoV 46 x 35° total, 42 x 27° instantaneous FoV, weight 342 lb, minimum collimation (focusing) distance 144 ft (centre field) and 33 ft (corners), and a viewing area with no image loss of 6.0 x 2.8 inches.

The MIDAS 26 with its additional viewing distance is better for applications involving larger crew compartments. MIDAS 20 has the advantages of lower cost and reduced weight and an eye point distance suitable for smaller crew compartment size.
Multiple MIDAS Units. Zero gap formats are available on all MIDAS systems, as are layouts with gaps between monitors in order to mimic cockpit window frames and/or to give more overall FoV.
Scan Capabilities. MIDAS units can be supplied to cover line scan rates in the range of 15 to 64 kHz (1000 lines at 50 Hz would be 50 kHz). MIDAS monitors generally operate in raster scan format only, although at customer request other units can be provided capable of accepting calligraphic as well as raster inputs.
Sales. MIDAS displays are used for a variety of simulation applications such as military and commercial pilot training, ship's bridge simulation, ground-based weapons simulation, vehicle driving simulation and research systems. By 1994, 65 sets of MIDAS systems involving over 130 monitors had been supplied to customers in Austria, European Space Agency, Germany, India, Japan, Switzerland, UK and USA.

Status

In production and available.

SEOS MIDAS 20 - two collimated monitors mounted side-by side. In this layout there is a gap between them (compare the picture on the other side). The monitor units can be seen on the top and the collimating mirror and beam-splitter units below. Compare with the optical diagram at the beginning of this section

SEOS MIDAS 26 - two collimated monitors mounted for zero gap in FoV. The monitor units can be seen on the top and the collimating mirror and beam-splitter units below

SEOS PANORAMA Back-Projected Collimated (BPC) Display System.

PANORAMA is a back-projected collimated display system which gives cross-cockpit viewing. See the definition of BPC in the glossary and also the diagrams of BPC systems at the beginning of this section on display systems.
Motion Platform Compatibility. Compatible.
Regulatory Approvals. US FAA and UK CAA.
Collimating Mirror. The collimating mirror consists of a continuous flexible plastic base coated with highly reflective aluminium and held at the correct double curvature by suction. A laser measurement system and an embedded microprocessor are used to continuously monitor and adjust the shape for correctness while the system is in operation.

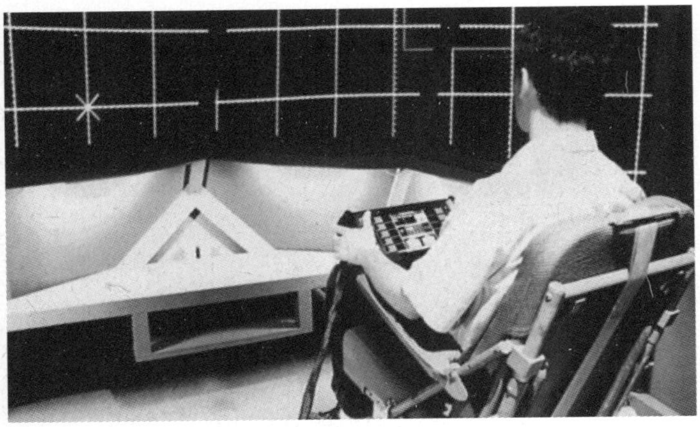

SEOS PANORAMA continuous FoV collimated system - Display calibration using the remote-control system and spherical reference grid

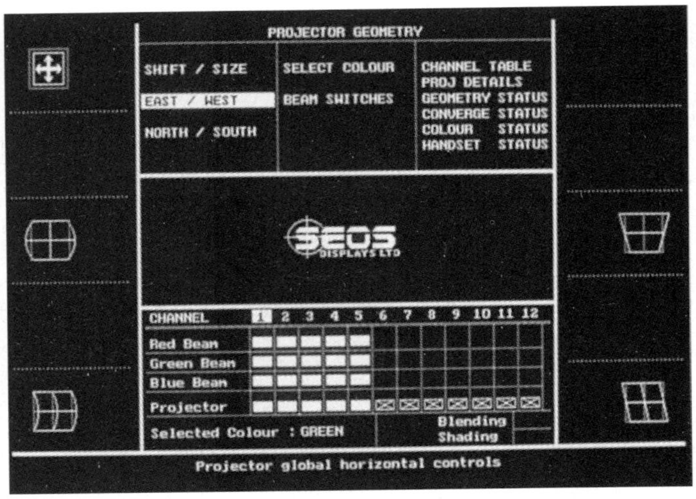

SEOS PANORAMA continuous FoV collimated system - Display graphics on remote-control system

FoV. The standard system gives a FoV of 150° x 40° in a three IG channel/three projector arrangement.
Projector Scan Modes. Although the standard PANORAMA SEOS-900 series of projectors accepts raster-only video signals with line scan rates between 15 and 64 kHz, other projectors can be offered with either a higher scan rate or to accept the raster/calligraphic video formats which are the output of CGI systems often used in civil flight simulators.
Screen Brightness. 6 Ft Lamberts in highlights, 3 Ft Lamberts average picture level.
Resolution. 3 arc minutes.
Contrast Ratio. 8:1.
Collimation/Focusing. Infinity down to 42 feet as required by application.
Weight. Mirror 1540 lb, intermediate screen 350 lb, 3 projectors 660 lb, total 2220 lb; extra structure weight depends on application.
Power Requirement. 3 kVA.
Display Interface Unit. A Digital Display Interface Unit provides soft edge blending, colour matching and intensity shading, and includes a remote-control handset. This is the master unit for system set up.
Spherical Reference Grid. A spherical reference grid projector gives a built-in optical test pattern for calibration purposes.
Sales. PANORAMA displays have been sold for aircraft simulators in the USA and for a museum display in Japan.

Status
In production and available.

SEOS Simulated Sight Systems

SEOS also makes a range of simulated sighting systems for simulations of AFVs and other devices. Because of the specialised nature of such devices on the equipment to be simulated, these are normally designed specially for the application. A typical AFV sight might have the following components:
Video Monitor. Frequently a colour shadow-mask monitor sized to the application. Monochrome and green screen displays can be provided for applications such as intensifiers (NVG/LLTV) and TI/FLIR.
Chassis and Controls. The chassis and controls will be modelled on the original equipment and may employ real or synthetic components.
Alphanumeric Displays and Graticules. Appropriate symbology and aiming graticules are provided as appropriate.
Sales. SEOS simulated sights have been sold to customers in Germany, Japan, UK, and the USA.

Status
In production and available.

Manufacturer
SEOS Displays Limited, Burgess Hill, West Sussex.
(For details see Directory of Suppliers)

SEOS Marder AFV Commander's sight supplied to KUKA Wehrtechnik GmbH

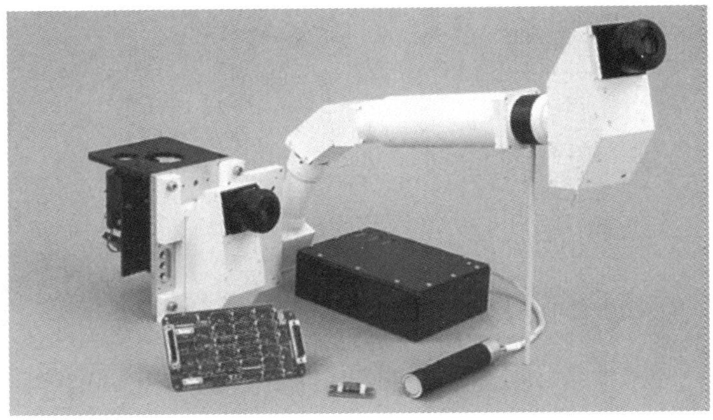

SEOS Gunner and Commander's sight supplied to ECC Simulation Ltd

Tector Display Systems

Tector Visual Systems has produced low-cost CGI systems and Collimated Monitor displays since the company was formed in 1974, and now also makes low-cost projected display systems. These can be configured to suit customer requirements.

Tector Display Monitor

The Tector range of display monitors includes both raster and raster/calligraphic scan models. Digital remote-control of alignment is standard on all models providing accurate alignment from the viewing position even with the monitor sited in the most inaccessible position. All monitors are fitted with a pattern generator to facilitate alignment, with linear scan monitors having a scan frequency selection. All are motion-system compatible.

General specifications for all models

Display: 660 mm diagonal shadow mask CRT; 0.51 or 0.37 mm dot pitch dependent on resolution requirements
Deflection: magnetic
Focus: electrostatic with dynamic focus
Scan frequency: 15 to 64 kHz, interlaced or non-interlaced
Colour registration: 0.75 mm in a circle equal to picture height, 1 mm elsewhere
Video bandwidth: medium resolution: 75 MHz, and high resolution: 150 MHz

Status

In production and widespread service.

Manufacturer

Tector Visual Systems, Newark.
(For details see Directory of Suppliers)

Tector 26 in (660 mm) collimated display system on a free-standing mounting

---

TTSL Collimated Display for Infantry Weapon Training Simulator (IWTS)

Development of the Thomson-CSF (formerly Link-Miles) Infantry Weapon Training Simulator (IWTS) began in early 1992.

The IWTS features a collimated wide-angle visual scene display with the detail and perspective to provide comprehensive training in target acquisition, weapon engagement over a wide arc, fire-control procedures, and shoot/no-shoot decisions. The system can also be used for anti-armour weapons and for mortar fire-control.

The wide-angle collimating mirror screen display is formed by moulding a thin plastic film coated in a reflective material over a suitably curved set of formers by a low-power vacuum pump. The resultant display ensures that all scenario and target images appear at the correct distance with the correct perspective for all trainees. The depth of field allows optical sights and image intensifiers to be used without modification.

The basic IWTS configuration is a five-lane system with a panoramic 72° horizontal field-of-view visual display. Two five-lane units can be combined, linked to one instructor station.

The IWTS is based on standard personal computers in a networked configuration (one for each firing position and one for the instructor). The system has facilities to enable instructors to prepare and present their own training exercises and scenarios.

Status

In production for Denmark and Germany.

Manufacturer

Thomson Training & Simulation Limited (TTSL), Crawley, West Sussex.
(For details see Directory of Suppliers)

TTSL Collimated Display for an Infantry Weapon Training Simulator (IWTS). A continuous scene can be portrayed when required. Note that all the screen can be viewed at once even from where the photograph was taken, unlike a collimated monitor system which has only a small viewing volume

---

TTSL AWARDS Collimated Display System

Current Situation. Thomson Training and Simulation Limited (TTSL) markets the ex-Link-Miles AWARDS display system.

Advanced Wide-Angle Reflective Display System (AWARDS).
AWARDS was Link-Miles' equivalent to Rediffusion's WIDE and uses glass mirrors compared to the flexible mirror surface used in WIDE. AWARDS has achieved a light-point brightness level of 8 Ft Lamberts and meets FAA Level D requirements.
FoV. Three projectors give a 150 × 40° FoV and AWARDS 200 extends the horizontal FoV to 200° using five projectors. The projectors use commercially available tubes with projection lenses permitting exposure of a large area of the cathode ray tube phosphor (nominally 127 mm diagonal) to provide high levels of brightness.

Status

Available.

Manufacturer

Thomson Training and Simulation Limited (TTSL), Crawley, West Sussex.
(For details see Directory of Suppliers)

TTSL WIDE Collimated Display System

Current Situation. Thomson Training and Simulation Limited (TTSL) continues to market the ex-Rediffusion WIDE product line of Back-Projected Collimated (BPC) cross-cockpit display systems which were the first of this type of display system in the market.

Wide-angle Infinity Display Equipment (WIDE)

In 1982 the first WIDE display system went into service and achieved a breakthrough in display technology by replacing conventional monitor-based displays with a collimated continuous-mirror system. WIDE was the first system to enable unlimited cross-cockpit viewing.
Screen Projection. Images are projected on a 20-ft (6.10 m) long screen situated in front of the simulator flight-deck. The screen has a convex shape in the vertical plane and this is calculated to give a distant focus (collimation) which is normally set at about 100 m.
Screen Construction. The screen is made of flexible material and is held in place by negative pressure from a suction pump.

FoV. 150° horizontally and 40° vertically with three projectors and CGI channels; 200° in the WIDE 2 version with five projectors and CGI channels.
SupraWIDE. SupraWIDE uses enhanced CRT technology with better picture quality and brightness.
Sales. By 1994 more than 220 WIDE systems had been sold.

Status
Available.

Manufacturer
Thomson Training and Simulation Limited (TTSL), Crawley, West Sussex.
(For details see Directory of Suppliers)

TTSL WIDE II Back-Projected Collimated (BPC) display - Note the rails at the top surrounding the projectors. The intermediate screen is not visible but is inside the conical cover, and the outside of the double-curvature mirror assembly can be seen in the foreground. This unit is fitted to the USAF C-141 WST

UNITED STATES OF AMERICA

CAE-Link ESPRIT Visual Display System

The CAE-Link ESPRIT (Eye-Slaved Projected Raster InseT) system is the first production application of an eye-slaved, area-of-interest (AOI) display where the high resolution (HR) AOI is always centred around the trainee's line-of-sight and moves as the eye moves. The background is filled in with a low resolution (LR) image. The system requires only two CGI channels in order to achieve a very large FoV, since high resolution imagery does not need to be generated in directions in which the user is not looking.
Projection. Projection can be on screens or on the inside of domes such as the 7.3 m diameter dome mounted on a motion platform used in the UK Harrier simulator made by TTSL (ex-Link-Miles). The HR AOI channel can be moved very rapidly to follow the user's eye movements since it is projected through a lightweight slaved mirror. It also has prediction functions built in to its drive laws which are designed to anticipate the eye movements of the user.
Resolution. Originally 3 arc minutes per optical line pair (OLP) in the slaved centre AOI channel (1.5 arc min per pixel), but CAE-Link now claims a potential of 1 arc minute per pixel (2 arcmin/OLP) in the centre AOI. The resolution of the human eye in its centre field (fovea) is about 0.5 arcmin/pixel. These resolutions in the display system are dependent on the cone angle of the AOI; the smaller the cone angle, the better the resolution, since the data of one CGI channel is presented in a smaller area. The UK MoD selected an AOI cone angle of about 16° which at the time corresponded to the 3 arcmin/OLP figure; a cone angle reduced to, say 10° would

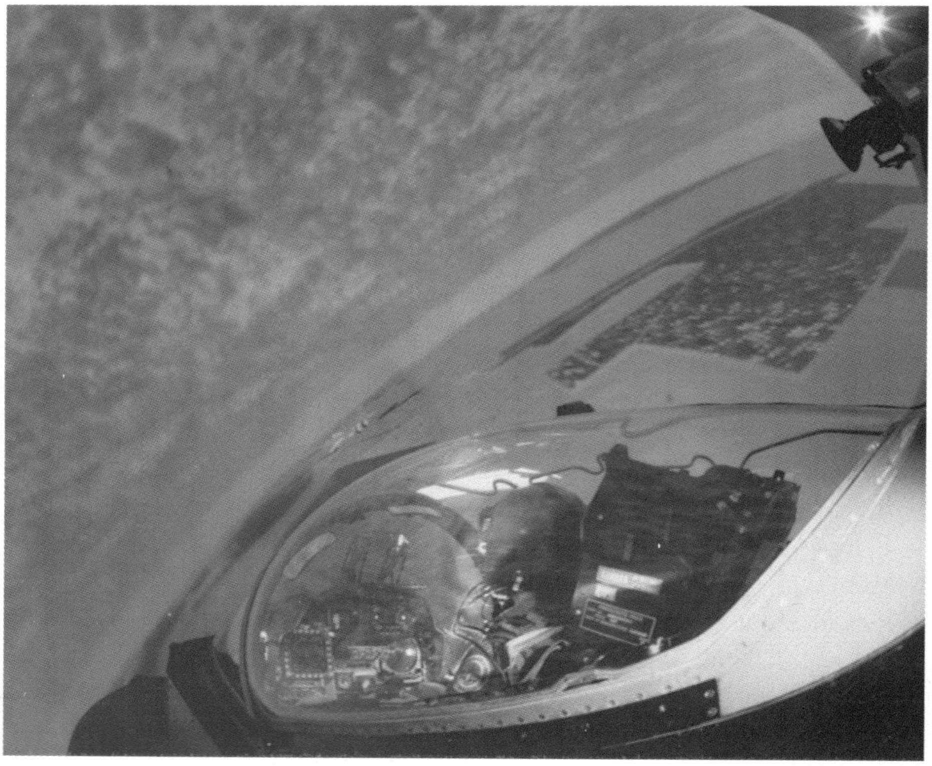

CAE-Link ESPRIT display system in use with a Royal Air Force Harrier GR5/7 mission simulator

increase the AOI resolution to about 1 arcmin/pixel for the same standard of optics and CGI data; the trade-off being a possible intrusive contrast between the high resolution AOI and the low resolution background and less margin to cope with any angular tracking predictions of the AOI slaving system.

Background Channel. The background channel has a much lower resolution of the order of 30 arcmin/OLP, but is only viewed by the user in the peripheral vision where the resolution of the human eye falls off with angle off the centre-of-look.

FoV. 130° vertical FoV. The baseline background projector gives a 270° horizontal FoV (±135°), but if the (low resolution) background is provided for greater angles, ESPRIT is capable of a full 360° FoV.

Features. Other features include:

Smooth electronic blending of the AOI/background images. A blending area of between 2 and 5° cone angle is used between the HR AOI and the LR background.

Distortion correction is included and also allows for parallax between the operator's eye point and the AOI projector.

Status

Available. The UK MoD selected the ESPRIT display system for the two Harrier GR5/7 mission simulators made by TTSL for the Royal Air Force at RAF Wittering.

Manufacturer

CAE-Link Corporation, Binghamton, New York.
(For details see Directory of Suppliers)

Contraves 360° Display System

Contraves manufactures a display projected on a patented high-reflectivity projection screen up to a FoV of 360° depending on the CGI channels and number of display projectors employed. It is used on Air Traffic and Fighter Control Simulators but could have other applications.

Manufacturer

Contraves-SSI, Tampa, Florida.
(For details see Directory of Suppliers)

Contraves ATC Tower Simulator (CATTS) showing four visual windows out of a possible 360° cover

Evans & Sutherland VistaView™ Head-Tracked Display System

The Evans & Sutherland (E&S) VistaView Head-Tracked Display System was introduced in 1992. A two-channel ESIG-3000 CGI system generates the imagery and projection is on a high-gain surface on the inside of a 24 ft/7.315 m diameter fibreglass dome.

Head-Tracking. The VistaView Projection System is head-tracked so that projected images follow wherever the user's head is pointing and are in the correct angular real-world orientation. This is achieved through a magnetically-coupled sensor attached temporarily to the user's own flying helmet and removed when the user leaves the simulator.

Projectors. The VistaView head-tracked projector system uses two light-valve projectors, one for each CGI channel. Its optical system projects and blends the two channels of imagery into a continuous viewed image.

System Brightness and Contrast. The combination of VistaView projection and the high-gain surface on the inside of the E&S projection dome give an image luminance of at least 5.0 Ft Lamberts over the whole FoR when measured at the pilot's eye-point. The minimum average contrast ratio in both projected images is 15:1 when measured directly ahead.

Image FoV. The background image has a FoV of 120° horizontal and 90° vertical (±60° H × ±45° V). Resolution is relatively low but matches that in the pilot's peripheral vision.

Centre High Resolution channel. The high resolution inset has a FoV of 40° horizontal by 30° vertical and is always centred in the middle of the background image.

Total Possible Field of Regard. Head-slaving allows the projected image to be seen by the pilot on the inside of the dome over the full Field of Regard of 360° (see the FoR angle plot diagram overleaf).

E&S VistaView Head-Tracked Display System - a closer view of the projectors mounted directly behind the pilot, giving minimum obscuration to the possible Field of Regard

E&S VistaView Head-Tracked Display System showing imagery produced when the pilot looks behind

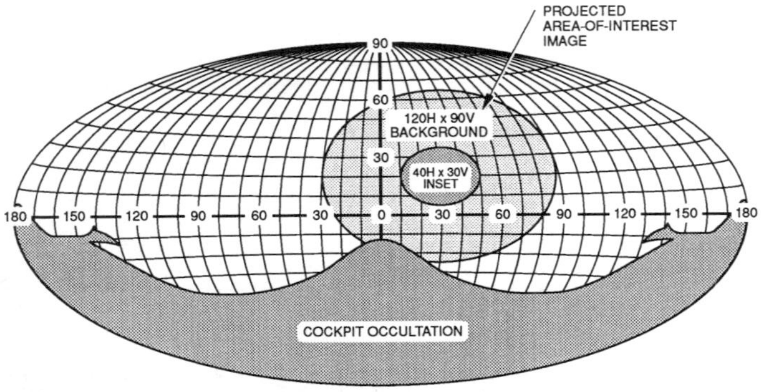

E&S VistaView Head-Tracked Display System - Field of Regard angular plot

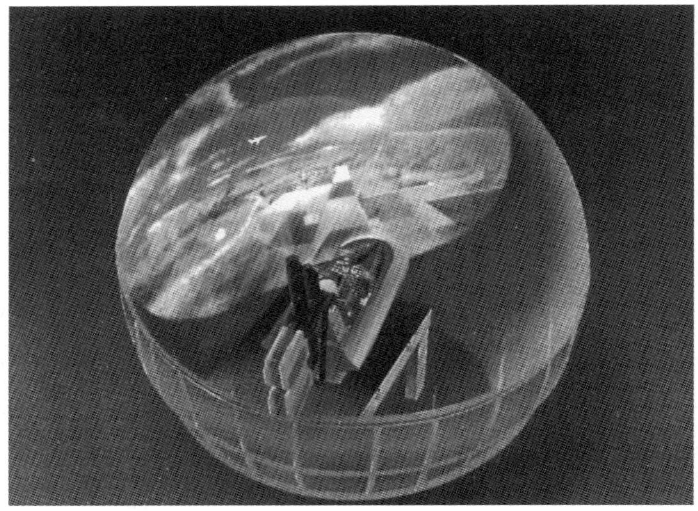

E&S VistaView Head-Tracked Display showing pilot looking left - note that no imagery is produced to the right which is outside the pilot's view

Extra Target Projectors. The system can be enhanced by the addition of target projectors to add images of aircraft and other objects for air-to-air combat. The E&S TargetView projector is available with a FoV of 2°.

Resolution.

Centre High Resolution Channel: 6.9 arcmin/Optical Line Pair (3.45 arcmin/pixel)

Background Channel: 20.6 arcmin/OLP (11.3 arcmin/pixel)

TargetView Projector: 1.5 arcmin/OLP (0.75 arcmin/pixel), equivalent to the centre field resolution of the human eye (6/6 vision or visual acuity 1.0).

Display Refresh Rate. 60 Hz standard, 50 Hz optional.

Slaving Follow-up Characteristics. Maximum servo performance within the FoR is a velocity of 20 rad/s and an acceleration of 40 rad/s/s (1146°/s and 2292°/s/s respectively).

Host Interface. Ethernet standard, SCRAMNET and others optional.

Power and Cooling. The system requires 13.2 kVA of electrical power and 34,000 BTU/hr cooling.

Status

Available. In use in F-16 simulators.

Manufacturer

Evans & Sutherland, Salt Lake City, Utah.
(For details see Directory of Suppliers)

FSI Vital Display Systems

FlightSafety International (FSI) acquired the Visual Simulation Systems (VSS) business unit of the McDonnell Douglas Corporation in February 1993. This included the St Charles, Missouri, facility and the Vital product line of simulation systems. This includes CGI systems and display systems such as conventional collimated monitors and the MultiView Back-Projected Collimated (BPC) cross-cockpit system described below. The first visual simulation system to be produced by McDonnell Douglas was Vital I, a laboratory prototype which appeared in 1969 and which was a forerunner in low-cost, CGI systems to rival the model board and travelling TV camera systems then in use for simulator visuals.

FSI MultiView™ Back-Projected Collimated (BPC) Cross-Cockpit Display System

MultiView was introduced in 1985 and is a BPC system allowing a continuous outside-world view for all crew members, regardless of their position in the cockpit.

Principle of Operation. See the definition of BPC in the glossary and also the diagrams of BPC systems at the beginning of this section on display systems.

Projectors. Vital SkyLight high resolution colour raster/calligraphic projectors are used. They employ commercially available 9 inch magnetic focus tubes.

Collimating Mirror. A one-piece mirror is made from 145 in-wide single-pass double-coated Mylar and held in place by pneumatic means.

A glass mirror is available as an alternative.

FoV. 40° vertical FoV with either 150, 180, 200 or 225° horizontally depending on the length of the mirror assembly and the number of CGI channels and projectors employed.

Channel Blending and Alignment. Digital edge blending is employed and automatic compensation and alignment reduce maintenance time. Automatic compensation produces repeatable setting of black level, white point, display response and linearity. This gives, amongst other things, accurate calibration of RVR.

Servicing. Modules can be changed in 10 minutes and CRTs in less than 30 minutes.

Eight-Channel USN System. The first MultiView system using imagery from the Vital VII CGI system was delivered to the US Navy in February 1989 for use with an MH-53 minesweeping helicopter operational flight trainer. This eight-channel system is at NAS Norfolk, Virginia.

FSI MultiView Back-Projected Collimated (BPC) Cross-Cockpit Display System - Note the outside curved casing for the Wraparound curved mirror assembly and the projectors on top of the cockpit

Weight and Size. A 150° MultiView system with a Mylar mirror weighs 2041 kg, and is 7.62 m wide, 4.12 m high and 5.64 m long.

Status

MultiView BPC display systems are in production and are in service with both military and commercial users. Known military applications of MultiView include simulators for the C-17, C-5B, P-3C and MH-53 and a system has been delivered to the Singapore Air Force.

Manufacturer

FSI Visual Simulation Systems Division (Ex-McDonnell Douglas Visual Simulation Systems), St Charles, Missouri.
(For details see Directory of Suppliers)

RGB Spectrum Display Systems

RGB Spectrum is a designer and manufacturer of video graphic and multimedia products for work stations and PCs. They make the SynchroMaster™ scan converter for use with displays including helmet and boom-mounted displays. Other products include ComputerWall™ multi-screen processor, the RGB/Videolink line of scan converters, the RGB/View line of video windowing systems, the Watchdog high resolution multi-window display systems, and the HRVM high scan rate signal mixer.

RGB Spectrum SynchroMaster 100.
SynchroMaster 100 is a field sequential colour scan converter designed for use with displays including helmet-mounted and boom-mounted displays. It was demonstrated at the SIGGRAPH Show held in Orlando, Florida, in July 1994, and has been available since August 1994.

HMD Compatibility. Most popular-use HMDs use small LCD panels for displaying images, whereas the more sophisticated HMDs used for military, medical, and industrial applications use lightweight miniature monitors. Such monitors offer higher resolution and wider FoV than the LCD systems, and to keep size and weight down, often utilise a single electron gun. The electron gun is fed signals from the red, green, and blue channels sequentially, with a synchronised mechanical or electronic shutter in front to provide the appropriate colour. These display systems require specially formatted field sequential colour input signals.

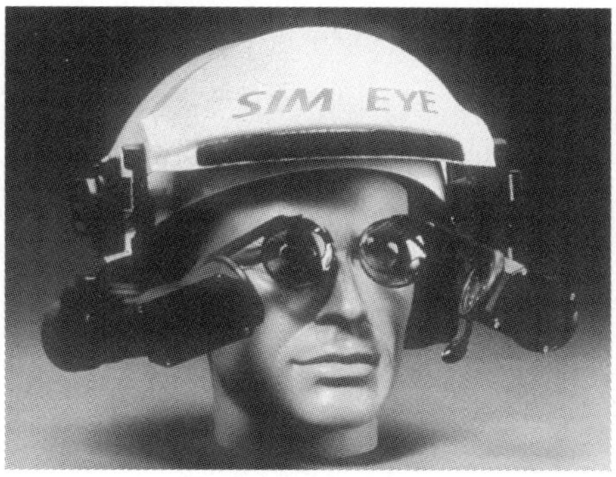

RGB Spectrum SynchroMaster 100 scan converter used with the Kaiser SimEye™ Helmet-Mounted Display

Scan Conversion. The SynchroMaster 100 converts the normal RGB parallel signals from computer work stations or CGI systems to the serial (field sequential colour) signals used by high-end display systems. It is compatible with systems from DEC, Evans & Sutherland, CAE-Link, General Electric, Hewlett-Packard, IBM, Silicon Graphics, Sun, and others. The photo shows the system in use on a SimEye™ HMD from Kaiser Electro-Optics, Inc.

Conversions in Both Directions. The SynchroMaster offers signal capability in both directions and can also convert field sequential colour signals to normal RGB for display on standard computer monitors. This is useful where there needs to be a monitor for an instructor's station or for viewing by others.

Status

Available.

Manufacturer

RGB Spectrum, Alameda, California.
(For details see Directory of Suppliers)

DOMES AND PARTIAL DOMES

This small item summarises the domes and partial domes that are available from a variety of manufacturers for use as part of wide-FoV projected display systems. Some are compatible with mounting on motion platforms, some are not, and some are large enough to have a complete motion platform and simulator cabin inside the projection dome itself. The word 'dome' is used loosely to refer to very wide FoV systems rather than partial spheroids. The SAM and AAA projection domes in particular are considerably flattened compared to a perfect sphere.

DOME AND PARTIAL DOME VISUAL DISPLAY SYSTEMS

Country	Maker (for details, see Directory of Suppliers)	Dome Radius or Focal Distance	Max FoV	Mounting possible on Motion Platform	Remarks
France	Sogitec	4 m radius	360°	No	Mirage Simulators
	Sogitec	4 m radius	360°	No	Puma Simulator mounted inside dome
	Thomson	4 m radius	360°	No	Air Combat Training
	Thomson	3 m radius partial dome	150° H × 105°	YES	On Super Puma Simulator
	Thomson	1 m radius mini dome	270°	No	For air combat training and research
Germany	Dornier, Friedrichshafen	20 m wide (10 m radius), 6.5 m high	360°	No	Stinger SAM Training
Russia	TsAGI	6 m radius	360°	No	Air Combat Simulation
Serbia	SDPR, Belgrade	8 m radius	360°	No	AAA training
Spain	CESELSA	3 m radius	360°	No	On AV-8B (Harrier) Simulator
	CESELSA	3 m radius half dome	170°	No	For Portuguese F-16 simulator
UK	BAe Defence	4.6 radius inflated dome	360°	No	2 used in Air Combat Simulator. Air pump needed to maintain dome shape
	DRA, Farnborough	4 m radius	360°	No	Air Combat Simulation
	Invertron	10 m radius wide, 5.5 m high	270°	No	SAM training
	Marconi, Cwmbran	12 m radius	360°	No	SAM & AAA Training
	Reflectone	6, 8 & 10 m radius	360°	No	SAM training
	TTSL (ex-Link-Miles)	3.68 m radius	360°	YES	On Harrier Simulators
	TTSL (ex-Rediffusion)	4 m radius partial dome	200° H × 60°	No	On Swiss and Malaysian Hawk Simulators, also USN Goshawk Simulators
USA	AAI	9.75 & 6 m radius	360°	No	SAM & AAA Training
	CAE-Link	3.7 m radius	360°	YES	On Turkish F-16 and Super Cobra Simulators
	Contraves - SSI	6.1 m radius	360°	No	For Air Combat Simulation
	Hughes Aircraft	6.1 m radius	360°	No	For Air Combat Simulation
	Hughes Training	3.7 m radius	360°	YES	On Osprey Simulator
	Lockheed Fort Worth	4 m radius partial dome	200°	No	F-16 Simulators
	Loral	3.7 m radius	360°	YES	For Gripen Simulator
	MDTS	4.6 m radius	360°	No	On A-6E Simulators
	MDTS	6.1 & 3.05 m radius	360°	No	On Air Combat simulator at MDTS St Louis facility
	MDTS	6.2 m radius	360°	No	On simulators for F-14, F-15 and others
	MDTS	4 m radius	360°	No	On AV-8B (Harrier) Simulators
	Perceptronics	6.4 m radius	240°	No	Used in the Avionics Situational Awareness Trainer (ASAT)
	SMI	3 m radius	360°	YES	On Sea Hawk Simulators
	USAF Materiel Command	6.1 m radius	360°	No	USAF Systems Command R&D Mission Simulator-1 at Wright-Patterson AFB

Table and Following Entries. The table is intended to be comprehensive. The entries following give photographs of various dome systems and further details on some others; they are a selection only and are not the full list.

FRANCE

Thomson Projection Domes

As well as making 4m radius domes for air combat simulators, TTS makes a motion-compatible half-dome which is fitted to a Super Puma simulator. It has also produced a mini-dome of only 1m radius for research and low-cost air combat simulations.

Manufacturer
TTS, Trappes.
(For details see Directory of Suppliers)

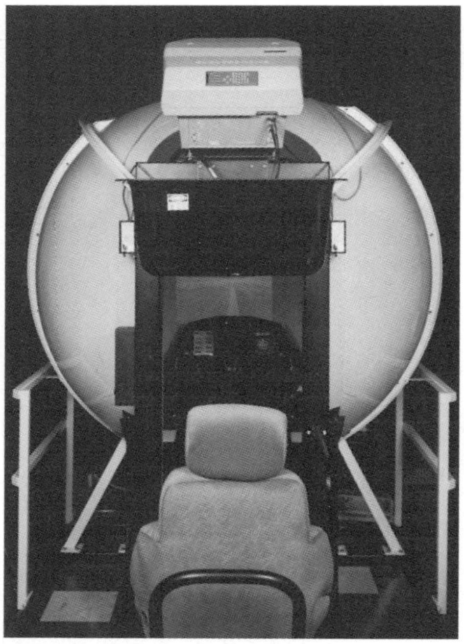

TTS COSMOS mini-sphere Air Combat Trainer

Thomson Super Puma simulator for the Swiss Air Force

GERMANY

Dornier Stinger Training Dome

The Dornier training dome diameter is 20 m with a height at the centre of 6.5 m. Trainees, consisting of one or two Stinger teams, are located at the centre of the dome. The dome trainers were designed primarily for use with shoulder-launched Stingers but the design accommodates expansion to other systems such as helicopter-launched Stinger. The dome is therefore provided with large access doors measuring 4.5 × 4 m.

Manufacturer
Dornier GmbH, Friedrichshafen.
(For details see Directory of Suppliers)

UNITED KINGDOM

Invertron Dome

Invertron makes a dome constructed of standard screen panels each covering a field-of-view of 20°. These are arranged in a circle, with extension panels around the top of the structure. The total usable screen area offered by the dome is 270° in azimuth and 115° in elevation (+90°, −15°). The azimuth restriction is due to the allowance made for an access doorway and the instructor's position. The dome was not designed to be weather-proof as it is intended to enclose the structure within a conventional building. This building can be supplied by the customer or by Invertron as part of a training package. The dimensions of the Invertron building housing the dome are 16 m long, 14 m wide and 6 m high (to the eaves). This building allows space for a small workshop or office area.

Manufacturer
Invertron Simulated Systems Limited, West Sussex.
(For details see Directory of Suppliers)

Marconi Air Defence Training Theatre

The Marconi (ex-Ferranti) Air Defence Tactical Training Theatre (ADT3-C) uses projected images on the inside of a dome of 12 m in radius and 10 m high. It is large enough to accommodate not only SAM systems such as Rapier but also substantially sized AAA guns such as the 35mm Oerlikon (see the picture in the Air Defence Training section).

Manufacturer
Marconi Simulation and Training, Cwmbran.
(For details see Directory of Suppliers)

Rapier SAM crew training in a Marconi Air Defence Tactical Training Theatre

RUKL Domes

Reflectone UK Limited (RUKL) makes a number of projection domes intended for the training of SAM operators. Domes of 6 m or 8 m radius are used for ground-mounted weapon systems, and a 10 m radius dome for multiple ground and vehicle-mounted systems. The dome versions can be installed in a building or housed inside a weather-proof outer dome. A wide-screen version is also available which consists of curved dome segments, designed for installation in drill halls, hangars or in purpose-built buildings.

Manufacturer
Reflectone UK Limited, Bristol.
(For details see Directory of Suppliers)

TTSL Motion-Compatible Projection Dome

TTSL (ex-Link-Miles element) has a dome projection system of 12.1 ft (3.7 m) radius which is stressed to be mounted on a motion platform. It is used on the two RAF Harrier flight simulators.

Manufacturer
TTSL, Crawley.
(For details see Directory of Suppliers)

TTSL motion-compatible dome fitted to GR5/7 simulators at RAF Wittering

UNITED STATES OF AMERICA

AAI Corporation 32 ft and 20 ft Radius Projection Domes

These domes are used in the MTS-II (Moving Target Simulator II) Redeye/Stinger simulator. Gunners can be trained on a wide variety of other weapon systems including Vulcan, Javelin, Chaparral and so forth. Two dome sizes are available with radii of 32 ft (9.75 m) or 20 ft (6.1 m). Vehicles can be driven in through the 4.5 × 3.5 m door in the side of the dome. A 5 m gantry mounts the visual projectors and allows vehicular weapon systems to be centred under the projectors while continuing to allow full weapon manoeuvrability.

Manufacturer
AAI Corporation, Maryland.
(For details see Directory of Suppliers)

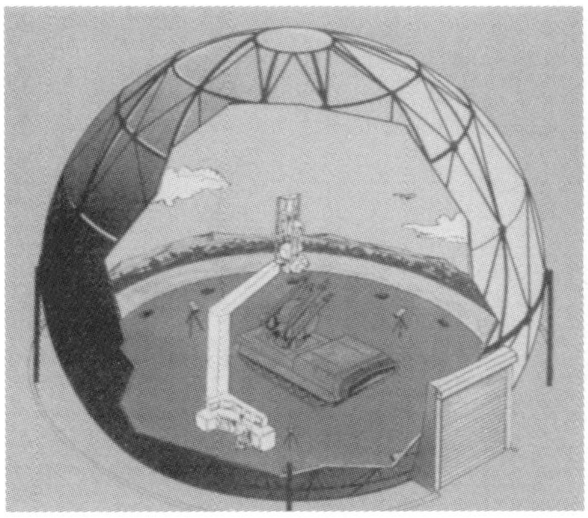

AAI Corporation MTS-II 32 ft (9.75 m) radius dome

AAI/SMI dome display

YUGOSLAVIA (SERBIA AND MONTENEGRO)

SDPR AAA Training Dome

The AAA training dome stands on a circular concrete foundation with a diameter of up to 14 m. The dome itself is 8 m in diameter. Erection time is 48 hours while 24 hours are needed for removal. Access doors are provided for weapons and other equipment. A three-phase 220 V 50 Hz electrical supply is required.

Contractor
Federal Directorate of Supply and Procurement (SDPR), Belgrade.
(For details see Directory of Suppliers)

MOTION CUEING SYSTEMS

This new section collates information on motion platforms and other motion cueing devices such as motion seats and G-Seats from all simulation areas, land, sea and air. It starts with some definitions and a table of the nomenclature used for the six degrees of freedom (6-DoF). A table follows giving a summary of the characteristics of motion platforms from various manufacturers. A short section on simulator-specific Motion Seats follows. In this edition the data is not complete and it is hoped to obtain more data for the 1995-96 edition.

Importance of Motion Cueing and low Latency of Simulator Motion Systems. As a change of position or a rotation is made in any of the 6 DoF, initial cues felt are those of acceleration which are quickly transmitted to the brain from organs such as the inner ear and the muscle and joint sensors, typically in less than 50 milliseconds; this is followed in the real world by a perception of change in the outside visual scene. The visual scene change is essentially one of detecting a change of perspective and/or displacement of objects such as the horizon. A displacement takes some time to build up to discernible amounts after the initial acceleration which caused it. A simulator should therefore reflect the real-world situation by ensuring that the motion latency is equal to or less than that of the visual system and not the other way round.

The rapid transmission of motion cues to the brain also illustrates the desirability of having cues of real motion in any simulation in which the vehicle to be simulated experiences accelerations, particularly where the accelerations are in response to control movements by a simulator trainee. It also illustrates that the motion platform has to be of low latency and be constantly monitored so that its performance and its relation to visual and other systems retains correct juxtaposition. It has been truly said that no motion is better than poor motion.

The rapid transmission of motion cues to the brain in the simulator allows realistic control movements to be made, particularly in situations critical for rapid control response such as failure cases and high-gain control tasks, or where outside visual cues are degraded such as in poor visibility, at night, or in the case of aircraft, when flying by instrument reference alone. It should be noted that in civil aircraft regulatory rules (see the introduction to the Flight Simulator section in this book), only a device with a motion platform is called a Flight Simulator; devices without motion platforms are known as Flight Training Devices and are allowed fewer training credits than a Full Flight Simulator.

Trade-offs with Motion Systems. Although in a capable mission simulator costing many millions of dollars, the motion platform itself is unlikely to be a major cost driver (see the notional prices in the tables that follow), the design of any other system mounted on a motion base has to be compatible with the maximum accelerations that will be experienced. This is a limitation particularly with wide FoV visual display systems. Many domes and partial domes are not compatible with motion platforms (see the table on domes at the end of the Visual Display Systems section). These non-motion domes can be built much more lightly and cheaply than those which are specifically designed to be mounted on a platform. When combined with G-Seat cueing, this is a commonly used compromise for fighter aircraft simulators. But G-Seats, however ingeniously designed, will never be able to replicate 6-DoF motion in the same way as a large-throw platform and so any training role which requires cues of real motion will continue to need one of the many good quality low latency platforms on the market. Finally, one solution used by Sogitec (France) in its Super Puma simulator, is to use a very large projection dome into which the whole simulator and its motion platform can be placed, thus avoiding any limitation of dome construction due to a motion platform. Since many of these large domes already exist and have been in use for other roles such as training for SAM and AAA crews, this solution may become more common in the future.

MOTION PLATFORMS

Synergistic - Definition. The word Synergy is from the Greek Syn, together, and Ergon, work, and means combined or co-ordinated action. A synergistic motion platform has 3 mounting points on the floor under the simulator, from each of which two jacks, displaced from each other at an angle, are connected to the simulator cockpit baseplate. By correct programming, all 6 DoF can be obtained from such a platform design. Nomenclature for the 6 DoF is tabulated below. It should be noted that, because of the disposition of the jacks, in a synergistic platform the amounts of movement possible will be greater if only one DoF is used at a time. However, frequently the movement of the vehicle being simulated requires, say, pitch, roll, heave and surge all to be applied together. Maximum displacement in a given DoF is reduced when, as is normal in real training operations, more than one DoF have to be used together.

Other Relevant Terms and Definitions

See the following in the Glossary of terms section at the beginning of this book:

Acceleration-Onset Cueing (principle used by Motion Platforms)

Anti-G Suit

g/G/Gz/G-Suit

Latency/Transport Delay

Sensory Threshold

Vestibular Apparatus (the human inner ear with its semicircular canals and otoliths)

CAE 6-DoF Synergistic Motion Platform

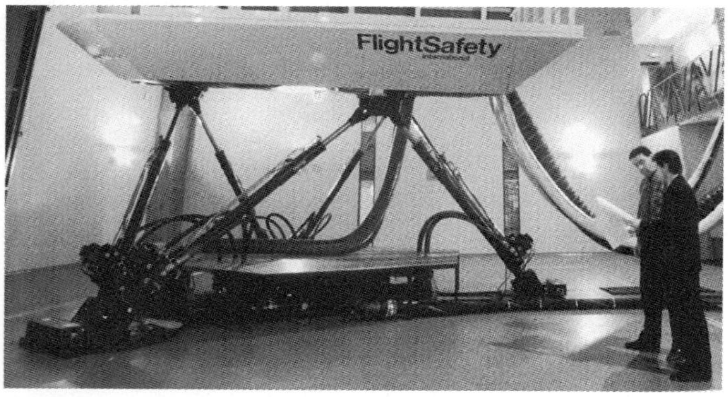

FSI 'Big Base' 6-DoF Synergistic Motion Platform - Note FSI's characteristic cover plate for the hydraulic accumulators between the floor mounting points

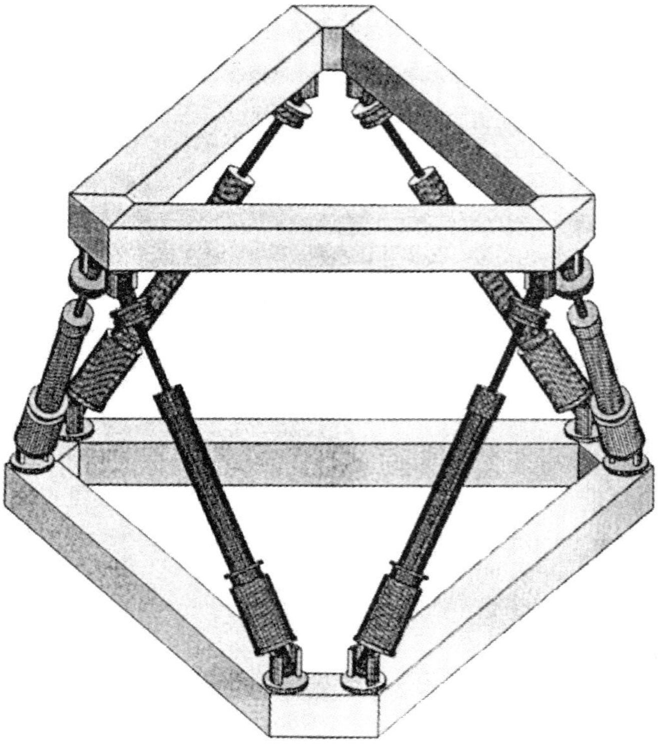

Diagram of synergistic 6-DoF platform showing the pairing of the jacks
(Diagram by permission of Fokker Control Systems)

Degrees of Freedom (DoF) Nomenclature

Type of Motion	Name of DoF (in Flight Simulator Terminology)	Description of motion	Axis of Rotation or Motion	Scientific Symbol for Acceleration
Angular Rates	Pitch Roll Yaw	Pitch Rate Roll Rate Yaw Rate	Lateral Longitudinal Normal (Vertical)	Ry Rx Rz
Linear Movement	Heave Sway Surge	Vertical (up and down) Lateral (side to side) Longitudinal (fore and aft)	Normal (Vertical) Lateral Longitudinal	Gz Gy Gx

TTS (France) 6-DoF synergistic platform on Super Puma simulator showing about 20° of nose-up pitch

3-DoF motion platform by SDPR of Belgrade - The two main beams produce heave and pitch and the separate small jack produces roll motion

Compacific 12-jack platform with jacks at rest

Compacific 12-jack platform with jacks extended

MOTION PLATFORMS - SUMMARY

These figures have been provided by manufacturers and are offered as an approximate comparison between the systems on the market. They are not contractual specifications. With synergistic platforms, throws will be reduced when more than one plane of movement is used at once. Heave acceleration is listed as a general indication of acceleration performance since to list accelerations in each DoF would take up too much space. Where a box is blank, the information is not known. If the box does not apply to that system, a dash symbol is used.

Maker (Details in Directory of Suppliers) **and Platform Title**	**Country**	**DoF**	**Main Principle of Operation H=Hydraulic; E = Electric; S = Synergistic**	**Max throws in training operation (individual movements, not combined)**						**Max Heave Acceleration in training use**			**Max Payload kg**	**Latency* ms**	**Remarks**
				Pitch °±	**Roll** °±	**Yaw** °±	**Heave** ±m	**Sway** ±m	**Surge** ±m	**± Gz**	**± Gz /sec**	**at Payload weight kg**			
ABC Dados	Brazil	6	H, S	25	20	20	0.86	0.86	0.86	0.4	0.8	10 886	10 886		
ABC Dados	Brazil	4	H	+17.5 −6.5	14	15	0.13	—	—	0.5		2722	2722		Front inverted Vee, rear vertical
ADE	India	6	H, S	25	20	20	0.762	1.02	1.02	0.9	3.0				
ARI	USA	6	H, S	+27 −25	20	20	0.86	0.8	0.86	1.0	32	13 400	13 400	100	About $800k All regulatory levels
CAE Series 800	Canada	6	H, S	+34 −37	31	34	1.3	1.67	1.75	1.0	30	15 422	15 876	60	Large throw platform
CAE Series 600	Canada	6	H, S	+32.5 −37.5	33	37.5	0.85	1.58	1.44	1.0	30	15 422	15 876	60	All regulatory levels
CAE 4-DoF	Canada	4	H, S	+11 −8	20	9	0.6	—	—	0.5	15	31 752	31 752	60	For large payloads such as evacuation trainers
CAE-Link	USA	6	H, S	25	23	29	0.9	1.14	1.09	1.1	50	10 900	16 300	25	All regulatory levels
CAE-Link Space Shuttle Motion Base	USA	6	H, S, with extra pitch rams	+90 −25	23	29	0.9	1.14	1.09	1.1	50	10 900	16 300	25	Standard CAE-Link 6-DoF platform plus extra pitch jacks to rotate the cockpit to 90° pitch to simulate the launch position of the Space Shuttle
Compacific	Singapore	6	H, S	21	33	25	+.82 −.72	1.28	0.82	0.9	15	4572	12 193		
Compacific Bisec	Singapore	6	H, S	+41 −45	27	32	+0.85 −0.60	0.76	+0.78 −0.65	0.8	15	500	500		12 jacks, multi-tier platform, cost about $600 000
Delft University	Netherlands	4	H	+40 −20	19	29	0.56	—	—	0.8	4	1500	2500	50	At NLR, used for research
DRA (Vickers)	UK	5	H	28	28	28	5.00	4.00	4.00	1.01		4000	4000	60	Either Sway or Surge, not both
FlightSafety "Big Base"	USA	6	H, S	+27 −30	29 5	32	2.05	1.07	+1.2 −.99	1.4	36	11 872	11 872	30	CG to left and rear of motion centroid
FlightSafety "Small Base"	USA	6	H, S	+27 −29	28 5	30.4	1.24	0.65	+0.80 −0.65	1.0	15	7258	7258	50	
FlytSim 4 DoF	UK	4	H	23	18	15	0.9	—	—	0.9	5	2500	3000	40	Compact platform, cost about £83 000. Option for either Sway or Yaw
FlytSim 3 DoF	UK	3	H	23	18	—	0.9	—	—	0.9	5	2500	3000	40	Compact platform, cost about £68 000. Same as the 4-DoF version but without the jacks for the 4th DoF
FlytSim 2 DoF	UK	2	H	14	10	—	—	—	—	—	—	—	2000	40	Compact platform, cost cost about £18 000
Fokker	Netherlands	6	E,S	20	21	30	0.71	0.52	0.57	1.5	7	2500	2500	50	MIL STD 1558 (compact)
Fokker	Netherlands	5	E	10	10	—	0.15	0.15	0.5	0.5	5	3000	3000	50	For Train (railway) simulation
Hydraudyne HSE-1-MS40	Netherlands	1	H	35	—	—	—	—	—	—	—	—			For submarine simulators
Hydraudyne HSE-2-MS10	Netherlands	2	H	+5 −20	5	—	—	—	—	—	—	—	5500	50	For Ship's Bridge simulations
Hydraudyne Universe Series	Netherlands	3	H	15	15	—	+0.16 −0.19	—	—	0.5	5	1800	1800	50	2 platforms in this series. Has a low height (2 ft) in settled position
Hydraudyne HSE-3-MS40	Netherlands	3	H	20	20	—	0.42	—	—	0.71					Typically for cabin crew trainers
Hydraudyne HSE-6-MS27 Series	Netherlands	6	H, S	21	22	27	+0.38 −0.36	0.47	+0.59 −0.45	1.0 to 0.82	15	1800	1800	42	3 platforms in this series with different heave accelerations. MIL STD 1558

Maker (Details in Directory of Suppliers) **and Platform Title**	**Country**	**DoF**	**Main Principle of Operation H=Hydraulic; E = Electric; S = Synergistic**	**Max throws in training operation (individual movements, not combined)**						**Max Heave Acceleration in training use**			**Max Payload kg**	**Latency* ms**	**Remarks**
				Pitch °±	**Roll** °±	**Yaw** °±	**Heave** ±m	**Sway** ±m	**Surge** ±m	**± Gz**	**± Gz /sec**	**at Payload weight kg**			
Hydraudyne HSE-6-MS60	Netherlands	6	H, S	25	26	30	+0.97 −0.84	1.02	+1.29 −0.99	0.82	10	9000	9000	42	2 platforms in this series. BITE, MIL STD 1558
Hydraudyne HSE-6-MS72	Netherlands	6	H, S	29	31	41	1.01	1.4	+1.7 -1.3	1.02	6.1	5000	5000	50	1.85 m jack stroke. NLR platform is this type.
McFadden 612AL	USA	6	H, S	18	19	21	0.18	0.18	0.23	0.5	1	907			
McFadden 611B	USA	6	H, S	+20 −18	19	23	+0.36 −0.33	+0.53 −0.41	0.46	1.0	5	3175	4536		
McFadden 624A	USA	6	H, S	20	19	22	0.28	0.46	+0.53 −0.41	0.8	1	3175		50	
McFadden 672B	USA	6	H, S	+30 −27	30	40	1.09	1.47	+1.78 −1.45	1.0	4	12 700			
Marconi (MTS)	UK	4	H	+17 −15	7.2	15	0.15	—	—	0.82		1000	1000		FlytSym platform modified to MTS specification and fitted to RAF Tucano sims
NASA Ames	USA	5	E-Mech				About 5 m								Large throw, similar to DRA Bedford UK Platform
Reflectone	USA	6	H, S	+45 −39	41	37	1.04	1.57	1.47	1.2	30	15 876	15 876	12.5	FAA level D and US MIL STD 1558
Reflectone	USA	4	E	10	10	10	0.61	—	—	0.8	4	1 814	1 814	12.5	
Reflectone	USA	3	H	24	22	—	0.71	—	—	1.5	12	15 876	15 876	12.5	
SDPR	Serbia	3	H			—		—	—						
Servos & Simulation Inc Model 700-2	USA	2	E-Mech	15	15	—	—	—	—	—		907	907		Used for USN gun training
Servos & Simulation Inc 3 Axis	USA	3	E-Mech	22	22	—	0.203	—	—	1		4082	4082		Used for USN gun training
Simtec	Germany	3	H	25	20	—	0.40	—	—	0.51	6	2500	2500	100	3 identical actuators
Simtec	Germany	6	H, S	25	30	35	0.43	0.59	+0.54 −0.74	0.51	7	2500	2500	100	0.8m stroke actuators
Simtec	Germany	6	H, S	30	25	40	0.15	0.25	0.25	3.0	12	500	500	80	Small 6-DoF
SIMULTEC	Romania	6	M, S	+33 −36	32	39	1.00	1.10	1.15	0.8	3	2500	4000	120	FAA Level B, Unit cost about US$200 000.
Sogitec	France	2	E	—	16	—	0.16	—	—	0.8	—	1300	1300	120	On Mirage 200D Sims
TTS	France	4	E	13	13	25	0.20	—	—	1.0	3.8	1000	1400	50	On Leclerc Tank driver trainer
TTS	France	5	E	35	23	39	0.45	0.5	—	0.55	4	1400	1800	50	On Leclerc Tank crew trainer
TTS	France	6	H, S	+28 −34	30	28	0.96	1.48	1.37	1.0	6	15 000	18 000	50	All regulatory levels
TTS (Ex Link-Miles)	UK	6	H, S	+31 −36	28	32	+0.81 −0.96	1.5	1.35	0.8	4	15 900	15 900	50	All regulatory levels
TTS (Ex Rediffusion)	UK	6	H, S	+32 −38	28	36	0.90	1.30	1.37	0.8	12	11 793	14 000	50	All regulatory levels
TTS (Ex Rediffusion)	UK	6	H, S	+29 −31	25	28	0.80	1.0	1.1	0.8	12	5500	6800	50	Lighter platform
TTS (Ex Rediffusion)	UK	3	H	18	22	—	0.60	—	—			6400	7300	120	Can also be used for entertainment simulators
TTS (Ex Rediffusion)	UK	3	H	23	23	—	0.8	—	—	0.7		2600	2600		
TTS (Burtek)	USA	2	H	+5	+8	—	—	—	—	—	—	—	42 670	High	Evacuation trainers (one at BA Heathrow)

*Latency is the time taken for a pilot's control input to produce a motion response clearly detectable by the pilot.

TTS Leclerc MBT Crew Training Simulator (left) and Driver Trainer (right), both on electrically actuated motion platforms

Hydraudyne special design of 6-DoF Synergistic Motion Platform installed at NLR Amsterdam for simulation research

Servos and Simulation Inc (S&SI) 3-DoF electric platform which can be used to simulate ship's motion for gunnery training

Servos and Simulation Inc (S&SI) Model 700-2 2-DoF electric platform which simulates ship's motion for gunnery training

MOTION SEATS

Motion Seats are specially designed simulator devices attached to crew seats designed to give cues of motion including vibration and acceleration. Principal acceleration cueing provided is normally for Gz (acceleration along the vertical axis) in which case the device is called a G-Seat. Some G-Seats produce cues in other axes such as through inflatable back-pads which give cues of longitudinal acceleration.

CANADA

CAE G-Seat

The CAE G-Seat has been available since the early 1980s, and has the following characteristics.

Components. A modified, safed real ejector seat is used as the basis.

A pneumatic G-Seat system is added to the ejection seat shell.

Anti-G suit facilities are provided.

Air bellows are operated by a hydraulic actuator

The system is digitally controlled; G-cue processing is integrated with the control loading system which has an iteration rate in excess of 2 kHz.

A Hydraulic Power Unit (HPU) provides for both control loading and pneumatic generation for the G-cueing system.

A Master Control Panel is used for control loading and G-cueing systems.

G-Cueing facilities

Eight Pneumatic Cushions. Four pneumatic cushions are fitted in the seat pan and four pneumatic cushions in the seat backrest. Pressure is adjustable from ±2 psi to ±5 psi and velocities up to 40 in/s can be achieved for a short period. Vibration and buffet cues are produced through short-term variations in seat cushion pressure and frequencies of between 0.25 and 20 Hz can be produced at amplitudes of up to 0.25 in.

Lap belt with variable tension. Designed to enhance cues of negative G, the lap belt can be symmetrically contracted (pulled in) by up to 1.5 inches. This increased lap belt tension simulates the crew member's body weight being thrown against the straps in real negative G conditions. Software limits the belt tension to 100 lb and a shear pin is also provided.

Anti-G suit pressure. The pressure gradient is adjustable to customer requirements but is normally supplied configured for zero pressure at 0 G (weightlessness) up to 5 psi at 10 simulated G. At normal 1 G conditions (for example, static on the ground) a pressure of 0.5 psi is applied in order to give some cueing at low G values. A blow-off valve operates at 7 psi.

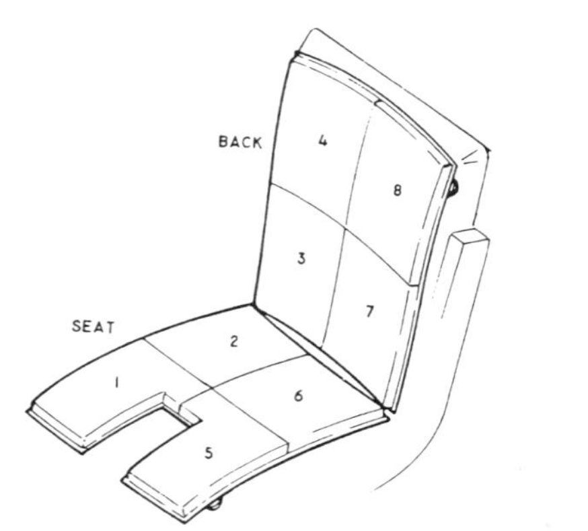

CAE (Montreal) G-Seat showing the layout of the eight inflatable pressure pads

Cues Available

The above facilities can be programmed at customer choice and are capable of giving the following cues:

Seat pan pressures on the buttocks are produced in response to simulated G by the four inflatable seat pan cushions and simulate the 'seat-of-the-pants' effects of Gz.

The pilot's anti-G suit is supplied with air pressure in the normal way as in the aircraft, the G-valve responding to computed G in accordance with a programmed schedule.

Cues of longitudinal acceleration are provided through the four inflatable cushions in the seat back. These provide pressure on the pilot's back in response to simulated acceleration along the fore-and-aft axis.

Buffet and vibration due to airbrakes, transonic effects, high G and so forth, is produced by variations of pressure in the seat and back cushions.

Roll Transients. Roll transients can be produced by pressurising either the left or the right pairs of seat cushions without pressurising the other.

Pitch Transients. Pitch transients can be produced by pressurising either the top or the bottom pair of seat back cushions without pressurising the other, and by pressurising the front or rear pair of seat pan cushions without the other pair.

Status
In service with the German Navy and German Air Force

Manufacturer
CAE, Montreal, and supply can also be through CAE Stolberg, Germany.
(For details see Directory of Suppliers)

FRANCE

Sogitec G-Seats

A Sogitec pneumatically operated simulator G-seat is fitted to many Mirage and other flight simulators. A new G-Seat is being developed for Rafale simulators. The basic system is described below.

Components
The system has the following components:

Mk 10 ejection seat modified with inflatable cushions and seat pan actuators.

Anti-G suit.

Control cabinet.

Pneumatic generator providing dry, filtered air at 7 bars.

Pneumatic rack containing the electropneumatic actuators.

Cues Produced
The following cues are produced:

Seat pan pressures on the buttocks and ischium bones (the posterior bones of the pelvic girdle) are produced in response to simulated G by two inflatable cushions (left and right) and two small pressure pads (also left and right) in the region of the ischium bones. These four pressure devices are programmed to simulate the 'seat-of-the-pants' effects of Gz, both positive and negative.

The pilot's anti-G suit is supplied with air pressure in the normal way as in the aircraft, the G valve responding to computed G.

Cues of longitudinal acceleration are provided through two inflatable cushions in the seat back. These provide pressure on the pilot's back in response to simulated acceleration along the fore-and-aft axis. The cushions are placed one above the other.

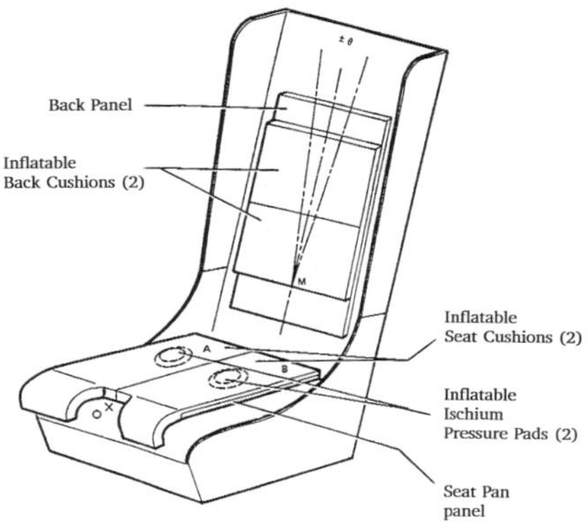

Sogitec G-Seat diagram showing the 4 pneumatic cushions and the small circular ischium pressure pads

Buffet and vibration due to airbrakes, transonic effects, high G and so forth, is produced by seat pan actuators which vibrate the seat pan.

Roll Transients. Roll transients can be produced by pressurising either the left or the right seat cushion without pressurising the other.

Pitch Transients. Pitch transients can be produced by pressurising either the top or the bottom seat back cushion without pressurising the other.

Other Specifications
Power Requirement. Less than 2 kVA.

Weights. Seat 150 kg, pneumatic module 60 kg, electronic control unit 300 kg.

Safety. Compliant with US MIL STD 454.

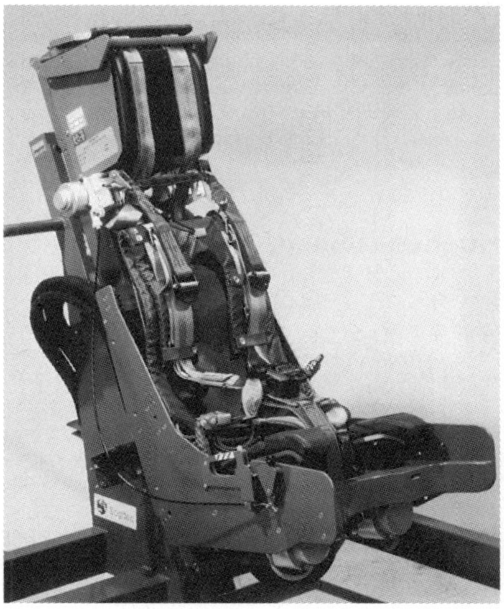

Sogitec G-Seat, photo of complete assembly

Status
All French Air Force simulators for the Alpha Jet, Mirage F1, Mirage 2000 are fitted. The system has also been installed in most of the Mirage F1 and Mirage 2000 simulators delivered abroad, and is fitted in the engineering simulator for the Italian/Brazilian AMX.

Manufacturer
Sogitec, Suresnes, France.
(For details see Directory of Suppliers)

NETHERLANDS

Hydraudyne Seat Shaker

Hydraudyne make an electrically-actuated seat with 25 cm of movement in sway, surge, or both. The seat has a maximum working load of 200 kg, and the device is described as a seat shaker.

Status
Available.

Manufacturer
Hydraudyne, Boxtel.
(For details see Directory of Suppliers)

UNITED KINGDOM

Cranfield G-Seats

Cranfield University Flight Systems and Measurement Laboratories (FSML) section has for many years modified ejection seats to provide G-Seat facilities for use in simulators. At first these were real ejection seats (made safe by removal of all explosive devices) but latterly replica seats have been used at substantially less cost. Standard ejection seat mountings are used to fit the seat to a simulator.

Cranfield Standard G-Seat
Components. In addition to the seat itself, the system comprises a hydraulic manifold block and an electronic control module. These may be mounted on the simulator chassis, typically below the cockpit floor.

Facilities. The following facilities are provided:

Seat Vibration and Buffet and, in response to simulated Gz:

Pressure under the thighs and buttocks is provided by variations in cushion feel

Variable tensioning of the shoulder and thigh straps

Changes in eye height are made by raising and lowering the seat pan.

Operation. A low friction hydraulic actuator is located between the seat vertical guide tubes and replaces the electrical actuator normally used by the pilot for seat height adjustment. This actuator is used for three purposes: seat height adjustment, buffet and vibration effects, and eye point raising and lowering.

A second actuator is coupled to a horseshoe-shaped frame inside the seat pan and this raises and lowers the horseshoe in order to vary the cushion feel as computed G is increased or decreased.

The harness straps are linked to the horseshoe fitting and so harness tension is varied continuously in proportion to horseshoe displacement.

Safety Devices. A digital controller monitors actuator performance and limits the velocities and accelerations of the actuators; the vertical velocity limit is 20 mm/s. If hydraulic or other power is lost or switched off, a safety lock limits the seat drop to less than 10 mm/min. Duplex isolating valves isolate the seat in the event of failures. Strap over-tensioning is prevented by levers and spring safety links.

Specification. Some specification details are as follows:

G Range. ±10 G scaled at ±1 V/G and 2.54 mm/g change of eye height.

Latency. Less than 50 msec response time.

Buffet/Vibration Simulation. Frequencies up to 35 Hz scaled at ±10 volts for ±2.5 mm displacement, reducing to 50% at 20 Hz and 25% at 30 Hz.

Services Required. Hydraulic and electrical supplies are needed. The hydraulic system operates in the range 69-103 bars.

Cranfield Multi-Axis G-Seat.
This design has the normal Cranfield Gz cueing systems but incorporates additional actuators to drive the following:

lateral seat cushion mechanism, moving the cushion sideways
lateral movement of the back rest
seat back pressure pad for fore-and-aft cueing.

A prototype multi-axis seat has been delivered to the UK DRA at Bedford where it is involved in trials to evaluate three-axis cueing.

Status
Standard model available, multi-axis seat in development. G-Seats to the Cranfield design have been fitted to many simulators produced by the ex-Rediffusion and ex Link-Miles companies, now both owned by Thomson Training and Simulation Limited (TTSL). Simulators equipped include the Sea Harrier simulator for the RN, Harrier GR5/7 and Tornado simulators for the RAF, and Hawk simulators for the Malaysian, South Korean and Swiss Air Forces.

Manufacturer
Cranfield University Flight Systems and Measurement Laboratories (FSML) Section.
(For details see Directory of Suppliers)

UNITED STATES OF AMERICA

CAE-Link G-Seat

CAE-Link designed a G-Seat for use with a specially designed simulator-specific anti-G suit which not only inflates with computed G but also has programmable pressure pads in various locations which can be inflated independently. The seat pan and seat back also have programmable pressure pads. The total number of independent pressure pads is believed to be about 20, and G-cueing is through a programme which is developed for optimum effect in the simulation concerned.

Status
In service with MBB in a research simulator at Ottobrun, Munich, Germany.

Manufacturer
CAE-Link, Binghamton, NY.
(For details see Directory of Suppliers)

Hughes G-Seats

HTI T-45 Goshawk OFTs
Hughes Training Inc (HTI) is supplying 22 operational flight trainers (OFTs, Device 2F138) and 10 instrument flight trainers (IFTs, Device 2F137) for the USN T-45 Goshawk. The OFT design is generally similar to that of the Rediffusion (ex-Hughes Rediffusion, now TTS and described under Thomson Training UK) Hawk simulators built for the Swiss Air Force and Malaysia. The T-45 OFT has the following motion cueing:
Motion Cueing. No motion platform is fitted, but pilots use their own anti-G suits in the simulator which allows realistic cueing for simulated G as the simulator inflates the suits appropriately. Also, a simulator-specific G-Seat is fitted which gives seat pressure with simulated G, seat pan lowering under high G (to model eye point lowering due to high G forces), and strap tightening and loosening in response to simulated G (modelling body pressures against the straps as G varies in the aircraft). The G-Seat is understood to be similar to the Cranfield, UK, design covered earlier in this section.

HTI F/A-18 OFTs
Hughes Training Inc (HTI) made 30 OFTs for the McDonnell Douglas F/A-18 and G-Seats were fitted, together with the facility to use the pilot's own anti-G suit in the simulator. The exact design of these G-Seats is not known but it is hoped to have more information in the next edition.

Status
In service with the USN (22) and Royal Australian Air Force (8)

Manufacturer
Hughes Training Inc (HTI), Herndon, Virginia.
(For details see Directory of Suppliers)

Loral G-Seats

Loral Defense Systems-Akron designed and produced the F-15 Operational Flight Trainer (OFT) in the 1970s.
Motion Cueing. The first seven F-15 OFTs were built with a 6-DoF motion system. However, the early 6-DoF platforms of the mid and late 1970s did not live up to their promise and had high latencies and few monitoring and diagnostic systems, sometimes in service becoming out of synchronisation with other simulator systems without this being detected. Consequently, in the seventh and subsequent simulators, motion cueing was confined to the use of the pilot's anti-G suit and a simulator-specific G-Seat.

Status
In service with the USAF.

Manufacturer
Loral Defense Systems-Akron, Akron, Ohio.
(For details see Directory of Suppliers).

Reflectone G-Seat

The Reflectone A-10 OFT is mounted on a fixed base, and a G-seat is used with pneumatic cushions to provide 'seat-of-the-pants' cues in response to pulling synthetic G in the simulator.

Status
In service with the USAF.

Manufacturer
Reflectone Inc, Tampa, Florida.
(For details see Directory of Suppliers)

UAV TARGETS FOR TRAINING

This section describes Unmanned Air Vehicles (UAVs) which are configured primarily to be targets for the training of SAM and AAA crews. Some of the more complex devices can also be used for Air-to-Air Gunnery and AAM Training. Targets towed by manned aircraft are not included.

The term UAV is the generic name for a range of air vehicles varying in complexity from small hand-launched air targets to large and sophisticated surveillance vehicles and armed Remotely Piloted Vehicles (RPVs) which can be ground-launched, ship-launched, or launched from aircraft. Unarmed UAVs are also sometimes called 'drones'. Also included are manned aircraft reconfigured for unmanned operation under remote-control, frequently obsolete aircraft now used as targets and also described as UnManned Aircraft (UMA).

UAV TARGETS FOR TRAINING

AUSTRALIA

ASTA Jindivik

Aerospace Technologies of Australia Limited (ASTA) make the Jindivik (Aborigine for 'the hunted one') small target aircraft. It was designed in 1948 at the former Government Aircraft Factories (GAF) and first flew, in piloted form (as the 'Pika') two years later. It has been continually improved and updated since that time and remains a standard weapons target for its principal customers, Australia and the United Kingdom.

Versions. The Mark 3A was first flown on 10 November 1961. A total of 147 were built, including 10 Mark 203A for the Royal Australian Navy. The Mark 3B is powered by a Viper Mark 201 engine. The Mark 3B first flew on 22 January 1970. There were 126 built, including 90 Mark 103B and Mark 103BL for the United Kingdom, 27 Mark 203B for the Royal Australian Navy, and 9 Mark 203B for the Weapon Research Establishment at Woomera. This version is intended for low level trials at speeds of up to 500 knots (925 km/h). It is equipped with a GEC Avionics L4 or L5 autopilot and the equipment incorporates printed circuit techniques.

An improved Mark 4A Jindivik entered production in March 1981; 15 were ordered by the United Kingdom, deliveries of which were completed by 1986. It has a rationalised electrical power supply system, integrated electronic systems, and increased speed, endurance and manoeuvring capability (up to more than 6 *g*).

For low altitude work, the standard span Mark 4A is fitted with Mark 9 wingtip pods each containing two cameras, a microwave reflector and a small amount of fuel. For high altitude work, also with Mark 9 wingtip pods, constant chord 1.02 m wing extension panels can be added outboard of the pods. For extra high altitude flying (with Mark 5 pods only), these panels can be replaced by 2.03 m panels, tapered on the leading edge. A ventral fin is fitted for high altitude and maximum turn rate configurations.

Construction. The airframe is a low/mid-wing monoplane with a dihedral of 2.5° and an incidence of 1°. The bonded multi-spar wing box has an integral trailing edge of aluminium alloy and a replaceable foam-filled aluminium alloy leading edge. The inter-spar torsion box forms the integral fuel tank. There are aluminium alloy monocoque flaps and ailerons with the ailerons fitted with inset geared tab and driven by twin-motor servos. The flaps are operated pneumatically. The fuselage is an aluminium alloy semi-monocoque and the front portion carries all the control equipment, autopilot and telemetry equipment. The pitot head is mounted on a nose probe. There is a moulded glass fibre honeycomb canopy which lifts off for access to equipment and also forms a ram-type air intake. The rear end of the front fuselage and the front end of centre fuselage form a bay for all mission equipment. The centre fuselage also houses a retractable landing skid while the removable rear fuselage carries engine and jet-pipe. The tailplane is a multi-spar light alloy unit with the elevators, with inset geared tabs, driven by a twin-motor servo actuator which is operable on one motor.

Engine. The Jindivik is powered by one Rolls-Royce Viper Mark 201 turbo engine rated at 12.36 kN. There is an engine relight capability in case of a flameout. The flexible rubber main fuselage tank has a capacity of 291 litres and the two integral wing tanks have a total capacity of 173 litres. The Mark 4A has an additional 91 litres fuselage tank. The Mark 9 pods each hold 86 litres. Total possible fuel capacity is 727 litres.

Take off and Landing. Take off is from a trolley steered by a gyro and a servo-controlled nosewheel which responds to signals from the ground controller. The aircraft/trolley combination accelerates under normal jet power with the flaps retracted and with the aircraft set at negative incidence. When unstick speed is reached, the aircraft flaps are lowered rapidly. The rotation of the aircraft initiates the trolley release system and the aircraft climbs away. When the aircraft is on the approach run the flaps and skid are selected down for landing. On touchdown, at approximately 120 knots (222 km/h), a 'sting' extended below the main skid rotates on impact and initiates rapid flap retraction.

ASTA Jindivik Mark 4A with Mark 9 wingtip pods

ASTA Jindivik

Fuel supply is terminated by radio command.

Radio Control. The radio control equipment comprises two UHF receivers and an auxiliary control unit. There is also a GEC Avionics flight control computer, with a gyroscope. Telemetry equipment consists of an Australian-designed transmitter and data unit.

Power Supplies. The Jindivik has a non-regenerative pneumatic system; air is stored at 138 bars which supplies air to the flaps and landing skid at a reduced pressure of 39.6 bars. Electrical power is derived from an engine-driven brushless DC generator, rated at 9 kW at 30 V DC. In the event of a generator failure, a 24 V battery provides limited power for essential control functions. Automatic orbit and/or destruct systems are provided, consistent with range safety requirements.

Signature Enhancements. Transponders and microwave reflectors are fitted for trials of active, semi-active or beam-riding missiles. Heat sources, including infrared flare packs mounted on the rear of the fuselage, can be fitted to provide low frequency infra-red output. Transponders in the X, S and C bands can be fitted for target acquisition and to enable Jindivik to be tracked to greater range. There is a provision for a recoverable towed target to be carried under each wing. These tow bodies can carry either active radar, in-flight commanded infra-red flares, or a forward-looking Luneberg lens. The latter can be towed at 15 to 150 m behind the aircraft; recovery is by an electric winch mounted in the centre of the fuselage. Other types of special tow bodies may also be carried. Cameras fitted with wide-angle lenses are carried in the wingtip pods, with an all-round viewing capability. To simulate different types of aircraft by varying the apparent radar cross-section, Plessey Microwave Limited developed a semi-active artificial radar target (SART Mark IV) for operation in a towed configuration with a Jindivik as the towing aircraft.

Specifications

Wing span:
(short span, low altitude) 6.32 m
(extended, high altitude) 7.92 m
(extended, extra high altitude) 9.78 m

Wing area:
(short span) 7.06 m^2
(extended, high altitude) 9.48 m^2
(extended, extra high altitude) 10.68 m^2

Length overall:
(including nose probe) 8.15 m
(excluding nose probe) 7.11 m

Weights:
(empty, equipped, minimum) 1315 kg
(maximum payload, short span) 249 kg
(maximum payload, extended span) 181 kg

Maximum take-off weight, short span: (Mark 9 wing pods) 1814 kg

Maximum take-off weight (high altitude): (Mark 9 wing pods) 1814 kg

Maximum take-off weight (extra high altitude): (Mark 5 wing pods) 1496 kg

Performance:
(max level speed at sea level, all configurations) 540 knots (1000 km/h)
(min operating height, short span, Mark 9 pods) 12 m
(max operational ceiling, short span, Mark 9 pods) 15 850 m
(max operational ceiling, high altitude, Mark 9 pods) 17 985 m
(max operational ceiling, extra high altitude, Mark 5 pods) 19 810 m

Time to max operational ceiling:
(short span, Mark 9 pods) 26 min
(high altitude, Mark 9 pods) 30 min
(extra high altitude, Mark 5 pods) 34 min

Typical max on-station endurance:
(short span, Mark 9 pods) 1 h 35 min
(high altitude, Mark 9 pods) 2 h
(extra high altitude, Mark 5 pods) 2 h 5 min

Max range:
(short span, Mark 9 pods) 825 nm (1529 km)
(high altitude, Mark 9 pods) 1125 nm (2085 km)
(extra high altitude, Mark 5 pods) 1160 nm (2150 km)

Status

In service with the Royal Australian Navy and with the United Kingdom. The main Jindivik programme ended in 1986 with the last United Kingdom Mark 4A delivery, but a Mark 4B version has been proposed.

Manufacturer

Aerospace Technologies of Australia Limited (ASTA), Port Melbourne, Victoria.
(For details see Directory of Suppliers)

BRAZIL

Aeromot K1 AM

This design is based on the imported Northrop KD2R-5 but using a Brazilian engine and equipment. Construction started in 1985.

Construction. The Aeromot K1 AM airframe is a tapered high-wing monoplane of duralumin construction, with oval section monocoque fuselage and conventional tail surfaces. The fuel tank, in the forward fuselage, is made of stainless steel. Wingtip radar reflector pods are of epoxy resin, reinforced with glass fibre. There is no landing gear.

Engine. Power is provided by a 65.6 kW (88 hp) Aeromot 11003 flat-four two-stroke engine, driving a two-blade fixed-pitch wooden propeller.

Operation. The Aeromot K1 AM is platform-launched, using a 12.75 kN solid propellant booster rocket motor. Recovery is made by an 11.58 m diameter parachute stowed in the fuselage immediately aft of the wings; deployment is initiated normally by the ground control operator, or automatically in the event of communications or engine failure. Recovery can be made on land or from water.

Guidance. Guidance is by radio command, transmitting to a digital autopilot via an onboard receiver/decoder to eliminate interference from other frequencies. The vertical gyro, distribution box, receiver computer, battery box and dual servos and their connecting cables are all waterproofed. The aerodynamic control surfaces are of conventional design. In the event of a communications failure the autopilot is programmed to continue flight until the link is restored, or to deploy the parachute if this is not achieved within a specified time.

Aeromot K1 AM target (Photo - Mário B. M. Vinagre)

Specifications

Wing span:
(without reflector pods) 3.502 m
(with reflector pods) 4.02 m
Wing area: 1.73 m^2
Length overall: 3.84 m
Fuselage (oval section):
(max depth) 0.346 m
(max width) 0.243 m
Propeller diameter: 1.12 m
Weights:
(basic operating weight empty) 130 kg
(fuel) 33 kg
(max launching weight) 163 kg
Performance (estimated, at max launching weight):
(never-exceed speed) 240 knots (444 km/h)
(max level speed at sea level) 194 knots (360 km/h)
(stalling speed) 68 knots (125 km/h)
(max rate of climb at sea level) 1000 m/min
(max operating height) 4000 m
(max range at sea level) 150 nm
(endurance at sea level) 50 mins

Status

10 K1 AMs ordered, including 6 pre-production examples, for the Brazilian Navy's Grupo de Alvos da Equadra (Fleet Target Group).

Manufacturer

Aeronaves E Motores SA (Aeromot), Porto Alegre.
(For details see Directory of Suppliers)

IGAE LAH/EB

Industria Gaucha de Aeronaves Especiais Ltda (IGAE) manufacture this design by Luis Arcari and Airton Hoch (the LAH in the name) for the Brazilian Army (Exército Brasileiro, the EB in the title).

Engine. It is powered by a nose-mounted OS MAX 2.6 kW (3.5 hp) model aircraft engine of Japanese manufacture driving a two-blade fixed-pitch 355 mm propeller.

Construction. The LAH/EB has a tricycle landing gear with a steerable nose wheel, no horizontal tail surfaces, and anti-vortex wingtips for increased landing stability.

Control. Control equipment installed to date has been Futaba radio control systems for line-of-sight operations.

Target Towing. The aircraft tows an all-aluminium dart-shaped target on a 100 m nylon cord. The target uses three equally spaced fins and has sufficient area for radar acquisition. It can be released in flight prior to landing.

Specifications

Wing span: 1.5 m
Length overall: 1.2 m
Weight: 4.5 kg
Speed: 250 km/h
Endurance: 45 min

Target length: 1.8 m
Target weight: 0.8 kg

IGAE LAH/EB air target

Status

In service with the Brazilian Army since 1984.

Manufacturer

Industria Gaucha de Aeronaves Especiais Ltda (IGAE), Porto Alegre.
(For details see Directory of Suppliers)

BULGARIA

Yastreb-2MB Air Target System

The Yastreb-2MB (Hawk) is a target system for SAM and AAA systems and is an updated version of the earlier Yastreb 2 and 2MA models. Over 1500 have been produced, mostly for the Former Soviet Union (FSU). Deliveries to the FSU were then suspended due to difficulties over payments. An improved version known as the Pelican has been developed.

The Yastreb-2MB system comprises the AD.06. YA2 target, the SKT.06 launcher and the SP.1B ground control station. There are two further components: the SP.1K preflight preparation station and the FS.1 target augmentation and towing devices.

The target is a mid-wing monoplane with a circular cross-section fuselage. A V-tail is used and the powerplant is a small single-cylinder petrol engine; a small booster rocket motor for launch is fitted to the rear of

Yastreb-2 air target system with accessories

Yastreb-2 target on launcher with booster rocket motor attached to rear fuselage. The circular guard in front of the nose is for safety during ground running of the engine and is removed before launch

the fuselage. The target can carry a single infra-red source under the port wing.

The complete system is operated by a crew of four. It takes 30 minutes to establish the ground control station, 25 minutes to prepare the target for flight, and a further 5 minutes for pre-launch operations. After launch, using the rocket booster, there are two possible control modes. One is direct remote-control by an operator and the other is automatic stabilisation in height, speed and heading, with an optional operator control in the horizontal plane. There is an automatic self-landing control in event of a system failure. Landing is by parachute.

There are two further developments of the Yastreb-2. The Yastreb-2MA has increased range and can carry electronic warfare equipment while the Yastreb-2MB has a more powerful engine providing a maximum speed of 124 knots (230 km/h).

Specifications

Wing span: 3.52 m
Length: 2.58 m
Height: 0.533 m
Fuselage diameter: 280 mm
Propeller diameter: 610 mm
Max take-off weight: 62.5 kg
Payload weight: 5 kg
Range: 15 km
Ceiling: 500-2400 m
Speed: 130-220 km/h
Endurance: 60 min

Status

In production. In service with Bulgaria and Russia.

Manufacturer

AVIOTECHNICA, Plovdiv.
Marketed by Metalchim Trading, Sofia.
(For details see Directory of Suppliers)

CANADA

Bristol Aerospace Scale Helicopter Target

This target is a one-fifth scale model representation of the Mil Mi-24 'Hind-D' attack helicopter, the first one having been flown in mid-1985. It was developed to simulate both attack and evasive manoeuvres by the Mi-24, but modular front and rear fuselage sections can be replaced with others to represent any modern single main rotor helicopter.

Construction. This target has a glass fibre fuselage shell with replaceable front and rear sections. It has two-blade main and tail rotors, with a simple but reliable drive train. The tricycle landing gear is non-retractable.

Engine. The target is powered by one 18.6 kW (25 hp) 342 Normalair-Garrett WAE 342 two-cylinder piston engine. There is enough fuel for 30 minutes operation initially.

Guidance. Guidance and control is via a PCM radio control system. Remote flight operations include jinking, hovering and 'S' turns in slow forward flight. There is a rate gyro to assist yaw control. Fully automatic flight is also possible.

Specifications

Main rotor diameter: 3.35 m
Tail rotor diameter: 0.66 m
Wing span: 1.37 m
Fuselage: (maximum width) 0.36 m
Height overall: 0.99 m
Wheel track: 0.55 m
Weights:
(max payload) 27.2 kg
(normal take-off weight, including 7.7 kg ballast) 47.5 kg
(max permissible take-off weight) 56.7 kg

Performance (required by specification):
(max level speed at 5 to 25 m above sea level) 60 kts (111 km/h)
(min ascent/descent velocity at 5 to 25 m above sea level) 55 m/min
(ceiling) 3050 m
(endurance) 1 h

Status

Ready for production.

Manufacturer

Bristol Aerospace Limited, Winnipeg, Manitoba.
(For details see Directory of Suppliers)

Bristol Robot-X

The Robot-X is intended primarily to simulate a low altitude anti-ship missile or attacking aircraft in order to exercise point defence missile systems such as Sea Sparrow. It can also be used as a target for aircraft, or as a low level air defence target for SAM and AAA crews. The first flight of the Robot-X took place on 4 November 1986 following initial design at the Defence Research Establishment at Suffield.

Construction. The airframe is a bullet-shaped body with sweptback rear-mounted wings, with endplate fins, and sweptback all-moving, nose-mounted foreplanes. It is of modular construction and made entirely of composite materials. The wings have a shockless symmetrical section, a 37° sweepback at quarter chord, and no anhedral or dihedral. Foreplanes and ailerons are actuated by Simmons Precision electric servos. There is no landing gear.

Engine. The Robot-X is powered by 19 2.75 in (70 mm) CRV-7 rocket motors fired sequentially to achieve over-the-radar-horizon ranges.

Robot-X built by Boeing of Canada
(Brian M. Service)

Launch and Recovery. The Robot-X is rocket-launched; normally it is surface-launched, but air or sea-borne launch is an option. There is a parachute recovery system.
Control. The Robot-X flies preprogrammed using a microprocessor-based onboard and three-axis digital autopilot.
Payload. The Robot-X has a payload capability for a miss-distance scoring system, a chaff dispenser, and infra-red flares or radar or visual augmentation devices.

Specifications
Wing span: 2.4 m
Wing aspect ratio: 4.6
Foreplane span: 1.097 m
Length overall: 3.41 m
Length of fuselage: 2.8 m
Body diameter: (max) 0.415 m
Wing area, gross: 1.25 m²
Foreplane area, total: 0.35 m²
Weights and loading:
(max payload) 22.7 kg
(max launching weight) 250 kg
(max wing loading) 200 kg/m²
Performance (surface launch):
(never-exceed speed) Mach 0.95
(max level speed at sea level) Mach 0.85
(operating altitude, surface launch) min 30 m; max 4575 m
(maximum range at sea level) 21 nm (40 km)
(endurance at sea level) 6 min
(maximum endurance) 30 min

Status
Ready for production.

Manufacturer
Bristol Aerospace Limited, Winnipeg, Manitoba.
(For details see Directory of Suppliers)

BSI TRAP artillery target rounds

BSI TRAP target rounds are special artillery rounds augmented to provide good radar cross-sections to enable them to be used as targets for SAM and AAA systems.
Weapons Fired From. Rounds are available for 3 in/50, 76 mm/62, 127 mm/38 and 127 mm/54 naval guns, and 155 mm NATO artillery. Also available are 76/62, 57 mm and 155 mm infra-red versions. X-band radar cross-section (RCS) echoes from 0.01 to 5 m² are possible. Other calibres, bands and RCS can be supplied. The Canadian 2.75 in/70 mm CRV-7 or compatible US/NATO rockets can also use the system.
Radar Characteristics. TRAP passive radar reflectors replicate the frontal radar echoes of typical air targets, both aircraft and missiles. The radar echo is approximately uniform over ±45° from the projectile axis. The echo falls off sharply after ±45° giving good Doppler fusing performance. Supersonic, transonic and subsonic presentations can be made.
Deployment. Extended range 155 mm TRAP projectiles are currently in use by NATO navies for surveillance and target tracking radar evaluation and for NATO Sea Sparrow live firing target exercises. The system is used for ship-to-ship and shore-to-ship firings, and over land ranges for low level air defence exercises.

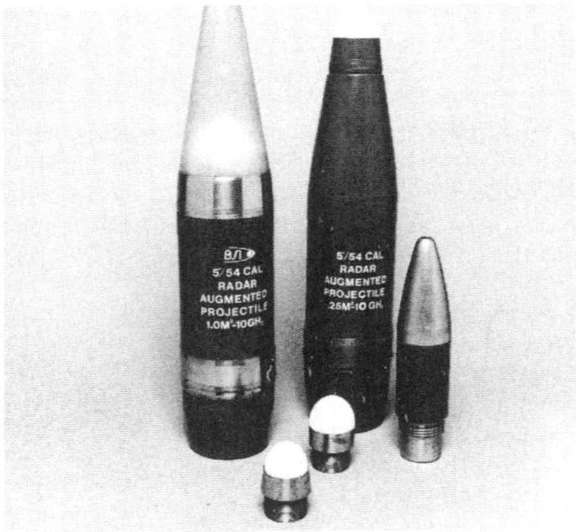

BSI TRAP radar-augmented projectiles for 127 mm/54 naval guns

Status
In production and service.

BSI TRAP radar-augmented projectile for 155 mm NATO artillery

Manufacturer
Ballistech Systems Inc (BSI), St Bruno, Québec.
(For details see Directory of Suppliers)

Targetair TATS Targets

The TATS 102 was developed to meet a CAF requirement for a low-cost, recoverable aerial target for low level air defence weapons training. It was designed for use with manportable SAM and for AAA training.
Construction. The TATS is a low-wing monoplane constructed of glass fibre, wood and foam plastic. The TATS 103 is an updated version of the TATS 102, with improved load carrying capabilities and an increased performance. The main visual difference is an increased wing span (3.6 m).
Engine. The TATS 102 is powered by one 13.8 kW (18.5 hp) 100 cc modified Yamaha piston engine, driving a two-blade propeller. Fuel capacity is 8.2 litres for the TATS 102 and 15 litres for the TATS 103; the latter provides the TATS 103 with an economical endurance of 2.5 hours. The TATS 103 is powered by a Yamaha KT100S 14.92 kW (20 hp) two-stroke engine, also driving a two-blade propeller.
Launch and Recovery. The TATS 102 and 103 are catapult-launched from a mechanical deadweight launcher. Recovery is by parachute or wheel/skid landing.
Guidance. Guidance is by radio command and control within visual range, using a binocular-based ground control system. Initially this restricted the maximum operating range to less than 1.6 nm (3 km); later the Canadian armed forces' CHAIR (control handling aid for increased range) system was used, which allowed visual control at ranges of up to 3.2 nm (6 km).
Signatures. Mission equipment comprises smoke and augmentation systems for visual tracking. For improved visual tracking the TATS 103 is fitted with a 110 000 candle power aircraft landing light. The TATS 102 can also be equipped with an infra-red target source, a 9 A electrical power source, and a DSQ-40, DSQ-41 or similar miss-distance scoring system.
Towing Targets. The TATS 103 can be used to tow the Targetair EX-10 and EX-11 tow target bodies.

Targetair TATS 102 aerial target

Specifications - TATS 103
Wing span: 3.6 m
Length overall: 2.6 m
Height overall: 0.285 m
Basic weight empty: 30.6 kg
Maximum take-off/launching weight: 45.1 kg
Performance:
(max speed in dive) 238 kts (442 km/h)
(max level speed) 129 kts (240 km/h)
(stalling speed) 30 kts (56 km/h)
(max rate of climb at sea level) 305 m/min
(take-off distance on wheels) 152 m
(take-off distance on launcher) 11.58 m
Landing distance:
(on wheels) 152-305 m
(on skids) 152 m
Radio range:
(whip antenna) 1.6 nm (2.9 km)
(linear amplifier and dipole antenna) 7.8 nm (14.5 km)
Endurance:
(at full throttle) 2 h
(at economy power setting) 2.5 h

Status
In service with the Canadian armed forces.

Manufacturers
Targetair Limited, Moncton, New Brunswick.
(For details see Directory of Suppliers)

CHINA, PEOPLE'S REPUBLIC

NAI Chang Kong 1C (Wide Blue Sky)

The Nanjing Aeronautical Institute (NAI) of the Nanjing University of Aeronautics and Astronautics (NUAA) finalised the design of a Chang Kong 1 (CK1) prototype in 1976. In 1977 NAI produced the CK1A version with underwing equipment pods, and in 1982 produced the CK1B variant which replaced these pods with non-jettisonable auxiliary fuel tanks.

The current standard version is the CK1C which has been available from 1984, and can be used as an aerial target for various types of missile. It has a capability for evasive manoeuvres at high bank angles. The CK1C is a mid-wing monoplane with constant chord wings, with unsymmetrical section and 2° anhedral. The fuselage is built in three sections, those at front (housing radio control, telemetry and electrical equipment) and rear (autopilot and flares) being of aluminium alloy. The central portion, made of steel sheet, forms an integral fuel tank. The CK1C has conventional rectangular tail surfaces; the tailplane is mounted near the base of the fin. There is no landing gear but the underside of the underslung engine is strengthened to absorb landing loads.

Engine. The CK1C is powered by one 25.5 kN turbojet. This is the Shenyang WP-6 (Wopen-6) from retired J-6 (MiG-19) fighters, without the afterburner section. The turbojet is mounted in a nacelle underslung beneath the centre of the fuselage. Auxiliary fuel is stored in underwing pods.

Launch. The CK1C is launched from a reusable trolley, upon which the target is mounted on three short guide rails and attached by a single connecting pin at the base of the engine nacelle. The complete ensemble accelerates along the runway under engine power, the connecting pin being withdrawn automatically by a pneumatic release system when the speed reaches 151-154 knots (280-285 km/h). The target then lifts off the trolley and enters the climbout phase. The trolley decelerates and is brought to a halt under radio command by a braking parachute and wheel brakes. The target can enter the firing area two or three times during a mission.

Recovery. If not shot down, the vehicle can be directed to a preselected landing site, where the engine is shut down at a predetermined speed and altitude and a glide landing is made. The engine nacelle is reinforced to absorb the landing impact, resulting in only minor damage which can be repaired easily before reuse.

Guidance. Guidance and control is provided by a four-channel autopilot (pitch, roll, yaw and altitude), located in the rear of the fuselage, which controls the flight in response to radio commands from the ground station. The autopilot incorporates the gyroscope, the directional gyro, a three-axis rate gyro, a programmer, an electrical actuator, an amplifier and a converter. After separation from the trolley, the first 85 seconds of the flight are programme-controlled. Thereafter the mission comes under preplanned radio command from the ground controller. Aerodynamic control is via ailerons, elevators and rudder.

Equipment Carried. Avionics and equipment include an onboard radio transponder for identification and tracking from the ground. Airborne radio equipment comprises receiver/decoder which enables up to 24 command signals to be conveyed to the autopilot and other components and equipment. A 52-channel telemetry system provides the ground controller with a continuous indication of altitude, speed, angle of bank, engine rpm and temperature, and other functions. Mission equipment includes: a miss-distance indicator with its associated antenna at the rear fin tip; five corner reflectors for radar signature augmentation; and three flares or smoke canisters (on the undersurface of each wing and on the rear edge of the engine nacelle fairing) to provide visual augmentation to aid tracking by ground-based optical aids. The main electrical power for avionics and equipment is provided by an engine-driven generator, with an alternator for AC power. An emergency battery supplies standby DC power.

CK1E Low-level Variant. The latest variant is the CK1E which is capable of operating at a height of 50 m AGL. The CK1E has a wing span of 6.88 m and a range of 600 km. The maximum speed range is between 840 and 870 km/h and it can be fitted with an infra-red pod. Power is supplied by a WP-6 turbojet providing 23.03 kN. The CK1E is used as a target for various kinds of guided missile.

Nanjing Aeronautical Institute Chang Kong 1C aerial targets

Specifications for CK1C Model

Wing span: 7.5 m
Wing area: 8.55 m^2
Length overall: 8.439 m
Body diameter (max): 0.55 m
Height overall: 2.955 m
Weights:
(weight empty) 1537 kg
(fuel): (fuselage) 600 kg,
(underwing tanks, total) 240 kg
(max launching weight) 2450 kg
Operating speed range: 458-491 knots (850-910 km/h)
Operating height range: 500-16 500 m
Range: 324-485 nm (600-900 km)
Endurance at low and medium altitude: 45-60 min

Status

In production.

Manufacturer

Nanjing Aeronautical Institute (NAI), RPV Department, Nanjing.
(For details see Directory of Suppliers)

NRIST YK-7

The YK-7 was developed by the Nanjing Research Institute on Simulation Technique (NRIST). First flight was in October 1986. It simulates an aircraft at one-fifth scale.

Design. The YK-7 is a reusable target in shoulder-wing monoplane form. Wings are of glass-reinforced plastic with a foam plastic core, and have a NASA GA(W)-1 aerofoil section (thickness/chord ratio 13 per cent), 5° dihedral, 0° incidence and 15° sweepback at quarter-chord. Conventional tail surfaces are of similar construction. The oval-section streamlined fuselage is a glass-reinforced plastic composites structure, with a landing skid under the forward portion.

Engine. The YK-7 is powered by one 3.7 kW (5 hp) single cylinder piston engine, fitted with a contactorless magneto, fuel pump and barostat, and driving a two-blade wooden propeller.

Launch and Recovery. It is launched from a bungee-powered rail launcher on a cradle which is jettisoned after take off. Recovery is by conventional skid landing or, optionally, by parachute; the latter can be commanded either manually, or automatically by a fail-safe function incorporated in the remote-control system. At the conclusion of the mission, the launcher can be folded to become the transport trailer for the air vehicle.

Two NRIST YK-7 aerial targets, one on its launch rail

Control. Remote-control is by a high power digital proportional PCM radio system; there is an optical system for out of sight operation. Aerodynamic control surfaces are conventional.

Specifications
Wing span: 2.115 m
Wing chord at root: 0.52 m
Wing chord at tip: 0.26 m
Wing aspect ratio: 5.4
Length overall: 2.05 m
Propeller diameter: 0.5 m
Wing area, gross: 0.825 m²
Weight empty: 12.5 kg
Max payload: 1 kg
Max launching weight: 14 kg
Max level speed: 105 knots (194 km/h)
Max rate of climb at sea level: 660 m/min
Operating height, minimum: 50 m
Operating height, maximum: 1000 m
Effective control range: 1.6-2.7 nm (3-5 km)
Endurance: 40 min

Status
In production.

Manufacturer
Nanjing Research Institute on Simulation Technique (NRIST), Nanjing.
(For details see Directory of Suppliers)

Type 130 Rocket IR Target System

The Type 130 rocket IR target system is also known as the TQ-4 Firefly. It is a training target for the HN-5A surface-to-air air defence missile system which is the Chinese equivalent of the Russian SA-7 'Grail' shoulder-launched SAM system.

Operation. The system fires a 130 mm rocket that emits visible and IR radiation and acts as a target for the SAM system. A single-barrel launch tube is mounted on an adjustable tripod. Firing is carried out using a control box and the target can follow a predetermined flight path to a height of 2000 m, with a flight path deviation of less than 120 m. The flight time of the target, during which it can be used for viable SAM training, is over 30 seconds.

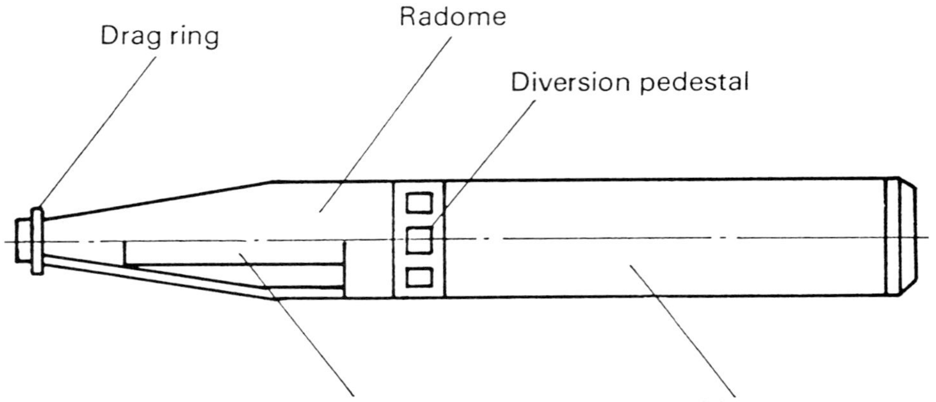

Type 130 IR target rocket - main components

Specifications
Diameter: 130 mm
Length: 1.02 m
Weight of target rocket: 33 kg
Velocity: 160 to 380 m/s
Altitude: 2000 m

Status
In production.

Marketing Agent
China Precision Machinery Import and Export Corporation, Beijing.
(For details see Directory of Suppliers)

YMF Air Targets

The Yuhe Machine Factory (YMF) in Nanjing manufactures a number of aerial targets which are described below.

YMF BJ7104
The BJ7104 is a recoverable air target designed by the Nanjing Research Institute for Simulation Technique (NRIST) and built by the Yuhe Machine Factory (YMF) in Nanjing. It is used primarily as a training target for small and medium-sized anti-aircraft artillery weapons. Series production began in 1973 and over 2500 have been built; production continues.

Airframe. The airframe is a mid-wing monoplane with a circular section fuselage and conventional tail surfaces. The wings have a 6° dihedral from the roots. There is a skid under the tail cone and each wingtip to minimise damage on landing. The design is similar to that of the CSIF B-2, except for the greater use of metal instead of composites.

The BJ7104 is powered by an 11.2 kW (15 hp) Yuhe Machine Factory (YH280) 280 cc four cylinder two-stroke engine, driving a two-blade wooden propeller. Launch is by rocket-assisted take off from a zero-length launcher; recovery is by skid landing.

Guidance is via a 22-channel radio control system. There is a provision for preprogrammed flights using microcomputer control. Aerodynamic control is provided by ailerons, elevators and a rudder. The BJ7104 is equipped with an onboard autopilot, a radio receiver and telemetry.

Specifications of BJ7104
Wing span: 2.7 m
Wing area: 1.14 m²
Length overall: 2.55 m
Fuselage, max diameter: 0.28 m
Height overall, including landing skid: 0.74 m
Weights:
(empty) 34 kg
(fuel) 8 kg
(payload) 10 kg
(max launching weight) 52 kg
Max level speed: 135 kts (250 km/h)
Max rate of climb at sea level: 600 m/min
Operating height range: 100-2000 m
Service ceiling: 6000 m
Control range: 8.1 nm (15 km)
Endurance: 1 h

YMF CSIF B-2 and B-2B
The YMF CSIF B-2 is a variant of the BJ7104 and differs chiefly in its engine. It has a GRP airframe and is fitted with conventional three-axis control surfaces. It can be launched from various kinds of surface using a zero-length launcher and rocket-assisted take off. A 3 W radio command guidance system is used with an autopilot, heading/azimuth/altitude/distance telemetry, and a number of safety devices. Frequencies used are 49.525 MHz (remote-control) and 129 MHz (telemetry). The B-2 is designed for weapon system training and gunnery practice.

Equipment. Operational equipment comprises two towed targets, plus four other devices which give a realistic representation of descending paratroops. The B-2 can also be modified to extend the range of applications, according to individual user requirements.

Engine. One 12 kW (16 hp) Huosai-16 flat-four two-stroke engine, with pull cord starting, driving a two-blade fixed-pitch wooden propeller. The fuel is a mixture of automobile gasoline and lubrication oil; the tank has a capacity of 8 litres.

Versions. The B-2B is the latest version of the B-2 and is powered by a 14.7 kW (19.7 hp) HS-350 petrol engine. Mission endurance is reduced to 45 minutes and overall dimensions are similar to those of the B-2.

YMF B-2B target following parachute recovery

YMF BJ7104 target, of which over 2500 have been built

Other Features. The radius of radio control is 30 km and operational ceiling 3000 m. The B-2B has a radar cross-section of 1 m^2.

Specifications of YMF CSIF B-2

Wing span: 2.695 m
Length overall: 2.544 m
Height overall: 0.6 m
Weights:
(empty) 53 kg
(effective payload) 3 kg
(fuel) 6 kg
(max launching weight) 62 kg
Performance:
(max level speed at 1000 m) 119-129 knots (220-240 km/h)
(max rate of climb at sea level) 480-600 m/min
(max controllable range) 10.75 nm (20 km)
(endurance) 1 h

YMF B-9 Navy Target

The B-9 Navy Target is a development of the B-2 and B-2B targets (see previous entry) adapted for use by naval units. The B-9 may be used to tow a target or used as a target itself when employed for gunnery purposes. It may also carry infra-red augmentation devices when employed as a target for surface-to-air missiles.
Design. The B-9 has an overall resemblance to the B-2 and B-2B but features a revised swept tail fin outline and a shoulder-mounted straight wing.
Operation. The B-9 can take off from a zero-length launcher mounted on a ship's deck with the assistance of a single booster rocket. Recovery is by parachute and the B-9 was designed to withstand numerous immersions in seawater.
Control. When airborne the B-9 is controlled by an autopilot and radio out to a radius of 15 km.
Engine. One 14.7 kW (19.7 hp) petrol engine.

Specifications of B-9 Navy Target

Wing span: 2.82 m
Length overall: 2.52 m
Weight at take off: 52 kg
Radio range: 15 km
Performance:
(max speed) 250 km/h
(max ceiling) 4000 m
(endurance) 1 h

Status

In production and in extensive service with the PLA. Over 2500 of the BJ7104 variants have been built.

Manufacturer

Yuhe Machine Factory (YMF), Nanjing.
(For details see Directory of Suppliers)

Marketing Agency

China National Aero-Technology Import and Export Corporation, Beijing.
(For details see Directory of Suppliers)

EGYPT

Kader TN-1A and TN-1B Targets

The Kader TN-1A is a simple, radio-controlled recoverable target for use by AAA systems. It is a high-wing monoplane with tapering wings and a long dorsal fin.

Kader TN-1A air target

The fuselage is flat-sided and narrow.
Engine. A two-stroke engine gives 1.34 kW (1.8 hp) and drives a two-bladed propeller with an aluminium spinner.
Undercarriage. A fixed tricycle undercarriage is fitted, allowing the use of any piece of flat ground from 15 to 20 m long.
Control. Control is line-of-sight using a form of model aircraft radio control system. Underwing pylons can carry smoke or infra-red tracking flares.
TN-1B Version. The Kader TN-1B is a modified version of the TN-1A. It has a larger modified wing (1.8 m span) and a more powerful 3 bhp engine which imparts better performance and manoeuvrability.

Specifications - TN-1B

Wing span: 1.8 m
Length: 1.7 m
Height: 0.5 m
Take-off weight: 8.5 kg
Payload: 1.5 kg
Max speed: 240 km/h
Max altitude: 2000 m
Maximum operating range: 3000 m
Endurance: 15 min

Status

Kader TN-1B in production for Egyptian Air Force.

Manufacturer

Kader Factory for Developed Industries, Cairo.
(For details see Directory of Suppliers)

FRANCE

Aérospatiale C.22 Target

The C.22 first flew on 6 June 1980 and is a variable speed target for air-to-air and surface-to-air weapon systems. Its dimensions, radar signature and high performance enable it to simulate combat aircraft flying at any altitude, and also sea-skimming missiles. Series production began in 1987 and it is in use on French firing ranges.
Versions. There are two versions: the C.22T operated with the fixed existing radio/radar installations of the CEL (Centre d'Essais des Landes), and the C.22L operated with the TTL (Telecommand, Telemetry and Localisation) system; the latter has been in service since 1986.
Design. The C.22 is a recoverable variable speed subsonic target. It has sweptback, moulded glass-reinforced plastic wings, with a light alloy main spar, a glass fibre skin and a foam core, a symmetrical profile (8 per cent thickness/chord ratio) and no control surfaces. The fuselage uses wound glass fibre on a central metal frame which is impregnated with epoxy resin and reinforced by metal inserts at attachment points. The nose cone and tail cone are made of glass-reinforced moulded plastic. The aluminium alloy tail fins have an 'X' configuration, with four control surfaces operated by Labinal electric actuators.
Engine. Power is provided by one 3.73 kN Microturbo TRI 60-2 turbojet, mounted in a pod on top of the fuselage, and two jettisonable solid propellant booster rockets (each providing 32 kN for 1 second) attached beneath wings on the sides of the fuselage. Fuel tanks for the TRI 60 are in the centre of the fuselage (capacity 240 litres). Oil capacity is 4 litres.
Launch and Recovery. The C.22 is launched by jettisonable rockets from a ground or ship base and there is a parachute system for recovery from land or sea. Inflatable airbags beneath the fuselage absorb the landing impact.
Control. The C.22 has a radio command digital guidance system. Flight control, based upon a mini-computer, permits complex manoeuvres at up to 6.5 *g*. Aerodynamic control is via four movable surfaces inset in cruciform tail fins indexed at 45° from the vertical axis.

Aérospatiale C.22L target at launch

Aérospatiale C.22L subsonic target

Payload. The operational equipment depends upon the weapon systems to be fired at the target. The nose compartment is available for equipment which could include a tow winch. The tail compartment payload includes a 20 kg smoke generator tank and recovery parachutes. The nose cone houses up to 60 kg of mission equipment; active or passive countermeasures; recording equipment for assessing the effectiveness of weapon systems during training; two 30 kg under-fuselage pods; or equipment for towing successively two 30 kg secondary targets at speeds of more than 485 knots (900 km/h), with 800 m of cable for each target.

Other Features. An onboard 28 V battery provides electrical power and an engine-driven generator is fitted. The TTL (Telecommand, Telemetry and Localisation) housed in a shelter constitutes, together with the C.22, a self-contained system which does not require any ground-based installation.

Specifications

Wing span: 2.5 m
Wing chord, at root: 0.92 m
Wing chord, at tip: 0.28 m
Span over tail fins: 1.4 m
Length overall: 5.25 m
Height overall: 1.15 m
Body diameter: 0.4 m
Wing area, gross: 1.6 m^2
Fin area, total: 0.6 m^2
Weight empty: 300 kg
Max fuel weight: 192 kg
Total internal/external mission load, incl towed targets: 130 kg
Max launching weight (excl boosters): 630 kg
Max speed (all altitudes): Mach 0.9
Max level speed at sea level: 600 kts (1112 km/h)
Max rate of climb at sea level: 4200 m/min
Time to 12 000 m: less than 6 min
Service ceiling: 14 000 m
Minimum operating height: 10 m
Range with max payload: 593 nm (1100 km)
Max endurance at 12 000 m: 2 h

Status

Currently in use on French firing ranges.

Manufacturer

Aérospatiale Missiles, Châtillon.
(For details see Directory of Suppliers)

Aeropar SOSI Aerial Target System

The SOSI is a remotely piloted training target designed to simulate an aerial threat for missile and gun air defence systems of all types.

Design. The SOSI airframe uses a delta wing layout and canard control surfaces. The forward fuselage is square in cross-section and contains the radio control system.

Engine. Power is provided by a 19.4 to 25.3 kW (26 to 34 bhp) twin cylinder, two-stroke propeller-driving engine located at the rear of the airframe. The engine can be started using an electrical starter.

Launch. A catapult is used which can be mounted on a trailer or fixed to a stationary mount on the ground or on a ship's deck.

Guidance. The SOSI can be tracked visually at ranges of up to 5 km or, using a radar system together with an autopilot, at ranges up to 30 km. The operator uses a dual frequency PCM command link for flight and altitude control. The vehicle is auto-stabilised in three axes by means of an autopilot. At the end of a mission the vehicle deploys a parachute for recovery, with a descent rate of 4 m/s. An undamaged SOSI can be recovered and relaunched within 30 minutes.

Payload. The SOSI has a payload capacity of 16 kg for 1 hour of flight and can fly at speeds up to 320 km/h at altitudes between 50 and 3000 m. Typical payloads could include remotely controlled smoke flares, a radar responder system using two 7 in (177.8 mm) Luneberg lenses, and infra-red flares for use with heat-seeking missiles. Other optional payloads include electronic, hit or telemetry transponders.

An aerial surveillance system variant of the SOSI is known as the AZI.

AEROPAR SOSI target on its launch catapult

Specifications

Wing span: 2.5 m
Length: 2.9 m
Height: 0.6 m
Wing area: 2 m^2
Weight empty: 40 kg
Max take-off weight: 70 kg
Fuel capacity: 15 l
Endurance: 1 h
Speed range: 95 to 320 km/h
PCM command link: 78/82 MHz
Power plant output: 19.4 to 25.3 kW
Max payload capacity: 16 kg
Weight of parachute: 3.5 kg

Status

In production.

Manufacturer

AEROPAR, Couhé.
(For details see Directory of Suppliers)

ALTEC Attacker Hand-launched Small Arms Aerial Target

The ALTEC Attacker Hand-launched Small Arms Aerial Target is part of a target system which comprises five Attacker target and one ground support equipment box and its contents.

Design. The Attacker target aircraft is a mid-wing monoplane with a rectangular cross-section fuselage and straight constant-chord wings. The fuselage is manufactured using glass-reinforced plastic, while the wings are vinyl-covered foam. The tail stabiliser fin is vinyl-covered balsa.

Engine. Power is provided by a 15 cc glow fuel engine delivering 1.65 kW; the engine is fitted with a silencer and drives a fibreglass-filled nylon propeller.

Guidance. Flight is controlled by a narrowband frequency modulated post-coded modulation VHF/UHF. The pilot's telecommand console has a two-axis gaitered joystick with a separate throttle control; it operates in the 35 to 40 MHz band. A dual-control connection is provided for training to enable an instructor to assume control when required. The telecommand range is 1500 m and the engine will shut down under the control of a failsafe system should the command link be lost.

Recovery. Recovery can be via a parachute or a belly landing. The aircraft can float should it descend on water.

ALTEC Attacker hand-launched small arms aerial target

Payload. Optional equipment can include two smoke or infra-red flares mounted on wing pods, a miss-distance indicator, and waterproofed electronics.
Support Equipment. Ground support equipment for a complete Attacker target system includes three aircraft support caddies, two hand-held telecommand transmitters, and a mains-powered charger unit. There is also a ground support equipment tray containing a fuel pump, a power panel, a set of Velcro straps, a 12 V 6 Ah battery for starting, a battery charger, a tool box, an engine starter and a starter lead.

Specifications
Wing span: 1.6 m
Length: 1.5 m
Weight empty: 4.22 kg
Fuel capacity: 0.57 l
Endurance: 25 min
Speed range: 56 to 176 km/h
Power plant output: 1.65 kW
Telecommand range: 1500 m

Status
In production.

Manufacturer
ALTEC Industries SA, Paris.
(For details see Directory of Suppliers)

ALTEC Vector Aerial Target System

The ALTEC Vector aerial target system was designed for short to medium range SAM systems.
Design. The Vector target aircraft is a high-winged monoplane with the slightly sweptback wings and stabiliser made from glass-reinforced plastic-covered PVC foam; the fuselage is made from fibreglass and Kevlar-reinforced plastics.
Engine. Power is provided by an 18.64 kW 290 cc two-stroke engine driving a hardwood propeller.
Launch. The aircraft is launched from an electrically actuated bungee-powered launching ramp 7.8 m long and carried on a single-axle trailer.
Guidance. Flight is controlled by a narrowband frequency modulated post-coded modulation VHF/UHF. The pilot's telecommand console has a two-axis gaitered joystick with a separate throttle control and an audio intercom; it operates in the 68 to 88 MHz band. A dual-control connection is provided for training to enable an instructor to assume control when required. The telecommand range is 10 km at 5 W transmitter output and 50 km at 10 W. The engine will shut down under the control of a failsafe system should the command link be lost.
Recovery. Recovery can be via a parachute or a belly landing. The aircraft can float should it descend on water and the electronics are waterproofed.
Payload. The aircraft can carry a payload of 6.3 kg. Underwing pods can carry smoke or infra-red flares, while other optional equipment includes a miss-distance indicator, target banners (with enhancement if required), a GPS tracking system, radar reflectors or other target enhancement devices, and various forms of telemetry.
Support Equipment. Ground support equipment includes an audio intercom head set, batteries, electronic support equipment, a general and an electronic tool box, an aerial, and a GRP repair kit.

ALTEC Vector medium-speed aerial target on its launcher

Specifications
Wing span: 2 m
Length: 2.38 m
Weight empty: 50.39 kg
Fuel capacity: 11 l
Endurance: 90 min
Speed range with payload: 90 to 320 km/h
Power plant output: 18.64 kW
Telecommand range: 10 km at 5 W; 50 km at 10 W
Payload: 6.3 kg

Status
In production.

Manufacturer
ALTEC Industries SA, Paris.
(For details see Directory of Suppliers)

GREECE

Hellenic Aerospace Aerial Target System

First shown publicly in October 1990, the Hellenic Aerospace Aerial Target System is configured as a scale model of an F-16 fighter but other configurations are possible.
Engine. A tail-mounted 28.88 kW NGL WAEL 342 cc two-stroke, twin cylinder engine.
Guidance. The command system uses a PCM command link and autostabiliser with an optical and/or radio tracking system.
Launch. The target can be launched from a low-maintenance catapult mounted on a light vehicle or naval vessel.
Recovery. Recovery can be by skid landing or by parachute under either manual control or automatically, using a fail-safe system.
Payload. Payloads include Luneberg lenses, corner reflectors, a miss-distance indicator system and infra-red flares.

Specifications
Wing span: 2.54 m
Length: 3.3 m
Height: 0.84 m
Max take-off weight: 72 kg
Useful load: 15 kg
Performance:
(max speed) 294 km/h
(economy speed) 175 km/h
(stalling speed) 75 km/h
(service ceiling) 3000 m
(range) 12 km
(endurance) 1 h 30 min

Status
In production. In service with the Greek armed forces.

Manufacturer
Hellenic Aerospace Industry Limited, Schimatari.
(For details see Directory of Suppliers)

Hellenic Aerospace Aerial Target
(Photo - K Munson)

INDIA

ADE Lakshya Air Target

The Lakshya is a reusable, ship- or ground-launched subsonic target, developed by the Aeronautical Development Establishment (ADE). It is used by all three Indian armed services and at the Indian Test Range Centre at Chandipur.

Propulsion. Production versions are powered by a Microturbo TRI-60-2 gas-turbine engine with rocket-assisted take off.

Design. The Lakshya is a low wing monoplane with cropped delta wings and tail surfaces, with the power pod suspended under the centre fuselage. Launch can be from the ground or from a ship using a zero length launcher.

Recovery. Recovery is on command over land or sea using a two-stage parachute system. In addition, onboard flight control electronics initiate recovery under certain specific system malfunction conditions. The target has a search beacon and flasher light for easy location and recovery. Each target is refurbished after recovery and can be reused up to ten times.

Guidance. Guidance is carried out by radio link. The flight controller is provided with all necessary flight information from sensors on the vehicle and he can command manoeuvres and recovery. Automatic altitude and direction hold modes are available. The launch phase is carried out under automatic control.

Payload. The Lakshya is designed to tow two HAL infra-red or radar-augmented targets, each on a cable up to 1100 m long. Acoustic or Doppler scoring systems provide accurate miss-distance information.

Specifications

Wing span: 3 m
Length overall: 5.929 m
Height (total): 1.64 m
Body diameter: 0.425 m
All-up weight at launch: 630 kg (with booster)
Weight empty: 320 kg
Fuel weight (usable): 175 kg
Speeds:
(max, at 300 m, clean) Mach 0.74
(one tow body in tow at 300 m) Mach 0.53
Operating altitude:
(clean) 9000 m
(one tow body in tow) 6000 m

Status

Entering service with the Indian armed forces.

Designer/Manufacturer

Designed by the Aeronautical Development Establishment (ADE) of the Ministry of Defence, Bangalore and manufactured by Hindustan Aeronautics Limited (HAL) at Bangalore.
(For details see Directory of Suppliers)

ADE Lakshya target - Note the underslung engine and the launch rocket under the tail

ADE Lakshya target with HAL target under the wing which can be trailed on a cable

ADE Ulka Missile Target (MT)

Ulka expendable air-launched target

The Ulka is an air-launched, expendable missile target for training SAM operators. Flight development and evaluation trials have been completed and the Ulka is in service at the Indian Test Range Centre at Chandipur.

Design. The Ulka uses modular construction and has a long cylindrical body, with a tapered nose cone having small movable canard surfaces near the tip, and rear-mounted delta wings with ailerons. A small 'arrowhead' end plate fin is located at each wingtip. There is a ventral fin under the rear fuselage.

Propulsion. The power plant consists of a single solid propellant rocket motor with two separate chambers arranged in tandem (a booster and a sustainer). The Ulka is available with a long-burn sustainer in place of the booster.

Launch. Ulka is air-launched from Canberra, Jaguar or other suitable aircraft. Two Ulkas can be carried, one under each wing. A typical Canberra launch takes place at an altitude of 800 m at speeds up to Mach 0.7. The Ulka's power plant is initiated 20 seconds after release. The Ulka can then be available to be tracked and intercepted for at least 100 seconds.

Guidance. Ulka is controlled by an onboard automatic flight control system. The ventral fin and two end fins provide directional stability. Lateral control is achieved by deflecting the ailerons in response to a sum of roll angle, roll rate and yaw angle errors. Longitudinal control of the target is by canard deflection in response to a sum of pitch angle and altitude error signals shaped through suitable control filters. Level flight is achieved using a barometric sensor and gyro stabilisation. An optional radar altimeter enables flights at down to 50 m above sea level. Directional orientation is obtained from the launch aircraft heading. Additional directional control is limited to 10° either side by roll-to-turn manoeuvres and can be effected by the range controller using a dedicated command link.

Payload. Mission equipment comprises an altitude sensor plus vertical, roll rate and yaw rate gyros. The Ulka has a single 30 V battery in the centre section. A Luneberg lens is fitted in the underbody pod. Optional equipment includes a fixed target/miss-distance indicator and a transponder/control display unit. A command destruction unit is provided.

Specifications

Wing span: 1.386 m
Length: 4.02 m
Weight: 360 kg
Engine weight: 220 kg
Speed range: Mach 0.7 to 1.4
Operating altitude: 50-13 000 m
Range: 35-70 km
Endurance:
(50 to 500 m) 150 s
(500 to 5000 m) 180 s
(over 5000 m) 300 s

Status

Understood to be in service in India.

Manufacturer

Aeronautical Development Establishment (ADE), Bangalore.
(For details see Directory of Suppliers)

IRAQ

SARAB-3 Air Target

The SARAB-3 has been referred to as both a target aircraft and a remotely piloted vehicle. It appears to be a similar design to the British TTL BTT-3 Banshee.

Design. The SARAB-3 is a low-wing monoplane with clipped delta wings and a sweptback fin. The entire airframe appears to be constructed from glass-reinforced plastics with some composite components. A payload bay is located just behind the nose. Payloads can include smoke, infra-red tracking flares and radar reflectors.

Propulsion. The power plant is located at the rear of the fuselage and is described as a WAE 342 cc air-cooled ultra-lightweight twin cylinder two-cycle unit developing 26 hp.

Launch and Recovery. Launching is carried out using a bungee catapult carried on a ground support trailer. Landing may be via a parachute or by belly landing on flat ground. The SARAB-3 can operate in wind conditions of up to 30 knots.

Guidance. The SARAB-3 is controlled by a transmitter/controller with a high power transmitter and RF attenuator at ranges of from 5 to 20 km. The flight can be monitored either by radar or a telemetry ground station; the latter also collects data such as battery voltage, flight path bearings and altitude.

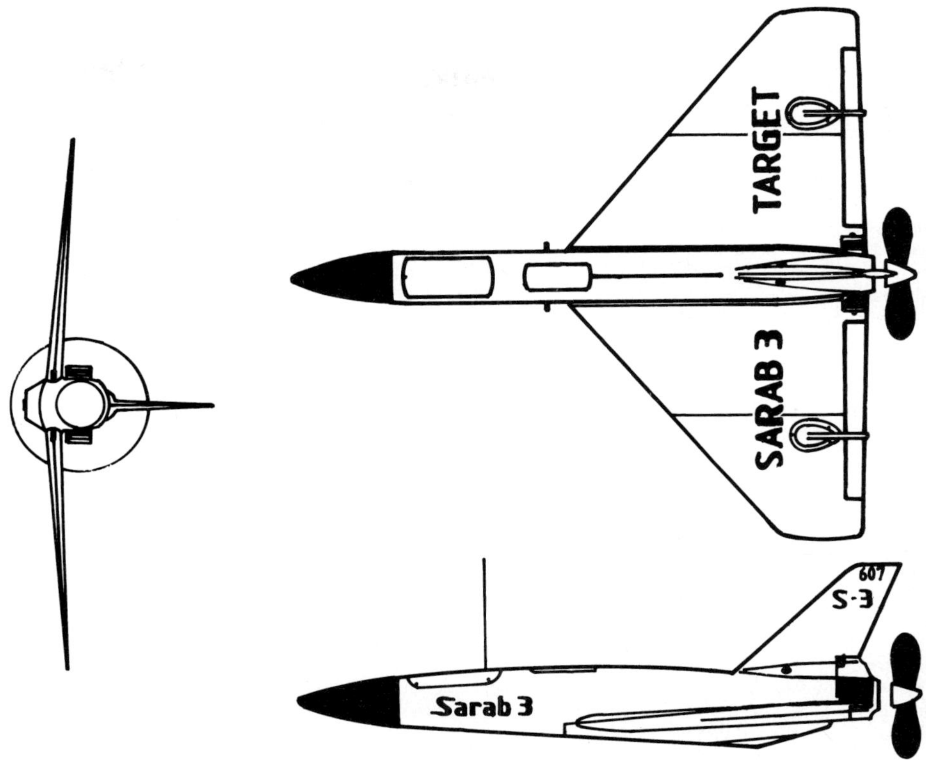

Iraqi SARAB-3 air target

Specifications

Wing span: 2.49 m
Length: 2.84 m
Height: 0.71 m
Wing area: 2.14 m^2
Weight:
(empty) 38.6 kg
(max take-off) 54.4 kg
Fuel capacity: 12 l
Speed range: 74 to 300 km/h
Launch speed: 27 m/s
Endurance: 1 h 15 min

Status

In service with the Iraqi Army and Navy.

Manufacturer

FSPF, Baghdad.
(For details see Directory of Suppliers)

ISRAEL

Tamnar Air Targets

Tamnar make a number of aerial targets which are described below.

TM-105 EDO

The name 'EDO' is derived from the initials of the Hebrew names of three soldiers killed in the Lebanon in 1982. Production began in December 1983.

Design. The EDO air defence target is a cantilever shoulder-wing monoplane. The wings have a birch plywood skin with a polystyrene core; the fuselage is of impact-absorbing glass fibre and epoxy; the tail unit is of balsa wood. A steel landing skid is standard; it can also be fitted with a tricycle landing gear.

Propulsion. The power plant is one 3 kW (4 hp) Quadra Q-50, Q-82 or 7.1 kW (9.5 hp) Q-100 single cylinder two-stroke engine, driving a Clark two-blade wooden propeller. The standard fuel capacity is 4.5 litres; an extended range tank is optional.

Launch and Recovery. Launch and recovery is by catapult from a Tamnar rail launcher; the target returns to land (if not destroyed during a mission) on an under-fuselage skid. Alternatively, it can be fitted with tricycle gear (jettisoned after take off) for conventional take off from any suitable 50 to 80 m length of roadway, track or grass. At sea, the EDO can be platform-launched from any vessel of patrol boat size or larger, and recovered in a retrieval net installed over the stern.

Guidance. The target is radio controlled by an operator located within the unit training area, using a Kraft 10 W transmitter able to control two targets simultaneously. During each exercise, instructions can be transmitted to both the air defence battery crew and the EDO operator. The commander thus remains in control throughout the training session, and can initiate or alter a situation within seconds.

Payload. Mission equipment includes a hit counter which records 'hits' at up to 6 m from the target and relays results in real time to the ground control display. For use as a target for IR-homing missiles, the target can be equipped with IR enhancement.

Tamnar TM-105 EDO air target

Specifications

Wing span: 2.5 m
Length overall: 2.31 m
Length of fuselage: 2.25 m
Propeller diameter:
(Q-50) 0.46 m
(Q-82) 0.51 m
(Q-100) 0.56 m
Weight empty: 12 kg
Max payload including fuel: 4.5 kg
Max take-off/launching weight: 18 kg
Max level speed: 162 kts (300 km/h)
Min flying speed: 19 kts (35 km/h)
Radio control range: 3.2 nm (6 km)
Endurance, standard fuel: 1 h 30 min
Endurance, optional fuel: 2 h

Tamnar MiG-27 and Su-25 Air Targets

Target models of the MiG-27 'Flogger' and Su-25 'Frogfoot' entered production in 1986, the latter (which is no longer being built) varying only slightly in appearance from the EDO (see previous entry).

The 'Flogger' target has a deeper fuselage, and resembles more closely the real aircraft. It is constructed, like the EDO, of impact-absorbing glass fibre, has a 7.1 kW (9.5 hp) Quadra Q-100 engine, and is similarly launched and recovered.

Specifications

Wing span: 3.3 m
Length overall: 3.05 m
Max take-off/launching weight: 18 kg
Endurance, standard fuel: 1 h
Endurance, optional fuel: 1 h 30 min

Status

In production. Over 800 EDOs ordered, in service with several nations including Israel, Chile, Honduras and Venezuela.

Tamnar MiG-27 'Flogger-D' scale target

Manufacturer

Tamnar Aviation Technology, Petah Tikva.
(For details see Directory of Suppliers)

ITALY

Meteor Mirach

Various types of Mirach aerial targets are currently available.

Mirach 10

The Mirach 10 target was originally developed during the late 1970s as a general-purpose surveillance UAV and is now available as a target for SAM and AAA training. It is a delta-winged vehicle powered by a single 20 kW aft-mounted air-cooled piston engine; an improved version has a 30 kW engine with commensurate improvements in performance. The Mirach 10 is launched from a pneumatic catapult launcher mounted either on land or on a ship and, once airborne, is controlled via a radio link. Also available is a portable ground control and tracking station based on the Global Positioning System which allows the control of missions beyond visual range. An optional ground receiving station can receive flight parameters such as engine rpm, height, and so on.

Mirach 70

The Mirach 70 is based on the airframe of the earlier Meteor 10. As well as the target role it can be used for electronic warfare or as a decoy. Power is provided by a single 52 kW piston engine. Guidance is via a radio command system. The Mirach 70 was produced under licence in Argentina by Quimar as the MQ-1 Chimango.

Mirach 100

The Mirach 100 is powered by a single 1.13 kN turbojet. It uses either ground control or automatic navigation with real-time data transmission. It is in production for the Italian armed forces and foreign countries (reportedly including Iraq and Libya), and was produced under licence in Argentina by Quimar as the MQ-2 Bigua. The one-way penetration range is 485 nm (900 km). The Agusta A 109/Meteor Mirach 100 combination, consisting of the transport, release and recovery by helicopter of one or two Mirach 100s has been delivered to various customers and is operational. The Mirach 100 is equipped with an automatic navigation subsystem called Sirah. With this subsystem installed, the vehicle can be programmed to loiter over a battlefield for surveillance purposes. It also has potential applications as a tactical cruise missile launched from an Aeritalia G222 'mother' aircraft carrying six Mirach 100s.

Mirach 100 production has exceeded 200 and a batch of 20 was ordered by the Greek Government for use as targets at the NATO range on Crete.

Mirach 10 target - Note the aircraft-style vertical stabiliser (fin)

Mirach ER

An extended-range Mirach 100ER, powered by a 1.78 kN NPT 401 turbojet and equipped with Pacific Aerosystem's Mizar avionic data link and ground equipment plus a British Aerospace MIRLS 4000 miniature infra-red line-scanner, was a candidate in the US Navy's MRR-RPV (medium range reconnaissance remotely piloted vehicle) competition.

Specifications

Wing span:
(Mirach 10) 2.6 m
(Mirach 70) 3.57 m
(Mirach 100) 1.804 m
Length overall:
(Mirach 10) 2.85 m
(Mirach 70) 3.66 m
(Mirach 100) 4.126 m
Height overall:
(Mirach 10) 0.8 m
(Mirach 70) 0.65 m
(Mirach 100) 0.804 m
Wing area, gross:
(Mirach 10) 2.3 m^2
(Mirach 70) 1.96 m^2
(Mirach 100) 0.82 m^2

Launch of a Mirach 70 target - Note the revised tail configuration with two vertical stabilisers on either side of the tailplane

Max launching weight, including booster where applicable:
(Mirach 10) 86.5 kg
(Mirach 70) 260 kg
(Mirach 100) 310 kg
Max level speed:
(Mirach 10) 62 kts (115 km/h)
(Mirach 70) 194 kts (360 km/h)
(Mirach 100) 458 kts (850 km/h)
Ceiling: (Mirach 100) 9000 m
Max endurance:
(Mirach 10) 1 h 40 min
(Mirach 70) 1 h
(Mirach 100) 1 h
Max altitude:
(Mirach 10) 8000 m
(Mirach 70) 4000 m
(Mirach 100) 9000 m

Status

Mirach 10 available.
Mirach 70 in production - licence production in Argentina has now ceased.
Mirach 100 in production, over 200 built - licence production in Argentina has now ceased.

Manufacturer

Meteor Costruzioni Aeronautiche Ed Elettroniche SpA, Gorizia.
(For details see Directory of Suppliers)

Launch of Mirach 100 target - Note the vee tail

JAPAN

Fuji/Teledyne Ryan BQM-34AJ Firebee 1

Under contract from the Japan Defence Agency, Fuji is building the Teledyne Ryan BQM-34A Firebee 1 subsonic target for use in the training of Tartar missile and gunnery crews, and for the evaluation of air-to-air missile systems and Japanese-built F-15J Eagle fighters. The first shipboard launch of a Fuji-built unit (known as the BQM-34AJ) was carried out in 1970.

Fuji-built BQM-34AJs include some fitted with the RALACS (radar altimeter low altitude control system) for simulating anti-shipping missiles.

For details of the Teledyne Ryan Firebee 1 refer to the entry under *United States* in this section.

Status

In production. A total of over 60 Fuji-built BQM-34AJs have been delivered to the Japanese Marine Self-Defence Force and 20 to the Japanese Army Self-Defence Force.

Manufacturer

Fuji Jukogyo Kabushiki Kaisha (Fuji Heavy Industries Limited), Tokyo.
(For details see Directory of Suppliers)

Fuji Model 820 - JASF J/AQM-1

The Model 820 expendable air-launched target was developed for the Japanese Self-Defence Force under a Japan Defence Agency (JDA) contract. Fuji was selected by the JDA in July 1983 as prime contractor. Production for the Japanese Air Self-Defence Force began in 1987. By 1994 a total of about 100 J/AQM-1s had been delivered to the Japanese Air Self-Defence Force and 20 to the Technical Research and Development Institute of the Japan Defence Agency.

Design. Construction is virtually all-metal (steel and aluminium). The airframe is a mid-wing monoplane, with clipped delta wings interchangeable port/starboard. It has four tail fins, of similar plan form, indexed in an 'X' configuration at 45° to vertical and horizontal axes. The target has a cylindrical metal body, with a glass fibre nose cone and tail cone. The engine is suspended on a centre line pylon beneath the rear of the fuselage. There is no landing gear.

Propulsion. The power plant is one 1.96 kN Mitsubishi Heavy Industries TJM3 turbojet in an underslung nacelle. A pressurised fuel tank in the centre of the fuselage holds 47 litres.

Launch. The Model 820-J/AQM-1 can be air-launched from an underwing pylon (one under each wing) of the McDonnell Douglas F-4EJ Phantom or F-15J/DJ Eagle.

Recovery. Non-recoverable.

Guidance. The target has a preprogrammed guidance system, which can be overridden by radio command from either the launch aircraft or a surface station. There is a digital flight control system, including the engine throttle control.

Payload. Mission equipment includes an X-band reflector in the fore and aft body compartments. Smoke or infra-red generators can be attached to the wingtips. A miss-distance indicator can be installed in the forward section of the fuselage.

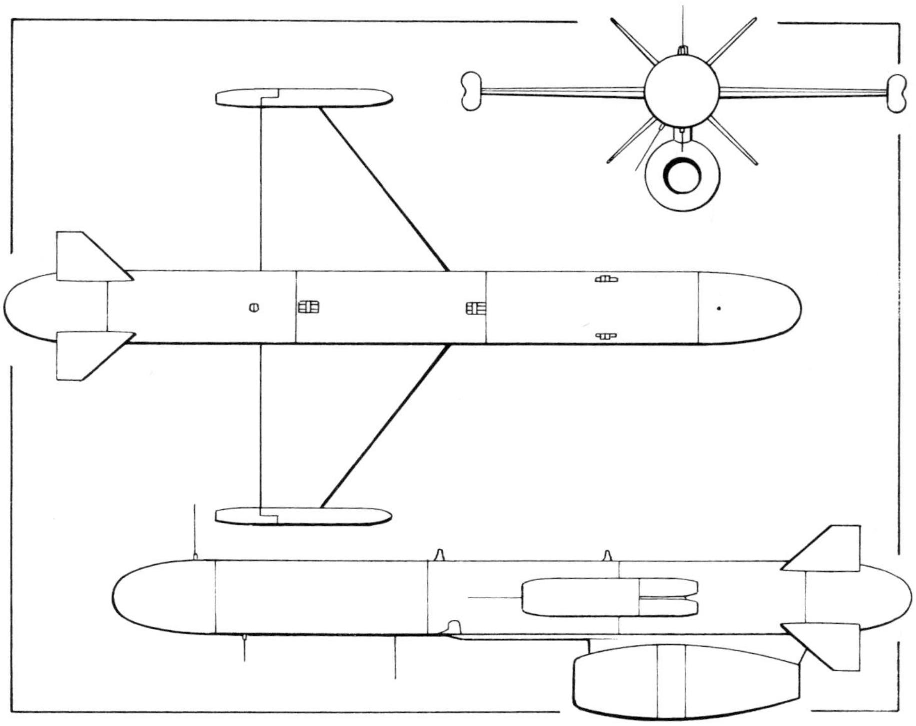

Fuji Model 820 J/AQM-1 air target

Specifications

Wing span over smoke generator pods: 2.07 m
Wing aspect ratio: 3.2
Length overall: 3.65 m
Height overall: 0.92 m
Body diameter (max): 0.35 m
Span over tail fins: 0.98 m
Wing area, gross: 1.2 m^2
Tail fins, total: 0.3 m^2
Weights:
(max mission payload) 42.4 kg
(fuel) 32 kg
(max launching weight) 235.5 kg
Max level speed at 9150 m: Mach 0.95
Max sustained turn at 9150 m: 3.2g
Operating height:
(min) 660 m
(max) 9150 m
Endurance at Mach 0.9 at 9150 m: 18 min

Status

In production.

Manufacturer

Fuji Jukogyo Kabushiki Kaisha (Fuji Heavy Industries Limited), Tokyo.
(For details see Directory of Suppliers)

Mitsubishi QF-104J Starfighter UMA Targets

Unmanned ex-Japanese Self-Defence Force F-104J Starfighters are used as QF-104 target UMAs for use in air-to-air combat training, and missile tests. Honeywell is the main contractor for the design, development and production of the control system.

The QF-104Js are operated from the Iwo Jima Island airbase.

Status

About 30 F-104s have been modified.

Manufacturer

Mitsubishi Jukogyo Kabushiki Kaisha (Mitsubishi Heavy Industries Limited), Tokyo.
(For details see Directory of Suppliers)

RUSSIA

Dan Aerial Target

The Dan (Russian for 'Tribute') aerial target is designed by the Sokol design bureau and intended for simulating cruise missiles and aircraft. The target is launched from a ground launcher unit with the assistance of a booster rocket and has a combined manual and programmed control system; preprogrammed flight paths are possible.

Components. The complete Dan system comprises the target, launcher, a loader transporter, an automatic ground monitoring system, an engine starter and other support equipment. Ground support equipment includes tracking and surveillance radars and a radio control station. Launch site preparation time takes 30 minutes and minimum turn-round time between flights is 2 hours. Each Dan target may be reused up to ten times.

Airframe. The airframe has a circular cross-section fuselage with centrally located straight wings set some two-thirds of the way towards the tail. The tail fin is swept and mounted over the above-fuselage housing containing the MD-120 turbojet engine. Recovery is by parachute.

Payload. Equipment carried can include miss-distance indication, radar reflectors and flares in wingtip housings.

Specifications

Wing span: 2.68 m
Length overall: 4.6 m
Height overall: 1.05 m
Fuselage diameter: 450 mm

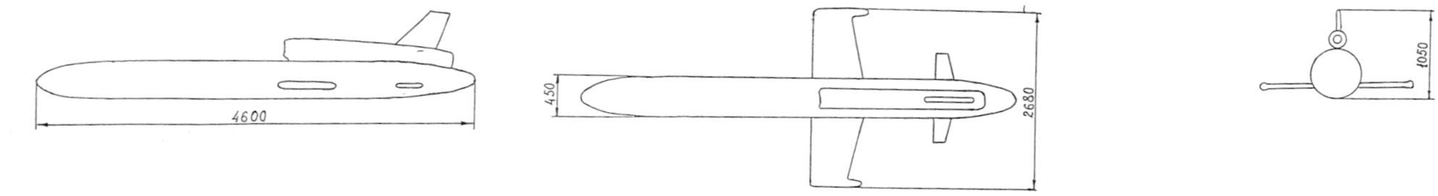

Drawing of Dan aerial target

Weight at take off: 345 kg
Operating speed range: 162-383 kts (300-710 km/h)
Max altitude: 9000 m
Endurance: 25-40 min

Status
In service.

Manufacturer
State manufacturers.

SOUTH AFRICA

Buzzard Low-speed Air Target

Buzzard low-speed target

The Buzzard low-speed air target system is for training SAM and AAA crews. There are three versions of the airframe, the basic Buzzard-3B, the enhanced Buzzard-3E, and a 60 per cent scaled-down training version, the Mini-Buzzard.

Buzzard-3B. The Buzzard-3B has a 10 km range through the use of a Semi-Remote-Control Ground Station.

Buzzard-3E. The Buzzard-3E has the same overall physical characteristics of the 3B but has a 30 km remote-control range through its Remote-Control Ground Station.

Mini-Buzzard. The Mini-Buzzard system is a 60 per cent scale version intended as a low cost training aid for Buzzard system pilots; it does not have the ground control stations and control is via a standard radio control box. The airframe has a wing span of 1.25 m, overall length is 1.38 m and endurance is 20 minutes. Control range is 1000 m.

System Components. A complete Buzzard-3 system comprises six airframes, a rail catapult launcher, a Semi-Remote Ground Control Station, a full Remote Ground Control Station, and a comprehensive support unit.

Airframe. The targets are small, delta-wing pusher propeller aircraft launched by a catapult launcher and controlled by radio or remote-control.

Payload. Each airframe can carry infra-red and smoke flares, Luneberg lens radar reflectors, radar corner reflectors and miss-distance indicators.

Ground Stations. The Semi-Remote-Control Ground Station (SRCS-3) is a manual optical tracking and radio control device which enables the operator to control the aircraft beyond normal visual range with the aid of binoculars and a radio control console, up to a range of 10 km. The station interfaces to the aircraft through a standard radio control uplink.

The Remote-Control Ground Station (RCGS-3) is a containerised station equipped with an operator display, a control panel, a ground control computer, an up/downlink transceiver (with antenna and mast) and cabling. The console allows the operator to adjust speed, heading, altitude and airspeed as well as enabling him to initiate alarm conditions, control speed and log events throughout a mission. Operator displays provide real-time information on position, flight path, speed, heading, altitude, airspeed, turn and climb rates and attitude.

Specifications of Buzzard 3

Wing span: 2.1 m
Length overall: 2.3 m
Height overall: 0.6 m
Wing area: 1.75 m^2
Weight empty: 50.7 kg
Fuel capacity: 20 l
Payload: 20 kg
Speed range: 60-180 kts (111-333 km/h)
Endurance: 1 h
Max altitude: 3000 m
Control range:
(Buzzard-3B) 10 km
(Buzzard-3E) 30 km
Launching speed: 34 m/s

Status
Development.

Marketing Agency
Denel (Pty) Limited, Hennopsmeer.
(For details see Directory of Suppliers)

IST LOCATS Air Target

The IST Low Cost Aerial Target System (LOCATS) air target training SAM and AAA crews.

System Components. The system includes a 4 × 4 configuration support and towing vehicle, a zero-length pneumatic trailer-mounted launcher, and three airframes. Containers are incorporated into the support vehicle and trailer to carry all ground support equipment, a communications system, spares and the airframes. The system is supported by a 'pilot' and two technicians who can prepare for a flight in less than one hour after arrival at a launch site.

Airframe. The LOCATS target airframe is a low wing tapered wing monoplane constructed using a modular glass fibre construction.

IST LOCATS air target

Propulsion. Power is provided by an 18 kW two-stroke petrol engine driving a 600 mm propeller.
Guidance. Control is via a 60 MHz nine-channel radio link with a range of 2.5 km, although this can be extended.
Fittings. A radar reflector is standard inside all fuselages but can be omitted on request. An integral parachute mechanism is used for recovery at the end of each mission.

Specifications
Wing span: 3.2 m
Length overall: 2.9 m
Height overall: 0.9 m
Propeller diameter: 600 mm
Weight empty: 42 kg
Max payload (inc fuel): 18 kg
Max launching weight: 60 kg
Max level speed: 162 kts (300 km/h)
Max dive speed (10°): 194 kts (360 km/h)
Launch speed: 58 kts (110 km/h)
Stalling speed: 47 kts (90 km/h)
Endurance: 60 min
Control range: 2.5 km (extendable)

Status
In production.

Manufacturer
Integrators of Systems Technology (Pty) Limited (IST), Waterkloof.
(For details see Directory of Suppliers)

Kentron Skua High-speed Target

The Kentron Skua is a high speed subsonic target designed to simulate an attack aircraft. It is part of a transportable system comprising a GCS-3 mobile ground control station, the targets themselves, an LRN-3 launcher unit and ground support equipment.
Flight Envelope. The Skua can fly at speeds up to Mach 0.86 at an altitude of 10 000 m.
Airframe. Skua is a high wing monoplane with slightly swept wings, twin tail fins and a circular cross-section fuselage.
Propulsion. An Apex TJ 450 turbojet engine is located under the rear of the fuselage, supplying up to 4.4 kN at sea level.
Towed Target. In order to reduce the possibility of the Skua being hit during a mission, a target can be towed from a 1500 m long cable. The tow target houses passive or active miss-distance measuring equipment and the associated telemetry; if required, these equipments can be carried inside the fuselage.
Launch and Recovery. Launch equipment consists of booster rockets, a zero length launching ramp and a control panel. Recovery is by means of a two-stage parachute system and air bags.
Guidance. The ground control station uses UHF telecommand and telemetry links. Controls and displays inside the station are used by the operator to control the target.

Specifications
Wing span: 3.57 m
Length overall: 6 m
Fuselage diameter: 500 mm
Weight (inc fuel and payload): 500-700 kg
Max payload: 100 kg
Max speed:
(sea level) Mach 0.75
(10 000 m) Mach 0.86
Stalling speed: 150 kts (278 km/h)
Max altitude: 12 000 m
Endurance (10 000 m and Mach 0.75): 1 h

Status
Development.

Manufacturer
Kentron, Hennopsmeer.
(For details see Directory of Suppliers)

Kentron Skua target launch

SWEDEN

Norabel RIPAN Air Target

The RIPAN is a small reusable target for training AAA crews.
Design. It is a cantilever shoulder-wing monoplane with a conventional tail unit. Wings are of symmetrical profile, and have a plywood skin with an expanded polystyrene core. Ailerons and tail surfaces are of balsa wood and the fuselage is of polystyrene. There is no landing gear.
Propulsion. The RIPAN is powered by one 1.9 kW (2.5 hp) 15 cc Super Tigre piston engine, driving a two-blade propeller. Fuel capacity is 0.45 litres.
Launch and Recovery. The RIPAN is hand-launched; recovery is by belly landing.
Guidance. Guidance is via radio command. Aerodynamic control is provided by full span ailerons, a rudder and one-piece elevator.
Payload. Mission equipment includes a three-element hit indicator system comprising a dorsally mounted acoustic sensor (microphone), a hit recorder and an indicator. The microphone can be preset for either 7.62 or 5.56 mm ammunition, and six sensitivity levels corresponding to radii from 2 to 12 m can be preselected to represent different hit zones or target sizes. The sound of a passing projectile is registered by microphone, and signals are processed electronically by a hit recorder. A sensor computer, in the centre fuselage, can be set at different numbers of registered hits (up to 15) required to produce a 'kill'. When the required number is reached, the computer signals to the adjacent indicator unit, which activates a pyrotechnic cartridge, producing a vivid flash that is visible up to at least 1000 m away, indicating to gunners that the target has been 'killed'. The hit recorder is powered by 8.4 V Ni/Cd batteries; the indicator uses a rechargeable cassette containing 10 disposable flash cartridges.

Norabel RIPAN target being hand-launched

Specifications
Wing span: 1.77 m
Length overall: 1.4 m
Height overall: 0.15 m
Payload: 1.5 kg
Max launching weight: 4 kg
Max level speed: 108 kts (200 km/h)
Stalling speed: 22 kts (40 km/h)
Range: 0.5 nm (1 km)
Endurance: 10-15 min

Status
In production.

Manufacturer
Norabel AB, Nora.
(For details see Directory of Suppliers)

SWITZERLAND

Farner Topaz

The Topaz AATS (anti-aircraft training system) is the current version of the *Kleinzieldrohne* (small target drone) 85 (KZD-85) which is produced by Farner Air Service AG for the Swiss Army. Farner produces the airframe, launcher and ground support equipment, and the remote-control system is produced by the Swiss Federal Aircraft Factory (FW). FW also acts as the interface to the military procurement authority.

General. The system is operated by one aircraft controller and one handler, and can be made ready for use within 20 minutes. Two or more target aircraft can be operated simultaneously, and the system was designed to work independently in the field for 12 hours per day, away from runways, roads and the general infrastructure. The current version is available in two airframe sizes.

Airframe. The target is a cantilever shoulder-wing monoplane with Kevlar-reinforced glass, reinforced plastic sandwich wings and tail, and a frames-and-stringers aluminium alloy fuselage, to the underside of which is bonded a foam rubber strip to cushion the landing impact. Wing and tail surfaces detach to simplify transportation. There is no landing gear.

Propulsion. The power plant is one 100 to 300 cc two-stroke engine, with electronic ignition, driving a two-blade propeller. Fuel capacity (20:1 petrol/oil mixture) is 2 litres.

Launch and landing. The target is catapult-launched from a trailer-mounted 6.5 m ramp. It is recovered by an 11 m diameter cruciform parachute or by belly landing on a ventral rubber skid.

Guidance. A PCM radio command guidance transceiver (output 1 W) is fitted for line-of-sight operation, powered by a 1.8 Ah rechargeable battery. If the command link is lost, the engine shuts down automatically after 0.5 second and the recovery parachute is deployed. Aerodynamic control is by conventional wing and tail control surfaces.

Versions. Topaz can be produced in a low-cost version as an expendable target for live anti-aircraft firing practice. In reusable form, it can carry a hit indicator and give active reflected signals of hits on a towed target. It can also use a miss-distance indicator with active signalling when Topaz itself is the target or it can emit a radar beam giving a passive reflected signal.

Farner Topaz gunnery target on launch rails

Specifications

Wing span:
(A version) 2.5 m
(B version) 3.5 m
Length overall:
(A version) 2 m
(B version) 2.4 m
Height overall: (A and B version) 0.62 m
Propeller diameter: (A and B version) 0.51 m
Wing area, gross (A version): 0.875 m^2
Weights:
(empty, equipped, A version) 20.9 kg
(fuel, A version) 1.6 kg
Max launching weight:
(A version) 23 kg
(B version) 30 kg
Max diving speed: 135 kts (250 km/h)
Max level speed: 108 kts (200 km/h)
Minimum flying speed: 43 kts (80 km/h)
Max operating altitude: 2500 m
Max range, line-of-sight: 1.1 nm (2 km)
Max range, with binoculars: 2.2 nm (4 km)
Max endurance: 30 min

Status

In service with the Swiss Army.

Manufacturer

Farner Air Service AG, Grenchen.
(For details see Directory of Suppliers)

UNITED ARAB EMIRATES

RPS Kestrel

Remote Piloted Systems (RPS) produces targets for small arms and AAA training. The Kestrel was designed for this purpose and a Kestrel II is under development. A larger turbojet version is planned for 1995.

Airframe. The Kestrel has a small straight-winged airframe.

Propulsion. 40 cc two-stroke petrol engine developing 4.4 hp.

Payload. Various radar reflectors and smoke systems can be carried.

System Components. A self-contained trailer holds five Kestrel airframes with spares, a fuel supply, and a generator for charging the electrical system. The aircraft launch ramp is mounted above the trailer and can be assembled by one person in 10 minutes.

Recovery. If the airframe has survived a live firing mission, it can be landed back at the launch site on its skid.

Specifications

Wing span: 2 m
Length: 1.6 m
Weight, empty: 10 kg
Speed range: 50 to 180 km/h
Minimum endurance (full throttle): 30 min
Operational range: limited only by operator's visual range with guidance system range up to 8 km
Airframe stress limits: +6 to −3 *g*

Status

Available.

Manufacturer

Remote Piloted Systems, Jebel Ali, Dubai.
(For details see Directory of Suppliers)

RPS Kestrel gunnery target

UNITED KINGDOM

BAJ Petrel

The BAJ Petrel is a solid propellant ballistic rocket in service with the Royal Navy for weapon trials and in-service practice with Sea Wolf and Sea Dart SAM systems. Its steep terminal dives closely resemble some ship threats. Well over 600 Petrel missiles have been fired.

Operation. The rocket is about 3 m long and 180 mm in diameter. It is fired from a free-standing launcher tube by small high-thrust booster motors which drag separate after about 0.25 second. The rocket accelerates for 30 seconds during main motor burn and then continues ballistically to impact. A payload bay allows various radar enhancement systems to be fitted. The launcher tube may be mounted on land or on a ship's deck. In the latter case the instant of firing is controlled by a gyro-stabilised reference unit.

Trajectory. Petrel trajectories are altered by selecting mass and drag prior to launch. It is possible to produce low altitude trajectories at Mach 2, medium altitude trajectories at Mach 2.5 and high altitude trajectories at Mach 3. Typical ranges are 10, 48 and 72 nm respectively, although these can be varied; maximum range is 100 nm.

Container Installation. The Petrel system, including all equipment needed for rocket preparation and launch, can be installed inside a 20 ft/6.096 m ISO container. The container is also used as a magazine and preparation room.

Multiple Launches. Petrels have been launched in pairs and a multi-launch system employing six or more launchers on a ship's deck is under development.

Status

In production. In service with the Royal Navy.

Manufacturer

BAJ Limited, Weston-super-Mare.
(For details see Directory of Suppliers)

FRL Falconet

The Falconet is made by Flight Refuelling Limited (FRL) and is a high performance jet-powered UAV designed to provide effective training by simulating the speed, manoeuvres and signature of modern fighter aircraft. It uses a robust modular construction designed to be assembled from kits by local technicians - a typical assembly time is quoted as 2 hours.

Construction. The airframe is largely of stressed skin aluminium construction with some composite components. The wings and underslung engine are bolted to the centre section which contains the fuel and smoke trail oil tanks.

Propulsion. Microturbo TJA24 turbojet supplying 147 kN static thrust, capable of speeds in excess of 700 km/h; the engine runs on standard aviation kerosene. The engine is started using compressed air from the Falconet ground start trolley; the trolley also provides fuel until immediately before launch, enabling the vehicle to take off with a full fuel tank.

Guidance. Manoeuvres are effected by aileron, elevator and engine thrust control - there is no rudder. External commands are detected by an antenna/receiver system and actioned by the onboard flight control computer. A telemetry transmitter relays data on air vehicle status back to the operations centre. An onboard transponder is fitted to assist radar tracking.

Launch. Falconet is able to operate using two methods of launch. For fixed base operation the vehicle launches under its own power without using booster rockets. This is achieved using a circular runway and launch trolley tethered to the centre. The circular runway permits take off in any wind direction at the cost of only 3 litres of fuel. For mobile or shipboard operation, Falconet can be fitted with booster rockets for zero-length launching.

Recovery. Falconet is recovered by parachute, descending to the ground vertically nose down under the main parachute. Ground impact is absorbed by a polystyrene nose cone. Recovery can be from land or sea.

Payload. Payloads can be carried either as underwing stores or within the fuselage. Internal payloads include smoke trail equipment, radar transponders, miss-distance indication equipment and passive radar signature enhancement. Wing-mounted payloads include thermal or visual wavelength flares, sub-targets, chaff and infra-red decoy dispensers, and special-to-mission pods. Sub-targets can be configured to suit a wide variety of weapon systems including HAWK, Roland, Crotale, Shahine II, Stinger, Redeye, Skyguard, close-in weapons systems and anti-aircraft guns.

FRL Falconet zero-length launch

Specifications

Wing span: 3 m
Wing area: 1.39 m^2
Length overall: 3.75 m
Body diameter, max: 0.39 m
Height overall: 1.52 m
Tailplane span: 1.39 m
Basic weight empty: 171 kg
Max fuel: 60 kg
Payload weight (at max fuel load): 23 kg
Max launching weight: 235 kg
Recovery weight: 175 kg
Max level speed (clean): 400 kts (740 km/h)
Ceiling: 3000 m
Range: 125 km
Typical endurance:
(clean) 60 min
(with sub-target) 42 min

Status

In service with the British Army at the Royal Artillery's Hebrides Range. The French DGA use the Falconet for Crotale, Roland and HAWK firings.

Manufacturer

Flight Refuelling Limited (FRL), Wimborne.
(For details see Directory of Suppliers)

FRL Multi-Role Towed Target (MRTT)

Flight Refuelling Limited (FRL) produce the Multi-Role Towed Target (MRTT) which is capable of simulating a sea-skimming missile as well as being used in other roles.

MRTT Mark 1. MRTT Mark 1 has a wingless lower section and is used for non sea-skimming roles.

MRTT Mark 2. MRTT Mark 2 has a radar altimeter and height control computer, specifically designed for sea-skimming and ditching prevention. A range of heights can be selected from 500 ft (152 m) down to 15 ft (4.5 m) ASL, with a height-keeping accuracy of ±7 ft (±2.1 m).

Towing. The MRTT may be towed behind a variety of aircraft at tow lengths up to 30 000 ft (9144 m), dependent upon aircraft type and the winch fitted.

Payloads. A variety of payloads is available, including miss-distance indication (acoustic or radar), IR and visual flares, Luneberg lens, and a laser reflecting panel.

Specifications

Weight: 45-60 kg
Max length: 3 m
Min length: 2.7 m
Overall height: 580 mm
Span overall: 600 mm
Minimum heights to minimise ditching probability:
(sea state 0-4) 5 m
(sea state 4-7) 15 m
(above sea state 7) 30 m
Design operating speed: 300 kts (550 km/h)
Min speed: 230 kts (400 km/h)
Max speed: 400 kts (740 km/h)
Max carriage altitude: 12 190 m

Status

In service with the Royal Navy and navies worldwide, and target towing contractors.

Manufacturer
Flight Refuelling Limited (FRL), Wimborne.
(For details see Directory of Suppliers)

FRL Multi-Role Towed Target (MRTT) Mark 2 in sea-skimming role

FRL Snipe Targets

The FRL Snipe product range consists of self-powered targets.

Snipe Mark 5

The Snipe Mark 5 is a monoplane design with a nose-mounted piston engine. A high-lift wing section is available for high altitude and/or high temperature operations.
Launch and Recovery. Launch is achieved by use of a bungee-operated launcher. Recovery is by parachute or skid landing.
Guidance. Using an autostabiliser and global positioning system tracking, the target can be flown out to a range of 30 km. The tracking system also provides the pilot with telemetry information.
Payload. The Snipe Mark 5 can deploy a number of payload options including various towed sleeves and sub-targets, miss-distance indication, Luneberg lens, and up to 16 smoke flares or eight infra-red flares.
Configuration. Can be configured to suit weapon systems including Javelin, Stinger, RBS 70, SA-7 'Grail', Crotale, Roland and air defence guns with calibres between 20 and 76.2 mm.

FRL Snipe Mark 5 target with wingtip-mounted Luneberg lens housings for radar cross-section enhancement

Specifications

- Snipe Mark 5
Wing span: 3 m; (high-lift variant) 3.2 m
Length: 2.7 m
Weight (less fuel and payload): 60 kg
Endurance (loiter): 100 min; (high-lift variant) 115 min
Max speed: 290 knots; (high-lift variant) 250 knots
Stalling speed: 85 knots; (high-lift variant) 70 knots

Snipe Mark 15

Construction. The Snipe Mark 15 is a small, piston-engined, hand-launched target for gun and missile training exercises or for training controllers for more expensive targets. Fuselage construction is of GRP and Kevlar. The wings have a foam core covered with marine ply.
Payload. The Snipe Mark 15 can carry four 45-s smoke flares and miss-distance indicator equipment.
Recovery. Recovery is by parachute or skid landing.

Specifications

- Snipe Mark 15
Wing span: 2.2 m
Length: 1.6 m
Weight (less fuel and payload): 11 kg
Endurance (loiter): 45 min
Max speed: 210 knots
Stalling speed: 32 knots

Status

In service.

Manufacturer

Flight Refuelling Limited, Wimborne.
(For details see Directory of Suppliers)

Shorts Skeet Mark 1

The Shorts Skeet is small high-wing monoplane of modular construction. The fuselage is made of metal and GRP. Wings, fin and tailplane are made of polyurethane foam-filled glass fibre.
Propulsion. The Skeet is powered by a 13.4 kW NGL WAM 274-6 274 cc two cylinder two-stroke engine.
Launch and Recovery. The target is launched pneumatically from a self-contained two-wheel trailer, having a capacity of eight launches from one charging, or from a launch rail mounted on a four-wheel truck. Recovery is by an Irvin(GB) cruciform parachute (in the forward fuselage), or by landing on the fuselage. The parachute can be deployed either by pilot demand, or automatically following failure of the engine, power supply or radio link. Launches can be carried out in crosswinds of up to 25 kts (46 km/h) and the aircraft can be operated from land or on board ship.
Guidance. Guidance is provided by an eight-channel radio command system consisting of a hand-held transmitter and 15 W amplifier in the 68 MHz band. The eight channels comprise pitch, roll, throttle, smoke generator firing, height hold, and three spare. Height hold uses a simple barometric device to maintain the aircraft at the desired altitude. A binocular sight allows the pilot to see and control the Skeet up to a range of 2.7 nm (5 km).
Payload. Mission equipment comprises an X-band reflector in the rear fuselage. Sixteen 40-second smoke generators are fitted in the landing skid. Two wing-mounted pods can carry, for example, Simflak for training, and the fuselage can accept either a miss-distance indicator or a transponder. An autopilot, complete with a heading device, is under development for out-of-sight operation. An automatic parachute release can be provided.

Specifications

Wing span: 3.35 m
Wing area: 1.53 m^2
Length overall, including MDI aerial: 2.72 m
Height overall: 0.45 m

Weights:
(basic aircraft, empty) 42 kg
(parachute pack) 4.1 kg
(fuel) 9.1 kg
(mission equipment, normal) 8 kg
(max launching weight) 68 kg
Max level speed: 130 kts (241 km/h)
Stalling speed: 50 kts (93 km/h)
Max in-sight range: 2.7 nm (5 km)
Max range under control: more than 5.4 nm (10 km)
Endurance at max speed: 1 h 15 min

Status
In service with the British Army and overseas.

Manufacturer
Short Brothers Plc (Defence Systems Division), Belfast.
(For details see Directory of Suppliers)

Shorts Skeet Mark 1 on launcher

Shorts Skeet Mark 2

The Shorts Skeet Mark 2 has the addition of a horizontal surface on top of the fin.
Propulsion. Initially a 18.6 kW NGL WAEL engine was fitted. This has been replaced by a Norton NR731 27 kW engine which provides a higher maximum speed and increased endurance.

Specifications
Dimensions: as for Mark 1
Weights:
(basic aircraft, empty) 46.65 kg
(parachute pack) 4.1 kg
(fuel) 9.1 kg
(mission equipment, normal) 9.1 kg
(max launching weight) 72.6 kg
Performance:
(max level speed, full throttle) 163 kts (302 km/h)
(level speed, half throttle) 149.6 kts (277 km/h)
(stalling speed) 50 kts (92.6 km/h)
Max in-sight range: 5 km
Max range under control: 10 km
Endurance: (half throttle) 93 min

Shorts Skeet Mark 2 with NGL engine

Status
An initial production contract was announced in June 1987. A £750 000 follow-on order for the United Kingdom Ministry of Defence was announced in June 1990.

Manufacturer
Short Brothers Plc (Defence Systems Division), Belfast.
(For details see Directory of Suppliers)

Skyleader MATS-A

The MATS-A aerial target is a small high-wing monoplane. It has a GRP fuselage and the wings have a polystyrene foam core with vinyl covering. Tail surfaces are of balsa wood. All flying surfaces are covered with vinyl.
Propulsion. The power plant is a 10 cc single cylinder piston engine, driving a two-blade propeller. The fuel mixture is methanol/castor oil.
Launch and Recovery. The MATS-A is hand-launched. Normal recovery is by belly landing on the fuselage undersurface; a parachute recovery system is optional.
Guidance. Guidance is by line-of-sight radio command system, using a hand-held transmitter. Conventional moving surfaces provide aerodynamic control.
Payload. Mission equipment comprises a four-channel superhet receiver, operating on 27 MHz bandwidth, within which 12 spot frequencies are available. A rechargeable Ni/Cd battery pack and three or four proportional feedback servos provide outputs to control surfaces. The transmitter power (1 W) is adequate for operating distance of up to 1.1 nm (2 km). Other frequencies are available up to 72 MHz.

Specifications
Wing span: 1.73 m
Length overall: 1.22 m
Max launching weight: 3.85 kg
Max level speed: 78 kts (145 km/h)
Operating range: approx 1.7 nm (3.2 km)
Endurance: 15-20 min

Status
In service in the British Army.

Manufacturer
Skyleader Radio Control Limited, Croydon.
(For details see Directory of Suppliers)

Tasuma Targets

Tasuma produce a number of Air Targets which are described below.

Tasuma Delta Target
The Tasuma Delta target is a delta-wing canard with twin fins. The complete airframe is constructed using epoxy composites and the skin is a moulded sandwich. The wings have a hollow core with carbon spars and the fuselage uses GRP reinforced with carbon/Kevlar.
Propulsion. Two alternative power plants can be used. One is a Tasuma 35 cc flat twin two-stroke power plant developing 11.9 kW (16 bhp) and the other is a Norton-Wankel unit developing 26.1 kW (35 bhp). The fuel tank is integral with the fuselage in the form of a fabricated rubber fuel cell.
Launch and Recovery. The launcher is a Tasuma unit using high-stretch elastic cords, although other existing launch units could be used. Recovery is by skid landing or by parachute.
Payload. For visual enhancement 16 smoke flares are carried in an underslung pod constructed of Derakane and Kevlar.

Specification of Tasuma Delta
Wing span: 2.5 m
Length: 3.1 m
Weight dry: 32 or 38 kg
Max take-off weight: 75 kg
Speed range: 45 to 180 kts (83 to 320 km/h)
Operating range: (with optics) 4 nm (7.4 km)
Endurance: 1 h plus

Tasuma AGT 30/40 Air Target
The Tasuma AGT 30/40 Air Target was at one time known as the Gyr (a species of falcon).
Design. It is a high-wing monoplane with the fuselage manufactured using epoxy resin composites with Kevlar reinforcement. The wing, tail and fin have a foam core with epoxy skins and carbon-fibre strengthening. The constant chord wings have an Eppler 374 section while the tail uses the NACA 0012 section.
Propulsion. The power plant is a two-stroke engine which may use a single or twin-cylinder and a capacity of from 100 to 200 cc, as required. The engine drives a two-blade fixed pitch wooden propeller.
Launch and Recovery. The aircraft is launched from a trailer-mounted rail assembly, energy being derived from stretching elastic cords of about 1 inch diameter. Recovery is by parachute or on an under-fuselage skid, depending on the terrain.
Guidance. Guidance and control is via a radio command link using a 10 W PCM transmitter. There is a full fail-safe mode with a sequencing system for firing flares. Optical sighting enhancement is available for the ground operator.
Payload. The AGT 30/40 can carry eight 45-second coloured smoke canisters under the fuselage or infrared emitters at the wing tips. Miss-distance indicators are optional.

Tasuma AGT 30/40 Air Target

Specifications of the AGT 30/40

Wing span: 2.75 m
Length: 2.2 m
Weight, empty: 18 kg
Launch weight: 27 kg
Max level speed: 180-240 km/h
Stalling speed: 70 km/h
Endurance: 30-60 min

Tasuma GT10 and MGT20

The Tasuma GT10 is designed for AAA training and the MGT20 is designed for SAM and AAA training. These models are smaller than the AGT 30/40 described earlier. The GT 10 has a span of 6 ft and the MGT 20 a span of 8 ft. The MGT 20 has an integral starter for the engine.

Tasuma MGT20 air target - the GT 10 is similar but slightly smaller

Tasuma Delta Target

Status

In production.

Manufacturer

Tasuma (UK) Limited, Verwood, Dorset.
(For details see Directory of Suppliers)

TTL BTT-1 Imp

The Target Technology Limited (TTL) Imp is a hand-launched target.

Design. The Imp is a mid-wing delta monoplane with a sweptback vertical tail and no horizontal tail surfaces. The fuselage is of glass fibre-reinforced resin, with additional strengthening; the wings are of plywood over a polystyrene foam core. All flying surfaces are further covered with a tough vinyl material, and primary components are interchangeable to simplify replacement.
Propulsion. The power plant is a 12.5 cc glow-plug engine, driving a two-blade propeller.
Launch and Recovery. The Imp is hand-launched. Recovery is by conventional skid landing or, optionally, by parachute. The latter can be commanded manually or automatically in the event of target damage or failure of the command link.
Guidance. Guidance and control is via a radio command system using a hand-held transmitter developed from a Skyleader four-channel digital system giving independent proportional control of the elevons, engine and (optionally) the recovery parachute. A variety of frequencies is available.

Two TTL BTT-1 Imp hand-launched targets

Specifications

Wing span: 1.83 m
Length overall: 1.09 m
Height overall: 0.43 m
Propeller diameter: 0.33 m
Weight empty: 4.5 kg
Fuel: 0.45 kg
Max launching weight: 5.9 kg
Max level speed: 82 kts (153 km/h)
Stalling speed: 13 kts (24 km/h)
Endurance: 30 min

Status

More than 500 have been produced for the Royal Navy, British Army, the Royal Air Force, and the armed forces of Denmark, Egypt, Indonesia, Norway, Oman, Qatar, Saudi Arabia, and Turkey.

Manufacturer

Target Technology Limited, Ashford, Kent.
(For details see Directory of Suppliers)

TTL BTT-3 Banshee

The Banshee is a low-cost target system designed primarily to simulate the threat of missiles (including sea-skimming weapons) and aircraft for training SAM and AAA crews. The Banshee first flew in 1983 and is in service with the armed forces of Abu Dhabi, Canada, Egypt, Indonesia, Oman, Qatar, Saudi Arabia, Turkey, and the UK.

Towing capability. The Banshee has been modified to be able to tow a target banner or target sleeve.
Variants. As the Vindicator, the Banshee was marketed in North America by Boeing of Canada (now Bristol Aerojet).

A payload-carrying version of the Banshee is known as the Spectre.

A reconnaissance version is known as Hawkeye.

The Iraqi SARAB-3 target aircraft (qv) appears to be similar to the Banshee.

The Series 300 has a more powerful engine coupled with improved aerodynamics. Improved command and control systems include a radar-controlled height lock developed by Target Technology Electronics Limited which allows accurate sea-skimming down to a height of 5 m. The Banshee series 300 is in service with the Royal Norwegian Navy as a target for the Sea Sparrow surface-to-air missile.

Design. Banshee is a low-wing monoplane with clipped delta-wings and a sweptback fin. There are no horizontal tail surfaces. The wings have 50° of sweepback and a dihedral of approximately 3°. The entire airframe is constructed from glass-reinforced plastics but the elevons are of injection-moulded epoxy composite material. Payload bays are situated in the detachable nose cone, and in the centre and rear fuselage. All aircraft subsystems are watertight for recovery from the sea. The series 300 uses identical construction methods but the vertical fin is detachable to facilitate entry into the rear payload bay.
Propulsion. The power plant is one 19.4 kW (26 hp) NGL WAEL 342 two-cylinder two-stroke. The series 300 uses a 26.1 kW (35 hp) Norton single-rotor rotating-piston engine. Both engines drive a two-blade laminated fixed-pitch pusher propeller with a spinner. Fuel is carried in an integral fuselage tank with a capacity of up to 15 litres.
Launch and Recovery. Launch is achieved with an elastic power band catapult launcher which is normally mounted on a ground support trailer, but which can be removed and used separately (on board ship for example). Recovery is by belly landing on land or water, or by deploying a cruciform parachute. An under-fuselage flare pod acts as a landing skid and mission payload bay. The parachute mode can be commanded manually or automatically by a fail-safe system in the event of

TTL BTT-3 Banshee target on a Pakistani Navy vessel

TTL BTT-3 Banshee target with its mobile ground command unit

in-flight damage or failure of the radio command link. Maximum turnround time on land or at sea is 30 minutes after a return to the launch site.

Guidance. A VHF radio command system and line-of-sight optical tracking incorporating gyro-stabilised optics, achieves ranges of up to 5.4 nm (10 km). Two-axis aerodynamic control is achieved with elevon control surfaces. Auto-stabilisation with a height lock facility is available as an option. This incorporates rate gyros controlling yaw, pitch and roll which can be switched on and off during flight. A radio tracking system is available for target control over ranges of up to 10.8 nm (20 km). Banshee can also be operated with the Vega tracking system for control at ranges of up to 24.3 nm (45 km). Onboard power is supplied by rechargeable battery packs. The power source can be supplemented by the use of a Thorn EMI/TTL 300 W alternator.

Payload. Alternative mission loads include a flare pod able to contain up to 24 smoke and/or infra-red flares, or a chaff dispenser. Corner array sets can be carried, consisting of an optimised triple array of reflective corners giving a spherical cardioid polar diagram with phase-interacted global fine structure and echo peaks tuned to customer requirement. Up to three Luneberg lenses can be carried, providing radar enhancement in a range of bandwidths and attitudes. Acoustic and Racal Doppler radar miss-distance indicating (MDI) systems can be carried. Banshee can also tow target banner and target sleeve which can be fitted with the MDI system. The banner and sleeve are towed from the rear of the strengthened ventral fin.

TTL BTT-3 Banshee target

Specifications

Wing span: 2.49 m
Length overall: 2.84 m
Height overall: 0.86 m
Body diameter (max): 0.305 m
Propeller diameter: 0.61 m
Wing area, gross: 2.14 m^2
Weights:
(empty) 38.5 kg
(max payload) 22.7 kg
(max launching weight) 72.6 kg
Launching speed: 52 kts (97 km/h)
Max level speed: 174 kts (322 km/h)
Stalling speed: 39 kts (73 km/h)
Max range:
(optical tracking system) 5.4 nm (10 km)
(radio tracking system) 10.8 nm (20 km)
(Vega tracking system) 24.3 nm (45 km)
Endurance at max speed: 1 h 15 min

Status

In production. In service with the British Army and the armed forces of Abu Dhabi, Canada, Denmark, Egypt, Indonesia, Oman, Pakistan, Qatar, Saudi Arabia and Turkey. The Banshee series 300 is in service with the Royal Norwegian Navy and some other NATO armed forces.

Manufacturer

Target Technology Limited, Ashford, Kent.
(For details see Directory of Suppliers)

TTL BTT-9

The TTL BTT-9 is modelled on the Su-25 'Frogfoot' aircraft and is produced in three different sizes.

Launch and Recovery. The two smaller versions of the BTT-9 are both capable of being hand-launched, although the TTL Imp launcher (see earlier entry) may also be used. The larger version of the BTT-9 uses the Banshee launcher (see earlier entry). All three versions are recovered by conventional skid landing or (optionally) by parachute.

Design. The BTT-9 is a shoulder wing monoplane with straight taper-wings. The fuselage is of glass-reinforced plastic construction with areas of local reinforcement. The two smaller versions have wings constructed of a hardwood veneer over a foam core while their control surfaces and empennage are of a lightweight hardwood construction, covered in a tough vinyl material. The flying surfaces and all control and tail surfaces of the larger version are of composite construction.

Propulsion. The three versions of the BTT-9 are powered by engines of different sizes. The smallest version is powered by one 15 cc two-stroke glow plug engine producing 1.82 kW (2.4 hp) and driving a two-blade propeller. The next largest version is powered by one 32 cc two-stroke petrol engine driving a fixed-pitch, two-blade propeller. The largest version is powered by one TTL P-100 two-stroke petrol engine producing 6.81 kW (9 hp) and driving a fixed-pitch two-blade propeller.

Guidance. For guidance and control there is a digital five-function command system giving independent proportional control of engine speed, aileron, elevator and (optionally) parachute and payload activation. A variety of frequencies is available as is an optional programmable PCM system.

Payload. The avionic systems and main control servos of all three versions are contained within a low-cost, splashproof module to facilitate easy replacement in the field or at sea. Mission payloads for the two smaller versions include a 4 in/102 mm Luneberg lens, four

TTL BTT-9 Frogfoot target after launch

Specifications

Type	A	B	C
Wingspan	2.16 m	2.47 m	2.77 m
Length (overall)	1.68 m	1.92 m	1.16 m
Height (overall)	0.5 m	0.57 m	0.64 m
Propeller diameter	0.33 m	0.46 m	0.56 m
Weight (dry)	5 kg	7 kg	20 kg
Fuel weight	0.5 kg	1 kg	3 kg
Max take-off weight	5.5 kg	9.5 kg	34 kg
Max speed	85 kts (157 km/h)	92 kts (170 km/h)	121 kts (224 km/h)
Endurance	27 min	45 min	65 min
Range (optical tracking)	3.2 nm (6 km)	3.2 nm (6 km)	3.2 nm (6 km)

smoke tracking flares, frequency specific corner arrays and an acoustic miss-distance indicator. The largest version can carry a 102 mm or 178 mm Luneberg lens, 10 smoke tracking flares and four infra-red flares, as well as frequency specific corner arrays and an acoustic miss-distance indicator.

Status
In production.

Manufacturer
Target Technology Limited, Ashford, Kent.
(For details see Directory of Suppliers)

TTL BRT-8 Dervish

TTL BRT-8 Dervish fitted to represent a Mil Mi-24 'Hind D'

The BRT-8 Dervish is a scale model helicopter. Possible uses vary from air defence target simulation to weapon system R&D and the carriage of small payloads. Deliveries of the first units commenced in mid-1989.

Design. The BRT-8 Dervish is a conventionally configured helicopter. Rotation of the main and tail rotors is clockwise and the tail rotor is elevated. The construction is semi- and full-monocoque and is entirely of aircraft grade aluminium, riveted and bolted. Various profiles are available using lightweight unstressed interchangeable panels.

Propulsion. The power plant is one water-cooled, twin cylinder two-stroke 242 cc petrol engine delivering 12.9 kW (17 hp). Transmission is by an internal toothed belt. Onboard starting is fitted as standard and a 12 V charging system is available.

Guidance. For control there is a damped, teetering two-bladed system with a Hiller control rotor. Control movement is effected by the use of electromechanical servos of various types.

Specifications
Main rotor diameter: 3.26 m
Tail rotor diameter: 720 mm
Height:
(at main rotor) 1.01 m
(at tail rotor) 1.25 m
Width at skids: 0.8 m
Main rotor area: 8.348 m^2
Tail rotor area: 0.407 m^2
Disk loading: 6.7 kg/m^2
Unladen weight: 56 kg
Payload: 20 kg
Speed: up to 43 kts (80 km/h)
Rate of climb: up 200 m/min
Endurance: 2 h 30 min

Status
In production.

Manufacturer
Target Technology Limited, Ashford, Kent.
(For details see Directory of Suppliers)

TTL UAV-Towed Targets

TTL targets which can be towed by UAVs with (from top) a towed banner, sleeve and target container. An acoustic miss-distance indicator is shown at the bottom

TTL produces a range of towed targets suitable for use with UAVs. They include a towed banner, a towed sleeve and a towed body.

Banner. The towed banner is a high visibility target for use with cannon-type weapon systems. It was designed to have a low initial cost and is non-recoverable.

Sleeve. The towed sleeve has a well-proven aerodynamic shape that is maintained by an integral flexible steel ring at the front opening. To enhance acquisition by radar the sleeve is completely covered with metallised threads woven into the surface.

Towed Bodies. A range of towed bodies is available to meet varying requirements. Most are able to carry Luneberg lenses, infra-red and smoke tracking flares. Also available is a towed acoustic miss-distance indicator for use with banners and sleeves.

Status
In production.

Manufacturer
Target Technology Limited, Ashford, Kent.
(For details see Directory of Suppliers)

UNITED STATES OF AMERICA

Beech UAV Targets

In addition to manufacturing piloted aircraft, Beech has been designing and producing UAV targets since 1955. Beech has built more than 7500 air targets, including more than 2200 MQM-61A Cardinals (now out of production), more than 4600 AQM-37s, and more than 1390 MQM-107s.

Beech AQM-37
Winner of a 1959 joint US Navy/US Air Force design competition, the AQM-37 (then designated XKD2B-1) is a supersonic, air-launched expendable target, powered by a liquid-propellant rocket motor. It entered US Navy service in 1963 and has been in continuous production ever since. Current US Navy launch aircraft are the A-4 Skyhawk, A-6 Intruder and F-4 Phantom. Versions produced for the US services are as follows:

AQM-37A. The original basic US Navy target version (Beech Model 1019). Became operational in 1963; a total of approximately 3950 had been delivered by 1992. Twenty Model 1019As were delivered in 1975-76 in a 'Sea Skipper' configuration, capable of flying at heights of between 10 and 300 m for weapon system evaluation. These targets had an autopilot incorporating a homing device for guidance, and a radar altimeter for height control. Ten other AQM-37As were modified in about 1981, under the name Challenger, to operate at 24 385 m altitude and Mach 3. These had enlarged tail surfaces, and a refined autopilot for higher speeds and altitudes; they exceeded the quoted altitude and speed, and also demonstrated end-of-flight dives simulating attack by air-to-surface missiles. The final US Navy AQM-37A version was the Beech Model 1104.

AQM-37C. This is the current basic US Navy target version (Beech Model 1104). The AQM-37C has payload space in the nose section for radar augmentation, scoring and active emitter payloads. It incorporates a digital autopilot with improved flight control features, in addition to on-command controlled dive angle and ground command of cruise heading control. A total of 735 has been ordered with deliveries beginning in 1986 and extending until July 1993.

In November 1988 the Brunswick Corporation Defense Division, at Costa Mesa, California, was awarded a $6 988 764 firm fixed-price contract to manufacture and deliver 10 AQM-37C missile targets in order to qualify as a second source for the AQM-37C.

Beech AQM-37C supersonic target

AQM-37C Variant. Eight were ordered, as the Beech Model 1105, for a 1984 evaluation resulting in the development and subsequent production of kits to modify standard AQM-37Cs into Variants, known as Extended Performance targets. The kits replace the vertical stabilisers, the wings and the radome and mid-nose sections with similar sections covered by insulating materials. These materials allow the performance of the Extended Performance versions of the AQM-37C to exceed Mach 4 at 30 480 m. Extended Performance targets have been successfully flown on test at Mach 4.5 and 34 140 m after launch from a US Navy F-4 Phantom.

The following description applies to the basic AQM-37C:

This air-launched expendable target is a mid-wing monoplane, of canard configuration, with slim-delta main-wings at the rear. Of double-wedge section and bonded aluminium honeycomb construction, these have 76° sweepback on the leading edges, 0° dihedral and incidence, and full span ailerons. The movable foreplanes are of modified double-wedge section. There is a fixed end plate vertical stabiliser at the tip of each main-wing. The centre fuselage is cylindrical with an ogival nose section and access panels in the aft nose and aft fuselage for access to the electronics and propulsion systems. There is an underbelly tunnel for the rocket engine cartridge-operated start valves, plumbing, battery, and antennas.

Propulsion. The power plant is a Harley Davidson LR64 P-4 two-chamber liquid propellant variable thrust rocket engine. Four preset thrust levels, ranging from 0.31 to 3.78 kN, are available from the booster and sustainer chambers. There are three stainless-steel propellant tanks, for the nitrogen pressurant, the mixed amine fuel (MAF-4) and IRFNA oxidiser.

Launch. The target is air-launched from a launcher adapted to A-4, A-6 or the F-4.

Recovery. Non-recoverable.

Guidance. Guidance and control is available preprogrammed or it can be supplemented with a command control guidance system responding to ground operator-generated commands. Flight is normally terminated by an automatic aerodynamic destruct system, which also operates from ground command if required by range safety.

The AQM-37C has a digital autopilot with improved flight control features which allows operations up to Mach 4 and 30 480 m. This is achieved by an optimum energy management programme and a command control capability which permits dual launch and in-flight control. Dives of 15 to 65°, and pull out from dives, can be initiated upon command from the ground.

Payload. Payloads can be installed in the AQM-37C to augment the radar cross-section capabilities of the target in four separate frequencies in addition to scoring and command/control kits.

Specifications

Wing span (over fins): 1 m
Foreplane span: 0.65 m
Length overall: 4.27 m
Height overall: 0.66 m
Body diameter (max): 0.33 m
Wing area, exposed: 0.87 m^2
Weight empty:
(Standard) 125.9 kg
(Extended Performance) 136 kg
Max payload weight: 12.2 kg
Max launching weight:
(Standard) 266.4 kg
(Extended Performance) 276.6 kg
Cruising height and speed, Standard: from Mach 0.87 (573 kts; 1062 km/h) at 305 m to Mach 3 (1719 kts; 3186 km/h) at 24 385 m
Cruising height and speed, Extended Performance: up to Mach 4 (2341 kts; 4339 km/h) at 30 480 m
Range, including glide to impact:
(Standard) more than 212 nm (393 km)
(Extended Performance) 229 nm (424 km)
Endurance, powered: 8 min

Status

In service with the US Navy and some other countries.

Manufacturer

Beech Aircraft Corporation, Wichita, Kansas.
(For details see Directory of Suppliers)

Beech Aircraft MQM-107 Streaker

The MQM-107 Streaker is the American military designation for the Beechcraft Model 1089. The Swedish Air Force designation is RB 06 and RB 06B Girun.

The Model 1089 was the winner of the US Army's VSTT (Variable Speed Training Target) competition with the award in April 1975 of the first instalment of a three-year production contract. Manufacture, under the designation MQM-107A, began in December 1975, and deliveries started in the following April. Contracts included design, development and delivery of the air vehicles, ground support equipment, spares and ancillaries, as well as operation and maintenance of the system by the company's subsidiary, Beech Aerospace Services Inc.

Role. The principal function of the MQM-107 is to provide a variety of threat simulations for training with missiles and automatic weapons. It is the primary subsonic missile training target for the US Army, and is also used in research and development testing of weapon systems. The Streaker is used extensively by the US Air Force for air-to-air combat training. It is designed to operate at altitudes from sea level up to 12 200 m and at speeds of up to Mach 0.8.

Versions. The following versions have been produced:

MQM-107A. The initial production version has a 2.85 kN Teledyne CAE J402-CA-700 turbojet and 5.13 m long fuselage as standard. Deliveries to the US Army, completed in early 1979, totalled 385. It was supplied also to the United Arab Emirates (with a TRI 60-2 engine), Jordan, the Republic of Korea, Sweden and Taiwan.

MQM-107B. Introduced in 1982, with the primary reason for redesignation being the installation of a 3.7 kN Microturbo North America TRI 60-2 Model 074 engine, permitting an increase in maximum operating speed and manoeuvring capability. Other changes include a longer fuselage with more volume available for payload, and systems improvements arising from experience with the MQM-107A. Deliveries to the US Army (139) and US Air Force (70) began in 1984 and were completed in 1986. Export customers include Egypt, Sweden and Taiwan.

MQM-107C. Similar to the MQM-107B, with the long fuselage, but powered by 2.85 kN Teledyne CAE J402-CA-700 turbojet. A total of 69 was delivered to the US Army in 1985. (This was a non-standard model, created to use up existing J402-CA-700 engines.)

MQM-107D. As for MQM-107C, but with a 4.27 kN Teledyne CAE 373-8 engine. The initial US Army contract was awarded in October 1985, with deliveries beginning in January 1987; follow-on contracts in March (for 108) and September 1986 (for 107, bringing total to 305) extended deliveries into 1990. Another US Army contract was awarded in late 1989 for 80 additional MQM-107D targets, but with a 4.27 kN Microturbo TRI 60-5 engine installed. The MQM-107D is used by both the US Army and the US Air Force.

MQM-107E. In February 1993 the US Army Missile Command issued a RFP for some 200 units of the improved-performance MQM-107E following qualification flight tests which took place at the White Sands Missile Range, New Mexico, during 1992. Data provided in the Specifications table refers to this version. Power is provided by a 4.72 kN Microturbo TRI 60-5 or Teledyne CAE 373-8B.

Exports. The international version of the Streaker is still designated MQM-107B but it contains all the improvements of the later versions used by the US Army except that the Microturbo TRI 60-2 is used to maintain commonality.

Design. The MQM-107 is a reusable variable speed target. Its all-aluminium airframe is a low-wing monoplane, with sweptback wings, fin, rudder, tailplane and elevators. The engine is suspended on a pylon beneath the centre of the mainly cylindrical fuselage. Modular design is used throughout, with flat section wing and tail surfaces of bonded honeycomb (fixed surfaces) or foam-filled aluminium (ailerons and elevators). The nose and tail cones are ogival in cross-section. Improved waterproofing is available on the MQM-107B and later models.

Propulsion. The power plant is a single turbojet (details provided under the model listings), pylon-mounted in an under-fuselage pod. The standard fuel capacity is 245 litres. Wing insert fuel tanks, available as a kit, can add a further 113 litres.

Launch and Recovery. The MQM-107B is surface-launched, using RATO booster, from a lightweight zero-length launcher. The booster is jettisoned approximately 2 s after ignition, after accelerating the target to approximately 220 kts (408 km/h). There is a command recovery system, using a 1.65 m diameter drogue and 15.24 m main parachutes stowed in the rear of the fuselage. A recovery cycle can be initiated by remote command, by a 6-second loss of the command carrier, or by an electrical power loss.

Guidance. Guidance and control systems provide for ground control with the trajectory preprogrammed for

Beech MQM-107E Streaker at launch

the first 20 seconds after RATO ignition. The flight controller is then in control and is provided with all pertinent flight information by radio link from sensors located in the air vehicle, and he can command vehicle manoeuvres and recovery. In flight, the guidance and control system stabilises automatically around the roll, yaw and pitch attitudes and provides an altitude hold mode. Flight control developments include a terrain-following guidance capability which has demonstrated extremely low altitude flight profiles. The MQM-107 has provisions for a high-*g* autopilot, to extend the manoeuvring and high-*g* envelope of the vehicle. The flight controller can select either constant airspeed or constant altitude high-*g* manoeuvres, and 6 *g* manoeuvres can be maintained during use with air-to-air or surface-to-air weapon systems. An extended payload section available on the long fuselage versions also provides easier access to the payload and vehicle electronics, and has specialised waterproofing provisions for seawater recoveries.

Payload. A total of 0.092 m^3 is available in the short fuselage version (0.136 m^3 on the B, C and D models) for mission equipment and core avionics. The vehicle's principal function is to tow a variety of targets for missile training and evaluation. Two radar, infra-red or visually augmented tow targets, or a 2.23 m banner with a corner reflector, can be carried on each mission and towed separately up to 2440 m behind the MQM-107. Wingtip and pod-mounted augmentation and scoring devices can be carried. Streaker serves as an aerial target for air defence systems such as Chaparral, HAWK, Improved HAWK, Patriot and Stinger, and the Vulcan air defence gun system. Air-to-air missiles such as the AIM-9 Sidewinder, AIM-7 Sparrow and AMRAAM can also use the MQM-107. The Streaker also serves as an aerial target for international air defence systems such as Crotale, Mistral, Rapier, Skyguard/Sparrow and Skyguard/35 mm. A 15-litre smoke/oil tank (26.5 litres in MQM-107B and subsequent models) is carried for visual augmentation. Typical external payloads can include pyrotechnic optical plume simulators (POPS), infra-red augmentor boom assemblies, point source and plume infra-red augmentation, flare/chaff dispensers, passive and active radar augmentation, bi-static radar reflector pods, gunnery tow banners, and miss-distance indicator score pods.

Specifications

- MQM-107E

Wing span: 3.02 m
Length overall: 5.51 m
Body diameter, max: 0.38 m
Height overall: 1.54 m
Tailplane span: 1.58 m
Mission equipment: 45 kg internal or 136 kg external
Max launching weight: 662 kg
Nominal launching weight: 454 kg
Operating speed range: 200-560 kts (370-1037 km/h)
Operating height range: sea level to 12 200 m
Endurance (max): 2 h 18 min

Status

In production. In service with Egypt, Jordan, South Korea, Sweden, Taiwan, United Arab Emirates, and the US Army and Air Force. Total deliveries of the MQM-107 (all versions) had exceeded 1390 by the beginning of 1992.

Manufacturer

Beech Aircraft Corporation, Wichita, Kansas.
(For details see Directory of Suppliers)

Continental RPVs Air Targets

In 1983 Continental RPVs became the contractor for the US Army National Training Center at Fort Irwin, California, supplying close air support system (CASS) aerial targets based on a scale representation of the MiG-27 'Flogger-D' combat aircraft. A further contract awarded in 1989 requires Continental RPVs to supply radio-controlled fixed- and rotary-wing aerial targets for continued support of the National Training Center and also for use by the US Army in Europe.

Aircraft Modelled. The most popular targets in the current Continental RPV family are one-fifth representations of the MiG-27, the Sukhoi Su-17 'Fitter-C' and Su-25 'Frogfoot', the Fairchild A-10, the General Dynamics F-16, the Vought A-7 and the Mil Mi-24 'Hind-D' in both autogyro and full helicopter forms. Other targets include the Bell AH-1, the F-111A, a one-half scale Cessna 150, and a 'self-propelled' aerial tow target. All these targets can reach an altitude of more than 3660 m.

Role. These scale aircraft are used primarily as part of a close air support training system (CASS) which includes one-fifth scale, remotely piloted airframes that carry MILES sensors and an onboard smoke generation system for MILES hit indication. The aircraft are also equipped with an Air-Ground Engagement System (AGES) to provide a 'shoot back' capability. These scale aircraft are integrated into laser-based, force-on-force engagement simulation training systems.

Design. Targets are manufactured in four sizes: one-ninth, one-seventh, one-fifth and one-half scale. All have fibreglass fuselages and fabricated-wood wings. The one-ninth targets run on nitromethanol fuel while the others use a petrol/oil mixture. Manoeuvrability is effected by elevator control and onboard payloads can include a gas or smoke grenade, pyrotechnics, an infra-red augmentation device, a cine-camera, a smoke hit indicator, a near-miss indicator, radar beacons and reflectors.

The aircraft have glass fibre fuselages and tail units with plywood wings. Standard control range is 3 km but this can be increased to 15 km by the introduction of an optical/video controller.

Launch and Recovery. All the targets are pneumatically launched by a non-hazardous zero-length pneumatic system that uses no pyrotechnics, JATO or explosive gas, and return for a skid or belly landing.

Specifications

- one-fifth MiG-27 'Flogger-D' aerial target

Wing span: 2.997 m
Weight empty: 19.05 kg
Payload: 13.6 kg
Fuel, standard: 0.9 kg
Speed range: 26 to 104 kts (48 to 193 km/h)
Endurance at max speed: (standard fuel) 30 min
Control range: (each controller) 8.1 nm (15 km)

Status

In production. In service with the US Army, Navy, Air Force, Marines and National Guard.

Manufacturer

Continental RPVs, Barstow, California.
(For details see Directory of Suppliers)

Continental Su-17 'Fitter' target

Continental F-16 target

Continental MiG-27 'Flogger-D' target

Continental A-10 target

Flight Systems QF-100 Super Sabre UAV Conversions

The conversion of F-100 aircraft as target UAVs is performed by Flight Systems Inc under contract from the US Air Force Sacramento Air Logistics Center, McClellan Air Force Base, California. This full-scale aerial target (FSAT) programme is for a multi-service target to provide air-to-air and ground-to-air missile evaluation and combat crew training. The QF-100 succeeded the F-102 target in these roles.

Programme. The programme was initially conducted by the Sperry Corporation and involved the delivery of 100 Super Sabres in four different configurations. Of these, configuration No.2, converted from the single seat F-100D, is the standard US Air Force target version. No.1 incorporates additional cockpit controls to permit evaluation of system performance from within the cockpit. No.3 is the same configuration as No.2, except that it is converted from the two-seat F-100F. No.4 incorporates a drone formation control system (DFCS) for multiple-target missions involving up to six QF-100s. The full-scale engineering development programme included DT & E (Development, Test and Evaluation) and IOT & E (Initial Operational Test and Evaluation), and was carried out at Sperry's facility at Phoenix-Litchfield Airport, Arizona. Conversions are now being made at FSI's Aircraft Modification and Test Center at Mojave, California.

Deliveries. Deliveries by Sperry began on 13 March 1981 to Tyndall Air Force Base, Florida, and the first unmanned flight was made on 19 November 1981. Development testing for the DFCS was carried out at Holloman Air Force Base, New Mexico; final test and evaluation took place during 1983 at Tyndall Air Force Base. The initial operational capability with Tactical Air Command was achieved in late 1983.

Follow-on Contract. In May 1984 the QF-100 follow-on production contract was awarded to Flight Systems, which delivered a total of 209 targets commencing from mid-1985.

Propulsion. The power plant is one Pratt & Whitney J57-P-21A turbojet, rated at 75.4 kN with afterburning.

Launch and Recovery. Launch and recovery is via normal runway take-off and landing. Automatic take-off and landing, utilising the IBM DFCS, has been demonstrated at Holloman Air Force Base, and will be installed on the Eglin Air Force Base Gulf range as part of the latter's drone control update system.

Guidance. Guidance and control is via dual Vega command guidance and telemetry systems. There is a fully redundant digital tracking and control system for command/telemetry link, used in conjunction with the AN/FPS-16 ground-based range radar. The simultaneous control of multiple targets has been demonstrated. The Honeywell digital flight control computer (FCC) system incorporates air data sensors, an SDP-175 processor, analogue/digital and digital/analogue converters, a power supply, and the necessary interface electronics. The FCC provides eight longitudinal/vertical and four lateral/directional modes, and interface between aircraft systems and command/telemetry system. Automatic modes are provided for take off, loss of command carrier, take off abort, and other safety modes. There are redundant power systems and dual autopilot channels. The digital FCC permits automatic checkout of many primary autopilot functions; it also provides a flexible system for incorporating target system functions, or for adaptation to other target programmes. Control is exercised by a mobile ground station for take off and recovery, and by a fixed ground site for guidance to, over and from the target range. The manoeuvre programmer can be preprogrammed for multiple manoeuvres, in any required sequence, and provides backup for the FCC system.

Payload. Mission equipment incorporates remotely operated smoke, braking and explosive destruct systems; the Digidops miss-distance scoring system is standard. Two types of scoring camera are operational, covering forward and aft areas to provide missile approach angle, velocity and miss-distance information. Manoeuvre destruct and explosive destruct systems are incorporated. The visual augmentation (smoke) system, is operable at any altitude or power setting. Radar and infra-red augmentation are not required, due to the size of the aircraft. An LVSS (laser vector scoring system), one of several systems tested in 1983, can provide a directional parameter to help further in evaluating missile performance. A DLQ-3B ECM pod and ALE-40 infra-red/chaff pod are incorporated to provide realistic evaluation of missile performance against anticipated countermeasures. A drone formation control system is incorporated to permit formation flying of two or more targets to provide a realistic challenge for missiles equipped with seeker heads.

Specifications

Wing span: 11.82 m
Length overall, including probe: 16.54 m
Height overall: 4.95 m
Wing area, gross: 35.79 m^2
Mission operational take-off weight: 14 060 kg
Max speed at altitude: Mach 1.3 (745 kts; 1381 km/h)
Operating height range: 60-15 240 m
Range, nominal (guidance radar range-limited): 120 nm (222 km)
Normal mission endurance: 40-55 min

Status

In service with the US Air Force.

Manufacturer

Conversions by Flight Systems Inc, Newport Beach, California.
(For details see Directory of Suppliers)

Honeywell QF-106 Full-Scale Aerial Target (FSAT)

In June 1986 the Defense Systems Division of the Sperry Corporation (now the Defense Avionics Systems Division of Honeywell Inc) was chosen by the US Air Force as prime contractor for the conversion of about 200 of the now-obsolete F-106 fighter aircraft. The target version is designated the QF-106. A new transponder was developed to work with the Gulf Range Drone Control System at Tyndall AFB, Florida, and the Drone Formation Control System at White Sands Missile Range, New Mexico. A signal processor/vehicle interface Line Replacement Unit was developed for use on both the QF-106 and MQM-107 sub-scale aerial target. Two identical digital flight control computers were developed for the primary and backup flight control systems.

Specifications

Wing span: 11.67 m
Length overall: 21.54 m
Height overall: 6.17 m
Max take-off weight: 16 062 kg
Max level speed at 11 000 m: 1324 kts (2455 km/h)
Service ceiling: 17 375 m
Range: 1000 nm (1850 km)

Status

Being supplied to the US Air Force. There are contract options for additional conversions.

Prime Contractor

Honeywell Defense Avionics Systems Division, Albuquerque, New Mexico.
(For details see Directory of Suppliers)

Northrop Basic Training Target (BTT)/MQM-3C and MQM-36

The US military designations of the Northrop Basic Training Target (BTT) are MQM-3C and MQM-36. This target, which first flew in July 1945, is in worldwide use, and was manufactured for the armed forces of 25 countries. More than 65 000 vehicles of this type were built. Also known as the Shelduck or KD2R-5, it has been used as a training device for anti-aircraft gunnery, as well as a target for surface-to-air missiles.

Northrop delivered a modified version of the BTT to the US Army. Known as the MQM-33C, it incorporates a Northrop-developed G-band command control system; 161 MQM-33Cs and 18 G-band ground control stations had been produced by early 1985.

The following description applies to the standard BTT:

Design. The BTT is a remotely controlled recoverable target. It is a high-wing monoplane of aluminium alloy construction. There is no dihedral. Wing incidence is 1° at the root and −2° at the tip. Ailerons and elevator are servo-operated by D-9 actuators.

Propulsion. The power plant is a 59.7 kW (80 hp) Northrop O-11-3 flat-four two-stroke engine, driving a two-blade fixed-pitch wooden propeller. There are steel integral fuel tanks in the mid-fuselage, with a capacity of 44 litres. The refuelling point is in the fuselage, forward of the wing.

Launch and Recovery. The BTT is surface-launched from land or ship, by either a rotary catapult or a zero-length launcher. Recovery is by parachute which is released by radio command. The engine is stopped

Northrop Basic Training Target (BTT)

and the parachute deploys automatically in the event of serious damage by gunfire, or any loss of radio control or electrical power.
Guidance. Guidance and control is via an AN/ARW-79 radio command guidance system, with automatic altitude hold. Visual or radar tracking (radar or FM type tracking systems or equivalent) can be employed.
Payload. Mission equipment includes a 28 V battery for all electrical power. Optional augmentation devices include two wingtip radar reflector pods, visual and infra-red flare kits, towed gunnery sleeves with real-time scoring, and six towed infra-red targets. An acoustic scoring system and auxiliary decoder can provide three spare command and control links in addition to the single spare channel of the basic system. An altitude hold control, a precision barometric sensor which detects variations in altitude and provides control corrections to secure constant flight altitude, can also be provided.

Specifications
Wing span: 3.5 m
Length overall: 3.85 m
Height overall: 0.76 m
Propeller diameter: 1.12 m
Wing area, gross: 1.74 m^2
Weight empty: 123 kg
Max launching weight: 181 kg
Max zero-fuel weight: 133 kg
Max landing weight: 154 kg
Max level speed at sea level and max cruising speed: 195 kts (360 km/h)
Stalling speed: 58 kts (108 km/h)
Max rate of climb at sea level: 1341 m/min
Service ceiling: more than 8230 m
Range at sea level with max fuel: 183 nm (338 km)
Average flight duration: 55 min

Status
Production complete. In service in 25 countries.

Manufacturer
Northrop Corporation, Hawthorne, California.
(For details see Directory of Suppliers)

Northrop Chukar MQM-74

The Chukar (pronounced 'chucker') is named after a bird, a rock partridge found in the Western USA. The Chukar was designated the MQM-74 by the DoD and has been used for a number of roles as well as a target for SAM and AAA training. In total 1800 MQM-74A were built, followed by the MQM-74B experimental model for a 500 kts (926 km/h) target. The MQM-74C is an improved version also called Chukar II. More than 1400 MQM-74Cs have been delivered to the US Navy since 1984. Other Chukar II customers included Germany, Greece, Iran, Italy, Japan, Netherlands, Saudi Arabia, Spain, and the UK. Japan's Chukar IIs can carry a Martin Marietta laser tracker pod at each wingtip.
Chukar II Variants. Modified versions of the MQM-74C were tested and operated as reconnaissance, electronic warfare and strike remotely piloted vehicles under the US Navy's Persistent Anti-Radiation Missile (PARM) and US Air Force's TED programmes. Under US Navy contract, Northrop has produced nearly 1000 of a version known as the BQM-74C (see next entry) for use by the US Navy as a cruise missile simulator and standard aerial training target, making more than 4300 MQM/BQM-74 Chukars manufactured in all.
Role. The basic Chukar II/MQM-74C target aircraft was designed to meet requirements for a small, lightweight target for anti-aircraft gunnery, surface-to-air and air-to-air missile training and weapons system evaluation.
NAMFI Deployment. Chukar II is in service at the NATO Missile Firing Installation (NAMFI) on the island of Crete, where it is used to train crews of AAA, SAM and air-to-air weapons. Meteor of Italy provides Chukar II services at the Salto di Quirra range in Sardinia.
Design. The Chukar II is a turbojet-powered recoverable target. It is a shoulder-wing monoplane, of aluminium construction. There is no dihedral. The detachable wings each have an electrically actuated aileron. The tapered circular-section body has an underslung air intake duct. There is an inverted 'Y' tail unit, with 30° of anhedral on the tailplane, electrically actuated elevators and a fixed vertical fin.
Propulsion. The power plant is a Williams International WR24-7 (J400-WR-401) turbojet, rated at 0.8 kN. The fuel tank is in the centre of the fuselage, having a capacity of 61.3 litres.
Launch and Recovery. Zero-length launching is accomplished by means of two Mark 91 Mod 0 JATO rockets and a ZL-5 launcher. Normal recovery is by automatic pull-up followed by the main parachute deployment, initiated automatically in emergencies such as the interruption of the continuous radio signal or loss of the parachute command channel. An alternative method consists of direct main parachute deployment and is initiated automatically on loss of electrical power. The main parachute is housed in the fuselage immediately aft of the wing, with an automatic disconnect on impact.
Guidance. Radio command is used for the guidance system. Out-of-sight control is by automatic stabilisation and command, with radar tracking; in-sight control uses visual acquisition aids. There is proportional feedback stabilisation and control system for pitch and bank. Engine throttle position, altitude hold initiation and recovery system initiation are controlled by audio tone signals. Components include a receiver, a decoder, an autopilot, aileron and elevator servos, and altitude hold and airspeed pressure transducers. The command control antenna is in the upper forward fuselage.
Payload. Target mission equipment uses electrical power from an engine-driven alternator passed through a rectifier-regulator. A 28 V nickel-cadmium battery secondary power source is used during the glide phase. Onboard acquisition and tracking aids include fore and aft Luneberg lenses for passive radar augmentation, four wingtip-mounted Mark 28 Mod 3 infra-red flares, pyrotechnic infra-red plume augmentors, active L-band augmentation, and a smoke system, designed to improve visual detection. The main payload compartment is in the front fuselage between control equipment bay and fuel tank. An improved manoeuvrability package (IMP), successfully flight tested in 1976, has a closed loop control device installed in the flight control system, enabling the Chukar II to perform constant *g* manoeuvres at any of five selected levels up to and including 6 *g*. A low-cost infra-red tow target, for use with the Chukar II system, is in production. One can be attached to each wingtip and towed approximately 30 m behind the Chukar II. A reusable active RF augmentation tow target system is in use for various RF guided missile systems. Production of the active tow targets began in 1979.

Northrop MQM-74C Chukar II target on the deck of a US Navy ship - note the power connections to the fuselage which indicate that it is being prepared to launch

Specifications
Wing span: 1.76 m
Length overall: 3.87 m
Body diameter: 0.35 m
Height overall: 0.71 m
Wing area, gross: 0.74 m^2
Weight empty: 128.4 kg
Fuel: 50.3 kg
Max launching weight:
(excluding boosters) 182 kg
(including boosters) 223 kg
Max level speed at 6100 m: 515 kts (954 km/h)
Max level speed at sea level: 475 kts (880 km/h)
Economic cruising speed at sea level: 250 knots (463 km/h)
Max rate of climb at sea level with full fuel: 1780 m/min
Service ceiling: 12 200 m
Range at max speed:
(at sea level) 205 nm (380 km)
(at 6100 m) 330 nm (611 km)
Range at economic cruising speed: (at sea level) 245 nm (454 km)
Average flight duration: more than 1 h

Status
Production complete. In service with the US Navy, other armed forces and other countries.

Manufacturer
Northrop Corporation Aircraft Division, Hawthorne, California.
(For details see Directory of Suppliers)

Northrop Chukar III BQM-74C/E

The BQM-74C is a US Navy target version of the MQM-74C Chukar II with added air-launch capability and preprogrammed flight profiles. It can be used as a cruise missile simulator, in training pilots for air-to-air combat, and as a target for SAM and AAA crews. The airframe is essentially similar to that of the MQM-74C, but with some components made of GRP.
Production. Since 1980 Northrop has produced more than 1000 BQM-74Cs for the US Navy, including 350 in 1985. A further 132 (including some BQM-74C/Recces) were delivered in 1986; 70 were delivered in 1987. A fixed-price contract placed by the US Naval Air Systems Command in June 1989 called for 129 BQM-74C to be produced by January 1992; the contract was worth $36 417 000. A further contract placed by the US Naval Air Systems Command in March 1990

called for 149 targets and support software and equipment with a contract value worth $38 200 000.

The 1990 contract marked the last BQM-74C as the US Navy changed to the BQM-74E, starting with the 1991 production of 200 units.

Payload. The BQM-74C target system includes pylon adaptors for air-launching from several types of aircraft, and payload kits for mobile sea range (MSR) operations. MSR enables fleet commanders to conduct realistic 'war games' on the open sea.

Role. The BQM-74C's role in these exercises is to simulate cruise missiles fired from multiple directions and altitudes. The BQM-74C's missions can be preprogrammed, and the target is capable of air- or surface-launch from beyond the horizon of target ships. It has been flown at an altitude of 3 m while simulating cruise missile flight profiles.

Description. The following description applies to the standard BQM-74C target version:

The BQM-74C is a radio-controlled or programmable, recoverable, multi-role target. It is a shoulder-wing monoplane of aluminium alloy and glass-reinforced plastic construction. The Northrop G-9224-080 wing section has an 8 per cent thickness/chord ratio. There is no dihedral. The non-swept wings each have an electrically actuated aileron. The semi-monocoque aluminium body houses all equipment, power plant and the fuel tank. The nose and tail skins are removable for access to equipment and the power plants. There is an underslung engine air intake. The inverted 'Y' aluminium tail unit comprises a fixed vertical fin, fixed tailplane halves (anhedral 30°) and two electrically actuated elevators.

Propulsion. The power plant is a Williams International WR24-7B (J400-WR-403) turbojet initially rated at 0.8 kN. The Chukar III/BQM-74E uses a Williams International WR24-8 (J400-WR-404) turbojet rated at 1.07 kN. The pressurised fuel tank in the centre fuselage section has a capacity of 61.3 litres.

Launch. The Chukar III/BQM-74E can be launched from ground or shipborne launcher by use of a zero-length launcher and a pair of JATOs. It can also be air-launched from the Grumman A-6E Intruder or McDonnell Douglas TA-4J Skyhawk (one under each wing); it is also compatible with the underwing launchers of the Lockheed DP-2 Neptune or DC-130A Hercules. The target can be recovered over land or water by the use of a parachute.

Northrop BQM-74C Chukar III in air-launched version

Guidance. Guidance and control can be either manual or automatic. In manual control an operator guides the target through use of a ground-based command and control station and an onboard receiver and transmitter. The operator controls the target pitch and bank angles, altitude and engine rpm, and can initiate altitude hold and parachute recovery. In automatic control, the target is preprogrammed and can fly a mission autonomously. Components include a Northrop digital avionics processor, vertical and yaw rate gyros, an AN/DKW-3 integrated target control system or any AN/DKW-4 control system, aileron and elevator servos, and an altitude hold pressure transducer. The BQM-74E programmed autonomous control is the same as that used for the BQM-74C.

Internal Power. Electrical power is provided by an engine-driven alternator through a rectifier/regulator. A 28 V Ni/Cd battery is used as a secondary power source. Onboard acquisition and tracking aids include fore and aft Luneberg lenses, passive radar augmentation, and a smoke system. The main payload compartment is in the forward fuselage section. Equipment includes a locator beacon, a radar altimeter, a seeker simulator (to duplicate cruise missile emissions), a radar transponder for IFF, and scoring equipment. Other provisions include a flight profile programmer with UHF command override; active J-band, B-band, L-band and X-band radar augmentation; and a Tacan receiver. The system also includes payload kits with flotation gear for mobile sea range (MSR) operations.

Specifications
(A) surface-launch; (B) air-launch
Wing span: 1.76 m
Length overall: 3.95 m
Body diameter (max): 0.36 m
Height overall: 0.72 m
Wing area, gross: 0.74 m^2
Weights:
(basic operating weight empty)
(A) 127 kg
(B) 133 kg
Max mission load: 78.5 kg
Fuel: 50.3 kg
Max launching weight:
(A) 233 kg
(B) 199 kg
Performance (at max launching weight (A) with J400-WR-402 engine, (B) with -404):
(max level speed at 6100 m)
(A) 500 kts (927 km/h)
(B) 530 kts (982 km/h)
Economic cruising speed:
(A) 300 kts (555 km/h)
Service ceiling:
(A) 9150 m
(B) 12 200 m
Max range: (A and B) 450 nm (833 km)

Status
In production. In service with the US Navy, Republic of China, France, India and Spain.

Manufacturer
Northrop Corporation Aircraft Division, Hawthorne, California.
(For details see Directory of Suppliers)

RS Systems Scaled Targets

RS Systems make a number of targets representing real aircraft at scales from one-fifth to one-ninth. Remotely actuated flares can be carried by the targets for IR tracking purposes.

RS One-Ninth Scale Targets
RS Systems produces one-ninth scale representations of two Russian and two Western aircraft: the MiG-27 'Flogger-D' and Su-25 'Frogfoot', the F-16 and AV-8B.

Propulsion. The power plant is one 1.5 kW (2 hp) 10 cc engine, mounted in the nose.

Payload. Four principal training payloads can be used: a Tactical Hit Indicator (THI), a Manual Hit Indicator (MHI), AIRMILES, or an infra-red source. When flown aboard one-ninth or one-fifth series targets, the THI automatically initiates up to 10 flash-bang-smoke 'target disabled' indications per flight, its sonic detection system providing users with an adjustable target size from 2 to 24 m and adjustable 'hits to disable' (1 to 15 rounds through the hit zone). Each time a preset 'disabled' criterion is met, the gunner is rewarded with an instant flash-bang-smoke feedback. The MHI can also provide up to 10 similar indications per flight, via push-button actuation at the target controller's transmitter. AIRMILES, which is compatible with the US Army MILES, enables air defence users to conduct training in virtually any open area, without live fire logistics and costs. It will signal 'target disabled' automatically with the same flash-bang-smoke indication as the THI and MHI, the 'disabled' criteria in its microprocessor varying according to lethality of the weapon fired against the target. AIRMILES has an optional eye-safe 'fire back' feature for use in 'cover and concealment' training by air defence units, using a laser gun mounted on the target and controlled by the target's pilot.

Specifications - one-ninth scale targets
Wing span:
(MiG-27, Su-25) 1.68 m
(F-16) 1.14m
(AV-8B) 1.6 m
Fuselage length:
(MiG-27, Su-25) 1.83 m
(F-16) 1.63 m
(AV-8B) 2.04 m
Max width:
(MiG-27, Su-25) 0.25 m
(F-16) 0.23 m
(AV-8B) 0.15 m
Max payload, MiG-27, Su-25: 2.3 kg
Max launching weight:
(MiG-27, Su-25) 3.61 kg
(F-16) 3.4 kg
(AV-8B) 4.5 kg
Max level speed (all): 78 kts (145 km/h)
Endurance (all): 12-24 min

RS Systems One-Seventh Scale Target
This aerial target is a representation of the Mil Mi-24 Hind-D battlefield helicopter. Hand-launched, it flies as an autogyro (ie the main rotor is not engine driven). The nose-mounted engine being offset to starboard driving a small tractor propeller. Lift is provided by the fixed wings and the freely-rotating main rotor. There is no tail rotor. Although it looks like a helicopter, it is controlled like a fixed-wing model and transition training for operators of fixed-wing aerial targets is minimal.

Propulsion. The power plant is one 17.7 cc methanol-fuelled standard glow plug engine, driving a tractor propeller.

Specifications - one-seventh scale target
Rotor diameter: 1.6 m
Wing span over rocket pods: 1.22 m

RS Systems one-ninth scale MiG-27 'Flogger' target

RS Systems one-ninth scale F-16

RS Systems target version of the Mil Mi-24 Hind-D

Fuselage:
(length) 2.16 m
(max width) 0.23 m
Max payload: more than 0.9 kg
Max launching weight: 4.5 kg
Max level speed: 43 kts (80 km/h)
Visual control range: 500 m
Endurance: 20 min

RS one-fifth scale targets

One-fifth scale targets representing the MiG-27 and Su-25 are available, being the replacement for RS Systems' earlier one-seventh scale series. The increased size, speed, range and endurance afford air defence trainees a low-cost realistic target that can be hand or mechanically launched from an unprepared training site.

Propulsion. The power plant is one 59 cc engine, mounted in the nose.
Payload. Mission equipment is as described for one-ninth scale targets (qv).

Specifications one-fifth scale targets

Wing span (both): 2.84 m
Fuselage length:
(MiG-27) 3.05 m
(Su-25) 2.78 m
Max width:
(MiG-27) 0.27 m
(Su-25) 0.38 m
Height overall:
(MiG-27) 0.81 m
(Su-25) 0.76 m
Max payload (both): 11.3 kg
Max launching weight (both): 14.5 kg
Max level speed: 95 kts (177 km/h)
Stalling speed: 22 kts (41 km/h)
Visual control range: 2.7 nm (5 km)
Endurance: 30 min

Status

In production.

Manufacturer

RS Systems, Beltsville, Maryland.
(For details see Directory of Suppliers)

Teledyne Ryan Model 124 Firebee 1

The Firebee 1 was developed as a joint US Army, Air Force and Navy project. A total of 1277 early Q-2A and KDA versions were built for all three US services and for the Royal Canadian Air Force.

Current Version. Development of the current BQM-34A improved Firebee for the US Air Force and US Navy began on 25 February 1958. The first production model flew on 25 January 1960.

By April 1990 a total of 6816 Firebee 1s had been produced, including the early Q-2C and KDA versions. The total includes Firebees supplied to NATO for use in a missile test programme and those supplied to the Japan Defense Agency to support the training of missile and gunnery crews in both surface-to-air and air-to-air operations. The latter are built by Fuji Heavy Industries and are known as the BQM-34AJ (see separate entry under *Japan* in this section).

Target Vulnerability. Since the Firebee has been in operational use, target presentations have been made to virtually every surface-to-air and air-to-air weapon in the US arsenal. To reduce the vulnerability of the target itself and decrease the number 'killed', extensive use has been made of targets towed behind the Firebees on cables or mounted on the wingtips. Such extra targets are IR or Radar enhanced as appropriate to the missile being used.

DoD Designations. The US Air Force designation for the Firebee 1 is BQM-34A. The US Army designation is MQM-34D. Current Firebee targets for the US Navy incorporate a Motorola Integrated Target Control System (ITCS) and have the designation BQM-34S. US Air Force BQM-34A targets have a Vega tracking and control system.

Design. The Firebee 1 is a cantilever mid-wing monoplane of aluminium alloy semi-monocoque construction. The constant chord wing has two spars and leading edge droop. There is no dihedral or incidence and sweepback is 45°. The single spar ailerons use electromechanical servo-actuators. The wingtips are detachable as there is provision for wingtip extensions. The body is tapered and has an oval cross-section, with a keel under the central portion to absorb landing impact. The horizontal tail surfaces are swept (45° at quarter-chord) and the multi-spar fin (swept at 48°) has a glass fibre tip housing the guidance and control antenna. There can be a ventral fin under the tail cone.

Propulsion. Various engines can be used. Firebee 1s in the current inventory are powered by one 7.56 kN Teledyne CAE J69-T-29, one 8.54 kN Teledyne CAE J69-T-41A, one 10.9 kN General Electric J85-GE-7, or one 11.8 kN J85-GE-100 turbojet. An integral fuel tank is located in the forward fuselage (capacity 378 litres) and there is provision for an auxiliary fuselage tank (capacity 68 litres).

Teledyne Ryan BQM-34A Firebee launch

Launch and Recovery. The Firebee 1 can be air-launched from a suitably modified aircraft or ground-launched using a 50.3 kN (nominal) JATO bottle. The US Navy has launched BQM-34Ss from ships under way at 15 knots (27.5 km/h). A two-stage parachute recovery system operates automatically in the event of loss of radio carrier wave from the remote-control station, engine failure, or on command by the operator. The recovery system incorporates a disconnect which releases the parachute from the target on contact with the ground or water.

Guidance. Remote-control methods include a choice of radar or UHF radio. The normal method is through the Motorola ITCS (Integrated Target Control System) for the US Navy BQM-34S and the Vega DTCS (Drone Tracking and Control System) for the US Army and Air Force. The target can be controlled from either a manned aircraft or from a surface station. Remote command includes the activation of special scoring and augmentation equipment carried in the aircraft. Basic commands consist primarily of on/off functions received by an onboard receiver unit and relayed to the appropriate subsystem. Other types of remote command and tracking systems can include a microwave command and guidance system to control the Firebee 1 beyond the line-of-sight from a ground station via an airborne relay station. The BQM-34A is equipped with a three-axis flight control system for tactical air combat simulation which gives the target an ability to perform 4, 5 or 6 *g* manoeuvres. Other systems include active and passive radar augmentation, wingtip-mounted infra-red sources and engine infra-red plume devices. The Radar Altimeter Low Altitude Control System (RALACS), when added to the Firebee 1's control system, permits precision low altitude flight down to 6 m over water and 30 m over land.

Internal Power. The Firebee 1 relies on electrical power derived from a 28 V, 200 A, DC engine-driven generator, with power for the control system vertical gyro furnished by a 400 Hz 115 V 250 W inverter. A 28 V lead-acid battery provides power for the electrical components of the recovery system and for control during the pre-landing glide phase.

Payload. Specialised mission equipment can include a wide range of 'building block' units that includes visual or radar-reflecting banner targets, radar or infra-red 'Towbee' towed targets or wingtip-mounted infra-red pods, two underwing drop tanks, wingtip tow launchers, camera pods, scoring equipment, flares or other forms of infra-red augmentation, or reflector pods for radar augmentation. The BQM-34A/S can be equipped with adjustable travelling wave tube or solid-state amplifiers for use as radar echo enhancers in a wide range of frequencies.

Specifications

Wing span: 3.93 m
Length overall: 6.98 m
Body diameter: 0.94 m
Height overall: 2.04 m
Wing area, gross: 3.34 m^2
Weights:
(empty) 680 kg
(basic, gross) 934 kg
(max launch) 1134 kg
Never-exceed speed: 635 kts (1176 km/h)
Max level speed at 1980 m: 600 kts (1112 km/h)
Max cruising speed at 15 240 m at 816 kg all-up weight: 547 kts (1015 km/h)
Stalling speed, power on, at 816 kg all-up weight: 101 kts (187 km/h)
Max rate of climb at sea level at 1000 kg all-up weight: 6707 m/min with J85-GE-7 turbojet
Operating height range: 3 to 18 300 m
Endurance at 15 240 m: 1 hour 15 min (including 3 minutes glide after fuel expended).
Max range: 692 nm (1282 km)
Flotation time with 25% fuel: 24 h

Status

In production. In service with the US Army, US Navy and US Air Force. Produced under licence in Japan (qv).

Manufacturer

Teledyne Ryan Aeronautical, San Diego, California.
(For details see Directory of Suppliers)

YUGOSLAVIA (SERBIA AND MONTENEGRO)

M-2M Radio-controlled Target

The M-2M radio-controlled target is for 20 mm light AAA training.

Design. The M-2M is a high-wing monoplane with straight wings and conventional tail surfaces. The fuselage is rectangular in cross-section but all surfaces are slightly rounded. A fixed tail wheel undercarriage is used.

Propulsion. Power is derived from a nose-mounted Aero 1000 MR-G engine driving a two-bladed propeller. The fuel tank is immediately behind the engine in the nose and contains a special fuel mixture supplied by the manufacturer.

Transportation. For transport, the wing is removable and is normally held in place by bungee cords. For transport, the aircraft is carried in a crate measuring 2.54 × 0.54 × 0.51 m. A second box contains the radio control equipment, spare parts, tools and accessories.

Launch and Recovery. The M-2M can take off from a circular area up to 200 m in diameter. The aircraft requires a minimum circle of 100 m within that area and for take off wind speed should not be more than 5 m/s. A grass surface can be used for take off and landing with the grass not more than 50 mm high.

Guidance. Once airborne, the aircraft is controlled by a Varioprop 12 radio controller with a maximum control range of 700 m. Maximum altitude is 500 m. It is possible to tow a small target sleeve from the M-2M.

Specifications

Wing span: 2.45 m
Length: 1.65 m
Weight: approx 5 kg
Max horizontal speed: approx 30 m/s
Take-off length: approx 30 m
Endurance: up to 1 h
Max climb angle: approx 20°
Max dive angle: 50°
Max operating altitude: 500 m

Status

In service with the armed forces.

Contractor

YUGOIMPORT, Belgrade.
(For details see Directory of Suppliers)

YUGO M-2M radio-controlled target

PRM-200 Programmable Rocket Target

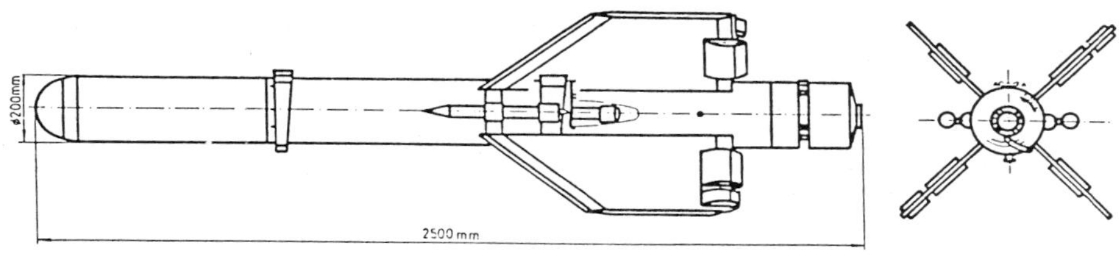

PRM-200 programmable rocket target

The PRM-200 programmable rocket target has a self-contained guidance system and once launched is non-recoverable.

Role. It is used as a target for radar or IR-guided AAM or medium-range SAM, and for radar-layed AAA.

Payload. It can be fitted with two optical and two infra-red or smoke flares to augment the infra-red signature. The radar signature is provided by a Luneberg lens in the forward and aft sections of the rocket target.

Launch. The PRM-200 can be launched from the Super Galeb trainer aircraft or similar aircraft using a launching kit. It is launched at speeds of Mach 0.5 to 0.8. After launch a solid booster rocket burns for 0.9 s, after which the solid sustainer rocket burns for 100 s. Maximum rocket speed is Mach 0.7. If the target is not hit, a programmer unit activates a self-destruct mechanism.

Guidance. The guidance system is self-contained, programmable and comprises a barometric altimeter and two free gyroscopes. Guidance instructions are programmed into the unit before fixing the PRM-200 onto the launcher. Two preprogrammed pitches or yaws can be entered.

Specifications

Fin span: 800 mm
Body diameter: 200 mm
Length overall: 2.5 m
Length as target: 2.244 m
Weight: 90 kg
Performance:
(max speed) Mach 0.7
(altitude range) 300-7000 m
Starter motor burn time: 0.9 s
Main motor burn time: 100 s
Target range:
(launch altitude 300 m) 22 km (±0.5 km)
(launch altitude 7000 m) 48 km (±2 km)

Status

In service with the armed forces.

Contractor

YUGOIMPORT, Belgrade.
(For details see Directory of Suppliers)

ADDENDA

The addenda section consists of items which were not able to meet the printing dates for the main sections, some of which are finalised well before publication date. In addition the new 'Miscellaneous Air Systems Trainers' section was finished too late for printing at the end of the Aviation Systems section, and for this year only appears here. In the next edition all addenda entries which are still relevant will be moved forward to the relevant specialist section.

Small Arms Training

INDIA

ZEN SATS-1 Trainer

Zen Technologies and Computers Ltd make the Zenteknix SATS-1 small arms trainer which allows aiming and firing practice at a monitor screen. Weapons such as rifles and revolvers can be used, and recoil simulation is provided. Sound of firing and battle can be provided, and any number of targets can be produced on the screen together with playback facilities and hit/miss calculations.

Status
Available.

Manufacturer
Zen Technologies and Computers Ltd, Secunderabad.
(For details see Directory of Suppliers)

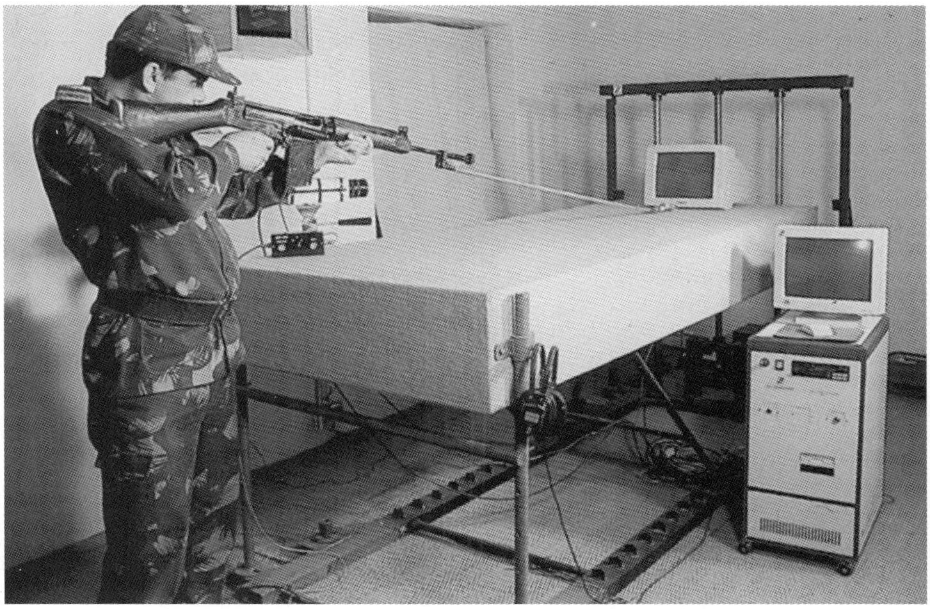

ZEN small arms trainer

Training Ammunition

INDIA

ZEN Training Hand Grenade

The Zenteknix ME-365 is a training hand grenade employing a small pyrotechnic cartridge instead of the main charge. The case can be re-used up to 100 times.

Status
Available.

Manufacturer
Zen Technologies and Computers Ltd, Secunderabad.
(For details see Directory of Suppliers)

Air Defence Training

UNITED KINGDOM

Inovex Multi-Role Crew Trainer (MRCT)

The MRCT is one of a family of air defence trainers designed by Inovex and its partner Simis Ltd. It consists of projectors, a wide FoV screen, and an Instructor Station. Air Defence weapons such as Rapier, Stinger, Starburst, Blowpipe, Javelin, SA-7, -8, -9, Chaparral, and AAA can be sited in view of the curved screen. Outside world visual scenes and aircraft targets are then projected on the screen and the weapons carry out their normal sighting, tracking and firing drills.

Status
Available.

Manufacturers
Inovex Technology, Winchester.
Simis Ltd, Godalming.
(For details see Directory of Suppliers)

Tank and Other Driving Simulators

SWITZERLAND

Oerlikon-Contraves ADAMS Driver Trainer

Oerlikon-Contraves make the Advanced Driving And Manoeuvring Simulator (ADAMS) which is designed for basic driver training for heavy trucks and special vehicles. It consists of a floor-mounted cubicle with a replica driver's cab and a visual display system with three collimated monitors. Image Generation is by CGI, and the driver's seat has a motion device mounted behind it and capable of giving small angular changes. ADAMS can be operated as a single unit or as fire training cubicle supervised by one IOS.

Status
Available.

Manufacturer
Oerlikon-Contraves, Zürich.
(For details see Directory of Suppliers)

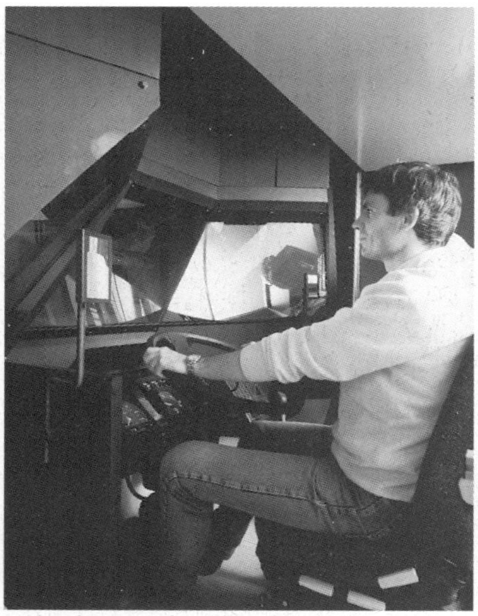

Oerlikon-Contraves ADAMS truck driving simulator

Visual display from Oerlikon-Contraves ADAMS truck driving simulator

Command and Staff Trainers

CANADA

CAE Tactical Management System (ITEMS)

CAE Electronics make the Intractive Tactical Environment Management System (ITEMS™) which is a software system that configures and runs a synthetic tactical environment, sometimes referred to as an 'electronic battlefield'. ITEMS allows the user to create a tactical scenario by creating scenario elements and using an off-line Database Management System (DBMS) to store the data until needed for real-time play.

Combat Decisions. The decision-making ability of

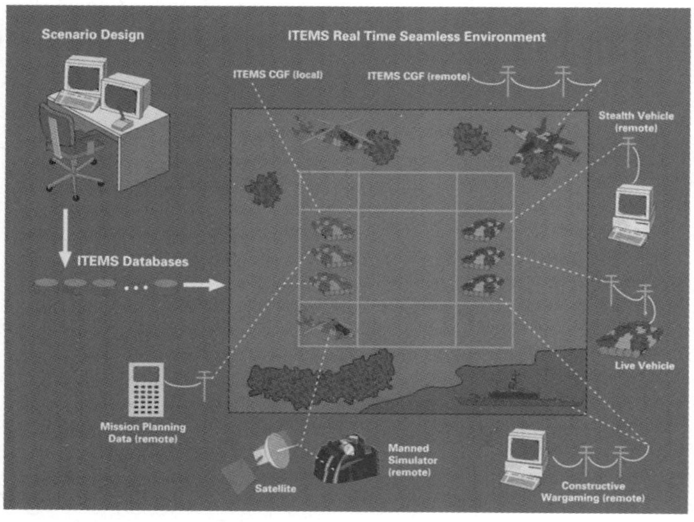

CAE ITEMS overall concept for a synthetic tactical environment or 'electronic battlefield'

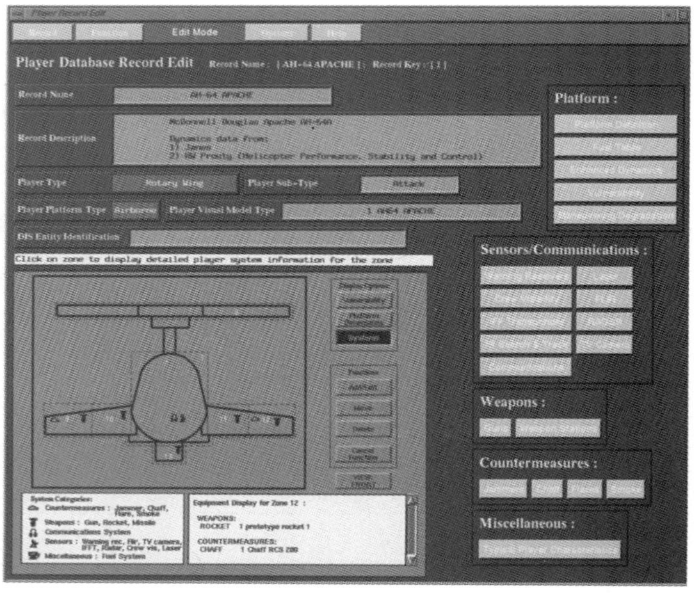

CAE ITEMS - Example of platform programme, in this case an AH-64 Apache helicopter

soldiers, commanders, drivers, weapon operators and pilots is controlled through a real-time rule-based system. Rules are written by the user and are executed in real-time play. The user can also take over a combat element manually, becoming an active participant in a battle scenario.

Networking. ITEMS can operate in a stand-alone mode or networked to other sites. It can be linked to simulators, computer programmes, and real units and vehicles. All of these units and facilities can be used interactively in a complex real-time battle scenario.

Real-World Database. ITEMS uses both terrain and cultural features, generated from sources such as DLMS DTED/DFAD data, and the later Project 2851/ MIL STD 1821 formats.

Visual Display. A map display and a 3-D view of the battle area are provided, selectable by the user.

Scenario Control. Pages presented on a control screen are used for scenario control. They have buttons and menu bars that can be used either with a mouse or touch-screen.

Data Recording. Any variable can be recorded for later analysis. Data reduction allows selective recording so that the user is not swamped with data. Continuous, triggered and one-shot data can be recorded. CAE also offers as an option their Relational DataBase Management System (RDBMS) which can be interrogated using industry-standard Structured Query Language (SQL).

Status
Available.

Manufacturer
CAE Electronics, Montreal.
(For details see Directory of Suppliers)

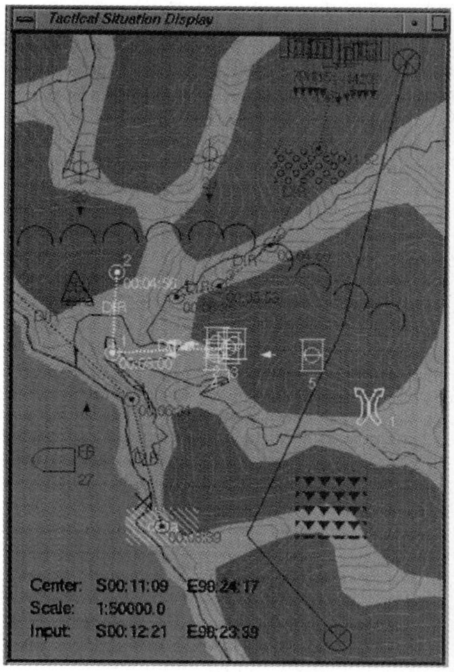

CAE ITEMS map display

CAE ITEMS 3-D display with combat in progress

Flight Simulators and Flight Training Devices

UNITED STATES OF AMERICA

ETC Simulators and Centrifuges

The Environmental Tectonics Corporation (ETC) manufacturer a number of training devices and simulators based on a cockpit module mounted on a rotating arm. The cockpit module is suspended in gymbals and has complete 360° freedom in pitch, roll, and yaw. Some examples of this product line follow:

ETC Advanced Spatial Disorientation Demonstrator (ASDD)

The ASDD is an ETC Model 2000 Gyrolab with cockpit controls and an outside-world visual system. It is designed to produce flight motion sensations to pilot trainees, including visual and motion perceptions which cause spatial disorientation.

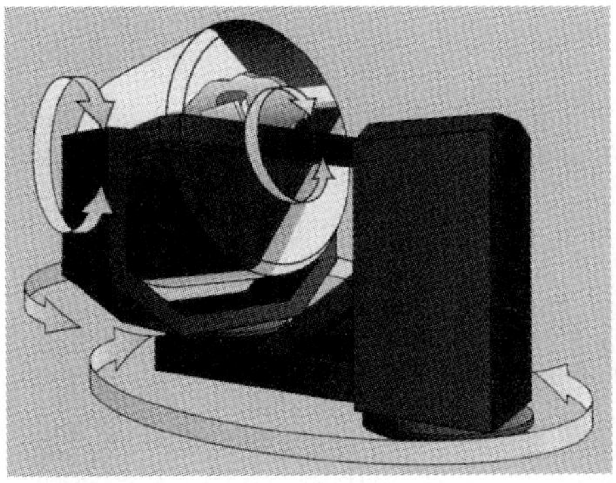

Drawing of ETC GYROSIM flight simulator showing centrifuge principle of sustained-G motion system

Cockpit. T-38, F-15 or F-16 are available, others on request.

Visual Image Generation. CGI with texture, shading and occulting, 3000 polygons at 30 Hz, DMA-derived real-world database. Day, night, cloud, rain, fog and lightning provided.

Visual Display System. FoV 120 × 53°, resolution 5.3 arcmin/pixel.

Motion System. ETC-patented all-electric system giving 360° in pitch, roll and yaw. The rotating centrifuge arm is capable of giving up to ±2.8 *g* at the cockpit.

IOS. 4 CRTs, profile and scenario selection, Record/Replay facility.

ETC Advanced Spatial Disorientation Demonstrator (ASDD) at USAF Brooks AFB

ETC GYROSIM Flight Simulator
The ETC GYROSIM™ flight simulator was developed from the earlier Gyrolab series. It is similar to the ASDD described earlier but has increased performance in G. It is capable of producing +4 Gz continuously and ±3 G at an onset rate of 4 G/second.

ETC G-FET High-G Flight Simulator
The ETC G-FET™ is a higher-G version of the GYROSIM. It is capable of producing a sustained 7.2 G and ±6 G at an onset of 6 G/second.

Status
Available. The ASDD is in service at the USAF School of Aviation Medcine at Brooks AFB, San Antonio, Texas.

Manufacturer
Environmental Tectonics Corporation, Southampton, Pennsylvania.
(For details see Directory of Suppliers)

USAF Materiel Command Simulators

In addition to the MS-1 mission simulator described in the main Flight Simulation section, the following devices are also operated or co-ordinated by the USAF Materiel Command's Aeronautical Systems Center (ASC) at Wright-Patterson AFB:

Air Mobility Command Simulator Upgrade Programme
This is a programme involving upgrades to allow simulators in Air Mobility Command (AMC) to be certificated to FAA simulator levels (see the tables at the beginning of the Flight Simulation section). This is in order to transfer an increased proportion of training from the aircraft themselves to the simulators. Aircraft concerned include KC-10, KC-135, C-5 and C-141.

Large Amplitude Research Simulator (LAMARS)
The Large Amplitude Multi-mode Aerospace Research Simulator (LAMARS) consists of a single-seat fighter cockpit with visual displays driven by CGI, mounted on a motion platform.
Motion Platform. The LAMARS motion system gives ±25° in pitch, roll and yaw, and ±10 ft (3.3 m) of heave and sway. It consists of a 30 ft long horizontal beam driven by hydraulic rams at the aft end of the beam; this provides the heave and sway motions. The cockpit section is mounted on the forward end of the beam and is mounted on gymbals in order that the rotational motions can be provided.
Cockpit. The cockpit is flexible and reconfigurable. About one hour is needed to change instrument panels and other components in order to change aircraft type. CRT displays are used extensively. The ejection seat is also modular and can be quickly changed.
Visual Display System. Several display systems may be used on LAMARS. The largest is a 20 ft diameter cockpit-mounted spherical screen. Four projectors are employed, generally three for outside-world CGI and one for targets. Maximum possible FoV is 266 × 108°. Two extra laser target projectors are also employed, so that three independent targets can be presented at once.
Image Generation. Martin Marietta Compu-Scene IV.
Location. Building 145, Area B, Wright Field, Wright-Patterson AFB.
Contractor. Northrop Corporation, California.
Cost. Contract cost was about $3 972 000.
Support Personnel. 14 support personnel are provided by EAI Services.

Wright Laboratory TACS/IMPACT Programmes
Simulators are used in the Tactical Aircraft Cockpit Simulator (TACS) and Integrated Mission Precision Attack Cockpit Technology (IMPACT) programmes. The TACS programme is for air-to-air combat and uses seven Silicon Graphics IRIS (Integrated Raster Imaging Systems) and six high resolution CRTs. The IMPACT programme also uses a pair of 10 inch square screens, and is configured with synthetic aperture radar for the air-to-ground role. The intention is both to learn from Desert Storm experiences and also to provide a single pilot crew with the same information and capability as a two-crew aircraft. Contact for these programmes is Captain Scott Vadrais on +1 513 255 2725.

Wright Laboratory Transport Aircraft Cockpit (TRAC) Simulator
The Transport Aircraft Cockpit (TRAC) simulator consists of a C-141 cockpit shell with four 21 inch and one 16 inch CRT displays for out-of-window and instruments.
Visual Image Generation. Silicon Graphics.
Visual Display System. Mitsubishi, Cypress, California.
Control loading. Servos and Simulation, Inc.
Controls. Centre stick, side stick or wheel (yoke).
Recording. 30 Hz.

Armstrong Laboratory SIRE facility
The Armstrong Laboratory has a Synthesised Immersion Research Environment (SIRE) consisting principally of a 40 ft-diameter half dome.
Visual Display System. 6 projectors, FoV 150 × 70°.
Image Generation. 6 channels of Onyx CGI.
Resolution. 1280 × 1024 pixels per channel, 60 Hz.
Programmes. Programmes include Fusion Interfaces for Tactical Environments (FITE), Brain-Actuated Control (BAC), Computerised Anthropomorphic Research and Data (CARD), bio-acoustics, data gloves, measures of workload, tactile displays, and situational awareness. A two-nation programme includes French-manufactured display control devices and will have a joint US/French cockpit.
Images. SIRE can portray anything that is generated on a GP graphics system. Therefore, scenes such as NASA-produced graphics of Mars terrain, the interiors of prototype cockpits and other vehicles, and mission rehearsal scenarios can all be displayed.
Contact. Captain Tracy O'Grady-Walsh, +1 513 255 2725.

Status
DoD R & D.

Location
USAF Wright-Patterson AFB, part of Materiel Command's Aeronautical Systems Center.
(For details see Directory of Suppliers)

Image Generation Systems

UNITED KINGDOM

RWS Reality

Real World Simulation's latest Reality CGI system is PC-based. A single circuit board gives a complete IG, and extra identical Reality boards can be added to provide more power per channel.
Multiple Channels. A single PC can house several channels each driving a separate display device. Each channel can comprise one or more Reality boards depending on power required. Channels can be synchronised to produce simultaneous updating.
Renderers. A Parallel Renderer divides the processing task evenly between processors, taking advantage of the parallel architecture of the system. A Reality scanline renderer is claimed to be unique and processes one line at a time, breaking up the line into spans between points where polygon edges cross the line. This technique is claimed to have advantages over conventional Z-buffer systems, particularly where polygon size increases, and also provides anti-aliasing.
Texture. Fully textured images are produced in real time. Textured polygons are available with lighting and fog effect, and may be mixed with flat shaded or smooth shaded polygons in the main database. Textures are MIP-mapped and map size is up to 1024 × 1024 pixels.
Overlays. A 16 colour full screen 2D overlay can super-impose menus, icons or animated graphics over the 3D background image. This is carried out in synchronisation with the 3D image.

Real World Simulation's Reality CGI showing a truck

Real World Simulation's Reality CGI showing a general driving scene

Real World Simulation's Reality PC-based CGI system, general view of capabilities showing textures and modelling

Specifications

Computer: Board format is a single-board PC/AT, processor is a Motorola 88110 RCIS, and the overlay processor is a TMS 34010
Channels: Unlimited, with up to 8 boards per channel and up to 32 processors per channel
Image Update Rate: Up to 60 Hz
Polygons: Up to 100 000 displayed polygons per second, or 3333 polygons at 30 Hz update, 1667 at 60 Hz
Latency: 50 ms at 30 Hz
Resolution: Up to 1024 × 768 (786 kpixels)
Data Compatibility: MultiGen™ compatible via database converter
Moving Model: Unlimited
Scene Management: LoD culling, bounding boxes
Hidden Surface Removal: Parallel Scanline algorithm
Colour: 24 true colour bits per pixel, 15 overlay colours plus transparency
Z-Depth Resolution: 32-bit floating point
Lighting and Time of Day: Ambient, diffuse and multi-directional lighting, real-time time of day
Weather Effects: Real-time visibility variation

Status

Available.

Manufacturer

Real World Simulation, St Albans.
(For details see Directory of Suppliers)

Visual Display Systems

BELGIUM

BARCO Projection Systems

The name BARCO stands for the Belgian American Radio Corporation, the company being founded in 1934 to produce radio receivers. In 1980 BARCO started making video projection systems, in 1983 producing data projectors capable of displaying computer images as well as video.

In 1987 they produced a graphics projector capable of reproducing images from work stations and CGI systems. BARCO projectors display video standards such as PAL, SECAM, NTSC 3.58/4.43, S-VHS(Y/C), and HDTV signals such as ACTV, IDTV, EDTV, HiVision, etc. They are compatible with a number of video sources such as VCR, laser-disc (video/HDTV), satellite receivers, direct input from video cameras, and so forth.

Three types of BARCO CRT projector, and an LCD light-valve projector which displays all colours through one lens

Status

Available and in extensive service.

Manufacturer

BARCO, Kuurne.
(For details see Directory of Suppliers)

A small section of back-projected display at the Stockholm control centre for Swedish Railways. 50 BARCO projectors are used, and the display panel is 131 ft long (40 m) (Photo: Studio toivo steen)

CANADA

Liquid Image MGR2 Head-Mounted Display

The MGR2 Head-Mounted Display (HMD) has a FoV of 84° Horizontal × 65° Vertical, and a resolution of 240 × 720 (172,800 pixels). Its display employs a full colour active matrix LCD. Its video signal can be composite or analogue RGB, and inputs can be in NTSC or PAL.

Status
Available.

Manufacturer
Liquid Image Corporation, Winnipeg.
(For details see Directory of Suppliers)

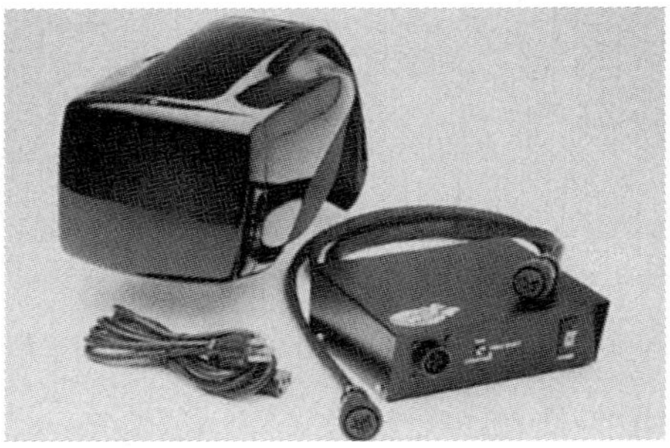

Liquid Image Head-Mounted Display Equipment

Liquid Image Head-Mounted Display with replica of scene displayed in the background

UNITED STATES OF AMERICA

Kaiser Helmet-Mounted Display

Kaiser Electro-Optics (KEO) manufacture the SimEye™ family of Helmet-Mounted Displays (HMD); Thomson Simulation and Training is system integrator and contributes technology such as HDTV and multiplexed fibre optic image transmission. The main components are an image collimation system and a head-tracker mounted on the helmet. An Image Generator such as a CGI system is needed and is connected to the helmet through fibre optic bundles (simpler images can be generated by sources mounted on the helmet if required).

Collimation System. This consists of two semi-transparent eye-glasses which use a patented cholesteric liquid crystal collimator which produces a direct view transmission efficiency greater than 20%, in order to see fixed objects such as controls and instrument panels.

CRT System. The two CRTs are ¾ inch in diameter and have a resolution greater than 1000 lines. To obtain colour, a switchable RGB colour filter is added in front of the CRT, and switching is provided by two LCD cells inserted between three polarising filters.

Fibre Optic Image Transmission. A technique known as dynamic multiplexing is used to reduce the unwanted texture effect of the individual fibres. This consists of rapidly moving the image at the entrance to the fibre optic bundle, and at the exit to the bundle moving the image in the opposite direction so that, overall, the image appears not to move. The unwanted texture effect of the fibres then moves rapidly with respect to the image and becomes less visible.

LCD Projector. An LCD Projector with CRT E-O addressing is used as an image source. It has a single exit lens, a luminosity over 1000 lumens, a resolution of 1280 × 1024 pixels, and geometric distortion correction.

Head Tracker. An electro-magnetic sensor is mounted on the helmet and is used in conjunction with a fixed E-M source over the pilot's head.

Eye-Slaving Oculometer. An oculometer may be added in order to slave any high-resolution area of image to the actual angle-of-look at any one time. This is achieved by using a low-intensity eye-safe IR source which, when reflected from the eye pupil, allows the angle-of-look to be sensed.

Luminous Intensity (Brightness). The image brightness is greater than 6 Ft Lamberts.

Resolution. 1 arcmin/pixel at a 25° circle of image, or 2 arcmin/pixel at a 50° image circle.

Instantaneous FoV. From 60° Horizontal × 40° to 120 × 80° depending on customer requirements.

Weight. Less than 2.5 kg overall helmet weight.

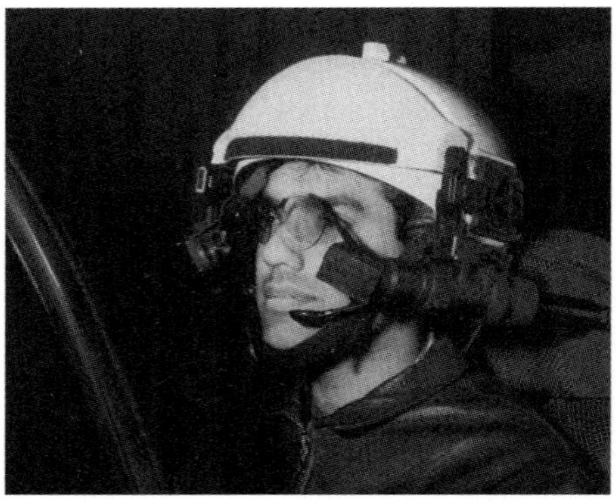

Kaiser Helmet-Mounted Display showing optical feed from fibre optic bundle

Status
Available.

Manufacturer
Kaiser Electro-Optics (KEO), Carlsbad, California.
Systems Integrator: Thomson Training and Simulation, Trappes, France.
(For details see Directory of Suppliers)

MISCELLANEOUS AIR SYSTEMS TRAINERS

ISRAEL

Tadiran Anti-Ballistic Missile (ABM) Simulator

Tadiran has developed a simulator for surface-to-surface missile (SSM) attacks and associated defence procedures. This includes the use of the Arrow anti-ballistic missile (ABM) as part of Ballistic Missile Defence (BMD). The simulator was built as part of the US Strategic Defense Initiative (SDI) and the programme allocated the equivalent of $35 million for development of the system.

Status
In production.

Manufacturer
Tadiran Limited, Holon.
(For details see Directory of Suppliers)

NETHERLANDS

Signaal Modular Radar Simulator

Hollandse Signaal not only manufactures radar systems but also provides a modular radar simulator for training purposes; it can also be used for testing. The characteristics of different radar types are stored on a 540 MByte hard disk which may be amended with details of specific radars. The simulators use Unix-based operating systems and characteristics fed into it include platform and antenna, radar/radar signal processing, target data, environment and effects.

Status
Available. In service.

Manufacturer
Hollandse Signaalapparaten BV, Hengelo Ov.
(For details see Directory of Suppliers)

NEW ZEALAND

Bullseye Television Bomb Scoring System

The Bullseye television automatic bomb scoring system was designed to provide instant target results of weapon impact or bursts. Fixed and portable systems are available.

Components. The system consists of a number of fixed cameras stationed on the weapons range and controlled by a computer at a base station. Views are recorded for verification and re-scoring purposes while freeze-frame and image processing facilities are available.

Uses. The system was originally designed for air weapons ranges but can be adapted for other purposes such as range and accuracy trials.

Results Interpretation. The programme contains procedures for printing and graphical display of summaries with a variety of coverage and formats. The former can include individual pilots, squadrons or total sorties while the latter can be in pilot or flight profile order. Scoring files are stored by date for analysis and by prior assignment of call signs, pilot numbers, tail numbers and profiles.

Components of the Bullseye television bomb scoring system

Status
In production.

Manufacturer
Marine-Air Systems Limited, Wellington.
(For details see Directory of Suppliers)

ROMANIA

SIMULTEC Radar Simulators

SIMULTEC of Bucharest make a system for the simulation of Airborne or Air Traffic radars. The system is based on a 486/66 PC using the C HLL operating under DOS. An instructor panel is also PC-based. Unit cost is about US$20 000.

Status
In service and available for export.

Manufacturer
SIMULTEC SA, Bucharest.
(For details see Directory of Suppliers)

UNITED KINGDOM

EASAMS Training Optimisation Programme (TOP)

EASAMS has developed a software tool for the UK MoD which helps in conducting Training Needs Analysis (TNA). Called the Training Optimisation Programme (TOP) it defines the most cost-effective training package to meet the requirements of any given situation.

Methodology. TOP takes into account the constraints and objectives of the overall training requirement. Factors such as Life-Cycle Costs (LCC), location of different training stages and weapon/training ranges, levels of initial and required trainee skills, training time and throughput of students, can be specified as limitations or targets. TOP will then indicate the best training solution for the given situation.

Changing Variables. Any solution may be highly dependent on some critical variables. TOP can be used for a sensitivity analysis involving changes in the training requirements relating to needs, device characteristics, constraints or objectives, in order that the final chosen solution is seen to be robust and not sensitive to small changes which may take place as the equipment programme progresses.

Use in Eurofighter Programme. TOP has been used for the TNA for aircrew training for Eurofighter 2000.

Status
Available.

Manufacturer
Easams Limited, Camberley.
(For details see Directory of Suppliers)

Graseby Dynamics MDI-RT269

The acoustic MDI System RT269 is a ground processing system which can provide real-time evaluation of air-to-ground gunnery. It consists of a readily identifiable aiming point and a butt to protect the microphone. The system will indicate the number of rounds which pass through an area which is set to be equivalent to a chosen target size.

Status
In service with the Royal Air Force and several overseas countries.

Manufacturer
Graseby Dynamics Limited, Watford.
(For details see Directory of Suppliers)

Pennant Computer-Aided Instruction

Computer-Aided Instruction (CAI) has been designed by Pennant International to be an instructor-led group teaching medium to complement traditional teaching methods.

Courseware. Instructional courseware can be developed in modular and lesson form, comprising two- or three-dimensional graphics, with or without animations. Courseware enables instructors to develop a subject systematically and at a pace to suit the students. CAI allows topics to be presented by computer graphics while colour-coded animation reinforces the learning process.

Total System. The Pennant CAI system consists not only of courseware but also a comprehensive training suite including a presentation control system, a display system and optional multi-media equipment such as a video recorder.

Classroom Control Console. A key feature of the system is the Classroom Control Console (CCC) which houses the hardware. It also enables the instructor to present material which can be interfaced with the monitor display by touch-screen, roller ball or mouse. The CCC can also be used as an instructor's station for part-task trainers, it can be networked with student work stations for self-paced instruction modules and it can operate with a Student Response Monitoring System (SRMS).

Status
CAI courseware packages have been supplied to the Royal Air Force, the Royal Navy and the British Army. Recent programmes were for the Chinook and Lynx helicopters.

Manufacturer
Pennant International Limited, Cheltenham.
(For details see Directory of Suppliers)

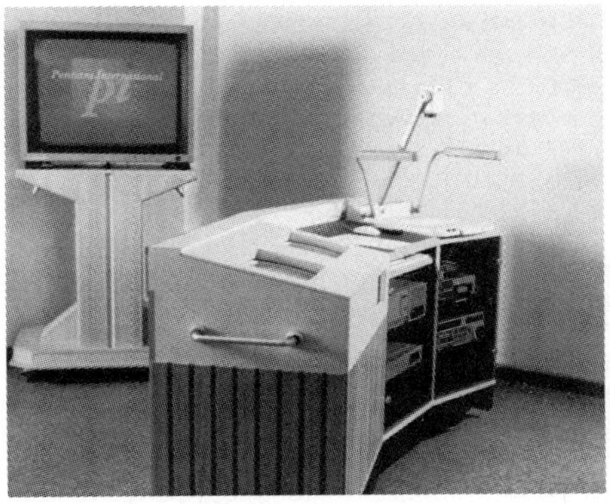

A Pennant International Classroom Control Console

SD-Scicon Scenario Generator for Air Combat Simulation

In January 1992 SD-Scicon delivered to the UK Defence Research Agency at Farnborough a real-time (20 Hz) intelligent opponent and scenario generator. This is designed to enhance a pilot's ability to fly in realistic scenarios and against realistic opponents in both fixed- and rotary-wing aircraft. Real-time line-of-sight calculation, player mobility, vulnerability and reliability are included.

Work Station. The work station-based system provides graphic displays and allows the incorporation and user-modification of opponent tactics using a knowledge-based framework.

Tactics. Tactics of both opposing aircraft and co-operating battlefield participants, such as armoured columns, are simulated. Subsystem behavioural characteristics, including detection, weapon/counter-measure deployment and networked communication, are also included.

Scenarios. Scenario definition is undertaken through a menu-driven interface and the operator can take the role of a variety of simulation players.

Status
In service

Manufacturer
SD-Scicon, Camberley.
(For details see Directory of Suppliers)

Signal Scenario Manager

Signal Computing produces an object-oriented software tool known as a Scenario Manager for the management of complex simulation applications. Designed to provide flexibility, the Scenario Manager uses object-oriented techniques and high level tools to assist in the configuration and control of the simulation environment.

Enhancements. Performance of the Scenario Manager can be further enhanced by the incorporation of SIgnal's Real-Time Expert System (SIRTES) to provide a high level method of modelling human or system responses to the simulation environment. SIRTES has been used to manage and to execute decisions about manoeuvres, tactics and weapon deployment of simulated aircraft.

Applications. The Scenario Manager is a generic tool for simulation applications and is equally suitable for aerospace, naval and land applications. It is particularly relevant for those involving the simulation of a large number of objects such as C³I or air traffic control trainers.

Status
In service with the UK Defence Research Agency, where it is used to manage up to 20 simulated aircraft simultaneously for an air combat research simulator.

Manufacturer
Signal Computing Limited, Guildford
(For details see Directory of Suppliers)

Simis VMP Mission Planning System

Simis have created, together with the Liquid Image Corporation of Winnipeg, Canada, a Virtual reality Mission Preview and rehearsal system (VMP). It consists of the Liquid Image Head-Mounted Display (HMD), a Polhemus 6-DoF head-tracker, a Pentium-processor PC, and a carrying case for all the equipment.

Software. The General Simulation Framework (GSF) is a set of tools and software components for development of real-time real-world imagery. Shape model development creates 3-D objects, and the terrain database is generated using a Cartesian-grid polygon model.

Status
Available.

Manufacturer
Simis Ltd, Godalming.
(For details see Directory of Suppliers)

TTSL Interactive Training Systems (ITS) Group

Thomson Training and Simulation Limited (TTSL, ex-Rediffusion element) has an Interactive Training Systems (ITS) Group based at Crawley. The group can assess training systems from analysis, design, development and implementation to the subsequent evaluation of the system in action. It uses desktop-based, interactive technology, and aims to provide inventive training solutions to meet the training needs of the aerospace industry in areas ranging from basic flight navigation to complex weapon system deployment.

Maintenance Improvement. ITS has also used simulation modelling and graphics to provide a training platform for diagnostic and fault-finding skills improvement in military maintenance.

Courseware Development. The provision of a courseware generating system to the Royal Air Force has led to improvements in courseware production rates, while allowing training staff to concentrate on the training material rather than courseware programming.

Lifetime Support. TTS is committed to the support of its training solutions throughout their lifetime.

Status
Available.

Manufacturer
Thomson Training and Simulation Interactive Training Systems Group, Crawley.
(For details see Directory of Suppliers)

UNITED STATES OF AMERICA

Applied Signal Technology 210 Multipath Simulator

The 210 Multipath Simulator, which is in a 7 in high rack mount chassis, simulates up to five multipath reflections and operates from 70 to 500 MHz in signal bandwidths up to 100 MHz. For each reflection there is independent multipath delay control, static attenuation control and independent dynamic attenuation control.

Status
In production.

Manufacturer
Applied Signal Technology Inc, Sunnyvale, California.
(For details see Directory of Suppliers)

CI Systems Infra-Red Simulation

CI produces IR scene simulators which can be integrated into hardware and software to enable electro-optic seekers and payloads to be tested in the laboratory for most conceivable battlefield scenarios. The unit weighs less than 60 kg, may be mounted on a CARCO S-495R-5D flight table, and combines at least three scenes.

Status
In production.

Manufacturer
CI Systems Inc, Agoura Hills, California.
(For details see Directory of Suppliers)

Fairchild Paveway II and Skipper Training Rounds

In May 1988 Fairchild-Weston Systems was awarded a $4.5 million contract by the US Naval Weapons Center for the development of a low-cost, mass-produced training round with a performance which will match that of the Paveway II series of laser-guided bombs (LGBs) and the AGM-123 Skipper missile.

Design. While retaining the basic features of the Paveway system (the seeker head, guidance unit, canard fins and wings) the training round is slimmer and lighter than the original weapons and weighs only 38.5 kg. This compares with the 500 lb (226 kg) Mark 82, the 1000 lb (453 kg) Mark 84 and the 2000 lb (907 kg) Mark 118 bombs normally converted to Paveway II guidance.

Components. The training round includes a pneumatic control actuator, a smoke-pyrotechnic marker and an ejector mechanism. Commercial components are used wherever possible to reduce expenditure, while alkaline batteries are used to give a long, two-year, shelf-life. The Skipper version will have a rocket motor and a new guidance system.

Guidance. The guidance unit will give the round the flight characteristics of the real missiles. The round will be guided to targets marked by air- or ground-based lasers.

Cost. The approximate unit cost of the Paveway training round is $1700, while that of the Skipper version will be $2900.

Status
In service with the US Navy and US Marine Corps. The Paveway version entered service in FY90 and the Skipper version in 1992.

Manufacturer
Fairchild-Weston Systems Inc, Syosset, New York.
(For details see Directory of Suppliers)

Litton Simulated Laser Target

Litton has modified its NT-90 laser to provide a simulated laser target. The system generates a diffuse laser spot at any specific location on a gunnery range where the device can be located to replace a laser main beam.

Operation. The system allows aircrew to be trained without the extensive range safety requirement normally associated with higher energies of laser target designation. The device simulates the laser signature of both ground and airborne designators and allows laser spot tracker-cued engagements with unguided ordnance and aircraft cannon for which the initial aim is by laser pointing.

Status

In production and in service.

Manufacturer

Litton Systems Inc, Orlando, Florida.
(For details see Directory of Suppliers)

Merit Battle Area Tactical Simulation (BATS)

The Merit Battle Area Tactical Simulation (BATS) combines a mission simulator with graphics displays to provide a system which will help plan, execute and analyse all types of air operation including air-to-air, air-to-ground and air defence. It uses Defense Mapping Agency (DMA) digital terrain data which is supplemented by graphics of maps and 3D views.

Flight Models. Three or six flight models are used to generate realistic flight trajectories and this allows the operator to simulate the effects of air- and ground-launched weapons, electronic warfare techniques, air- and ground-based sensors and terrain following. The 'aircraft' can be controlled by an automatic flight controller, an on-line man-in-the-loop joystick or by external programmes.

Graphics. The graphics system interface shows command, control communications and intelligence links between fire-control centres and firing sites, the intervisibility of aircraft and ground radars and all planned flight paths. The BATS system provides strip chart and histogram plots of selected mission data. Threat-tracking states are displayed during mission playback.

Status

In service and in production.

Manufacturer

Merit Technology Inc, Plano, Texas.
(For details see Directory of Suppliers)

Motorola Integrated Range Instrumentation Systems

Motorola Government and Systems Technology Group produces a number of combat systems for aircraft use.

Motorola Integrated Target Control System (ITCS)

ITCS consists of a modular group of subsystems which are designed to provide remote-control of air target vehicles and pilotless aircraft. The subsystems include ground control stations, target vehicle subsystems and ground support equipment to test and to maintain them.

The system remotely controls randomly spaced target vehicles by means of time-division multiplexing of the target command telemetry information by individual stations operating on a common frequency pair. Compatibilities include range and angle tracking, Global Positioning System (GPS) tracking, reception and display of individual target telemetry data and target control hand-over between one control station and another. These functions are performed at both radio frequency line-of-sight range (250 nm/465 km) and over-the-horizon ranges (350 nm/650 km) through relay links.

Deployment. Operational US Ranges using ITCS include the Pacific Missile Test Center, Port Mugu, California; Naval Air Warfare Center, China Lake, California; Atlantic Fleet Weapons Training Facility, Roosevelt Roads, Puerto Rico and the Pacific Missile Range Facility, Barking Sands, Hawaii.

Non Co-operative Aerial Vector Scorer (NAVS)

NAVS is a system which is installed in a drone or target vehicle and used for testing missile effectiveness. It incorporates a high resolution multistatic impulse radar, transmitting a narrow impulse and having four fixed narrow-range gates in the receiver. The use of precisely positioned, narrow, fixed-range gates reduces ground clutter and multipath effects.

As a missile travels through a specific range gate, the system measures the instant at which fragments from the missile body enter the range gate. Because the most prominent fragments will be the missile nose and tail fins these measurements allow a separate estimation and reconstruction of both the nose and tail trajectory. The measurements are then telemetered to the ground for processing.

Missile End Game Scoring (MEGS)

Missile End Game Scoring (MEGS) is a co-operative vector scoring system which consists of a missile transponder, the drone target subsystem, ground station and ground support equipment. It has been developed by Motorola and the US Air Force to determine automatically the miss-distance in three dimensions from the target.

The measurement is computed using RF phase history which is transmitted to the ground station via standard telemetry links. The full missile trajectory relative to the target is reconstructed through the scoring volume and is graphically displayed in three-dimensional representation.

Status

In service and in production.

Manufacturer

Motorola Government and Systems Technology Group, Scottsdale (Phoenix), Arizona.
(For details see Directory of Suppliers)

USAF ASC Dynamic Environment Simulator (Centrifuge)

The USAF Aeronautical Systems Center (ASC) has a centrifuge, operated by the Combined Stress Branch at the Armstrong Laboratory, Wright-Patterson AFB. The centrifuge is known as the Dynamic Environment Simulator (DES). It has a rotating mass of 180 US tons and can reach 20 G at the end of the main rotating arm. It can simulate heat at the same time as acceleration, and tumbling motions can be produced because the cabin in which the subject is located is mounted in a fork arm and can be rotated separately from the main rotating arm.

Status

Military R & D use.
Location: USAF Wright-Patterson AFB.

Directory of Suppliers

This Directory of Suppliers of equipment and services in the Training and Simulation field is in alphabetical order of the first main name of the organisation concerned, ignoring any initials or first names of persons (unless the main name itself consists of initials). Where there could be ambiguity in the main name, both names are listed and cross-referencing is used. Telephone and Fax numbers are the international ones used from other countries and the country code is prefixed with a '+' symbol. Where, for inside-country calls, other numbers are added (such as the extra '0' in the UK), these are shown in brackets and should be omitted for international calls.

Special note for UK numbers - Changeover on 16 April 1995.

For UK numbers, note that from August 1994 an extra '1' is added to the regional code; the old numbers may be used until 16 April 1995 after which only those with the extra '1' will work. In this directory the new numbers with the additional '1' are given.

AAI Corporation, PO Box 126, Hunt Valley, Maryland 21030-0126, USA.
Tel: +1 410 628 3900/3184. Telex: 8-7849.
Fax: +1 410 628 3465.
President and CEO of AAI Corp: Richard R Erkeneff
President, AAI Training & Test Systems: Maurice P Ranc
Vice President, AAI Training & Test Systems: J Richard Weger
Director of Marketing, Domestic: E P (Bud) Miller, Tel: 410 628 8499
Director of Marketing, International: Richard A Mosher, Tel: 410 628 3177
Marketing Contact: Sharon C Reid
The AAI Corporation is a wholly owned subsidiary of the United Industrial Corporation. It has about 1700 employees and a shop floor area of about 275 000 ft^2 (25 500 m^2). AAI has three primary business units, Vehicle and Ordnance Systems, Training and Test Systems, and Transportation.

Aardvark Clear Mine Limited, Shevock Farm, Insch, Aberdeenshire, UK.
Tel: +44 (0) 1464 20122/3/4. Telex: 73509 AARD G.
Fax: +44 (0) 1464 20985.

AB TELEANALYS, Gelbgjutarevägen 4, Box 1212, S-171 23 Solna, Sweden.
Tel: +46 8 83 17 48.
Fax: +46 8 82 94 15.

ABC Dados Informática SA, Avenida Heitor Villa Lobos 2176, CEP 12243, São José dos Campos, SP, Brazil 12245-280.
Tel: +55 123 21 75 22. Telex: 123 3325.
Fax: +55 123 21 17 89.
President: José Gabriel Correia David
Chief Executive: Junzo Habiro
Technical Director: A Guilherne de A Lorenzi
Marketing Contact: Maria Virgínia Cellet
ABC Dados is a member of the ABC Algar Group, and was previously known as ABC Teleinformática and ABC Systemas Electrônicos. The flight simulation division started operations in 1984, at that time based on a technical partnership with GMI. ABC Dados has 140 employees and a shop floor area of 4500 m^2.

AEL Industries Inc, 305 Richardson Road, Lansdale, Pennsylvania 19446-1485, USA.
Tel: +1 215 822 7980/2929.
Fax: +1 215 822 7993/9165.

Aeronautical Development Establishment (ADE), Defence Research and Development Organisation (DRDO), Ministry of Defence, C V Raman Nagar, Bangalore 560 093, India.
Tel: +91 080 558 3404. Telex: 0845-8461 ADE IN
Fax: +91 080 558 3188.
Director: Dr K G Narayanan.
ADE was started as a laboratory in 1959 under the DRDO. Its facilities include aerodynamics, structures, flight control systems, telecommand and telemetry, mathematical modelling, computing, simulation engineering, avionics and HUDs, E-O systems, and prototype engineering. It has made Hovercraft, CPTs, flight simulators, a reuseable rocket pod, an air-launched expendable drone, and target drones. ADE is also concerned with the development of the Light Combat Aircraft (LCA) for the Indian Air Force, particularly the FBW control system and the core avionics. ADE Bangalore has about 1500 employees.

Aeronautical Development Establishment (ADE), Ministry of Defence, Jeevanbimanagar, Bangalore 560 075, India.
Tel: +91 0812 543404. Telex: 0845-8461 ADE IN
Fax: +91 0812 543188.

Aeronaves E Motores SA (Aeromot), Aeroporto Internacional, Salgado Filho, PO Box 8031, 90201 Porto Alegre, RS, Brazil.
Tel: +55 0512 42 3 344. Telex: (051) 1991 AEMT BR.

AEROPAR, Aérodrome de Couhé-Brux, BP 30, F-86700 Couhé, France.
Tel: +33 (1) 16 49 59 17 18. Telex: AEROPAR 793261.

Aerospace Technologies of Australia Limited (ASTA), 226 Lorimer Street, Port Melbourne 3207, Victoria, Australia.
Tel: +61 3 646 3011. Telex: AA 34851.
Fax: +61 3 646 4381.

Aérospatiale Missiles, 2 à 18 rue Béranger, F-92322 Châtillon Cedex, France.
Tel: +33 (1) 47 46 21 21. Telex: AISPA 250881 F.
Fax: +33 (1) 47 46 33 19.

AGARD, 7 rue Ancelle, F-92200 Neuilly-sur-Seine, France.
Tel: +33 (1) 47 38 57 00.
AGARD stands for the Advisory Group for Aerospace R&D, and is a NATO group. About every 5 years it holds a conference on military flight simulation to which attendance is restricted to NATO affiliates (including companies involved in flight simulation).

Agusta Sistemi Srl, Via Isonzo 33, I-21049 Tradate, Italy.
Chairman: Ing A Caporaletti.
Managing Director: Ing V Consani.
Technical Director: Ing A Ceriotti.
Marketing Director: Ing D Vaccari.
Marketing Contact: Ing P Della Rosa.
Tel: +39 331 815 111. Telex: 350 405.
Fax: +39 331 843 910.
Agusta Sistemi was merged with Agusta on 12 February 1994. Agusta itself is part of the Finmeccanica SpA group. Agusta Sistemi is located about 40 km north of Milan at Tradate, and is involved in avionics, mission systems, simulation and training systems. Agusta Sistemi has about 250 employees and a shop floor area of 9000 m^2.

Airshows DownUnder
P O Box 1254, Geelong, Victoria 3220, Australia.
Tel: +61 52 24 20 20
Fax: +61 52 21 38 78/24 11 44

Air Target Sweden AB, Box 3006, S-171 03 Solna, Sweden.
Tel: +46 8 730 22 33.
Fax: +46 8 730 34 24.
Chairman and Marketing Director: Bo Eklund
Technical Director: Robert Stake
Marketing Contact: Christer Nyberg
Air Target makes ground-launched air targets and miss distance indicators, and has about 30 employees. Devices delivered exceed 50 000 to 28 countries.

AirteQ Limited (and AirteQ Training & Simulation Limited), 72 Tanners Drive, Blakelands, Milton Keynes, Buckinghamshire MK14 5BP, UK.
Tel: +44 (0) 1908 614 365.
Fax: +44 (0) 1908 613 167.
Chairman: N Randall
Managing Director: Phil Colley
Technical Director: David Fairhurst
Marketing Director: Mike Boseley
AirteQ is part of the multimedia division within the Air Group, and has 21 employees and a shop floor area of about 2000 m^2. It produces firearms training systems, target identification trainers and HGV driver trainers. Over 200 devices have been delivered.

Alenia Elsag Sistemi Navali, Via S Alessandro 28/30, I-00131, Rome, Italy.
Tel: +39 6 419 0631/2/3. Telex: 621276 SESN I.
Fax: +39 6 419 0710.

Alenia Defence Systems, Simulators and RPVs Division, 34077 Ronchi dei Legionari, Goriza, Italy.
Tel: +39 481 4791. Telex: 460 369.
Fax: +39 481 474 321/776 202.

ALTEC Industries SA, 77 rue de Lille, F-75009 Paris, France.
Tel: +33 (1) 47 70 48 01. Telex: 649601.
Fax: +33 (1) 48 24 85 00.

Alliant Techsystems Inc, Marine Systems, 600 Second Street North East, Hopkins, Minnesota 55343, USA.
Tel: +1 612 931 5295.
TWX: 910 7563 437.

AlliedSignal Aerospace Canada, Bendix Avelex Inc, 200 Laurentian Boulevard, Montreal, Québec, Canada H4M 2L5.
Tel: +1 514 744 2811.
Fax: +1 514 342 3795.

AlliedSignal Ocean Systems, 15825 Roxford Street, Sylmar, California 91342-3597, USA.
Tel: +1 818 367 0111.
Fax: +1 818 367 0403.

AMES, see NASA

Anaren Microwave Limited, Frimley Business Park, Frimley, Camberley, Surrey GU16 5SG, UK.
Tel: +44 (0) 1276 692 829.
Fax: +44 (0) 1276 682 635.

Applied Signal Technology Inc, 160 Sobrante Way, Sunnyvale, California 94086, USA.
Tel: +1 408 522 3421.
Fax: +1 408 738 1928.

ARI, ARL, see Aviation Resources Inc.

Armament Systems and Procedures (ASP) Inc, Box 1794, Appleton, Wisconsin 54913, USA.
Tel: +1 414 735 6242.
Fax: +1 414 735 6245.
President: Donald Roegner
Chairman and CEO: Dr Kevin Parsons
Technical Director: Jerome Webber
Marketing Director: Kathy Bruns
Marketing Contact: Kathy Faker
ASP was founded in 1978, employs about 85 people and has a shop floor area of about 14 000 ft^2 (1300 m^2). It produces weapon systems for US law enforcement agencies including training and simulation equipment. Product lines include training batons, the Red Gun™ line of training revolvers, pistols, and shotguns, and the Red range of training radios, flashlights and knives. Over 62 000 devices have been delivered.

Armscor, see Denel.

Armi Adler, Via delle Fornaci 4, I-17025 Loano (SV), Italy.
Tel: +39 19 669 294.
Fax: +39 19 669 294.

ATC Flight Simulator Company, 1650 19th Street, Santa Monica, California 90404, USA.
Tel: +1 213 453 3557. Telex: 284687.
Fax: +1 213 453 5276.

Atlantis Aerospace Corporation, 1 Kenview Boulevard, Brampton, Ontario, Canada L6T 5E6.
Tel: +1 905 792 1981. Telex: 06-986766.
Fax: +1 905 792 7251.
Managing Director: Dale Simmons
Manager, Marketing: Chris Lehman
Marketing Contact: Ms Patricia Takeuchi
Atlantis was formed in 1978 and has about 220 employees and a shop floor area of about 225 000 ft^2 (20 900 m^2). It specialises in simulation systems, electronic test equipment and instrumentation and control systems. Simulation currently accounts for some 80% of sales.

Atlantis Aerospace Corporation, 951 Rontree Dairy Road, Woodbridge, Ontario, Canada L4L 4E4.
Tel: +1 905 851 8531. Telex: 06 986766.
Fax: +1 905 851 7251.

Atlas Elektronik GmbH, Sebaldsbrücker Heerstrasse 235, D-28305 Bremen 44, Germany.
Tel: +49 421 457 33 26. Telex: 2 457 4641 ae d.
Fax: +49 421 457 37 25.
Marketing Manager, Simulation Systems Division: Rainer Huelsz
Atlas Elektronik is part of the Electronics and Systems Technology Division of the Bremer Vulkan Group, together with STN (qv). The Vulkan Group consists of about 30 companies grouped into six functional divisions and deals with shipbuilding, maritime electronics and systems technology, employing about 15 000 people. Atlas Elektronik deals with hydroacoustics, E-O devices, signal and data processing, simulators for ship handling and submarines, and complete sonar systems for ships including minesweepers. Atlas also has an army division which produces fire-control equipment, air defence equipment and measuring and test equipment.

ATS Aerospace Inc, 1250 Marie-Victorin, St-Bruno, Québec, Canada J3V 6B8.
Tel: +1 514 441 9000.
Fax: +1 514 441 6789.
Aero Mechanical Systems: Louis Roy
Marketing Co-ordination: Carmelita Tartaglia
ATS Aerospace Inc has a staff of about 85, divided between two divisions, the Aerospace division and the Ballistech Systems (BSI) division. The Aerospace division, amongst other activities, makes a product line of Air Traffic Control trainers and simulation systems. BSI produces aerial targets and also has an UNDerwater EXplosion (UNDEX) test facility at Bedford, Québec.

Australian Defence Industries Limited (ADI), PO Box 985, Level 22, Plaza 2, Grosvenor & Grafton Streets, Bondi Junction, NSW 2022, Australia.
Tel: +61 2 365 9320/9300.
Fax: +61 2 369 2168/2404.
Chairman: Russel Fynmore
Managing Director: Ken Harris
Marketing Director: Les Targ
Chief General Manager, Training Systems Division: Neville Rogers
In 1989, ADI took over the functions of the Australian DoD Production Office and is organised as a normal commercial venture. It has about 4500 employees, and makes live firing training systems and targets, amongst other products. ADI has offices outside Australia in Abu Dhabi, Berlin, and Kuala Lumpur. Its divisional structure is as follows: Ammunition Division, Weapons and Engineering, Naval Engineering, Clothing, Electronics, Training Systems, Industrial Decontamination, Logistic Support, Consulting, Property, and Minehunter Projects Division.

Aviation Software, 16 Chemin Colladon, CH-1209 Geneva, Switzerland.
Tel: +41 22 798 1365/5707.
Fax: +41 22 788 2072.
Contact: John Plaistowe

Aviation Resources Inc (ARI), 3100 North Hemlock Circle, Broken Arrow, Oklahoma 74012, USA.
Tel: +1 918 258 7000. Telex WU 797 834.
Fax: +1 918 251 9002.
President: George W Moody.
Vice-President, Operations: Phil M Dooley.
Vice-President, Technical Marketing: V (Prab) Prabhu.
ARI has a 40 000 ft^2 facility at Broken Arrow near Tulsa, and can assemble up to four flight simulators concurrently. ARI has over 100 employees, and manufactures full flight simulators, CPTs, maintenance trainers, cabin trainers, door trainers, lavatory fire trainers, cabin service and galley trainers, and CBT and classroom training systems. ARI also upgrades, modifies and relocates existing training equipment. It has a UK affiliate company, Aviation Resources Limited (ARL).

Aviotechnica Ltd, H. Est 'Trakia', PO Box 423, 4000 Plovdiv, Bulgaria.
Tel: +359 32 83 10 40, and 23 86 90. Telex: 44639.
Fax: +359 32 83 10 80.
President: Dobri Dobrev
Marketing by Metalchim Trading, 21 Gavril Genov Street, 1000 Sofia, Bulgaria.
Tel: +359 02 65 70 68. Telex: 24533 POB 839.
Fax: +359 02 87 76 60.

Aware Simulation Limited, Laser House, 132/140 Goswell Road, London EC1V 7LE, UK.
Tel: +44 (0) 171 253 40 14.
Fax: +44 (0) 171 250 41 43.
Managing Director: Geoffrey D Hawkins
Technical Director: Paul A Shepard
Marketing Contact: G D Hawkins
Aware Simulation was formed after a management buy-out of Walmore Defence Systems in 1992.
Aware Simulation specialises in RF stimulation devices for sea, land and air use. It also makes portable RF emitters, power meters, and classroom EW trainers. Its customers include the Royal Navy, British Army, RAF, and other navies. It has a shop floor area of about 3500 ft^2 (325 m^2).

Aydin Corporation, Radar and EW Division, 32 Great Oaks Boulevard, PO Box 1371, San Jose, California 95119, USA.
Tel: 408 629 1100. Telex: 348393.

BAeSEMA Limited, Marine Division, PO Box 5, Filton, Bristol BS12 7QW, UK.
Tel: +44 (0) 117 9 368 670.
Fax: +44 (0) 117 9 368 925.
Chairman: John Weston
Managing Director: Alex Dorrian
Technical Director: Ray Foulkes
Business Group Manager, Mine Warfare: P J (Jim) Elsworth
Marketing Director: Alan Nicholson-Florence
Marketing Contact: A A Gomm
BAeSEMA is a joint venture between British Aerospace and the French SEMA group. It has about 1400 employees.

BAeSEMA Limited, Biwater House, Portsmouth Road, Esher, Surrey KT10 9SJ, UK.
Tel: +44 (0) 1372 466 660.
Fax: +44 (0) 1372 466 566.

Baghdad FSPF, Iraq.
Tel: + 964 55 68 330. Telex: 214162 ALFA IK.
Fax: +964 53 84 634.
No information received recently due to the international situation

Ballistech Systems Inc (BSI). A division of ATS Aerospace Inc (qv).

Elsag Bailey SpA, Marine Business Unit, Via Puccini 2, I-16154 Genoa, Italy.
Tel: +39 10 65 831. Telex: 275 531.

BAJ Limited, Banwell, Weston-super-Mare, Avon BS24 8PD, UK.
Tel: +44 (0) 1934 82 22 51. Telex: 44259 BAJBAN G.
Fax: +44 (0) 1934 82 39 97.

Ball Systems Engineering Division, 5580 Morehouse Drive, San Diego, California 92121-1709, USA.
Tel: +1 619 457 5550.
Fax: +1 619 457 5410.

Ball Systems Technology Limited, 15/16 Mercia Business Village, Westwood Business Park, Warwick University, Coventry CV4 8HX, UK.
Tel: +44 (0) 1203 421212.
Fax: +44 (0) 1203 421500.

BARCO Projection Systems,
Noordlaan 5, B-8520 Kuurne, Belgium
Tel: +32 56 36 82 11
Fax: +32 56 35 16 51

BBN, see Bolt Beranek and Newman Inc

Beech Aircraft Corporation, PO Box 85, Wichita, Kansas 67201-0085, USA.
Tel: +1 316 733 0022. Telex: RCA 2036603 Beech UR.
Fax: +1 316 733 5664.

Bell Aerospace Textron, PO Box 1, Buffalo, New York, 14240-0001, USA.
Tel: +1 716 297 1000.
Fax: +1 716 696 6981.

Bharat Electronics Limited, Trade Centre, 116/2 Race Course Road, Bangalore-560 001, India.
Tel: +91 812 267 322. Telex: 0845-2477 BE IN.
Fax: +91 812 258 410 & 265 657.

BMY Combat Systems, PO Box 15512, York, Pennsylvania 17405, USA.
Tel: +1 717 225 4781.
Fax: +1 717 225 4615.

Boeing Defense & Space Group, Missiles & Space Division, Simulators & Training Systems, PO Box 240002, M/S JM-00, Huntsville, Alabama 35824-6402, USA.
Tel: +1 205 461 3453.
Fax: +1 205 461 3453.

Boeing Canada Technology Limited, 99 Murray Park Road, Winnipeg, Manitoba, Canada, R3J 3M6.
Tel: +1 204 888 2300. Telex: 07 57309.
Fax: +1 204 888 2951.

Bofors AB, S-691 80 Karlskoga, Sweden.
Tel: +46 586 860 18; Telex: 73 210 bofors s
Fax: +46 586 581 45
Advertising Manager: Kjell Mohlin

Bolt Beranek and Newman Inc, 10 Moulton Street, Cambridge, Massachusetts 02138, USA.
Tel: +1 617 873 3000/3008.
Fax: +1 617 873 4315/3776.
Bolt Beranek and Newman (BBN) has sold the simulation part of its business to Loral, which also has an office in Moulton Street, Cambridge, Mass (qv under Loral)

Bristol Aerospace Limited, PO Box 874, Winnipeg, Manitoba, Canada R3C 2S4.
Tel: +1 204 775 8331. Telex: 07 57774.
Fax: +1 204 885 3195.
Bristol Aerospace used to be Boeing Canada.

Britanite SA Indústrias Quimicas, Rodovia BR-116, Km 71, Quatro Barras, Paraná, Brazil.
Tel: +55 41 772 1211. Telex: (041) 2193 BRIQ BR.
Fax: +55 41 772 2031.

British Aerospace Australia, PO Box 180, Salisbury, South Australia 5108
Tel: +61 8 343 82 11
Fax: +61 8 349 66 29
Chairman: Ted Harris
Chief Engineer: Alastair Hodgson
Technical Director: Roman Olesnicky
Marketing Director: Alf Jaugietis; contact at Moore St, Canberra, see below.
BAeA has about 800 employees, and is a subsidiary of BAe plc. It was formed in 1953 and is involved in Rapier support, FLIR, light armoured vehicles, link 11 communications and various training facilities.

British Aerospace Australia, Fifth Floor, 12 Moore Street, Canberra ACT, Australia.
Tel: +61 6 257 66 66

British Aerospace Defence Limited, Military Aircraft Division, Warton Aerodrome, Preston, Lancashire PR4 1AX, UK.
Tel: +44 (0) 1772 852 697, 854 621, and 633 333.
Telex: 67 627.
Fax: +44 (0) 1772 855 279.
Head of Flight Simulation Research: Peter Beckett.
Director of Sales and Marketing Support: Peter Anstiss.

British Aerospace Simulation Limited (BAeSL), see Reflectone UK Limited (RUKL).

British Manufacture and Research Company Limited (BMARC), Springfield Road, Grantham, Lincolnshire NG31 7JB, UK.
Tel: +44 (0) 1476 65 577. Telex: 37 635.
Fax: +44 (0) 1476 65 614.

Brocks Explosives Limited, Sanquhar, Dumfries and Galloway DG4 6JP, UK.
Tel: +44 (0) 1659 50 531.
Fax: +44 (0) 1659 50 526.

Browning SA, Police Department, Parc Industriel des Hauts-Sarts, 3e Avenue, B-4400 Herstal, Belgium.
Tel: +32 41 646 555. Telex: 41223 fabna b.
Fax: +32 41 481 490.

Burtek Inc, part of Thomson Training and Simulation; see under Thomson

BVR Technologies Limited, 1 Korazin Street, Gyvataim 53583, Israel.
Tel: +972 03 571 5671.
Fax: +972 03 571 5668.
Chairman: Yaron Sheinman.
Chief Executive: Zvi Kanor.
Technical Director: Aviv Tzidon.
Marketing Director and Contact for Enquiries: Avi Zamir.
BVR has about 100 employees and a shop floor area of 2000 m^2. The company was founded in 1986 by two ex-Israeli Air Force pilots. It makes HOTSHOT combat trainers for aircraft, the AMOS pilot screening device, sensor trainers, ACMI pods, data link systems, mission planning and war game systems, and is involved in studies and simulations.

CAA (UK), see Civil Aviation Authority

CACI International Inc, 1700 North Moore Street, Arlington, Virginia 22209, USA.
Tel: +1 703 875 2919
Fax: +1 703 875 2919

CAE Electronics Ltd, 8585 Côte de Liesse, PO Box 1800, Saint-Laurent, Québec, Canada H4L 4X4.
Tel: +1 514 341 6780. Telex: 05 824 856.
Fax: +1 514 341 7699.
President: N. Byran Cavadias
Director, Business Planning and RAeS liaison: Brian P Hampson.
Head of Engineering: Jack Shlien.
Visual Systems: Bob Barette.
Marketing Contacts: Andrew Morris, Max Rutherford
Saint-Laurent is a suburb of Montreal close to the airport and this site is often referred to as CAE Montreal. CAE Electronics is a member of the Aerospace and Electronics Group of CAE Inc. It has a factory floor area which includes over 20 bays for accommodating full-size flight simulators in build or under test.

CAE Electronics GmbH, Postfach 1220, D-52 201 Stolberg, Germany
Tel: +49 24 02 1060. Telex: 832 220.
Fax: +49 24 02 106 270.
Civil Marketing: Rob Bakker.
Military Marketing: Wolf Winkler.
CAE Electronics GmbH is a member of the Aerospace and Electronics Group of CAE Inc and maintains simulators for, amongst others, the German military. They built for the Luftwaffe an advanced Tornado Simulator Testbed including the CAE (Montreal) 6-DoF motion platform and the CAE FOHMD display system; the image generator being the Evans and Sutherland ESIG 1000 with a centre (AOI) channel and a background channel. The Stolberg facility also manufactures aircrew selection systems and PTTs. It employs 480 people and has a factory floor area of 12 000 m^2 with three full size bays for the construction of flight simulators.

CAE-Link Corporation, PO Box 1237, Binghamton, New York 13902-1237, USA.
Tel: +1 607 721 5465/5650.
Fax: +1 607 771 8477 or 721 5574.
e-mail: RGROW@Link.com
President: George G Houser
Vice-President, Operations: Howard Marshall
Vice-President, Engineering: Gary Nesta
Vice-President, Marketing: Albert H Gagne
Director, US Sales: P H Mallory
Director, Sales outside the US: F J Donohue
Director, Marketing Communications: Robert Grow.
CAE-Link is a member of the Aerospace and Electronics Group of CAE Inc of Toronto, Canada. It specialises in training systems and simulators and is the largest simulator and training manufacturing company in the USA. It employs about 3400 people and has a factory floor area of 7630 m^2. The company history goes back to 1929 when Edwin A Link invented ground-based flight simulation with his famous 'blue box'.

CAE-Link Corporation, 2224 Bay Area Boulevard, Houston, Texas 77058, USA
Tel: +1 713 280 4001
Fax: +1 713 280 4402

CAE-Link Silver Spring Operation - now moved to Binghamton address given above.

Caltronic Systems Limited, Portsmouth Enterprise Centre, Quartermaine Road, Portsmouth, Hampshire PO3 5QT, UK.
Tel: +44 (0) 1705 662 996.
Fax: +44 (0) 1705 673 647.

Caswell International Corporation, 1221 Marshall Street NE, Minneapolis, Minnesota 55413-1035, USA.
Tel: +1 612 379 2000. Telex: 910 576 1306.
Fax: +1 612 379 2367.

CCI, Omark Industries, Sporting Equipment Division, PO Box 856 Lewiston, Idaho 83501, USA.
Tel: +1 208 746 2351.
Fax: +1 208 746 2915.

Cegelec Soprano, Parc Technologique, Isle d'Abeau, 40 rue Condorcet, F-38090 Vaulx Milieu, France.
Tel: +33 74 822 727. Telex: 308 039 F.
Fax: +33 74 822 704.

CelsiusTech Systems AB (and CelsiusTech Electronics AB), Nettovägen 6, Jakobsberg, S-175 88 Järfälla, Sweden.
Tel: +46 8 580 853 85 or 855 03 or 840 00.
Fax: +46 8 580 872 16/38 or 322 44.
President: Per Ove Morberg
Vice-President, Engineering: Dr Jaak Urmi
Vice-President, Marketing: Jan G Linder
Marketing Contacts: Guy Hörnfeldt, Kent Javestad
CelsiusTech used to be known as NobelTech until 1994. It is part of the Celsius group which also includes companies such as Bofors, Kockums, FFV Aerotech and Telub. The overall group employs 16 000 personnel. CelsiusTech Systems makes naval and land-based command and weapon control systems, and Air Traffic Control systems; it has a workforce of about 1000 and a shop floor area of about 474 000 ft^2 (44 000 m^2). CelsiusTech Electronics also has a workforce of about 1000 and makes EW, comms, fire-control, optronics, radar and munition electronics, and microwave products.

Central Military Bureau of Designs and Technology, Warsaw, Poland. See under WCBKT which are its initials in Polish. However, marketing is through the Cenzin Company whose details are given below.

Cenzin Company Limited, 2 Frascati Street, PL-00-489 Warsaw, Poland.
Tel: +48 29 67 38. Telex: 814505 pl
Fax: +48 62 86 356.
Deputy Commercial Director: Wlodzimierz Strzelczyk MScEng
The Cenzin Company is a 'Foreign Trade Enterprise' and markets equipment on behalf of organisations such as OBRUM and WCBKT which are also listed in this Directory.

CEI Technologies Pte Limited, 249 Jalan Boon Lay, Singapore 2261, Republic of Singapore.
Tel: +65 264 1803/265 3226/261 3755. Telex: RS 55 992 CEIPL.
Fax: +65 264 2687.
Chairman: Madam Ho Ching
General Manager: Toh Kim Huat
Technical Director: Mark Chan
Marketing Director: Khoo Swee Peng
CEI Technologies used to be known as Chartered Electronics Industries until 1992. It has about 170 employees and a shop floor area of about 20 000 ft^2 (1860 m^2). It has delivered about 2500 battlefield engagement systems and also manufactures hand grenade simulators and dummy rifles.
Marketing is by Unicorn International Pte Limited, 11-01/02 Singapore Technologies Building, Singapore 0208, Republic of Singapore.

CESELSA, Simulation and Avionics, C/La Granja 84, E-28100 Alcobendas, Madrid, Spain.
Tel: +34 (1) 623 70 00. Telex: 23962 CSEL-E.
Fax: +34 (1) 623 73 52.

CESELSA SA, Aeronautical Application Systems Branch, C/Mar Egeo, 4 Poligno Industrial No 1, E-28850 San Fernando de Henares, Madrid, Spain.
Tel: +34 (1) 396 72 78/29.
Fax: +34 (1) 396 73 52.
Chairman: Jose Antonio Perez Nievas
Managing Director: Humberto Figarola
Branch Director: Cristobal Martin Rico
Branch Marketing Director: Juan Izquierdo
Branch Marketing Contact: Jose Fernandez-Savater
PR Contact: A Martinez Palazon.
CESELSA is part of the Grupo (Group) Indra and has about 300 employees; the Indra group has about 4000 employees in all. CESELSA has links to other companies such as AISA, ASDL, ELT SA, EMAC, ENOSA, Giraviona Dorand, GYCONOSA, SEAS, and SAINSEL. It deals with defence-related technologies, and its Aeronautical Application Systems Branch makes flight training devices and simulators, and ATE systems.

CGA Alcatel Thomson, 33 rue Emeriau, F-75725 Paris Cedex 15, France.
Tel: +33 (1) 45 71 11 55.

CGA Cegelec, Le Plessis-Pâté, F-91220 Brétigny-sur-Orge Cedex, France.
Tel: +33 (1) 69 88 52 00. Telex: 601 262F.
Fax: +33 (1) 60 84 82 30.

Chemring Limited, Alchem Works, Fratton Trading Estate, Portsmouth, Hampshire, PO4 8SX, UK.
Tel: +44 (0) 1705 735 457. Telex: 86 242 ALCHEM G.
Fax: +44 (0) 1705 817 509.
Group Chairman: P Billington
Group Managing Director: D Evans
Chemring Ltd Managing Director: I D Pollicott
Business Development Director: B Butters
Sales Director: G P Hopkinson
Business Development Manager: P J Ankers
Chemring Limited is wholly owned by the Chemring Group plc and has about 150 employees and a shop floor area of about 20 000 ft^2 (1860 m^2). It was founded in 1941 as a chemical plating company and made, amongst other things, silver-coated nylon mesh for use in radar reflectors, and chaff packs. In 1953 it amalgamated with the Automatic Light Controlling Company and moved to Portsmouth. It acquired the Benders Company in 1974, an organisation specialising in making chaff, which provided the original 'window' for the RAF during WW2. In 1986 it acquired Pains-Wessex pyrotechnics. As well as coated radar-reflective materials, it now makes AFV target systems and small arms targets, amongst other products.

China, People's Republic
There are several agencies concerned with simulation and training. Those headed 'China' appear below in alphabetical order. See also under Nanjing.

China National Aero-Technology Import and Export Corporation, No 5 Liangguochang Road, East City District, PO Box 647, Beijing, People's Republic of China.
Tel: +86 (1) 445 831. Telex: 22318 AEROT CN.
Fax: +86 (1) 445 831.

China National Electronics Import & Export Corporation (CEIEC), PO Box 140, Electronics Building, A23 Fuxing Road, Beijing, People's Republic of China.
Tel: +86 (1) 821 2361.
Fax: +86 (1) 821 2352.

China Precision Machinery Import and Export Corporation, PO Box 845, No 17 Wenchang Hutong, Xidan, Beijing, People's Republic of China.
Tel: +86 (1) 895 012. Telex: 22484 CPMC CN.
Fax: +86 (1) 895 012.

China Xinxing Corporation, The 2nd Business Department, 26-312 Fu Xing Road, Beijing, China.
Tel: +86 (1) 810 121. Telex: 22062 CXXC CN.
Fax: +86 (1) 810 121. Cable: 3298 BEIJING.

CI Systems Inc, 5137 Clareton Drive, Suite 220, Agoura Hills, California 91301-4526 USA.
Tel: +1 818 865 0402.
Fax: +1 818 865 0403.

Cincinnati Electronics, 2630 Glendale-Milford Road, PO Box 3187, Cincinnati, Ohio 45241, USA.
Tel: +1 513 733 6100. Telex: 214452.
Fax: +1 513 733 6855.

Civil Aviation Authority (CAA), Operator Training Standards (OTS), Aviation House, Gatwick Airport South, West Sussex RH6 0YR, UK.
Tel: +44 (0) 1293 573 487.
Fax: +44 (0) 1293 573 999/991.
Principal Training Inspector and Head of OTS Section: Captain T (Paddy) Carver FRAeS.
Technical Manager, Flight Simulation: D R (Don) Irving BSc CEng MRAeS.
The CAA is the national civil aviation regulatory authority for the UK. Captain Carver is also the Chairman of the JAA Simulator (JARSIM) Working Group.

Comator Process AB, Rundgången 10, Box 12050, S-254 52 Helsingborg, Sweden.
Tel: +46 42 18 51 00. Telex: 72864.
Fax: +46 42 15 02 30.
Marketing Director: Joakim Gustafsson
The ESIM artillery trainer is marketed outside Sweden by CelsiusTech Systems AB, but all other products are marketed directly by Comator.

Comet GmbH Pyrotechnik Apparatebau, Postfach (PO Box) 100267, Vieländer Weg 147, D-27574 Bremerhaven 1, Germany.
Tel: +49 471 39 30. Telex: 238 731.
Fax: +49 471 393 94.
Chairman: Dieter Ritzman
Marketing Director: Wolfgang Kowalski
Comet specialises in tank and artillery hit and impact indications, and also makes signal and illumination ammunition. It has about 340 employees and a shop floor area of about 180 000 ft^2 (16 730 m^2).

Compacific Engineering Pte Ltd, PO Box 331, Singapore 9161.
Tel: +65 265 95 96.
Fax: +65 265 61 70.
Chairman: Chew Yam Poey.
Contact: Cheah Ghee Wai.
Compacific makes, amongst other things, 6-DoF synergistic motion platforms.

Comptek Federal Systems Inc, 2732 Transit Road, West Seneca, New York 14224, USA.
Tel: +1 716 677 4070 Ext 519.
Fax: +1 716 677 0014.
President and CEO: John Sciuto
Vice-President, Engineering: Russel Hulsing
Vice-President, Marketing & Business Development: Dr Joseph Sciuto
Director Training and Simulation and Marketing Contact: Tim Kriebel
Founded in 1969, Comptek is involved in EW systems and has about 650 employees and a shop floor area of about 118 000 ft^2 (10 960 m^2).

Computer Sciences Corporation, Integrated Systems Division, Training System Center, 813 Diligence Drive, Newport News, Virginia 23606, USA.
Tel: +1 804 873 1024.
Fax: +1 804 873 1454.

Computer Sciences Corporation, Defense Systems Division, 304 West Route 38, Box N, Moorestown, New Jersey 08057, USA.
Tel: +1 609 234 1166
Fax: +1 609 234 1166

Computing Devices Canada Limited, 3785 Richmond Road, Nepean, Ontario, Canada
Mailing address: PO Box 8508, Ottawa, Ontario K1G 3M9, Canada.
Tel: +1 613 596 7000. Telex: 053 4139.
Fax: +1 613 820 5081.
President: Jack Fawcett
Marketing Director: Paul Coderre
Information Contact: Ruth McMahon +1 613 576 7059
Computing Devices Canada has about 1000 employees and a shop floor area of about 425 000 ft² (39 484 m²).

Continental RPVs, 34924 Victor Street, Barstow, California 92311, USA.
Tel: +1 619 252 4741.
Fax: +1 619 252 7074.

Contraves-SSI, 5902 Breckenridge Parkway, Tampa, Florida 33610-4233, USA.
Tel: +1 813 628 6100.
Fax: +1 813 621 4814/3280.
President: George Dorkhom
Managing Director: Roger Charbonneau
Technical Director: Dr Ed Freeman
Marketing Director: Dave Teichman
Manager, Business Development: Frank J Zinghini, Ext 5039
Contraves Inc Simulation and Systems Integration (SSI) Division is solely concerned with design and production of simulation and training systems, and has been functioning in this capacity for 30 years under previous names of Hydrosystem and Gould Inc. The company was acquired by Contraves in January 1990. SSI is a division of Contraves Inc of Pittsburgh, a wholly owned but independent US corporation which is a subsidiary of Oerlikon-Contraves AG of Zurich, Switzerland, part of the Oerlikon-Bührle group. Contraves-SSI has about 600 employees and a shop floor area of 300 000 ft² (27 870 m²). Product lines include FTDs and flight simulators, marine training systems for ships and submarines, ATC training systems, and fire-fighting trainers.

Cranfield University, Bedford MK43 0AL, UK.
Flight Simulation Courses: Professor David J Allerton (Professor of Avionics).
Course Bookings: Miss A L Roff.
Tel: +44 (0) 1234 750 111 and 1234 754 700.
Fax: +44 (0) 1234 751 206 and 1234 750 083.
e-mail: dja@uk.ac.cranfield
G-Seats for fighter and helicopter simulators: FSML Section, D A (David) Williams.
Tel: +44 (0) 1234 754 047.
Fax: +44 (0) 1234 750 878.
Cranfield is an aeronautically orientated university whose title was at one time the College of Aeronautics. It has an airfield on site which is used by its own flight test and engineering facilities, the International Test Pilots' School (ITPS), and lodger organisations dealing with flying training and sales of fixed-wing and rotary-wing aircraft. Cranfield University deals with, amongst other things, flight simulation theory, hardware and software. As well as the University function, some divisions carry out aircraft modifications and other commercial engineering which, on the simulation side includes the Cranfield designs of G-Seat based either on modifications to real seats provided by simulator manufacturers, or on more cheaply produced replica seats. Seats include ejector seats as well as helicopter designs. Flight Systems and Measurement Laboratories (FSML) engineering facilities have a shop floor area of 580 m² and employ 24 people, but staff numbers involved in the aerospace area total about 350 and a hangar/shop floor area of about 5000 m² is available.

Crisis Simulations Limited, 958A Barrington Street, Halifax, Nova Scotia, Canada B3H 2P7.
Tel: +1 902 420 9775.
Fax: +1 902 422 0571.

Croatia, see LAIR Croatia.

Cross Systems Inc, 8601 Dunwoody Place, PO Box 1284, Building 100, Atlanta, Georgia 30350, USA.
Tel: +1 404 998 8585.
Fax: +1 404 998 8163.

CTA Incorporated, Simulation Systems Division (SSD), 5670 Greenwood Plaza Boulevard, Suite 2000, Englewood, Colorado 80111, USA
Tel: +1 303 889 1200.
Fax: +1 303 889 1296.
President and CEO: C E Velez.
Managing and Technical Director: B A Clausen.
Marketing Director and Contact: Thomas Donohue.
Contact: Diane Herwitz.
The company's Simulation Systems Division (SSD) at Englewood has 140 employees and a 45 000 ft² shop floor area. CTA Incorporated has group headquarters in Rockville, Maryland, and was founded in 1979; the overall group has 1200 employees. The group has annual revenues exceeding $130 million. Main products are space products and development, simulation systems, avionics, C3I and air traffic systems.

Cubic Defense Systems, 9333 Balboa Avenue, San Diego, California 92123, USA.
Tel: +1 619 277 6780.
Fax: +1 619 277 1878.

Custom Training Aids (CTA) Inc, PO Box 759, Swansea, South Carolina 29160, USA.
Tel: +1 803 794 3219.
Fax: +1 803 796 1810.

Daedalean Inc, 9017 Red Branch Road, Columbia, Maryland 21045, USA.
Tel: +1 301 964 6700.
Fax: +1 301 964 6700.

Dassault Electronique, 55 quai Marcel Dassault, F-92214 Saint-Cloud, France.
Tel: +33 (1) 49 11 86 53.
Fax: +33 (1) 46 02 57 58.

Data Sciences Limited (DSL), Meudon Avenue, Farnborough, Hampshire GU14 7NB, UK.
Tel: +44 (0) 1252 544 321. Telex: 858228.
Fax: +44 (0) 1252 513 739.

DATAMAT Ingegneria dei Sistemi SpA, Via Simone Martini 126, I-00143 Rome, Italy.
Tel: +39 6 50 451. Telex: 613 436.
Fax: +39 6 50 45 305.

Datron Systems, 200 Los Angeles Avenue, Simi Valley, California 93065, USA.
Tel: +1 805 584 1717.
Fax: +1 805 584 1717.

DCN, 2 rue Royale, BP 1, F-75200 Paris Naval, France.
Tel: +33 (1) 42 60 33 30. Telex: 270734 SCO DCN.
Fax: +33 (1) 45 54 06 89.
DCN stands for Direction des Constructions Navales.

Defense Mapping Agency (US), DMA Support Center (Attn PMA), Washington DC 203 15-0010, USA

Defense Mapping Agency Systems Support Center, 12100 Sunset Hills Road, 200 Reston, Virginia 22090-3207, USA
Tel: +1 703 756 9329
Fax: +1 703 285 9396
The US Defense Mapping Agency (DMA) is, amongst other things, involved in the multinational DLMS programme which involves digitising terrain data (DTED) and feature data (DFAD) to a number of levels of fidelity. It is also involved in the successor system MIL STD 1821/Project 2851 which has a more sophisticated architecture than DLMS. These programmes involved in the digitisation of real-world data are not only for the construction of databases for simulators and training aids, but also for the long-term replacement of paper maps with digital data, and with changes of mapping and navigational data to worldwide geographic reference systems such as the WGS 84 Geodetic System, and associated navigational aids such as the satellite-based US GPS which gives accurate ground positions from simple receivers anywhere in the world.

Delex Systems Inc, 1953 Gallows Road, Suite 700, Vienna, Virginia 22182-3991, USA.
Tel: +1 703 734 8300. Telex: 901 154.
Fax:+1 703 734 9303 or +1 703 893 5338.
President: Jack Piper
Vice-President, Operations: James Wigge
Vice-President, Business Development: Doug Greenlaw
Marketing Contact: Rick Wasser
Delex has about 250 employees.

Denel (Pty) Limited, PO Box 8322, Hennopsmeer 0046, South Africa.
Tel: +27 12 428 0911/0639.
Fax: +27 12 347 0300 and 428 0947.
Communication Manager: Eben Mouton
Denel (part of Armscor) consists of 21 divisions in the following six categories: Aerospace, Mechanical & Electronic Systems, Informatics and Properties, Manufacturing, Business Development, and International Marketing.

Deutsche Aerospace AG (DASA), Defense and Civil Applications Group, Postfach 44 03 20, D-8000 München 44, Germany.
Tel: +49 89 38 19 98 42.
Fax: +49 89 38 19 98 46.

Deutsche Aerospace AG (DASA), Simulation Department of the Military Aircraft Division, D-81663 München 44, Germany.
Tel: +49 89 60 72 57 14.
Fax: +49 89 60 72 49 36.

Diagnostic Retrieval Systems (DRS) Inc, 16 Thornton Road, Oakland, New Jersey 07436, USA.
Tel: +1 201 337 3800/2603.
Fax: +1 201 337 4775/3314.
President: David E Gross
CEO: Leonard Newman
President, DRS Electronic Systems Group: Paul Casner
Vice-President, Engineering: Eric Hanson
Senior Vice-President, Business Development: Carl Boecher
Marketing Contact: Ronald Freitag
Asst Vice-President, Company Communication: Patricia Williamson
DRS was formed in 1968 and has about 550 employees and a shop floor area of about 10 000 ft² (930 m²). DRS makes, amongst other things, sonar training sets and a periscope image generator.

DIEHL GmbH & Co, Elektronik und Luftfahrgeräte, Stephanstrasse 49, D-8500 Nürnberg 30, Germany.
Tel: +49 911 947 2843. Telex: 623 942.
Fax: +49 911 947 3680.

Diemaco, 1036 Wilson Avenue, Kitchener, Ontario, Canada N2C 1J3.
Tel: +1 519 893 68 40.
Fax: +1 519 893 31 44.
Chairman and Managing Director: Henrik Noesgaard
Technical Director: Gord Brumby
Marketing Director: Maurice Clermont
Marketing Contact: Philip C O'Dell
Diemaco makes training rifles and blank training adaptors. It has about 100 employees and a shop floor area of about 55 000 ft² (5100 m²).

DIXI SA, Usine No 1, 35 rue de la Côte, CH-2400 Le Locle, Switzerland.

DLR (Deutsche Forschungsanstalt für Luft- und Raumfahrt e. V), Institut für Flugmechanik, Braunschweig, Germany Tel: +49 531 2951
Fax: +49 531 2951
Contacts: Dr Gary J Gray, Dr Wolfgang von Grünhagen

Dornier GmbH, PO Box 1420, D-7990 Friedrichshafen 1, Germany.
Tel: +49 7545 82 965. Telex: 734209-0.
Fax: +49 7545 85 583.

Dornier Luftfahrt GmbH, Naval Training, Support, and Training and Special Products Divisions, Aumunder Feldstrasse 54, D-2820 Bremen 70, Germany.
Tel: +49 421 658 430. Telex: 245 176.
Fax: +49 421 658 4318.

Dornier Luftfahrt GmbH, International Logistics Division, Oberpfaffenhofen-Flugplatz, Postfach (PO Box) 1103, D-82230 Wessling, Germany.
Tel: +49 815 30 30 29 92 and 49 40.
Fax: +49 815 30 30 49 54.
President: H Kränzle
Managing Director: Dr K Bauer
Technical Director: K Höpfl
Marketing Director: H-J von Scheven
Marketing enquiries: K Höpfl, H-J von Scheven
Dornier International Logistics is part of Dornier Luftfahrt GmbH, a corporate unit of Deutsche Aerospace.

Doron Precision Systems Inc, 174 Court Street, PO Box 400, Binghamton, New York 13902-0400, USA.
Tel: +1 607 772 1610.
Fax: +1 607 772 6760.
President: Karl J Hirschman
Chairman: Carl J Wenzinger
Executive Vice-President: W Ralph Price
Vice-President, Engineering: Kenneth R Kittinger
Marketing Director, International: Owen W Murray
Marketing Director, Domestic: David E Lindsey
Doron has about 100 employees and a shop floor area of about 47 000 ft² (4370 m²). It makes driving trainers for cars, vans, trucks and buses. Over 3000 systems have been delivered to over 50 countries.

DRA, UK. Defence Research Agency of the UK Ministry of Defence (MoD)
Headquarters: DRA, Farnborough, Hants GU14 6TD, UK
Tel: +44 (0) 1252 392 000
Fax: +44 (0) 1252 370 753
Chief Executive: J A R Chisholm
Managing Director, Programmes: Dr R H Warren
Managing Director, Operations: P D Ewins
New Business Manager, Flight Simulation: Tel: +44 (0) 1234 225 224; Fax +44 (0) 1234 225 409.
The DRA was formed in 1990 by the amalgamation of the

UK MoD Research Establishments. Previously these had been separate organisations namely, the Admiralty Research Establishment (ARE) with its HQ at Portsmouth, the Admiralty Underwater Weapons Research Establishment (AUWE) with its HQ at Portland, the Royal Aircraft Establishment (RAE) with its HQ at Farnborough, the Royal Armament Research and Development Establishment (RARDE) with its HQ at Fort Halstead, and the Royal Signals Research Establishment (RSRE) with its HQ at Malvern. Since the merger, rationalisation of facilities has taken place and a more commercially orientated structure established. Facilities may also be used by industry and other outside bodies at negotiated rates subject to being compatible with the planned programme of research and associated trials, and appropriate security and other clearances. The DRA is organised into Business Sectors as follows: Above Water Systems, Aircraft Systems, Command and Information Systems, Chemical and Electronics, Electronics, Fighting Vehicles and Systems, Quality Assurance and Technical Services, Operational Studies, Underwater Systems, and Weapon Systems. The Divisions listed below are part of the Flight Systems Department of the Aircraft Systems Business Sector.

DRA Flight Dynamics and Simulation Department, DRA Bedford, Bedfordshire MK41 6AE, UK
Tel: +44 (0) 1234 270 077
Fax: +44 (0) 1234 225 409
Department Manager: Peter England
Simulator Policy: Barry Tomlinson
Motion Systems: John Hall
Manager, Advanced Flight Simulator: Barry Moulang
Business Manager, Flight Simulation: Tel: +44 (0) 1234 225 369; Fax +44 (0) 1234 225 409.
The department is responsible, amongst other things, for the basic technology of flight simulation and its application in training and to validation of flight models. See the entry for DRA Flight Simulation in the Flight Simulator section of the book

DRA Systems Integration Division, DRA Farnborough, Hants GU14 6TD, UK
Tel: +44 (0) 1252 393 061
Fax: +44 (0) 1252 394 700
Head of Division: Geoff Butler
Head of Simulation: Derek Barter
Business Manager, Flight Simulation: Tel: +44 (0) 1234 225 224; Fax +44 (0) 1234 225 409.
Manager, Flight Systems Department and Point of Contact: Tel: +44 (0) 1252 392 400; Fax: +44 (0) 1252 376 392.
The Systems Integration Division is responsible, amongst other things, for the use of flight simulation techniques in the integration of systems in flight.

DRA Battle Group Wargaming and Simulation Section, Operational Studies Business Sector, DRA Fort Halstead, Sevenoaks, Kent TN14 7BP
Tel: +44 (0) 1959 514 924
Fax: +44 (0) 1959 514 924

DRA Parallel Processing and Military Simulation Section, E 214 Building, DRA Malvern, St Andrews Road, Malvern, Worcestershire WR14 3PS, UK.
Tel: +44 (0) 1684 895 121.
Fax: +44 (0) 1684 894 389.
e-mail: jbgr@dra.hmg.gb.
On-line information also from the UK Simulation Interoperability Working Group gopher database URL ref: gopher://siwg.dra.hmg.gb/
Contact: Dr J B G Roberts.

DRA Partnership in Technology Programme, DRA Farnborough, Hants GU14 6TD, UK.
Tel: +44 (0) 1252 39 45 17.
Fax: +44 (0) 1252 39 46 50.
Programme Manager: Colonel Roy Giles
This programme is the DRA Interface with TsAGI (qv) in Russia. The programme manager is a Russian speaker and has the appropriate contacts in TsAGI and other organisations.

DSD Digital Systems and Design Limited, 18 Shakespeare Business Centre, Hathaway Close, Eastleigh, SO5 4SR, UK.
Tel: +44 (0) 1703 620 499. Telex: 477575.
Fax: +44 (0) 1703 629 164.

DST Deutsche System-Technik GmbH, Hans-Bredow-Strasse 20, D-2800 Bremen 44, Germany.
Tel: +49 421 42 87-0.
Fax: +49 421 40 46 60.

DST Deutsche System-Technik GmbH, Postfach 6420, Edisonstrasse 3, D-2300 Kiel 14, Germany.
Tel: +49 431 71 09-1.
Fax: +49 431 713 084.

Dundridge College Limited, Harberton, Totnes, Devon TQ9 7PP, England.
Tel: +44 (0) 1803 865 118
Fax: +44 (0) 1803 863 195
Dundridge College runs courses for air traffic controllers and is associated with Marconi (ex-Ferranti) Simulation and Training.

DynaLantic Corporation, 100 Jefryn Boulevard, Deer Park, New York 11729-5759, USA.
Tel: +1 516 242 4300.
Fax: +1 516 242 4306.
President and Chief Executive: Paul J Patin
Vice-President, Engineering: Gerald St George
Vice-President, Marketing: Robert Catron
DynaLantic was founded in 1984 and specialises in simulation and training devices. It has about 50 employees and a shop floor area of about 25 000 ft² (2323 m²). It has produced maintenance trainers, ordnance trainers, propulsion trainers, a helicopter flight simulator and a ship control trainer.

Dynamit Nobel AG, Defence Division, Kaiserstrasse 1, PO Box 1261, D-53 839 Troisdorf, Germany.
Tel: +49 224 189 1298. Telex: 885 666 DN D or 889 660-44.
Fax: +49 224 189 1542/1669 or 224 185 4585.
Chairman and CEO: Dr Axel Homburg
Managing Director: Dr Norbert Wedel
Marketing Manager: Erich Tappert
Marketing Contact: Franz Wolfgarten
The Dynamit Nobel group has about 11 000 employees.

EASAMS Limited, Lyon Way, Frimley, Camberley, Surrey GU16 5EX, UK.
Tel: +44 (0) 1276 63 377. Telex: 858115.
Fax: +44 (0) 1276 683 468.

ECA, 17 avenue du Château, F-92194 Meudon Cedex, France.
Tel: +33 (1) 42 14 86 59. Telex: 614 416F.
Fax: +33 (1) 41 14 86 64.
Chairman: J P Gorgé
Business Development Manager: C Vigne
Marketing Director: P Corman
Marketing Contact: Caroline Detalle
ECA was founded in 1936 and specialises in remotely operated vehicles. It has about 180 employees.

ECC International Corporation (Marketing and Executive Offices), 175 Strafford Avenue, Wayne, Pennsylvania 19087-3377, USA.
Tel: +1 610 687 2600.
Fax: +1 610 254 9268.
President and CEO: George W Murphy
Executive Vice-President: George L Frye, Jr
Vice-President, Engineering: Ajit Hirani
Vice-President, Manufacturing: Spencer C Whitehead
Vice-President, Instructional Systems Development: Nicholas A Sieko
Vice-President, Marketing: Patrick M Donohue

ECC International Corporation (Systems Design and Development Center), 2882 South Tampa Avenue, Orlando, Florida 32809, USA.
Tel: +1 407 859 74 10.
Fax: +1 407 851 18 71.
ECC has about 900 employees and a shop floor area of about 500 000 ft² (46 450 m²) at the Orlando site.

ECC Simulation Limited, Kingston Wharf, Brighton Road, Shoreham-by-Sea, West Sussex BN43 6RN, UK.
Tel: +44 (0) 1273 870 486.
Fax: +44 (0) 1273 592 122.
Marketing Support Manager: Mike Wright
ECC Simulation Ltd is a wholly owned subsidiary of the ECC International Corporation of Wayne, Pennsylvania. ECC Simulation Ltd makes and supports training systems for the UK and European markets.

EDS (Electronic Data Systems Limited), 1 Bartley Way, Bartley Wood, Hook, Hants RG27 9XA, UK
Tel: +44 (0) 1256 74 20 00
Fax: +44 (0) 1256 74 27 00

Kurt Eichweber Präzionsgerätewerk
This company is understood to be no longer trading. Any further information on it and its products would be welcome. Its last known address was: Postfach 50 04 23, Leverkusenstrasse 13, D-2000 Hamburg 50, Germany.
Tel: (49) 40 85 39 60. Telex: 214348.

Elbit Computers Limited, POB 5390, Haifa 31053, Israel.
Tel: +972 4 315 315, 556 677. Telex: 46 774.
Fax: +972 4 550 002.

Elsag Bailey (Genoa, Italy), see Bailey

Electronics and Space Corporation (ESCO), 8100 West Florissant Avenue, St Louis, Missouri 63136, USA.
Tel: +1 314 553 2000/4444/4678/3334.
Fax: +1 314 553 4122/4698/4321.
President: G A Potthoff
Senior Vice-President and General Manager: D M Purdy
Vice-President Engineering: V V Reznack
Vice-President Business Development: M Murphy
Manager, Trade Shows: Donald C Gasch
ESCO is a subsidiary of the ESCO Electronics Corporation and has about 1000 employees and a shop floor area of about 500 000 ft² (46 450 m²). It is also sometimes abbreviated to E&S, but should not be confused with Evans and Sutherland which manufactures the ESIG range of CGI systems in Salt Lake City. ESCO manufactures defence and surveillance products including airborne and ground radars, ELINT equipment, sighting and guidance systems, and ATE for avionics and vehicles. In training equipment ESCO makes a target simulator for the ground-launched TOW ATM, a turret trainer for the M901A1, and the Helitow gunner trainer.

Electronintorg Limited, 24/2 ul. Usievicha, 125315 Moscow, Russia.
Tel: +7 095 151 06 40. Telex: 411326.
Fax: +7 095 151 54 41.

Elesco Oy, Luomannotko 4, PO Box 128, FIN-02201, Espoo, Finland.
Tel: +358 420 8600.
Fax: +358 420 8610.

Elisra Electronic Systems Limited, 48 Mivtza Kadesh Street, Bene Beraq IL-51203, Israel.
Tel: +972 3 754 5111.
Fax: +972 3 754 5468.

EMS Development Corporation, 100 Sea Lane, Farmingdale, New York 11735, USA.
Tel: +1 510 293 7900.
TWX: 510 244 6480.

Engineering Computer Optecnomics Inc (ECO), 1356 Cape St Clair Road, Annapolis, Maryland 21401, USA.
Tel: +1 410 757 32 45.
Fax: +1 410 757 82 65.
President: Virgil F Keith
Vice-President: Dr Stan Finger
Technical Director: Jon Porricelli
Marketing Director: Colleen Purcell
ECO has about 25 employees and a shop floor area of about 8000 ft² (740 m²).

Environmental Tectonics Corporation (ETC), 125 James Way, County Line Industrial Park, Southampton, Pennsylvania 18966-3877, USA.
Tel: +1 215 355 91 00. Telex: RCA244926 ETC UR
Fax: +1 215 357 40 00.
ETC was founded in 1969 by William F Mitchell, the present President and CEO. It makes aircrew training devices, flight simulators, environmental test chambers and equipment, and gas and steam sterilisers. The Simulators Division is located at Orlando, Florida.

Ericsson Radar Electronics AB, Airborne Electronics Division, S-164 84 Stockholm, Sweden.
Tel: +46 8 757 3000. Telex: 19784
Fax: +46 8 752 8967 and +46 31 67 17 03
Chairman: Lars Ramquist
Managing Director: Bengt Halse
Marketing Director: Lars-Olof Linders
PR Manager: Lars Köhler
Ericsson Radar Electronics AB is a wholly owned subsidiary of Telefonaktiebolaget LM Ericsson which has some 70 000 employees and operates in about 100 countries, concentrating on telecommunications. Ericsson Radar Electronics AB employs about 2700 people and has been producing radar equipment since 1956. It produces a number of radars for ships, artillery location and for aircraft such as the Gripen and for AEW purposes. It is associated with Rodale Electronics Inc of Garden City, New York (qv in this directory), a company specialising in EW training equipment.

Esperanza y Cia SA (ECIA), Avda. de Xemein, Marquina, E-48270 Vizcaya, Spain.
Tel: +34 4 616 7616/7626. Telex: 34 725 ECMA E.
Fax: +34 4 616 7878.

Etienne Lacroix SA, 6 boulevard de Joffrery, F-31600 Muret, France.
Tel: +33 61 566 500. Telex: 531 478 F.
Fax: +33 61 524 277.

European Aviation Consultants (EAC), 16 Scoular Drive, North Seaton Village, Northumberland NE63 9SE, UK
Tel: +44 (0) 1670 811 811.
Fax: +44 (0) 1670 811 811.
Contact: Frode Iversen.
EAC, amongst other things, organise conferences on aviation training matters.

Evans and Sutherland, Simulation Division, 600 Komas Drive, Salt Lake City, Utah 84 108, USA.
Tel: +1 801 582 5847.
Fax: +1 801 582 5848 and 581 9852.
President: Rod Rougelot
Evans and Sutherland (E&S) develops and manufactures

Computer Generated Image equipment in the ESIG series (E&S Image Generator). Previous products included the SP and CT series (Special Product, Continuous Tone); the SP-X system was particularly popular. From 1973 to July 1994, E&S had an agreement with the Rediffusion company of the UK (subsequently Hughes Rediffusion and then Thomson Training; see the entry under Thomson) that E&S CGI systems would be marketed on Rediffusion civil simulators in the worldwide civil market and the UK military market. Since the development of Thomson's own SPACE CGI system and the termination of the agreement with E&S, E&S is now an independent supplier of CGI systems to the world market. It should be noted that Thomson has stated that it will continue to fit E&S CGI systems to Thomson-built simulators at customer request.

FAA, see under Federal Aviation Administration.

Fairchild-Weston Systems Inc, 300 Robbins Lane, Syosset, New York 11791, USA.
Tel: +1 516 349 2200.

Farner Air Service AG, Grenchen Airport, CH-2540 Grenchen, Switzerland.
Tel: 065 525341. Telex: 934565.

Federal Aviation Administration (FAA), National Simulation Program, AFS-205, PO Box 20636, Atlanta, Georgia 30320, USA.
Tel: +1 404 305 6100.
Fax: +1 404 305 6118.
Managers, National Simulation Program: Ed Cook, Paul A Ray.
The FAA is the national civil aviation regulatory authority for the USA.

Federal Directorate of Supply and Procurement (SDPR), Belgrade, Serbia.
Title changed to YUGOIMPORT-SDPR Ltd, and address and telephone numbers changed. See YUGOIMPORT.

Ferranti International Simulation and Training, Cwmbran, UK;
Ferranti International Simulation and Training, Cheadle Heath, UK;
Ferranti Technologies Inc, Lancaster, Pennsylvania, USA;
Titles changed from Ferranti to Marconi on the acquisition by GEC of Ferranti's remaining defence interests. Other details remain as before. Now listed under Marconi.

FIAT Componenti e Impianti per l'Energia e l'Industria SpA, Unitá Tecnologico Industriale SEPA, Corso Guilio Cesare 300, I-10154 Turin, Italy.
Tel: +39 11 26 821. Telex: 221 527.
Fax: +39 11 242 0372.

Firearms Training Systems Inc, International Division, 110 Technology Parkway, Norcross, Georgia 30092, USA.
Tel: +1 404 813 0180
Fax: +1 404 813 0741

Firearms Training Systems Inc, 7340 McGinnis Ferry Road, Suwanee, Georgia 30174, USA.
Tel: +1 404 813 0180.
Fax: +1 404 813 0741.

Flight Refuelling Limited (FRL), Brook Road, Wimborne, Dorset BH21 2BJ, UK.
Tel: +44 (0) 1202 88 21 21. Telex: 41247.
Fax: +44 (0) 1202 88 00 96.

FlightSafety International (FSI)
FSI, Marine Air Terminal, La Guardia Airport, Flushing, New York 11371-1061, USA.
Tel: +1 718 565 4100/4120/4124.
Fax: +1 718 565 4134
President: Albert L Ueltschi
Executive Vice-President: Bruce N Whitman
Vice-President, Marketing: James S Waugh
FlightSafety International (FSI) was founded in 1951 by Mr Al Ueltschi, a former PanAm pilot. Its activities revolve about training. It has a fleet of over 150 training devices including 135 FAA-certified simulators, claimed to be the world's largest simulator fleet. It trains some 40 000 pilots and maintainers annually, and has a staff of about 1300 of whom over 1000 are instructors. It also trains in the marine and power industries, under MarineSafety International at King's Point, New York (also at Rotterdam, Holland), and PowerSafety International at Homer City, Pennsylvania (in association with the Babcock and Wilcox Company). See below for details of MarineSafety and PowerSafety

FSI had revenues of $297 million in 1993, split between training revenues of $245 million and $54 million for product sales. The company has three divisions, the Instructional Systems Division at Hurst, Texas; Simulation Systems Division at Broken Arrow near Tulsa, Oklahoma; and Visual Simulation Systems Division at St Charles, Missouri.

The company has Flight Training Academies at Fort Worth, Texas; Vero Beach, Florida; and Lakeland, Florida. It has learning centres for pilots at College Park and Savannah, Georgia; Bethany, Oklahoma; Erlanger, Kentucky; Daleville and Dothan, Alabama; Fort Worth, Houston, and San Antonio, Texas; New Castle, Delaware; La Guardia, New York; Long Beach, California; Miami and West Palm Beach, Florida; Montreal and Toronto, Canada; Paris le Bourget, France; St Louis, Missouri; Salt Lake City, Utah; Seattle, Washington State; Moonachie, New Jersey; Toledo, Ohio; Tucson, Arizona; Wichita, Kansas (4 centres at Wichita for Beech, Cessna, Citation and Learjet aircraft). Maintenance learning centres are at Savannah and there are 2 centres at Wichita for Beech and Cessna aircraft.

FSI Simulation Systems Division (SSD), 2700N Hemlock Circle, Broken Arrow, Oklahoma 74012, USA
Tel: +1 918 251 0500
Fax: +1 918 258 0306
Vice-President, Simulation Systems: Dennis Gulasy
Chief Hardware Engineer: Fred Narrin
Chief Software Engineer: Ken Allen
Marketing: Paul Hancock
Manager, Product Information: John Slish.
Broken Arrow is near Tulsa. SSD was purchased in 1978 from Atkins and Merrill and now has about 435 employees and a shop floor area of about 110 000 ft² (10 220 m²).

FSI Visual Simulation Systems Division, 2590 North Highway 94, St Charles, Missouri 63301, USA
Tel: +1 314 925 8400/8500/8411
Fax: +1 314 925 8444
Marketing: Nancy Ferebee.
FlightSafety International acquired the Visual Simulation Systems (VSS) business unit of the McDonnell Douglas Corporation on 1 February 93. This included its St Charles, Missouri facility and its Vital product line of simulation systems which includes CGI systems and display systems such as the MultiView collimated cross-cockpit system.

FSI Subsidiary: See under MarineSafety International
FSI Associate: PowerSafety International, Mike Wida Industrial Park, 325 Ferguson Road, Homer City, Pennsylvania 15748, USA
Tel: +1 412 479 3585
Fax: +1 412 479 3911

Flight Systems Inc, 1901 Dove Street, PO Box 2400, Newport Beach, California 92660, USA.
Tel: +1 714 833 9661

FlytSim Training Limited, 53-57 Queens Road, High Wycombe, Bucks HP13 6AH, UK
Tel: +44 (0) 1494 45 95 45. Telex 849 462 TELFAC G
Fax: +44 (0) 1494 45 95 47
Chairman: D A (Dennis) Jeffries
Managing Director: L J (Les) Burnett
Technical Director: C Eaton
Marketing Contact: L J (Les) Burnett
FlytSim was founded in 1972 and markets a range of low-cost motion platforms giving 2 DoF, 3 DoF and 4 DoF. It also manufactures flight simulators and flight training devices to order. It operates a DC-9 flight simulator at its site at High Wycombe for British Midland Airways and other DC-9 operators. It has 10 employees and a shop floor area of 4500 ft². A related company is Simtech Limited (a UK company not related to SIMTECH of Israel) which makes simulator instruments and subassemblies, and also simulators for other vehicles such as cars, buses and trains. Simtech operates from the High Wycombe address given above.

FMC Corporation, see United Defense LP.

FN Herstal, Fabrique Nationale Nouvelle Herstal SA, Voie de Liège 33, B-4040 Herstal, Belgium.
Tel: +32 41 408 111. Telex: 41 223 fabna
Fax: +32 41 408 679.

Fokker Control Systems, Fokker Aircraft BV, PO Box 7600, NL-1117 ZJ Schiphol, Netherlands.
Tel: +31 20 605 2653
Fax: +31 20 605 2524
Commercial and General Manager: Aart J Hemmink

Ford Aerospace, Sunnydale Division, 1260 Crossman Avenue, PO Box 1198, Sunnyvale, California 94089, USA.
Tel: +1 408 743 3957.

Frasca International Inc (FII), 906 East Airport Road, Urbana, Illinois 61801, USA.
Tel: +1 217 344 9200.
Fax: +1 217 344 9207.
President: Rudy Frasca
Vice-President Operations: John Frasca
Technical Director: Marty Hendrikson
Marketing Director: Charlie Douglas
Marketing Enquiries: Joel Prichard
The Frasca company was founded in 1958 by Rudy Frasca after service in the USN and research in Flight Simulation at the University of Illinois. Frasca International Inc (FII) has made over 1400 training devices and simulators for customers in 60 countries. The company is located at Frasca Field GA airfield and has about 100 employees and a shop floor area of 6500 m².

FSPF, Baghdad, see Baghdad

Fuji Jukogyo Kabushiki Kaisha (Fuji Heavy Industries (FHI) Limited), Aerospace Division, Subaru Building, 7-21 l-chome, Nishi-Shinjuku, Shinjuku-ku, Tokyo 160, Japan.
Tel: +81 03 347 2533/2525. Telex: 03-232-2268/2588 FUJI J.
President: Isamu Kawai
Managing Director Aerospace Division: Yasuyuki Kogure
General Manager Utsunomia Plant: Tsutoo Ono
General Manager, Marketing Department: Kisaburo Wani
The FHI group employs about 14 800 people and has a shop floor area of about 1 329 400 ft² (123 509 m²).

Gajon Institute of Technology, Via Piaggio 9, I-16136 Genoa, Italy.
Tel: +39 (10) 224 641. Telex: 281 229.
Fax: +39 (10) 212 671.

GEC Avionics Limited, Sonar Systems Division, 26-28 Hyde Way, Welwyn Garden City, Hertfordshire AL7 3BD, UK.
Tel: +44 (0) 1707 328 511. Telex: 269 404.
Fax: +44 (0) 1707 327 822.

GEC-Marconi Avionics Limited, Displays Division, 1 South Gyle Crescent, Edinburgh EH12 9HQ, UK.
Tel: +44 (0) 131 316 4545. Telex: 727 101.
Fax: +44 (0) 131 314 8237.

GEC Avionics Limited, Maritime Aircraft Systems Division, Airport Works, Rochester, Kent ME1 2XX, UK.
Tel: +44 (0) 1634 816 124. Telex: 96 333.
Fax: +44 (0) 1634 816 579.

GEC-Marconi Avionics Limited, Sensors Division, Christopher Martin Road, Basildon, Essex SS14 3EL, UK.
Tel: +44 (0) 1268 522 822.
Fax: +44 (0) 1268 883 140.

GEC-Marconi Underwater Weapons Division, Elettra Avenue, Waterlooville, Hampshire PO7 7XS, UK.
Tel: +44 (0) 1705 264 466. Telex: 869 233.
Fax: +44 (0) 1705 260 246.

General Dynamics, Fort Worth, see Lockheed Fort Worth Company.

General Electric Company (GE USA, not GEC UK), Simulation and Control Systems Department, Daytona Beach (Makers of Compu-Scene series of CGI image generation systems), now under Martin Marietta (Lockheed Martin).

Giat Industries, 2 rue d'Alsace-Lorraine, BP 1450, F-65014 Tarbes Cedex, France.
Tel: +33 62 386 270. Telex: ARSEN 531 581 F.
Fax: +33 62 386 151.

Giat Industries, 13 route de la Minière, Satory, F-78034 Versailles Cedex, France.
Tel: +33 30 97 37 37.
Fax: +33 30 97 39 00.

GIFAS (Groupement des Industries Fançaises Aéronautiques et Spatiales), 4, rue Galilee, 75782 Paris Cedex 16, France.
Tel: +33 (1) 44 43 17 00/50 Telex: 645615F
Fax: +33 (1) 40 70 91 41
Press and Communication Manager: Patrick Guerin
GIFAS is the French Aerospace industry trade association and, amongst other activities, is the organiser of the Paris Air Show which takes place every two years, normally in June. The next shows are scheduled for 1995 and 1997.

Giravions Dorand Industries, 5 rue de Jean-Macé, BP 30, F-92151 Suresnes Cedex, France.
Tel: +33 (1) 46 972 600. Telex: VERSUR 612 961 F.
Fax: +33 (1) 42 045 820.
Giravions Dorand is linked to CESELSA and the Indra Group, see under CESELSA

Glock GmbH, Hausfeldstrasse 17, A-2232 Deutsch-Wagram, Austria.
Tel: 2247 24 60. Telex: 133307 GLOCK A.
Fax: 2247 24 60 12.

Graseby Dynamics Limited, Ionics Division, 459 Park Avenue, Bushey, Hertfordshire WD2 2BW, UK.
Tel: +1 (0) 1923 238 483. Telex: 923 010.
Fax: +1 (0) 1923 221 361.
Chairman: Chris P Fowler
Managing Director: Robert C Hood
Technical Director: Peter J Day

Marketing Director: Mike Booth
Marketing Contact: Ravi Mawkin
Publicity Manager: Monica Ridley.
Graseby has about 220 employees and a shop floor area of about 960 000 ft^2 (8900 m^2).

Gunpell & Co Limited, Wildacres, Reading Road North, Fleet, Hampshire GU13 8HS, UK.
Tel: +44 (0) 1252 617 590.
Fax: +44 (0) 1252 815 560.
Managing Director: John A Vertanness
Gunpell was formed in 1985 and makes bullet traps and target systems for pistols and rifles.

Haley & Weller Limited, Wilne, Draycott, Derbyshire DE72 3QJ, UK.
Tel: +44 (0) 1332 872 475. Telex: 378215 HALWEL G.
Fax: +44 (0) 1332 873 046.

Harris Corporation, Government Systems Division, 1025 West NASA Boulevard, Melbourne, Florida 32919, USA.
Tel: +1 305 727 9100. Telex: 803719.
TWX: 510 959 6010.

Haven Automation Limited, Cwmdu Industrial Estate, Gendros, Swansea, SA5 5LQ, UK.
Tel: +44 (0) 1792 588 722. Telex: 48479.
Fax: +44 (0) 1792 582 624.

Hawker De Havilland Limited, Systems Division, PO Box 580, Salisbury, South Australia 5108, Australia.
Tel: +61 (8) 258 3422. Telex: 89706.
Fax: +61 (8) 281 2529.

Hawker De Havilland Victoria Limited, 304 Lorimer Street, Port Melbourne, Victoria, Australia 3207
Tel: +61 3 647 6587
Fax: +61 3 646 6310
Contact: P Middleton
Hawker de Havilland (HDH) is Australia's oldest and largest aerospace company, having been founded in 1927. In addition to the Hawker Pilot Trainer, its Victoria division also manufactures cabin service trainers, emergency evacuation trainers and door trainers, all for the larger transport aircraft.

Hayes International Corporation, PO Box 707, Leeds, Alabama 35094-0707, USA.
Tel: +1 205 699 6151. Telex: 78 2550.
Fax: +1 205 699 8999.

Heckler and Koch GmbH, D-7238, Oberndorf-Neckar, Germany.
Tel: +49 742 3791. Telex: 760 313 HUKO D.
Fax: +49 742 379 2539.

Hellenic Aerospace Industry Limited, PO Box 23, Tanagara, GR-320 09 Schimatari, Greece.
Tel: +30 262 52 272.
Fax: +30 262 52 170.

D Hendry Electronics Limited, 2 Fitzalan Road, Arundel, West Sussex BN18 9JS, UK.
Tel: +44 (0) 1903 882 255. Telex: 877 159.
Fax: +44 (0) 1903 884 252 and 883 716.
Chairman and Managing Director: D J Hendry
Technical Director and Marketing Contact: P J King
The company was formed in 1979 and specialises in small arms trainers. It has delivered some 350 training devices. It has about 15 employees and a shop floor area of about 7000 ft^2 (650 m^2).

Hercules Defense Electronics Systems Inc, PO Box 4648, Clearwater, Florida 34618-4648, USA.
Tel: +1 813 572 1900.

Hollandse Signaalapparaten BV, PO Box 42, NL-7550 GD Hengelo Ov, Netherlands.
Tel: +31 74 488 111. Telex: 44310.
Fax: +31 74 425 936.

Honeywell Geschäftsbereich Sondertechnik GmbH, Honeywellstrasse, D-6457 Maintal 1, Germany.
Tel: +49 6181 4011. Telex: 4184 820.
Fax: +49 6181 4011.

Honeywell, Defense Avionics Systems Division, 9201 San Mateo Boulevard NE, Albuquerque, New Mexico 87113, USA.
Tel: +1 505 828 5571. Telex: 501 386.
Fax: +1 505 828 5500.

Hordaland Mek. Verkstad A/S, Hardangerv. 23, Box 23, N-5701 Voss, Norway.
Tel: +47 55 11 422. Telex: 42522 hmvvs n.
Fax: +47 55 10 065.

Hughes Aircraft Company, Ground Systems Group, 1901 Malvern Street, PO Box 3310, Fullerton, California 92634, USA.
Tel: +1 714 871 3232. Telex: 685504.
Fax: +1 714 871 3232.

Hughes Aircraft Company, Product Support Division, Training and Support Systems Group, Building 6 MB/A20, PO Box 1011, 1240 Rosecrans Avenue, Manhattan Beach, California 90266-8511, USA.
Tel: +1 213 536 5564/5569.
Fax: +1 213 536 5635.

Hughes Rediffusion Simulation Limited (HRSL, obsolete title), now part of Thomson Training and Simulation; see Thomson-CSF

Hughes Training Inc (HTI), 5301 East River Road, Minneapolis, Minnesota 55421, USA.
Tel: +1 612 572 7700.
Fax: +1 612 572 7732.

Hughes Training Inc (HTI), 13775 McLean Road, Herndon, Virginia 22071, USA.
Tel: +1 703 481 4500.
Fax: +1 703 689 0018.

Hughes Training Inc (HTI), 1200 East San Bernardino Road, West Covina, California 91790-1098, USA.
Tel: +1 818 331 0011. Telex: 670 452.
Fax: +1 818 915 9250.

Hughes Training Inc (HTI), 2200 Arlington Downs Road, Arlington, Texas 76011, USA.
Tel: +1 817 695 2000.
Fax: +1 817 695 2222/2690.

Hughes UK Limited (HUKL), ATC Simulation and Training, Jenner Road, Crawley, West Sussex RH10 2GA, UK
Tel: +44 (0) 1293 401 914 and 900
Fax: +44 (0) 1293 401 901
General Manager: Tom Howard-Jones
Operations Manager: Jerry Richold
Technical Manager: David Tomlinson
Marketing Manager: Ron Mahendran
Sales Manager: Paul Thompson
HUKL ATC Simulation and Training is a Hughes Aircraft Company in the Hughes UK Systems Division.

Hydraudyne Systems and Engineering BV, Kruisbroeksestraat 1a, PO Box 32, 5280 AA Boxtel, The Netherlands
Tel: +31 (0) 41 16 51 951
Fax: +31 (0) 41 16 51 453
Contact: Huub Vermeulen
The Hydraudyne Group is an affiliate of Mannesmann Rexroth, and manufactures, amongst other things, a range of hydraulic motion platforms for ship, submarine, AFV and aircraft simulation. Hydraudyne was founded in 1954 and specialises in hydraulic systems from small units to docks capable of lifting whole ships, and also in electrically actuated systems. They have fitted large hydraulic units to semi-submersible drilling rigs for heave compensation, and even jacked up the 40 000 ton Ekofisk platform in the North Sea when it suffered from subsidence.

I. C. Sim, of Irvine, California. Acquired by CTA Inc. See CTA.

IDEX (International Defence EXhibition), PO Box 309-40, Abu Dhabi, United Arab Emirates.
Tel: +971 2 336 864
Fax: +971 2 332 393

IDEX Foreign Trading, Contracting and Engineering Company Limited, PO Box 24, H-1251 Budapest, Hungary.
Tel: +36 (1) 150 090. Telex: 22-4541 IDEX H.
Fax: +36 (1) 351 393.
Note that IDEX stands for InDustrialEXport. The company acts as agent for manufacturers such as Struktura Szerezesi Vallalat.

I/ITSEC (Inter-Service/Industry Training Systems and Education Conference)
Organiser: 1994 - National Security Industrial Association (NSIA)
1995 - American Defense Preparedness Association (ADPA)

I/ITSEC: NSIA, Director National Activities, Col Bruce Green USA (Ret), 1025 Connecticut Avenue NW, Suite 300, Washington DC 20036, USA
Tel: +1 202 775 14 40
Fax: +1 202 775 13 09

I/ITSEC: ADPA, Director Technical Services (Contact for I/ITSEC 1995), Captain Nelson Jackson USN (Retd), 2101 Wilson Boulevard, Suite 400, Arlington, Virginia 22201, USA
Tel: +1 703 247 25 75
Fax: +1 703 522 18 85

I/ITSEC: Publicity and Information Office: I/ITSEC, PO Box 2509, Goldenrod, Florida 32733-2509, USA
Tel: +1 407 671 3258
Fax: +1 407 380 5690 (address to: Allen Collier, I/ITSEC Publicity Office)

I/ITSEC: Stand and Hotel Bookings: JMK Associates, PO Box 4313, Enterprise, Florida 32725, USA
Tel: +1 407 322 7000
Fax: +1 407 322 7119
The I/ITSEC conference and exhibition is held annually in the USA in late November or early December, and is the largest world event dedicated to training and simulation. In 1994 the venue is the Marriott Orlando World Center and in 1995 will be the Convention Centre, Albuquerque, New Mexico 12-15 November 1995, thereafter returning to the Marriott Orlando World Center 28 November - 1 December 1996.

INDEP, INDustrias E Participacòes de Defesa SA, Rua Fernando Palha, P-1802 Lisboa Codex, Portugal.
Tel: +351 (1) 858 4371. Telex: 12 514 indfbp p.
Fax: +351 (1) 858 2330.

Indian Head Division, Naval Surface Warfare Center (IHD/NSWC), Weapons Simulation Department, Indian Head, Maryland 20640-5000, USA.
Tel: +1 301 743 4583/4360 and 753 9098/9.
Fax: +1 301 743 4299 and 753 9164.

Industria Gaucha de Aeronaves Especiais Ltda (IGAE), Rua Padre Henrique Koelher 87, Bairro Anchieta, Porto Alegre, RS, 90200, Brazil.
Tel: +55 051 342 6199.

Industrialexport, Hungary, see IDEX entry above.

Initiative Computing AG, Zürichstrasse 25, CH-8185 Winkel-Rüti, Switzerland.
Tel: + 41 (1) 861 05 61.
Fax: +41 (1) 861 05 63.
Contact: S Vogt.

Innovative Projects Limited, Forum Stirling Road, Chichester, West Sussex PO19 2EN, UK.
Tel: +44 (0) 1243 787 932.
Fax: +44 (0) 1243 786 930.

Innotech Aviation Limited, 595 Stuart Graham Boulevard, Dorval, Québec, Canada H4Y 1E3.
Tel: +1 514 636 8484.
Fax: +1 514 636 8887.
Executive Vice-President and COO: Stephen K Plummer
Technical Director: Flo Ciobotaru
Marketing Contact: Kevin J Harris
Innotech is a member of the Innotech/Execaire Group, a wholly owned subsidiary of the IMP group based in Halifax, Nova Scotia. Innotech has about 320 employees and a shop floor area of about 35 000 ft^2. Innotech/Execaire has its headquarters in Montreal and has facilities in Toronto, Ottawa, Vancouver, and Burlington (Vermont, USA).

Inovex Technology, 33 St Thomas Street, Winchester, Hampshire, S023 9HJ, UK.
Tel: +44 (0) 1962 866 586
Fax: +44 (0) 1962 841 665
Contact: Bob D'Silva, Commercial Director

Institute for Combat Arms and Tactics, 2715 Broadbent Parkway NE, Suite J, Albuquerque, New Mexico 87107, USA.
Tel: +1 505 344 3887.

Institutul de Aviatie Bucuresti, Bucuresti 77538, Sector 6, Bd. Pacii 220, Romania.
Tel: 45 25 52/89 69 55. Telex: 11 907; 11 460.
See also Simultec, Bucharest.

Integrators of System Technology (Pty) Limited (IST), 221 Garstfontein Road, Menlyn, Pretoria, Republic of South Africa.
Tel: +27 012 348 8705. Telex: 3-22077 SA.
Fax: +27 012 348 9472.

Integrators of Systems Technology (Pty) Limited (IST), Private Bag 95355, Waterkloof 0145, South Africa.
Tel: 27 12 348 8705. Telex: 3 22077 SA.
Fax: 27 12 348 9472.

Intellisys, 4641 Crossroads Park Drive, Liverpool, New York 13088, USA.
Tel: +1 315 453 2300.

Interatom GmbH, Friedrich-Ebert-Strasse, D-5060 Bergisch Gladbach 1, Germany.
Tel: +49 022 0484-2400. Telex: 8 878 457/8 878 492 iagld.
Fax: +49 022 0484-2400.

International Civil Aviation Organisation (ICAO), 1000 Sherbrooke Street West, Montreal, Quebéc, Canada H3A 2R2.
Tel: +1 514 285 8220
Fax: +1 514 286 6376
Chief, Public Relations: Hutton G Archer

Intertechnik, PO Box 100, Industriezeile 56, A-4040 Linz, Austria.
Tel: +43 (0) 732 78 92. Telex: 22 15 22 his a.
Fax: +43 (0) 732 78 92/13.

Invertron Simulated Systems Limited, Albert Drive, Burgess Hill, West Sussex RH15 9TN, UK.
Tel: +44 (0) 1444 247 535. Telex: 877766 INTRON G.
Fax: +44 (0) 1444 244 895.
Sales and Marketing Manager: Paul Matthews
Marketing Executive: Bruce Lutterloch

Israel Aircraft Industries Limited, Ben Gurion International Airport, Lod, IL-70100, Israel.
Tel: +972 3 971 3111. Telex: 371-114.
Fax: +972 3 971 3111.

Istec Inc, 1810 Highway 6 North, Hamilton, Ontario, Canada L9J 1H2.
Tel: +1 416 529 5132. Telex: 061 8548.
Istec was founded in 1974 with the purchase of assets and rights for the WESCAM camera system from Westinghouse of Canada. In 1993 a separate operating division was created called Wescam, see the entry under this name.

ITEC (International Training Equipment Conference Ltd), 51 Market Place, Warminster, Wilts BA12 9AZ, UK.
Tel: +44 (0) 1985 84 61 81
Fax: +44 (0) 1985 84 61 63
ITEC are conference organisers and, amongst other activities, are responsible for the European training conference and exhibition held annually about April, and for other conferences such as the Asian Civil Aviation Conferences held in Singapore. See also ITSEC for conferences in the USA.

ITSEC, see I/ITSEC

IVEX Corporation, 4357 J Park Drive, Norcross, Georgia 30093, USA.
also at: 4355 International Boulevard, Norcross, Georgia 30093, USA.
Tel: +1 404 564 1148. Telex: 6503144415 MCIUW.
Fax: +1 404 381 0622.
President: John Klipper
Chief Executive and CEO: Peter Sherlock
Vice-President, Marketing: Paul Kingston
Marketing Enquiries: Schelly Weedman
IVEX was formed in 1983 and produces a range of image generators employing a unique principle involving database storage on video disk rather than conventional computer hard disk.

Jet Propulsion Laboratory, California Institute of Technology, 4800 Oak Grove Drive, Pasadena, California 91109, USA.
Tel: +1 818 354 4321.
Fax: +1 818 354 4321.

Joint Aviation Authorities (JAA), Saturnusstraat 8-10, PO Box 3000, NL-2130 KA Hoofddorp, Netherlands.
Tel: +31 25 03 79 700
Fax: +31 25 03 21 714
Secretary General: Ronald Ashford.
Operations Director: Richard Yates
Chairman JAA Simulation (JARSIM) Working Group: Capt Paddy Carver, UK (See CAA entry).
The JAA is the European multinational airworthiness and civil aviation regulatory authority. The 23 nations which currently constitute the JAA are all the European Community (EC) nations plus others: Austria, Belgium, Cyprus, Denmark, Finland, France, Germany, Greece, Iceland, Ireland, Italy, Luxembourg, Malta, Monaco, Netherlands, Norway, Poland, Portugal, Slovenia, Spain, Sweden, Switzerland, UK. The US FAA has observer status. Joint certifications, accepted by all the JAA nations, have been achieved for aircraft from Airbus, Canada, Russia, Sweden, UK, and the USA; and for engines from France, the UK and USA. Since January 1992, JAA regulations have been law (rather than recommended practice) in the states of the European Union. The current JARSIM requirements use the technical standards of the US FAA Levels A and B (see table 1) for JARSIM Levels A and B; levels C and D use the technical standards from the RAeS/ICAO standard. See the introduction to the Flight Simulator section.

Kader Factory for Developed Industries, El Tayaran Street, Nasr City, Cairo, Egypt.
Tel: +20 60 43 19/24. Telex: 92136 KADFA UN.

Kaiser Electro-Opt.cs, 2752 Loker Avenue West, Carlsbad, California 92714, USA.
Tel: +1 619 438 9255
Fax: +1 619 438 6875
President: J T (Jeny) Carollo
Marketing Manager: J W Shunk

Keltron, Keltron House, Vellayambalam, Trivandrum 695 033, India.
Tel: +91 471 60 621. Telex: 045 3273 KEDC IN.

Kentron, PO Box 7412, Hennopsmeer 0046, South Africa.
Tel: +27 12 671 1239.
Fax: +27 12 671 1779.

Kintex, PO Box 209, 66 James Baucher Boulevard, 1407 Sofia, Bulgaria.
Tel: +359 66 23 11. Telex: 22 471/23 243.
Fax: +359 65 81 91/01
Contact: Stamenkov Chalev
Kintex is an export agency working on behalf of Bulgarian manufacturers.

Kollmorgen Corporation, Electro-Optical Division, 347 King Street, Northampton, Massachusetts 01060, USA.
Tel: +1 413 583 2330. Telex: 955409
Fax: +1 413 586 1234
Marketing Contact: Susan Eddy

Krauss-Maffei AG, Ordnance Division, Krauss-Maffei-Strasse 2, D-8000 Munich 50, Germany.
Tel: +49 89 88 99-0. Telex: 523 163-31.
Fax: +49 89 88 99-0.

Kruithoorn, see NWM De Kruithoorn BV

Krupp Fördertechnik GmbH, Franz-Schubert-Strasse 1-3, D-47226 Duisburg, Germany.
Tel: +49 (0) 203 65 78-0. Telex: 855 486.
Fax: +49 (0) 203 65 78-2890.

Lacroix, Muret, France; see Etienne Lacroix SA.

LAIR Croatia, Bijenička cesta.54, 4100 Zagreb, PO Box 1016, Croatia.
Tel: +38 541 27 27 59. Telex: 21383.
Fax: +38 541 42 48 69.
Chairman: Dr Antun Persin
LAIR is understood to mean Laser and Atomic R&D and is associated with the Auder Boskovic Institute. It has about 60 employees, a shop floor area of about 500 m², and has delivered about 1700 training devices to various customers.

Leentjens-Boës, Rue en Bois 210, B-4040 Herstal, Belgium.
Tel: +32 41 64 13 98. Telex: 41754 LEEMAR.
Fax: +32 41 48 05 84.

Le Proton LIMA Sdn. Bhd, Suite 804, 8th Floor, Ming Building, Jalan Bukit Nanas, 50250 Kuala Lumpur, Malaysia.
Tel: +60 3 238 5857/5866.
Fax: +60 3 238 6272.
Organiser of the Langkawi International Maritime & Aerospace (LIMA) Exhibition and Conference, to be held 5 - 10 December 1995.

Link-Miles Ltd (obsolete title), now part of Thomson Training and Simulation; see Thomson

Liquid Image Corporation, 659 Century Street, Winnipeg, Manitoba, Canada R3H OL9.
Tel: +1 204 775 2633.
Internet: HAVELK@MBNET. MB. CA.
Fax: +1 204 772 0239.
Contact: Tony Havelka

Litton Applied Technology, 4747 Hellyer Avenue, PO Box 7012, San Jose, California 95150-7012, USA.
Tel: +1 408 365 4747. TWX: 910 339 9271.

Litton Data Systems, 29851 Agoura Road, PO Box 6008, Agoura Hills, California 91376-6008, USA.
Tel: +1 818 707 4946.
Fax: +1 818 706 5939.

Litton Systems Canada Limited, 25 Cityview Drive, Rexdale, Ontario, Canada M9W 5A7.
Tel: +1 416 249 1231. Telex: 06-989406.

Litton Systems Inc, Laser Systems Division, PO Box 547300, Orlando, Florida 32854-7300, USA.
Tel: +1 407 295 4010.
President: Gregory Misiak
Technical Director: William Wilson
Marketing Director: James Cole
Marketing Contact: Steve Gulch
2787 South Orange Blossom Trail, Apopka, Florida 32703, USA
Tel: +1 407 297 4405
The Laser System Division of Litton Systems Inc has a staff of about 400 and a shop floor area of about 140 000 ft². The division was founded in 1969 as International Laser Systems. It was acquired in 1979 by Martin Marietta and in 1983 by Litton Systems Inc. A new plant was occupied in 1986. The division has made, amongst other things, over 30 training devices, and also makes simulated laser targets.

Lockheed, Lockheed Martin. It was announced in September 1994 that Lockheed and Martin Marietta were to merge, subject to approval by the US Government from the standpoint of competition and monopoly. As we go to press, it appears likely that the merger will be approved early in 1995 and the new organisation will be known as Lockheed Martin. Combined aerospace sales of the two organisations in 1993 were $21.8bn, second only to Boeing at $25bn.

Lockheed (Lockheed Martin) Fort Worth Company, PO Box 748, Fort Worth, Texas 76101, USA.
Tel: +1 817 763 7524.
Fax: +1 714 620 7511.
Telex: 758 231.
TWX: 710 4531 089.
The Tactical Military Aircraft Division of General Dynamics, producer of the F-16 aircraft, was acquired by Lockheed in December 1992 and renamed the Lockheed Fort Worth Company. Its new title under Lockheed's proposed merger with Martin Marietta is not yet known.

Lockheed Sanders, Daniel Webster Highway South, PO Box 0868, Nashua, New Hampshire 03061-0868, USA.
Tel: +1 603 885 2816/5777.
Fax: +1 603 885 2816.

Locus Inc, PO Box 0740, State College, Pennsylvania 16804-0740, USA.
Tel: +1 814 466 6275.
Fax: +1 814 466 3341.

Logicon, Tactical and Training Systems Division, PO Box 85158, San Diego, California 92138-5158, USA.
Tel: +1 619 455 1330. TWX: 910 337 1780.
Fax: +1 619 455 5524

Loral Control Systems, Kennedy Drive, Archibald, Pennsylvania 18403, USA.
Tel: +1 717 876 1500.
Fax: +1 717 876 1500.

Loral Defense Systems-Akron, 1210 Massillon Road, Akron, Ohio 44315, USA.
Tel: +1 216 796 9687/2762.
Fax: +1 216 796 8974.

Loral (East Coast Facility), 50 Moulton Street, Cambridge, Massachusetts 02138, USA.
Tel: +1 617 873 3577
Fax: +1 617 873 3577
Manager: J C Williams
This facility includes the simulation interests of Bolt Beranek and Newman (BBN), sold to Loral in 1993.

Loral Electro-Optical Systems (LEOS), 300 North Halstead Street, PO Box 7101, Pasadena, California 91107, USA.
Tel: +1 818 351 5555.
Fax: +1 818 351 5081/2315.
President: Robert W Mueller
Vice-President, Training and Simulation: Steve Gideon
Vice-President, Business Development: Gary W Morton
LEOS has about 800 employees and a shop floor area of about 209 000 ft² (19 440 m²). It makes the MILES series of laser engagement systems, of which over 250 000 vehicle systems and 700 000 man systems are in service. LEOS also makes IR and R+E-O countermeasures systems.

Loral Europe Limited, 71 Buckingham Avenue, Slough, Berks SL1 4PN, UK.
Tel: +44 (0) 1753 696 488
Fax: +44 (0) 1753 696 218

Loral Hycor, 10 Gill Street, Woburn, Massachusetts 01801-1788, USA.
Tel: +1 617 935 59 50.
Fax: +1 617 932 37 64.
President: L James Larson
Director, Programs & Engineering: T Richard Ellis
Marketing Director: Laurence D White
The Loral Corporation acquired Hycor Inc in 1985. It makes decoys and decoy dispensing systems for ships and aircraft. Loral Hycor has about 225 employees and a shop floor area of about 40 000 ft² (3700 m²).

Loral Librascope, 833 Sonora Avenue, Glendale, California 91201, USA.
Tel: +1 818 244 6541. Telex: 674912.
Fax: +1 818 244 6541

Loral Solartron, Pindar Road, Hoddesdon, Hertfordshire EN11 0DB, UK.
Tel: +44 (0) 1992 470 011.
Fax: +44 (0) 1992 470 485 (General).
+44 (0) 1992 470 486 (Purchasing).
+44 (0) 1992 470 492 (Engineering).
Chairman: Bernard Schwartz
Managing Director: Jeffrey Kaye
Technical Director: Dr Mark Skipper

Marketing Director: John Mottram
Marketing Enquiries: Alexander Hooper
Loral Solartron is owned by the Loral Corporation and has about 120 employees and a shop floor area of 11 500 ft². It works closely with the Training and Simulation division of Loral Electro-Optical Systems (LEOS) in California. The company name was Solartron Defence Systems and before that, Weston Simfire which made the SIMFIRE product line of laser Weapon Effect Simulators (WES). About 7000 systems have been sold.

Loral Vought Systems, PO Box 650 003, Dallas, Texas 75 265-0003, USA.
Tel: +1 214 603 9040/1615
Fax: +1 214 603 1009.
Manager, Communications: Craig Vanbebber

Loral (West Coast Facility), 13810 SE Eastgate Way, Suite 500, Bellevue, WA 98005, USA.
Tel: +1 206 746 6800.
Fax: +1 206 746 1335.

Losfeld-Industries, 15 rue Thiébault, F-94220 Charenton, France.
Tel: +33 (1) 43 68 10 31. Telex: 262 053 F.
Fax: +33 (1) 43 68 10 31

LTV Missiles and Electronics Group, Sierra Research Division, 247 Cayuga Road, PO Box 222, Buffalo, New York 14225, USA.
Tel: +1 716 631 6200. Telex: 6854308.
Fax: +1 716 631 6200.

McDonnell Douglas Training Systems (MDTS), PO Box 516, St Louis, Missouri 63166, USA.
Tel: +1 314 232 0232. Telex: 44857.
McDonnell Douglas Training Systems (MDTS) was formerly the American Airlines Training Corporation which was purchased by McDD in August 1986. MDTS develops flight simulators for the McDonnell Aircraft (McAir) Company and other manufacturers.

McDonnell Douglas Electronic Systems Company, Visual Simulation Systems, 2600 North Third Street, MC 500 2442, St Charles, Missouri 63301, USA.
Tel: +1 314 925 4533. Telex: 447369.
Fax: +1 314 925 4119.

McFadden Systems Inc, 11835 East Smith Avenue, Santa Fe Springs, California 90670, USA.
Tel: +1 310 948 44 14
Fax: +1 310 948 12 91
Sales and marketing: Michele Grantham
McFadden Systems produces servo-mechanisms and makes, amongst other things, a range of hydraulic motion platforms.

McQueen Limited, Target Division, Nether Road, Galashiels, Scotland TD1 3HE, UK.
Tel: +44 (0) 1896 4866.
Fax: +44 (0) 1896 50 774.
Chairman: M M Gray OBE
General Manager: G Moncur
McQueen was founded in 1846 and has been producing targets for the British Army since 1890. Computer software production started in 1985. Targets are produced not only for the military but also the police and various rifle and small arms shooting associations. The Target Division has about 250 employees and a shop floor area of about 70 000 ft² (6500 m²).

Marconi Radar and Control Systems Limited (MRCS), New Parks, Leicester LE3 1UF, UK.
Tel: +44 (0) 116 287 14 81. Telex: 34551 gecnpl g.
Fax: +44 (0) 116 287 17 46.
Manager, Training Systems Group: Alan Jones

Marconi Radar Systems Limited (MRSL), Writtle Road, Chelmsford, Essex CM1 3BN, UK.
Tel: +44 (0) 245 267 111. Telex: 99 108.
Fax: +44 (0) 245 357 927.

Marconi Simulation and Training (MST), The John Sutcliffe Building, Fulmar Way, Donibristle Industrial Park, nr Dunfermline, Fife KY11 5JX, UK.
Tel: +44 (0) 383 821 921.
Fax: +44 (0) 383 824 227.
The Donibristle site is the HQ of Marconi Simulation and Training (MST), and the group now includes the ex-Ferranti Simulation sites at Cheadle Heath (Stockport) and Cwmbran (Wales), details of which are given below.

Marconi Simulation and Training (MST), Ty Coch Way, Cwmbran, Gwent NP44 7XX, UK.
Tel: +44 (0) 1633 871 111.
Fax: +44 (0) 1633 835 022.
Retitled from Ferranti on GEC's acquisition of Ferranti's remaining defence interests. See also Dundridge College for ATCO training. Headquarters of MST are at the Donibristle (Scotland) site shown above.

Marconi Simulation and Training (MST), Bird Hall Lane, Cheadle Heath, Stockport, Cheshire SK3 0XQ, UK.
Tel: +44 (0) 161 428 0771. Telex: 666803.
Fax: +44 (0) 161 428 9121.
Retitled from Ferranti on GEC's acquisition of Ferranti's remaining defence interests. Headquarters of MST are at the Donibristle (Scotland) site shown above.

Marconi Technologies Inc, 3725 Electronics Way, PO Box 3025, Lancaster, Pennsylvania 17604-3025, USA.
Tel: +1 717 285 7151. Telex: 173244 FTI-UT.
Fax: +1 717 285 3468.
Retitled from Ferranti on GEC's acquisition of Ferranti's remaining defence interests.

Marine-Air Systems Limited, PO Box 30-248, 24 Bridge Street, Lower Hutt, Wellington, New Zealand.
Tel: +64 4 692 170. Telex: NZ 31 412.
Fax: +64 4 661 247.

MarineSafety International (MSI), National Maritime Research Center, CAORF Building, US Merchant Marine Academy, Steamboat Road, King's Point, New York 11024, USA.
Tel: +1 516 773 5866/5602
Fax: +1 516 773 5604
Managing Director: Dr F Eugene Guest
Technical Director: Steven Fisch
Marketing Director: Thomas Garrigan
Marketing Contact: Douglas Ullman
MarineSafety International (MSI) is a wholly owned subsidiary of FlightSafety International (FSI) and was formed in 1976 to provide simulator-based training for shipping companies. It has facilities at King's Point and La Guardia NY, Newport Rhode Island, San Diego California, and Rotterdam in Holland. It has about 75 employees.

MarineSafety International, Newport Simulator Training/ Research Complex, 344 Aquidneck Avenue, Middletown, Rhode Island 02840, USA.
Tel: +1 401 849 0222.
Fax: +1 401 849 9264.

MarineSafety Rotterdam BV, Wilhelminakade 701, PO Box 51290, NL-3077 GG Rotterdam, Netherlands
Tel: +31 10 486 6654
Fax: +31 10 484 6071
MarineSafety is a subsidiary of FlightSafety International (FSI). For more company information see the entry under FlightSafety.

Maritime Simulation Centre the Netherlands, Van Uvenweg 9, PO Box 90, NL-6700 AB Wageningen, Netherlands.
Tel: +31 837 079 911.
Fax: +31 837 079 999.

Martin Marietta Corporation (Lockheed Martin), PO Box 1260, Denver, Colorado 80201, USA.
Tel: +1 309 977 5364; TWX: 910 935 0718.
Fax: +1 309 977 5364
See also Lockheed Martin for the proposed merger with Lockheed.

Martin Marietta Corporation (Lockheed Martin), Simulation and Automated Systems Division, PO Box 2825, Daytona Beach, Florida 32115, USA.
Tel: +1 904 239 2906. Telex: 56 6556.
Fax: +1 904 239 2176.
Note that Martin Marietta acquired this facility during 1993 from the General Electric (GE) Company. It produces, amongst other things, the Compu-Scene family of high-fidelity CGI systems. It is understood that some parts of this activity are being transferred to the Orlando site. See also Lockheed Martin for the proposed merger with Lockheed.

Martin Marietta Electronic Systems (Lockheed Martin), PO Box 55 5837, MP-416, Florida 32855-5837, USA.
Tel: +1 407 356 1647. Telex: 564414.
Fax: +1 407 356 1647.

Martin Marietta Information Systems Company (Lockheed Martin), 12506 Lake Underhill Road, Orlando, Florida 32825, USA.
Tel: +1 407 356 6645
Fax: +1 407 356 0919
See also under Lockheed Martin for the proposed merger with Lockheed.

Mauser-Werke Oberndorf GmbH, Teckstrasse 11, D-7238 Oberndorf Neckar, Germany.
Tel: +49 742 370 0. Telex: 742 3161.
Fax: +49 742 370 670.

MECAR SA, Rue Grinfaux 50, B-7181 Petit-Roeulx-les-Nivelles, Belgium.
Tel: +32 67 217 795. Telex: 574 38 mecar b.
Fax: +32 67 216 307.

Merit Technology Inc, 5068 West Plano Parkway, Plano, Texas 75075-5009, USA.
Tel: +1 214 248 2502.
Fax: +1 214 733 4788.

Messerschmidt-Bölkow-Blohm GmbH (MBB), Unternehmensbereich Flugzeuge, Training Systems, PO Box 80 11 60, D-8000 Munich 80, Germany.
Tel: +49 (0) 89 607 2 48 60. Telex: 5287-134 mbb d.
Fax: +49 (0) 89 607 2 93 52.

Metalchim Trading, see AVIOTECHNICA, Sofia, Bulgaria.

Meteor Costruzioni Aeronautiche Ed Elettroniche SpA, Via Mario Stoppani 21, Zona Industriale di Soleschiano, I-34077 Ronchi dei Legionari, Gorizia, Italy.
Tel: +39 481 47 81 11 and 47 93 04. Telex: 460 288 Metmon I.
Fax: +39 481 77 62 02, 47 93 00 and 47 83 13.
Chairman: Guiseppe Tonutti
Managing and Marketing Director: Filippo Lanzarotto
Technical Director: Luigi De Giorgio
Marketing Contact: Antonio Greco
Meteor CAE is part of the Finmeccanica group and has about 330 employees and a shop floor area of about 20 500 m². In the past it has made light aircraft under licence and gliders for the Italian armed services. This developed into the manufacture of RPVs under licence from Aerospatiale, Beechcraft, Northrop and Canadair. It now makes its own range of Mirach air targets.

Metric Systems Corporation, Electronics Division, 645 Anchors Street, PO Box 9990, Fort Walton Beach, Florida 32548-9990, USA.
Tel: +1 904 863 6200. Telex: 702447.
Fax: +1 904 863 6371.
President: Coy J Scribner
Senior Vice-President and Director, Electronics Division: Harry V Fowler
Director, Electronics Engineering Department: George D Lee
Director, International Marketing; Stanley F Wilusz
Metric has about 750 employees and a shop floor area of about 130 000 ft² (12 100 m²).

Micro Processor Systems A/S, Baneveien 32, N-3600 Kongsberg, Norway.
Tel: +47 32 735 766.
Fax: +47 32 736 965.

Miltra Engineering Limited, Production and Administration, Bentalls, Pipps Hill Industrial Estate, Basildon, Essex SS14 3BY, UK.
Tel: +44 (0) 1268 526 756.
Fax: +44 (0) 1268 531 485.

Miltra Engineering Limited, Research and Marketing, 207 Chester Road, Watford, Hertfordshire WD1 7RH, UK.
Tel: +44 (0) 1923 818 342.
Fax: +44 (0) 1923 818 342.

Miltrain Limited, 64A High Street, Godalming, Surrey GU7 1DU, UK.
Tel: +44 (0) 1483 426 664. Telex: 858 893 FLETEL G.
Fax: +44 (0) 1483 415 953.

Mitsubishi Jukogyo Kabushiki Kaisha (Mitsubishi Heavy Industries Limited), 5-1, Marunouchi 2-chome, Chiyoda-ku, Tokyo 100, Japan.
Tel: +81 (03) 212 3111. Telex: J22282 and J22443.

Mitsubishi Precision Company Limited, 3-13-16 Mita, Minato-ku, Tokyo, Japan 108.
Tel: (03) 3453 6421.
Fax: (03) 3453 6434.

ML Wallop Defence Systems Limited, Arkay House, Weyhill Road, Andover, Hampshire SP10 3NR, UK.
Tel: +44 (0) 1264 333 322. Telex: 47 692 WALIND G.
Fax: +44 (0) 1264 332 186.

Motorola Inc, Government and Systems Technology Group (GSTG), 8220 E Roosevelt Street, Scottsdale, Arizona 85257, USA.
Tel: +1 602 441 7737/7357.
Fax: +1 602 441 7749.
Scottsdale is a suburb of Phoenix, Arizona.
General Manager GSTG: David Wolfe.
Marketing Director GSTG: Robert Bigony.

MS Instruments plc, Electron House, 33-49 Farwig Lane, Bromley, Kent BR1 3RE, UK.
Tel: +44 (0) 181 290 0200.
Fax: +44 (0) 181 464 6596.
Chairman: N N Graham-Maw
Managing Director: Dr B D MaCarthy
Technical Director: A H Philbrook
Marketing Director: P A H Etherington
MS Instruments was formed in 1972 primarily for the design and manufacture of ballistic instrumentation.

Subsequently it has expanded into training target systems and project management. It has delivered some 100 training systems. It has about 10 employees and a shop floor area of about 16 150 ft² (1500 m²).

MSG Marine- und Sondertechnik GmbH, now renamed STN (qv)

MSI-Defence Systems Limited, 10 Cambridge Road, Granby Industrial Estate, Weymouth, Dorset DT4 9XA, UK.
Tel: +44 (0) 1305 760 111.
Fax: +44 (0) 1305 760 222.

MSI Defence Systems Limited, Salhouse Road, Norwich, Norfolk NR7 9AY, UK.
Tel: +44 (0) 1603 484 065. Telex: 975067.
Fax: +44 (0) 1603 415 649.

Nanjing Research Institute on Simulation Technique (NRIST), 2 Huangpu Road, PO Box 1607, Nanjing 210016, Jiangsu Province, People's Republic of China.
Tel: +86 25 648 061; 25 647 765; 664 80 61; 448 27 84.
Cable: 0319 Nanjing.
Fax: +86 25 445 28 94.
Chairman: Si Baokang
NRIST was established in 1979 and has developed some 76 types of training equipment. Having been developed by NRIST, equipment is produced in state factories and then goes into service with the People's Liberation Army (PLA). NRIST has about 200 employees and a shop floor area of about 8500 m².

Nanjing University of Aeronautics and Astronautics (NUAA), Nanjing Aeronautical Institute (NAI), RPV Department, 29 Yu Dao Jei, Nanjing 20016, Jiangsu Province, People's Republic of China.
Tel: +86 0 25 44 94 514. Telex: 34 155 NAINJ CN.
Fax: +86 0 25 44 10 724.

NASA (National Aeronautics and Space Administration), Simulation Division, Ames Research Centre, Moffett Field, California 94035, USA
Tel: +1 415 604 5000.
Fax: +1 415 604 5000.

NATO: NATO has a number of organisations related to training matters. Its Aerospace Group for R&D (AGARD) monitors training technology and organises seminars on matters such as low flying and other training by simulation. See the entry under AGARD.
On 8 December 1993, a statement by ministers of the NATO Eurogroup (NATO nations less Canada, Iceland and the USA) announced that the former EURONATO-TRAINING sub-group on joint training, had become a full NATO group with the title NATOTRAINING.
Contact: The Secretary, NATOTRAINING Group, NATO Headquarters, Brussels, Belgium.
Tel: +32 2 728 41 11
Fax: +32 2 728 45 79

Naval Air Warfare Center Training Systems Division (NAWCTSD), 12350 Research Parkway, Orlando, Florida 32826-3224.
Tel: +1 407 380 8150.
Commanding Officer: Captain Christopher L Addison, USN
Director, R&D: H C Okraski.
Head, Human Systems Integration Division: W A Rizzio.
Head, Advanced Simulation Concepts Division: P Little.
On 1 October 1993 the US Navy Naval Training Systems Center (NTSC) was retitled the Training Systems Division (TSD) of the Naval Air Warfare Centre (NAWC). It has responsibilities for simulation and training which vary from R&D to acquisition of systems and life-cycle support. An unclassified Project Summary of its R&D activities is published annually.

Naval Surface Warfare Center, Indian Head Division; see Indian Head.

Nederlandsche Instrumenten Compagnie 'NEDINSCO' BV, Molensingel 17, PO Box 40, NL-5900 AA Venlo, Netherlands.
Tel: +31 077 558 777.
Fax: +31 077 518 787.

Nico-Pyrotechnik, Hanns-Jürgen Diederichs GmbH & Co KG, Bei der Feuerwerkerei 4, Postfach 1227, D-22946 Trittau, Germany.
Tel: +49 415 4805-0. Telex: 2189 413 nico d.
Fax: +49 415 42 451.
Chairman and Managing Director: Hanns-Jürgen Diederichs
Technical Director: Willi Lübbers
Marketing Director: Hans G Morotini
Nico has been making pyrotechnics at Trittau near Hamburg since 1949. It makes flares, signals, alarm and simulation devices, screening and signal smokes and training systems for anti-tank weapons, hand grenades, mines, mortars, training ammunition, and weapon effect signature simulators. Nico has about 300 employees and a shop floor area of about 180 000 m². It has subsidiaries in Austria, Canada, Germany, Singapore and the USA, and has joint ventures with other companies in Singapore and South Korea.

The Nitor Group Ltd, Nitor House, PO Box 1, Godalming, Surrey GU7 1UJ, UK.
Tel: +44 (0) 1483 426 299. Telex: 94011723 NITR G.
Fax: +44 (0) 1486 85 953.
The Nitor Group specialises in training and operational complexes and ranges for the military, special forces, police and commercial markets. It is the holding company for five subsidiaries, Sector, OTEC Support Services Ltd, OTEC Special Products Ltd, UK Training Ltd, and Quest Research Ltd. Sector is an engineering consultancy, OTEC Support Services is for project management and contracting, OTEC Special Products supplies and installs training devices and equipment, UK Training offers training support and Quest undertakes research in audio, visual, software, and systems integration matters.

NLR, Antony Fokkerweg 2, NL-1059 CM Amsterdam, Netherlands.
Tel: +31 20 511 31 13
Fax: +31 20 511 32
Head of Flight Division: J T M van Doorn
Man-Machine Integration: P G A M Jorna
Flight Mechanics: W P de Boer
Helicopters: L T Renirie
Air Traffic Management: J Brogan
Flight Simulation : W G Vermeulen
Transport & Environmental Studies: G Bekebrede
NLR stands for the Nationaal Lucht- en Ruimtevaart Laboratorium or the National Aerospace Laboratory (NLR) of the Netherlands. NLR has a total staff of about 900, and carries out theoretical and experimental research for Dutch and foreign aircraft industries, civil and military aircraft operators and organisations concerned with aviation and space flight. About one third of employees are graduates from universities and one third from technical colleges. NLR has laboratories in Amsterdam and Noordoostpolder, and is organised into the following divisions: Fluid Dynamics, Informatics, Flight, Structures and Materials and Space.

The Flight Simulation Department is responsible for the basic technology of flight simulation, its application in training research, performing experiments for customers and for use of flight simulation techniques in the integration of systems in flight.

Nobel, see Dynamit Nobel AG

NobelTech Systems of Järfälla, Sweden; see CelsiusTech

Noptel Oy, Teknologiantie 2, FIN-90570 Oulu, Finland
Tel: +358 81 551 43 51; Telex: 32 272 outec sf
Fax: +358 81 36 41 01
President and Managing Director: Dr Juhani Heinula
R&D Manager: Markku Koskinen
Marketing Director and contact for Marketing Enquiries: Martti Säkelä
Noptel Oy was founded in 1982 by a university research group. It employs about 30 people and has a shop floor area of about 600 m². It specialises in marksmanship training devices and has sold its products to 40 countries.

Norabel AB, PO Box 803, S-713 00 Nora, Sweden.
Tel: +46 587 114 50. Telex: 73 451 norabel s.
Fax: +46 587 141 55.

Norcontrol Systems AS, Bekkajordet 8A, PO Box 1024, N-3194 Horten, Norway.
Tel: +47 33 41436. Telex: 70218 NCNOR N.
Fax: +47 33 43 410.
Norcontrol is wholly owned by Norsk Forsvarsteknologi A/S (NFT), the largest manufacturer of defence, aerospace and maritime products in Norway. NFT is owned by the government of Norway through the Ministry of Industry.

Norden Service Company Inc, 200 Professional Drive, Gaithersburg, Maryland 20879, USA.
Tel: +1 301 921 8100.

Northbend Instrumentation (Pty) Limited, 12 Middle Street, Bellville, 7530, Republic of South Africa.
Tel: +27 21 949 0123.
Fax: +27 21 948 0151.

Northrop Corporation, Aircraft Division, One Northrop Avenue, Hawthorne, California 90250-3277, USA.
Tel: +1 310 332 1000. Telex: 181861 NORTHR HWTH A.
Fax: +1 310 332 3396.

NTSC - On 1 October 1993 the US Navy Naval Training Systems Center (NTSC) was retitled the Training Systems Division (TSD) of the Naval Air Warfare Center (NAWC). See Naval Air Warfare Center.

NWM De Kruithoorn BV, Poeldonkweg, PO Box 1050, NL-5200 BC's-Hertogenbosch, Netherlands.
Tel: +31 73 810 222. Telex: 50 229.
Fax: +31 73 810 310.

Oak Switch Systems Inc, PO Box 517, Crystal Lake, Illinois 60014, USA.
Tel: +1 815 459 5000.
Fax: +1 815 459 5011.

OBRUM, Research and Development Centre of Mechanical Appliances, ul. Toszecka 102, PL-44-101 Gliwice, Poland.
Tel: +48 032 31 72 41. Telex: 036 197.
Fax: +48 032 31 58 87.
General Director: Henryk Knapczyk
Technical Directors: Jan Wodjcik and Wladyslaw Mazurkiewicz
The Initials OBRUM stand for Ośrodek Badawczo-Rozwojowy Urzadzeń Mechanicznych. OBRUM was established in 1969 and has about 450 employees. It not only makes military devices but also hydraulic power packs, mining machines, asphalt milling machines and devices for dealing with organic waste. Other civil products are made to order.
The Polish Foreign Trade Enterprise, used for marketing, is the CENZIN Co, for details see CENZIN in this Directory.

Oerlikon-Contraves AG, Schaffhauserstrasse 580, CH-8052 Zürich, Switzerland.
Tel: +41 (1) 306 22 11.
Fax: +41 (1) 301 34 66.

Oerlikon-Contraves AG, Birchstrasse 155, CH-8050 Zürich, Switzerland.
Tel: +41 (1) 316 22 11.
Fax: +41 (1) 316 42 02 and 311 41 22.

Officine Galileo SpA, Via A. Einstein 35, I-50013 Campi Bisenzio, Firenze, Italy.
Tel: +39 55 89 501. Telex: 570 126 GALILE I.
Fax: +39 55 895 0600.

Officine Panerai SpA, 2 Piazza Galileo Ferraris, I-50131 Florence, Italy.
Tel: +39 055 579 304. Telex: 573144 PANFIRI.

OGMA (Oficinas Gerais de Material Aeronáutico), P-2615 Alverca, Portugal.
Tel: +351 (1) 258 1000. Telex: 14 479.
Fax: +351 (1) 258 1000.

Oscmar Products Limited, 27-29 Shaddock Street, PO Box 6008, Wellesley Street, Auckland, New Zealand.
Tel: +64 9 370 211. Telex: NZ60646 AISINS.
Fax: +64 9 370 211.

OTEC Special Products Limited (also OTEC Support Services Ltd), Nitor House, PO Box 1, Godalming, Surrey GU7 1UJ, UK.
Tel: +44 (0) 1483 426 299. Telex: 94011723 NITR G.
Fax: +44 (0) 1486 85 953.
OTEC is part of the Nitor Group.

Ośrodek Badawczo-Rozwojowy Sprzętu Mechanicznego, ul. Kochanowskiego 30, 33-100 Tarnow, Poland.
Tel: +48 14 221 466. Telex: 066351 A-E.
Fax: +48 14 221 466.

Pains-Wessex, see Schermuly (Pains-Wessex Limited)

Paramax, Shipborne and Ground Systems Group, Marcus Avenue, Great Neck, New York 11020, USA.
Tel: +1 516 574 0111. Telex: 277 259.
Fax: +1 516 574 3318.

Pennant International Limited, Pennant Court, Staverton Technology Park, Cheltenham, Gloucestershire, GL51 6TL, UK.
Tel: +44 (0) 1452 714 914.
Fax: +44 (0) 1452 714 920.

Perceptronics Inc, Training and Simulation Systems Division, 21135 Erwin Street, Box 4198, Woodland Hills, California 91367, USA.
Tel: +1 818 884 3485.
Fax: +1 818 340 6067.

Phoenix Simulation, Southfield House, 11 Liverpool Gardens, Worthing, West Sussex BN11 1RY, UK.
Tel: +44 (0) 1903 821 412. Telex: 878665 CENRES G.
Fax: +44 (0) 1903 238 885.

Piepenbrock Pyrotechnik GmbH, Ruhweg 21, D-67307 Göllheim, Germany.
Tel: +49 635 140 20. Telex: 451 231.
Fax: +49 635 140 226.

Plasticas Oramil (PO) SA, División AM, Apdo Correos 192, San Sebastian, Spain.
Tel: +34 43 216710. Telex: 38 063 ORAM E.
Tel: +34 43 216710.

Polytronic International Limited, Pilatusstrasse 1125, CH-5630 Muri, Switzerland.
Tel: +41 57 45 11 45. Telex: 828 080 poly ch.
Fax: +41 57 44 16 55.
Chairman: Claude Thalmann
Technical Director: Arnold Jost
Marketing Director: Heinrich Schweizer
Polytronic International is wholly owned by Polytronic Holdings Ltd and was founded in 1966 with the invention of its electronic hit scoring system. It specialises in equipment for shooting ranges and more than 2000 ranges in 80 countries have been provided. It employs about 50 people and has a shop floor area of about 3500 m^2.

PowerSafety International, an affiliate of FlightSafety International (FSI), see entry under FlightSafety.

Primary Image Limited, Lever House, 9 St James's Road, Kingston-upon-Thames, Surrey KT1 2BA, UK.
Tel: +44 (0) 181 546 4908. Telex: 928893.
Fax: +44 (0) 181 549 8035.
Managing Director: Alan Davenport
Director: Geoff Thiel
Technical Director: Paul Pontin
Marketing Manager: Chris Bratchell
Primary Image used to be known as Akebia. It has about 50 employees and a shop floor area of about 10 000 ft^2 (830 m^2).

Primary Image Inc, 12424 Research Parkway Suite 101, Orlando, Florida 32826, USA
Tel: (1) 407 658 0557
Fax: (1) 407 658 0478

PRONIT Plastic Works, ul Zakladowa 7, 26-940 Pionki, Poland
Tel: +48 48 124 723 and 122 082. Telex: 067 2232.
Fax: +48 48 125 534 and 122 046.
Managing Director: Wieslaw Kiepiel
Technical Director: Henryk Lecki
Marketing Director: Jerzy Kustrzepa
The PRONIT company is also known as Zaklady Tworzyw Sztucznych and has about 2500 employees. It was established in 1925 as a state production plant for propellants and explosives. It now produces a range of chemical and defence products. The PRONIT group includes PRONICAL Ltd, POL-PLAST Ltd, and FAM-SIMADEX Ltd.

Racal Communications Limited, PO Box 112, 472 Basingstoke Road, Reading, Berkshire RG2 0QF, UK.
Tel: +44 (0) 1734 875 181. Telex: 848011.
Fax: +44 (0) 1734 866 014.

Radio Holland Group, Electronics Systems, Defence, Postbox 9094, NL-1006 AB Amsterdam, Netherlands.
Tel: +31 20 101 972. Telex: 13 166.
Fax: +31 20 106 826.

RAeS, see Royal Aeronautical Society

Rafael is the Israel Armament Development Authority. The first address given is the Corporate HQ

Rafael, PO Box 2082, IL-31021 Haifa, Israel.
Tel: +972 04 706 965. Telex: 471508 VERED IL.
Fax: +972 04 794 657.

Rafael Electronic Systems Division (ESD), PO Box 2250/80, IL-31021 Haifa, Israel.
Tel: +972 4 794 784 and 792 002/3. Telex: 471508 VERED IL.
Fax: +972 4 794 093.
Chairman: Itzhak Gat
Marketing Director: Dr Joram Walfisch
Marketing Contact: Naftali Benad
Rafael ESD has about 800 employees.

RamTech Enterprises Inc, 2401 West Grand Avenue, Yoakum, Texas 77995, USA.
Tel: +1 512 293 2154.
Fax: +1 512 293 2154.

Theodor Rapp GmbH & Co KG, Alte Hausacher Strasse 2-4, Postfach 1320, D-7612 Haslach iK, Germany.
Tel: +49 783 2703-0. Telex: 752 322 rapp.
Fax: +49 783 2703-22.

Raufoss A/S, PO Box 2, N-2831 Raufoss, Norway.
Tel: +47 615 20 00. Telex: 71 144 RA N.
Fax: +47 615 20 01.
President: Bjarne Gravdahl
Executive Vice-President: Johs Norheim
Technical Directors: Einar Willassen, Erland Paulsrud
Marketing Director: Per Lillehaug
Raufoss A/S (RA) has about 2300 employees and a shop floor area of about 101 600 m^2. Its Defence Technology division has 1100 employees and a shop floor area of about 56 400 m^2. Raufoss was established in 1896 and became a company with the Norwegian Government as the major shareholder in 1990. RA has two divisions, Defence Technology and Automotive; the Defence Technology Division is further divided into Aerospace, Ammunition, and Metal Products Groups

Raytheon Company, Submarine Signal Division, 1847 West Main Road, Portsmouth, Rhode Island 02871-1087, USA.
Tel: +1 401 847 8000. Telex: 716 817 111.

Real World Simulation, The Annex, New Barns Mill, Cotton Mill Lane, St Albans, Herts AL1 2HA, UK.
Tel: +44 (0) 1727 840 312.
Fax: +44 (0) 1727 840 651.
The company used to be known as Real World Graphics and has moved from Hertford to St Albans.

Rediffusion (obsolete title), now part of Thomson Training and Simulation; see Thomson-CSF

Reflectone Inc, 4908 Tampa West Boulevard, Tampa, Florida 33634-2481, USA.
Tel: +1 813 885 7481, 887 1436/1449, +1 800 237 8414.
Fax: +1 813 887 1177/1439, 887 1419.
President and CEO: Richard G Snyder
Vice-President Marketing and Military Products: Frank T Tobin.
Director, Business Development: Bill Yarnall.
Reflectone employs just over 1000 people and has a shop floor area of about 23 765 m^2. The company was founded in 1939 by Luther Simjian and claims to be the second oldest simulator company in the world. In 1980 the company relocated from Stamford, Connecticut to Tampa, Florida. In 1987 British Aerospace bought 53% of the company stock and continues to hold this proportion in 1994. In 1993, Reflectone assumed responsibility for the British Aerospace Dulles Flight Training centre. Also in 1993 Reflectone acquired British Aerospace Simulation Limited (BAeSL) from BAe and renamed it Reflectone UK Limited (RUKL); see next entry.

Reflectone UK Limited (RUKL), PO Box 77, Building 2041, Southmead Road, Filton, Bristol BS99 7AR, UK.
Tel: +44 (0) 11 79 36 41 71 and 41 29
Fax: +44 (0) 11 79 36 44 69
President: F T Tobin
General Manager: J R Paulson
Marketing Manager: Peter Sanders
Following a restructuring of BAe's training and simulation interests in 1993, British Aerospace Simulation Ltd (BAeSL) became a wholly owned subsidiary of Reflectone Inc of Tampa, Florida (see entry above), and was renamed Reflectone UK Ltd (RUKL). RUKL has about 60 employees and a shop floor area of about 23 000 ft^2 (2140 m^2).

Remote Piloted Systems (RPS), PO Box 16991, Jebel Ali, Dubai, United Arab Emirates.
Tel: +971 84 35 976.
Fax: +971 84 35 221.
Chairman: Michael K Garry
Managing Director: Charles H Haslam
Technical Director: John Martin
Marketing Director: D E Graham
RPS was formed in March 1991 in the Jebel Ali free zone. It produces the Kestrel air target which was first exhibited at the Dubai 1991 air show. RPS has 10 employees and a shop floor area of about 540 m^2.

Republic Electronics, 575 Broad Hollow Road, Melville, New York 11747, USA.
Tel: +1 516 249 1414. Telex: 96 1460.

Republic Electronics Company, 200 Wireless Boulevard, Hauppauge, New York 11788, USA.
Tel: +1 516 273 9800. Telex: 961460.

Research and Development Centre of Mechanical Appliances, Poland; see OBRUM

RGB Spectrum, 950 Marina Village Parkway, Alameda, California 94501, USA
Tel: +1 510 814 7000
Fax: +1 510 814 7026
Contact: Kathy Kahn
RGB Spectrum is a designer and manufacturer of video graphic and multimedia products for work stations and PCs. It makes the SynchroMaster™ scan converter for use with displays including helmet and boom-mounted displays. Other products include ComputerWall™ multi-screen processor, the RGB/Videolink line of scan converters, the RGB/View line of video windowing systems, the Watchdog high resolution multi-window display systems, and the HRVM high scan rate signal mixer.

Rheinmetall GmbH, Pempelfuststrasse 1663, D-4030 Ratingen, Germany.
Tel: +49 2102 90-0.
Fax: +49 2102 90-0.

Richmond Electronics & Engineering International Limited, Armtec Estate, North Lopham, Norfolk IP22 2LR, UK.
Tel: +44 (0) 1379 888 221. Telex: 975 630 RICHMD G.
Fax: +44 (0) 1379 888 134.

Rockwell International, Autonetics Electronics Systems, Autonetics Marine Systems Division, PO Box 4921, 031-BA86, 3370 Miraloma Avenue, Anaheim, California 92803-4921, USA.
Tel: +1 714 762 3327.
Fax: +1 714 762 0146.

Rodale Electronics Inc, 603 Chestnut Street, Garden City, New York 11530, USA.
Tel: +1 516 222 0050.
Fax: +1 516 222 0223.
Chairman: John B Clement
Technical Director: Vincent Maida
Marketing Director: Tommy Kahlin
Marketing Contact: Marguerite Fullman

RO El TV Elektronika OOUR '7 Oktobar' NIŠ, Bulevar V. Vlahovića 80-82, Serbia.
Telephone and Fax numbers not known, information welcome.

de Rotterdamsche Droogdok Maatschappij BV, 21 Heijplaatstraat, PO Box 913, NL-3000 AX Rotterdam, Netherlands.
Tel: +31 10 487 2034. Telex: 20753.
Fax: +31 10 487 2913.

Royal Aeronautical Society, Flight Simulation Group, 4 Hamilton Place, London W1V 0BQ.
Tel: +44 (0) 171 499 3515.
Fax: +44 (0) 171 499 6230.
Director RAeS: R J Kennett.
Chairman, Flight Simulation Group: 1994, Brian Laven (Thomson Training and Simulation); 1995, Professor David Allerton (Cranfield University).
Secretary, Flight Simulation Group: Jane Lewis (RAeS Staff).
Chairman, International Regulations Working Group: Brian Hampson (CAE Electronics).
The RAeS was founded in 1866. Its Flight Simulation Group (FSG) normally organises two conferences on simulation subjects each year in May and November, held at RAeS headquarters in London. A Position Paper on Flight Simulation was published by the RAeS in 1994 and copies may be obtained from the above address.

Royal Ordnance plc, Guns & Vehicles Division, Kings Meadow Road, Nottingham NG2 1EQ, UK.
Tel: +44 (0) 115 986 3341.
Fax: +44 (0) 115 986 1436.

Royal Ordnance plc, Ammunition Division, Euxton Lane, Chorley, Lancashire PR7 6AD, UK.
Tel: +44 (0) 1257 265 511. Telex: 677 495.
Fax: +44 (0) 1257 260 614.

Royal Ordnance Simulation (ROSIM), Control Systems and Fuzes Division, Blackburn, Lancashire BB1 2LE, UK.
Tel: +44 (0) 1254 55 131. Telex: 63 121.
Fax: +44 (0) 1254 667 884.

RS Systems, 5301 Holland Drive, Beltsville, Maryland 20705-2383, USA.
Tel: +1 301 474 2900. Telex: WSH 904059.

Ruggieri, Direction Commerciale Technologie, 86 avenue de Saint-Ouen, F-75018 Paris, France.
Tel: +33 1 46 271 208. Telex: 283 581 F.
Fax: +33 1 42 266 465.

RUSSIA, Central Aerodynamic Institute, see under TsAGI.

RX Products Limited, Pontygwindy Industrial Estate, Caerphilly, Mid Glamorgan CF8 3HU, UK.
Tel: +44 (0) 1222 882 126/886 869.
Fax: +44 (0) 1222 883 226.

Saab Training Systems AB, S-561 85 Huskvarna, Sweden.
Tel: +46 36 194 600. Telex: 70045 saabjkg s.
Fax: +46 36 143 285.

Saab Training Systems AB, Mobile Data Systems, Box 1234, S-171 23 Solna, Sweden.
Tel: +46 8 272 680.
Fax: +46 8 274 260.

Saab-Scania AB, Saab Military Aircraft, S-581 88 Linköping, Sweden.
Tel: +46 13 180 000. Telex: 50040.
Fax: +46 13 181 802.

SABRE Computers International Limited, Process House, 43 Selsdon Road, South Croydon, Surrey CR2 6PY, UK.
Tel: 081 681 8241. Telex: 927290.
Fax: 081 688 1848.

SAINSEL, Pol Indal Ctra Amaralla, C/Manuel Velasco Pando s/n, E-41007 Seville, Spain.
Tel: +34 54 57 3691. Telex: SNCD-E 73 222
Fax: +34 54 57 9362.
SAINSEL is linked to CESELSA and the Indra Group, see CESELSA

SAINSEL, Aravaca 24, E-28040 Madrid, Spain.
Tel: +34 1 53 51 500. Telex: 47755 SNCE.
Fax: +34 1 53 44 529.

Salzburg Technology Centre, Jakob-Haringer Strasse 1-6, A-5020 Salzburg, Austria
Tel: +43 662 45 48 88
Fax: +43 662 45 48 89
The full title is Salzburger Technologiezentrumgesellschaft mbH. They are organising a conference on civil flight simulation in May 1995. Contact Toni Eiser at the above address.

SBAC (Society of British Aircraft Constructors), 29 King Street, St James's, London SW1Y 6RD, England.
Tel +44 (0) 171 839 3231.
Fax: +44 (0) 171 930 3577.
The Society of British Aircraft Constructors (SBAC) is the trade association of the British Aerospace industry. It represents some 200 companies in the aerospace business sector, and so is also a focus for aerospace matters within the UK. Amongst other activities, SBAC is the organiser of the Farnborough Air Show and Aerospace Exhibition, held every two years in or about the first week in September. The next events are scheduled for the years 1996 and 1998, and many companies involved in training and simulation are exhibitors.

Schermuly (Pains-Wessex Limited), High Post, Salisbury, Wiltshire SP4 6AS, UK.
Tel: +44 (0) 1722 411 611. Telex: 474 86 PW SCH G.
Fax: +44 (0) 1722 412 121.

Scovati Italia, Via Cernaia 16, I-10122 Turin, Italy.
Tel: +39 011 562 0702.
Fax: +39 011 562 0703.

SD-Scicon, Consultancy Group, Pembroke House, Pembroke Broadway, Camberley, Surrey GU15 3XD, UK.
Tel: +44 (0) 1276 686 200.
Fax: +44 (0) 1276 61 292.

SD-Scicon, see EDS (Electronic Data Systems Ltd)

SDPR, Belgrade, Serbia. See YUGOIMPORT

SEOS Displays Inc, PO Box 11526 Daytona Beach, Florida 32120, USA.
Tel: +1 904 672 0202.
Fax: +1 904 672 5108.

SEOS Displays Limited, Marchants Way, Burgess Hill, West Sussex RH15 8QY, UK.
Tel: +44 (0) 1444 870 888.
Fax: +44 (0) 1444 870 777.
Managing Director: Owen J Wynn
Technical Director: Stephen Elmer
Marketing: J B (Barry) Swainson
Business Development Manager: Ms Sarah Cornish
Contact for enquiries: Mrs Julie Lutterloch, J B (Barry) Swainson
SEOS Displays was started in 1984 by Owen Wynn and Stephen Elmer. SEOS manufactures a number of visual display systems including the MIDAS range of collimated monitors, the PRODAS range of direct screen projection systems of various FoV, the PANORAMA back-projected collimated cross-cockpit display, laser target projectors, helmet-mounted displays, and various simulated weapon and observation sights for gunners and AFV commanders. It has about 60 employees and a shop floor area of about 10 850 ft² (1010 m²).

Serbia, see YUGOIMPORT-SDPR Ltd, Belgrade and RO El TV Elektronika, Vlahovića.

Servos and Simulation Inc, 620 Douglas Avenue, Suite 1308, Alamonte Springs, Florida 32714, USA.
Tel: +1 407 862 2158
Fax: +1 407 862 2166
President: E Bruce Baker
Chief Engineer: Paul Kenard
Director of Marketing: Ronald E Handy
Alamonte Springs is a suburb of Orlando. Servos and Simulation Inc (S&SI) was founded by E Bruce Baker in 1981 for the purpose of high technology engineering. It specialises in the design and production of computer-controlled servo mechanisms, particularly for the aerospace industry. In the simulation field, S&SI makes electric control loaders for simulators and electric motion platforms.

SHAPE Technical Centre, Oude Waalsdorperweg 61, PO Box 174, The Hague, Netherlands
Tel: +31 70 314 2100
Fax: +31 70 314 2111

Ship Analytics Inc, 183 Prov-New London Turnpike, North Stonington, Connecticut 06359-0250, USA.
Tel: +1 203 535 3092. Telex: 643 732.
Fax: +1 203 535 0560.
Marketing Co-ordinator: Elizabeth C Casteel
Ship Analytics is involved in training systems and simulation for ships both commercial and military, and provides services such as analysis of channel and harbour dredging, ship casualty and damage control stability. The company is also involved in training, contingency planning, and containment and remedial measures for oil spills. Its simulation activities include ship handling simulators, and trainers for radar, sonar, ECM, submarine, cargo handling, dredging, engine room, and CBT packages.

Short Brothers plc, Defence Systems Division, Montgomery Road, Castlereagh, Belfast BT6 9HBN, UK.
Tel: +44 (0) 1232 458 444. Telex: 747 087.
Fax: +44 (0) 1232 705 293.

Short Brothers plc, Defence Systems Division, Alanbrooke Road, Castlereagh, Belfast BT6 9HB, UK.
Tel: +44 (0) 1232 465 406.
Fax: +44 (0) 1232 733 863.

Sichuan University, Chengdu, Sichuan, 610064, China.
Tel: 583875-2564.

Sierra Technologies Inc, Sierra Research Division, 485 Cayuga Road, Buffalo, New York 14225, USA.
Tel: +1 716 631 6343.
Fax: +1 716 631 7849.

Signaal Special Products, Regulusweg 15, PO Box 16060, NL-2500 AB, The Hague, Netherlands.
Tel: +31 70 348 29 92. Telex: 31 164.
Fax: +31 70 348 29 38.

Signaal, Hengelo Ov, Netherlands, see Hollandse Signaalapparaten BV.

Signal Computing Limited, 20 Nugent Road, The Surrey Research Park, Guildford, Surrey GU2 5AF, UK.
Tel: +44 (0) 1483 579 900.
Fax: +44 (0) 1483 62 836.

SimCom International Inc, 7500 Municipal Drive, Orlando, Florida 32 819, USA.
Tel: +1 800 272 0211 and +1 407 345 0511.
Fax: +1 407 345 9450.
President: Walter (Wally) W David
Vice-President, Engineering: Bill Conrad.
Vice-President, Training: Dick Powers.
Vice-President, Sales: Sandy Scott.
SimCom operates flight training centres and manufactures its own fixed-base flight training devices based on real aircraft cockpit sections and a wide FoV Visual System.

Simis Limited, Simis House, 90 Meadrow, Godalming, Surrey GU7 3HY, UK.
Tel: +44 (0) 1483 41 43 44
Fax: +44 (0) 1483 41 42 87
Contact: Chris Tubbs

Simtec Simulation Technology GmbH, Forststrasse 33, D-38108 Braunschweig, Germany.
Tel: +49 531 35 00 91/92.
Fax: +49 531 35 03 53.
President: Dipl-Ing Bemd Kaufmann
Chief Executive: Dipl-Ing Volker Brandt
Contact for Enquiries: Dipl-Ing Bemd Kaufmann
Simtec is a member of the Avionic Centre Braunschweig (Brunswick). Co-operation exists with the Zentrum für Luft- und Raumfahrttechnik (ZLR) at the Technical University of Braunschweig, and the Deutsche Forschungsanstalt für Luft- und Raumfahrt (DLR), the German Aerospace Research Institute. The Luftfahrt Bundesampt (German Civil Aviation Authority) is also located at Braunschweig. Simtec manufactures flight simulator components and also whole flight simulators for the smaller class of aircraft. It employs CGI systems and display systems from other manufacturers but makes its own 6-DoF synergistic compact motion platform, and all other components. It employs 30 people and has a factory floor area of 1500 m².

SIMTECH Advanced Training and Simulation Systems Limited, 5 Hamerkava Street, PO Box 150, Holon 58101, Israel.
Tel: +972 3 557 7469.
Fax: +972 3 556 4471.

SimuFlite Training, DFW Airport, PO Box 619 119, 2929 West Airfield Drive, DFW Airport, Dallas, Texas 75261, USA.
Tel: +1 214 456 8000/8094/8118, and 800 527 2463. Telex: 735 037.
Fax: +1 214 456 8303/8383
President: Charles C Carson II
Technical Director: Thomas M Ferranti
Managing Director, Sales: Dale M Martin
SimuFlite was launched in 1981 with the support of the Singer company. SimuFlite is now an independent company operating training facilities at DFW, Totowa NJ, Marietta Georgia and Tampa Florida. Simulator training is offered on the C-130 Hercules, Boeing 727 and 747, MD11, Learjet 25/35/36/55, Citation, Falcon, Gulfstream, Hawker, King Air amongst others. SimuFlite has about 200 employees and a floor space of 164 000 ft² (15 240 m²) at DFW alone.

Simulaser Corporation, 15251 E. Don Julian Road, City of Industry, California 91746, USA.
Tel: +1 818 369 1355.
Tlx: +1 910 584 4890.

SIMULTEC SA, Platforma Magurele CP 24, Bucharest, Romania.
Tel: +40 1 780 3725. Telex 11 460.
Fax: +40 1 312 2465.
General Manager; Dr Olivotto Constantin.

Simunition Division, SNC Industrial Technologies Inc, 5 Montée des Arsenaux, Le Gardeur, Québec, Canada J5Z 2P4.
Tel: +1 514 581 3080/5999. Telex: 05-24 642
Fax: +1 514 581 0231.
Co-ordinator, International Sales: Lise Laberge
Simunition produces small arms training ammunition and SNC produces larger calibres.

Sindel srl, Via Divisione Acquie 66f, I-16166 Genoa, Italy.
Tel: +39 10 330 891. Telex: 275 851.

Singapore Technologies, Chartered Industries of Singapore.
Marketing by: Unicorn International, 3 Lim Teck Road, 11-01/02, Singapore Technologies Building, Singapore 0208.
Tel: 2254788. Telex: RS 39882.
Fax: 2248862.

SINTRO Electronics AG, Hauptstrasse 21, CH-3800 Interlaken, Switzerland.
Tel: +41 36 22 22 22.
Fax: +41 36 22 01 02.
Chairman: Oskar Schärz
Managing and Marketing Director: Fritz Dürst
SINTRO was founded in 1967 and specialises in training simulators. Its product lines include simulators for aircraft (2 made), air defence systems (24 made), artillery (22), tanks (58), and vehicles (11), a total of 117 devices by mid-1994. Sintro employs about 60 people and has a shop floor area of about 5500 m².

Sippican Inc, 7 Barnabas Road, PO Box 861, Marion, Massachusetts 02738-1499, USA.
Tel: +1 508 748 1160/3626.
Fax: +1 508 748 2091.

SIUS AG, Im Langhag 1, CH-8307 Effretikon, Switzerland.
Tel: +41 52 32 91 91.
Fax: +41 52 32 12 14.
Contact: H R Walti
The full company name is SIUS-ASCOR.

Skyleader Radio Control Limited, Airport House, Purley Way, Croydon CR0 0XZ, UK.
Tel: +44 (0) 181 686 6688 or 0700
Telex: 265871 MONREF G.

SNC Industrial Technologies Inc, see Simunition.

Société Française de Matériels d'Armement (SOFMA), 17 boulevard Malesherbes, F-75008 Paris, France.
Tel: +33 (1) 42 65 97 10. Telex: SOFMA 280 566F.
Fax: +33 (1) 42 65 10 06.

Society of British Aircraft Constructors, see SBAC.

Sogitec Industries SA, 4 rue Marcel Monge, Immeuble Nobel, F-92158 Suresnes Cedex, France.
Tel: +33 (1) 41 18 57 00. Telex: 616 664F.
Fax: +33 (1) 41 18 57 18.
Président Directeur Général: Yves Fouché
Financial Director: Jean-Pierre Taquet
Director, Electronic Division: Jean Mayou
Director, Communications Division: Christopher Barre
Director, Marketing: Joël Clairon
Marketing Contact: Jean-Claude de Laforcade
Public Relations: Bruno le Cornec.
Sogitec is a subsidiary of Dassault Aviation specialising in

Simulation and Integrated Logistics. The company has an Electronics Division and a Communications Division, and has facilities at Suresnes (Paris area), Rennes (Brittany), and Bordeaux. Sogitec has about 400 employees and a factory floor area of 11 000 m².

US Office: Sogitec, Suite 300, 205 Robin Road, Paramus, New Jersey 07652, USA.
Tel: +1 207 261 61 22.
Fax: +1 207 261 52 58.

Spartanics Limited, 3605 Edison Place, Rolling Meadows, Illinois 60008-1077, USA.
Tel: +1 708 394 5700. Telex: 240 329.
Fax: +1 708 394 0409.
President: William N Knotts
Chairman: Samuel P Willits
Vice-President, Engineering: Thomas Kleeman
Spartanics was founded in 1961 and makes marksmanship training devices for civil and military use. It employs about 75 people.

Solartron, see Loral Solartron

Sperry, see Honeywell Limited, Sperry Aerospace Division,

Star Technologies Inc, PO Box 13951, Research Triangle Park, North Carolina 27709, USA.
Tel: +1 919 361 3800.

STN Systemtechnik Nord GmbH, Hünefeldstrasse 1-5, D-28199 Bremen 1, Germany.
Tel: +49 (0) 421 538 4848 and 02, also +49 40 88 25-0. Telex: 245 825.
Fax: +49 (0) 421 538 5461 and 3320, and also +49 40 88 25 4022.
Chairman: Hans E W Hoffmann
Managing Director: Klaus-Dieter Gerster
Technical Director: Helmut Meyer-Baumeister
Marketing Director: Helmut Füssinger
STN is part of the Electronics and Systems Technology Division of the Bremer Vulkan Group, together with Atlas Elektronik (qv). The Vulkan Group consists of about 30 companies grouped into six functional divisions and deals with shipbuilding, maritime electronics and systems technology, employing about 15 000 people. STN employs about 2600 people and has subsidiaries of STF at Rostock, DEBEG (Hamburg), W G Schulz (Hamburg) and Eurodrone (Paris).

Strachan & Henshaw (S&H), PO Box 103, Ashton Works, Ashton Vale Road, Bristol, BS99 7TJ, UK.
Tel: +44 (0) 117 966 4677. Telex: 44170.
Fax: +44 (0) 117 966 4259.
Chairman: Alistair Fleming
Managing Director: Ken Grove
Marketing Manager: Mike Bravery, Ext 2281
S&H is a subsidiary of the Weir Group plc and has about 950 employees and a shop floor area of about 45 000 m². It manufactures Weapon Handling and Discharge Equipment (WHDE) trainers for submarines such as the Swiftsure, Trafalgar, Upholder, and Vanguard classes for the RN, and the Collins class for the RAN.

Struktura Szerezesi Vallalat. See IDEX (InDustrialEXport), Budapest, Hungary

Swiss Federal Armament Works, Allmendstrasse 86, CH-3602 Thun, Switzerland.
Tel: +41 33 281 111. Telex: 921 256.
Fax: +41 33 282 047.

Symtron Systems Inc, 17-01 Pollitt Drive, Fair Lawn, New Jersey, 07410, USA.
Tel: +1 201 794 0200.

Synertech (Pty) Limited, PO Box 420, Halfway House, 1685, Republic of South Africa.
Tel: +27 11 313 9000. Telex: 4 27859 SA.
Fax: +27 11 313 9100.

Syscon Corporation, 1000 Thomas Jefferson Street North West, Washington DC 20007, USA.
Tel: +1 202 342 4000.
Fax: +1 202 342 4449.

Systems and Simulation Inc, 5902 Breckenridge Parkway, Tampa, Florida 33610-4233, USA.
Tel: +1 813 628 6100. Telex: 810 876 0809.
Fax: +1 813 621 7763.

Systems Management Incorporated (SMI), PO Box 126, Hunt Valley, Maryland 21030-0126, USA.
Tel: +1 410 628 3000.
Fax: +1 410 628 3465.
SMI is a wholly owned subsidiary of the AAI Corporation, also of Hunt Valley, Maryland.

Tadiran Limited, 29 Hamerkava Street, PO Box 150, Holon 58101, Israel.
Tel: +972 3 557 32 45.
Fax: +972 3 557 32 91.
Marketing Director: Ms Margalit Degani.

Tamnar Aviation Technology, 51 Hailanot Street, Gat Rimon, Petah Tikva 49920, Israel.
Tel: +972 03 93 25 461.
Fax: +972 03 93 07 045.

Targetair Limited, RR4, D-268 Moncton, New Brunswick, Canada E1C 8J8.
Tel: +1 506 857 2141.
Fax: +1 506 853 9182.

Target Technology Limited, Unit 3, Brunswick Industrial Centre, Ashford, Kent TN24 9QP, UK.
Tel: +44 (0) 1233 639 762. Telex: 965058 BANSHEE G.
Fax: +44 (0) 1233 624 883.

Tasuma (UK) Limited, Unit 3, 27 Blackmoor Road, Ebblake Industrial Park, Verwood, Dorset BH21 6BE, UK.
Tel: +44 (0) 1202 82 54 89.
Fax: +44 (0) 1202 82 41 16.

Technical Applications and Service Company (TAS), 200 Professional Drive, Gaithersburg, Maryland 20879, USA.
Tel: +1 301 921 8100.
Fax: +1 301 977 6158.

Tector Visual Systems, Operational Displays Division, Woodhill Road, Collingham, Newark, Nottinghamshire NG23 7NR, UK.
Tel: +44 (0) 1636 892 246.
Fax: +44 (0) 1636 893 317.
Chairman: A V Tushingham.
Managing Director: Alan L Leonard.
Technical Director: N R S Nesbitt
Marketing: Alan L Leonard.
Tector was formed in 1974 and launched its low-cost OPDIS day/dusk/night image generator in 1976. Tector is a division of Graff Electronic Machines Ltd and employs about 25 staff and has a shop floor area of about 32 300 ft² (3000 m²). It makes a range of visual system products for simulation of land, sea and air systems equipment, and is also involved in displays and audio training systems. Its visual system products include the VISTAR CGI system, and display systems such as collimated monitors and screen projection systems.

Teklogic (Pty) Limited, PO Box 420, Halfway House, 1685, Republic of South Africa.
Tel: +27 11 313 9000. Telex: 427859.
Fax: +27 11 313 9100.

Teledyne Brown Engineering (TBE), Cummings Research Park, Huntsville, Alabama 35807-7007, USA.
Tel: +1 205 726 1426. TWX 810 726 2103.
Fax: +1 205 726 1507.

Teledyne Ryan Aeronautical, 2701 Harbor Drive, PO Box 80311, San Diego, California 92138-9012, USA.
Tel: +1 619 291 7311. Telex: 68-31061.

Theissen Fassaden- und Maschinenbau GmbH, PO Box 104942, D-40231 Düsseldorf, Germany.
Tel: +49 (0) 211 2106-0. Telex: 858 2323.
Fax: +49 (0) 211 2106-233.

Theissen Fassaden- und Maschinenbau GmbH, PO Box 150425, D-40081 Düsseldorf, Germany.
Tel: +49 (0) 211 92 107-0. Telex: 858 2323.
Fax: +49 (0) 211 97 504-33.

Thomson-CSF SA, 173 boulevard Haussmann, F-75008 Paris, France
The parts of the Thomson-CSF organisation which are relevant to Training and Simulation are listed below:

FRANCE: Thomson Training and Simulation (TTS), 3 avenue Albert Einstein, BP 116-78192 Trappes Cedex, France
Tel: +33 (1) 30 69 41 00
Fax: +33 (1) 30 69 44 44 or 62

Thomson Training and Simulation (TTS), 26 chaussée Jules César, Z. I. Les Beaux Soleils, Osny - BP 226, F-95523 Cergy Pontoise Cedex, France.
Tel: +33 (1) 34 22 82 00.
Fax: +33 (1) 34 22 87 00.

Thomson Sintra, ASM (Activités Sous-Marinés), Departement de Sophia, 525 route des Dolines, BP 157, F-06903 Sophia Antipolis Cedex, France.
Tel: +33 92 96 41 21. Telex: Thom 616 780F/Thomsi X 470 111F.
Fax: +33 92 96 39 50.

UK: Thomson Training and Simulation Limited (TTSL), Gatwick Road, Crawley, West Sussex RH10 2RL, UK.
Tel: +44 (0) 1293 562 822. Telex: 87 327.
Fax: +44 (0) 1293 563 366.
This facility is the ex-Rediffusion site. The Lancing facility (ex Link-Miles) is being closed and all UK resources will be based at Crawley.

USA: Thomson Training and Simulation (TTS), 7041 East 15th Street, PO Box 1677, Tulsa, Oklahoma 74101A, USA
Tel: +1 918 836 4621
Fax: +1 918 834 3768
This facility is the ex-Burtek company.

USA: Thomson Training and Simulation Inc, Box 210 929, 2000 East Lemar Boulevard, Suite 300, Arlington, Texas 76006-7337, USA
Tel: +1 817 543 5802
Fax: +1 817 543 5813
This facility is the ex-Rediffusion Inc company.

Thomson Training and Simulation (TTS) Executives:
President and Chairman: Louis Le Portz
Managing Director, Civil Business: Louis Le Portz
Managing Director, Military and Energy Business: Dominique Vernay
Technical Director, Civil Business: Eric Le Mer
Technical Director, Military and Energy Business: Bob Ison (Crawley address)
Marketing Director, Civil Business: Gilbert Dangleterre (Crawley address)
Marketing Director, Military and Energy Business: Jean-Marie Huet
Marketing Contact, Civil Business: John Lillicrap (Crawley address)
Marketing Contact, Military and Energy Business: Jean-Claude Canioni
Communications Contact: Brigitte Porêe (Trappes address)
John Barrett (Crawley address)
Thomson-CSF created Thomson Training and Simulation (TTS) in May 1994, having added the purchase of Hughes Rediffusion (HRSL) in late 1993 to the previous purchase of Link-Miles in 1990. The previous Link-Miles site at Lancing is being closed and all UK assets are being centred on the ex-Rediffusion site at Crawley. The Crawley site is the headquarters for civil business and the Cergy Pontoise site is the HQ for military and energy business. TTS employs 2600 people at its sites in France, the UK and the USA and has a total shop floor area of 21 500 m². It has an annual turnover of $US500 million, and has some 50 years of experience in the training and simulation area. Its products include simulators for military and civil aircraft, airline cabin crew trainers, CGI and visual display systems, and simulators for land force equipment, ships and submarines, nuclear and conventional power plants, and for the entertainment market. TTS is the second largest industrial group in the world specialising in Training and Simulation equipment, after the CAE group, and has the second largest world market share, also after the CAE group.

THORN Automation, PO Box 4, Rugeley, Staffordshire WS15 1DR, UK.
Tel: +44 (0) 1889 585 151.
Fax: +44 (0) 1889 578 209.
General Manager: J C Sheldon
Technical Director: J P Wardale
Marketing Director: C C T Millard
Marketing Contact: K A Mortimer
THORN Automation is a division of THORN EMI Electronics and has about 430 employees. It manufactures naval systems such as data integration and distribution, navigational databus systems, plasma displays, power conversion and control systems, degaussing ranges and magnetic sensors, and sonar.

THORN EMI Electronics Limited, Command Systems Division, Wookey Hole Road, Wells, Somerset BA5 1AA, UK.
Tel: +44 (0) 1749 672 081. Telex: 44254.
Fax: +44 (0) 1749 670 671.

THORN EMI Electronics Limited, Defence Group, 120 Blyth Road, Hayes, Middlesex UB3 1HY, UK
Tel: +44 (0) 181 910 8000. Telex 22 417
Fax: +44 (0) 181 910 8361
PR Manager: Mrs Penny Wreford
THORN EMI Defence Group makes, amongst other things, weapon ranges and telemetry systems.

THORN EMI Electronics Limited, Electronic Systems Division, Longford Unit, Victoria Road, Feltham, Middlesex TW13 7DZ, UK.
Tel: +44 (0) 181 890 3600.
Fax: +44 (0) 181 751 0207.

THORN Security Limited, Electronic Systems Group, Brymers Avenue, Fortuneswell, Portland, Dorset DT5 1JT
Tel: +44 (0) 1305 860 922
Fax: +44 (0) 1305 860 912
Business Manager: Brian Thomas

TIRAX AG, Bernstrasse 390, CH-8953 Dietikon, Switzerland.
Tel: +41 7 42 17 17.
Fax: +41 7 42 17 18.
Chairman and Managing Director: Carl Leutwyler
Technical Director: Rolf Frey
Marketing Director: H R Holzer
TIRAX has about 20 employees and a shop floor area of about 950 m^2. The company makes combat simulators and targets, and has delivered over 12 000 devices.

TNO Physics and Electronics Laboratory, PO Box 96 864, Oude Waalsdorperweg 63, NL-2509 JG, The Hague, Netherlands.
Tel: +31 70 326 42 21.
Fax: +31 70 328 09 61.

TNO-IWECO (Institute of Mechanical Engineering), PO Box 29, 2600 AA Delft, Leeghwaterstraat 5, NL-2628 CA Delft, Netherlands.
Tel: +31 15 60 86 08. Telex: 38192.

Tracor Flight Systems Inc, Simulation Systems, 1241 East Dyer Road, Santa Ana, California 92705, USA.
Tel: +1 714 662 0333.
Fax: +1 714 662 1030.

Training Systems Associates (TSA), 78 Buckingham Gate, London SW1E 6PE, UK
Tel: +44 (0) 19 56 24 44 42
Fax: +44 (0) 17 12 89 86 08
United States Office: TQI Inc, 7301 Forest Avenue, Suite 250, Richmond, Virginia 23226, USA
Tel: +1 804 673 22 53
Fax: +1 804 282 30 66
Principal Consultants: R M (Bob) Prothero
Dr Susan B Hardwicke
Training Systems Associates (TSA) specialises in Training Needs Analyses, the design of training systems, contractor training systems, and the application of TQM (Total Quality Management) in the training sector. TSA has experience in the land and air systems areas, and works with US and UK companies. TSA offers TQM courses to the training, defence and other industries.

Trenton Technical Services Limited, Unit 6, 28 Botley Road, Hedge End, Southampton SO3 3HE, UK.
Tel: +44 (0) 1489 781 182. Telex: 86646 VYNER G.
Fax: +44 (0) 1489 781 182.

TsAGI (Central AeroHydrodynamic Institute), 1 Zhukovsky Street, Zhukovsky 140160, Russia
Tel: +7 095 556 40 54 and 41 59. Telex: 412573 ZAKAT SU
Fax: +7 095 556 43 37 & 271 00 19
Director: Professor German I Zagainov
Deputy Director: Valery Suhanov
General Contact: International Business Department (IBD)
Chief of IBD: Elena Tolchennikova
Chief of Section, Flight Simulation and Handling Qualities: Dr Victor V Rodchenko
TsAGI stands for Tsentral'ny Aerogidrodinamicheski Institut (Central AeroHydrodynamic Institute). The Flight Dynamics and Control Systems Department is responsible, amongst other things, for the design of training simulators.

TVI Corporation, 10700 Hanna Street, Beltsville, Maryland 20705, USA.
Tel: +1 301 595 5252.

UEC Projects (Pty) Limited, PO Box 432, Mount Edgecombe, 4300, Natal, South Africa.
Tel: +27 031 593 242. Telex: 6-20821 SA.
Fax: +27 031 591 802.

Unicorn International, 3 Lim Teck Road, 11-01/02, Singapore Technologies Building, Singapore 0208.
Tel: 2254788. Telex: RS 39882.
Fax: 2248862.

Unisys Corporation, Naval Systems Division (Also Surveillance and Fire-Control System Division), 365 Lakeville Road, Great Neck, New York 11020, USA.
Tel: +1 516 574 0111/3259. Telex: 960 167.
Fax: +1 516 574 1967.
Marketing Manager: Stephen H Young

Unisys Corporation, Shipboard & Ground Systems Group, 4000 South Memorial Parkway, Huntsville, Alabama 35802-4304, USA.
Tel: +1 205 880 5500.

United Defense LP, Armament Systems Division, 4800 East River Road, PO Box 59043, Minneapolis, Minnesota 55421-1498, USA.
Tel: +1 612 572 6517/9201.
Fax: +1 612 571 8025.
President and CEO: T W Rabault
Marketing Communications: Richard C Snider
United Defense LP divisions are Ground Systems, Defense Systems International, Armament Systems, Combat Systems, and Steel Products.

USAF Aeronautical Systems Center (ASC), USAF Materiel Command, ASC/CC Building 14, 1865 4th Street, Suite 12, Wright-Patterson AFB, Ohio 45433-7126, USA.
Tel: +1 513 225 33 34
Fax: +1 513 476 40 22
Commanding Officer, ASC: Lt Gen James A Fain, Jr
Director, Office of Public Relations: Lt Col Patrick C Mullaney, 1865 4th Street, Suite 15, Wright-Patterson AFB, Ohio 45433-7129, USA.
The USAF Aeronautical Systems Center (ASC) is responsible for R&D, Test and Evaluation, and support of aeronautical systems and equipment for the USAF. It has a work force of about 11 000 and manages about one-fifth of the USAF budget. As part of this task ASC is involved in some flight simulation and human factors activities which are described in the main sections of this book.

US Defense Mapping Agency, see under D for Defense.

US Department of Transportation, National Highway Traffic Safety Administration, 400 Seventh Street, SW, Washington DC 20590.

USN, see item under Naval Air Warfare Center Training Systems Division (NAWCTSD), Orlando, Florida.

Utilex (UK) Limited, 5 Copthorne Avenue, London SW12 0JZ, UK.
Tel: +44 (0) 181 674 7989.

Van Halteren Metaal BV, Marconiweg 2, NL-3752 LS Bunschoten, Netherlands.
Tel: +31 3499 83 124.
Fax: +31 3499 83 674.

Van Rietschoten en Houwens Defensiesystemen BV, Sluisjesdijk 155, PO Box 5054, NL-3008 AB Rotterdam, Netherlands.
Tel: +31 10 48 71 911.
Fax: +31 10 48 71 702.

Veda Incorporated, 6 Pecan Court, Suite A, California, Maryland 20619, USA.
Tel: +1 301 737 1555

Vickers Defence Systems, Scotswood Road, Newcastle upon Tyne, NE99 1BX, UK.
Tel: +44 1091 273 8888. Telex: 53 104.
Fax: +44 1091 273 2324.
Contact: Stuart K LaMoury

Vickers Shipbuilding and Engineering Limited, Barrow-in-Furness, Cumbria LA14 1AF, UK.
Tel: +44 (0) 1229 23 366. Telex: 65411.
Fax: +44 (0) 1229 35 281.

Vickers Shipbuilding and Engineering Limited, Combat Systems, Barrow-in-Furness, Cumbria LA14 1AF, UK.
Tel: +44 (0) 1299 873 982.
Fax: +44 (0) 1229 873 846.

Virtual Prototypes Inc, 701 Palomar Airport Road, Suite 300, Carlsbad, California 92009, USA.
Tel: +1 619 931 4707.
Fax: +1 619 929 9669.

Virtual Prototypes Inc, 4700 de la Savanne, Suite 300, Montreal, Québec, Canada H4P 1T7.
Tel: +1 514 341 3874 and 483 4712.
Fax: +1 514 341 8018 and 483 1231.

Virtual Prototypes, 1 bis rue du Petit Clamart F-78144 Velizy Cedex, France
Tel: +33 (1) 41 07 97 37.
Fax: +33 (1) 41 07 97 47.

Vitro Services Corporation, Industrial Park, 557 Mary Esther Cutoff, Fort Walton Beach, Florida 32548, USA.
Tel: +1 904 244 7711/7795.
Fax: +1 904 244 1183.
President and Chief Executive: Richard L Manley.
Technical Director: Ronald M Van Oeveren.
Marketing Director: James L Atkinson.
Marketing Contact: Tony Gardecki.
Vitro Services Corporation is wholly owned by Tracor Inc of Austin, Texas. Vitro's activities in weapon ranges date back to 1946 when the company designed and constructed the first missile firing range in the USA. Vitro designs and constructs test and training ranges, and provides logistic support, operations and maintenance for them.

The Electronics Systems Division (ESD) was established in 1962 and is the manufacturing arm of Vitro. It employs 1720 people and has a shop floor area of 57 500 ft^2. It manufactures Range Tracking and Surveillance Radars, Optical Tracking Systems, Range Control and Display Systems, and Threat Radar Simulators. The latter are described in the Aviation Systems section under EW Trainers for Air Systems.

Vitro, 45 West Gude Drive, Rockville, Maryland 20850-1160, USA.
Tel: +1 301 738 4300.
Fax: +1 301 738 4323.

Vosper Naval Systems Pte Limited, 232 Tanjong Rhu Road, Singapore 1543.
Tel: +65 344 9388. Telex: 21 219.
Fax: +65 344 6642.

Vosper Thornycroft (UK) Limited, Controls Unit, Cow Lane, Wymering, Portsmouth, Hampshire PO6 3TR, UK.
Tel: +44 (0) 1705 373 511. Telex: 86 528.
Fax: +44 (0) 1705 326 885.

Vosper Thornycroft (UK) Limited, Victoria Road, Woolston, Southampton, Hampshire SO9 5GR, UK.
Tel: +44 (0) 1703 445 144. Telex: 47682.
Fax: +44 (0) 1703 421 539.

Walmore Defence Systems Limited, London; see Aware Simulation Limited.

WCBKT, Central Military Bureau of Designs and Technology, Bemowo, 00-908 Warsaw-49, Poland.
Tel: +48 22 36 94 91
Fax: +48 22 36 41 62
WCBKT stands for Wojskowe Centralne Biuro Konstrukcyjno-Technologiczne.
Note that marketing is through the Cenzin Company whose details are given earlier.

Wegmann & Co GmbH, August-Bode-Strasse 1, D-34127 Kassel, Germany.
Tel: +49 561 105-0 or +49 561 105 23 19. Telex: 99 859.
Fax: +49 561 105 22 08 or +49 561 105 27 11.
Chairman and Managing Director: Dr Manfred Bode
Directors: Wolfgang Bode, Jürgen Bode, Dr Werner F Zimni
Chief Executive: Detlev Jaritz
Technical Director: Josef Klebing
Marketing Director: Dr Günter Hönig
Wegmann & Co has been active in railroad engineering since 1882, and in defence-related products since early this century. It produces tank turrets, artillery, AAA and rocket launch systems (eg MRLS), and training systems for crew and maintenance personnel, and has 1050 personnel. The Wegmann Group employs over 5000 people and includes the Gebr Bode Company of Kassel, FAC of Munster, the Hofmann Group of Pfungstadt, Schleifring-und Apparatebau of Fürstenfeldbruck, BWG of Butzbach, and Wegmann Industrietechnik GmbH of Kassel.

J L Weibull AB, Möllegården, PO Box 43, S-232 02 Åkarp, Sweden.
Tel: +46 40 465 080. Telex: 33 159.
Fax: +46 40 461 677.

Wescam Inc, 1810 Highway 6 North, Hamilton, Ontario, Canada L9J 1H2.
Tel: +1 416 529 5132. Telex: 061 8548.
Chairman: Mark Chamberlain
Chief Executive: Alastair Jenkins
Technical Director: Paul Jennison
Marketing Director: John Bastedo
Marketing Contact: Brian Klodt
Wescam is a division of Istec Inc and has about 80 employees. It makes air-to-ground surveillance and identification systems including the WESCAM camera and the stabilised WESCAM gimbal system.

Westinghouse Canada Inc, Government Projects Division, Information and Defence Technology Division, PO Box 5009, Burlington, Ontario L7R 4B3, Canada.
Tel: +1 416 333 6006. Telex: 061 8401.

Westinghouse Electric Corporation, Naval Systems Division, 18901 Euclid Avenue, Cleveland, Ohio 44117, USA.
Tel: +1 216 692 5126.
Fax: +1 216 692 6993.
Group President: Richard Linder
General Manager, Naval Systems Division: D Wayne Snodgrass
Technical Director: Zoltan Dudevszky
Operations Manager: Elmer Guiher
The Naval Systems Division is part of the Electronic Systems group of the Westinghouse Electric Corporation. The Electronic Systems group has its headquarters in Baltimore, Maryland. The Naval Systems Division has about 800 employees and a shop floor area of about 430 000 ft^2 (40 000 m^2). It used to be Gould Inc Ocean Systems Division until 1988 when acquired by Westinghouse.

Westland System Assessment Limited, Telec House, Goldcroft, Yeovil, Somerset BA21 4DQ, UK.
Tel: +44 (0) 1935 24 545. Telex: 46542 W-SYST.
Fax: +44 (0) 1935 32 864.

Whitehead SpA, Via di Levante 48, I-57128 Livorno, Italy.
Tel: +39 586 840 111. Telex: 500 192.
Fax: +39 586 854 060.

Whittaker Electronic Systems, 1785 Voyager Avenue, PO Box 8000, Simi Valley, California 93063-8000, USA.
Tel: +1 805 584 8200. Telex: 651329.
TWX: 301 993 6698.

Yuhe Machine Factory (YMF), 2 Jie Fang Road, Nanjing, Jiangsu Province, People's Republic of China.
Tel: +86 42133 and 46462. Cables: 1630 Nanjing.

YUGOIMPORT-SDPR Ltd, Bulevar umetnosti 2, 11070 Novi Beograd (Belgrade), Serbia.
Tel: +38 111 22 - 22 743, - 23 697, - 21 284. Fax: +38 111 32 48 - 791/738/748. Telex: 71000 JU SDPR

Zaklady Tworzyw Sztucznych, see PRONIT.

Zen Technologies and Computers Limited, Plot No 44, Vikrampuri, Secunderabad 500 003, India.
Tel: +91 40 84 33 07
Fax: +91 40 84 87 68
Director: Ashok Atluri

Zhukovsky, Russia, see TsAGI.

Manufacturers Index

Equipment Index

D

E

J

K

L

M

U

V

W

Y

Z